COMPREHENSIVE BIOTECHNOLOGY

IN 4 VOLUMES

COMPREHENSIVE BIOTECHNOLOGY

The Principles, Applications and Regulations of Biotechnology in Industry, Agriculture and Medicine

EDITOR-IN-CHIEF

MURRAY MOO-YOUNG

University of Waterloo, Ontario, Canada

Volume 1

The Principles of Biotechnology: Scientific Fundamentals

VOLUME EDITORS

ALAN T. BULL

University of Kent at Canterbury, UK

and

HOWARD DALTON

University of Warwick, UK

PERGAMON PRESS

OXFORD · NEW YORK · TORONTO · SYDNEY · FRANKFURT

U.K.	Pergamon Press Ltd., Headington Hill Hall, Oxford OX3 0BW, England
U.S.A.	Pergamon Press Inc., Maxwell House, Fairview Park, Elmsford, New York 10523, U.S.A.
CANADA	Pergamon Press Canada Ltd., Suite 104, 150 Consumers Road, Willowdale, Ontario M2J 1P9, Canada
AUSTRALIA	Pergamon Press (Aust.) Pty. Ltd., P.O. Box 544, Potts Point, N.S.W. 2011, Australia
FEDERAL REPUBLIC OF GERMANY	Pergamon Press GmbH, Hammerweg 6, D-6242 Kronberg-Taunus, Federal Republic of Germany

First edition 1985

Library of Congress Cataloging in Publication Data

Main entry under title:
Comprehensive biotechnology.
Includes bibliographies and index.
Contents: v. 1. The principles of biotechnology – scientific fundamentals / volume editors, Alan T. Bull, Howard Dalton –
v. 2. The principles of biotechnology – engineering considerations / volume editors, A.E. Humphrey, Charles L. Cooney – [etc.] –
v. 4. The practice of biotechnology – speciality products and service activities / volume editors, C.W. Robinson, John A. Howell.
1. Biotechnology. I. Moo-Young, Murray.
TP248.2.C66   1985   660′.6   85–6509

British Library Cataloguing in Publication Data

Comprehensive biotechnology: the principles, applications and regulations of biotechnology in industry, agriculture and medicine.
1. Biotechnology
I. Title   II. Moo-Young, Murray
660′.6   TP248.3

ISBN 0–08–032509–2 (vol. 1)
ISBN 0–08–026204–X (4–vol. set)

Printed in Great Britain by A. Wheaton & Co. Ltd., Exeter

Contents

Preface

In his recent book, entitled 'Megatrends', internationally-celebrated futurist John Naisbitt observed that recent history has taken industrialized civilizations through a series of technology-based eras: from the chemical age (plastics) to an atomic age (nuclear energy) and a microelectronics age (computers) and now we are at the beginning of an age based on biotechnology. Biotechnology deals with the use of microbial, plant or animal cells to produce a wide variety of goods and services. As such, it has ancient roots in the agricultural and brewing arts. However, recent developments in genetic manipulative techniques and remarkable advances in bioreactor design and computer-aided process control have founded a 'new biotechnology' which considerably extends the present range of technical possibilities and is expected to revolutionize many facets of industrial, agricultural and medical practices.

Biotechnology has evolved as an ill-defined field from inter-related activities in the biological, chemical and engineering sciences. Inevitably, its literature is widely scattered among many specialist publications. There is an obvious need for a comprehensive treatment of the basic principles, methods and applications of biotechnology as an integrated multidisciplinary subject. *Comprehensive Biotechnology* fulfils this need. It delineates and collates all aspects of the subject and is intended to be the standard reference work in the field.

In the preparation of this work, the following conditions were imposed. (1) Because of the rapid advances in the field, it was decided that the work would be comprehensive but concise enough to enable completion within a set of four volumes published simultaneously rather than a more encyclopedic series covering a period of years to complete. In addition, supplementary volumes will be published as appropriate and the work will be updated regularly via *Biotechnology Advances*, a review journal, also published by Pergamon Press with the same executive editor. (2) Because of the multidisciplinary nature of biotechnology, a multi-authored work having an international team of experts was required. In addition, a distinguished group of editors was established to handle specific sections of the four volumes. As a result, this work has 10 editors and over 250 authors representing 15 countries. (3) Again, because of the multidisciplinary nature of the work, it was virtually impossible to use a completely uniform system of nomenclature for symbols. However, provisional guidelines on a more unified nomenclature of certain key variables, as provided by IUPAC, was recommended. (4) According to our definition, aspects of biomedical engineering (such as biomechanics in the development of prosthetic devices) and food engineering (such as product formulations) are not included in this work. (5) Since the work is intended to be useful to both beginners as well as veterans in the field, basic elementary material as well as advanced specialist aspects are covered. For convenience, a glossary of terms is supplied. (6) Since each of the four volumes is expected to be fairly self-contained, a certain degree of duplication of material, especially of basic principles, is inevitable. (7) Because of space constraints, a value judgement was made on the relative importance of topics in terms of their actual rather than potential commercial significance. For example, 'agricultural biotechnology' is given relatively less space compared to 'industrial biotechnology', the current raison d'être of biotechnology as a major force in the manufacture of goods and services. (8) Finally, a delicate balance of material was required in order to meet the objective of providing a comprehensive and stimulating coverage of important practical aspects as well as the intellectual appeal of the field. Readers may wish to use this work for initial information before possibly delving deeper into the literature as a result of the critical discussions and wide range of references provided in it.

Comprehensive Biotechnology is aimed at a wide range of user needs. Students, teachers, researchers, administrators and others in academia, industry and government are addressed. The requirements of the following groups have been given particular consideration: (1) chemists, especially biochemists, who require information on the chemical characteristics of enzymes, metabolic processes, products and raw materials, and on the basic mechanisms and analytical techniques involved in biotechnological transformations; (2) biologists, especially microbiologists and molecular biologists, who require information on the biological characteristics of living organisms involved in biotechnology and the development of new life forms by genetic engineering techniques; (3) health scientists, especially nutritionists and toxicologists, who require information on biohazards and containment techniques, and on the quality of products and by-products of biotechnological processes, including the pharmaceutical, food and beverage industries; (4) chemical engineers, especially biochemical engineers, who require information on mass and energy balances and rates of processes, including fermentations, product recovery and feedstock pretreatment, and the equipment for carrying out these processes; (5) civil engineers, especially environmental engineers, who require information on biological waste treatment methods and equipment, and on contamination potentials of the air, water and land within the ecosystem, by industrial and domestic effluents; (6) other engineers, especially agricultural and biomedical engineers, who require information on advances in the relevant sciences that could significantly affect the future practice of their professions; (7) administrators, particularly executives and legal advisors, who require information on national and international governmental regulations and guidelines on patents, environmental pollution, external aid programs and the control of raw materials and the marketing of products.

No work of this magnitude could have been accomplished without suitable assistance. For guidance on the master plan, I am indebted to the International Advisory Board (J. D. Bu'Lock, T. K. Ghose, G. Hamer, J. M. Lebault, P. Linko, C. Rolz, H. Sahm, B. Sikyta and H. Taguchi). For structuring details of the various sections, the invaluable assistance of the section editors is gratefully acknowledged, especially Alan Bull, Charles Cooney, Harvey Blanch and Campbell Robinson, who also acted as coordinators for each of the four volumes. For the individual chapters, the 250 authors are to be commended for their hard work and patience during the two years of preparation of the work. For checking the hundreds of literature references cited in the various chapters, the many graduate students are thanked for a tedious but important task well done. A special note of thanks is due to Jonathan and Arlene Lamptey, who acted as editorial assistants in many diverse ways. At Pergamon Press, I wish to thank Don Crawley for originally suggesting this project and Colin Drayton for managing it. Finally, I am pleased to note the favourable evaluations of the work by two distinguished authorities, Sir William Henderson and Nobel Laureate Donald Glaser, who provided a foreword and a guest editorial, respectively, to the treatise.

MURRAY MOO-YOUNG
Waterloo, Canada
December 1984

Foreword

This very comprehensive reference work on biotechnology is published ten years after the call by the National Academy of Sciences of the United States of America for a voluntary worldwide moratorium to be placed on certain areas of genetic engineering research thought to be of potential hazard. The first priority then became the evaluation of the conjectural risks and the development of guidelines for the continuation of the research within a degree of containment. There had hardly been a more rapid response to this type of situation than that of the British Advisory Board for the Research Councils. The expression of concern by Professor Paul Berg and the committee under his chairmanship, and the call for the moratorium, was published in *Nature* on 19 July 1974. The Advisory Board agreed at their meeting on the 26 July to establish a Working Party with the following terms of reference:

> 'To assess the potential benefits and potential hazards of the techniques which allow the experimental manipulation of the genetic composition of micro-organisms, and to report to the Advisory Board for the Research Councils.'

Because of the conviction of those concerned that recombinant DNA techniques could lead to great benefits, the word order used throughout the report of the Working Party (Chairman, Lord Ashby) always put 'benefits' before 'hazards'. The implementation of the recommendations led to the development of codes of practice. This was followed by the establishment of the Genetic Manipulation Advisory Group as a standing central advisory authority operating within the framework of the Health and Safety at Work, *etc.* Act 1974 and, later, more specifically within the framework of the Health and Safety (Genetic Manipulation) Regulations 1978. Similar moves took place in many other countries but the other most prominent and important activity was that of the US National Institutes of Health. This resulted in the adoption by most countries of the NIH or the UK guidelines, or the use of practices based on both.

The significant consequence of the debates, the discussions and of the recommendations that emerged during these early years of this decade (1974–1984) was that research continued, expanded and progressed under increasingly less restriction at such a pace that now makes it possible and necessary to devote the first Section of Volume 1 of this work to genetic engineering. Many chapters of the subsequent Sections and Volumes are of direct relevance to the application of genetic engineering.

The reason for identifying today's genetic engineering for first mention in this foreword is its novelty. It was being conceived barely more than ten years ago. Ten years by most standards is a short time. Although in the biological context it represents at least 10^4 generation times of the most vigorous viruses, it is less than one of man even for the most precocious. The current developments in biotechnology, whether they be in recombinant DNA, monoclonal antibodies, immobilized enzymes, *etc.* are mostly directed towards producing a better product, or a better process. This is commendable and is supportable by the ensuing potential commercial benefits. The newer challenge is the application of the new biotechnology to achieve what previously could scarcely have been contemplated. Limited biological sources of hormones, growth regulators, *etc.* are being, and will be increasingly, replaced by the use of transformed microorganisms, providing a vastly increased scale of production. Complete safety of vaccines by the absence of ineffectively inactivated virus is one of the great advantages of the genetically engineered antigen. This is quite apart from the ability to prepare products for which, at present, there is a technical difficulty or which is economically not feasible by standard methods.

A combination of advances in recombinant DNA research, molecular biology and in blastomere manipulation has provided the technology to insert genetic material into the totipotent animal cell. The restriction on the application of this technology for improved animal production is the lack of knowledge on the genetic control of desirable biological characteristics for transfer from one breed line to another.

There are probably greater potential benefits to be won in the cultivation of the domesticated plants than in the production of the domesticated animals. In both cases, the objectives are to increase the plant's or the animal's resistance to the prejudicial components of its environment and to increase the yield, quality and desired composition of the marketable commodities. These include the leaf, the tuber, the grain, the berry, the fruit or the milk, meat and other products of animal origin. This is not taking into account the other valuable products of horticulture, of oil or wax palms, rubber trees and forestry in general. Genetic engineering should be able to provide short-cuts to reach objectives attainable by traditional procedures, for example by by-passing the sequential stages of a traditional plant breeding programme by the transfer of the genetic material in one step. Examples of desirable objectives are better to meet user specifications with regard to yield, quality, biochemical composition, disease and pest resistance, cold tolerance, drought resistance, nitrogen fixation, *etc.* One of the constraints in this work in plants is the scarcity of vectors compared with the many available for the transformation of microorganisms. The highest research priority on the plant side is to determine by one means or another how to increase the efficiency of photosynthesis. The photosynthetic efficiency of temperate crop plants is no more than 2–2.5% in terms of conversion of intercepted solar energy. These plants possess the C_3 metabolic pathway with the energy loss of photorespiration. Tropical species of plants with the C_4 metabolic pathway have a higher efficiency of photosynthesis in that they do not photorespire. One approach for the breeder of C_3 plants is to endow them with a C_4 metabolism. If this transformation is ever to be achieved, it is most likely to be by genetic engineering. Such an advance has obvious advantages with regard, say, to increased wheat production for the ever-increasing human population. Nitrogen fixation as an agricultural application of biotechnology is given prominence in Section 1 of Volume 4. Much knowledge has been acquired about the chemistry and the biology of the fixation of atmospheric nitrogen. This provides a solid foundation from which to attempt to exploit the potential for transfer of nitrogen-fixing genes to crop plants or to the symbiotic organisms in their root systems. If plants could be provided with their own capability for nitrogen fixation, the energy equation might not be too favourable in the case of high yielding varieties. Without an increase in the efficiency of photosynthesis, any new property so harnessed would have to be at the expense of the energy requirements of existing characteristics such as yield.

Enzymes have been used for centuries in the processing of food and in the making of beverages. The increasing availability of enzymes for research, development and industrial use combined with systems for their immobilization, or for the immobilization of cells for the utilization of their enzymes, is greatly expanding the possibilities for their exploitation. Such is the power of the new biotechnology that it will be possible to produce the most suitable enzymes for the required reaction with the specific substrate. An increasing understanding at the molecular level of enzyme degradation will make it possible for custom-built enzymes to have greater stability than those isolated from natural sources.

The final section of Volume 4 deals with waste and its management. This increasingly voluminous by-product of our society can no longer be effectively dealt with by the largely empirical means that continue to be practised. Biological processes are indispensable components in the treatment of many wastes. The new biotechnology provides the opportunity for moving from empiricism to processes dependent upon the use of complex biological reactions based on the selection or the construction of the most appropriate cells or their enzymes.

The very comprehensive coverage of biotechnology provided by this four-volume work of reference reflects that biotechnology is the integration of molecular biology, microbiology, biochemistry, cell biology, chemical engineering and environmental engineering for application to manufacturing and servicing industries. Viruses, bacteria, yeasts, fungi, algae, the cells and tissues of higher plants and animals, or their enzymes, can provide the means for the improvement of existing industrial processes and can provide the starting points for new industries, for the manufacture of novel products and for improved processes for management of the environment.

Sir William Henderson, frs
Formerly of the *Agricultural Research Council*
and *Celltech Ltd., London, UK*

Guest Editorial

Since 1950, the new science of molecular biology has produced a remarkable outpouring of new ideas and powerful techniques. From this revolution has sprung a new discipline called genetic engineering, which gives us the power to alter living organisms for important purposes in medicine, agriculture and industry. The resulting biotechnologies span the range from the ancient arts of fermentation to the most esoteric use of gene splicing and monoclonal antibodies. With unprecedented speed, new scientific findings are translated into industrial processes, sometimes even before the scientific findings have been published. In earlier times there was a more or less one-way flow of new discoveries and techniques from scientific institutions to industrial organizations where they were exploited to make useful products. In the burgeoning biotechnology industry, however, developments are so rapid that there is a close intimacy between science and technology which blurs the boundaries between them. Modern industrial laboratories are staffed with sophisticated scientists and equipped with modern facilities so that they frequently produce new scientific discoveries in areas that were previously the exclusive province of universities and research institutes, and universities not infrequently develop inventions and processes of industrial value in biotechnology and other fields as well.

Even the traditional flow of new ideas from science to application is no longer so clear. In many applications, process engineers may find that the most economical and efficient process design requires an organism with new properties or an enzyme of previously unknown stability. These requirements often motivate scientists to try to find in nature, or to produce through genetic engineering or other techniques of molecular biology, novel organisms or molecules particularly suited for the requirements of production. A recent study done for the United States Congress* concluded that "in the next decade, competitive advantage in areas related to biotechnology may depend as much on developments in bioprocess engineering as on innovations in genetics, immunology, and other areas of basic science."

These volumes bring together for the first time in one unified publication the scientific and engineering principles on which the multidisciplinary field of biotechnology is based. Following accounts of the scientific principles is a large set of illustrations of the diverse applications of these principles in the practice of biotechnology. Finally, there are sections dealing with important regulatory aspects of the potential hazards of the growing field and of the need for promoting biotechnology in developing countries.

Comprehensive Biotechnology has been produced by a team of some of the world's foremost experts in various aspects of biotechnology and will be an invaluable resource for those wishing to build bridges between 'academic' and 'commercial' biotechnology, the ultimate form of any technology.

Donald A. Glaser
University of California, Berkeley
and *Cetus Corp., Emeryville, CA, USA*

*"Commercial Biotechnology: An International Analysis," Office of Technology Assessment Report, U.S. Congress, Pergamon Press, Oxford, 1984.

Executive Summary

In this work, biotechnology is interpreted in a fairly broad context: the evaluation and use of biological agents and materials in the production of goods and services for industry, trade and commerce. The underlying scientific fundamentals, engineering considerations and governmental regulations dealing with the development and applications of biotechnological processes and products for industrial, agricultural and medical uses are addressed. In short, a comprehensive but concise treatment of the principles and practice of biotechnology as it is currently viewed is presented. An outline of the main topics in the four volumes is given in Figure 1.

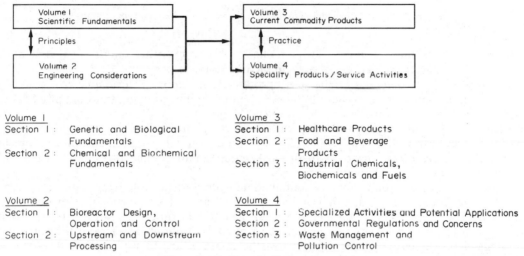

Volume I
| Section 1 : | Genetic and Biological Fundamentals |
| Section 2 : | Chemical and Biochemical Fundamentals |

Volume 2
| Section 1 : | Bioreactor Design, Operation and Control |
| Section 2 : | Upstream and Downstream Processing |

Volume 3
Section 1 :	Healthcare Products
Section 2 :	Food and Beverage Products
Section 3 :	Industrial Chemicals, Biochemicals and Fuels

Volume 4
Section 1 :	Specialized Activities and Potential Applications
Section 2 :	Governmental Regulations and Concerns
Section 3 :	Waste Management and Pollution Control

Figure 1 Outline of main topics covered

As depicted in Figure 2, it is first recognized that biotechnology is a multidisciplinary field having its roots in the biological, chemical and engineering sciences leading to a host of specialities, *e.g.* molecular genetics, microbial physiology, biochemical engineering. As shown in Figure 3, this is followed by a description of technical developments and commercial implementation,

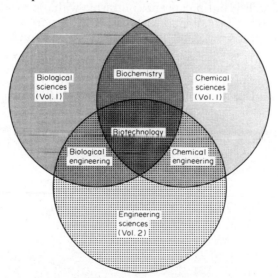

Figure 2 Multidisciplinary nature of biotechnology

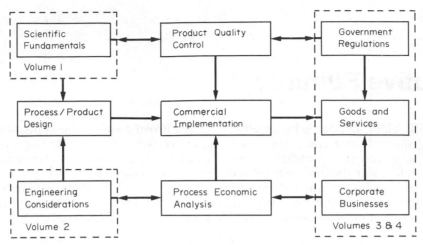

Figure 3   Interrelationships between biotechnology principles and applications

the ultimate form of any technology, which takes into account other important factors such as socio-economic and geopolitical constraints in the marketplace.

There are two main divisions of the subject matter: a pedagogical academic coverage of the disciplinary underpinnings of the field (Volumes 1 and 2) followed by a utilitarian practical view of the various commercial processes and products (Volumes 3 and 4). In the integration of these two areas, other common factors dealing with product quality, process economics and government policies are introduced at appropriate points throughout all four volumes. Since biotechnological advances are often ahead of theoretical understanding, some process descriptions are primarily based on empirical knowledge.

The four volumes are relatively self-contained according to the following criteria. Volume 1 delineates and integrates the unifying multidisciplinary principles in terms of relevant scientific fundamentals. Volume 2 delineates and integrates the unifying multidisciplinary principles of biotechnology in terms of relevant engineering fundamentals. Volume 3 describes the various biotechnological processes which are involved in the manufacture of bulk commodity products. Volume 4 describes various specialized services, potential applications of biotechnology and related government concerns. In each volume, a glossary of terms and nomenclature guideline are included to assist the beginner and the non-specialist.

This work takes into account the relative importance of the various topics, primarily in terms of current practice. Thus, bulk commodity products of the manufacturing industries (Volume 3) are accorded more space compared to less major ones and for potential applications (part of Volume 4). This proportional space distribution may be contrasted with the expectations generated by the recent news media 'biohype'. For example, virtually no treatment of 'biochips' is presented. In addition, since the vast majority of commercial ventures involve microbial cells and cell-derived enzymes, relatively little coverage is given to the possible use of whole plant or animal cells in the manufacturing industries. As future significant areas of biotechnology develop, supplementary volumes of this work are planned to cover them. In the meantime, on-going progress and trends will be covered in Pergamon's complementary review journal, *Biotechnology Advances*.

M. Moo-Young
University of Waterloo, Canada

Contributors to Volume 1

Dr C. A. Boulton
Department of Biochemistry, University of Hull, Hull HU6 7RX, UK

Dr M. L. Britz
CSIRO, Clayton, Victoria, Australia

Professor C. M. Brown
Department of Brewing & Biological Sciences, Heriot-Watt University, Chambers Street, Edinburgh EH1 1HX, UK

Professor A. T. Bull
Biological Laboratory, University of Kent at Canterbury, Canterbury, Kent CT2 7NJ, UK

Dr P. H. Calcott
NCD Fermentation, Monsanto Co., 800 N Lindberg Boulevard, St Louis, MO 63167, USA

Dr M. Chandler
Centre de Recherches de Biochemie et Génétiques Cellulaire du CNRS, 118 route de Narbonne, 31077 Toulouse Cedex, France

Dr B. W. Churchill
The Upjohn Company, Kalamazoo, MI 49001, USA

Dr G. Churchward
Département de Biologie Moleculaire, Université de Genève, 30 quai Ernest-Ansermet, CH-1211 Genève 4, Switzerland

Mr K. Corbett
Beecham Pharmaceuticals, Clarendon Road, Worthing, West Sussex BN14 8QH, UK

Dr A. Cornish-Bowden
Department of Biochemistry, University of Birmingham, PO Box 363, Birmingham B15 2TT, UK

Professor S. Dagley
Department of Biochemistry, College of Biological Sciences, University of Minnesota, St Paul, MN 55108, USA

Professor H. Dalton
Department of Biological Sciences, University of Warwick, Coventry CV4 7AL, UK

Professor A. M. Demain
Department of Nutrition & Food Sciences, Massachusetts Institute of Technology, Cambridge, MA 02139, USA

Miss A. Dietz
The Upjohn Company, Kalamazoo, MI 49001, USA

Dr L. Dijkhuizen
Department of Microbiology, University of Groningen, Kerklaan 30, 9751 NN Haren, Netherlands

Dr C. S. Dow
Department of Biological Sciences, University of Warwick, Coventry CV4 7AL, UK

Dr G. M. Dunn
Fermentech Ltd, Heriot-Watt, Riccarton, Currie EH14 4AS, Scotland, UK

Dr R. G. Forage
Biotechnology Australia Pty Ltd, PO Box 20, Roseville, New South Wales 2096, Australia

Professor M. W. Fowler
Wolfson Institute of Biotechnology, University of Sheffield, Sheffield S10 2TN, UK

Dr B. A. Haddock
Bioscot Ltd, King's Building, West Mains Road, Edinburgh EH9 3JF, UK

Dr B. G. Hall
Biological Sciences Group, College of Liberal Arts & Sciences, University of Connecticut, Storrs, CT 06268, USA

Professor W. Harder
Department of Microbiology, University of Groningen, Kerklaan 30, 9751 NN Haren, Netherlands

Dr D. E. F. Harrison
Biotechnology Australia Pty Ltd, PO Box 20, Roseville, New South Wales 2096, Australia

Professor G. Holt
School of Biotechnology, Faculty of Engineering and Science, Polytechnic of Central London, 115 New Cavendish Street, London W1M 8JS, UK

Dr I. S. Hunter
Pfizer Central Research, Sandwich, Kent CT13 9NJ, UK

Dr C. W. Jones
Department of Biochemistry, University of Leicester, University Road, Leicester LE1 7RH, UK

Professor N. F. Millis
Department of Microbiology, University of Melbourne, Royal Parade, Parkville, Victoria 3052, Australia

Professor J. G. Morris
Department of Botany & Microbiology, University College of Wales, Penglais, Aberystwyth, Dyfed SY23 3DA, UK

Professor D. P. Nagle, Jr
Department of Botany and Microbiology, University of Oklahoma, Norman, OK 73019, USA

Dr O. M. Neijssel
Laboratorium voor Microbiologie, Universiteit van Amsterdam, Nieuwe Achtergracht 127, 1018 WS Amsterdam, Netherlands

Dr S. G. Oliver
Department of Biochemistry, University of Manchester Institute of Science and Technology, PO Box 88, Manchester M60 1QD, UK

Dr M. A. Payton
Biogen SA, PO Box 1211, Geneva, Switzerland

Dr D. E. Pitt
Biotechnology Australia Pty Ltd, PO Box 20, Roseville, New South Wales 2096, Australia

Professor J. R. Postgate, FRS
AFRC Unit of Nitrogen Fixation, University of Sussex, Falmer, Brighton BN1 9RQ, UK

Dr F. G. Priest
Department of Brewing and Biological Sciences, Heriot-Watt University, Chambers Street, Edinburgh EH1 1HX, UK

Professor C. Ratledge
Department of Biochemistry, University of Hull, Hull HU6 7RX, UK

Dr G. Saunders
School of Biotechnology, Faculty of Engineering and Science, Polytechnic of Central London, 115 New Cavendish Street, London W1M 8JS, UK

Professor Dr H. G. Schlegel
Institut für Mikrobiologie der Universität Göttingen, Grisebachstrasse 8, D-3400 Göttingen, Federal Republic of Germany

Dr K. Schneider
Institut für Mikrobiologie der Universität Göttingen, Grisebachstrasse 8, D-3400 Göttingen, Federal Republic of Germany

Professor J. H. Slater
Department of Applied Biology, University of Wales Institute of Science and Technology, King Edward VII Avenue, Cardiff CF1 3NU, UK

Professor R. E. Spier
Department of Microbiology, University of Surrey, Guildford, Surrey GU2 5XH, UK

Dr D. I. Stirling
Celanese Research Corporation, 86 Morris Avenue, Summit, NJ 07901, USA

Professor A. H. Stouthamer
Biological Laboratory, Vrije Universiteit, Postbus 7161, de Boelelaan 1087, 1007 MC Amsterdam, Netherlands

Dr K. E. Suckling
Department of Biochemistry, University of Edinburgh Medical School, Edinburgh EH8 9XD, UK

Professor D. W. Tempest
Laboratorium voor Microbiolgie, Universiteit van Amsterdam, Nieuwe Achtergracht 127, 1018 WS Amsterdam, Netherlands

Professor A. P. J. Trinci
Department of Botany, University of Manchester, Manchester M13 9PL, UK

Dr H. W. van Verseveld
Biological Laboratory, Vrije Universiteit, Postbus 7161, de Boelelaan 1087, 1007 MC Amsterdam, Netherlands

Professor R. S. Wolfe
Department of Microbiology, University of Illinois, Urbana, IL 61801, USA

Contents of All Volumes

Section 2   Governmental Regulations and Concerns

Section 3   Waste Management and Pollution Control

SECTION 1

GENETIC AND BIOLOGICAL FUNDAMENTALS

1
Introduction

A. T. BULL
University of Kent at Canterbury, UK

1.1 INTRODUCTION

Successful development of biotechnology is dependent on advances in the fundamental sciences which underpin it; empiricism and insufficient attention to scientific principles can only produce less than optimal performance. Over the past few years various attempts have been made to define biotechnology but I shall retain that proposed by my colleagues and me for the OECD: 'the application of scientific and engineering principles to the processing of materials by biological agents to provide goods and services' (Bull *et al.*, 1982). These scientific and engineering principles embrace a broad spectrum of disciplines but foremost amongst the sciences are microbiology, biochemistry and genetics, the chief elements of which make up the subject of this first volume of *Comprehensive Biotechnology*. Research achievements have been such that 'over the last 30 years or so, our knowledge of microorganisms has increased enormously and with it a growing appreciation of their potentials and how best they may be realized. We are, therefore, at a point in the history of microbiology where it is possible to indicate, with some degree of accuracy, whether or not a given product could be produced microbiologically and, moreover, what type of organism would have to be used; where it might be found if it is not yet known; and under what conditions it must be grown to express its desirable properties. We can also say how the productivity of a microorganism might be improved: by genetic manipulation; by judicious choice of culture conditions; by addition, or subtraction, of a particular nutrient. This rational approach to (biotechnology) is now with us and is ripe for exploitation.' (Bull *et al.*, 1979). Microbiology is a fundamental component of biotechnology and in Section 1 are highlighted some features which have a major influence on its development. The stuff of biotechnology is the exploitation of biocatalysts. When developing a process or product it is necessary to select the most advantageous biocatalyst, optimize its structure, its properties and the environment within which it will be required to operate.

1.1.1 Biocatalysts — Organisms

Biotechnology is much concerned by the selection of appropriate biocatalysts. The focus in Section 1 is upon microorganisms and animal and plant cells where selection for desirable attributes constitutes a very critical feature of exploratory research. It is very noticeable that, compared with the large number of described species, those that have been developed as biocatalysts are remarkably few. Microbiologists and geneticists have concentrated on a handful of organisms while traditional use and know-how on the one hand and experimental opportunism on the other have only encouraged this narrow focus. The upsurge of the 'new biotechnology' has significantly changed the situation because of the need to select a wide range of microbial and cultured cell types for diverse activities such as pest control, mineral processing, health care and food production.

The search for and the discovery of novelty is a recurring goal of biotechnology and at best two, complementary, approaches are commonly made. One, the selective isolation of microorganisms, is based on an improved knowledge of microorganisms in their natural environments and has benefitted immeasurably from the emergence of microbial ecology as a defined and well-subscribed sector of microbiology. The other approach focuses on the products of microorganisms and cells and their anticipated utility in proscribed target areas, in other words the familiar screening strategy. With regard to selective isolation, continuous-flow enrichment cultivation provides a powerful means of obtaining microorganisms possessing desired properties in a rational and reproducible manner, while innovative work on microbial enzyme inhibitors has demonstrated the value of new thinking in devising novel screening procedures. The range of novel microbial products emanating from the enzyme inhibitor screen is impressive and includes new antibiotics and anticancer agents and candidate compounds for the treatment of, for example, thromboses, obesity and hypertension. Comprehensive accounts of search and discovery strategies appear in recent proceedings of the Society for General Microbiology (Bu'Lock *et al.*, 1982; Nisbet and Winstanley, 1983).

It is sometimes forgotten or under-emphasized that organism selection for a biotechnological purpose depends on a range of criteria that are relevant to the optimization of the product or the process. Among such criteria are found the following, and ultimate selection will often be made on the basis of best compromise: (1) nutritional characteristics, (2) temperature characteristics, (3) type of process, (4) reaction of organism to equipment, (5) organism stability (phenotypic and genotypic), (6) amenability to genetic manipulation, (7) product yield, (8) productivity, and (9) product recovery (Bull *et al.*, 1979).

1.1.2 Physiology — Optimization

Microbial physiology might be defined as the study of relationships between the metabolic capabilities and the environment within which the organism exists in either a growing or a non-growing state. Microorganisms are extremely adaptable and they are unrivalled in their capacity to respond to major fluctuations in the environment. This adaptability is founded on a highly developed physiology that enables microorganisms to change structurally and functionally in reponse to environmental change and, thereby, to maintain either totally or partially their growth potential (Bull and Brown, 1979; Tempest and Neijssel, 1978). Such phenotypic variability is both exploitable and a potential source of problems in biotechnology. Clearly careful manipulation of environmental conditions (process optimization) produces fermentations having precisely defined properties, but during process scale-up, if the environmental conditions are not exactly reproduced, phenotypic variability may result in the loss or diminution of the desired property.

Continuous-flow cultures and fermentations possess all the advantages that continuous industrial operations have over batch equivalents, namely ease of control and uniformity, reduced reactor volumes and operational down-time, and, most significantly, increased productivity. Moreover the chemostat type of continuous-flow culture is the ideal system for exploring the phenotypic variability of microorganisms and hence for making optimization studies. Of course, chemostat culture exacerbates the problem of organism instability because of the strong and continuous selective pressure that is imposed on the population and this is a problem that is usually heightened in genetically engineered strains.

The physiology of slowly growing or non-growing microorganisms has been a relatively neglected area but understanding becomes essential to the design and operation of the new generations of bioreactors based on immobilized cells. Thus, two high priority research targets must be strain stability and metabolic stability under non-growth circumstances.

1.1.3 Genetics

The impact of modern genetics on biotechnological innovation has been singular and the two extensive chapters on the subject in this section expose the technology and reveal the impact. The most spectacular advances have led to the breaching of the species barrier by *in vitro* recombinant

DNA procedures and by cell fusion, and to the tailoring of protein properties *via* the rapidly evolving technology of protein engineering. In the past, genetic manipulation has centred predominantly on product yield improvement; the 'new genetics' opens the way to novel products and processes to an extent that was inconceivable a decade or so ago.

Just as industrial microbial physiologists and biochemists have worked with a relatively small portfolio of organisms, detailed genetic understanding is restricted to a few genera or even species. For example, the development of suitable host–vector systems is urgently required for many industrial microorganisms and, although progress is being made on several fronts, the frequent lack of fundamental work on the genetics of such organisms inevitably hampers the speed with which the new technology can be integrated into biotechnological operations.

1.1.4   Integration

While biotechnological innovation depends on advances in the underpinning scientific and engineering disciplines, successful commercialization is totally reliant on an effective and interactive relationship between these disciplines. It may be thought that such truisms need no further iteration. However, the experiences of many companies founded on the technical wizardry of the 'new biotechnology' suggest that this is not always the case and that the lesson is learnt only after a certain amount of pain. Since this first volume of *Comprehensive Biotechnology* is concerned with scientific fundamentals, it is worth repeating, in conclusion, that the exploitation of basic science is not simply a one way process. Biotechnology has and will continue to act as a major stimulus for fundamental studies on microorganisms, animals and plants and probably lead research into areas that otherwise would be neglected or even unperceived.

1.2   REFERENCES

Bull, A. T. and C. M. Brown (1979). Continuous culture applications to microbial biochemistry. In *International Review of Biochemistry*, ed. J. R. Quayle, vol. 21, pp. 177–226. University Park Press, Baltimore, MD.

Bull, A. T., D. C. Ellwood and C. Ratledge (1979). The changing scene in microbial technology. In *Microbial Technology*, ed. A. T. Bull, D. C. Ellwood and C. Ratledge, pp. 1–27. Cambridge University Press, Cambridge.

Bull, A. T., G. Holt and M. D. Lilly (1982). *Biotechnology. International Trends and Perspectives*. Organisation for Economic Co-operation and Development, Paris.

Bu'Lock, J. D., L. J. Nisbet and D. J. Winstanley (eds.) (1982). *Bioactive Microbial Products: Search and Discovery*. Academic, London.

Nisbet, L. J. and D. J. Winstanley (eds.) (1983). *Bioactive Microbial Products 2: Development and Production*. Academic, London.

Tempest, D. W. and O. M. Neijssel (1978). Eco-physiological aspects of microbial growth in aerobic, nutrient-limited environments. *Adv. Microb. Ecol.*, **2**, 105–153.

2

The Organisms of Biotechnology

N. F. MILLIS
University of Melbourne, Australia

2.1 CELLS — THE BASIC UNITS

Biotechnology is based on the catalytic activities of cells, either intact cells or extracts from them. This chapter will present an overview of the cellular catalysts. All cells have certain characteristics in common. They have a semipermeable osmotic barrier (cytoplasmic membrane), their genetic information is coded in double-stranded deoxyribonucleic acid (DNA), and they contain ribonucleic acids (RNA) which are concerned with the transcription and translation of the genetic information coded in the DNA into the proteins which are characteristic for each type of cell. The viruses will be considered in Section 2.3 although they are not cells. All cells have properties in common, but the forms in which cells occur in nature, and their other properties, are very diverse.

2.2 TYPES OF MICROORGANISM

Individual cells are rarely greater than 1 mm in diameter and many are only 1 nm in diameter. Unicellular organisms (Protista) are thus often referred to as microorganisms, and while this term is in common use, it is not a true taxonomic grouping.

Microorganisms include the smallest entities, the viruses, which can be seen only with an electron microscope, as well as those seen by eye and with the light microscope. The protists include the single-celled bacteria, protozoa, algae and fungi, and some multicellular species of algae and fungi which are large and conspicuous (seaweeds and mushrooms).

Despite their small size and superficial similarity, the protists embrace species which are fundamentally different at the intracellular level. From studies of the form of their nuclear apparatus and the complexity of their intracellular organelles, the protists are divided into two large classes, the prokaryotes and the eukaryotes. The bacteria (including the cyanobacteria or blue-green algae) have a simpler internal organization and constitute the prokaryotes; the protozoa, algae and fungi, which have a more complex organization, are placed with the multicellular, highly differentiated plants and animals in the eukaryotes. Some of the major differences among these cells are summarized in Table 1.

Table 1 Differentiating Properties of the Protists

Characteristic	Prokaryotes	Eukaryotes
Genome		
No. of DNA molecules	One[a]	More than one
DNA in organelles	No	Yes
DNA observed as chromosomes	No	Yes
Nuclear membrane	No	Yes
Mitotic and meiotic division of the nucleus	No	Yes
Formation of partial diploid	Yes	No
Organelles		
Mitochondria	No	Yes
Endoplasmic reticulum	No	Yes
Golgi apparatus	No	Yes
Photosynthetic apparatus	Chlorosomes	Chloroplasts
Flagella	Single protein, simple structure	Complex structure, with microtubules
Spores	Endospores[b]	Endo- and exo-spores
Heat resistance	High	Low

[a] Bacteria may contain extra, small DNA molecules (plasmids) in addition to the genome proper.
[b] A few groups produce spores by budding or make chlamydospores and these have low heat resistance.

By contrast, the viruses have no cytoplasm or cytoplasmic membrane; they are obligate parasites whose replication inside a specific host is achieved using the biosynthetic system of the host under the direction of the viral nucleic acid. The genome of an infective virus particle (virion) may be DNA or RNA, never both; but the nucleic acid may be either single- or double-stranded. The genome is enclosed within a protein coat (capsid) made of identical subunits, and in some viruses this is covered by an outer envelope of lipoprotein derived from the host cell membrane. Viruses with no capsid or envelope (viroids) also occur. Some large virions contain several hundred proteins, many of which are associated with viral synthesis, but the small virions contain very few proteins and their nucleic acid codes for only two or three proteins. Such virions are about 20 nm in diameter, whereas the largest virions are about 300 nm across. Many virions appear as geometrically symmetrical solids; others appear as filaments or helices. The viruses infecting bacteria (bacteriophages or phages) include some with more complex morphology, for example part of the virion is concerned with recognition of the host surface and part with injecting DNA into the host. An excellent introduction to virology is offered by Primrose and Dimmock (1980).

2.3 VIRUSES

The viruses are classified largely on morphology. The properties used include: the type of nucleic acid, the number of protein subunits (capsomers) in the capsid, the type of symmetry or form of arrangement of the capsomers, the presence or absence of an envelope around the capsid, the host cell infected, and the disease characteristics produced (Matthews, 1982).

Viral pathogens of man and animals are of little significance to biotechnology except for the production of vaccines for which the virus is grown in tissue culture (see Section 2.5.4.1). A comprehensive treatment of viruses pathogenic for animals is given in White and Fenner (1985).

Biological control of insect pests by viruses is an attractive prospect because viruses can be highly specific and are non-polluting. However, control by this method has been achieved for only a few pests (Burges and Hussey, 1971; Tinsley, 1978). Generally the virus must be grown in whole insects, which restricts the volume of production. The challenge is to develop appropriate insect tissue culture lines so that processes can be scaled up. It may then be possible to expand the range of pests controlled by viral pathogens.

Bacteriophages have been found in almost every bacterial species examined (Adams, 1959), and a single bacterial species may be attacked by several types of bacteriophage. For example, *Escherichia coli* is most commonly attacked by phages containing double-stranded DNA, but phages containing single-stranded DNA or single-stranded RNA also occur. In electron microscope studies some phages appear tadpole-shaped with a polyhedral 'head' enclosing the nucleic acid and a 'tail' of helical protein. Other phages have a polyhedral head and no tail, and one group of single-stranded DNA phages is filamentous. The phage tail is associated with phage attachment to the specific receptor on the bacterial cell. The receptor may be in the lipoprotein outer layer, or in the lipopolysaccharide layer of the cell, or on the flagella or pili (appendages to the cell). Some phage tails have a base plate with fine fibres attached and these recognize the host receptors. The tail of the T-even phages of *E. coli* is contractile. After adsorption the core of the tail penetrates to the cell membrane, then the outer protein of the tail contracts and the DNA is injected into the cell, leaving the head and tail outside. The mechanism of penetration of the nucleic acid of other phages is unknown. In a lytic cycle the nucleic acid of the phage may be immediately translated into messenger RNA (m-RNA) coding for viral products which are synthesized by the host ribosomes. Depending on the phage, the DNA of the host may be hydrolyzed by deoxyribonuclease coded by the virus, or it may remain intact to provide for the synthesis of viral proteins. The synthesis of the capsid subunits begins after the synthesis of the viral nucleic acid is complete. The nucleic acid condenses into a polyhedron around which the subunits are packed. In phages with tails, the tail components are made and assembled, then the completed tails and heads are joined into infective virions. Virally-coded lysozyme is then produced which attacks the host cell wall to release mature phages (from 10 to over 100 per bacterium). The number of phage particles can be measured by spreading appropriate dilutions of phage and susceptible host on solid media. On average, each phage gives rise to a focus of lysis seen as a clearing or plaque in the lawn of cell growth. During phage replication it is possible for some bacterial DNA to be packaged into the head of the phage. These phages (called transducing phages) cannot replicate, but are infective, and thus provide a means of transferring DNA from one host to another.

Another association between bacteriophage and host may occur. This is called lysogeny. Here, viral DNA is incorporated into the bacterial genome and is replicated stably with it. Viral DNA codes for a repressor protein which prevents the expression of the genes associated with viral replication. At low frequency, however, viral DNA is excised from the host genome and a cycle of lytic replication begins.

Bacteriophages are thus important in two aspects of biotechnology. They can lyse the process organism causing serious economic loss (cheese starters and other bacterial fermentations), but they are extremely useful in genetic studies, both for mapping the position of genes and for the construction of novel strains either by transduction or recombination. In recombination, phages are used as vectors to introduce foreign DNA into cells.

2.4 PROKARYOTES

This section will consider briefly the general properties of the prokaryotes. It will make no attempt to consider the distinguishing properties of the major taxonomic groups; these are set out by Buchanan and Gibbons (1974). A two-volume handbook by Starr *et al.* (1981) should be consulted for an excellent comprehensive discussion of the habitats, isolation and identification of the diverse groups of which the prokaryotes are composed.

As was indicated in Section 2.2, the prokaryotes are the smallest cells, but within the group the sizes of particular species range from 0.01×0.15–0.3 nm to as large as 25 nm; filaments up to 1.5 cm (*e.g.* streptomycetes, cyanobacteria) may also occur. The cells of many bacteria are regular cylinders, spheres or spirals, but various distortions from these basic forms occur. They generally divide by simple fission (see Trinci and Oliver, Volume 1, Chapter 10). The shape is conferred by the rigid outer layer (cell envelope) which overlies the cell protoplast. The envelope of some bacteria consists largely of peptidoglycan (murein) (50–80%), along with polysaccharides and teichoic acids in a single layer 20–80 nm thick. In other bacteria the murein layer is thin (2–3 nm) and is covered by multilayered unit membranes of lipopolysaccharide and protein 8–10 nm thick. Cells rich in murein stain Gram-positive, whereas cells poor in murein stain Gram-negative. This stain is a most useful distinguishing character in taxonomy and in practical microbiology, as many other properties are associated with this difference.

Two groups of bacteria contain no murein: the mycoplasmas, or L-forms, which are obligately intracellular, and the archaebacteria (methane producers, some halobacteria and acidothermophiles) which differ from other bacteria in other properties also (types of RNA and coenzymes, and components of the cell membrane).

The cell protoplast, which lies below the cell envelope, is bounded by a semipermeable cytoplasmic membrane (10 nm) enclosing the cell contents. This is a double-layered structure of phospholipid and protein. The region between the cell envelope and the membrane is called the periplasmic 'space', but in fact it is a site of intense enzymic activity.

Exterior to the cell envelope various projections (pili and fimbriae), surface slime layers and holdfasts may be formed. The two latter types generally contain polysaccharide and assist the non-specific attachment of the cell to surfaces, a very important property for survival in environments low in nutrients. In the case of pathogens of animals and plants the cells often colonize particular tissues preferentially (diphtheria bacilli in the throat, gonococci in the urethra, rhizobia in the roots of particular legumes) and this specificity is thought to reside in the structure of the surface of the bacteria. In the case of *E. coli* K12, strains containing the sex factor (F) produce sex pili which react specifically with the surface of F^- strains and allow the formation of mating pairs and the subsequent unilateral transfer of DNA from F^+ to F^- cells.

Adsorption of cells to surfaces is an important property in process microbiology. It is used to advantage in the formation of stable flocs between organisms (as occurs in the activated sludge process), for the rapid production of methane in a process where the sewage flows up through flocculated sludge, and in beer production with bottom fermenting yeasts. Slime formation may be a troublesome feature if the process organism forms thick films on reactor surfaces, or reduces the rate of mass transfer, *e.g.* of oxygen, in the fermentation broth. On the other hand valuable processes are based on the accumulation of bacterial polysaccharides. The earliest was the production of dextran as a plasma substitute by *Leuconostoc*. More recently the plant pathogen *Xanthomonas campestris* has been exploited for the production of xanthan (a polymer of glucose, glucuronic acid and mannose) which is used in the food industry. Alginate made by *Azotobacter* is also used as a food gel.

Whereas morphological differences in the prokaryotes are not great, their physiological diversity is spectacular, and they colonize an extremely wide range of habitats. Bacteria are found in nature at pH values from 1 to 11, and at temperatures from 0 to 92 °C at atmospheric pressure and of over 200 °C at 10 000 m in ocean deeps. They occur in both aerobic and anaerobic environments and in situations where the water activity ranges from 1.0 to as low as 0.75, although most bacteria do not tolerate water activities below 0.91.

The substrates they can use for growth are similarly diverse. Their growth capabilities range from autotrophism to obligate, intracellular parasitism. Among the autotrophs, the phototrophic bacteria obtain carbon for growth from carbon dioxide using the energy from light, and use either reduced inorganic sulfur compounds or organic carbon as the source of reducing power under anaerobic conditions. All other growth requirements are obtained from inorganic salts. Photosynthesis in the cyanobacteria is aerobic and water provides the reducing power as it does in eukaryotes. The chemoautotrophs fix carbon dioxide using the energy captured from the oxidation of specific inorganic molecules. Among the heterotrophs, some require only inorganic salts and a single source of organic carbon for growth, whereas others have complex growth requirements and need, in addition, a full complement of amino acids, vitamins, growth factors, purines and pyrimidines. In the extreme, a few heterotrophs are obligate intracellular parasites and can only be grown in whole animals or in tissue culture. Carbon is supplied for most industrial heterotrophic organisms as a crude carbohydrate, with nitrogen supplied as ammonium ion and protein-rich by-products. For a full discussion of microbial nutrition see Dunn, Volume 1, Chapter 7.

The organic molecules which can be used by heterotrophs extend from simple molecules like methane, hexoses and amino acids, to high molecular weight polymers such as starch, cellulose, lipids and proteins; straight, branched, aromatic, cyclic and polycyclic organic compounds can also be assimilated. Some of these compounds are attacked only slowly by quite restricted groups of organisms. To treat industrial wastewater containing recalcitrant molecules, many of which are toxic, attention must be given to developing special consortia of organisms able to oxidize these compounds (see Bull, Volume 1, Chapter 15). However, there are limits to the microbial appetite. Synthetic chemicals, such as highly chlorinated or brominated ring compounds, and highly polymerized plastics, such as PVC, nylon and polystyrene, cannot be attacked.

For centuries cottage fermentations depended upon mixed cultures dominated by restricted groups of bacteria. Modifications of these processes remain an important section of the fermentation industry today. Lactic cocci and lactobacilli are used for making cheese, fish and vegetable sauces, and for pickling vegetables, fish, meat and sausages; acetobacters oxidize ethanol for vinegar; a consortium of anaerobes converts organic complexes to methane; and thiobacilli solubilize metals from sulfide ores. Processes developed more recently use pure cultures of bacteria to produce lactic, oxogluconic and various amino acids, antibiotics, biomass, purine nucleotides, solvents, polysaccharides, enzymes, vitamin B_{12} and polyols, and to carry out important molecular conversions, such as the bioconversions of steroids and antibiotics.

Of the highly-branched filamentous prokaryotes, one genus (*Streptomyces*) has over 1000 species. Most are saprophytes in soils and have simple growth requirements. On solid media, their filaments (hyphae) grow as a ramified mat (mycelium) with aerial hyphae bearing spores (conidia) which are resistant to drying but not to heat. The hyphae are about 1–4 nm in diameter with walls like those of Gram-positive bacteria. Cross walls are not common. The hyphae are multinucleate and the nuclear material lies free in the cytoplasm; plasmids have also been demonstrated. The morphology, the colour and arrangement of the spores, the colour of the aerial mycelium, the gross morphology of the colonies, and physiological properties have been used in taxonomy, but a generally accepted classification has yet to emerge. *Streptomyces* provide many therapeutically useful antibiotics (*e.g.* streptomycin, chloramphenicol, erythromycin, kanamycin, lincomycin, nystatin, neomycin, rifamycin and the tetracyclines) and some species are used in bioconversions of steroids and antibiotics. Species of a related genus, *Micromonospora*, produce gentamicin.

The fermentation products derived from prokaryotes fall into a number of categories: single cell protein or biomass, end-products (*e.g.* solvents and acids), primary metabolites (*e.g.* amino acids, enzymes and nucleotides), and secondary metabolites (*e.g.* antibiotics, pigments and polysaccharides). Table 2 lists examples of organisms grown for biomass, and Table 3 shows some of the prokaryotes which are used to produce end-products or metabolites.

Table 2   Production of Single Cell Protein (SCP) or Biomass[a]

Substrate	Organism
Bacteria	
Hexose	*Rhizobium*
Methane	*Methylococcus capsulatus*, *Pseudomonas extorquens*
Methanol	*M. capsulatus*, *Methylophilus methylotrophus*, *Pseudomonas* AM1, *Methylosinus*
Soya bean	*Bacillus natto*
Carbon dioxide	*Spirulina maxima*
Fungi	
Alkanes	*Candida tropicalis*, *Candida lipolytica*
Carob bean	*Aspergillis niger* M1
Cellulose	*Agaricus bisporus*, *Volvariella volvacea*, *Lentinus edodes*
Coffee wastes	*Aspergillus oryzae*
Glucose	*Saccharomyces cerevisiae*, *Candida utilis*
Methanol	*Hansenula*, *Candida*, *Kloeckera*
Peanut	*Neurospora* (ontjom)
Pentose	*C. utilis*
Rice	*Rhizopus* (tempeh)
Starch	*Paecilomyces varioti* (Pekilo), *Endomycopsis fibuliger* + *C. utilis* (Symba), *Fusarium graminearum* (Rank–Hovis)
Whey	*Torulopsis cremoris*, *Kluveromyces fragilis*
Algae	
Carbon dioxide	*Chlorella*, *Scenedesmus*, *Dunaliella*

[a] Collated from Rose, 1979b; Tannenbaum and Wang, 1975.

In developing strains for use in industrial processes, fundamental knowledge of the genetics of *E. coli* was applied to the breeding of production strains in which the normal controls on metabolism were circumvented. This allowed spectacular improvements in the yield of amino acids, antibiotics and enzymes. When biomass was required, selection was directed towards high yields, rapid growth rates and appropriate chemical composition. At present, considerable effort is being directed to developing novel processes using strains constructed by recombinant DNA methodology (see Chandler and Churchward, Volume 1, Chapter 6). By this technique, foreign DNA is introduced into a bacterial or tissue culture cell. The host cell may then produce the protein coded by the introduced DNA. Human and animal hormones and immunogenic proteins of viruses are examples of potentially useful products which are now being investigated.

In all industrial processes the physical form of the cells is important and this can be influenced by the inoculum (for example the number of growing points in the inoculum of a *Streptomyces* fermentation). The composition of the medium can influence the metabolism of the cells directly through nutrition, or indirectly by altering the form of growth and so influencing factors like the efficiency of aeration. This latter is particularly important when growing filamentous organisms. Details of the ways in which these objectives are achieved appear in later chapters of this treatise.

Table 3   Products Made by Prokaryotes[a]

Product	Organism	Product	Organism
Acids (organic)		Glucose isomerase	Streptomyces, B. coagulans,
Acetic	Acetobacter aceti		Actinoplanes missouriensis
Gluconic	Pseudomonas ovalis	Lactase	E. coli
Lactic	Lactobacillus delbrueckii	Penicillin acylase	E. coli, B. megaterium
2-Oxogluconic	Pseudomonas fluorescens	Penicillinase	B. subtilis, B. licheniformis
5-Oxogluconic	Gluconobacter (syn.	Protease	B. licheniformis
	Acetobacter, Acetomonas)	Protease (alkaline)	B. subtilis, B. licheniformis
	suboxydans	Protease (neutral)	B. subtilis, B. amylo-
Acids (amino)			liquifaciens
Glutamic		Pullulanase	Enterobacter (Klebsiella)
Isoleucine			aerogenes
Leucine	Corynebacterium glutamicum	Methane	Methanobacterium,
Lysine	(syn. Brevibacterium flavum)		Methanococcus,
Serine			Methanogenium,
Threonine			Methanosarcina,
Valine			Methanospirillum
Antibiotics from Bacillus		Mineral recovery	Thiobacillus, Ferrobacillus
Bacitracin	B. licheniformis	Nucleotides	
Colistin	B. polymyxa	Guanylic acid	Bacillus subtilis,
Gramicidin S	B. brevis	Inosinic acid	Brevibacterium ammonia-
Polymyxin	B. polymyxa	Xanthilic acid	genes
Tyrocidine	B. brevis	Polysaccharides	
Antibiotics from		Curdlan	Alcaligenes faecalis var.
Streptomyces			myxogenes
Amphotericin B	S. nodosus	Dextran	Leuconostoc mesenteroides
Avermectins	S. avermitilis	Xanthan	Xanthomonas campestris
Blasticidins	S. griseochromogenes	Zanflo	Erwinia tahitica
Candidin	S. viridoflavus	Solvents/fuels	
Chloramphenicol	S. venezuelae	Acetone	Clostridium acetobutylicum
Cyclohexamide	S. griseus	Butanol	Clostridium acetobutylicum
Erythromycins	S. erythreus	Ethanol	Zymomonas mobilis
Kanamycin	S. kanamyceticus	Steroid conversions	
Lincomycin	S. lincolnensis	Dehydrogenation at	
Monensin	S. cinnamonensis	1,2	Arthrobacter (Coryne-
Neomycins	S. fradiae		bacterium) simplex, Bacillus
Novobiocin	S. spheroides		sphaericus
Nystatin	S. noursei	Hydroxylation at	
Oleandomycin	S. antibioticus	11,16	Streptomyces roseo-
Spectinomycin	S. spectabilis		chromogenes
Streptomycin	S. griseus	Side-chain cleavage	Mycobacterium, Nocardia
Tetracyclines	S. aureofaciens, S. rimosus	Vaccines	Bacillus anthracis, strain S
Tobramycin	S. hachijoensis		Bacteroides nodosus
Tylosin	S. fradiae		Bordetella pertussis
Vancomycin	S. orientalis		Brucella abortus, strain 19
Antibiotics from other			Clostridium spp.
prokaryotes			(C. botulinum, C. chauvoei,
Gentamicin	Micromonospora purpurea		C. oedematiens,
Rifamycins	Nocardia mediterranei		C. perfringens, C. septique,
Enzymes			C. tetani)
Amylase	B. subtilis, B. licheniformis		Corynebacterium diphtheriae
Asparaginase	Escherichia coli, Erwinia		Mycobacterium tuberculosis,
	carotovora		strain BCG
Catalase	Micrococcus lysodeikticus		Salmonella spp. (S. typhi,
Fumarase	Brevibacterium ammonia-		S. paratyphi A, B)
	genes		Vibrio cholerae
β-Glucanase	B. subtilis, B. circulans,		Yersinia pestis
	Arthrobacter		

[a] Collated from Berkeley *et al.*, 1979; Perlman, 1979; Rose, 1978, 1979a; Taylor and Richardson, 1979.

2.5   EUKARYOTES

Table 1 lists some of the important properties of the eukaryotic microorganisms. Whereas these properties are common to many of the eukaryotes, in other characteristics they are a very diverse group. There are three major subdivisions of the eukaryotic microorganisms: algae, protozoa and fungi. The algae and protozoa are of relatively little significance to biotechnology, but the fungi are used to make a number of valuable fermentation products.

2.5.1   Algae

Many algae are unicellular, but multicellular plant-like forms occur, particularly in marine waters. All are photosynthetic with a variety of photosynthetic pigments contained within true laminated chloroplasts. The major properties which divide the algae into six groups are shown in Table 4. Details of the classification and properties of the algae can be obtained from Bold and Wayne (1978) and Rosowski and Parker (1982).

Algae are not generally cultivated industrially in closed reactors. Some single cell protein is produced from treated wastewater using green unicellular algae, such as *Chlorella* and *Scenedesmus*, grown in shallow lagoons or ditches. *Dunaliella*, which is rich in β-carotene, has been grown on brackish wastewater.

Extracts of marine algae provide useful gelling agents; agar (a sulfate-containing galactan) is derived from red algae (*Gelidium*, *Gracilaria*, *Eucheuma*) and alginic acid (β-1,4 glycoside of D-mannuronic acid) from brown algae (*Laminaria*, *Macrocystis*, *Ascophyllum*). These products, are, however, obtained by harvesting naturally-occurring seaweeds. The seaweeds traditionally eaten in the Far East like the Japanese nori (*Porphyra*) and konbu (*Laminaria*) are also gathered from the ocean rather than cultivated.

2.5.2   Protozoa

The protozoa are essentially single-celled eukaryotes, typically motile and non-photosynthetic, ranging in size from 1 nm to 50 mm, but with most between 5 and 250 nm. They are generally uninucleate; sexual and asexual reproduction occur, depending on the group. Some are colonial and many form stalks at some stage of growth (Curds and Ogden, 1978). The earlier classifications provided for four main groups: Mastigophora (flagellates), Rhizopoda (amoeboid forms), Sporozoa (sporozoite parasites) and Ciliata (ciliates). Levine *et al.* (1980), however, have revised this classification and proposed seven major groups.

Important diseases (malaria, amoebic dysentery, sleeping sickness) are caused by protozoa, but most species are saprophytes in soil and water and are of little significance to biotechnology. However, in the treatment of wastewater, the protozoa (particularly the attached ciliates and amoeboid forms) associated with the solid surfaces in trickling filter beds and the flocs in activated sludge plants are important in predating the bacteria in the water. They lower the BOD, COD, suspended solids, numbers of bacteria and organic nitrogen in the final effluent (Curds, 1982). They are essential to obtaining effluents of high quality.

2.5.3   Fungi

The fungi are heterotrophs whose usual habitat is the soil, but many species cause serious diseases of plants and a few are pathogens of animals. They are an important source of solvents, organic acids, antibiotics, single cell protein and enzymes. They may cause serious economic loss through spoilage of foodstuffs, paper, wood, fabrics and many manufactured goods.

Like the streptomycetes, the fungi have branched, filamentous rigid hyphae which ramify to form a mycelium, but the diameters of fungal hyphae are generally larger (2 – 18 nm) and the composition of the hyphal wall is entirely different. The walls contain chitin, cellulose, glucan or mixtures of these occurring in microfibrils; yeast cell walls are composed of glucans and mannans. Cross walls are prominent in the hyphae of three of the major fungal groups (ascomycetes, basidiomycetes and the fungi imperfecti), but there are no cross walls in two groups (oomycetes and zygomycetes). The hyphae of all groups are multinucleate with the genetic material contained in a nuclear membrane. The cells contain mitochondria and an endoplasmic reticulum. Spores are produced by sexual and asexual processes, depending on the species, and some are asporogenous. Asexual spores may be formed in a number of ways: by fragmentation of the hyphae (arthrospores), by budding from special conidiophores (conidia), by formation within a single sac, or sporangium, at the end of a hypha (sporangiospores), or within sporangia at the ends of branched hyphae (zoospores). Sexual spores may also be formed in different ways. Ascospores (usually

Table 4   The Properties of the Six Major Divisions of the Algae[a]

Division	Common name	Habitat	General structural arrangement	Pigments contained	Selected reserve materials	Motility	Method of reproduction
Chlorophycophyta	Green algae	Fresh water and moist environments	Unicellular to multicellular	Chlorophylls *a* and *b*, carotenes, xanthophylls	Starch, oils	Mostly non-motile	Asexual by multiple fission; spores sexual
Chrysophycophyta	Golden algae (includes diatoms)	Fresh and salt water	Mainly unicellular	Chlorophylls *a* and *c*, special carotenoids, xanthophylls	Oils, leucosin, chrysolaminarin, oils	Unique movement with diatoms; others utilize flagella	Asexual and sexual
Euglenophycophyta	Euglenoids[b]	Fresh water	Unicellular	Chlorophylls *a* and *b*, carotenes, xanthophylls	Fats, paramylum	Motile by means of flagella	Asexual only by binary fission
Phaeophycophyta	Brown algae	Salt water (cool environment)	Multicellular	Chlorophylls *a* and *c*, special carotenoids xanthophylls	Fats, laminarin	Motile	Asexual by motile zoospores; sexual by motile sex cells (gametes)
Pyrrophycophyta	Dinoflagellates	Fresh and salt water	Unicellular	Chlorophylls *a* and *c*, carotenes, xanthophylls	Starch, oils	Motile	Asexual; sexual rare
Rhodophycophyta	Red algae	Salt water (warm environment)	Multicellular	Chlorophyll *a*, phycobillins, carotenes, phycoerythrin, xanthophylls	Starch, oils	Non-motile	Asexual by spores; sexual by sex cells (gametes)

[a] Reproduced with permission from Wistreich and Lechtman, 1980.
[b] These microorganisms possess characteristics of both animals and plants. Euglenoids seem intermediate between algae and protozoa.
[c] One genus, *Gonyaulax*, occurs in algal blooms referred to as the 'red tide'.

eight) are formed inside a sac or ascus, basidiospores (usually four) are borne on a club-shaped basidium, oospores (usually one to twenty per oogonium) are formed by fertilization of a special egg cell, and zygospores result from the fusion of pairs of cells to form thick-walled single spores.

Fungal classification (>50 000 species) is based largely on the morphology of the mycelium and the spores, particularly the sexual spores; unfortunately, the sexual stage is not always readily observed. Alexopoulos and Mims (1979) have published an introductory text and Booth (1978) has summarized the major distinguishing characteristics of the principal groups of fungi, following the classification of Ainsworth (1973). Onions *et al.* (1981) have produced simplified keys to the species of economic importance and they also list references to monographs on particular groups. Species useful in biotechnology appear in all major groups of fungi, but no discussion of their taxonomy can be attempted here.

Fungi are, generally speaking, strict aerobes. Hence processes based on them must be adequately aerated. This was achieved traditionally in Asia by growing the organisms on solid substrates (rice, soya beans, wheat) to make foods such as natto, soya sauce, miso and tempeh and to provide amylases (as koji) for use in other fermentations. Edible fungi like *Agaricus*, *Volvariella* and *Pleurotus* were grown on the surface of wood or in shallow composts. In the West, the curd of certain cheeses was ripened with *Penicillium*.

In early industrial processes based on fungi, citric and gluconic acids were made by growing moulds (*Penicillium* or *Aspergillus*) on the surface of still liquid media. The discovery of penicillin led to the development of submerged culture techniques which are now standard practice in producing antibiotics with fungi or streptomycetes. The characteristics of the mycelium (physical and chemical) are critical to obtaining high yields of antibiotics; these can be controlled by the type of inoculum, the composition of the medium, the management of nutrient additions and the design of the vessel. The best conditions will differ for each case, but for many it is desirable to achieve a highly-branched mycelium and to maintain a low concentration of available carbohydrate in the medium.

Yeasts (*e.g. Saccharomyces cerevisiae*) are true ascomycetes but are unusual fungi in a number of ways. They grow as single cells, multiply by budding and can metabolize aerobically and anaerobically. Alcoholic beverages and power alcohol are based on anaerobic fermentation, whereas aerobic growth is used to produce Bakers' yeast and single cell protein.

Although penicillin was the first important antibiotic made industrially, the antibiotic industry is now dominated by streptomycetes rather than fungi. However, large amounts of cephalosporin, fusidic acid and griseofulvin are made by fungi. Table 5 lists examples of these and other products made on an industrial scale by fungi

2.5.4 Tissue Cultures

Plant and animal cells were first grown *in vitro* about the turn of the century, and remained a laboratory research technique until the late 1940s when the large-scale production of viral vaccines was developed. Now, in addition to vaccines, plant and animal cells can be grown on a large scale to accumulate cell metabolites and for the bioconversion of substrates like steroids and alkaloids. Tissue cultures of cells from reptiles, fish, birds, amphibia and insects have also been developed. See Fowler and Spier (Volume 1, Chapter 16) for a comprehensive discussion of cell and tissue culture.

2.5.4.1 Animal cells

Animal cells can be grown as discrete cells, pieces of tissue, or whole or parts of organs. Primary cell lines are prepared directly from animals; these cells are diploid and exhibit contact inhibition, that is the cells stop growing when they touch. Primary cells usually do not survive repeated passages in culture, although the life in culture of different types of cells varies from days to several months. After repeated passaging, some primary cell lines undergo a change and survive to become established or transformed cell lines. These cells are aneuploid, multiply more rapidly and to higher cell densities than primary cell lines, lack the property of contact inhibition and can be cultured indefinitely. Cell lines derived directly from tumours are already aneuploid and transformed. Transformed cell lines usually consist of one type of cell, but lines containing more than one type do occur. Cloned lines, that is lines derived from a single cell, have also been developed. Standard cell lines derived from various mammals (human, hamster, monkey, mouse, rat, cat,

Table 5 Products Made by Eukaryotes[a]

Product	Organism
Acids	
Citric	*Aspergillus niger*
Fumaric	*Rhizopus nigricus*
Gluconic	*A. niger*
Itaconic	*A. terreus*
Alkaloids	
Ergot alkaloids	*Claviceps purpurea, Clavicaps paspali*
Antibiotics	
Cephalosporins (C and P)	*Cephalosporium acremonium*
Fusidic acid	*Fusidium coccineum*
Griseofulvin	*Penicillium*
Penicillins	*P. chrosogenum*
Zeranol	*Gibberella zeae* (*Fusarium graminearum*)
Enzymes	
L-Amino acid acylase	*A. oryzae*
α-Amylase	*A. oryzae, A. niger, Rhizopus*
Catalase	*A. oryzae, A. niger*
Cellulase	*Trichoderma reesei, T. koningii, A. oryzae, A. niger*
Dextranase	*P. funiculosum*
β-Glucanase	*A. oryzae, A. niger*
Glucoamylase (amyloglucosidase)	*A. oryzae, A. niger, Endomyces fibuliger, Rhizopus*
Glucose oxidase	*A. oryzae, A. niger, Penicillium*
Hemicellulase	*A. oryzae, A. niger, T. reeseii*
Invertase	*Saccharomyces cereviseae*
Lactase	*A. oryzae, A. niger, Kluyveromyces fragilis, Kluyveromyces lactis*
Lipase	*A. oryzae, A. niger, Mucor miehei, Rhizopus*
Naringinase	*A. oryzae, A. niger*
Pectinase	*A. oryzae, A. niger, T. reesei, Rhizopus*
Proteases	*Rhizopus*
Acid	*A. oryzae, A. niger*
Alkaline	*A. oryzae*
Neutral	*A. niger, A. oryzae*
Rennets	*M. meihei, Endothia parasitica, Mucor pusillus*
Gibberellins	*Gibberella fujikuroi* (*Fusarium moniliforme*)
Polysaccharide	
Pullulan	*Aureobasidium* (*Pullularia*) *pullulans*
Solvent/fuel	
Ethanol	*Saccharomyces cerevisiae*
Steroid conversions	
Dehydrogenation at 1,2	*Cylindrocarpon radicola, Fusarium javanicum, Septomyxa affinis*
Hydroxylation at 6,7,9,11,15	*Curvularia lunata, Rhizopus nigricans, A. ochraceus, A. niger, Fusarium, Mucor griseocyanas*
Side-chain cleavage	*P. lilacinum, A. flavus, Fusarium*
Vitamin	
Riboflavin	*Ashbya gossypii*

[a] Collated from Berkeley *et al.*, 1979; Fogarty and Kelly, 1979; Perlman, 1979; Rose, 1978, 1970a; Taylor and Richardson, 1979; Vezina *et al.*, 1971.

dog and pig), fish, frogs and birds can be obtained from Type Culture Collections (see Brown, Volume 1, Chapter 3). Cell lines can be stored conveniently in liquid nitrogen.

Tissue culture monolayers are an indispensable tool in the isolation and identification of animal viruses. Infection by some viruses causes readily recognized changes in cell lines, so-called cytopathic effects. They may cause total destruction of the cells, sublethal damage to all cells, foci of destruction (plaques), or distortion of the cells without stopping their metabolism; on the other hand, some viruses produce no cytopathic effects.

The media first developed for culturing animal cells were very complex, containing buffered mineral salts, glucose, vitamins, amino acids, growth factors, foetal extracts and serum. Totally defined media have been developed for some cell lines, but tissue media remain complex and often include foetal extracts and usually serum. Cell lines derived from widely different mammals or from different organs of the same host will grow in virtually the same medium.

Antibiotics are usually incorporated in culture media to control microbial contamination, but for some large-scale processes, antibiotics may interfere with the isolation and purification of the process metabolites, and then contamination must be controlled by very strict attention to aseptic techniques and to the design of equipment.

Animal cells lack the tough outer walls of microbial and plant cells and, depending on the type of cell, must be grown on flat glass or plastic surfaces as monolayers, or as suspended cells in slowly-rotated containers, or in gently-agitated media.

Kruse and Patterson (1973) and Paul (1975) provide details of the techniques and media for cell culture and Feder and Tolbert (1983) describe modifications of conventional fermenters to allow the large-scale culturing of mammalian cells.

Monoclonal antibodies have assumed great importance in diagnosis. Hybridoma cell lines (constructed by fusing a myeloma cell to a cell secreting a single antibody) can be used to produce a specific antibody *in vitro* indefinitely (Fathman and Fitch, 1982). Unfortunately, the titres of antibodies obtained in tissue culture are often low, and so the antibodies in these culture supernatants must be concentrated for use in diagnosis. High titres can also be obtained directly by growing the hybridomas in the peritoneum of animals such as mice, and harvesting the peritoneal fluid.

The production of viral vaccines and interferon are at present the main commercial processes using animal cell lines (Table 6), although research into new products made by tissue cells is active at present.

Table 6 Products Made with Animal Cell Lines[a]

Product	Cell line
Angiogenic factors[b]	Human tumour cells
Interferon	Human leucocytes
Urokinase[b]	Human kidney
Viral vaccines	
Canine distemper	Dog kidney
Foot and mouth	
disease	Bovine kidney
Influenza	Chick embryo fluids
Measles	Chick embryo fluids
Mumps	Chick embryo fluids
Newcastle disease	Chick embryo fluids
Rabies	Duck embryo fluids
Rubella	Duck embryo fluids
Yellow fever	Chick embryo fluids

[a] Collated from Stones, 1981.
[b] Not in commercial production.

2.5.4.2 Plant cells

The tissues of a very wide range of plants can be grown on solid media (callus) or in suspension. The techniques and media used to grow plant cells are similar to those for animal cells, except that light is provided and foetal extracts and serum are replaced with plant extracts like coconut milk and plant auxins (cytokinins, and indoleacetic, naphthaleneacetic and 2,4-dichlorophenoxyacetic acids). Carbohydrate must be added to the medium as photosynthesis by cells in culture is less efficient than in whole plants (Staba, 1980). Plant cells range in diameter from 20 – 150 nm; they have cellulosic cell walls of high tensile strength but relatively low shear resistance. However, as they have a slower metabolic rate than aerobic microorganisms, reactors to grow plant cells are not required to achieve very high rates of oxygen transfer, although mixing must be adequate to prevent cells sedimenting.

The first cultures of plant cells were explants (tissues excised and grown in culture medium). Tissues rich in parenchyma cells, for example segments of seedlings, are especially suitable as explants; those from the tips of roots and shoots may retain their structure in culture, and given the right conditions may reorganize into entire plants again; those from stems, rhizomes, tubers and roots tend to give suspensions of cells in culture which do not redifferentiate easily. Tissue taken from different parts of the same plant ultimately yield parenchyma cell lines with similar metabolic capabilities.

Suspension cultures can be developed by germinating seeds aseptically, placing pieces of the young shoots and roots on solid media containing auxins and incubating them until callus tissue forms. The callus develops as a mixture of parenchymatous cells. On repeated subculture, the cells of the callus tend to grow faster, have less complex growth requirements, become less able to differentiate and may change in chromosome number. Pieces of callus can be dispersed in liquid medium and thereafter grown as a cell suspension. Other cultures can be made by excising mononucleate pollen and growing this as a haploid callus, or as homozygous haploid tissue cells from which, under special conditions, a homozygous haploid plant can be derived. Plant cell protoplasts can be prepared by enzymically removing the outer cell wall under conditions which protect the protoplast against osmotic shock. Protoplast fusion and genetic manipulation of protoplasts offer great promise in plant breeding, for example crosses can be achieved between plants which are otherwise infertile or nearly so.

Plant tissues can now be grown on a large scale (up to 20 000 l) and this makes industrial processes based on plant cells technically possible (Fowler, 1982). Nickell (1980) has listed some of the many potentially useful products detected in plant tissue cultures. She groups these under 28 headings, each of which contains numerous examples of different plant tissues known to accumulate compounds belonging to each class. This review should be consulted for references. There are no commercial processes based on plant tissues at present, as many of the products or bioconversions are more economically achieved with microorganisms. However, pharmacologically-active agents which are produced by plants would seem to offer the most promising area for future exploitation, especially if mutation and selection programmes and recombinant DNA methodology can be successfully applied to plant tissues as they have been to microorganisms. Bioconversion of precursors of drugs (alkaloids and steroids) is also a potentially interesting field.

2.6 REFERENCES

Adams, M. H. (1959). *Bacteriophages*. Interscience, New York.

Ainsworth, G. C. (1973). Introduction and keys to higher taxa. In *The Fungi*, ed. G. C. Ainsworth, F. K. Sparrow and A. S. Sussman, vol. 4A, pp. 1–7. Academic, New York.

Alexopoulos, C. J. and C. W. Mims (1979). *Introductory Mycology*, 3rd edn. Wiley, New York.

Berkeley, R. C. W., G. W. Gooday and D. C. Ellwood (eds.) (1979). *Microbial Polysaccharides and Polysaccharases, Special Publications of the Society for General Microbiology 3*. Academic, London.

Bold, H. C. and M. J. Wynne (1983). *Introduction to the Algae*, 2nd edn. Prentice-Hall, Englewood Cliffs, NJ.

Booth, C. (1978). Form and Function — II. Fungi. In *Essays in Microbiology*, ed. J. R. Norris and M. H. Richmond, pp. 3/1–3/32. Wiley, Chichester.

Buchanan, R. E. and N. E. Gibbons (eds.) (1974). *Bergey's Manual of Determinative Bacteriology*, 8th edn. Williams and Wilkins, Baltimore.

Burges, H. D. and N. W. Hussey (eds.) (1971). *Microbial Control of Insects and Mites*. Academic, New York.

Curds, C. R. (1982). The ecology and role of protozoa in aerobic sewage treatment processes. *Annu. Rev. Microbiol.*, **36**, 27–46.

Curds, C. R. and C. G. Ogden (1978). Form and Function – IV. Protozoa. In *Essays in Microbiology*, ed. J. R. Norris and M. H. Richmond, pp. 5/1–5/32. Wiley, New York.

Fathman, C. G. and F. W. Fitch (eds.) (1982). *Isolation, Characterization and Utilization of T Lymphocyte Clones*. Academic, New York.

Feder, J. and W. R. Tolbert (1983). The large-scale cultivation of mammalian cells. *Sci. Am.*, **248** (1), 24–31.

Fogarty, W. M. and C. T. Kelly (1979). Starch-degrading enzymes of microbial origin. *Prog. Ind. Microbiol.*, **15**, 87–150.

Fowler, M. W. (1982). The large scale cultivation of plant cells. *Prog. Ind. Microbiol.*, **16**, 207–229.

Kruse, P. F., Jr. and M. K. Patterson, Jr. (eds.) (1973). *Tissue Culture Methods and Applications*. Academic, New York.

Levine, N. D. *et al.* (1980). A newly revised classification of the protozoa. *J. Protozool.*, **27**, 37–58.

Matthews, R. E. F. (1982). Classification and nomenclature of viruses. Fourth report of the International Committee on Taxonomy of Viruses. *Intervirology*, **17**, 1–200.

Nickell, L. G. (1980). Products. In *Plant Tissue Culture as a Source of Biochemicals*, ed. E. J. Staba, pp. 235–269. CRC Press, Florida.

Onions, A. H. S., D. Allsopp and H. O. W. Eggins (1981). *Smith's Introduction to Industrial Mycology*, 7th edn. Edward Arnold, London.

Paul, J. (1975). *Cell and Tissue Culture*, 5th edn. Livingstone, Edinburgh.

Perlman, D. (1979). Microbial production of antibiotics. In *Microbial Technology*, ed. H. J. Peppler and D. Perlman, 2nd edn., vol. 1, pp. 241–280. Academic, New York.

Primrose, S. B. and N. J. Dimmock (1980). *Introduction to Modern Virology*, 2nd edn., vol. 2. Blackwell, Oxford.

Rose, A. H. (ed.) (1978). *Primary Products of Metabolism, Economic Microbiology*, vol. 2. Academic, London.

Rose, A. H. (ed.) (1979a). *Secondary Products of Metabolism, Economic Microbiology*, vol. 3. Academic, London.

Rose, A. H. (ed.) (1979b). *Microbial Biomass, Economic Microbiology*, vol. 4. Academic, London.

Rosowski, J. R. and B. C. Parker (1982). *Selected Papers in Phycology II*. Phycological Society of America, Lawrence, KS.

Staba, E. J. (1980). *Plant Tissue Culture as a Source of Biochemicals*. CRC Press, Florida.

Starr, M. P., H. Stolp, H. G. Truper, A. Balows and H. G. Schlegel (eds.) (1981). *The Prokaryotes*, vols. 1 and 2. Springer-Verlag, Berlin.

Stones, P. B. (1981). Viral vaccines. In *Essays in Applied Microbiology*, ed. J. R. Norris and M. H. Richmond, pp. 10/1–10/32. Wiley, New York.

Tannenbaum, S. R. and D. I. C. Wang (eds.) (1975). *Single-Cell Protein II*. MIT Press, Cambridge, MA.

Taylor, M. J. and T. Richardson (1979). Applications of microbial enzymes in food systems and in biotechnology. *Adv. Appl. Microbiol.*, **25**, 7–35.

Tinsley, T. W. (1978). Use of insect pathogenic viruses as pesticidal agents. *Perspect. Virol.*, **10**, 199–210.

Vezina, C., S. N. Sehgal, K. Singh and D. Kluepfel (1971). Microbial aromatization of steroids. *Prog. Ind. Microbiol.*, **10**, 1–47.

White, D. O. and F. Fenner (1985). *Medical Virology*, 3rd edn. Academic, New York.

Wistreich, G. A. and M. D. Lechtman (1980). *Microbiology*, 3rd edn., p. 199. Glenco, Encino, CA.

3

Isolation Methods for Microorganisms

C. M. BROWN
Heriot-Watt University, Edinburgh, UK

3.1 INTRODUCTION

The sources of microorganisms available to the biotechnologist are many and varied and on many occasions it may be sufficient to obtain cultures from one of the many culture collections. Large collections are held by national organizations (Table 1) and in addition many industrial companies maintain their own (often private) collections which are maintained through a continuous programme of enrichment and isolation. Clearly private companies will carefully secure organisms of great commercial value for their own exploitation, and some which have been constructed genetically to carry out specific biosynthetic, biocatalytic or degradation reactions may be protected by patents. In effect this means that a large percentage of isolated organisms are not available for general exploitation. In addition, cultures maintained for long periods may lose some of the characteristics which distinguished them on isolation. For this latter reason alone, it is often beneficial to return to natural habitats for fresh isolates when contemplating a new process. In most instances, cultures of single organisms (pure cultures) are required and these are selected for varying properties including (a) ability to grow on a particular substrate, (b) ability to produce a satisfactory growth yield in a short time period, (c) ability to produce some primary or secondary product reproducibly, (d) ability to carry out a specific chemical transformation and (e) genetic stability. In addition there are instances when a (stable) mixture of organisms can provide a more effective process than can a single organism and specialized methods are required for their production. Clearly many processes will in future be based on the transfer by recombinant DNA methods of a particular property to a suitable host organism (such as *Escherichia coli*,

Bacillus subtilis or *Saccharomyces cerevisiae*) and this is dealt with elsewhere (see Chandler and Churchward, Volume 1, Chapter 6).

Table 1 Sources of Industrially Important Microorganisms[a]

ACC	Akers Culture Collection of Imperial Chemical Industries Ltd., Alderley Park, Macclesfield, Cheshire, UK
AMC	Walter Reed Army Medical Center (formerly Army Medical School and Army Medical Department Research and Graduate School), Washington, DC, USA
AMIF	American Meat Institute Foundation, Chicago, IL, USA
AMNH	American Museum of Natural History, New York, NY, USA
ARL	*see* ACC
ATCC	American Type Culture Collection, Rockville, MD, USA
ATU	*see* FAT
BBL	Baltimore Biological Laboratory, Baltimore, MD, USA
BKM	All-Union Collection of Microorganisms, USSR
BRL	*see* ACC
BU	*see* BUCSAV
BUCSAV	Biologicky Ustav, Ceskoslovenska Akademie Ved, Prague, Czechoslovakia
CASE	Case Laboratories, Chicago, IL, USA
CBS	Centraalbureau voor Schimmelcultures, Baarn, Netherlands
CCAP	Cambridge Collection of Algae and Protozoa, Cambridge, UK
CCEB	Culture Collection of Entomogenous Bacteria, Prague, Czechoslovakia
CCF	Culture Collection of Fungi, Department of Botany, Charles University, Prague, Czechoslovakia
CCM	Czechoslovak Collection of Microorganisms, J E Purkyne University, Brno, Czechoslovakia
CCTM	Centre de Collection de Type Microbienne, Lille, France
CDA	Canadian Department of Agriculture, Ottawa, Canada
CDC	Centre for Disease Control, Atlanta, GA, USA
CI	Carnegie Institute, Cold Spring Harbor, Long Island, NY, USA
CIP	Collection of the Institut Pasteur, Paris, France
CLMR	Central Laboratory, South Manchurian Railway Co. Ltd.
CMI	Commonwealth Mycological Institute, Kew, UK (strain numbers formerly prefixed, IMI, Imperial Mycological Institute)
CSIRO	Commonwealth Science and Industrial Research Organization, Sydney, Australia
CU	*see* CCAP
DAOM	Plant Research Institute, Department of Agriculture, Mycology, Ottawa, Canada
DIFCO	Difco Laboratories, Detroit, MI, USA
DSM	Deutsche Sammlung von Mikroorganismen, Göttingen, West Germany
EPA	Environmental Protection Agency, Washington, DC, USA
ETH	Eidgenosische Technische Hochschule, Zurich, Switzerland
FAT	Faculty of Agriculture, Tokyo University, Japan
FDA	Food and Drug Administration, Washington, DC, USA
FERM	Fermentation Research Institute, Agency of Industrial Science and Technology, Chiba, Japan
FGSC	Fungal Genetic Stock Center, Humboldt State College, Arcata, CA, USA
FI	Farmitalia SpA, Milan, Italy
FRR	Division of Food Research, Food Research Laboratory, CSIRO, Sydney, Australia
GRIF	Government Research Institute of Formosa
HACC	Hindustan Antibiotics Ltd, Pimpri, Poona, India
HUT	Hiroshima University, Faculty of Engineering, Hiroshima, Japan
IAM	Institute of Applied Microbiology, University of Tokyo, Japan
IAUR	Instituto de Antibioticos da Universidade de Recife, Brazil
IAW	Institute of Antibiotics, Warsaw, Poland
ICI	Imperial Chemical Industries Ltd., Butterwick Research Laboratories, Welwyn, UK
ICPB	International Collection of Phytopathogenic Bacteria, University of California, Davis, CA, USA
HEM	Institute of Epidemiology and Microbiology, Prague, Czechoslovakia
IFM	Institute of Food Microbiology, Chiba University, Chiba, Japan
IFO	Institute of Fermentation, Osaka, Japan
IHM	Instituto de Higiene Experimental, Montevideo, Uruguay
IMCAS	Institute of Microbiology, Czechoslovak Academy of Sciences, Prague, Czechoslovakia
IMI	*see* CMI
IMRU	Institute of Microbiology, Rutgers—The State University, New Brunswick, NJ, USA
IMUR	Instituto de Micologia, Universidade de Recife, Brazil
INA	Institute for New Antibiotics, Moscow, USSR
INMI	Institute for Microbiology, USSR Academy of Sciences, Moscow, USSR
IOC	Instituto Oswaldo Cruz, Rio de Janeiro, Brazil
IP	Institut Pasteur, Paris, France
IPV	Instituto de Patologia Vegetale, Milan, Italy
ISC	International Salmonella Center
ITCC	Indian Type Culture Collection, New Delhi, India
ITCCF	Indian Type Culture Collection of Fungi, New Delhi, India
IU	*see* UTEX
KCC	Kaken Chemical Company Ltd., Tokyo, Japan

MDB	*see* CCM
MATHU	*see* MTU
MTU	Faculty of Medicine, University of Tokyo, Japan
NADC	National Animal Disease Center, Ames, IA, USA
NADL	*see* NADC
NBL	Naval Biological Laboratory, Oakland, CA, USA
NCA	National Canners' Association, Washington, DC, USA
NCAIA	National Center for Antibiotics and Insulin Analysis, FDA, Washington, DC, USA
NCDC	*see* CDC
NCDO	National Collection of Dairy Organisms, Reading, UK
NCIB	National Collection of Industrial Bacteria, Aberdeen, UK
NCIM	National Collection of Industrial Microorganisms, National Chemical Laboratory, Poona, India
NCMB	National Collection of Marine Bacteria, Aberdeen, UK
NCPPB	National Collection of Plant Pathogenic Bacteria, Harpenden, UK
NCTC	National Collection of Type Cultures, London, UK
NCYC	National Collection of Yeast Cultures, Food Research Institute, Norwich, Norfolk, UK
NDRC	National Defense Research Committee, Washington, DC, USA
NEA	Nobel Explosives Co., Ardeer, UK
NI	Nagao Institute, Tokyo, Japan
NIH	National Institute of Health, Bethesda, MD, USA
NIHJ	National Institute of Health, Japan
NIRD	National Institute for Research in Dairying, Reading, UK
NRC	*see* NRCC
NRCC	National Research Council, Ottawa, Canada
NRRL	Northern Utilization Research and Development Division, US Department of Agriculture, Peoria, IL, USA
OUT	Osaka University, Faculty of Engineering, Osaka, Japan
PD	Plantenziektenkundige Dienst, Wageningen, Netherlands
PRL	Prairie Regional Laboratory, Saskatoon, Canada
PSA	Progetto Sistematica Actinomiceti, Instituto 'P. Stazzi', Milan, Italy
QM	Quartermaster Research and Development Center, US Army, Natick, MA, USA
RIA	*see* USSR RIA
SAUG	Sammlung von Algenkulturen, University of Göttingen, West Germany
SMG	*see* DSM
TMC	Trudeau Mycobacterial Culture Collection, Trudeau Institute, Saranac Lake, NY, USA
TRTC	Department of Botany, University of Toronto, Toronto, Canada
UAMII	University of Alberta Mold Herbarium and Culture Collection, Alberta, Canada
UC	Upjohn Company, Kalamazoo, MI, USA
U.Md.	University of Maryland, College Park, MD, USA
USDA	United States Department of Agriculture
USDI	United States Department of the Interior
USPHS	United States Public Health Service
USSR RIA	USSR Research Institute for Antibiotics, Moscow, USSR
UTEX	University of Texas, Culture Collection of Algae, Austin, TX, USA
UTMC	University of Texas, Myxomycete Collection, Austin, TX, USA
VPI	Virginia Polytechnic Institute and State University, Blacksburg, VA, USA
WB	University of Wisconsin, Bacteriology Department, Madison, WI, USA
WHO	World Health Organization
WLRI	Warner Lambert Research Institute, Morris Plains, NJ, USA
WRAIR	Walter Reed Army Institute Research, Washington, DC, USA
WRRL	Western Utilization Research and Development Division, US Department of Agriculture, Albany, CA, USA

[a] Information from Atkinson and Mavituna (1983).

The ultimate sources of microorganisms for industrial use are water, soil, plant materials, sewage, manure, spoiled foods, *etc.*, where they exist in complex mixtures from which isolation must be attempted. It may be possible to isolate the required organisms by a direct procedure such as plating out a suitably diluted sample. Alternatively some enrichment stage may be necessary in which the numbers of organisms with the required properties are increased (by growth) relative to those associated with them. A range of such enrichment methods may be used. The aim of this chapter is to outline the more common enrichment and isolation procedures available (Figure 1).

3.2 SOURCES OF ORGANISMS AND SOME SAMPLING STRATEGIES

The choice of inocula for some isolations may be straightforward if the sought-after properties are expressed during microbial growth. For example, anaerobic sewage sludge and flocs from an aerated sludge plant are rich sources of organisms with varied biodegradative properties, agricultural soils treated with pesticides might be expected to contain organisms well adapted to the

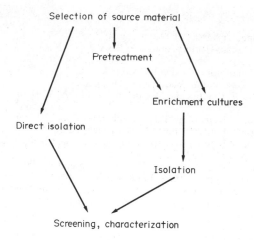

Figure 1 An outline of some common isolation procedures.

degradation of these chemicals, fermentative yeasts are found in abundance on grape must, thermophiles abound in hot springs and rotting compost, while animal and human faeces are brimfull of the enterobacteriaceae. Non growth-related properties, however, present problems in the design of isolation strategies and in the case of the antibiotics industry, organisms have been isolated from a wide range of habitats and secondary metabolite production demonstrated only after extensive screening of the isolates. The choice of natural materials (soils are the most used) is often based on the assumption that samples from widely different locations are more likely to yield novel isolates which in turn might produce novel metabolites. In addition attention is often directed towards relatively extreme environments (such as sea water and sediments and salt-rich and acid soils) in the search for new organisms (Williams and Wellington, 1982). There will continue to be many examples in which novel activities will become apparent only after extensive and time consuming screening trials (see, for example, Nakayama, 1982).

The chemical composition of a sample from any microbial habitat will begin to change the moment that sample is taken and such chemical change will clearly influence the survival opportunities of the organisms present. The golden rule must be to process samples in the laboratory as rapidly as possible and if there is likely to be a significant delay then thought must be given to provide effective storage conditions. Maintenance under conditions similar to those of the natural habitat, cooling to just above freezing or rapid freezing in liquid nitrogen are some alternatives.

3.3 DIRECT ISOLATION METHODS

3.3.1 Pretreatment of Samples

The purpose of the pretreatment of samples is to increase the relative numbers of desired organisms without involving selective growth (enrichment) procedures. The most common pretreatments of samples are temperature variations and concentration techniques.

Heat treatments have been used frequently to decrease the numbers of vegetative bacteria prior to the isolation of endospore-forming bacteria and actinomycetes (Williams and Wellington, 1982). For example, the heating of water samples to 80 °C for 2–5 min is a common practice in water testing laboratories intent on enumerating and isolating the endospore-forming *Clostridium perfringens*. Rowbotham and Cross (1977) reported that heating samples of water, soil, dung, *etc.* for 6 min at 55 °C before plating out increased the incidence of *Rhodococcus coprophilus* and *Micromonospora* spp. while Nonomura and Ohara (1969, 1971) described the isolation of *Microbispora* spp. and *Actinomadura* from soil preheated to 100 °C for 1 h. This selective effect on the survival of actinomycetes is not understood but it is apparent that many actinomycete propagules (*i.e.* both spores and hyphal fragments) are more resistant to elevated temperatures than are the vegetative cells of Gram-negative bacteria. Conversely the maintenance of samples at normal laboratory temperatures may decrease the numbers or even eliminate the presence of psychrophilic bacteria.

Concentration techniques are applied to samples containing relatively sparse microbial populations and are often used in conjunction with highly selective growth media. Filtration through a cellulose acetate (or nitrate) membrane of known pore size (usually 0.45 μm) was described by Windle-Taylor *et al.* (1953) as a method of concentrating bacteria from water and sewage effluents but it has much wider applications. A special advantage is that the filter may be incubated directly on a pad saturated with nutrient medium or on a nutrient-medium agar surface. Watson (1965) described the first isolation of the nitrifying organism *Nitrosomonas oceanus* from membrane-filtered sea water. Other reports of the successful isolation of bacteria from water using this method include those of Canale-Parola *et al.* (1967) for *Spirillum* spp., Mouraret and Baldensperber (1977) for thiobacilli, Burman *et al.* (1969) for *Micromonospora* spp. and other actinomycetes, and Al-Diwany *et al.* (1978) for *Thermoactinomyces* endospores. Miller (1967) used filters of pore size 0.8, 1.2 and 3.0 μm to remove zoospores of the Mastigomycotina from large volumes of water. The filters containing the fungal propagules were then incubated on suitable selective media for isolation purposes. Okami and Okazaki (1972) concentrated *Streptomyces* spp. from sea water and sediment by centrifugation and continuous flow centrifugation was used as a stage in fungal isolation from water by Fuller and Poynton (1964), Miller (1967) and Jones (1971).

Spores of thermophilic actinomycetes in plant materials may be removed by agitation in air, for example in a sedimentation chamber or wind tunnel (Gregory and Lacey 1963, Lacey 1971, Lacey and Dutkiewicz 1976), and the air impacted onto plates of isolation media in an air sampler (Andersen, 1958). This procedure lowers the relative numbers of non-sporing organisms present on the isolation plates.

Samples of plant and animal tissues are often required to be finely divided before isolation is attempted and a variety of commercial blenders and macerators are available for this purpose. Harley and Waid (1955) reported that fungal propagules could be removed from plant roots by efficient washing in sterile water. Similar methods are also used extensively to remove fungal spores from soil and decomposing leaf litter (Parkinson 1967, Hayes 1979). Both the washings and the remaining soil particles may then be plated out, the particles being more likely to contain hyphal fragments (see below). Discrete pieces of plant material have also been used in this way to isolate surface attached fungi and yeasts. This is a standard method used in the examination of storage fungi on stored grains and malt (Flannigan *et al.*, 1982). Organisms from the plant phylloplane may also be isolated using a sticky tape method in which a commercial sticky tape is used to transfer organisms from the leaf surface to the surface of a nutrient-medium agar.

Some microbial cells and spores are sufficiently large to be isolated by micromanipulation methods. This is widely applied to the separation of individual ascospores from the ascosporogenous yeasts and moulds, and a more simple method using sterile capillary tubes has been used to obtain pure cultures from the individual zoospores of aquatic phycomycetes (Parkinson, 1982). Warcup (1955) has described the direct isolation of fungal hyphal fragments from soil particles using extremely fine forceps and needles.

3.3.2 Dilution and Incubation of Samples

Direct methods for the isolation of bacteria and fungi often involve the preparation of a (serial) dilution series followed by plating out onto or into nutrient-medium agar. The choice of diluent is of critical importance, especially at the more extreme dilutions, if maximum organism recovery is to be achieved. NaCl (0.9% w/v) is often used but the use of phosphate buffered-NaCl or 1/4 strength Ringer's solution is to be preferred (Cruikshank, 1965). While more expensive, the use of a liquid nutrient medium as diluent is worthy of consideration. The temperature of the diluent should be that to be used for subsequent incubation. Special care is required with anaerobic bacteria and it is usual to incorporate a reducing agent and a redox indicator (*e.g.* 0.0001% w/v resazurin) into diluents used for the isolation of these organisms.

While 1–2% (w/v) agar is commonly used to gel nutrient media it may be toxic to some organisms, while others show agarolytic activity. Silica gel (Skerman, 1967) is recommended for the isolation of autotrophs while polyacrylamide or gelatin provide other alternatives for agarolytic organisms.

The isolation of anaerobic bacteria may be carried out using the roll-tube technique of van Niel (1931) in which molten nutrient-medium agar in tubes is used as diluent. The more sophisticated Hungate method (Hungate, 1950, 1969), employing the flushing of tubes with oxygen-free gas, is essential for the isolation of organisms such as methanogenic bacteria which display extreme oxygen sensitivity. Self-contained anaerobic 'work stations' are now commercially available while glove bag enclosures provide a useful alternative (Leftley and Vance, 1979). The use of aerobic/anaerobic incubation conditions for organism selection was discussed by Herbert (1982).

The incubation temperature used has a major bearing on organism growth and hence the opportunity for successful isolation. A temperature of 10 °C is used for psychrophiles while thermophiles are usually isolated at temperatures in excess of 45–55 °C. Mesophilic bacteria are usually grown at 30–35 °C, while for actinomycetes and fungi 25–30 °C is more common. The time of incubation required clearly depends upon the nature of the organism and the growth medium involved, in addition to the temperature employed.

Phototrophic organisms require incubation under suitable conditions of lighting. These were considered by Herbert (1982).

3.3.3 Media Considerations

Isolation media (and those used for enrichments) may be composed entirely of chemically defined constituents—*defined* or *synthetic media*—or may contain ingredients from animal, plant, soil, microbial sources, *etc.* containing a range of undefined constituents. Such *complex media* are used widely for the direct isolation of heterotrophic bacteria and fungi. Commercial manufacturers such as Difco and Oxoid produce a wide range of both general purpose and selective media and publish analytical data (*e.g.* Difco Manual 1966, Oxoid Manual 1979) which are a useful source of information on the overall chemical nature of these products. These preparations are used widely to isolate organisms of significance in food and public health laboratories. Common ingredients include enzymic and acid hydrolysates of proteins (generally known as 'peptones') derived from meat, plant resources (*e.g.* soya) and milk, along with yeast extract, malt extract, blood products, *etc.* The most common complex media include nutrient agar, plate count agar, malt extract agar, blood agar, marine agar, casein starch agar, soil extract agar and tryptone soy agar. There are an almost infinite number of variations possible in the composition of both defined and complex media. In addition to the Difco and Oxoid Manuals, the following papers give useful examples of the range of media used successfully for the isolation of bacteria and fungi: Harrigan and McChance (1966), Skerman (1967), Norris and Ribbons (1970), Veldkamp (1970), Booth (1971), Parkinson *et al.* (1971), Johnson and Curl (1972), Rodina (1972), Collins and Lyne (1976), Stanier *et al.* (1977), Herbert (1982), Williams and Wellington (1982), Atkinson and Mavituna (1983). The following paragraphs illustrate some of the variations employed.

Many organisms have complex nutritional requirements which can be satisfied by the inclusion of, for example, yeast extract or a peptone in the medium. Stanier *et al.* (1977) quote the example of the lactic acid bacterium *Leuconostoc mesenteroides* which grows satisfactorily in a medium containing a mineral base plus glucose and yeast extract. A defined medium supporting the growth of this organism contained, in addition to the mineral base, sodium acetate, nineteen amino acids, four purine and pyrimidine bases and ten vitamins. Some organisms have even more exacting requirements, for example isolation media for *Lactobacillus* spp. are often supplemented with tomato juice to supply further vitamins, *etc.*, the isolation of some rumen anaerobes requires the addition of rumen fluid, some animal pathogens, *e.g. Haemophilus* spp., require ingredients from whole blood, and many media require supplementation with vitamin B_{12}. At the other extreme, chemolithotrophic bacteria such as the thiobacilli and nitrifying bacteria require only simple salts media. There is also evidence that media with low nutrient concentrations can favour the growth of some groups of microorganisms; examples include the mineral salts, sodium propionate, thiamine agar (M3) described by Rowbotham and Cross (1977) for the isolation of *Rhodococcus* and *Micromonospora* and the low nutrient sea water medium of Zobell (Difco Manual 1966) used extensively for the isolation of marine bacteria. Between these extremes a range of synthetic and 'semisynthetic' media are in general use, for example the addition of low concentrations of peptone and yeast extract to a defined medium may avoid the requirement for amino acid and vitamin supplements for less exacting organisms. Clearly the absence of specific nutrients makes a medium *selective* in favour of those organisms lacking such specific requirements. Media may also be made selective by including substrates used only or mainly by a particular group of organisms, for example chitin-containing media are often used for the isolation of actinomycetes while a medium containing lysine as sole nitrogen source is used widely in the brewing industry to select for yeasts other than *Saccharomyces* spp.

Selective media often also contain specific inhibitors to discriminate against unwanted isolates.

Examples include the use of bile salts in media for the isolation of *Escherichia coli* and the use of cetrimide to favour the growth of *Pseudomonas* spp. (Brown and Lowbury, 1965). It is also common practice to include antibacterial agents in media designed for the isolation of actinomycetes and fungi. Thus, the selective medium of Orchard and Goodfellow (1971) for the isolation of *Nocardia* spp. contained cycloheximide and mycostatin (to inhibit yeasts and moulds) together with the antibacterial agents chlorotetracycline and methacycline. Other examples of selective media for isolation of actinomycetes are given by Williams and Wellington (1982). Similarly Parkinson (1982) indicated that when isolating fungi from soils containing high numbers of bacteria it is useful to add crystal violet, rose bengal, sodium propionate or sodium deoxycholate to restrict bacterial growth (see also Papvizas and Davey, 1959). The addition of penicillin G, streptomycin and aureomycin is also commonly employed for the same purpose. Conversely the antifungal agent cycloheximide (actidione) is used extensively for the isolation of bacteria present alongside large numbers of yeasts and moulds. The salinity and osmolarity of media may have a profound influence on organism selection, for example many marine bacteria have an obligate requirement for at least 0.2M NaCl (McLeod, 1965) while osmophilic yeasts such as *Saccharomyces rouxii* tolerate high sucrose concentrations in growth media (Phaff *et al.*, 1978).

In addition to nutrient composition and the presence or absence of inhibitors, the pH has a marked selective effect in isolation media. Thus a pH of 5.0 may encourage the growth of many soil fungi (Parkinson, 1982) while a pH of 4.0 is suitable for many fermentative yeasts (Phaff *et al.*, 1976). Most actinomycetes are neutrophiles and grow best in isolation media of pH between 6.7 and 7.5 although acidophilic (pH 4.5–5.0) actinomycetes have been recognized (Williams *et al.*, 1971; Khan and Williams, 1975). Bacteria in general have a broad pH spectrum for growth (pH 4.0–9.0, Thimann, 1964) although particular organisms grow optimally in a range of about 2 pH units. The acid thiobacilli and some lactic acid bacteria can grow at pH less than 3 while some *Bacillus* spp., *Micrococcus* spp. and *Pseudomonas* spp. grow well at pH 10–11 (see Stanier *et al.*, 1977; Herbert, 1982).

The requirements for suitable aerobic/anaerobic incubation conditions were outlined above. It is also important to determine the requirements for electron acceptors for anaerobic bacteria. For fermentative organisms the provision of fermentable sugar (glucose or sucrose) or amino acid mixture is usually sufficient. There may also be requirements for nitrate, sulfate and CO_2/acetate for nitrate reducing, sulfate reducing and methanogenic bacteria respectively and the presence of these potential electron acceptors in anaerobic cultures serve as powerful selective agents.

3.4 ENRICHMENT CULTURE METHODS

3.4.1 Baiting Methods

The baiting of samples with a range of sterile solid substrates is a technique used commonly to enrich for actinomycetes and fungi prior to their isolation by plating methods or by micromanipulation.

The paraffin rod technique was first used by Gordon and Hagen (1936) to isolate acid-fast actinomycetes from soil and has since been employed widely to isolate *Nocardia* spp. (Karup and Schmidt, 1971). Some genera within the Actinoplanaceae produce motile spores and a variety of baits including pollen have been used in their isolation from soils and water (Couch, 1950; Kuznetsov, 1969). Snakeskin and human hair have also been employed in the enrichment /isolation of *Ampullariella* spp. (Booth, 1971) and *Pilimelia* spp. (Kane, 1966) respectively from soil. Parkinson (1982) has described the use of a number of sterile baits including conifer (usually *Pinus*) pollen, cellulose (usually wettable cellophane), boiled hemp seeds, boiled grass leaves and dead insects to enrich for fungi present in soils and water. In general the baits are added to Petri dishes half filled with sterile water (or water sample) and inoculated as required with soil, organic matter, *etc.* Samples are withdrawn at regular intervals for up to 2–4 weeks and the baits examined microscopically for the appearance of rhizoids, mycelia and sporangia. Alternatively, small mesh boxes containing bait may be suspended *in situ* and returned to the laboratory for examination. Isolation then involves either removal with a sterile capillary tube or plating onto a suitable nutrient agar. Some baits may be used selectively, for example hemp seed and grass leaves attract the motile spores of a wide range of Mastigomycotina, *Pinus* pollen is especially useful in attracting zoospores of the Chytridiales (Parkinson, 1982; Gaertner, 1968) and Fell *et al.* (1960) used cores of banana stalks in the isolation of marine yeasts from sea water. Blocks of wood submerged in water and then incubated for several weeks in the laboratory have been used to isolate lignicolous ascomycetous fungi (Jones, 1971). Fruiting bodies develop on the wood and ascospores removed from these may be removed and streaked on to a suitable nutrient agar.

3.4.2   General Chemical Enrichment

The general principles of the use of enrichment cultures in the isolation of bacteria were outlined elegantly by Stanier *et al.* (1977) and embellished by Veldkamp (1970) and Nakayama (1982). In addition the recent review of Herbert (1982) includes details of enrichment procedures and media for autotrophic and phototrophic bacteria.

When a mixture of microorganisms is introduced into a liquid medium there is competition for the available nutrients between the members of the developing population. These conditions tend to select for those organisms able to sustain the highest growth rate and these will predominate. The selectivity of enrichment cultures is determined by the chemical composition of the medium used together with other factors such as medium pH and ionic strength, conditions of illumination and aeration, and the incubation temperature. In addition the chemical composition of a traditional batch (closed) culture alters with microbial growth and the predominant organisms often vary as a function of incubation time. The use of the alternative open (continuous flow) culture systems is considered in Section 3.4.3. Comments on the influence of nutrient composition, incubation details, presence of growth inhibitors, *etc.* on cultures used for direct isolation procedures (Section 3.3) apply equally to enrichment cultures. These latter allow the development of organisms present perhaps only in low numbers and therefore likely to be overgrown in attempts at direct isolation. Stanier *et al.* (1977) considered enrichment cultures to be one of the most powerful techniques available to the microbiologist and these methods are used extensively as an important stage in bacterial isolation. Williams and Wellington (1982) commented, however, that this technique has been used only infrequently for the isolation of actinomycetes with common exceptions being only the paraffin baiting method and the use of chitin in the isolation of *Streptomyces* from soil (Williams and Mayfield, 1971).

The following comments are based on the details given in the texts of Veldkamp (1970) and Stanier *et al.* (1977). In the enrichment of fermentative organisms, much depends on the nature of the carbon and energy source used. Sugars are often preferred but other organic compounds of similar oxidation level are also suitable. Enrichments are often carried out anaerobically both to prevent competition from aerobes and to provide growth conditions conducive to the development of obligate anaerobes. Nitrate addition (as nitrogen source) is avoided in order to prevent the development of denitrifying bacteria. $CaCO_3$ addition for pH control is useful to counter excess acid production. Denitrifying bacteria develop well in media containing acetate, butyrate, methanol or ethanol and supplemented with nitrate, while sulfate reducing bacteria are enriched in media containing lactate or malate and supplemented with sulfate. The use of non-fermentable substrates often results in the enrichment of obligate aerobes, such as *Pseudomonas* spp. Specific substrates are also useful, for example the inclusion of sodium benzoate as sole carbon source often leads to the enrichment of organisms using that substrate such as *Pseudomonas putida*. In principle this strategy has wide application but has not proved entirely successful in many instances when complex substrates are employed (see Bull, 1980). This is discussed further in Section 3.4.3.

Complex media are used extensively in the enrichment of nutritionally fastidious organisms such as the lactic acid bacteria. These organisms produce large concentrations of lactic acid and are resistant to it. A poorly-buffered medium containing glucose and yeast extract, inoculated with raw milk or sewage and incubated anaerobically will often first show the growth of *Enterobacter* and *Escherichia* spp. As incubation proceeds, however, lactic acid production usually increases leading to the predominance of lactic acid bacteria and the cessation of growth and even the death of the competing bacteria. Table 2, modified from Stanier *et al.* (1977), contains a number of further examples of the enrichment of heterotrophic bacteria in complex media.

3.4.3   Specialized Enrichment Systems and their Applications

One of the earliest enrichment systems to be described is the Winogradsky column (Winogradsky, 1888; Aaronson, 1970; Veldkamp, 1970; Herbert, 1982). This consists of a glass or plastic cylinder containing at its base a sample of soil or sediment supplemented with $CaCO_3$ and (usually) strips of filter paper as substrate. This is then covered with a column of water and the

Table 2   Complex Media for Enrichment of Heterotrophic Bacteria[a]

Additions (g l^{-1})	Special environmental conditions	Preferred choice of inoculum	Organisms enriched
None	pH 7.0; aerobic	Soil	Aerobic amino acid oxidizers
None	pH 7.0; aerobic	Pasteurized soil	*Bacillus* spp.
None	pH 7.0; anaerobic	Pasteurized soil	Amino acid-fermenting clostridia
Urea, 50.0	pH 8.5; aerobic	Pasteurized soil	Alkali-tolerant urea-decomposing bacilli (*Bacillus pasteurii*)
Glucose, 20.0	pH 2.0–3.0; anaerobic	Soil	Anaerobic *Sarcina* spp.
Glucose, 20.0	pH 6.5; anaerobic	Plant materials, milk	Lactic acid bacteria
Glucose, 20.0; CaCO$_3$, 20.0	pH 7.0; aerobic or anaerobic	Soil or sewage	Enteric bacteria
Glucose, 20.0; CaCO$_3$, 20.0	pH 7.0; anaerobic	Pasteurized soil	Sugar-fermenting clostridia
Sodium lactate, 20.0	pH 7.0; anaerobic	Swiss cheese	Propionic acid bacteria
Ethanol, 40.0	pH 6.0; aerobic	Fruits, unpasteurized beer	Acetic acid bacteria

[a] Information from Stanier *et al.*, 1977.

cylinder covered with a lid to prevent water evaporation but to allow entry of air. The column may be illuminated for the enrichment of phototrophs. A gradient of nutrients and metabolic end-products develops as a result of fermentative metabolism in the sediment. Thus a typical column will be anaerobic and nutrient rich at the base and aerobic and nutrient poor at the surface. Herbert (1982) described such an arrangement and commented on the enrichment of fermentative bacteria, thiobacilli, *Desulfovibrio* spp. and a range of phototrophic bacteria. Wimpenny (1982) has described a gradient system (the 'gradostat') consisting of a series of linked continuous flow cultures (see below) which has a high potential as an enrichment system.

The commonly used dilution-plating method for the direct isolation of fungi from soils, detritus, *etc.* (see above and Park, 1972) is thought to select for organisms present as spores rather than as active hyphae (Warcup, 1957). Attempts to isolate the latter usually involve an immersion method in which the enrichment medium (agar) is placed in soil in such a manner that it is separated from the soil particles by a small air gap (Chesters, 1940; Mueller and Durrell, 1957; Luttrell, 1967). Mueller and Durrell (1957) used plastic centrifuge tubes with holes bored in the walls at intervals and then covered with plastic tape. The tubes are then filled with medium and before insertion into soil the plastic tape is pierced opposite the holes in the tube walls. After 5–7 days the tubes are removed from the soil, the agar examined and the fungi present subcultured.

In the batch or closed culture system used traditionally for enrichments, the concentrations of nutrients employed are usually high initially in order to produce large populations of organisms. Those organisms able to grow fastest at any one time in a changing environment predominate and selection is therefore governed by the maximum specific growth rates of the species involved. This type of enrichment is time dependent with the likelihood of a species succession. In contrast the open (continuous) culture enrichment system is based (in its most simple forms) on the provision of a constant chemical environment and one in which the nutrient concentrations involved are very low. This simple system is a well-mixed homogeneous culture usually functioning as a chemostat but more complex systems including solid substrates and other heterogeneous cultures have been used in many laboratories. As discussed below, the soil percolation column is an example of such a heterogeneous continuous enrichment culture.

The theory of continuous cultivation has been detailed in many reviews since the principles were expounded by Monod (1950) and Novick and Szilard (1950) and made comprehensible by Herbert *et al.* (1956). The chemostat exploits the fact that at a constant temperature and pH the specific growth rate of a microorganism (μ) is dependent on the concentration of a growth limiting substrate (s) in the culture medium. Defined media may be constructed with known limiting substrates (Evans *et al.*, 1970), while in complex media the nature of the growth limitation is often obscure. The Monod equation describes the relationship between μ and s as $\mu = \mu_m[s/(K_s+s)]$, where μ_m is the growth rate achieved at saturating values of s (those usually found in batch cultures) and K_s is a saturation constant with values in the order milligrams per litre for carbohydrates and micrograms per litre for amino acids (Pirt, 1975). In a chemostat one nutrient in the incoming medium is usually maintained at a low relative concentration (see Evans *et al.*, 1970); s then is fixed by the rate of addition of fresh medium (the dilution rate) and in turn controls μ at some point on the μ/s curve (see Figures 2 and 3). The behaviour of mixed cultures in a chemostat has been described by many authors including Powell (1958), Veldkamp (1970),

Jannasch and Mateles (1974), Brown *et al.* (1978), and Bull and Brown (1980). Such systems are highly selective and in a mixed population competing for a single limiting nutrient then the outcome is determined by the μ/s relationships of the organisms involved. Figure 2 shows a culture of organism A with a higher μ_m and lower K_s than organism B, A will outgrow B at any value of μ imposed by the dilution rate and will predominate both in batch or continuous culture enrichments in the medium involved. In contrast Figure 3 shows organism A with an organism C with a lower μ_m and K_s which results in crossing μ/s curves. Thus in a batch culture A would be likely to predominate. In continuous culture A would predominate at high growth rates (high s) and C would predominate at low growth rates (low s). The outcome of a continuous flow competition experiment enrichment is therefore dependent on the dilution rate (D) applied and it is possible to use such cultures to enrich for organisms never likely to predominate in batch culture systems. Examples are to be found in the papers of Jannasch (1967), Harder and Veldkamp (1971), Veldkamp and Jannasch (1972), Veldkamp and Kuenen (1973), Veldkamp (1976), and Jannasch and Mateles (1974).

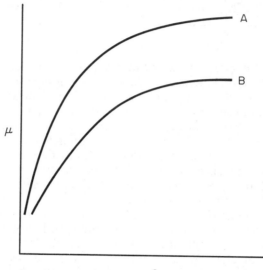

Figure 2 The relationship between μ and s; organism A will outgrow organism B at all growth rates.

Jannasch (1965, 1967) pioneered the use of continuous culture enrichment methods and suggested the following advantages. (1) No succession of species occurs and if there if no wall growth or interaction between members of the microbial population then the predominance of one species increases with time. (2) The growth advantages of the successful competitor are determined by the growth parameters (μ_m/K_s) of the organisms concerned on the particular substrate(s) used. If these parameters are known their enrichments should be reproducible. (3) Enrichments may be carried out in the presence of extremely low concentrations of a limiting nutrient and therefore at low population densities.

In practice such continuous flow procedures seldom, if ever, lead to the enrichment of only one organism and while single organisms may predominate they are usually associated with others in apparently stable associations. A number of examples will serve to illustrate this point and all are dependent on the fact that the mixture of organisms grows more effectively under the imposed conditions than does any single organism. The general course of such enrichments is illustrated in Table 3. These data were obtained during the enrichment of bacteria from offshore marine sediment and water (Brown and Wardell, 1983). The initially complex mixture of organisms resolved into a stable consortium of two organisms after 4–5 days. It is interesting to note that neither organism enriched (*i.e.* *Pseudomonas* A or B) was able to grow alone on an agar medium containing low concentrations of benzoate as sole source of carbon and energy.

3.4.3.1 Enrichments from sea water

Jannasch (1967) described a number of enrichments from sea water and demonstrated that a high degree of reproducibility could be achieved within a range of dilution rates and limiting substrate concentrations. Using these methods he isolated a number of *Vibrio*, *Spirillum*, *Micrococ-*

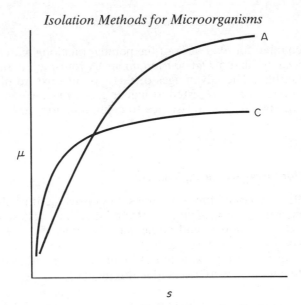

Figure 3 The relationship between μ and s; organism A will outgrow organism C at high growth rates; C will outgrow A in continuous culture at low growth rates.

Table 3 Continuous Flow Enrichment Culture in Benzoate-limited Artificial Sea Water Medium[a]

Day	1		2		3		4		5		6		7		8		9	Organism
	—	10	75	70	78	78	72	76	95	87	100	77	78	86	81	86	73	*Pseudomonas* sp. A
	33	40	15	6	22	21	28	22	4	13	—	23	22	14	19	14	27	*Pseudomonas* sp. B
	22	10	5	24	1	1	1	2	—	—	—	—	—	—	—	—	—	*Pseudomonas* sp.
	—	—		5	—	1	—											—	*Pseudomonas* sp.
	22	40																	Unknown
	23	—																	Others
	1×10^2		2×10^3		2×10^4		5×10^4		2×10^4		1×10^4		5×10^5		2×10^5		5×10^5	Total population

[a] Information from Brown and Wardell, 1983.

cus and *Pseudomonas* species. He especially noted that some organisms showed unstable growth parameters when cultures were maintained at nutrient concentrations higher than those present in the enrichment. This phenomenon may be common and is likely to be due to selection in batch cultures of mutants with an increased μ_m. Brown *et al.* (1978) described the enrichment of marine bacteria with a high affinity for nitrate (as nitrogen source) using an artificial sea water medium maintained at 10 °C and used at a dilution rate of 0.02 h^{-1}. Such low growth rates (Jannasch used dilution rates from 0.03–0.02 h^{-1}) are thought to be in the same range as those of the organisms in their natural growth environment. Brown and Wardell (1983) have described general methods and equipment for the enrichment and isolation of bacteria able to degrade both simple and complex substrates (such as quaternary biocides) from marine and estuarine waters and sediments.

3.4.3.2 Enrichments for biomass production

Harrison *et al.* (1976) described in detail the enrichment of bacteria for single cell protein production from methane or methanol as carbon and energy source. Since many organisms which grew well in a batch system did not adapt readily to a continuous flow culture (which was the system of choice for production purposes) then continuous flow enrichment methods were developed in order to obtain new isolates. Enrichments were run for over one thousand hours with methane or methanol as sole carbon source and on no occasion did a single organism culture emerge. Stable cultures were obtained in which one organism predominated but which also contained five to ten other microbial species most of which could not utilize the supplied carbon source. The predominant C(1)-utilizing organisms were isolated but in monoculture did not grow with the same vigour as in the mixture. What was selected was a microbial consortium in which interspecies interactions were of prime significance in culture stability, yield and the growth rate obtained. Harrison *et al.* (1976) described a methane-utilizing mixture in which the sole methane-oxidizing organism accounted for over 90% of the population along with a *Hyphomicrobium* sp. (4–9%), an *Acinetobacter* sp. and a *Flavobacterium* sp. A four-membered mixture enriched on methanol was also described. The advantages of this type of mixture of organisms specifically enriched for biomass production were listed as higher growth rate (than monocultures), higher yield coefficient, greater stability and resistance to contamination, and reduced foaming problems.

3.4.3.3   Enrichments for nitrate-reducing bacteria

Dunn *et al*. (1978, 1980) described the continuous flow enrichment of nitrate-reducing bacteria from estuarine water and sediments using media of varying salinity. These media contained either glycerol or acetate as carbon sources and ammonia as nitrogen source together with nitrate. Enrichments were carried out first aerobically to select against obligate anaerobes and then anaerobically for facultative anaerobes able to reduce nitrate. A number of fermentative bacteria able to reduce nitrate to ammonia were isolated in this manner.

3.4.3.4   Enrichments in complex media

Continuous flow cultures were used by Ellwood and Hunter (1976) for enrichment of anaerobic bacteria from dental plaque. A glucose-limited complex medium was employed and cultures were allowed to develop under batch conditions overnight before medium flow was commenced. Complex mixtures of organisms developed with *Streptomyces milleri* as the predominant organism. The isolation of an 'unusual' lactobacillus was also recorded. A further example is provided by Marsh *et al*. (1983).

3.4.3.5   Biodegradation

Microorganisms play an active role in the biodegradation of complex organic compounds including xenobiotics. Some compounds disappear very rapidly from soil while others may persist for long periods of time and studies on biodegradative processes have been carried out in many laboratories (see Bull, 1980; Harder, 1981; Slater and Bull, 1982). Modern studies usually involve the enrichment of populations able to carry out specific transformations and individual isolates are required in order to determine physiological and biochemical details of the biodegradation pathways.

The sources of organisms for these enrichments may be soil or water previously treated with the chemical in question or samples from an industrial wastewater plant involved in the processing of a specific chemical or structural analogue. In such instances some 'natural' enrichment has already occurred and isolation may be possible by direct plating or after a further laboratory enrichment, usually employing the organic substrate concerned as sole source of carbon. One clear problem may arise in that some chemicals may act mainly or solely as a source of nitrogen or phosphorus in which case some additional carbon source must be included in the enrichment/ isolation medium. Harder (1981) quotes as an example the herbicide amitrole (3-amino-1,2,4-triazole) which is degraded rapidly in fertile soil. Isolation of amitrole-degrading bacteria was possible only by using sucrose as an added source of carbon and amitrole as nitrogen source. With this type of example in mind it has become common practice to conduct two parallel enrichments, one containing the organic compound only and a second which is augmented with a readily metabolized carbon and energy source such as succinate or sucrose.

While many genera of microorganisms which degrade xenobiotics as growth substrates have been isolated (see Higgins and Burns, 1975; Alexander, 1977), attempts at isolation using novel substrates are not always successful even from an enrichment culture in which the particular chemical is actively utilized. Of many reasons for this, the following are the most likely: some compounds may be partly or completely degraded by cooxidation (*i.e.* another substrate is also required; see Dalton and Stirling, 1982), degradation may require interfaces and/or gradients, and the compound in question may be degraded only by a microbial mixture or consortium with no single organism possessing all the required characteristics. A number of examples are discussed in the articles of Bull (1980), Harder (1981) and Slater and Bull (1982).

3.4.3.6 Heterogeneous continuous flow systems

The gradostat (Lovitt and Wimpenny, 1981) was mentioned earlier as an example of a system employing nutrient gradients and which has a high potential as an enrichment system. The soil percolation (perfusion) system is an older example of a heterogeneous continuous flow system described by Lees and Quastel (1946) and improved by Jeffrys and Smith (1951). The equipment consists essentially of a column packed with soil through which the enrichment medium is circulated by applying either a gentle suction at the outlet or an inlet pressure of air or inert gas. The enrichment medium may be supplemented with specific substrates, inhibitors, *etc.* and may be used for both aerobic and anaerobic organisms.

3.5 REFERENCES

Aaronson, S. (1970). *Experimental Microbial Ecology*. Academic, New York.

Al-Diwany, L. J., B. A. Unsworth and T. Cross (1978). A comparison of membrane filters for counting *Thermoactinomyces* endospores in spore suspensions and river water. *J. Appl. Bacteriol.*, **45**, 249–258.

Alexander, M. (1977). *Introduction to Soil Microbiology*, 2nd edn. Wiley, New York.

Andersen, A. A. (1958). New sampler for the collection, sizing and enumeration of viable airborne particles. *J. Bacteriol.*, **47**, 471–484.

Atkinson, B. and F. Mavituna (1983). *Biochemical Engineering and Biotechnology Handbook*. Macmillan, London.

Booth, C. (1971). Fungal culture media. In *Methods in Microbiology*, ed. J. R. Norris and D. W. Ribbons, vol. 3, pp. 49–94. Academic, London.

Brown, C. M. and J. N. Wardell (1983). Bacterial processes in marine sediments. In *Experimental Biology at Sea*, ed. A. MacDonald. Academic, London.

Brown, C. M., D. C. Ellwood and J. R. Hunter (1978). Enrichments in a chemostat. In *Techniques for The Study of Mixed Populations*, ed. W. D. Lovelock and R. Davies, pp. 213–222. Academic, New York.

Brown, V. I. and E. J. L. Lowbury (1965). Use of an improved certimide agar medium and other culture methods for *Pseudomonas aeruginosa. J. Clin. Pathol.*, **18**, 752–756.

Bull, A. T. (1980). Biodegradation: some attitudes and strategies of microorganisms and microbiologists. In *Contemporary Microbial Ecology*, ed. D. C. Ellwood, J. N. Hedger, M. J. Latham, J. M. Lynch and J. H. Slater, pp. 107–136. Academic, London.

Bull, A. T. and C. M. Brown (1980). Continuous culture applications to microbial biochemistry. In *Microbial Biochemistry*, ed. J. R. Quayle, pp. 177–226, University Park Press, Baltimore.

Burman, N. P., C. W. Oliver and J. K. Stevens (1969). Membrane filtration techniques for the isolation from water of coli-aerogenes, *Escherichia coli*, faecal streptococci, *Clostridium perfringens*, actinomycetes and microfungi. In *Isolation Methods for Microbiologists*, ed. D. A. Shapton and G. W. Gould, pp. 127–134. Academic, London.

Canale-Parola, E., S. C. Holt and Z. Udris (1967). Isolation of free living anaerobic spirochaetes. *Arch. Mikrobiol.*, **59**, 41–48.

Chesters, C. G. C. (1940). A method for isolating soil fungi. *Trans. Br. Mycol. Soc.*, **24**, 352–355.

Collins, C. H. and P. M. Lyne (1976). *Microbiological Methods*. University Park Press, Baltimore.

Couch, J. N. (1950). *Actinoplanes*, a new genus of the Actinomycetales. *J. Elisha Mitchell Sci. Soc.*, **66**, 87–92.

Cruikshank, B. (1965). *Medical Microbiology*. Livingstone, Edinburgh.

Dalton, H. and D. I. Stirling (1982). Co-metabolism. *Philos. Trans. R. Soc. London, Ser. B*, **297**, 481–496.

Difco Manual (1966) Difco Laboratories. Detroit, MI.

Dunn, G. M., R. A. Herbert and C. M. Brown (1978). Physiology of denitrifying bacteria from tidal mudflats in the River Tay. In *Physiology and Behaviour of Marine Organisms*, ed. D. S. McLusky and A. J. Berry, pp. 135–140. Pergamon, Oxford.

Dunn, G. M., J. N. Wardell, R. A. Herbert and C. M. Brown (1980). Enrichment, enumeration and characterisation of nitrate reducing bacteria present in sediments of the River Tay. *Proc. R. Soc. Edinburgh, Ser. B*, **78**, 47–56.

Ellwood, D. C. and J. R. Hunter (1976). The mouth as a chemostat. In *Continuous Culture 6: Applications and New Fields*, ed. A. C. R. Dean, D. C. Ellwood, C. G. T. Evans and J. Melling, pp. 270–282. Ellis Horwood, Chichester.

Evans, C. T. G., D. Herbert and D. W. Tempest (1970). The continuous culture of microorganisms. 2. Construction of a chemostat. In *Methods in Microbiology*, ed. J. R. Norris and D. Ribbons, vol. 2, pp. 277–327. Academic, London.

Fell, J. W., D. G. Ahearn, S. P. Myers and F. J. Roth (1960). Isolation of yeasts from Biscayne Bay, Florida and adjacent benthic areas. *Limnol. Oceanogr.*, **5**, 366–371.

Flannigan, B., R. N. Okagbue, R. Khalid and C. K. Teoh (1982). *Brew. Distilling Int.*, **12**, 31–37.

Fuller, M. S. and R. O. Poyton (1964). A new technique for the isolation of aquatic fungi. *Bioscience*, **14**, 45–46.

Gaertner, A. (1968). Eine Methode des quantitation Nachweises niederer, mit Pollen Köderbarer Pilze im Meerwasser und im Sediment. *Veroeff. Inst. Meeresforsch. Bremerhaven, Sonderband*, **3**, 75–92.

Gordon, R. E. and W. A. Hagen (1936). A study of some acid fast actinomycetes from soil with special reference to pathogenicity to animals. *J. Infect. Dis.*, **59**, 200–206.

Gregory, P. H. and M. E. Lacey (1963). Mycological examination of dust from mouldy hay associated with farmer's lung disease. *J. Gen. Microbiol.*, **30**, 75–88.

Harder, W. (1981). Enrichment and characterization of degrading organisms. In *Microbial Degradation of Xenobiotics and Recalcitrant Compounds*, ed. T. Leisinger, A. M. Cook, R. Hunter and J. Neusch, pp. 77–96. Academic, London.

Harder, W. and H. Veldkamp (1971). Competition of marine psychrophilic bacteria at low temperatures. *Antonie van Leeuwenhoek*, **37**, 51–57.

Harley, J. L. and J. S. Waid (1955). A method for studying active mycelia on living roots and other surfaces in the soil. *Trans. Br. Mycol. Soc.*, **38**, 104–118.

Harrigan, W. F. and M. E. McChance (1966). *Laboratory Methods in Microbiology*. Academic, London.

Harrison, D. E. F., T. G. Wilkinson, S. J. Wren and J. H. Harwood (1976). Mixed bacterial cultures as a basis for continuous production of SCP from C_1 compounds. In *Continuous Culture 6: Application and New Fields*, ed. A. C. R. Dean, D. C. Ellwood, C. T. G. Evans and J. Melling, pp. 122–134. Ellis Horwood, Chichester.

Hayes, A. J. (1979). The microbiology of plant litter decomposition. *Sci. Prog.*, **66**, 25–42.

Herbert, D., R. Elsworth and R. C. Telling (1956). The continuous culture of bacteria: a theoretical and experimental study. *J. Gen. Microbiol.*, **14**, 601–622.

Herbert, R. A. (1982). Procedures for the isolation, cultivation and identification of bacteria. In *Experimental Microbial Ecology*, ed. R. G. Burns and J. H. Slater, pp. 3–21. Blackwell, Oxford.

Higgins, I. J. and R. G. Burns (1975). *The Chemistry and Microbiology of Pollution*. Academic, New York.

Hungate, R. E. (1950). The anaerobic mesophilic cellulolytic bacteria. *Bacteriol. Rev.*, **14**, 1–49.

Hungate, R. E. (1969). A roll tube method for cultivation of strict anaerobes. In *Methods in Microbiology*, ed. J. R. Norris and D. W. Ribbons, vol. 3B, pp. 117–132. Academic, London.

Jannasch, H. W. (1965). Continuous culture in microbial ecology. *Lab. Pract.*, **14**, 1162–1163.

Jannasch, H. W. (1967). Enrichments of aquatic bacteria in continuous culture. *Arch. Mikrobiol.*, **59**, 165–169.

Jannasch, H. W. and R. I. Mateles (1974). Experimental bacterial ecology studied in continuous culture. *Adv. Microb. Physiol.*, **11**, 165–212.

Jeffrys, E. G. and W. K. Smith (1951). A new type of soil percolator. *Proc. Soc. Appl. Bacteriol.*, **14**, 169–171.

Johnson, L. F. and E. A. Curl (1972). *Methods for Research on the Ecology of Soil-Borne Plant Pathogens*. Burgess, Minnesota.

Jones, E. B. G. (1971). Aquatic fungi. In *Methods in Microbiology*, ed. C. Booth, vol. 4, pp. 335–365. Academic, London.

Kane, W. D. (1966). A new genus of *Actinoplanaceae*, *Pilimelia terevasa* and *P. annulata*. *J. Elisha Mitchell Sci. Soc.*, **82**, 220–230.

Karup, P. V. and J. A. Schmidt (1971). Isolation of *Nocardia* from soil by a modified paraffin bait method. *Mycologia*, **63**, 175–177.

Khan, M. R. and S. T. Williams (1975). Studies on the ecology of actinomycetes in soil. VIII. Distribution and characteristics of acidophilic actinomycetes. *Soil Biol. Biochem.*, **7**, 345–348.

Kuznetsov, V. D. (1969). Cultivation of members of the family *Actinoplanaceae* on plant pollen. *Mikrobiologia*, **38**, 118–120.

Lacey, J. (1971). The microbiology of moist barley storage in unsealed silos. *Ann. Appl. Biol.*, **69**, 187–212.

Lacey, J. and J. Dutkiewicz (1976). Isolation of actinomycetes and fungi from mouldy hay using a sedimentation chamber. *J. Appl. Bacteriol.*, **41**, 315–319.

Lees, H. and J. H. Quastel (1964). Biochemistry of nitrification in soil. 1. Kinetics of, and effects of poisons on, soil nitrification as studied by a soil perfusion technique. *Biochem J.*, **40**, 803–810.

Leftley, J. and I. Vance (1979). An inflatable anaerobic glove bag. In *Cold Tolerant Microbes in Spoilage and the Environment*, ed. A. D. Russell and R. Fuller, pp. 51–57. Academic, London.

Lovitt, R. W. and J. W. T. Wimpenny (1981). The gradostat, a bidirectional compound chemostat, and its application in microbiological research. *J. Gen. Microbiol.*, **127**, 261–268.

Luttrell, E. S. (1967). A strip bait for studying the growth of fungi in soil and aerial habitats. *Phytopathology.*, **57**, 1266–1267.

Marsh, P. D., J. R. Hunter, G. H. Bowden, I. Hamilton, A. S. McKee, J. M. Hardie and D. C. Ellwood (1983). The influence of growth rate and nutrient limitation on the microbial composition and biochemical properties of a mixed culture of oral bacteria grown in a chemostat. *J. Gen. Microbiol.*, **129**, 755–770.

McLeod, R. A. (1965). The question of the existence of specific marine bacteria. *Bacteriol. Rev.*, **29**, 9–23.

Miller, C. E. (1967). Isolation and pure culture of aquatic phycomycetes by membrane filtration. *Mycologia*, **59**, 524–527.

Monod, J. (1950). La technique de culture continuée. Théorie et application. *Ann. Inst. Pasteur, Paris*, **79**, 390–410.

Mouraret, M. and J. Baldensperber (1977). Use of membrane filters for the enumeration of autotrophic thiobacilli. *Microb. Ecol.*, **3**, 245–359.

Mueller, K. E. and L. W. Durrell (1957). Sample tubes for soil fungi. *Phytopathology*, **47**, 243.

Nakayama, N. (1982). Sources of industrial microorganisms. In *Biotechnology: Microbial Fundamentals*, ed. H.-J. Rehm and G. Ree, pp. 355–410. Verlag Chemie Weinheim, Basel.

Nonomura, H. and Y. Ohara (1969). Distribution of actinomycetes in soil: VI. A culture method effective for both preferential isolation and enumeration of *Microbispora* and *Streptosporangium* strains in soil. *J. Ferment. Technol.*, **47**, 463–469.

Nonomura, H. and Y. Ohara (1971). Distribution of actinomycetes in soil. IX. New species of the genera *Microbispora* and *Microtetraspora* and their isolation method. *J. Ferment. Technol.*, **49**, 887–894.

Norris, J. R. and D. W. Ribbons (1970). *Methods in Microbiology*, vol. 3A. Academic, London.

Novick, A. and L. Szilard (1950). Description of the chemostat. *Sciences, NY*, **112**, 715–716.

Okami, Y and T. Okazaki (1972). Studies on marine microorganisms. I. Isolation from the Japan Sea. *J. Antibiot.*, **25**, 456–460.

Orchard, J. A. and M. Goodfellow (1974). The selective isolation of *Nocardia* from soil using antibiotics. *J. Gen. Microbiol.*, **85**, 160–162.

Oxoid Manual (1979). 4th edn. Oxoid Ltd., London.

Papvizas, G. C. and C. B. Davey (1959). Evaluation of various media and anti-microbial agents for isolation of soil fungi. *Soil Sci.*, **88**, 112–117.

Park, D. (1972). Methods of detecting fungi in organic detritus in water. *Trans. Br. Mycol. Soc.*, **58**, 281–290.

Parkinson, D. (1967). New methods for qualitative and quantitative study of fungi in the rhizosphere. *Pedologie*, **7**, 146–154.

Parkinson, D. (1982). Procedures for the isolation, cultivation and identification of fungi. In *Experimental Microbial Ecology*, ed. R. G. Burns and J. H. Slater, pp. 22–30. Blackwell, Oxford.

Parkinson, D., T. R. G. Gray and S. T. Williams (1971). *Methods for Studying the Ecology of Soil Microorganisms*. Blackwell, Oxford.

Phaff, H. J., M. W. Miller and E. M. Mrak (1978). *The Life of Yeasts*, 2nd ed. Harvard University Press, Cambridge, MA.

Pirt, S. J. (1975). *Principles of Microbe and Cell Cultivation*. Blackwell, Oxford.

Powell, E. O. (1958). Criteria for the growth of contaminants and mutants in continuous culture. *J. Gen. Microbiol.*, **18**, 259–268.

Rodina, A. G. (1972). *Methods in Aquatic Microbiology*. University Park Press, Baltimore.

Rowbotham, T. J. and T. Cross (1977). Ecology of *Rhodococcus coprophilus* and associated actinomycetes in fresh water and agricultural habitats. *J. Gen. Microbiol.*, **100**, 231–240.

Skerman, V. D. B. (1967). *A Guide to the Identification of Genera of Bacteria*, 2nd. edn. Williams and Wilkins, Baltimore.

Slater, J. H. and A T. Bull (1982). Environmental microbiology: biodegradation. *Philos. Trans. R. Soc. London, Ser. B*, **297**, 575–596.

Stanier, R. Y., E. A. Adelberg and J. L. Ingraham (1977). *General Microbiology*, 4th edn. Macmillan, London.

Thimann, K. V. (1964). *Das Leben der Bakterien*. Fischer, Jena.

van Niel, C. B. (1931). On the morphology and physiology of the purple and green sulphur bacteria. *Arch. Mikrobiol.*, **3**, 1–112.

Veldkamp, H. (1970). Enrichment cultures of prokaryotic microorganisms. In *Methods in Microbiology*, ed. J. R. Norris and D. W. Ribbons, vol. 3A, pp. 303–361. Academic, London.

Veldkamp, H. (1976). Mixed culture studies with the chemostat. In *Continuous Culture 6: Applications and New Fields*, ed. A. C. R. Dean, D. C. Ellwood, C. T. G. Evans and J. Melling, pp. 315–328. Ellis Horwood, Chichester.

Veldkamp, H. and H. W. Jannasch (1972). Mixed culture studies with the chemostat. *J. Appl. Chem. Biotechnol.*, **22**, 105–123.

Veldkamp, H. and J. G. Kuenen (1973). The chemostat as a model system for ecological studies. *Bull. Ecol. Res. Commun.*, *NFR*, **17**, 247–256.

Warcup, J. H. (1955). The soil plate method for isolation of fungi from soil. *Nature (London)*, **116**, 117.

Warcup, J. H. (1957). Studies on the occurrence and activity of fungi in a wheat field soil. *Trans. Br. Mycol. Soc.*, **40**, 237–262.

Watson, S. W. (1965). Characteristics of a marine nitrifying bacterium, *Nitrocystis oceanus* sp. nov. *Limnol. Oceanogr.*, **10**, 274–289.

Williams, S. T. and C. I. Mayfield (1971). Studies on the ecology of actinomycetes in soil. III. The behaviour of streptomyces in acid soil. *Soil Biol. Biochem.*, **3**, 197–208.

Williams, S. T. and E. M. H. Wellington (1982). Principles and problems of selective isolation of microbes. In *Bioactive Microbial Products: Search and Discovery*, ed. J. D. Bu'Lock, L. J. Nisbet and D. J. Winstanley, pp. 9 26. Academic, London.

Williams, S. T., F. L. Davies, C. I. Mayfield and M. R. Khan (1971). Studies on the ecology of actinomycetes in soil. The pH requirements of streptomycetes from two acid soils. *Soil Biol. Biochem.*, **3**, 187–195.

Wimpenny, J. W. T. (1982). Responses of microorganisms to physical and chemical gradients. *Philos. Trans. R. Soc. London, Ser. B*, **297**, 497–515.

Windle-Taylor, E., N. P. Burman and C. W. Oliver (1953). Use of the membrane filter in the bacteriological examination of water. *J. Appl. Chem.*, **3**, 233–239.

Winogradsky, S. (1888). Beitrage zur Morphologie und Physiologie der Bakterium. *Pflanzenforschung*, **1**, 1–120.

4

Culture Preservation and Stability

A. DIETZ and B. W. CHURCHILL
The Upjohn Company, Kalamazoo, MI, USA

4.1   INTRODUCTION

Stability of microorganisms preserved for scientific and practical studies and use may be effected by various methods. Selection of a preservation method and the stability of desired properties of a microorganism are correlated with that microorganism's identity, properties, use, restoration and growth conditions. Record keeping is an integral part of the system. In this chapter we shall address determinants and procedures considered in the preservation of industrial microorganisms according to interest in and/or application of one or more of their properties.

Quality control is accepted as essential to the production of manufactured consumer goods. Reference to the term in the laboratory in relation to microorganisms is generally relegated to cultures used in microbial product evaluation. In the past decade the American Type Culture Collection, the British Federation for Culture Collections, the United States Federation for Culture Collections and the World Federation for Culture Collections have sponsored or participated in sessions at national and international meetings on the value to the scientific and industrial communities of authenticated, stable cultures.

4.2   PROCEDURES PRIOR TO SELECTING A PRESERVATION METHOD

One frequently hears it said that all cultures are lyophilized or frozen or stored on slants. What is not said is why, for example, they were lyophilized and not frozen, which properties of the culture affected the culture conditions prior to preservation, which conditions were employed to prepare the cultures for lyophilization, and which lyophilization procedures were used. In selecting a preservation method, consideration of the object of the project, records of previous treatment and lineage, and notation or reported characteristics of the culture are recommended.

4.2.1   Object of Preservation

Cultures are preserved for many reasons including demonstration of specific properties, reference, research, assay, phage host, plasmid host, comparison studies, type strains, production of fermentation products, starter cultures (cheese, bakery products), enzymes, insecticides and herbicides, and waste decomposition products.

The preservation of production cultures has the following goals: (1) to preserve the culture as near to initial productivity as possible, (2) preservation of the genetic purity of the culture without loss of any biochemical properties, and (3) preservation in such a way that they can be transported and handled with ease. The latter can sometimes be a major factor in the selection of a method of culture preservation.

In all cases the object is to have a source of cultures of known characteristics that can be stored for an unspecified time without loss of the characteristics.

4.2.2   Good Record Keeping of Previous Treatment and Lineage

Valuable information for a culture user is found in records tracing the lineage of a culture. Use of the culture for a specific test or production of a product is not guaranteed without information detailing cultivation and maintenance. Examples of this are given by Haynes *et al.* (1961) for bacteria whose spores are used to combat the Japanese beetle, by Sourek (1974) for preservation of pathogenic bacteria in the Czechoslovak National Collection of Type Cultures, and by Snyder (1981) for quality control in the clinical microbiology laboratory. One of us has also emphasized the role of record keeping in two recent publications (Dietz, 1981, 1982). The previous maintenance conditions of a culture may be very different from those to be employed by the culture recipient, therefore it is useful to record the culture source, the optimum growth conditions, medium, incubation temperature, length of incubation, incubation in light or dark, any special gaseous requirements, state of medium (liquid, solid, semisolid), aeration, *etc.*, as well as preservation conditions (preservation process, storage and storage temperature) as discussed in Heckly (1978) and Hesseltine and Haynes (1974).

The early establishment of a system for recording the previous treatment and lineage of production cultures is very important since the record may warn of, or elucidate, rundown in, for examples, antibiotic production.

4.2.3 Notation of Reported Characteristics of a Culture

Receipt or possession of a culture as a member of a certain genus and as being a certain species of that genus does not automatically guarantee that the culture should bear that specific epithet. It is important that both the supplier and the recipient observe the same characteristics for a culture. A culture received from a culture collection by a commercial transaction should have the properties cited for it in a scientific journal or in an issued patent. The recipient is requested to check for these properties upon receipt or within a specified time after receipt. The American Type Culture Collection in Rockville, MD, USA (Daggett *et al.*, 1982) repeatedly stresses quality control of its cultures and correctly disclaims responsibility for cultures bearing ATCC® numbers that have been distributed by others and for which the original properties are often lost. Mutagenized strains, phage hosts, and strains with inserted plasmids may carry the specific epithet of the Type Strain but differ significantly in macro- and micro-scopic morphology. The treatment of a strain and the resultant appearance of that strain should be noted to avoid taxonomic chaos. In noting reported characteristics, specific conditions under which the characteristics are observed must be given. The characteristics may be discerned under other conditions but reference points are essential to careful studies.

Notation of reported characteristics is equally important in laboratories using in-house isolates for potential product and/or product processes. Users of the culture should be able to confirm that the culture has the pertinent morphological and fermentative characteristics reported when the culture first became of interest. Such notations are imperative to the user of test strains. When large lots of a given strain are prepared, selective testing of samples can result in assurance that no mixup has occurred in the processing.

In selecting a preservation method, the maintenance of the significant properties of a culture and the quality of the culture upon restoration are considered. Notations made about a culture before preservation and after restoration enable the preserver to judge the reliability of the method chosen.

Notations of physical and biochemical characteristics of production cultures are very important and are discussed by Bader *et al.* (1981), Frazier and Westhoff (1979) and Mlall (1975). Culture back mutations, loss of biochemical properties, and even phenotypic traits may warn of unstable traits in a production strain.

4.3 DETERMINANTS FOR CULTURE IDENTITY, CHARACTERISTICS AND PURITY

Culture identity may be made simply on the basis of the same appearance as when received when grown on a specified medium, on the basis of exhibition of certain macro- and micro-scopic properties, or on the basis of an extensive evaluation using many tests. In all cases standardized test procedures must be employed. The selection of tests is determined by the class to which the culture belongs. In this presentation we are referring to prokaryotic and eukaryotic microorganisms. Selective tests are used for culture identity, characterization, and purity determinants. Commercially available test systems (API, Minitek; reference to such systems does not mean endorsement by The Upjohn Company) may be used for identification of a number of genera of prokaryotes and for some eukaryotes. For others, classical systems used by actinomycetologists, bacteriologists, mycoplasmologists, mycologists and zymologists must be applied. Examples are given in Barnett *et al.* (1983) and Starr *et al.* (1981).

4.3.1 Authenticated Cultures — Confirmation of Stated Traits

Authenticated cultures, *i.e.* cultures purchased from a collection devoted to the preservation of reference cultures, are guaranteed by that collection to have the properties cited in published descriptions of the cultures. The recipient should have no difficulty verifying stated traits of these cultures.

4.3.1.1   Morphological

The culture recipient and user are concerned with authenticity of the culture being used. The most practical verification of an authenticated strain is by a morphologic check. This is done using growth conditions specific for the strain being checked. Morphological traits such as degree of pigment production and zones of hydrolysis of starch in an agar medium should be recorded for the culture so that each time a slant is inoculated with a production culture these macromorphological traits can be checked.

Verification of reported traits of surface color and texture, reverse color, and pigment production and growth at a specific temperature are recommended. One familiar with microbes with certain traits may rely on this information alone.

Verification of microscopic properties is a must for many organisms which do not exhibit strong macromorphologic properties. Gram stain, acid-fast stain, flagella and other stains may be required for confirmation of properties of many unicellular forms. Spore and bud production are also important. For filamentous forms the type of spore formation and fragmentation or not of mycelium are significant. Observation of surface detail of spores may also be needed.

4.3.1.2   Biochemical

Reactions on various differential media and chemotaxonomic tests must be performed if more detailed differentiation is required. This is particularly true for test organisms. Many diagnostic kits are available making such tests rapid and reliable. These kits are limited to specific prokaryotic groups and yeasts (API, Minitek). Tests for biochemical differentiation of actinomycetes, filamentous fungi and most yeasts are significant and time-consuming (Barnett *et al.*, 1983; Starr *et al.*, 1981).

The level of antibiotic biosynthesis by a production culture should be determined each time a culture is retrieved from storage. Care must be taken to retrieve a culture each time by the same method. A complete record should be kept of the exact fermentation conditions (media ingredients, method used to prepare medium, sterilization time and temperature, seed growth conditions and conditions used to obtain optimum titers) used to obtain high antibiotic titers.

4.3.1.3   Physiological

Included here are verification of product potential, reproducibility of assays or any test dealing with the function and activity of the culture.

4.3.2   Research and Development Strains

Research and development strains are usually new isolates for which characterization is incomplete. Minimal observations of properties will have been made. Additional tests are not economically feasible until product potential is recognized.

4.3.2.1   Elimination of 'leaky' mutants

'Leaky' mutants must be eliminated since they are not reliable in recombination or nutritional studies.

4.3.2.2   Assurance of auxotrophic traits (elimination of mixed 'genetic bag')

We have found that a mixed 'genetic bag' can be eliminated by grinding colonies in a tissue homogenizer and plating out on agar to select genetically pure clones. The testing of cultures for purity of auxotrophic characteristics can be accomplished by replica plating the strain to a selective medium for demonstration of auxotrophic traits.

4.3.2.3 *Selective pressure for maintaining specific culture traits*

Many methods of applying selective pressure to maintain specific culture traits can be designed for each improved yielding clone. One effective tool is the challenging of an antibiotic-producing culture with relatively high levels of that antibiotic.

4.3.2.4 *A complete medium for the expression of all clones of a genetically altered population*

We have designed a medium for the expression of all clones of a genetically altered population by adding high nucleic acid yeast, casein hydrolysate and corn-steep liquor.

4.4 REVIEW OF METHODS FOR CULTURE STORAGE

Published culture storage methods have been cited and analyzed by Heckly (1978) and Dietz (1981, 1982). There are probably many methods used by individuals with small collections that are very efficacious but which have not been brought to the attention of most workers.

The methods used must ensure preservation of culture traits essential to the use of the restored culture. The method used in a small laboratory is not necessarily one that will be efficient and reliable for preserving a large collection of diverse strains, for preserving strains which may be undergoing research evaluation, for test strains, or for production.

4.4.1 Long-term Storage

Stock culture collections, especially those housing patent cultures, must ensure preservation of specified properties of the preserved cultures for at least 30 years according to patent treaties (Saliwanchik, 1982). The method selected and its reliability are strengthened by records pertaining to culture handling prior to institution of the preserved state. This includes recording the age of the culture at preservation as well as growth conditions. An example of the value of this is in the study of cellular vulnerability at various stages in the cell cycle in *Saccharomyces cerevisiae* (Cottrell, 1981).

It is essential that production cultures be prepared for preservation at the physiological state at which optimal survival will be obtained. It is necessary to examine cultures frequently until the incubation period which gives optimum sporulation has been determined. Streptomycetes are generally incubated from a week to 10 days to give good sporulation but mature spores are usually assured if incubation is extended to two weeks. Non-sporulating cultures have been preserved adequately by growing vegetative mycelia in a broth medium. In order to greatly increase the number of growth centers, we frequently grind the vegetative mycelia in a tissue homogenizer prior to lyophilization or storage in the vapor phase in a liquid nitrogen refrigerator.

The long-term methods of choice for the storage of production cultures vary from one to another. However, there seems to be little doubt that lyophilization and soil tube storage are the most cost efficient and require the least maintenance for long-term storage (Hatt, 1980). Most antibiotic biosynthetic strains can be reliably stored over a long period in either lyophilized tubes or in glass ampoules sealed with a gas–oxygen flame and stored in a liquid nitrogen refrigerator. The necessity of developing an enduring labeling system for long-term storage is apparent. A lyophilized tube after being sealed in glass is coated with a good grade finger nail polish which preferably is somewhat opaque. The lyophilized tube or ampoule is then labeled by writing the desired label on the nail polish surface with India ink. Lyophilized tubes labeled in this way have endured storage in a 3–4 °C refrigerator for 30 years. Recently ampoules with silk screen labels have become available. The importance of developing a precise inventory system for long-term storage has dictated the use of a more complete label than would otherwise be required. The distribution of lyophilized tubes with labels containing research notebook number, initials of research scientist, page and entry number of that research notebook page, as well as antibiotic or other product code designation, culture number and tube number, give a fail-proof system. The tubes are distributed in ascending order so that the remaining number of tubes in a lyophilized lot is always easily determined. The storage facility must always be equipped with an alarm system to warn of temperature fluctuation and a locked door since production stocks are priceless.

4.4.1.1 Cost efficiency

Cost of preparation, preservation and housing cultures is a consideration in the method selection. The method of choice for culture collections housing reference strains and patent cultures is lyophilization (Hatt, 1980; Hesseltine and Haynes, 1974). Storage of lyophiles may be at room temperature (if the temperature is kept constant at 24–26 °C) or under refrigeration (4 °C).

4.4.1.2 Minimal maintenance

Constant culture transfer is costly both in time and supplies. Storage states that retain the desired culture properties with minimal attention are refrigerated sterile soil stocks, lyophile preparations, and preparations stored in liquid nitrogen. Such states are effective for most prokaryotes and eukaryotes. Assurance of constant temperature and refrigeration states is required. If liquid nitrogen is used there must be assurance of a constant supply. Back-up power sources must be available in the event of a sudden failure. The frequent, such as once a week, transfer of important cultures is very time consuming. In addition biosynthetic potential is frequently lost by serial transferring.

The simple altering of a refrigerator to include sliding drawers in a built in cabinet gives the reliability of a mechanical refrigerator with no maintenance cost. Our refrigerator has maintained a temperature of 4 ± 0.5 °C over a period of 30 years. Thousands of lyophilized tubes of fermentation and production strains are stored in that refrigerator and can be easily retreived from the sliding drawers which are labeled with the culture designations. Even though the unit is equipped with a sensor for a change of 1 °C temperature, the alarm has never recorded a malfunction. The combination lock on the refrigerator door gives good security for the cultures in storage.

4.4.1.3 Reliability

The three conditions of storage cited above, sterile soil, lyophilization and liquid nitrogen, are discussed in Antheunisse *et al.* (1981), Dietz (1975, 1982), Hill (1981), Hossack (1972), Nei and Oshiro (1979), Pridham *et al.* (1973), Sokolski (1975) and Trejo (1975).

4.4.1.4 Endurance of label

All stored cultures must be labeled. Even if only one culture is important enough to be stored it should be identified; it may be a potent pathogen which should not be handled carelessly. An emergency may occur and someone other than the curator may have to retrieve it. There should be no mistaking the stored culture from some similarly stored material. Refrigeration tape, or direct marking with cold-tolerant, non-smearing ink are recommended for refrigerated samples. Painting the sample vial with a good grade nail polish, letting it harden (preferably overnight) and labeling with a Sanford Sharpie® or CMS™ (Curtis Matheson Scientific Inc.) ink pen is very good for ultra-cold storage. Alternatively, vials with an etched labeling spot may be used.

4.4.1.5 Precise inventory system

Stored cultures should be coded so that retrieval is simple. Alphabetical or numerical systems are recommended. Numerical systems are favored by experienced curators. For storing different types of organisms a combination alphabetical–numerical system is recommended so that like classes are grouped together. For example, in the Upjohn stock culture collection numerical sequences for actinomycetes, bacteria and fungi are such that no confusion of members of these groups occurs in storage, retrieval, or record files.

4.4.2 Short-term Storage

Short-term storage implies use of the stored material within a specified time. However, the stability of the culture properties in this period is as vital as in long-term storage.

It is essential to have a convenient, simple method of preparing samples for short-term storage for potential production isolates. A convenient method is to use sterile 2 or 3 g screw-cap vials for the storage of 3–5 ml aliquots of vegetative seed as tissue homogenizer ground suspensions from agar surface or spore suspension in a liquid nitrogen freezer. There is a rapid turnover of the short-term storage so that the storage space is economical per group of samples. The labeling of the screw-cap vials is easily accomplished since these are now equipped with a rectangular silk screen area for labeling. This method has been found to be very reliable as well as easy to use since the productivity of antibiotic-producing isolates so stored is retained for several months. The suspensions in the vials can be thawed, sampled and refrozen with a minimum of effort and can be later retrieved for retesting or recloning.

4.4.2.1 Ease of sample preparation

Short-term storage frequently involves large volumes of cells for test purposes. Growth of the material for storage, preparation of the growth for storage, and choice of preservation state are preferably those that entail minimal effort, for example subculturing cells in liquid media, harvesting the cells and dispensing them in preservation vials while maintaining them in the log phase. Screw-cap or snap-top vials, filter paper discs in plates or vials may all be seeded rapidly so that growth beyond the log phase does not occur.

4.4.2.2 Label reliability

Labels must not smudge or come off. There are usually volumes of cells of different microbes stored in a container. Storage containers as well as the vials must be well marked and not subject to smudging from handling.

4.4.2.3 Economic aspects

Storage containers that are temperature controlled may be as simple as a household type refrigerator. After the initial investment in a storage unit, costs for storage should be only those for materials, manpower, and utilities, all of which will be applied to other projects as well as to the storage project.

Cultures native to a particular climate may be stored at the temperature of the isolation source. This is particularly important to workers in developing countries and to those in certain areas of developed countries.

4.4.2.4 Reliability

Cultures must be stable for the use period of the preserved material. Examples are weekly subculturing for some assays, refrigerated frozen cultures as agar plugs, or shake flask seed dispensed in inoculum size aliquots, or spore suspensions in capped vials.

4.4.2.5 Ease of retrieval

Again, a labeling/coding system must be devised so that no costly mixups occur. It is desirable to use the same size vial and storage container for all samples so that a uniform system can be employed.

4.4.2.6 Rapid retrieval

Removal from the storage container and restoration of the stored material should be accomplished in as short a time interval as possible, particularly in clinical or industrial operations where results are needed for evaluation and determination of patient treatment, continuation of a research experiment, evaluation of a product, or production of a product.

4.5 SELECTION OF MAINTENANCE CONDITIONS AND PROCEDURES FOR IMPLEMENTATION, BASED ON CULTURE USE

Cultures deposited in a collection should come to the curator of the collection with specific information, *i.e.* growth medium and optimum incubation temperature, reason for deposit, and ideal age of culture at storage.

It is recognized that certain details may not be disclosed by depositors to outside collections and that such information may not always be provided to in-house collections. The individual charged with maintaining the culture should be familiar with the characteristics and taxonomy of cultures maintained so that a rational selection of means of handling the culture for preservation and restoration may be made.

4.5.1 Long-term Storage

Optimum conditions must be selected and no universal condition is employed even by such well-known collections as ATCC® (Hatt, 1980) and NRRL (Hesseltine and Haynes, 1974). Cultures used for production of fermentation products in our manufacturing plant are each studied extensively for the optimum maintenance conditions to retain their highest level of productivity. Most cultures are stored in liquid nitrogen freezers although occasionally microorganisms are tested which cannot tolerate the freezing process used for nitrogen frozen inoculum. Recently some plant pathogen microorganisms were studied which could not tolerate liquid nitrogen freezing or lyophilization and they were stored in a frozen state in a mechanical freezer (Churchill, 1982). Lyophilized tubes of some cultures are prepared since they are much more readily transported from one manufacturing plant to another or liquid nitrogen may not be available at a particular plant location. In addition, long-term storage of production cultures is usually by means of lyophilization. It is imperative to maintain those isolates which have lost their ability to produce conidia through mutation and selection in a vegetative state in liquid nitrogen frozen ampoules which have been sealed with a gas–oxygen torch for production working stocks.

In preparing production cultures for lyophilization, we routinely add 10% dried skim milk (Difco) suspension with the spores, vegetative growth or tissue homogenizer ground growth from shake flasks or agar surface. Many additives to enhance the lyophilization process have been suggested by various investigators with the oldest being horse serum, but the most popular today is skim milk. Many penetrating and non-penetrating agents which reportedly protect freeze sensitive microorganisms have been added to suspensions to be lyophilized with good success. Production cultures which have auxotrophic requirements usually are lyophilized with an enrichment of the auxotrophic requirement to the suspension to be lyophilized. This results in a substantially higher percent viability of the cells in most instances.

The lyophilization process requires great care in order to obtain high percentage survival of the cells upon restoration. Our sterile non-absorbent cotton plugged tubes are prepared for lyophilization by placing 0.2 ml of the skim milk–spore or cell suspension in each tube with a sterile syringe and needle assembled, slowly cooling the suspension in the tubes to −30 °C, and placing the 100 mm × 8 mm O.D. tubes plugged with cotton in a lightly vacuum greased connector in a lyophilizer. The ends of the tubes containing the frozen skim milk culture suspension are covered with styrofoam boots so that the frozen suspension will not melt during the freeze drying process. The vacuum valves are turned on and the operational level of vacuum to the individual tubes is maintained at 5–10 μm with a direct drive vacuum pump. A lot of 48 tubes in the manifold of a conventional lyophilizer is dry in about 2 h. The Pyrex tubes are then sealed with a gas–oxygen torch when the vacuum in the tubes is at least 10 μm. The sealed tubes are tested for leaks after 24 h with a Tesla Coil. Tubes without leaks contain a blue 'vacuum glow' when the Tesla Coil is placed against the non-filled end of the tube which shows that the tube still has an acceptable vacuum. The viability of spores or vegetative fragments of hyphae of several antibiotic production *Streptomyces* sp. strains after lyophilization by the above process was in the range of 90–95%. Comparison lyophilization runs made when the vacuum was purposely not as good, and consequently the drying time required was 6–7 h, had a respective range of viabilities of only 9–10%. In order to be assured that the production culture will maintain its expected level of production of the desired antibiotic or other fermentation product, the viability of the lyophilized culture should be as close to 100% as is possible.

Filamentous forms may be fragmented and stored in sterile soil (sandy-loam). The soil in tubes plugged with gauze-enclosed non-absorbent cotton is inoculated with sterilized distilled water suspensions of 10–14 day agar slant growth. The soil is saturated with culture suspension. Soil stocks are air dried at 24 °C for about one month. The tubes when dry should be tapped gently to loosen the soil; the tubes are then stored at 4 °C.

4.5.1.1 Analytical organisms

In our facilities long-term preservation of analytical organisms is the responsibility of the central culture collection. The satellite or user collections request a new subculture only if a problem develops in its use. The central collection is responsible for maintaining the culture in such a state that the desired use properties are retained. The subject is discussed by Ceyle (1976).

4.5.1.2 Comparison strains

Such strains are stored (as the name implies) for comparative purposes.

For assay purposes a decision may be made to use an organism or a different strain of the currently used assay organism. It will be desirable to use strains stored under the same conditions for such an evaluation. Therefore, if obtained from an outside source or from strain selection programs in-house, the new strains will be stored under the same conditions as the present strain and revived from such a storage state.

Metabolite strains naturally have potential for a product-oriented company and the valuable metabolite potential must be retained. This topic is covered in Miall (1975) and Perlman and Kikuchi (1977).

Pathogenic strains are usually significant *in vivo* test strains and retention of the pathogenicity feature is required. A sample of the preserved lot must be evaluated for this feature as soon as possible after preservation.

Strains are preserved for comparitive taxonomic studies with newly isolated or acquired or altered strains, *i.e.* type cultures are needed for comparison purposes. All the properties cited in a scientific publication and/or a patent must be retained.

4.5.1.3 Manufacturing plant cultures

Cultures used for production of fermentation products in the manufacturing plant are each studied extensively for the optimum maintenance conditions to retain their highest level of productivity. Many cultures are stored in liquid nitrogen freezers although occasionally cultures are tested which cannot stand the freezing process used for liquid nitrogen frozen inoculum such as described in Churchill (1982). The majority of production cultures are maintained in liquid nitrogen freezers unless they are to be transported to plants at some distance from the location where they are maintained. Lyophilized tubes of some strains are prepared since they can be much more readily, safely, and economically transported to other plants. It is imperative to maintain those cultures which no longer form conidia due to mutation and selection in a vegetative state in sealed ampoules in liquid nitrogen.

4.5.2 Short-term Storage

Many cultures are stored on a short-term basis for evaluation of certain useful and/or interesting properties; others are expected to be used over a limited time span. The storage state for these latter cultures may or may not be the same as those for long-term storage.

Clones from populations for culture improvement studies are routinely removed from agar surfaces, ground in a tissue homogenizer and stored in screw-capped vials. The use of a slow cooling rate of 1 °C min^{-1} has been recommended by numerous investigators in order to retain maximum viability and production of desirable metabolites such as antibiotics (Trejo, 1975). A large variety of slow controlled-rate cooling units are available which will hold from a few to many ampoules or vials. Very simple units such as styrofoam blocks with holes in them can be used to simulate slow cooling and can be used to determine the importance of this factor in particular culture preservation problems. Microorganisms which are very sensitive to freeze damage are usually protected with a penetrating agent such as DMSO or glycerol, or a non-penetrating agent such as dextran, glucose, lactose, malt extract, mannitol, polyvinylpyrrolidone, polyglycol, sorbitol or sucrose. Frequently a combination of two or more of these agents give even more protection to the microorganism upon freezing for storage in liquid nitrogen.

4.5.2.1   New metabolite producers for investigative studies

Such cultures may be stored on refrigerated slants or plates, as frozen (deep-freeze or liquid nitrogen gas-phase) agar plugs or shake flask growth. If it is economically feasible the frozen state may be the best, as it will be possible to make a smooth transition to long-term storage in the same state (Rinfret and LaSalle, 1975).

4.5.2.2   Clones from populations for improved metabolite producers

These may be stored in the same manner as the metabolite producer.

4.5.2.3   'Working stocks' of analytical organisms

These organisms may be subcultured weekly, or daily in some cases, or stored in large working lots in a refrigerated or frozen state. Alternatively, commercial preparations may be obtained from, for example, Difco or ATCC.

4.6   CULTURE RESTORATION AND GROWTH CONSIDERATIONS

Requirements for restoration of a stored culture may differ from the growth requirements for actively growing cultures. The methods used for restoration and subsequent growth of cultures for antibiotic or other desired metabolite production have frequently been demonstrated to be critical to obtain maximum productivity. Some general guidelines can be outlined for most microorganisms but each culture should be tested under a variety of restoration and growth conditions in order to select optimum conditions.

4.6.1   Restoration

A number of factors merit consideration in the restoration process. Extreme care as well as standard methods must be taken in culture restoration, especially from lyophilized tubes that will be used for metabolite production. It has frequently been demonstrated that the more nearly conditions that were used for growth of the production cultures to be lyophilized are repeated when the culture is restored, the better will be the viability. The preparation of liquid nitrogen frozen ampoules for restoration is usually not as critical as is the rehydration of lyophilized tubes. Production frozen ampoules are thawed rapidly, in contrast to the very slow cooling during their preparation, in order to retain high viability and productivity. Again, the medium used for restoration from sealed ampoules is best if the same osmotic and nutrient composition and medium strength is used as was used to grow the culture for dispensing in the ampoules. Assay cultures used in control, which are stored in the liquid phase of liquid nitrogen, are rapid-frozen and rapid-thawed. Cultures preserved in sterile soil may be restored by aseptic transfer of an inoculating loop of the soil to the appropriate growth medium in plates, tubes, or shake flasks.

4.6.1.1   Concentration of inocula

The inocula used may be based on viable cell count (in the case of unicellular forms), on the volume of thawed frozen material, on the size of transfer loop of material, or on the volume of suspension of rehydrated lyophilized sharp frozen material. For example, we have found that 0.2 ml of a thawed suspension of cells frozen in the gas-phase of liquid nitrogen gives good growth in 16–18 h in 10 ml of broth culture in a test tube and in a 100 ml broth culture in a 500 ml shake flask if the appropriate aerobic or anaerobic growth conditions and temperature state are employed.

The restoration of lyophilized tubes is found to be most successful, in terms of percent survival, if during restoration the concentration of spores, cells or vegetative hyphae are at the same concentration as they were in the suspension milk before lyophilization. Frequently the vegetative growth or spore suspension is placed in ampoules before liquid nitrogen storage has been concentrated. This concentration is either by centrifugation or gravity settling before the ampoules are filled. This can be diluted when it is used as primary inoculum for the manufacturing process. During the restoration of liquid nitrogen frozen ampoules, we seldom observe any difference between diluted or concentrated growth. We have found that regardless of whether a microorganism to be used in the manufacturing plant is derived from a lyophilized tube, liquid nitrogen frozen ampoule, deep freeze stored ampoule or a soil tube, it must have optimum concentration for restoration on an agar surface. That concentration must be such that confluent growth will occur over the entire agar surface. The use of too low a concentration either on agar or in nutrient broth results in an excessive number of generations of cells. Productivity of the antibiotic may be greatly decreased in subsequent fermentations.

4.6.1.2 Nutrition

The nutritional broth used during restoration is predominantly the nutrient medium used for vegetative growth of the culture. There is a similar requirement for rehydration and growth regardless of whether a culture is being restored from a lyophilized or soil tube. It is not unusual to increase the survival of spores or vegetative propagules by at least 10–25% by replacing distilled water with an acceptable nutrient broth for restoration. The nutrient medium used for the restoration of cultures from sealed ampoules is best if the nutrient medium composition and medium strength is the same as was used for initial culture growth.

4.6.1.3 Osmotic (rehydration)

The placing of a lyophilized dry pellicle, which is void of any moisture, in distilled water for rehydration will usually result in the destruction of many/most of the cells because of the too rapid hydration resulting in burst cells. Even the use of isotonic solution will improve the restoration viability because of the increased osmotic strength during rehydration. Similar results are experienced when the culture is restored from dry soil culture stocks. The use, during restoration, of nutrient medium with the same or similar osmotic strength to that used in initially growing the culture gives a greatly increased percent survival from soil or lyophilized cultures.

4.6.1.4 Temperature (rehydration and/or rate of melting)

The temperature at which a liquid nitrogen frozen ampoule is thawed is not critical as long as the frozen ampoule is thawed very rapidly. Occasionally microorganisms are found which need not be thawed rapidly in liquid nitrogen frozen ampoules in order to give high survival and subsequent optimum antibiotic production. This is the exception rather than a general rule. Many investigators have used a system of placing a rack of frozen ampoules in an agitated water bath at approximately 40 °C. Usually the rack is agitated slightly to cause mixing in the ampoules during thawing. Cultures thawed rapidly not only have a greater percent survival but we have found that most antibiotic-producing organisms retain their maximum productivity when they are melted rapidly. In the case of lyophilized cultures, experience has shown that the first step is to place the sealed glass tube at ambient temperature for a period of at least 30 min to allow equilibration in warming from 4 to about 26 °C. The temperature of the nutrient broth used during the rehydration and restoration of lyophilized or otherwise dried culture stocks can greatly affect the rate of rehydration (Churchill, 1982). The use of a temperature substantially lower than optimum for growth is not always necessary during rehydration. Great care taken to maintain a high enough osmotic strength during rehydration will result in a very slow rehydration of the cells or spores.

This slow rehydration results in maximum survival of the previously dried spores without a lowering of the temperature during rehydration.

4.6.2 Growth

The conditions of nutrition, temperature and dissolved gases that are optimum for a particular culture during the growth phase after restoration cannot be predicted even within a group of closely related microorganisms. The growth phase can be started much more easily and quickly with liquid nitrogen frozen ampoules regardless of whether the organism has been stored as cells, spores or vegetative mycelia. The growth stage of previously dried cultures from lyophilized, soil, or other dried viable organisms may require conditions for growth drastically different to those used for rehydration.

4.6.2.1   Requirements

In general, one has most assurance of success in establishing good growth with the use of a balanced nutrient medium as soon after rehydration as is feasible. It is sometimes found that the initial growth after rehydration has an extended lag period. This lag can be substantially reduced if a growth medium is initially employed which is of the same composition as that which gives optimum growth but which is of decreased strength of 25–50% of the original medium. Care must be taken if an organism has auxotrophic or special nutrient requirements to ensure that a sufficient level of the required nutrients are included in the growth medium. Such organisms are often extremely sensitive at this stage to medium deficiencies. We have found that a change in ability to produce enzymes often occurs under storage conditions which can only be restored after a period of active growth. An example is the loss of amylase production in some cultures after liquid nitrogen storage. The initial growth after storage is much more rapid if glucose is added to the nutrient medium even though the organisms are routinely grown on starch.

4.6.2.2   Temperature

The preferred growth temperature after rehydration which gives the optimum growth rate is usually identical to the temperature used in the manufacturing plant for seed growth. Frequently the temperature required for rapid growth of antibiotic-producing microorganisms is not the same as that used for the maximum antibiotics production phase in the fermentation stage. The optimum incubation temperature for growth of a liquid nutrient medium is frequently lower than that which gives optimum growth on nutrient agar media.

4.6.2.3   Aeration (including dissolved gases)

We routinely use low and high aeration rates in parallel in growing cultures in shake flasks after the cultures have been retrieved from lyophilized tubes, liquid nitrogen frozen ampoules, soil tubes or deep freeze frozen ampoules (Churchill, 1982). We have observed with some cultures that growth is more rapid after retrieval if a somewhat different aeration rate is maintained than is used when the culture is being routinely grown. Only rarely have we found isolates which require either an oxygen enriched gas stream or higher levels of carbon dioxide in air to give optimum growth.

4.6.2.4   Duration

The duration of growth on a nutrient agar should be such that mature spores are produced which usually requires from 7–14 days. Vegetative growth after rehydration in shake flasks usually requires from 2–4 days until the peak growth is obtained. Antibiotic-producing cultures should usually be transferred to secondary seed media before the stationary phase is reached which is usually accompanied by fragmenting and autolysis of vegetative growth.

4.6.3 Verification of Purity

The purity of the production culture after retrieval is usually checked microscopically as well as by transfer to a phenol red or thioglycollate broth for the presence of contaminating bacteria. Occasionally serial dilutions on nutrient agar will demonstrate morphological variants in the culture. Frequently cultures which are to be used in the manufacturing plant for metabolite production are tested in shake flask fermentations for authentication of the expected productivity.

4.7 REFERENCES

Antheunisse, J. J. W. de Bruin-Tol and M. E. van der Pol-Van Soest (1981). Survival of microorganisms after drying and storage. *Antonie van Leeuwenhoek*, **47**, 539–545.

Bader, F. G., M. D. Young and L. L. Kempe (1981). Lincomycin production and culture stability in continuous production. *Abstr. Pap. Am. Chem. Soc. 182 Meeting*, MICR 49.

Barnett, J. A., R. W. Payne and D. Yarrow (1983). *Yeasts, Characteristics and Identification*, Cambridge University Press, London.

Ceyle, M. B. (1976). Reproducibility of control strains for antibiotic susceptibility testing. *Antimicrob. Agents Chemother.*, **10**, 436–440.

Churchill, B. W. (1982). Mass production of microorganisms for biological control. In *Biological Control of Weeds with Plant Pathogens*, ed. R. Charudattan and H. Walker, pp. 139–156. Wiley, New York.

Cottrell, S. F. (1981). Yeast freeze–thaw survival rates as a function of different stages in the cell cycle. *Cryobiology*, **18**, 506–510.

Daggett, P.-M., R. L. Gherna, P. Pienta, W. Nierman, S.-C. Jong, H.-ti Hsu, B. Brandon and M. T. Alexander (1982). *American Type Culture Collection Catalogue of Strains. I*, 15th edn. American Type Culture Collection, Rockville, MD.

Dietz, A. (1975). Nitrogen preservation of stock cultures of unicellular and filamentous microorganisms. In *Round Table Conference on the Cryogenic Preservation of Cell Cultures*, ed. A. P. Rinfret and B. LaSalle, pp. 22–36. National Academy of Sciences, Washington, DC.

Dietz, A. (1981). Pure culture methods for industrial microorganisms. In *Biotechnology*, ed. H.-J. Rehm and G. Reed, vol. 1, pp. 412–434. Verlag Chemie, Weinheim.

Dietz, A. (1982). Culture preservation and instability. In *Bioactive Microbial Products: Search and Discovery*, ed. J. D. Bu'Lock, L. J. Nisbet and D. J. Winstanley, pp. 27–35. Academic, New York.

Frazier, W. C. and D. C. Westhoff (1979). Production of cultures for food fermentations. *Food Microbiology*, 3rd edn., chap. 20, pp. 331–340. McGraw Hill, New York.

Hatt, H. (ed.) (1980) *American Type Culture Collection Methods. I. Laboratory Manual on Preservation, Freezing and Freeze-Drying as Applied to Algae, Bacteria, Fungi and Protozoa.* American Type Culture Collection, Rockville, MD.

Haynes, W. C., G. St. Julian, Jr., M. C. Shekleton, H. L. Hall and H. Tashiro (1961). Preservation of infectious milky disease bacteria by lyophilization. *J. Insect Pathol.*, **3**, 55–61.

Heckly, R. J. (1978). Preservation of microorganisms. *Adv. Appl. Microbiol.*, **24**, 1–53.

Hesseltine, C. W. and W. C. Haynes (1974). Sources and management of microorganisms for the development of a fermentation industry. *Agriculture Handbook No. 440*, Supt. of Doc., US Government Printing Office, Washington, DC.

Hill, L. R. (1981). Preservation of microorganisms. In *Essays in Applied Microbiology*, ed. J. R. Norris and M. H. Richmond, pp. 2/1–30. Wiley, Chichester.

Hossack, D. J. N. (1972). Liquid nitrogen frozen inocula — a year's experience. *Proc. Soc. Analt. Chem.*, **9**, 35–38.

Miall, L. M. (1975). Historical development of the fungal fermentation industry. In *The Filamentous Fungi*, ed. J. E. Smith and D. R. Berry, vol. 1, chap. 6, pp. 104–121. Edward Arnold, London.

Nei, T. and R. Oshiro (1979). Effects of freezing and drying on genetic characteristics of yeast cells. *Cryo-Lett.*, **1**, 24–27.

Perlman, D. and M. Kikuchi (1977). Culture maintenance. *Annu. Rep. Ferment. Processes*, **1**, 41–48.

Pridham, T. G., A. J. Lyons and B. Phrompatima (1973). Viability of *Actinomycetales* stored in soil. *Appl. Microbiol.*, **26**, 441–442.

Rinfret, A. P. and B. LaSalle (1975). *Round Table Conference on the Cryogenic Preservation of Cell Cultures*. National Academy of Sciences, Washington, DC.

Saliwanchik, R. (1982). *Legal Protection for Microbiological and Genetic Engineering Inventions*. Addison-Wesley, Reading, MA.

Snyder, J. W. (1981) Quality control in clinical microbiology. *Species*, **5**, 13–21.

Sokolski, W. T. (1975). Preservation of cultures in routine microbiological assay operations. In *Round Table Conference on the Cryogenic Preservation of Cell Cultures*, ed. A. P. Rinfret and B. LaSalle, pp. 50–59. National Academy of Sciences, Washington, DC.

Sourek, J. (1974). Long-term preservation by freeze-drying of pathogenic bacteria of the Czechoslovak National Collection of type cultures. *Int. J. Syst. Bacteriol.*, **24**, 358–365.

Starr, M. P., H. Stolp, H. G. Truper, A. Balows and H. G. Schlegel (1981). *The Procaryotes*. Springer-Verlag, New York.

Trejo, W. H. (1975). Some considerations in the cryogenic preservation of cultures. In *Round Table Conference on the Cryogenic Preservation of Cell Cultures*, ed. A. P. Rinfret and B. LaSalle, pp. 9–15. National Academy of Sciences, Washington, DC.

5

Genetic Modification of Industrial Microorganisms

G. Holt and G. Saunders
The Polytechnic of Central London, UK

5.1 INTRODUCTION

The life and activities of microbial cells have developed as a 'compromise' (Tempest *et al.*, 1983) dictated by conditions of environmental and nutritional constraint and of competition to which they have been routinely exposed in their natural ecosystems. Although, when transferred to laboratory culture, the conditions are usually far removed from those prevailing in the natural environment, regulatory mechanisms still operate to prevent 'wasteful' expenditure of energy. Thus, it is not surprising that microbial metabolites of possible industrial interest are normally produced initially at very low levels. If the product turns out to be of commercial value, programmes of media development and strain improvement are initiated.

In Section 5.2 we describe how, in the light of knowledge gained over a number of years, particularly with respect to the role of DNA repair mechanisms in mutagenesis, it is now possible to exploit mutagenesis more rationally than in the past. Mutagenesis can create genotypes different from the progenitor strain by altering the genes in a microorganism, whereas recombination can rearrange genes or parts of genes and bring together in one organism genetic information from two or more parental strains. A range of recombination procedures are now available and will be discussed in Section 5.3. It should also be noted that successive steps of mutagenesis have often led to the accumulation of cryptic deleterious mutations in industrial production strains (see, for example, Macdonald and Holt, 1976) which only a process of recombination offers real hope of removing. The ideal strain improvement programme as well as the search for new bioactive metabolites will, therefore, aim to exploit the advantages of both mutation and recombination (see Section 5.4).

In discussing these various aspects of modifying genetically industrial microorganisms we have chosen to base most of our examples on those organisms producing secondary metabolites, notably antibiotics, although the principles are generally applicable to any industrial microorganism used to produce either primary or secondary metabolites.

5.2 MUTATION

The artificial induction of mutation was first demonstrated in *Drosophila* by Muller (1927) using X-rays. Subsequently, ionizing radiations in general were shown to be mutagenic. In 1932, Promptov, using the same organism, induced mutation with UV light and Auerbach and Robson (1946) described the first chemical mutagen, nitrogen mustard, again in *Drosophila*. Since this time an enormous number of agents, both physical and chemical, have been shown to induce mutations in organisms from viruses to man. In many cases, although the primary molecular change induced in DNA is known, the process between this step and the appearance of a mutant clone is highly complex.

Auerbach contributed greatly to our understanding of mutagenesis and there is no better way to introduce this section than to quote from her classical paper of 1967 which followed one of the same title published 20 years earlier (Auerbach *et al.*, 1947) 'A chemical change in DNA is necessary but not a sufficient condition for the production of an observable mutation. Intercalated between this primary change and the emergence of a population of cells with a new hereditary property is a whole series of cellular events, including a variety of repair mechanisms, transcription and translation of the new information, and growth of the mutant cell into a mutant population, often in the face of severe competition from non-mutant cells. These events act as so many sieves that screen out a proportion of potential mutations for realization. The study of mutation as a cellular process has hardly begun, but it already shows the importance of these cellular events for the numbers and types of mutations produced. In addition to its theoretical interest this approach is the only one likely to lead to the production of direct mutations for "mutation breeding".'

5.2.1 DNA Repair Mechanisms

As stated by Auerbach, one of the most important cellular events in this context is DNA repair (Saunders *et al.*, 1982). DNA repair mechanisms, including replication-associated processes, have been demonstrated in a wide range of microorganisms and broadly consist of two types, those which restore the original base sequence of the DNA with fidelity ('error-proof') and those which correct damage but in so doing are liable to make errors leading to changes in base sequence ('error-prone'). Such pathways may be constitutive or inducible. Cells often differ in which type of pathway operates on its damaged DNA and this can significantly alter the yield of mutants from any particular mutagenic treatment.

The first mutant deficient in DNA repair was isolated by Hill in 1958 and since then DNA repair processes have been extensively studied in the prokaryote *Escherichia coli* (Witkin, 1976). Similarities in repair of DNA damage between this organism and others have been found to exist (Howard-Flanders, 1973). Investigations centred initially on the repair of thymine dimers induced by UV irradiation although repair of distinct types of damage is now receiving greater attention. However, an initial consideration of the different repair pathways which may operate to reverse the mutation process initiated by induction of a thymine dimer in DNA will serve as a model in illustrating the basic principles upon which one can seek to increase mutant yield.

There are at least four known major types of enzymatic process which may be involved in the repair of UV-induced damage. These are shown in Table 1 together with the major gene loci associated with them. The simplest form, photoreactivation, involves breakage of thymine dimers by light in the presence of the enzyme photolyase, a process specific for repair of UV damage. This and the second method, excision repair, act prior to replication. Excision repair has been well characterized biochemically (Grossman *et al.*, 1975) and with minor variations, has been shown to act on other types of damage. It is defined as a four-step process which consists of: (i) recognition of damage and endonuclease incision, (ii) excision of the damage, (iii) resynthesis of the excised portion of the DNA and (iv) ligation of free ends (for review see Hanawalt and Setlow, 1975).

Table 1 Major Pathways and Genes Involved in Repair of UV Damage and Mutagenesis in *E. coli*

System	Gene symbol	Wild-type gene function
A. ERROR-PROOF REPAIR PATHWAYS		
(i) Photorepair		
	phr^+	Structural gene for photoreactivating enzyme.
(ii) Excision repair		
	$uvr\,A^+$ }	UV endonuclease, nicks DNA at site of
	$uvr\,B^+$ }	damage.
	$uvr\,C^+$	Acts together with $uvr\,A^+$ and B^+.
	$pol\,A^+$	Structural gene for main DNA polymerase and exonuclease.
	lig^+	Structural gene for DNA ligase catalyzing the sealing of broken DNA strands.
(iii) Post-replication/recombination repair		
	$rec\,A^+$	Promotes synapsis and strand exchange.
	$rec\,B^+$	Structural gene for exonuclease V.
	$rec\,C^+$	Governs $rec\,BC^+$ recombination pathway.
B. ERROR-PRONE OR SOS REPAIR PATHWAY		
	$rec\,A^+$	Structural gene for protein X, activated by DNA damage to a specific protease capable of cleaving $lex\,A^+$ repressor protein.
	$lex\,A^+$	Codes for protein repressing SOS response.
	$umu\,C^+$	Essential for mutagenesis by UV.

Although prereplication repair processes can reverse a substantial number of the potentially mutagenic lesions initially induced in the DNA structure, some do persist to replication. At this point in the cell cycle the proof-reading activities of the polymerases of *E. coli* prevent the insertion of any bases opposite a 'non-coding' lesion, such as a thymine dimer, leading to the stalling of replication and restarting further along the DNA molecule (Rupp and Howard-Flanders, 1968). This produces a series of gaps in the DNA which the cell seals by employing a recombination type of repair (Rupp *et al.*, 1971). Basically this process entails sister chromosome exchange between the damaged strand and its undamaged copy. A number of theoretical models of this process exist and it has been shown to be dependent on several gene products and to operate by more than one pathway (Youngs and Smith, 1976).

The processes of photoreactivation, excision repair and postreplication recombination repair are error-free and almost certainly act to reduce mutant yield although there is clear evidence to suggest that a low level of mistakes does occur with, for example, excision repair (Bridges and Mottershead, 1971). However, it is now known that the majority of UV-induced mutations arise through an inducible repair activity termed 'SOS' repair (Radman, 1977).

The ability of *E. coli* or its bacteriophage to be mutated by UV and many chemical agents can be blocked by mutation at three bacterial loci, $rec\,A^+$ (Witkin, 1969), $lex\,A^+$ (Witkin, 1967) and $umu\,C^+$ (Kato and Shinoura, 1977). The requirement for mutagenesis of the products of the wild-type genes $rec\,A^+$ and $lex\,A^+$ (formerly $exr\,A^+$) has been known for some time (for review see Howard-Flanders, 1981) and on the basis of the *E. coli* model Bridges (1976) has classified mutagens simply into two types. Firstly, there are those which will mutate bacteria of the genotype $rec\,A\,lex\,A$ (*i.e.* lacking functional wild-type gene products of these loci) and which include ethyl methanesulfonate (EMS) and *N*-methyl-*N'*-nitro-*N*-nitrosoguanidine (NTG). In the case of such mutagens it is likely that mutation occurs as a result of direct errors at replication (*i.e.* 'mispairing') and failure of a repair system which detects and corrects mismatched base pairs. In fact the exact mechanism by which NTG produces mutations remains controversial. Several theories, including miscoding involving preferential action at the replication fork and direct interaction with the replication enzyme complex, have been postulated (see Kimball, 1978).

The second type is represented by those which will only mutate $rec\,A^+\,lex\,A^+$ strains and which includes UV, ionizing radiations, methyl methanesulfonate, 7-bromomethylbenzanthracene, nitrogen mustard, mitomycin C, captan (the fungicide), dichlorvos (the insecticide), a number of nitrofurans, nitroquinoline l-oxide and psoralens in the presence of near-UV light (360 nm). In this case mutations are said to arise by a process of 'misrepairing'.

The single division of mutagens into these two types is important in ensuring that in strain development programmes where more than one mutagenic line is employed mutagens are chosen for efficiency but also to represent these different modes of action. However, there is great complexity and variety in cellular responses induced by DNA-damaging agents. The SOS response of *E. coli* entails derepression of at least 12 different operons. The expression of the so-called 'SOS regulon' is controlled by the *lex A*$^+$ and *rec A*$^+$ genes. During normal growth the *din* genes (damage inducible) of the SOS regulon would be repressed by *lex A*$^+$ protein which would be maintained at a constant level by the autoregulatory repression which *lex A*$^+$ protein exerts on its own synthesis. It appears that single-stranded DNA, unusually long lengths of which are likely to be present after DNA damage, is the signal that activates *rec A*$^+$ protein into a specific protease capable of cleaving the *lex A*$^+$ repressor protein and this induces the SOS error-prone repair response. Little *et al.* (1981) showed *in vitro* that homogeneous *rec A*$^+$ protein cleaves purified *lex A*$^+$ protein into two fragments that cannot repress transcription from the *rec A*$^+$ promoter. The *rec A*$^+$ protein can also promote two central steps of genetic recombination *in vitro*, namely synapsis (the pairing of homologous molecules and even a single strand with homologous duplex DNA in the absence of a free end in any of the strands) and strand exchange, which produces long heteroduplex joints containing paired strands from each of two DNA molecules (DasGupta *et al.*, 1981). The *rec A*$^+$ protein can also act as a single-stranded DNA-dependent ATPase and a DNA unwinding or annealing enzyme.

The inducible *umu C*$^+$ gene product of *E. coli* does not have the range of pleiotropic effects of *rec A*$^+$ and *lex A*$^+$ but is absolutely required for UV and some chemical mutagenesis. It specifies two proteins of molecular weight 45 000 and 16 000 and is probably the key participant in the processing of DNA damage induced by UV that results in mutation (Bagg *et al.*, 1981).

Thus, it can be seen from the example of UV mutagenesis in *E. coli* that a complex system has evolved in that organism to deal with DNA damage. Mechanisms have also evolved to cope with DNA damage caused by other mutagens and indeed in some cases the same repair pathway and genes are involved. In the case of excision systems these are usually specific for certain classes of mutagen. For example, UV excision repair deficient strains of *E. coli* are normal for repair of damage by ionizing radiation (Kimball, 1980). Similarly, monofunctional alkylating agents exemplified by *N*-methyl-*N'*-nitro-*N*-nitrosoguanidine (NTG) and methyl methanesulfonate (MMS) have their own specific excision systems (Strauss *et al.*, 1980). The major difference appears to reside in actual recognition and subsequent endonuclease incision at the site of damage, the more subtle base changes induced by NTG and in some cases X-rays, for example, apparently not being recognised by UV endonuclease which picks up rather more bulky, helix distorting damage.

However, not all bulky chemical adducts are repaired in the same way. A recent study of Tang *et al.* (1982) showed that DNA damage caused by the carcinogen *N*-acetoxy-2-acetylaminofluorene requires all three *uvr*$^+$ genes (A, B and C) as does the correction of induced pyrimidine dimers, whereas damage caused by *N*-hydroxyaminofluorene seemed only to require the *uvr C*$^+$ gene. The authors suggested that the *uvr C*$^+$ gene product is required for nucleotide excision repair of a wide range of DNA adducts while *uvr A*$^+$ and *uvr B*$^+$ gene products may be required only for those which produce a great deal of distortion of the DNA helix.

In addition to an excision system as outlined above for UV damage, the minor base alterations produced by alkylating agents may be repaired by a direct 'replacement base' procedure not actually involving strand excision (Linn *et al.*, 1978). Either way these processes act to reduce mutant yield.

Recombination repair may also operate on damage induced by other mutagens. Evidence exists, for example, that, after treatment with some alkylating agents, DNA is of a lower than expected molecular weight immediately after replication but with continued incubation returns to normal size (Kimball *et al.*, 1978).

Just as photoreactivation is specific for thymine dimer damage, certain repair processes have been found to be peculiar to damage induced by monofunctional alkylating agents (Strauss *et al.*, 1980). Discovery of one such system followed from investigations of an 'adaptive' response whereby pretreatment with low doses of monofunctional alkylating agents leads to a decrease in both the lethal and mutagenic effects of a subsequent challenge dose. The principal premutagenic lesion produced by methylating agents in DNA is O^6-methylguanine (m^{6}G). The adaptive response to mutagenesis involves the induction of a 'suicidal' DNA methyl transferase which transfers the methyl group from m^{6}G residues to a cysteine residue in the enzyme and thereby inactivates it. Residues of m^{6}G which have not been repaired before passing through the replication fork will pair with thymine with high frequency. This base pair is then subject to an enzymic 'mismatch correction system' capable of recognizing and correcting mismatched base pairs.

Mutants affected in the system which removes mismatched bases from DNA have also been discovered. For example, strains carrying the *dam* mutation lack the major DNA adenine methylase (Marinus and Morris, 1973), are hypersensitive to the effects of base analogue mutagens such as 2-aminopurine and NTG and are defective in mismatch correction (Glickman *et al.*, 1978; Karran and Marinus, 1982). It was also shown that phenotypic reversion of sensitivity to 2-aminopurine by the introduction of a second site *mut* mutation into a *dam* strain is accompanied by an increased frequency of both spontaneous and base analogue-induced mutagenesis presumably due to the absence of a mismatch repair system (Rydberg, 1978).

In summary, investigation of the mutation process in prokaryotes has clearly demonstrated that DNA repair mechanisms play a fundamental role in mutation induction by the majority of commonly utilized mutagens. In *E. coli* alone over 50 genes affecting DNA repair, replication and recombination have been recognized. Similar evidence is now accumulating in the study of eukaryotic DNA repair. By appropriate genetic impairment of error-proof repair within the cell mutant yields can be increased significantly by channelling a larger amount of the initial damage through error-prone pathways. Table 2 shows some examples where this has been achieved in both prokaryotic and eukaryotic organisms. Table 3 lists a number of common ways in which repair deficient mutants have been successfully isolated.

Table 2 Examples of Microorganisms in which Induced Mutation Frequency is Increased as a Consequence of Genetic Impairments in DNA Repair

Organism	Mutation or mutant designation	Effect on mutation induction in the mutant cells in comparison with wild type by physical and chemical agents[a]
PROKARYOTES		
Bacillus subtilis	*hcr*	Increased mutation with FUV, 4NQ and AF2 (Tanooka, 1977)
Escherichia coli	*uvr A*	Increased mutation with ICR-191 (Newton *et al.*, 1972)
Micrococcus radiodurans	*mtc A*	Increased mutation with MMS, EMS and NTG (Tempest and Moseley, 1980)
Streptomyces clavuligerus	CL105	Increased mutation with FUV (Saunders and Holt, 1982)
	CL89	Increased mutation with 4NQ (Saunders and Holt, unpublished)
EUKARYOTES		
Aspergillus nidulans	*uvs D53*	Mutated by benomyl whereas wild type is not (Kappas and Bridges, 1981)
Dictyostelium discoideum	*rad C35*	Increased mutation with FUV (Deering and Sheely, 1977)
Neurospora crassa	*uvs −2*	Increased FUV, γ-ray, NTG and 4NQ induced mutation frequency (De Serres, 1980)
Saccharomyces cerevisiae	*uvs − 9*	Increased mutation with FUV (Resnick, 1969)
MAN	*Xeroderma pigmentosa*	Increased mutation with UV (Arlett and Lehmann, 1978)

[a] Abbreviations: FUV—far UV light; 4NQ—4-nitroquinoline 1-oxide; MMS—methyl methanesulfonate; EMS—ethyl methanesulfonate; NTG—*N*-methyl-*N*-nitro-*N*-nitrosoguanidine.

Extensive information is also available, mostly from experiments with *E. coli*, on chemicals which affect DNA repair in different ways (see Table 4). Thus, by environmental conditions alone it is also possible to increase the yield of mutants obtained from a mutagen-treated population of cells by, for example, inhibiting error-proof repair pathways while leaving the error-prone pathways operational. Useful synergistic interactions increasing mutation frequency of various chemicals (*e.g.* caffeine and isoniazid) as well as physical mutagens have been demonstrated (Saunders *et al.*, 1982).

5.2.2 Mutagen Specificity

As pointed out by Auerbach (1966) the term 'mutagen specificity' has been used to denote several different phenomena but is usually taken to mean that under the influence of certain mutagenic factors particular types of mutation preferentially arise. The first examples of this were published in 1953 both from bacteria (*E. coli*; Demerec) and from fungi (*N. crassa*; Kolmark). Strains carrying double auxotrophic markers were treated with different mutagens and different relative frequencies for reversion to prototrophy at each locus were obtained depending on the mutagens employed. Later (Paterson, 1974; Kilby, 1974; Auerbach, 1976), it was shown that such relative frequencies of reversion in *Neurospora crassa* could be modified dramatically by treatment conditions such as temperature, ionic environment, liquid holding, presence of light and also by the cytoplasmic background of the strain used.

Table 3 Some Methods by Which Repair-deficient Mutants Have Been Successfully Isolated

Screening/selection basis	Organism	Mutant designation	Effect on induced mutation frequency[a]
(a) Sensitivity to radiations			
UV	*Streptomyces clavuligerus*	CL105	Increased UV induced mutation (Saunders and Holt, 1982)
		CL89	Shows a detectable level of 4NQ induced mutation (Saunders and Holt, 1982)
(b) Sensitivity to chemicals			
Mitomycin C	*Micrococcus radiodurans*	*mtc A*	Increased EMS, MMS, NTG and nitrous acid induced mutation (Tempest *et al.*, 1980)
Aminopurine	*Escherichia coli*	*dam*	Increased base analogue induced mutation (Glickman *et al.*, 1978)
(c) Resistance to chemicals			
Aphidicolin	Chinese Hamster V79		Elevated spontaneous mutation frequency (Liu *et al.*, 1983)
Cycloheximide	*Schizophyllum commune*	Various	Sensitive to UV (Shneyour *et al.*, 1978)
Cycloheximide	*Saccharomyces cerevisiae*	Various	Sensitive to UV (Shneyour *et al.*, 1978)
(d) Strains lacking nuclease haloes on plates containing DNA	*Neurospora crassa*	*nuh-4*	Elevated spontaneous mutation frequency (Fraser *et al.*, 1980)
(e) Selection on plates containing *p*-aminobenzoic acid	*Neurospora crassa*	Various	Sensitive to UV and representing a discrete type of UV sensitivity in this organism (Schroeder and Olsen, 1980)

[a] Abbreviations: see Table 2.

Table 4 Some Chemicals Affecting DNA Repair in a Range of Microorganisms and their Effect on Mutation Frequency

Chemical	Repair target and effect on mutation frequency
Acriflavine	Inhibits excision repair and enhances UV induced mutation frequency (Doudney *et al.*, 1964).
Antipain	Inhibits error-prone repair and diminishes UV, EMS, MMS and NTG induced mutation frequency (Ichikawa-Ryo and Kondo, 1980).
Caffeine	Inhibits excision repair (Fong and Bockrath, 1979), under correct conditions enhances UV induced mutation frequency (Saunders and Holt, 1982), but can have antimutagenic effect (Ichikawa-Ryo and Kondo, 1980). In addition caffeine inhibits post-replication repair and enhances mutation induction by alkylating agents in Chinese Hamster cells (Roberts *et al.*, 1974).
Chloramphenicol	Inhibits error-prone repair and reduces UV induced mutagenesis (Sedgewick, 1975). Inhibits inducible error-proof repair and enhances NTG induced mutagenesis (Sklar, 1978).
Cobalt chloride	Corrects error-proneness of DNA polymerase III. Inhibits induced mutation by NTG and γ-rays in *E. coli* (Kada *et al.*, 1979) and inhibits spontaneous mutation frequency in *Bacillus subtilis* (Inoue *et al.*, 1981) and *S. clavuligerus* (Coleman and Holt, unpublished).
Cysteine arabinose	Inhibits excision repair in human fibroblasts (Johnson *et al.*, 1982). Effects on mutation frequency not assessed.
Isoniazid	Inhibits post-replication repair in mammalian cells (Klamerth, 1978) and enhances UV induced mutagenesis in *Streptomyces clavuligerus* (Saunders *et al.*, 1982).
Methylpurines/ xanthines	Inhibit excision repair and enhance radiation induced mutation frequency (Sideropoulos *et al.*, 1968; Donesan and Shankel, 1964).
Novobiocin	Inhibits UV repair in mammalian cells (Collins and Johnson, 1979). Effects on mutation frequency not assessed (Collins and Johnson, 1979).
Quinines	Inhibit excision repair and increase UV induced mutation frequency (Sideropoulos *et al.*, 1980).
Sodium arsenite	Inhibits excision repair and enhances UV induced mutagenesis (Rossman, 1981)
Sodium bisulfite	Inhibits excision repair and enhances UV induced mutation frequency (Mallon and Rossman, 1981).

Results from *Aspergillus nidulans* serve well to illustrate the importance of mutagen specificity. In this fungus two systems have been employed by Alderson and coworkers (Alderson and Clarke, 1966; Alderson and Hartley, 1969; Alderson and Scott, 1970; Scott and Alderson, 1971) for examining the simultaneous assessment of mutation at more than one locus. Reverse (suppressor) mutation of the *meth G1* locus gives recognizably different phenotypes at five gene loci and forward mutation to 2-thioxanthine resistance enables easy phenotypic recognition of alterations at at least eight gene loci. Using a range of mutagens it was shown that only in the case of 8-

methoxypsoralen in the presence of near-UV light was a random distribution of mutations among the loci achieved. Cove (1976) has also used a system in *A. nidulans* capable of simultaneous examination of mutations but involving at least nine different loci (selection of mutants resistant to chlorate). The nature of the nitrogen source used in the medium for isolating the mutants was found to affect markedly the relative frequencies of the mutant classes obtained.

Despite the fact that the cellular environment can modify specificity, the phenomenon of mutagen specificity can also be exerted at the molecular level within a gene. For example, the frequency of UV-induced mutations at specific nucleotides along the *lac I* gene in *E. coli* varies by as much as eighty-fold and the spectrum of mutations to amber, ochre and UGA chain-terminating codons include five hot spots accounting for over 30% of these induced nonsense mutations (Coulondre and Miller, 1977; Miller and Schneissner, 1979). Such specificity might also be expected to operate within different genes.

These examples serve to emphasize the need to consider all aspects of treatment and post-treatment conditions in the adoption of a procedure for mutagenesis.

5.2.3 Survival Curves and Optimum Conditions for the Use of a Mutagen and Expression of Mutations

A survival curve is obtained by treating a population with increasing doses of mutagen and plotting the surviving fraction as a function of dose. This relationship is usually exponential in nature and the 'hit' or 'target theory' was developed in the 1930s by Lea (see Lea, 1955) to explain the effects of ionizing radiation on cells. In its simplest form the theory states that if a number of targets is exposed to a dose sufficient to hit all of them, some targets will not be hit at all, others once, twice and so on as predicted by the Poisson distribution:

$$P_x(m) = (e^{-m}m^x)/x! \text{ or, when } m = 1, P_x = e^{-1}/x!$$

which gives the probability (P) that the target will be hit x times when the mean number of hits per target is one. Thus, 37% of the targets will be hit once ($e^{-1}/1! = 0.37$) and 37% will not be hit at all ($e^{-1}/0!$). Thus, for a dose of D_0, or D_{37} as it is sometimes designated, sufficient to produce one hit per target, 37% of the original population of organisms will survive. The target theory can be applied to most of the physical mutagens with simple biological systems such as a population of bacteria or fungal spores. The D_0 dose varies for different biological systems for the same mutagen and although the target theory is not of general application it does often provide an indication of the dose at which one can expect surviving cells to be 'hit' approximately once giving survivors more likely to be mutated once only.

Of course a better guide would always include a plot of the induced frequency of the desired mutation against dose and survival (*vide infra*). However, even in this case it is necessary to use a mutagenic test system (*e.g.* forward mutation to chemical resistance) which is usually unrelated to the desired mutation (*e.g.* increasing antibiotic yield). It is, however, always necessary to use some sort of mutagenic test system in the industrial organism in which mutation and selection will be made since there are many examples where published rationales for mutagenesis in one organism do not work in another. A good example of this is NTG where in *E. coli* optimum mutagenesis is achieved at pH 6 whilst in *Streptomyces coelicolor* the optimum is achieved at pH 9.

In industrial programmes of strain improvement, in order to save time, effort and hence expense, conditions of mutagenesis should be chosen to give the highest possible frequency of mutants per survivor. Thus it is necessary to plot induced mutants against survival (*i.e.* lethal hit) and not simply against dose. The advantages of this plot were described by Munson and Goodhead (1977) and demonstrated by Normansell and Holt (1979) to be useful when comparing the induction of deletions by ionizing radiation in *Aspergillus nidulans*. In this work X-rays were shown to yield the highest number of deletants per rad but 15 MeV electrons produced the most deletants per survivor. The additional advantage of this type of plot is that it allows direct comparison of mutagenicity of different mutagens, both physical and chemical, which do not have the same units of dose and therefore are not comparable on a mutation frequency/dose plot. It should be noted that another approach to comparing different mutagens by using mutant yield data has been described by Eckardt and Haynes (1980) based on the number of mutants per cell treated.

These then are the principles to be applied to the general choice of mutagenic treatment conditions. However, the situation is far more complex in the case of mutations for yield enhancement. Early work in this area (*e.g.* Alikhanian, 1962) led to a general conclusion that the optimum dose for antibiotic titre-increasing mutations in industrial production strains was always lower than for titre-decreasing and morphological mutations. This is not surprising since production strains with high yields will have resulted from mutation at a large number of loci selected sequentially for an optimum but balanced regulation of gene products. Clearly, high doses which lead to higher probabilities of multiple hits in surviving cells are more likely to unbalance the system by introducing multiple deleterious mutations instead of building on the mutations present. This again underlies the value of the D_0 kinetic approach.

The need for expression of new mutations is often neglected in mutagenic procedures. In some organisms expression periods for particular mutations are not required. In others it is possible to fail to obtain mutants of a particular type unless suitable conditions for expression are provided. For example in our laboratory the detection of acriflavine-resistant mutants after UV treatment of *Streptomyces clavuligerus* requires an expression period of 16–18 h prior to selection with acriflavine. Similarly, Lemontt and Lair (1982) have described the effects of expression time on the yield of candidine-resistant mutants in yeast treated with both chemicals and radiations.

Expression effects on mutant yield obviously are not dependent solely on the time allowed for the cell prior to selection. Both the environment and genetic background of the cell can also significantly affect mutant yield. For example, no revertants of a histidine auxotrophic strain of *Bacillus subtilis* were obtained unless high levels of threonine were present in the plating medium (Corran, 1969). Also, an easily revertible tryptophan allele in *E. coli* could not be reverted when an additional adenine requirement was induced in the strain (Chopra, 1967). Many other such effects have been observed (see Auerbach, 1976) but the important point is to take into account the need for an expression period in designing any mutagenic rationale.

5.2.4 Site Specific Mutagenesis

As more information becomes available on genes and gene products concerned in industrially important biosynthetic pathways, recombinant DNA technology and directed mutagenesis will often be the tools of choice to change specifically particular DNA sequences in a predetermined way. These techniques have been described by Timmis (1981) and by Chandler and Churchwood in Volume 1, Chapter 6. The introduction of a single base change at or near restriction enzyme sites using such techniques provides an excellent example. In the presence of ethidium bromide, some restriction enzymes nick DNA in one strand rather than cutting both strands. This nick can then be extended to a gap by exonuclease treatment and base transitions introduced into the remaining single-stranded DNA by bisulfite treatment (Shortle and Nathans, 1978). Bisulfite specifically deaminates cytosine to uracil in single-stranded nucleic acids (Hayatsu, 1976), resulting in a transition mutation during repair. Unfortunately, restriction sites are not always conveniently placed. However, such a limitation has been overcome by utilizing the fact that in the presence of *E. coli rec* A^+ protein, small single-stranded DNA will efficiently displace a complementary sequence in a super-coiled double-stranded DNA molecule forming a D-loop. This displaced single-stranded loop is then available for mutagenesis using nuclease S_1 and bisulfite (Shortle *et al.*, 1980).

Winter *et al.* (1982) have recently provided a good example of site directed mutagenesis which gives a general method for systematic replacement of amino acids in an enzyme in order to analyze enzyme structure and activity. However, even without knowledge of genes and gene products being available other approaches towards directing mutagenesis are possible. Clear potential exists, for example, in utilizing the unique action of NTG at the replication fork for sequential mutagenesis in synchronized cultures (Godfrey, 1974) and comutation (Cerda-Olmedo and Ruiz-Vasquez, 1979). Such techniques should be widely applicable in prokaryotic organisms and its use in eukaryotes seems promising (Dawes *et al.*, 1977). In addition, transcription of the genes involved in production of particular metabolites often occurs at specific stages in growth and hence mutagenesis of chemostat-grown cultures can be investigated.

5.2.5 Applications of Mutation to Antibiotic-producing Microorganisms

Induced mutation followed by selection has been widely and successfully utilized for yield enhancement. Thus, penicillin titres in *Penicillium chrysogenum* have been increased from 100 to over 50 000 units ml^{-1} (Queener and Swartz, 1979). Similarly, some streptomycete products are produced at levels of 20 g l^{-1} or more (Hopwood and Chater, 1980).

Most published work on the genetics of antibiotic production in fungi has involved penicillin production in *P. chrysogenum* but studies have also been undertaken using *Aspergillus nidulans* as a model because the formal genetics of this mould is well established (Macdonald and Holt, 1976; Macdonald, 1983). In this organism single step mutations leading to overproduction of penicillin (*pen* mutations) and impairment of yield (*npe* mutations) have been identified (Holt *et al.*, 1976). A standard and successful method of investigating biosynthetic pathways is to use mutants unable to synthesize the product of the pathway. Such mutants isolated in *A. nidulans*, *P. chrysogenum* and *Cephalosporium acremonium* have proved useful in studies of β-lactam biosynthesis (Normansell *et al.*, 1979; Makins *et al.*, 1981). In particular the application of genetic complementation tests to divide large numbers of mutants, often into a small number of complementation groups all of which would be expected to be blocked at different biochemical steps, provides an indication of the number of different genetic loci involved in the pathway. This leads to an enormous economy of effort as a result of the need to study fewer isolates in detail. In some organisms where diffusion of intermediates can take place, this type of analysis can be done by looking for cosynthesis in mixed cultures involving pairs of mutants. The impermeability of fungal cell walls to penicillin intermediates has been attributed as a cause of failure to observe penicillin production when pairs of non-producing mutants were grown together (Bonner, 1947; Nash *et al.*, 1974; Macdonald and Holt, 1976). For this reason the analysis of Normansell *et al.* (1979) with *npe* mutants of *P. chrysogenum* employed diploid complementation tests. However, Makins *et al.* (1980) have overcome the impermeability obstacle by using osmotically fragile mycelia, prepared by the digestion of cell walls under osmotically stabilizing conditions, and incubating the resulting (partially) wall-less cells in the presence of inhibitors of cell wall synthesis to prevent wall regeneration. The technique has also been exploited with regard to intergeneric cosynthesis (Makins *et al.*, 1981).

A knowledge of the steps in the pathway enables specific mutants to be sought. A good example is provided by sulfur metabolism with respect to β-lactam synthesis. Penicillin-producing strains of *P. chrysogenum* can obtain sulfur for antibiotic synthesis very efficiently by the sulfate reduction pathway from inorganic sulfur whereas in *C. acremonium* the sulfur of cephalosporin normally comes from methionine by reverse transulfuration. Treichler *et al.* (1979) were able to isolate specifically blocked mutants of *C. acremonium* which resembled *P. chrysogenum* in their ability to utilize inorganic sulfate for antibiotic synthesis.

Blocked mutants of antibiotic pathways may produce new compounds as a result of 'shunt metabolism' of accumulated intermediates. Mutants of *C. acremonium* impaired in cephalosporin C formation have been shown to produce MTC, *i.e.* 7-(D-5-amino-5-carboxy-*N*-valeramido)-3-methylthiomethyl-3-cephem-4-carboxylic acid (Kanzaki *et al.*, 1974, 1976).

The use of blocked mutants in a general scheme (mutational biosynthesis or mutasynthesis) has been proposed by Fleck (1979). Fleck exemplified the system by reference to a culture of *Streptomyces violaceus* which produces violamycin composed of four moieties A, B, C and D (A, aglycone; B, rhodosamine; C, 2-deoxy-L-fucose; D, rhodinose) from simple carbon and nitrogen sources. A mutant is obtained that does not produce the antibiotic because it cannot synthesize the aglycone (component A). When analogue A' (the mutasynthon) is added to the media instead of A the mutant will produce a new anthracycline A'BCD (mutasynthetic compound) if A' can penetrate the cell, if the incorporating enzyme is not too specific and if the new anthracycline has antibiotic activity. Fleck (1979) listed a number of examples in actinomycetes where this approach yielded new antibiotics. It has also been demonstrated for β-lactam antibiotics where the side-chain of all hydrophilic penicillins and cephalosporins is, or is derived from, L-α-aminoadipic acid. A block in the synthesis of α-aminoadipate gave a mutant which was a lysine auxotroph and which was unable to produce antibiotic unless aminoadipate was added. A range of α-aminoadipate analogues was added and one, L-S-carboxymethylcysteine, led Troonen *et al.* (1976) to the discovery of a new antibiotic called RIT-2214 (6-D-{[(2-amino-2-carboxy)ethylthio]acetamido}penicillanic acid).

The special role that mutants impaired in antibiotic production can play in strain improvement programmes has been highlighted by Normansell (1982). Reversion of such mutants (which nearly always results from a second site suppressor mutation) has often been shown to give increased titre mutants (Dulaney and Dulaney, 1967).

Table 5 gives some examples where mutagenesis has been employed to (a) increase yield, (b) alter the spectrum of metabolites, and (c) produce mutasynthetic compounds.

A major problem associated with successive mutagenesis in industrial strain improvement programmes is that of strain degeneration. The desired stability and production characteristics of the strain are genetically controlled. The danger is that under a given set of environmental conditions a spontaneous variant might become better adapted and dominate the culture, in almost every

Table 5 Some Examples where Mutagenesis has been Employed to Increase Yield or Alter the Spectrum of Metabolites or Produce Mutasynthetic Compounds

	Organism	Product	Phenotypic change[a]	Reference
Increased yield	Streptomyces aureofaciens	Chlortetracycline	—	Valerianov et al. (1981)
	Streptomyces rimosus	Oxytetracycline	—	Mindlin and Alikhanian (1958)
	Aspergillus nidulans	Penicillin	—	Ditchburn et al. (1976)
	Cephalosporium acremonium	Cephalosporin C	—	Elander (1975)
	Penicillium chrysogenum	Penicillin	—	Elander (1976)
Altered spectrum	Streptomyces noursei	Cycloheximide	Antiphenol[−] Fungicidin[−]	Spizek et al. (1965)
	Streptomyces aureofaciens	Tetracycline	Aureovocin[−]	Blaumauerova et al. (1972)
	Penicillium chrysogenum	Penicillin	Chrysogenin[−]	Backus and Stauffer (1955)
	Streptomyces peucitius	Daunomycin	Adriamycin[+]	Arcamone et al. (1969)
	Bacillus vitellinus	Butirosin A	Butirosin A derivatives[+]	Hideo et al. (1982)
	Micromonospora sagamiensis	Streptomycin	Antibiotic SU-2	Hiroshi et al. (1982)
Mutasynthesis	Streptomyces fradiae	Neomycin	Hybrimycins	Shier et al. (1969)
	Micromonospora purpurea	Gentamycin	2-Deoxygentamycin	Rosi et al. (1977)
	Streptomyces niveus	Novobiocin	Cholornovobiocin	Sebek (1976)
	Micromonospora sagamiensis	Sagamycin	2-Hydroxysagamycin	Kase et al. (1982)

[a] + indicates production; − indicates loss of product.

case leading to a reduction in yield. The stability of industrial microorganisms has been discussed in a number of symposia (see, for example, Johnston, 1975; Kirsop, 1980) and Nisbet (1980) has suggested ways of preventing strain degeneration in antibiotic-producing microorganisms by both physiological and genetic control. The former included optimum preservation and storage conditions and the use of media for preventing culture variation whilst the latter recommended selection of genetic construction of antidegenerate strains. Although antimutator compounds such as manganese(II) in fungi (Sermonti and Morpurgo, 1979) and cobalt chloride in bacteria have been reported, this is an area where much valuable work relating to particular products has been undertaken, but little published.

When the genes coding for particular proteins or pathways are carried on plasmids, as is the case for some industrial organisms (see Chandler and Churchward, Volume 1, Chapter 6), the problem of strain stability may be exacerbated. The ability of the host cells to maintain the plasmids unchanged through several growth cycles may be affected by the genetic characteristics of the host cells, the culture conditions, the copy number of the plasmid and the gene(s) carried on the plasmid. The plasmid may be lost totally or undergo segregation or other rearrangements with the loss of significant gene regions. With multicopy plasmids, the copy number may change, thus reducing the predicted gene amplification. It has been assumed sometimes that plasmid stability is related directly to size, with small plasmids being less liable to spontaneous loss. That this simplistic interpretation is by no means true is obvious from recent studies on large plasmids in a variety of organisms, including *Pseudomonas* and *Rhizobium*, which are highly stable, in contrast to the situation with some small *E. coli* plasmids involved in, for example, somatostatin or insulin production (Old and Primrose, 1982).

5.3 RECOMBINATION

Recombination is usually classified into two types (see Glass, 1982; Whitehouse, 1982); legitimate (or homologous) and illegitimate (or non-homologous). As with studies of mutation, much of our fundamental knowledge comes from studies in *E. coli* where homologous recombination requires fairly extensive homology to act efficiently and is mediated by the *rec A*[+] system (see Section 5.2.1). On the other hand non-homologous recombination is *rec A*[+] independent and aspects of the exploitation of this system with specific reference to transposons will be discussed in Section 5.3.7. There is a wide range of natural and artificial methods available to bring about recombination and the most important ones are discussed below.

5.3.1 Protoplast Fusion

Natural methods of gene exchange have been replaced for the most part by protoplast fusion techniques (Alfoldi, 1982). Protoplast fusion overcomes the problem of low recombination frequencies typically obtained with natural systems and further may allow recombination between divergent species to occur. It has been found, for example, in *Streptomyces* species that recombination frequencies of 10–20% are attainable (Hopwood and Wright, 1976). These frequencies can be further increased by UV irradiation of the fused protoplasts. Such crosses can also be polarized by UV irradiation of one of the parental protoplast populations to a very low (approximately 1%) survival level (Hopwood, 1981a).

Protoplast fusion is universally applicable and reported in a wide range of bacteria, fungi, yeasts and algae (Alfoldi, 1982). Protoplasts which totally lack a cell wall can be successfully regenerated on a suitably osmotically buffered medium. There are many similar features of all protoplast formation and fusion processes designed for the various microorganisms to which this technique has been applied, although some differences exist here and there. For example, the optimum polyethylene glycol concentration for *Streptomyces* species is 40% v/v (Ochi *et al.*, 1979) whilst for fungi it is 25% v/v (Ferenczy *et al.*, 1976). The efficiency of protoplast regeneration varies also from one organism to another. Regeneration frequencies of 100% have been reported (Gabor and Hotchkiss, 1979) but it can be as low as 1 in 1000 (Szvoboda *et al.*, 1980). Recombinant formation is a sensitive process in the systems which have been studied in detail. Media composition for protoplast formation and growth phase of cells/mycelium prior to protoplast formation influences the yield of recombinants obtained (Gabor and Hotchkiss, 1979; Foder and Alfoldi, 1976; Baltz, 1978). Recently Vidoli *et al.* (1982) reported an interesting technique using encapsulation of protoplast mixtures in metal ion gelled polysaccharide matrices which apparently gave rise to a considerable increase in numbers of recombinants recovered in comparison with other procedures.

Although technical problems may be encountered within a particular system, protoplast fusion represents a dramatic advance towards widespread application of recombination to achieve *inter alia* mapping of genes, intraspecific hybrid formation, yield improvement, and novel product formation. The advantages and uses of protoplast fusion as genetic tools are outlined in Table 6.

Table 6 Advantages of Protoplasts and Protoplast Fusions in Genetic Manipulation

1. Protoplast formation and fusion are universally applicable.
2. Recombinants can be obtained in microorganisms in which no natural form of gene transfer has been demonstrated.
3. Both parents appear to play an equal role in recombinant formation circumventing any mating type polarity which may normally exist. It should be noted, however, that protoplast fusions can be polarized by UV irradiation or SDS treatment to select against one parental population if desired.
4. Protoplast fusions between more than two sets of parental strains can be utilized to generate multiparental recombinants.
5. Protoplasts can be transformed by plasmid and actinophage DNA in *Streptomyces*. Plasmid DNA can similarly transform yeast protoplasts and those of *Neurospora crassa*.
6. Protoplasts can be fused with a wide range of other cells and cellular organelles to promote interspecific hybrid formation or uptake of organelles such as mitochondria. Fusion of mitochondria with yeast protoplasts and subsequent expression of mitochondrial genes, has been demonstrated.
7. Protoplasts can be fused with liposomes, which act as 'carriers' allowing delivery of biological macromolecules, biosynthetic intermediates and cellular organelles into the cell.

5.3.2 Conjugation and Natural Plasmids

The classical experiments in *E. coli* genetics were based on the one-way conjugal transfer of DNA. It soon became apparent that this process relied on physical contact between cells and that not all strains of *E. coli.* could mate to give recombinants. Subsequently, male donor strains F$^+$, Hfr and F' (all containing the sex factor/plasmid F) and female recipient cells, F$^-$ (lacking the sex plasmid) were recognized. For details of the process the reader is referred to reviews by Clark and Warren (1979) and Glass (1982).

Other conjugative plasmids have been discovered in both Gram-negative and Gram-positive bacteria, although not all plasmids are self-transmissible (*i.e.* act as sex factors). Some of the plasmids, for example RP4 of *Pseudomonas* species, have broad host ranges making it possible to transfer genes very widely. In *E. coli*, plasmids can often be integrated into the chromosome either by a *rec A*$^+$-dependent or *rec A*$^+$-independent method of recombination (Cullum and Broda, 1979) which means that plasmid and/or chromosomal genes can be conjugally transferred. Conjugal transfer of plasmid DNA is normally followed by its recircularization and reestablishment as a replicon, whereas to be inherited, transfer of chromosomal DNA must be followed by *rec A*$^+$-dependent integration into the recipient chromosome. Although most of the studies of conjugation have been done with the Enterobacteriaceae, members of this group are not important eubacterial antibiotic producers. In this connection the most important genera are *Bacillus* and *Pseudomonas*. In the former, most of the genetic studies depend on transformation and transduction but well developed conjugation systems do exist in the latter with many instances of plasmid control of catabolic functions (Clarke, 1978). However, little genetic work on antibiotic-producing strains has been undertaken.

Most actinomycetes produce antibiotics (Hopwood and Merrick, 1977) and in the case of the best characterized streptomycete system, *Streptomyces coelicolor*, early experiments on recombination relied on mixing two strains differing in nutritional requirements on a complete medium and allowing growth, followed by spreading spores on a selective medium to detect recombinants. Subsequent work, notably by Hopwood and his colleagues at the John Innes Institute, England, demonstrated that gene transfer was mediated by two sex factors or fertility plasmids, SCP 1 (Hopwood and Wright, 1973) and SCP 2 (Bibb *et al.*, 1977). Although there are many differences in the actual process, overall it is possible to compare the various types of *S. coelicolor* strains with counterparts in the *E. coli* system and this is shown in Table 7. Soon after the demonstration of conjugative plasmids in *S. coelicolor*, investigations of at least two industrially important species of *Streptomyces* gave genetic evidence for the presence of similar plasmids (Friend *et al.*, 1978; Lomorskaya *et al.*, 1977).

Table 7 A Comparison of Broad Categories of *E. coli* Strains with Counterparts in *S. ceolicolor* with Respect to Conjugative Plasmids

Designants of E. coli *strains*	Original nomenclature used by Hopwood for S. coelicolor *strains*	Current nomenclature of S. coelicolor strains with respect to SCP1 plasmid	Plasmid status
F$^-$	UF ultrafertility	SCP 1$^-$	Lacks plasmid
F$^+$	IF initial fertility	SCP 1$^+$	Possesses autonomous plasmid
Hfr	NF normal fertility	SCP 1	Possesses integrated plasmid
F'	—	SCP 1'	Possesses plasmid with integrated chromosomal genes

The useful general property of many conjugative *Streptomyces* plasmids is their ability to produce 'pocks' (circular zones of inhibited or retarded growth) when plasmid-containing strains are grown in contact with a strain lacking the corresponding plasmid. The production of 'pocks' allows easy detection of single transformants in large populations of non-transformed cells and is useful in the isolation of plasmid clones (Bibb *et al.*, 1978).

5.3.3 Transformation

Transformation requires the uptake and subsequent expression by cells of exogenously supplied DNA. As a means of bringing about gene exchange, transformation is usually the preferred method (Old and Primrose, 1982). Transformation by both plasmid and phage DNA has been demonstrated in a wide range of microorganisms including *Streptomyces coelicolor* and other streptomycetes (Thompson *et al.*, 1982), *Saccharomyces cerevisiae* (Struhl *et al.*, 1979), *Bacillus subtilis* (Chang and Cohen, 1979), *Neurospora crassa* (Case *et al.*, 1979) and *Aspergillus nidulans* (Tilburn *et al.*, 1983).

Particular mention should also be made of the use of liposomes (Makins and Holt, 1982a) which can be employed to bring about recombination even in those industrially important microbes where extensive genetic information is lacking. The application of liposomes in a variety of fields has been described by Makins and Holt (1982b) who have indicated the importance of liposome encapsulation as an adjunct to gene cloning techniques. Recently, the value of liposomes in promoting transfection frequencies in *Streptomyces* has also been demonstrated (Rodicio and Chater, 1982) and in analyzing a number of closely linked markers in *Streptomyces* (Makins, Wright and Hopwood, personal communication). It has also been reported that liposome-encapsulated DNA of *Neurospora crassa* can bring about transformation of protoplasts (Radford *et al.*, 1981).

5.3.4 Transduction

The virus-mediated transfer of genetic information is known as transduction. Bacterial viruses or bacteriophages can be divided into two types: virulent or temperate. In the former case phages infect and destroy bacteria by causing lysis of the host cell. On the other hand, phages may be carried within a bacterium without causing immediate lysis. These are known as temperate phages and the bacterial strains are called lysogenic since the phage can be induced by various treatments to multiply rapidly, leading to lysis of the host cell.

Zinder and Lederberg (1952) first demonstrated that phage P22 of *Salmonella typhimurium* was able to transport genetic information from one cell to another and bring about transduction. Within a few years of this discovery transduction was shown to occur in *E. coli*. In generalized transduction the virus can be temperate or virulent and random DNA fragments are occasionally packaged within viral particles during lytic growth. Because such fragments will not normally carry replication origins, their inheritance depends on integration (Sherratt, 1981). In 1956 Morse *et al.* discovered a case where transducing activity by phage λ of *E. coli* was restricted to the locus controlling galactose utilization. This, in contrast to generalized transduction, has been described as specialized transduction. Here viral DNA covalently linked to non-viral DNA is packaged within a single viral particle. Because the viral heads are only big enough to accommodate the viral genome, specialized transducing particles must normally contain a deleted viral genome and the virus must become covalently attached to the sequences to be transduced. For a fuller discussion of this and ways in which this restriction has been overcome in the case of λ, see Sherratt (1981) and Murray (1977).

Transduction has been a powerful and effective tool in the study of gene organization and regulation in eubacteria. Generalized transduction has been widely used for genetic mapping and strain construction while specialized transduction, in addition, can be used for complementation analysis, gene enrichment, and purification followed by expression studies. Also a general method for introducing mutations into any gene cloned in a λ vector is described by Chandler and Churchward in Volume 1, Chapter 6.

Outside the eubacterial systems the use of phages as recombinogenic tools has been very limited. Transduction was reported in *Streptomyces olivaceous* by Chater and Carter in 1978 and Stuttard (1979) has demonstrated transduction of auxotrophic markers in a chloramphenicol-producing strain of *Streptomyces venezualae*. *Streptomyces* phages and their potential as cloning vectors have been discussed recently by Chater *et al.* (1982). There are also a number of cases where viruses (mycophages) have been shown to be associated with industrially important fungi. For example, the presence of RNA (35 nm polyhedral) virus-like particles in penicillin-producing *P. chrysogenum* was first reported by Banks *et al.* (1969) and Normansell and Holt (1978) demonstrated the presence of such (35 nm polyhedral) viral particles in penicillin-producing *P. chrysogenum* and in a series of derived blocked mutants representative of each of the identified complementation groups controlling penicillin biosynthesis. The interest in mycophages for genetic manipulation was particularly awakened by a publication of Tikhonenko *et al.* (1974) suggesting that a new type of DNA virus (of the PBV series) found in *P. chrysogenum* was capable of replicating in *E. coli*. However, since that time no further information has been published.

5.3.5 Sexuality and Parasexuality in Fungi

Studies of fungal life cycles show that most fungi exist in the haploid state, and the diploid phase, when it occurs, is normally of relatively short duration. As far as sexual, meiotic recombination is concerned, techniques of genetic analysis devised for higher diploid organisms are particularly easy to apply. The interpretation is often easier since meiosis usually occurs within recognized structures and the four products of meiosis can often be isolated together as a tetrad (ordered or unordered depending on the organism). Furthermore, segregants are usually haploid and their phenotype can be directly assessed. These are the principal reasons why fungal genetics has made significant contributions to our ideas on mechanisms of recombination (Burnett, 1975; Whitehouse, 1982). Major contributions in this area have come from studies in both the ascomycetes (*e.g. Saccharomyces cerevisiae, Neurospora crassa* and *Aspergillus nidulans*) as well as the basidiomycetes (*e.g. Ustilago maydis* and *Coprinus* species).

The term parasexuality was first proposed by Pontecorvo (1954) for those processes other than standard sexual reproduction which result in recombination. In eukaryotes recombination of genetic information originating from different individuals requires four steps: (1) the introduction of two genomes (or parts of genomes) into the same cell; (2) their association in the same nucleus; (3) crossing over leading to intrachromosomal recombination; and (4) reduction whereby the amount of genetic information per nucleus is reduced to the haploid level. Interchromosomal recombination typically occurs during the reduction step through independent separation of non-homologous chromosomes (Caten, 1981). The essential differences between the sexual and parasexual cycles are shown in Table 8. Further details of genetic analysis in fungi can be found in Pontecorvo *et al.* (1953), Roper (1966), Macdonald and Holt (1976) and Caten (1981). In addition, Fincham *et al.* (1979) give a general review of control of gene expression in fungi which has recently been supplemented by an excellent article by Arst (1981) in the same field.

Table 8 A General Comparison of the Main Features of the Sexual Cycle and the Parasexual Cycle[a]

Sexual	*Parasexual*
1. Highly regulated procedure.	Sequence of uncoordinated, fortuitous events.
2. Nuclear fusion regulated by specific mating type factors or through morphological differentiation. High proportion of compatible haploid nuclei fuse in specialized structures.	Rare nuclear fusion in somatic cells, fusions between nuclei of like genotype predominate over nuclear fusions of different genotype. Heterozygotes selected by appropriate markers; homozygotes, not detected.
3. Zygote persists one nuclear generation only.	'Zygote' may persist through many mitotic divisions.
4. Recombination at meiosis: crossing-over, at 4-strand stage, in all chromosome pairs.	Recombination by rare 'accidents' of mitosis: (a) mitotic crossing-over at 4-strand stage, usually one exchange in a single chromosome arm, (b) mitotic chromosome non-disjunction, independent of crossing-over.
5. Segregation through ordered processes of two successive nuclear divisions.	Segregation by several successive mitoses from aneuploid nuclei.
6. Products of meiosis readily recognized and isolated.	Recombinants, occurring among vegetative cells, recognized only by use of selective markers.

[a] Modified from Roper (1966).

5.3.6 Recombinant DNA Technology

Gene cloning techniques extend both the range and selectivity of recombination. Barriers to the successful application of these procedures within any one system have been discussed by Bull *et al.* (1982). Major requirements are the availability of a cloning vector together with a suitable transformation system. In addition to the recombinational possibilities, gene cloning has led to a revolution in mutagenic methods allowing, in appropriate cases, up to 100% mutant recovery. For a more complete discussion of gene cloning see Chandler and Churchward, Volume 1, Chapter 6.

It is clear that the majority of techniques associated with genetic engineering have been developed within the *E. coli* system. It is essential to develop similar cloning systems for use in organisms of industrial importance (Bull *et al.*, 1982). An excellent start has been made with respect to the industrially important genus *Streptomyces* where both low and high copy plasmids are available as vectors (Chater *et al.*, 1982). Just as the use of antibiotic-resistant markers on plasmids together with insertional inactivation of one of the markers has been successfully employed in the now classical gene cloning experiment of *E. coli* (*e.g.* pBR322 bearing resistance to tetracycline and ampicillin) it has been possible to prepare similarly marked plasmids in *Streptomyces*. For example, Thompson *et al.* (1980), using several combinations of restriction enzymes and plasmid SLP1.2, cloned genes for resistance to neomycin mediated by either an acetyltransferase or a phosphotransferase from total *S. fradiae* DNA and thiostrepton resistance mediated by a ribosomal RNA pentose methylase from *S. azureus* to produce the plasmid pIJ41. With this plasmid there is a unique *Bam H1* site within the neomycin phosphotransferase gene which is suitable for clone recognition by insertional inactivation.

More recently, Katz *et al.* (1983) have reported the construction of pIJ702, a multicopy, broad host range plasmid carrying genes for thiostrepton resistance and tyrosinase production. With this plasmid transformants are initially selected as thiostrepton resistant. Colonies harbouring recombinant plasmids with inserts in the tyrosinase gene can be easily recognized by virtue of their white colour on agar containing tyrosine. An inactive tyrosinase gene is unable to direct the production of melanin which is responsible for the normal black colony pigmentation.

Already, using such cloning vectors, at least three enzymes involved in antibiotic biosynthesis have been cloned (Hopwood *et al.*, 1983) and a cloned streptomycete gene sequenced (Thompson and Gray, 1983).

Although streptomycete phages have not been exploited to the same extent as plasmids, the potential of two heteroimmune temperate phages ϕ C31 and R4 as cloning vectors has been described and a shuttle vector composed of DNA from ϕ C31 and pBR322 is available (Chater *et al.*, 1982, 1984). Together with the plasmid vectors there now exists a wide choice of low, medium and high copy number vectors, some of which are able to accommodate several tens of kilobases of DNA.

Not unexpectedly current vectors available for the filamentous fungi lag far behind the level of those constructed in prokaryotic systems. Rudimentary cloning vectors, usually based on relief of an auxotrophic requirement, have been constructed for *Neurospora crassa* (Stohl and Lambowitz, 1983) and *Aspergillus nidulans* (Ballance *et al.*, 1983).

Undoubtedly much effort is currently being directed at developing vectors for industrially important species such as *Penicillium chrysogenum*. In our own laboratories an analysis of the different species of nucleic acids (including nuclear, mitochondrial and ribosomal) in this organism has been made with a view to moving further along the path of vector construction (see Saunders *et al.*, 1984a, 1984b; Smith *et al.*, 1984; Smith and Holt, 1984).

Mutations which render the cell non-mutable by UV and ionizing radiation (see Section 5.2.1, and Saunders and Holt, 1982) have been identified in our laboratories working with *Streptomyces* and this type of mutation, giving a phenotype analogous to that of *rec* in *E. coli*, may well prove useful *inter alia* in overcoming plasmid instability allowing homologous genes from one organism to be cloned on a plasmid in another. In addition, one can envisage using the DNA repair mutations to establish a 'maxi-cell' system for characterization of products expressed by cloned fragments (Sancar *et al.*, 1979). In *E. coli* this system relies on the preferential degradation of chromosomal DNA following UV irradiation of a mutant carrying the gene *rec A uvr A*. With an appropriate dose of UV, plasmid DNA remains undamaged and the resulting 'maxi-cells' will direct the synthesis of plasmid-encoded proteins. Further, UV irradiated cells with a genotype *uvr A rec A*$^+$ can be applied to study phage encoded proteins (Murialdo and Siminovitch, 1972).

5.3.7 Transposable Elements

Illegitimate recombination can best be represented by the transposable drug resistance elements or transposons which have many potential areas of application in both the field of mutation and recombination. Transposons are discrete genetic elements which have the property of randomly inserting into other genes causing inactivation of that gene (Kleckner *et al.*, 1977; Kleckner, 1981). Associated with this property is usually a selectable drug resistance phenotype which enables one to mark any gene inactivated by the *rec A*$^+$ independent insertional event. This ability to 'tag' a gene of interest by insertion in or near it can be of great use in subsequent recombination studies and also in the application of site-directed mutagenic procedures. Additionally, the insertion of a transposon can provide useful restriction targets within or near a site and also regions of homology for *rec A*$^+$ mediated recombination. After insertion into a gene, transposons often exhibit negative regulator effects on genes distal to the site of insertion. Such effects are useful in determining the size of transcription units for example. Positive effects on gene expression have also been reported (Nevers and Saedler, 1977). Transposons can also cause genetic rearrangements such as deletions and inversions near the site of insertion.

Several possible protocols may be employed for transposon mutagenesis. The most commonly used involve infecting cells containing the target DNA with a replicon containing the transposon. Selection is then made against maintenance of the replicon either by virtue of some incompatibility system or by blocking replicon replication. For a more detailed description of the protocol used in practical applications of transposons, readers should consult Kleckner *et al.* (1977) and

Kleckner (1981) and for an outline of the genetics and biochemistry of transposition Calos and Miller (1980).

So far the study of transposons has been confined mainly to the prokaryote *E. coli*. Indications of similar genetic instabilities have been reported in other organisms, for example in yeast (Cameron *et al.*, 1979; Szostack and Wu, 1979), and more recently in the filamentous fungus *Ascobolus immersus* (Decaris, 1981). The promise of transposons may soon be realized in the streptomycetes. Although no transposon-like elements have so far been discovered in *Streptomyces* the construction of a bifunctional replicon (Suarez and Chater, 1980) has opened the way for the introduction of the *E. coli* transposable elements into these organisms. In addition it might be possible to construct transposon-like structures employing suitably engineered streptomycete phages. A recent article by Skotnicki *et al.* (1982) amply illustrates with a practically useful example involving the ethanol producer *Zymomonas mobilis* rapid advances likely to be made by the use of such elements.

The discovery of the *copia* elements in *Drosophila* and similar elements in maize demonstrates that transposable elements probably also exist widely in eukaryotes. Recently, with reference to *Drosophila*, Finnegan (1983) has discussed the intriguing possibility of an evolutionary link between a transposable element able to rearrange sequences within genomes and a retrovirus able to rearrange sequences between genomes.

5.3.8 Applications of Recombination to Antibiotic-producing Microorganisms

Recombination as a technique for improving yield has always been the poor relation to mutagenesis and current yields of industrially important products are due almost entirely to mutation and selection of the producing microorganism, accompanied of course by media development. In an excellent review of genetic recombination and strain improvement, Hopwood (1977) has clearly established the case for recombination by a consideration of the number of genotypes (2^n) possible by back-crossing a production strain obtained after n rounds of mutagenesis and selection in a single lineage. This can be further extended by considering the combination by protoplast fusion of two or more strains in the same experiment derived from multiline development programmes. Some examples of strain improvement and isolation of new products by recombination are shown in Table 9.

Table 9 Examples of the Use of Inter or Intra Specific Recombination to Improve Strains and Generate Novel Products

Organism	Product	Genetic manipulation	Strain improvement/ novel product	Reference
Aspergillus nidulans	Penicillin	Sexual recombi- nation	Improved yield	Holt and Macdonald (1968b)
Aspergillus nidulans/ Aspergillus rugulosus	Penicillin	Protoplast fusion	Brown pigment	Kevei and Peberdy (1979)
Cephalosporium acremonium	Cephalosporin C	Protoplast fusion	Improved yield, growth rate and sporulation	Ball and Hamlyn (1982)
Nocardia mediterranei	Rifamycin	Conjugation	Range of novel ansamycins	Schupp *et al.* (1981)
Penicillium chrysogenum	Penicillin	Parasexual recom- bination	Increased yield	Queener and Baltz (1979)
		Protoplast fusion	Increased yield reduced level of *p*-hydroxy- penicillin in final fermentation	Pesti *et al.* (1981) Elander (1981)
Streptomyces hygroscopicus/S. violaceous	Turimycin/ violamycin	Conjugation	Iremycin	Fleck (1979)

The need to consider recombination early in a strain development programme when natural isolates with different yields are available has been urged by Macdonald and Holt (1976) who demonstrated the attractiveness of using recombination in this way in their model system of penicillin production by *Aspergillus nidulans* (Holt and Macdonald, 1968b). These workers have also demonstrated the use of sexual and parasexual analysis to map genes involved in increasing (*pen* mutations) and decreasing (*npe* mutations) penicillin yields on five of the eight known chromosomes of *Aspergillus nidulans* (Figure 1, Makins *et al.*, 1983; Macdonald, 1983). An interesting extension of this model system at present under investigation (O'Donnell, Upshall and Macdonald, personal communiation) is to utilize the mapping knowledge to produce disomic strains of *A. nidulans* with duplications of each of the known *pen* genes with a view to increasing penicillin yield further. Similar mapping studies utilizing the parasexual cycle have been undertaken in *P. chrysogenum* where mutations blocking particular steps in the penicillin biosynthetic pathway have been located on different haploidization groups (probably chromosomes) (Normansell *et al.*, 1979; Makins *et al.*, 1981). It is expected that in the near future this type of mapping knowledge will be important in the design of experiments involving newly developed cloning vectors together with improved biochemical detection methods, particularly HPLC (see, for example, Neuss *et al.*, 1982 and Rogers *et al.*, 1983a, 1983b) for the further analysis and exploitation of β-lactam biosynthetic pathways.

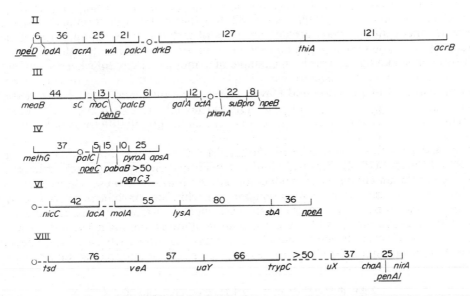

Figure 1   Genetic map of 5 of the 8 chromosomes of *Aspergillus nidulans*. Mutations concerned with the biosynthesis and overproduction of penicillin are underlined and a few other genetic markers are shown to allow the location of these penicillin mutations.

Explanation of symbols: O, centromere; *npeA*, *npeB*, *npeC*, *npeD*, impairing penicillin titre; *penA1*, *penB2*, *penC3*, increasing penicillin titre; *acrA*, *acrB*, acriflavine resistant; *actA*, actidione resistant; *apsA*, anucleate primary sterigmata; *chaA*, chartreuse coloured conidia; *drkB*, dark coloured conidia; *galA*, galactose non-utilizing; *iodA*, iodoacetate resistance; *lacA*, lactose non-utilizing; *lysA*, lysine requiring; *meaB*, methylammonium resistant; *methG*, methionine requiring; *moC*, morphologically abnormal; *molA*, molybdate resistant; *nicC*, nicotinic acid requiring; *nirA*, nitrite non-utilizing; *pabaB*, p-aminobenzoic acid requiring; *palC*, alkaline phosphatase deficient; *palcA*, *palcB*, acid and alkaline phosphatase deficient; *phenA*, phenylalanine requiring; *pyroA*, pyridoxin requiring; *sbA*, sorbitol non-utilizing; *sC*, reduced inorganic sulfur requiring; *suBpro*, suppressor of proline requirement; *thiA*, thiamine requiring; *trypC*, tryptophan requiring; *tsD*, temperature sensitive; *uX*, urea non-utilizing; *uaY*, uric acid non-utilizing; *veA*, velvet morphology; *wA*, white coloured conidia

Notwithstanding the availability of a wide range of naturally occurring mechanisms for recombination, the technique of choice will often be protoplast fusion which can overcome incompatibility between different strains, species and genera and which has many advantages including that of giving enhanced recombination frequencies (Hopwood, 1981; Holt and Saunders, 1983). Indeed, it seems that the optimism shared by geneticists involved in this field is beginning to bear fruit. Steadily over the last few years reports of successful and practically useful recombination have appeared in the literature. Elander (1981) has described recombination of two desirable characteristics found in separate mutant strains of *Penicillium chrysogenum* into one strain by protoplast fusion. Genetic linkage data are available, for example in *Streptomyces fradiae* (Baltz, 1978), and work in eukaryotes (Gunge and Tamaru, 1978) suggests that protoplast fusion offers no insurmountable barrier to genetic analysis in these systems. It has become clear in several studies that regenerated clones can subsequently segregate to yield a number of individual colonies of varied phenotype. This has been shown to occur, for example in *Bacillus megaterium* (Fodor and Alfoldi, 1978) and *Streptomyces coelicolor* (Hopwood and Wright, 1976). A further interesting feature of protoplasts, particularly in the streptomycetes, is that plasmid loss (or curing) can often be brought about by regenerating colonies after protoplast formation (Furiumai *et al.*, 1982). Finally, reference to the list of microorganisms to which protoplast fusion has been applied recently produced by Alfoldi (1982) shows an expanding number and illustrates the universal applicability of this technique to prokaryotes and eukaryotes alike.

5.4 GENETICS AND SCREENING

Considerable effort in the biotechnological industries is devoted to screening both for new products and improved yield (Nisbet, 1982). Screening may be divided into two basic types: (a) the non-selective random screen where all isolates are individually tested for the desired activity and (b) selective or rational screens.

Traditionally, random screens were carried out using shake flasks but miniaturization including both aerated liquid culture and agar disc techniques (Ditchburn *et al.*, 1974; Ball and McGonagle, 1978; Chang and Elander, 1979; Trilli *et al.*, 1982) together with automation and microprocessor technology advances allows much greater numbers of isolates to be tested. In practice, these newer methods are often used as a primary screen or preselection to identify those isolates worth detailed testing in shake flasks. This is an example of a 'multilevel screen' where a primary screen is used to test large numbers of isolates using single replicates and fewer strains are passed to the next level of testing. Calam (1964) and Davies (1964) have discussed the use of statistical principles to optimize the process of strain selection and favour the use of such multilevel screens together with 'rapid recycling', a technique used to enrich the population for isolates having small yield increases.

Many rational screens, some specifically targeted for the discovery of new compounds, have been described (Nisbet, 1982). For strain improvement this selective approach often arises from the knowledge of the biochemistry of product formation. Some of the best examples are provided by the amino acid industry where normal regulatory mechanisms have been overcome by the use of toxic amino acid analogues. Table 10 lists a number of instances where resistance to toxic metabolites has led to improved yield. The application of this technique to secondary metabolites has been less successful mainly due to the difficulty of finding sufficiently toxic analogues of intermediates although within industrial companies ways have been found to overcome this problem.

Table 10   Selection of Deregulated Over-producing Strains

Organism	Product	Selection basis[a]	Reference
Aspergillus niger	Citric acid	Citric acid as sole carbon source	Zhu *et al.* (1981)
Bacillus licheniformis	Bacitracin	Aurolysin[r]	Lukin *et al.* (1981)
Brevibacterium lactofermentum	Lysine	2-Aminoethylcysteine	Tosaka and Takunami (1978)
Candida pelliculosa	Lysine	S-(β-aminoethyl)-L-cysteine[r]	Takenouchi *et al.* (1979)
Cephalosporium acremonium	Cephalosporin C	Mercury (II) chloride[r]	Chang and Elander (1979)
Corynebacterium glutamicum	Histidine	Histidine analogue[r]	Araki *et al.* (1974)
Micromonospora species	Lincomycin	Lincomycin[r]	Ivanitskaya *et al.* (1981)
Pseudomonas fluorescens	Pyrollnitrins	Fluorotryptophan[r]	Elander *et al.* (1971)
Streptomyces aureofaciens	Chlortetracycline	Chlortetracycline[r]	Valerianov *et al.* (1981)
Streptomyces olivoreticuli	Olivomycin	Olivomycin[r]	L'vova *et al.* (1981)

[a] Abbreviations: [r]—resistance.

With reference to the search for new antibiotics Hopwood (1981b) has listed four main ways in which these are, or might be, obtained: (a) chemical modification of existing compounds—the semisynthetic approach; (b) isolation of wild microorganisms and their culture under suitable (varied) conditions leading to maximal gene expression; (c) mutations of a single strain, or recombination between strains, to achieve expression of 'silent' genes; and (d) recombination between two or more strains to generate a 'hybrid' antibiotic. The first method usually follows from the initial identification of the structure of an antibiotic shown to have commercial possibilities but does not require novel microbial strains. However, there is no doubt that chemical modification of structure could be far more powerful if the biosynthetic pathway and hence all intermediates were known. In addition, such knowledge would enable the design of specific screens for use in programmes designed to increase yield. It is in this connection that experiments are under way in a number of laboratories to amplify selectively genes governing the biosynthesis of antibiotics, for example β-lactam genes from *Streptomyces clavuligerus* using mutants impaired in β-lactam biosynthesis similar to those isolated by Normansell *et al.* (1979) as a detection system in transformation experiments. This should help to overcome a major problem associated with studies of β-lactam antibiotics—the difficulty of isolating pathway enzymes to enable a good enzymological study to be made. Once genes of enzymes are characterized other new techniques of molecular biology, including site specific mutagenesis and monoclonal antibody preparation for purification, may be applied. Recently, Baltz (1982) has outlined a useful model for the application of genetic engineering methods to tylosin biosynthesis, fully complementing these techniques with known genetic and biochemical features of the synthesis of this antibiotic.

Turning to the second broad category, the screening of large numbers of wild isolates, sometimes of the same species, takes account of the fact that a particular desired activity and its yield are often properties of a single strain which are not necessarily shared by other members of the same species. For example, the diversity in penicillin yield shown by wild-type isolates of *Aspergillus nidulans* was demonstrated by Holt and Macdonald (1968a). The same pattern, with a number of natural non-producing isolates was also found for wild-type isolates classified as members of the *Penicillium chrysogenum/notatum* groups (Blake and Holt, unpublished data) and for a number of dermatophytic species (Youssef *et al.*, 1976). Strains of *A. nidulans* are known to have the structural genetic information to synthesize a wide variety of different antibiotics and the spectrum of biologically active compounds depends on growth and media conditions employed (Cole *et al.*, 1976a; Cole and Holt, unpublished data). In some cases synthesis of particular metabolites has been shown to be related to specific genetic differences. For example, a number of natural heterokaryon compatible (h-c) groups have been recognized in *A. nidulans* (Jinks *et al.*, 1966) and no strains belonging to h-c groups f and g can produce penicillin (Cole *et al.*, 1976b) whereas cordycepin is only synthesized by members of h–c groups B and L.

Possibilities in the third area involving mutation of one strain can be exemplified by 'mutasynthesis' (Fleck, 1979, and Section 5.2.5) and although its potential has been much publicized (Demain, 1981) the discovery of new products of commercial importance by this approach is awaited.

Meiotic recombination between different h-c groups in *A. nidulans* as a model for the isolation of novel antibiotics has been attempted in our laboratories and those of Macdonald (Middleton *et al.*, 1978; Macdonald, unpublished data). The experiments so far have demonstrated that recombination is most effective in increasing the yield of an antibiotic to a detectable level. However, once the antibiotic is detected and identified a closer scrutiny has always revealed other natural isolates capable of producing that same antibiotic. Even so this approach enabled us to discover from *A. nidulans* an interesting but commercially useless case of a metabolite (a high molecular weight glycoprotein) which acted synergistically with sterigmatocystin to inhibit several Gram-positive bacteria (Perry *et al.*, 1982b). Neither compound alone showed any activity. As pointed out by Hopwood (1981a), random generalized recombination between strains using protoplast fusion is probably a good strategy for novel product formation.

Finally, truly hybrid antibiotics, compounds whose structures depend on the activity of gene products from two or more different strains, will most probably be best achieved by the use of gene cloning techniques (Hopwood, 1981b). Here it is important to remember that microorganisms from nature are the product of evolution. If during this process a new biosynthetic activity were to arise which was selectively harmful to the organisms it would be unlikely to be retained. However, by genetic manipulation (*e.g.* protoplast fusion or 'shot gun' cloning of complete gene libraries) it may be possible to generate recombinant microorganisms in the laboratory with the ability of producing novel compounds. Experiments are in progress in several laboratories to test this hypothesis using highly specific screens for bioactive metabolites. As cloning vectors are produced for a wider range of organisms than have been available hitherto, one of the many benefits will be that even largely empirical programmes for isolating new antibiotics should prove fruitful.

In addition to using genetics to produce novelty in organisms to be screened it is important to consider modification of the test organism used in bioassays. For example, Imada *et al.* (1981) have described the value of supersensitive mutants of *E. coli* and *Pseudomonas aeruginosa* in the isolation of new β-lactams.

5.5 CONCLUSION

In this article we have attempted to show how an appreciation of the fundamental knowledge of mutation and recombination can lead to a more rational approach to the modification of industrial microorganisms. The specific approach and the methods to employ in achieving particular objectives may differ widely but will increasingly involve both mutagenic and recombinational procedures. In this connection we would like to endorse the view of Bull *et al.* (1982) that recombinant DNA technology represents only an additional (albeit exceedingly powerful) tool available to the geneticist. For ingenious and profitable exploitation of all the techniques, as we said in a recent lecture at the Second Symposium on Industrial Microbiology held in Milan 1982 (Holt and Saunders, 1983), there are lessons to be learnt from Michaelangelo, who, given an exceptional creative talent, achieved success by combining hard work with a knowledge of the capabilities and limitations of the tools and the materials.

5.6 REFERENCES

Alderson, T. and A. M. Clark (1966). Interlocus specificity for chemical mutagens in *Aspergillus nidulans*. *Nature (London)*, **210**, 593–595.

Alderson, T. and J. Hartley (1969). Specificity for spontaneous and induced forward mutation at several gene loci in *Aspergillus nidulans*. *Mutat. Res.*, **8**, 255–264.

Alderson, T. and B. R. Scott (1970). The photosensitizing effect of 8-methoxypsoralen on the inactivation and mutation of *Aspergillus* conidia by near ultraviolet light. *Mutat. Res.* **9**, 569–578.

Alfoldi, L. (1982). Fusion of microbial protoplasts: problems and perspectives. In *Genetic Engineering of Microorganisms for Chemicals*, ed. A. Hollaender, pp. 59–71. Plenum, New York.

Alikhanian, J. I. (1962). Induced mutagenesis in the selection of microorganisms. *Adv. Appl. Microbiol.*, **4**, 1–50.

Araki, K., S. Shimojo and K. Nakayama (1974). Histidine production by *Corynebacterium glutamicum* mutants, multiresistant to analogues of histidine, tryptophan, purine and pyrimidine. *Agric. Biol. Chem.*, **38**, 837–846.

Arcamone, F., G. Cassinelli, G. Fantini, A. Grein, P. Orezzi, C. Pol and C. Spalla (1969). Adriamycin, 14-hydroxydaunomycin, a new antitumor antibiotic from *S. peucetius* var. *caesius*. *Biotechnol. Bioeng.*, **11**, 1101–1110.

Arlett, C. R. and A. R. Lehman (1978). Human disorders showing increased sensitivity to the induction of genetic damage. *Annu. Rev. Genet.*, **12**, 95–115.

Arst, H. N., Jr. (1981). Aspects of the control of gene expression in fungi. In *Genetics as a Tool in Microbiology, Society for General Microbiology Symposium No. 31.*, ed. S. W. Glover and D. A. Hopwood, pp. 131–163. Cambridge University Press, Cambridge.

Auerbach, C. (1966). The role of mutagen specificity in mutation breeding. *Sov. Genet.*, **2**, 1–5.

Auerbach, C. (1967). The chemical production of mutations. *Science*, **158**, 1141–1147.

Auerbach, C. (1976). *Mutation Research: Problems, Results, Perspectives*. Chapman and Hall, London.

Auerbach, C. and J. M. Robson (1946). Chemical production of mutations. *Nature (London)*, **157**, 302.

Auerbach, C., J. M. Robson and J. G. Carr (1947). The chemical production of mutations. *Science*, **105**, 243.

Backus, M. P. and J. F. Stauffer (1955). The production and selection of a family of strains in *Penicillium chrysogenum*. *Mycologia*, **47**, 429–463.

Bagg, A., C. J. Denyon and G. C. Walker (1981). Inducibility of gene product required for UV and chemical mutagenesis in *E. coli*. *Proc. Natl. Acad. Sci. USA*, **78**, 5749–5753.

Ball, C. and P. F. Hamlyn (1982). Genetic recombination studies with *Cephalosporium acremonium* related to the production of the industrially important antibiotic cephalosporin C. *Rev. Bras. Genet.*, **5** (1), 1–14.

Ball, C. and M. P. McGonagle (1978). Development and evaluation of a potency index screen for detecting mutants of *Penicillium chrysogenum* having increased penicillin yield. *J. Appl. Bacteriol*, **45**, 67–74.

Ballance, D. J., F. P. Buxton and G. Turner (1983). Transformation of *Aspergillus nidulans* by the orotidine-5'-phosphate-decarboxylase gene of *Neurospora crassa*. *Biochem. Biophys. Res. Commun.*, **112**, 284–289.

Baltz, R. H. (1978). Genetic recombination of *Streptomyces fradiae* by protoplast fusion and cell regeneration. *J. Gen. Microbiol.*, **107**, 93–102.

Baltz, R. H. (1982). Genetics and biochemistry of tylosin production. In *Genetic Engineering of Microorganisms for Chemicals*, ed. A. Hollaender, pp. 431–444. Plenum, New York.

Banks, G. T., K. W. Buck, E. B. Chain, F. Himmelweit, J. E. Marks, J. M. Tyler, M. Hollings, F. T. Last and O. M. Stone (1969). Viruses in fungi and interferon stimulation. *Nature (London)*, **218**, 542–545.

Bibb, M. J., R. F. Freeman and D. A. Hopwood (1977). Physical and chemical characterisation of a second sex factor SCP 2 for *Streptomyces coelicolor*. *Mol. Gen. Genet.*, **154**, 155–166.

Bibb, M. J., J. M. Ward and D. A. Hopwood (1978). Transformation of plasmid DNA into *Streptomyces* at high frequency. *Nature (London)*, **274**, 398–400.

Blaumaurova, M., Z. Hostalek and Z. Vanek. (1972). Biosynthesis of tetracyclines; problems and perspectives of genetic analysis. In *Fermentation Technology Today*, ed. G. Terui, pp. 223–232. Society of Fermentation Technology, Japan.

Bonner, D. (1947). Studies on the biosynthesis of penicillin. *Arch. Biochem.*, **13**, 1–9.

Bridges, B. A. (1976). Mutation Induction. In *Second International Symposium on the Genetics of Industrial Microorganisms*, ed. K. D. Macdonald, pp. 7–14. Academic, New York.

Bridges, B. A. and R. Mottershead (1971). *Rec A*[+]-dependent mutagenesis occurring before replication in UV and γ-irradiated *Escherichia coli*. *Mutat. Res.*, **13**, 1–18.

Bull, A. T., G. Holt and M. D. Lilly (1982). *Biotechnology—International Trends and Perspectives*. OECD, Paris.

Burnett, J. H. (1975). *Mycogenetics*. Wiley, New York.

Calam, C. T. (1964). The selection, improvement and preservation of microorganisms. *Prog. Ind. Microbiol.*, **5**, 1–53.

Calos, M. P. and J. H. Miller (1980). Transposable elements. *Cell*, **20**, 579–595.

Cameron, J. R., E. Y. Loh and R. W. Davis (1979). Evidence for transposition of dispersed repetitive DNA families in yeast. *Cell*, **16**, 739–751.

Case, M. E., M. Schweizer, S. R. Kushner and N. H. Giles (1979). Efficient transformation of *Neurospora crassa* by utilising hybrid plasmid DNA. *Proc. Natl. Acad. Sci. USA*, **76**, 5259–5263.

Caten, C. E. (1981). Parasexual processes in fungi. In *The Fungal Nucleus*, ed. K. Gull and S. G. Oliver, pp. 191–214. Cambridge University Press, Cambridge.

Cerda-Olmedo, E. and R. Ruiz-Vazquez (1979). Nitrosoguanidine mutagenesis. In *Third International Symposium on the Genetics of Industrial Microorganisms*, ed. O. K. Sebek and A. I. Laskin, pp. 15–20. American Society for Microbiology, Washington, DC.

Chang, L. T. and R. P. Elander (1979). Rational selection for improved Cephalosporin C production in strains of *Acremonium chrysogenum*. *Dev. Ind. Microbiol.*, **20**, 367–379.

Chang, S. and S. N. Cohen (1979). High frequency transformation of *Bacillus subtilis* protoplasts by plasmid DNA. *Mol. Gen. Genet.*, **168**, 111–115.

Chater, K. F. and A. T. Carter (1978). Restriction of a bacteriophage in *Streptomyces albus* P (CMI 52766) by endonuclease *Sal* PI. *J. Gen. Microbiol.*, **109**, 181–185.

Chater, K. F., D. A. Hopwood, T. Keiser and C. J. Thompson (1982). Gene cloning in *Streptomyces*. *Curr. Top. Microbiol. Immunol.*, **96**, 69–95.

Chater, K., M. Bibb, C. Bruton, D. A. Hopwood, G. Janssen, F. Malpartida and C. Smith (1984). Dissecting the *Streptomyces* genome. *Biochem. Soc. Trans.*, **12**, 584–586.

Chopra, V. L. (1967). Gene controlled change in mutational stability of a tryptophanless mutant of *E. coli* WP2. *Mutat. Res.*, **4**, 382–384.

Clark, A. J. and G. J. Warren (1979). Conjugal transmission of plasmids. *Annu. Rev. Genet.*, **13**, 99–125.

Clarke, P. H. (1978). *Experimental Evolution in the Bacteria*, vol. 6, pp. 137–219. Academic, London.

Cole, D. S., G. Holt and K. D. Macdonald (1976a). Antibiotic production by wild type and mutant strains of *Aspergillus nidulans*. In *Abstracts of the 5th International Fermentation Symposium, Berlin*, ed. H. Dallweg, p. 191. Westbreuz Drukerei Verlag, Berlin.

Cole, D. S., G. Holt and K. D. Macdonald (1976b). Relationship of the genetic determination of impaired penicillin production in naturally occurring strains to that of induced mutants of *Aspergillus nidulans*. *J. Gen. Microbiol.*, **96**, 423–426.

Collins, A. R. S. and R. T. Johnson (1979). Repair and survival after UV in quiescent and proliferating *Microtus agrestis* cells. *J. Cell Physiol.*, **99**, 125–138.

Corran, J. (1969). Analysis of an apparent case of gene controlled mutational stability: the auxotrophic preemption of a specific growth requirement. *Mutat. Res.*, **7**, 287–295.

Coulondre, C. and J. H. Miller (1977). Genetic studies of the *lac* repressor. III. Additional correlation of mutational sites with specific amino acid residues. *J. Mol. Biol.*, **117**, 525–567.

Cove, D. J. (1976). Chlorate toxicity in *Aspergillus nidulans*: the selection and characterisation of chlorate resistant mutants. *Heredity*, **36**, 191–203.

Cullum, J. and P. Broda (1979). Chromosome transfer and Hfr formation by F in *rec*+ and *rec* A strains of *E. coli* K12. *Plasmid*, **2**, 358.

DasGupta, C., A. M. Wu, R. Kahn, R. P. Cunningham and C. M. Redding (1981). Concerted strand exchange and formation of Holliday structures by *E. coli rec* A protein. *Cell*, **25**, 507–516.

Davies, O. L. (1964). Screening for improved mutants in antibiotic research. *Biometrics*, Sept., 576–591.

Dawes, I. W., D. A. Mackinnon, D. E. Ball, I. D. Hardie, F. M. Ross and F. Macdonald (1977). Identifying sites of simultaneous DNA replication in eukaryotes by NTG multiple mutagenesis. *Mol. Gen. Genet.*, **152**, 53–57.

Decaris, G. (1981). Intragenic location of an unstable insertion element within gene b5 of the fungus *Ascobolus immersus*. *Mol. Gen. Genet.*, **184**, 434–439.

Deering, R. A. and M. Sheely (1977). Mutation induction in vegetative and developing cells of *Dictyostelium discoideum*. In *Development and Differentiation in the Cellular Slime Molds*, ed. P. Cappucinelli and J. M. Ashworth, pp. 63–68. Elsevier/North Holland Biomedical Press, Amsterdam.

Demain, A. L. (1981). Production of new antibiotics by directed biosynthesis and by the use of mutants. In *Future of Antibiotherapy and Antibiotic Research*, ed. P. Ninet and P. Bost, pp. 417–435. Academic, New York.

Demerec, M. (1953). Reaction of genes of *E. coli* to certain mutagens. *Symp. Soc. Exp. Biol.*, **7**, 43–54.

De Serres, F. J. (1980). Mutation induction in repair deficient strains of *Neurospora crassa*. In *DNA Repair and Mutagenesis in Eukaryotes*, ed. W. M. Generoso, M. D. Shelby and F. J. De Serres, pp. 75–84. Plenum, New York.

Ditchburn, P., B. Giddings and K. D. Macdonald (1974). Rapid screening for the isolation of mutants of *Aspergillus nidulans* with increased penicillin yield. *J. Appl. Bacteriol.*, **37**, 515–523.

Ditchburn, P., G. Holt and K. D. Macdonald (1976). The genetic location of mutations increasing penicillin yield in *Aspergillus nidulans*. In *Proceedings of the Second International Symposium on the Genetics of Industrial Microorganisms*, ed. K. D. Macdonald, pp. 213–227. Academic, London.

Donesan, I. N. and D. M. Shankel (1964). Mutational synergisms between radiations and methylated purines in *Escherichia coli*. *J. Bacteriol.*, **87**, 61–67.

Doudney, C. O., B. F. White and B. J. Bruce (1964). Acriflavin modification of nucleic acid formation, mutation induction and survival, in ultra-violet light exposed bacteria. *Biochem. Biophys. Res. Commun.*, **15**, 70–75.

Dulaney, E. L. and D. A. Dulaney (1967). Mutant populations of *Streptomyces viridifaciens*. *Trans. NY Acad. Sci.*, **29**, 782–799.

Eckhardt, F. and R. H. Haynes (1980). Quantitative measures of mutagenicity and mutability based on mutant yield data. *Mutat. Res.*, **74**, 439–458.

Elander, R. P. (1975). Genetic aspects of cephalosporin and cephamycin producing microorganisms. *Dev. Ind. Microbiol.*, **16**, 356–374.

Elander, R. P. (1976). Mutation to increased product formation in antibiotic producing microorganisms. In *Microbiology 1976*, ed. D. Schlessinger, pp. 517–521. American Society for Microbiology, Washington, DC.

Elander, R. P. (1981). Strain improvement programmes in antibiotic producing microorganisms; present and future strategies. In *Advances in Biotechnology*, ed. M. Moo-Young., C. W. Robinson and C. Vezina, vol. 1, pp. 3–8. Pergamon, Oxford.

Elander, R. P., J. A. Mabe, R. L. Hamill and M. Gorman (1971). Biosynthesis of pyrollnitrins by analogue resistant mutants of *Pseudomonas fluorescens*. *Folia Microbiol.*, **16**, 157–165.

Ferenczy, L., F. Kevei and M. Szegedi (1976). Fusion of fungal protoplasts induced by polyethylene glycol. In *Microbial and Plant Protoplasts*, ed. J. F. Peberdy and A. H. Rose, pp. 213–218. Academic, London.

Fincham, J. R. S., P. R. Day and A. Radford (1979). *Fungal Genetics*, 4th edn. Blackwell, Oxford.

Finnegan, D., Jr. (1983). Retroviruses and transposable elements—which came first? *Nature* (*London*), **302**, 105–106.

Fleck, W. F. (1979). Genetic approaches to new streptomycete products. In *Genetics of Industrial Microorganisms*, ed. O. K. Sebek and A. I. Laskin, pp. 117–122. American Society for Microbiology, Washington, DC.

Fodor, K. and L. Alfoldi (1976). Fusion of protoplasts of *Bacillus megaterium*. *Proc. Natl. Acad. Sci. USA*, **73**, 2147–2150.

Fong, K. and R. C. Bockrath (1979). Inhibition of deoxyribonucleic acid repair in *Escherichia coli* by caffeine and acriflavine after ultra-violet light irradiation. *J. Bacteriol.*, **139**, 671–674.

Fraser, M. T., T. Y. K. Chow and E. Kafer (1980). Nucleases and their control in wild type and *nuh* mutants of *Neurospora*. In *DNA Repair and Mutagenesis in Eukaryotes*, ed. W. M. Generoso, M. D. Shelby and F. J. De Serres, pp. 63–74. Plenum, New York.

Friend, E. F., M. Warren and D. A. Hopwood (1978). Genetic evidence for a plasmid controlling fertility in an industrial strain of *Streptomyces rimosus*. *J. Gen. Microbiol.*, **106**, 201–206.

Furumai, T., K. Takeda and M. Okanishi (1982). Function of plasmids in the production of aureothricin. I. Elimination of plasmids and alteration of phenotypes caused by protoplast regeneration in *Streptomyces kasagaensis*. *J. Antibiot.*, **35**, 1367–1373.

Gabor, M. H. and R. D. Hotchkiss (1979). Parameters governing bacterial regeneration and genetic recombination after fusion of *Bacillus subtilis* protoplasts. *J. Bacteriol.*, **137**, 1346–1353.

Glass, R. E. (1982). *Gene function*: E. coli *and its heritable elements*. Croom Helm, London.

Glickman, B., P. Van den Elsen and M. Radman (1978). Induced mutagenesis in *dam⁻* mutants of *Escherichia coli*: a role for 6-methyladenine residues in mutation. *Mol. Gen. Genet.*, **163**, 307–312.

Godfrey, O. W. (1974). Directed mutagenesis in *Streptomyces lipmanii*. *Can. J. Microbiol.*, **20**, 1479–1485.

Grossman, L., A. Braun, R. Feldberg and J. Mahler (1975). Enzymatic repair of DNA. *Annu. Rev. Biochem.*, **44**, 19–43.

Gunge, N. and A. Tamaru (1978). Genetic analysis of products of protoplast fusion in *Saccharomyces cerevisiae*. *Jpn. J. Genet.*, **53**, 41–50.

Hanawalt, P. C. and R. B. Setlow (eds.) (1975). *Molecular Mechanisms for DNA Repair*. Plenum, New York.

Hayatsu, H. (1976). Bisulphite modification of nucleic acids and their constituents. *Prog. Nucleic Acid Res.*, **16**, 75–124.

Hideo, S., K. Nakahama, I. Nogami, M. Kida and M. Yoneda (1982). Accumulation of diphosphorylated butirosin A derivatives by phosphatase negative mutants of *Bacillus vitellinus*. *Agric. Biol. Chem.*, **46**, 1599–1612.

Hill, R. F. (1958). A radiation sensitive mutant of *E. coli*. *Biochim. Biophys. Acta*, **30**, 636–637.

Hiroshi, K., S. Kitamura and K. Nakayama (1982). Production of antibiotic SU-2 complex by a 2-deoxystreptidine idiotroph of *Micromonospora sagamunsis*. *J. Antibiot.*, **35**, 385–390.

Holt, G., G. F. St. L. Edwards and K. D. Macdonald (1976). The genetics of mutants impaired in the biosynthesis of penicillin. In *Second International Symposium on the Genetics of Industrial Microorganisms*, ed. K. D. Macdonald, pp. 199–212. Academic, London.

Holt, G. and K. D. Macdonald (1968a). Penicillin production and its mode of inheritance in *Aspergillus nidulans*. *Antonie van Leeuwenhoek*, **34**, 409–416.

Holt, G. and K. D. Macdonald (1968b). Isolation of strains with increased penicillin yield after hybridisation in *Aspergillus nidulans*. *Nature* (*London*), **219**, 616–617.

Holt, G. and G. Saunders (1983). Genetics, strain selection and improvement with particular respect to antibiotic producing microorganisms. *Riv. Biol.*, **76**, 183–193.

Hopwood, D. A. (1977). Genetic recombination and strain improvement. *Dev. Ind. Microbiol.*, **18**, 9–21.

Hopwood, D. A. (1978). The opening address. In *Proceedings of the Third International Symposium on the Genetics of Industrial Microorganisms*, ed. O. K. Sebek and A. I. Laskin, pp. 3–8. American Society for Microbiology, Washington, DC.

Hopwood, D. A. (1981a). Genetic studies with bacterial protoplasts. *Annu. Rev. Microbiol.*, **35**, 237–273.

Hopwood, D. A. (1981b). Genetic studies of antibiotics and other secondary metabolites. In *Genetics as a Tool in Microbiology, Society for General Microbiology Symposium No. 31*, ed. S. W. Glover and D. A. Hopwood, pp. 187–218. Cambridge University Press, Cambridge.

Hopwood, D. A. and K. F. Chater (1980). Fresh approaches to antibiotic production. *Philos. Trans. R. Soc. London, Ser. B*, **290**, 313–328.

Hopwood, D. A. and H. M. Wright (1973). A plasmid of *Streptomyces coelicolor* carrying a chromosomal locus and its inter-specific transfer. *J. Gen. Microbiol.*, **79**, 331–342.

Hopwood, D. A. and M. J. Merrick (1977). Genetics of antibiotic production. *Bacteriol. Rev.*, **41**, 595–635.

Hopwood, D. A. and H. M. Wright (1976). Bacterial protoplast fusion: recombination in fused protoplasts of *Streptomyces coelicolor*. *Mol. Gen. Genet.*, **162**, 307–317.

Hopwood, D. A., M. J. Bibb, C. J. Bruton, K. F. Chater, J. S. Feitelson and J. A. Gil (1983). Cloning *Streptomyces* genes for antibiotic production. *Trends Biotechnol.*, **1** (2), 40–48.

Howard-Flanders, P. (1968). DNA Repair. *Annu. Rev. Biochem.*, **37**, 175–200.

Howard-Flanders, P. (1973). DNA repair and recombination. *Br. Med. Bull.*, **29**, 226–235.

Howard-Flanders, P. (1981). Inducible repair of DNA. *Sci. Am.*, **245** (5), 56–64.

Ichikawa-Ryo, H. and S. Kondo (1980). Differential antimutagenic effects of caffeine and the protease inhibitor antipain in *Escherichia coli*. *Mutat. Res.*, **72**, 311–322.

Imada, A., K. Kitano, K. Kintaka, M. Muroi and M. Asai (1981). Sulfazecin and isosulfazecin, novel β-lactam antibiotics of bacterial origin. *Nature* (*London*), **289**, 590–591.

Imray, F. P. and D. G. Macphee (1976). Spontaneous and induced mutability of frameshift strains of *Salmonella typhimurium* carrying *uvr* and *pol A* mutations. *Mutat. Res.*, **34**, 35–42.

Inoue, T., Y. Ohta, Y. Sadaie and T. Kada (1981). Effect of cobaltous chloride on spontaneous mutation induction in a *Bacillus subtilis* mutator strain. *Mutat. Res.*, **91**, 41–45.

Ivanitskaya, L. P., M. V. Bibikova, M. N. Gromova, Y. V. Zhadanovich, I. M. Vagina and E. N. Ishatov (1981). Selective media with lincomycin for directed screening of antibiotic producing microorganisms. *Antibiotiki (Moscow)*, **26**, 83–86.

Jinks, J. L., C. E. Caten, G. Simchen and J. H. Croft (1966). Heterokaryon incompatibility and variation in wild populations of *Aspergillus nidulans*. *Heredity*, **21**, 227–239.

Johnson, R. T., A. R. S. Collins, C. S. Docones and S. Squires (1982). DNA synthesis inhibitors and the analysis of UV repair. In *Progress in Mutation Research*, ed. A. T. Najarajan, pp. 357–373. Elsevier/North Holland Biomedical Press, Amsterdam.

Johnston, J. R. (1975). Strain improvement and strain stability in filamentous fungi. In *The Filamentous Fungi*, ed. J. Smith and D. R. Berry, pp. 59–78. Edward Arnold, London.

Kada, T., T. Inoue, A. Yokoiyama and L. B. Russel (1979). Combined genetic effects of chemicals and radiations. In *Radiation Research, Proceedings of the 6th International Congress on Radiation Research*, ed. S. Okada, M. Imamura, T. Terashima and H. Yamaguchi, pp. 711–720. Japanese Association for Radiation Research, Tokyo.

Kappas, A. and B. A. Bridges (1981). Induction of point mutations by benomyl in DNA-repair deficient *Aspergillus nidulans*. *Mutat. Res.*, **91**, 115–118.

Karran, P. and M. G. Marinus (1982). Mismatch correction at O^6-methylguanine residues in *E. coli* DNA. *Nature (London)*, **296**, 868–869.

Kanzaki, T., T. Fukita, K. Kitano, K. Katamoto, K. Nara and Y. Nakao (1976). Occurrence of a novel cephalosporin compound in the culture broth of a *Cephalosporium acremonium* mutant. *J. Ferment. Technol.*, **54**, 720–725.

Kanzaki, T., T. Fukita, H. Shirafuji, Y. Fujisawa and K. Kitano (1974). Occurrence of a 3-methylthiomethylcephem derivative in a culture broth of a *Cephalosporium* mutant. *J. Antibiot.*, **27**, 361–362.

Kato, T. and Y. Shinoura (1977). Isolation and characterisation of mutants of *Escherichia coli* deficient in induction of mutations by ultra-violet light. *Mol. Gen. Genet.*, **156**, 121–131.

Katz, E., C. J. Thompson and D. A. Hopwood (1983). Cloning and expression of the tyrosinase gene from *Steptomyces antibioticus* in *Streptomyces lividans*. *J. Gen. Microbiol.*, **129**, 2703–2714.

Kevei, F. and J. F. Peberdy (1979). Induced segregation in interspecific hybrids of *Aspergillus nidulans* and *Aspergillus rugulosus* obtained by protoplast fusion. *Mol. Gen. Genet.*, **170**, 213–218.

Kilby, B. J. (1974). Determinants of the mutagenic specificity of chemical and physical agents in microorganisms. *Radiat. Res.*, **59**, 966–975.

Kimball, R. F. (1978). The relation of repair phenomena to mutation induction in bacteria. *Mutat. Res.*, **55**, 85–120.

Kimball, R. F. (1980). Relationship between repair processes and mutation induction in bacteria. In *DNA Repair and Mutagenesis in Eukaryotes*, ed. W. M. Generoso, M. D. Shelby and F. J. De Serres, pp. 1–23. Plenum, New York.

Kimball, R. F., S. W. Perdue and M. E. Boling (1978). The role of prereplication and postreplication processes in mutation induction in *Haemophilus influenzae* by N-methyl-N'-nitro-N-nitrosoguanidine. *Mutat. Res.*, **52**, 57–72.

Kirsop, B. E. (ed.) (1980). *The Stability of Industrial Organisms*. The Commonwealth Mycological Institute, Kew, Surrey.

Kitamura, S., H. Kase, Y. Odakura, T. Iida, K. Shirahata and K. Nakayama (1982). 2-Hydroxysagamycin: a new antibiotic produced by mutational biosynthesis of *Micromonospora sagamiensis*. *J. Antibiot.*, **35**, 94–98.

Klamerth, O. L. (1978). Inhibition of post-replication repair by isonicotinic acid hydrazide. *Mutat. Res.*, **50**, 251–261.

Kleckner, N. (1981). Transposable elements in prokaryotes. *Annu. Rev. Genet.*, **15**, 295–341.

Kleckner, N., J. Roth and D. Botstein (1977). Genetic engineering *in vivo* using translocatable drug resistance elements. *J. Mol. Biol.*, **116**, 125–159.

Kolmark, G. (1953). Differential response to mutagens as studied by the *Neurospora* reverse mutation test. *Hereditas*, **39**, 270–276.

Krivonokov, S. V. and V. A. Novitskaja (1981). A protein connected with the integrity of the *rec F* gene in *E. coli* K12. *Mol. Gen. Genet.*, **187**, 302–304.

Lea, D. E. (1955). *Actions of Radiations on Living Cells*, 2nd edn. Cambridge University Press, Cambridge.

Lemontt, J. F. and S. V. Lair (1982). Plate assay for chemical and radiation induced mutagenesis of *can 1* in yeast as a function of post treatment DNA replication: the effect of *rad 6-1*. *Mutat. Res.*, **93**, 339–352.

Linn, S., U. Kuhnlein and W. A. Deutsch (1978). Enzymes from human fibroblasts for the repair of AP DNA. In *DNA Repair Mechanisms*, ed. P. C. Hanawait, E. C. Friedberg and C. F. Fox, pp. 199–203. Academic, New York.

Little, J. W., D. W. Mount and C. R. Yanisch-Perron (1981). Purified *lex*A protein is a repressor of the *rec*A and *lex*A genes. *Proc. Natl. Acad. Sci. USA*, **78**, 4199–4203.

Lomovskaya, N. D., T. A. Volykova and N. M. Mkrturian (1977). Construction and properties of hybrids obtained in interspecific crosses between *Streptomyces coelicolor* A3(2) and *Streptomyces griseus* Kr. 15. *J. Gen. Microbiol.*, **98**, 187–198.

Lukin, A. A., V. I. Zveningordskii and V. G. Zhdanov (1981). Changes in bacitracin synthesis levels in autolysin resistant strains of *Bacillus licheniformis*. *Antibiotiki (Moscow)*, **26**, 87–89.

L'vova, N. A. and O. A. Lapchinskaya (1981). Selection of productive variants from cultures of olivomycin producing *Streptomyces olivoreticuli* 16749. *Antibiotiki (Moscow)*, **26**, 890–893.

Macdonald, K. D. (1983). Fungal genetics and antibiotic production. In *Biochemistry and Genetic Regulation of Commercially Important Antibiotics*, ed. L. C. Vining. Addison-Wesley, Reading, MA.

Macdonald, K. D. and G. Holt (1976). Genetics of the biosynthesis and overproduction of penicillin. *Sci. Prog.*, **63**, 547–573.

Makins, J. F., A. E. Allsop, G. Holt and K. D. Macdonald (1981). Intergeneric cosynthesis of penicillin by strains of *Penicillium chrysogenum*, *P. chrysogenum/notatum* and *Aspergillus nidulans*. *J. Gen. Microbiol.*, **122**, 339–343.

Makins, J. F. and G. Holt (1982a). Liposome mediated transformation of Streptomycetes by chromosomal DNA. *Nature (London)*, **293**, 671–673.

Makins, J. F. and G. Holt (1982b). Liposome–protoplast interactions. *J. Chem. Technol. Biotechnol.*, **32**, 347–353.

Makins, J. F., G. Holt and K. D. Macdonald (1980). Cosynthesis of penicillin following treatments of mutants of *Aspergillus nidulans* impaired in antibiotic production with lytic enzymes. *J. Gen. Microbiol.*, **119**, 397–404.

Makins, J. F., G. Holt and K. D. Macdonald (1983). The genetic location of three mutations impairing penicillin production in *Aspergillus nidulans*. *J. Gen. Microbiol.*, **124**, 3027–3033.

Mallon, R. G. and T. G. Rossman (1981) Bisulphite (sulphur dioxide) as a co-mutagen in *E. coli* and in Chinese hamster cells. *Mutat. Res.*, **88**, 125–133.

Marinus, M. G. and N. R. Morris (1973). Isolation of deoxyribonucleic acid methylase mutants of *Escherichia coli* K12. *J. Bacteriol.*, **114**, 1143–1150.

McCormick, J. R. D., N. O. Sjolander, U. Hirsh, E. R. Jensen and A. P. Doershuk (1957). A new family of antibiotics: the demethyltetracyclines. *J. Am. Chem. Soc.*, **79**, 4561–4563.

Middleton, A. J., D. S. Cole and K. D. Macdonald (1978). A hydroxamic acid from *Aspergillus nidulans* with antibiotic activity against *Proteus* species. *J. Antibiot.*, **31**, 1110–1115.

Miller, J. and U. Schneissner (1979). Genetic studies of the *lac* repressor. X—Analysis of missense mutations in the *lac* I gene. *J. Mol. Biol.*, **131**, 223–248.

Mindlin, S. Z. and S. I. Alikhanian (1958). Studies on ultraviolet ray induced variability and on selection of *Actinomyces rimosus*. *Antibiotiki*, **3**, 18–21.

Mirscher, L. A., J. H. Martin, P. A. Miller, P. Shu and N. Bohonos (1966). 5-Hydro-7-chlortetracyclin. *J. Am. Chem. Soc.*, **88**, 3647–3648.

Morse, M. L., E. M. Lederberg and J. Lederberg (1956). Transductional heterogenotes in *E. coli*. *Genetics*, **41**, 758.

Muller, H. J. (1927). Artificial transmutation of the gene. *Science*, **66**, 84–87.

Munson, R. J. and D. T. Goodhead (1977). The relation between induced mutation frequency and cell survival—a theoretical approach and an examination of experimental data for eukaryotes. *Mutat. Res.*, **42**, 145–160.

Murialdo, H. and L. Siminovitch (1972). The morphogenesis of bacteriophage lambda IV. *Virology*, **48**, 785–823.

Murray, K. (1977). Application of bacteriophage λ in recombinant DNA research. In *Molecular Cloning of Recombinant DNA, Miami Winter Symposia 13*, ed. W. A. Scott and R. Werner, pp. 133–154. Academic, New York.

Nakayama, K., K. Araki, H. Hagino, H. Kase and H. Yoshida (1976). Amino acid fermentations using regulatory mutants of *Corynebacterium glutamicum*. In *Proceedings of the Second Symposium on the Genetics of Industrial Microorganisms*, ed. K. D. Macdonald, pp. 437–449. Academic, New York.

Nash, C. H., N. de la Higuerea, N. Neuss and P. A. Lemke (1974). Applications of biochemical genetics to the biosynthesis of β-lactam antibiotics. *Dev. Ind. Microbiol.*, **15**, 114–123.

Neuss, N., D. M. Berry, J. Kupka, A. L. Demain, S. W. Queener, D. C. Duckworth and L. L. Huckstep (1982). High performance liquid chromatography (HPLC) of natural products. V. The use of HPLC in the cell free biosynthetic conversion of α-aminoadipylcysteinylvaline (LLD) into isopenicillin N. *J. Antibiot.*, **35**, 580–584.

Nevers, P. and H. Saedler (1977). Transposable genetic elements as agents of gene instability and chromosomal rearrangements. *Nature (London)*, **268**, 109–115.

Nevers, P. and H. C. Spatz (1975). *Escherichia coli* mutants *uvr D* and *uvr E* deficient in gene conversion. *Mol. Gen. Genet.*, **139**, 233–243.

Newton, A., D. Masys, E. Leonardi and D. Wygal (1972). Association of induced frameshift mutagenesis and DNA replication in *E. coli*. *Nature (London), New Biology*, **236**, 19–22.

Nisbet, L. J. (1980). Strain degeneration in antibiotic producing Actinomycetes. In *The Stability of Industrial Organisms*, ed. B. E. Kirsop, pp. 39–52. The Commonwealth Mycological Institute, Kew, Surrey.

Nisbet, L. J. (1982). Current strategies in the search for bioactive microbial metabolites. *J. Chem. Technol. Biotechnol.*, **32**, 251–270.

Normansell, I. D. (1982). Strain improvement in antibiotic producing microorganisms. *J. Chem. Technol. Biotechnol.*, **32**, 296–303.

Normansell, I. D. and G. Holt (1978). Viruses in strains of *Penicillium chrysogenum* impaired in penicillin biosynthesis. *Mycovirus Newsletter*, **6**, 15.

Normansell, P. J. M., I. D. Normansell and G. Holt (1979). Genetic and biochemical studies of mutants of *Penicillium chrysogenum* impaired in penicillin production. *J. Gen. Microbiol.*, **112**, 113–126.

Normansell, I. D. and G. Holt (1979). The ability of ionizing radiations of different LET to induce chromosomal deletions in *Aspergillus nidulans*. *Mutat. Res.*, **59**, 167–177.

Ochi, K., M. J. M. Hitchcock and B. Katz (1979). High frequency fusion of *Streptomyces parvulus* or *Streptomyces antibioticus* protoplasts induced by polyethylene glycol. *J. Bacteriol.*, **139**, 983–992.

Old, R. W. and S. B. Primrose (1982). *Principles of Gene Manipulation, an Introduction to Genetic Engineering*, 2nd edn. Blackwell, Oxford.

Paterson, H. F. (1974). Investigations into a reversal of diepoxybutane specificity in *Neurospora crassa*. *Mutat. Res.*, **25**, 411–413.

Perry, M. J., M. W. Adlard and G. Holt (1982a). The isolation of a fungal metabolite which exhibits antimicrobial synergy with sterigmatocystin. *J. Appl. Bacteriol.*, **52**, 83–89.

Perry, M. J., M. W. Adlard and G. Holt (1982b). The effect of the synergy between sterigmatocystin and a fungal metabolite on *Bacillus subtilis*. *J. Appl. Bacteriol.*, **52**, 91–96.

Pesti, M., E. Konszky, J. Erdei, K. Polya and L. Ferenczy (1980). Penicillin production of *Penicillium chrysogenum* strains obtained by protoplast fusion. *Acta Microbiol. Acad. Sci. Hung.*, **27**, 249–250.

Pontecorvo, G. (1954). Mitotic recombination in the genetic system of filamentous fungi. *Caryologia (Florence)*, **6**, Suppl., 192–200.

Pontecorvo, G., J. A. Roper, L. M. Hermmans, K. D. Macdonald and A. W. J. Bufton (1953). The genetics of *Aspergillus nidulans*. *Adv. Genet.*, **5**, 141–238.

Promptov, A. N. (1932). The effect of short ultraviolet rays on the appearance of heriditary variations in *Drosophila melanogaster*. *J. Genet.*, **26**, 56.

Queener, S. W. and R. H. Baltz (1979). Genetics of industrial microorganisms. In *Annual Reports on Fermentation Processes*, ed. D. Perlman, vol. 3, pp. 5–45. Academic, New York.

Queener, S. and R. Swartz (1979). Penicillins: biosynthetic and semisynthetic. In *Economic Microbiology*, ed. A. H. Rose, vol. 3, pp. 35–122. Academic, London.

Radford, A., S. Pope, A. Sazci, M. J. Fraser and J. H. Parish (1981). Liposome mediated genetic transformation of *Neurospora crassa*. *Mol. Gen. Genet.*, **184**, 567–569.

Radman, M. (1977). Inducible pathways in DNA repair, mutagenesis and carcinogenesis. *Biochem. Soc. Trans.*, **5**, 1194–1196.

Resnick, M. A. and J. K. Setlow (1972). Repair of pyrimidine dimer damage induced in yeast by ultraviolet light. *J. Bacteriol.*, **109**, 979–986.

Roberts. J. J., J. E. Sturrock and K. Ward (1974). The enhancement by caffeine of alkylation induced cell death, mutations and chromosomal aberrations in Chinese Hamster cells as a result of inhibition of post-replication DNA repair. *Mutat. Res.*, **26**, 129–135.

Rodicio, M. R. and K. F. Chater (1982). Small DNA free liposomes stimulate transfection of Streptomycete protoplasts. *J. Bacteriol.*, **151**, 1078–1085.

Rogers, M., M. W. Adlard, G. Saunders and G. Holt (1983a). High performance liquid chromatographic determination of β-lactam antibiotics using fluorescence detection following post-column derivatization. *J. Chromatogr.*, **257**, 91–100.

Rogers, M., M. W. Adlard, G. Saunders and G. Holt (1983b). High performance liquid chromatographic determination of penicillins following derivatization to mercury stabilised penicillenic acids. *J. Liq. Chromatogr.*, **6**, 2019–2031.

Roper, J. A. (1966). Mechanisms of inheritance. 3. The parasexual cycle. In *The Fungi, an Advanced Treatise*, ed. G. C. Ainsworth and A. S. Sussman, pp. 589–617. Academic, London.

Rosi, D., W. A. Gross and S. J. Daum (1977). Mutational biosynthesis by idiotroph of *Micromonospora purpurea*: 1. Conversion of aminocyclitol to new aminoglycoside antibiotic. *J. Antibiot.*, **30**, 88–97.

Rossman, T. G. (1981). Enhancement of UV mutagenesis by low concentrations of arsenite in *E. coli. Mutat. Res.*, **91**, 207–211.

Rupp, W. D. and P. Howard-Flanders (1968). Discontinuities in the DNA synthesised in an incision defective strain of *Escherichia coli* following ultraviolet irradiation. *J. Mol. Biol.*, **31**, 294–304.

Rupp, W. D., C. E. Wilde, D. L. Reno and P. Howard-Flanders (1971). Exchanges between DNA strands of ultraviolet irradiated *Escherichia coli. J. Mol. Biol.*, **61**, 25–44.

Rydberg, B. (1978). Bromouracil mutagenesis and mismatch repair in mutator strains of *Escherichia coli. Mutat. Res.*, **52**, 11–24.

Sancar, A., A. M. Hack and W. D. Rupp (1979). Simple method for identification of plasmid coded proteins. *J. Bacteriol.*, **137**, 692–693.

Saunders, G. and G. Holt (1982). Far ultraviolet light sensitive derivatives of *Streptomyces clavuligerus. J. Gen. Microbiol.*, **128**, 381–385.

Saunders, G., A. E. Allsop and G. Holt (1982). Modern developments in mutagenesis. *J. Chem. Technol. Biotechnol.*, **32**, 354–364.

Saunders, G., M. E. Rogers, M. W. Adlard and G. Holt (1984a). Chromatographic resolution of nucleic acids extracted from *P. chrysogenum. Mol. Gen. Genet.*, **194**, 343–348.

Saunders, G., M. E. Rogers, M. W. Adlard and G. Holt (1984b). Chromatographic resolution of nucleic acids: application to organisms of industrial importance. *Trans. Biochem. Soc.*, **12**, 694–695.

Schroeder, A. L. and L. D. Olson (1980). Mutagen sensitive mutants in *Neurospora*. In *DNA Repair and Mutagenesis in Eukaryotes*, ed. W. M. Generoso, M. D. Shelby and F. J. De Serres, pp. 55–62. Plenum, New York.

Schupp, T., P. Traxler and J. A. L. Auder (1981). New rifamycins produced by a recombinant strain of *N. mediterranei. J. Antibiot.*, **34**, 965–970.

Scott, B. R. and T. Alderson (1971). The random (non-specific) forward mutational responses of gene loci in *Aspergillus* conidia after photosensitization to near ultraviolet light (365 nm) by methoxypsoralen. *Mutat. Res.*, **12**, 29–34.

Sebek, O. K. (1976). Use of mutants for the synthesis of new antibiotics. In *Microbiology 1976*, ed. D. Schlessinger, pp. 522–525. American Society for Microbiology, Washington, DC.

Sedgewick, S. G. (1975). Inducible error prone repair in *Escherichia coli. Proc. Natl. Acad. Sci. USA*, **72**, 2753–2757.

Sermonti, G., and G. Morpurgo (1959). Action of manganous chloride on induced somatic segregation in *Penicillium chrysogenum* diploids. *Genetics*, **44**, 437–447.

Sherrat, D. (1981). *In vivo* genetic manipulation in bacteria. In *Genetics as a Tool in Microbiology*, ed. S. W. Glover and D. A. Hopwood, pp. 35–47. Cambridge University Press, Cambridge.

Shier, W. T., K. L. Rinehart, Jr. and D. Gottlieb (1969). Preparation of four new antibiotics from a mutant of *Streptomyces fradiae. Proc. Natl. Acad. Sci. USA*, **63**, 198–204.

Shneyour, J., J. Stamberg, P. Hundert, R. Werzberger and Y. Koltin (1978). Selection with cycloheximide of metabolic and UV sensitive mutants of *Schizophyllum commune* and *Saccharomyces cerevisiae. Mutat. Res.*, **49**, 195–201.

Shortle, D., D. Koshland, G. M. Weinstock and D. Boslein (1980). Segment directed mutagenesis; construction *in vitro* of point mutations limited to a small predetermined region of a circular DNA molecule. *Proc. Natl. Acad. Sci. USA*, **77**, 5375–5379.

Shortle, D. and D. Nathans (1978). Local mutagenesis: a method for generating viral mutants with base substitutions in preselected regions of the viral genome. *Proc. Natl. Acad. Sci. USA*, **75**, 2170–2174.

Sideropoulos, A. S. and D. M. Shankel (1968). Mechanism of caffeine enhancement of mutations induced by sublethal ultraviolet dosages. *J. Bacteriol.*, **96**, 198–204.

Sideropoulos, A. S., S. M. Specht and M. T. Jones (1980). Feasibility of testing DNA repair inhibitors for mutagenicity by a sample method. *Mutat. Res.*, **74**, 95–105.

Sklar, R. (1978). Enhancement of nitrosoguanidine mutagenesis by chloramphenicol in *Escherichia coli* K12. *J. Bacteriol.*, **136**, 460–462.

Skotnicki, M. L., K. J. Lee, D. E. Tribe and P. L. Rogers (1982). Genetic alteration of *Zymomonas mobilis* for ethanol production. In *Genetic Engineering of Microorganisms for Chemicals*, ed. A. Hollaender, pp. 271–290. Academic, New York.

Spizek, J., I. Malek, L. Dolezilova, M. Vondrack and Z. Vanek (1965). Metabolites of *Streptomyces noursei* IV. Formation of secondary metabolites by producing mutants. *Folia Microbiol.*, **10**, 259–262.

Smith, T. M., G. Saunders, L. M. Stacey and G. Holt (1984). Restriction endonucleases map of mitochondrial DNA from *Penicillium chrysogenum. J. Biotechnol.*, **1**, 37–46.

Smith, T. M. and G. Holt (1984). Cloning of DNA from *P. chrysogenum. Trans. Biochem. Soc.*, **12**, 645–646.

Steinborn, G. (1978). *Uvr* mutants of *Escherichia coli* K12 deficient in UV mutagenesis. *Mol. Gen. Genet.*, **165**, 87–93.

Stohl, L. L. and Lambowitz, A. M. (1983). Construction of a shuttle vector for the filamentous fungus *N. crassa. Proc. Natl. Acad. Sci. USA*, **80**, 1058–1062.

Strauss, B., K. N. Ayres, K. Bose, P. Moore, R. Sklar and K. Tatsumi (1980). Role of cellular systems in modifying the response to chemical mutagens. In *DNA Repair and Mutagenesis in Eukaryotes*, ed. W. M. Generoso, M. D. Shelby and F. J. De Serres, pp. 25–48. Plenum, New York.

Struhl, K., D. T. Stinchcomb, S. Scherer and R. W. Davis (1979). High frequency transformation of yeast: autonomous replication of hybrid DNA molecules. *Proc. Natl. Acad. Sci. USA*, **76**, 1035–1039.

Stuttard, C. (1979). Transduction of auxotrophic markers in a chloramphenicol producing strain of *Streptomyces*. *J. Gen. Microbiol.*, **110**, 479–482.

Suarez, J. E. and K. F. Chater (1980). DNA cloning in *Streptomyces*: a bi-functional replicon comprising pBR322 inserted into a *Streptomyces* phage. *Nature (London)*, **286**, 527–529.

Szostak, J. W. and R. Wu (1979). Insertion of a genetic marker into the ribosomal DNA of yeast. *Plasmid*, **2**, 536–554.

Szvoboda, G., T. Lang, I. Gado, G. Ambrus, C. Kari, K. Fodor and L. Alfoldi (1980). Fusion of *Micromonospora* protoplasts. In *Advances in Protoplast Research*, ed. L. Ferenczy and G. L. Farkas, pp. 235–240. Pergamon, Oxford.

Takenouchi, E., T. Yamamoto, D. K. Nikolova, H. Tanaka and K. Soda (1979). Lysine production by δ-(β-aminoethyl)-L-cysteine resistant mutants of *Candida peliculosa*. *Agric. Biol. Chem.*, **43**, 727–734.

Tang, M., M. W. Lieberman and C. M. King (1982). *Uvr* genes function differently in repair of acetylaminofluorene and aminofluorene DNA adducts. *Nature (London)*, **299**, 646–648.

Tanooka, H. (1977). Development and application of *Bacillus subtilis* test systems for mutagens involving DNA repair deficiency and repressible auxotrophic mutations. *Mutat. Res.*, **42**, 19–32.

Tempest, P. R. and B. E. B. Moseley (1980). Defective excision repair in a mutant of *Micrococcus radiodurans* hypermutable by some monofunctional alkylating agents. *Mol. Gen. Genet.*, **179**, 191–199.

Tempest, D. W., O. M. Neijssel and W. Zevenboom (1983). Properties and performances of microorganisms in laboratory culture; their relevance to growth in natural ecosystems. In *Microbes in their Natural Environment, Society for General Microbiology Symposium No. 34*, ed. J. H. Slater, R. Whittenbury and J. W. Wimpenny, pp. 119–152. Cambridge University Press, Cambridge.

Thompson, C. J., J. M. Ward and D. A. Hopwood (1980). DNA cloning in *Streptomyces*: resistance genes from antibiotic producing species. *Nature (London)*, **286**, 525–527.

Thompson, C. J., J. M. Ward and D. A. Hopwood (1982). Cloning of antibiotic resistance and nutritional genes in Streptomycetes. *J. Bacteriol.*, **151**, 668–677.

Thompson, C. J. and Gray, G. S. (1983). Nucleotide sequence of a streptomycete aminoglycoside phosphotransferase gene and its relationship to phosphototransferases encoded by resistance plasmids. *Proc. Natl. Acad. Sci. USA*, **80**, 5190–5194.

Tikhonenko, T. I., G. A. Velikodvorskaya, A. F. Bobkova, Y. E. Bartosherich, E. P. Lebed, N. M. Chaplygina and T. S. Maksimova (1974). New fungal virus capable of reproducing in bacteria. *Nature (London)*, **249**, 454–456.

Tilburn, J., C. Scazzochio, G. Taylor, R. A. Lockington and J. O. Zabicki-Sissman (1983). Transformation by integration in *Aspergillus nidulans*. In *Abstracts of the 198th Meeting of the Genetical Society*, p. 10.

Timmis, K. N. (1981). Gene manipulation *in vitro*. In *Genetics as a Tool in Microbiology*, ed. S. W. Glover and D. A. Hopwood, pp. 49–109. Cambridge University Press, Cambridge.

Tosaka, O. and K. Takinami (1978). Pathway and regulation of lysine biosynthesis in *Brevibacterium lactofermentum*. *Agric. Biol. Chem.*, **42**, 95–100.

Triechler, H. J., M. Liersch and J. Nuesch (1978). Genetics and biochemistry of Cephalosporin biosynthesis. In *Antibiotics and Other Secondary Metabolites*, ed. R. Hutter, T. Leisinger, J. Nuesch and W. Wehrli, pp. 97–104. Academic, New York.

Triechler, H. J., M. Liersch, J. Nuesch and H. Bobeli (1979). The role of sulphur metabolism in Cephalosporin C and penicillin biosynthesis. In *Genetics of Industrial Microorganisms*, ed. O. K. Sebek and A. I. Laskin, pp. 97–104. American Society for Microbiology, Washington, DC.

Trilli, A., I. Costanzi, F. Lamenna and N. Di Dio (1982). Development of the agar disc method for the rapid screening of strains with increased productivity. *J. Chem. Technol. Biotechnol.*, **32**, 281–291.

Troonen, H. P., P. Roelants and B. Boon (1976). RIT 2214, a new biosynthetic penicillin produced by a mutant of *Cephalosporium acremonium*. *J. Antibiot.*, **29**, 1258–1267.

Valerianov, Z., T. Todonov, M. Naydenova, S. Kozarova, D. Minchov and V. Gancheva (1981). UV irradiation of *Streptomyces aureofaciens* and isolation of chlortetracycline highly productive mutants. *Acta Microbiol. Bulg.*, **8**, 25–29.

Vidoli, R., H. Yamazaki, A. Nasim and I. A. Veliky (1982). A novel procedure for the recovery of hybrid products from protoplast fusion. *Biotechnol. Lett.*, **4**, 781–784.

Whitehouse, H. L. K. (1982). *Genetic Recombination—Understanding the Mechanisms*. Wiley, London.

Winter, G., A. R. Fersht, J. J. Wilkinson, M. Zoller and M. Smith (1982). Redesigning enzyme structure by site directed mutagenesis: tyrosyl tRNA synthetase and ATP binding. *Nature (London)*, **299**, 756–758.

Witkin, E. M. (1967). Mutation proof and mutation prone modes of survival in derivatives of *Escherichia coli* B differing in sensitivity to ultra-violet light. *Brookhaven Symp. Biol.*, **29**, 17–55.

Witkin, E. M. (1969). The role of DNA repair and recombination in mutagenesis. *Proc. XII Congr. Genet.*, **3**, 225–245.

Witkin, E. M. (1976). Ultraviolet mutagenesis and inducible DNA repair in *Escherichia coli*. *Bacteriol Rev.*, **40**, 869–907.

Youngs, D. A. and K. C. Smith (1976b). Genetic control of multiple pathways of post-replicational repair in *uvr* B strains of *Escherichia coli* K12. *J. Bacteriol.*, **125**, 102–110.

Youssef, N., C. H. E. Wyborn, G. Holt, W. C. Noble and T. M. Clayton (1976). Antibiotic production by dematophyte fungi. *J. Gen. Microbiol.*, **105**, 105–111.

Zhu, H., Q. Hou, Y. Yang, A. Zheng, G. Wang and X. Manfu (1981). Direct fermentation of citric acid from highly concentrated sweet potato mash. The selection of *Aspergillus niger* 5016. *Acta Microbiol. Sin.*, **21**, 363–366.

Zinder, M. D. and J. Lederberg (1952). Genetic exchange in *Salmonella*. *J. Bacteriol.*, **64**, 679.

6

In Vitro Recombinant DNA Technology

G. CHURCHWARD and M. CHANDLER
University of Geneva, Switzerland

6.1 INTRODUCTION

The past 10 years have seen a dramatic acceleration in our understanding of the organization and expression of both prokaryotic and eukaryotic genomes. These advances have been due in large part to the availability of a range of techniques known collectively as recombinant DNA technology.

The most important contribution to this rapid growth has been the development of methods for obtaining specific genes in high yield and in a pure form. The magnitude of the problem confronting the molecular biologist wishing to study a specific gene which forms part of a complex genome can best be understood by considering that the human β-globin gene (1.6 kb), for example, represents only 0.00005% of the human genome (2.9×10^6 kb). Previously, detailed study of gene expression at the molecular level was limited to those genes that could be propagated as part of a viral genome or a bacterial plasmid because these autonomously replicating molecules carry relatively few genes and can be purified easily, in quantity, away from the genomic DNA of the host. The discovery of enzymes able to cleave a DNA molecule at specific sites (restriction endonucleases) or to join covalently double-stranded DNA molecules (DNA ligases) *in vitro*, together with a detailed knowledge of the organization of bacterial viruses and plasmids (particularly those of *Escherichia coli*), provided a means of purification and propagation of genes from any source. Bacterial viruses and plasmids can be harnessed as vectors for foreign DNA fragments by covalent linkage of the fragment to the vector at a predetermined position defined by a restriction endonuclease cleavage site. Provided that the insertion of the DNA does not interrupt any essential vector functions, the hybrid molecule can be propagated and easily isolated.

The basic gene enrichment procedure, known as molecular cloning, together with the many complementary methods allowing rapid DNA sequencing and *in vitro* manipulation of the cloned segment, has permitted the purification and study of the structure and the expression of genes obtained from a wide variety of organisms. The subsequent development of host–vector systems for organisms other than *E. coli* and of 'shuttle' vectors (bifunctional replicons able to propagate in several hosts) has introduced a large degree of flexibility into the study of gene structure and function. Not only can a gene be studied in a heterologous system, for example mammalian genes in *E. coli*, but it is now possible to study many genes in an homologous system by reintroduction of a cloned gene into cells of the original organism. This flexibility has profoundly changed the approach of the geneticist. In addition to isolating mutants by screening for a given phenotype and subsequently locating these mutations within a specific gene by employing suitable genetic crosses, recombinant DNA technology has provided the means to introduce predetermined changes in a given gene and to observe the effects of such changes on gene expression. This type of approach has been called 'reversed genetics'.

Although the full potential of these methods has yet to be realized, they have already been applied with success to a wide range of problems of economic, environmental and medical importance.

In this review, we propose to outline briefly the concepts involved in recombinant DNA technology and to present the reader with a comprehensive, although not exhaustive, discussion of many of the techniques available for the cloning and manipulation of genes. Where possible we have cited review articles rather than the original papers. A list of several important articles, some of which include detailed experimental protocols, will be found at the end of this chapter.

We have divided gene cloning into five different stages: (a) generation of DNA fragments for cloning, (b) covalent linkage of the fragments to a vector plasmid or bacteriophage, (c) introduction of this hybrid molecule into a suitable host, (d) detection of the desired clone, and (e) characterization of the purified cloned fragment. Subsequent operations may involve: (f) modification of the cloned piece by removal of inessential DNA regions and mutagenesis, and (g) recloning into a vector to study and optimize expression of the gene in its original host or in other secondary host organisms.

6.2 GENERATION AND CLONING OF DNA FRAGMENTS

6.2.1 Fragmentation of DNA

One of the first problems encountered prior to cloning a given gene is to separate the gene physically from the rest of the DNA molecule of which it forms a part. This is normally accomplished either by fragmentation of the DNA with restriction enzymes or by synthesis of a double-stranded DNA copy of partially purified mRNA (cDNA).

6.2.1.1 *Class II restriction enzymes*

Fragmentation of a genome for cloning is most conveniently carried out by the use of class II restriction endonucleases of which over 200 are known (Roberts, 1983). Each enzyme recognizes a specific nucleotide sequence and cleaves both strands of the DNA. A specific recognition sequence may consist of six, five or four nucleotides which on average would be expected to occur once every 4096 (1 in 4^6; infrequent cutters), 1024 (1 in 4^5) and 256 (1 in 4^4; frequent cutters) base pairs respectively in DNA with a 50% G + C content.

Many such enzymes recognize palindromic sequences and cleave each strand of the DNA, several base pairs from the axis of symmetry generating protruding 5' or 3' complementary, single-stranded extensions (Figure 1, Table 1). This complementarity facilitates the eventual insertion of the fragment into a vector molecule cleaved with the same enzyme, since the extremities of the vector and the fragment are able to hybridize and form metastable structures prior to their covalent linkage (Mertz and Davis, 1972; Section 6.2.2). The role of homology in cloning using this type of enzyme in general precludes the direct insertion of a fragment generated by one enzyme into a vector cleaved with another. However, exceptions exist, for example in the case of the enzymes *Sal*1 and *Xho*1, and *Bam*HI and *Bgl*II, where the extensions (but not the entire recognition sequences) are identical (Table 1).

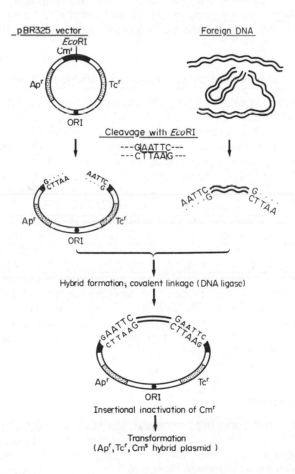

Figure 1 Cloning foreign DNA into an *E. coli* plasmid vector using insertional inactivation. Following cleavage of pBR325 vector and foreign DNA with a restriction endonuclease, *Eco*RI, the DNAs are ligated together. Insertion of DNA into the vector inactivates the gene specifying chloramphenicol resistance (Cm^r). The desired clones may be recovered after transformation as colonies that are resistant to ampicillin (Ap^r) or tetracycline (Tc^r), but sensitive to chloramphenicol

Other class II restriction enzymes cleave both DNA strands at the axis of symmetry of the recognition site and thus generate fragments which carry neither 5' nor 3' extensions. These 'flush' or 'blunt' ended fragments can also be cloned directly (Sgaramella *et al.*, 1970). Since no single-

Table 1   Recognition Sites of Some Restriction Endonucleases Used in Cloning

		5′ Extensions		*3′ Extensions*		*Blunt ends*
(A)	BamHI	–G|G A T C C– –C C T A G|G–	KpnI	–G G T A C|C– –C|C A T G G–	PvuII	–C A G|C T G– –G T C|G A C–
	BglII	–A|G A T C T– –T C T A G|A–	PstI	–C T G C A|G– –G|A C G T C–	SmaI	–C C C|G G G– –G G G|C C C–
	EcoRI	–G|A A T T C– –C T T A A|G–				
(B)	BamHI/BglII hybrid site			–G G A T C A– –G C T A G T–		Cleaved by *Sau*3A –G A T C– –C T A G–

stranded extensions are involved, any 'blunt-ended' fragment can be inserted into any 'blunt-end' site in the vector allowing greater flexibility.

When a fragment produced by digestion of DNA with a restriction enzyme is inserted into a vector digested with the same enzyme, a recognition sequence is regenerated at each junction between vector and inserted DNA. This can facilitate the recovery of fragments for subsequent manipulation. If different enzymes are used to produce the fragment and cleave the vector, hybrid recognition sequences are generated which are usually not susceptible to cleavage by either enzyme. However, they may be cleaved by a third enzyme (Table 1B).

6.2.1.2   *Random DNA fragments and the generation of genomic libraries*

If the genome of interest is small, for example a virus or plasmid, or if the gene can be enriched in the population, direct cloning using class II restriction enzymes should be the method of choice. For larger genomes, however, it is usual to generate a genomic library (a collection of clones which together include most or all of the genome), and to screen amongst groups of these clones for the presence of the desired gene (Maniatis *et al.*, 1978). For this purpose fragmentation of the DNA should be random. This can be achieved by shearing to generate fragments of a suitable length. However, partial digestion of the genome with a restriction enzyme which cleaves frequently is normally employed (*e.g.* Ish-Horowicz and Burke, 1981). For example, DNA which has been partially digested with *Sau*3A can be inserted into a vector cleaved with *Bam*HI, making use of the homologous 5′ extensions generated by these two enzymes (see Table 1). The number of independent clones required to ensure the appearance of a given gene in the library is determined by the average length of the cloned fragments, the length of the desired sequence and the total length of the genome (Carbon *et al.*, 1977). With fragments of average size 20 kb, about 900 clones are required to ensure that a given *E. coli* gene (genome size 4000 kb) is present with 99% probability. This increases to 3100 in the case of yeast (13 500 kb), 6.7×10^5 for the human genome (2.9×10^6 kb) and 2.4×10^7 for the South American lungfish (1.02×10^8 kb). For a detailed discussion of the theoretical and practical problems in constructing genomic libraries the reader is referred to the paper by Seed *et al.* (1982).

6.2.1.3   *Enrichment for specific DNA sequences*

In certain circumstances it is possible to obtain an initial enrichment of the desired DNA fragment prior to cloning. Where the fragment can be defined by hybridization with purified or enriched mRNA or other specific probes, an initial enrichment can be obtained by fractionation of the population of fragments according to their size. The most convenient methods of fractionation are agarose or acrylamide gel electrophoresis (see Southern, 1979) followed by extraction of the DNA from the gel matrix (Smith, H. O., 1980; Yang *et al.*, 1979) or sedimentation in sucrose density gradients.

6.2.1.4   *Synthesis of cDNA*

For the isolation of a gene from large, complex eukaryotic genomes, a second method is frequently employed. This involves the synthesis of a double-stranded cDNA copy from isolated mRNA molecules. The power of this technique is that only a fraction of the genome (that fraction which is transcribed into mRNA) is copied. The resulting cDNA clones can subsequently be used as probes to identify genomic fragments contained in a genomic library. The coding region of many eukaryotic structural genes is frequently interrupted by introns (intervening sequences) which are removed from the mature RNA by splicing (*e.g.* Breathnach and Chambon, 1981). Since a primary bacterial host will not possess splicing systems, cloning of cDNA copies of mature mRNA is the only way to obtain expression of such genes in bacteria. A comparison of cDNA clones with their genomic counterparts also provides important information concerning RNA processing.

Schemes for purification of bulk eukaryotic mRNA often make use of the ability of polyade-nylated 3′ ends to hybridize with, and be reversibly retained by, oligodeoxythymidilate residues immobilized on a solid matrix (Aviv and Leder, 1972). Species such as ribosomal RNA which do not contain large tracts of adenine residues cannot hybridize with the immobilized oligo(dT) and are therefore not retained by the matrix (Figure 2). The polyadenylated mRNA is then eluted from the matrix. If a suitable assay exists for the gene product, an additional purifica-tion can be carried out at this stage by fractionating the purified mRNA according to size on a sucrose gradient and testing the capability of each fraction to direct the synthesis of the gene product (*e.g.* Buell *et al.*, 1978; Gray *et al.*, 1982; Section 6.4.1.2). These methods do not generally result in sufficient purification to make cloning cDNA copies of rare mRNA species a straightforward task. If suitable antibodies are available they can be used to precipitate poly-somes to purify a specific mRNA. This can greatly enrich for rare mRNA molecules (Korman *et al.*, 1982).

Double-stranded DNA copies are synthesized in several steps from the collection of purified mRNA molecules (Figure 2). DNA synthesis is carried out by use of avian myeloblastosis virus reverse transcriptase (AMVrt) and is primed by an oligo(dT) residue which specifically hybridizes with the 3′ polyadenylated end of the mRNA. After hydrolysis of the mRNA template, con-ditions are employed which lead to the formation of small 'hairpins' at the 3′ end of the newly synthesized DNA. The small hairpin is used to prime DNA synthesis by AMVrt (Rougeon and Mach, 1976) or by *Escherichia coli* DNA polymerase I (Efstratiadis *et al.*, 1976) using the newly synthesized cDNA copy as template. Both strands of the resulting double-stranded cDNA copy are therefore covalently joined at one end. Prior to cloning, this covalent linkage must be broken. The most common method has been the use of the single strand specific endonuclease S1 (Figure 2). Self-priming employing hairpin structures, followed by S1 treatment, can seriously reduce the probability of obtaining full length cDNA, particularly if the mRNA is large. This problem can be circumvented by extending the 3′ end of the initial single-stranded cDNA copy using dCTP and terminal deoxynucleotidyl transferase (Section 6.2.3), removal of the mRNA template by hydrolysis, and priming of complementary DNA synthesis with oligo(dG) which has been hybri-dized to the 3′ oligo(dC) extremity (Figure 2, Land *et al.*, 1981). More recently, an extremely efficient system has been developed which uses a specialized vector plasmid (Okayama and Berg, 1982). With this system, the cDNA is synthesized directly into the vector and cDNA copies of RNA molecules capable of encoding proteins in excess of 200 000 D have been cloned.

6.2.1.5 *Chemical synthesis of DNA*

Although the methods for generating DNA fragments described above are those most com-monly used, an increasingly important method for generating specific DNA sequences is that of chemical synthesis. Solid phase synthetic methods can give a chain growth rate of up to one resi-due per 30 min and generate tetradecamers with a yield in excess of 70% (Alvarado-Urbina *et al.*, 1981). Rapid development in this area has resulted in the production of an enormous variety of DNA sequences. Short, double-stranded oligonucleotide linkers encoding a restriction enzyme recognition sequence can be used to modify the extremities of DNA fragments prior to cloning (Section 6.2.3). Single-stranded oligonucleotides can be used to introduce short deletions or specific base changes into any DNA fragment (Section 6.5.1) and are of increasing importance in DNA sequencing techniques (Anderson *et al.*, 1980).

A very important use of chemically synthesized oligonucleotides is in the screening of cDNA libraries for specific genes (Section 6.4.1). If the partial amino acid sequence of a protein is known, a synthetic oligonucleotide can sometimes be synthesized, permitting detection of the cloned gene by nucleic acid hybridization. Often this is the only feasible approach to cloning a particular gene since other methods of detection are too laborious.

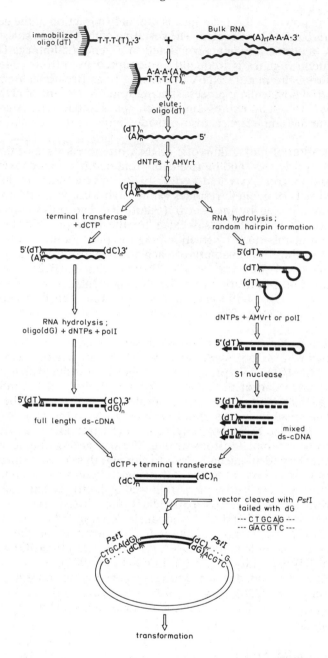

Figure 2 Synthesis and cloning of cDNA. The steps in the synthesis of cDNA copies of a population of polyadenylated mRNA molecules and the subsequent complementary strand synthesis are shown. Two alternatives for complementary strand priming are demonstrated: either by tailing with (dc) and terminal transferase, and using oligo(dG) as primer, or by hairpin formation and subsequent S1 nuclease digestion. The double-stranded cDNA is cloned into a vector by tailing, although addition of synthetic linkers is a frequently used alternative

6.2.2   Covalent Linkage of DNA Fragments to Vector Molecules

In order to purify and amplify a given DNA fragment, it must be inserted and covalently joined to a suitable vector molecule. In a ligation reaction both inter- and intra-molecular ligation may occur and often a large proportion of the products are reconstituted vector molecules that do not contain an inserted DNA fragment. In many cases, it is thus important to ensure that the majority of resulting molecules do indeed carry an inserted segment of foreign DNA since this greatly reduces the effort involved in the subsequent identification of the hybrid molecule.

6.2.2.1 *Ligation to vector molecules*

Covalent linkage of DNA fragments to a suitable vector molecule is generally accomplished by treatment of a DNA mixture with T4 or, less frequently, *E. coli* DNA ligase. These enzymes use the high energy pyrophosphate linkage of a nucleotide cofactor (ATP for T4; NAD for *E. coli*) to generate a phosphodiester bond between 5'-phosphoryl and 3'-hydroxyl termini of DNA fragments (*e.g.* Higgins and Cozzarelli, 1980). Since nearly all class II restriction enzymes generate 5'-phosphoryl and 3'-hydroxyl ends on cleavage of DNA, fragments generated by these enzymes are excellent substrates for ligation.

Both T4 and *E. coli* ligase can efficiently join molecules carrying complementary extensions but T4 DNA ligase is also able to join fragments lacking single-stranded terminal extensions. The latter reaction requires high concentrations of DNA extremities and of ligase (Sgaramella and Khorana, 1972) and is preferentially inhibited by high concentrations of ATP (Ferretti and Sgaramella, 1981). *E. coli* ligase is unable to catalyze this type of reaction.

Optimal ligation conditions depend not only on the type of DNA extremity involved but also on the type of vector employed and on the sizes of both vector and fragment. In the case of ligation of DNA fragments to a plasmid vector, reaction conditions should favour an initial joining of vector to fragments and subsequently a circularization of the hybrid molecule (Figure 1). In the case of cloning using a linear vector such as phage λ, three fragments must be joined: the two arms of λ and the foreign DNA fragment. Intramolecular ligation (circularization) or intermolecular ligation can be favoured in a particular reaction by varying the overall concentration of DNA and the relative concentration of fragment and vector (Dugaiczyk *et al.*, 1975).

Two reaction parameters are important: the effective concentration of one end of a DNA fragment in the immediate neighbourhood of the other (j), which depends on the length of the fragment and is independent of the DNA concentration; and the concentration of the DNA termini in the reaction mixture (i) (Dugaiczyk *et al.*, 1975). Circularization is more likely at low DNA concentrations where j for a given molecule is greater than the sum of the concentration of all ligatable ends in the mixture ($j > \Sigma i$). Intermolecular ligation and thus the formation of hybrid molecules are favoured at high DNA concentrations when $j < \Sigma i$. Optimal conditions for insertion of foreign DNA into a λ vector have been determined empirically to require a ratio of vector ends to fragment ends of about 2:1 and a DNA concentration sufficiently high to ensure that i for each DNA species is greater than j (Maniatis *et al.*, 1978). For cloning using a plasmid vector, conditions that result in the insertion of single DNA fragments into the vector molecule are normally chosen empirically.

6.2.2.2 *Methods favouring formation of hybrid DNA molecules*

Methods have been developed which favour intermolecular ligation and thus increase the probability of obtaining an insert into a vector molecule. In certain cases, where insertion results in the destruction of a unique restriction site in the vector and no site is present in the inserted DNA, the ligation mixture can simply be redigested with the enzyme prior to introduction of the DNA into the host organism. In the second method, the vector and the DNA to be cloned are both cleaved with two restriction endonucleases which each generate single strand extensions of different sequence. Recovery of reconstituted vector molecules is frequently reduced since recircularization requires the formation of a dimer molecule. In *E. coli*, for example, head to head dimer molecules normally cannot be propagated.

The third method exploits the requirement for 5'-phosphoryl and 3'-hydroxyl groups in the ligase-catalyzed reaction. After cleavage with a restriction enzyme the vector molecule is treated with alkaline phosphatase to remove the 5'-phosphoryl groups (Table 2). This precludes both circularization of the vector and the formation of oligomeric vector molecules, and ensures the insertion of 5'-phosphorylated foreign DNA (Ullrich *et al.*, 1977). Ligation results in the covalent linkage of the 5'-phosphorylated extremities of the fragment to the vector. Covalent linkage of the 3' extremities to the dephosphorylated 5' extremities of the vector presumably occurs *in vivo* following the introduction of the molecule into the host cell.

A fourth method which ensures covalent linkage of the vector to foreign DNA is that of homopolymer tailing and is described below.

The 'selection' procedures described above operate at the level of the ligation reaction and select against the reconstitution of the vector molecule. Additional or alternative methods which

Table 2   Enzymes Used in the Manipulation of DNA

Class II restriction enzymes (*e.g. Eco*RI)	5'-G.A.A.T.T. C-3' 3'-C.T.T. A.A.G-5'	$-G^{OH}$ $^{P}A.A.T.T.C-$ $-C.T.T.A.A._P$ $_{HO}G-$	Site specific cleavage
T4 DNA ligase	$-(N\ )^{OH}$ $+$ $^{P}(N\ )-$ $-(N')_P$ $_{HO}(N')-$	$-(N\ )\cdot(N\ )-$ $-(N')\cdot(N')-$	Covalent linkage
Alkaline phosphatase	$-(N\ )^{OH\ or\ P}$ $-(N')_P$	$-(N\ )^{OH\ or\ P}$ $-(N')_{OH}$	(i) Vector manipulation prior to ligation (ii) End-labelling together with poly-nucleotide kinase
DNA polymerase I (a) 5'–3' polymerase (b) 5'–3' exonuclease (c) 5'–3' exonuclease	$-N\ ^{OH}$ $+$ dNTP $-N'.N'.N'.N'_{P\ or\ OH}$ $^{P}N\ .N\ .N\ .N-$ $N'.N'.N'.N'-$ $_{HO\ or\ P}$ $-N\ .N\ .N\ .N\ ^{OH}$ $-N'.N'.N'.N'_{P\ or\ OH}$	$-N\ .N\ .N\ .N\ ^{OH}$ $-N'.N'.N'.N'\cdot_{P\ or\ OH}$ ^{P}N $_{HO\ or\ P}N'.N'.N'.N'-$ $-N\ ^{OH}$ $-N'.N'.N'.N'_{P\ or\ OH}$	(a) + (b) nick translation for *in vitro* labelling
Klenow fragment	Activities (a) and (c) of DNA polymerase I		(i) End-labelling (ii) Removal of 3' extensions (iii) Conversion of 5' extensions to blunt ends
Polynucleotide kinase	$-(N\ )^{P\ or\ OH}$ $-(N')_{OH}$	$-(N\ )^{OH}$ $-(N')_P$	End-labelling
Terminal transferase	$-(N\ )^{OH}$ $+$ dNTP $-(N')_{P\ or\ OH}$	$-(N\ )\cdot(N\ )^{OH}$ $-(N')_{P\ or\ OH}$	(i) End-labelling (ii) Tailing
Exonuclease III	$-N\ .N\ .N\ .N\ ^{P\ or\ OH}$ $-N'.N'.N'.N'_{P\ or\ OH}$	$-N\ ^{OH}$ $-N'.N'.N'.N'_{P\ or\ OH}$	

select against the propagation of reconstituted vector molecules are also available. These are described in Section 6.3.

6.2.3   Modification of DNA Extremities

In many situations, cloning a given segment of DNA can be greatly facilitated by modification of its extremities. For a variety of reasons it might be desirable to generate cohesive ends, change one type of cohesive end for another or generate blunt ends from cohesive ends.

In cloning random, sheared DNA fragments or double-stranded cDNA into plasmid vectors, modification frequently involves single-strand extension of the polynucleotide chain from the 3' extremity using terminal nucleotidyl transferase (tailing; *e.g.* Nelson and Brutlag, 1980). This enzyme catalyzes the addition of deoxynucleotide 5'-monophosphates and does not require a template. Tailing of the molecule is carried out using a single deoxynucleotide triphosphate to generate a homopolymer at the 3' end. Vector DNA carrying 3' extensions of the complementary homopolymer is synthesized and can then be annealed to the fragment to form non-covalently linked circular molecules. The annealed molecules, while less efficient in establishing themselves than covalently closed circular molecules, can nevertheless be introduced into a suitable host and undergo covalent linkage within the host cell (Wensink *et al.*, 1974). If a suitable choice of homopolymer and vector cleavage site is made, it is possible to regenerate the original recognition sequence at the joints between vector and foreign DNA. This approach is important in the eventual recovery of the cloned fragment. The most common procedure has been to use the restriction enzyme *Pst*I to cleave the vector. Tailing of the vector with dGTP and the fragment with dCTP usually results in regeneration of the *Pst*I recognition sequence at each junction of the hybrid molecule (Figure 2; Villa-Komaroff *et al.*, 1978). If dA:dT homopolymers are employed, conditions of partial denaturation can be found in which the cloned fragment can be isolated by specific cleavage of the polydA:dT duplex regions with endonuclease S1 (Hofstetter *et al.*, 1976). A considerable advantage of the tailing technique is that it prevents recircularization of the vector molecule and ensures the formation of hybrid molecules.

Several additional methods have been developed for modifying DNA extremities. Both 5′ and 3′ cohesive ends can be converted to blunt ends to increase their flexibility in cloning: 5′ extensions can be converted by 'filling in' the duplex strand from its 3′ recessed end using T4 DNA polymerase or *E. coli* DNA polymerase I (Klenow fragment); 5′ and 3′ extensions may be removed by controlled digestion with S1 endonuclease, and 3′ extensions by the 3′–5′ exonuclease activity of the Klenow fragment (Table 2).

Perhaps the most important and most flexible technique available for modifying DNA extremities is the use of oligonucleotide linkers which contain recognition sequences for one or several restriction endonucleases. The most common type of linker is a duplex molecule which can be attached to the extremities of a DNA molecule by blunt end ligation. Once attached, the modified DNA segment is subjected to digestion by the restriction endonuclease for which the linker carries a recognition sequence. This generates specific cohesive ends which enable the fragment, carrying its new extremities, to be ligated into a suitable site in a vector molecule. It is, of course, essential that the inserted piece does not carry a site susceptible to the enzyme or, if present, that the site can be protected (*e.g.* using *Eco*RI methylase together with *Eco*RI). This technique has been used to modify the ends of randomly sheared DNA or double-stranded cDNA in preparation for cloning (Maniatis *et al.*, 1978). It can be used to aid manipulation of cloned DNA fragments and for subsequent mutagenesis (Section 6.5.1). The number of these specific oligonucleotides available at present is large enough to allow a wide choice of extremities and thus allows the investigator to optimize the use of available vector molecules.

6.2.4 Isolation of Recombinant Molecules and Interspecies DNA Transfer

Following ligation, the desired hybrid molecules must be purified from the ligation mixture and propagated. This is accomplished in a first stage by introduction of the mixture into a suitable recipient organism. Since the introduction and maintenance of a given molecule in a host cell requires that the molecule is able to replicate, only those molecules containing the vector will be isolated. If the gene of interest has no selectable character, the resulting clones must be screened further (Section 6.4.1).

Several host–vector systems are at present available. These include various bacterial species, fungi and higher eukaryotes (Section 6.3.4). While certain of these are relatively efficient, and have been used to isolate cloned genes directly, the organism of choice in an initial cloning is generally *Escherichia coli*. Not only have very efficient methods been developed for the introduction of cloned DNA fragments into this host, but the enormous body of information on the genetics of *E. coli* and its bacteriophages and plasmids has resulted in the construction of a wide range of vector molecules. These allow convenient isolation of large quantities of the fragment of interest and subsequent *in vivo* and *in vitro* manipulation of the gene. *E. coli* also has the advantage that problems of restriction, the degradation of unmodified foreign DNA by host-specific enzymes, can be avoided by the use of restriction-deficient mutants as recipient cells. Restriction is often encountered in bacteria and can reduce the probability of introducing a foreign DNA molecule by several orders of magnitude. It should be noted that restriction-deficient mutants of other bacteria, in particular the pseudomonads (Bagdasarian and Timmis, 1982) and *B. subtilis*, are available.

Ligated DNA may be introduced into a cell in two ways: in its native state, in which case the process is referred to as transformation (for plasmid and chromosomal DNA) and transfection (for viral DNA), or, where a viral vector is employed, the DNA may be packaged *in vitro* into mature virus particles and used to infect the host (see also Holt and Saunders, Volume 1, Chapter 5).

6.2.4.1 *Transformation and transfection*

In *E. coli*, the efficiency of transformation or transfection can be as high as 10^8 transformants per microgram of vector DNA (see Maniatis *et al.*, 1982). The process requires prior treatment of the cells with $CaCl_2$. The treated cells are mixed with the sample and DNA uptake proceeds for about one hour at 4 °C. The final yield of transformants is significantly increased in most cases by a short heat shock (Cohen *et al.*, 1972; but see also Norgard *et al.*, 1978). Following a period of growth, to allow expression of a vector-associated marker, the cells are plated on a selective medium.

Similar protocols have been used with some success for other Gram-negative bacteria, in particular the pseudomonads (*e.g.* Bagdasarian and Timmis, 1982) and *Klebsiella* (Cannon *et al.*, 1977). In certain cases, addition of RbCl with the $CaCl_2$ has been reported to increase the transformation frequency (Norgard *et al.*, 1978; Bagdasarian and Timmis, 1982; Hanahan, 1983).

Certain Gram-negative bacteria such as *Agrobacterium tumefaciens* seem not to be susceptible to $CaCl_2$-mediated DNA uptake. Since these organisms have a unique relationship with the plants they infect and may prove a powerful tool in introducing cloned DNA into plant cells using their natural plasmid transfer systems (Section 6.3.4), it is important to develop methods for the introduction of cloned DNA into these cells. A treatment which has been used for *A. tumefaciens* is to freeze and thaw the recipient cells. The DNA presumably enters the damaged cells which are then allowed to regenerate (Holsters *et al.*, 1978). This process is inefficient and thus precludes the use of *A. tumefaciens* as a primary host for cloning. Introduction of cloned DNA fragments can be accomplished by cloning using a broad host range plasmid in *E. coli* and subsequent transfer of the plasmid, using natural conjugation and mobilization systems into *A. tumefaciens*. This cross-species transfer has been applied to several Gram-negative bacterial species (Section 6.3.4).

A more general technique which has been applied with success to a variety of organisms including Gram-positive bacteria, fungi and plants, involves the use of polyethylene glycol (PEG) to promote DNA uptake. Here it is necessary to remove all or part of the cell wall to generate protoplasts or spheroplasts prior to PEG treatment. This can, in general, be accomplished by treating the cells with enzymes such as lysozyme, lysostaphin (Gram-positive bacteria), zymolase or snail gut helicase (yeast) which degrade the cell wall (see Hopwood, 1981; Ferenczy, 1981; and Hicks *et al.*, 1978). DNA uptake is promoted by a short exposure to PEG in the presence of $CaCl_2$ and the spheroplasts are then allowed to regenerate. The time of exposure, concentration and molecular weight of the PEG are all critical in optimizing transformation (or transfection) frequencies (Hopwood, 1981). PEG is known to provoke protoplast aggregation and fusion. Indeed, in the case of *Saccharomyces cerevisiae*, it appears that protoplast fusion may be intimately associated with DNA uptake (see Hicks *et al.*, 1978).

With this type of technique, transformation frequencies approaching 10^7 per microgram of vector DNA have been reported for both *B. subtilis* and *Streptomyces* (Chang and Cohen, 1979; see Hopwood, 1981 and Chater *et al.*, 1982) using plasmids of between 5 and 30 kb. For yeast, values of 10^4 per microgram of DNA have been obtained (Kingsman *et al.*, 1979).

A modification of this procedure, which has been reported to increase transformation frequencies in *Streptomyces* (Makins and Holt, 1982), plants (Lurquin, 1979) and animal cells, is to add the transforming DNA to a dried phospholipid film to generate artificial phospholipid vesicles or liposomes. The liposomes are then mixed with cell protoplasts and treated with PEG (see Fraley and Papahadjopoulos, 1982 for an extensive discussion).

Transformation and transfection techniques have also been developed for animal cells. Two methods are currently employed for DNA delivery to large numbers of cells. The first involves the direct addition of a calcium phosphate/DNA precipitate to growing cells (Wigler *et al.*, 1979). In the second, used primarily for transfection with SV40-based vectors (Section 6.3.4), DNA uptake is promoted by DEAE–dextran in the culture medium (*e.g.* Sompayrac and Danna, 1981). A third technique, but one limited to the delivery of DNA to a small number of cells, is microinjection. In some cases microinjection has proved to be an extremely efficient means of introducing cloned DNA into animal cells (see Colbère-Garapin *et al.*, 1982).

An efficient method for introducing cloned DNA into a given host is essential in the primary isolation of a rare DNA segment. Less efficient systems can be tolerated when the cloned DNA is to be introduced in a pure form into a secondary host after initial isolation and propagation in a primary host. In addition to transformation, other methods of interspecies DNA transfer have been developed. Potentially the most general technique involves PEG-induced protoplast fusion directly between bacterial cells carrying the cloned DNA and the recipient cells. This technique has been used to transfer DNA cloned and propagated in *E. coli* into yeast (Kingsman *et al.*, 1979) and animal cells (*e.g.* Rassoulzadegan *et al.*, 1982). Very high efficiency fusion between various cell types has also been obtained by electro-pulse techniques (Zimmermann *et al.*, 1981). These techniques may prove valuable in the future.

6.2.4.2 In vitro packaging

One problem encountered in the use of transformation and transfection techniques is that while the frequency per microgram of vector DNA can be quite high, the overall efficiency of introducing and establishing a viable hybrid molecule is quite low (*e.g.* 10^{-4} transformants per viable cell in the case of *E. coli*). The use of cosmids and vectors based on the bacteriophage λ (Section 6.3.2) allows an alternative procedure to be employed. DNA in the ligation mixture can be packaged directly into phage particles *in vitro*. Once the phage particles are assembled, they may be used to infect a suitable *E. coli* host. Packaging followed by infection can give a frequency two orders of magnitude higher than transfection using native DNA (Enquist and Sternberg, 1979). The packaging reaction employs a mixture of complementing bacterial lysates obtained from two λ lysogens in which each λ prophage carries a mutation in a different gene involved in phage assembly. When provided with a suitable DNA preparation, the mixed lysate is capable of assembling phage particles *in vitro*. In order to optimize the packaging reaction and to reduce the probability that phage present in the bacterial lysates will contribute to the final *in vitro* packaged phage particles, mutant prophages are used and the lysates can be irradiated with UV light prior to mixing with the ligated DNA (Enquist and Sternberg, 1979; Hohn, 1979; Collins, 1979).

6.3 CLONING VECTORS

The choice of vector is an important consideration in designing experimental strategies for cloning. The number and variety of available vectors, however, preclude a comprehensive discussion here. The following is, therefore, simply intended to demonstrate the useful features of the various types of vector available. For more detailed information, the reviews of Bernard and Helinski (1980), Timmis (1981) and Williams and Blattner (1980) should be consulted.

DNA molecules to be used as cloning vectors should be relatively small, autonomous replicons which are physically and genetically well characterized. They must carry an easily selectable character for introduction into a host and should be simple to isolate in large quantities. Furthermore, insertion of a foreign DNA fragment into the vector must not disrupt its replication functions or the expression of the selectable character. Additional properties which are desirable but not essential include: the presence of single recognition sequences for the restriction enzymes to be used in cloning, the availability of a means to identify recombinant molecules, and the presence of suitable vector-associated promoters which will allow transcription of the cloned fragment. The presence of vector-associated promoters is important in cases where the endogenous promoter is not recognized by the primary host or where it has been separated from the gene during cloning. Both bacterial viruses and plasmids have been modified to fulfil some or all of these criteria and, in certain cases, analogous replicons from other organisms have also been successfully harnessed.

Initial cloning is usually carried out using *E. coli* vectors together with an *E. coli* host. These vectors are of three basic types: plasmids, bacteriophages and cosmid vectors which are maintained as plasmids but which can be packaged into phage particles *in vitro* and then used to infect the primary host. Each type of vector has specific properties which lend themselves to different experimental problems.

Many of these vectors have been modified to permit more detailed analysis of a DNA fragment or the isolation of DNA with particular functions. Vectors are available which allow optimization of gene expression, facilitate DNA sequencing or permit the isolation of transcription and translation signals, or signal sequences involved in replication.

6.3.1 Plasmid Vectors

The plasmid vectors most widely used in *E. coli* are derived from the small, multicopy plasmids pMB9, p15A and colE1. They possess the important property of continued replication in the absence of protein synthesis, in contrast to the host chromosome. Treatment of the host cells with chloramphenicol, or other antibiotics which inhibit protein synthesis, results in a large increase in the number of copies of the plasmid and a concomitant increase in the proportion of plasmid DNA in the culture (amplification; Clewell and Helinski, 1972). A very significant advantage in the use of pMB9 derivatives such as pBR322 and pBR325 (Bolivar *et al.*, 1977a, 1977b; Bolivar, 1978) is that the entire nucleotide sequence of these plasmids is known (Sutcliffe, 1979; Prentki *et al.*, 1981). This together with the precise location of the plasmid promoters (West and Rodriguez, 1980; Stüber and Bujard, 1981; Queen and Rosenberg, 1981) greatly facilitates physical studies of cloned DNA fragments. Extensive knowledge of the organization of pBR322 and pBR325 has also led to the generation of deletion derivatives which exhibit even higher copy numbers (Twigg and Sheratt, 1980; Soberon *et al.*, 1980; Covarrubias and Bolivar, 1982).

A second type of plasmid vector is based on low copy number plasmids such as F and pSC101. These differ from those described above in being unable to undergo amplification in the absence of host protein synthesis. This kind of plasmid vector is useful in cases where the product of the cloned gene might be deleterious to the host cell, since the gene will be maintained in fewer copies per cell than when cloned in a multicopy plasmid, and might be expected to produce less product.

A third type of vector has been constructed from 'runaway' mutants of the low copy number plasmid R1 (Uhlin *et al.*, 1979; Bittner and Vapnek, 1981). At low temperature these plasmids exhibit the wild-type copy number while at elevated temperatures replication is no longer controlled and a large amplification of plasmid DNA occurs. This eventually results in cell death. The use of such a vector should also allow the cloning and expression of genes which are normally deleterious to the host.

All these plasmids, however, have relatively restricted host ranges. Other plasmids with broader host ranges allowing eventual propagation in a variety of bacterial species have been developed as vectors and are described below.

The size of DNA fragments that can be cloned into plasmid vectors is limited only by the reduction in transformation frequency associated with increased plasmid size (see Hanahan, 1983).

Many plasmid vectors have been designed to allow easy detection of hybrid molecules. This is important since the methods available for ensuring the formation of hybrid molecules at the ligation step in the cloning procedure (Section 6.2.2.2) are often only partially effective. Commonly these vectors specify resistance to at least two antibiotics. The resistance genes ideally carry unique sites for restriction enzymes commonly used in cloning. One of these characters is employed to select for the introduction of the plasmid into the host cell. The other can be employed to detect hybrid molecules since cloning within an antibiotic resistance gene generally inactivates the gene. Following ligation and transformation, clones are normally selected by resistance to one antibiotic and then screened for sensitivity to the second antibiotic by replica plating. Clones sensitive to the second antibiotic should all carry recombinant DNA molecules in which the resistance gene has been interrupted by insertion of the cloned piece. This procedure is known as insertional inactivation (see Timmis, 1981). In several cases screening can be performed directly on the primary transformants by use of indicator plates. Thus inactivation of the chloramphenicol transacetylase gene (Figure 1) can be scored by including crystal violet in the selective plates (Proctor and Rownd, 1982) and clones in which the β-lactamase gene has been interrupted can be scored using the β-lactam nitrocephin or starch/iodine (O'Callaghan *et al.*, 1972; Boyko and Ganschow, 1982).

An analogous method based on the interruption of a cloned β-galactosidase gene has also been developed, using a derivative of pBR322 (pUR222; Rüther *et al.*, 1981) which carries a cloned fragment of the *lacZ* gene of *E. coli*. The cloned *lacZ* fragment carries several unique restriction sites at its 5′ extremity. Certain *lacZ*⁻ host cells are complemented by the protein fragment synthesized by the plasmid to produce active β-galactosidase and form blue colonies on selective media containing the chromogenic substrate 5-bromo-4-chloroindoyl-β-D-galactosidase (X-gal). Insertion of DNA into the *lacZ* gene fragment stops synthesis of the protein fragment and, in such a host background, results in the production of white colonies.

Inactivation of an antibiotic resistance gene can also be used to enrich for recombinant clones *prior* to plating the transformation mixture. If the action of the antibiotic is bacteriostatic, *i.e.* prevents the growth of the cells (*e.g.* tetracycline), the transformation mixture can be treated with this antibiotic and subsequently with a second antibiotic whose action is bacteriocidal but which kills only growing cells (*e.g.* ampicillin or cycloserine). Cells carrying clones in which a tetracycline resistance gene has been inactivated will not grow in the presence of the drug. Those clones which have retained tetracycline resistance will continue to grow and will be killed by the action of the second antibiotic, resulting in an overall enrichment for recombinant clones (Bolivar *et al.*, 1977a). Methods also exist for the direct selection of inactivation of both tetracycline (Malloy and Nunn, 1981) and kanamycin resistance genes (Slutsky *et al.*, 1980).

Frequently, complementation between different cloned fragments must be examined. This cannot be easily achieved when the two fragments are cloned in identical plasmid vectors. In the absence of a selection for each plasmid, two plasmids with the same replication system will not be stably maintained but will segregate, giving rise to a mixed population of cells that contain one or the other plasmid. This phenomenon is known as incompatibility. Two derivatives of p15A, pACYC177 and pACYC184, have been constructed and are compatible with pMB9 derived plasmids (Chang and Cohen, 1978). Other vectors are also available which circumvent this problem.

6.3.2 Vectors Derived from Bacteriophage λ

6.3.2.1 *Phage vectors*

The attraction of bacteriophage λ as a vector results from the fact that 40% of the phage genome (20 kb) is inessential for phage growth, that reconstituted phage particles can be efficiently prepared *in vitro* and that once formed, recombinant molecules can be propagated as phage particles *in vivo*. This allows the cloning of large fragments of foreign DNA and efficient introduction of the hybrid molecules into the primary *E. coli* host, and facilitates the isolation and handling of large numbers of independent clones. This is particularly important in the construction of genomic libraries of higher eukaryotic cells which may require the propagation of at least 10^7 independent clones. In addition, size constraints imposed on the phage genome in both *in vivo* and *in vitro* packaging to produce infectious phage particles (see below) present a powerful selective technique for ensuring the isolation of recombinant phages. These properties have made λ-based vectors important in the construction of genomic libraries. An additional advantage of λ vectors stems from the fact that most of the inessential DNA is located in a single region of the genome whose transcription is directed by a strong phage promoter P_L. Foreign DNA cloned in the inessential region would therefore be expected to undergo transcription from the phage promoter. Finally, phage vectors are ideal for cloning genes whose continuous expression adversely affects the growth of the host cells because the cloned fragment is not continuously propagated within the host cell.

Two basic types of λ vector have been developed: (i) insertion vectors which carry a deletion in the phage genome and allow the insertion of foreign DNA into a unique restriction site without the concomitant removal of phage DNA; and (ii) replacement vectors in which an inessential region of the phage genome, flanked by suitable restriction sites, can be replaced by the foreign DNA fragment. Useful variants of the replacement vectors have been constructed in which the replaceable fragment carries genes whose inactivation or loss permits the direct selection or screening of the recombinant clones. An extensive list of λ vectors can be found in the article by Williams and Blattner (1980).

For both types of vector, the final recombinant genome must be between 39 and 52 kb (78% and 105% of the wild-type λ genome) if they are to be packaged into infectious particles. Insertion vectors must therefore be at least 39 kb in length to maintain their viability. This places an upper limit of about 12 kb on the size of foreign DNA fragments which can be inserted. Replacement vectors have a larger capacity because the entire inessential region can be replaced, allowing the cloning of fragments up to 22 kb.

A disadvantage in the use of λ, as with other vectors, is that a substantial proportion of vector molecules can be reconstituted during ligation. With λ replacement vectors, this problem can be avoided in two ways. Since λ has a linear genome, the two fragments carrying essential functions can be purified away from the replaceable fragment by physical methods, such as sucrose gradient centrifugation, prior to ligation to the foreign DNA. The purified fragments rather than the entire phage can then be employed in ligation. Purification of the vector fragments can be avoided if the replaceable fragment carries functions that prevent plaque formation (lytic growth) of the reconstituted vector on particular host strains. Only recombinant phages of the correct size and in which the fragment has been replaced will form plaques, making this type of λ vector the most useful for constructing genomic libraries.

The observation that wild-type λ phages are unable to form plaques on bacteria lysogenic for phage P2 ('Spi$^+$': sensitivity to P2 inhibition) permits a powerful selection of this type. The inhibition of plaque formation is due to the expression of the λ *red* and *gam* genes. These functions are located in the inessential region of λ, and vectors have been constructed where the genes are replaced during the cloning procedure (Loenen and Brammer, 1980). Recombinant phages that have lost the *red* and *gam* genes are selected on a P2 lysogenic strain.

Several types of vector have been developed which allow direct screening for recombinant phages and are useful for cloning specific DNA fragments. Many of these vectors synthesize β-galactosidase, and form blue plaques on X-gal media. Cloning DNA fragments into the vector prevents the synthesis of β-galactosidase and recombinant phages form colourless plaques on lacZ bacteria which can be easily detected. Many different screening and selection procedures can be used with λ vectors and are described in detail by Williams and Blattner (1980).

The use of λ vectors generally entails the use of the λ *in vitro* packaging system and the delivery of the hybrid molecules by infection of the primary *E. coli* host. The λ packaging system normally requires concatamers of phage DNA as a substrate. Such concatamers are a normal product of phage replication in the host cell. A specific 12 bp sequence (the *cos* site) is recognized by the packaging machinery and DNA is presumably drawn into the phage head until a second *cos* site is reached. The recognition of the second site results in cleavage of the λ DNA. The smallest DNA molecule which can be packaged will therefore contain two *cos* sites separated by approximately one genome's length of DNA. The *cos* sites must occur in the correct orientation. For this reason, any ligation mixture in which the delivery to the primary *E. coli* host relies on *in vitro* packaging must be sufficiently concentrated to generate concatameric molecules. This constraint applies both to λ phage vectors and to cosmids (see below).

6.3.2.2 Cosmid vectors

Cosmids are plasmids that can be packaged into λ phage particles by virtue of the presence of a cloned λ *cos* site (Collins, 1979; Hohn and Hinnen, 1980). Any plasmid molecule that contains a *cos* site can be packed into a phage particle provided that it is large enough to overcome the size constraints of the λ packaging system. Thus cosmids have advantages of both plasmid and bacteriophage vectors since they can be delivered to the host by the more efficient infection procedures rather than by transformation. Since phage heads will accept about 50 kb of DNA, small cosmid vectors can be used to clone very large fragments of DNA. For example, a 5 kb cosmid (approximately the size of many of the useful plasmid vectors described above) cannot be packaged as a monomeric unit unless it contains between 40 and 45 kb of additional DNA. This is a significant increase in capacity compared to λ phage vectors and reduces the number of clones required to generate a genomic library. Recently, an elegant system of cosmid cloning has been described by Ish-Horowicz and Burke (1981). Normally, ligation produces a random mixture of different polymers of vector and insert DNA. Only a small fraction of these have the correct structure to be packaged. By consecutive use of alkaline phosphatase (Table 2; Section 6.2.2.2) and suitable restriction enzymes for which the vector carries unique sites, the only ligation products generated using this technique are hybrid molecules which carry two *cos* sites and a single inserted foreign DNA fragment.

6.3.3 Special Purpose Cloning Vectors

6.3.3.1 Expression vectors

Once a gene has been cloned and identified, subsequent steps will probably involve its recloning into a secondary vector so that its transcription is directed by a strong vector-associated promoter. The *E. coli lac*, *trp* and the λ P_L promoters have all been employed for this purpose. Various vectors have been developed in which it is possible to insert DNA fragments downstream from these promoters (Mercereau-Puijalon *et al.*, 1978; Hallewell and Emtage, 1980; Remaut *et al.*, 1981). An additional problem in maximizing expression of cloned genes in *E. coli* which is frequently encountered with genes from a heterologous source is that the gene carries no translation start signal which can be efficiently recognized by the *E. coli* translation system. This problem may arise for heterologous genes cloned into any host. Thus even though the gene can be transcribed from a promoter within the *E. coli* vector, the resulting mRNA is poorly translated and little or no protein product will be synthesized. In such cases alternative strategies must be employed. One is to fuse the gene to the amino-terminal portion of a vector gene that is efficiently translated in the host. Both translation and transcription start signals are thus provided by the vector-associated gene. The fusion must be engineered in the correct reading frame and preferably as few amino acids as possible deleted from, or added to, the cloned gene. Suitable vectors have been developed to enable fusion of a cloned gene to β-galactosidase (Charnay *et al.*, 1978), anthranilate synthetase (Tacon *et al.*, 1980) and β-lactamase (Talmadge and Gilbert,

1981). The β-lactamase fusion vectors are also useful in cases where secretion of the product by the *E. coli* host is desirable (Section 6.5.2).

An alternative strategy is to couple the coding region of the cloned gene to a DNA segment carrying both a strong promoter and a ribosomal binding site (Guarente *et al.*, 1980; Panayotatos and Truong, 1981; see Section 6.5.2).

6.3.3.2 Single-stranded phage vectors

An interesting class of *E. coli* vector has been derived from the single-stranded DNA phages fd and M13 (see Barnes, 1980). These phages do not kill the host cell but are continuously secreted into the culture medium allowing simple isolation of DNA from the phage. Cloning into these vectors requires the isolation of double-stranded replication intermediates from the host cell. Various antibiotic resistance genes have been cloned into the vectors to allow simple selection for their introduction into the host cell. The *lac* gene has also been used for detection of recombinant molecules in the same way as with other vectors.

Because isolation of the phage DNA is so simple, use of these vectors is an excellent method for the preparation and purification of single strands of cloned DNA for sequencing studies, mutagenesis (Section 6.5.1) and analysis of replication properties (Ray *et al.*, 1982). The most common use of single-stranded phage vectors is in sequencing large regions of DNA. Short fragments of DNA are cloned at random. Sequencing is carried out by annealing a single-stranded fragment complementary to the phage. This is derived from either a purified phage restriction fragment or a synthetic oligonucleotide. The annealed fragment then serves as a primer in the chain termination sequencing method (Sanger *et al.*, 1977; see Smith, A. J. H., 1980). The overall sequence can be compiled from a series of overlapping clones using established computer scanning techniques. Sequencing of both strands of a given fragment can be simply accomplished by cloning the fragment in both orientations within the vector.

Fragments of DNA cloned in single-stranded vectors should be ideal hybridization probes for colony or plaque screening, or for hybridization to 'Southern' or 'Northern' blots (Sections 6.4.1 and 6.4.2.2), since there is no complementary probe strand to compete in the hybridization reactions. This competition can, for example, make it impossible to detect rare mRNA species in such experiments. It has been difficult to obtain sufficiently high specific activities of radioactive labelling of phage DNA, but techniques are now available that circumvent this problem (Brown *et al.*, 1982).

6.3.3.3 Plasmid vectors for subcloning and sequencing

A major advantage in the use of single-stranded phage vectors in DNA sequencing is that the DNA needs no extensive purification. Recently, two plasmid vectors have also been developed which permit DNA sequencing without extensive purification of radioactively labelled DNA. One, pHP34 (Prentki and Krisch, 1982), derived from pBR322, allows the cloning of any blunt end DNA fragment with the possibility of its easy recovery. Fragments cloned in this vector can be sequenced directly by the chemical method of Maxam and Gilbert (1980). The second plasmid, pUR222 (Rüther *et al.*, 1981; Section 6.3.1) has similar properties. An improved version of pUR222, pUR250, permits sequencing of both strands of the cloned fragment (Rüther, 1982). These plasmids are ideal vectors for the rapid sequencing protocol described by Guo and Wu (1982) which uses the chain terminator method rather than the chemical method.

6.3.3.4 Vectors for the detection of transcription and translation signals

Two series of *E. coli* plasmid vectors have been described that permit the detection of promoters and transcription termination signals (Casabadan *et al.*, 1980; McKenny *et al.*, 1981). Both types of vector contain genes that are not transcribed and therefore not expressed unless a promoter is cloned into a suitable site on the plasmid. One carries the gene for β-galactosidase, and the other a gene encoding galactokinase. The levels of expression of both these genes affect the colour of colonies growing on indicator media facilitating the isolation and subsequent mutational analysis of promoter sequences. A plasmid that expresses β-galactosidase or galactokinase can be used to isolate and analyze sequences involved in the termination of transcription. Apart from their utility in the analysis of transcription in *E. coli*, these plasmids can also be used to isolate DNA sequences from other organisms that function as promoters in *E. coli* (*e.g.* Casabadan and Cohen, 1980).

The vectors facilitate the analysis of the control of expression of those genes whose products cannot be easily assayed. Not only do the β-galactosidase vectors allow analysis of transcriptional control (by isolating promoters and terminators) but they also permit the fusion of coding sequences in a cloned DNA segment with a DNA sequence encoding β-galactosidase. This results in the production of a hybrid protein that retains β-galactosidase activity and therefore allows investigation of regulation at the translational level.

6.3.4   Vector Systems for Organisms other than *E. coli*

Establishment and maintenance of cloned genes in organisms other than *E. coli* represents an important aspect of recombinant DNA technology. The ability to introduce genes into an homologous host allows the study of their expression in a natural environment. Moreover, the availability of a range of host–vector systems permits the study of a given gene in a variety of species. Such studies should provide essential information on gene expression and aid in the optimization of production of specific gene products.

Techniques which allow the introduction of a gene by integration into the host genome following transformation have been available for several years both for bacteria and higher organisms. While these methods can be useful in strain improvement they have limited utility. In cases where there is no direct selection, the gene of interest must be delivered to the host in a relatively pure form, preferably following an initial cloning in a more suitable host–vector system, together with DNA containing a selectable genetic marker to select transformed clones (*e.g.* Wigler *et al.*, 1979; Colbère-Garapin *et al.*, 1982). A disadvantage of this method is that once integrated, the recovery of the gene is not always simple. In addition, integration is not the method of choice where high yield of a gene product is required since, in general, the integrated gene will be present in only a few copies per cell.

In view of these considerations it is frequently more useful if the cloned DNA can be introduced into the host cell as part of an autonomously replicating molecule. Since initial cloning is usually performed in an organism such as *E. coli* where well-developed vector systems are available, a useful feature is to design a vector so that it can be propagated in more than one host. This is generally obtained by the *in vitro* fusion of two replicons (see below). While not yet as sophisticated as those of *E. coli*, vector systems exist for a variety of organisms and are being rapidly improved.

For several Gram-negative bacterial species, a fruitful approach has been the use of broad host range plasmids. Two types of plasmid vector are being employed at present: (i) those based on the small high copy number plasmid RSF1010 (Bagdasarian *et al.*, 1981) and (ii) those based on low copy number plasmids of the P incompatibility group such as RP4 and RK2 (Kahn *et al.*, 1979). A significant advantage of RSF1010-based vectors is their apparent ability to undergo spontaneous amplification as the host cells reach stationary phase (J. Frey and M. Bagdasarian, personal communication). This facilitates isolation of large quantities of cloned DNA. In addition, while RSF1010 is itself non-transmissible, it can be readily mobilized into other cells in the presence of a second, conjugal plasmid such as RP4 (Willetts and Crowther, 1981). These vectors are therefore more versatile than the colE1 related vectors of *E. coli*. They can be used for direct cloning in a variety of species, or, where the transformation system in a particular species is inefficient, cloning can be carried out in *E. coli* and the resulting plasmids susbsequently mobilized into the species of choice (Bagdasarian and Timmis, 1982).

Various derivative plasmids have been constructed, each carrying a selectable antibiotic resistance marker and single sites for several restriction endonucleases. In many cases, recombinant plasmids can be recognized by insertional inactivation, as for *E. coli* vectors (Section 6.3.1). Cosmid vectors allowing for efficient cloning and introduction into *E. coli* using the *in vitro* packaging system of bacteriophage λ have also been constructed (Bagdasarian *et al.*, 1981; J. Frey, personal communication).

Vectors based on the *inc*P group plasmids have provided another useful system. Since these relatively large plasmids specify their own conjugation system, direct cloning into such plasmids allows simple *in vivo* transfer into a secondary bacterial host. In addition, binary vector systems composed of two derivatives of the plasmid have been developed. A mini-plasmid, deleted for the transfer functions of the parent, is used as a vector. Transfer to an alternative host can be accomplished by the presence of a second plasmid or phage carrying the cloned transfer functions in the same cell (Ditta *et al.*, 1980). Cosmid derivatives of the *inc*P plasmids have been constructed and used with success (Klee *et al.*, 1982).

Progress has also been made in the development of vectors for Gram-positive bacteria. Several low and high copy number plasmids which replicate in both *Staphylococcus aureus* and *Bacillus subtilis* have been modified for use as vectors. As for many of the vectors discussed so far, antibiotic resistance is generally employed as the vector-associated marker. Some vectors carry cloned *Bacillus* genes which can be selected by complementation of a suitable auxotrophic marker in the recipient strain (see Kreft and Hughes, 1982 and Ehrlich *et al.*, 1982). Insertional inactivation can be used in many cases to screen for recombinant plasmids. Certain of these plasmids can be amplified 4–8 fold by treatment with hydroxyurea and in the case of one vector, pUB110, amplification to about 10^3 copies per cell can be achieved by propagation at the non-permissive temperature in strains carrying a temperature-sensitive mutation in chromosomal DNA replication (Ehrlich *et al.*, 1982).

One important problem encountered in the present *B. subtilis* cloning systems is the relatively high instability of the vectors and inserts. This instability derives from two sources: (i) segregation of the plasmids during growth of the cells (a problem which may be overcome by suitable selective pressure), and (ii) deletion or other rearrangements of the cloned DNA. The reasons for the molecular instability are at present unclear. One way in which this phenomenon can be partially overcome, at least operationally, is by the use of shuttle vectors. These are replicons able to propagate in two alternatives hosts. Vectors able to replicate in both *B. subtilis* and *E. coli* have been constructed by *in vitro* fusion of a *B. subtilis* plasmid with the well-characterized *E. coli* plasmid pBR322. They allow the initial cloning and manipulation to be carried out in *E. coli*, where they are stable, and can subsequently be introduced into a suitable *B. subtilis* strain. This type of vector may be superseded, however, by the discovery of small plasmid derivatives which are able to replicate efficiently in *Streptococcus pneumoniae*, *B. subtilis* and *E. coli* (Barany *et al.*, 1982). One shuttle vector has been constructed for DNA sequencing. This vector is a hybrid between a *B. subtilis* plasmid and the single strand DNA phage, M13, of *E. coli*. The hybrid replicates as a double-stranded plasmid molecule in *B. subtilis* and can be packaged as a single-stranded phage in *E. coli* (see Kreft and Hughes, 1982). While plasmid vectors for *Bacillus* are relatively well advanced, little use has been made as yet of *Bacillus* phages as vector molecules.

For the industrially important genus *Streptomyces*, both low and high copy number plasmids are available as vectors. Certain of these have a relatively broad host range (Chater *et al.*, 1982). Derivative plasmids have been constructed which allow direct selection for an antibiotic resistance marker and in certain cases allow screening for recombinant molecules by insertional inactivation. While *Streptomyces* phages have not been exploited to the same extent as plasmid vectors, use has been made of a shuttle vector composed of the broad host range *Streptomyces* phage, ϕC31, and pBR322. When grown as a phage in *Streptomyces*, the packaging constraints of ϕC31 can be exploited, in the same way as those of coliphage λ, to generate deletions (Chater *et al.*, 1982).

Amongst the lower eukaryotes, the yeast *S. cerevisiae* has the most developed cloning system. Two classes of autonomous vector have been developed. They are based either on the 2 μm plasmid (see Hollenberg, 1982 and Hinnen and Meyhack, 1982) or on autonomously replicating segments of chromosomal DNA (*ars*). Many of the autonomous vectors also contain an *E. coli* vector plasmid such as pBR322. Cosmid derivatives of such shuttle vectors are available (Hohn and Hinnen, 1980).

Like many *B. subtilis* vectors, yeast vectors based on the 2 μm plasmid are unstable and undergo molecular rearrangements and segregation. Their degree of stability and copy number vary according to the particular vector, the inserted sequence and the presence of endogenous 2 μm plasmids in the recipient cells. The frequency of transformation is, in general, higher than with other types of vector, presumably reflecting relatively high copy numbers (the parent has a copy number of about 50–100 per cell). The other class of high frequency transforming vectors is based on autonomously replicating yeast genomic sequences (*ars*) cloned in an *E. coli* plasmid vector. The plasmids are quite unstable and segregate rapidly in yeast but can be stabilized by the insertion of a yeast centromere. They then behave as a chromosome during meiosis and mitosis (see Hinnen and Meyhack, 1982).

Most yeast vectors employ a yeast biosynthetic marker for selection purposes. Recipients are, therefore, limited to those strains which carry the relevant auxotrophic mutation. More recently, use has been made of bacterial chloramphenicol and gentamycin G418 resistance as selective markers (Hollenberg, 1982; Jimenez and Davies, 1980).

An expression vector based on the 2 μm plasmid has been developed that carries the control region of the yeast phosphoglycerate kinase gene. This has been shown to promote high level expression of human α-interferon in yeast (Tuite *et al.*, 1982).

Ars sequences from other organisms have been isolated by cloning into yeast and may prove useful as a basis of vector construction in their homologous host although initial results with an *ars* locus derived from *N. crassa*, indicate that the 'plasmid' integrates when reintroduced into its parent organism (Vapnek and Case, 1982).

For mammalian cells, the most widely used autonomously replicating vectors are based on the papova virus SV40. This is perhaps the best characterized of all the animal viruses and has many useful features (see Elder *et al.*, 1981; Hamer, 1980; Gruss and Khoury, 1982). Its genome is composed of a small circular DNA molecule. It can be propagated by lytic infection of permissive cells such as African Green Monkey Kidney (AGMK) cells to yield up to 10^5 copies per cell (Hamer, 1980). In non-permissive cells, SV40 undergoes integration into the host genome thus maintaining a low copy number of any cloned gene (see Elder *et al.*, 1981). The virus can be recovered subsequently by fusion of the 'transformed' non-permissive cells with a permissive cell line (Hamer, 1980). In AGMK cells, the lytic cycle occurs in two phases which involve initial expression of the early functions followed by high level expression of the late functions. In non-permissive cells only early functions are expressed. The regions coding for these functions are located in opposite halves and transcribed in opposite directions on the circular genome. Vectors have been constructed in which the cloned DNA can replace either the early or late regions. As long as the hybrid molecule represents between 70% and 105% of the viral genome in length, it can be isolated and maintained as a viral stock by growth with a complementing helper virus. For production of encapsidated virus, the present vectors are limited to the inclusion of 2.5 kb of cloned DNA. Where it is unnecessary to obtain amplification of a viral stock or reinfection of a secondary host, other derivative vectors are available. These include both defective SV40 molecules which can be propagated in cells engineered to produce constitutively the SV40 T-antigen necessary for replication of the virus, and shuttle vectors composed of part of the SV40 genome combined with an *E. coli* plasmid such as pBR322 (see Elder *et al.*, 1981). Such molecules can also be combined with selectable genetic markers. Examples include the thymidine kinase gene of *Herpes simplex* virus and bacterial neomycin phosphotransferase which specifies resistance to the antibiotic G418 (see Colbère-Garapin *et al.*, 1982).

The use of SV40 vectors to express cloned eukaryotic genes is well advanced. Many of the controlling elements involved in viral gene expression have been precisely located on the SV40 genome and have been successfully harnessed for the expression of foreign genes (Gruss and Khoury, 1982).

Recently an autonomously replicating vector based on bovine papilloma virus has been developed and used to promote expression of cloned human β-globin and β-interferon in mouse cells (Dimaio *et al.*, 1982; Zinn *et al.*, 1982). At the time of writing, other vector systems based on adeno and murine sarcoma viruses are being developed (see Gluzman, 1983).

An alternative to viral vectors for introducing cloned genes into eukaryotic hosts is to use a transposable element. Rubin and Spradling (1982) have shown that it is possible to introduce exogenous DNA into *Drosophila* embryos using the P element as a vector.

Of all the important groups of organisms which have received attention to date, cloning systems in plants are the least well developed. There are, however, several systems under investigation.

Only two classes of the known plant viruses, the gemini viruses and the caulimoviruses, have a DNA genome and are, therefore, potential vectors. The small double-stranded circular DNA genome of cauliflower mosaic virus (CMV) has received the most attention among these (Hohn *et al.*, 1982) although host range is restricted to members of the Cruciferae.

Another potential class of autonomous vectors comprises sequences equivalent to the genomic *ars* sequences of yeast (see above). Such elements, able to replicate in yeast, have been isolated from *Zea* maize (Stinchcomb *et al.*, 1980) but it is as yet unclear whether they are able to replicate autonomously in cells cultured from the parental plant. *Chlamydomonas rheinhardii* genomic segments, able to replicate in the parental cells, have recently been isolated (J.-D. Rochaix, personal communication) and may provide the basis for a useful vector system.

Perhaps the most unusual method for introducing foreign DNA into plants (both dicotyledons and gymnosperms; see Nester and Kosuge, 1981) stems from the observation that segments (T-DNA) of certain plasmids carried by the genus *Agrobacterium* are transferred from the bacterium into the plant cell where they become integrated into the plant genome. The best studied of these are the Ti (tumour inducing) plasmids of *Agrobacterium tumefaciens*, the causative agent of Crown Gall disease. Similar observations have been made in the case of a plasmid carried by *A. rhizogenes* and which induces 'hairy root' disease. While many problems remain to be solved before this integrating vector can be used as a vehicle for routine genetic manipulation in plants, several successful attempts have been made to introduce foreign DNA cloned into the integrating sequences. Cloning into these plasmids is complicated by their large size. A system has been described in which DNA is first cloned into a suitable shuttle vector, and then subsequently inserted into the Ti plasmid by recombination *in vivo* (Leemans *et al.*, 1982). This has been used to promote the expression of chloramphenicol acyltransferase in plant cells (Herrera-Estrella *et al.*, 1983). Ti plasmid DNA may be introduced into plants cells by the PEG/Ca^{2+} method (Krens *et al.*, 1982), or by the use of liposomes (see Fraley and Papahadjopoulos, 1982), and the T-DNA undergoes integration and is expressed.

6.4   DETECTION AND ANALYSIS OF CLONES

Following their introduction into a host cell, hybrid molecules must be detected and characterized. Detection may be accomplished by selection or screening of transformant clones by physical, chemical or biological methods. Their subsequent characterization requires efficient methods for the isolation of recombinant DNA molecules, for their physical analysis, and in many cases to examine their ability to direct the synthesis of a given gene product.

6.4.1   Screening Recombinant Clones

The simplest procedure for clone identification is direct selection. This is normally only possible when the cloned gene is expressed and complements a defect in the new host. Since each complementation is rarely possible, indirect methods must generally be used.

As described in Section 6.3, many vector systems have been designed to ensure, as far as is practicable, that only recombinant clones are recovered. Once isolated, a set of recombinant clones can sometimes be screened and the desired clone identified simply by size or restriction pattern. More usually, alternative methods are required. Screening procedures based on nucleic acid homology, *in vitro* translation, and immunodetection of a protein product have been developed and successfully employed.

6.4.1.1   Nucleic acid homology

Screening using nucleic acid homology requires that a specific DNA or RNA probe, complementary to the desired DNA fragment, is available. This need not necessarily be highly purified provided that the mixture gives the required specificity of hybridization. The probe may be a radioactive DNA fragment, mRNA labelled *in vitro* (Maniatis *et al.*, 1976) or a radioactive cDNA copy of the mRNA (Rougeon *et al.*, 1975). Labelled DNA probes can be prepared by 'nick translation' (Rigby *et al.*, 1977) using *E. coli* DNA polymerase I or by 'end labelling' (Table 2). Where no such probe can be made an alternative approach is to synthesize chemically a complementary oligonucleotide. This requires that some part of the protein or nucleotide sequence of the gene is already known. Mixtures of oligonucleotides that derive from partial protein sequence data have been used to screen cDNA libraries and show great reliability in the detection of rare clones (*e.g.* Edlund *et al.*, 1983).

Such probes can be used to screen clones by hybridization, either to bacterial colonies (Grunstein and Hogness, 1975), or to phage plaques (Benton and Davis, 1977). The colonies or plaques are replicated from a Petri dish onto nitrocellulose or paper filters and treated with alkali, which in colony hybridization lyses the bacteria, and in both cases denatures the DNA. The DNA is then baked onto the filter. Following hybridization with the probe, the filters are autoradiographed and compared with the original plate to detect those clones exhibiting homology. Under optimal conditions, initial screening can be performed at high density (10^5 colonies per plate; Hanahan and Meselson, 1980). Examples of the methods used to maintain and screen large numbers of clones are given by Maniatis *et al.* (1978) and Gergen *et al.* (1979).

One important use of this technique is in the screening of cDNA clones. While initially this was performed using purified radioactive mRNA or its cDNA copy, in most cases such a purified probe is not available. An alternative method ('plus and minus' screening) is to isolate mRNA from tissues which express the gene of interest and from tissues which do not. Duplicate filters can be prepared from each Petri dish and hybridized with either radioactive mRNA or cDNA synthesized from one or other of the probe preparations. Clones that give an increased response to the 'plus' preparation can then be analyzed further (*e.g.* Gray *et al.*, 1982). The second important use of this type of screening is in the isolation of genomic DNA clones. Using appropriate probes, overlapping cloned segments of genomic DNA can be isolated (walking; W. Bender, P. Spierer and D. Hogness, personal communication) allowing the structural analysis of extended regions of chromosomal DNA.

6.4.1.2   Translation in vitro

For screening cDNA libraries two methods that rely on the *in vitro* translation of RNA have been developed. In one, hybrid arrest translation (Paterson *et al.*, 1977), cloned DNA is hybridized to the bulk mRNA used to generate the cDNA and the preparation is then used to direct the synthesis of proteins in an appropriate *in vitro* system. Cloned DNA will titrate homologous mRNA from the preparation by duplex formation and thus prevent the synthesis of the protein product. The second technique relies on the hybridization of the cloned DNA with its homologous RNA to select the mRNA of interest. The RNA can then be recovered and used to direct an *in vitro* translation reaction (Ricciardi *et al.*, 1979; Parnes *et al.*, 1981). DNA is prepared from pools of recombinant clones, denatured and immobilized on filters (Kafatos *et al.*, 1979). The filters are then hybridized with the RNA which is then eluted and translated. This second technique has the advantage that a positive result is obtained. Both techniques require that the protein product of interest can be detected by size, activity or antigenicity.

6.4.1.3   Immunological screening

Methods analogous to those for plaque and colony hybridization have been developed to allow detection by antibody binding of clones that synthesize a specific protein product. Broome and Gilbert (1978) describe a technique where PVC sheets, to which antibody has been bound, are applied directly to plaques or lysed colonies on a Petri dish. The protein product binds specifically to the immobilized antibody and the sheets are exposed to a second, radioactive preparation of the antibody which binds to the immobilized antigenic protein product. Autoradiography can then be used to locate the positive clones on the original plate. An alternative sandwich technique uses radioactive protein A from *S. aureus* to detect antibody binding (Erlich *et al.*, 1978).

The use of antibodies to screen transformed colonies implies that at least part of the cloned gene is expressed and produces an antigenic protein fragment. To accomplish this, cDNA (or fragmented genomic DNA) may be cloned directly into an expression vector. Transformed colonies can then be replica plated onto nitrocellulose filters and screened for antibody binding (Helfman *et al.*, 1983). An alternative is to clone into a specialized vector that contains the *lac* promoter and β-galactosidase gene with a frameshift mutation that prevents the synthesis of active β-galactosidase. Selection is made for the insertion of DNA fragments that restore the correct reading frame by selecting for Lac[+] colonies. These colonies can then be screened by an antibody sandwich technique (Rüther *et al.*, 1982).

6.4.2   Characterization of Cloned DNA

6.4.2.1   Isolation of cloned DNA

The method of choice for isolating cloned DNA will be determined by the nature of the vector molecule and of the host. It will also depend on whether DNA is to be isolated for analytical or for more extensive studies.

It is often more convenient to screen directly DNA isolated from many individual transfor-

result in the preferential expression of proteins encoded by the cloned DNA fragment and the vector. Normally, detection is carried out by size fractionation of radioactively labelled proteins on denaturing polyacrylamide gels. This may be coupled with immunological or enzymatic detection of the protein if necessary (Dottin *et al.*, 1979).

The most versatile technique is to use recombinant DNA to direct protein synthesis in a coupled *in vitro* transcription/translation system (Zubay, 1973). Here, exogenous DNA is added to a bacterial cell extract together with radioactive amino acids. The procedure has several advantages. The extract is stable and a single extract can be used with different DNA templates. With suitable modifications (Pratt *et al.*, 1981), linear restriction fragments rather than entire plasmids can be used as templates, facilitating the physical localization of start and stop signals for transcription and translation. The extracts can be prepared from many *E. coli* strains, permitting the use of different genetic backgrounds to analyze the effects of various host functions on expression. In addition, specific negative regulators of transcription and translation should not interfere with the analysis since they are usually diluted in the extract. The major disadvantage is that a variable proportion of the product may be composed of prematurely terminated protein molecules. This can reduce the ability to detect both large proteins, which will have a greater probability of premature termination, and small proteins, which will be obscured by the presence of the protein fragments. The three current *in vivo* methods do not suffer from this problem.

The use of minicells has been the most commonly employed *in vivo* technique. They are small, DNA-less cells that are produced by certain mutants of *E. coli* and other bacterial species (see Frazer and Curtiss, 1975 for a review), and which can readily be separated from the parent cells by centrifugation of sucrose gradients. If the strain carries a plasmid, the plasmid can segregate into the minicells and which support limited synthesis of plasmid proteins when incubated in the presence of exogenous amino acids. Minicells will also support protein synthesis after phage infection (Reeve, 1979). In at least one case, this system has been shown to process correctly *E. coli* membrane proteins (Clement *et al.*, 1982).

An analogous *in vivo* system relies on the preferential degradation of chromosomal DNA following UV irradiation of a *recA uvrA* mutant of *E. coli* (Sancar *et al.*, 1979). Damaged DNA cannot be repaired and is degraded. With the correct dose of UV, plasmid DNA remains undamaged and the resulting 'maxicells' will direct the synthesis of plasmid-encoded proteins. The maxicell system requires fewer manipulations than the minicell system, but for either a separate transformation and preparation must be performed for each plasmid to be analyzed. UV irradiated $uvrA^-$ $recA^+$ cells can be used to study the synthesis of phage-encoded proteins. A much higher UV dose, sufficient to block all transcription in the cells, is given prior to infection (Murialdo and Siminovitch, 1972). With λ vectors many proteins are synthesized and complicate detection of the cloned gene product. It is possible to repress the synthesis of the vector proteins by using lysogenic cells that contain a non-inducible prophage which produces the λ *ind*$^-$ repressor. Comparison of proteins synthesized in lysogenic and non-lysogenic cells can be used to determine whether or not transcription of a cloned gene can initiate in the cloned DNA segment.

Provided that the cloned gene is strongly expressed, an alternative *in vivo* system can be used. This relies on the selective amplification of colE1-type plasmid vectors in the presence of antibiotics that inhibit protein synthesis. After amplification the antibiotic is removed and the cells are incubated with amino acids. Under these conditions the majority of host proteins are poorly expressed while plasmid products, due to the high copy number of the gene, can be detected (Neidhardt *et al.*, 1980). The advantage of this technique is that no special strains or additional transformation steps are required.

Although no analogous methods exist for examining protein synthesis directly from cloned DNA fragments in eukaryotic organisms, *Xenopus* oocytes will efficiently translate injected mRNA molecules. In addition, *in vitro* translation systems based on extracts of wheat germ or reticulocytes are widely used. The *Xenopus* system will correctly modify proteins and can export heterologous secreted proteins (Colman *et al.*, 1981). A recent report (Contreras *et al.*, 1982) that capped mRNA molecules can be synthesized directly from plasmid DNA in an *in vitro* reaction using *E. coli* RNA polymerase, and that these RNAs are translated efficiently in all three eukaryotic systems, promises to facilitate the screening of cDNA libraries. Protein products may be identified by immunoprecipitation followed by gel electrophoresis. Alternatively, proteins may be transferred to nitrocellulose filters after electrophoresis ('Western blotting'; Towbin *et al.*, 1979; Bowen *et al.*, 1980) and exposed to radioactive antiserum.

Transcription products of cloned DNA can be characterized by methods analogous to those used for the characterization of translation products. Minicells will incorporate exogenous labelled uridine into RNA. The RNA products can be extracted and fractionated by gel electrophoresis (see Frazer and Curtiss, 1975). Similar experiments can be performed *in vivo* with maxicells and with phage infected, UV irradiated cells (Lund *et al.*, 1976) or *in vitro* using either purified bacterial RNA polymerase, or eukaryotic cell extracts (Manley *et al.*, 1980; Weil *et al.*, 1979).

For mapping *in vivo* transcription products, a modification of the Southern transfer technique can be employed (Southern blotting; Southern, 1979). Radioactive RNA from the original organism is hybridized to DNA of a clone which has been digested with various restriction enzymes, fractionated by gel electrophoresis and transferred to a nitrocellulose filter. This will locate the transcript relative to the restriction map of the cloned fragment. To identify various RNA products, for example primary and processed transcripts, unlabelled RNA can be fractionated by gel electrophoresis and transferred to chemically treated paper or nitrocellulose ('Northern blotting', Alwine *et al.*, 1979; Thomas, 1980) in an analogous manner. Radioactively labelled subclones of the DNA fragment can be used as hybridization probes.

To obtain more precise information as to the location of the ends of the transcript, a technique known as S1 mapping can be used (Berk and Sharp, 1977). RNA is extracted from cells and hybridized to specific, radioactively labelled, denatured DNA restriction fragments. The hybridization is performed under conditions where only DNA–RNA hybrids would be expected to form and the resulting hybrids are treated with S1 nuclease which degrades single-stranded nucleic acid. Single-stranded radioactive DNA is thus degraded but hybridized DNA is protected. The size of the protected DNA is then determined by electrophoresis on denaturing gels. With appropriate DNA probes and electrophoresis conditions, the method can be used to determine the 5' and 3' ends of a transcript, often with an accuracy of one or two bases. An alternative, used to map 5' ends of transcripts, is to hybridize a labelled restriction fragment to the RNA. The fragment is chosen so that the 5' end of the RNA remains single stranded. The hybridized DNA fragment is then used to prime DNA synthesis with AMVrt (Proudfoot *et al.*, 1980). Synthesis proceeds to the 5' end of the RNA molecule and the size of the elongated DNA fragment is then determined by electrophoresis and autoradiography.

Promoter sites, as well as the direction of transcription, can be determined by electron microscopy. Complexes between RNA polymerase and DNA, formed by incubation of DNA with purified enzyme, can be visualized directly (Williams, 1977). The direction of transcription can be detected by measurement of the RNA chain length and the position of the growing point in preparations isolated at different times from an *in vitro* transcription reaction (Brack, 1981; Stuber and Bujard, 1981). For an alternative method using gel electrophoresis rather than electron microscopy see Chelm and Geiduschek (1979).

6.5 MANIPULATION OF CLONED GENES *IN VITRO*

6.5.1 Mutagenesis

The development of *in vitro* methods to make defined alterations in cloned DNA sequences not only provides a greater degree of control than the use of *in vivo* methods, but, more importantly, it enables mutations to be generated and identified in the absence of any selective system. The types of mutation which can be introduced *in vitro* include all those which occur *in vivo*: rearrangements, particularly deletion and insertion of DNA segments, and the alteration of specific base pairs (point mutations). The techniques can be used both in localizing a given gene within a large region of cloned DNA and in generating specific changes within a gene in order to study its expression or the function of its product. In the former case, mutations are generated at random throughout the segment and are generally deletion or insertion mutations whilst, in the latter, they are directed to a specific region or site. Many of these methods have been extensively reviewed recently (Shortle *et al.*, 1981).

6.5.1.1 *Generation of deletions and insertions*

The deletion and insertion of fragments in cloned DNA is important in defining specific genes, in analyzing their expression and in modifying the fragment for subsequent manipulation by, for example, the introduction or removal of restriction sites. While powerful *in vivo* techniques employing transposable elements are available and are often simpler to use (see Guyer, 1978;

Kleckner *et al.*, 1977), the advantage of *in vitro* manipulation is that the investigator retains a greater degree of control in site specificity.

The simplest method for generating deletions is by the use of restriction enzymes to remove a segment of DNA (Figure 4). This, of course, requires that suitable sites are present in the region of interest and that only a few such sites occur per molecule (Lai and Nathans, 1974). Methods for modification of DNA extremities (Section 6.2.3) can be conveniently employed to enable rejoining of ends generated by two different restriction endonucleases.

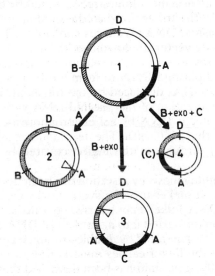

Figure 4 Construction of deletions *in vitro*. Deletion of DNA sequences in a plasmid molecule by restriction enzyme (A) cleavage and religation (i), or by restriction enzyme (B) cleavage followed by exonuclease digestion prior to ligation (ii). The first procedure results in deletion of DNA between the cleavage sites, and the second in deletions extending symmetrically from the original cleavage site. Cleavage of the DNA with a second restriction enzyme (C) prior to ligation, after cleavage with B and digestion with endonuclease (iii), results in deletions extending for a variable distance from a fixed site. Under appropriate conditions a cleavage site (C) is regenerated at the site of the deletion

More versatile *in vitro* methods for deletion formation rely on controlled nuclease digestion from a target site generated in a closed circular plasmid molecule. This target site can be a double-strand break, introduced by use of a restriction enzyme, or a single-strand break, introduced by an endonuclease such as DNaseI. In all cases, it is usually desirable to generate a single target site in each molecule. This can be achieved by partial cleavage with restriction enzymes for which there are two or more sites per molecule or by limited digestion with DNaseI. In the case of DNaseI, digestion in the presence of Mn^{2+} rather than Mg^{2+} results in double-strand cleavage instead of the introduction of single-strand breaks (Shenk *et al.*, 1976). Partial cleavage can often be conveniently controlled by the addition of ethidium bromide to the reaction. This limits cleavage of supercoiled molecules by DNaseI to a single nick and some restriction enzymes to a single, double-strand break (Greenfield *et al.*, 1975; Parker *et al.*, 1977; Shortle and Nathans, 1978).

The nature of the deletion generated depends upon both the exonuclease used and the target site. If a double-strand break is introduced into a plasmid, a deletion extending in both directions from the target site can be generated by the 3'–5' single-strand activity of exonuclease III, followed by digestion of the resulting single-stranded regions with the nuclease S1 (Heffron *et al.*, 1977; Figure 4). If the target is a single-stranded break, this procedure will result in the formation of a deletion extending only in one direction, since only one DNA strand is digested by exonuclease III. It may be more convenient to use the nuclease *Bal*31 (Legerski *et al.*, 1978) since this nuclease has a double-strand exonuclease activity. It also possesses a single-strand endonuclease activity and is apparently capable of cleaving DNA opposite a single-strand break. Treatment of DNA with *Bal*31 should thus always produce deletions extending in both directions from the target site. If a suitable restriction site exists in the molecule, this can be used to define one end point of a deletion by digestion of exonuclease-treated DNA prior to ligation

(Figure 4). For ligation to occur, the ends of the DNA must usually be treated to generate flush ends (Section 6.2.3) and in some cases the restriction site can be regenerated and used in further manipulation (Panayotatos and Truong, 1981).

Where deletions must be directed to a small region that carries no appropriate restriction site, two elegant methods involving the use of small DNA fragments have been developed. (i) A synthetic oligonucleotide carrying the flanking sequences of the desired deletion is hybridized to a preparation of single-stranded circular plasmid DNA and is used as a primer for synthesis of the complementary strand. The resulting double-stranded plasmid DNA, now carrying a deletion in the newly synthesized strand, is treated with DNA ligase, introduced into a suitable bacterial host by transformation and the transformant colonies are screened by hybridization using the original oligonucleotide as a probe. With stringent hybridization conditions, the probe will only form stable hybrids with mutant plasmid DNA (Wallace *et al.*, 1980). The procedure can be simplified by use of a single-stranded phage vector, thus avoiding the necessity for preparing single-stranded plasmid DNA. (ii) A small restriction fragment, homologous to the region to be mutagenized, is hybridized to double-stranded plasmid DNA to form a displacement loop. The single-stranded region of the circular plasmid DNA, displaced by the fragment, is sensitive to S1 nuclease and digestion finally results in linearization of the plasmid DNA with the loss of several base pairs. Following purification of the linear DNA by gel electrophoresis, it is recircularized with DNA ligase and transformed into the host cells. The technique results in the formation of small deletions (on average 10 bp) throughout the region covered by the single-stranded fragment (Green and Tibbets, 1980).

A powerful technique for mutagenesis by insertion has been developed by Heffron and collaborators (Heffron *et al.*, 1978). It makes use of a synthetic linker, in this case carrying an *Eco*RI cleavage site, although any DNA linker may be used. Normally, a linker is chosen that would generate a frameshift mutation if inserted into a segment of DNA encoding a protein. The linker is ligated to linear plasmid DNA generated by double-strand cleavage with DNaseI. This method can be made more efficient if the linker carries a marker allowing direct selection for plasmids carrying the insertion. Recently, a plasmid has been described that carries a DNA fragment specifying resistance to streptomycin and spectinomycin that is flanked at both ends by cleavage sites for *Eco*RI and *Sma*I (Prentki and Krisch, 1982). The *Sma*I fragment, which has blunt ends, can be used directly for insertion mutagenesis, but results in a large (3.5 kb) insertion. Alternatively, the fragment can be generated by *Eco*RI digestion, and the extremities converted to blunt ends (Section 6.2.3). Mutant plasmids can then be redigested with *Sma*I to delete the drug resistance gene and recircularized leaving an insertion of 14 bp containing a *Sma*I cleavage site. A similar fragment, carrying transcription termination sequences at each end, has recently been constructed (P. Prentki and H. Krisch, personal communication).

6.5.1.2 *Point mutations*

Single base pair changes can be generated in cloned fragments, either by treating the DNA *in vitro* with chemical mutagens, or by *in vitro* repair synthesis using base analogues or an error-prone DNA polymerase. With these methods any base pair can be replaced by any other, often at high frequency. This is an important consideration if labour intensive methods such as DNA sequencing are to be used to identify clones containing the desired mutations.

Exposure of DNA *in vitro* to either hydroxylamine or bisulfite results in transitions that replace G:C by A:T base pairs. Since hydroxylamine reacts equally well with single- or double-stranded DNA it can only be used for generalized mutagenesis unless a small restriction fragment is purified, mutagenized and recloned. Bisulfite, in contrast, reacts much more rapidly with single-stranded DNA than with double-stranded DNA (see Shortle *et al.*, 1981). Because of this specificity mutations can be limited to a small region of a DNA molecule. Short single-stranded stretches of DNA suitable for bisulfite mutagenesis can be produced by use of a restriction enzyme to introduce a nick into a plasmid from which a small gap of a few nucleotides is then generated by exonuclease digestion. After reaction with bisulfite, the gap is repaired (Shortle *et al.*, 1981). The advantage of this strategy is that frequently the restriction site, if it contains a C residue, is destroyed. This greatly facilitates the identification of mutant clones if no direct selection is available. There are also techniques which expose a number of sites within a larger defined region of the plasmid to the mutagen. DNA containing large single-stranded regions can be prepared: by reannealing purified single plasmid DNA strands after cleavage with different combinations of restriction enzymes and strand separation (Giza *et al.*, 1981); by use of a single-stranded phage vector (Weiher and Schaller, 1982); or by formation of a displacement loop (Everett and Chambon, 1982). These procedures have the disadvantage that multiple mutations can occur within the single-stranded region. A modification of the displacement loop technique that avoids this problem has been reported (Shortle *et al.*, 1981).

Transitions that result in the replacement of A:T by G:C base pairs can be induced by the use of N^4-hydroxy dCTP in place of dTTP during nick translation with *E. coli* DNA polymerase I (Müller *et al.*, 1978). Since the analogue will pair either with G or A residues, it should also be possible to replace G:C by A:T base pairs if dCTP rather than dTTP is replaced in the reaction.

Certain DNA polymerases, for example AMVrt, will misincorporate nucleotides at reasonably high frequencies *in vitro* and can be used to generate any base substitution. DNA with a single-stranded region produced by any of the methods described above can be repaired in the presence of only three dNTPs. This results in the misincorporation of one of them in place of the missing nucleotide (Shortle *et al.*, 1982). A similar method has been described that involves copying of a single-strand phage template with AMVrt, using an annealed restriction fragment as primer (Zakour and Locb, 1982). This method has been adapted and successfully used to generate single mutations in defined regions of plasmid DNA (G. Cesareni, personal communication).

Defined point mutations can be introduced into DNA segments cloned in single-stranded phages, or plasmids, using synthctic oligonucleotides (Zoller and Smith, 1982; Dalbadie-McFarland *et al.*, 1982). The advantage of these techniques is that any particular base in a DNA sequence can be altered, since there is no need for an appropriate restriction site, which is normally required to expose a specific site for chemical mutagenesis, or to generate defined primers for repair synthesis.

Although it is not always possible to select directly a mutation in a specific segment of DNA, the choice of a particular vector may permit an indirect selection procedure to be used (Traboni *et al.*, 1982). A short DNA segment can be cloned into a vector at a site where insertion disrupts the synthesis of active β-galactosidase, due either to the introduction of a frameshift, or to the presence of translation termination codons in the cloned fragment. Any of the standard procedures for generating mutations in *E. coli* can be used to produce mutations in the cloned fragment that restore the synthesis of active β-galactosidase. These are then easily detected on media containing X-gal.

The methods for directed mutagenesis described here are often most convenient if applied to short segments of cloned DNA coding for only part of a gene. This means that the mutant DNA must be substituted into the wild-type gene for further analysis. This can be very difficult to do using *in vitro* techniques, and methods that rely on *in vivo* recombination to introduce a mutation generated *in vitro* into a gene are being developed. For example, mutations induced in a cloned fragment of T4 DNA have been efficiently introduced into the phage genome by growing phages on a pool of bacteria transformed with mutagenized plasmid DNA (Volker and Showe, 1980; Figure 5). A more general technique would be to mutagenize a small DNA segment which has been inserted in a plasmid vector. The mutation can then be introduced into the entire gene, cloned in a λ phage vector, by *in vivo* recombination. As shown in Figure 5, *in vivo* recombination results in a mixture of recombinant phages carrying mutant and wild-type alleles of the cloned gene. For this approach to be useful, it must be possible to distinguish readily between the two alleles, for example by screening plaques by hybridization with an oligonucleotide used to generate the mutant allele.

6.5.2 Efficient Expression of Cloned Genes

Since the ultimate goal of many gene cloning experiments is to obtain the maximum yield of the cloned product, the actual cloning and characterization of the gene may represent only the first of many steps. The problems involved in obtaining large yields of an active product are so many and varied as to be beyond the scope of this review. However, the first manipulations, involving *in vitro* recombinant DNA technology, are normally to ensure that the gene is expressed at a high level.

The transcription and translation signals that must be provided will obviously depend upon the host organism that is chosen. These constructions are not necessarily straightforward. For translation in a prokaryotic host, for example, a site that allows ribosome binding must be present, correctly positioned with respect to the initiation codon of the cloned gene. This positioning can

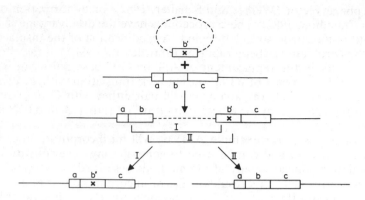

Figure 5 Reconstitution of mutagenized genes by *in vivo* recombination. Transfer of a mutation from a cloned segment to an entire gene. A segment of DNA is cloned into a plasmid and mutagenized (b′). The resulting mutation may be placed in the original gene by *in vivo* recombination. To reconstitute the gene a second recombination event is required, which results in the formation of either a mutant gene, or reconstitution of the original wild-type genes

be crucial in determining the level of translation of a protein (Gold *et al.*, 1981). Similar considerations apply to the expression of genes cloned in vectors that are propagated in eukaryotic hosts. Efficient expression of genes cloned in SV40 vectors, for example, requires that the mRNA transcript of the cloned segment is correctly spliced, capped and polyadenylated (see Hamer, 1980; Elder *et al.*, 1981) and therefore the appropriate signals for all these events must be correctly positioned in the final construction.

6.5.2.1   *Constructions that maximize expression*

The route followed in the construction of a recombinant DNA molecule that directs the high level expression of a cloned gene product will be different in each particular case. Usually, high level expression is first attempted in a prokaryotic host. The simplest method is to use an expression vector (Section 6.3.3.1) but cloning in many of those that are available frequently results in the addition of extraneous amino acids to the protein product, which might affect its function. In order to avoid this, Guarente *et al.* (1980) have made use of a three-step process that is applicable to any gene cloned in an *E. coli* plasmid vector.

First a hybrid gene is constructed which encodes a protein consisting of the amino-terminal end of the cloned gene product, fused to a portion of β-galactosidase, which retains its enzymatic activity. This is then linked to DNA containing the *lac* promoter and ribosome binding site, although other promoters could be used. This step results in the generation of a series of plasmids with different spacing between the ribosome binding site and the initiation codon of the cloned gene. Clones are examined for the level of β-galactosidase activity by simply comparing the intensity of colouration of colonies on X-gal medium. Those that produce the maximum level of β-galactosidase are picked, DNA is prepared, the β-galactosidase DNA segment is excised from the plasmid and the cloned gene is reconstituted. The advantage of this method is the easy selection of clones that direct the synthesis of high levels of the cloned gene product.

Another method, that also avoids the addition of extraneous amino acids, is to construct expression vectors that permit the insertion of DNA immediately adjacent to a promoter, ribosome binding site and initiation codon (Panayatatos and Truong, 1981).

Alternatively, a synthetic oligonucleotide containing appropriate translation initiation signals may be employed to modify the 5′ end of the cloned gene (see Yelverton *et al.*, 1983).

6.5.2.2   *Secretion of cloned products*

The manipulations described in the preceding section can be carried one step further to provide the signals necessary for secretion of the cloned gene product from the host organism. This may facilitate recovery of active protein made from the cloned gene. Nearly all secreted proteins are synthesized as preproteins and contain a signal sequence of additional amino acids at the amino-terminal end. This signal sequence is cleaved off during secretion through the cell membrane

(Blobel and Dobberstein, 1975). Some eukaryotic signal sequences, such as that of rat proinsulin, are recognized by *E. coli*, and the product of the cloned eukaryotic gene is secreted into the periplasmic space between the inner and outer cell membranes (Talmadge *et al.*, 1980b). The preprotein is correctly processed to yield a protein with no extraneous amino acids (Talmadge *et al.*, 1980a). After selective release of periplasmic proteins by osmotic shock, the proinsulin has already been separated from the bulk of the host proteins. At the time of writing, secretion vectors are available for both *E. coli* and *B. subtilis* (Talmadge and Gilbert, 1980; Palva *et al.*, 1982).

Secretion may also have another advantage in that the secreted proteins are protected from intracellular proteases by the cell membrane (Talmadge and Gilbert, 1982). In the case of *B. subtilis*, which has no outer membrane, proteins are secreted directly into the culture medium (Palva *et al.*, 1981). Since many bacterial strains produce extracellular proteases, the recovery of large yields of exogenous proteins is not yet a simple task.

Human interferons have been purified from culture media following expression in yeast (Hitzeman *et al.*, 1983). However, two major forms of the interferons were found, one identical to that produced by human cells and the other with three additional amino acids at the *N*-terminal end. Whether this reflects a difference in the mechanism of secretion between yeast and human cells is not yet clear. A considerable proportion of the interferon is not secreted. The cellular location of the non-secreted interferon has not been determined, but it is processed, raising the possibility that the non-secreted interferon is entering various organelles.

If the product of a cloned gene is an integral membrane protein, purification of the protein can present formidable problems. A recent report using influenza virus haemagglutinin as a model suggests that by suitable manipulation of clones such proteins can be secreted into the culture medium rather than being sequestered in the cell membranes of the host organism (Gething and Sambrook, 1982). Many such membrane proteins contain a very hydrophobic segment that serves to anchor the protein in the membrane. In the case of the influenza virus haemagglutinin this segment can be deleted and following expression from an SV40 vector in CV-1 cells, a truncated protein product that is glycosylated and retains normal haemagglutinin antigenic properties is secreted into the culture medium.

6.6 USEFUL SOURCES OF MORE DETAILED INFORMATION

Several collections of review articles contain much more detailed information on both the principles and practice of recombinant DNA technology. These include the series 'Genetic Engineering Principles and Practice' edited by J. K. Setlow and A. Hollaender, Plenum Press, New York and volumes 65 and 68 of 'Methods in Enzymology'. For an extensive review of gene cloning in organisms other than *E. coli* the reader is referred to 'Current Topics in Microbiology and Immunology', volume 96. Two manuals published by Cold Spring Harbor Laboratory, 'Advanced Bacterial Genetics' and especially 'Molecular Cloning' provide detailed experimental protocols.

ACKNOWLEDGEMENTS

During the writing of this review we were supported by grant No. 3.169.81 from the Swiss National Fund to Lucien Caro. We thank Marc Ballivet, Julian Davies and Gerald Selzer for actually reading the manuscript and suggesting many improvements; Otto Jenni for drawing the figures and Brigitte Brun, Jacqueline Snowdon and especially Monique Visini for endlessly retyping the manuscript.

6.7 REFERENCES

Alvarado-Urbina, G., G. M. Sathe, W.-C. Liu, M. F. Gillen, P. D. Duck, R. Bender and K. K. Ogilvie (1981). Automated synthesis of gene fragments. *Science*, **214**, 270–274.
Alwine, J. C., D. J. Kemp, B. A. Parker, J. Reiser, J. Renart, G. R. Stark and G. M. Wahl (1979). Detection of specific RNAs or specific fragments of DNA by fractionation in gels and transfer to diazobenzyloxymethyl paper. In *Methods in Enzymology*, ed. R. Wu, vol. 68, pp. 220–244. Academic, New York.
Anderson, S., M. J. Gait, L. Mayol and I. G. Young (1980). A short primer for sequencing DNA cloned in the single-stranded phage vector M13mp2. *Nucleic Acids Res.*, **8**, 1731–1743.
Aviv, H. and P. Leder (1972). Purification of biologically active globin messenger RNA by chromatography on oligothymidylic acid cellulose. *Proc. Natl. Acad. Sci. USA*, **69**, 1408–1412.

Bagdasarian, M., R. Lurz, B. Rückert, F. C. H. Franklin, M. M. Bagdasarian, J. Frey and K. Timmis (1981). Specific broad host range, high copy number, RSF1010-derived vectors, and a host–vector system for gene cloning in *Pseudomonas*. *Gene*, **16**, 237–247.

Bagdasarian, M. and K. N. Timmis (1982). Host vector systems for gene cloning in *Pseudomonas*. *Curr. Top. Microbiol. Immunol.*, **96**, 47–67.

Barany, F., J. D. Boeke and A. Tomasz (1982). Staphylococcal plasmids that replicate and express erythromycin resistance in both *Streptococcus pneumoniae* and *Escherichia coli*. *Proc. Natl. Acad. Sci. USA*, **79**, 2991–2995.

Barnes, W. M. (1976). Plasmid detection and sizing in single colony lysates. *Science*, **195**, 393–394.

Barnes, W. M. (1980). DNA cloning with single-stranded phage vector. In *Genetic Engineering: Principles and Methods*, ed. J. K. Setlow and A. Hollaender, vol. 2. Plenum, New York.

Benton, W. D. and R. W. Davis (1977). Screening λgt recombinant clones by hybridisation to single plaques *in situ*. *Science*, **196**, 180–182.

Berk, A. J. and P. A. Sharp (1977). Sizing and mapping of early adenovirus mRNA's by gel electrophoresis of S1 endonuclease-digested hybrids. *Cell*, **12**, 721–732.

Bernard, H. U. and D. R. Helinski (1980). Bacterial plasmid cloning vehicles. In *Genetic Engineering: Principles and Methods*, ed. J. K. Setlow and A. Hollaender, vol. 2, pp. 133–168. Plenum, New York.

Bibb, M. J., R. F. Freeman and D. A. Hopwood (1977). Physical and genetical characterization of a second sex factor, ScP2, for *Streptomyces coelicolor* A3 (2). *Mol. Gen. Genet.*, **154**, 155–166.

Birnboim, H. C. and J. Doly (1979). A rapid alkaline extraction procedure for screening recombinant plasmid DNA. *Nucleic Acids Res.*, **7**, 1513–1523.

Bittner, M. and D. Vapnek (1981). Versatile cloning vectors derived from the run-away replication plasmid pKN402. *Gene*, **15**, 319–329.

Blobel, G. and D. Dobberstein (1975). Transfer of proteins across membranes. I. Presence of proteolytically processed and unprocessed nascent immunoglobulin light chains on membrane-bound ribosomes or murine myeloma. *J. Cell Biol.*, **67**, 835–851.

Bolivar, F. (1978). Construction and characterisation of new cloning vehicles. III. Derivatives of plasmid pBR322 carrying unique EcoRI sites for selection of EcoRI generated recombinant DNA molecules. *Gene*, **4**, 121–136.

Bolivar, F., R. L. Rodriguez, M. C. Betlach and H. W. Boyer (1977a). Construction and characterisation of new cloning vehicles. I. Ampicillin resistant derivatives of the plasmid pMB1. *Gene*, **2**, 75–93.

Bolivar, F., R. L. Rodriguez, P. J. Greene, M. C. Betlach, H. L. Heyneker, H. W. Boyer, J. H. Crosa and J. Falkow (1977b). Construction and characterisation of new cloning vehicles. II. A multipurpose cloning system. *Gene*, **2**, 95–113.

Bowen, B., J. Steinberg, U. K. Laemmli and H. Weintraub (1980). The detection of DNA-binding proteins by protein blotting. *Nucleic Acids Res.* **8**, 1–20.

Boyko, W. L. and R. E. Ganschow (1982). Rapid identification of *Escherichia coli* transformed by pBR322 carrying inserts at the *Pst* I site. *Anal. Biochem.*, **122**, 85–88.

Brack, C. (1981). DNA electron microscopy. *CRC Crit. Rev. Biochem.*, **10**, 113–169.

Breathnach, R. and P. Chambon (1981). Organization and expression of eucaryotic split genes coding for proteins. *Annu. Rev. Biochem.*, **50**, 349–383.

Broome, S. and W. Gilbert (1978). Immunological screening method to detect specific translation products. *Proc. Natl. Acad. Sci. USA*, **75**, 2746–2749.

Brown, D. M., J. Frampton, P. Goelet and J. Karn (1982). Sensitive detection of RNA using strand-specific M13 probes. *Gene*, **20**, 139–144.

Buell, G. N., M. P. Wickens, F. Payvar and R. T. Schimke (1978). Synthesis of full length cDNAs from four partially purified oviduct mRNAs. *J. Biol. Chem.*, **253**, 2471–2482.

Cannon, F. C., E. E. Riedel and F. M. Ausubel (1977). Recombinant plasmid that carries part of the nitrogen fixation gene cluster of *Klebsiella pneumoniae*. *Proc. Natl. Acad. Sci. USA*, **74**, 2963–2967.

Carbon, J., L. Clarke, C. Ilgen and B. Ratzkin (1977). The construction and use of hybrid plasmid gene banks in *Escherichia coli*. In *Recombinant Molecules: Impact on Science and Society*, ed. R. F. Beers, Jr. and E. G. Bassett, pp. 355–378. Raven, New York.

Casadaban, M. J. and S. N. Cohen (1980). Analysis of gene control signals by DNA fusion and cloning in *Escherichia coli*. *J. Mol. Biol.*, **138**, 179–207.

Casadaban, M. J., J. Chou and S. N. Cohen (1980). *In vitro* gene fusions that join an enzymatically active β-galactosidase segment to amino terminal fragments of exogenous proteins: *Escherichia coli* plasmid vectors for the detection and cloning of translational initiation signals. *J. Bacteriol.*, **143**, 971–980.

Chang, A. C. Y. and S. Cohen (1978). Construction and characterisation of amplifiable multicopy DNA cloning vehicles derived from the P15A cryptic miniplasmid. *J. Bacteriol.*, **134**, 1141–1156.

Chang, S. and S. N. Cohen (1979). High frequency transformation of *Bacillus subtilis* protoplasts by plasmid DNA. *Mol. Gen. Genet.*, **168**, 111–115.

Charnay, P., M. Perricaudet, F. Galibert and P. Tiollais (1978) Bacteriophage lambda and plasmid vectors, allowing fusion of cloned genes in each of the three transitional phases. *Nucleic Acids Res.*, **12**, 4479–4494.

Chater, K. F., D. A. Hopwood, T. Kieser and C. J. Thompson (1982). Gene cloning in Streptomyces. *Curr. Top. Microbiol. Immunol.*, **96**, 69–95.

Chelm, B. K. and E. P. Geiduschek (1979). Gel electrophoretic separation of transcription complexes: an assay for RNA polymerase selectivity and a method for promoter mapping. *Nucleic Acids Res.*, **7**, 1851–1865.

Clement, J.-M., S. Perrin and J. Hedgpeth (1982). Analysis of λ receptor and β-lactamase synthesis and export using cloned genes in a minicell system. *Mol. Gen. Genet.*, **185**, 302–310.

Clewell, D. B. and D. R. Helinski (1969). Supercoiled circular DNA–protein complex in *E. coli*: purification and induced conversion to an open circular form. *Proc. Natl. Acad. Sci. USA*, **62**, 1159–1166.

Clewell, D. B. and D. R. Helinski (1972). Effect of growth conditions on the formation of the relaxation complex supercoiled colEl deoxyribonucleic acid and protein in *Escherichia coli*. *J. Bacteriol.*, **110**, 1135–1146.

Cohen, S. N., A. C. Y. Chang and L. Hsu (1972). Non-chromosomal antibiotic resistance in bacteria: genetic transformation of *E. coli* by R factor DNA. *Proc. Natl. Acad. Sci. USA*, **69**, 2110–2114.

Colbère-Garapin, F., A. Garapin and P. Kourilsky (1982). Selectable markers for the transfer of genes into mammalian cells. *Curr. Top. Microbiol. Immunol.*, **96**, 145–157.

Collins, J. (1979). *Escherichia coli* plasmids packageable *in vitro* in λ bacteriophage particles. In *Methods in Enzymology*, ed. R. Wu, vol. 68, pp. 309–325. Academic, New York.

Colman, A., C. D. Lane, R. Craig, A. Boolton, T. Mohun and J. Morser (1981). The influence of topology and glycosylation on the fate of heterologous secretory proteins made in *Xenopus* oocytes. *Eur. J. Biochem.*, **113**, 339.

Contreras, R., H. Cheroutre, W. Degrave and W. Fiers (1982). Simple efficient *in vitro* synthesis of capped RNA useful for direct expression of cloned eukaryotic clones. *Nucleic Acids Res.*, **10**, 6353–6362.

Covarrubias, L. and F. Bolivar (1982). Construction and characterisation of new cloning vehicles V1. Plasmid pBR329, a new derivative of pBR328 lacking the 482-base-pair inverted duplication. *Gene*, **17**, 79–89.

Dalbadie-McFarland, G., L. W. Cohen, A. D. Riggs, C. Morin, K. Itakura and J. H. Richards (1982). Oligonucleotide-directed mutagenesis as a general and powerful method for studies of protein function. *Proc. Natl. Acad. Sci. USA*, **79**, 6409–6413.

Davis, R. W., M. Thomas, J. Cameron, P. Thomas, S. Scherer and R. A. Padgett (1980). Rapid DNA isolation for enzymatic and hybridization analysis. In *Methods in Enzymology*, ed. L. Grossman and K. Moldave, vol. 65, pp. 404–411. Academic, New York.

Dimaio, D., R. Treisman and T. Maniatis (1982). Bovine papilloma virus vector that propagates as a plasmid in both mouse and bacterial cells. *Proc. Natl. Acad. Sci. USA*, **79**, 4030–4034.

Ditta, G., S. Stanfield, D. Corbin and D. R. Helinski (1980). Broad host range DNA cloning system for Gram-negative bacteria: construction of a gene bank of *Rhizobium meliloti*. *Proc. Natl. Acad. Sci. USA*, **77**, 7347–7351.

Dottin, R. P., R. E. Manrow, B. R. Fishel, S. L. Aukerman and J. L. Culleton (1979). Localization of enzymes in denaturing polyacrylamide gels. In *Methods in Enzymology*, ed. R. Wu, vol. 68, pp. 513–526. Academic, New York.

Dugaiczyk, A., H. W. Boyer and H. M. Goodman (1975). Ligation of EcoRI endonuclease-generated DNA fragments into linear and circular structures. *J. Mol. Biol.*, **96**, 171–184.

Edlund, T., T. N., M. Ranby, L. O. Heden, G. Palm, E. Holmgren and S. Joselmson (1983). Isolation of cDNA sequences coding for a part of human tissue plasminogen activator. *Proc. Natl. Acad. Sci. USA*, **80**, 349–352.

Efstratiadis, A., F. C. Kafatos, A. M. Maxam and T. Maniatis (1976). Enzymatic *in vitro* synthesis of globin genes. *Cell*, **7**, 279–288.

Ehrlich, S. D., B. Niaudet and B. Michel (1982). Use of plasmids from *Staphylococcus aureus* for cloning of DNA in *Bacillus subtilis*. *Curr. Top. Microbiol. Immunol.*, **96**, 19–29.

Elder, J. T., R. A. Spritz and S. M. Weissman (1981). Simian virus 40 as a eukaryotic cloning vehicle. *Annu. Rev. Genet.*, **15**, 295–340.

Enquist, L. and N. Sternberg (1979). *In vitro* packaging of λ*Dam* vectors and their use in cloning DNA fragments. In *Methods in Enzymology*, ed. R. Wu, vol. 68, pp. 281–298. Academic, New York.

Erlich, H. A., S. N. Cohen and H. O. McDevitt (1978). Immunological detection and characterization of products, translated from cloned DNA fragments. In *Methods in Enzymology*, ed. R. Wu, vol. 68, pp. 443–453. Academic, New York.

Everett, R. D. and P. Chambon (1982). A rapid and efficient method for the region- and strand-specific mutagenesis of cloned DNA. *EMBO J.*, **1**, 433–437.

Ferenczy, L. (1981). Microbial protoplast fusion. In *Genetics as a Tool in Microbiology. Society for General Microbiology Symposium 31*, ed. S. W. Glover and D. A. Hopwood, pp. 1–34. Cambridge University Press, Cambridge.

Ferretti, L. and V. Sgaramella (1981). Specific and reversible inhibition of the blunt endjoining activity of the T4 DNA ligase. *Nucleic Acids Res.*, **9**, 3695–3705.

Fraley, R. and D. Papahadjopoulos (1982). Liposomes: the development of a new carrier system for introducing nucleic acids into plant and animal cells. *Curr. Top. Microbiol. Immunol.*, **96**, 171–191.

Frazer, A. C. and R. Curtiss (1975). Production, properties and utility of bacterial minicells. *Curr. Top. Microbiol. Immunol.*, **69**, 1–84.

Gergen, J. P., R. H. Stern and P. Wensink (1979). Filter replicas and permanent collections of recombinant DNA plasmids. *Nucleic Acids Res.*, **7**, 2115–2136.

Gething, M.-J. and J. Sambrook (1982). Construction of influenza haemagglutinin genes that code for intracellular and secreted forms of the protein. *Nature (London)*, **300**, 598–603.

Giza, P. E., D. M. Schmit and B. L. Murr (1981). Region and strand-specific mutagenesis of a recombinant plasmid. *Gene*, **15**, 331–342.

Gluzman, Y. (ed.) (1983). *Eukaryotic Viral Vectors*. Cold Spring Harbor Laboratory, NY.

Gold, L., D. Pribnow, T. Schneider, S. Shinedling, B. S. Singer and G. Stormo (1981). Translational initiation in prokaryotes. *Annu. Rev. Microbiol.*, **35**, 365–403.

Gray, P. W., D. W. Leung, D. Pennica, E. Yelverton, R. Najarian, C. C. Simonsen, R. Derynck, P. J. Sherwood, D. M. Wallace, S. L. B. Eber, A. D. Levinson and D. V. Goeddel (1982). Expression of human immune interferon cDNA in *E. coli* and monkey cells. *Nature (London)*, **295**, 503–508.

Green, C. and C. Tibbets (1980). Targeted deletion of sequences from cloned circular DNA. *Proc. Natl. Acad. Sci. USA*, **77**, 2455–2459.

Greenfield, L., L. Simpson and D. Kaplan (1975). Conversion of closed circular DNA molecules to single molecules by digestion with DNAseI in the presence of ethidium bromide. *Biochim. Biophys. Acta*, **407**, 365–375.

Grosjean, H. and W. Fiers (1982). Preferential codon usage in prokaryotic genes: the optimal codon–anticodon interaction energy and the selective codon usage in efficiently expressed genes. *Gene*, **18**, 199–209.

Grunstein, M. and D. Hogness (1975). Colony hybridisation: a method for the isolation of cloned DNAs that contain a specific gene. *Proc. Natl. Acad. Sci. USA*, **72**, 3961–3965.

Gruss, P. and G. Khoury (1982). Gene transfer into mammalian cells: use of viral vectors to investigate regulatory signals for the expression of eukaryotic genes. *Curr. Top. Microbiol. Immunol.*, **96**, 159–170.

Gryczan, T. J., S. Contente and D. Dubnau (1978). Characterization of *Staphylococcus aureus* plasmids introduced by transformation into *Bacillus subtilis*. *J. Bacteriol.*, **134**, 318–329.

Guarente, L., T. M. Roberts and M. Ptashne (1980). A technique for expressing eukaryotic genes in bacteria. *Science*, **209**, 1428–1430.

Guo, L. H. and R. Wu (1982). New rapid methods for DNA sequencing based on endonuclease III digestion followed by repair synthesis. *Nucleic Acids Res.*, **10**, 2065–2084.

Guyer, M. S. (1978). The sequence of F is an insertion sequence. *J. Mol. Biol.*, **126**, 347–365.

Hallewell, R. A. and S. Emtage (1980). Plasmid vectors containing the tryptophan operon promoter suitable for efficient regulated expression of foreign genes. *Gene*, **9**, 27–47.

Hamer, D. H. (1980). DNA cloning in mammalian cells with SV40 vectors. In *Genetic Engineering: Principles and Methods*, ed. J. K. Setlow and A. Hollaender, vol. 2, pp. 83–101. Plenum, New York.

Hanahan, D. (1983). Studies on transformation of *Escherichia coli* with plasmids. *J. Mol. Biol.*, **166**, 557–580.

Hanahan, D. and M. Meselson (1980). Plasmid screening at high colony density. *Gene*, **10**, 63–67.

Hansen, J. B. and R. H. Olsen (1978). Isolation of large bacterial plasmids and characterization of the P2 incompatibility group plasmids pMG1 and pMG5. *J. Bacteriol.*, **135**, 227–238.

Heffron, F., P. Bedinger, J. J. Champoux and S. Falkow (1977). Deletions affecting the transposition of an antibiotic resistance gene. *Proc. Natl. Acad. Sci. USA*, **74**, 702–706.

Heffron, F., M. So and B. J. McCarthy (1978). *In vitro* mutagenesis of circular DNA molecules using synthetic restriction sites. *Proc. Natl. Acad. Sci. USA*, **75**, 6012–6016.

Helfman, D. M., J. R. Feramisco, J. C. Fiddes, G. P. Thomas and S. H. Hughes (1982). Identification of clones that encode chicken tropomyosin by direct immunogical screening of a cDNA expression library. *Proc. Natl. Acad. Sci. USA*, **80**, 31–35.

Herrera-Estrella, L., A. Depicker, M. van Montagu and J. Schell (1983). Expression of chimaeric genes transferred into plant cells using a Ti-plasmid-derived vector. *Nature (London)*, **303**, 209–213.

Hicks, J. B., A. Hinnen and G. R. Fink (1978). Properties of yeast transformation. *Cold Spring Harbor Symp. Quant. Biol.*, **43**, 1305–1313.

Higgins, N. B. and N. R. Cozzarelli (1980). DNA joining enzymes: a review. In *Methods in Enzymology*, ed. R. Wu, vol. 68, pp. 50–71. Academic, New York.

Hinnen, A. and B. Meyhack (1982). Vectors for cloning in yeast. *Curr. Top. Microbiol. Immunol.*, **96**, 101–117.

Hirt, B. (1967). Selective extraction of polyoma DNA from infective mouse cell cultures. *J. Mol. Biol.*, **26**, 365–369.

Hitzeman, R. A., D. W. Leung, L. J. Perry, W. T. Kohr, H. L. Levine and D. V. Goeddel (1983). Secretion of human interferon by yeast. *Science*, **219**, 620–625.

Hofstetter, H., A. Schambock, J. van den Berg and C. Weissmann (1976). Specific excision of the inserted DNA segment from hybrid plasmids constructed by the poly(dA)-poly(dT) method. *Biochim. Biophys. Acta*, **454**, 587–591.

Hohn, B. (1979). In vitro packaging of λ and cosmid DNA. In *Methods in Enzymology*, ed. R. Wu, vol. 68, pp. 299–309. Academic, New York.

Hohn, B. and A. Hinnen (1980). Cloning with cosmids in *E. coli* and yeast. In *Genetic Engineering: Principles and Methods*, ed. J. K. Setlow and A. Hollaender, vol. 2. Plenum, New York.

Hohn, T., K. Richards and G. Lebeurier (1982). Cauliflower mosaic virus on its way to becoming a useful plant vector. *Curr. Top. Microbiol. Immunol.*, **96**, 193–236.

Hollenberg, C. P. (1982). Cloning with 2 μm DNA vectors and the expression of foreign genes in *Saccharomyces cerevisiae*. *Curr. Top. Microbiol. Immunol.*, **96**, 120–144.

Holsters, M., B. Silva, F. Van Vliet, J. P. Hernalsteens, G. Genetello, M. Van Montagu and J. Schell (1978). *In vivo* transfer of the Ti plasmid of *Agrobacterium tumefaciens* to *Escherichia coli*. *Mol. Gen. Genet.*, **163**, 335–338.

Hopwood, D. A. (1981). Genetic studies with bacterial protoplasts. *Annu. Rev. Microbiol.*, **35**, 237–272.

Ish-Horowicz, D. and J. F. Burke (1981). Rapid and efficient cosmid cloning. *Nucleic Acids Res.*, **9**, 2989–2998.

Jimenez, A. and J. Davies (1980). Expression of a transposable antibiotic element in *Saccharomyces cerevisiae*: a potential selection for eukaryotic cloning vectors. *Nature (London)*, **287**, 869–871.

Kado, C. I. and S. T. Liu (1981). Rapid procedure for detection and isolation of large and small plasmids. *J. Bacteriol.*, **145**, 1365–1373.

Kafatos, F. C., C. W. Jones and A. Efstratiadis (1979). Determination of nucleic acid sequence homologies and relative concentrations by a dot hybridization procedure. *Nucleic Acids Res.*, **7**, 1541–1552.

Kahn, M., R. Kolter, C. Thomas, D. Figurski, R. Meyer, E. Remaut and D. Helinski (1979). Plasmid cloning vehicles derived from plasmids ColE1, F, R6K and RK2. In *Methods in Enzymology*, ed. R. Wu, vol. 68, pp. 268–280. Academic, New York.

Kingsman, A. J., L. Clarke, R. K. Mortimer and J. Carbon (1979). Replication in *Saccharomyces cerevisiae* of plasmid pBR313 carrying DNA from the yeast trpl region. *Gene*, **7**, 141–152.

Kleckner, N., J. Roth and D. Botstein (1977). Genetic engineering *in vivo* using translocatable drug-resistance elements. New methods in bacterial genetics. *J. Mol. Biol.*, **116**, 125–159.

Klee, H. J., M. P. Gordon and E. W. Nester (1982). Complementation analysis of *Agrobacterium tumefaciens* Ti plasmid mutations affecting oncogenicity. *J. Bacteriol.*, **150**, 327–331.

Klein, R. D., E. Selsing and R. D. Wells (1980). A rapid microscale technique for isolation of recombinant plasmid DNA suitable for restriction enzyme analysis. *Plasmid*, **3**, 88–91.

Korman, A. J., P. J. Knudsen, J. F. Kaufman and J. L. Strominger (1982). cDNA clones for the heavy chain of HLA-DR antigens obtained after immunoprecipitation of polysomes by monoclonal antibody. *Proc. Natl. Acad. Sci. USA*, **79**, 1844–1848.

Kreft, J. and C. Hughes (1982). Cloning vectors derived from plasmids and phage of *Bacillus*. *Curr. Top. Microbiol. Immunol.*, **96**, 1–17.

Krens, F. A., L. Molendijk, G. J. Wullems and R. A. Schilperoot (1982). *In vitro* transformation of plant protoplasts with Ti-plasmid DNA. *Nature (London)*, **296**, 72–74.

Lai, C.-J. and D. Nathans (1974). Deletion mutants of SV40 generated by enzymatic excision of DNA segments from the viral genome. *J. Mol. Biol.*, **89**, 179–193.

Land, H., M. Grez, H. Hauser, W. Lindenmaier and G. Schütz (1981). 5′-Terminal sequences of eukaryotic mRNA can be cloned with high efficiency. *Nucleic Acids Res.*, **9**, 2251–2266.

Leemans, J., C. Shaw, R. Deblaere, H. de Greve, J. P. Hernalsteens, M. Maes, M. van Montague and J. Schell (1982).

Site specific mutagenesis of *Agrobacterium* Ti plasmids and transfer of genes to plant cells. *J. Mol. Appl. Genet.*, **1**, 149–162.

Legerski, R. J., J. L. Hodnett and H. B. Gray (1978). Extracellular nucleases of *Pseudomonas* Bal31. III. Use of the double-strand deoxyriboexonuclease activity as the basis of a convenient method of the mapping of fragments of DNA produced by cleavage with restriction enzymes. *Nucleic Acids Res.*, **5**, 1445–1464.

Loenen, W. A. M. and W. J. Brammer (1980). A bacteriophage lambda vector for cloning large DNA fragments made with several restriction enzymes. *Gene*, **10**, 249–259.

Lovett, P. S. and K. M. Keggins (1979). *Bacillus subtilis* as a host for molecular cloning. In *Methods in Enzymology*, ed. R. Wu, vol. 68, pp. 342–357. Academic, New York.

Lund, E., J. E. Dahlberg, L. Lindahl, S. R. Jaskunas, P. P. Dennis and M. Nomura (1976). Transfer RNA genes between 16 and 23S rRNA transcription units of *E. coli*. *Cell*, **7**, 165–177.

Lurquin, P. F. (1979). Entrapment of plasmid DNA by liposomes and their interactions with plant protoplasts. *Nucleic Acids Res.*, **6**, 2773–2784.

Makins, J. F. and G. Holt (1982). Liposome-mediated transformation of streptomyces by chromosomal DNA. *Nature (London)*, **293**, 671–673.

Malloy, J. R. and W. D. Nunn (1981). Selection for loss of tetracycline resistance by *Escherichia coli*. *J. Bacteriol.*, **145**, 1110–1112.

Maniatis, T., S. G. Kee, A. Efstratiadis and F. C. Kaftos (1976). Amplification and characterization of a β-globin gene synthesized *in vitro*. *Cell*, **8**, 163–182.

Maniatis, T., R. C. Hardison, E. Lacy, J. Lauer, C. O'Connell and D. Quon (1978). The isolation of structural genes from libraries of eukaryotic DNA. *Cell*, **15**, 687–701.

Maniatis, T., E. F. Fritsch and J. Sambrook (1982). *Molecular cloning: a laboratory manual*. Cold Spring Harbor Laboratory, N.Y.

Manley, J. E., A. Fire, A. Cano, P. A. Sharp and M. L. Gefter (1980). DNA-dependent transcription of adenovirus genes in a soluble system. *Proc. Natl. Acad. Sci. USA*, **77**, 3855–3859.

Maxam, A. M. and W. Gilbert (1980). Sequencing end labelled DNA with base-specific chemical cleavages. In *Methods in Enzymology*, ed. L. Grossman and K. Moldave, vol. 65. Academic, New York.

McKenney, K., H. Shimatake, D. Court, U. Schmeissner, C. Brady and M. Rosenberg (1981). A system to study promotor and terminator signals recognised by *Escherichia coli* RNA polymerase. In *Gene Amplification and Analysis*, ed. J. C. Chirikjian and T. S. Papos, vol. II. Elsevier/North Holland Biomedical Press, Amsterdam.

Mercereau-Puijalon, O., A. Royal, B. Cami, A. Garapin, A. Krust, F. Gannon and P. Kourilsky (1978). Synthesis of an ovalbumin-like protein by *Escherichia coli* K12 harbouring a recombinant plasmid. *Nature (London)*, **275**, 505–510.

Mertz, J. E. and R. W. Davis (1972). Cleavage of DNA by RI restriction endonuclease generates cohesive ends. *Proc. Natl. Acad. Sci. USA*, **68**, 3370–3374.

Muller, W., H. Weber, F. Meyer and C. Weissman (1978). Site directed mutagenesis of DNA: generation of point mutations in cloned β-globin complementary DNA at positions corresponding to amino acids 121 and 123. *J. Mol. Biol.*, **124**, 343–358.

Murialdo, O. H. and L. Siminovitch (1972). The morphogenesis of bacteriophage lambda. IV. Identification of gene products and control of the expression of morphogenetic information. *Virology*, **48**, 785–823.

Neidhardt, F. C., R. Wirth, M. W. Smith and R. V. Bogelen (1980). Selective synthesis of plasmid-coded proteins by *Escherichia coli* during recovery from chloramphenicol treatment. *J. Bacteriol.*, **143**, 535.

Nelson, T. and D. Brutlag (1980). Addition of homopolymers to the 3'-ends of duplex DNA with terminal transferase. In *Methods in Enzymology*, ed. R. Wu, vol. 68, pp. 41–50. Academic, New York.

Nester, E. W. and T. Kosuge (1981). Plasmids specifying plant hyperplasias. *Annu. Rev. Microbiol.*, **35**, 531–565.

Norgard, M. V., K. Keem and J. J. Monahan (1978). Factors affecting the transformation of *Escherichia coli* strain 1776 by pBR322 plasmid DNA. *Gene*, **3**, 279–292.

O'Callaghan, C. H., A. Morris, S. M. Kirby and A. H. Shinger (1972). Novel method for detection of β-lactamases by using a chromogenic cephalosporin substrate. *Antimicrob. Agents Chemother.*, **1**, 283–288.

Okayama, H. and P. Berg (1982). High efficiency cloning of full length cDNA. *Mol. Cell. Biol.*, **2**.

Palva, I., R. F. Pettersson, N. Kalkkinen, P. Lehtovarra, M. Sarvas, H. Söderlund, P. Takinen and L. Kääriäinen (1981). Nucleotide sequence of the promoter and NH_2-terminal signal peptide region of the α-amylase gene from *Bacillus amyloliquefaciens*. *Gene*, **15**, 43–51.

Palva, I., M. Sarvas, P. Lehtovaara, M. Sibakov and L. Kääriäinen (1982). Secretion of *Escherichia coli* β-lactamase from *Bacillus subtilis* by the aid of α-amylase signal sequence. *Proc. Natl. Acad. Sci. USA*, **79**, 5582–5586.

Panayotatos, N. and K. Truong (1981). Specific deletions of DNA sequences between preselected bases. *Nucleic Acids Res.*, **1**, 5679–5688.

Parker, R. C., R. M. Watson and J. Vinograd (1977). Mapping of closed circular DNAs by cleavage with restriction endonucleases and calibration by agarose gel electrophoresis. *Proc. Natl. Acad. Sci. USA*, **74**, 851–855.

Parnes, J. E., B. Velan, A. Felsenfeld, L. Ramanathan, U. Ferrini, E. Appell and J. G. Seidman (1981). Mouse β-microglobulin cDNA clones: a screening procedure for cDNA clones corresponding to rare mRNAs. *Proc. Natl. Acad. Sci. USA*, **78**, 2253–2257.

Paterson B. M., B. E. Roberts and E. L. Kuff (1977). Structural gene identification and mapping by DNA:mRNA hybrid arrested cell-free translation. *Proc. Natl. Acad. Sci. USA*, **74**, 4370–4374.

Pratt, J. M., G. M. Boulnois, V. Darby, E. Orr, E. Wahle and I. B. Holland (1981). Identification of gene products programmed by restriction endonuclease DNA fragments using an *E. coli in vitro* system. *Nucleic Acids Res.*, **9**, 4459–4474.

Prentki, P., F. Karch, S. Ilda and J. Meyer (1981). The plasmid cloning vector pBR325 contains a 482 base pair-long inverted duplication. *Gene*, **14**, 289–299.

Prentki, P. and H. M. Krisch (1982). A modified pBR322 vector with improved properties for the cloning, recovery and sequencing of blunt ended DNA fragments. *Gene*, **17**, 189–196.

Procter, G. N. and R. H. Rownd (1982). Rosanilins: indicator dyes for chloramphenicol resistant enterobacteria containing chloramphenicol acetyltransferase. *J. Bacteriol.*, **150**, 1375–1382.

Proudfoot, N. J., M. H. H. Shander, J. L. Manley, M. L. Gefter and T. Maniatis (1980). Structure and *in vitro* transcription of human globin genes. *Science*, **209**, 1329–1336.

Queen, C. and M. Rosenberg (1981). A promotor of pBR322 activated by cAMP receptor protein. *Nucleic Acids Res.*, **9**, 3365–3377.

Rassoulzadegan, M., B. Bineruy and F. Cuzin (1982). High frequency of gene transfer after fusion between bacteria and eukaryotic cells. *Nature (London)*, **285**, 257–259.

Ray, D. S., J. C. Hines, M. H. Kim, R. Imber and N. Nomura (1982). M13 vectors for selective cloning of sequences specifying initiation of DNA synthesis of single-stranded templates. *Gene*, **18**, 231–238.

Reeve, J. N. (1979). Use of minicells for bacteriophage-directed polypeptide synthesis. In *Methods in Enzymology*, ed. R. Wu, vol. 68, pp. 493–502. Academic, New York.

Remaut, E., P. Stanssens and W. Fiers (1981). Plasmid vectors for high efficiency expression controlled by the P_L promotor of coliphage lambda. *Gene*, **15**, 81–93.

Ricciardi, R. P., T. S. Miller and B. E. Roberts (1979). Purification and mapping of specific mRNAs by hybridization selection and cell-free translation. *Proc. Natl. Acad. Sci. USA*, **76**, 4927–4931.

Rigby, P. W. J., M. Dieckman, C. Rhodes and P. Berg (1977). Labelling deoxyribonucleic acid to high specific activity *in vitro* by nick translation with DNA polymerase I. *J. Mol. Biol.*, **113**, 237–251.

Roberts, R. J. (1983). Restriction and modification enzymes and their recognition sequences. *Nucleic Acids Res.*, **11**, 135–167.

Rougeon, F., P. Kourilsky and B. Mach (1975). Insertion of a rabbit *β*-globin gene sequence into an *E. coli* plasmid. *Nucleic Acids Res.*, **2**, 2365–2378.

Rougeon, F. and B. Mach (1976). Stepwise biosynthesis *in vitro* of globin genes from globin mRNA by DNA polymerase of avian myeloblastosis virus. *Proc. Natl. Acad. Sci. USA*, **73**, 3418–3422.

Rubin, G. M. and A. C. Spradling (1982). Genetic transformation of *Drosophila* with transposable element vectors. *Science*, **218**, 348–353.

Rüther, U. (1982). pUR250 allows rapid chemical sequencing of both DNA strands of its inserts. *Nucleic Acids Res.*, **10**, 5765–5772.

Rüther, U., M. Koenen, K. Otto and B. Müller-Hill (1981). pUR222, a vector for cloning and rapid chemical sequencing of DNA. *Nucleic Acids Res.*, **9**, 4087–4098.

Rüther, V., M. Koenen, A. E. Sippel and B. Müller-Hill (1982). Exon cloning: immunoenzymatic detection of exons of the chicken lysozyme gene. *Proc. Natl. Acad. Sci. USA*, **79**, 6852–6855.

Sancar, A., A. M. Hack and W. D. Rupp (1979). Simple method for identification of plasmid coded protein. *J. Bacteriol.*, **137**, 692–693.

Sanger, F., S. Nicklen and A. R. Coulson (1977). DNA sequencing with chain-terminating inhibitors. *Proc. Natl. Acad. Sci. USA*, **12**, 5463–5467.

Schreier, P. H. and R. Cortese (1979). A fast and simple method for sequencing DNA cloned in the single-stranded bacteriophage M13. *J. Mol. Biol.*, **129**, 169–172.

Seed, B., R. C. Parker and N. Davidson (1982). Representation of DNA sequences in recombinant DNA libraries prepared by restriction enzyme partial digestion. *Gene*, **19**, 201–209.

Sgaramella, V., J. H. van de Sande and H. G. Khorana (1970). Studies on polynucleotides: C. A novel joining reaction catalyzed by the T4-polynucleotide ligase. *Proc. Natl. Acad. Sci. USA*, **67**, 1468–1475.

Sgaramella, V. and H. G. Khorana (1972). Studies on polynucleotides: CXVI. A further study of the T4 ligase catalysed joining of DNA at base-paired ends. *J. Mol. Biol.*, **72**, 493–502.

Shenk, T., J. Carbon and P. Berg (1976). Construction and analysis of viable deletion mutants of SV40. *J. Virol.*, **18**, 664–671.

Shortle, D. and D. Nathans (1978). Local mutagenesis: a method for generating viral mutants with base substitutions in preselected regions of the viral genome. *Proc. Natl. Acad. Sci. USA*, **75**, 2170–2174.

Shortle, D., D. Dimaio and D. Nathans (1981). Directed mutagenesis. *Annu. Rev. Biochem.*, **15**, 265–297.

Shortle, D., P. Grisafi, J. J. Bencovic and D. Botstein (1982). Gap misrepair mutagenesis: efficient site directed induction of transition, transversion and frameshift mutations *in vitro*. *Proc. Natl. Acad. Sci. USA*, **79**, 1588–1592.

Slutsky, A. M., P. M. Rabinovich, L. Z. Yakubov, I. V. Sineokaya, A. I. Stepanov and V. K. Gordeyev (1980). Direct selection of DNA inserts in plasmid gene of kanamycin resistance. *Mol. Gen. Genet.*, **180**, 487–489.

Smith, A. J. H. (1980). DNA sequence analysis by primed synthesis. In *Methods in Enzymology*, ed. L. Grossman and K. Moldave, vol. 65, pp. 560–579. Academic, New York.

Smith, G. E. and M. D. Summers (1980). The bidirectional transfer of DNA and RNA to nitrocellulose or diazobenzyloxymethyl-paper. *Anal. Biochem.*, **109**, 123–129.

Smith, H. O. (1980). Recovery of DNA from gels. In *Methods in Enzymology*, ed. L. Grossman and K. Moldave, vol. 65, pp. 371–380. Academic, New York.

Smith, H. O. and M. L. Birnsteil (1976). A simple method for DNA restriction site mapping. *Nucleic Acids Res.*, **3**, 2387–2398.

Soberon, X., L. Covarrubias and F. Bolivar (1980). Construction and characterization of new cloning vehicles. IV. Deletion derivatives of pBR322 and pBR325. *Gene*, **9**, 282–305.

Sompayrac, L. M. and K. J. Danna (1981). Efficient infection of monkey cells with DNA of simian virus 40. *Proc. Natl. Acad. Sci. USA*, **78**, 7575–7578.

Southern, E. (1979). Gel electrophoresis of restriction fragments. In *Methods in Enzymology*, ed. R. Wu, vol. 68, pp. 152–176. Academic, New York.

Stinchcomb, D. T., M. Thomas, J. Kelly, E. Selker and R. W. Davies (1980). Eukaryotic DNA segments capable of autonomous replication in yeast. *Proc. Natl. Acad. Sci. USA*, **77**, 4559–4563.

Stüber, D. and H. Bujard (1981). Organisation of transcriptional signals in plasmids pBR322 and pACYC184. *Proc. Natl. Acad. Sci. USA*, **78**, 167–171.

Sutcliffe, J. G. (1979). Complete nucleotide sequence of the *Escherichia coli* plasmid pBR322. *Cold Spring Harbor Symp. Quant. Biol.*, **43**, 77–90.

Tacon, W., N. Carey and S. Emtage (1980). The construction and characterisation of plasmid vectors suitable for the expression of all DNA phases under the control of the *E. coli* tryptophan promotor. *Mol. Gen. Genet.*, **177**, 427–438.

Talmadge, K. and W. Gilbert (1981). Construction of plasmid vectors with unique PstI cloning sites in a signal sequence coding region. *Gene*, **12**, 235–241.

Talmadge, K. and W. Gilbert (1982). Cellular location affects protein stability in *Escherichia coli*. *Proc. Natl. Acad. Sci. USA*, **79**, 1830–1833.

Talmadge, K., J. Kaufman and W. Gilbert (1980a). Bacteria mature preproinsulin to proinsulin. *Proc. Natl. Acad. Sci. USA*, **77**, 3988–3992.

Talmadge, K., S. Stahl and W. Gilbert (1980b). Eukaryotic signal sequence transports insulin antigen in *Escherichia coli*. *Proc. Natl. Acad. Sci. USA*, **77**, 3369–3373.

Telford, J., P. Boseley, W. Schaffner and M. Birnstiel (1976). Novel screening procedure for recombinant plasmids. *Science*, **195**, 391–392.

Thomas, P. S. (1980). Hybridisation of denatured RNA and small DNA fragments transferred to nitrocellulose. *Proc. Natl. Acad. Sci. USA*, **77**, 5201–5205.

Timmis, K. N. (1981). Gene manipulation *in vitro*. In *Genetics as a Tool in Microbiology, Society for General Microbiology Symposium 31*, ed. S. W. Glover and D. A. Hopwood. Cambridge University Press, Cambridge.

Towbin, H., T. Staehelin and J. Gordon (1979). Electrophoretic transfer of proteins from polyacrylamide gels to nitrocellulose sheets: procedure and some applications. *Proc. Natl. Acad. Sci. USA*, **76**, 4350–4354.

Traboni, G., G. Ciliberto and R. Cortese (1982). A novel method for site directed mutagenesis: its application to an eukaryotic tRNA^{PRO} gene promotor. *EMBO J.*, **1**, 415–420.

Tuite, M. F., M. J. Dobson, N. A. Roberts, R. M. King, D. C. Burke, S. M. Kingsman and A. J. Kingsman (1982). Regulated high efficiency expression of human interferon-alpha in *Saccharomyces cerevisiae*. *EMBO J.*, **1**, 603–608.

Twigg, A. J. and D. Sherratt (1980). Trans-complementable copy-number mutants of plasmid colE1. *Nature (London)*, **283**, 216–218.

Uhlin, B. E., S. Molin, P. Gustafsson and K. Nordström (1979). Plasmids with temperature dependent copy number for amplification of cloned genes and their products. *Gene*, **6**, 91–106.

Ullrich, A., J. Shine, J. Chirgwin, R. Pictet, E. Tischer, W. J. Rutter and H. M. Goodman (1977). Rat insulin genes: construction of plasmids containing the coding sequences. *Science*, **196**, 1313–1319.

van Montagu, M. and J. Schell (1982). The Ti plasmids of *Agrobacterium. Curr. Top. Microbiol. Immunol.*, **96**, 237–254.

Vapnek, D. and M. Case (1982). Gene cloning in *Neurospora crassa*. *Curr. Top. Microbiol. Immunol.*, **96**, 98–100.

Villa-Komaroff, L., A. Efstratiadis, S. Broome, P. Lomdico, R. Tizard, S. P. Naber, W. L. Chick and W. Gilbert (1978). A bacterial clone synthesizing proinsulin. *Proc. Natl. Acad. Sci. USA*, **75**, 3727–3731.

Völker, T. A. and M. K. Showe (1980). Induction of mutations in specific genes of bacteriophage T4 using cloned restriction fragments and marker rescue. *Mol. Gen. Genet.*, **177**, 447–452.

Wallace, R. B., P. F. Johnson, S. Tanaky, M. Schöld, K. Itakura and J. Abelson (1980). Directed deletion of a yeast transfer RNA intervening sequence. *Science*, **209**, 1396–1400.

Weiher, H. and H. Schaller (1982). Segment specific mutagenesis and extensive mutagenesis of a *lac* promotor/operator element. *Proc. Natl. Acad. Sci. USA*, **79**, 1408–1412.

Weil, P. A., D. S. Lose, J. Segall and R. G. Roeder (1979). Selective and accurate initiation of transcription at the Ad2 major late promoter in a soluble system dependent on purified RNA polymerase II and DNA. *Cell*, **18**, 469–484.

Wensink, P. C., D. J. Finegan, J. E. Donelson and D. S. Hogness (1974). A system for mapping DNA sequences in the chromosomes of *Drosophila melanogaster*. *Cell*, **3**, 315–325.

West, R. W. and R. L. Rodriguez (1980). Construction and characterization of *E. coli* promotor probe plasmid vectors. II. RNA polymerase binding studies on antibiotic resistance promotors. *Gene*, **9**, 175–193.

Wigler, M., A. Pellicer, S. Silverstein, R. Axel, G. Urlamb and L. Chasin (1979). DNA-mediated transfer of the adenine phosphoribosyl transferase locus into mammalian cells. *Proc. Natl. Acad. Sci. USA*, **76**, 1373–1376.

Willets, N. and C. Crowther (1981). Mobilization of the non-conjugative Inc and plasmid RSF1010. *Genet. Res.*, **37**, 311–316.

Williams, R. C. (1977). Use of polylysine for adsorption of nucleic acids and enzymes to electron microscope specimen films. *Proc. Natl. Acad. Sci. USA*, **74**, 2311–2316.

Williams, W. and F. R. Blattner (1980). Bacteriophage lambda vector for DNA cloning. In *Genetic Engineering: Principals and Methods*, ed. J. K. Setlow and A. Hollaender, vol. 2, pp. 133–168. Plenum, New York.

Yank. R. C., J. Lis and R. Wu. (1979). Elution of DNA from agarose gels after electrophoresis. In *Methods in Enzymology*, ed. R. Wu, vol. 68, pp. 176–182. Academic, New York.

Yelverton, E., S. Norton, J. F. Obijeski and D. V. Goeddel (1983). Rabies virus glycoprotein analogs: biosynthesis in *Escherichia coli*. *Science*, **219**, 614–620.

Zakour, P. A. and L. A. Loeb (1982). Site-specific mutagenesis by error-directed DNA synthesis. *Nature (London)*, **295**, 708–710.

Zasloff, M., G. D. Ginder and G. Felsenfeld (1978). A new method for the purification and identification of covalently closed circular DNA molecules. *Nucleic Acids Res.*, **5**, 1139–1152.

Zimmerman, U., P. Schreurich, G. Pilwat and R. Benz (1981). Cells with manipulated functions: new perspectives for cell biology, medicine and technology. *Angew. Chem.*, **20**, 325–344.

Zinn, K., P. Mellon, M. Ptashne and T. Maniatis (1982). Regulated expression of an extrachromosomal human β-interferon gene in mouse cells. *Proc. Natl. Acad. Sci. USA*, **79**, 4897–4901.

Zoller, J. J. and M. Smith (1982). Oligonucleotide-directed mutagenesis using M13-derived vectors: an efficient and general procedure for the production of point mutations in any fragment of DNA. *Nucleic Acids Res.*, **10**, 6487–6500.

Zubay, G. (1973). *In vitro* synthesis of proteins in microbial systems. *Annu. Rev. Genet.*, **7**, 267–287.

7

Nutritional Requirements of Microorganisms

G. M. DUNN
University of Kent at Canterbury, UK*

7.1 INTRODUCTION

Although the microbial world was discovered as long ago as the seventeenth century by van Leeuwenhoek, the first great advance in the development of microbiology as a science did not occur until the mid-nineteenth centry with the actual cultivation of microorganisms. Implicit in this cultivation is an understanding of microbial nutrition. Our knowledge of the growth requirements of microorganisms is still not complete and in a few instances formulation of growth media has not advanced significantly since the pioneering work of Pasteur and his contemporaries.

However, with the current interest in expanding the industrial exploitation of microbes it is

* Now at Fermentech Ltd.

likely that more attention will be directed towards media design. In the interests of profitability it is essential to supply necessary nutrients to ensure optimal growth of an organism, or to maximize product yield, at minimum cost. This can only be achieved by an understanding of what microorganisms require as nutrients, and in some instances, what is not required. The aim of this chapter is to illustrate the principles of microbial nutrition rather than provide a catalogue of compounds and formulae. Such reviews are available and the reader will be referred to them. Particular emphasis will be placed on those microbes that have current or potential applications to biotechnology (which may include most microbes!) and these will be used as examples to illustrate particular points arising within the general principles of microbial nutrition. It will be convenient to subdivide our consideration of nutrition into a number of main sections.

7.2 BACTERIA AND FUNGI

Of the 100 or more elements that appear in the periodic table, some 35–40 have been demonstrated as being essential nutrients (or claimed as such). Although the majority of these are metals, six non-metals (C, O, H, N, P, S) and two metals (K, Mg) comprise an average 98% of the dry weight of bacteria and fungi. It is usual to speak of these elements collectively as *macronutrients* (required concentration in growth media greater than 10^{-4} mol l^{-1}). All other elemental nutrients are termed *micronutrients* or trace elements and are usually required at concentrations less than 10^{-4} mol l^{-1}. Carbon and general nutrition of fungi have been reviewed by Lily (1965) and Perlman (1965), of bacteria by Pirt (1975), of yeasts by Suomalainen and Oura (1971), and general inorganic nutrition by Hutner (1972).

7.2.1 Macronutrients

The eight macronutrients and their physiological functions are outlined in Table 1. The total weight of all microorganisms comprises 80–90% water, with C, O, H and N, the main constituents of cellular material, comprising 90–95% of the dry weight.

Table 1 The Eight Macronutrient Elements, some of their Physiological Functions and Growth Requirements

Element	Physiological function	Required concentration (mol l^{-1})
Carbon	Constituent of organic cellular material. Often the energy source.	$>10^{-2}$
Nitrogen	Constituent of proteins, nucleic acids and coenzymes.	10^{-3}
Hydrogen	Organic cellular material and water.	—
Oxygen	Organic cellular material and water. Required for aerobic respiration.	—
Sulfur	Constituent of proteins and certain coenzymes.	10^{-4}
Phosphorus	Constituent of nucleic acids, phospholipids, nucleotides and certain coenzymes.	10^{-4} to 10^{-3}
Potassium	Principal inorganic cation in the cell and cofactor for some enzymes.	10^{-4} to 10^{-3}
Magnesium	Cofactor for many enzymes, chlorophylls (photosynthetic microbes) and present in cell walls and membranes.	10^{-4} to 10^{-3}

7.2.1.1 Carbon

The biochemistry of this planet is based on carbon. Traditionally, microorganisms are divided into categories on the basis of their carbon nutrition, *i.e.* heterotrophs and autotrophs. Heterotrophs obtain cell carbon from organic compounds and this group includes all fungi and most bacteria and protozoa. Some algae can grow as heterotrophs under certain conditions. Autotrophs obtain cell carbon from CO_2 (carbon fixation). This group comprises the algae and a few bacteria. These groups are not mutually exclusive since certain bacteria and algae show both types of carbon nutrition. Microbes that show concomitant autotrophy and heterotrophy are termed mixotrophs (Rittenberg, 1969). They will grow as autotrophs, but autotrophic growth is invariably stimulated by certain organic carbon compounds.

In addition to its role as a major constituent of cellular material, the carbon source is frequently the energy source. Although this topic is treated in detail by Stouthamer in Volume 1, Chapter 12, it may be useful to show here how microbes can be assigned to nutritional categories on the basis of their use of carbon as an energy source. All organisms require an energy source for cell synthesis and two types are available, namely light energy and chemical oxidation. On the basis of these two criteria combined with the trophic state outlined above, microbes can be classified into four groupings (Table 2). CHEMOLITHOTROPHS depend on oxidation–reduction reactions for energy and use inorganic electron donors, *e.g.* H_2 (hydrogen bacteria and methanogenic bacteria), reduced sulfur compounds (*Thiobacillus* spp.), or reduced nitrogen compounds (nitrifying bacteria). These bacteria can grow autotrophically but growth of some is stimulated by organic carbon and hence these are mixotrophs. Nearly all chemolithotrophs require oxygen as an electron acceptor. The methanogens are an example of anaerobic chemolithotrophs which use H_2 as the electron donor and CO_2 as the electron acceptor. These too are mixotrophs since they can utilize a restricted range of organic carbon compounds, *e.g.* formate and acetate. CHEMOORGANOTROPHS depend on oxidation–reduction reactions for energy and use organic electron donors. These include most bacteria and protozoa, and all fungi. Oxygen is the commonest electron acceptor, but in its absence NO_3^-, SO_4^{2-} and CO_2 can be utilized by certain groups of bacteria. Where organic compounds serve as both electron donors and acceptors, this is fermentation *sensu stricto*.

Table 2 Classification of Microorganisms on the Basis of Energy
Source and Electron Donor Requirements

	Electron donor	
Energy source	*Inorganic*	*Organic*
Chemical (chemosynthetic)	Chemolithotroph	Chemoorganotroph
Light (photosynthetic)	Photolithotroph	Photoorganotroph

PHOTOLITHOTROPHS depend on radiant energy and use inorganic electron donors. Examples here are the algae, which use water as the electron donor, and most photosynthetic bacteria, which use other reduced inorganic compounds, *e.g.* H_2 and H_2S. Some of the algae can also grow heterotrophically. PHOTOORGANOTROPHS depend on radiant energy and use organic carbon sources. This mode of nutrition is restricted to the purple non-sulfur photosynthetic bacteria (Athiorhodaceae) and a few algae. It is apparent that these are not rigid compartments since many overlaps occur. However, it does provide a useful starting point in determining a particular organism's nutritional requirements.

As a source of cellular material carbon forms approximately 50% of the dry weight of bacteria and fungi. The range of carbon compounds that can be utilized is vast, particularly for bacteria. The versatility of the microbial world is illustrated by the fact that all biologically synthesized carbon compounds are biodegradable, as are many xenobiotics, although the breakdown of the latter is often very slow. Some microorganisms, *e.g.* certain *Pseudomonas* spp. can utilize over 90 organic compounds as their sole source of carbon and energy. Others, *e.g.* the methanogens, can only utilize a very restricted number (Taylor, 1982). Carbohydrates, particularly glucose, are very popular choices for growth media, but, depending upon the particular nutritional versatility of the organism under study, suitable carbon sources also include carbohydrate derivatives, fatty acids, alcohols, organic acids, hydrocarbons and organic nitrogenous compounds. Some commonly utilized carbon and nitrogen feedstocks for the fermentation industry are given in Table 3. Autotrophs require a source of CO_2, but all organisms have a growth requirement for this compound since it is an essential metabolic intermediate. Actively growing cells produce sufficient CO_2 as a result of their own metabolism, but problems can occur during the lag phase of growth when a certain amount is required as a 'primer'. This is particularly a problem with anaerobes because such cultures are often sparged with an inert gas (*e.g.* N_2) to remove oxygen, which also removes CO_2. A small enrichment of the gas phase with CO_2 or addition of the HCO_3^- ions will often overcome this effect. Aerobic cultures obtain CO_2 from the air but it is poorly retained in low pH fermentations, *e.g.* many fungal cultures. Problems here may be overcome by use of certain CO_2-sparing compounds, *e.g.* Krebs Cycle acids and related compounds (Hutner, 1972; Pirt, 1975). Some animal pathogens require high CO_2 concentrations in the culture atmosphere which must be supplied by enrichment (5–10%). The methanogens have an absolute requirement for CO_2 as a terminal electron acceptor.

Table 3 Some Carbon and Nitrogen Sources Utilized by the Fermentation Industry

Carbon sources	Nitrogen sources
Starch waste (maize and potato)	Soya meal
Mollases (cane and beet)	Yeast extract
Whey	Distillers solubles
n-Alkanes	Cotton seed extract
Gas oil	Dried blood
Sulfite waste liquor	Corn-steep liquor
Domestic sewage	Fish solubles and meal
Cellulose waste	Ground-nut meal
Carob bean	

7.2.1.2 Nitrogen

Nitrogen comprises 8–14% of the dry weight of bacteria and fungi. A wide range of inorganic and organic nitrogen compounds can be utilized to satisfy the requirement for this element. Nitrogen metabolism by microorganisms has been recently reviewed by Payne (1980).

It is likely that all bacteria and fungi can utilize ammonia (as the NH_4^+ ion), which reflects its central role in nitrogen metabolism as the form in which nitrogen is incorporated into organic cell components. Failure to demonstrate growth with NH_4^+ is usually caused by the rapid decrease in pH produced when NH_4^+ compounds are utilized. It is frequently the preferred nitrogen source in many laboratory and industrial scale fermentations, where it is added as $(NH_4)_2SO_4$ or NH_4Cl. Many microbes can utilize ammonia as the sole source of nitrogen. Nitrate can be used by many filamentous fungi, and some bacteria and yeasts. In addition, many bacteria can utilize NO_3^- as an alternative electron acceptor to O_2. The product of this reaction is frequently the toxic NO_2^- ion which can lead to growth inhibition. However, it is an important reaction in the treatment of sewage effluents where high NO_3^- levels can pose a disposal problem. The chemolithotrophic nitrifying bacteria utilize NH_4^+ (producing NO_2^-) or NO_2^- (producing NO_3^-) as electron donors. This too has potential importance in effluent treatment. Nitrogen gas can be 'fixed' by some prokaryotic microbes. Although growth is often slow due to the high energy requirements for N_2 fixation, there is current interest in the production of NH_3 from sunlight using N_2-fixing blue-green algae.

Many organic nitrogen sources can be utilized if the microbe under consideration is capable of breaking down these compounds into smaller units that can be transported into the cell, *e.g.* amino acids and NH_4^+. Many filamentous fungi and bacteria can break down proteins and peptides, this latter source also being utilized by yeasts. Amides (notably urea) and amino acids can be used in many cases, with amino acids being particularly good sources of nitrogen giving increased growth when compared to NH_4^+ in many instances. Growth on amino acids relieves the cell of many biosynthetic activities leading to more efficient growth. However, economic considerations invariably make the use of NH_4^+ salts more attractive in many industrial processes. Additionally, some essential amino acids can often act as growth inhibitors. This is often linked to transport into the cell since many of these compounds share a common permease with resulting antagonism of uptake. Similarly, uptake of amino acids and NH_4^+ ions is often mutually inhibitory. Nitrogen sources that are broken down slowly into repressive amino acids and NH_4^+ are very useful in this context (Drew and Demain, 1977). Examples of commonly utilized nitrogen sources are given in Table 2 and the nutrient profiles of many of these can be found in Atkinson and Mavituna (1983). Many of the so-called growth factors are organic nitrogen compounds (see Section 7.2.3). Cyanide can be used by a restricted number of bacteria, which has great potential in the treatment of effluents containing this extremely hazardous chemical species.

7.2.1.3 Hydrogen

All organic compounds of the cell contain hydrogen, derived invariably from the carbon source. It is also present in cell water. The hydrogen bacteria and some methanogens can utilize H_2 gas as an energy source.

7.2.1.4 Oxygen

Oxygen is present in all organic cell components and in cell water. The carbon source usually provides oxygen as well, but molecular O_2 is required for those organisms that can utilize hydrocarbons. The O_2 is incorporated into these compounds by an oxygenase and is also required for aerobic degradation of aromatic carbon compounds. Molecular O_2 is also required as a terminal electron acceptor in aerobic metabolism.

7.2.1.5 Phosphorus

Inorganic PO_4^{3-} is the usual source of this element for microbial nutrition. These compounds are pH buffers and are often added in excess of the growth requirement to fulfil this role. Organic sources, *e.g.* glycerophosphate, are an alternative supply. Much of this element is bound up in RNA with the result that demand increases with the specific growth rate. It is also an important structural component (as teichoic acids) in the cell wall of Gram-positive bacteria.

Phosphorus is a key element is many industrial microbial processes because of its effect on the formation of many secondary metabolites (Demain, 1972). Many antibiotic fermentations are sensitive to inorganic PO_4^{3-} and it is an important regulator of carbohydrate and lipid metabolism. It also regulates the production and excretion of many organic acids of commercial importance, *e.g.* citric acid. Whereas the growth requirement is met for different microbes in the range 0.3 to 300 mmol l^{-1}, this is often high enough to repress secondary metabolite formation. The average highest level for the maximum production of these metabolites is 1 mmol l^{-1}, with levels greater than 10 mmol l^{-1} being strongly inhibitory.

7.2.1.6 Sulfur

Most bacteria and fungi can utilize inorganic SO_4^{2-} to satisfy the sulfur requirement. A useful source for media is $(NH_4)_2SO_4$ which can often be used concomitantly as the nitrogen source. An important exception to this is the methanogenic bacteria. These bacteria cannot utilize SO_4^{2-} and in some cases it is inhibitory. In defined media S^{2-} is usually the best source for these latter organisms. However, S^{2-} is often toxic at high concentrations and will also precipitate many metal ions which can lead to trace element deficiencies. It is also the inorganic electron donor of choice for culturing many of the photosynthetic bacteria. Sodium sulfide is the usual salt used which, because it is also a strong reducing agent, contributes to the lowering of redox potential required to grow these strict anaerobes.

The uptake and incorporation of SO_4^{2-} into cell material is an energy-requiring process. Since much cell sulfur is present in protein in the form of the sulfur-containing amino acids L-cystine, L-cysteine and L-methionine, stimulation of growth is often apparent if these are used in place of SO_4^{2-}. For industrial scale processes the high cost of these amino acids precludes their use in fermentation media. Sulfur is also present in some coenzymes, *e.g.* thiamine, biotin and ferredoxin.

Certain autotrophs, *e.g.* many *Thiobacillus* spp., use reduced inorganic sulfur compounds as electron donors. These can be supplied as sulfur, tetrathionate and thiosulfate. These organisms are important in metal winning by leaching low grade ores (see Ralph, Volume 4, Chapter 12).

7.2.1.7 Potassium

Potassium is the principal inorganic cation in the cell. Although much of it is bound up in the ribosomes of prokaryotic microbes (Tempest, 1969), it is a cofactor of some enzymes, is required in carbohydrate metabolism and is involved in many transport processes. In respect of the latter ionic balances of K^+ and other inorganic cations (notably Na^+) may be very important regulators of growth since cells tend to actively take up K^+ and Mg^{2+} and exclude Na^+ and Ca^{2+}. Although

Na^+, Rb^+ and NH_4^+ can partially replace the K^+ requirement in some microbes, it is usual to add an inorganic K salt, *e.g.* K_2SO_4, $KHPO_4$ or KH_2PO_4, to satisfy this requirement.

7.2.1.8 Magnesium

The magnesium requirement of bacteria and fungi is principally the result of a specific requirement of the ribosomes (Tempest, 1969). It also functions as an enzyme cofactor and is present in cell walls and membranes. It is usually supplied as $MgSO_4 \cdot 7H_2O$.

7.2.2 Micronutrients

7.2.2.1 Growth requirements

Consideration of the micronutrients (more commonly termed trace elements) brings us into one of the more uncertain areas of microbial nutrition. It is relatively easy to demonstrate an absolute requirement for the macronutrients, but how can this be shown for a nutrient that can satisfy a growth requirement at a concentration of less than 10^{-9} mol l^{-1}? The requirement for trace elements is usually satisfied at the laboratory stage by using standard formulations of these nutrients that have been 'handed down' from earlier workers. The mixtures are often peculiar to each research group and have often been formulated by a review of some relevant literature and attempts made to include 'a bit of everything'. This approach has obviously been successful where laboratory scale culture is concerned, but to the industrial microbiologist this *ad hoc* approach has the major disadvantage that many trace elements have profound effects on the production of many commercially important primary and secondary metabolites (Weinberg, 1970). Weinberg's review lists the concentrations of trace elements important in the control of such metabolites and reference will be made to the more important of these later in this section.

Trace element nutrients can be classified under four headings (Table 4). It is immediately apparent that most of them can be classified under group 4, *i.e.* rarely required (or claimed to be). Attempts to demonstrate a growth requirement usually follow two lines of investigation. The first of these is removal of a trace element from the medium. If the test organism does not grow, varying amounts of the element are added and the growth response should show a proportional increase. Complete removal is often extremely difficult and is complicated by a number of factors. The problem that causes most difficulty is the purity of the other medium components. Perusal of the declaration of analysis on even 'analytical grade' reagents will show that a macronutrient source, *e.g.* $(NH_4)_2SO_4$, contains sufficient micronutrients (notably Cu^{2+} and Na^+) as contaminants to satisfy a number of growth requirements. In industrial scale processes where much cruder nutrient sources are often used, lack of many of these micronutrients is clearly not going to be a problem. Similarly, the growth vessel can release significant amounts of an essential nutrient, *e.g.* Zn^{2+}, from laboratory glassware and many metal ions from the construction materials of a fermenter (Co^{2+}, Ni^{2+}, *etc.*). The second line of investigation is destructive analysis of the organism by ashing. Unfortunately, microorganisms are very indiscriminate about the uptake of many elements, particularly many metal ions. Indeed, this bioaccumulation has great potential in the removal and recovery of certain valuable and toxic metals from solutions and effluents.

Table 4   The Trace Elements required for Bacteria and Fungi

---

1. Trace elements required by all bacteria and fungi:
   Mn, Zn, Fe[a].
2. Trace elements required by many bacteria and fungi under specific growth conditions:
   Cu, Co, Mo, Ca.
3. Trace elements required by some bacteria and fungi under specific growth conditions:
   Na, Cl, Ni, Se.
4. Trace elements which are rarely required (or claimed as such):
   B, Al, Si, Cr, As, V, Sn, Be, F, Sc, Ti, Ga, Ge, Br, Zr, W, Li, I.

---

[a] With the possible exception of *Lactobacillus* spp.

Fortunately, most of the foregoing is academic since fermentation feedstocks usually contain sufficient trace elements as contaminants. More importantly, stimulation of growth or product formation (entirely different from an absolute requirement) is often the main consideration and this is discussed in the next section.

7.2.2.2 *Effects of trace elements*

Although lack of an essential trace element will sometimes prevent growth completely, deficiency is usually manifest in batch culture by an increase in the lag phase and a decrease in subsequent specific growth rate and yield when compared with a complete medium (Figure 1a). This may be due, in part at least, to partial replacement by other trace elements. In continuous culture (Figure 1b) the growth yield varies inversely with growth rate although the critical dilution rate is not altered (Pirt, 1975).

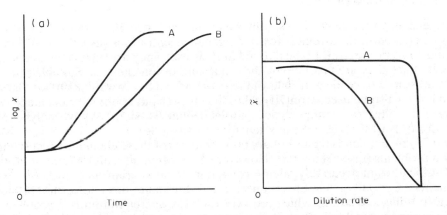

Figure 1 Effects of trace element deficiency on growth of microorganisms. (a) Batch culture and (b) chemostat culture: x = biomass in batch; $\bar{x}$ = steady state biomass in chemostat. Line A, optimum amount of trace element; line B, deficiency of trace element. (Drawn from Pirt (1975) with the permission of Blackwell Scientific Publications)

(i) Fe, Zn and Mn

Iron is present in cytochromes and ferredoxins, and all three elements are enzyme cofactors. In addition, Fe and Mn are the most important of the trace elements in the regulation of secondary metabolism and also play a role in excretion of primary metabolites. Deficiencies of both these elements are required for excretion of citric acid by *Aspergillus niger*. Iron deficiency is required for the excretion of riboflavin, and its concentration affects penicillin production by *Penicillium chrysogenum*. Zinc is inhibitory to this latter fermentation, to griseofulvin production by *P. griseofulvum* and to aflatoxin production by *A. flavus*. Conversely, *Streptomyces* cultures require five times the growth requirement of Fe for streptomycin production.

(ii) Cu, Co, Mo and Ca

Copper is present in certain respiratory chain components and enzymes and so is probably required by all aerobes; deficiency stimulates penicillin and citric acid production. Cobalt is present in corrinoid compounds, for example vitamin B_{12} which is synthesized by many prokaryotes. Corrinoids are found in many methanogenic bacteria growing on H_2 and CO_2 as carbon and energy sources. Molybdenum is a cofactor in nitrate reductase and nitrogenase and so is required for growth on NO_3^- or N_2 gas as the sole nitrogen source. It is also required by methanogens. Calcium is present in bacterial spores and is required as a cofactor (and for stability) of α-amylase and many proteases. It is usually expelled from the cytoplasm and can have an important structural role in cell wall proteins and extracellular polymer materials, capsules, *etc.*

(iii) Na, Cl, Ni and Se

Trace amounts of Na are probably required for some bacteria, growth and methane formation in certain methanogens being one such example. A trace requirement is easily satisfied since Na is a very common contaminant of other media components. Massive amounts of Na^+ are often added to growth media since Na salts are often the most convenient (and cheapest) source of some of the macronutrients, *e.g.* $NaHPO_4$, Na_2SO_4. Large amounts of Na are often inadvertently added as the titrant (NaOH) for pH control of cultures. This may have profound effects on the ionic balance of the cell and its environment. Also, energy is required to expel Na^+ from the cell together with a concomitant increase in the K^+ requirement. It is probably wise to omit much of this Na^+ from growth media except where a definite requirement for large amounts exists, *e.g.* with halobacteria and marine microbes which also require large amounts of Cl^- ions. Chloride is required for certain Cl-containing fungal antibiotics, *e.g.* griseofulvin and chlor-tetracyclines. It must be omitted from tetracycline fermentations to prevent conversion to chlor-tetracycline. Chloride, like Na^+ and Ca^{2+}, is usually expelled from the cytoplasm. Pirt (1975) has suggested that a large part of the maintenance energy ration is required to maintain such concentration gradients. As a result, it should be supplied as a trace element only and not added in large amounts with macronutrient sources such as NH_4Cl, KCl, *etc.* Nickel is required for methanogens and Se is required in formate metabolism of some organisms.

7.2.2.3 Addition of trace elements

The trace element requirement is usually small, in the range 10^{-9} mol l^{-1} to 10^{-6} mol l^{-1}. Many of these elements are extremely toxic at higher concentrations (usually $> 10^{-4}$ mol l^{-1}, see Figure 2) and so some care must be exercised in their use. They are all usually added as soluble inorganic salts since they are taken up by the cell as ions or ion chelates. Suitable salts are chlorides and nitrates since trace element salts of these ions are all freely water soluble. The sulfates of Zn, Cu and Ni can also be used. Iron(III) chloride may precipitate out in acidic media where iron(III) citrate is often a better choice. Molybdenum is supplied as molybdenate (*e.g.* $(NH_4)_6Mo_7O_{24}$ or Na_2MoO_4) and Se as selenate (*e.g.* Na_2O_4Se).

Hutner (1972) recommends the use of dry powder mixes of these elements because many cause precipitation problems (discussed later). However, this is often difficult for laboratory scale work because of difficulties in accurately adding nanogram or microgram quantities of dry powders. These elements are usually made up as concentrated stock solutions and certain of them can be combined to produce 'cocktails' which are very useful. Coprecipitation is a problem here and usually only experience will tell which elements can be combined in this way. Iron(III) salts are a particular problem and often have to be added as dry powders.

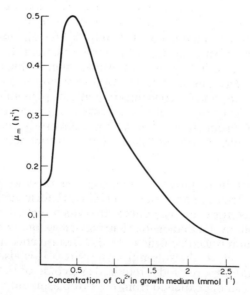

Figure 2 Maximum specific growth rate, μ_m, as a function of Cu^{2+} concentration for an *Acinetobacter* sp. growing in a glycerol/mineral salts medium

7.2.2.4 Chelation

Problems caused by precipitation of trace elements occur as a result of mixing incompatible medium components and can often involve some of the macronutrients, particularly Mg^{2+} and

PO_4^{3-}. As all Na and K salts are water soluble, such problems are usually associated with non-alkali metal components, although certain complex salts containing alkali metals are only sparingly soluble, *e.g.* magnesium sodium phosphate. Precipitated media components are in most cases unavailable to the cell and so deficiencies can occur. If this precipitate is of one of the trace elements, it may not be visible to the naked eye due to the low initial concentration of the nutrient and so possible problems can go undetected. Iron is particularly troublesome in this context.

Conversely, there is also a need to prevent toxicity problems due to accidental oversupply as contaminants of other components or leaching from growth vessel walls. There is, therefore, a need to control the concentrations of trace elements and this is achieved by chelation. Chelating compounds act as metal ion buffers maintaining a constant supply at physiological concentrations to the growing cells. These compounds contain groupings (termed ligands) which bind reversibly to metals forming soluble complexes. Examples of biologically important ligand binding groups are carboxylates ($-CO_2H$), amines ($-NH_2$) and mercapto groups ($-SH$). Compounds which contain these ligands are polybasic acids (*e.g.* citric acid, polyphosphates and ethylenediaminetetra-acetic acid), amino acids (*e.g.* histidine, tyrosine and cysteine), peptides and proteins.

A quantitative measure of the extent to which a metal ion will combine with a particular ligand is given by the formation or stability constant, K. For a metal ion M and monodentate ligand L a series of stepwise equilibria will be set up as follows provided precipitation does not occur (Hughes 1981):

$$M + L \rightleftharpoons ML \qquad K_1 = [ML]/[M][L]$$
$$ML + L \rightleftharpoons ML_2 \qquad K_2 = [ML_2]/[ML][L]$$
$$ML_{(N-1)} + L \rightleftharpoons ML_N \qquad K_N = [ML_N]/[ML_{(N-1)}][L]$$

The square brackets denote reactant concentration. These concentrations strictly should be expressed as activities since pH and ionic strength of the growth medium will modify absolute concentrations of the reactants. Formation constants are usually quoted as log K; the greater the value of log K, the greater is the affinity of the ligand for a particular metal ion. The measurement and use of formation constants and tables of values of log K for a number of biologically important chelating agents are given by Pirt (1975) and Hughes (1981).

In complex media addition of chelating agents should not be required since many of the components, especially amino acids, proteins and peptides, will chelate trace elements. For chemically defined media the chelating agent should satisfy two main criteria; it must be non-toxic and not be metabolized. EDTA (supplied as the disodium salt) is one of the more popular choices but at recommended levels (1 mol EDTA mol^{-1} Mg and other non-alkali metals; Pirt, 1975) it can be inhibitory, especially for yeasts. Metal ion-dependent exoenzymes are inhibited by this compound and it has also been suggested that it removes metal ion components of the cell wall (Ca^{2+}, Mg^{2+}, Zn^{2+}) leading to a loss of the structural integrity of the cell wall. However, it is probably resistant to metabolic attack. Citrate is another popular chelating agent, but it is metabolized by many bacteria and fungi. This is often only problematical in carbon-limited media. Although it should be possible to calculate the concentration of chelating agent required by stoichiometry, it is often necessary to derive values empirically in order to achieve a balance between toxicity and sufficient chelation to prevent trace element deficiencies.

The values of log K for certain chelators and metals are very high, *e.g.* EDTA/Fe^{3+}, log K = 25.1 at 25 °C. As very few chelating agents can permeate the plasma membrane, microorganisms often produce endogenous chelators which will compete effectively enabling uptake. The best studied of these are the siderophores which are produced by many bacteria for uptake of Fe^{3+}. Formation constants for certain siderophore/Fe^{3+} chelates are very high (log K = 30).

7.2.3 Growth Factors

Not all microorganisms can synthesize an entire complement of cell components from the compounds discussed so far. Some require an external supply of preformed organic molecules which form the building blocks of cell components. Such compounds are called growth factors. They stimulate the growth of all organisms since they decrease the biosynthetic load on the cell and certain bacteria and fungi (notably yeasts) have an absolute requirement for a number of them. Requirements, function and structure of growth factors required by microorganisms have been reviewed by Spector (1956), for fungi by Fries (1965), for yeasts by Suomalainen and Oura (1971), for bacteria by Guirard and Snell (1962), and for bacteria and yeasts by Koser (1968). It will be useful to split these compounds into three broad groupings; vitamins, amino acids and miscellaneous growth factors.

7.2.3.1 Vitamins

These are compounds which are components (or precursors of these components) of coenzymes or enzyme prosthetic groups. Examples of commonly required vitamins are *p*-aminobenzoic acid, thiamine (vitamin B_1), riboflavine (B_2), nicotinic acid (B_3), pantothenic acid (B_5), pyridoxine (B_6), biotin, cyanocobalamin (B_{12}), folic acid, lipoic acid, inositol, vitamin K, mevalonic acid and haemin. Any growth requirement is usually met at very low concentrations (10^{-12} mol l^{-1} to 10^{-6} mol l^{-1}) for most vitamins. This requirement may increase under conditions of stress, *e.g.* growth at temperatures above optimum, or anaerobic growth of yeasts. Estimates of requirements can be complicated by nutrient carry-over on subculturing. Certain structural components, *e.g.* inositol, which is a component of phospholipids, are usually required at the higher end of the range quoted above. Since they are components of higher plants, and residues of these are often used as fermentation feedstocks, there is usually no need to supplement the medium further with vitamins. Some feedstocks can be combined to provide a range of vitamins, *e.g.* production of Bakers' yeast from molasses using cane molasses (rich in biotin) and beet molasses (which contains B group vitamins). Pure vitamins are sometimes used since there is a large scale production of these compounds by the fermentation industry. Similarly, extracts of other microorganisms are obvious vitamin sources, yeast extract being particularly useful.

The growth yields from vitamins are still not fully detailed but culture under growth-limiting concentrations is important in a number of fermentation processes. An example is glutamic acid production by *Corynebacterium* spp. which is stimulated by biotin deficiency. Lack of this vitamin alters the cell membrane rendering it 'leaky' to glutamic acid which accumulates in the medium and not intracellularly where it would cause feedback inhibition of its production.

7.2.3.2 Amino acids

These have been considered in Section 7.2.1.2. These are usually structural components and concentrations of 10^{-6} mol l^{-1} to 10^{-3} mol l^{-1} are needed to satisfy any requirement.

7.2.3.3 Miscellaneous growth factors

Fatty acids, *e.g.* oleic acid, are required by certain bacteria, notably *Lactobacillus* spp. (Spector, 1956). They may be supplied as Tween 80. Sterols are not generally required for bacteria and fungi, although certain yeasts show a requirement for fatty acids and sterols when grown anaerobically, *e.g. Saccharomyces cerevisiae*.

7.3 ALGAE

The elemental needs of the algae are, in general, the same as those for bacteria and yeasts. The trace element requirement of eukaryotic algae shows more similarity to that of higher plants. This section will also include the photosynthetic bacteria. Much useful information on the nutrition and culturing of algae is to be found in Stewart (1974), Lewin (1962), Stein (1973) and Droop (1969), and of photosynthetic bacteria in Carr (1969).

7.3.1 Macronutrients

7.3.1.1 Carbon, oxygen and hydrogen

Most photosynthetic microbes grow in aquatic habitats and utilize CO_2 or one of its ionization products in solution as the principal carbon source. CO_2 dissolves in water to form carbonic acid (H_2CO_3) which, being a weak acid, is only slightly ionized:

$$CO_2 + H_2O \rightleftharpoons H_2CO_3$$
$$H_2CO_3 \rightleftharpoons HCO_3^- + H^+(aq) \qquad pK_1 = 6.37 \text{ at } 25\,°C$$
$$HCO_3^- \rightleftharpoons H^+(aq) + CO_3^{2-} \qquad pK_2 = 10.25 \text{ at } 25\,°C$$

The first dissociation is the greater and at physiological pH the dominant species is HCO_3^-. Its concentration can be determined from the equation

$$\log [HCO_3^-] = pH - pK_1 + \log \alpha \frac{[CO_2]}{2240}$$

where $[HCO_3^-]$ = bicarbonate concentration (mol l^{-1}) in equilibrium with the gas phase; pK_1 and α are temperature-dependent constants that can be found in critical tables; and $[CO_2]$ = percentage concentration of CO_2 gas in equilibrium with the culture solution (= 0.03% in normal air). The major species present at various pH values are: below pH 5, CO_2 gas, pH 7–10, HCO_3^-, and above pH 10.5, CO_3^{2-}. Carbon dioxide can be supplied as HCO_3^- or CO_3^{2-} depending upon medium pH. Air enriched with CO_2 gas can also be used. A 5% enrichment is often used although 100% CO_2 gas has been used for a *Chlorella* sp. (Pirt, 1982). Oxygen for cell material is obtained from this source and also from water which also supplies hydrogen. In the algae, water is the photosynthetic electron donor. The majority of photosynthetic bacteria are anaerobes and utilize reduced sulfur compounds as electron donors.

Hetcrotrophy is widespread in algae, with some achlorotic groups being obligately so. Depending upon species, acetate and/or glucose are useful carbon sources for growth. Such heterotrophic potential is often confined to growth in the dark, but some species can photoassimilate organic carbon, particularly under CO_2 limitation. Green and purple sulfur photosynthetic bacteria (Chlorobiaceae and Thiorhodaceae respectively) are photolithotrophs but can photoassimilate a very restricted range of organic carbon compounds, *e.g.* acetate, fatty acids, pyruvate and dicarboxylic acids. The purple non-sulfur bacteria are photoorganotrophs that can utilize a wide range of organic compounds when growing anaerobically in the light. Some can grow as aerobic heterotrophs in the dark.

7.3.1.2 Nitrogen

Ammonium ion and NO_3^- can be used by most algae and photosynthetic bacteria. Preferential utilization of NH_4^+ occurs if they are provided together. Nitrogen fixation is widespread in the prokaryotic algae. Certain organic nitrogen sources can be utilized by many algae, *e.g.* urea, other amides and amino acids. These are poor nitrogen sources for prokaryotic algae due to permeability problems.

7.3.1.3 Phosphorus and sulfur

Inorganic PO_4^{3-} is usually supplied to meet the phosphorus requirement of algae, although certain groups can utilize organic sources. Most photosynthetic microbes can utilize SO_4^{2-} as the sole source of sulfur. The photosynthetic bacteria require reduced sulfur compounds as electron donors.

7.3.1.4 Potassium and magnesium

Magnesium, which can be supplied as inorganic Mg^{2+}, is an important nutrient since it is a component of chlorophyll. Potassium and magnesium requirements are similar to those of bacteria and fungi, but marine algae often require a wide range of inorganic ions for successful growth. Such requirements can be met by use of artificial sea water formulations. These can be made up from components or supplied as commercial sea water mixes (Stein, 1973).

7.3.2 Micronutrients

All of the micronutrients described previously for bacteria and fungi are required with the possible exception of Se. In addition Si (for diatoms), I, V and B are required by some eukaryotic algae. The diatoms have silica frustules and this structural role is reflected in a relatively high requirement (10^{-4} mol l^{-1}). Sodium is required by certain prokaryotic algae and for all marine algae which reflects its presence as a major ionic component of sea water.

7.3.3 Growth Factors

Biotin, thiamine and cyanocobalamin (B_{12}) are required by many algae, these three vitamins being the only ones of general importance in algal nutrition. Green and purple sulfur photosynthetic bacteria require vitamin B_{12}. Purple non-sulfur bacteria in general have more complex requirements including biotin, thiamine and nicotinic acid which can be supplied as yeast extract.

7.4 PROTOZOA

The nutritional requirements of protozoa are still poorly understood when compared to other microbes, presumably because of their rather restricted biosynthetic capabilities. Most non-photosynthetic free-living protozoa feed by ingesting other microorganisms and organic detritus which presents problems when trying to elucidate their minimal needs. Parasitic protozoa have very complicated requirements reflecting a nutrient-rich growth habitat that provides many preformed macromolecules and biosynthetic intermediates.

Early attempts at culturing protozoa were hampered by two main problems. All free-living forms have associated microflora and microfauna, some of which are true symbionts rather than casual contaminants, resulting in cross-feeding. This problem has been largely overcome by use of antimicrobial compounds so that axenic culture is now possible. Secondly, the complex substances required for growth are often difficult to fractionate and are in any case contaminated with other nutrients. However, many protozoa, including certain parasites, can now be cultured in truly defined media. In general the phytoflagellates are much less nutritionally exacting than other groups.

Carbon sources utilized include carbohydrates (notably glucose), organic acids, *e.g.* acetate and citrate, amines and glycerophosphate. Only the phytoflagellates can utilize inorganic nitrogen sources; all other groups require organic nitrogen. Many amino acids are required, *e.g.* all ciliates require at least 10, with others in addition depending upon the species. Purines and pyrimidines are often needed, the ciliates and zooflagellates showing an absolute requirement. Inorganic ion requirements are similar to other microbes with Na^+ often being required in amounts which render it a macronutrient.

Growth factor requirements are complex for all protozoa with the exception of phytoflagellates. There is an almost universal requirement for most of the water soluble vitamins, with a few groups requiring fatty acids, *e.g.* lipoic acid, and sterols. Haem is required by many parasites, particularly the trypanosomes. The space required to detail the complex nutrition of these organisms is beyond the scope of this chapter and the reader is referred to Kidder (1967) and Gutteridge and Coombs (1977).

7.5 ANIMAL AND PLANT CELLS AND TISSUES

This topic is covered by Fowler and Spier in Volume 1, Chapter 16.

7.6 DESIGN OF CULTURE MEDIA

Although this topic is covered by Corbett in Volume 1, Chapter 8, it may be useful to raise a few points here with particular reference to media optimization. Stoichiometric determinations of yield from the carbon and energy and macronutrient sources can often be used as a starting point for medium design, but for the other nutrients empirical determinations are usually required. Such experiments are usually performed by use of a large number of shake flasks containing permutations of various nutrients until particular growth limiting concentrations have been identified.

Mateles and Battat (1974) describe an elegant use for continuous culture in media optimization for a single cell protein organism, *Pseudomonas* C, growing on methanol as the sole source of carbon and energy. All other media components were supplied as inorganic salts. The experimental design is summarized in Figure 3. High biomass concentration was the most important consideration and so steady state conditions were initially achieved with methanol as the growth limiting nutrient. The concentration of methanol was increased in the medium reservoir until analysis of the spent medium indicated it was no longer growth limiting. Various components of the nutrient medium were then injected individually into the growth vessel to produce higher concentrations than in the medium reservoir. Monitoring of the biomass and residual methanol indicated whether any particular nutrient stimulated biomass production and methanol utilization. If any stimulation by a particular nutrient was observed, its concentration in the medium reservoir was increased to the same level. This was repeated by increments until the nutrient ceased to limit growth and the other nutrients were then tested sequentially until all had ceased to limit growth. There is no requirement to achieve a steady state between additions since any increase in biomass coupled with a decrease in residual methanol can be interpreted as an indication of growth limitation by the nutrient producing this effect. As a result it is possible to screen all nutrients rapidly at various concentrations.

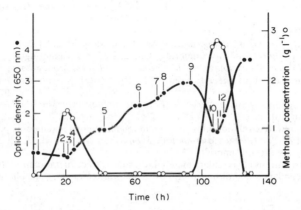

Figure 3 Changes in organism and methanol concentrations during growth of *Pseudomonas* C in a chemostat. *Pseudomonas* C was grown for 20 h in a chemostat with MSM medium containing methanol (1 g l^{-1}), $(NH_4)_2SO_4$ (1 g l^{-1}), as nitrogen source, and $CuSO_4 \cdot 5H_2O$ (10 μg l^{-1}) at 35 °C. Dilution rate was 0.35 h^{-1}, with stirring of 600 rpm and aeration of 0.5 volume volume^{-1} min^{-1}. OD of the bacterial suspension was measured at 650 nm, and the methanol concentration was that in the growth vessel (S). Numbers indicate additions as follows: 1, methanol (1 g l^{-1}) added to the medium in the reservoir (S_0 methanol = 2 g l^{-1}); 2, $MnSO_4 \cdot 5H_2O$ (10 μg l^{-1}) added to the growth vessel; 3, $CuSO_4 \cdot 5H_2O$ (10 μg l^{-1}) added to the growth vessel; 4, $CuSO_4 \cdot 5H_2O$ (70 μg l^{-1}) added to the medium in the reservoir; 5, methanol (1 g l^{-1}) added to the medium in the reservoir (S_0 methanol = 3 g l^{-1}); 6, methanol (1 g l^{-1}) added to the medium in the reservoir (S_0 methanol = 4 g l^{-1}); 7, $(NH_4)_2SO_4$ (0.5 g l^{-1}) added to the growth vessel; 8, $(NH_4)_2SO_4$ (1.5 g l^{-1}) added to the medium in the reservoir; 9, methanol (1 g l^{-1}) added to the medium in the reservoir (S_0 methanol = 5 g l^{-1}); 10, $FeSO_4 \cdot 7H_2O$ (1 mg l^{-1}) added to the growth vessel; 11, $MgSO_4 \cdot 7H_2O$ (0.1 g l^{-1}) added to the growth vessel; 12, $MgSO_4 \cdot 7H_2O$ (0.1 g l^{-1}) added to the medium in the reservoir. The same procedure was continued until S_0 was 10 g l^{-1}. (Drawn from Mateles and Battat (1974) with the permission of the American Society for Microbiology)

Although this work was performed using a methanol–mineral salts medium, growth factor requirements can also be determined in this way. This technique could also be applied to determination of the toxic threshold concentration of the various heavy metal micronutrients which could indicate the margin between adequacy and toxicity. This information would be useful at the production plant stage where media components are measured out with rather less precision, often by untrained personnel, than at the laboratory stage. The great advantage of studies of this nature is that it permits the formulation of an optimal medium, tailor-made for a particular process and organism, rather than having to rely on general principles of microbial nutrition to design media.

7.7 REFERENCES

Atkinson, B. and F. Mavituna (1983). *Biochemical Engineering and Biotechnology Handbook*. Macmillan, London.
Carr, N. G. (1969). In *Methods in Microbiology*, ed. J. R. Norris and D. W. Ribbons, vol. 3B, chap. 2, pp. 53–77. Academic, New York.
Demain, A. L. (1972). Cellular and environmental factors affecting the synthesis and excretion of metabolites. *J. Appl. Chem. Biotechnol.*, **22**, 345–362.
Drew, S. W. and A. L. Demain (1977). Effects of primary metabolites on secondary metabolism. *Annu. Rev. Microbiol.*, **31**, 343–356.
Droop, M. R. (1969). In *Methods in Microbiology*, ed. J. R. Norris and D. W. Ribbons, vol. 3B, chap. 11, pp. 269–313. Academic, New York.
Fries. N. (1965). In *The Fungi*, ed. G. C. Ainsworth and A. S. Sussman, vol. 1, chap. 19, pp. 491–523. Academic, New York.

Guirard, B. M. and E. E. Snell (1962). In *The Bacteria*, ed. I. C. Gunsalus and R. Y. Stanier, vol. 4, chap. 2, pp. 33–94. Academic, New York.

Gutteridge, W. E. and G. H. Coombes (1977). *Biochemistry of Parasitic Protozoa*. Macmillan, London.

Hughes, M. N. (1981). *The Inorganic Chemistry of Biological Processes*, 2nd edn. Wiley, Chichester.

Hutner, S. H. (1972). Inorganic nutrition. *Annu. Rev. Microbiol.*, **26**, 313–346.

Kidder, G. W. (ed.) (1967). *Chemical Zoology*, vol. 1. Academic, New York.

Koser, S. A. (1968). *Vitamin Requirements of Bacteria and Yeasts*. Charles C. Thomas, Springfield, IL.

Lewin, R. A. (ed.), (1962). *Physiology and Biochemistry of Algae*. Academic, New York.

Lily, V. G. (1965). In *The Fungi*, ed. G. C. Ainsworth and A. S. Sussman, vol. 1, chap. 8, pp. 163–177. Academic, New York.

Mateles, R. I. and E. Battat (1974). Continuous culture used for media optimization. *Appl. Microbiol.*, **28**, 901–905.

Payne, J. W. (ed.) (1980). *Microorganisms and Nitrogen Sources*. Wiley, Chichester.

Perlman, D. (1965). In *The Fungi*, ed. G. C. Ainsworth and A. S. Sussman, vol. 1, chap. 18, pp. 479–489. Academic, New York.

Pirt, S. J. (1975). *Principles of Microbe and Cell Cultivation*. Blackwell, Oxford.

Pirt, S. J. (1982). Microbial photosynthesis in the harnessing of solar energy. *J. Chem. Technol. Biotechnol.*, **32**, 198–202.

Rittenberg, S. C. (1969). The role of exogenous organic matter in the physiology of chemolithotrophic bacteria. *Adv. Microb. Physiol.*, **3**, 156–196.

Spector, W. S. (ed.) (1956). *Handbook of Biological Data*. W. B. Saunders, Philadelphia.

Stein, J. R. (ed.) (1973). *Phycological Methods*. Cambridge University Press, Cambridge.

Stewart, W. D. P. (ed.) (1974). *Algal Physiology and Biochemistry. Botanical Monographs*, vol. 10. Blackwell, Oxford.

Suomalainen, H. and E. Oura (1971). In *The Yeasts*, ed. A. H. Rose and J. S. Harrison, vol. 2, chap. 2, pp. 3–74. Academic, New York.

Taylor, G. T. (1982). The methanogenic bacteria. In *Progress in Industrial Microbiology*, ed. M. J. Bull, vol. 16, pp. 231–329. Elsevier, Amsterdam.

Tempest, D. W. (1969). Quantitative relationships between inorganic cations and anionic polymers in growing bacteria. In *Microbial Growth, Society for General Microbiology Symposium No. 19*, ed. P. M. Meadow and S. J. Pirt, pp. 87–111. Cambridge University Press, Cambridge.

Weinberg, E. D. (1970). Biosynthesis of secondary metabolites: role of trace metals. *Adv. Microb. Physiol.*, **4**, 1–44.

Weinberg, E. D. (1974). Secondary metabolism: control by temperature and inorganic phosphate. *Dev. Ind. Microbiol.*, **15**, 70–81.

8
Design, Preparation and Sterilization of Fermentation Media

K. CORBETT
Beecham Pharmaceuticals, Worthing, Sussex, UK

8.1 INTRODUCTION

The use of microorganisms to produce foods and beverages stretches back to antiquity and records exist which suggest that the Sumarians used fermentation processes around 6000 years ago. Throughout the ensuing centuries most civilizations have developed microbiological processes, large numbers of which have survived until the present day. With the increase in scientific knowledge and the pressure of increasing industrialization, new processes were developed in the first half of this century for the production of speciality chemicals to meet particular needs, *e.g.* citric acid, acetone and butanol. The current boom in biotechnology was catalyzed by the discovery and successful large scale production of antibiotics and has been fuelled by the exciting advances in gene-modifying techniques.

The materials and media used in the traditional processes have been developed over hundreds of years by a slow evolutionary process. They originated as the small scale home production of food and beverages which formed part of the diet of the producer's family. In recent times, methods have been refined and adapted to the large scale, mainly using traditional raw materials but improved process control has resulted in more consistent products. Unlike these traditional products the more recent biotechnological processes have no established raw materials as their starting point, therefore it has been necessary to develop suitable media for the exploitation of individual scientific discoveries. These processes have flourished in a hot-house atmosphere of commercial competition which has led to intense pressure to improve productivity and reduce costs. It is well to remember that the aim of modern biotechnology is the production of materials required by society at a cost which allows the producer to make a profit. If this last criterion is not satisfied the particular process becomes a modern fossil.

The prime requirement of a biotechnological process is a high-yielding organism, but optimum culture conditions must be developed for all stages of the process if maximum yields are to be obtained. Pirt (1975) has listed five criteria for optimal growth of microorganisms, namely, an energy source, nutrients to provide essential materials for growth, absence of inhibitors, a viable inoculum and suitable physicochemical conditions. Of these parameters, medium design and preparation govern the first three and together with physical processes contribute significantly to the fifth parameter. However, for the practice of modern biotechnology, designing media to satisfy the requirements of the organism in line with these criteria only constitutes part of the story. Manufacturers must strive to improve the producer organism; strains that were considered to be high-yielding in the penicillin G processes of the 1950s would be totally unacceptable today. A whole family of specialized media are required for various stages of strain improvement, culture maintenance and inoculum preparation. Even for the seed and final stages of the process the above parameters must be tempered by economic considerations and availability of materials in quantities necessary to conduct a large scale commercial process. Furthermore, the nature of the ultimate product will affect the design philosophy of the final stage medium; whereas one must maintain a high growth rate throughout a single cell protein process, it is necessary to limit growth rates in the latter stages of secondary metabolite producing processes.

Medium development is, therefore, a many faceted, complex operation critically affecting the success of the overall process. Having designed a series of satisfactory media it is important to ensure that the advantages endowed on the process are not dissipated by careless preparation or sterilization. The key to success in these areas is consistency of operation to ensure that each and every batch of medium is as similar as it is possible to achieve. This may appear to be a simple criterion to satisfy under laboratory conditions using highly trained staff but it requires considerable attention on a large-scale production plant using unskilled operators on a 24 h shift-work basis. Moreover, one must ensure that the requirements of regulatory authorities are satisfied when designing the operating system for medium preparation and sterilization on the production scale.

8.2 MEDIUM DESIGN

A fermentation process consists of a number of operations and stages for which a whole family of inter-related media must be developed. Before attempting to design any media one must make the objective of the specific stage of the process absolutely clear. Table 1 lists the various stages which would be involved in the maintenance of a successful process. At least one medium would need to be designed specifically to achieve the objectives inherent in each step and in certain cases, *e.g.* selection in liquid media, several different media would be used to enhance the chances of identifying an improved strain. A wide variety of objectives is apparent, ranging from a requirement for optimum growth, *e.g.* growth of colonies on agar surfaces following mutation, to the inhibition of growth of all organisms except ones which have specific characteristics and even here one might well obtain sparse growth, the aim being to distinguish between those organisms which will grow and those that will not. Obviously a wide range of media are necessary to meet the varied objectives shown in Table 1.

The mutation step would be achieved by application of chemical or physical methods capable of damaging the genetic material. Obviously in the case of chemical damage a suitable chemical, *e.g.* NTG or MMS, would be incorporated into the medium to bring about this effect. Even when physical damage is the objective, *e.g.* using UV light, it may be necessary to use a chemical adjunct to inhibit repair of the damaged DNA, *e.g.* nalidixic acid or caffeine. These are examples of one specific chemical being added to the medium to obtain a very specific effect and further examples may be found in the design of media for specific selection techniques. Having carried out a mutation step it may be necessary to select putative mutants having characteristics which have been shown to correlate with improved productivity of the desired product. In such cases inhibitory substances would be added to the medium to prevent the growth of all but those organisms resistant to the added compound. Examples of the application of this technique are the addition of high concentrations of toxic amino acid analogues in the selection of high-yielding amino acid producers and the addition of high levels of the required antibiotic or a precursor, *e.g.* phenylacetic acid, in the search for better penicillin G producers.

Interspersed with these operations aimed at modifying and inhibiting growth, there would be stages in which it would be necessary to promote growth. Random selection of mutants requires solid media which encourage the growth of all organisms which survive the mutation stage. Likewise the productivity testing stage would, at least initially, support active growth, although depending on the product, it may be necessary to restrict total growth if one is dealing with secondary metabolite production. Since it is difficult to reproduce large-scale fermentation conditions exactly on a small scale, particularly in shaken flasks, it may be necessary to develop a number of media to enhance the chances of selecting high-producing mutants. It is customary to try to use a shaken flask selection medium which is as close to the large-scale production medium as possible. However, because of the differences in scale, the limiting condition in the shaken

Table 1   Objectives Guiding the Selection of Media for Each Stage of an
Industrial Fermentation Process

Operation	Stage	Objective
Genome modification	Mutation	Damage to the genetic material
	Protoplast formation	Produce maximum numbers of viable protoplasts
	Protoplast storage	Store protoplasts in an undamaged condition
	Protoplast fusion	Enhance the chance of protoplasts fusing with production of viable progeny
	Regenerate cells from protoplasts	Regenerate the maximum number of whole cells from fused protoplasts
	Recombination methods	Enhance the entry of vectors bearing cloned genes with maximum cell viability
Mutant selection	Random growth	Encourage the growth of all surviving organism
	Selective growth	Allow growth of organism with specific characteristics
Productivity testing	Shaken flask testing	Identify any strain with improved ability to produce the desired product
	Fermenter testing	Develop optimum conditions for selected strains
Preservation	Preservation	Prepare improved strains for long term storage with retention of maximum viability and productivity
Inoculum preparation	Vegetative inoculum	Rapid growth to give sufficient cells to inoculate seed stage
	Spore formation	Production of maximum numbers of spores
	Spore harvest	Collect spores, evenly dispersed to give maximum number of potential growth centres
	Spore storage	Short term storage of spores with retention of maximum viability and productivity
Seed stage	Seed stage	Rapid production of a large cell mass in the optimum metabolic state for inoculation of the final stage fermenter
Final stage	Final stage	Maximum production of desired product at minimum cost

flask may be different to that pertaining to the large scale, thus it may be necessary to design media specifically for the small scale which allows the mutants to express their full potential, *e.g.* if oxygen transfer is a limiting factor in shaken flasks it may be necessary to use a more dilute medium than that used on the large scale. Also, in the shaken flask system it is not possible to monitor and control conditions continuously throughout the fermentation, as is the case in large fermenters, therefore, for the selection step the medium may have to be designed specifically to overcome this problem. Two examples are found in the selection of organisms for penicillin G and clavulanic acid production. In the former it is necessary to maintain a continuous but limited supply of glucose throughout the fermentation. In large fermenters a small amount of glucose is added initially to the medium, the concentration monitored and the correct level maintained by slowly feeding glucose throughout the fermentation. In shaken flasks the limited glucose supply stems from the slow hydrolysis of lactose which is incorporated in the medium, thus achieving a similar end result. Control of pH is important in clavulanic acid production. Additions of acid or alkali are easy to perform in a fermenter but in numerous shaken flasks it may be necessary to add a buffering agent to achieve the desired effect (Aharonowitz and Demain, 1977).

The preservation and inoculum preparation stages require quite different media from any so far described. Preservation of organisms for long term storage is a necessary part of any industrial process in order to ensure that there are adequate stocks of the producer organism for a long running production operation. It is essential to design conditions which retain both viability and high productivity of the organism. Various freezing or drying techniques are the basis of such systems, *e.g.* lyophilization or freezing in liquid nitrogen, but the medium from which the organism is dried or frozen may have a crucial influence on the success of the operation (Lapage *et al.*, 1970; Dietz and Churchill, Volume 1, Chapter 4).

Preparation of the inoculum for the large scale may be in the form of vegetative cells or spores. A straightforward growth medium would be used in the former case but specialized media are required for the preparation of a spore inoculum. Spores would usually be produced on a solid surface which could be agar or a particulate material soaked in a suitable medium, *e.g.* rice grains or perlite. The medium should contain all the factors necessary to support a good cover of growth over the whole surface before sporulation is induced. Having obtained luxuriant sporulation, a medium is required to remove spores from the solid surface and from vegetative material. It must also prevent aggregation of spores without causing damage or allowing germination to occur. Buffered salt solutions containing a surfactant are often used but the latter material must be very carefully chosen to avoid spore damage, particularly if the spores are stored in this medium for any length of time.

All the operations considered so far are conducted on the laboratory scale. In some cases the amount of medium required would be very small, of the order of a few litres per year, but the importance of these stages to the continuing successful commercial exploitation of the process is beyond doubt. Unless improved high-yielding mutants can be produced the process will stagnate and commercial competitors will move ahead; the number of companies manufacturing penicillin has fallen dramatically since the 1950s. A large part of the success of the operation will depend on devising a successful medium for each stage of the process. The major factors in designing media for these stages are scientific knowledge and ingenuity. Unlike the two large scale stages, cost is relatively unimportant; the maximum amount of medium used at any one laboratory stage would not exceed 50 l per week. Even if relatively expensive chemicals are involved the cost is negligible when compared to the cost of materials for a large fermenter.

Most commercial processes operate as a batch or a fed batch system in stirred tank fermenters between 10 m^3 and 200 m^3 capacity. However, the advent of large scale production of single cell protein has led to the development of novel fermenters with capacities up to 4000 m^3 (Schreier, 1977). Typically, the production medium is a complex heterogeneous mixture with a dry solids content in excess of 10% of the volume. Even for the traditional stirred tank fermenter, several tons of raw materials are required for each production batch, *e.g.* approximately 10–15 tonnes of corn-steep liquor and a similar amount of glucose syrup would be required for each batch of medium in a 200 m^3 penicillin G fermenter. While it is still necessary to satisfy the nutritional requirements of the organism it is essential on this scale to design a medium with constituents which are cheap and readily available throughout the year. It is also desirable that they are consistent, stable and easy to store and handle.

The first large-scale step in the process is usually the seed stage fermentation which is designed to produce a large inoculum for the production vessel. There is no requirement for final product formation therefore the medium may be designed purely to satisfy the growth requirements of the organism. Rapid growth is also a requirement for the first phase of the production stage fermentation. If single cell protein is the product, a high growth rate will be required throughout the final stage but if a secondary metabolite is desired, *e.g.* an antibiotic, then growth must be limited before it will be formed.

In order to obtain rapid growth in both seed and final stages, the medium must contain sources of energy, carbon, nitrogen and phosphate, trace elements and any specific growth factors which the organism itself cannot manufacture (see Dunn, Volume 1, Chapter 7). A single carbohydrate material may act as both carbon and energy sources although a second material, usually a lipid, may be required and the carbon skeletons of nitrogenous organic compounds may contribute to both requirements. As much as 15% of the biomass dry weight may be composed of nitrogen, therefore the medium must provide at least this amount of a suitable material. Although many industrial organisms utilize inorganic nitrogen, growth and productivity is invariably stimulated by the addition of suitable organic nitrogenous materials. The requirement for phosphate is usually supplied as either the sodium or potassium salt of orthophosphoric acid. As a consequence of using crude technical grade materials, the required trace elements may be supplied automatically. It is common practice to add growth factors as part of a complex mixture, *e.g.* yeast extract, distillers solubles, so that the nature, much less the amount required, is imperfectly understood. In order to obtain good growth it is necessary to design a balanced medium containing appropriate nutrients in proportions specifically selected for the particular organism concerned.

It is easier to design media for good initial growth than for the subsequent step, maximum product accretion. This is particularly the case where a secondary metabolite (idiolite) is required. Microorganisms only appear to be capable of producing such compounds when the specific growth rate falls below a given value (Bu'Lock, 1975). Therefore the key to good secondary metabolite production is the development of a final-stage medium which will become deficient in one or more nutrients after an initial rapid growth phase. Exhaustion of any nutrient makes balanced growth impossible and causes biochemical differentiation (Bu'Lock, 1974; Bu'Lock *et al.*, 1974).

Phosphate is the nutrient most widely studied for its effect on secondary metabolite production. Whereas vegetative growth is possible in the range 0.1–500 mM, secondary metabolite production is generally inhibited at phosphate levels greater than 10 mM (Weinberg, 1978). The biosynthesis of many important antibiotics, particularly those produced by streptomycetes, is controlled by phosphate. Weinberg (1978) lists more than 30 such compounds and Martin (1977) has reviewed the regulation of antibiotics formed by bacteria. Secondary metabolites produced by fungi may also be controlled by this nutrient, *e.g.* citric acid (Marchesini, 1966), bikiverin (Brewer *et al.*, 1973) and ergot alkaloids (Amici *et al.*, 1967). This latter fermentation gives a good demonstration of biochemical differentiation in response to phosphate depletion, since not only is ergot alkaloid production induced but there is also a change in cell type from sphaecelial to sclerotial form which is accompanied by an increase and a change in the composition of intracellular lipids (Mantle *et al.*, 1969). For phosphate-controlled fermentations it is necessary to establish the optimum concentration of phosphate which will give adequate growth to provide the maximum production of required product. Such optima can only be established empirically using samples of the technical grade nutrients which will be used on the production scale, since they may contain significant quantities of both inorganic and organic phosphate.

As with phosphate, it may be necessary to deplete the medium of a specific trace element before certain compounds are produced. Maximum production of citric and itaconic acids only occurs if the medium is deficient in iron (Lockwood, 1979). When an impure carbohydrate source is used, *e.g.* molasses, it is necessary to remove the trace elements in order to stimulate synthesis. Both ion exchange resins and ferricyanide treatment have been used for this purpose. Similarly, several organisms only overproduce riboflavin when the iron content of the medium is restricted (Straube and Fritsche, 1973; Perlman, 1979).

Early medium development studies of the penicillin process showed that oligosaccharides or polysaccharides supported better production than glucose (Soltero and Johnson, 1953). The inhibitory effect of glucose is due to catabolite repression and is evident in the production of a number of other antibiotics (Martin and Demain, 1980). Similarly, citrate and glycerol inhibit the production of novobiocin (Kominek, 1972) and cephamycin (Aharonowitz and Demain, 1978) respectively. A more complicated situation exists with ergot alkaloid biosynthesis by *Claviceps*. The medium must contain either a sugar or a polyol, plus a member of the TCA cycle or a related compound, in the ratio of approximately 10:1, *e.g.* sucrose and citric acid for ergotamine production (Amici *et al.*, 1967). There is no diauxic effect but rather a simultaneous metabolism of the two components (Spalla *et al.*, 1978). However, if the TCA cycle component is missing there is massive growth and little alkaloid production. In all the above cases it is necessary to design the medium so that growth is restricted by depletion of a nutrient or by carefully controlled supply during the production phase.

Addition of specific organic compounds to stimulate productivity has met with limited success and is usually the result of serendipity in finding a suitable material. Precursor addition is an obvious possibility although it is necessary to have a detailed knowledge of the biosynthetic sequence. Penicillin fermentations provide the best known case of precursor addition but even in this instance the initial finding was due to chance. Use of a corn-steep liquor medium to replace Czapek–Dox medium not only provided suitable nitrogenous nutrients but also stimulated the synthesis of penicillin G by supplying a source of the phenylacetic acid side chain (Moyer and Coghill, 1946). This discovery led to the rapid commercial development of the penicillin G process. A whole range of side chain precursors has since been shown to be incorporated with resultant directed synthesis of the final product (Cole, 1966). Not all precursors are as useful as phenylacetic acid since they may be more expensive than the desired product, *e.g.* cysteine is a costly precursor of the penicillin nucleus and fails to stimulate increased productivity in a complex production medium.

The literature contains references to a number of instances of compounds other than precursors being added to stimulate productivity. Methionine and norleucine have been shown to increase cephalosporin synthesis (Demain *et al.*, 1963) and thiosulfate was reported to have a similar effect on penicillin production (Hockenhull *et al.*, 1955). These compounds are not incorporated into the antibiotic but exert a regulatory influence on productivity. Apart from established literature reports there are many industrial rumours of beneficial effects of particular compounds. Every manufacturer must at some time have looked for the magical properties of chelating agents, surfactants and exotic oils!

A well known property of protein solutions is their propensity to foam when agitated, an effect which is exacerbated if air is blown through the solution. As we have seen, the seed and final stage media of a production process usually contain proteins and they may be both agitated and aerated. If the medium is allowed to foam it will occupy a larger volume than a non-foaming medium therefore the effective working volume of the fermenter is reduced. It may be necessary to lower aeration rate and/or agitation rate, thus reducing the oxygen transfer rate with possible limitation to growth and productivity. If excessive foam formation occurs, it may enter the fermenter vent system thus posing a possible contamination risk. Control of foam formation is essential, therefore, and it is common practice to add a chemical agent for this purpose. The most commonly used materials are fatty acids and derivatives, polyglycols, higher alcohols and silicones. A large variety are available as proprietary preparations but great care is necessary in selecting one which is suitable, since many of these compounds may inhibit growth or spore germination and all will degas the liquid and reduce oxygen availability. The degree of foam control exhibited by different compounds and preparations varies widely. Some will reduce foam for a short period but will allow it to reform after a relatively short time, whereas others will suppress foam formation for long periods. The nature of the fermentation governs the type of antifoam to be used. In cases where foam formation only occurs for a short period of time it may be convenient to add small amounts of short acting defoamers until the danger has passed. In cases where foaming is a constant threat a longer acting defoamer would be necessary. Inevitably, more efficacious materials are more expensive therefore it is necessary to balance the overall cost of multiple additions of cheap material against the lower utilization of an expensive compound. Since each type of fermentation behaves differently, the type of antifoam and its method of use must be specifically selected for individual circumstances.

Selection of suitable media for production scale processes depends on factors additional to optimizing productivity of the organism. Cost and availability of a regular supply are the two critical factors affecting the suitability of any material. Additionally one must take into account consistency and stability of the material, ease of handling during transportation, storage and medium preparation, and also approval by regulatory authorities.

Many materials used in production media are of animal or vegetable origin, therefore like all such commodities they are subject to world price fluctuations depending on many uncontrollable factors (Figure 1). Since the final stage medium is one of the major cost centres of the process, along with energy utilization, these fluctuations can have a dramatic effect on profitability, therefore it is advisable to investigate alternative materials. Also it is necessary to seek alternative suppliers of individual materials since equipment failure, an accident, a strike or bankruptcy of the sole supplier can have disastrous effects on the production schedule. Alternative supplies and materials must be tested to ensure they are suitable since even simple materials, *e.g.* chalk, made by different processes may result in unequal productivity (Corbett, 1980). It is necessary to maintain close liaison with raw material manufacturers so that one is aware of the range of products available and also the possible effects of changes in manufacturing methods on their composition. Having selected suitable materials it is necessary to balance them in correct proportions to obtain maximum productivity. In order to cover comprehensively the number of possibilities with the least time and effort it is usual to employ statistical methods, *e.g.* EVOP techniques (Box and Draper, 1969; Chatfield, 1970). Examples of the application of such techniques to medium design may be found in reports by Kupletskaya *et al.* (1969) and McDaniel *et al.* (1976). Since the medium must be specifically adapted to each and every mutant, medium development can never be finalized as long as the strain improvement programme is successful.

8.3 MEDIUM PREPARATION

The guiding principle governing medium preparation is a consistent operating procedure so that, having established the optimum medium, each batch will be as similar as possible to this ideal. There is little point spending time and effort by a highly qualified medium development

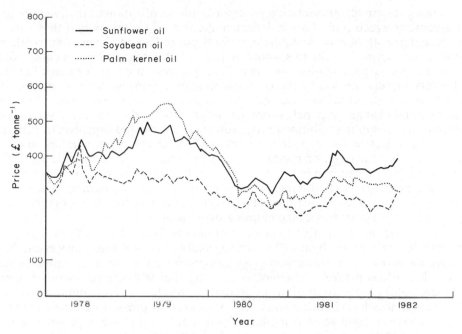

Figure 1 Fluctuation of the price of some vegetable oils for the period 1978–1982

team to establish the best conditions only to throw away the advantages by careless preparation. It is not an area which requires erudite scientific thought but it does require careful technical planning and supervision. It is essential to devise storage, handling and recording procedures which reduce the possibility of error to the minimum.

Storage and handling of bulk materials is a subject in its own right. In order that powders remain free-flowing they must be kept in a dry condition otherwise samples such as phosphate will become rock-like and others, like proteins, form glutinous lumps. Not only do lumpy materials make batching difficult, they also make subsequent sterilization unpredictable. Many bulk liquids contain a high solids content and they must be kept warm to prevent them solidifying, *e.g.* glucose syrups. If such a material solidifies in tanks and pipes it is expensive and time consuming to return it to the liquid state. However, the storage temperature must not be allowed to go too high otherwise there may be degradation of the material with ensuing impairment of the fermentation. Handling bulk materials poses problems of health and safety for the operators. One must provide them with protective clothing combined with adequate instruction and training, particularly for handling corrosive materials, *e.g.* acids and alkalis, and dusty powders which may pose fire and allergy risks.

Whereas one can reasonably expect skilled technical laboratory staff to prepare media reproducibly with limited supervision, the same cannot always be true for the production plant. The use of unskilled labour to prepare media often under hot, humid conditions in the middle of the night requires that each batch of medium is accompanied by explicit, unambiguous instructions, coupled with careful supervision. Operators must be trained to follow the instructions implicitly since not only will the wrong medium be produced if incorrect weights or volumes are used, but also unwanted reactions may arise if ingredients are added in the wrong order, *e.g.* media for penicillin production may contain both ammonium sulfate and chalk which must not be mixed together before other ingredients are added otherwise ammonia may be evolved. It is common practice to prepare the media in a separate tank for subsequent transfer to the fermenter. It is essential to ensure that the medium is suitably agitated so that it is not allowed to separate into phases, otherwise solid or lipid materials may be left behind causing distortion of the medium in the fermenter.

Although medium preparation may appear to be a trivial step in the process, the correct composition of every batch of medium is fundamental to achieving consistently high-yielding fermentations which are essential for a successful commercial operation.

8.4 STERILIZATION

The preliminary step in all fermentation processes is the sterilization of the medium to remove unwanted organisms which could have a detrimental effect on productivity if they were allowed to flourish. Rapid growth of contaminating organisms can reduce the level of nutrients available to the producer organism. Usually this would be the carbohydrate and nitrogenous sources but even expensive secondary metabolite precursors, *e.g.* phenylacetic acid for penicillin G production, can be very rapidly utilized by the contaminating organism, in this case by *Paecilomyces*. The growth of an aerobic organism can limit the amount of oxygen available and may result in other environmental changes, *e.g.* pH, which are detrimental to the process. Excretion of extraneous materials may result in inhibition of the growth of the selected organism or contamination of the required product at the end of the fermentation. If the contaminating material is an enzyme, it may destroy the required product, *e.g.* a β-lactamase producer can destroy all the product of a penicillin G fermentation in a matter of hours. Finally, the contaminating organism may produce such a mass of growth that it increases the viscosity of the medium to such an extent that agitation is impaired or efficiency of the harvest equipment is reduced. This is particularly the case if the unwanted organism has a filamentous growth form.

Theoretically, it should be possible to prepare sterile media by filtration, radiation, ultrasonic treatment, chemical treatment, heat or by various combinations of these processes. In practice the large volumes involved in production processes preclude the use of radiation or ultrasonic methods. The hazardous nature of chemical agents together with the necessity of removing or destroying them before inoculation of the desired organism, has severely limited their use. Only in special circumstances has chemical treatment been used to prevent excessive damage to the medium, *e.g.* growth of food yeast is inhibited by denaturation of whey proteins, therefore to allow growth in a medium containing whey solids the medium may be sterilized by hydrogen peroxide at 60 °C (Bechtle, 1974). In cases where a totally soluble medium is used, particularly if it contains heat labile materials, as may be the case for the growth of animal cells in deep culture, the medium may be sterilized by filtration. Similarly, soluble feeds for fed batch processes or water for making medium up to a desired volume may be filter sterilized. A greater risk of contamination is associated with this sterilization method than with the most commonly used procedure, the use of heat.

The universal method for sterilizing heterogeneous industrial media is steam heating under pressure. In the presence of moisture it is possible to sterilize both the vessel, associated equipment and the medium at a temperature of 120 °C whereas temperatures of approximately 400 °C would be required if dry heat were used. If the medium contains only vegetative cells, they can be killed at lower temperatures, but where spores are present the higher temperature is required. Conglomerations of materials and the presence of oil droplets are common in industrial media and they reduce the efficiency of sterilization by providing an environment which is resistant to the penetration of moisture. Relative humidities as low as 20% have been reported with such materials, resulting in kill rates which are 2 to 3 orders of magnitude lower than in the aqueous phase. The destruction of spores which are encapsulated in such a hydrophobic environment may depend on the size of the particle, the distance of the spore from the surface, the thermoconductivity of the material, and permeability to steam and water. Media containing such particulate materials require more extensive heating than totally soluble media. It is often necessary to determine empirically suitable times for the sterilization of heterogeneous media; a theoretical treatment of sterilization under these conditions may be found in Carslaw and Jaeger (1959).

Heating complex chemical mixtures to temperatures above 120 °C for periods exceeding 20 min can cause destruction of medium constituents either by direct thermal degradation or by unwanted chemical reactions. The best known case is the destruction of glucose by the Maillard or browning reaction when heating in the presence of amines. Table 2 shows the effect of extending the sterilization time period on glucose concentrations when a CSL medium was sterilized after glucose addition. Not only is glucose destroyed by the reaction but toxic products are formed which interfere with the subsequent fermentation. In such cases it is usual to sterilize the glucose solution separately and add it to the remainder of the medium after it has also been sterilized.

The conditions prevailing during sterilization can affect not only the stability of components but also their form and availability to the organism. Minor medium components, *e.g.* trace elements, may exert an influence on the solubility of major components, *e.g.* the nitrogen source (Table 3). From the same table it can be seen that both temperature and length of heating also affect the solubility of medium components. The degree of solubility may have a profound

Table 2 Effect of Sterilization Time on Glucose Concentration
and Antibiotic Accretion Rate

Time at 121 °C (min)	Amount of added glucose remaining (%)	Relative accretion rate
60	35	90
40	46	92
30	64	100

influence on the subsequent fermentation pattern and ultimately on the productivity of the process. Although sterility is of paramount importance, it is wise to investigate a range of sterilization conditions for their effect on productivity.

Table 3 Effect of Sterilization Conditions on the Soluble
Nitrogen Content of a Medium

	Sterilization Conditions		
pH	Time at 121 °C (min)	Addition of trace elements	Total nitrogen in soluble form (%)
6.4	20	Yes	33
6.4	40	Yes	42
7.0	20	Yes	67
7.0	40	Yes	68
6.4	20	No	45

The thermal death of microorganisms follows first order reaction kinetics, that is the rate of reaction (the death rate) is directly proportional to the concentration of the reacting substance. This may be expressed mathematically by equation (1).

$$-\frac{dc}{dt} = kc \tag{1}$$

In equation (1) c = concentration of reacting substance; t = time; and k = specific reaction rate.

The specific reaction rate, k, is a constant at a defined temperature for a specific reaction. The above equation may be integrated to give equation (2).

$$c = c_0 e^{-kt} \tag{2}$$

In equation (2) c_0 is the initial concentration and c is the concentration at time t. It can be seen that the quantity of the reacting material falls off exponentially and reaches zero only after infinite time.

The effect of temperature on the reaction rate constant is expressed by the Arrhenius equations (3) and (4).

$$\ln k = A - E/RT \tag{3}$$

$$k = A e^{-E/RT} \tag{4}$$

A = Arrhenius constant; E = activation energy; R = Boltzmann constant; and T = absolute temperature.

If one plots ln k against $1/T$ (Figure 2) the slope of the line is equal to $-E/R$ and the intercept is equal to the constant A. Also, since R is a constant it is possible to calculate E, the activation energy. Therefore, the slope of the line is dependent on the activation energy, *i.e.* the higher the activation energy the steeper the slope therefore the faster the rate of reaction at higher temperature.

The term sterilization implies the death of all contaminating microorganisms. However, we have seen that for a first order reaction to go to completion, *i.e.* to ensure the death of all microorganisms, it would be necessary to heat for an infinite time. The major penalty of heating large volumes of medium for long periods are loss of nutrients and high costs due to excessive use of steam, therefore it is necessary to use a criterion other than absolute sterility. One must compromise and accept the probability of a certain percentage of contaminations. It is usual to aim for a figure of 1/100 or 1/1000 contaminations on a commercial plant.

Obviously, one is considering the survival of some organisms in the whole mass of broth, there-

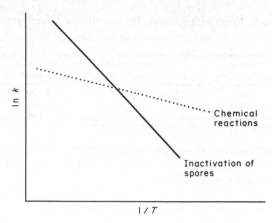

Figure 2 Effect of temperature on the rate constants for chemical reactions and inactivation of spores

fore in applying first order kinetics to sterilization it is usual to consider the whole volume as unit volume and the total number of organisms or spores rather than a concentration. Equation (2) now becomes equation (5).

$$N = N_0 e^{-kt} \tag{5}$$

where N_0 = number of organisms present at time zero, and N = number of organisms present at time t.

If one accepts a probability of contaminated fermentations of 1/100 or 1/1000 one can assign a value to N of 10^{-2} or 10^{-3}. It is now possible to determine a factor equivalent to the heat treatment necessary to bring about reduction from N_0 to N, *i.e.* 10^{-2} or 10^{-3}. This factor is the Del Factor or sterilization criterion and is defined as equation (6).

$$\text{Del Factor} = \nabla = \ln N_0/N \tag{6}$$

However, equation (7) can be derived from equation (5).

$$\ln N_0/N = kt \tag{7}$$

Therefore, from equation (4), equation (8) is obtained.

$$kt = At e^{-E/RT} = \nabla \tag{8}$$

Knowing the activation energy E and the Arrhenius constant for any organism it is possible to calculate the time at a set temperature which is necessary to achieve the requisite degree of killing during batch sterilization. However, the unsterilized medium will contain both vegetative cells and spores of many different organisms, therefore it is difficult to calculate a specific value. In practice it is usual to take the values for *Bacillus stearothermophilus* which has the most heat resistant spores known, and to base the calculation on these values (E = 287.2 kJ mol^{-1}; A = 7.94 × 10^{38} s^{-1}). In this way a certain amount of overkill is built into the system and acts as an extra safety factor.

Since the amount of heat treatment required to kill off an acceptable ratio of the organisms depends on the total number initially present, large volumes of media require longer sterilization times. Also, the larger the volume the longer will be the times for heating and cooling and it is not unusual to have large volumes of media above 100 °C for several hours (Figure 3). The two drawbacks to such long heating and cooling times are the high cost of steam and cooling water and the danger of destruction of labile compounds. The rate constant for the destruction of medium components follows first order kinetics exactly as described above. However, the activation energies for chemical reaction are in the range 40–130 kJ mol^{-1} (*e.g.* hydrolysis of casein, 86.1 kJ mol^{-1}, destruction of riboflavin, 85.7 kJ mol^{-1}), whereas the activation energies for the heat destruction of spores are in the range 200–300 kJ mol^{-1} (*e.g. B. stearothermophilus* 287.2 kJ mol^{-1}). Increasing the temperature causes a greater increase in the rate of those reactions which have high activation energies (Figure 2), therefore raising the temperature will increase the rate of spore destruction compared with the degradation of medium components.

The practical application of this phenomenon is the continuous medium sterilizer, which comprises a series of continuous flow heat exchangers and a lagged serpentine holding coil (Figure 4).

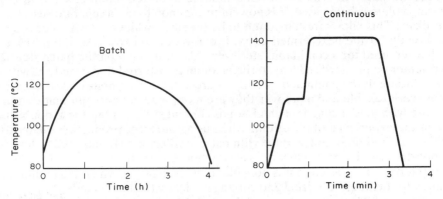

Figure 3 Temperature profiles of batch and continuous medium sterilizations

Traditional plate heat exchangers are unsuitable for this type of equipment because failure of the gaskets which separate the plates can allow sterile and non-sterile streams to mix. Modern continuous sterilizers use double spiral heat exchangers in which the two streams of liquid or liquid and steam are separated by a continuous stainless steel division. Initially water is cycled through the system to establish operating temperatures and to presterilize the equipment. Cold medium is introduced into the first heat exchanger, the economizer, which allows the incoming liquid to receive heat from the outgoing sterile stream. In this way the incoming liquid is preheated and the outgoing steam is cooled with considerable saving in both heating and cooling costs compared with a batch sterilizer; savings in excess of 60% are possible in a well designed and carefully operated system. The medium passes to the main heat exchanger in which high pressure steam is used to attain the desired temperature, 140–160 °C. The medium is passed through the lagged serpentine coil which holds it at the required temperature for the requisite time, 1–4 min, itself determined by the length of the coil and the flow rate of the medium. Hot medium is passed through the outward spiral of the economizer and finally through a water cooled exchanger before going to the presterilized fermenter. A typical temperature profile for a continuously sterilized medium is shown in Figure 3 where it can be compared with a corresponding profile for a batch sterilized medium.

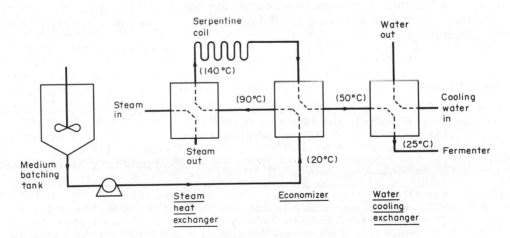

Figure 4 Continuous medium sterilizer

Despite the advantages of improved nutrient stability, continuous medium sterilization has not been widely used on a large scale. The disadvantages of the method have been considered to be high initial capital cost, complexity, unreliability, doubts regarding use with medium containing a high solid content and the risk of increased contamination because medium and vessel are sterilized separately, necessitating the transfer of sterile medium to the vessel over long distances. However, the dramatic increase in energy costs has led to a greater emphasis on reducing running costs and improving the efficiency of vessel utilization leading to a reappraisal of these disadvantages. In order to improve vessel turnround time between fermentations and to cope with the ever increasing volumes being used, it is now quite common to use a separate vessel specifically as a batch sterilizer. This allows fresh medium to be prepared while the fermenter is being emptied and cleaned ready for the next fermentation. Provision of an extra vessel requires a capital cost similar to that required for a continuous sterilizer. Also use of a separate batch sterilizer requires that the fermenter be presterilized and sterile medium transferred to it exactly as with continuous sterilization. Although the continuous sterilizers are still more complicated than batch sterilizers they are now very reliable and moreover they are more amenable to automatic control. Doubts about their use with media containing high solids content still exist but these doubts have proved groundless in the experience of the author when using different production media. It is claimed that shorter turnround times are possible with batch sterilization (Banks, 1979), however in practice this is not the case. Large production scale continuous sterilizers can produce sterile media at rates up to 50 tonnes h^{-1} therefore it is possible to fill even large fermenters in under 5 h which is the time taken for heating, sterilizing and cooling media by the batch method.

In these days of high energy costs and increased competitiveness, continuous sterilization offers an alternative to batch sterilization. The lower heating and cooling costs coupled with good reliability and turnround times offer a cheaper alternative to the batch method. Moreover because degradation of heat sensitive nutrients is reduced it may be possible to increase productivity or at least to obtain the same productivity with lower levels of nutrients, thus reducing medium costs. In addition to reducing costs, continuous medium sterilization offers the advantage that scaling up operations are easier, because the long heat up and cool down times associated with large volumes are eliminated. All in all, the advantages are now in favour of continuous medium sterilization which is now the first choice method for large scale fermentations. For a further discussion of media sterilization see Cooney, Volume 2, Chapter 18.

8.5 SUMMARY

It is necessary to design the optimum medium for each step of the process from mutant selection to the final stage fermentation. Having established the best media it is important that they are carefully and accurately prepared. Traditionally, batch sterilization has been used to remove unwanted organisms from fermentation media but this can lead to excessive nutrient destruction and is more expensive to operate than the alternative procedure of continuous sterilization which is gaining favour throughout the industry. While productivity plays the crucial role, cost is a major factor in selecting the medium composition since the success of any fermentation process must ultimately depend on its economic viability in relation to other methods of manufacturing the same product.

8.6 REFERENCES

Aharonowitz, Y. and A. L. Demain (1977). Influence of inorganic phosphate and organic buffers on cephalosporin production by *Streptomyces clavuligerus*. *Arch. Mikrobiol.*, **115**, 169–173.
Aharonowitz, Y. and A. L. Demain (1978). Carbon catabolite regulation of cephalosporin production in *Streptomyces clavuligerus*. *Antimicrob. Agents Chemother.*, **14**, 159–164.
Amici, A. M., A. Minghetti, T. Scotti, C. Spalla and L. Tognoli (1967). Ergotamine production in submerged culture and physiology of *Claviceps purpurea*. *Appl. Microbiol.*, **15**, 597–602.
Banks, G. T. (1979). Scale-up of fermentation processes. In *Topics in Enzyme and Fermentation Biotechnology*, ed. A. Wiseman, vol. 3, pp. 170–266.
Bechtle, R. M. (1974). Conversion of whey solids to an edible yeast cell mass. *US Pat.* 3 818 109.
Box, G. E. P. and N. R. Draper (1969). *Evolutionary Operations*. Wiley, London.
Brewer, D., G. P. Arsenault, J. L. C. Wright and L. C. Vining (1973). Production of bikaverin by *Fusarium oxysporium* and its identity with lycopersin. *J. Antibiot.*, **26**, 778–783.
Bu'Lock, J. D. (1974). Secondary metabolism of microorganisms. In *Industrial Aspects of Biochemistry*, ed. B. Spencer, pp. 335–346. North Holland/American Elsevier, Amsterdam.
Bu'Lock, J. D. (1975). Secondary metabolism in fungi and its relationship to growth and development. In *The Filamentous Fungi*, ed. J. E. Smith and D. R. Berry, vol. 1, pp. 33–50. Edward Arnold, London.
Bu'Lock, J. D., R. W. Detroy, Z. Hostalek and A. Munim-al-Shakarchi (1974). Regulation of secondary biosynthesis in Gibberella. *Trans. Brit. Mycol. Soc.* **62**, 377–389.
Carslaw, H. S. and J. C. Jaeger (1959). *Conduction of heat in solids*. Clarendon Press, Oxford.
Chatfield, C. (1970). *Statistics for Technology*. Chivers–Penguin, London.
Cole, M. (1966). Microbial synthesis of penicillins. *Process Biochem.* **1**, 334–338.

Corbett, K. (1980). Preparation, sterilisation and design of media. In *Fungal Biotechnology*, ed. J. E. Smith, D. R. Berry and B. Kristiansen, pp. 25–41. Academic, London.

Demain, A. L., J. F. Newkirk and D. Hendlin (1963). Effect of methionine, norleucine and lysine derivatives on cephalosporin C formation in chemically defined media. *J. Bacteriol.*, **85**, 339–344.

Hockenhull, D. J. D., G. D. Wilkin and A. R. J. Quilter (1955). Penicillin. *Br. Pat.*, 730 185.

Kominek, L. A. (1972). Biosynthesis of novobiocin by *Streptomyces niveus*. *Antimicrob. Agents Chemother.*, **1**, 123–134.

Kupletskaya, M. B., V. N. Maksimov and T. B. Kasatkina (1969). Selection of a medium for the biosynthesis of gramicidin C by the method of steepest ascent. *Appl. Biochem. Microbiol.*, **5**, 541–548.

Lapage, S. P., J. E. Shelton and T. G. Mitchell (1970). Media for the maintenance and preservation of bacteria. In *Methods in Microbiology*, ed. J. R. Norris and D. W. Ribbons, vol. 3A, pp. 1–133. Academic, London.

Lockwood, L. B. (1979). Production of organic acids by fermentation. In *Microbial Technology*, ed. H. J. Peppler and D. Perlman, vol. I, pp. 353–387. Academic, New York.

McDaniel, L. E., E. G. Bailey, S. Ethiraj and H. P. Andrews (1976). Application of response surface optimisation techniques to polyene macrolide fermentation studies in shaken flasks. *Dev. Ind. Microbiol.*, **17**, 91–98.

Mantle, P. G., L. J. Morris and S. W. Hall (1969). Fatty acid composition of sphaecelial and sclerotial growth forms of *Claviceps purpurea* in relation to the production of ergoline alkaloids in culture. *Trans. Br. Mycol. Soc.*, **53**, 441–447.

Marchesini, A. (1966). Effect of chelating agents on the solubility of soil phosphorus. IV. Relationship between phosphorus availability and growth of *Aspergillus niger*. *Ann. Microbiol. Enzymol.*, **16**, 147–151.

Martin, J. F. (1977). Control of antibiotic synthesis by phosphate. *Adv. Biochem. Eng.*, **6**, 105–127.

Martin, J. F. and A. L. Demain (1978). Fungal development and metabolite formation. In *The Filamentous Fungi*, ed. J. E. Smith and D. R. Berry, vol. III, pp. 426–450. Edward Arnold, London.

Martin, J. F. and A. L. Demain (1980). Control of antibiotic biosynthesis. *Microbiol. Rev.*, **44**, 230–251.

Moyer, A. J. and R. D. Coghill (1946). Penicillin. X. The effect of phenylacetic acid on penicillin production. *J. Bacteriol.*, **53**, 329–341.

Perlman, D. (1979). Microbial processes for riboflavin production. In *Microbial Technology*, ed. H. J. Peppler and D. Perlman, vol. I, pp. 521–527. Academic, New York.

Pirt, S. J. (1975). *Principles of Microbe and Cell Culture*, Blackwell, Oxford.

Schreier, K. (1977). The 4000 m^3 fermenter for petro-protein production *Abstracts of the 4th FEMS Symposium, Vienna.*, p. 1331.

Soltero, F. V. and M. J. Johnson (1953). Effect of the carbohydrate nutrition on penicillin production by *Penicillium chrysogenum* Q-176. *Appl. Microbiol.*, **1**, 52–57.

Spalla, C., S. Filippini and A. Grein (1978). A hypothesis on the regulation mechanisms governing the biosynthesis of alkaloids in *Claviceps*. *Folia Microbiol.*, **23**, 505–508.

Straube, G. and W. Fritsche (1973). Influence of iron concentration and temperature on growth and riboflavin overproduction of *Candida guilliermondii*. *Biotechnology and Bioengineering, Symposium No. 4*, pp. 225–231.

Weinberg, D. (1978). Secondary metabolism; regulation by phosphate and trace elements. *Folia Microbiol.*, **23**, 496–504.

9

Nutrient Uptake and Assimilation

I. S. HUNTER
Pfizer Central Research, Sandwich, Kent, UK

9.1 INTRODUCTION

All living cells take up nutrients and make products. Many aspects of biotechnology are concerned with the use of cells to make commercial products. For example: (1) Cells may be grown for *biomass, e.g.* single cell protein; new cells are the major product of this fermentation. (2) The commercial product may be a *macromolecular constituent* of the cell, *e.g.* enzymes. (3) A *metabolite* may be the major product. Metabolites may be classified as 'primary metabolites', *e.g.* ethanol, organic acids, or 'secondary metabolites', *e.g.* antibiotics.

In each case there is flow of nutrients into the cell, conversion of the nutrients into cellular intermediates, and conversion of these intermediates into products. Specific details of the biochemical pathways which carry out the interconversions are considered elsewhere in standard texts. The mechanisms by which the nutrients are taken up, and the fluxes involved in their assimilation are the subject of this chapter.

9.2 NUTRIENT UPTAKE

There are two major problems facing a cell which wishes to take up nutrients. (1) The cell membrane is hydrophobic; most nutrients (present in aqueous environments) are hydrophilic. (2) The nutrients are usually present in the environment in vanishingly small concentrations. (The laboratory situation is the major exception to this.) The nutrient may have to be taken into the cell 'uphill' against a concentration gradient; free energy will have to be spent on its accumulation.

It is convenient to classify mechanisms of nutrient uptake by the methods the cells use to overcome these two problems. The examples quoted are invariably for uptake of carbon sources or amino acids since they have been studied most. The subject of nutrient uptake has been reviewed extensively. The contributions by Hamilton (1975), and Hamilton and Booth (1980) on bacterial uptake and by Eddy (1982) on yeast uptake are particularly noteworthy.

9.2.1   Simple Diffusion

If a nutrient molecule is sufficiently lipophilic or small enough there is no problem. The cellular membrane does not present a barrier to the molecule (Figure 1) which enters the cell down its electrochemical potential (*i.e.* concentration) gradient. Examples of this type are poorly described; it is much more difficult to prove that an uptake system does not exist for a particular nutrient than to prove the converse, for example a carrier molecule has yet to be identified for acetate, although *E. coli* can grow on acetate readily. It is likely that many organic compounds may enter by this route.

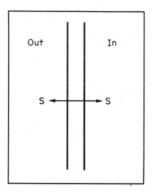

Figure 1   Uptake of nutrient S by simple diffusion

It has been suggested that solute uptake previously assigned to a passive diffusion mechanism was, in some instances, due to recruitment of an alternative transport system (reviewed in Payne, 1980), for example Cowell (1974) assigned 10% of the measured uptake of glycylglycine by *E. coli* to passive diffusion; this uptake was probably due to the oligopeptide transport system.

The subject of passive diffusion in yeast has been discussed in a particularly searching review (Fiechter *et al.*, 1981). The plasma membrane allows limited diffusion of some solutes. Linear sugars, *e.g.* D-ribose, penetrate well, but hexoses are sterically incapable of entering passively. In *Saccharomyces cerevisiae*, entry of xylitol, ribitol, mannitol and sorbitol by facilitated diffusion through the plasma membrane has been demonstrated (Canh *et al.*, 1975).

It will be apparent that translocation of nutrient molecules by this mechanism can only occur 'downhill', that is for uptake to occur the concentration of the nutrient outside the cell has to be greater than or equal to that inside the cell.

9.2.2   Transport Systems

The word 'permease' is often used erroneously to describe a system responsible for uptake of a nutrient. The term was coined (*e.g.* Cohen and Monod, 1957) when one protein species was assumed to be responsible for translocation of the nutrient. Subsequent reports (reviewed in Harold, 1972; Hamilton, 1975) have shown that, in general, this was an oversimplification; in some cases, many gene products may be involved as in maltose transport (see Section 9.2.2.2(ii)). The term 'transport system' is used here since it is a general one which assumes nothing about the number of protein species or the mechanism of translocation.

9.2.2.1   Facilitated diffusion

Facilitated diffusion is the simplest type of transport system. A hydrophilic nutrient molecule is taken up through the cell membrane (which would otherwise be impermeable to the nutrient) by interacting in some way with a membrane protein (or proteins) which facilitates its entry. As in simple diffusion, no energy is expended on transport. The nutrient enters the cell down its electrochemical potential gradient, therefore the intracellular concentration of nutrient taken up by such a system is always less than (or at equilibrium with) the extracellular concentration (Figure 2). So far, only uncharged solutes have been shown to be taken up by this mechanism. Although a charged molecule could be taken up by facilitated diffusion, it is possible that it would not be recognized as such because the charged 'back potential' would limit initial entry and would oppose further entry by the solute.

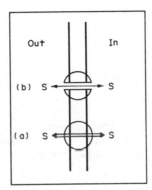

Figure 2 Schematic diagram for uptake of nutrient S by facilitated diffusion, drawn to accommodate either (a) the protein translocator or (b) pore models

Few transport systems of this type have been characterized. The classical example is glycerol uptake. Transport of glycerol in microorganisms appears to occur universally by this mechanism (see Lin, 1976, for an extensive review). The best-characterized system is *E. coli*, where Hayashi and Lin (1965) originally showed that ^{14}C-glycerol was taken up but not accumulated in glycerokinase-negative mutants. The facilitator function was subsequently characterized by osmotic techniques (Sanno *et al.*, 1968). Originally it was thought that the gene product of the GlpT locus was the facilitator protein (Richey and Lin, 1972). Considerable confusion arose subsequently about the role of GlpT and associated proteins. However, the controversy has been cleared up recently (Larson *et al.*, 1982); the facilitator locus is now recognized as GlpF. Kinetic analysis of glycerol uptake has suggested that the solute may enter through a pore in the cytoplasmic membrane in which GlpF is a component (Heller *et al.*, 1980). *Klebsiella pneumoniae* has a system for glycerol dissimilation similar to *E. coli* (Jin *et al.*, 1982).

In *S. cerevisiae*, Romano (1982) has interpreted data on 6-deoxyglucose uptake in terms of facilitated diffusion for glucose. Earlier, Fuhrmann *et al.* (1976) had found several solutes which entered yeast by facilitated diffusion.

The maltose channel of the λ receptor and porin can be envisaged as facilitated diffusion systems (Wandersmann *et al.*, 1979) but these proteins mediate entry only across the outer membrane of the bacterial cell. The facilitators such as GlpF described above catalyze uptake across the inner membrane of the bacteria which is the true diffusion barrier.

Under certain conditions, protein components of the more complex group translocation mechanism may transport nutrients on their own, without energy expenditure in the facilitated diffusion mode. They are discussed in Section 9.2.2.2(ii).

9.2.2.2 *Active transport*

In active transport, the nutrient molecule is taken up through the membrane into the cell in a chemically unaltered state. However, unlike facilitated diffusion, the intracellular concentration of the nutrient may be higher than that outside the cell. Energy has to be expended to translocate nutrients into the cell against a concentration gradient. Active transport systems can be classified according to the source of free energy used to transport their substrates against a concentration gradient.

(i)   Active transport linked to cotransport of an ion

The nutrient is transported actively against its chemical potential either by fluxing or counter-fluxing with another species which is itself moving down its electrochemical potential (concentration) gradient.

Ions which are cotransported in the same direction as the nutrient are said to *symport* (Figure 3a) with the nutrient: those which move in the opposite direction are said to *antiport* (Figure 3b). A charged nutrient may enter on it own, the energy coming from electrostatic interactions with the membrane potential of the cell; in that case the nutrient is said to *uniport* (Figure 3c). To understand this mechanism fully, one has to become familiar with the concept of ion gradients across cellular membranes as the link between electron transport and oxidative phosphorylation/nutrient uptake.

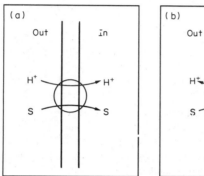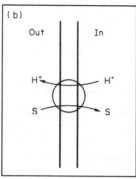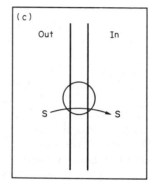

Figure 3   Schematic diagram for uptake of solute S by *symport* (Figure 3a), *antiport* (Figure 3b), and *uniport* (Figure 3c); the illustration is for a proton as the cotransported species

It is now generally accepted that the hypothesis of Mitchell (1970, 1977) best explains the situation. Electron carriers of the electron transport chain are organized vectorially within the membrane (Figure 4). Passage of electrons along the chain to the terminal acceptor (which is oxygen if the cells are growing aerobically) results in protons being pumped out across the membrane. The proton gradient or *proton motive force* (PMF) can then drive ATP synthesis *and/or* nutrient uptake.

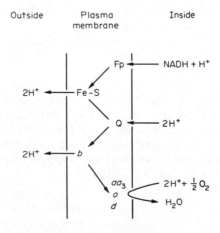

Figure 4   Electron transport from NADH to oxygen in *Escherichia coli*

The proton motive force (Δp) therefore has two components: a concentration factor (ΔpH) which is the pH gradient across the membrane, and a charge component ($\Delta \psi$) which is the membrane potential. The PMF is described thermodynamically by equation (1).

$$\Delta p = \Delta \psi - Z\Delta pH \tag{1}$$

In equation (1) Δp and $\Delta \psi$ are expressed in millivolts. Z is a constant (59 at 25 °C) to convert pH units to millivolts.

The proton gradient may be converted into gradients of other ions by carriers which exchange protons for inorganic ions. The existence of Na^+/H^+ antiports has been demonstrated in *E. coli* (West and Mitchell, 1974).

(*a*) *Active transport linked to proton cotransport.* Of all the classes of transport system, this type has been studied most intensively. Perhaps the best understood system of the class is the β-galactoside (*lac*) transport system. The pioneering work on *lac* transport initiated by Monod has been continued by many others. Uptake of thiomethylgalactoside (TMG) by *E. coli* under anaerobic conditions required an intact membrane which was impermeable to protons (Pavlasova and Harold, 1969). Adding uncouplers (which made the membrane permeable to protons) affected transport without inhibiting the generation or utilization of ATP, which was produced by glycolysis under these anaerobic conditions. Direct involvement of ATP in this type of mechanism was, therefore, unlikely. Until then it had been thought that ATP was always involved directly in active transport. It was West (1970) who first demonstrated cotransport of lactose with protons. More rigorous (West and Mitchell, 1972) and quantitative experiments (West and Mitchell, 1973) established that the proton:lactose stoichiometry was 1:1 and that transport involved a net charge displacement. A mutant which could no longer *accumulate* solute against a concentration gradient was shown to have lost the ability to cotransport protons with solute (West and Wilson, 1973). Thus, the obligatory role between accumulative uptake and proton translocation was indicated.

A theoretical prediction of the chemiosmotic mechanism for active transport has been confirmed experimentally for resting cells of *Staphylococcus aureus* (Niven and Hamilton, 1974). Amino acids with different charges were used as models for nutrient uptake and energy-depleted cells were manipulated to generate pH gradients (or, in separate experiments, to have membrane potentials) similar to those of metabolically-active bacteria. The results, considered in terms of the gradients present in metabolically-active bacteria, could be interpreted as follows. (1) Lysine, which is positively charged, was taken up by the transport system by a uniport mechanism (Figure 5a). Uptake was driven only by the membrane potential ($\Delta \psi$) component of the PMF. The net charge of the carrier/lysine complex was positive, so transport can be envisaged as being driven electrostatically by the membrane potential by repulsion from the outside (positive) and attraction to the inside (negative). (2) Uptake of isoleucine, which is itself neutral, occurred by proton symport (Figure 5b). The net charge on the carrier/isoleucine/H^+ complex was positive. Hence transport could be driven by both components of the proton motive force: electrostatically by the membrane potential and by the pH gradient as the proton returned through the membrane down its own electrochemical potential gradient from the outside where the proton concentration was high, to the inside where the proton concentration was lower. In teleological terms, the isoleucine 'hitched a ride' on the transport carrier which was being driven by the return of the proton. (3) Uptake of glutamate which is negatively charged was by proton symport. The carrier/glutamate/H^+ complex was electrically neutral (Figure 5c), therefore the membrane potential had no influence on it. Glutamate transport was driven solely by the ΔpH.

Similar conclusions were drawn later for carbohydrate transport. Transport of galactose, which is uncharged and therefore similar to isoleucine, was driven by both the pH gradient and membrane potential (Flagg and Wilson, 1976) while uptake of 2-oxo-3-deoxygluconate (negatively charged and therefore similar to glutamate) was driven by the membrane potential alone (Lagarde and Haddock, 1977).

These qualitative demonstrations were a major factor in verifying the chemiosmotic mechanism for nutrient uptake in bacteria. Work in Hamilton's laboratory has gone on to attempt to *quantitate* the relationship between the PMF and amount of solute accumulated (Hamilton and Booth, 1980).

The study and elucidation of the mechanism of proton-linked solute transport was assisted greatly by the development of membrane vesicles (reviewed by Kaback, 1976). In principle, these vesicles are the ideal experimental system on which to study proton-coupled transport. Since the vesicles are devoid of intracellular enzymes, there is no interference from subsequent metabolism of the solute after it has been transported. However, the preparation of the vesicles often results in the presence of atypical membranes within the total population; during preparation some proteins are translocated from the inner to the outer membrane surface (Adler and Rosen, 1977). Membrane-bound NADH oxidase and ATPase are translocated whereas lactate dehydrogenase is not. It is possible the membrane carriers under investigation will also be translocated during vesicle preparation. The current state of the art on vesicles has been put succinctly by Hamilton

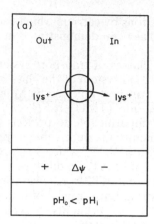
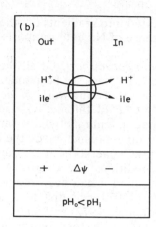
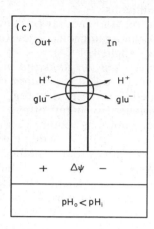

Figure 5   Uptake of lysine (Figure 5a), isoleucine (Figure 5b) and glutamate (Figure 5c) in *S. aureus* (Niven and Hamilton, 1974)

and Booth (1980): 'Whilst vesicles have played a significant role in studies of the mechanism of membrane transport, they should only be used with a clear vision of their shortcomings.'

The availability of a good genetic system in *E. coli* has allowed mutants to be used to investigate the mechanism of transport. The mutants of West and Wilson (1973) which no longer cotransported protons with lactose were discussed above. Loss of proton symport converted the active transport system into one with the characteristics of facilitated diffusion.

During prolonged slow growth in the chemostat, mutants of *E. coli* were isolated (Collins *et al.* 1976) which were adventitiously capable of higher accumulation of amino acids. This was thought to occur by a change in the amino acid:proton stoichiometry from 1:1 to 1:2 and eventually 1:4, although integer values were never achieved. The inference here was that, in the nutrient-limiting environment of the chemostat, increasing the stoichiometry of the transport system conferred a selective growth advantage.

Many studies on proton-linked transport have quoted integer values for proton cotransport and indeed many investigators seem to have gone to great lengths to obtain integer stoichiometries. This is envisaged, conceptually, in terms of protonation and deprotonation reactions of the carrier. However, Lagarde (1976) has presented a convincing criticism of this misconception using a non-equilibrium thermodynamic approach; the values need not be integers. The glucose 6-phosphate:H^+ stoichiometry was shown to increase at alkaline pH (Le Blanc *et al.*, 1980). In some instances the change in stoichiometry can be rationalized, for example in *E. coli* membrane vesicles at high pH, electroneutral Na^+/H^+ antiport became an electrogenic system when more than one H^+ was exchanged for each Na^+ (Schuldiner and Fishkes, 1978). The result was an apparent change in stoichiometry. In the yeast *Rhodotorula gracilis*, two carriers were found to operate simultaneously, one carrier operated by the H^+ route and the other not. A change in their relative activities appeared as a variable stoichiometry (Hauer and Hofer, 1982).

The discussion so far has been restricted mainly to *E. coli*. Many nutrients in this organism are taken up by proton symport (tabulated by Harold, 1977), but other species and genera use proton cotransport, *e.g. Streptococcus faecalis* (Ashgar *et al.*, 1973), *S. lactis* (Kashket and Wilson, 1974), *Clostridium pasteuranium* (Booth and Morris, 1975), *Mycobacterium phlei* (Hinds and Brodie, 1974) and *Chlorella vulgaris* (Tanner, 1974 for a review). Transport in yeast has been shown to occur by this mechanism (Seaston *et al.*, 1976; review by Eddy, 1978, 1982).

(*b*) *Active transport linked to movement of other ion species.* There is ample evidence in higher eukaryotic systems for cotransport of solutes with other inorganic ions. In microorganisms too, examples have been discovered. The ion gradient used for transport is the result of conversion of the proton gradient, usually *via* antiport systems described in Section 9.2.2.2(i). The ion gradient is a '*secondary*' gradient, the proton gradient being the '*primary*' event. Solute uptake in this context is said to be '*secondary transport*'.

Na^+ and K^+ ions have been shown to be required for transport of glutamate in *E. coli* (Halpern *et al.*, 1973) and of γ-aminobutyrate in a marine pseudomonad (Drapeau *et al.*, 1966). Na^+ was required for transport of melibiose in *S. typhimurium* (Stock and Roseman, 1971). In *E. coli*, Na^+ is involved in both transport and binding of melibiose (Cohn and Kaback, 1980). Na^+

is also required for transport of glutamate in *B. licheniformis* (MacLeod *et al.*, 1973); the transport of citrate in *K. aerogenes* has been shown to require K^+ (Eagon *et al.*, 1972); Mg^{2+} has been shown to be required for the transport of citrate in *B. subtilis* (Willecke *et al.*, 1973).

(ii)   Active transport involving periplasmic binding proteins

Activities of some transport systems are reduced drastically by osmotic shock (Heppel *et al.*, 1972) due to loss of binding proteins from the periplasmic space. Berger (1973) using an ATPase negative mutant of *E. coli*, showed that the energy requirement for transport of glutamine (a shock-sensitive system) could be satisfied only by ATP or a 'high-energy' compound related to ATP, whereas the energy for transport of proline (a shock-resistant system) came from an 'energized membrane state', now recognized as the proton motive force. The mutant could not generate ATP from electron transport *via* the proton motive force because it was ATPase-negative. ATP could only be generated by substrate-level phosphorylation, mainly from glycolysis. So, the PMF and ATP synthesis were separated in this strain. In the conclusive experiment, the mutant was first starved of endogenous energy reserves and then lactate was added which resulted in the generation of a PMF. Only proline uptake was stimulated; glutamate was not taken up. Addition of glucose, which resulted in the generation of both a PMF from NADH oxidation *and* ATP from glycolysis stimulated both glutamate and proline uptake. Berger and Heppel (1974) subsequently showed with the ATPase mutant that when a PMF was generated using ascorbate–phenazine methosulfate as the electron donor, proline transport occurred but glutamate transport was not stimulated. Hong *et al.* (1979) have implicated acetyl phosphate as the source of the high-energy compound, but except for that report the nature of the compound has remained elusive.

A large number of transport systems of this type have been identified, *e.g.* diaminopimelic acid, arginine, histidine, ornithine (Berger and Heppel, 1974), isoleucine (Kobayashi *et al.*, 1974), glycylglycine (Cowell, 1974), glutamate (Miner and Frank, 1974), ribose (Curtis, 1974), glycerol 3-phosphate (Silhavy *et al.*, 1976) and galactose *via* the Mgl system (Wilson, 1974).

The maltose transport system is probably the most intensively studied of this class. It transports maltodextrins as well as maltose. The periplasmic binding protein MalE was identified as being essential for the uptake of maltose (Kellerman and Szmelcman, 1974). The system is complex, with other proteins involved: MalF is membrane-bound (Shuman *et al.*, 1980); MalK and MalG are also essential. The receptor for phage λ located on the outer membrane was identified as an element of the uptake mechanism (Szmelcman *et al.*, 1976) and subsequently shown (Wandersman *et al.*, 1979) to be an essential component only when the molecule is of size greater than maltotriose. The λ receptor is thought to form a pore in the lipoprotein of the outer membrane which facilitates the entry of the maltodextrins into the periplasmic space. Maltose is small enough to pass through the outer membrane passively on its own. The example of maltose has shown that periplasmic binding systems are complex in molecular identity and architecture. As other systems are submitted to scrutiny they may prove to be equally complex.

The periplasmic binding proteins are also important as receptors in the chemotactic response of bacteria (see Koshland, 1977 for a review). Mutants which lacked the periplasmic binding protein for maltose (Hazelbauer, 1975) and galactose (Ordal and Adler, 1974) failed to exhibit a chemotactic response to them. However, not all chemotactic receptors are binding proteins, *e.g.* the enzymes II of the mannose and glucose PT systems (Section 9.2.2.2(iii)(a)) are responsible for the chemotactic response to these solutes (Adler and Epstein, 1974).

(iii)   Group translocation

Since the initial discovery of the phosphotransferase (PT) group translocation system (Kundig *et al.*, 1964), PT systems have been studied intensively, particularly in *E. coli* and *S. typhimurium*. Uptake of solute by these systems is accompanied by phosphorylation. Thus, the solute enters the cytoplasm in a chemically-altered state; by contrast to the systems discussed earlier, the substrate and product are chemically as well as vectorially distinct.

At least four proteins are required to effect translocation and phosphorylation. The identification of the components, the specificities of the enzymes and the elucidation of their mechanisms of action have been reviewed (Hamilton, 1975; Kornberg and Jones-Mortimer, 1977). The phosphate group is supplied by phosphoenolpyruvate (Figure 6). A cytoplasmic enzyme, E-I, catalyzes the transfer of the phosphate group from phosphoenolpyruvate to HPr, a small, heat-stable, soluble protein. This part of the mechanism is common, irrespective of the PT substrate to be translocated (equation 2).

$$PEP + HPr \xrightarrow{\text{E-I} + Mg^{2+}} pyruvate + HPr\text{-}P \qquad (2)$$

Since the proteins are cytoplasmic, their action can be studied in detail, for example in the presence of Mg^{2+} and PEP the E-I dimerizes to its active form (Misset *et al.*, 1980).

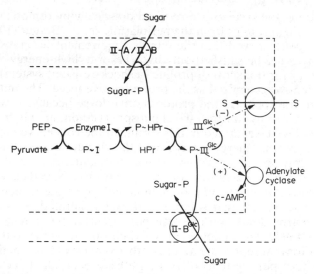

Figure 6   The glucose PT system of *Salmonella typhimurium* (Nelson *et al.*, 1982)

The next step in the mechanism involves proteins which are sugar-specific and within the membrane, making them more difficult to study carefully. However, the discovery of sugar phosphate exchange by membranes, in which a sugar is phosphorylated at the expense of dephosphorylation of a second sugar phosphate species has been used as a tool to look at membrane interactions (equation 3) (Saier *et al.*, 1977).

$$\text{sugar-1} + \text{sugar phosphate-2} \rightleftharpoons \text{sugar phosphate-1} + \text{sugar-2} \tag{3}$$

This phenomenon has been named exchange group phosphorylation (Raphaeli and Saier, 1978) and is best assayed by radiolabelling one of the sugars.

The phosphate group is passed to a membrane-bound protein (IIB), which is responsible for phosphorylation of the substrate. The transfer of phosphate to protein IIB is associated either with phosphorylation of a membrane-bound protein (IIA) or of a cytoplasmic enzyme (E-III), in which case the membrane-bound component is called IIB′ (Figure 6). In *E. coli*, the E-III/ IIB′$_{glucose}$ route is the main one for glucose uptake. The route is synonymous with the genetic locus PtsG. The molecular nature of PtsG mutants has yet to be described, but they are likely to be impaired in the IIB′$_{glucose}$ protein. Although blocked in the major glucose uptake system, PtsG mutants can still take up glucose but at a slower rate. This occurs *via* the IIA/IIB$_{glucose}$ system. The genetic locus PtsM (Jones-Mortimer and Kornberg, 1976) is probably synonymous with one or other of the proteins IIA/IIB$_{glucose}$. The PtsG system has a high affinity for glucose ($K_m <$ 10 μM) whereas the PtsM system has a lower affinity ($K_m \sim 1$ mM); the presence of two transport systems with differing affinities for one substrate will be discussed later (Section 9.2.3).

Other solutes taken up by PT systems include mannitol, sorbitol, galactitol, fructose and *N*-acetylglucosamine. The membrane-bound enzymes-II are unique for each substrate; their structural genes are found within the operons for the metabolism of the respective sugars, whereas the proteins common to all substrates (E-I, HPr) are found in a separate operon (Raphaeli and Saier, 1978). The PT system has only been found to transport sugars to date, with one notable exception: AMP has been reported (Yagil and Beacham, 1975) to be translocated by a phosphotransferase system in which the AMP is first dephosphorylated to adenosine in the periplasm then rephosphorylated by the PT system. Lactose is taken up by a PT system in *S. typhimurium* but is transported by proton symport in *E. coli*, therefore one cannot generalize on the mode of uptake for one particular compound amongst the bacterial species.

The PT mechanism for sugar uptake is widely distributed amongst the bacteria but not amongst the fungi. Romano *et al.* (1970, 1979) surveyed a large number of bacterial species and concluded that PT activity was usually present in facultative anaerobes: non PT-containing species were usually obligate aerobes. There have been reports of investigations of PT activity in *Bacillus*

(Saheb, 1972), *Klebsiella* (O'Brien *et al.*, 1980), *Streptococcus* (Vadebancoeur and Trahan, 1982) and *Pseudomonas* (Tsay *et al.*, 1971). The study of the genetics of the fructose PT system in *Ps. aeruginosa* has just begun (Roehl and Phibbs, 1982).

Currently, attempts are being made to purify the components of the PT system to investigate their enzymology thoroughly. Several approaches have been taken. Proteins have been purified from the cells directly, *e.g.* the E-III$_{glc}$ of *S. typhimurium* (Scholte *et al.*, 1981), and from mini-cells in the case of the E-II$_{mannitol}$ of *S. typhimurium* (Jacobsen *et al.*, 1979). The cloning of the genes for E-I and HPr has been reported recently (Lee *et al.*, 1982; Bitoun *et al.*, 1983). The glucose PT system is subject to intense interest because it mediates control over carbon dissimilation by the cell (Section 9.2.2.2(iii)(b)) and this has an inevitable influence on assimilation. With reagent quantities of the purified proteins available, advances should soon be made on the understanding of the molecular interactions of the system.

(*a*) *Facilitated diffusion mediated by proteins of the group translocation system*. It has been reported that for *E. coli* (Kornberg and Riordan, 1976) and *S. typhimurium* (Postma, 1976) some of the PT proteins may operate independently, without phosphorylation, to facilitate the entry of galactose into the cell. This unusual situation was recognized in strains which lacked the major uptake systems for galactose (GalP$^-$, Mgl$^-$). There was no *active* uptake of galactose; in galactokinase-negative mutants, the substrate was taken up only until it equilibrated with the external galactose. The novel transport activity had a poor affinity (high K_m) for galactose. In *E. coli*, the protein responsible for uptake was identified genetically as being close to the PtsG locus, and tentatively designated *tgl* (Bourd *et al.*, 1975; Kornberg and Jones-Mortimer, 1977).

These earlier papers were criticized subsequently by Postma and Stock (1980) for not establishing that all residual PT activity was absent, or for not checking that an alternative system was present. They claimed that the PT system cannot operate as a facilitator. The criticism of Kornberg and Riordan's work seems particularly unjustified. However, recently another elegant demonstration of non-phosphorylative uptake by a PT system *has* been reported (Postma, 1981): mutants deleted in E-I and HPr are incapable of using glucose since the phosphorylative pathway is missing. Glucose-positive revertants of the mutants were shown to be altered in PtsG. The apparent K_m for glucose had increased 1000-fold in the revertant. When E-I and HPr activities were restored by transduction the revertant was unable to transport glucose with concomitant phosphorylation. Thus, the mutation of the revertant allowed PtsG to operate in a new mode by facilitated diffusion to the exclusion of the phosphorylative step. These mutants should be useful in examining the structure–activity role of the proteins at the molecular level.

(*b*) *Regulation of transport mediated by the glucose phosphotransferase system*. The phenomenon of glucose dominance over lactose utilization (Monod, 1942) is often the subject of undergraduate practical classes. Not only does glucose repress the synthesis of β-galactosidase, the first enzyme of lactose utilization, but it also inhibits the uptake of lactose, a proton-linked system (McGinnis and Paigen, 1969). This simple experiment has been extended to show that glucose metabolism also dominates uptake of glycerol (a facilitated diffusion system) and expression of the enzymes of glycerol dissimilation (Edgar *et al.*, 1973), uptake *via* shock-sensitive systems, *e.g.* galactose (Adhya and Echols, 1966), and even uptake by other group translocation systems, *e.g.* fructose (Amaral and Kornberg, 1975). Glucose dominance manifests itself in four separate ways (listed below), some of which are more important than others.

(1) Competition exists amongst the membrane-bound carrier proteins for supply of phosphate from HPr-P. By virtue of its kinetic constants, the E-III$_{glucose}$ protein competes very effectively for HPr-P against the other IIA proteins. Therefore if glucose and fructose are available to the cell, the glucose will be used in preference since the E-III$_{glucose}$ (*i.e.* the PtsG system) is more efficient at capturing the phosphate group of the HPr-P than the E-IIA$_{fructose}$.

(2) Sugar phosphates (*e.g.* glucose 6-phosphate and fructose 6-phosphate; Kornberg, 1973) can sterically inhibit the operation of other transport systems. Since sugar phosphates are the products of uptake by the PT system, the result is dominance by the group translocation system.

(3) A protein component of the PT system may interact with a non-PT transport system to inhibit its activity. This phenomenon has been called 'inducer exclusion'. The topic has been reviewed by Dills *et al.* (1980). Non-phosphorylated E-III$_{glucose}$ is believed to inhibit the second transport system in a manner which is yet to be elucidated (Saier *et al.*, 1978). Presumably a complex is formed between E-III and the transport system which renders the latter catalytically inactive. In support of this hypothesis, Nelson *et al.* (1982) have shown in *S. typhimurium* that lactose transport (a PT-sensitive system) is no longer sensitive to inducer exclusion when it is induced at high level in a mutant. The E-III protein is titrated out by the excess of PT-sensitive protein. So, the membrane contains both the inhibited complex with E-III bound to it, and free lactose trans-

port protein which is catalytically active. The relative amounts of E-III and the lactose carrier, together with the phosphorylation state of the E-III all contribute to the final effect.

(4) The first three methods of regulation described all involve inhibition by the glucose PT system of pre-existing transport activity. The fourth method involves modulation of the *synthesis* of the second transport activity. Glucose repression of the synthesis of operons coding for the proteins of uptake and dissimilation of other carbon sources (Magasanik, 1961) was shown to be reversed by the addition of cyclic 3′,5′-AMP (cAMP) to the cells (Perlman and Pastan, 1968). The direct effect of glucose on inhibition of cAMP synthesis in whole cells (Peterkofsky and Gazdar, 1974) and subsequently *in vitro* on toluenized cell preparations (Harwood and Peterkofsky, 1975) established conclusively the relationship between glucose entry and decreased cAMP levels resulting in repression of cAMP-sensitive operons. The mechanism by which this is manifest has been the subject of great debate. The reduced level of cAMP within the cell was shown to be due to decreased activity of adenylate cyclase (Harwood *et al.*, 1976). It is now believed that adenylate cyclase is stimulated by phosphorylated E-III (RPr in the model of Saier and Stiles, 1975). Consequently, when the PtsG system is working, phosphorylated E-III usually available to stimulate the adenylate cyclase will be diverted to PtsG. The net effect is to reduce the activity of the adenylate cyclase and the level of cAMP is reduced as a result.

Finally, lest it be thought that the glucose PT system of the enteric bacteria is itself immune from regulation, it appears that an elevated redox potential can inhibit the system: the membrane-bound E-II loses the ability to bind the sugar substrate (Robillard and Konings, 1981).

9.2.3 Redundancy of Transport Systems

The presence within the membrane of two or more distinct transport systems for the same substrate appears at first to be wasteful in terms of protein synthesis; *E. coli*, for example, has two specific galactose transport systems, GalP and Mgl (Henderson *et al.*, 1977). The GalP system is proton-linked and has a lower affinity for galactose than the Mgl system which operates by a shock-sensitive mechanism. Similarly, in *Ps. aeruginosa* branched-chain amino acids are transported by two system: LIV I, a high affinity shock-sensitive complex, and LIV II, an ion driven system of lower affinity (Hoshino, 1980). Amongst these and other similar reports, a pattern has emerged. The shock-sensitive systems have a high affinity for their substrate but a lower velocity of uptake. By contrast, the ion-driven systems generally have a poorer affinity but have higher velocities. Although it could never be proven, the redundancy of transport systems may be an evolutionary quirk to ensure adaptability to environmental conditions. If the nutrient is in short supply, the high affinity periplasmic system captures it efficiently. If the nutrient is present in excess, a fast ion-driven system will supply nutrients for rapid growth, effectively competing against other organisms present in the environment. In some fermentation situations, the presence of one transport system and the absence of the other can improve the efficiency of growth (Hunter *et al.*, 1981).

9.3 ASSIMILATION

With the nutrient taken up into the cytoplasm, the next problem facing the cell is to assimilate it and incorporate it into new cell material or products. The step-by-step enzymatic routes for assimilation are amply discussed in most biochemistry textbooks. The aim here will be to give an overview of the pathways taken as the nutrients are assimilated. It is convenient to consider assimilation in elemental terms, *e.g.* carbon, nitrogen, oxygen. This fundamental approach not only simplifies the exposition of the problem, but has direct relevance to growth in simple defined media.

With the availability of analytical techniques of increasing sophistication, the current trend in the fermentation industry is to place greater emphasis on process monitoring; the use of a defined medium reduces the number of analytical parameters which have to be measured. In addition, the successful implementation of computer control of a fermentation process is greatly assisted by simplification of the medium. In theory, at least, many fermentation processes could be run on a medium containing a carbon source, inorganic salts and trace metals. In practice, this is not plausible because either (1) a complex medium is available at less cost or (2) the cells to be grown in the fermenter are fastidious and require certain essential additions. The latter restriction is likely to be removed in the future by the application of recombinant DNA technology to introduce new genes into the cells which now would allow the synthesis of these essential components *in vivo*. With regard to the media cost, the increased productivity of a defined medium process which would almost certainly result from better process control using the computer may offset the higher cost of defined media.

The elemental composition of cells is reasonably constant. Although the composition of the macromolecules of the cell may change depending on the growth conditions (Herbert, 1961), the routes taken to make the monomers of the cell are usually the same, for example nucleotide synthesis for RNA or DNA proceeds by the same biochemical reactions irrespective of the growth conditions of the cell. By examination of a cell's average monomer composition (Morowitz, 1968), it is possible to gain an insight into the assimilation routes which will be taken to supply monomers to the cell from the biochemical compounds generated by intermediary metabolism. This approach is similar to that considered by Stouthamer (1973).

9.3.1 Assimilation of Carbon

All of the major macromolecular components of the cell (*e.g.* protein, nucleic acid, lipid, carbohydrate) contain carbon. The assimilation of carbon into the monomers of which the macromolecules are made can be thought of as a supply and demand phenomenon. If, for example, there is insufficient free alanine in the cell, the biosynthesis of alanine proceeds by transamination of pyruvate (equation 4).

$$\text{pyruvate} + NH_3 \rightleftharpoons \text{alanine} \tag{4}$$

One pyruvate molecule will be removed from the cellular pool to make an alanine molecule. Similarly, if AMP is required for RNA biosynthesis or as a precursor to supplement the ATP pool of the cell, the purine biosynthetic pathway will need a supply of carbon to make AMP. The following carbon-containing compounds are required as precursors to synthesize AMP (equation 5).

$$\text{glycine} + \text{phosphoribosyl pyrophosphate} + 2\,\text{formate} + CO_2 \rightarrow \text{AMP} \tag{5}$$

Glucose 6-phosphate is the source of phosphoribosyl pyrophosphate *via* the pentose phosphate pathway, and glycine is synthesized from 3-phosphoglycerate by transamination. So, the carbon requirement for AMP synthesis can be summarized as equation (6).

$$\text{3-phosphoglycerate} + \text{glucose 6-phosphate} + 2\,\text{formate} + CO_2 \rightarrow \text{AMP} \tag{6}$$

A balance sheet can be drawn up for the major components of the cell (Table 1). Clearly, except for some of the carbon atoms in the nucleotides, most of the carbon required by the cell can be assimilated through eight biochemical intermediates: glucose 6-phosphate, triose phosphate, 3-phosphoglycerate, phosphoenolpyruvate, pyruvate, acetyl-CoA, oxalacetate and 2-oxoglutarate. Figure 7 describes the fluxes for assimilation of carbon from glucose. The central role of the Embden–Meyerhof pathway (glycolysis) is evident from inspection of Figure 7. The modulation of activity of the Embden–Meyerhof pathway is reviewed in most biochemical texts.

9.3.2 Assimilation of Nitrogen

Whereas the assimilation of carbon into cells proceeds in parallel through the eight intermediates described above, the cellular assimilation of nitrogen involves only two key compounds, glutamate and glutamine. The subject of nitrogen assimilation has been reviewed extensively in the past (Tyler, 1978; Wheelis, 1975; Tempest *et al.*, 1973).

Ammonia is incorporated into glutamate by glutamate dehydrogenase or glutamine synthetase (Figure 8). When amino acids are the supply of nitrogen to the cell, the amino acid is degraded to glutamate by the classical amino acid degradation pathway or nitrogen is passed to glutamate in transamination reactions. Glutamate is the primary donor for the incorporation of amino groups into amino acids in transamination reactions. This is the main supply of nitrogen to the biosynthetic pathways (equation 7).

$$\text{glutamate} + NADP^+ + RCOCO_2H \rightleftharpoons \text{2-oxoglutarate} + NADPH + RCHNH_2CO_2H \tag{7}$$

Glutamine serves as the donor to, for example, tryptophan, histidine, asparagine, and also purine nucleotides to satisfy the requirement for nitrogen within the monomer skeletons. Therefore all

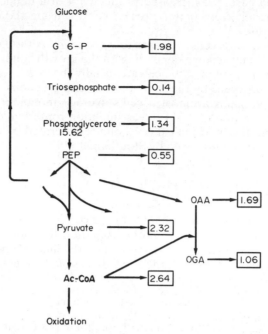

Figure 7 Flux of glucose to the seven key intermediates of carbon assimilation. Abbreviations: Ac-CoA, acetyl-CoA; OGA, 2-oxoglutarate; OAA, oxalacetate; PEP, phosphoenolpyruvate. The numbers in the boxes are the amounts of each intermediate required to make 1 g cells

Table 1 Metabolic Precursors Required to Supply the Monomers of Cells

Monomer	Precursors	Monomer	Precursors
Alanine	Pyruvate	Proline	2-Oxoglutarate
Arginine	2-Oxoglutarate	Serine	3-Phosphoglycerate
Aspartate	Oxalacetate	Threonine	Oxalacetate
Asparagine	Oxalacetate	Tryptophan	Glucose 6-phosphate, 3-phosphoglycerate, pyruvate
Cysteine	3-Phosphoglycerate	Tyrosine	Glucose 6-phosphate, phosphoenolpyruvate
Glutamate	2-Oxoglutarate	Valine	Pyruvate
Glutamine	2-Oxoglutarate	AMP/dAMP	Glucose 6-phosphate, 3-phosphoglycerate
Glycine	3-Phosphoglycerate	GMP/dGMP	Glucose 6-phosphate, 3-phosphoglycerate
Histidine	Glucose 6-phosphate	CMP/dCMP	Glucose 6-phosphate, oxalacetate
Isoleucine	Pyruvate, oxalacetate	UMP	Glucose 6-phosphate, oxalacetate
Leucine	Pyruvate, acetyl-CoA	dTMP	Glucose 6-phosphate, oxalacetate
Lysine	Pyruvate, oxalacetate	Fatty acid	Acetyl-CoA
Methionine	Oxalacetate	Glycerophosphate	Triose phosphate
Phenylalanine	Glucose 6-phosphate, phosphoenolpyruvate	Carbohydrate	Glucose 6-phosphate

cellular nitrogen has to pass through glutamine, glutamate or both in order to be assimilated, unless amino acids are supplied in which case they are incorporated directly.

The usual nitrogen source in a well-controlled fermentation process is ammonia. Two possibilities exist for the incorporation of ammonia into glutamate. Ammonia may be incorporated into glutamate by the reductive amination of 2-oxoglutarate mediated by the enzyme glutamate dehydrogenase (EC. 1.4.1.4.; equation 8).

$$2\text{-oxoglutarate} + NADPH^+ + NH_3 \rightleftharpoons \text{glutamate} + NADP^+ \tag{8}$$

Historically, this enzyme was thought to be the primary pathway for assimilation of ammonia. In fact, the enzyme has a poor affinity ($K_m \sim 1$ mM) for ammonia and in *K. aerogenes* is repressed during growth on a simple glucose/ammonia medium (Meers *et al.*, 1970). Glutamate dehydrogenase-negative mutants of *E. coli* (Berberich, 1972) and *K. aerogenes* (Brenchley and Magasanik,

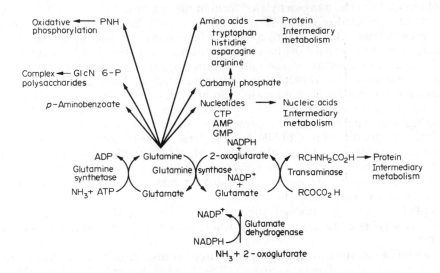

Figure 8 Pathways of ammonia assimilation in the enteric bacteria for the production of glutamate and glutamine and some of the roles of these compounds in intermediary metabolism (Tyler, 1978)

1974) do no require glutamate, suggesting the existence of an alternative pathway for glutamate biosynthesis in these organisms.

The alternative route for ammonia assimilation is by glutamine synthetase (EC. 6.3.1.2.) which, in an ATP-dependent reaction, catalyzes the production of glutamine from ammonia and glutamate (equation 9).

$$NH_3 + \text{glutamate} + ATP \rightarrow \text{glutamine} + ADP \tag{9}$$

This is the main route for ammonia assimilation and is highly regulated (see Section 9.3.2.1). Nitrogen incorporated in this way can be assimilated immediately into the skeletons of precursors of amino acids and nucleic acids for which glutamine is a primary donor. However, in order to supply the nitrogen requirement for the amino groups of the amino acids, glutamine has to be converted into glutamate, the substrate for the transamination reactions. Glutamate synthase (GOGAT; EC. 2.6.1.53) catalyzes the formation of two molecules of glutamate from glutamine and 2-oxoglutarate (equation 10).

$$\text{glutamine} + \text{2-oxoglutarate} + NADPH \rightleftharpoons 2\,\text{glutamate} + NADP^+ \tag{10}$$

9.3.2.1 Control of nitrogen assimilation

On the assumption that assimilative pathways are best controlled at the early (or first) steps, the unique role of glutamine synthetase in nitrogen assimilation makes it a prime candidate for regulation. The flux of nitrogen passing through glutamine synthetase is regulated both by changes in the *amount* of glutamine synthetase protein made and by modulation of the *activity* of the enzyme molecule.

In enteric bacteria the amount of glutamine synthetase made depends on the status of the nitrogen source available. In ammonia-excess, synthesis of the protein is repressed. Ammonia can be taken up by the more energetically-favourable glutamate dehydrogenase route. During ammonia-limitation, synthesis of the protein increased many-fold (Ginsberg and Stadtman, 1973; Brenchley *et al.*, 1975) until the protein was of the order of 1% of the cellular protein.

Once the protein is synthesized, the activity may be modulated in three ways which are interrelated: (1) activation by the presence of divalent cations, particularly Mn^{2+}; (2) feedback inhibition of the enzyme by some of the products to which nitrogen is assimilated from glutamate, *e.g.* CTP, glucosamine, carbamyl phosphate, and some of the amino acids; and (3) covalent modification of the enzyme by adenylation. Each glutamine synthetase molecule is composed of 12 identical subunits. Within the duodecameric quaternary structure, each subunit may be adenylated or not. The state of adenylation may therefore vary between no subunits adenylated or all 12 subunits adenylated. The highly-adenylated form is the most inactive whereas the totally-deadenylated form has the highest activity. Partially-adenylated glutamine synthetase is more sensitive to feedback inhibition than the non-adenylated form of the enzyme.

Attachment and removal of the AMP units in the adenylation mechanism is catalyzed by an adenyltransferase whose activity in turn is affected by several metabolites, particularly glutamine and 2-oxoglutarate. The activity of the adenyltransferase is controlled by a small protein, P_{II}, which itself exists in two forms: in the native form P_{II} interacts with the adenyltransferase to enhance the adenylation state of glutamine synthetase whereas, in an uridylated form, P_{II} stimulates deadenylation of glutamine synthetase. The uridylation of P_{II} is catalyzed by a uridyltransferase whose activity is stimulated by 2-oxoglutarate and ATP, and inhibited by glutamine or inorganic phosphate. Therefore, in the presence of high concentrations of 2-oxoglutarate, the uridyltransferase is stimulated to add UMP to P_{II}. Uridylated P_{II} interacts with the adenyltransferase to favour deadenylation of the glutamine synthetase in which form the glutamine synthetase is most active. Conversely, when the level of glutamine is high, P_{II} exists in its native form: this stimulates the adenyltransferase to adenylate glutamine synthetase, making it catalytically-inactive. The glutamine synthetases of *E. coli*, *S. typhimurium* and *K. aerogenes* are regulated by adenylation (Tyler, 1978). By contrast, the enzymes of Gram-positive bacilli are not apparently regulated in this way, nor are those of fungi, whereas that of *Streptomyces cattleya* is (Streicher and Tyler, 1981).

Nitrogen assimilation from other nitrogen sources is also regulated. It was shown that the enzymes of histidine and proline dissimilation (Prival and Magasanik, 1971) and the enzymes for nitrogen fixation (reviewed by Brill, 1975) were repressed when ammonia was present. A correlation existed between the levels of the dissimilation enzymes and the activity of glutamine synthetase (Prival *et al.*, 1973). It was proposed that under conditions of nitrogen starvation when glutamine synthetase was totally deadenylated and highly active catalytically, the enzyme also acted on DNA to stimulate transcription of the operons for nitrogen dissimilation. This model has been consistent with the experimental data of the last 10 years. Recently, however, fine structure genetic mapping has invalidated this model which had grown to become a central dogma in cellular regulation. Glutamine synthetase is no longer considered to be the regulator protein (reviewed by Merrick, 1982). Two cistrons (NtrB and NtrC) which lie adjacent to the structural gene for glutamine synthetase (McFarland *et al.*, 1981) regulate the transcription of the nitrogen dissimilation genes *and* of glutamine synthetase. Transcription may be activated or repressed, with NtrC likely to be the DNA-binding protein. A locus, NtrA, distant on the genetic map from that for NtrB and NtrC, produces an activator of NtrC, further complicating the issue.

9.3.3   Assimilation of Other Elements

Sulfur is assimilated through cysteine in most bacteria. In a well-defined process sulfur is usually supplied as sulfate. Activation of sulfate has been reviewed by Peck (1974). Sulfate is assimilated in a series of reductive reactions of which 3'-phospho-5'-adenyl sulfate (PAPS) is the main intermediate. PAPS is the donor of active sulfate in the cell. Assimilation proceeds through sulfite reductase to cysteine.

Hydrogen is usually assimilated with other elements but, apart from this route, the pyridine nucleotide dehydrogenases can add hydrogen. Similarly, the hydratases are capable of adding hydrogen across a double bond.

Oxygen, like hydrogen, is assimilated along with other elements. Hydration across a double bond, followed by dehydrogenation, formally assimilates oxygen into the cell, although this is a quantitatively minor route. Molecular oxygen may be assimilated directly in oxygenase reactions, but this is also a minor route.

Phosphorus is assimilated in the ATP-dependent reactions of intermediary metabolism.

Most biochemistry texts deal with the general details of the above reactions (*e.g.* Metzler, 1977). Little attention has been paid to the assimilation of these elements in the literature.

9.4   SUMMARY

Nutrients are taken up into microorganisms by a variety of mechanisms, each with different characteristics tailored to a specific need of the cell. Control of uptake of a substrate by other transport systems is often demonstrated. In contrast to the diversity of uptake mechanisms, assimilation of nutrients once inside the cell proceeds in a unified pattern which can be rationalized in terms of supply and demand. When considering the use of cells to make biotechnological products, the assimilation of carbon and nitrogen is of prime importance.

9.5 REFERENCES

Adhya, S. and H. Echols (1966). Glucose effect and the galactose enzymes of *Escherichia coli*: correlation between glucose inhibition of induction and inducer transport. *J. Bacteriol.*, **92**, 601–608.

Adler, J. and W. Epstein (1974). Phosphotransferase-system enzymes as chemoreceptors for certain sugars in *Escherichia coli* chemotaxis. *Proc. Natl. Acad. Sci. USA*, **71**, 2895–2899.

Adler, L. W. and B. P. Rosen (1977). Functional mosaicism of membrane proteins in vesicles of *Escherichia coli*. *J. Bacteriol.*, **229**, 959–966.

Amaral, D. and H. L. Kornberg (1975). Regulation of fructose uptake by glucose in *Escherichia coli*. *J. Gen. Microbiol.*, **90**, 157–168.

Ashgar, S. S., E. Levin and F. M. Harold (1973). Accumulation of neutral amino acids by *Streptococcus faecalis*. *J. Biol. Chem.*, **248**, 5225–5233.

Berberich, M. A. (1972). A glutamine-dependent phenotype in *E. coli* K12: the result of two mutations. *Biochem. Biophys. Res. Commun.*, **47**, 1498–1503.

Berger, E. A. (1973). Different mechanisms of energy coupling for the active transport of proline and glutamine in *Escherichia coli*. *Proc. Natl. Acad. Sci. USA*, **70**, 1514–1518.

Berger, E. A. and L. A. Heppel (1974). Different mechanisms of energy coupling for the shock-sensitive and shock-resistant amino acid permeases of *Escherichia coli*. *J. Biol. Chem.*, **249**, 7747–7755.

Bitoun, R., H. de Rouse, A. Touti-Schwartz and A. Danchin (1983). The phosphoenolpyruvate-dependent carbohydrate phosphotransferase system of *Escherichia coli*: cloning of the *pts III crr* region and studies with *pts-luc* operon fusion. *FEMS Microbiol. Lett.*, **16**, 163–167.

Booth, I. R. and J. G. Morris (1975). Proton-motive force in the obligately anaerobic bacterium *Clostridium pasteurianum*: a role in galactose and gluconate uptake. *FEBS Lett.*, **59**, 153–157.

Bourd, G. I., R. S. Erlagaeva, T. N. Balshakova and V. N. Gershanovitch (1975). Glucose catabolite repression in *Escherichia coli* K12 mutants defective in methyl-α-D-glucoside transport. *Eur. J. Biochem.*, **53**, 419–427.

Brenchley, J. E. and B. Magasanik (1974). Mutants of *Klebsiella aerogenes* lacking glutamate dehydrogenase. *J. Bacteriol.*, **117**, 544–550.

Brenchley, J. E., L. G. Baker and L. G. Patil (1975). Regulation of the ammonia assimilation enzymes in *Salmonella typhimurium*. *J. Bacteriol.*, **124**, 182–189.

Brill, W. J. (1975) Regulation and genetics of bacterial nitrogen fixation. *Annu. Rev. Microbiol.*, **29**, 109–129.

Canh, D. S., J. Horak, A. Kotyk and L. Rohvara (1975). Transport of acyclic polyols in *Sacharomyces cerevisiae*. *Folia Microbiol.*, **20**, 320–325.

Cohen, G. N. and J. Monod (1957). Bacterial permeases. *Bacteriol. Rev.*, **21**, 169–194.

Cohn, D. E. and H. R. Kaback (1980). Mechanism of the melibiose porter in membrane vesicles of *Escherichia coli*. *Biochemistry*, **19**, 4237–4243.

Collins, S. H., A. W. Jarvis, R. J. Lindsay and W. A. Hamilton (1976). Proton movements coupled to lactate and alanine transport in *Escherichia coli*: isolation of mutants with altered stoichiometry in alanine transport. *J. Bacteriol.*, **126**, 1232–1244.

Cowell, J. L. (1974). Energetics of glycl glycine transport in *Escherichia coli*. *J. Bacteriol.*, **120**, 139–146.

Curtis, S. J. (1974). Mechanism of energy coupling for transport of D-ribose in *Escherichia coli*. *J. Bacteriol.*, **120**, 295–303.

Dills, S. S., A. Apperson, M. R. Schmidt and M. H. Saier (1980). Carbohydrate transport in bacteria. *Microbiol. Rev.*, **44**, 385–418.

Drapeau, G. R., T. I. Matula and R. A. MacLeod (1966). Nutrition and metabolism of marine bacteria. XV. Relation of Na$^+$ activated transport to the Na$^+$ requirement of a marine pseudomonad for growth. *J. Bacteriol.*, **92**, 63–71.

Eagon, R. J. and L. S. Wilkerson (1972). A potassium-dependent citric acid transport system in *Aerobacter aerogenes*. *Biochem. Biophys. Res. Commun.*, **45**, 1944–1950.

Eddy, A. A. (1978). Proton-dependent solute transport in microorganisms. *Curr. Top. Membr. Transp.*, **10**, 279–360.

Eddy, A. A. (1982). Mechanism of solute transport in selected eukaryotic microorganisms. *Adv. Microb. Physiol.*, **23**, 1–78.

Edgar, W., I. S. Forrest, W. H. Holms and B. Jasani (1972). The control of glycerol utilization by glucose metabolism. *Biochem. J.*, **127**, 59p.

Fiechter, A., G. F. Fuhrmann and O. Kapelli (1981). Regulation of glucose metabolism in growing yeast cells. *Adv. Microb. Physiol.*, **22**, 123–183.

Flagg, J. L. and T. H. Wilson (1976). Galactoside accumulation by *Escherichia coli*, driven by a pH gradient. *J. Bacteriol.*, **125**, 1235–1236.

Fuhrmann, G. F., C. Boehm and A. P. R. Theuvenet (1976). Sugar transport and potassium permeability in yeast plasma membrane vesicles. *Biochem. Biophys. Acta*, **433**, 583–596.

Ginsberg, A. and E. R. Stadtman (1973). In *The Enzymes of Glutamine Metabolism*, ed. S. Pruisner and E. R. Stadtman, pp. 9–42. Academic, London.

Halpern, Y. S., H. Barash, H. B. S. Dover and K. Druck (1973). Sodium and potassium requirements for active transport of glutamate by *Escherichia coli* K12. *J. Bacteriol.*, **114**, 53–58.

Hamilton, W. A. (1975). Energy coupling in microbial transport. *Adv. Microb. Physiol.*, **12**, 1–53.

Hamilton, W. A. and I. R. Booth (1980). In *Microorganisms and Nitrogen Sources*, ed. J. W. Payne, pp. 171–207. Wiley, Chichester.

Harold, F. M. (1972). Conservation and transformation of energy by bacterial membranes. *Bacteriol. Rev.*, **36**, 172–230.

Harold, F. M. (1977). Mechanisms of energy transduction in bacteria. *Curr. Top. Bioenerg.*, **6**, 84–151.

Harwood, J. P. and A. Peterkofsky (1975). Glucose-sensitive adenylate cyclase in toluene-treated cells of *Escherichia coli* B. *J. Biol. Chem.*, **250**, 4656–4662.

Harwood, J. P., C. Gazdar, C. Prasad, A. Peterkofsky, S. J. Curtis and W. Epstein (1976). Involvement of the glucose enzymes II of the sugar phosphotransferase system in regulation of adenylate cyclase by glucose in *Escherichia coli*. *J. Biol. Chem.*, **251**, 2462–2468.

Hauer, R. and M. Hofer (1978). Evidence for interactions between the energy dependent transport of sugars and membrane potential in the yeast *Rhodotorula gracilis*. *J. Membr. Biol.*, **43**, 335–349.

Hayashi, S. and E. C. C. Lin (1965). Product induction of glycerol kinase in *Escherichia coli*. *J. Mol. Biol.*, **14**, 515–521.

Hazelbauer, G. L. (1975). Maltose chemoreceptor of *Escherichia coli*. *J. Bacteriol.*, **122**, 206–214.

Heller, K. B., E. C. C. Lin and T. H. Wilson (1980). Substrate specificity and transport properties of the glycerol facilitator of *Escherichia coli*. *J. Bacteriol.*, **144**, 274–278.

Henderson, P. J. F., R. A. Giddens and M. C. Jones-Mortimer (1977). Transport of galactose, glucose and their molecular analogues by *Escherichia coli* K12. *Biochem. J.*, **162**, 309–320.

Heppel, L. A., B. P. Rosen, I. Frieberg, E. A. Berger and J. H. Weiner (1972). In *The Molecular Basis of Biological Transport*. *Miami Winter Symposia*, vol. 3, pp. 133–156. Academic, New York.

Herbert, D. (1961). The chemical composition of microorganisms as a function of their environment. *Symp. Soc. Gen. Microbiol.*, **11**, 391–416.

Hinds, T. R. and A. F. Brodie (1974). Relationship of a proton gradient to the active transport of proline with membrane vesicles of *Mycobacterium phlei*. *Proc. Natl. Acad. Sci. USA*, **71**, 1202–1206.

Hong, J. S., A. G. Hunt, P. S. Masters and M. A. Lieberman (1979). Requirement of acetyl-phosphate for the binding protein dependent transport systems in *Escherichia coli*. *Proc. Natl. Acad. Sci. USA*, **76**, 1213–1217.

Hoshino, T. (1980). Transport systems for branched chain amino acids in *Pseudomonas aeruginosa*. *J. Bacteriol.*, **139**, 705–712.

Hunter, I. S., P. J. F. Henderson and M. C. Jones-Mortimer (1981). Influence of transport system on growth of galactose by *Escherichia coli* K12. *Proceedings of the 2nd European Congress of Biotechnology*, p. 23.

Jacobsen, G. R., C. A. Lee and M. H. Saier (1979). Purification of the mannitol-specific enzyme II of the *Escherichia coli* PEP: sugar phosphotransferase system. *J. Biol. Chem.*, **254**, 249–252.

Jin, R. Z., R. G. Forage and E. C. C. Lin (1982). Glycerol kinase as a substitute for dihydroxyacetone kinase in a mutant of *Klebsiella pneumoniae*. *J. Bacteriol.*, **152**, 1303–1307.

Jones-Mortimer, M. C. and H. L. Kornberg (1976). Order of genes adjacent to *ptsX* on the *E. coli* genome. *Proc. R. Soc. London, Ser. B*, **193**, 313–315.

Kaback, H. R. (1976). Molecular biology and energetics of membrane transport. *J. Cell Physiol.*, **89**, 575–594.

Kashket, E. and T. H. Wilson (1974). Proton motive force in fermenting *Streptococcus lactis* 7962 in relation to sugar accumulation. *Biochem. Biophys. Res. Commun.*, **59**, 879–886.

Kellerman, O. and S. Szmelcman (1974). Active transport of maltose in *Escherichia coli* K12: involvement of a 'periplasmic' maltose binding protein. *Eur. J. Biochem.*, **47**, 139–149.

Kobayashi, M., E. Kin and Y. Anraku (1974). Transport of sugars and amino acids in bacteria: sources of energy and energy coupling reactions of the active transport systems for isoleucine and proline in *E. coli*. *J. Biochem. (Tokyo)*, **76**, 251–261.

Kornberg, H. L. (1973). Carbohydrate transport by micro-organisms. *Proc. R. Soc. London., Ser. B*, **183**, 105–123.

Kornberg, H. L. and M. C. Jones-Mortimer (1977). The phosphotransferase system as a site of cellular control. *Symp. Soc. Gen. Microbiol.*, **27**, 217–240.

Kornberg, H. L. and C. Riordan (1976). Uptake of galactose in *Escherichia coli* by facilitated diffusion. *J. Gen. Microbiol.*, **94**, 75–89.

Koshland, D. E. (1977). Bacterial chemotaxis and some enzymes in energy metabolism. *Symp. Soc. Gen. Microbiol.*, **27**, 317–331.

Kundig, W., S. Ghosh and S. Roseman (1964). Phosphate bound to histidine in a protein as an intermediate in a novel phosphotransferase system. *Proc. Natl. Acad. Sci. USA*, **52**, 1067–1074.

Lagarde, A. L. (1976). A non-equilibrium thermodynamic analysis of active transport within the framework of the chemiosmotic theory. *Biochem. Biophys. Acta*, **426**, 198–217.

Lagarde, A. L. and B. A. Haddock (1977). Proton uptake linked to the 3-deoxy-2-oxo-D-gluconate transport system of *Escherichia coli*. *Biochem. J.*, **162**, 183–187.

Larson, T. S., G. Schumacher and W. Boos (1982). Identification of the GlpT-encoded Sn-glycerol 3-phosphate permease of *Escherichia coli*, an oligomeric integral membrane protein. *J. Bacteriol.*, **152**, 1008–1021.

LeBlanc, G., G. Rimon and H. R. Kaback (1980). Glucose 6-phosphate transport in membrane vesicles isolated from *Escherichia coli*: effect of imposed electrical potential and pH gradient. *Biochemistry*, **19**, 2522–2528.

Lee, L. G., P. Britton, F. Parra, A. Boronat and H. Kornberg (1982). Expression of the *ptsH*⁺ gene of *Escherichia coli* cloned on plasmid pBR322: a convenient means for obtaining the histidine-containing protein. *FEBS Lett.*, **149**, 288–292.

Lin, E. C. C. (1976). Glycerol dissimilation and its regulation in bacteria. *Annu. Rev. Microbiol.*, **30**, 535–578.

Macfarland, N., L. McCarter, S. Artz and S. Kustu (1981). Nitrogen regulatory locus 'glnR' of enteric bacteria is composed of cistrons *ntr* B and *ntr* C: identification of their protein products. *Proc. Natl. Acad. Sci. USA*, **78**, 2135–2139.

MacLeod, R. A., P. Thurman and H. J. Rogers (1973). Comparative transport activity of intact cells, membrane vesicles and mesosomes of *Bacillus licheniformis*. *J. Bacteriol.*, **113**, 329–340.

Magasanik, B. (1961). Catabolite repression. *Cold Spring Harbor Symp. Quant. Biol.*, **26**, 249–256.

McGinnis, J. F. and K. Paigen (1969). Catabolite inhibition: a general phenomenon in the control of carbohydrate utilization. *J. Bacteriol.*, **100**, 902–913.

Meers, J. L., D. W. Tempest and C. M. Brown (1970). 'Glutamine (amide): 2-oxoglutamate amino transferase oxidoreductase (NADP)', an enzyme invoked in synthesis of glutamate by some bacteria. *J. Gen. Microbiol.*, **64**, 187–194.

Merrick, M. J. (1982). A model for nitrogen control. *Nature (London)*, **297**, 362–363.

Metzler, D. E. (1977). *Biochemistry, the Chemical Reactions of Living Cells*. Academic, New York.

Miner, K. M. and L. Frank (1974). Sodium-stimulated glutamate transport in osmotically-shocked cells and membrane vesicles of *Escherichia coli*. *J. Bacteriol.*, **117**, 1093–1098.

Misset, O., M. Brouwer and G. T. Robillard (1980). *Escherichia coli* PEP-dependent phosphotransferase. Evidence that the dimer is the active form of enzyme I. *Biochemistry*, **19**, 883–890.

Mitchell, P. (1970). Membranes of cells and organelles: morphology, transport and metabolism. *Symp. Soc. Gen. Microbiol.*, **20**, 121–166.

Mitchell, P. (1977). From energetic abstraction to biochemical mechanism. *Symp. Soc. Gen. Microbiol.*, **27**, 382–423.

Monod, J. (1942). *Recherches sur la Croissances de Cultures Bactériennes*. Herman, Paris.

Morowitz, H. J. (1968). *Energy Flow in Biology*. Academic, New York.

Nelson, S. O., B. J. Scholte and P.W. Postma (1982). Phosphoenolpyruvate sugar phosphotransferase system mediated regulation of carbohydrate metabolism in *Salmonella typhimurium*. *J. Bacteriol.*, **150**, 604–615.

Niven, D. F. and W. A. Hamilton (1974). Mechanisms of energy coupling to the transport of amino acids by *Staphylococcus aureus*. *Eur. J. Biochem.*, **44**, 517–522.

O'Brien, R. W., O. M. Niejssel and D. W. Tempest (1980). Glucose phosphoenolpyruvate phosphotransferase activity and glucose uptake rate of *Klebsiella aerogenes* growing in chemostat culture. *J. Gen. Microbiol.*, **116**, 305–314.

Ordal, G. A. and J. Adler (1974). Isolation and complementation of mutants in galactose taxis and transport. *J. Bacteriol.*, **117**, 509–516.

Pavlasova, E. and F. M. Harold (1969). Energy coupling in the transport of β-galactosides by *Escherichia coli*. Effect of proton conductors. *J. Bacteriol.*, **98**, 198–204.

Payne, J. W. (1980). In *Microorganisms and Nitrogen Sources*, ed. J. W. Payne, p. 360. Wiley, Chichester.

Peck, H. D. (1974). Biological transformations of inorganic sulfur compounds. *Abstr. Pap. Am. Chem. Soc.*, **169**, 26–27.

Perlman, R. L. and I. Pastan (1968). Cyclic 3′, 5′-AMP: stimulation of galactosidase and tryptophanase synthesis in *E. coli*. *Biochem. Biophys. Res. Commun.*, **30**, 656–664.

Peterkofsky, A. and C. Gazdar (1974). Glucose inhibition of adenylate cyclase in intact cells of *Escherichia coli*. B. *Proc. Natl. Acad. Sci. USA*, **71**, 2324–2328.

Postma, P. W. (1976). Involvement of the phosphotransferase system in galactose transport in *Salmonella typhimurium*. *FEBS Lett.*, **61**, 49–53.

Postma, P. W. (1981). Involvement of the phosphotransferase system in galactose transport in *Salmonella typhimurium*. Defective enzyme IIB of the phosphoenolpyruvate : sugar phosphotransferase system leading to uncoupling of transport and phosphorylation in *Salmonella typhimurium*. *J. Bacteriol.*, **147**, 382–389.

Postma, P. W. and J. B. Stock (1980). Enzymes II of the phosphotransferase system do not catalyse sugar transport in the absence of phosphorylation. *J. Bacteriol.*, **141**, 476–484.

Prival, M. J. and B. Magasanik (1971). Resistance to catabolite repression of histidase and proline oxidase during nitrogen-limited growth of *Klebsiella aerogenes*. *J. Biol. Chem.*, **246**, 6288–6296.

Prival, M. J., J. E. Brenchley and B. Magasanik (1973). Glutamine synthetase and the regulation of histidase formation in *Klebsiella aerogenes*. *J. Biol. Chem.*, **248**, 4334–4344.

Raphaeli, A. W. and M. H. Saier (1978). Kinetic analysis of the sugar phosphate : sugar transphosphorylation reaction catalysed by the glucose enzyme II complex of the bacterial phosphotransferase system. *J. Biol. Chem.*, **253**, 7595–7597.

Richey, D. P. and E. C. C. Lin (1972). Phosphorylation of glycerol in *Staphylococcus aureus*. *J. Bacteriol.*, **114**, 880–881.

Robbilard, G. T. and W. T. Konings (1981). Physical mechanism for the regulation of PEP-dependent glucose transport activity in *Escherichia coli*. *Biochemistry*, **20**, 5025–5032.

Roehl, R. A. and P. V. Phibbs (1982). Characterization and genetic mapping of fructose phosphotransferase mutations in *Pseudomonas aeruginosa*. *J. Bacteriol.*, **149**, 897–905.

Romano, A. H. (1982). Facilitated diffusion of 6-deoxyglucose in Bakers' yeast: evidence against phosphorylation-associated transport of glucose. *J. Bacteriol.*, **152**, 1295–1297.

Romano, A. H., S. J. Eiserhard, S. L. Dingle and T. D. McDowell (1970). Distribution of the phosphoenolpyruvate : glucose phosphotransferase system in bacteria. *J. Bacteriol.*, **104**, 808–813.

Romano, A. H., J. B. Trifone and M. Brustolon (1979). Distribution of the phosphoenolpyruvate : glucose phosphotransferase system in fermentative bacteria. *J. Bacteriol.*, **139**, 93–97.

Saheb, S. A. (1972). Permeation du glycerol et sporulation chez *Bacillus subtilis*. *Can. J. Microbiol.*, **18**, 1307–1313.

Saier, M. H. and C. D. Stiles (1975). In *Molecular Dynamics in Biological Membranes*, pp. 99–105. Springer Verlag, Berlin.

Saier, M. H., B. U. C. Feucht and S. Roseman (1977). Sugar phosphate : sugar transphosphorylation and exchange group translocation catalyzed by the enzyme II complexes of the bacterial PEP : sugar phosphotransferase system. *J. Biol. Chem.*, **252**, 8899–8907.

Saier, M. H., L. S. Stroud, L. S. Massman, J. J. Judice, M. J. Newman and B. U. Feucht (1978). Permease-specific mutations in *Salmonella typhimurium* and *Escherichia coli* that release glycerol, maltose, melibiose and lactose transport systems from regulation by the phosphoenolpyruvate : sugar phosphotransferase system. *J. Bacteriol.*, **133**, 1358–1367.

Sanno, Y., T. H. Wilson and E. C. C. Lin (1968). Control of permeation to glycerol in cells of *Escherichia coli*. *Biochem. Biophys. Res. Commun.*, **32**, 344–349.

Scholte, B. J., A. R. Schiutema and P. W. Postma (1981). Isolation of III$_{glc}$ of the phosphoenolpyruvate-dependent glucose phosphotransferase system of *Salmonella typhimurium*. *J. Bacteriol.*, **148**, 257–264.

Schuldiner, S. and H. Fishkes (1978). Sodium–proton antiport in isolated membrane vesicles of *Escherichia coli*. *Biochemistry*, **17**, 706–711.

Seaston, A., G. Carr and A. A. Eddy (1976). The concentration of glycerol by preparations of the yeast *Saccharomyces carlsbergensis* depleted of adenosine triphosphate. *Biochem. J.*, **154**, 669–676.

Shuman, H. A., T. S. Silhavy and J. R. Beckwith (1980). Labelling of proteins with β-galactosidase by gene fusion. Identification of a cytoplasmic membrane component of the *Escherichia coli* maltose transport system. *J. Biol. Chem.*, **255**, 168–174.

Silhavy, T. J., I. Hartig-Beecken and W. Boos (1976). Periplasmic protein related to the sn-glycerol 3-phosphate transport system of *Escherichia coli*. *J. Bacteriol.*, **126**, 951–958.

Stock, J. and S. Roseman (1971). A sodium-dependent sugar cotransport system in bacteria. *Biochem. Biophys. Res. Commun.*, **44**, 132–138.

Stouthamer, A. H. (1973). A theoretical study of the amount of ATP required for synthesis of microbial cell material. *Antonie van Leeuwenhoek*, **39**, 540–565.

Streicher, S. L. and B. Tyler (1981). Regulation of glutamine synthetase activity by adenylation in the Gram-positive bacterium *Streptomyces cattleya. Proc. Natl. Acad. Sci. USA*, **78**, 229–233.

Szmelcman, S., M. Schwarz, T. J. Silhavy and W. Boos (1976). Maltose transport in *Escherichia coli* K12. *Eur. J. Biochem.*, **65**, 13–19.

Tanner, W. (1974). Energy-coupled sugar transport in *Chlorella. Biochem. Soc. Trans.*, **2**, 793–797.

Tempest, D. W., J. L. Meers and C. M. Brown (1973). In *The Enzymes of Glutamine Metabolism*, ed. S. Prusiner and E. R. Stadtman, pp. 167–182. Academic, New York.

Tsay, S. S., K. K. Brown and E. T. Gaudy (1971). Transport of glycerol by *Pseudomonas aeruginosa. J. Bacteriol.*, **108**, 82–88.

Tyler, B. (1978). Regulation of the assimilation of nitrogen compounds. *Annu. Rev. Biochem.*, **47**, 1127–1162.

Vadebancoeur, C. and L. Trahan (1982). Glucose transport in *Streptococcus salivarus*. Evidence for the presence of a distinct PEP : glucose phosphotransferase system which catalyzed phosphorylation of α-methylglucoside. *Can. J. Microbiol.*, **28**, 190–197.

Wandersmann, C., M. Schwarz and T. Ferenci (1979). *Escherichia coli* mutants impaired in maltodextrin transport. *J. Bacteriol.*, **140**, 1–13.

West, I. C. (1970). Lactose transport coupled to proton movements in *Escherichia coli. Biochem. Biophys. Res. Commun.*, **41**, 655–661.

West, I. C. and P. Mitchell (1972). Proton-coupled β-galactoside translocation in non-metabolizing *Escherichia coli. J. Bioenerg.*, **3**, 445–662.

West, I. C. and P. Mitchell (1973). Stoichiometry of lactose-H^+ symport across the plasma membrane of *Escherichia coli. Biochem. J.*, **132**, 587–592.

West. I. C. and P. Mitchell (1974). Proton/sodium antiport in *Escherichia coli. Biochem. J.*, **144**, 87–90.

West, I. C. and T. H. Wilson (1973). Galactoside transport dissociated from proton movements in *Escherichia coli. Biochem. Biophys. Res. Commun.*, **50**, 551–558.

Wheelis, M. L. (1975). The genetics of dissimilatory pathways of *Pseudomonas. Annu. Rev. Microbiol.*, **29**, 505–524.

Willecke, K., E. M. Gries and P. Oehr (1973). Coupled transport of nitrate and magnesium in *Bacillus subtilis. J. Biol. Chem.*, **248**, 807–814.

Wilson, D. B. (1974). Source of energy for the *Escherichia coli* galactoside transport systems induced by galactose. *J. Bacteriol.*, **120**, 866–871.

Yagil, E. and I. R. Beacham (1975). Uptake of adenosine 5-monophosphate by *Escherichia coli. J. Bacteriol.*, **121**, 401–405.

10

Modes of Growth of Bacteria and Fungi

S. G. OLIVER
University of Manchester Institute of Science and Technology, UK
and
A. P. J. TRINCI
University of Manchester, UK

10.1 INTRODUCTION

In this chapter we describe the growth of individual organisms and colonies of bacteria, yeasts and moulds. Filamentous organisms, unlike unicellular organisms, are able to increase their size indefinitely without altering the ratio between protoplasmic volume and surface area, and such organisms are usually well adapted to colonize solid substrates, including surfaces in fermenter vessels. Prokaryotic actinomycetes and eukaryotic moulds have remarkably similar morphologies and modes of growth, and it has been suggested that these organisms represent an example of convergent evolution. However, actinomycete hyphae never attain the extension rates typically observed in fungi (up to 100 μm min^{-1}), possibly because they form complete septa and lack a vesicular system of wall growth. The similarities between the morphology and growth of mycelia of actinomycetes and moulds will be described in this chapter and a description will also be given of the modes of growth of cocci, rods and stalked bacteria and of budding yeasts. Liquid cultures of unicellular microorganisms exhibit low viscosities and more or less Newtonian flow behaviour. In contrast, liquid cultures of disperse, filamentous mycelia exhibit high viscosities and pseudo-plastic flow behaviour, and since the oxygen transfer coefficient falls with increasing viscosity, such cultures may rapidly become oxygen limited. Cultures of moulds and actinomycetes which form spherical colonies (pellets) in liquid media have a similar rheology to unicellular organisms. However, biomass at the centre of these pellets is often starved of oxygen and may autolyze. Thus, the mode of growth of a microorganism has significant implication for its behaviour in submerged culture and hence for microbial technology.

10.2 GROWTH OF UNICELLULAR ORGANISMS

10.2.1 Cocci

More is known about the mode of growth of individual cells of the Gram-positive cocci than about any other class of organisms. There are two reasons for this. First, the sphere is the simplest possible cell shape and the increase in surface area which occurs during growth and division is almost entirely due to the formation of the septum which separates the two daughter cells. Second, the Gram-positive wall is much simpler than that of Gram-negative bacteria (40% of its mass consists of peptidoglycan) and in the cocci there is little or no turnover of wall components during growth and division (Higgins and Shockman, 1971; Boothby *et al.*, 1973). This latter fact enables the sites of intercalation of new wall material to be identified by radioisotopic labelling and by fluorescent antibody or bacteriophage binding techniques.

The most complete picture of the growth and division of cocci has come from the study of *Streptococcus faecalis*, mainly by Shockman, Higgins and their collaborators. Streptococci divide in the same plane in successive cell divisions and therefore form chains of cells during rapid growth. This contrasts with the micrococci in which each successive plane of division is at right-angles to the previous one (see below), thus forming clusters or packets of cells. The fact that the principal site of wall growth in the cocci is the septum means that 44% of the total wall volume is assembled in this region (Rogers *et al.*, 1980).

The first successful application of the fluorescent antibody labelling technique was the study by Cole and Hahn (1962) on *Streptococcus pyogenes*. They prepared an antibody against the Group A antigen of this organism (the M protein) and rendered it fluorescent by coupling the antibody to fluorescein isothiocyanate. Two types of experiment were performed, equivalent to the 'pulse-chase' and 'chase-pulse' regimes used in radioisotopic labelling. In the first, *S. pyogenes* was grown in a medium containing the fluorescent antibody and then the organism was harvested, washed and resuspended in a medium containing an antibody without the fluorescent tag. Fluorescence microscopy revealed that new (non-fluorescent) wall appeared as a well-defined band around the equator of the coccus cell. The reverse ('chase-pulse') experiment, in which cells were switched from a medium containing a non-fluorescent antibody to one containing a fluorescent antibody, gave similar results.

In spite of the advantages of *Streptococcus* for labelling studies, the electron microscope has probably been the most important tool in deciphering this organism's mode of growth and division. This has involved the use of computer techniques to make three-dimensional reconstructions of the *Streptococcus* cell from exact axial and longitudinal sections (Higgins, 1976; Higgins and Shockman, 1976). The following is a summary of the findings of these analyses (see Figure 1).

Streptococcus has a raised band of wall material on its surface at the junction between new and old wall. A 'new-born' *Streptococcus* cell has one hemisphere of old wall (synthesized in the previous generation) and one of new wall (synthesized in the generation which led to the cell division which produced it). The raised band marks the approximate position where the septum or cross-wall will be formed. As the septum grows in towards the centre of the cell, a notch appears in the cell surfaces dividing the wall band into two bands which begin to move apart. At completion of cell division they will again occupy an equatorial position in the two daughter cells at the junction between old and new wall.

The septum is laid down between the two layers of a double membrane invagination. Wall precursors are thought to be associated with this septal membrane and are incorporated into the septum itself at its leading edge. The growth of the cross-wall necessitates the further invagination

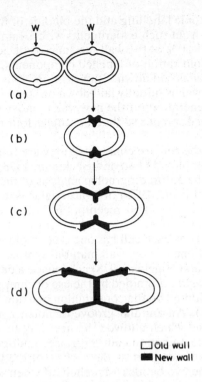

Figure 1 Growth of the Gram-positive coccus *Streptococcus faecalis*. Cell wall extension is almost entirely associated with septation which is initiated at the wall band (w). (Redrawn from Rogers *et al.*, 1980)

and growth of the membrane. Following the division of the wall bands extension of the septum ceases and most wall growth occurs by the lateral extension of the region between the two bands. When most of the two new wall hemispheres has been constructed in this manner, the invagination of the septum continues to separate the cytoplasms of the two daughter cells. During this final period of growth the septum peels into two layers which form the poles of the daughter cocci.

Although there is little wall turnover in *Streptococcus*, the organism does possess an autolytic enzyme, *N*-acetylmuramınıdase. This enzyme is located in the wall where it is present in both latent and active forms, conversion from one to the other being effected by proteases. The wall-bound active autolysin preferentially hydrolyzes newly synthesized wall material. In contrast, the artificial activation of autolysin by trypsin treatment results in the random hydrolysis of the entire wall.

10.2.2 Gram-positive Rods

Rod-shaped bacteria grow and divide by increasing their length until it is twice that of a 'new-born' cell and the laying down of a septum half-way between the two poles of the parent cell. The septum separates the two daughter cells each of which contains a nucleoid. The study of the mechanism of rod growth therefore involves investigation of the three processes of elongation, nuclear segregation and septation. These investigations are more likely to be successful with Gram-positive than Gram-negative rods since the former have simpler walls. More than 40% of their mass consists of peptidoglycan whereas less than 10% of the Gram-negative wall is made up of that polymer.

It is evident that the increase in cell volume of a bacillus cannot entirely be due to the growth of the septum as it is in the coccus. Indeed, only 15% of the increase in wall volume is due to the formation of septa in *Bacillus subtilis* and the majority of length extension occurs in regions distinct from the area of septum formation. The problem, therefore, is to determine whether length extension occurs by the intercalation of new material in distinct zones or bands or whether the wall surface increases by a process of diffuse intercalation. It should be possible to distinguish between these alternatives by the use of similar methods to those employed for the study of the growth of cocci, *i.e.* radioisotopic labelling and the binding of fluorescent antibodies or of phage particles. However, the data from such experiments with Gram-positive rods is difficult to interpret for two reasons. First, 30–50% of the wall substance is turned over per generation (Mauck *et al.*, 1971) leading to the randomization of labelled components. The second complication results from the way in which new wall is laid down. Studies on *Bacillus subtilis* by Pooley (1976a, 1976b) have demonstrated that new wall is initially laid down on the inner surface close to the cell membrane. It takes more than a generation for the new wall to move to the cell surface. As it moves to the periphery, the wall material spreads so that it occupies four times its original surface area two generations after its synthesis.

It is clear that the poles of the cell are conserved; they are relatively resistant to autolysis once the process of septation is completed (Fan and Beckman, 1973). Fluorescent antibody labelling (Hughes and Stokes, 1971), the autoradiographic analysis of the incorporation of labelled diaminopimelic acid (De Chastellier *et al.*, 1975) or *N*-acetylglucosamine (Pooley *et al.*, 1978) and the development of new bacteriophage SP50 receptors following phosphate starvation (Archibald and Coapes, 1976) all show that new wall material is incorporated at the site of septum formation but not at the existing poles. Thus each cell has one pole of old wall material (it may be from one to many generations old) and one of new wall material synthesized during the generation which gave rise to that cell. Burdett and Higgins (1978) have made a detailed electron microscopic study of septum formation in *Bacillus subtilis* aided by the use of computer reconstructions. They found that septum formation was initiated by the construction of a new cross-wall under the pre-existing cylindrical wall (see Figure 2). An annular groove or notch appeared in the overlying old wall beneath which a split appeared which allowed the new wall to separate into two layers. During notching two raised bands appeared on the outer surface, analogous to those previously seen in *S. faecalis*. These bands were pushed apart as the surface of the two nascent poles enlarged. The area of outer wall including the two bands is sloughed off when septation is complete.

The details of the topography of wall extension in Gram-positive rods is far from clear. Most of the early studies concluded that the cylindrical wall grew by a process of diffuse intercalation (De Chastellier *et al.*, 1975; Archibald and Coapes, 1976). However, the high rate of wall turnover probably invalidates this conclusion. Indeed, it is difficult to imagine how the process of cell wall enlargement can be coordinated with chromosome segregation and cell division if new wall material is incorporated at random. The problem of wall turnover causing the dispersion of incorporated material has been solved by the use of mutants deficient in autolytic enzymes. The earliest study of this sort was performed using fluorescent antibody labelling of the peptidoglycan of a *lyt*⁻ mutant of *Bacillus licheniformis* (Hughes and Stokes, 1971). The 'chasing' of uniformly labelled cells with non-labelled antibody revealed non-fluorescent (new) wall material half-way along the length of the cell at the site of cross-wall material. Other non-fluorescent sites were seen halfway between the cross-wall and the poles of the cell.

Sargent (1975) proposed a model in which it was envisaged that there were a limited number of growth zones in a *Bacillus* cell and that the appearance of new zones was dependent on DNA synthesis. New growth zones were said to appear at the junction of new and old wall and to operate to cause a symmetrical increase in cell length. Once length extension had ceased a septum could be formed from the central growth zone. This model has now been essentially substantiated. Pooley *et al.* (1978) used a low turnover-rate mutant of *B. subtilis* in an autoradiographic study of *N*-acetylglucosamine incorporation. As in previous studies they found that the cross-wall was a major site of incorporation and that the septa were conserved. They allowed an extensive period (five generations) of 'chase' with non-labelled *N*-acetylglucosamine in order to permit labelled material incorporated on the inner surface of the wall to reach the outer surface. They then found that the silver grains in their autoradiographs which were not in the septal regions were, nevertheless, distributed non-randomly down the length of the cells. This demonstrated that there were distinct zones of incorporation involved in length extension. A similar finding has been made by Howard *et al.* (1982) using ferritin-labelled antibody against the T-layer protein of *Bacillus sphaericus*. The T-layer is the outermost layer of the cell wall and, unlike the peptidoglycan, is subject to almost no turnover. The fluorescent labelling pattern showed that a major site of T-layer deposition is the new cross-wall and that there was little or no new T-layer inserted at the pre-existing poles. In addition to insertion at the site of incipient cell division, minor bands of new T-layer material were found at a number of distinct sites along the length of the cylindrical wall.

The septum delimits the two daughter cells once the parental cell has doubled in size. This septum must be formed at a site equidistant from the two poles and it must also be ensured that each of the daughters contains a nucleoid. The problem of nuclear segregation in bacteria was first

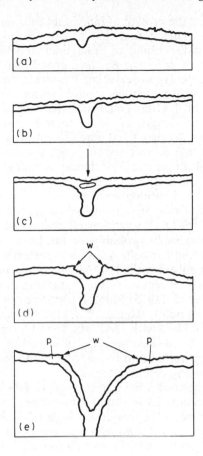

Figure 2 Growth of the septum in the Gram-positive rod *Bacillus subtilis*. The cross-wall is first seen as a V-shaped indentation in the parent wall (a). A split (arrowed in c) appears later at the base of the septum, producing the wall bands (w). These wall bands are located at the junction between poles (p) and the cylindrical part of the wall. They are displaced by enlargement of the poles. (Redrawn from the electron micrographs of Burdett and Higgins, 1978)

addressed by Jacob *et al.* (1963). These workers envisaged that the bacterial chromosome was attached to the cell membrane *via* its origin of replication. Since the origin, by definition, replicates first, then the two daughter chromosomes would automatically be dragged towards opposite poles of the cell as the membrane extended. The model predicts that the distance between the two bacterial nucleoids should be directly proportional to cell length. Sargent (1974, 1975) has examined this relationship in *B. subtilis* and his results do not conform to the predictions of the Jacob *et al.* (1963) model. Rather, the nucleoids appear to 'jump' from the centre of the cell to sites 0.25 of a cell length from the poles. The new attachment sites correspond to cell wall growth zones and represent the junctions between new and old walls. The daughter chromosomes are attached to these sites *via* their replication termini rather than their origins. There is now compelling evidence that genes associated with the chromosome terminus of *B. subtilis* are complexed with the cell membrane (Yamaguchi and Yoshikawa, 1973; Sargent and Bennett, 1982).

10.2.3 Gram-negative Rods

More is known about the genetics and molecular biology of *Escherichia coli* than about any other organism. Nevertheless, our understanding of the mechanisms of cell growth and division in *E. coli* has lagged behind that in *Bacillus* species, mainly because of the greater complexity of the Gram-negative wall. The peptidoglycan portion of the Gram-negative wall is a very thin layer which turns over slowly, if at all (Rogers *et al.*, 1980). It should therefore be quite amenable to analysis. The outer membrane presents more difficulties since its components may be mobile and could therefore be inserted at a specific site and then become randomized by migration.

One of the most direct studies of the growth and division of *E. coli* was performed by Donachie and Begg (1970) who measured the growth of individual cells on an agar surface relative to latex beads embedded in the agar. They found that slowly growing cells (on a minimal medium) increased their length in a unidirectional or polar manner while fast growing cells (on a rich medium) grew bidirectionally. *E. coli*, like *Bacillus*, grows by increasing in length but not width (Marr *et al.*, 1966). From this early observation Donachie and Begg evolved their 'unit cell' hypothesis (Donachie *et al.*, 1976) which has aroused some controversy. This states that a unit cell of *E. coli* has a length of 1.4 μm (=L) and that cells of length < $2L$ (2.8 μm) grow unidirectionally until they reach length $2L$ when their rate of cell extension doubles due to bidirectional growth. They then divide a fixed time (20 min) later. The model would explain why faster-growing cells are larger than slower-growing ones since a greater increase in length would occur during this fixed interval. In practice this 'linear doubling' model for the rate of cell extension is difficult to distinguish from truly exponential growth.

Studies on the extension of the peptidoglycan layer of *E. coli* have mainly used ^{3}H-diaminopimelic acid (^{3}H-DAP) as a specific wall precursor. In an early study of this sort Van Turbergen and Setlow (1961) concluded that ^{3}H-DAP was inserted at random into the *E. coli* wall. However, their light microscope autoradiographs probably did not have sufficient resolution to distinguish localized growth zones. Ryter and her colleagues have performed EM autoradiography on murein sacculi isolated from cells pulsed with ^{3}H-DAP. In wild-type cells Ryter *et al.* (1973) found grains concentrated over the septal region, but in mutants temperature-sensitive for septum formation (*fts* mutants) Schwarz *et al.* (1975) found second regions of DAP incorporation half-way between the septum and the cell poles. Koppes *et al.* (1978) made an EM autoradiographic study of osmium-fixed whole cells labelled with DAP and found a similar pattern of incorporation to that of Schwarz *et al.* (1975). All of these workers observed small cells labelled at one pole but assumed this was due to incorporation into the septum in the previous generation rather than to the polar growth pattern of the unit cell hypothesis.

Both bacteriophages and antibodies have been used to label specific protein or lipopolysaccharide components of the Gram-negative outer membrane. The synthesis of porins, the major proteins of the outer membrane, is repressed by high salt concentrations. Smit and Nikaido (1978) derepressed porin synthesis in *Salmonella typhimurium* by a high to low salt switch. They followed the incorporation of porins into the membrane using ferritin-conjugated antibody and electron microscopic observation of freeze-etched preparations. The technique produced electron micrographs of high quality and clearly demonstrated that these proteins intercalate into the outer membrane in a random manner. Similar results were obtained by Begg (1978) using *Escherichia coli* and a ^{3}H-histidine labelling procedure.

The extension of the lipopolysaccharide component of the outer membrane of *E. coli* (Beachey and Cole, 1966) and *Salmonella typhimurium* (May, 1963) has been studied using immunofluorescent and immunoelectron microscopic techniques. In both organisms incorporation of material along the length of the cell showed a diffuse pattern. However, in *E. coli* it was obvious that material was preferentially incorporated at the septum. Once again, small cells labelled at one pole were assumed to be new-born and to have inherited a labelled septum from their parents.

The phage T6 receptor is an *E. coli* outer membrane protein encoded by the *tsx* gene. A temperature-sensitive amber suppressor (Begg and Donachie, 1973, 1977) and also zygotic induction (Leal and Marcovich, 1975) have been used to specifically switch on the expression of this gene and follow the incorporation of the *tsx* protein into the outer membrane. The two groups come to opposite conclusions. Leal and Marcovich (1975) found that T6 receptors were incorporated at the septal region. Begg and Donachie (1977) found that in the 30–42 °C switch (where receptors are diluted out) new material was incorporated at the poles. They also supplied data from synchronous cultures to support their contention that small cells grow unidirectionally rather than the polarity of labelling reflecting the inheritance of septal material from the parent. In neither case can growth be said to be normal. In one the outgrowth of a zygote is observed while in Begg and Donachie's experiment cells were growing in the presence of low concentrations of penicillin.

Ryter *et al.* (1975) observed the incorporation of the phage λ receptor (maltose permease) into the *E. coli* outer membrane following its induction with cyclic AMP and maltose. They found a high density of labelling with λ particles over the septum. In small, aseptate cells the phage particles labelled one pole which they assumed to represent an inherited septum.

The data relating to the growth and division of *E. coli* are thus as complex as its cell wall. It is evident that, as with the Gram-positive bacteria, the septal region is a major site for the intercalation of new material into the cell surface structures. The studies on peptidoglycan synthesis using ^{3}H-DAP suggest a similar pattern of growth to the Gram-positive rods with new material being incorporated at the septum at the sites of future cell divisions. A polar mode of growth for small cells cannot, however, be excluded and careful studies need to be made on the growth of synchronously dividing cells for a period of more than one generation. The recent development of nonstressful techniques for generating such cultures (Gordon and Elliott, 1977) should allow the necessary experimentation.

10.2.4 The Prosthecate or Stalked Bacteria

These bacteria have a more complex morphology than is typical for prokaryotes. Their cells have a long narrow extension known as a prostheca which is bounded by a cell wall and membrane which are contiguous with those of the rest of the cell. There are two classes of prosthecate bacteria. In one, the prostheca has a reproductive role and daughter cells are produced as buds from its end. The best-known members of this class are *Hyphomicrobium* and *Rhodomicrobium*. *Caulobacter* is the best-known member of the second class of prosthecate bacteria in which the prostheca has no reproductive role and daughter cells are budded from the pole of the mother cell distal to the prostheca.

Growth is polar in both classes and permits the distinction to be made between a mother cell and its daughter. As in the budding yeasts (see below) an individual mother cell has a limited reproductive capacity. In a culture of *Rhodomicrobium vannielii* undergoing branched development, for instance, no more than four daughter cells are produced from each mother (Whittenbury and Dow, 1977). The mother cell can therefore be considered as going through an ageing process analogous to that of human cells in tissue culture (Hayflick and Moorhead, 1961). In the prosthecate bacteria this may, in part, be a reflection of the fact that new cell wall material is incorporated only into the daughter cell (Dow and Whittenbury, 1979). A detailed study of wall biogenesis in these bacteria has not been made, however.

The life cycle of *Rhodomicrobium* is shown in Figure 3. It should be noted that its asymmetric or polar mode of growth permits differentiation from a stalked cell into a swarmer cell to occur as part of the cell cycle. This developmental step is not obligatory however, and the stalked cell has a 'choice' as to whether it produces another stalked cell or a swarmer. The production of the latter is triggered by such environmental stimuli as high CO_2 concentrations and low light intensity. The motile swarmer stage therefore probably represents a means of escape when conditions in the *Rhodomicrobium* microcolony become too crowded. The swarmer cell, which in *Rhodomicrobium* has numerous, peritrichous flagella, is solely a dispersal device. It has no reproductive capacity itself and must undergo the transition to a stalked cell before it can generate buds. Dow and Whittenbury (1979) have described this transition as the 'maturation' phase of the *Rhodomicrobium* life cycle, while the cyclic generation of new stalked cells is called the 'reproductive' phase (Figure 3). The molecular mechanisms which underlie the developmental switch between these two phases are not well understood.

The cell cycle of *Caulobacter* (see Figure 4) has been subjected to a more detailed molecular analysis than has that of *Rhodomicrobium* or *Hyphomicrobium*. Differentiation is an obligate part of the cycle: mother cells only give rise to swarmer cell daughters which in turn must change into stalked cells before they themselves can reproduce. As in *Rhodomicrobium*, the *Caulobacter* swarmer is solely a dispersal phase; it can neither duplicate its DNA nor divide. The swarmer cell enters the maturation phase by shedding its flagellum and becoming non-motile. DNA synthesis and prostheca formation are then initiated coordinately.

This prostheca is formed from the site of attachment of the flagellum, which is pushed out as an appendage on the tip of the stalk and is eventually shed into the medium (Shapiro *et al.*, 1971). The stalk grows by the introduction of new material at its base (Schmidt and Stanier, 1966). The distal portion of the stalk is inert and no wall synthesis occurs within it, as is evidenced by its insensitivity to penicillin. The stalk becomes longer with each succeeding cell cycle and crossbands, sometimes called querbalken (bulkheads), are laid down within it. The cross-bands are at least partly made of peptidoglycan since they are sensitive to lysozyme (Schmidt, 1973). They have a complex structure, consisting of concentric rings which stain alternately darkly and lightly with phosphotungstic acid (Jones and Schmidt, 1973). The cross-bands do not necessarily have a hole at their centre and they may contain membrane attachment sites. Both the length of the stalk and the number of cross-bands which it contains have been used as morphological markers of cell age but this interpretation has been challenged. Phosphate limitation results in an acceleration in the rate of stalk extension (Schmidt and Stanier, 1966).

The transition from the motile swarmer to the non-motile stalked stage is marked by the loss of sensitivity to phages ϕCb5 and ϕCbk. The former, which is an RNA phage, binds to the pili of

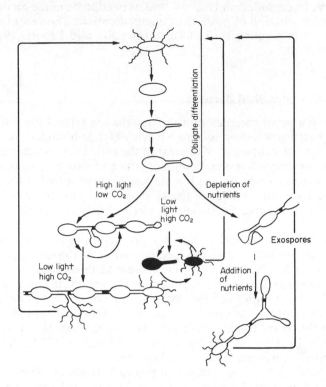

Figure 3 Life cycle of *Rhodomicrobium*. The stalked cell is the only reproductive phase of the cycle. At high light intensities and low CO_2 concentrations stalked cells bud to produce more stalked cells. In low light and high CO_2 progeny cells are of the dispersal 'swarmer' type. In starvation conditions exospores are formed. (Redrawn from Dow and Whittenbury, 1979)

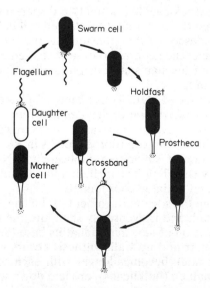

Figure 4 *Caulobacter* cell cycle. (Redrawn from Dow and Whittenbury, 1979)

the swarmer cell while the latter DNA phage has a receptor on the swarmer cell which cannot be correlated with either pili or flagellum. The stalked cell divides asymmetrically at a division furrow which is produced early in the cycle. Wall growth occurs in both the mother and daughter cell. There has been little study of the biogenesis of the daughter cell wall, although its chemistry is well understood. The *Caulobacter* wall is typically Gram-negative except that it lacks the 'lipid A' component of the water-soluble lipopolysaccharide (D. Button and R. D. Bevill, quoted in

Shapiro *et al.*, 1971). About three-quarters of the way through the cell cycle coordinate formation of several surface structures on the daughter cell occurs: the flagellum, several pili and the ϕCbk receptor. When these and DNA replication and segregation are complete the daughter cell is released into the medium to swim away.

The most puzzling thing about the prosthecate bacteria is the purpose of the prostheca itself. In *Caulobacter* it terminates in a holdfast which can mediate adhesion either to surfaces or to the prosthecae of other cells to form rosettes. It may be that this helps to maintain a colony of *Caulobacter* in a favourable ecological niche. However, there is no holdfast associated with the reproductive prostheca of *Rhodomicrobium* and *Hyphomicrobium* and in the latter some strains have holdfasts on the pole opposite to the flagellum of the motile cell (Moore, 1981). It has been suggested that the prostheca serves to separate the cellular processes of mother and daughter cell, but it does not appear to provide a barrier to cytoplasmic exchange. Prosthecae may serve to increase the buoyancy of the cell in its aquatic habitat or to increase the surface area available for solute transport. The extension of the prostheca of *Caulobacter* under conditions of phosphate limitation offers some support for the latter contention.

10.2.5   Budding Yeasts (*Saccharomyces*)

The growth of the budding yeasts is, like that of the budding bacteria, polar or asymmetric. New cell wall material is incorporated into the growing bud and not into its mother cell (Cabib *et al.*, 1981) and, following cell division, mother and daughter cells can be distinguished morphologically. The new-born daughter cell is usually slightly smaller than its mother and must grow to full size before it can itself initiate bud formation (see Section 10.3). The site of separation between mother and daughter cell is marked by a bud scar on the surface of the mother. These bud scars may be stained by fluorescent dyes such as primulin, brightener or calcofluor which bind to the chitin which is found in the bud scar. A new bud cannot be initiated from a preexisting scar (Barton, 1950) and thus a mother cell gains a new scar with each passing generation. Micromanipulatin techniques have been used to demonstrate that a mother cell may only produce a finite number of buds, *ca.* 24 in a diploid strain (Mortimer and Johnston, 1959) and 34 in a brewing strain (Johnston, 1966). Although it has been suggested that this represents a system for studying cellular ageing (Müller, 1971) this seems unlikely, especially as the last-born bud from a given mother is able to go through the full number of generations itself (Johnston, 1966).

The duplication of the nuclear spindle plaque is coincident with the initiation of bud formation in *Saccharomyces cerevisiae* (Byers and Goetsch, 1975). The double plaque is always oriented towards the bud and microtubules extend from it into the bud. Vesicles are found near the free ends of these microtubules and it is thought that these are involved in the transport of enzymes or precursors required for the growth of the bud or septum.

Septum formation in *Saccharomyces cerevisiae* has been studied in great detail by Cabib and his coworkers (Cabib *et al.*, 1973; Cabib, 1976). An electron microscopic examination of bud formation has shown that an electron-translucent ring encircles the base of the bud soon after its initiation. When the bud is almost full grown this ring advances inwards until it closes the gap between the mother and daughter cells (Figure 5). The two cytoplasms are now separated by a thin disc which Cabib has termed the primary septum; it is composed of chitin. This primary septum is thickened by the deposition of secondary septa on either side of it which are composed of mannan. One of the secondary septa forms part of the daughter cell wall while the primary septum remains with the mother cell as the bud scar.

The formation of the primary septum can be prevented by the action of polyoxin D, an antibiotic which inhibits chitin synthesis. Yeast cells treated with polyoxin D continue the process of bud enlargement, showing that septum formation is not required for this. However, in the absence of a completed septum, cell division cannot occur and yeasts treated with polyoxin D eventually lyse.

The synthesis of the chitin of the primary septum is determined by chitin synthetase, an enzyme which exists in the plasma membrane of yeast as an inactive zymogen (Duran and Cabib, 1978). The zymogen is converted into the active enzyme by an activating factor present in yeast vacuoles (Cabib *et al.*, 1973). It is envisaged that the migration of vesicles to the neck of the bud permits the localized activation of the zymogen (see Figure 6). This process is prevented elsewhere by a cytoplasmic inhibitor (Cabib and Farkas, 1971).

The chitin synthetase zymogen can be activated by trypsin *in vitro* (Cabib and Farkas, 1971) and Ulane and Cabib (1976) have purified the activating factor from *Saccharomyces* and shown it

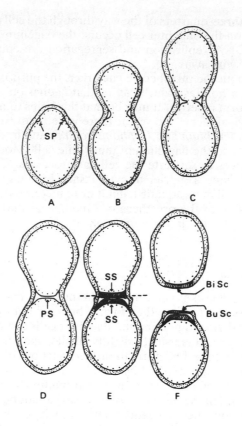

Figure 5 Budding and septum formation in *Saccharomyces cerevisiae*. SP, septal primordia; PS, primary septum; SS, secondary septum; BiSc, birth scar; BuSc, bud scar. (Redrawn from Cabib, 1976)

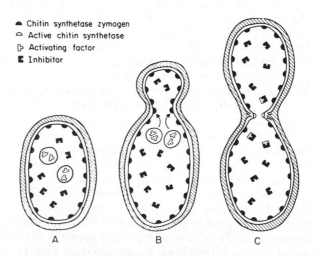

Figure 6 Cabib's model for the spatial regulation of chitin synthesis in yeast. (Redrawn from Cabib and Farkas, 1971)

to be identical with protease B, one of the principal vacuolar proteases. However, mutants deficient in protease B have subsequently been isolated and shown to be proficient in septum formation (Zubenko *et al.*, 1979; Wolf and Ehmann, 1978). Wolf and Ehmann (1981) have gone on to show that mutants deficient in both carboxypeptidases Y and S can also produce normal septa. The presence of four novel carboxypeptidases was revealed in the Y/S deficient mutants and it may be that one of these enzymes is involved in the activation of the zymogen, or at least can substitute for the action of protease B. Until this matter is resolved, Cabib's elegant scheme for the regulation of chitin deposition must retain the status of hypothesis.

10.3 THE CELL CYCLE

The cell cycle has been extensively reviewed (Mitchison, 1971; Lloyd *et al.*, 1982; Elliott and McLaughlin, 1983) and will not be dealt with at length here. Rather, we will confine our remarks to some general points about the timing of DNA replication in the cycle of prokaryotic and eukaryotic cells and about fixed and variable intervals in the cell cycle. The period of DNA replication has been regarded as an important landmark event in the cell cycle ever since studies moved away from purely cytological observations.

In eukaryotic cells the period of nuclear DNA replication, the S-phase, occurs within the period designated interphase by cytologists. S-phase is preceded by G1-phase which is the phase in which non-growing cells arrest under conditions which permit protein synthesis (Unger and Hartwell, 1976). There is increasing evidence that there are important distinctions between the G1-phase which actively growing cells pass through and the resting or G_0 state in which stationary phase and other starved cells arrest (Boucherie, personal communication; Ludwig, personal communication). In particular, G_0 yeast cells synthesize a specific group of proteins, including the so-called 'heat shock' proteins, not found in G1 cells (Boucherie, personal communication). S-phase is followed by a second gap, G2, which itself precedes the M-phase of mitotic nuclear division. In a eukaryotic organism, then, nuclear DNA replication is confined to a particular proportion of the cycle and it is thus ensured that all chromosomes have been replicated before their segregation is attempted in M-phase. The position and length of S-phase is characteristic for the cycles of particular species and examples for *Saccharomyces cerevisiae* and *Schizosaccharomyces pombe* are given in Figure 7.

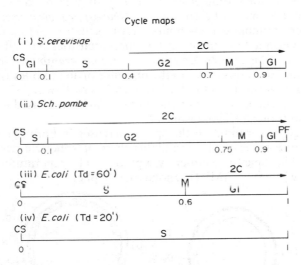

Figure 7 Cell cycle maps for two eukaryotic microbes (*S. cerevisiae* and *S. pombe*) and the prokaryote, *E. coli*. CS, cell separation; G1, gap 1; S, period of DNA synthesis; G2, gap 2; M, mitosis or genome segregation; PF, cell plate (septum) formation; C, one genome complement of DNA; 2C, two genomes complement of DNA; Td, doubling time

Growth rate has an effect on both the absolute and relative length of the different phases of the eukaryotic cell cycle. A batch growth analysis on *S. cerevisiae* was performed by Barford and Hall (1976), a variety of carbon sources being used to set a range of growth rates. They discovered that the S- and M-phases occupied the same length of time at all growth rates and that the *S. cerevisiae* cell cycle adjusted to different growth rates mainly by altering the length of G1. Similar conclusions were reached using a carbon-limited chemostat by Jagadish and Carter (1978) who found that the length of G1 contracted to accommodate higher growth rates but that the time required for the completion of S + G2 + M was relatively constant. A very different conclusion was reached by Rivin and Fangman (1980) who used different nitrogen sources to vary the rate of growth of yeast batch cultures. They found that, over a three-fold range of growth rates, the S-phase varied in absolute length but occupied a constant proportion (*ca.* 50%) of the cycle. These apparent contradictions emphasize the fact that both the nutrient environment and the growth rate may have separate effects on the control of growth and that the chemostat should be used to dissect them (see below).

The division of the cell cycle into G1, S, G2 and M phases is meaningless for prokaryotic organisms. The bacterial genome is organized as a single, circular chromosome which can be segregated into daughter cells by the extension of the cell membrane (Jacob *et al.*, 1963). There is only one origin of replication on the bacterial chromosome but DNA replication can be reinitiated while the current round of replication is still in progress. The replication of the chromosome is completed in a time period *C* which is constant over a wide range of growth rates (Helmstetter and Cooper, 1968; Cooper and Helmstetter, 1968). Cell division is triggered by the completion of a round of DNA replication and occurs at a fixed interval *D* after the termination event. The ability to sustain a number of rounds of DNA replication at the same time enables bacteria to grow with doubling times less than the DNA replication time *C*. At fast growth rates, initiation of DNA replication can take place in one cell cycle to generate daughter molecules which will not be segregated until the end of the next cell cycle. The requirement for the completion of a round of DNA replication to trigger cell division means, however, that bacteria cannot divide with a generation time of less than *D*. For *E. coli* B/r, $C \simeq 40$ min and $D \simeq 20$ min (Cooper and Helmstetter, 1968). This strain, therefore, has a minimum generation time of 20 min. At generation times $\leqslant 40$ min it carries out DNA replication throughout its cell cycle while at slower growth rates there are gap periods in which no DNA synthesis occurs (Figure 7).

10.4 GROWTH OF FILAMENTOUS ORGANISMS

10.4.1 Germination of Fungal Spores

Some fungal spores exhibit constitutive dormancy and therefore do not germinate readily when seeded onto nutrient media. This dormancy can usually be broken experimentally by a heat shock, *e.g.* 3 min at 50 °C for sporangiospores of *Phycomyces blakesleeanus*, or by a chemical treatment, *e.g.* furfural treatment of ascospores of *Neurospora tetrasperma* (Cotter, 1981). Most fungal spores, however, germinate a few hours after inoculation onto a nutrient medium. Some 'non-dormant' spores germinate normally at low spore densities but fail to germinate or germinate slowly when seeded at high spore densities. In *Puccinia graminis tritici*, this self-inhibition of germination at high spore densities is due to the presence of an inhibitor (methyl *cis*-ferulate) and uredospores do not form germ-tubes until it has been leached away (Macko, 1981). In contrast, self-inhibition of spore germination in *Geotrichum candidum* is caused by an inadequate supply of carbon dioxide (Robinson and Thompson, 1982).

During the first stage of germination the spore increases in both diameter (Gull and Trinci, 1971) and biomass (Van Etten *et al.*, 1977) and this phase of 'spherical growth' (Bartnicki-Garcia, 1981) can be prolonged in some fungi either by supraoptimal temperatures (Anderson and Smith, 1972) or by treatment with cyclic AMP (Paznokas and Sypherd, 1975).

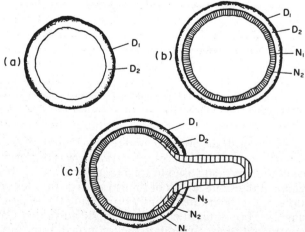

Figure 8   Changes in wall structure during germination of spores of *Botrytis cinerea*. (a) Dormant conidium, (b) swollen conidium, 4 to 6 h after inoculation, and (c) conidium with germ-tube, approximately 10 h after inoculation. D_1 and D_2 represent wall layers in the dormant spore; N_1, N_2 and N_3 represent wall layers formed during germination. (Redrawn from Gull and Trinci, 1971)

During spherical growth, new wall layers are formed in the spore and as the spore increases in diameter the wall of the dormant spore becomes stretched and may eventually rupture (Van Etten *et al.*, 1977; Bartnicki-Garcia, 1981). Gull and Trinci (1971) found that the spore wall of *Botrytis cinerea* increased in thickness from 263 to 339 nm during germination, and, as in other fungi, the germ-tube was continuous with a wall layer formed *de novo* during spherical growth (Figure 8). There is a uniformly dispersed pattern of wall synthesis during spherical growth of sporangiospores of *Mucor rouxii* but wall synthesis becomes polarized prior to germ-tube emergence and germ-tube growth is associated with a transfer of wall synthesizing activity from the swollen spore to the emerging germ-tube (Bartnicki-Garcia and Lippman, 1977). Membrane bound vesicles have been implicated in wall synthesis in fungi and vesicles may also contain enzymes capable of lyzing wall polymers (Rosenberger, 1979). The production of germ-tubes from fungal spores and cysts (Grove, 1972; Grove and Bracker, 1978) and the formation of hyphal initials from regenerating fungal protoplasts (Van der Valk and Wessels, 1976) have been correlated with vesicle accumulation at the site of hyphal emergence, suggesting that the transition from isotropic to anisotropic wall growth involves a change in the pattern of vesicle migration. Although this transition is crucial to many aspects of hyphal morphogenesis, little is known about how it is regulated. De Vries and Wessels (1982) have shown that the point at which hyphae emerge from regenerating protoplasts of *Schizophyllum commune* is influenced by external electrical fields and that outgrowths are preferentially oriented towards the anode. They suggest that the relatively weak polarization of hyphal outgrowths at a low field strength (0.7 mV cell^{-1}) could be explained in terms of a lateral migration of charged particles within the lipid bilayer of the protoplasmic membrane, whilst the stronger polarization observed at higher field strengths could involve the induction of a current within the protoplast which causes negatively charged vesicles to migrate to the anode-facing side. However, De Vries and Wessels were unable to detect any redistribution of chitin synthetase activity prior to hyphal emergence. If a difference in membrane potential between apical and subapical parts of a hypha (Slayman and Slayman, 1962) is involved in maintaining hyphal polarity, it is not upset by external electrical fields of the kind employed by De Vries and Wessels.

10.4.2 Hyphal Morphology

A mycelium of a fungus or an actinomycete consists of a system of branching hyphae which are either aseptate, *e.g. Phycomyces blakesleeanus*, or septate, *e.g.* actinomycetes and fungi belonging to the Ascomycotina, Basidiomycotina and Deuteromycotina. Mycelial hyphae of fungi usually have a diameter of about 3–10 μm but may have a diameter of up to 100 μm, *e.g.* the water mould *Saprolegnia*. Hyphae of actinomycetes are much narrower than those of fungi and usually have a diameter of only 0.5–1.5 μm (Kalakoutskii and Agree, 1976). Mycelia formed by *Coprinus sterquilinus* (a mould) and *Streptomyces coelicolor* (an actinomycete) are illustrated in Figure 9; the hyphae of both mycelia branch monopodially and execute 'avoidance' reactions. The similarity between the morphology of these prokaryotic and eukaryotic mycelia is remarkable and supports the hypothesis of convergent evolution. However, not all actinomycetes form stable mycelia when grown on solid media, *e.g. Nocardia corallina* forms a mycelium which subsequently fragments (*cf.* the mould *Geotrichum candidum*) and members of the Mycobacteriaceae and Actinomycetaceae either do not form mycelia or form transient mycelia (Kalakoutskii and Agree, 1976).

A septate hypha of a mould or an actinomycete is composed of an apical compartment and a series of intercalary compartments (Figure 10). Septa of fungi belonging to the Ascomycotina and most members of the Deuteromycotina usually have a single central pore bordered by a number of spherical, Woronin bodies or, more rarely, hexagonal crystals (Gull, 1978). These pores are wide enough (50–500 nm in diameter) to allow the passage of cytoplasm, nuclei, mitochondria and other organelles from one compartment to the next. However, septal pores may become plugged following hyphal damage or during mycelial differentiation (Trinci and Collinge, 1973a). The most complex type of fungal septum is found in the Basidiomycotina; it is characterized by a swelling around the central pore (the dolipore) and is bounded on either side by a hemispherical, membranous, perforated cap (the parenthosome); nuclei are unable to traverse this type of septum. Mucoraceous fungi, *e.g. Mucor hiemalis*, and actinomycetes form septa which lack pores, whilst *Geotrichum candidum* has a septum which is perforated by numerous micropores 8.5–9.5 nm wide.

The presence of septa and their structure have considerable significance for hyphal extension and branching. Unplugged septa of the type found in the Ascomycotina and Basidiomycotina allow intercalary compartments to contribute to apical extension but this cannot happen in organisms which have complete septa. Septa also influence the location of branch initials (Trinci, 1979).

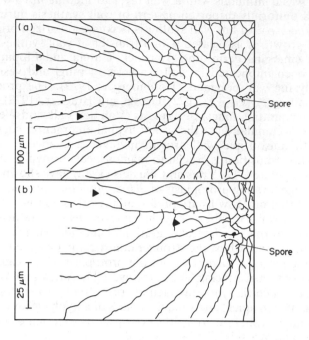

Figure 9 Mycelia of (a) *Coprinus sterquilinus* (redrawn from Buller, 1931) and (b) *Streptomyces coelicolor* (redrawn from Hopwood, 1960). The arrows indicate 'avoidance' reactions

10.4.3 Growth of Individual Hyphae

A mycelial hypha only extends in length at its tip (the extension zone) although apical growth is supported by protoplasm in the non-extending part of the hypha (the peripheral growth zone).

10.4.3.1 The extension zone

Mycelial hyphae of fungi have extension zones which vary from about 2–35 μm in length. When a hypha grows at a linear rate, the length and shape of this zone remain constant and there is a linear relationship between extension zone length and the rate of hyphal extension (Steele and Trinci, 1975a). The shape of the extension zone resembles a half ellipsoid of revolution and expansion of its surface area at any point (P) is related to the cotangent of the angle (α in Figure 10) at the base of the extension zone between P and its longitudinal axis (Trinci and Saunders, 1977). This model of tip growth, which is based upon one proposed by Green (1974), predicts that wall growth (specific rate of area expansion) is maximal at the tip and decreases to zero at the base of the extension zone; since the extension zone wall does not vary appreciably in thickness, the specific rate of wall synthesis at any point in the extension zone must equal the specific rate of area expansion. Predictions of wall growth generated by the cotangent model are qualitatively similar to experimental observations. Since the shape and length of the extension zone remain constant, wall rigidification must move forward along a hypha at exactly the same rate as hyphal extension (Robertson, 1965). Little is known about this rigidification process except that it occurs very rapidly and is influenced by temperature (Robertson, 1958; Steele and Trinci, 1975a). Wessels and Sietsma (1981) have suggested that in *Schizophyllum commune* plastic wall at the hyphal tip becomes transformed into relatively rigid wall at the base of the extension zone by the formation of β-glucan cross-links between chitin microfibrils. Their model of tip growth, unlike that of Bartnicki-Garcia (1973), does not include a requirement for autolysins (enzymes which lyze wall polymers).

Saunders and Trinci (1979) showed that it is not possible to account for tip shape if one assumes that the wall is a rigid layer, with the expansion of each small area of the surface due to the intersusception of new wall material in that area itself. The rate of supply of wall material does, how-

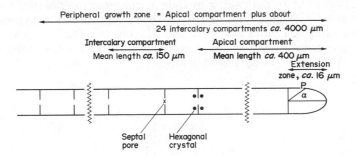

Figure 10   Extension zone and peripheral growth zone of a leading hypha of *Neurospora crassa spco* 9 from the margin of a mature colony

ever, determine the ratio of the specific growth rate in the longitudinal and circumferential directions (the allometric coefficient). Saunders and Trinci suggest that the 'elastic' wall of the extension zone adjusts itself in response to turgor pressure so as to adopt a shape which minimizes surface energy; they thus believe that tip shape is largely determined by wall elasticity. Koch (1982) also suggests that tip shape is determined by surface tension effects. He compares a hyphal tip with a molten glass bubble blown under conditions such that the pressure is constant throughout the tip and that surface tension is very low at the apex and increases towards infinity with distance from the tip. Koch developed a formula which enables the rate of wall synthesis at any point in the extension zone to be calculated from its distance from the hyphal axis and the slope of the tip at that point. Growth rates predicted by Koch are qualitatively similar to experimental observations.

The extension zone of a fungal hypha increases in surface area as vesicles (100–400 nm in diameter) and microvesicles (30–100 nm in diameter) fuse with protoplasmic membrane, releasing their contents (precursors of wall polymers and/or enzymes involved in synthesizing wall polymers) into the expanding wall; this vesicular system provides a convenient way of integrating membrane and wall growth. In some fungi, a large proportion of the volume (up to 80%) of the tip is occupied by vesicles but their concentration decreases steeply towards the base of the extension zone where they only occupy about 5% of the hyphal volume (Collinge and Trinci, 1974). Vesicles are produced in the non-extending part of the hypha by dictyosomes (in Oomycetes) or by the smooth endomembrane system (in other fungi).

Microtubules (Howard and Aist, 1977, 1979; Howard, 1981) and microfilaments (Howard, 1981) have been observed in the tips of hyphae of *Fusarium acuminatum* adjacent to the protoplasmic membrane and parallel to the long axis of the hypha, and it has been suggested that the former structures may be involved in guiding or transporting vesicles to the tip. However, the distribution of vesicles and growth of germlings of *Uromyces phaseoli* var. *vignae* remained normal in the apparent absence of microtubules (Herr and Heath, 1982).

Bartnicki-Garcia *et al.* (1978) have isolated vesicles (chitosomes) from fungi which are similar in size (40–70 nm in diameter) to microvesicles. These vesicles usually contain an inactive (zymogen) form of chitin synthetase which can be activated by proteolytic enzymes. When 'activated' chitosomes are incubated in a mixture containing UDP-*N*-acetylglucosamine and Mg^{2+}, they form microfibrils which are chemically and morphologically indistinguishable from chitin microfibrils found *in vivo*. During chitin synthesis a fibroid structure is formed within the chitosome; the 'membrane' of the chitosome is then opened or shed and a slender microfibril, which is continuous with the fibroid particle, emerges. Chitin synthetase is usually present in a predominantly active form in the hyphal tip and in a predominantly zymogenic form in the rest of the hypha (Archer, 1977; Isaac *et al.*, 1978); it is thought that active chitin synthetase is located in the protoplasmic membrane where it receives substrates from the cytoplasm and releases chitin chains into the expanding wall.

Preparations containing vesicles have also been obtained from fungi which can synthesize $\beta1$–3, $\beta1$–4, $\beta1$–6 and $\alpha1$–4 linkages from UDP-glucose; these linkages are commonly found in wall polymers (Gooday and Trinci, 1980). The possession of a vesicular system enables fungi to synthesize their extension zone walls very rapidly, and its presence is probably one of the main reasons why fungal hyphae are able to attain very high extension rates (up to 100 μm min^{-1}). Besides vesicles and ribosomes, the extension zone of a fungal hypha usually contains endoplasmic reticulum and mitochondria; nuclei, storage products and vacuoles are never present.

Compared with the fungi, few studies have been made of the growth of actinomycete hyphae. Braña *et al*. (1982) have shown that wall growth in *Streptomyces antibioticus*, as in fungi, preferentially occurs at the tip (Figure 11), and it has been reported that hyphae of *Streptomyces hygroscopicus* have extension zones which are about 20 μm long (Kalakoutskii and Agree, 1976). Actinomycete hyphae have a typical prokaryotic type of cellular organization and their growth does not involve the vesicular system found in fungi.

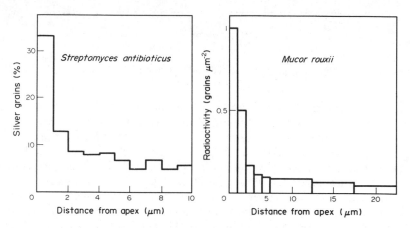

Figure 11 Wall synthesis in hyphae of *Streptomyces antibioticus* (redrawn from Braña *et al*., 1982) and *Mucor rouxii* (redrawn from Bartnicki-Garcia and Lippman, 1969) incubated in *N*-acetyl-D-glucosamine for 1 min

10.4.3.2 Cytology of the non-extending part of fungal hyphae

The first nucleus in a hypha is usually situated some distance from the tip and its position varies during hyphal growth (Trinci, 1979). In *Basidiobolus ranarum* (Robinow, 1963) and *Uromyces phaseoli* var. *vignae* (Heath and Heath, 1978), protoplasm migrates towards the tip, leaving behind a hypha which is virtually devoid of protoplasm. In other fungi the protoplasm becomes progressively more vacuolated with distance from the tip. Organelles such as mitochondria and nuclei maintain characteristic positions during hyphal extension, suggesting that there is some cytoplasmic organization of cell components which may involve microtubules (Heath and Heath, 1978; Herr and Heath, 1982).

10.4.3.3 The peripheral growth zone

During linear growth, tip extension of a hypha is supported by a volume of protoplasm which remains constant as the hypha moves forward; the hyphal region involved in tip growth is called the peripheral growth zone (Figure 10) and the linear growth rate (K_r) of a hypha is a function of the length of this zone (w) and the organism's specific growth rate (μ). Thus:

$$K_r = \mu w \tag{1}$$

In fungi and actinomycetes which have complete septa the peripheral growth zone must be restricted to the apical compartment. It is not clear, however, how these hyphae maintain linear growth despite appreciable changes in apical compartment length following septation. However, it may be significant that the volume of protoplasm in germlings of *Uromyces phaseoli* var. *vignae* remains virtually constant and migrates forward at the same rate as tip extension (Heath and Heath, 1978). In contrast, the peripheral growth zone of *Neurospora crassa* consists of a large number of intercalary compartments in addition to the apical compartment (Figure 10) and in this fungus, cytoplasm, vacuoles and nuclei migrate through septal pores towards the hyphal tip.

Any change in the rate of extension of a hypha must be associated with a change in peripheral growth zone length, a change in specific growth rate or a change in both (equation 1). Further, hyphal extension or colony radial growth rate can only be used to assess the effect of a particular variable, *e.g.* temperature, on fungal growth provided that the variable concerned does not affect peripheral growth zone length. The validity of hyphal extension and colony radial growth rate as parameters of fungal growth is discussed by Bull and Trinci (1977).

10.4.4 Growth of Mycelia

When cultured under unrestricted conditions (all nutrients present in excess and in the absence of inhibitory substances, Righelato, 1975), a fungal spore germinates to form a mycelium whose total hyphal length and number of branches increase exponentially at a specific rate characteristic of the organism and the conditions; the mean length of hypha associated with each tip is a constant called the hyphal growth unit (Trinci, 1974). At any given time, some hyphae in such a mycelium will be extending at the linear rate characteristic of the organism and the conditions, whilst the remainder will be accelerating towards this rate (Trinci, 1974). Steele and Trinci (1975b) showed the mean hyphal extension rate (E) of a fungal mycelium is also a constant and is a function of the length of the organism's hyphal growth unit (G) and its specific growth rate (μ). Thus:

$$E = G\mu \qquad (2)$$

In contrast, germ-tubes formed by *Candida albicans* and actinomycetes (Schuhmann and Bergter, 1976; Allan and Prosser, 1983) initially grow at a linear rate and thus the growth of mycelia of these organisms is biphasic (Figure 12). However, the general pattern of growth of mycelia of moulds and actinomycetes is very similar (Figures 11 and 12), and Schuhmann and Bergter (1976) and Gow and Gooday (1982) have shown that equation (2) also describes the growth kinetics of *Streptomyces hygroscopicus* and *Candida albicans*.

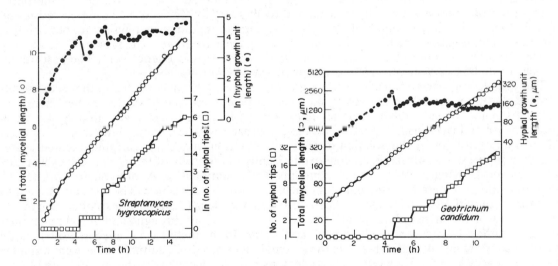

Figure 12 Growth of mycelia of *Streptomyces hygroscopicus* (redrawn from data of Schuhmann and Bergter, 1976) and *Geotrichum candidum* (redrawn from Trinci, 1974) on solid media

Steele and Trinci (1975b) and Gow and Gooday (1982) observed differences between the dimensions of hyphae grown under unrestricted conditions and those formed at the margin of mature colonies; this variation in hyphal morphology is reflected in differences in peripheral growth zone length and extension rate.

10.5 YEAST–MYCELIAL DIMORPHISM

A number of yeasts form pseudomycelia in late stationary phase and this can affect the mass transfer characteristics of a fermentation and the recovery of organisms or product. A few organisms, however, have two distinct modes of vegetative growth, as true yeasts or true mycelia. The dimorphic fungi which have been most thoroughly studied are *Candida albicans* and a number of species of *Mucor* (Sypherd *et al.*, 1978; Stewart and Rogers, 1983).

The ability to switch from a yeast to a mycelial form is important to the pathogenicity of *Candida albicans*. A classic method of triggering the yeast–mycelial transition in the laboratory is to inoculate yeast into a medium containing serum. The other is to dilute stationary (but not exponential) phase yeast cells into fresh medium and at the same time to heat shock from 25 to 37 or 40 °C. Such temperature shocks are known to induce an increased rate of synthesis of a specific group of proteins, the 'heat shock' proteins, in a number of organisms including the yeast *Saccharomyces cerevisiae* (McAlister and Finkelstein, 1980; Miller *et al.*, 1980).

A number of regimes have been used to induce the yeast–mycelial transition in *Mucor* (Sypherd *et al.*, 1978). The most common method is to change the atmosphere in which the culture is growing from CO_2 to air. However, mycelial growth may be attained in 30% CO_2/70% N_2 or even in 100% N_2 if the glucose concentration in the medium is kept low (Bartnicki-Garcia, 1968). Correspondingly high concentrations of hexose sugars maintain *Mucor genevensis* in the yeast phase even under aerobic conditions (Rogers *et al.*, 1974). Mooney and Sypherd (1976) have demonstrated that *M. racemosus* may be maintained in either the yeast or mycelial phase when growing under N_2 on 2% glucose depending on the flow rate of the gas. This indicates that some volatile compound must be produced which stimulates either yeast or hyphal growth.

Although the phenomenon of yeast–mycelial dimorphism has been known for more than a century, there is still no clear idea as to the mechanism of the change in morphology. The yeast cell wall of *Mucor* is much richer in both mannose and protein than is the hyphal wall (Bartnicki-Garcia and Nickerson, 1961); however, there are no striking differences in the pattern of individual proteins synthesized by the two cell types (Sypherd *et al.*, 1979). The search for regulatory molecules involved in morphogenesis has relied, almost exclusively, on correlative evidence. Thus changes in the levels of cyclic 3',5'-adenosine monophosphate (cAMP, Larsen and Sypherd, 1974) in the rates of RNA, protein and polyamine synthesis (Orlowski and Sypherd, 1977, 1978a, 1978b; Larsen and Sypherd, 1979; Sypherd *et al.*, 1979), in the activity of various enzymes (Paznokas and Sypherd, 1977; Peters and Sypherd, 1979; Inderlied *et al.*, 1980), and in the phosphorylation of ribosomal proteins (Larsen and Sypherd, 1979) have all been implicated in the morphological switch. A number of changes in addition to that of growth mode are occurring at the same time, notably changes in nutritional status and in growth rate. The problem is to identify which metabolic or molecular changes are specific to the morphogenetic event. The best way to do this would be to isolate mutants specifically blocked in morphogenesis.

The genetic approach has made little progress to date and this may be due to the fact that differences between growth in the yeast and mycelial phases may be of degree rather than kind (Thompson and Wheals, 1980). Thus any mutation which blocks mycelial growth may also block yeast growth and *vice versa*. The most profitable approach may be to control the rate of synthesis of individual proteins by the isolation of nonsense mutations in a strain carrying a temperature-sensitive suppressor. This has been employed recently in the study of the *E. coli* cell cycle (Lutkenhaus and Donachie, 1979). The *coy* mutants (for *c*onditional *y*east) of Sypherd *et al.* (1979) require methionine, not for growth, but for the yeast–mycelial transition. The lesion is thought to be associated with polyamine metabolism. Orlowski and Ross (1981) have exploited this mutation to study the changes which occur when two cultures are put through an identical nutritional switch in either the presence or absence of methionine. In the first case a change in growth mode accompanies the nutritional shift and in the second it does not. Both cultures underwent a similar increase in growth rate. They were able to demonstrate that a change in the specific rate of protein synthesis was a consequence of the change in the growth rate, and not the morphology of the culture. However, a fall in the intracellular level of cAMP only occurred in cells undergoing the morphological transition. These data indicate that the genetic approach should eventually solve the problem of the mechanism of yeast–mycelial dimorphism.

10.6 COLONY GROWTH

10.6.1 Growth of Colonies on Solid Media

Pirt (1967) considered the formation of a microbial colony on a solid medium and predicted that (1) cells would initially grow exponentially; (2) a gradient of nutrient concentration would be formed in the medium beneath the colony; (3) eventually the concentration of some 'growth-limiting' nutrient would fall virtually to zero in the medium under the centre of the colony and that these central cells would stop growing; (4) growth of the colony would eventually be restricted to a peripheral annulus and that the width of this zone would remain constant; and (5) the colony would eventually expand at a linear rate.

The initial exponential phase of microbial growth on solid medium has been confirmed experimentally for some unicellular (Wimpenny and Lewis, 1977; Wimpenny, 1979) and filamentous (Schuhmann and Bergter, 1976) bacteria, and for moulds (Trinci, 1974). Rieck *et al.* (1973) have confirmed that gradients in glucose concentration are formed beneath bacterial colonies, and various workers have shown that cells at the margin of fungal and bacterial colonies grow faster than cells in other parts of the colony (Reyrolle and Letellier, 1979; Wimpenny, 1979; Galun, 1972). However, as shown in Figure 13, growth does occur in the centre of both bacterial and fungal colonies even when these colonies are expanding at a linear rate. Wimpenny (1979) suggests that cells at the centre of most bacterial colonies grow slowly at a diffusion-limited rate and showed that bacterial colonies are made up of a peripheral region whose height rises relatively steeply and a central region whose height rises much less steeply (Figure 13). Reyrolle and Letellier (1979) used autoradiography to estimate the relative growth rates of cells within bacterial colonies, and their results suggest that only cells in the top two thirds (*ca.* 70 μm) of 34 h colonies of *Pseudomonas putida* (an obligate aerobe) receive sufficient oxygen for growth. Similarly, Wimpenny and Parr (1979) found very much higher specific activities of oxidative enzymes in the top 120 μm of colonies of *Enterobacter cloacae* than in the basal 580 μm. Thus, growth of cells within a microbial colony is influenced by diffusion of oxygen into the colony from above and diffusion of nutrients from below.

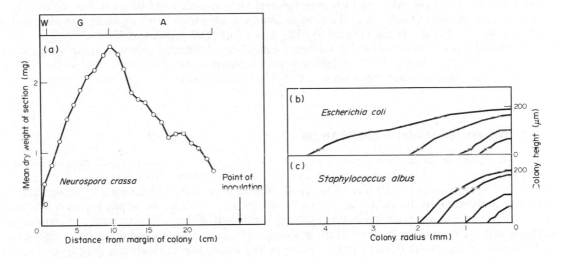

Figure 13 (a) Distribution of biomass in a colony of *Neurospora crassa* cultured in a growth tube on solid medium; W, peripheral growth zone, G, internal growth zone, and A, autolysis zone (redrawn from Gillie, 1968). (b) and (c) Profiles of colonies of *Escherichia coli* and *Staphylococcus albus* at various times after inoculation (redrawn from Wimpenny, 1979)

As predicted by Pirt, colonies of yeasts (Gray and Kirwan, 1974), non-motile bacteria (Pirt, 1967), gliding bacteria (Burchard, 1974) and fungi increase in radius at a linear rate (Table 1). Although the colonies of most fungi and some bacteria (Burchard, 1974) expand at this linear rate until they have completely colonized the medium, colonies of most unicellular microorganisms and some fungi only grow at a linear rate for a time, after which expansion slows and may eventually stop (Gray and Kirwan, 1974; Palumbo *et al.*, 1971). Cooper *et al.* (1968) found that during this deceleration phase, the area of the colony increased linearly with time, indicating a decrease in peripheral growth zone width. This decrease in the rate of colony expansion may be correlated with diffusion of secondary metabolites from the centre to the periphery of the colony and with a reduction in nutrient concentration at the colony margin (Rieck *et al.*, 1973).

Fungi and actinomycetes usually branch monopodially and this is a very economical and efficient branching pattern for substrate colonization. Leading hyphae at the margin of fungal and actinomycete colonies grow approximately parallel to one another, and at approximately the same distance apart (Figure 9), so that new leading hyphae are added as the colony expands. Moulds, unlike unicellular, non-motile bacteria, can regulate their hyphal density in a manner

Table 1   Growth of Bacterial and Fungal Colonies on Solid Medium

Organism	Incubation temperature	Width of peripheral growth zone (w, μm)	Specific growth rate[b] (μ, h⁻¹)	Colony radial growth rate (K_r, μm⁻¹)	Reference
Bacteria					
Escherichia coli	25	91.4[a]	0.28	18	Pirt (1967)
Streptococcus faecalis	25	ND	0.65	18	Pirt (1967)
Pseudomonas fluorescens	26	ND	0.59	29	Palumbo *et al*. (1971)
Myxococcus xanthus F By (motile strain)	30	ND	0.21	192	Burchard (1974)
Myxococcus xanthus NM (non-motile strain)	30	ND	ND	20	Burchard (1974)
Streptomyces coelicolor	30	ND	0.32	22	Allan and Prosser (1983)
Fungi					
Candida albicans (mycelial form)	37	119[a]	0.39	46	Gow and Gooday (1982)
Penicillium chrysogenum	25	496	0.16	76	Trinci (1971)
Neurospora crassa SYR-17-3A	25	6800	0.26	2152	Trinci (1973)

[a] Estimated values; ND = not determined.   [b] Usually determined in shake flask culture.

which ensures spread of the colony even on media containing low concentrations of nutrients (Bull and Trinci, 1977). The mechanism involved in regulating hyphal density is not fully understood but involves an 'avoidance' reaction.

Fungi usually form colonies with wider peripheral growth zones and faster radial growth rates than bacterial colonies (Table 1), and although actinomycetes form mycelia which are morphologically similar to fungal mycelia (Figure 9), they never form fast growing colonies. Thus, the success of moulds as colonizers of solid subtrates cannot be attributed solely to their filamentous morphology. It may, however, be significant that actinomycetes have complete septa, lack a vesicular growth system and form a very small hyphal growth unit.

10.6.2   Growth of Colonies in Liquid Media

When a mould or an actinomycete forms disperse, filamentous mycelia in submerged culture, all hyphae are exposed to the medium and growth is exponential. However, when these organisms form large colonies (pellets) in submerged culture, growth is not exponential because nutrients cannot diffuse into the pellet fast enough to maintain growth of the whole biomass. Instead, when a pellet attains a certain diameter, the peripheral shell of the pellet continues to grow exponentially but the remaining biomass stops growing and may eventually autolyze, forming a space at the centre of the pellet (Clark, 1962). When pellets attain this critical size, they subsequently increase in diameter at a linear rate (Trinci, 1970) and their growth is described by cube root kinetics (Marshall and Alexander, 1960; equation 3).

$$x^{1/3} = x_0^{1/3} + kt \tag{3}$$

In equation (3) k is a constant. Pirt (1966) suggested that for an obligately aerobic organism, oxygen was the nutrient most likely to limit growth and he predicted that the region of the pellet exhibiting growth rates between zero and the maximum value would be quite narrow. The thickness of the shell of the pellet which grows exponentially will depend upon the depth to which medium can penetrate the pellet, and this in turn is influenced by the texture of the pellet. If the margin of a pellet is densely packed with hyphae, nutrients will only be able to penetrate the pellet by simple diffusion and in this case Pirt (1966) predicted that the growing shell would be only 77 μm wide; this value should be compared with the depth to which oxygen can apparently penetrate bacterial colonies on solid media. In contrast, the medium will be able to penetrate more deeply into pellets which have a loose texture and therefore these pellets will have wider, exponentially growing shells. Yoshida *et al.* (1968) suggested that oxygen penetrated the peripheral 4 mm of pellets of *Lentinus edodes* but this work has been criticized by Metz and Kossen (1977). In contrast, the results of Kobayashi *et al.* (1973) suggest that the central biomass of pellets of *Aspergillus niger* becomes oxygen limited before they attain a radius of 250 μm.

van Suijdam *et al.* (1982) have formulated an unstructured model to describe pellet growth in submerged culture; this model integrates the kinetics of hyphal growth with the physical mechanisms of mass transfer in the pellets and fermenter. In contrast to previous workers van Suijdam *et al.* have incorporated the phenomenon of autolysis in their model, but even this model fails to consider the effects on growth which result from the diffusion of secondary metabolites and products of autolysis from the core to the periphery of the pellet and then into the medium.

In some fungi a pellet may be formed from a single spore, but in most fungi pellets are formed from a number of spores which aggregate prior to germ-tube emergence (Galbraith and Smith, 1969). Hyphae (Burkholder and Sinnot, 1945) and even small pellets (Clark, 1962) may also aggregate to form large pellets. This aggregation phenomenon explains why only a small number of pellets are formed even when cultures are inoculated with very large numbers of spores (Trinci, 1970). Spore aggregation can be reduced by incorporating anionic polymers such as carbopol-934 (carboxypolymethylene) or Junlon-111 (a cross-linked acrylic resin) in the medium (Trinci, 1983). Elmayergi *et al.* (1973) suggest that carbopol is effective because its ionized carboxyl groups induce electrostatic repulsion between spores and found that carbopol was adsorbed on the surfaces of spores and hyphae during growth but was released at the end of the fermentation.

Cultures of mycelial pellets exhibit low viscosities and more or less Newtonian flow behaviour. In contrast, cultures containing disperse, filamentous mycelia exhibit high viscosities and pseudoplastic flow behaviour, and since the oxygen transfer coefficient falls with increasing viscosity, such cultures may rapidly become oxygen limited. Metz (1976) found that the viscosity of cultures of *Penicillium chrysogenum* was related to biomass concentration and hyphal growth unit length (equation 4).

$$\eta_\infty = (K_2 \times G^{0.6})^2 \tag{4}$$

In equation (4) η_∞ = viscosity at infinitely high shear rate, K_2 is a constant, x = biomass concentration and G = hyphal growth unit length. Thus, mycelial morphology has a significant influence on culture viscosity.

10.7 EFFECT OF GROWTH RATE AND OTHER VARIABLES ON CELL COMPOSITION AND MORPHOLOGY

10.7.1 Unicellular Organisms

At extremely low growth rates *Aerobacter aerogenes* shows a dramatic change in morphology producing long filamentous cells (Tempest *et al.*, 1967). At these very low growth rates (less than 6% of the maximum specific growth rate, μ_{max}) an increasing proportion of the cells in the culture are in a non-growing state (Tempest *et al.*, 1967; Pirt, 1972). This means that there is a minimum growth rate (μ_{min}) as well as a maximum (μ_{max}) for a given organism. In the case of *A. aerogenes* μ_{min} was found to be 0.009 h^{-1} (equivalent to a doubling time of 77 h) at 37 °C. Determination of μ_{min} is complicated, however, by an increase in cell lysis at low growth rates and 'cryptic' growth on the lysis products (Drozd *et al.*, 1978). This phenomenon will also lead to an overestimate of the organism's maintenance coefficient (Pirt, 1965) at these low growth rates.

A number of organisms undergo quite specific changes in morphology, determined by their growth rate. Luscombe and Gray (1971) demonstrated that, at any given temperature, there was a specific value of D (dilution rate) above which *Arthrobacter* would grow in the rod form and below which it grew as a coccus. They suggested that cocci could be considered as rods of greatly reduced length and pointed out that at normal soil temperatures *Arthrobacter* would grow as a coccus. It seems likely that the observation that *Nocardia corallina* grows as a coccus on minimal media and as a rod on complex media (Heinzen and Ensign, 1975) also can be attributed to the effect of growth rate. However, this observation again emphasizes the importance of using the chemostat to distinguish between nutritional and growth rate effects.

The study of the effect of growth rate on the cellular proportions of DNA, RNA and protein and on the rates of synthesis of these macromolecules has obvious implications for growth control (Nierlich, 1974). The protein and DNA content of bacterial cells increases in direct proportion to the growth rate (μ) whereas RNA content increases as μ^2 (Maaløe and Kjelgaard, 1966; Nierlich, 1972; Rosset *et al*, 1964, 1966; Schaechter *et al.*, 1958). The syntheses of the different RNA species do not increase coordinately with increasing growth rate. The rate of rRNA synthesis increases to a greater extent than either tRNA or mRNA synthesis and accounts for the geometric relationship between RNA content and growth rate.

The disproportionate increase in rRNA content reflects the fact that it is the number of ribosomes in a bacterial cell which determines its rate of protein synthesis and, therefore, its growth rate (Brachet, 1960). Bacterial ribosomes work at their maximum efficiency at all growth rates (Dennis and Bremer, 1973; Schaechter *et al.*, 1958). This means that during a shift-up in growth rate (due, for instance, to enrichment of the nutrient medium) there is a lag before the net rate of protein synthesis may increase (Dennis and Bremer, 1973; Neidhardt, 1964; Nierlich, 1974). During this lag there is a rapid increase in the rate of synthesis of both rRNA and ribosomal proteins in order to supply the larger number of ribosomes that are required to sustain an increased rate of protein synthesis.

Among eukaryotic microorganisms these growth rate relationships have been studied most extensively using *Saccharomyces cerevisiae*, *Schizosaccharomyces pombe* and *Neurospora crassa* (Oliver, 1981). In *S. cerevisiae* cellular DNA content is constant at all growth rates (Waldron and Lacroute, 1975); this probably reflects the fact that DNA replication in eukaryotes is confined to a portion of the cycle time, the S-phase (see Section 10.3). The protein content of yeast cells varied directly with growth rate, but the pattern for RNA content was more complex. As in the bacteria, the response of rRNA and tRNA synthesis to increasing growth rate was non-coordinate. Furthermore, at slow growth rates ($\mu < 0.14$ h^{-1}) cellular RNA content was constant and did not fall with decreasing growth rate. A marked difference between yeast and bacteria is that the efficiency of yeast ribosomes is not constant but increases with increasing growth rate (Boehlke and Friesen, 1975; Waldron and Lacroute, 1975; Bull and Trinci, 1977).

This last fact has important consequences for the response of a eukaryotic microorganism, such as yeast, to a shift-up. It means that there is spare translational capacity at lower growth rates and that cells may increase their rate of protein synthesis immediately following the nutritional shift (Ludwig *et al.*, 1977). After a short lag there is a large shift in the rate of synthesis of rRNA, that of tRNA increasing later and to a lesser extent (Ludwig *et al.*, 1977; Waldron, 1977). The rate of DNA synthesis does not increase until the cells achieve their new rate of balanced growth.

The mean cell volume of both the budding yeast *Saccharomyces cerevisiae* (Jagadish *et al.*, 1977; Lord and Wheals, 1980; McMurrough and Rose, 1967; Mor and Fiechter, 1968; Thompson and Wheals, 1980; Tyson *et al.*, 1979) and the fission yeast *Schizosaccharomyces pombe* (Fantes and Nurse, 1977) increases as an exponential function of the growth rate. In budding yeast the new-born daughter cell is always smaller than its mother (Lord and Wheals, 1980) and this asymmetry of cell size at division increases with decreasing growth rate (Jagadish *et al.*, 1977; Lord and Wheals, 1980; Thompson and Wheals, 1980). The new-born daughter cell must grow to a certain critical size (which is itself a function of growth rate) before it can begin a new cell cycle (Jagadish *et al.*, 1977; Lord and Wheals, 1980; Nurse 1981; Thompson and Wheals, 1980; Tyson *et al.*, 1979). This means that daughter cells have a longer generation time than do their mothers and that the lower the growth rate the larger will be the proportion of daughter cells in a culture of *S. cerevisiae* (Jagadish *et al.*, 1977).

Cell size also plays a critical role in controlling cell division in *Schizosaccharomyces pombe*. Under most conditions this control is operated at the level of nuclear division (Fantes and Nurse, 1977); successful nuclear division is, of course, an absolute prerequisite for cell division. Fantes and Nurse (1977) studied the effects of shifts in growth rate on cell size in *S. pombe*. A shift-down resulted in accelerated cell division and a reduction in cell size. During a shift-up *ca.* 20% of cells divided at their original length. These were, presumably, those already committed to division at the time of the shift. Cell division then halted for a period before resuming at a faster rate, cell length at division then being greater and characteristic for the new, higher, growth rate.

The use of chemostat culture has revealed that at slow growth rates ($\mu < 0.15$ h^{-1}) a proportion of *S. cerevisiae* cells are not actively dividing but are arrested in the G_0 or stationary phase state (Bujega *et al.*, 1982). These cells are phase-light and resistant to wall-degrading enzymes, whereas actively dividing cells appear dark under the phase contrast microscope and are readily lyzed by such enzymes. The non-dividing cells represented 30–40% of the culture at growth rates of $\mu = 0.05$–0.06 h^{-1}). At very low growth rates ($\mu = 0.03$ h^{-1}) a proportion of *S. cerevisiae* cells become filamentous (Walmsley and Oliver, unpublished), a phenomenon which is also shown at the fastest growth rates ($\mu \geqslant 0.3$ h^{-1}). In the latter case this has been attributed to a failure in cell separation (Thompson and Wheals, 1980). Low growth rates trigger more specific differentiation events in *Schizosaccharomyces pombe*. McDonald *et al.* (1982) have found that at $\mu = 0.06$ h^{-1} and under conditions of low aeration, a carbon-limited (but not a nitrogen-limited) chemostat culture of *S. pombe* will conjugate and sporulate.

10.7.2 Fungi and Actinomycetes

When the hyphal growth unit hypothesis was proposed (Caldwell and Trinci, 1973; Trinci, 1974) it was assumed that the mean radius of hyphae remained constant when mycelia grew exponentially under unrestricted conditions. It was concluded therefore that hyphal length could be used as a valid parameter of hyphal volume and that growth of a mycelium involved the duplication of a growth unit whose length (G) and volume remained constant (Trinci, 1974). However, Robinson and Smith (1979) have shown that when *Geotrichum candidum* is grown in chemostat (restricted) culture, hyphal diameter increases with increase in dilution rate, although the volume of the growth unit does not vary appreciably with dilution rate or with temperature (Figure 14); the growth unit of *G. candidum* had a volume of about 1500 μm^3 at 25 °C and about 1470 μm^3 at 30 °C. Similarly, Riesenberg and Bergter (1979) have shown that the mean diameter of hyphae of *Streptomyces hygroscopicus* increases with increase in dilution rate (Figure 14), but in this species hyphal growth unit volume also varies with dilution rate (from about 1.07 μm^3 at 1.20 h^{-1} to about 2.30 μm^3 at 0.418 h^{-1}).

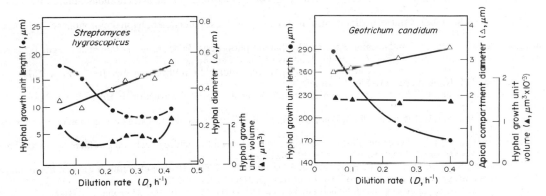

Figure 14 Effect of dilution rate on hyphal growth unit length, hyphal growth unit volume and hyphal diameter of mycelia of *Streptomyces hygroscopicus* (redrawn from data of Riesenberg and Bergter, 1979) and *Geotrichum candidum* (redrawn from Robinson and Smith, 1979)

Morrison and Righelato (1974) and Miles and Trinci (1983) found that at a stirrer speed of 1000 rev min^{-1} the length of the hyphal growth unit of *Penicillium chrysogenum* decreased significantly with increase in dilution rate. A similar result was obtained by Robinson and Smith (1979) with *G. candidum* and Riesenberg and Bergter (1979) with *Streptomyces hygroscopicus* (Figure 14). In contrast, when *P. chrysogenum* was grown at a stirrer speed of 500 rev min^{-1} hyphal growth unit length increased with increase in dilution rate (Miles and Trinci, 1983). van Suijdam and Metz (1981) also found that stirrer speed had a significant effect on mycelial morphology; they grew *P. chrysogenum* at a dilution rate of 0.055 h^{-1} and found that hyphal growth unit length typically increased from 54 $\pm$ 20 μm (mean and standard deviation) at 1000 rev min^{-1} to 89 $\pm$ 28 μm at 450 rev min^{-1}. Thus, mycelial morphology is affected both by dilution rate and shear stress. If it is assumed that the volume (V_g) of the growth unit of a fungus remains constant during mycelial growth in a chemostat, it follows that hyphal growth unit length (G) will be inversely related to the square of hyphal radius (r; equation 5).

$$G = V_g/(\pi r^2) \tag{5}$$

The implication of equation (5) is that relatively small changes in hyphal radius will have appreciable effects on hyphal growth unit length. Equation (2) can be rewritten as equation (6).

$$E = [V_g/(\pi r^2)]\mu \tag{6}$$

If V_g and r remain constant, changes in μ will only affect branching frequency (G) if E is not linearly related to μ, *i.e.* if the ratio E/μ is not constant. Trinci (1973, 1974) showed that the length of the hyphal growth unit of batch cultured *Neurospora crassa spco* 1 did not vary significantly with temperature and that E did indeed vary directly with μ. In contrast, L-sorbose inhibited hyphal extension of *N. crassa spco* 1 but did not affect its specific growth rate; L-sorbose consequently causes a change in the ratio E/μ which results in the formation of densely branched mycelia (Trinci and Collinge, 1973b).

Steady-state chemostat cultures of fungi can only be obtained provided that mycelia fragment. This fragmentation may occur spontaneously (Caldwell and Trinci, 1973) or may result from shear stress imposed by the impeller. van Suijdam and Metz (1981) found that shear stress only had a limited influence on the length of hyphae of *P. chrysogenum*. However, shearing of mycelia occurred when *Trichoderma aureoviride* was grown in chemostat cultures at stirrer speeds of *ca.* 1100 rev min^{-1} (Pitt and Bull, 1982). This shearing resulted in the formation of large numbers of non-growing, sheared tips.

10.8 REFERENCES

Allan, E. J. and J. I. Prosser (1983). Mycelial growth and branching of *Streptomyces coelicolor* on solid medium. *J. Gen. Microbiol.*, **129**, 2029–2036.

Anderson, J. G. and J. E. Smith (1972). The effects of elevated temperatures on spore swelling and germination in *Aspergillus niger*. *Can. J. Microbiol.*, **18**, 289–297.

Archer, D. B. (1977). Chitin biosynthesis in protoplasts and subcellular fractions of *Aspergillus fumigatus*. *Biochem. J.*, **164**, 653–658.

Archibald, A. P. and H. E. Coapes (1976). Bacteriophage SP50 as a marker for cell wall growth in *Bacillus subtilis*. *J. Bacteriol.*, **125**, 1195–1206.

Barford, J. P. and R. J. Hall (1976). Estimation of the length of cell cycle phases from asynchronous cultures of *Saccharomyces cerevisiae*. *Exp. Cell Res.*, **102**, 276–284.

Bartnicki-Garcia, S. (1968). Control of dimorphism in *Mucor* by hexoses. Induction of hyphal morphogenesis. *J. Bacteriol.*, **96**, 1586–1594.

Bartnicki-Garcia, S. (1973). Fundamental aspects of hyphal morphogenesis. In *Microbial Differentiation*, ed. J. M. Ashworth and J. E. Smith, pp. 245–267. Cambridge University Press, Cambridge.

Bartnicki-Garcia, S. (1981). Cell wall construction during spore germination in Phycomycetes. In *The Fungal Spore*, ed. G. Turian and H. R. Hohl, pp. 533–556. Academic, London.

Bartnicki-Garcia, S. and E. Lippman (1969). Fungal morphogenesis: cell wall construction in *Mucor rouxii*. *Science*, **165**, 302–303.

Bartnicki-Garcia, S. and E. Lippman (1977). Polarization of cell wall synthesis during spore germination of *Mucor rouxii*. *Exp. Mycol.*, **1**, 230–240.

Bartnicki-Garcia, S. and W. Nickerson (1961). Isolation, composition and structure of cell walls of filamentous and yeast-like forms of *Mucor rouxii*. *Biochem. Biophys. Acta*, **58**, 102–119.

Bartnicki-Garcia, S., C. E. Bracker, E. Reyes and J. Ruiz-Herrera (1978). Isolation of chitosomes from taxonomically diverse fungi and synthesis of chitin microfibrils *in vitro*. *Exp. Mycol.*, **2**, 173–192.

Barton, A. A. (1950). Some aspects of cell division in *Saccharomyces cerevisiae*. *J. Gen. Microbiol.*, **4**, 84–86.

Beachey, E. H. and R. M. Cole (1966). Cell wall replication in *Escherichia coli*, studied by immunofluorescence and immunoelectron microscopy. *J. Bacteriol.*, **92**, 1245–1251.

Begg, K. J. (1978). Cell surface growth in *Escherichia coli*: distribution of matrix protein. *J. Bacteriol.*, **135**, 307–310.

Begg, K. J. and W. D. Donachie (1973). Topography of outer membrane growth in *E. coli*. *Nature (London), New Biol.*, **245**, 38–39.

Begg, K. J. and W. D. Donachie (1977). Growth of the *Escherichia coli* cell surface. *J. Bacteriol.*, **129**, 1524–1536.

Boehlke, K. W. and J. D. Friesen (1975). Cellular content of ribonucleic acid and protein in *Saccharomyces cerevisiae* as a function of growth rate. *J. Bacteriol.*, **121**, 429–439.

Boothby, D., L. Daneo-Moore, M. L. Higgins, J. Coyette and G. D. Shockman (1973). Turnover of bacterial cell wall peptidoglycans. *J. Biol. Chem.*, **248**, 2161–2169.

Brachet, J. (1960). *The Biological Role of Ribonucleic Acids*. Elsevier, Amsterdam.

Braña, A. F., M. B. Manzanal and C. Hardsson (1982). Mode of cell wall growth in *Streptomyces antibioticus*. *FEMS Microbiol. Lett.*, **13**, 231–235.

Bujega, V. C., J. R. Piggott and B. L. A. Carter (1982). Differentiation of *Saccharomyces cerevisiae* at slow growth rates in glucose-limited chemostat culture. *J. Gen. Microbiol.*, **128**, 2707–2714.

Bull, A. T. and A. P. J. Trinci (1977). The physiology and metabolic control of fungal growth. *Adv. Microb. Physiol.*, **50**, 1–84.

Buller, A. H. R. (1931). *Researches in Fungi*, vol. 4. Longmans Green, London.

Burchard, R. P. (1974). Growth of surface colonies of the gliding bacterium *Myxococcus xanthus*. *Arch. Mikrobiol.*, **96**, 247–254.

Burdett, I. D. J. and M. L. Higgins (1978). Study of pole assembly in *Bacillus subtilis* by computer reconstruction of septal growth zones seen in central longitudinal thin sections. *J. Bacteriol.*, **133**, 959–971.

Burkholder, P. R. and E. W. Sinnot (1945). Morphogenesis of fungal colonies in submerged shaken cultures. *Am. J. Bot.*, **32**, 424–431.

Byers, B. and L. Goetsch (1975). Behaviour of spindles and spindle plaques in the cell cycle and conjugation of *Saccharomyces cerevisiae*. *J. Bacteriol.*, **124**, 511–523.

Cabib, E. (1976). The yeast primary septum: a journey in three-dimensional biochemistry. *Trends Biochem. Sci.*, **1**, 275–277.

Cabib, E. and V. Farkas (1971). Control of morphogenesis: enzymatic mechanism for the initiation of septum formation in yeast. *Proc. Natl. Acad. Sci. USA*, **68**, 2052–2056.

Cabib, E., R. Ulane and B. Bowers (1973). Yeast chitin synthetase. Separation of the zymogen from its activating factor and recovery of the latter in the vacuole fraction. *J. Biol. Chem.*, **248**, 1451–1458.

Cabib, E., M. Shematek, J. A. Bratz and H. Kawai (1981). Biosynthesis of the major structural polysaccharide of the yeast cell wall, β-(1, 3)-glucan, and its regulation. In *Microbiology—1981*, ed. D. Schlessinger, pp. 235–237. American Society for Microbiology, Washington, DC.

Caldwell, I. Y. and A. P. J. Trinci (1973). The growth unit of the mould *Geotrichum candidum*. *Arch. Mikrobiol.*, **88**, 1–10.

Clark, D. A. (1962). Submerged citric acid fermentation of ferrocyanide-treated beet molasses: morphology of pellets of *Aspergillus niger. Can. J. Microbiol.*, **8**, 133–136.

Cole, R. M. and J. J. Hahn (1962). Cell wall replication in *Streptococcus pyogenes. Science*, **135**, 722–724.

Collinge, A. J. and A. P. J. Trinci (1974). Hyphal tips of wild-type and spreading colonial mutants of *Neurospora crassa. Arch. Mikrobiol.*, **99**, 353–368.

Cooper, A. L., A. C. R. Dean and C. H. Hinshelwood (1968). Factors affecting the growth of bacterial colonies on agar plates. *Proc. R. Soc. London, Ser. B*, **171**, 175–199.

Cooper, S. and C. E. Helmstetter (1968). Chromosome replication and the division cycle of *Escherichia coli* B/r. *J. Mol. Biol.*, **31**, 519–540.

Cotter, D. A. (1981). Spore activation. In *The Fungal Spore*, ed. G. Turian and H. R. Hohl, pp. 385–411. Academic, London.

De Chastellier, C., R. Hellio and A. Ryter (1975). Study of cell wall growth in *Bacillus megaterium* by high resolution autoradiography. *J. Bacteriol.*, **123**, 1184–1196.

Dennis, P. P. and H. Bremer (1973). Regulation of ribonucleic acid synthesis in *Escherichia coli* B/r: an analysis of a 'shift-up'. *J. Mol. Biol.*, **75**, 145–159.

De Vries, S. C. and J. G. H. Wessels (1982). Polarized outgrowth of hyphae by constant electrical fields during reversion of *Schizophyllum commune* protoplasts. *Exp. Mycol.*, **6**, 95–98.

Donachie, W. D. and K. J. Begg (1970). Growth of the bacterial cell. *Nature (London)*, **227**, 1220–1224.

Donachie, W. D., K. J. Begg and M. Vicente (1976). Cell length, cell growth and cell division. *Nature (London)*, **264**, 328–333.

Dow, C. S. and R. Whittenbury (1979). Prosthecate bacteria. In *Developmental Biology of Prokaryotes*, ed. J. H. Parish, pp. 139–165. Blackwell, Oxford.

Drozd, J. W., J. D. Linton, J. Downs and R. J. Stephenson (1978). An *in situ* assessment of the specific lysis rate in continuous cultures of *Methylococcus* sp. (NCIB11083) grown on methane. *FEMS Microbiol. Lett.*, **4**, 311–314.

Duran, A. and E. Cabib (1978). Solubilization and partial purification of yeast chitin synthetase. *J. Biol. Chem.*, **253**, 4419–4425.

Elliott, S. G. and C. S. McLaughlin (1983). The yeast cell cycle: coordination of growth and division rates. *Prog. Nucleic Acid Res. Mol. Biol.*, **28**, 143–176.

Elmayergi, H., J. M. Scharer and M. Moo-Young (1973). Effect of polymer additives on fermentation parameters in a culture of *A. niger. Biotechnol. Bioeng.*, **15**, 845–859.

Fan, D. P. and B. E. Beckman (1973). Structural differences between walls from hemispherical caps and partial septa of *Bacillus subtilis. J. Bacteriol.*, **114**, 790–797.

Fantes, P. and P. Nurse (1977). Control of cell size at division in fission yeast by a growth-modulated size control over nuclear division. *Exp. Cell Res.*, **107**, 377–386.

Galbraith, J. C. and J. E. Smith (1969). Filamentous growth of *Aspergillus niger* in submerged shake culture. *Trans. Br. Mycol. Soc.*, **52**, 237–246.

Galun, E. (1972). Morphogenesis in *Trichoderma*: autoradiography of intact colonies labelled by [^{3}H]N-acetylglucosamine as a marker of new cell wall biosynthesis. *Arch. Mikrobiol.*, **86**, 305–314.

Gooday, G. W. and A. P. J. Trinci (1980). Wall structure and biosynthesis in fungi. In *The Eukaryotic Microbial Cell*, ed. G. W. Gooday, D. Lloyd and A. P. J. Trinci, pp. 207–251. Cambridge University Press, Cambridge.

Gordon, C. N. and S. G. Elliott (1977). Fractionation of *Saccharomyces cerevisiae* cell population by centrifugal elutriation. *J. Bacteriol.*, **129**, 97–100.

Gow, N. A. R. and G. W. Gooday (1982). Growth kinetics and morphology of colonies of the filamentous form of *Candida albicans. J. Gen. Microbiol.*, **128**, 2187–2194.

Gray, B. F. and N. A. Kirwan (1974). Growth rates of yeast colonies on solid media. *Biophys. Chem.*, **1**, 204–213.

Green, P. B. (1974). Morphogenesis of the cell and organ axis—biophysical models. *Brookhaven Symp. Biol.*, **25**, 166–190.

Grove, S. N. (1972). Apical vesicles in germinating conidia of *Aspergillus parasiticus. Mycologia*, **64**, 638–641.

Grove, S. N. and C. E. Bracker (1978). Protoplasmic changes during zoospore encystment and cyst germination in *Pythium aphanidermatum. Exp. Mycol.*, **2**, 51–98.

Gull, K. (1978). Form and function of septa in filamentous fungi. In *The Filamentous Fungi*, ed. J. E. Smith and D. R. Berry, vol. 3, pp. 78–93. Edward Arnold, London.

Gull, K. and A. P. J. Trinci (1971). Fine structure of spore germination in *Botrytis cinerea. J. Gen. Microbiol.*, **68**, 207–220.

Hayflick, L. and P. S. Moorhead (1961). The serial cultivation of human diploid cell strains. *Exp. Cell Res.*, **25**, 585–621.

Heath, I. B. and M. C. Heath (1978). Microtubules and organelle movements in the rust fungus *Uromyces phaseoli* var. *vignae. Cytobiologie*, **16**, 393–411.

Heinzen, R. J. and J. C. Ensign (1975). Effect of growth substrates on morphology of *Nocardia corallina. Arch. Mikrobiol.*, **103**, 209–217.

Helmstetter, C. E. and S. Cooper (1968). DNA synthesis during the cell cycle of rapidly growing *Escherichia coli* B/r. *J. Mol. Biol.*, **31**, 507–518.

Herr, F. B. and M. C. Heath (1982). The effects of antimicrotubule agents on organelle positioning in the cowpea rust fungus, *Uromyces phaseoli* var. *vignae. Exp. Mycol.*, **6**, 15–24.

Higgins, M. L. (1976). Three-dimensional reconstruction of whole cells of *Streptococcus faecalis* from thin sections of cells. *J. Bacteriol.*, **127**, 1337–1345.

Higgins, M. L. and G. D. Shockman (1971). Procaryotic cell division with respect to wall and membranes. *CRC Crit. Rev. Microbiol.*, **1**, 29–72.

Higgins, M. L. and G. D. Shockman (1976). Study of a cycle of cell wall assembly in *Streptococcus faecalis* by three-dimensional reconstructions of thin sections of cells. *J. Bacteriol.*, **127**, 1346–1358.

Hopwood, D. A. (1960). Phase-contrast observations on *Streptomyces coelicolor. J. Gen. Microbiol.*, **22**, 295–302.

Howard, L. V., D. D. Dalton and W. K. McCaubrey, Jr. (1982). Expansion of the tetragonally arrayed cell wall protein layer during growth of *Bacillus sphaericus. J. Bacteriol.*, **149**, 748–757.

Howard, R. J. (1981). Ultrastructural analysis of hyphal tip cell growth in fungi: Spitzenkörper, cytoskeleton and endo-membranes after freeze-substitution. *J. Cell Sci.*, **48**, 89–104.

Howard, R. J. and J. R. Aist (1977). Effects of MBC on hyphal tip organization, growth and mitosis of *Fusarium acuminatum*, and their antagonism by D_2O. *Protoplasma*, **92**, 195–210.

Howard, R. J. and J. R. Aist (1979). Hyphal tip cell ultrastructure of the fungus *Fusarium*: improved preservation by freeze-substitution. *J. Ultrastruct. Res.*, **66**, 224–234.

Hughes, R. C. and E. Stokes (1971). Cell wall growth in *Bacillus licheniformis* followed by immunofluorescence with mucopeptide-specific antiserum. *J. Bacteriol.*, **106**, 694–696.

Inderlied, C. B., R. L. Cihlar and P. S. Sypherd (1980). Regulation of ornithine decarboxylase during morphogenesis of *Mucor racemosus*. *J. Bacteriol.*, **141**, 699–706.

Isaac, S., N. S. Ryder and J. F. Peberdy (1978). Distribution and activation of chitin synthase in protoplast fractions released during the lytic digestion of *Aspergillus nidulans* hyphae. *J. Gen. Microbiol.*, **105**, 45–50.

Jacob, F., S. Brenner and F. Cuzin (1963). On the regulation of DNA replication in bacteria. *Cold Spring Harbor Symp. Quant. Biol.*, **28**, 329–347.

Jagadish, M. N. and B. L. A. Carter (1978). Effect of temperature and nutritional condition on the mitotic cell cycle of *Saccharomyces cerevisiae*. *J. Cell Sci.*, **31**, 71–78.

Jagadish, M. N., A. Lorincz and B. L. A. Carter (1977). Cell size and cell division in yeast cultured at different growth rates. *FEMS Microbiol. Lett.*, **2**, 235–237.

Johnston, J. R. (1966). Reproductive capacity and mode of death in yeast cells. *Antonie van Leeuwenhoek*, **32**, 94–98.

Jones, H. C. and J. M. Schmidt (1973). Ultrastructural studies of crossbands occurring in stalks of *Caulobacter crescentus*. *J. Bacteriol.*, **116**, 466–470.

Kalakoutskii, L. V. and N. S. Agree (1976). Comparative aspects of development and differentiation in Actinomycetes. *Bacteriol. Rev.*, **40**, 450–524.

Kobayashi, T., G. Van Dedem and M. Moo-Young (1973). Oxygen transfer into mycelial pellets. *Biotechnol. Bioeng.*, **15**, 27–45.

Koch, A. L. (1982). The shape of the hyphal tips of fungi. *J. Gen. Microbiol.*, **128**, 947–951.

Koppes, L. J. H., N. Overbeeke and N. Nanninga (1978). DNA replication pattern and cell wall growth in *Escherichia coli* PAT84. *J. Bacteriol.*, **133**, 1053–1061.

Larsen, A. D. and P. S. Sypherd (1974). Cyclic adenosine 3′, 5′-monophosphate and morphogenesis in *Mucor racemosus*. *J. Bacteriol.*, **117**, 432–438.

Larsen, A. D. and P. S. Sypherd (1979). Ribosomal proteins of the dimorphic fungus *Mucor racemosus*. *Mol. Gen. Genet.*, **175**, 99–109.

Leal, J. and H. Marcovich (1975). Electron microscopic observation of the tsx^s gene expression in *Escherichia coli* K12F⁻ cells after conjugation with HFr bacteria. *Mol. Gen. Genet.*, **139**, 203–212.

Lloyd, D., R. K. Poole and S. W. Edwards (1982). *The Cell Division Cycle. Temporal Organization and Control of Cellular Growth and Reproduction*. Academic, London.

Lord, P. G. and A. E. Wheals (1980). Asymmetrical division of *Saccharomyces cerevisiae*. *J. Bacteriol.*, **142**, 808–818.

Ludwig, J. R., II, S. G. Oliver and C. S. McLaughlin (1977). The regulation of RNA synthesis in yeast. II. Amino acids shift-up experiments. *Mol. Gen. Genet..*, **158**, 117–122.

Luscombe, B. M. and T. R. G. Gray (1971). Effect of varying growth rate on the morphology of *Arthrobacter*. *J. Gen. Microbiol.*, **69**, 433–434.

Lutkenhaus, J. F. and W. D. Donachie (1979). Identification of the *ftsA* gene product. *J. Bacteriol.*, **137**, 1088–1094.

Maaløe, O. and N. O. Kjelgaard (1966). *Control of Macromolecular Biosynthesis*. W. D. Benjamin, New York.

Macko, V. (1981). Inhibitors and stimulants of spore germination and infection structure formation in fungi. In *The Fungal Spore*, ed. G. Turian and H. R. Hohl, pp. 565–584. Academic, London.

Marr, A. G., R. J. Harvey and W. C. Trentini (1966). Growth and division in *Escherichia coli*. *J. Bacteriol.*, **91**, 2388–2389.

Marshall, K. C. and M. Alexander (1960). Growth characteristics of fungi and actinomycetes. *J. Bacteriol.*, **80**, 412–416.

Mauck, J., L. Chan and L. Glaser (1971). Turnover of the cell wall of Gram-positive bacteria. *J. Biol. Chem.*, **246**, 1820–1827.

May, J. W. (1963). The distribution of cell-wall label during growth and division of *Salmonella typhimurium*. *Exp. Cell Res.*, **31**, 217–220.

McAlister, L. and D. B. Finkelstein (1980). Heat shock proteins and thermal resistance in yeast. *Biochem. Biophys. Res. Commun.*, **93**, 819–824.

McDonald, I. J., G. B. Calleja and B. F. Johnson (1982). Conjugation in chemostat cultures of *Schizosaccharomyces pombe*. *J. Gen. Microbiol.*, **128**, 1981–1987.

McMurrough, I. and A. H. Rose (1967). Effect of growth rate and substrate limitation on the composition and structure of the cell wall of *Saccharomyces cerevisiae*. *Biochem. J.*, **105**, 189–203.

Metz, B. (1976). From pulp to pellet. *Ph.D. thesis*, Technical University of Delft, The Netherlands.

Metz, B. and N. W. F. Kossen (1977). The growth of molds in the form of pellets—a literature review. *Biotechnol. Bioeng.*, **19**, 781–799.

Miles, E. A. and A. P. J. Trinci (1983). Effect of pH and temperature on the morphology of batch and chemostat cultures of *Penicillium chrysogenum*. *Trans. Br. Mycol. Soc.*, **81**, 193–200.

Miller, M. J., N. Xuong and E. P. Geiduschek (1980). A response of protein synthesis to temperature shift in yeast. *Proc. Natl. Acad. Sci. USA*, **76**, 5222–5225.

Mitchison, J. M. (1971). *The Biology of the Cell Cycle*. Cambridge University Press, Cambridge.

Mooney, D. T. and P. S. Sypherd (1976). Volatile factor involved in the dimorphism of *Mucor racemosus*. *J. Bacteriol.*, **126**, 1266–1270.

Moore, R. L. (1981). The biology of *Hyphomicrobium* and other prosthecate, budding bacteria. *Annu. Rev. Microbiol.*, **35**, 567–594.

Mor, J. R. and A. Fiechter (1968). Continuous cultivation of *Saccharomyces cerevisiae*. I. Growth on ethanol under steady state conditions. *Biotechnol. Bioeng.*, **10**, 159–176.

Morrison, K. B. and R. C. Righelato (1974). The relationship between hyphal branching, specific growth rate and colony radial growth rate in *Penicillium chrysogenum. J. Gen. Microbiol.*, **81**, 517–520.

Mortimer, A. K. and J. R. Johnston (1959). Life span of individual yeast cells. *Nature (London)*, **183**, 1751–1752.

Müller, I. (1971). Experiments on ageing in single cells of *Saccharomyces cerevisiae. Arch. Mikrobiol.*, **77**, 20–25.

Neidhardt, F. C. (1964). The regulation of RNA synthesis in bacteria. *Prog. Nucleic Acid Res. Mol. Biol.*, **3**, 145–181.

Nierlich, D. P. (1972). Regulation of ribonucleic acid synthesis in growing bacterial cells. 1. Control over total rate of RNA synthesis. *J. Mol. Biol.*, **72**, 751–764.

Nierlich, D. P. (1974). Regulation of bacterial growth. *Science*, **184**, 1043–1050.

Nurse, P. (1981). Genetic control of the yeast cell cycle: a reappraisal of 'start'. In *The Fungal Nucleus*, ed. K. Gull and S. G. Oliver, pp. 331–345. Cambridge University Press, Cambridge.

Oliver, S. G. (1981). Coordination of transcription with translation in yeast. In *The Fungal Nucleus*, ed. K. Gull and S. G. Oliver, pp. 315–330. Cambridge University Press, Cambridge.

Orlowski, M. and J. F. Ross (1981). Relationship of internal cyclic AMP levels, rates of protein synthesis and *Mucor* dimorphism. *Arch. Mikrobiol.*, **129**, 353–356.

Orlowski, M. and P. S. Sypherd (1977). Protein synthesis during morphogenesis of *Mucor racemosus. J. Bacteriol.*, **132**, 209–218.

Orlowski, M. and P. S. Sypherd (1978a). Regulation of ribonucleic acid synthesis during morphogenesis of *Mucor racemosus. Arch. Mikrobiol.*, **119**, 145–152.

Orlowski, M. and P. S. Sypherd (1978b). Regulation of translation rate during morphogenesis of the fungus *Mucor. Biochemistry*, **17**, 569–575.

Palumbo, S. A., M. G. Johnson, V. T. Rieck and L. D. Witter (1971). Growth measurements on surface colonies of bacteria. *J. Gen. Microbiol.*, **66**, 137–143.

Paznokas, J. L. and P. S. Sypherd (1975). Respiratory capacity, cyclic adenosine 3′,5′-monophosphates, and morphogenesis in *Mucor racemosus. J. Bacteriol.*, **124**, 134–139.

Paznokas, J. L. and P. S. Sypherd (1977). Pyruvate kinase isozymes of *Mucor racemosus*: control of synthesis by glucose. *J. Bacteriol.*, **130**, 661–666.

Peters, J. and P. S. Sypherd (1979). Morphology-associated expression of nicotinamide adenine dinucleotide-dependent glutamate dehydrogenase in *Mucor racemosus. J. Bacteriol.*, **137**, 1134–1139.

Pirt, S. J. (1965). The maintenance energy of bacteria in growing cultures. *Proc. R. Soc. London, Ser. B.*, **163**, 224–231.

Pirt, S. J. (1966). A theory of the mode of growth of fungi in the form of pellets in submerged culture. *Proc. R. Soc. London, Ser. B.*, **166**, 369–373.

Pirt, S. J. (1967). A kinetic study of the mode of growth of surface colonies of bacteria and fungi. *J. Gen. Microbiol.*, **47**, 181–197.

Pirt, S. J. (1972). Prospects and problems in continuous flow culture of microorganisms. *J. Appl. Chem. Biotechnol.*, **22**, 55–64.

Pitt, D. E. and A. T. Bull (1982). Use of autoradiography to quantify aspects of fungal growth and starvation in submerged liquid culture. *Trans. Br. Mycol. Soc.*, **78**, 97–104.

Pooley, H. M. (1976a). Turnover and spreading of old wall during surface growth of *Bacillus subtilis. J. Bacteriol.*, **125**, 1127–1138.

Pooley, H. M. (1976b). Layered distribution, according to age, within the cell wall of *Bacillus subtilis. J. Bacteriol.*, **125**, 1139–1147.

Pooley, H. M., J.-M. Schaleppi and D. Karamata (1978). Localised insertion of new cell wall in *Bacillus subtilis. Nature (London)*, **274**, 264–266.

Reyrolle, J. and F. Letellier (1979). Autoradiographic study of the localization and evolution of growth zones in bacterial colonies. *J. Gen. Microbiol.*, **111**, 399–406.

Rieck, V. T., S. A. Palumbo and L. D. Witter (1973). Glucose availability and the growth rate of colonies of *Pseudomonas fluorescens. J. Gen. Microbiol.*, **74**, 1–8.

Riesenberger, D. and F. Bergter (1979). Dependence of macromolecular composition and morphology of *Streptomyces hygroscopicus* on specific growth rate. *Z. Allg. Microbiol.*, **19**, 415–430.

Righelato, R. C. (1975). Growth kinetics of mycelial fungi. In *The Filamentous Fungi*, ed. J. E. Smith and D. R. Berry, vol. 1, pp. 77–103. Edward Arnold, London.

Rivin, C. J. and W. L. Fangman (1980). Cell cycle phase expansion in nitrogen-limited cultures of *Saccharomyces cerevisiae. J. Cell Biol.*, **85**, 96–107.

Robertson, N. F. (1958). Observations on the effect of water on the hyphal apices of *Fusarium oxysporum. Ann. Bot. (London)*, **32**, 279–291.

Robertson, N. F. (1965). The fungal hypha. *Trans. Br. Mycol. Soc.*, **48**, 1–8.

Robinow, C. F. (1963). Observations on cell growth, mitosis and division in the fungus *Basidiobolus ranarum. J. Cell Biol.*, **17**, 123–152.

Robinson, P. M. and J. M. Smith (1979). Development of cells and hyphae of *Geotrichum candidum* in chemostat and batch culture. *Trans. Br. Mycol. Soc.*, **72**, 39–47.

Robinson, P. M. and L. A. A. Thompson (1982). Volatile promoter of germination and hyphal extension produced by *Geotrichum candidum. Trans. Br. Mycol. Soc.*, **78**, 353–355.

Rogers, H. J., J. B. Ward and H. R. Perkins (1980). *Microbiol Cell Walls and Membranes*. Chapman and Hall, London.

Rogers, P. J., G. D. Clark-Walker and P. R. Stewart (1974). Effects of oxygen and glucose on energy metabolism and dimorphism in *Mucor genevensis* grown in continuous culture: reversibility of yeast–mycelial conversions. *J. Bacteriol.*, **119**, 282–293.

Rosenberger, R. F. (1979). Endogenous lytic enzymes and wall metabolism. In *Fungal Walls and Hyphal Growth*, ed. J. H. Burnett and A. P. J. Trinci, pp. 265–277. Cambridge University Press, Cambridge.

Rosset, R., J. Julien and R. Monier (1966). Ribonucleic acid composition of bacteria as a function of growth rate. *J. Mol. Biol.*, **18**, 308–320.

Rosset, R., R. Monier and J. Julien (1964). RNA composition of *E. coli* as a function of growth rate. *Biochem. Biophys. Res. Commun.*, **15**, 392–437.

Ryter, A., Y. Hirota and U. Schwarz (1973). Process of cellular division in *Escherichia coli*, growth pattern of *E. coli* murein. *J. Mol. Biol.*, **78**, 185–195.

Ryter, A., H. Shuman and M. Schwartz (1975). Integration of the receptor for bacteriophage lambda in the outer membrane of *Escherichia coli*: coupling with cell division. *J. Bacteriol.*, **122**, 295–301.

Sargent, M. G. (1974). Nuclear segregation in *Bacillus subtilis*. *Nature (London)*, **250**, 252–254.

Sargent, M. G. (1975). Control of cell length in *Bacillus subtilis*. *J. Bacteriol*, **123**, 7–19.

Sargent, M. G. and M. F. Bennett (1982). Attachment of the chromosomal terminus of *Bacillus subtilis* to a fast-sedimenting particle. *J. Bacteriol.*, **150**, 623–632.

Saunders, P. T. and A. P. J. Trinci (1979). Determination of tip shape in fungal hyphae. *J. Gen. Microbiol.*, **110**, 469–473.

Schaechter, M., O. Maaløe and N. O. Kjelgaard (1958). Dependency on medium and temperature of cell size and chemical composition during balanced growth of *Salmonella typhimurium*. *J. Gen. Microbiol.*, **19**, 592–606.

Schmidt, J. M. (1973). Effect of lysozyme on crossbands in stalks of *Caulobacter crescentus*. *Arch. Mikrobiol.*, **89**, 33–40.

Schmidt, J. M. and R. Y. Stanier (1966). The development of cellular stalks in bacteria. *J. Cell Biol.*, **28**, 423–426.

Schuhmann, E. and F. Bergter (1976). Mikroskopische untersuchungen zur wachstumskinetik von *Streptomyces hygroscopicus*. *Z. Allg. Microbiol.*, **16**, 201–215.

Schwarz, U., R. Ryter, A. Rambach, R. Hellio and Y. Hirota (1975). Process of cellular division in *Escherichia coli*: differentiation of growth zones in sacculus. *J. Mol. Biol.*, **98**, 749–759.

Shapiro, L., N. Agabian-Keshishian and I. Bendis (1971). Bacterial differentiation. *Science*, **173**, 884–892.

Slayman, C. L. and C. W. Slayman (1962). Measurement of membrane potentials in *Neurospora*. *Science*, **136**, 876–877.

Smit, J. and H. Nikaido (1978). Outer membrane of Gram-negative bacteria. XVIII. Electron microscopic studies on porin insertion sites and growth of the cell surface in *Salmonella typhimurium*. *J. Bacteriol.*, **135**, 687–702.

Steele, G. C. and A. P. J. Trinci (1975a). The extension zone of mycelial hyphae. *New Phytol.*, **75**, 583–587.

Steele, G. C. and A. P. J. Trinci (1975b). Morphology and growth kinetics of hyphae of differentiated and undifferentiated mycelia of *Neurospora crassa*. *J. Gen. Microbiol.*, **91**, 362–368.

Stewart, P. R. and P. J. Rogers (1983). Fungal dimorphism. In *Fungal Differentiation: A Contemporary Synthesis*, ed. J. E. Smith, pp. 267–313. Dekker, New York.

Sypherd, P. S., P. T. Boogia and J. L. Paznokas (1978). Biochemistry of dimorphism in the fungus *Mucor*. *Adv. Microb. Physiol.*, **18**, 67–104.

Sypherd, P. S., M. Orlowski and J. Peters (1979). Models of fungal dimorphism: control of dimorphism in *Mucor racemosus*. In *Microbiology 1979*, ed. D. Schlessinger, pp. 224–227. American Society for Microbiology, Washington, DC.

Tempest, D. W., D. Herbert and P. J. Phipps (1967). Studies on the growth of *Aerobacter aerogenes* at low dilution rates in a chemostat. In *Microbial Physiology and Continuous Culture*, ed. E. O. Powell, C. G. T. Evans, R. E. Strange and D. W. Tempest, pp. 240–254. HMSO, London.

Thompson, P. W. and A. E. Wheals (1980). Asymmetrical division of *Saccharomyces cerevisiae* in a glucose-limited chemostat culture. *J. Gen. Microbiol.*, **121**, 401–409.

Trinci, A. P. J. (1970). Kinetics of the growth of mycelial pellets of *Aspergillus nidulans*. *Arch. Mikrobiol.*, **73**, 353–367.

Trinci, A. P. J. (1971). Influence of the peripheral growth zone on the radial growth rate of fungal colonies. *J. Gen. Microbiol.*, **67**, 325–344.

Trinci, A. P. J. (1973). The hyphal growth unit of wild type and spreading colonial mutants of *Neurospora crassa*. *Arch. Mikrobiol.*, **91**, 127–136.

Trinci, A. P. J. (1974). A study of the kinetics of hyphal extension and branch initiation of fungal mycelia. *J. Gen. Microbiol.*, **81**, 225–236.

Trinci, A. P. J. (1979). The duplication cycle and branching in fungi. In *Fungal Walls and Hyphal Growth*, ed. J. H. Burnett and A. P. J. Trinci, pp. 319–358. Cambridge University Press, Cambridge.

Trinci, A. P. J. (1983). Effect of Junlon on the morphology of *Aspergillus niger* and its use in making turbidity measurements of fungal growth. *Trans. Br. Mycol. Soc.*, **77**, 20–24.

Trinci, A. P. J. and A. J. Collinge (1973a). Structure and plugging of septa of wild type and spreading colonial mutants of *Neurospora crassa*. *Arch. Mikrobiol.*, **91**, 355–364.

Trinci, A. P. J. and A. J. Collinge (1973b). Influence of L-sorbose on the growth and morphology of *Neurospora crassa*. *J. Gen. Microbiol.*, **78**, 179–192.

Trinci, A. P. J. and P. T. Saunders (1977). Tip growth of fungal hyphae. *J. Gen. Microbiol.*, **103**, 243–248.

Tyson, C. B., P. G. Lord and A. E. Wheals (1979). Dependency of size of *Saccharomyces cerevisiae* cells on growth rate. *J. Bacteriol.*, **38**, 92–98.

Ulane, R. E. and E. Cabib (1976). The activating system of chitin synthetase from *Saccharomyces cerevisiae*. *J. Biol. Chem.*, **251**, 3367–3374.

Unger, M. W. and L. H. Hartwell (1976). Control of cell division in *Saccharomyces cerevisiae* by methionyl-tRNA. *Proc. Natl. Acad. Sci. USA*, **73**, 1664–1668.

Van der Valk, P. and J. G. H. Wessels (1976). Ultrastructure and localization of wall polymers during regeneration and reversion of protoplasts of *Schizophyllum commune*. *Protoplasma*, **90**, 65–87.

Van Etten, J. L., L. D. Dunkle and S. N. Freer (1977). Germination of *Rhizopus stolonifer* sporangiospores. In *Eukaryotic Microbes as Model Development Systems*, ed. D. M. O'Day and P. A. Horgen, pp. 372–374. Dekker, New York.

van Suijdam, J. C. and B. Metz (1981). Influence of engineering variants upon the morphology of filamentous moulds. *Biotechnol. Bioeng.*, **23**, 111–148.

van Suijdam, J. C., H. Hols and N. W. F. Kossen (1982). Unstructured model for growth of mycelial pellets in submerged cultures. *Biotechnol. Bioeng.*, **24**, 177–191.

Van Turbergen, R. P. and R. B. Setlow (1961). Quantitative autoradiographic studies on exponentially growing cultures of *Escherichia coli*—the distribution of parental DNA, RNA, protein and cell wall among progeny cells. *Biophys. J.*, **1**, 589–625.

Waldron, C. (1977). Synthesis of ribosomal and transfer ribonucleic acids in yeast during a nutritional shift-up. *J. Gen. Microbiol.*, **96**, 215–221.

Waldron, C. and F. Lacroute (1975). Effect of growth rate on the amounts of ribosomal and transfer ribonucleic acids in yeast. *J. Bacteriol.*, **122**, 855–865.

Wessels, J. G. H. and J. H. Sietsma (1981). Cell wall synthesis and hyphal morphogenesis. A new model for apical growth. In *Cell Walls*, ed. D. G. Robinson and H. Quader, pp. 135–142. Wissenschaftliche Verlagsgesellschaft, Stuttgart.

Whittenbury, R. and C. S. Dow (1977). Morphogenesis and differentiation in *Rhodomicrobium vanielli* and other budding and prosthecate bacteria. *Bacteriol. Rev.*, **41**, 754–808.

Wimpenny, J. W. T. (1979). Growth and form of bacterial colonies. *J. Gen. Microbiol.*, **114**, 483–486.

Wimpenny, J. W. T. and M. W. A. Lewis (1977). Growth and respiration of bacterial colonies. *J. Gen. Microbiol.*, **103**, 9–18.

Wimpenny, J. W. T. and J. A. Parr (1979). Biochemical differentiation in large colonies of *Enterobacter cloacae*. *J. Gen. Microbiol.*, **114**, 487–489.

Wolf, D. H. and C. Ehmann (1978). Isolation of yeast mutants lacking protease B activity. *FEBS Lett.*, **92**, 121–124.

Wolf, D. H. and C. Ehmann (1981). Carboxypeptidase S- and carboxypeptidase Y-deficient mutants of *Saccharomyces cerevisiae*. *J. Bacteriol.*, **147**, 418–426.

Yamaguchi, K. and H. Yoshikawa (1973). Topography of chromosome–membrane junction in *Bacillus subtilis*. *Nature* (*London*), *New Biol.*, **244**, 204–206.

Yoshida, T., H. Taguchi and S. Teramoto (1968). Studies on submerged culture of basidiomycetes. IV. Distribution of respiration and other metabolic activities in pellets of Shiitake (*Lentinus edodes*). *J. Ferment. Technol.*, **46**, 119–124.

Zubenko, G. S., A. P. Mitchell and E. W. Jones (1979). Septum formation, cell division and sporulation in mutants of yeast deficient in protease B. *Proc. Natl. Acad. Sci. USA*, **76**, 2395–2399.

11

Microbial Growth Dynamics

J. H. SLATER
University of Wales Institute of Science and Technology, Cardiff, UK

11.1 MICROBIAL GROWTH IN UNLIMITED ENVIRONMENTS

The *basic growth equations* are simply derived by considering the processes involved when a microorganism grows in ideal conditions with all the essential requirements for balanced growth present in unlimited quantities. The nutritional conditions need not be identical, except for the requirement for unrestricted supplies of appropriate nutrients. Thus, for example, a proto-trophic, aerobic heterotroph requires a utilizable carbon and energy source, a simple combined nitrogen supply, and provision of the appropriate amounts of all the elements required to synthesize more microbial cell material. This organism would also need unlimited quantities of oxygen to serve as the terminal electron acceptor for the electrons freed during metabolism. A lysine auxotroph of the same heterotroph would additionally need an unrestricted external supply of this amino acid. The same basic equations could also be derived for a photolithotroph using light as its energy source and carbon dioxide as its starting point for all cellular biosyntheses. Similarly, organisms growing anaerobically with sulfate or nitrate as terminal electron acceptors would grow in the same way leading to the derivation of identical basic growth equations.

For conceptual purposes it is convenient to formulate the basic growth equations with reference to a microorganism which replicates by a binary fission process, that is a cell increases in size until a point is reached when the parental cell splits into two identical daughter cells (see Oliver and Trinci, Volume 1, Chapter 10). However, as we shall see, the basic growth equations can apply equally well to growth processes measured in terms of the increase in cellular material or *biomass*. Consequently the basic growth equations could also be derived by considering other types of microbes where population growth is not dependent on binary fission mechanisms, that is the same basic growth equations apply to yeasts growing by budding mechanisms, or the extension of the length of fungal hyphae, or more complex growth processes involving multicellular forms.

Whatever may be the physiological method of growth or the mode of organization and structure of the organism, under ideal conditions with unlimited resources available, the pattern of population development will be the same. It follows that in all cases the kinetic analysis presented below must assume that the ideal, uniform starting conditions are maintained and that the products of metabolism do not change in any way the prevailing physicochemical conditions.

Analysis of a growing population at any point in time reveals that on an average basis all cells process the available nutrients, produce new biomass and increase in cell size at exactly the same rate. Examination of just one of the cells within the total population shows that after a period of time equivalent to that required for the cell to increase in size by a factor of exactly two, the cell divides to produce two daughter cells identical in composition to each other and to their parent cell. The two new cells are the same size as each other and so the amount of biomass per cell is exactly halved, generating two daughter cells which are the same size as the previous parent. Since there is a maintained, favourable environment, both daughter cells repeat the growth and division process, each generating two new cells. Thus for microbes which grow by binary fission, growth is essentially a cyclical process composed of two main stages during which: (i) a recently divided, small cell increases in size as a result of biosynthetic activity; and (ii) a cell which has exactly doubled in size then divides into two equally sized progeny.

For all unicellular microorganisms growth is characterized by cell size increase, cell division, and separation of the divided parent cell. This applies to both the binary fission and budding type mechanisms and means that population growth is represented by an increase in the number of cells (per unit volume of population) and an increase in biomass (per unit volume of population). On this basis the basic growth equations may be derived in two different, but obviously related, ways.

11.1.1 Basic Growth Equation from Cell Number Increase

The cyclical growth processes described above and repeated from generation to succeeding generation may be represented as in Table 1, where x is the initial population size, n is the nth generation and a is the multiple constant at the nth generation (see also Figure 1). By definition the time taken to go from one generation $(n-1)$ to the next generation (n) is constant under stable environmental conditions and so the rate of increase in cell number in the population increases with time. In fact the population progresses exponentially and the rate of acceleration in the population size is governed by two major factors. Firstly, the rate is controlled by the intrinsic nature of the organism, that is its basic genetic structure and potential which determine its physiological characteristics. Secondly, the rate of population increase is substantially modified by extrinsic factors due to the organism's growth environment. So in 'rich' environments containing many preformed cellular components, such as amino acids and nucleic acids, the rate of increase is more rapid than in a 'poor' environment in which the growing organism has to synthesize all its cellular requirements, devoting proportionally much more of the available resources and energy to these activities. In both rich and poor environments the multiplication factor per generation and the rate of increase of cell numbers per generation are constant but the time taken to complete each generation varies. It is much shorter in a rich environment compared with a poor environment.

Thus the rate of cell number increase with time (compared to the preceding considerations which have been based on a generation basis) is a variable dictated by the type of organism and the nature of its growth environment. Under constant environmental conditions the time taken to complete each generation and double the size of the population is a characteristic constant known

Table 1 Cyclical Growth Processes

| | Number of complete cell cycles *or* number of generations | | | | | | | |
	0	1	2	3	4	5	6	......*n*
Number of individual cells in the population at the beginning of each generation	x	$2x$	$4x$	$8x$	$16x$	$32x$	$64x$	......*ax*
	2^0x	2^1x	2^2x	2^3x	2^4x	2^5x	2^6x	......2^nx
Population size when: $x = 1$ *organism*	1	2	4	8	16	32	64	......*a*
$x = 1 \times 10^6$ *organisms*	1×10^6	2×10^6	4×10^6	8×10^6	1.6×10^7	3.2×10^7	6.4×10^7	......$a \times 10^6$
Rate of cell number increase per generation	—	x	$2x$	$4x$	$8x$	$16x$	$32x$	......$(a - b)x$
Multiplication factor per generation	2	2	2	2	2	2	2	2

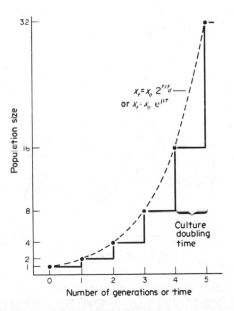

Figure 1 Exponential growth in a closed, unlimited environment. The smooth curve is described by the basic growth equations (1) and (7) and is the form of growth obtained if the population size is large. The stepwise line is obtained in a synchronized population and seen if the population size is measured in terms of cell number (Slater, 1979)

as the *culture doubling time* t_d (units: time, conventionally hours although sometimes given as minutes, or days for very slow growing populations). Theoretically the culture doubling time should be the same as the individual *generation time* for a single cell growing and producing two daughter cells. However, in a growing population there is considerable variation in individual generation times, principally due to the fact that when cell division occurs it is not rigidly related to parental cell size and that the two progeny are not of exactly equal size, that is there is a distribution of different generation times within a population. In principle it is possible to measure a statistically significant number of individual generation times and calculate an average value, the so-called *mean generation time*, which should be equivalent to the culture doubling time. In practice this is never carried out since it is much simpler technically to measure an overall population parameter, such as cell number per unit volume of population, and use the change in this parameter with time to describe the progress of the culture. Accordingly, the term culture doubling time is much preferred to others, such as mean generation time.

It is also important to recognize that most microbial cultures contain very large numbers of cells per unit volume. Thus not only is there an 'averaging effect' due to the variations in individual cell generation times, but there is also a distribution of cells at various stages of the cell cycle, that is some fraction of the population will be small cells which are the recent product of a cell division. Another fraction will contain large cells on the point of division after completing their period of cell growth and doubling in size (Figure 2). At one instant in time only a small fraction of the population will be dividing to increase cell number, followed by a different fraction which is proportionally equal to the size of the first fraction. This repeated process measured at a population (culture) level generates a typical smooth exponential curve as illustrated in Figure 1. If it was technically possible to follow the development of a population initiated by a single cell, then the stepwise line would be observed and maintained for a limited number of generations until the averaging effects discussed above became significant. It is possible to regulate a population in such a way that all the individual cells are maintained at the same stage of cell cycle development (Figure 3). This may be achieved by various physicochemical treatments such as temperature shocks and the intermittent use of DNA replication inhibition (see Mitchison, 1971). These cultures are known as *synchronous populations* and at a population level show the stepwise increases depicted for the single cell in Figure 1.

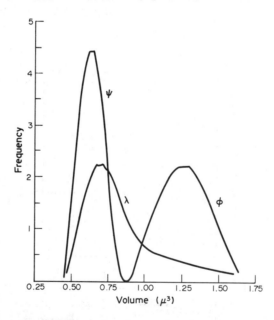

Figure 2 Frequency functions of the distribution of volume of newly formed cells (ψ), extant cells (λ), and dividing cells (ϕ) of *Escherichia coli* strain B/r/1 growing in glucose minimal medium at 30 °C (Marr *et al.*, 1969)

Inspection of Table 1 and Figure 1 reveals that since the multiplication factor between succeeding generations is constant, then the size of the population in the nth generation depends on the size of the population in the $(n-1)$th generation. This exponential progression may be formulated by saying that the size of the population after a period of growth x_t depends on the initial population size x_0 and the period of time for growth t (or alternatively the number of elapsed generations). Thus the progression in Table 1 may be written as equation (1), where $n = t/t_d$

$$x_t = x_0 2^{t/t_d} \tag{1}$$

When $t = t_d$, that is the period of growth time is equal to the culture doubling time and, by defini-

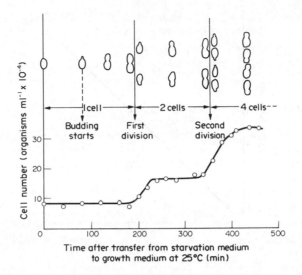

Figure 3 Growth of a synchronized yeast culture (*Saccharomyces cerevisiae*) on inoculation into synthetic medium (Williamson and Scopes, 1961)

tion, a single generation time elapses and so $n = t_d/t_d = 1$, then equation (1) becomes equation (2), describing as expected a doubling in the population size over a single generation.

$$x_t = 2x_0 \tag{2}$$

Similarly, if the growth time t is three times the culture doubling time, then $n = 3t_d/t_d = 3$, giving equation (3).

$$x_t = 2^3 x_0 = 8x_0 \tag{3}$$

Equation (1) has the general form of an exponential function and characterizes the fundamental pattern of growth of a microbial population in an optimal, unlimited growth environment. For practical purposes fitting an exponential curve to experimental data is difficult and it is therefore more convenient to derive a linear form of the exponential growth equation by taking the natural logarithms of each side of equation (1), giving equations (4) and (5).

$$\ln x_t = \ln x_0 + \frac{t}{t_d} \ln 2 \tag{4}$$

$$= \ln x_0 + 0.693 \frac{t}{t_d} \tag{5}$$

This graphically takes the form of Figure 4 when $\ln x$ is plotted against t. The intercept on the abscissa shows the initial population size and the slope of the line is $0.693/t_d$, from which it is possible to calculate the culture doubling time.

11.1.2 Basic Growth Equation from Incremental Increase in the Population over a Small Growth Time

The alternative derivation of the basic growth equation starts from the position that in a heterogeneous population (that is a mixture of cells at different stages in their cell cycle and with variations in individual generation times) at a particular instant in time a small fraction of the population is in the act of cell division thereby contributing to an increase in the population size. Thus, if we consider an initial population size x_0 existing at time $t = 0$, then after a small interval of time δt a small fraction of the organisms present within x_0 complete their cell cycles, divide and increase the population size by δx. It has already been stated that the magnitude of δx (assuming that δt remains constant) depends on the size of x_0. Thus the rate of change of the population

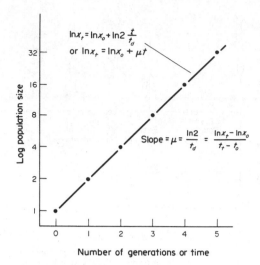

Figure 4 Exponential growth in a closed, unlimited environment. This is the linear representation of exponential growth derived from equations (4) and (8) (Slater, 1979)

size, or more conveniently the rate of growth, is directly proportional to the initial population size. This is expressed in equation (6).

$$\frac{dx}{dt} \propto x_0$$

$$\frac{dx}{dt} = \mu x_0 \tag{6}$$

In equation (6) μ is a proportionality constant known as the *specific growth rate* (units: reciprocal time, conventionally hours $^{-1}$ but see below). This constant is sometimes known as the *instantaneous growth rate* or *intrinsic growth rate* and may be given the symbol r, largely through the influence of plant and animal ecologists. The constant μ does, however, have a precise biological meaning since it is a measure of the number of new individuals produced by a given number of existing individuals in a fixed period of growth time. Strictly, therefore, μ has units of (number of new individuals) (number of individuals)$^{-1}$ hours $^{-1}$, which can be simplified to the more usual reciprocal time units. Clearly the larger the numerical value the greater is the rate of production of new individuals within the population. For the same reasons which were discussed for the culture doubling time, the exact value of μ depends on the species of microorganism in question and the influence of the environmental conditions on the rate of growth. In an optimum environment with excess nutrient supply, the rate of growth is maximized and only subject to the limitations imposed by the genetic composition and potential of the organism in question. If the rate of growth is maximized then so too is μ which becomes known as the *maximum specific growth rate* μ_{max}. In many closed (batch) cultures the ideal conditions which have been assumed in deriving the basic growth equations are to all intents and purposes realized and growth proceeds at μ_{max}. A fuller appreciation of μ and μ_{max} is provided when the Monod equation is discussed in Section 11.2.3.

By integration from $t = 0$ to $t = t$, equation (6) has the solution:

$$x_t = x_0 e^{\mu t} \tag{7}$$

which is analogous to equation (1) describing an exponential curve (Figure 1). In a similar fashion, therefore, equation (7) has a linear form:

$$\ln x_t = \ln x_0 + \mu t \tag{8}$$

which also gives a straight line (Figure 4). Since equations (5) and (8) describe the same line, then it follows that the slopes are equivalent and by comparison between the two equations:

$$\mu = \frac{0.693}{t_d} \tag{9}$$

Alternatively, from equation (8), if $t = t_d$, then $x_t = 2x_0$ and so:

$$\ln 2x_0 = \ln x_0 + \mu t_d$$

$$\mu = \frac{\ln 2}{t_d} = \frac{0.693}{t_d} \tag{10}$$

Thus the specific growth rate is inversely proportional to the culture doubling time.

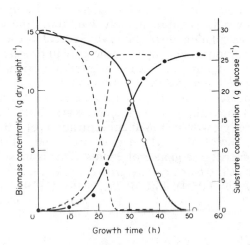

Figure 5 The modelling of the growth of *Aspergillus nidulans* in a closed culture with glucose as the limiting substrate ($S_R = 30$ g glucose l^{-1}). The observed biomass concentrations (g dry weight l^{-1}) are shown by ● and the observed glucose concentrations (g glucose l^{-1}) by ○. The logistic model (solid lines) was fitted using $\mu_{max} = 0.2$ h^{-1} and $x_f = 13.0$ g dry weight l^{-1}. The saturation model (broken lines) was fitted using the following constants: $\mu_{max} = 0.2$ h^{-1}, $K_s = 0.072$ g glucose h^{-1} and $Y = 0.43$ g dry weight g^{-1} glucose. (Data from Carter and Bull, 1969, and modelled according to Bull *et al.*, 1975)

11.1.3 Basic Growth Equations

The two preceding sections have derived the basic growth equations (1) and (7) in two related ways. They have been formulated on a basis of considering populations as numbers of individual cells and rates of increase as increases in the number of cells in the population. However, there are many different ways of determining population size (see Jones, 1979; Burns and Slater, 1982): these include total and viable counts, biomass weight, ATP content, lipopolysaccharide content, or nucleic acid content. Thus x in the preceding equations can be used to represent any microbial parameter which is proportional to population size. It is crucially important, however, to recognize that many cellular components, such as DNA or protein, vary substantially under different conditions and these variations do not necessarily reflect proportional changes in the rate of growth. Under optimal conditions where μ is constant these variations reach steady state values and therefore do not influence calculations of μ or t_d. In most practical situations μ is not constant and cellular compositions and cell sizes vary substantially as a result (see, for example, Herbert, 1961; Mandelstam *et al.*, 1982). In practice it is preferable to use some parameter for x which reflects the complete composition of the cell and normally this is biomass in terms of grams dry weight per unit volume. This is particularly appropriate, since, as we shall see, other processes involved in growth, such as substrate utilization, use equivalent units. Comparisons, therefore, between substrate use, biomass production and growth yields are standardized and more readily understandable.

11.2 MICROBIAL GROWTH IN LIMITED ENVIRONMENTS

To simplify the derivation of the basic growth equations it was assumed that unlimited quantities of all the growth resources were available. Such conditions might exist for short periods of growth (for example early stages of a batch culture) but in most natural habitats resource limitation is the rule rather than the exception. Furthermore, since the biosphere is ultimately a closed system, unlimited resource availability cannot occur. In the case of microorganisms this is just as well since the growth potential of a single microbe is stupendous. Many microbes have the potential to grow with doubling times in the order of 0.33 h and, if a population under unlimited conditions was initiated by a single cell, then the population x_t after $t = 48$ h would be (substituting in equation 1):

$$x_t = 1.2^{48/0.33} = 2.33 \times 10^{43} \text{ individuals}$$

Each cell typically has a dry weight of 10^{-12} g which gives a final weight of 2.23×10^{28} kg for the population after 48 h. This weight is approximately 4000 times the weight of the Earth and is clearly preposterous. This calculation does, however, reveal the extraordinary potential for growth under appropriate conditions. Fortunately, the growth process is self-limiting, frequently as the result of exhausting the supply of an essential growth nutrient (Figure 5). Other factors, however, can cause or contribute towards the termination of growth; for example, end products of metabolism can be inhibitory. Yeasts fermenting glucose anaerobically produce ethanol which at levels of about 10% (v/v) cause the yeast to stop growing. Sulfate-reducing bacteria produce hydrogen sulfide which causes growth to stop at concentrations of 1.0 mM unless it is volatilized or used by other organisms. Microbial growth in all environments, if started with a small enough population, follows a characteristic sequence of events. This is most clearly seen in laboratory batch cultures and is known as the *batch culture growth cycle* (Figure 6). The various phases are described in more detail by Bull (1974), Pirt (1975), Slater (1979), and Pritchard and Tempest (1982).

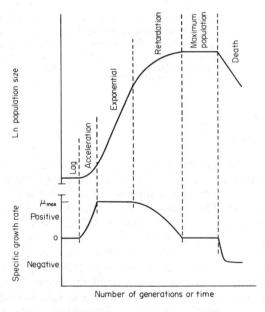

Figure 6 The phases of growth occurring during a closed culture growth cycle in a limited environment, showing the changes in the organism specific growth rate (Bull, 1974)

11.2.1 Growth Limitation by Substrate Exhaustion

Depletion of the *growth-limiting substrate* restricts the final size of the population and indeed the two parameters may be shown to be proportional to each other (Figure 7):

$$x_f \propto S_R \tag{11}$$

where S_R is the initial concentration of the growth-limiting substrate and x_f is the final population size. Thus:

$$x_f = YS_R \tag{12}$$

where Y is a proportionality constant known as the *observed growth yield*. In its simplest form it is defined as that weight of new biomass produced as a result of utilizing a unit amount of the growth-limiting substrate. If x_f is measured as grams of biomass per litre, then the observed growth yield has units of grams of biomass produced per gram of substrate utilized.

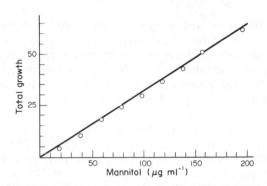

Figure 7 Total growth of *Escherichia coli* in synthetic medium with the organic source (mannitol) as the limiting factor. Ordinate: arbitrary units, where one unit is equivalent to 0.8 μg dry weight ml^{-1} (Monod, 1949)

The yield term indicates how much of the available substrate is used for new biomass production and also the amount used for energy generation to drive the biosynthetic reactions. The yield term is therefore a variable depending on the balance between these two processes. Clearly the greater the efficiency in energy generation the less the amount of available substrate which has to be dissimilated and so the greater the amount available for new biomass production. This can be well illustrated by comparing the yields of a facultative anaerobe growing either aerobically or anaerobically. *Streptococcus faecalis* using glucose aerobically produces a Y value of 0.32 g biomass (g glucose used) $^{-1}$, whereas anaerobically Y is 0.12 g biomass (g glucose used) $^{-1}$. The reasons for the lower anaerobic growth yield are quite clear. A greater fraction of the available glucose has to be metabolized to generate an equivalent amount of energy since under anaerobic conditions only substrate level phosphorylation can lead to ATP production. Furthermore, some of the glucose (or derivative molecules) has to act as terminal electron acceptors during fermentation and this further reduces the available carbon for biomass production.

Similarly, the yield value depends on the nature of the substrate since this determines at what point it is fed into central metabolic pathways. This in turn dictates the amount of energy which may be generated per substrate molecule. For example, for *Klebsiella aerogenes* grown aerobically on glucose, $Y = 0.39$ g biomass (g glucose used) $^{-1}$, whereas on succinate under aerobic conditions $Y = 0.23$ g biomass (g substrate used) $^{-1}$.

Many different ways of expressing the observed growth yield have been proposed in efforts to find a standard term which gives an adequate basis for comparison between different organisms, different growth substrates and different growth conditions. These are discussed in detail by Stouthamer and van Verseveld in Volume 1, Chapter 12.

Y_{sub} is the *molar growth yield* with units of grams of biomass per mole of substrate consumed. The subscript indicates the substrate for which the yield has been determined. The molar growth yield is widely used since it more directly relates the efficiency of substrate conversion to biomass and is more closely akin to the conventional way of expressing substrate concentrations. It has serious shortcomings since it does not standardize with respect to parameters more directly concerned with biomass production (that is the carbon content) or with the substrate energy content (that is electron availability for energy production). Y_{oxygen} is the molar growth yield with units of grams of biomass per mole of oxygen consumed, and is widely used for aerobic organisms. It is based on the reasonable assumption that the amount of oxygen consumed is directly related to the amount of energy generated and so to the amount of growth substrate used for energy-generating purposes. The difficulty is that the relationship between ATP production and oxygen consumption is not necessarily constant and does not, for example, allow for uncoupling between oxidative phosphorylation and substrate used in energy-generating steps. It is also substantially influenced by oxygen availability.

Molar growth yields based on the gram carbon atom content of substrate and biomass are not widely used but do directly relate the amount of carbon used for biosynthesis to the total carbon availability. Y_{carbon} has units of grams of cell carbon per gram of substrate carbon consumed. When expressed as a percentage this yield term is sometimes known as the *carbon conversion efficiency* (*CC*) where:

$$CC = Y \times (\% \text{ carbon in biomass}/\% \text{ carbon in growth-limiting substrate}) \times 100 \qquad (13)$$

This term effectively standardizes with respect to unit carbon thereby making comparisons between different carbon sources valid. The disadvantage is that the amount of energy generated is not directly related to the carbon content but is dependent on the redox potential of the substrate, that is organisms need to catabolize a small proportion of more reduced substrates compared to more oxidized substrates in order to generate the same amount of useful cell energy.

The molar growth yield for ATP, Y_{ATP}, has been widely used to overcome this difficulty and relate biomass production to the use of moles of ATP. Y_{ATP} has units of grams of biomass per mole of ATP generated and theoretically ought to give reliable estimates of the observed growth yield since it is based directly on the amount of energy (ATP and other high energy compounds, such as NADH and NADPH) available for biosynthetic purposes. For anaerobic organisms generating ATP by substrate level phosphorylation, it is possible to calculate accurately the amount of ATP generated from the stoichiometry of substrate fermentation. For many organisms constant Y_{ATP} values of 10.5 g biomass (mol ATP)$^{-1}$ have been calculated. For aerobic organisms, however, the amount of ATP produced cannot be calculated readily and is subject to the same sort of errors as were given for Y_{oxygen} values. In the past Y_{ATP} has been elevated to the level of a mystical biological constant but it is now known that many factors, such as environmental conditions, substantially influence ATP yields and so Y_{ATP} values.

Calculation and expression of yield values is fraught with difficulty and the reader is referred to the following articles for more detailed discussions of these problems (Payne, 1970; Harrison, 1978; Stouthamer, 1977; Stouthamer and van Verseveld, Volume 1, Chapter 12).

11.2.2   Variation in the Observed Growth Yield

The previous discussion has recognized difficulties in providing a standard growth yield term. It was assumed, however, that for a given organism the observed growth yield for a given substrate would be constant. This is not the case and it is now known to be substantially influenced by factors such as the growth rate of the organism. The observed growth yield shows a much lower value than expected when populations are grown slowly under carbon limited conditions. Pirt (1965) suggested that this variation is due to a fraction of the growth substrate being oxidized to produce energy which is *not* used for biomass production. This fraction of cellular energy is used for maintaining solute gradients, motility, maintenance of internal pH, turnover of macromolecules and transport mechanisms within the cell. This group of growth disassociated, energy-requiring processes are collectively known as *maintenance functions* and the energy required to drive them as the *maintenance energy*. It is normal to express the maintenance energy coefficient *m* in terms of the amount of substrate used per unit amount of biomass in unit time (units: grams of substrate used per gram of biomass per hour; see Section 11.4.2.1 for the determination of the maintenance coefficient). It is a useful concept and the models proposed accounting for the observed variability of the observed growth yield fit reasonably well. However, it must be recognized that little quantitative information is available on the energetic requirements of the individual maintenance functions. It is clear that environmental conditions and growth rate influence the magnitude of the maintenance energy requirements (see Stouthamer and van Verseveld, Volume 1, Chapter 12).

11.2.3   Influence of the Growth-limiting Substrate on Growth Rate

Section 11.2.1 demonstrated the effect of the initial growth-limiting substrate concentration on the final population size. However, prior to reaching the maximum population phase there is a stage during which substrate depletion causes a restriction in the specific growth rate (Figures 5 and 8). Monod was the first to establish a relationship between growth-limiting substrate concentration and specific growth rate by recognizing that it is a rectangular hyperbola and closely resembles that between the velocity of an enzyme-catalyzed reaction and substrate concentration. This was shown to conform to the Michaelis–Menten equation:

$$v = \frac{VS_0}{S_0 + K_m} \tag{14}$$

where v is the reaction velocity, V is the maximum velocity attained at saturating substrate concentration, S_0 is a subsaturating concentration and K_m is the value of the substrate concentration which gives a reaction velocity equal to $V/2$. Monod deduced that for microbial cultures:

$$\mu = \mu_{max} \frac{s}{K_s + s} \tag{15}$$

where s is the growth-limiting substrate concentration, μ is the specific growth rate, μ_{max} is the maximum specific growth rate and K_s is a constant known as the *saturation constant* and is equivalent to the value of s which gives a growth rate equal to $\mu_{max}/2$ (units: as for substrate concentration and normally grams per litre). The saturation constant is a measure of the affinity the organism has for the growth-limiting substrate; the smaller the K_s value, the higher the substrate affinity and the greater the capacity to grow rapidly at low growth-limiting substrate concentrations. The saturation constant attains particular importance in open growth systems (see Section 11.4) and in the outcome of competition between different organisms (see Section 11.5). In closed culture systems the saturation constant has little significance since most of the growth occurs when substrates are in excess conditions, when $s \gg K_s$, and so equation (15) reduces to $\mu = \mu_{max}$. For most organisms the K_s values for carbohydrate substrates are in the order of milligrams per litre and for other compounds, such as amino acids, approximately micrograms per litre. For unusual or unnatural substrates the organism's affinity may be poor and so the K_s term will have a measurable effect. It is important to recognize that many organisms inhabit extremely nutrient-deficient environments and have evolved exceptionally high affinity mechanisms.

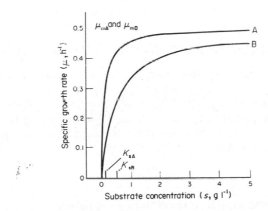

Figure 8 The relationship between the specific growth rate and the growth-limiting substrate concentration. The Monod relationship, equation (15), is shown for two organisms: A, $\mu_{max} = 0.5$ h^{-1}, $K_s = 0.1$ g substrate l^{-1} and B, $\mu_{max} = 0.5$ h^{-1}, $K_s = 0.5$ g substrate l^{-1} (Slater, 1979)

In theory the K_s value may be determined from a series of batch cultures differing in their initial growth-limiting substrate concentration. The initial growth rate of the different cultures is determined and these values plotted against their corresponding growth-limiting substrate concentration (Figure 8). In practice this procedure is extremely difficult since in order to measure submaximal growth rates, very low growth-limiting substrate concentrations have to be used. This means that, firstly, the amount of growth which can be observed is very limited and, secondly, the smallest fraction of biomass increase represents a substantial decrease in the available substrate which reduces the growth rate still further. It is, therefore, difficult to achieve an adequate period of growth when the specific growth rate is constant and can be measured.

Curve-fitting a hyperbolic function to data of the sort shown in Figure 8 is unreliable and better estimates of K_s (and indeed μ_{max}) values can be obtained from suitable rearrangements of equation (15) with linear functions, such as the familiar Lineweaver–Burk rearrangement used in enzyme kinetics (equation 16).

$$\frac{1}{\mu} = \frac{K_s}{s\mu_{\max}} + \frac{1}{\mu_{\max}}$$

(16)

Plotting $1/\mu$ against $1/s$ yields a linear relationship (Figure 9) intercepting the abscissa at $1/\mu_{\max}$ and the ordinance at $-1/K_s$

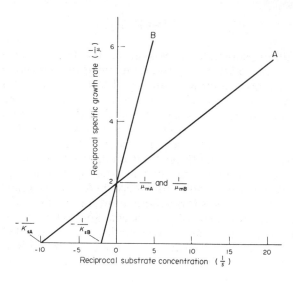

Figure 9   The linear form of the Monod equation (16) for the same organisms as in Figure 8 (Slater, 1979)

11.2.4   Deviation of the Monod Equation at High Substrate Concentrations

It is frequently observed at high substrate concentrations that there is a decline in the specific growth rate rather than an asymptotic approach to $\mu_{\max}$ (Figure 10); that is at high concentrations the growth substrate also expresses growth inhibiting properties. Many modifications of the Monod function have been proposed (see, for example, Dijkhuizen and Harder, 1975) to include an inhibition constant K_i. These modifications have been arrived at empirically and the most widely used is the Haldane equation.

$$\mu = \frac{\mu_{\max}s}{(K_s + s)(1 - s/K_i)}$$

(16a)

One of the difficulties with these models is determining the theoretical $\mu_{\max}$ value since there are no growth conditions under which it can be directly measured.

11.2.5   Basic Growth-limiting Substrate Equation

From equations (11) and (12) it is clear that in a small time interval, δt, the population increases in size by a small amount, δx, as a consequence of using a small amount of the growth-limiting substrate $-\delta s$. Thus:

$$-\frac{dx}{ds} = Y$$

(17)

that is the rate of change of biomass with respect to used substrate depends on the observed growth yield. Hence the rate of change of the growth-limiting substrate may be calculated by substituting equation (6) for dx and rearranging:

$$-\frac{ds}{dt} = \frac{\mu x}{Y}$$

(18)

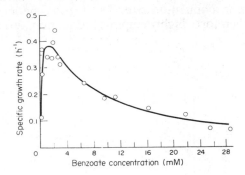

Figure 10 Inhibition of specific growth rate of *Klebsiella aerogenes* growing on sodium benzoate (Edwards, 1970)

The negative sign indicates a decrease in substrate concentration with time. The constants μ and Y may be replaced by another constant q, the *specific metabolic rate* or *metabolic quotient*:

$$- \frac{ds}{dt} = qx \tag{19}$$

The specific metabolic rate defines the rate of uptake of the growth-limiting substrate by a unit amount of biomass per unit time (units: grams of growth-limiting substrate per gram of biomass per hour). It must be noted, however, that q can vary widely since, as has already been discussed, μ and Y are subject to environmental variation (see Section 11.4.2.1).

11.3 MODELLING MICROBIAL GROWTH IN LIMITED ENVIRONMENTS

The two basic parameters involved in a growing microbial population, namely the increase in cell number or biomass concentration and the decrease in substrates required for growth (particularly the growth-limiting substrate), may be modelled in many ways. For example, it is possible to generate polynomial expressions which adequately fit the observed data, although these models are often unsatisfactory since the relevance, in a biological sense, of the constants used is not clear. The models are widely used to describe the characteristic sigmoidal curves of the batch culture growth cycle (Figures 5 and 6).

11.3.1 The Logistic Equation

It has been known for many years (Pearl and Reed, 1920; Verhulst, 1839) that the classical s-shaped curves generated when a population grows may be empirically described by equation 20.

$$\frac{dx}{dt} = \mu_{max} x \left[1 - \frac{x}{x_f} \right] \tag{20}$$

The variables μ_{max}, x and x_f are the same terms as previously described. This equation describes the change in population density from the moment growth begins (the beginning of the exponential phase) to the end of the maximum population phase. Equation (20) differs from equation (6) by the term $[1-(x/x_f)]$ which provides the 'restriction' factor absent from the basic growth equation (6), and enables a reduction in the rate of change of biomass concentration to a zero value to be modelled. At the beginning of the batch growth cycle when the biomass concentration is small, all nutrients are in excess and conditions are essentially those required for unlimited exponential growth; then x/x_f is small and the term $[1-(x/x_f)]$ tends towards one and equation (20) reduces to equation (6). As x increases, x/x_f tends towards unity and $[1-(x/x_f)]$ becomes significantly less than one. Since μ_{max} is a constant this reducing factor effectively decreases the value of μ_{max} thereby restricting the rate of increase in the population size with time. By the time the population reaches the maximum population phase, $x = x_f$ and so $[1-(x/x_f)] = 0$ and $dx/dt = 0$ which is characteristic of the maximum population phase. This simple model frequently describes population growth but it must be recognized that there are likely to be many factors other than x_f which contribute to the limitation in population size. The logistic model in no way identifies these unknown variables and is, therefore, highly empirical. Equation (20) has been used to model the data shown in Figure 11.

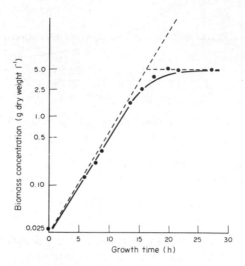

Figure 11 The modelling of the growth of *Geotrichum candidum* in a closed culture with glucose as the limiting substrate ($S_R = 10$ g l^{-1}). The observed biomass concentrations (g dry weight l^{-1}) are shown by ●. The logistic model (——) was fitted using $\mu_{max} = 0.353$ h^{-1} and $x_f = 4.9$ g dry weight l^{-1}. The saturation model (- - -) was fitted using the following constants: $\mu_{max} = 0.353$ h^{-1}, $K_s = 0.024$ g glucose l^{-1} and $Y = 0.49$ g dry weight g^{-1} glucose. The continuation line (- - -) indicates continued exponential growth in an unlimited environment. (Data from Trinci, 1971, and modelled according to Bull *et al.*, 1975)

The equation may be used to determine the actual growth rate at a particular population density using equation (21).

$$\mu = \mu_{max}\left[1 - \frac{x}{x_f}\right] \tag{21}$$

Substituting for μ in equation (18) gives equation (22).

$$-\frac{ds}{dt} = \frac{\mu_{max}\left[1 - (x/x_f)\right]}{Y}\,x \tag{22}$$

Thus by an iterative computation using equations (20) and (22) it is possible to calculate the increase in biomass concentration and the associated decrease in growth-limiting substrate concentration.

11.3.2   The Saturation Model

The same stages of growth may be described by a model based on the Monod equation to modify the constant μ_{max} value used in equation (6). This involves substituting equation (15) in equation (6) to give equation (23).

$$\frac{dx}{dt} = \frac{\mu_{max}\,sx}{(K_s + s)} \tag{23}$$

During the initial stages of growth when $s \gg K_s$, $s/(K_s + s)$ equals unity and so equation (23) reduces to the basic exponential growth equation for unlimited environments. As s tends towards K_s, $s/(K_s + s)$ tends towards zero, reducing the value of μ_{max} and restricting the rate of change of biomass with time. Equation (23) has been used to model the data shown in Figure 11.

11.4 MICROBIAL GROWTH IN OPEN ENVIRONMENTS

Open growth systems differ from closed growth systems in that there is a continuous input of growth substrates and removal of waste products, cells and unused substrates. These systems are known as continuous-flow cultures and many different forms have now been described and used (for examples see Slater and Hardman, 1982). Continuous-flow cultures enable the exponential growth phase to be prolonged indefinitely, establishing steady state conditions. Unlike batch cultures, where to a large extent the behaviour of cells at a particular point in time is influenced by the culture's previous history, steady state conditions remove the influence of transient conditions. There are substantial, additional advantages to open growth: specific growth rate may be directly controlled by external factors, continued substrate-limited growth may be established, submaximal growth rates can be imposed, and biomass concentration may be set independently of the growth rate. These advantages are discussed in more detail by Pirt (1972, 1975) and Bull (1974). The most widely used continuous culture system is the *chemostat* which is characterized by growth control exercised through a growth-limiting substrate.

11.4.1 Chemostat Growth Kinetics

The major feature of a chemostat is a culture vessel containing a fixed volume, V, of growing culture (units: volume). For both aerobic and anaerobic systems agitation is needed to maintain a homogeneous culture which, in the case of aerobic cultures, also ensures adequate oxygen transfer (Figure 12). A defined growth medium containing one of the constituent substrates S_R at a concentration which is known to be growth-limiting is pumped into the growth vessel at a constant rate F (units: volume time $^{-1}$). By mechanisms such as a weir overflow or another pumping system, culture, unused substrates and products are removed at exactly the same flow rate F. It is assumed that in an adequately mixed culture the incoming fresh medium is instantaneously mixed with the resident culture preventing temporary, heterogeneous conditions being established. Within the culture vessel a certain biomass concentration, x, is set up (and this also gives the biomass concentration in the effluent culture) as a result of growth, thereby reducing the initial concentration of the growth-limiting substrate, S_R, to a value s.

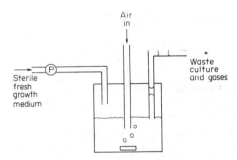

Figure 12 Schematic diagram of a simple continuous-flow culture system (Slater and Hardman, 1982)

11.4.1.1 The dilution rate

In a chemostat the concentration of the growth-limiting substrate clearly depends on the rate at which the organisms use it and on its rate of supply. There is an additional factor which depends on the rate of dilution as fresh substrate is dispersed throughout the growing culture. Thus the growth-limiting substrate concentration depends on a ratio, known as the *dilution rate*, D, of the flow rate through the system F and the volume of the culture in the vessel, V (equation 24).

$$D = F/V \qquad (24)$$

The dilution rate has units of reciprocal time, normally hours $^{-1}$, and is a measure of the number of volume changes achieved in unit time. Thus a dilution rate of 0.5 h $^{-1}$ means that in one hour a volume of fresh medium enters (and an equal volume of spent culture leaves) which is equivalent to half the constant culture volume. The dilution rate, therefore, has units which are the same as the specific growth rate and it will be shown later that under steady state conditions the dilution rate is equal to the culture specific growth rate. For the moment, however, the important point to recognize is that D is easily controlled by the experimenter since the culture volume may be fixed and the flow rate controlled to desired values. The reciprocal of the dilution rate is known as the *mean residence time*, θ (units: time, usually hours), and denotes the average time a cell remains within the culture vessel.

11.4.1.2   *The dilution rate and biomass concentration*

If it is assumed for a moment that the organisms are non-growing, then the rate of change of the organism concentration in a system with dilution rate D is given by:

$$\frac{dx}{dt} = -Dx$$

and by integration:

$$x_t = x_0 e^{-Dt} \tag{25}$$

This equation describes exponential decay and is analogous to radioactive element decay and constitutes a measure of the culture *washout*.

However, in a growing culture an additional factor influencing the rate of change of biomass concentration is a growth term, such that:

The overall rate of change of biomass concentration in the culture vessel	=	The rate of biomass production (growth)	−	The rate of biomass removal (washout)
$\dfrac{dx}{dt}$	=	μx	−	Dx
$\dfrac{dx}{dt}$	=	$x(\mu - D)$		$\tag{26}$

Substituting μ for equation (15) gives equation (27).

$$\frac{dx}{dt} = x\left[\frac{\mu_{max}s}{K_s + s} - D\right] \tag{27}$$

Equation (26) predicts three general cases for the overall rate of change of biomass concentration. (i) If $\mu > D$, dx/dt is positive; the biomass concentration increases and the rate of biomass production is greater than the rate of culture washout. (ii) If $\mu < D$, dx/dt is negative; the biomass concentration declines and the growth rate is less than the washout rate. (iii) If $\mu = D$, dx/dt is zero and the rate of biomass production balances the rate of culture washout. Under these conditions the culture is said to be in a *steady state* and, as we shall see (Section 11.4.1.6), this is the preferred, stable state and in time all chemostat cultures will reach a steady state provided that $D < \mu_{max}$ and the environmental conditions are kept constant.

11.4.1.3   *The dilution rate and growth-limiting substrate concentration*

A similar balanced equation may be derived to describe the change of growth-limiting substrate such that:

The overall rate of change of growth-limiting substrate concentration in the culture vessel	=	The rate of supply of substrate (input)	−	The rate of substrate removal (wash out)	−	The rate of organism use for growth (growth)
$\dfrac{ds}{dt}$	=	DS_R	−	Ds	−	$\dfrac{\mu x}{Y}$

The first two terms are rates dependent on the dilution rate whilst the growth use term was derived previously (equation 18), and

$$\frac{ds}{dt} = D(S_R - s) - \frac{\mu x}{Y} \tag{28}$$

Inspection of equation (28) shows that the two terms $(S_R - s)$ and x/Y are equivalent and both measure the proportion of substrate used for growth, s_g, giving equation (29).

$$\frac{ds}{dt} = Ds_g - \mu s_g \tag{29}$$

Three general cases may be considered. (i) If $D > \mu$, ds/dt is positive; the growth-limiting substrate concentration increases and the rate of supply of growth-limiting substrate is less than the rate of use by x. (ii) If $D < \mu$, ds/dt is negative and the overall growth-limiting substrate concentration declines. (iii) If $D = \mu$, ds/dt is zero and the growth-limiting substrate concentration in the culture vessel reaches a steady state. This occurs at the same time as the value for $dx/dt = 0$.

11.4.1.4 Biomass and growth-limiting substrate concentrations in the steady state

As we have stressed, the value of chemostat systems is the ability to establish steady state conditions where $ds/dt = 0$ and $dx/dt = 0$. These values are attained for unique values of x and s for a given value of D and constant environmental conditions. Furthermore as shown in Section 11.4.1.6 this is a self-balancing system and the culture adjusts itself to maintain the steady state condition. In the steady state the unique values of x and s are given the symbols $\bar{x}$ and $\bar{s}$ respectively.

By substituting in equation (27) and then rearranging, equation (30) is obtained.

$$0 = \bar{x}\left[\frac{\mu_{max}\bar{s}}{(K_s + \bar{s})} - D\right]$$
$$D = \frac{\mu_{max}\bar{s}}{(K_s + \bar{s})}$$
$$\bar{s} = \frac{DK_s}{[\mu_{max} - D]} \tag{30}$$

Similarly, equation (28) leads to equation (31).

$$0 = D(S_R - \bar{s}) - (\mu x/Y)$$
$$\bar{x} = Y(S_R - \bar{s}) \tag{31}$$

Substituting equation (30) gives equation (32).

$$\bar{x} = Y\left[S_R - \frac{DK_s}{(\mu_{max} - D)}\right] \tag{32}$$

Equations (30) and (31) enable the steady state biomass concentration and growth-limiting substrate concentration to be calculated provided that three basic growth parameters, K_s, μ_{max} and Y, are known. Since for a given organism under constant conditions these parameters are constant, then the unused growth-limiting substrate concentration in the culture vessel depends solely on the imposed dilution rate: it is even independent of the initial substrate concentration S_R. On the other hand the biomass concentration depends on D and S_R. The relationships between x, s, the organism constants and the dilution rate are demonstrated in Figure 13. As D tends towards μ_{max} then, from equation (30), s increases and tends towards its maximum value, namely S_R. Similarly, from equation (31), x tends towards zero as s tends towards S_R. Indeed, a combination of equations (30) and (31) shows that there is an upper limit to the dilution rate above which a steady state culture cannot be achieved. This is because the organism has a maximum specific growth rate which cannot be exceeded. So if $D > \mu_{max}$, then from equation (26), dx/dt must be negative. As shown in Figure 13 the maximum dilution rate attainable is slightly less than the value for μ_{max} and is dependent on the value of S_R. This value of the maximum dilution rate which can be obtained is known as the *critical dilution rate* and is obtained when s equals S_R.

It is possible to calculate the various basic growth parameters from measurements of x and s,

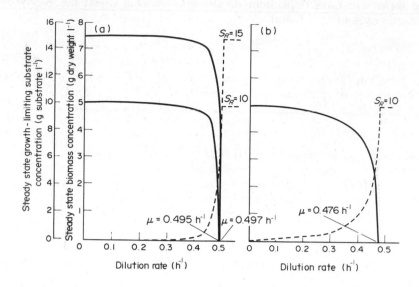

Figure 13 Theoretical steady state biomass and growth-limiting substrate concentrations for two organisms, A and B, in chemostat culture. In (a) organism A ($\mu_{max} = 0.5$ h^{-1}, $K_s = 0.1$ g substrate l^{-1} and $Y = 0.5$ g biomass g^{-1} substrate) is growing in two different cultures with varying concentrations of the growth-limiting substrate. When $S_R = 10$ g substrate l^{-1}, the lower line results and $D_{crit} = 0.495$ h^{-1}. When $S_R = 15$ g substrate l^{-1}, the upper line results and $D_{crit} = 0.497$ h^{-1}. In (b) organism B ($\mu_{max} = 0.5$ h^{-1}, $K_s = 0.5$ g substrate l^{-1} and $Y = 0.5$ g biomass g^{-1} substrate), differing from A only in the affinity for the limiting substrate, is growing with an $S_R = 10$ g substrate l^{-1} and $D_{crit} = 0.476$ h^{-1} (Slater, 1979)

especially K_s. However, as can be seen from Figure 13, the lower the K_s value, the smaller the unused growth-limiting substrate concentration over a wide range of dilution rates. Accordingly it is necessary to have reliable and highly quantitative methods for analyzing low growth-limiting substrate concentrations.

11.4.1.5 *Determination of μ_{max} from washout kinetics*

As noted in the previous section, if $D > \mu_{max}$, then dx/dt is negative and the biomass concentration must decrease. Under these conditions *washout* is said to occur. Thus as s tends towards S_R at the critical dilution rate, then:

$$D_{crit} = \frac{\mu_{max} S_R}{(K_s + S_R)} \tag{33}$$

If $K_s \ll S_R$, as is normally the case, then $D_{crit} = \mu_{max}$. It is important to note that D_{crit} is not a constant, unlike μ_{max}, and depends on the organism and also the value of S_R (see Figure 13). During washout from a culture vessel, so long as x tends towards 0 and s tends towards S_R, then μ tends towards μ_{max}. Thus:

$$\frac{dx}{dt} = x(\mu_{max} - D) \tag{34}$$

When D is fixed at a value $D > \mu_{max}$, then equation (34) has the solution:

$$x_t = x_0 \exp\left[(\mu_{max} - D)t\right]$$

and

$$\ln x_t = \ln x_0 + (\mu_{max} - D)t \tag{35}$$

From data of the type shown in Figure 14, μ_{max} can be calculated as follows:

$$\mu_{max} = \frac{\ln x_t - \ln x_0}{t} + D \tag{36}$$

It is important to note that this method assumes that μ_{max} is attained immediately the dilution rate is stepped up to $D > \mu_{max}$. This is not the case and in shift up experiments of this sort a period of time equal to the C and D time for the cell cycle must elapse before the new, higher growth rate is established (Mandelstam *et al.*, 1982; Oliver and Trinci, Volume 1, Chapter 10). This may be clearly seen in washout experiments with the cyanobcaterium *Anacystis nidulans* (Karagouni and Slater, 1978).

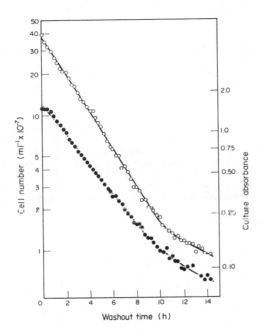

Figure 14 Washout curves for *Anacystis nidulans* in terms of cell number (○) and absorbance (●) for a light-limited culture grown at an initial dilution rate of 0.03 h^{-1}. At $t = 0$ the dilution rate was instantaneously increased to 0.37 h^{-1}
(Karagouni and Slater, 1978)

11.4.1.6 Establishing and maintaining the steady state

As we have noted several times, a particular attribute of chemostat kinetics is the ability to maintain steady state culture conditions with unique values of x and s for a given value of D. The mechanism is as follows: consider a culture growing at D, with an initial, small biomass concentration x_1 ($x_1 \ll \bar{x}_1$ for D_1) and a high growth-limiting substrate concentration s_1 ($s_1 \gg \bar{s}_1$ for D_1) (see Figure 15). Under these conditions the growth-limiting substrate concentration is much greater than that value required to maintain $\mu_1 = D_1$. Thus with $s_1 > \bar{s}_1$ then $\mu_1 > D_1$ and as we have seen in Sections 11.4.1.2 and 11.4.1.3, dx/dt is positive and ds/dt is negative. As growth proceeds x_1 approaches $\bar{x}_1$ and s_1 declines towards $\bar{s}_1$ until they equal the steady state values when $\mu_1 = D_1$.

The reverse situation may be described in establishing a new steady state at D_2 starting from a steady state culture at D_1 (Figure 15). If D_1 is instantaneously increased to D_2 the immediate conditions are that s_2 ($= s_1$) $< \bar{s}_2$ and x_2 ($= x_1$) $> \bar{x}_2$. Thus the growth-limiting substrate concentration s_2 is unable to support a growth rate μ_2 equal to D_2. Thus it must be the case that $\mu_2 < D_2$ and so dx/dt is negative and ds/dt is positive. The biomass concentration declines as the growth-limiting substrate concentration increases until the new steady state conditions are established.

It is clear, therefore, that at a fixed D, minor changes in x or s result in appropriate changes in dx/dt and ds/dt such that the required steady state values are restored.

11.4.2 Deviations from Theoretical Chemostat Kinetics

In many instances the observed growth kinetics in a chemostat fit well with the above theory and, incidentally, provide good support for the Monod model of growth kinetics (equation 15).

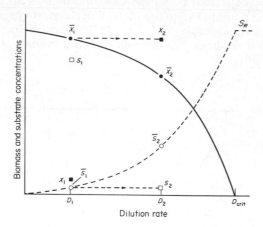

Figure 15 Establishing and maintaining the steady state chemostat culture. (a) From an initial inoculation of the culture vessel at a dilution rate of D_1: ■ = initial biomass concentration x_1; □ = initial growth-limited substrate concentration s_1; ● = steady state biomass concentration $\bar{x}_1$; and ○ = steady state growth-limiting substrate concentration $\bar{s}_1$. (b) After a change in the dilution rate from D_1 to D_2: ■ = initial biomass concentration x_2; □ = initial growth-limiting substrate concentration s_2; ● = steady state biomass concentration $\bar{x}_2$ at D_2; and ○ = steady state growth-limiting substrate concentration $\bar{s}_2$ at D_2 (Slater, 1979)

However, it is important to note that many additional factors influence the observed values for x and s (Figure 16).

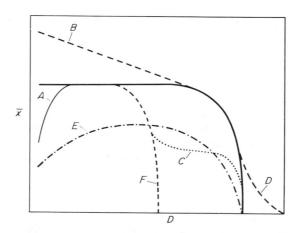

Figure 16 A schematic representation of experimental departures from the simple chemostat theory. The solid line indicates the theoretical variation of the steady-state organism concentration $\bar{x}$ with dilution rate D. The curves A to F indicate various observed non-ideal behaviour: A, maintenance energy; B, synthesis of polymeric reserves; C, switch in metabolism to that which is energetically less efficient; D, imperfect mixing; E, substrate toxicity; F, second substrate becomes growth limiting (Bull, 1974)

11.4.2.1 *Influence of variation in the observed growth yield*

In Section 11.2.2 it was shown that the observed growth yield is not a constant (equation 12). If it is assumed that the quantity of the substrate used for maintenance purposes is constant and independent of changes in the dilution rate, then the proportion of substrate available for biomass production declines and so x declines (Figure 16A). Thus the overall rate of substrate utilization may be expressed as:

The overall rate of substrate utilization	=	The rate of substrate use for biomass production	+	The rate of substrate use for maintenance purposes

$$\frac{ds}{dt} = \left(\frac{ds}{dt}\right)_{growth} + \left(\frac{ds}{dt}\right)_{maintenance} \qquad (37)$$

Now equation (19) gives the overall rate of substrate utilization and may be rewritten to distinguish the rates of substrate use for growth and maintenance purposes:

$$\frac{ds}{dt} = qx = q_{growth}x + q_{maintenance}x \qquad (38)$$

where $q_{growth} = \mu/Y_G$, where Y_G is the true or maximum growth yield, and $q_{maintenance} = m$, where m = the maintenance coefficient. Thus:

$$\frac{\mu x}{Y} = \frac{\mu x}{Y_G} + mx$$

and rearranging:

$$\frac{1}{Y} = \frac{1}{Y_G} + \frac{m}{\mu} \qquad (39)$$

Thus deviations in the observed growth yield with respect to different steady state cultures at different dilution rates enable the values for the constants Y_G and m to be determined. From Figure 17, which shows $1/Y$ plotted against $1/D(= 1/\mu)$, the intercept on the abscissa gives $1/Y_G$ and the slope of the line yields m. Typically, values for m are in the order of 10–100 mg carbon substrate used (g biomass)$^{-1}$ h^{-1}.

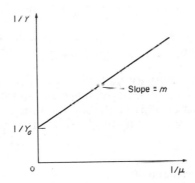

Figure 17 Determination of the true growth yield (Y_G) and maintenance coefficient (m) (Pirt, 1975)

11.5 MICROBIAL COMPETITION

There are many types of interaction between two or more microbial populations (Slater and Bull, 1978; Bushell and Slater, 1981; Bull and Slater, 1982; Bull, Volume 1, Chapter 15), but from the point of view of the industrial exploitation of microorganisms, competition is the most significant. The problem stems from the behaviour of contaminants in pure culture fermentations (Powell, 1958) and ranges to the behaviour of industrial fermentations with unstable recombinant DNA culture (Ollis, 1982). Moreover, competition is the most important interaction between populations in Nature, and understanding the kinetic principles is important in primary studies involving the enrichment and isolation of particular microorganisms.

Competition between two or more microorganisms occurs when the component populations are restricted, either in terms of their growth rates or final population sizes, as the result of a common dependence on an external factor. Competition can, within the terms of this widely accepted definition, occur either in closed culture where growth is ultimately limited by the availability of a particular growth resource, or in open culture where growth is continuously limited. Bazin (1981) has erroneously attacked the view that competition can occur in batch culture, misunderstanding the substantial significance of competitive events which influence final population sizes. Ecologically, and in many industrially important fermentations, this type of competition is important and is entirely consistent with the above definition.

Many mathematical models have been proposed (see, for example, Fredrickson, 1977; Bazin, 1981; Ollis, 1982) and the models given below are presented simply to illustrate the basic principles. The following analyses consider two populations, x_a and x_b, growing in an environment with a single growth-limiting substrate, and assume that apart from the competition interaction there are no other interactions between the two populations.

11.5.1 Competition in Closed Environments

In this situation, normally occurring for batch culture, the two populations are present in an environment in which initially all the substrates are present at excess concentrations. Under these conditions, growth of both populations proceeds *via* the basic growth equation (6):

$$\frac{dx_a}{dt} = \mu^a_{max} t \quad \text{and} \quad \frac{dx_b}{dt} = \mu^b_{max} t$$

where μ^a_{max} and μ^b_{max} are the maximum specific growth rates for organisms A and B. Thus:

$$\ln x_{at} - \ln x_{a0} = \mu^a_{max} t \tag{40}$$

$$\ln x_{bt} - \ln x_{b0} = \mu^b_{max} t \tag{41}$$

and rearranging these two equations

$$\mu^a_{max} - \mu^b_{max} = \frac{\ln R_t - \ln R_0}{t} \tag{42}$$

where R_t = ratio of population A to population B after time t for growth, and R_0 = the initial ratio of population A to population B.

As expected the greater the difference between the maximum specific growth rates of the two populations, the greater the difference in the final population sizes once the limiting nutrient has been exhausted (Figure 18).

Equation (42) is probably satisfactory for the situations where maximum growth rates are sustained for long periods in batch culture. Clearly, however, appropriate terms which give a variable μ [namely the Monod equation (15) or the logistic function (20)] in conjunction with an appropriate rate of change of substrate term, may be substituted for μ_{max} terms.

11.5.2 Competition in Open Environments

Open growth systems illustrate particularly competition characteristics dependent on different growth rates due to the availability of a single growth-limiting substrate. As previously shown (Sections 11.4.1.3 and 11.4.1.4) an organism's growth rate is controlled by the concentration of the growth-limiting substrate. Clearly in a mixed culture the organism able to sustain the highest growth rate will increase its population size more rapidly than any of the competing populations (assuming that there are no other stabilizing interactions occurring between the different populations). It also inevitably follows that in an open growth system all uncompetitive populations will eventually be eliminated, the rate of elimination being dependent on the differences in growth rate.

In a chemostat at a dilution rate D, organism A establishes a steady state population such that (from equation 30):

$$\mu_a = D = \frac{\mu^a_{max} s}{K^a_s + s} \tag{43}$$

If a mutant should arise or a contaminating organism enter the growth vessel, then its chance of establishing itself as a successful, competitive population depends on its growth rate potential at a

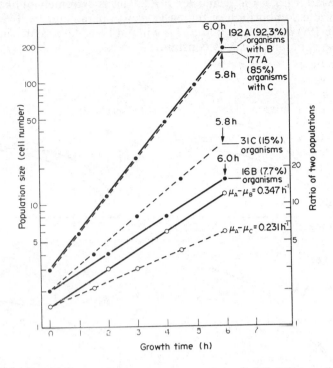

Figure 18 Competition between two pairs of microorganisms, A and B or A and C, in a closed culture system. For both mixed cultures the starting population sizes were 3:2, A to B or A to C. (●——●) growth curves for A and B together where $\mu_{max}^a = 0.693$ h^{-1} and $\mu_{max}^b = 0.347$ h^{-1}, (○——○) logarithm of the ratio of population A to population B; (●----●) growth curves for A and C together where $\mu_{max}^a = 0.6593$ h^{-1} and $\mu_{max}^c = 0.462$ h^{-1}: (○----○) logarithm of the ratio of population A to population C (Slater and Bull, 1978)

substrate concentration s which is initially dictated by the major growth characteristics (μ_{max} and K_s^a of organism A. Initially the growth of population B is given by

$$\frac{dx_b}{dt} = (\mu_b - D)x_b \tag{44}$$

and

$$\frac{dx_b}{dt} = \left[\frac{\mu_{max}^b s}{K_s^b + s} - D\right]x_b \tag{45}$$

There are three basic cases to be considered in assessing whether dx_b/dt will be positive (B outcompeting A), negative (A outcompeting B), or zero (B and A coexisting).

(i) For population x_b to increase in size and replace population x_a then dx_b/dt will be positive when $\mu_b > D$. Thus by comparing equations (43) and (45) this will occur when:

$$\frac{\mu_{max}^b s}{K_s^b + s} > \frac{\mu_{max}^a s}{K_s^a + s}$$

Since, initially, s is fixed, this can occur if either $\mu_{max}^b > \mu_{max}^a$ (Figure 19a), or $K_s^b < K_s^a$ (Figure 19b). It must be noted, however, that it is the combination of μ_{max}^b and K_s^b which must be greater than the equivalent combination of μ_{max}^a and K_s^a and it is entirely possible that $\mu_{max}^b > \mu_{max}^a$ but $K_s^b > K_s^a$ (Figure 19c). The salient factor is that at a given value of s, $\mu_b > \mu_a$.

(ii) For population x_b to be uncompetitive relative to x_a, the reverse situation pertains with:

$$\frac{\mu_{max}^b s}{K_s^b + s} < \frac{\mu_{max}^a s}{K_s^a + s}$$

and this occurs if either $\mu_{max}^b < \mu_{max}^a$ (Figure 19d) or $K_s^b > K_s^a$ (Figure 19e).

(iii) It is theoretically possible that the second population x_b could have an identical growth rate to the initial population x_a, that is

$$\frac{\mu_{max}^b s}{K_s^b + s} = \frac{\mu_{max}^a s}{K_s^a + s}$$

This would be achieved if $\mu_{max}^b = \mu_{max}^a$ and $K_s^b = K_s^a$. This is extremely unlikely but it is conceivable that an appropriate combination of the basic growth parameters will yield a substrate concentration s_1 (Figure 19f) at which $\mu_b = \mu_a$. This will not be a stable state since it would be very difficult practically to maintain s_1 exactly constant.

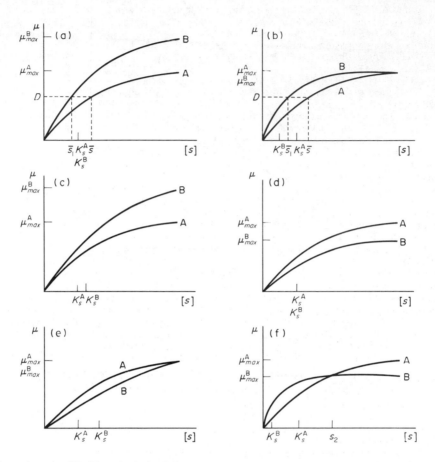

Figure 19 The various possible Monod relationships between two organisms, A and B, used to predict the outcome of free competition between them under substrate limited growth (Slater and Bull, 1978)

11.6 CONCLUSIONS

This chapter has provided the basic outlines to the important principles of microbial growth in batch and continuous systems. The literature abounds with many variations and extensions to the widely used Monod kinetics. Other important systems such as light-limited kinetics, or concepts such as intracellular accumulation of growth-limiting nutrients, have not been considered here. There are speculations and some evidence that the maintenance coefficient is not a constant (see Stouthamer and van Verseveld, Volume 1, Chapter 12 for discussion) with some evidence that it varies in response to organism growth rate. The kinetics of many other important microbial interactions have not been considered, nor has the kinetic behaviour of organisms growing as either micro- or macro-films on surfaces. In all these instances the fundamentals will be the same, being modified by additional variables which reflect the peculiar circumstances of the organism's habitat or growth system.

11.7 REFERENCES

Bazin, M. J. (1981). Mixed culture kinetics. In *Mixed Culture Fermentations*, ed. M. E. Bushell and J. H. Slater, pp. 25–51. Academic, London.

Bull, A. T. (1974). Microbial growth. In *Companion to Biochemisty*, ed. A. T. Bull, J. R. Lagnado, J. O. Thomas and K. F. Tipton, pp. 415–442. Longman, London.

Bull, A. T., M. E. Bushell, T. G. Mason and J. H. Slater (1975). Growth of filamentous fungi in batch culture: a comparison of the Monod and logistic models. *Proc. Soc. Gen. Microbiol.*, **3**, 62–63.

Bull, A. T. and J. H. Slater (1982). *Microbial Interactions and Communities*, vol. 1. Academic, London.

Burns, R. G. and J. H. Slater (1982). *Experimental Microbial Ecology*. Blackwell, Oxford.

Bushell, M. E. and J. H. Slater (1981). *Mixed Culture Fermentations*. Academic, London.

Carter, B. L. A. and A. T. Bull (1969). Studies of fungal growth and intermediary carbon metabolism under steady and non-steady state conditions. *Biotechnol. Bioeng.*, **11**, 785–804.

Dijkhuizen, L. and W. Harder (1975). Substrate inhibition in *Pseudomonas oxalaticus*, OX1: a formate using extended cultures. *Antonie van Leeuwenhoek*, **41**, 135–146.

Edwards, V. H. (1970). The influence of high substrate concentrations on microbial kinetics. *Biotechnol. Bioeng.*, **12**, 69–712.

Fredrickson, A. G. (1977). Behaviour of mixed cultures of microorganisms. *Annu. Rev. Microbiol.*, **31**, 63–87.

Harrison, D. E. F. (1978). Efficiency of microbial growth. In *Companion to Microbiology*, ed. A. T. Bull and P. M. Meadow, pp. 155–179. Longman, London.

Herbert, D. (1961). The chemical composition of microorganisms as a function of their environment. In *Microbial Reaction to the Environment*, ed. G. G. Meynell and H. Gooder, pp. 391–416. Cambridge University Press, London.

Jones, J. G. (1979). *A Guide to Methods for Estimating Microbial Numbers and Biomass in Fresh Water*. Freshwater Biological Association. Scientific Publication Number 39, Ambleside, Cumbria.

Karagouni, A. D. and J. H. Slater (1978). Growth of the blue-green alga *Anacystis nidulans* during washout from light- and carbon dioxide-limited chemostats. *FEMS Microbiol. Lett.*, **4**, 295–299.

Mandelstam, J., K. McQuillen and I. Davies (1982). *Biochemistry of Bacterial Growth*. Blackwell, Oxford.

Marr, A. G., P. R. Painter and E. H. Nilson (1969). Growth and division of individual bacteria. In *Microbial Growth*, ed. S. J. Pirt and P. M. Meadow, pp. 237–261. Cambridge University Press, Cambridge.

Mitchison, J. M. (1971). *The Biology of the Cell Cycle*. Cambridge University Press, Cambridge.

Monod, J. (1949). The growth of bacterial cultures. *Annu. Rev. Microbiol.*, **3**, 371–394.

Ollis, D. F. (1982). Industrial fermentations with (unstable) recombinant cultures. *Philos. Trans. R. Soc. London, Ser. B.*, **297**, 617–629.

Payne, W. J. (1970). Energy yields and growth of heterotrophs. *Annu. Rev. Microbiol.*, **24**, 17–90.

Pearl, R. and L. J. Reed (1920). On the rate of growth of the population of the United States since 1790 and its mathematical representation. *Proc. Natl. Acad. Sci. USA*.

Pirt, S. J. (1965). Maintenance energy of bacteria in growing cultures. *Proc. R. Soc. London, Ser. B.*, **163**, 224–231.

Pirt, S. J. (1972). Prospects and problems in continuous flow culture of microorganisms. *J. Appl. Chem. Biotechnol.*, **22**, 55–64.

Pirt, S. J. (1975). *Principles of Microbe and Cell Cultivation*. Blackwell, Oxford.

Powell, E. O. (1958). Criteria for growth of contaminants and mutants in continuous culture. *J. Gen. Microbiol.*, **18**, 249–268.

Pritchard, R. H. and D. W. Tempest (1982). Growth: cells and populations. In *Biochemistry of Bacterial Growth*, ed. J. Mandlestam, K. McQuillen and I. Dawes, pp. 99–123. Blackwell, Oxford.

Slater, J. H. (1979). Population and community dynamics. In *Microbial Ecology: A Conceptual Approach*, Blackwell, Oxford.

Slater, J. H. and A. T. Bull (1978). Interactions between microbial populations. In *Companion to Microbiology*, ed. A. T. Bull and P. M. Meadow, pp. 181–206. Longman, London.

Slater, J. H. and D. J. Hardman (1982). Microbial ecology in the laboratory: experimental systems. In *Experimental Microbial Ecology*, ed. R. G. Burns and J. H. Slater, pp. 225–274. Blackwell, Oxford.

Stouthamer, A. H. (1977). Energetic aspects of the growth of microorganisms. In *Microbial Energetics*, ed. B. A. Haddock and W. A. Hamilton, pp. 285–315. Cambridge University Press, Cambridge.

Trinci, A. P. J. (1971). Influence of the width of the peripheral growth zone on the radial growth rate of fungal colonies on solid medium. *J. Gen. Microbiol.*, **57**, 325–344.

Verhulst, P. F. (1839). Notice sur la que la population suit dans son accroissement. *Corr. Math. Phys. Publ. Para. A*, **10**, 113–121.

Williamson, D. H. and A. W. Scopes (1961). Synchronization of division in cultures of *Saccharomyces cerevisiae* by control of the environment. In *Microbial Reaction to the Environment*, ed. G. G. Meynell and H. Gooder. Cambridge University Press, Cambridge.

12
Stoichiometry of Microbial Growth

A. H. STOUTHAMER and H. W. VAN VERSEVELD
Vrije Universiteit, Amsterdam, The Netherlands

12.1 INTRODUCTION

Studies on the energetics of microbial growth play a central role in microbial physiology. Yield studies in microorganisms have the purpose of finding the relationship between substrate utilization, ATP generation, the formation of new cell material and the formation of products. During growth fixed relationships exist between the yield of biomass and product and the consumption of substrate, oxygen and nitrogen source, and between oxygen consumption and heat production. Part of the energy derived from the catabolism of the substrate is used for maintenance purposes. In this chapter the theory which gives the relationship between these parameters will be given and the methods for their determination will be discussed. Three classes of low molecular weight fermentation products are distinguished. The energetics of each class of product have their own characteristics, which lead to different strategies to optimize product formation. With the equations for oxygen uptake and heat production an optimal utilization of the possibilities of fermentation equipment used in industrial fermentations can be obtained. Although great advances in our knowledge of the energetics of microbial growth and product formation have been achieved during the last years, a large number of gaps are still present.

215

12.2 GROWTH YIELDS AND MATERIAL BALANCES

During heterotrophic growth the utilization of a certain amount of substrate will give rise to the formation of a certain amount of new cell material. This amount is defined as the coefficient Y_{sub}. Under these conditions part of the substrate will be catabolized to yield ATP and this ATP is utilized in the formation of new cell material. Under anaerobic conditions the substrate is converted to a number of fermentation products. All microbiologists are familiar with the fact that under anaerobic conditions there should be a correct carbon and redox balance. Under aerobic conditions the efficiency of growth is characterized by Y_{sub} and Y_{O_2}. It must be emphasized that these parameters are not independent. It is surprising that this point is not fully appreciated by microbiologists. It has been pointed out (Stouthamer, 1977a, 1979) that the relationship between these parameters is absent in many of the published experimental yield data, which means that not enough attention was given to obtaining a good mass balance, or that the authors did not realize that in the experimental determination of these parameters large errors can be made. It seems appropriate therefore to treat this point first. The relation between the utilization of substrate, nitrogen source and oxygen and the formation of biomass, product and CO_2 is obtained by elemental balance equations. The first authors who presented a generalized treatment were Minkevich and Eroshin (1973). Further developments have been discussed in an extensive recent review by Roels (1980a).

In these treatments an empirical formula of cell material is used: $CH_pO_nN_g$. For a large number of organisms it has been shown that $CH_{1.8}O_{0.5}N_{0.2}$ is a good average (Roels, 1980a). All substrates and products are converted to formulae which contain one mole of carbon. This leads to the balance equation (1) for growth:

$$CH_mO_l \text{ (substrate)} + a NH_3 + b O_2 \rightarrow y_c\, CH_pO_nN_g \text{ (biomass)} +$$
$$z\, CH_rO_sN_t \text{ (product)} + c H_2O + d CO_2 \tag{1}$$

y_c and z are the fractions of the substrate converted into biomass and product respectively.

A carbon balance may be written utilizing:

$$y_c + z + d = 1 \tag{2}$$

Consequently the CO_2 production is given by equation (3):

$$d = 1 - y_c - z \tag{3}$$

The nitrogen balance is given by equation (4):

$$a = y_c g + zt \tag{4}$$

Further balances may be obtained by introducing the concept of reductance degree or equivalents of available electrons that may be transferred to oxygen. The reductance degree for organic substrate, biomass and product respectively were defined by Minkevich and Eroshin (1973) as follows:

$$\gamma_s = 4 + m - 2l \tag{5}$$
$$\gamma_b = 4 + p - 2n - 3g \tag{6}$$
$$\gamma_p = 4 + r - 2s - 3t \tag{7}$$

where γ is the number of equivalents of available electrons per gram atom carbon, based on carbon = 4, hydrogen = 1, oxygen = -2 and nitrogen = -3. A valence of -3 is used with nitrogen because this is the reductance degree of nitrogen in biomass and ammonia. The available electron balance based on equation (1) gives:

$$\gamma_s - 4b = y_c \gamma_b + z \gamma_p \tag{8}$$

Consequently the oxygen consumption is given by (9):

$$b = (\gamma_s - y_c \gamma_b - z \gamma_p)/4 \tag{9}$$

These basic equations can be used in a variety of ways as follows.

(1) In the case of growth on a substrate of known degree of reduction and known absence of product formation the equation can be used to estimate oxygen consumption if the substrate yield factor is known.

(2) In the case of measured oxygen yield and measured substrate yield the relationship can be used to check the consistency of the data. If there is a discrepancy and, of course, positive indications that no measuring errors are present, the deviations may indicate the presence of undetected other carbon sources or products. If the amount of carbon converted to product is known, for example from a total organic carbon measurement of the broth filtrate, the degree of reduction of the product can be estimated. This can provide indications of the chemical nature of the product.

(3) It is evident from (1) that the whole system can be derived by measuring two flows. In many cases the consumption of substrate and oxygen and the production of CO_2 and biomass are measured. In this case the system obtained by application of equation (1) is overestimated. This makes an improvement of the method for statistic analysis of growth yields and maintenance parameters by de Kwaadsteniet *et al.* (1976) possible. In all measurements errors are made and the measured flows are not equal to the real ones. A treatment has been developed to find the most probable values of the errors and this treatment may be used to correct the raw experimental data (de Kok and Roels, 1980). This has been applied to growth of *Saccharomyces cerevisiae* in a chemostat with glucose or glucose–ethanol mixtures as growth-limiting substrate (Bonnet *et al.*, 1980; Dekker *et al.*, 1981; Geurts *et al.*, 1980). A somewhat different method was proposed by Solomon *et al.* (1982). In this method the consistency of the measured variables is checked and then the parameters are estimated from different sets of data. By this method a maximum likelihood estimator of biomass energetic yield and maintenance parameters can be determined.

Elemental balance equations also have consequences for anaerobic yield studies. Their influence can be demonstrated by considering anaerobic growth of *Klebsiella aerogenes* with gluconate. With resting cells the fermentation balance is 1 gluconate $\rightarrow$ 1.5 acetate + 0.5 ethanol + 2 formate. With growing cells the balance is: 1 gluconate + 0.174 NH_3 + 0.04 H_2O $\rightarrow$ 0.87 $CH_{1.8}O_{0.5}N_{0.2}$ + 1.40 acetate + 0.31 ethanol + 1.71 formate (Stouthamer, unpublished results). Thus 14.5% of the C is incorporated into biomass. With growing cells the acetate/ethanol ratio is much higher than with resting cells. The high acetate/ethanol ratio for growing cells is due to the difference in the degree of reduction of gluconate ($\gamma_s = 3.67$) and biomass ($\gamma_b = 4.2$). Many times this effect is overlooked in anaerobic yield studies and in the construction of fermentation balances.

12.3 RELATION BETWEEN ATP PRODUCTION AND GROWTH YIELDS, Y_{ATP}

The experimental determination of the ATP requirement for the formation of microbial cell material started with the classic work of Bauchop and Elsden (1960). These authors concluded that the amount of growth of a microorganism was directly proportional to the amount of ATP that could be obtained from the degradation of the energy source in the medium. They introduced the coefficient Y_{ATP}. It is evident that Y_{ATP} can be determined only if the ATP yield during the degradation of the energy source is known. Only under anaerobic conditions can the ATP yield for substrate breakdown be calculated exactly, because the catabolic pathways for anaerobic breakdown of substrates are known. However, during the last years it has been proposed that during anaerobic breakdown of substrates useful energy in addition to ATP may be produced. Therefore it seems useful at this moment to consider energy production and utilization during growth in more detail, before starting a discussion on the various Y_{ATP} values found. A scheme for generation and utilization of ATP is given in Figure 1. It is generally accepted that transmembrane gradients as proposed by Mitchell (1966, 1970) play a central role in energy production and utilization. Electron-transport to oxygen or alternative electron-acceptors yields a proton motive force. This proton motive force, $\Delta\mu_{H^+}$, is the driving force for ATP synthesis, transport processes and transhydrogenation (Mitchell 1966, 1970; Harold, 1977). Under anaerobic conditions ATP is produced by substrate-level phosphorylation. The proton motive force, needed to drive transport processes, is obtained by hydrolysis of ATP or in some cases by reduction of fumarate (Harold, 1977; Riebeling *et al.*, 1975; Gutowski and Rosenberger, 1976; Boonstra *et al.*, 1978). From Figure 1 it is clear that a number of processes directly yield a proton motive force, which is partly used for processes other than ATP formation. Consequently, in yield studies, energy-requiring or -producing processes, in which ATP is not directly involved, can be expressed in ATP equivalents. The total ATP yield during substrate breakdown is then the sum of actual ATP produced plus the ATP equivalents.

It has been pointed out that under anaerobic conditions another likely mechanism to generate an electrochemical potential could be the excretion of protons in symport with metabolic end products (Michels *et al.*, 1979). In a theoretical study it was calculated that (based on an H^+/ATP

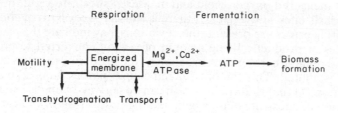

Figure 1   Scheme for the generation and utilization of ATP

ratio of 2) the excretion of lactate formed from glucose could yield an equivalent of 0.6 mol of ATP in the initial phase. This would indicate that excretion of lactate produced by homolactic fermentation of glucose can increase the theoretical ATP yield by about 30% (Michels *et al.*, 1979) to 2.6 mol per mol glucose fermented. When lactate accumulates in the medium the energy yield obtained by lactate–proton efflux decreases rapidly. The energy gain by lactate efflux is also dependent on the extracellular pH. At low pH no energy can be gained by efflux of metabolites. In a study in which molar growth yields were measured for lactose-limited chemostat cultures of *Streptococcus cremoris* in the presence of low and high external lactate concentrations, it was concluded that the net energy gain from lactate efflux amounted to 12% (Otto *et al.*, 1980a). In mixed cultures the molar growth yield of *S. cremoris* on lactose was increased by lactate consumption by *Pseudomonas stutzeri* (Otto *et al.*, 1980b). The lactate consumption of *P. stutzeri* under anaerobic conditions in the presence of nitrate kept the ATP gain of lactate efflux by *S. cremoris* at the maximal level. In *S. cremoris*, lactate efflux could potentiate the uptake of leucine, histidine and tetraphenylphosphonium (Otto *et al.*, 1980a; Otto *et al.*, 1982). Therefore it seems that the generation of a $\Delta\mu_{H^+}$ by lactate efflux in *S. cremoris* is well documented. It is not yet clear, however, how large this contribution is to the establishment and maintenance of the $\Delta\mu_{H^+}$ in growing cells. It has been demonstrated using ATPase negative mutants of *Escherichia coli* that 51% of the total energy production is used for membrane energization (Stouthamer and Bettenhausen, 1977; Stouthamer, 1979). Therefore it is likely that ATP hydrolysis must be involved in the generation and maintenance of the $\Delta\mu_{H^+}$ in *S. cremoris*. It is not yet known whether efflux of metabolites other than lactate can generate an electrical potential. Some of these products (formate, acetate, propionate and butyrate) are weak acids, which can cross the membrane as undissociated acid. It has even been demonstrated that accumulation of these products during fermentation has an uncoupling effect (Linton *et al.*, 1981; de Vries *et al.*, 1980; Hueting and Tempest, 1977). More work is necessary in this area. The influence of energy generation coupled to metabolite efflux on the interpretation of earlier yield studies is not clear. Most of these studies were performed in batch cultures in which large amounts of metabolites accumulated and in which the pH was not controlled. Therefore it is assumed that the influence of energy generation by metabolite efflux on Y_{ATP} determinations has been small.

In previous reviews Y_{ATP} values have been listed for a large number of microorganisms growing with a variety of substrates (Stouthamer, 1977a, 1978, 1979; Forrest and Walker, 1971; Payne, 1970). From these data the following conclusions can be drawn.

(a) Y_{ATP} is not a constant for different microorganisms. For batch cultures Y_{ATP} can vary from 4.7 for *Zymomonas mobilis* (Bélaich *et al.*, 1972) to 20.9 for *Lactobacillus casei* (de Vries *et al.*, 1970). This (Stouthamer and Bettenhaussen, 1973; Stouthamer, 1977a, 1979) is in contradiction with earlier conclusions that Y_{ATP} is a well-defined biological constant with a value of 10.5 which can be used to predict yields of organisms (Bauchop and Elsden, 1960; Forrest and Walker, 1971).

(b) The highest value for Y_{ATP} for any one organism is obtained for growth in a complex medium (Stouthamer, 1977a, 1979).

(c) The same Y_{ATP} value is found for growth of an organism on many different substrates. However, in most cases only growth of various sugars is compared for which the theoretical ATP requirement for cell formation is the same (see Section 12.6). In other cases growth on citrate has been studied in addition to growth on sugars. In these cases complex media were used for which the calculated ATP requirement is not very much different from that for growth on glucose in mineral salts medium (see Section 12.6). For the determination of Y_{ATP}, citrate, tartrate and pyruvate are very suitable growth substrates. The ATP yield is 1 mol per mol substrate. This may be very helpful for the elucidation of ATP yields in organisms with very complex fermentation pathways for other compounds.

(d) The experimental Y_{ATP} values are always much smaller than those based on the calculation of the theoretical ATP requirement for the formation of cell material (see Section 12.6). The difference between experimental and calculated Y_{ATP} values can partly be explained by non-equilibrium thermodynamics (*e.g.* see Westerhoff *et al.*, 1982), which explains that the actual P/O ratio is lower than the mechanistic stoichiometry. In aerobic growth studies experimental Y_{ATP} values are calculated using the mechanistic stoichiometry of oxidative phosphorylation, so these values have to be underestimates.

12.4 INFLUENCE OF GROWTH RATE AND MAINTENANCE ENERGY ON Y_{ATP}: ANAEROBIC CHEMOSTAT CULTURES

It is well known from chemostat experiments that Y and Y_{ATP} values are dependent on the specific growth rate (μ). This is generally ascribed to the occurrence of maintenance energy. Maintenance energy is required for protein synthesis to replace thermally denatured proteins, reuptake of metabolites that escaped through leakage across the cytoplasmic membrane, the preservation of the correct intracellular pH, *etc.* However, loss of the proton motive force due to proton permeability of the cytoplasmic membrane or ATP consumption in futile cycles also contribute to the maintenance energy. An equation that relates the molar growth yield and the specific growth rate has been derived by Pirt (1965). In this derivation it is assumed that during growth the consumption of the energy source is partly growth-dependent and partly growth-independent. This leads to equation (10):

$$1/Y_{sub} = [1/Y_{sub}^{max}] + [m_s/\mu] \tag{10}$$

The maintenance coefficient m_s is normally determined by studying the influence of μ on Y in chemostat cultures. A double reciprocal plot of estimated Y_{sub} values against the experimental μ values yields a straight line in which the intercept is the reciprocal of Y_{sub}^{max} and the slope the maintenance coefficient. A fairly large number of data points are necessary to obtain a reliable estimate of Y_{sub}^{max} and m_s. Consequently this is a very time-consuming exercise. It is also possible to determine these parameters in fed batch cultures. If a constant rate of substrate addition a is used and x_t is the biomass concentration at time t, the increase in biomass is given by equation (11):

$$\frac{dx}{dt} = (a - m_s x_t) Y_{sub}^{max} \tag{11}$$

If m is independent of μ it is very easy to determine m_s and Y_{sub}^{max} in a single experiment (van Verseveld, 1979, Esener *et al.*, 1981). In such experiments μ decreases exponentially. Care must be taken that in those experiments values of μ which are too small are not included, since at these values aberrations occur (see later).

In a number of cases a double reciprocal plot of Y_{sub} values against μ did not yield a straight line (de Vries *et al.*, 1970; Pirt, 1965; Watson, 1970). This was shown to be due to the influence of μ on the fermentation pattern and on the ATP yield of the organism (de Vries *et al.*, 1970; Watson, 1970). Therefore Y_{sub} and Y_{sub}^{max} were replaced by Y_{ATP} and Y_{ATP}^{max}. A double reciprocal plot of Y_{ATP} against μ did indeed yield a straight line with *L. casei* (de Vries *et al.*, 1970) and *S. cerevisiae* (Watson, 1970). By multiplication of equation (10) by μ, equation (12) was obtained:

$$q_{ATP} = \mu/Y_{ATP} = (\mu/Y_{ATP}^{max}) + m_e \tag{12}$$

The results of a number of determinations of Y_{ATP}^{max} and m_e are shown in Table 1.

The following conclusions can be drawn. (a) The data for *A. aerogenes* indicate that Y_{ATP}^{max} and m_e are dependent on the growth condition. The same conclusion has been reached for m_e from the results of aerobic yield studies with *E. coli* (Hempling and Mainzer, 1975; Farmer and Jones, 1976). (b) The Y_{ATP}^{max} value for growth of *A. aerogenes* with citrate is much lower than for growth with glucose. This is in accordance with theoretical calculations of the ATP requirement for the formation of cell material from various substrates (see Section 12.6.1). (c) The Y_{ATP}^{max} values for *L. casei* and *C. acetobutylicum* are close to the theoretical amount of cell material that can be formed per mole of ATP. In all other cases the experimental Y_{ATP}^{max} values are much smaller (about 50%) than the amount of cell material that theoretically can be formed per mole of ATP, therefore the concept of maintenance energy offers no explanation *per se* for the discrepancy between theoretical and experimental growth yields (Stouthamer and Bettenhaussen, 1975;

Table 1   Growth parameters of some organisms growing anaerobically in a chemostat

Organism	Growth-limiting factor	m_e	Y_{ATP}^{max}	References
Lactobacillus casei	Glucose	1.5	24.3	de Vries *et al.* (1970)
Aerobacter aerogenes	Glucose[a]	6.8	14.0	Stouthamer and Bettenhaussen
	Glucose[b]	2.3	17.6	(1973, 1975), Stouthamer (1977a)
	Tryptophan	38.7	25.4	
	Citrate	2.2	9.0	
Escherichia coli	Glucose	18.9	10.3	Hempfling and Mainzer (1975)
		6.9	8.5	Stouthamer and Bettenhaussen (1977)
Saccharomyces cerevisiae	Glucose	0.5	11.0	Watson (1970)
		0.25	13.0	Rogers and Stewart (1974)
Saccharomyces cerevisiae (petite)	Glucose	0.7	11.3	
Candida parapsilosis	Glucose	0.2	12.5	
Clostridium acetobutylicum	Glucose	—	23.8	Bahl *et al.* (1982)
Streptococcus cremoris	Lactose[c]	2.3	12.6	Otto *et al.* (1980a)

[a] Minimal medium.   [b] Complex medium.   [c] In the presence of a high extracellular lactate concentration.

Stouthamer, 1979). The same conclusion can be reached for aerobic yield studies. (d) The Y_{ATP}^{max} for tryptophan-limited growth of *A. aerogenes* is very high. However the m_e value is also very high. Therefore at a high μ value (0.65 h^{-1}) the Y_{ATP} values for glucose- and tryptophan-limited cultures are about the same. The very high m_e value was the reason for concluding that m_e also included an energy component that was required for processes other than those of true maintenance (Stouthamer and Bettenhaussen, 1975). It is very likely that in this case m_e is not independent of μ and in fact decreases with increasing μ values. Again similar conclusions have been reached for aerobic yield studies (Neijssel and Tempest, 1976; Nagai and Aiba, 1972). These phenomena which occur when the carbon source is in excess will be discussed in more detail in Section 12.7. The maintenance coefficient is of the utmost importance in industrial fermentations. The yield of product is highly dependent on the maintenance coefficient. This aspect will be discussed in Section 12.10.

The determination of m_e and Y_{ATP}^{max} is not very accurate. With the aid of a stochastic model 95% confidence limits can be determined (de Kwaadsteniet *et al.*, 1976). In experiments of this kind generally no points or only some are measured at μ values of about 0.1 h^{-1} or lower. Then extrapolation to $\mu = 0$ is performed. Of course there is no guarantee that in this region the effects suggested by the extrapolation procedure will really occur. In fact there is strong evidence that this is not the case. Very low growth rates can be studied in a recycling fermenter. Anaerobic growth of *E. coli* in a recycling fermenter has been studied by Chesbro *et al.* (1979). If equation (10) is valid it is expected that growth will asymptotically approach a limiting mass determined by the glucose provision rate. Instead, three phases could be distinguished. After the fermenter is seeded, growth follows the usual exponential path (phase 1). During continuous cultivation $Y_{glu}^{max} = 28.9$. When the culture becomes glucose-limited the growth rate dx/dt becomes constant and a linear function of the glucose provision rate (phase 2). In this phase $Y_{glu} = 16$. The length of this second phase was dependent upon the strain used. In wild type *E. coli*, at a critical specific growth rate (0.013 h^{-1}) phase 2 ends abruptly and phase 3 commences. In phase 3 $Y_{glu} = 6$; no experiment carried out yielded a termination of this phase. The fermentation pattern indicated that the pathway of glucose fermentation in the three phases was the same. The results show that in phase 2 and phase 3 constant but different fractions of the metabolic energy are used for maintenance purposes, although both x and μ change considerably during the experiment. Consequently the growth behaviour of *E. coli* B is not described by simple mass transfer analysis using the maintenance energy concept as it is currently accepted. In later work it was shown that guanosine 3',3', 5',5'-tetraphosphate (ppGpp) accumulation commenced at the start of phase 2 and reached a maximum level at the start of phase 3 concurrent with a curbing of RNA accumulation (Arbige and Chesbro, 1982a). The length of phase 2 was influenced by mutations, which affect the synthesis or the degradation of ppGpp. Therefore, it seems that these changes from one phase to another are due to a change from one metabolic state to another. These phenomena seem to be very general. They have also been observed for anaerobic cultures of *Bacillus polymyxa* (Arbige and Chesbro, 1982b) and for aerobic cultures (cultures in the recycling fermenter or fed batch cultures) of *E. coli*, *Paracoccus denitrificans* and *Rhizobium leguminosarum* (van Verseveld *et al.*, 1984; Stam *et al.*, 1983).

These observations, which are not yet completely understood, will have a very great influence on our ideas about the energetics of growth of microorganisms. More work is necessary in this area. It is already clear, that when on the basis of these newer findings a plot of q *versus* μ must be given, the most likely one is shown in Figure 2. It is evident that when only a limited number of these data points in the parts A and B are available, the line obtained by normal regression analysis will not differ very much from that drawn in part C. These phenomena are very important for our understanding of survival of bacteria in natural environments.

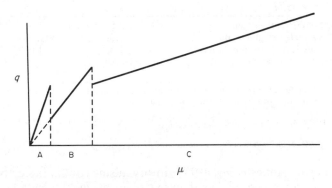

Figure 2 Probable course of q_{sub} *versus* μ: part A is identical with phase 3, part B with phase 2 and part C is the normal q *versus* μ curve for chemostat cultures

12.5 AEROBIC YIELD STUDIES AND THE INFLUENCE OF THE EFFICIENCY OF OXIDATIVE PHOSPHORYLATION ON GROWTH YIELDS

In the presence of external hydrogen acceptors the energy source may be degraded further than under anaerobic conditions. Concomitantly the ATP yield in the presence of hydrogen acceptors will be larger than in its absence. This is illustrated very nicely by the difference between the anaerobic and aerobic Y_{glu} values for *A. aerogenes* which were found to be 26.1 and 72.7 respectively (Hadjipetrou *et al.*, 1964). In aerobic yield studies, in addition to Y_{sub}, Y_{O_2} is usually measured. The coefficient Y_{O_2} is taken as a measure of that part of the carbon source that is completely oxidized. In continuous cultures $Y_{O_2}^{max}$ is used in many cases; it is determined by equation (13)·

$$1/Y_{O_2} = m_0/\mu + 1/Y_{O_2}^{max} \tag{13}$$

Equation (13) is a modified form of equation (10).

The total ATP yield during growth in the presence of hydrogen acceptors cannot be determined easily, since the $P/2e^-$ ratio for phosphorylation coupled to electron transport to the hydrogen acceptor is not known. The $P/2e^-$ ratio has a very strong influence on Y_{sub} and Y_{O_2} as shown by a theoretical study (Stouthamer, 1978a). The problem in the interpretation of aerobic yield studies is illustrated in Table 2. The method of the calculation (Stouthamer, 1977b, 1979) is a more microscopic approach than that given in Section 12.2. From the growth equation and Y_{glu}^{max} the O_2 uptake, CO_2 evolution and $Y_{O_2}^{max}$ can be calculated and thus can be compared with the experimental values. It is evident that there is a very good agreement between the calculated and the observed parameters. This is a further illustration that Y_{sub}^{max} and $Y_{O_2}^{max}$ are not independent parameters as outlined in Section 12.2. Assuming an unbranched respiratory chain and 2 or 3 phosphorylation sites the total ATP production can be calculated. In this way we arrive at Y_{ATP}^{max} values of 11.0 and 7.6 respectively when 2 or 3 sites are present. Consequently either Y_{ATP}^{max} must be known to determine the $P/2e^-$ ratio or conversely the $P/2e^-$ ratio must be known to determine Y_{ATP}^{max}.

During the last years the progress in the determination of the $P/2e^-$ ratio has been very slow. According to the chemiosmotic theory protons are extruded during respiration. The H^+/O ratio is a measure of the efficiency of oxidative phosphorylation. In microorganisms there is an enormous diversity in the composition of the respiratory chain (Jones, 1977; Haddock and Jones, 1977; Stouthamer, 1978a). It has been demonstrated that there is a relation between the complexity of the composition of the respiratory chain, the H^+/O ratio and the Y_{O_2} value for a

222

Genetic and Biological Fundamentals

Table 2 Calculation of growth parameters for gluconate-limited growth of *Paracoccus denitrificans* with nitrate as nitrogen source[a]

	Calculated	Observed
O_2 uptake (mol/mol gluconate)	1.43	1.52
CO_2 evolution (mol/mol gluconate)	3.26	3.03
$Y_{O_2}^{max}$	48.4	45.5

ATP formation	Sites for oxidative phosphorylation	
	2 sites	3 sites
1.77 NADH/mol gluconate	3.54	5.31
1.09 $FADH_2$/mol gluconate	1.09	2.18
Substrate level phosphorylation	1.63	1.63
Total ATP formation	6.26	9.12
Y_{ATP}^{max}	11.0	7.6

[a] Data from Stouthamer (1977b, 1979). For the calculation of the ATP yield 1 ATP is used per site (compare Table 4). The growth equation is:

$$C_6H_{12}O_7 + 1.4\,HNO_3 + 6.85\,\text{'}H_2\text{'} \rightarrow C_6H_{10.84}N_{1.4}O_{3.07} + 8.13\,H_2O$$

$Y_{glu}^{max} = 69.2$. Dissimilated: 54.4%; assimilated 45.6%.

number of microorganisms (Jones *et al.*, 1977). In organisms containing cytochrome *c* both the H^+/O ratio and the Y_{O_2} value were higher than in organisms lacking cytochrome *c*. These data indicate that the presence of cytochrome *c* leads to an increase in the ATP production. This is nicely demonstrated for *Pseudomonas* AM1 where for growth with succinate $Y_{suc} = 50.1$ and $Y_{O_2} = 44.1$ were found for the wild type and $Y_{suc} = 42.0$ and $Y_{O_2} = 27.6$ for a cytochrome *c* deficient mutant (Keevil and Anthony, 1979). According to the chemiosmotic theory the relation between the H^+/O ratio and the $P/2e^-$ ratio is given by equation (14):

$$H^+/O = 2(P/2e^-) \qquad (14)$$

However, during the last few years the validity of this equation has been questioned both for mitochondria (Wikstrøm and Krab, 1980) and bacteria (Stouthamer, 1977a; Kell *et al.*, 1978). For mitochondria there is wide disagreement about the precise numbers of protons translocated by the various segments of the respiratory chain. It may be concluded that proton translocation does not occur by the arrangement of respiratory carriers in proton-translocating loops, but is due to the presence of conformational proton pumps (Wikstrøm and Krab, 1980). Due to these newer findings the definition of sites of oxidative phosphorylation has become obsolete.

These newer ideas have not yet led to a new vision on the mechanism of energy conservation in microorganisms. Some of the uncertainties which are introduced by these points are as follows.

(a) In *E. coli*, *K. pneumoniae* and *K. aerogenes* the H^+/O ratio is 4 (Jones, 1977; Stouthamer, 1978a), indicating that there are two proton-translocating segments in the respiratory chain. With H^+/ATP ratios of 2 and 3 the $P/2e^-$ ratios for oxidation of NADH are 2.0 and 1.33 respectively. Consequently it is no longer necessary that $P/2e^-$ ratios are integral numbers. From yield studies a $P/2e^-$ ratio of 1.2–1.4 was calculated for *K. aerogenes* (Stouthamer and Bettenhaussen, 1975; Stouthamer; 1977a) and a $P/2e^-$ ratio of about 1 for *E. coli* (Hempfling and Mainzer, 1975). These data could correspond with $H^+/ATP = 3$ and $P/2e^- = 1.33$. Very recently evidence has been presented which indicates that in growing *E. coli* the H^+/ATP ratio is indeed 3 (Kashket, 1982).

(b) For mitochondria it has been suggested that there is a difference in charge separation across the membrane (which is proportional to the efficiency of oxidative phosphorylation) for the classical sites. Stoichiometrics of 2–3, 2 and 4 charges are reported for sites I, II and III respectively (Wikstrøm and Krab, 1980). This implies that the three sites have different $P/2e^-$ ratios. The same has been proposed for *P. denitrificans* (van Verseveld *et al.*, 1980; Boogerd *et al.*, 1981).

In previous reviews a number of methods have been described to determine directly the $P/2e^-$ ratio from yield studies (Stouthamer, 1977a, 1980). These methods are as follows. (i) Deviation from linearity; this method has been applied to glucose-limited growth of *K. aerogenes*. At a μ value of about 0.5 h^{-1} the organisms start to excrete acetate (Stouthamer and Bettenhaussen, 1975). At the same time the rate of oxygen uptake decreases and the rate of glucose consumption increases. By comparing the parts of the curves before and after $\mu = 0.5$ h^{-1} one has enough information for an independent resolution of all the growth parameters. One finds a value of $Y_{ATP}^{max} = 12.4$ and $P/2e^- = 1.4$ (Stouthamer, 1977a). The Y_{ATP}^{max} compares favorably with the anaerobic value, which was about 14. (ii) Comparison of aerobic and anaerobic growth; in this case it is stated $(Y_{ATP}^{max})_{aer} = (Y_{ATP}^{max})_{anaer}$ or $(Y_{ATP})_{aer} = (Y_{ATP})_{anaer}$. This method has been applied to *K. aerogenes* (Stouthamer and Bettenhaussen, 1975), *E. coli* (Hempfling and Mainzer, 1975) and various other organisms. (iii) Sulfate-limited growth; it has been observed in *Candida utilis* (Haddock and Garland, 1971), in *Paracoccus denitrificans* (Meijer *et al.* 1977) and various other microorganisms that site 1 is lost by sulfate-limited growth in the chemostat. By comparison of growth parameters for succinate- and sulfate-limited growth it is possible to estimate a $P/2e^-$ ratio. (iv) Growth on mixed substrates; this possibility results from the fact that for each substrate the proportion of substrate-level phosphorylation and oxidative phosphorylation during complete oxidation is different (Table 3). This approach has been used in *P. denitrificans* (van Verseveld *et al.*, 1979, van Verseveld and Stouthamer 1980) and in *S. cerevisiae* (Bonnet *et al.*, 1980; Geurts *et al.*, 1980).

Table 3 Production of NADH, FADH$_2$, XH$_2$ and ATP by substrate phosphorylation during complete oxidation of a number of substrates[a]

		Number of molecules		
Substrate	*NADH*	*FADH$_2$*	*PQQH$_2$*	*ATP*
Mannitol	11	2	—	3
Glucose	10	2	—	3
Acetate	3	1	—	−1
Ethanol	5	1	—	−1
Methanol	2	—	1	0

[a] The oxidation of mannitol and glucose is assumed to occur by the Entner–Doudoroff system. PQQH$_2$ is pyrroloquinolinequinone which is the cofactor of methanol dehydrogenase.

All these methods have a number of uncertainties and therefore as many as possible should be applied to a single organism. Furthermore the data then obtained must be in accordance with the biochemical data on respiratory chain composition, H$^+$/O ratio and H$^+$/ATP ratio. At this moment this situation is only approached in *P. denitrificans*. A scheme for proton pumping in aerobic and anaerobic cells of *P. denitrificans* is shown in Figure 3. At the first site two to three protons are extruded. The second site is split into two parts; two protons are released from ubiquinol (IIa) and two are extruded at the cytochrome *b* level (IIb). A similar scheme for II$^+$ ejection at site II in the mitochondrial respiratory chain has recently been proposed (Alexandre *et al.*, 1980). Two protons are pumped by the cytochrome oxidase (van Verseveld *et al.*, 1981). The charge separation for NADH oxidation at sites I, II and III are 2–3, 2 and 4 respectively. Oxidation of NADH *via* cytochrome *o* gives a total charge separation of 6–7. In cells grown heterotrophically cytochrome *o* is the main oxidase (Meijer *et al.*, 1977; van Verseveld and Stouthamer, 1978). In accordance with this view the molar growth yield on glycerol of a cytochrome *c* deficient mutant is only slightly smaller than that of the wild type (Willison and Haddock, 1981). Electrons from methanol enter the respiratory chain at the level of cytochrome *c*. In cells grown on methanol or on a mixed substrate of mannitol and methanol, electron transport mainly occurs *via* cytochrome *aa$_3$* (van Verseveld and Stouthamer, 1978a; van Verseveld *et al.*, 1979). During sulfate-limited growth phosphorylation site I is lost (Meijer *et al.*, 1977), therefore the functioning of the phosphorylation sites is strongly dependent on the growth conditions. Under anaerobic conditions (Figure 3b) cytochrome *aa$_3$* is not formed. Electron transport to nitrite and nitrous oxide involves *c*-type cytochromes, that to nitrate a special *b*-type cytochrome (Stouthamer *et al.*, 1982). However, the reduction of nitrite and nitrous oxide occur at the periplasmic side of the membrane, whereas nitrate is reduced at the cytoplasmic side of the membrane, therefore the charge separations for the oxidation of NADH and FADH$_2$ with all nitrogenous oxides are 4–5 and 2 respectively, although the pathway of electron transport to nitrate differs from that to nitrite and nitrous oxide. With these data it is possible to calculate overall $P/2e^-$ ratios for the oxidation of various substrates under conditions with a different number of phosphorylation sites. With these $P/2e^-$ ratios Y_{ATP}^{max} values can then be calculated (Table 4). For one substrate the same value for Y_{ATP}^{max} is found for the various growth conditions in which the $P/2e^-$ ratio is different. In fact the values obtained fit better than in previous calculations in which one ATP per phosphorylation site was used. The value found for anaerobic nitrite-limited growth with succinate is too low. In recent experiments higher $Y_{nitrite}$ values are found (Boogerd *et al.*, 1984). In these experiments Y_{ATP} values are consistent with the values of other experiments mentioned in Table 4. A possible explanation is that in earlier experiments part of the nitrite is reduced to nitrous oxide instead of to nitrogen.

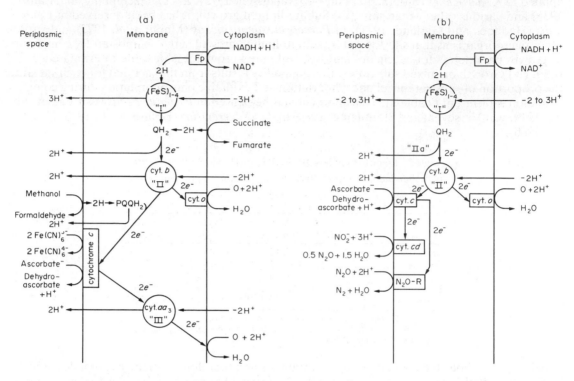

Figure 3 Simplified schemes for proton pumping and electron transport in aerobically (a) and anaerobically (b) grown *P. denitrificans*. Abbreviations: Fp, flavoprotein; FeS, iron–sulfur center; Q, ubiquinone; cyt, cytochrome; PQQ, pyrroloquinolinequinone. I, IIa, IIb and III are the traditional sites of energy conservation. (Data from van Verseveld *et al.* (1981) and Boogerd *et al.* (1981). Reproduced with the permission of Elsevier Biomedical Press)

Table 4 Influence of the number of phosphorylation sites on molar growth yields of *Paracoccus denitrificans*[a]

Substrate	Hydrogen acceptor	Sites	Overall $P/2e^-$	Y_{sub}^{max}	$Y_{el\ acc}^{max}$	Y_{el}^{max}	Y_{ATP}^{max}
Gluconate	O_2	I, II	2.15	80.2	44.7	11.2	8.7
Gluconate	O_2	II	1.33	65.7	31.3	7.8	8.1
Gluconate	NO_3^-	I, II	1.48	72.0	35.2	7.0	8.0
Mannitol	O_2	I, II	2.18	95.3	37.0	9.2	8.2
Mannitol	O_2	I, II, III	2.85	109.4	53.8	13.5	8.8
Succinate	O_2	I, II	2.05	39.6	34.2	8.5	7.6
Succinate	O_2	II	1.33	37.5	21.0	5.2	7.0
Succinate	NO_2^-	I, II	1.38	38.4	13.6	4.5	6.1

[a] The overall $P/2e^-$ ratio was calculated using charge separations of 3, 4 and 2 for sites I, II and III respectively, for electron transfer from NADH to O_2 and charge transfers of 5 and 2 for electron transfer from NADH and $FADH_2$ respectively to nitrogenous oxides. Furthermore, it was assumed that a charge separation of 3 is needed for the synthesis of 1 ATP. For the calculation of the overall $P/2e^-$ ratio the number of molecules of NADH and $FADH_2$ formed per mole of substrate is taken. The experimental data are from Meijer *et al.*, 1977; van Verseveld *et al.*, 1977, 1979. As an example the calculation of the overall $P/2e^-$ for succinate with two phosphorylation sites will be given. Succinate yields 5 NADH + 2 $FADH_2$. The overall $P/2e^-$ is thus $[(5 \times 7) + (2 \times 4)]/(7 \times 3) = 2.05$.

12.6 THEORETICAL CALCULATIONS ON THE ATP REQUIREMENT FOR THE FORMATION OF MICROBIAL BIOMASS

12.6.1 Influence of the Carbon Source and Complexity of the Medium

In theoretical calculations of the ATP requirement for the formation of cell material, the macromolecular composition of the cells is taken as a base. Subsequently the ATP requirement for the formation of each cell constituent is calculated. Such calculations have been performed by several authors. Stouthamer (1973) used the detailed analysis of the cell composition of *Escherichia coli* by Morowitz (1968) as a base. The ATP requirement for the formation of cell material of this composition from glucose and preformed monomers (amino acids and nucleic acid bases) and from glucose and simple inorganic salts is given in Table 5. The results show that, theoretically, 31.9 g cells can be formed per mol of ATP for growth with glucose and preformed monomers and 28.8 for growth with glucose and inorganic salts. The difference between these values is relatively small, which is due to a small ATP requirement for monomer synthesis from glucose, as shown in Table 5. With other carbon sources the situation is completely different. The ATP requirement for the formation of cell material from pyruvate and preformed monomers and from pyruvate and inorganic salts is included in Table 5 for comparison. The amount of cell material formed per mol of ATP from pyruvate and inorganic salts is much smaller than the amount formed from glucose and inorganic salts. This difference is due to a larger ATP requirement for monomer formation and for transport processes during growth on pyruvate than during growth on glucose. Furthermore, during growth on pyruvate the effect of supplementation of the medium with amino acids and nucleic acid bases is much more pronounced than during growth on glucose. The results in Table 5 demonstrate that, theoretically, 21 g of cell material can be formed per mol of ATP from pyruvate and preformed monomers and 13.5 g of cell material from pyruvate and inorganic salts. The ATP requirement for the formation of cell material in a mineral salts medium with various carbon sources is given in Table 6. Again it is evident that the nature of the carbon source has a very profound influence on the ATP requirement for the formation of cell material. For growth on acetate, theoretically only 10 g of cell material can be formed per mol of ATP. For autotrophic growth this value is only 6.5 g (Harder and van Dijken, 1976), which is due to the very large ATP requirement for monomer formation under these circumstances.

Table 5 ATP requirement for the formation of microbial cells from glucose pyruvate in the presence or absence of amino acids and nucleic acid bases[a]

Macromolecule	Amount of macromolecule (g/100 g cells)	ATP requirement ($10^4 \times$ mol/g cells formed)			
		A	B	C	D
Polysaccharide	16.6				
G6P formation		10.3	10.3	61.5	61.5
Polymerization		10.3	10.3	10.3	10.3
Protein	52.4				
Amino acid formation		0	13.5	0	148
Polymerization		191.4	191.4	191.4	191.4
Lipid	9.4	1.4	1.4	27	27
RNA	15.7				
Nucleoside monophosphate formation		14.8	34.5	37	62
Polymerization		9.2	9.2	9.2	9.2
DNA	3.2				
Deoxynucleoside monophosphate formation		3.8	8.6	8	14
Polymerization		1.9	1.9	1.9	1.9
Turnover mRNA		13.9	13.9	13.9	13.9
Total		257.0	295.0	369.2	539.2
Transport of					
Carbon source		0	0	58.1	148
Amino acids		47.8	0	47.8	0
Ammonium ions		0	42.4	0	42.4
Potassium ions		1.9	1.9	1.9	1.9
Phosphate		7.7	7.7	7.7	7.7
Total ATP requirement		314.5	347.1	475.7	740.2
Gram cells per mole ATP		31.9	28.8	21.0	13.5

[a] Data from Stouthamer (1973, 1979). Media: A glucose, amino acids and nucleic acid bases; B glucose and inorganic salts; C pyruvate, amino acids and nucleic acid bases; D pyruvate and inorganic salts.

These calculations offer an explanation for two experimental observations. First, for aerobic batch cultures the Y_{O_2} values are smaller for growth on simple compounds than for growth on glucose (Stouthamer, 1978). A Y_{O_2} value of 63.8 for growth of *Aerobacter aerogenes* on glucose and a value of 7.8 for growth of *Pseudomonas oxalaticus* on formate may be mentioned as extremes. Many $Y_{O_2}^{max}$ values for chemostat cultures of a number of organisms growing on various substrates have been determined. From these studies it is well documented that the $Y_{O_2}^{max}$

Table 6   ATP requirement for the formation of cell material in mineral
salts medium containing various carbon sources[a]

Material synthesized	ATP requirement ($10^4 \times$ mol/g cells)			
	Lactate	Malate	Acetate	CO_2
Polysaccharide	71	51	92	195
Protein	339	285	427	907
Lipid	27	25	50	172
RNA	85	70	101	⎫
				⎬ 212
DNA	16	13	19	⎭
Transport	200	200	306	52
Total	738	644	995	1538
Gram cells per mole of ATP	13.4	15.4	10.0	6.5

[a] The data are simplified from Stouthamer (1973) and Harder and van Dijken (1977).
The composition of the microbial cells is as given in Table 5.

values for growth on simple compounds are smaller than for growth on glucose (Stouthamer, 1977a). Second, supplementation of a glucose–mineral salts medium with amino acids has scarcely any influence on the Y_{O_2} value. On the other hand, supplementation with amino acids strongly increases the Y_{O_2} value for cells growing on ethanol or acetate. This difference is explained by the higher ATP requirement for amino acid formation during growth on simple compounds than for formation during growth on glucose (compare Table 5). Similarly the results of the determination of Y_{ATP}^{max} values for anaerobic chemostat cultures (Section 12.4) show that the ATP requirement for the formation of cell material from citrate is larger than for its formation from glucose (Stouthamer, 1977a, 1979).

There are a number of uncertainties in these calculations, especially in the amount of ATP required for transport processes. The data in Table 5 indicate that on theoretical grounds 15 or 27% of the total ATP requirement for the formation of cell material during growth on mineral salts and either glucose or pyruvate, respectively, is used for transport processes. It has been stressed before that these estimates are approximations. However, it seems a safe conclusion that the ATP requirement for transport processes during growth on glucose is much smaller than the requirement during growth on pyruvate. Recently, experimental evidence for the influence of the ATP requirement for transport processes on molar growth yields has been presented by Dijkhuizen *et al.* (1977). They observed molar growth yields of 3.8 g mol^{-1} and 3.4 g mol^{-1} for growth of *P. oxalaticus* with oxalate and formate, respectively. This was rather unexpected, because formate is assimilated by the ribulose diphosphate cycle and oxalate by the glycerate pathway. Therefore, the ATP requirement for the formation of cell material from formate is much higher than that from oxalate. However, it was estimated that about 50% of the potentially available energy in oxalate is required for its active transport across the cell membrane, whereas translocation of formate requires approximately 25% of the energy available in the formate. This difference explains why the molar growth yields on oxalate and formate are not very different.

12.6.2   Influence of Cell Composition

The influence of cell composition on the ATP requirement for the formation of cell material is relatively small; in order to evaluate this, the ATP requirement for the formation of cellular macromolecules is given in Table 7. It is evident from these data that the ATP requirement for the formation of protein, RNA, and DNA is larger than the ATP requirement for the formation of polysaccharide and lipid. The composition of *A. aerogenes* differs considerably from that of *E. coli*, which was taken as the basis for the calculations in the previous sections. A high protein content (75%) and a low polysaccharide content have been reported. On the basis of these data it can be calculated that, theoretically, about 25 g of cells per mol of ATP can be formed, a value that does not differ greatly from the value of 28.8 (Table 5, column B) calculated for the composition of *E. coli* as reported by Morowitz (1968). It is well known that the composition of microbial cells is influenced by the growth rate and the growth conditions. The influence of these changes on the ATP requirement for the formation of cell material is extremely small, unless large amounts of storage material are formed.

The influence of growth conditions on the formation of energy storage compounds has been described extensively in a recent review by Dawes and Senior (1973). The ATP requirement for

Table 7 ATP requirement for the formation of
cellular macromolecules in a glucose–inorganic
salts medium[a]

Macromolecule	ATP requirement ($10^4 \times$ mol/g macromolecule)
Polysaccharide	123.6
Protein	391.1
Lipid	14.8
RNA	373.2
DNA	330.0

[a] Results from Stouthamer (1977a).

the formation of storage materials has been treated by Stouthamer (1977a, 1979). When large amounts of storage materials are formed, the ATP requirement for the formation of cell material is much smaller than in the absence of the formation of storage material. Indeed it has been observed that molar growth yields are higher under conditions in which storage materials are formed than under conditions in which they are not.

12.6.3 Influence of the Nitrogen Source

There is a large ATP requirement for nitrogen fixation. In cell-free extracts, utilization of 12–15 mol of ATP per mol of nitrogen fixed has been reported (Zumft and Mortenson, 1975). Theoretical calculations of the amount of cell material that can be formed per mol of ATP for various assumed ATP requirements per mol of ammonia have been published by Stouthamer (1977b). During nitrogen fixation the assimilation of the ammonia formed occurs by the glutamine synthetase/glutamate synthase pathway (Dalton, 1979). Assimilation of ammonia by this pathway requires 1 mol of ATP, whereas ammonia assimilation by glutamate dehydrogenase, which occurs during growth with excess ammonia, does not require ATP. If this is taken into account, the results described in Table 8 are obtained (Stouthamer 1979). During ammonia-limited growth, or during growth with nitrate, the assimilation of ammonia also occurs by the glutamine synthetase/glutamate synthase system. Theoretically, for ammonia-limited growth or growth with nitrate in a mineral salts medium, 23.1 g of cell material can be formed per mol of ATP, whereas 28.8 g of cell material can be formed with excess ammonia (Table 8).

Table 8 Influence of the nitrogen source on the theoretical amount of cell
material formed per mole of ATP[a]

Source	ATP requirement/mol N_2 fixed	Cell material (g/mol ATP)
NH_3	—	28.8
NO_3^-	—	23.1
N_2	12	11.1
	18	8.7
	24	7.1
	30	6.0

[a] For growth with molecular nitrogen various ATP: N_2 ratios are used. The value of 12 for the ATP : N_2 ratio is the minimal ratio that is possible based on a requirement of 4 mol ATP per electron pair transferred in the nitrogenase reaction (Zumft and Mortenson, 1975; Dalton, 1979). Data from Stouthamer (1979).

It is evident that the nature of the nitrogen source in the medium has a very drastic influence on the ATP requirement for the formation of cell material. The amount of cell material that can be formed per mol of ATP is much lower during growth with molecular nitrogen than with ammonia. The experimentally observed molar growth yields are in accordance with these theoretical calculations.

12.6.4 Influence of the Carbon Assimilation Pathway of the Growth Substrate

The calculations given above indicate that the pathway of nitrogen assimilation has a profound influence on the ATP requirement for the formation of cell material. A similar influence is exerted by the carbon assimilation pathway. Different assimilation pathways have been found for growth with methane and methanol. The theoretical amounts of cell material formed per mol of ATP growth with methane or methanol using the different carbon assimilation pathways are shown in Table 9. It is evident that the influence of the carbon assimilation pathway on the ATP requirement for the formation of cell material is very great. Several attempts have been made to predict molar growth yields of methylotrophs growing on methane and methanol (Harder and van Dijken, 1977; Stouthamer, 1977b; Anthony, 1978. In agreement with these theoretical calculations, it has been observed that the molar growth yield for methanol varies from 15.7 to 17.3 for a number of organisms using the ribulose monophosphate cycle and from 9.8 to 13.1 for a number of organisms using the serine pathway (Goldberg *et al.*, 1976). However a high $Y_{methanol}$ of 13.4 was found in *Paracoccus denitrificans*, which uses the ribulose bisphosphate cycle for carbon assimilation (van Verseveld and Stouthamer, 1978b).

Table 9   Theoretical amount of cell material formed per mole of ATP for microorganisms using different carbon assimilation pathways for growth on methane and methanol[a]

Assimilation pathway	Cell material (g/mol ATP)
Ribulose monophosphate cycle, fructose diphosphate aldolase variant	27.3
Ribulose monophosphate cycle, 2-keto-3-deoxy-6-phosphogluconate aldolase variant	19.4
Serine pathway	12.5
Ribulose bisphosphate cycle	6.5

[a] Data from Harder and van Dijken (1976).

12.7   ENERGY-DISSIPATING MECHANISMS DURING GROWTH WITH EXCESS CARBON AND ENERGY SOURCE

Under a number of growth conditions the growth yield is much lower than expected on the basis of the ATP yield. Senez (1962) concluded that under these conditions growth and energy production are uncoupled and he introduced the term 'uncoupled growth'. Under these conditions there is a discrepancy between the rate of ATP production by catabolism and the rate of ATP utilization by anabolism. The capacities of bacteria to regulate cellular processes are evidently insufficient to regulate the rate of catabolism exactly to the needs of anabolism. This subject has been reviewed by Stouthamer (1979). It has been concluded that in microorganisms there is a great variety in the extent to which they can regulate the rate of ATP production to the needs of anabolism.

The following mechanisms exist for the adjustment of the rate of ATP production. (a) Excretion of products; this has been called 'overflow metabolism' (Neijssel and Tempest, 1975). This possibility does not exist in all organisms. (b) Deletion of sites of oxidative phosphorylation; in Section 12.5 it has already been mentioned that in sulfate-limited cultures of a number of organisms site I phosphorylation is lost. In cultures of *Azotobacter vinelandii* growing with molecular nitrogen as nitrogen source with a high dissolved oxygen tension, a sharp decrease in efficiency of energy conservation at site I was observed (Yates and Jones, 1974). (c) Branching of the respiratory chain; the respiratory chain of *A. vinelandii* has a highly branched structure and the cytochrome $b \rightarrow d$ pathway is non-phosphorylating (Yates and Jones, 1974). In nitrogen-fixing cultures under high aeration the cytochrome d content is highly increased and therefore a larger part of the electrons follow the non-phosphorylating branch. These mechanisms are not sufficient, however, to adapt the rate of ATP formation to the needs of anabolism under all conditions, so energy-dissipating mechanisms must occur. That this is indeed the case was demonstrated in the experiments of van der Beek and Stouthamer (1973) who measured simultaneously respiration and the ATP, ADP and AMP content of *K. aerogenes*. It was demonstrated that respiration continued after ATP had reached its maximal and ADP and AMP their minimal levels in the cell. The following possible energy-dissipating mechanisms have been considered (Stouthamer, 1979). (d) Dissipation of the energized membrane state; this may occur by an increase in the proton-conducting capacity of the membrane and by futile ion cycles. (e) Wastage of ATP by an ATPase. (f) Wastage of ATP in the formation of biomass; two possibilities may be mentioned, namely the formation of guanosine tetra- and penta-phosphates and futile cycles. The formation and their importance has been discussed earlier (Stouthamer, 1979; Arbige and Chesbro, 1982). The importance of futile cycles has been discussed (Tempest and Neijssel 1980; Westerhoff *et al.*, 1982).

These mechanisms offer an explanation for the observation that in aerobic yield studies the Y_{O_2} values for carbon-limited cultures are much higher than for carbon-sufficient cultures (Neijssel and Tempest, 1976), especially at low μ values. At μ_{max} the Y_{O_2} values are about the same. This indicates that in these cases the maintenance coefficient is dependent on the specific growth rate (see also Section 12.4). The consequence is that at low μ values the amount of energy spilled in carbon-sufficient cultures is much larger than at high μ values.

Energy-dissipating mechanisms must occur during growth with highly reduced compounds (see Section 12.8). Furthermore the existence of energy-dissipating mechanisms is necessary to allow the production of intermediates of primary metabolism with microorganisms (see Section 12.10).

12.8 INFLUENCE OF THE DEGREE OF REDUCTION OF THE GROWTH SUBSTRATE

During growth with a number of highly reduced compounds the ATP production may be too high in comparison with the ATP required for the biosynthesis of cell material (Neijssel and Tempest, 1975; Linton and Stephenson, 1978; Roels, 1980b; Erickson and Hess, 1981). This can easily be demonstrated with the use of elemental balances. The oxygen uptake per carbon equivalent of substrate is given by equation (9). If the amount of substrate-level phosphorylation is low in comparison with the amount of oxidative phosphorylation the ATP production per mole of substrate consumed can be approximated by equation (15).

$$\frac{\text{ATP production}}{\text{carbon equivalent substrate}} = \frac{(\gamma_s - y_c\gamma_b - z\gamma_p)(P/2e^-)2}{4} \tag{15}$$

y_{ATP} is the growth yield in carbon equivalents biomass per mol ATP. The growth yield is given by the equation:

$$y_c = (\text{ATP production/carbon equivalent substrate})\, y_{ATP}$$

Consequently:

$$y_c = [(\gamma_s - y_c\gamma_b - z\gamma_p)(P/2e^-)y_{ATP}]/2 \tag{16}$$

From equation (16)

$$y_c = \frac{(\gamma_s - z\gamma_n)P/2e^- \cdot y_{ATP}}{2 + \gamma_b \cdot P/2e^- \cdot y_{ATP}} \tag{17}$$

It is evident that $y_c \leqslant 1$ if no substantial fixation of CO_2 occurs, which is true for heterotrophic organisms. This means that

$$\gamma_s \leqslant \frac{2}{(P/2e^-)y_{ATP}} + \gamma_b + z\gamma_p \tag{18}$$

If γ_s is higher than this value growth will not be energy-limited but carbon-limited. An influence of the degree of reduction on molar growth yields has been reported in several studies. Figure 4 is a reproduction of the results presented by Linton and Stephenson (1978). In this study the critical value of γ_s is placed at $\gamma_s = 4$. Below this value y_c was linearly dependent on γ_s. Above this value y_c was independent of γ_s. Payne (1970) has introduced the concept of $y_{av.e^-}$ (= dry weight of organisms per electron available in the substrate for transfer to oxygen). An average value of 3.07 g for $y_{av.e^-}$ was found for growth of a large number of organisms on a variety of substrates (Payne, 1970). However, lower values were obtained for growth on ethanol ($\gamma_s = 6$) and hydrocarbons (Payne and Wiebe, 1978; Roels, 1980a). These results indicate that during growth of organisms with these substrates the efficiency of energy generation must be adapted to the possibilities of ATP consumption in biomass formation. This can be accomplished by a decrease in $P/2e^-$ ratio (deletion of one or more phosphorylation sites) or by the occurrence of energy-spilling mechanisms. Equation (18) shows that the critical value of γ_s is dependent on the $P/2e^-$ ratio and y_{ATP}. Both are characteristic properties of an organism during growth under certain conditions, therefore it is impossible to give a fixed critical value of γ_s. This is illustrated by the observation that in *P. denitrificans* growth with mannitol ($\gamma_s = 4.33$) or methanol ($\gamma_s = 6$) is certainly energy-limited (van Verseveld and Stouthamer, 1976, 1978a, b; Meijer *et al.*, 1977; van Verseveld *et al.*,

1979). The latter is due to the utilization of the Calvin cycle for the formation of cell material during growth on methanol (Cox and Quayle, 1975). Although during growth with methanol three phosphorylation sites are operative (van Verseveld and Stouthamer, 1978a, 1978b; van Verseveld *et al.*, 1979), growth will be energy-limited because of the high energy demand for biomass synthesis (Section 12.6.4).

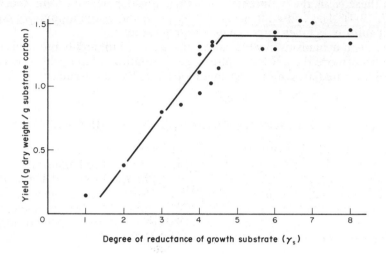

Figure 4   Influence of the degree of reductance of the growth substrate (γ_s) on the growth yield

Equation (18) is obtained by the application of macroscopic principles. In this way a number of simplifications have been made. If growth with ethanol as carbon and energy source is considered this approach can be refined by taking substrate-level phosphorylation and the involvement of the glyoxylate cycle in substrate assimilation into account. If this is done it can be calculated that growth will be carbon-limited at y_{ATP} values above 6.3 and 4.2 when respectively two or three phosphorylation sites are functional (A. H. Stouthamer, unpublished results). The theoretical Y_{ATP}^{max} for growth with ethanol is 13.5 (Stouthamer, 1973 and unpublished results). Since the experimental Y_{ATP}^{max} values are about 50% of the theoretical ones, carbon-limitation will occur with both two and three phosphorylation sites.

It must be realized that y_{ATP} in equation (18) is dependent on μ (equation 13). The strength of the influence of μ on y_{ATP} depends on the maintenance coefficient (Roels, 1980b). This dependency implies that at low μ values growth with ethanol can be energy-limited and at higher μ values growth can be carbon-limited. With two phosphorylation sites the range of μ values at which growth is energy-limited is much longer than with three phosphorylation sites. In accordance with these theoretical considerations, the curves relating q_{sub} and q_{oxygen} with μ for ethanol-limited growth of *Acinetobacter calcoaceticus* are not straight lines (Du Preez *et al.*, 1981).

The considerations above have important implications for the production of single cell protein from highly reduced compounds such as methanol, ethanol and hydrocarbons. In general not all the energy available in these substrates can be used in the formation of biomass. The utilization of mixed substrates might be an advantage in this respect. In mixtures of, for example, glucose and methanol, the glucose can be used as a carbon source, the conversion of which into biomass is characterized by a high theoretical Y_{ATP}^{max}. The reduced substrate can be used to yield additional energy to assimilate a larger part of the glucose. A procedure for the calculation of the mixing ratio which gives the greatest advantages has been developed (van Verseveld *et al.*, 1979; van Verseveld and Stouthamer, 1980).

12.9   HEAT PRODUCTION

The heat production during growth is of great practical importance as this quantity may determine the productivity of a fermenter in cases where the cooling capacity of the system is limited. The methods used to calculate the heat production during microbial growth have been treated extensively by Roels (1980a). During aerobic growth there is a strict proportionality between oxygen uptake and heat production. The oxygen uptake during aerobic growth is given by equation (9). The heat produced during the conversion of 1 mol of substrate (Q) is thus given by equation (19):

$$Q = A(\gamma_s - y_c\gamma_b - z\gamma_p)/4 \tag{19}$$

in which A is a proportionality constant. For growth with ammonia or nitrogen the value of A has been calculated to be 460 kJ mol^{-1} O$_2$ (Minkevich and Eroshin, 1973; Payne and Wiebe, 1978; Roels, 1980a) for the average composition of biomass. In a fermenter with a biomass content of x g biomass l^{-1} and a substrate conversion rate q_{sub} and a volume V the heat production (j_q) will be given by equation (20).

$$j_q = xq_{sub}V115(\gamma_s - y_c\gamma_b - z\gamma_p) \tag{20}$$

Equation (20) is also valid for growth in complex media and when considerable amounts of product are formed. Heat production and oxygen consumption are proportional as long as all components present obey the relationships that their heat of combustion is proportional to the oxygen consumption on combustion which is known to hold for most compounds commonly encountered in fermentation systems.

Much less is known about heat production during anaerobic growth. As a comparison Roels (1980a) mentions a heat production per unit biomass produced of 334 and 90 kJ/C mol biomass produced during aerobic and anaerobic growth of *S. cerevisiae* respectively. Consequently anaerobic growth results in a smaller heat production than aerobic growth, even if the heat production is expressed per unit biomass produced. This conclusion is expected to be of a fairly general validity.

12.10 THE STOICHIOMETRY OF PRODUCT FORMATION

The formation of products has been treated already in Section 12.2, equation (1). For modelling product formation one can use in the first instance equations (1)–(9) derived from elemental balancing. However, a number of kinetic equations are required in addition. For this purpose modified forms of equations (10)–(12) can be used. In addition for the study of the kinetics of product formation it is necessary to know the relation between the rate of product formation (q_{prod} = mole of product produced g^{-1} h^{-1}) and the specific growth rate. For this relation Erickson uses equation (21):

$$q_{prod} = \alpha + \beta\mu \tag{21}$$

in which α and β are constants.

Roels and Kossen (1978) have proposed equation (22):

$$\frac{q_{prod}}{q_{prod,\,max}} = \frac{\varepsilon\mu(\mu_{max})^{-1}}{1 + (\varepsilon - 1)\mu(\mu_{max})^{-1}} \tag{22}$$

in which $q_{prod,\,max}$ is the maximal rate of product formation and ε is a constant. Since this constant can have all values a large number of different relationships between q_{prod} and μ can exist. As an example, polysaccharide production in *Rhizobium trifolii* can be considered. The utilization of substrate for growth, maintenance and product formation is as outlined in Figure 5. At low growth rates no polysaccharide is formed and the amount produced increases with increasing values of μ. As a consequence the molar growth yield decreases at increasing μ values. If q_{sub} is extrapolated to zero growth rate a negative maintenance coefficient results (de Hollander *et al.*, 1979). It is clear that a more complex treatment is necessary to find the Y_{glu}^{max} and m_s values. For that purpose it is necessary to correct q_{sub} for the amount of substrate converted into polysaccharide. An analysis of the carbon and electron balance according to equations (2) and (8) gave satisfactory results (Erickson and Hess, 1981). The relation between q_{prod} and μ must be established in each case. Later on the relationship between the specific rate of penicillin production and μ will be treated.

It must be realized that various classes of low molecular weight fermentation products can be considered. The relation between substrate consumption, energy production, maintenance requirement and biomass and product yield is different for these classes. The various classes are listed in Table 10 (Stouthamer, 1978b; Roels and Kossen, 1978).

End products of energy-producing pathways are thought to be formed mainly during growth which is characterized by strong coupling between growth and energy production. Part of these

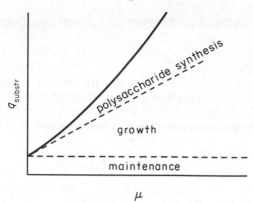

Figure 5 Utilization of substrate for growth, maintenance and polysaccharide production as a function of μ in chemostat cultures of _Rhizobium trifolii_

Table 10 Characterization of low molecular weight fermentation products and their production process

Class of metabolite	Examples	Coupling of growth to energy production
End products of energy metabolism	Ethanol, acetic acid, gluconic acid, lactic acid, acetone, butanol	Coupled
Intermediates of primary metabolism	Amino acids, citric acid, nucleotides	Uncoupled
Products of secondary metabolism	Antibiotics, _e.g._ penicillin	Coupled

processes are carried out under anaerobic conditions when molar growth yields are low, or in rich media, in which the carbohydrate is mainly used as an energy source (Stouthamer, 1977a). The production of acetic and gluconic acids is carried out under aerobic conditions by acetic acid bacteria, which are characterized by a low efficiency of energy conversion (Stouthamer, 1969; Olijve and Kok, 1979). It is evident that for the formation of these products no free choice of the carbon source as starting material is possible. The yield of product can be achieved by performing the production process at a temperature above the optimum, which causes an increase in m_e. This might also be accomplished in a continuous process with feedback of biomass. This has two advantages: an increase in the rate of product formation and a higher yield due to a larger maintenance energy demand by the high concentration of biomass. One way to perform such a process is in an upflow reactor with sedimenting microorganisms, another is the use of a recycling fermenter.

The formation of large amounts of products of intermediary metabolism occurs mostly with organisms which carry mutations or which grow under conditions of nutritional stress. Both approaches lead to abnormal regulation of metabolic pathways and result in an overproduction of the desired metabolite. As an example, the formation of lysine is treated; 4 mol of reduced nicotinamide adenine dinucleotide are formed whereas 1 mol of ATP is consumed per mol of lysine formed. It is thus evident that lysine formation is associated with a net production of ATP. In these production processes sugar concentrations as high as 1 M are used. For an experimental Y_{ATP} value of 12 for coupled growth, the biomass concentration would be about 65 g l^{-1} for two phosphorylation sites. These concentrations are much higher than the actual ones. This example shows that a very strong uncoupling between growth and energy production must occur under these circumstances. The example given above is not an isolated case, but represents the general case for the production of metabolites of intermediary metabolism. A very important conclusion is that such production processes are made possible by the occurrence of energy-spilling mechanisms. In these processes the carbon source can be chosen more freely than in the case of end-products of energy-generating pathways. The only essential condition is that the conversion of the carbon source to the precursors of the desirable metabolite is sufficiently rapid.

The formation of secondary metabolites has not been thought to be associated with growth. In general, the production process is divided into a phase of rapid growth (trophophase), in which no secondary metabolites are formed, followed by a subsequent production stage (idiophase) (Demain, 1971), but it is questionable whether this division is always so sharp. As an example, penicillin fermentation will be considered (Stouthamer, 1978b, Heijnen *et al.*, 1979). The formation of penicillin was studied in continuous cultures. It was found that the specific rate of its production (q_{pen} = units per milligram per hour) is independent of the growth rate above a critical value. This critical value can be estimated to be between 0.006 and 0.014 h^{-1}. When the specific growth rate was lower than the critical one, Pirt and Righelato (1967) observed a decaying penicillin synthesis rate with time. It has been assumed that a stable specific rate of penicillin synthesis is possible at specific growth rates lower than the critical value, however it decreases linearly with decreasing specific growth rate if it is lower than the critical one (Heijnen *et al.*, 1979). For the critical growth rate a value of 0.01 h^{-1} seems reasonable. Consequently equation (23) can be derived:

$$q_{pen} = q_{pen}^{max} \text{ for } \mu \geqslant 0.01 \text{ h}^{-1} \tag{23a}$$

$$q_{pen} = q_{pen}^{max} \frac{\mu}{0.01} \tag{23b}$$

Penicillin production is mostly carried out in fed batch cultures. Glucose (amongst other substances) is fed to the fermenter. The glucose balance is given by a modification of equation (11) as in equation (24):

$$dx/dt = (a - m_s x - bq_p)Y_{glu}^{max} \tag{24}$$

in which b = the amount of glucose needed for the formation of 1 mol penicillin. This amount was calculated by Cooney and Avecedo (1977) to be 1.78. The side chain is supposed to be supplied fully by precursor added to the medium. In the model of the penicillin fermentation elemental balances were used in addition. The course of the fermentation was calculated using the following parameters: $Y_{glu}^{max} = 90$ g mol^{-1}; $m_s = 3.4 \times 10^{-3}$ mol g^{-1} h^{-1}; $q_{pen}^{max} = 8$ U mg^{-1} h^{-1}; initial sugar concentration = 10 g l^{-1}; initial mycelium concentration = 1 g l^{-1}; $\mu_{max} = 0.12$ h^{-1}; glucose feed rate = 0.01 mol l^{-1} h^{-1}, growth rate period 200 h. Very large mycelium concentrations are reached (about 50 g l^{-1}) and therefore the amount of substrate used for maintenance purposes is very high. At the end of the fermentation (Figure 6) about 85% of the glucose added is used for maintenance purposes. It is therefore understandable that small variations in m_s have a very high influence on the penicillin yield. An increase in m_s from 3.4 × 10^{-3} mol g^{-1} h^{-1} to 4.1 × 10^{-3} g^{-1} h^{-1} gives a 15% decrease in penicillin yield. This variation is much smaller than the 95% confidence limits normally determined for maintenance coefficients. Due to this factor the temperature of the fermentation process has to be chosen and maintained very carefully. On the other hand the influence of Y_{glu}^{max} on the penicillin yield is negligible. It is also possible to calculate the influence of variable glucose feed rates. The example given in Figure 7 demonstrates that the influence of variation of the glucose feed rate on the penicillin production is tremendous. In the example given in Figure 7 the total amount of glucose given in the total cultivation period is the same for the three feeding schemes. This approach may help to optimize the production process. It is also possible to make calculations of the oxygen uptake and the heat production during the process. Application of the model allows the optimal use of a given fermenter with a fixed capacity for oxygen transfer and cooling capacity. The model given above fits the real production process very well. The formation of penicillin is also different from the formation of other types of fermentation products with respect to the influence of the carbon source. Carbohydrates are the preferred carbon sources for the formation of penicillin. The formation of penicillin from simpler substrates requires too much energy and will therefore result in low yields.

12.11 SUMMARY AND GENERAL CONCLUSIONS

(1) In many studies no attention has been given to the relationship between measured substrate consumption, oxygen consumption and product formation during microbial growth. These parameters are strictly interdependent and results from yield studies will certainly improve if accurate mass- and energy-balances are used.

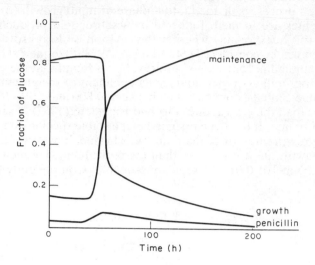

Figure 6 Glucose utilization for growth, maintenance and penicillin synthesis during the course of the penicillin fermentation using a constant glucose feed rate of 1000 mol l⁻¹. (Reproduced from Heijnen *et al.* (1979) with the permission of Wiley, New York)

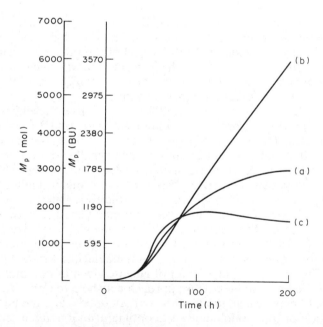

Figure 7 Influence of various glucose feed schemes on the penicillin concentration. (a) Constant 0.01 mol l⁻¹ h⁻¹, (b) 0.005 + 0.000015 × time, (c) 0.015 − 0.00005 × time. Penicillin concentration expressed as M_p = total penicillin amount in fermenter (broth weight after inoculation, 10⁵ kg) and in BU = billion units in the fermenter. (Reproduced from Heijnen *et al.* (1979) with the permission of Wiley, New York)

(2) Experiments using a recycling fermenter with 100% biomass feedback shed some doubt on the concept of maintenance as a constant need per gram biomass per hour, independent of the specific growth rate. Very accurate measurements in chemostat cultures at specific growth rates lower than 0.02 h⁻¹ and in recycling fermenters with a high biomass concentration are necessary for a final opinion about the real maintenance demands. Determination of the physiological parameters regulating the flow of carbon- and energy-source to maintenance processes (*e.g.* ppGpp) is of the utmost importance.

(3) Results presented in this chapter indicate the existence of two kinds of aerobic microorganisms: (a) organisms exhibiting a high efficiency of energy conservation, but low yields on ATP; (b) organisms with a high biomass yield per mole ATP, but a low efficiency of energy conservation. More microscopic research on the efficiency of oxidative phosphorylation and the efficiency of utilization of the proton motive force for the synthesis of ATP in microorganisms is necessary for a full understanding.

(4) Theoretical calculations on the ATP requirement for biomass formation show that Y_{ATP} is not a biological constant, but depends on the carbon substrate, the presence of preformed monomers, the used anabolic and catabolic pathways, the nitrogen source and the efficiency of oxidative phosphorylation. In general there is a good correspondence between the theoretical calculations and the growth yields observed experimentally. However, measured Y_{ATP} values are mostly 50–60% of the theoretical calculated ones. A possible explanation for the differences between theoretical and experimental Y_{ATP} values is given by non-equilibrium thermodynamics (see, *e.g.*, Westerhoff *et al.*, 1983; van Verseveld, 1981), which indicates that in many microorganisms growing aerobically optimization of the net rate of ATP synthesis has taken place. This is equal to an apparent efficiency of 50%.

(5) Different strategies are necessary for optimization of the rate of synthesis and the yield of products, depending on their status as end-products or intermediary products of primary metabolism or products of secondary metabolism. When biomass itself is the desired product (*e.g.* single cell proteins) an increase in yield can be reached by using a cheap second carbon source (*e.g.* methanol) that is used only as replacing energy source for the more expensive carbon source (*e.g.* glucose). In this case glucose will be anabolized more efficiently.

12.12   NOMENCLATURE

Symbol	Specification	Units
Y_{sub}	Molar growth yield for substrate	g dry weight per mol substrate
Y_{O_2}	Growth yield per mol O_2	g dry weight per mol O_2
Y_{ATP}	Growth yield per mol ATP	g dry weight per mol ATP
Y_{sub}^{max}	Molar growth yield for substrate corrected for maintenance energy	g dry weight per mol substrate
$Y_{O_2}^{max}$	Growth yield per mol O_2 corrected for maintenance energy	g dry weight per mol O_2
Y_{ATP}^{max}	Molar growth yield for ATP corrected for maintenance energy	g dry weight per mol ATP
$Y_{el\ acc}^{max}$	Growth yield per mol electron acceptor (oxygen, nitrate, nitrite) corrected for maintenance energy	g dry weight per mol electron acceptor
Y_{el}^{max}	Growth yield per g equivalent of electrons transferred to oxygen, nitrate or nitrite corrected for energy of maintenance	g dry weight per g equivalent of electrons
y_c	Growth yield in C-moles	C-moles dry weight per C-mole substrate
y_{ATP}	Growth yield in C-moles per mol ATP	C-moles dry weight per mol ATP
m_s	Maintenance coefficient	mol substrate per g dry weight h^{-1}
m_{O_2}	Maintenance coefficient	mol O_2 per g dry weight h^{-1}
m_e	Maintenance coefficient	mol ATP per g dry weight h^{-1}
μ	Specific growth rate	h^{-1}
q_{sub}	Specific rate of substrate consumption	mol substrate per g dry weight h^{-1}
q_{ATP}	Specific rate of ATP production	mol ATP per g dry weight h^{-1}
$P/2e^-$	Moles of ATP formed per electron pair transferred	mol ATP per electron pair
H^+/O	Number of protons ejected per atom O	number of H^+ per atom O
γ_s	Reductance degree of substrate	number of electrons per C-mol substrate
γ_b	Reductance degree of biomass	number of electrons per C-mol biomass
γ_p	Reductance degree of product	number of electrons per C-mol product

12.13   REFERENCES

Alexandre, A., F. Galiazzo and A. L. Lehninger (1980). On the location of the H^+-extruding steps in site 2 of the mitochondrial electron transport chain. *J. Biol. Chem.*, **255**, 10721–10730.

Anthony, C. (1978). The prediction of growth yields in methylotrophs. *J. Gen. Microbiol.*, **104**, 91–104.

Arbige, M. and W. R. Chesbro (1982a). Rel A and related loci are growth rate determinants for *Escherichia coli* in a recycling fermentor. *J. Gen. Microbiol.*, **128**, 693–703.

Arbige, M. and W. R. Chesbro (1982b). Very slow growth of *Bacillus polymyxa*: stringent response and maintenance energy. *Arch. Microbiol.*, **132**, 338–344.

Bahl, H., W. Andersch, K. Braun and G. Gottschalk (1982). Effect of pH and butyrate concentration on the production of acetone and butanol by *Clostridium acetobutylicum* grown in continuous culture. *Eur. J. Appl. Microbiol. Biotechnol.*, **14**, 17–20.

Bauchop, T. and S. R. Elsden (1960). The growth of microorganisms in relation to their energy supply. *J. Gen. Microbiol.*, **23**, 457–469.

Belaich, J. P., A. Belaich and P. Simonpietri (1972). Uncoupling in bacterial growth: effect of pantothenate starvation on growth of *Zymomonas mobilis*. *J. Gen. Microbiol.*, **70**, 179–185.

Bonnet, J. A. B. A. F., H. E. de Kok and J. A. Roels (1980). The growth of *Saccharomyces cerevisiae* CBS426 on mixtures of glucose and ethanol: a model. *Antonie van Leeuwenhoek*, **46**, 565–576.

Boogerd, F. C., H. W. van Verseveld and A. H. Stouthamer (1981). Respiration-driven proton translocation with nitrate and nitrous oxide in *Paracoccus denitrificans*. *Biochim. Biophys. Acta*, **638**, 181–191.

Boogerd, F. C., H. W. van Verseveld, D. Torenvliet, M. Braster and A. H. Stouthamer (1984). Reconsideration of the efficiency of energy transduction in *Paracoccus denitrificans* during growth under a variety of culture conditions. A new approach for the calculation of $P/2e^-$ ratios. *Arch. Microbiol.*, **139**, 344–350.

Boonstra, J., J. A. Downie and W. N. Konings (1978). Energy supply for active transport in anaerobically grown *Escherichia coli*. *J. Bacteriol.*, **136**, 844–853.

Chesbro, W., T. Evans and R. Eifert (1979). Very slow growth of *Escherichia coli*. *J. Bacteriol.*, **139**, 625–638.

Cooney, C. L. and F. Avecedo (1977). Theoretical conversion yields for penicillin synthesis. *Biotechnol. Bioeng.*, **19**, 1449–1462.

Cox, R. B. and J. R. Quayle (1975). The autotrophic growth of *Micrococcus denitrificans* on methanol. *Biochem. J.*, **150**, 569–571.

Dalton, H. (1979). In *Microbial Biochemistry*, ed. J. R. Quayle, pp., 227–266. University Park Press, Baltimore.

Dawes, E. A. and P. J. Senior (1973). The role and regulation of energy reserve polymers in micro-organisms. *Adv. Microb. Physiol.*, **10**, 135–266.

Dekker, J. G. J., H. E. de Kok and J. A. Roels (1981). Energetics of *Saccharomyces cerevisiae* CBS426: comparison of anaerobic and aerobic glucose limitation. *Biotechnol. Bioeng.*, **23**, 1023–1035.

Demain, A. L. (1971). Overproduction of microbial metabolites and enzymes due to alteration of regulation. *Adv. Biochem. Eng.*, **1**, 113–142.

de Hollander, J. A., C. W. Bettenhaussen and A. H. Stouthamer (1979). Growth yields, polysaccharide formation and energy conservation in chemostat cultures of *Rhizobium trifolii*. *Antonie van Leeuwenhoek*, **45**, 401–415.

de Kok, H. E. and J. A. Roels (1980). Method for the statistical treatment of elemental and energy balances with application to steady state continuous-culture growth of *Saccharomyces cerevisiae* CBS426 in the respiratory region. *Biotechnol. Bioeng.*, **22**, 1097–1104.

de Kwaadsteniet, J. W., J. C. Jager and A. H. Stouthamer (1976). A quantitative description of heterotrophic growth in micro-organisms. *J. Theor. Biol.*, **57**, 103–120.

de Vries, W., W. M. C. Kapteyn, E. G. van der Beek and A. H. Stouthamer (1970). Molar growth yields and fermentation balances of *Lactobacillus casei* L3 in batch cultures and in continuous cultures. *J. Gen. Microbiol.*, **63**, 333–345.

de Vries, W., H. G. D. Niekus, M. Boellaard and A. H. Stouthamer (1980). Growth yields and energy generation by *Campylobacter sputorum* subspecies *bubulus* during growth in continuous culture with different hydrogen acceptors. *Arch. Microbiol.*, **124**, 221–227.

Dijkhuizen, L., M. Wiersma and W. Harder (1977). Energy production and growth of *Pseudomonas oxalaticus* OX1 on oxalate and formate. *Arch. Microbiol.*, **115**, 229–236.

Du Preez, J. C., D. F. Toerien and P. M. Lategan (1981). Growth parameters of *Acinetobacter calcoaceticus* on acetate and ethanol. *Eur. J. Appl. Microbiol. Biotechnol.*, **13**, 45–53.

Erickson, L. E. and J. L. Hess (1981). Analysis of growth and polysaccharide yields in chemostat cultures of *Rhizobium trifolii*. *Ann. NY Acad. Sci.*, **369**, 81–90.

Esener, A. A., J. A. Roels and N. W. F. Kossen (1981). Fed. batch culture: modelling and applications in the study of microbial energetics. *Biotechnol. Bioeng.*, **22**, 1851–1871.

Farmer, I. S. and C. W. Jones (1976). The energetics of *Escherichia coli* during aerobic growth in continuous culture. *Eur. J. Biochem.*, **67**, 115–122.

Forrest, W. W., and D. J. Walker (1971). The generation and utilization of energy during growth. *Adv. Microb. Physiol.*, **5**, 213–274.

Geurts, Th. G. E., H. E. de Kok and J. A. Roels (1980). A quantitative description of the growth of *Saccharomyces cerevisiae* CBS426 on a mixed substrate of glucose and ethanol. *Biotechnol. Bioeng.*, **22**, 2031–2043.

Goldberg, I., J. S. Rock, A. Ben-Bassat and R. I. Mateles (1976). Bacterial yields on methanol, methylamine, formaldehyde and formate. *Biotechnol. Bioeng.*, **18**, 1657–1668.

Gutowski, S. J. and H. Rosenberg (1976). Effects of dicyclohexylcarbodiimide on proton translocation coupled to fumarate reduction in anaerobically grown cells of *Escherichia coli* K12. *Biochem. J.*, **160**, 813–816.

Haddock, B. A. and P. B. Garland (1971). Effect of sulfate-limited growth on mitochondrial electron transfer and energy conservation between reduced nicotinamide adenine dinucleotide and the cytochromes in *Torulopsis utilis*. *Biochem. J.*, **124**, 155–170.

Haddock, B. A. and C. W. Jones (1977). Bacterial respiration. *Bacteriol. Rev.*, **41**, 47–99.

Hadjipetrou, L. P., J. P. Gerrits, F. A. G. Teulings and A. H. Stouthamer (1964). Relation between energy production and growth of *Aerobacter aerogenes*. *J. Gen. Microbiol.*, **36**, 139–150.

Harder, W. and J. P. van Dijken (1977). In *Microbial Production and Utilization of Gases* (H_2, CH_4, CO), ed. H. G. Schlegel, N. Pfennig and G. Gottschalk, pp. 403–418. E. Goltze Verlag, Göttingen.

Harold, F. M. (1977). Membranes and energy transduction in bacteria. *Curr. Top. Bioenerg.*, **6**, 83–149.

Heijnen, J. J., J. A. Roels and A. H. Stouthamer (1979). Application of balancing methods in modelling the penicillin fermentation. *Biotechnol. Bioeng.*, **21**, 2175–2201.

Hempfling, W. P. and S. E. Mainzer (1975). Effects of varying the carbon source limiting growth on yield and maintenance characteristics of *Escherichia coli* in continuous culture. *J. Bacteriol.*, **123**, 1076–1087.

Hueting, S. and D. W. Tempest (1977). Influence of acetate on the growth of *Candida utilis* in continuous culture. *Arch. Microbiol.*, **115**, 73–78.

Jones, C. W. (1977). Aerobic respiratory systems in bacteria. *Symp. Soc. Microbiol.*, **27**, 23–59.

Jones, C. W., J. M. Brice and C. Edwards (1977). The effect of respiratory chain composition on the growth efficiency of aerobic bacteria. *Arch. Microbiol.*, **115**, 85–93.

Kashket, E. R. (1982). Stoichiometry of the H^+-ATPase of growing and resting aerobic *Escherichia coli*. *Biochemistry*, **21**, 5534–5538.

Keevil, C. W. and C. Anthony (1979). Effect of growth conditions on the involvement of cytochrome *c* in electron transport, proton translocation and ATP synthesis in the facultative methylotroph *Pseudomonas* AMI. *Biochem. J.*, **182**, 71–79.

Kell, D. B., P. John and S. J. Ferguson (1978). The proton motive force in phosphorylating membrane vesicles from *Paracoccus denitrificans*. Magnitude, sites of generation and comparison with the phosphorylation potential. *Biochem. J.*, **174**, 257–266.

Linton, J. D. and R. J. Stephenson (1978). A preliminary study on growth yields in relation to the carbon and energy content of various organic growth substrates. *FEMS Microbiol. Lett.*, **3**, 95–98.

Linton, J. D., K. Griffiths and M. Gregory (1981). The effect of mixtures of glucose and formate on the cell yield and respiration of a chemostat culture of *Beneckea natriegens*. *Arch. Microbiol.*, **129**, 119–122.

Meijer, E. M., H. W. van Verseveld, E. G. van der Beek and A. H. Stouthamer (1977). Energy conservation during aerobic growth in *Paracoccus denitrificans*. *Arch. Microbiol.*, **112**, 25–34.

Michels, P. A. M., J. P. J. Michels, J. Boonstra and W. N. Konings (1979). Generation of an electrochemical proton gradient in bacteria by the excretion of metabolic end products. *FEMS Microbiol. Lett.*, **5**, 357–364.

Minkevich, J. G. and V. K. Eroshin (1973). Productivity and heat generation of fermentation and oxygen limitation. *Folia Microbiologia*, **18**, 376–386.

Mitchell, P. (1966). In *Chemiosmotic Coupling and Energy Transduction*. Glynn Research Ltd., Bodmin, England.

Mitchell, P. (1970). Membranes of cells and organelles: morphology, transport and metabolism. *Symp. Soc. Gen. Microbiol.*, **20**, 121–166.

Morowitz, H. J. (1968). In *Energy Flow in Biology*, Academic, New York.

Nagai, S. and S. Aiba (1972). Reassessment of maintenance and energy uncoupling in the growth of *Azotobacter vinelandii*. *J. Gen. Microbiol.*, **73**, 531–538.

Neijssel, O. M. and D. W. Tempest (1975). The regulation of carbohydrate metabolism in *Klebsiella aerogenes* NCTC 418 organisms, growing in chemostat culture. *Arch. Microbiol.*, **106**, 251–258.

Neijssel, O. M. and D. W. Tempest (1976). Bioenergetic aspects of aerobic growth of *Klebsiella aerogenes* NCTC 418 in carbon-limited and carbon-sufficient chemostat culture. *Arch. Microbiol.*, **107**, 215–221.

Olijve, W. and J. J. Kok (1979). An analysis of the growth of *Gluconobacter oxydans* in chemostat cultures. *Arch. Microbiol.*, **121**, 291–297.

Otto, R., A. S. M. Sonnenburg, H. Veldkamp and W. N. Konings (1980a). Generation of an electrochemical proton gradient in *Streptococcus cremoris* by lactate efflux. *Proc. Nat. Acad. Sci. USA*, **77**, 5502–5506.

Otto, R., J. Hugenholtz, W. N. Konings and H. Veldkamp (1980b). Increase of molar growth yield of *Streptococcus cremoris* for lactose as a consequence of lactate consumption by *Pseudomonas stutzeri*. *FEMS Microbiol. Lett.*, **9**, 85–88.

Otto, R., R. G. Lageveen, H. Veldkamp and W. N. Konings (1982). Lactate efflux-induced electrical potential in membrane vesicles of *Streptococcus cremoris*. *J. Bacteriol.*, **149**, 733–738.

Payne, W. J. (1970). Energy yields and growth of heterotrophs. *Annu. Rev. Microbiol.*, **24**, 17–90.

Payne, W. J. and W. J. Wiebe (1978). Growth yield and efficiency in chemosynthetic micro-organisms. *Annu. Rev. Microbiol.*, **32**, 155–183.

Pirt, S. J. (1965). The maintenance energy of bacteria in growing cultures. *Proc. R. Soc. London, Ser. B*, **163**, 224–231.

Pirt, S. J. and R. C. Righelato (1967). Effect of growth rate on the synthesis of penicillin by *Penicillium chrysogenum* in batch and chemostat cultures. *Appl. Microbiol.*, **15**, 1284–1290.

Riebeling, V., R. K. Thauer and K. Jungermann (1975). The internal-alkaline pH gradient, sensitive to uncoupler and ATPase inhibitor in growing *Clostridium pasteurianum*. *Eur. J. Biochem.*, **55**, 445–453.

Roels, J. A. (1980a). Application of macroscopic principles to microbial metabolism. *Biotechnol. Bioeng.*, **22**, 2457–2514.

Roels, J. A. (1980b). Simple model for the energetics of growth on substrates with different degrees of reduction. *Biotechnol. Bioeng.*, **22**, 33–53.

Roels, J. A. and N. W. F. Kossen (1978). On the modelling of microbial metabolism. *Prog. Ind. Microbiol.*, **14**, 95–203.

Rogers, P. J. and P. R. Stewart (1974). Energetic efficiency and maintenance energy characteristics of *Saccharomyces cerevisiae* (wild type and petite) and *Candida parapsilosis* grown aerobically and micro-aerobically in continuous culture. *Arch. Microbiol.*, **99**, 25–46.

Senez, J. C. (1962). Some considerations on the energetics of bacterial growth. *Bacteriol. Rev.*, **26**, 95–107.

Solomon, B. O., L. E. Erickson, J. E. Hess and S. S. Yang (1982). Maximum likelihood estimation of growth yields. *Biotechnol. Bioeng.*, **24**, 633–649.

Stam, H., H. W. van Verseveld and A. H. Stouthamer (1983). Derepression of nitrogenase in chemostat cultures of the fast growing *Rhizobium leguminosarum*. *Arch. Microbiol.*, **135**, 199–204.

Stouthamer, A. H. (1973). A theoretical study on the amount of ATP required for synthesis of microbial cell material. *Antonie van Leeuwenhoek*, **39**, 545–565.

Stouthamer, A. H. (1977a). Energetic aspects of the growth of micro-organisms. *Symp. Soc. Gen. Microbiol.*, **28**, 285–315.

Stouthamer, A. H. (1977b). Theoretical calculations on the influence of the inorganic nitrogen source on parameters for aerobic growth of micro-organisms. *Antonie van Leeuwenhoek*, **43**, 351–367.

Stouthamer, A. H. (1978a). In *The Bacteria*, ed. L. N. Ornston and J. R. Sokatch, vol. 6, chap. 6, pp. 389–462. Academic, New York.

Stouthamer, A. H. (1978b). In *Genetics of Industrial Micro-organisms*, ed. O. K. Sebek and A. I. Laskin, pp. 70–76. American Society for Microbiology, Washington, DC.

Stouthamer, A. H. (1979). In *Microbial Biochemistry*, ed. J. R. Quayle, chap. 1, pp. 1–47. University Park Press, Baltimore.

Stouthamer, A. H. (1980). Energetic regulation of microbial growth. *Vierteljahrschr. Naturforsch. Gesellsch. Zürich*, **125**, 43–60.

Stouthamer, A. H. and C. W. Bettenhaussen (1973). Utilization of energy for growth and maintenance in continuous and batch cultures of micro-organisms. *Biochim. Biophys. Acta*, **301**, 53–70.

Stouthamer, A. H. and C. W. Bettenhaussen (1975). Determination of the efficiency of oxidative phosphorylation in continuous cultures of *Aerobacter aerogenes*. *Arch. Microbiol.*, **102**, 187–192.

Stouthamer, A. H. and C. W. Bettenhaussen (1977). A continuous culture study of an ATPase-negative mutant of *Escherichia coli*. *Arch. Microbiol.*, **113**, 185–189.

Stouthamer, A. H., F. C. Boogerd and H. W. van Verseveld (1982). The bioenergetics of denitrification. *Antonie van Leeuwenhoek*, **48**, 545–553.

Tempest, D. W. and O. M. Neijssel (1980). In *Diversity of Bacterial Respiratory Systems*, ed. C. J. Knowles, vol. 1, chap. 1, pp. 1–31. CRC Press, Boca Raton, FL, USA.

Thauer, R. K., K. Jungermann and K. Dekker (1977). Energy conservation in chemotrophic anaerobic bacteria. *Bacteriol. Rev.*, **41**, 100–180.

van der Beek, E. G. and A. H. Stouthamer (1973). Oxidative phophorylation in intact bacteria. *Arch. Mikrobiol.*, **89**, 327–339.

van Verseveld, H. W. (1981). Proton translocation and oxidative phosphorylation in *Paracoccus denitrificans*. *Antonie van Leeuwenhoek*, **47**, 178–180.

van Verseveld, H. W. and A. H. Stouthamer (1976). Oxidative phosphorylation in *Micrococcus denitrificans*. Calculation of the P/O ratio in growing cells. *Arch. Microbiol.*, **107**, 241–247.

van Verseveld, H. W. and A. H. Stouthamer (1978a). Electron transport chain and coupled oxidative phosphorylation in methanol-grown *Paracoccus denitrificans*. *Arch. Microbiol.*, **118**, 13–20.

van Verseveld, H. W. and A. H. Stouthamer (1978b). Growth yields and the efficiency of oxidative phosphorylation during autotrophic growth of *Paracoccus denitrificans* on methanol and formate. *Arch. Microbiol.*, **118**, 21–26.

van Verseveld, H. W., J. P. Boon and A. H. Stouthamer (1979). Growth yields and the efficiency of oxidative phosphorylation of *Paracoccus denitrificans* during two-(carbon) substrate-limited growth. *Arch. Microbiol.*, **121**, 213–223.

van Verseveld, H. W. and A. H. Stouthamer (1980). Two-(carbon) substrate-limited growth of *Paracoccus denitrificans*. A direct method to determine the P/O ratio in growing cells. *FEMS Microbiol. Lett.*, **7**, 207–211.

van Verseveld, H. W., K. Krab and A. H. Stouthamer (1981). Proton pump coupled to cytochrome c oxidase in *Paracoccus denitrificans*. *Biochim. Biophys. Acta*, **635**, 525–534.

van Verseveld, H. W., W. R. Chesbro, M. Braster and A. H. Stouthamer (1984). Eubacteria have three growth modes keyed by nutrient flow. Consequences for the concept of maintenance and maximal growth yield. *Arch Microbiol.*, **137**, 176–184.

Watson, T. G. (1970). Effect of sodium chloride on steady-state growth and metabolism of *Saccharomyces cerevisiae*. *J. Gen. Microbiol.*, **64**, 91–99.

Westerhoff, H. V., J. S. Lolkema, R. Otto and K. J. Hellingwerf (1982). Thermodynamics of growth. Non-equilibrium thermodynamics of bacterial growth: The phenomenological and the mosaic approach. *Biochim. Biophys. Acta*, **683**, 181–220.

Wikström, M. and K. Krab (1980). Respiration-linked H^+ translocation in mitochondria. Stoichiometry and mechanism. *Curr. Top. Bioenerg.*, **10**, 51–101.

Willison, J. C. and B. A. Haddock (1981). The efficiency of energy conservation in cytochrome c-deficient mutants of *Paracoccus denitrificans*. *FEMS Microbiol. Lett.*, **10**, 53–57.

Yates, M. G. and C. W. Jones (1974). Respiration and nitrogen fixation in *Azotobacter*. *Adv. Microb. Physiol.*, **11**, 97–135.

Zumft, W. G. and L. E. Mortenson (1975). The nitrogen-fixing complex of bacteria. *Biochim. Biophys. Acta*, **416**, 1–52.

13

Ageing and Death in Microbes

J. R. POSTGATE
University of Sussex, Brighton, UK
and
P. H. CALCOTT
Monsanto Company, St. Louis, MO, USA

13.1 BASIC PRINCIPLES

13.1.1 Death of Microbes

It is a truism that unicellular fissile prokaryotes have no physiological equivalent of the natural death of higher organisms (Postgate, 1976). At fission, all identifiable properties of the parent organism are evenly distributed between the two daughters, so the parent does not die but becomes two individuals of equal youth. In contrast, filamentous, budding, stalked or otherwise differentiated microbes often have a discernible mother–daughter relationship and may show true ageing (Yanagita, 1977). This always implies a commitment to death; a good example is a multiscarred yeast which has budded off so many progeny that it can produce no more and eventually dies (Beran, 1968; Sando, 1977).

Death of a unicellular microbe is only recognizable retrospectively, by tests in which the cell either grows or does not. Examples of such tests are the conventional viable counts in Petri dishes or the more direct viability determination using slide culture (Postgate, 1969); essentially the organisms are placed on a medium judged to be optimal for growth and the question is asked whether they grow or not. Short-cuts to the assessments of the ability of cells to multiply, such as dye-uptake, ATP content, respiration or other enzyme activity, are generally unreliable as estimates of viability (Postgate, 1967, 1969). A problem of principle therefore arises: since the viability of a cell can only be assessed after transfer to a recovery medium and incubation, was a dead cell indeed dead at the time of such transfer or did it die during the recovery procedure? The phenomenon of substrate-accelerated death (see below) provides a case of the latter kind.

Instances of suspended animation (*e.g.* as frozen or lyophilized yet viable cells) are common among prokaryotes. However, near-semantic problems which arise in discussions of death in differentiated eukaryotes, for example those concerned with the survival of organs (kidneys, grafted tissue, *etc.*) or cells (tissue cultures, *etc.*) after death of the whole organism, do not arise in the present context.

13.1.2   Ageing of Microbes

The period between its origin and division into two daughters comprises the lifetime of a fissile prokaryote. A freely-growing population shows a skewed age distribution throughout this period (Powell, 1956; Sargent, 1978) paralleled by changes in enzyme constitution which can be studied using synchronized cultures (Mitchison, 1971). This is the 'natural' ageing of a microbe; it is progressive, irreversible and, as far as is known, it bears no analogy to the ageing of complex metazoa. Ageing of a different kind occurs when growth and multiplication are prevented by a stress; such physiological in addition to temporal ageing is by far the most important practical aspect of ageing and death in microbes. Stresses which may initiate such ageing include chilling, starvation, irradiation, immobilization, and chemical or osmotic stress.

Ageing in higher organisms leads to death by way of a senescent state, in which the organism is physiologically a living individual but its reproductive potential is low or zero. Comparable senescent states probably exist among ageing fissile prokaryotes and certainly occur in differentiated microbes.

13.1.3   Viability among Microbes

The term viability has a precise meaning in microbiology: it applies to microbial populations and is the ratio of the number of organisms capable of multiplying on a recovery medium to the total number of organisms present. A viable cell is an organism capable of multiplication in such circumstances; a senescent cell is effectively non-viable but it may possess all other definitive properties of a living organism. Moribund populations often contain a mixture of viable cells, senescent cells and dead cells. The latter two subclasses of a moribund population are operationally dead. However, the existence of a senescent state implies that ageing is a progressive process and that, after a certain stage has been reached, the reproductive machinery of the cell is irrecoverably damaged although other vital physiological processes persist.

13.1.4   Survival of Populations: Cryptic Growth

An important distinction must be made between clonal survival, the survival of a population as a whole, and the survival of the individuals within that population. Consider a population of coliform bacteria starved of a nitrogen source but with otherwise adequate supplies of nutrient. After a period during which all the population survives, a proportion will die. They will then release protein, amino acids and other nitrogen metabolites which will serve as nitrogen sources for survival and multiplication of their surviving cousins. Thus the clone survives at the expense of some of its constituents. This process has been referred to as 'regrowth' and 'cell turnover' but the term most widely use is 'cryptic growth'. Examples were discussed by Postgate and Hunter (1963a): the number of divisions supported by the death of one cell (the cryptic growth factor) can range from less than one for carbon-limited populations to more than three for Mg^{2+}-limited populations of *Klebsiella aerogenes*. Cryptic growth has often been overlooked by research workers in the interpretation of data from such experiments as Warburg manometry, enzyme induction, prolonged microbicide assays; it can usually be allowed for by incorporating carefully chosen microbistats into the test system (Postgate, 1969).

13.1.5   Injury among Microbes

Stresses such as chilling, freezing followed by thawing, exposure to sublethal irradiation or disinfectant, mild heat shock or even an abrupt surge of a growth-limiting substrate can render a proportion of a microbial population fragile: they will die under mild stress more rapidly than they otherwise would have done. Such fragile organisms can be regarded as injured. In some cases, recovery from such injury can be demonstrated if the organisms are incubated in 'resting' conditions; an example is presented in Figure 1. Two classes of injury are recognized. Some stresses (*e.g.* irradiation or detergent treatment) cause *structural injury* in the sense that damage is done to the cells' structure (*e.g.* DNA or membranes) and a proportion of the population cannot repair this damage in time to avoid death. 'Fragile' mutants, which die more readily than the wild type, are examples of relative stable structural injury. Other stresses (*e.g.* sublethal microbicide treatment or freeze-thawing) lead to *metabolic injury*, a physiological state, as yet imperfectly understood, in which the organisms' metabolic pathways are forced into an unbalanced condition so that a proportion of the population is unable to multiply except in special media. Most commonly 'rich' media give higher counts than 'minimal' media with metabolically injured populations, no doubt by promoting recovery from injury.

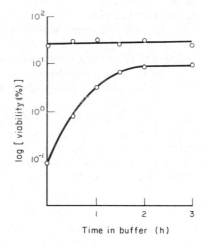

Time in buffer (h)

Figure 1 Injury as a result of osmotic shock and recovery therefrom: mercury-resistant *Pseudomonas aeruginosa* were subjected to an osmotic stress which lowered viability to about 20% assayed on conventional nutrient agar (upper graph). When assayed on $HgCl_2$-containing medium, osmotic stress had induced sensitivity to Hg^{2+} toxicity from which most of the population recovered gradually over 2 h at 37 °C in magnesium–potassium phosphate buffer (Calcott, 1981).

13.1.6 Stress and Survival

In this chapter we shall be concerned almost entirely with ageing and death resulting from stress. Powerful stresses, such as exposure to lethal concentrations of disinfectant or sterilizing temperature, induce death of vegetative microbes which is so rapid that no useful distinction between ageing and death can be made. Less powerful stresses rarely cause immediate death; they usually initiate an ageing process which leads to death of the population in a time-dependent manner. The nature of the survival pattern for a population of microbes subject to a mild stress is determined by two major factors: the physiological status of the population and the nature of the stress. These factors tend to modify the logarithmic death curve which would be expected for a population in its death phase, at least as far as the majority of the population is concerned, and a variety of death curves may be obtained (Postgate and Hunter, 1962). In the majority of practical situations both factors are complex and even in controlled laboratory situations it is often difficult to analyse the precise physiological and environmental status of a stressed population. The nature of the problem will be illustrated in the following two sections.

13.1.6.1 *The physiological status of the population*

Consider a population of a simple fissile bacterium growing in a minimal medium in a chemostat at a moderate growth rate, the population density being limited by the concentration of glucose in the medium. Such a population will, if stressed (*i.e.* starved, subjected to mild heat shock or freezing) show certain responses reflected in a decreased viability which would be observed after that stress. A population growing at the same rate but with its density determined at a similar level but by the concentration of ammonium ions (the nitrogen source) in the medium would show quite different responses to such stresses; it would, in most circumstances, be less fragile. Yet another population, limited to a similar density by the concentration of magnesium ions in the medium, would be considerably more fragile. These principles illustrate the profound effect that nutritional status can have on the physiological status of vegetative cells (for further discussion see Dawes, 1976; Strange, 1976). More complex organisms such as *Arthrobacter globiformis* undergo morphological changes in response to changes in nutritional status (Luscombe and Gray, 1974) and can become more resistant to certain stresses; others form relatively stress-resistant cysts (*e.g. Azotobacter*) and the extreme case is represented by the numerous prokaryotic groups which sporulate: spore formation is generally highly dependent on nutritional status and spores are usually, though not always, highly resistant to a variety of stresses.

These considerations apply to the physiological status of microbes obtained from chemostats, in which the nutritional status can be clearly defined. Populations from batch or surface cultures, or from the natural environment, present a considerably more complicated problem because the nutritional status of the population is rarely defined. Normally such populations are of mixed nutritional status: for example, if the culture is of an aerobe growing in stagnant broth, most of the population is oxygen-limited except that within a few microns of the liquid surface, which will be limited by a component of the medium which is rapidly metabolized but relatively slowly diffusible, often the carbon/energy source. As another example, a well-stirred, well-aerated batch culture of an aerobe, growing at its maximum growth rate (μ_{max}) will be limited by whichever intrinsic metabolic pathway is rate-limiting and determines μ_{max}. This pathway is rarely known (although in diazotrophs there is good evidence that N_2 fixation itself is often the rate-limiting process, see Postgate, 1982). In principle, populations undergoing logarithmic growth are limited by intrinsic metabolic pathways rather than extrinsic factors, although external factors such as temperature, pH and O_2 regime may influence which of several intrinsic pathways is limiting.

13.1.6.2 Overt and actual stresses

Microorganisms age and die because environmental stress leads to delay in, or cessation of, growth. Seemingly simple stresses are often complex when analysed logically. Freezing and thawing, for example, can be analysed in the following manner. The initial stress is cold shock, to which fast-growing populations are hypersensitive (MacLeod and Calcott, 1976). As the menstruum freezes, solutes become concentrated around the cell generating a plasmolytic osmotic stress. With actual freezing, mechanical penetration of the cell by ice crystals may cause structural damage. If a period of cold storage ensues, devitrification of vitrified zones of the frozen environment will also cause mechanical damage. Thawing will lead to a second osmotic stress as a result of dilution of the solutes which earlier became concentrated around the cell; if these were osmotically active molecules of 'medium' molecular weight (*e.g.* sucrose, glucose) which had concentrated inside the cell wall, transient osmotic pressures may build up during thawing, yet before the solutes have diffused out again, such that the cell wall may become disrupted (Record and Taylor, 1953, 1960). Postgate (1976) analysed the preparation of cells for experiments on starvation into at least five subsidiary stresses involved in transferring actively growing cells to the starvation environment; he concluded that the most scientifically sound system for studying ageing and death during starvation would be chemostats of known nutritional status running so slowly that the death rate of the cells made a significant contribution to the steady state kinetics of the system as a whole. Changes in population density and in minutiae of the chemical environment may introduce unsuspected stresses. The essential message of these considerations, which were amplified by Postgate (1976) and Strange (1976), is that (a) an apparently simple stress may well be complex, and (b) removal of the population from a stress and its transfer to a 'recovery' environment may well entrain new stresses.

13.2 STARVATION

When a bacterial population is subjected to conditions in which growth is prevented by depriving it of one or more nutrient, its metabolism changes significantly from that seen in growing organisms (Dawes, 1976; Trinci and Thurston, 1976). Among the first events seen are a cessation of net synthesis of protein, RNA, wall, lipid and DNA, although turnover of most of these polymers continues (see Dawes, 1976; Strange, 1976). Net degradation of these molecules results, by the process called endogenous metabolism. In most of the species of bacteria examined, RNA appears to be the polymer that is most readily degraded during starvation, and the content of ribosomal RNA per cell decreases. The free pools of amino acids, adenine nucleotides and probably mRNA, may decrease initially but soon stabilize; they are probably turned over rapidly. Protein, which must be degraded during this initial period of endogenous metabolism to sustain the amino acid pool and adjust the cell's metabolism to its new situation, does not usually show any gross decrease in concentration until the concentrations of more readily metabolized materials have declined significantly. DNA, as one would imagine, is usually one of the last polymers to be broken down; a decreased level is rarely observed unless the population is of low viability.

Viability is lost some time after the onset of endogenous metabolism, the time depending on the organism and its nutritional status. The cells remain osmotically intact, metabolizing entities after loss of viability (Postgate and Hunter, 1962) and no quantitative correlations between levels of cell polymers and viability have been obtained. Bacteria in the period between loss of viability and total loss of metabolic function may correspond to senescent forms of more complex organisms.

Magnesium plays an important role in survival of many stresses, including starvation. Many experiments on starvation can be interpreted in terms of the effects of magnesium deprivation or loss even though other nutrients may have been removed (Dawes, 1976). Addition of magnesium to a starvation environment usually prolongs the life of the population.

The dependence of the survival potential of a strain on its physiological status, that is on the composition, aeration, temperature, *etc.* of the growth medium used prior to starvation, can be rationalized in terms of the type and quantity of endogenous reserve material laid down, the synthesis of which is determined by both genetic and environmental factors. Endogenous reserve materials such as polyphosphates, polysaccharides (often polyglucose) and lipids (such as polyhydroxybutyric acid, PHB) presumably serve to spare other more critical cell components and generate energy to be used for maintenance. Dawes (1976) pointed out that the presence of a rapidly oxidizable energy-rich polymer does not constitute evidence that the material is an endogenous reserve, but evidence from many sources indicates that the polyglucose reserves of enteric bacteria, streptococci and clostridia can constitute endogenous reserves which aid survival of starvation.

Some bacteria accumulate more than one reserve material. For instance, a *Pseudomonas* species isolated from activated sludge can accumulate glycogen, PHB and lipid reserves; the relative concentrations depend on the growth medium (Zevenhuizen and Ebbink, 1974). In this case, PHB is synthesized first and is broken down to yield glycogen and lipid. In general, the cultures that degraded their reserves more slowly died less rapidly.

Bacteria which are adapted to survive long periods of starvation, such as *Arthrobacter* or *Nocardia*, tend to metabolize their endogenous reserves much more slowly than more starvation-sensitive bacteria such as *Escherichia coli* and *Klebsiella aerogenes*. While the factors that govern rate of endogenous metabolism have not been elucidated, the glycogen of *E. coli* and *K. aerogenes* tends to have a chain length of 12–15 residues, whereas that of *Arthrobacter* has shorter chains of 7–9 residues with a high degree of branching. A lower level of debranching enzymes such as α-1,6-glucosidase may slow endogenous metabolism and prolong survival (Dawes, 1976).

During the initial period of rapid endogenous metabolism, when viability is maintained, energy becomes available for 'maintenance' (see Stouthamer, Volume 1, Chapter 12). The maintenance energy of bacteria can be calculated by a variety of techniques, but no conclusive data have been presented to demonstrate which of the events requiring maintenance energy specifically maintain cell viability (Dawes, 1976; Strange, 1976). Potentially this energy could be used to keep polymer turnover at a minimal level, for DNA repair and maintenance of nutrient gradients across the plasma membrane. Konings and Veldkamp (1981) proposed that the last of these is the critical factor for viability maintenance. Since nutrient gradients depend on the existence of the proton motive force (PMF) which is made up of H^+ concentration and electrical gradient components, they proposed that the death phase is triggered by a decrease of the PMF to zero. This attractive hypothesis has not been tested quantitatively.

Some understanding of the physiology of starvation is now available and a role for certain intracellular control mechanisms is becoming evident at the molecular level. In bacteria such as *E. coli*, *Peptococcus* and *Chromatium*, transition from growth to non-growth situations leads to a decrease in the adenylate energy charge (ATP + 0.5 ADP)/(ATP + ADP + AMP) from near 1 to between 0.75 and 0.5 (Atkinson, 1971; Dawes, 1976; Knowles, 1977) but it then stabilizes. After a period of survival, viability may drop dramatically, accompanied in *E. coli* by a sudden decrease in the adenylate charge to values less than 0.5. These data indicate a correlation between viability and energy charge but a causal relation has not been proved.

13.2.1   Substrate-accelerated Death (SAD)

This is a phenomenon which has been useful in furthering our understanding of starvation at the molecular level. It takes the following form: the substrate which limited growth of a culture of bacteria can appear to accelerate the death of the population when it is subjected to starvation in the presence of that nutrient (Dawes, 1976; Strange, 1976; Calcott and Calvert, 1981). The traumatic substrate is usually the carbon-energy source, but SAD due to the inorganic nitrogen source has been reported. An important feature of SAD is that the traumatic substrate must be present in the medium used to assess viability. For example, a lactose-limited population of *K. aerogenes* will show rapid death when starved in non-nutrient buffer containing lactose, compared with its death rate in plain buffer, provided viability is assessed on a lactose medium. Assessment of viability on a glycerol medium, for example, indicates no difference in death rate. Thus the organisms being starved in the presence of the traumatic substrate are not dying; they die on the recovery medium. Consistent with this fact, they retain intact osmotic barriers (Postgate and Hunter, 1964) and synthesize protein, RNA and DNA (P. H. Calcott, unpublished). In the case of lactose-accelerated death, the mechanism seems to be the following. The traumatic substrate, lactose, causes the induced lactose permease to be suddenly saturated, so that the sugar is transported rapidly into the cell. This rapid flux of lactose switches off the enzyme adenylate cyclase, the enzyme which catalyses the conversion of ATP to 3′,5′-cyclic AMP (Peterkofsky, 1977; Peterkofsky and Gazdar, 1979). The cells have an enzyme, cyclic phosphodiesterase, which catalyses the conversion of cyclic AMP to 5′-AMP, as well as an active cyclic AMP excretion mechanism; these continue to function, disrupting the cell's metabolic control so that a rapid decrease in the intracellular level of the cyclic nucleotide occurs. This decrease in cyclic AMP concentration prevents the organism from signalling, *via* synthesis of guanosine tetraphosphate, that the starvation medium is unfavourable for growth (an unbalanced C/N ratio; Bennett and Calcott, 1982). Somehow, this predisposes the cell to an inability to divide on a plating medium which would normally be favourable. SAD can be prevented in several ways, all consistent with the mechanism just outlined: by addition of exogenous cyclic AMP to the starvation or plating media, by the inclusion of magnesium in the starvation medium (which allows a higher internal cyclic AMP level to be maintained, which in turn allows guanosine tetraphosphate to be synthesized), by addition of guanosine tetraphosphate to the starvation medium (giving an external signal) or, of course, by plating the traumatized population on a medium containing a nontraumatic substrate.

Cyclic AMP is in fact excreted, although to a lesser extent, by starving populations of *K. aerogenes* which are not being subject to SAD (Calcott *et al.*, 1972). During starvation of *E. coli* for carbon or nitrogen sources, there is also a rapid synthesis of polyphosphorylated nucleotides (Cashel, 1975; Silverman and Atherly, 1979). Matin (1982) and Calcott (1982a) have proposed that total cellular enzyme levels could be due to the interplay between the concentrations of various small molecules such as cyclic AMP, cyclic GMP and guanosine tetraphosphate in the cell. Any imbalance between these should result in a disruption of the total cell control which might lead to inability to grow, *i.e.* to death.

13.3   METABOLIC AND STRUCTURAL INJURY

Bacteria subject to stress often pass through fragile or injured states prior to death. Indeed, populations subject to SAD are in an injured state, being predisposed to death on media containing the traumatic substrate. Two such states, referred to as metabolic and structural injury, have been the focus of many studies recently because many organisms in the environment, in water supplies, in the gut, in frozen and freeze-dried foods are in an injured state (Ray and Speck, 1973; MacLeod and Calcott, 1976; Calcott, 1978; Russell and Andrew, 1983). These injured organisms are often not detected by enumeration methods used in conventional public health practice so, since they are viable in special conditions, the safety of a product or process has not been fully evaluated (see Russell and Andrew, 1983).

Metabolic injury, which has been studied predominantly in frozen-thawed populations, is a repairable, transient inability to grow in the absence of complex nutrients, often amino acid or peptide in nature. The traumatized population can recover in a simple salts solution, given an appropriate temperature and time. Exposure to sublethal concentrations of disinfectants can lead to comparable metabolic injury. The mechanism of injury or the repair process involved is unclear; a transient repression of control systems involving small molecules, such as the synthesis and degradation of cyclic nucleotide and guanosine tetraphosphate, may be involved (P. H. Calcott and M. Thomas, unpublished observation). A mechanistic similarity to SAD has been noted.

Structural injury has been observed in mildly heated, chilled, frozen-thawed, frozen-stored, freeze-dried and dried organisms (Adams, 1978; Calcott, 1978; Ray and Speck, 1973; Strange and Cox, 1976). It involves a transient, repairable susceptibility to an agent or agents that are innocuous to unstressed populations. Membrane damage can be detected in these populations since stressed organisms are sensitive to high NaCl concentrations; wall injury in Gram-negative bacteria can be detected as hypersensitivity to detergents in the plating medium. In the case of the outer membrane, the sites of damage include the lipopolysaccharide, protein and lipid, as well as possibly the linkage between them (Bennett *et al.*, 1981).

Damage to ribosomes has been observed in heated bacteria (Pierson *et al.*, 1978) and DNA is also a target. Evidence has accrued to show that mild stresses such as freeze-thawing or heating induce both single-stranded and double-stranded breaks into the DNA (Alur and Grecz, 1975; Alur *et al.*, 1977; MacLeod and Calcott, 1976). This damage can be minimized with the inclusion of cryoprotective agents in the freezing menstruum, and can be repaired if the cell contains an active complement of DNA excision repair and recombinant repair genes. Such damage can be lethal if not repaired, for example in repair-deficient mutants of *E. coli* (Calcott and Thomas, 1981) and of *Pseudomonas aeruginosa* (Williams and Calcott, 1982). Mutants may also arise (Postgate and Hunter, 1963b; Calcott and Gargett, 1981). In at least one case a plasmid carrying DNA repair genes can confer cryoresistance on certain cryosensitive strains (Williams and Calcott, 1982). The damage to the DNA is not confined to the chromosome; plasmid DNA can be cured by freeze-thaw (P. H. Calcott, D. Wood and L. Anderson, unpublished experiments). Although evidence is not available it is likely that starvation could result in DNA damage which might not be repaired effectively (Dawes, 1976).

These studies serve to illustrate that improvident handling and cryopreservation can cause genetic changes in bacteria. Plasmids constructed *in vivo* or *in vitro* can be relatively unstable and may be lost completely or may dissociate into smaller more stable forms when the host strain is stressed (*e.g.* Skotnicki and Rolfe, 1977) with a concomitant loss of gene function. At a more general level, they indicate that, in the assessment of bacterial viability, the scientist must take into account metabolic and structural injury by a careful choice of plating medium.

13.4 THYMINE-LESS DEATH

The phenomenon called thymine-less death seems superficially akin to substrate-accelerated death. It occurs in bacteria that require thymine for growth and DNA replication, either because of mutation or inhibitor action; these, when starved of thymine, attempt to grow but then die (Stacey, 1976). An early hypothesis explained the phenomenon in terms of unbalanced growth, since the cells still make protein, RNA, walls and membranes but fail to accomplish more than one round of DNA replication. During thymine-less death, single- and double-strand breaks occur in the DNA, possibly caused by an activated nuclease attempting a repair process. A simplistic view would be that DNA damage at thymine-containing codons occurs during replication, but damage could also occur during transcription (Stacey, 1976). Thymine starvation also triggers prophage induction and colicin synthesis, but this must be a secondary effect because *rec*A strains are still sensitive to thymine-less death although they can neither induce prophage nor synthesize colicins (Stacey, 1976). This phenomenon has not been fully elucidated.

13.5 SURVIVAL OF SLOWLY GROWING BACTERIA

Klebsiella aerogenes grown at different growth rates in a chemostat showed a faster death rate from starvation the more slowly it was grown. This generalization was true of all nutritional status studied, indicating that growth rate plays an important role in determining survival potential. Thus it is possible to grow continuous cultures of *K. aerogenes* so slowly that the death rate makes a significant contribution to the steady state parameters of the population. Postgate and Hunter (1962; see also Tempest *et al.*, 1967, and Postgate, 1973) grew *K. aerogenes* at generation times greater than 20 h using chemostat cultures with four types of growth-limiting nutrient (N, C, S, Mg). Depending on the limiting nutrient, steady state populations with viabilities down to 40% were obtained. In such a population it is self-evident that the normal relationship between doubling time of the cells and dilution rate of the culture does not apply: the viable cells are multiplying faster than they would if some of their siblings were not dead. Tempest *et al.* (1967) observed that the doubling time of the survivors tended to a constant value. They proposed that for these organisms a minimum doubling time exists which they termed μ_{min}, and that growth of a population at a slower rate would lead to death of some cells so as to maintain a μ_{min} in the surviving portion. The concept of μ_{min} was new and is not readily understood; it is not simply a matter of energy of maintenance because ammonia- and carbon-limited populations both gave the same μ_{min}. As pointed out by Calcott (1982b), the observation of a μ_{min} could be due to operational artefacts of sheer stress by the culture stirrer or of a non-optimal plating medium. These ideas have not been tested. Assuming that the moribund populations are not operational artefacts, it would be useful to know whether the dead and living portions are similar in terms of cell composition (DNA, RNA and protein), macromolecular synthesis potential and whether members of the dead portion of the population have intact cell control mechanisms. Such studies require effective experimental means of separating the live and dead components of the populations. Tempest *et al.* (1967) recorded morphological aberrants among such moribund populations of *K. aerogenes*; comparable changes occur when *E. coli* grows at these low rates: its division processes change from a true symmetric binary fission to a polar growth with a symmetric type division. Thus when growing slowly (and incidentally at rates in its natural habitat, the gut), it resembles other so-called budding bacteria that give rise to a daughter cell and retain the mother cell (Dow and Whittenbury, 1981).

It is relevant that most natural microbial environments resemble a slow-growing, moribund chemostat rather than a conventional laboratory batch culture, which may in part account for the low viabilities such populations often show (Postgate, 1973). However, the correlation between a low growth rate and a high death rate is not universally true: autochthonous microbes are often starvation-resistant and tend to have naturally slow growth rates (Dawes, 1976; Gray, 1976).

Some biotechnological processes involve the use of immobilized microbes. In such systems multiplication is restricted and they might approximate to slow-growing, moribund chemostat systems which would soon become of low viability. This would be unimportant if the required metabolic function were retained and the fact that metabolic function can be retained in such systems for long periods (*e.g.* Chibata, 1979) is remarkable. Data on the relationship of viability to function in immobilized biotechnological systems are not readily available.

13.6   DIFFERENTIATION AND SURVIVAL

Many microbes respond in more complex ways to starvation than do coliform bacteria. *Arthrobacter crystallopoietes* is a well known example of nutrient-dependent dimorphism: a rod form appears in rich media and a coccus form, with the higher surface to volume ratio, predominates in nutrient-poor situations. Chemostat culture shows the dimorphic change to be triggered by growth rate; the coccus is favoured at low growth rates (Luscombe and Gray, 1974). The precise role of the dimorphism in survival of the organism is not clear (Gray, 1976).

There are other examples of nutrient-induced polymorphism described in detail by Dow and Whittenbury (1981). Under very dilute nutrient concentrations, the prosthecae of *Ancalomicrobium* sp. are fully expressed while under conditions of higher nutrient levels the structures are absent. These structures might play significant roles in scavenging nutrients by increasing the surface to volume ratio. This might allow a slow but definite growth rate which would favour survival. It is a pity that chemostat culture studies have not been done with this organism.

Certain budding and prosthecate bacteria show assymetric division with clear mother and daughter cells being formed (see Trinci and Oliver, Volume 1, Chapter 10). When *Caulobacter* divides it forms a cross-wall in the mother cell stalk. After subsequent division another cross-wall is produced. Thus it is possible to determine the age (or more correctly the number of divisions that the mother cell has undergone) from the number of cross-walls. Presumably there is a finite number of cross-walls that can ensue and consequently the mother cell will eventually enter an aged state. This is best demonstrated in *Rhodomicrobium vannielli*, where only four cross-walls can form, after which the mother cell cannot divide (Whittenbury and Dow, 1977). As was mentioned at the opening of this chapter, differentiation of this kind implies a commitment to senescence and death in the species. The morphological scarring of budding yeasts, which was also mentioned, leads to senescence and death for similar reasons: it imposes a restriction on further budding.

In the cases described so far, the cells change morphology or structure to delay or prevent death, but the cells remain biochemically active. More sophisticated microbes switch from a biochemically active to a non-active form as cysts or spores. These 'resting stages' are more resistant to environmental stresses such as heating, freezing, drying and starvation than the original vegetative cells. The physiology of cyst formation has been relatively little studied (*e.g.* Sadoff, 1975) and the resistance to stress provided by encystment appears to be modest. However, spore formation can lead to dramatic stress tolerance and has been studied in great detail by numerous workers. Sporulation as a topic cannot be surveyed here (Gould and Hurst, 1969). For present purposes we recall that conversion from vegetative cell to spore requires a considerable commitment by the cell (Freese, 1981; Setlow, 1981). The organism must recognize the environmental trigger, often a depletion of key nutrients in the medium, and subject itself to a dramatic metamorphosis (Freese, 1981; Setlow, 1981; Ellar, 1978). These events consume energy which must be fuelled by reserve polymers, polysaccharide and protein, since the environment is often exhausted. While the cellular events that lead to sporulation are not fully understood, sporulation seems to be associated with the synthesis of highly phosphorylated nucleotides such as guanosine penta- and tetra-phosphates, adenine-containing nucleotides and certain antibiotics. Spores can, like vegetative cells, display both metabolic and structural injury (*e.g.* on heating, Adams, 1978; see also Russell and Andrew, 1983).

Sporulation is probably the most sophisticated means available to microbes of preserving viability in stress conditions. Once the cell has sporulated and lain dormant for a period of time, it needs to be able to detect external changes that indicate the environment has become favourable for growth. In most cases low molecular weight physiological molecules such as glucose and alanine are important although pressure, chemicals such as KBr and heat shock can cause germination.

13.7 REFERENCES

Adams, D. M. (1978). Heat injury of bacterial spores. *Adv. Appl. Microbiol.*, **23**, 245–261.

Alur, M. D. and N. Grecz (1975). Mechanism of injury of *Escherichia coli* by freezing and thawing. *Biochem. Biophys. Res. Commun.*, **62**, 108–112.

Alur, M. D., N. F. Lewis and N. Grecz (1977). Evidence of chemical protection against freeze-induced DNA breakage. *FEMS Microbiol. Lett.*, **1**, 367–369.

Atkinson, D.E. (1971). Adenine nucleotides as stoichiometric coupling agents in metabolism and as regulatory modifiers: the adenylate energy charge. In *Metabolic Pathways*, ed. H. J. Vogel, vol. 5, pp. 1–21. Academic, New York.

Bennett, G. M., A. Seaver and P. H. Calcott (1981). Effect of defined lipopolysaccharide core defects on resistance of *Salmonella typhimurium* to freezing and thawing and other stresses. *Appl. Environ. Microbiol.*, **42**, 843–849.

Bennett, G. M. and P. H. Calcott (1982). *Abstracts of the ASM Annual Meeting*, Atlanta, I43, p. 101.

Beran, K. (1968). Budding of yeast cells, their scars and ageing. *Adv. Microb. Physiol.*, **2**, 143–172.

Calcott, P. H. (1978). *Freezing and Thawing Microbes*. Meadowfield Press, Sheldon, Co. Durham, UK.

Calcott, P. II. (1981). Transient loss of plasmid-mediated mercuric ion resistance after stress in *Pseudomonas aeruginosa*. *Appl. Environ. Microbiol.*, **41**, 1348–1354.

Calcott, P. H. (1982a). Cyclic AMP and cyclic GMP control of synthesis of constitutive enzymes in *Escherichia coli*. *J. Gen. Microbiol.*, **128**, 705–712.

Calcott, P. H. (1982b). Continuous culture: where it came from and where it's going. In *Continuous Culture of Cells*, ed. P. H. Calcott, pp. 1–11. CRC Press, Boca Raton, FL.

Calcott, P. H. and T. J. Calvert (1981). Characterization of 3′,5′-cyclic AMP phosphodiesterase in *Klebsiella aerogenes* and its role in substrate-accelerated death. *J. Gen. Microbiol.*, **122**, 313–321.

Calcott, P. H. and A. M. Gargett (1981). Mutagenicity of freezing and thawing to *Escherichia coli*. *FEMS Microbiol. Lett.*, **10**, 151–155.

Calcott, P. H. and M. Thomas (1981). Sensitivity of DNA repair deficient mutants of *Escherichia coli* to freezing and thawing. *FEMS Microbiol. Lett.*, **12**, 117–120.

Calcott, P. H., W. Montague and J. R. Postgate (1972). The levels of cyclic AMP during substrate-accelerated death. *J. Gen. Microbiol.*, **73**, 197–200.

Cashel, M. (1975). Regulation of bacterial ppGpp and pppGpp. *Annu. Rev. Microbiol.*, **29**, 301–318.

Chibata, I. (1979). Immobilized microbial cells with polyacrylamide gel and carrageenan and their industrial applications. In *Immobilized Microbial Cells*, ed. K. Venkatasubramanian, *Symp. Am. Chem. Soc.*, **106**, 187–202.

Dawes, E.A. (1976). Endogenous metabolism of the survival of starved microbes. In *The Survival of Vegetative Microbes*, ed. T. R. G. Gray and J. R. Postgate, *Symp. Soc. Gen. Microbiol.*, **26**, 19–53.

Dow, C. S. and R. Whittenbury (1981). Prokaryote form and function. In *Contemporary Microbial Ecology*, ed. D. C. Ellwood, J. N. Hedger, M. J. Latham, J. M. Lynch and J. H. Slater, pp. 391–417. Academic, New York.

Ellar, D. J. (1978). Spore specific structures and their function. In *Relations between Structure and Function in the Prokaryotic Cell*, ed. R. Y. Stanier, H. J. Rogers and B. J. Ward, *Symp. Soc. Gen. Microbiol.*, **28**, 295–325.

Freese, E. (1981). Initiation of bacterial sporulation. In *Sporulation and Germination*, ed. H. S. Levinson, A. L. Sonenshein and D. J. Tippe, pp. 1–2. The American Society for Microbiology, Washington, DC.

Gould, G.W. and A. Hurst (eds.) (1969). *The Bacterial Spore*. Academic, New York.

Gray, T. R. G. (1976). Survival of vegetative microbes in soil. In *The Survival of Vegetative Microbes*, ed. T. R. G. Gray and J. R. Postgate, *Symp. Soc. Gen. Microbiol.*, **26**, 327–364.

Knowles, C. J. (1977). Microbial metabolic regulation by adenine nucleotide pools. In *Microbial Energetics*, ed. B. A. Haddock and W. A. Hamilton, *Symp. Soc. Gen. Microbiol.*, **27**, 241–283.

Konings, W. N. and H. Veldkamp (1981). Phenotypic responses to environmental change. In *Contemporary Microbial Ecology*, ed. D. C. Ellwood, J. N. Hedger, M. J. Latham, J. M. Lynch and J. M. Slater, pp. 161–191. Academic, New York.

Luscombe, B. M. and T. R. G. Gray (1974). Characteristics of *Arthrobacter* grown in continuous culture. *J. Gen. Microbiol.*, **82**, 213–222.

MacLeod, R. A. and P. H. Calcott (1976). Cold shock and freezing damage to microbes. In *The Survival of Vegetative Microbes*, ed. T. R. G. Gray and J. R. Postgate, *Symp. Soc. Gen. Microbiol.*, **26**, 81–109.

Matin, A. (1982). Microbial regulatory mechanisms as studied in continuous culture. In *Continuous Culture of Cells*, ed. P. H. Calcott, pp. 69–97. CRC Press, Boca Raton, FL.

Mitchison, J. M. (1971). *The Biology of the Cell Cycle*. Cambridge University Press, Cambridge.

Peterkofsky, A. (1977). Regulation of *Escherichia coli* adenylate cyclase by phosphorylation–dephosphorylation. *Trends Biochem. Sci.*, **2**, 12–14.

Peterkofsky, A. and C. Gazdar (1979). *Escherichia coli* adenylate cyclase complex: regulation by the proton electrochemical gradient. *Proc. Natl. Acad. Sci. USA*, **76**, 1099–1103.

Pierson, M. D., R. F. Gomez and S. E. Martin (1978). The involvement of nucleic acids in bacterial injury. *Adv. Appl. Microbiol.*, **23**, 263–285.

Postgate, J. R. (1967). Viability measurements and the survival of microbes under minimum stress. *Adv. Microb. Physiol.*, **1**, 2–23.

Postgate, J. R. (1969). Viable counts and viability. *Methods Microbiol.*, **1**, 611–628.

Postgate, J. R. (1973). The viability of very slow-growing populations: a model for the natural eco-system. *Bull. Ecol. Res. Commun. (Stockholm)*, **17**, 287–292.

Postgate, J. R. (1976). Death in macrobes and microbes. In *The Survival of Vegetative Microbes*, ed. T. R. G. Gray and J. R. Postgate, *Symp. Soc. Gen. Microbiol.*, **26**, 1–18.

Postgate, J. R. (1982). *The Fundamentals of Nitrogen Fixation*. Cambridge University Press, Cambridge.

Postgate, J. R. and J. R. Hunter (1962). The survival of starved bacteria. *J. Gen. Microbiol.*, **29**, 233–267; corrigenda: (1964). *ibid.*, **34**, 473.

Postgate, J. R. and J. R. Hunter (1963a). The survival of starved bacteria. *J. Appl. Bacteriol.*, **26**, 295–306.

Postgate, J. R. and J. R. Hunter (1963b). Metabolic injury in frozen bacteria. *J. Appl. Bacteriol.*, **26**, 405–414.

Postgate, J. R. and J. R. Hunter (1964). Accelerated death of *Aerobacter aerogenes* starved in the presence of growth-limiting substrates. *J. Gen. Microbiol.*, **34**, 459–473.

Powell, E. O. (1956). Growth rate and generation time of bacteria with special reference to continuous culture. *J. Gen. Microbiol.*, **15**, 492–511.

Ray, B. and M. D. Speck (1973). Freeze-injury in bacteria. *CRC Crit. Rev. Clin. Lab. Sci.*, **4**, 161–213.

Record, B. R. and R. Taylor (1953). Some factors influencing the survival of *Bacterium coli* on freeze-drying. *J. Gen. Microbiol.*, **9**, 475–484.

Record, B. R. and R. Taylor (1960). Survival of bacteria on drying in sugar/protein mixtures. *Nature (London)*, **185**, 944.

Russell, A. D. and M. H. E. Andrew (eds.) (1983). Revival of injured microbes. *10th Symposium of the Society for Applied Bacteriology*. Academic, New York.

Sadoff, H. L. (1975). Encystment and germination in *Azotobacter vinelandii*. *Bacteriol. Rev.*, **39**, 516–539.

Sando, N. (1977). Ageing and sporulating ability in yeast cells. In *Growth and Differentiation in Micro-organisms*, ed. T. Ishikawa, Y. Maruyama and H. Matsumiya, pp. 151–164. University Park Press, Baltimore.

Sargent, M. G. (1978). Surface extension and the cell cycle in prokaryotes. *Adv. Microb. Physiol.*, **18**, 106–176.

Setlow, P. (1981). Autochemistry of bacterial forespore development and spore germination. In *Sporulation and Germination*, ed. H. S. Levinson, A. L. Sonenshein and D. J. Tippe, pp. 13–28. The American Society for Microbiology, Washington, DC.

Silverman, R. H. and A. G. Atherly (1979). The search for guanosine tetra-phosphate (ppGpp) and other unusual nucleotides in eukaryotes. *Microbiol. Rev.*, **43**, 27–41.

Skotnicki, M. L. and B. G. Rolfe (1977). Stepwise selection of defective nitrogen-fixing phenotypes in *Escherichia coli* K-12 by dimethyl sulfoxide. *J. Bacteriol.*, **130**, 939–942.

Stacey, K. A. (1976). The consequences of thymine starvation. In *The Survival of Vegetative Microbes*, ed. T. R. G. Gray and J. R. Postgate, *Symp. Soc. Gen. Microbiol.*, **26**, 365–382.

Strange, R. E. (1976). *Microbial Response to Mild Stress*. Meadowfield Press, Sheldon, Co. Durham, UK.

Strange, R. E. and C. S. Cox (1976). Survival of dried and air-borne bacteria. In *The Survival of Vegetative Microbes*, ed. T. R. G. Gray and J. R. Postgate, *Symp. Soc. Gen. Microbiol.*, **26**, 111–154.

Tempest, D. W., D. Herbert and P. J. Phipps (1967). Studies on the growth of *Aerobacter aerogenes* at low dilution rates in a chemostat. In *Microbial Physiology and Continuous Culture*, ed. E. O. Powell, C. G. T. Evans, R. E. Strange and D. W. Tempest, *Proceedings of the 3rd International Symposium*, pp. 240–254. HMSO, London.

Trinci, A. P. J. and C. F. Thurston (1976). Transition to the non-growing state in eukaryotic micro-organisms. In *The Survival of Vegetative Microbes*, ed. T. R. G. Gray and J. R. Postgate, *Symp. Soc. Gen. Microbiol.*, **26**, 55–80.

Whittenbury, R. and C. S. Dow (1977). Morphogenesis and differentiation in *Rhodomicrobium vannielii* and other budding and prosthecate bacteria. *Bacteriol. Rev.*, **41**, 754–808.

Williams, D. L. and P. H. Calcott (1982). Role of DNA repair genes and an R-plasmid in conferring cryoresistance on *Pseudomonas aeruginosa*. *J. Gen. Microbiol.*, **128**, 215–218.

Yanagita, T. (1977). Cellular age in micro-organisms. In *Growth and Differentiation in Micro-organisms*, ed. T. Ishikawa, Y. Maruyama and H. Matsumiya, pp. 1–36. University Park Press, Baltimore.

Zevenhuizen, L. P. T. M. and A. G. Ebbink (1974). Interrelations between glycogen, poly-β-hydroxybutyric acid and lipids during accumulation and subsequent utilization in a *Pseudomonas*. *Antonie van Leeuwenhoek*, **40**, 103–120.

14

Effect of Environment on Microbial Activity

R. G. FORAGE, D. E. F. HARRISON and D. E. PITT
Biotechnology Australia Pty. Ltd., Roseville, NSW, Australia

14.1   INTRODUCTION

Unlike the cells of higher organisms, those of microorganisms are not naturally maintained in controlled environments but are exposed completely to environmental influences. They must be able to tolerate and react to a wide range of environmental changes. To attempt to catalogue such changes in a chapter such as this would clearly be fatuous. Rather, we will examine in general the underlying effects of some of the main environmental influences on microbial physiology.

One problem with trying to identify the response of a microorganism to a change in a single environmental parameter is that rarely can one alter only one aspect of the environment independently. For instance, a change in pH may also affect the osmolarity of the solution, the CO_2 partial pressure and perhaps O_2 solubility. Thus any discussion of the influence of environmental parameters must be a gross over-simplification of the actual situation.

14.2   MECHANISMS OF MICROORGANISM RESPONSE TO THE ENVIRONMENT

Before considering individual environmental influences in turn, it is as well to consider the ways in which an organism can be expected to react to environmental changes. As in sophisticated man-made control systems, microorganisms demonstrate control at several levels from the extremely rapid but superficial reactions to slower but more profound adaptations to prolonged environmental changes. The levels of control can be categorized as described in the following sections.

14.2.1   Primary Response due to Direct Chemical or Physicochemical Effects

These are the inevitable and immediate consequence of a chemical or physical change. Examples would be the alterations in rates of reaction brought about by changes in substrate or product levels, the effect of temperature changes on enzyme rates, and the effect of osmotic changes on solute movements across membranes.

14.2.2   Enzyme Inhibition and Stimulation

Rapid and fine control of metabolism is exerted by regulation of enzyme reaction rates through negative or positive feedback control mechanisms. These often rely on allosteric control by substrates, products or coenzymes which bind to specialized regulating sites on enzymes. An example of such control is the Pasteur effect in yeast which is complex but depends largely on the regulation of the key glycolytic enzyme, phosphofructokinase by the adenine nucleotides; a high 'energy charge', *i.e.* high levels of ATP relative to ADP, inhibits the enzyme but activation occurs when the 'energy charge' is low.

Responses of these mechanisms tend to be rapid, being completed in a few seconds or minutes, and represent the fine tuning of metabolism.

14.2.3   Induction and Repression of Protein Synthesis

Microorganisms are remarkable in their ability to undergo quite radical reorganisation of metabolism in response to environmental changes. This is achieved by switching on and off synthesis of specific enzymes and by protein turnover. The control is at the level of transcription of DNA and has been studied in more detail than any other type of regulatory mechanism. The time scale for response can vary greatly depending on the complexity of the enzyme system involved and on the growth rate of the organism. Potentially, transcription can be switched on within a few seconds of the appropriate stimulus, but there is often a built-in delay in the regulatory system. For example, in recent studies on the induction of citrate permease in *Klebsiella aerogenes*, it was found that there was a delay of about 10 min after initial exposure to citrate before uptake was stimulated but subsequent exposures led to an immediate response (Ashby and Harrison, 1980). The reversal of stimulation tends to take longer as protein turnover often proceeds at a relatively slow rate. This was also illustrated for the case of citrate permease in *K. aerogenes*.

Control by transcription no doubt represents a 'coarse' control system and the changes brought about are often profound. Response times may be anything from a few minutes to hours.

14.2.4   Changes in Cell Morphology

Many microorganisms undergo morphological changes in response to environmental changes. Often resting forms or spores may be formed in response to adverse variations in the environment. Fungi demonstrate a very great variety of morphological adaptations, including the formations of complex fruiting bodies, in response to appropriate stimuli. But many bacteria and fungi demonstrate smaller morphological changes: some yeasts switch to a filamentous mode of growth under certain conditions and even *Escherichia coli* can be induced to grow in semi-filamentous form. Such changes usually occur gradually and require several hours or longer to complete.

14.2.5   Changes in Genotype

Undoubtedly the rapid proliferation rate of microorganisms is utilized so that selection of new genotypes forms part of their response to environmental changes. Any population of microorganisms may contain a range of genotypes and selection of a new, predominant form begins whenever the environment is changed. This fact is often overlooked in studies using continuous culture although the technique was originally developed by Novick and Szilard (1950) to study mutant selection. Plasmids obviously play an important role in bacteria in extending the genotype variation and are especially prevalent in organisms such as *Pseudomonas* spp. which show a large degree of adaptation to the environment. The rate of response due to selection of genotype will depend on the mutation rate and selection pressure and this has been quantified by authors such as Kubitschek (1970). Generally the response will be very slow, take many generations to complete and will not be rapidly reversible.

In the discussions that follow it should be borne in mind that all the above levels of response can, and are likely to, be induced by changes in most environmental parameters.

14.3   DISSOLVED OXYGEN

14.3.1   Cell Interactions with Oxygen

14.3.1.1   Respiration

By far the largest utilization of oxygen in aerobic organisms is for respiration. Oxygen uptake in this case involves no incorporation of molecular oxygen into the cell or metabolites; oxygen acts as the terminal electron acceptor for the oxidative reactions which provide energy for cellular activities. The terminal electron donors to oxygen (terminal oxidases) are most commonly cytochromes. There are other oxidases such as flavoprotein-linked oxidases which usually play a very minor role, if any, in respiration.

The terminal cytochrome oxidase is part of an electron transport system (ETS) of electron carriers situated in a membrane which serves to link oxidation reactions to ATP generation. In eukaryotic cells the ETS resides in the mitochondrial membrane and is highly conservative, being the same for almost all eukaryotes. Between prokaryotes, however, there is considerable variation in the nature of the ETS and the terminal cytochrome oxidase. Furthermore, individual species of bacteria often possess several types of terminal oxidase, often as many as three, and the relative amounts of these may vary with the environmental conditions, especially with oxygen availability (Table 1). There is evidence that changes in the terminal oxidase are accompanied by variation in the overall energy conservation efficiency of respiration in some organisms. For example, in *Azotobacter* cytochromes a_1 and o linked respiration is thought to have a P/O ratio of 3 while cytochrome d linked respiration may have a P/O ratio of only 2 (Ackrell and Jones, 1971). Thus, respiration in bacteria may be influenced both qualitatively and quantitatively in response to environmental changes.

Table 1   Examples of Cytochrome Oxidases and their Relation to O_2 Status in Bacteria

Organism	Cytochrome oxidases	Comments	References
Klebsiella aerogenes	a_1, d, o	a_1 and d maximum under O_2-limitation	Harrison, 1972b
Micrococcus parainfluenzae	aa_3, o	aa_3 maximum under excess O_2 o maximum under O_2-limitation	Sapshead and Wimpenny, 1970
Azotobacter vinelandii	a_1, d, o	All maximum under O_2-limitation	Akrell and Jones, 1971
Beneckea natriegens	CO-binding c, o	o maximum under O_2-limitation	Linton *et al.*, 1975

14.3.1.2   Oxygen incorporation

Although respiration does not involve any incorporation of molecular oxygen into biomass, there is a range of enzymes in microorganisms which catalyse the fixing of oxygen into organic molecules. These enzymes are broadly termed oxygenases. Either one or both of the atoms of an O_2 molecule may be incorporated by oxygenase reactions. Dioxygenases catalyse reactions in which both atoms of the oxygen molecule are incorporated into an organic molecule. Where only one of the atoms of oxygen is incorporated (monooxygenase) the other atom is usually reduced to water. This type of reaction requires an appropriate electron donor such as coenzymes NADH or NADPH, or tetrafolic acid or ascorbic acid. The requirement of monooxygenases for an electron donor is physiologically important as the consumption of the reduced coenzymes or substrate without the formation of high energy phosphate bonds is a net consumption of energy compared with the case of dioxygenases where there is no net energy consumption and with cytochrome oxidase systems which are associated with the generation of ATP by oxidative phosphorylation.

With the vast majority of bacteria, an extremely small amount of the cellular oxygen, if any, is derived from molecular oxygen; rather it comes from water and carbon dioxide. However, with organisms growing on highly reduced substrates, such as hydrocarbons, as sole carbon sources, the primary attack on the substrate is usually *via* an oxygenase reaction and involves the incorporation of molecular oxygen into the molecule. In such organisms a significant proportion of the total cell oxygen may be derived from molecular oxygen. For instance, it can be calculated that for bacteria growing on methane as the sole carbon source, up to 30% of the cellular oxygen may be derived from molecular oxygen through the action of methane monooxygenase. Oxygenase reactions may be physiologically very significant even though only a very small proportion of the total cell oxygen is derived from molecular oxygen. In yeasts and fungi, oxygen is incorporated into unsaturated fatty acids and sterols by means of oxygenase reactions. Although the oxygen incorporated represents a minute part of the total cell mass, these fatty acids and sterols are essential for growth and the inability of the organism to form unsaturated fatty acids under anaerobic conditions creates an absolute requirement for small amounts of oxygen in the absence of an external supply of the essential fatty acid or sterol.

14.3.1.3 Oxygen as an inhibitor

For all organisms, including obligate aerobes, oxygen is to some extent toxic. However, the degree of susceptibility to oxygen toxicity varies enormously from the obligate anaerobes at one extreme, such as some members of the genus *Clostridium* and methanogenic bacteria which require highly reducing atmospheres for growth, through microaerophilic organisms which are inhibited by oxygen in air-saturated liquids while requiring oxygen in small amounts for growth, to facultative and obligate aerobes, some of which can tolerate oxygen tensions up to five times that of air-saturated liquids. The mechanism for oxygen toxicity which has gained widespread support in recent years and for which there is considerable experimental evidence, is through the formation of superoxide radicals (O_2^-) which are destructive to many aspects of cell metabolism (Gregory *et al.*, 1973). Oxygen radicals are formed as products during the reduction of oxygen by flavoprotein oxidase. Oxygen tolerance is conferred on aerobic organisms by the enzyme superoxide dismutase which removes the oxygen radical. Gregory and Fridovich (1973) found that conditions of growth under high oxygen tension which enhanced superoxide dismutase content in *Escherichia coli* corresponded also with increased resistance to hyperbaric oxygen in this organism.

It was also shown that *E. coli* possesses two distinct superoxide dismutase enzymes. One, which is present in cells grown at low dissolved oxygen tension, is not induced by oxygen and is iron-containing and another, which is absent in anaerobically grown cells, is induced by oxygen and contains manganese. The iron-containing superoxide dismutase protects against exogenous superoxide radicals, while the manganese-containing enzyme which is inducible by oxygen, protects against endogenous superoxide radicals formed under hyperbaric oxygen tensions.

Although there is strong evidence that superoxide plays a major role in the toxic effect of hyperbaric oxygen, possibly under some circumstances other mechanisms are more important. Thus, oxygen toxicity has been attributed to oxidation of thiol groups, enzyme inactivation and lipid peroxidation. Also, prolonged exposure to oxygen pressures of ten atmospheres have been reported to have a mutagenic effect.

14.3.1.4 Oxygen as an enzyme regulator

The level at which dissolved oxygen is effective as an inducer of enzymes can be very low indeed, much lower than that at which respiration rate is affected. *Escherichia coli* grown under white spot grade nitrogen in chemostat culture where the dissolved oxygen tension was immeasurably low (< 100 Pa) still possesses a fully functional electron transport system and a high potential respiration rate (Harrison and Loveless, 1971a). If, however, strict precautions are taken to exclude all oxygen, then the respiration rate is negligible and the terminal oxidase generation is repressed. At the other end of the scale, high oxygen concentration may also act as an inducer of some enzymes, *e.g.* superoxide dismutase. The precise mechanism by which oxygen acts to induce or repress enzymes is not known. Probably in some cases it is the intracellular redox potential which serves to change the concentration of regulator molecules which then react with the DNA bases. This has been discussed in some detail by Cole (1976). In yeast, molecular oxygen is required for a synthetic step in cytochrome generation (Chen and Charalampous, 1969).

14.3.2 Measurement of Dissolved Oxygen

Modern membrane oxygen electrodes enable the accurate determination of dissolved oxygen over long periods of time in microbial cultures at oxygen tensions above about 1% of air saturation. However, for most probe designs, the calibration becomes unreliable at oxygen tensions below about 100 Pa. Unfortunately, it is at this latter level that the most interesting effects of oxygen on cells occurs so there is still a requirement for the development of probes to monitor very low oxygen tensions. The design of oxygen electrodes for microbial cultures has been reviewed by

Harrison (1976a). Membrane electrodes respond to the equivalent partial pressure or activity of oxygen rather than to concentration:

$$T_1 = fc \tag{1}$$

where T_1 is the oxygen tension or activity in the liquid (Pa), c is the oxygen concentration in the liquid (g l^{-1}) and f is an activity coefficient (Pa^{-1} l). In fact, dissolved oxygen activity or tension (DOT) is a more meaningful parameter than concentration in relation to microbial response; the rate of transfer of oxygen to the cell is a direct function of activity rather than concentration.

14.3.3 Generalized Response to DOT

Early studies on the response of microorganisms to dissolved oxygen tension concurred that there was a general response by all microorganisms: above a certain low value, termed the critical DOT, changes in oxygen tension had no effect on the respiration rate of the microorganisms, but decreasing DOT below the critical value resulted in a decrease in respiration rate. The more reliable measurements of critical DOT indicate that it is very low, less than 5% of the air saturated value, but there is considerable variation in observed critical DOT values (Harrison, 1972a). This generalized model of a critical DOT above which respiration rate is insensitive to changes is still a useful one and applicable to most microorganisms and cell tissue culture lines. However, the response demonstrated by microorganisms, especially actively growing cultures, may be much more complex than the simple model would predict.

Variations in the critical DOT are found even in the same species, but these can perhaps be rationalized in the light of studies on respiration rate. Studies on the inhibition of respiration rate of *Beneckea natriegens* by Linton *et al.* (1976) demonstrated that the terminal cytochrome oxidase was rate limiting for respiration rate only when the organisms were respiring at their maximum rate. Respiration rates below this maximum could be obtained by limiting the substrate supply, lowering the temperature of the culture, or by harvesting the cells late in the growth cycle. Plotting the respiration (q_{O_2}) against DOT would give a family of curves as shown in Figure 1. From this it can be seen that the critical DOT will depend upon the rate limiting factor and the maximum respiration rate, the highest value for the critical DOT being obtained when the cytochrome oxidase is rate limiting.

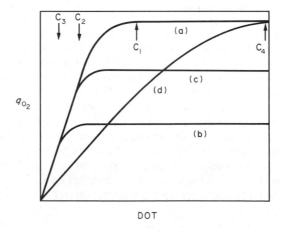

Figure 1 Theoretical response of respiration rate (q_{O_2}) to changes in dissolved oxygen tension (DOT). Curve (a): maximum q_{O_2} is limited only by the terminal oxidase. Curves (b) and (c): maximum q_{O_2} is limited by reductant supply, *e.g.* by substrate levels. Curve (d): a diffusion gradient exists between bulk liquid and cell membrane. C_1, C_2, C_3 and C_4 indicate the respective critical DOT values for curves (a), (b), (c) and (d)

The affinity of cytochrome oxidase for oxygen is very great. Thus, the critical DOT can be expected to be below about 300 Pa. However, many values quoted in the literature are well above this. There are two main causes for the overestimation of the critical DOT. Dissolved oxygen probes often overestimate DOT particularly at low concentrations (Harrison, 1976a) or respiration rate may be diffusion limited.

14.3.4   Diffusion Limitation

If there is a diffusion gradient between the bulk liquid or the membrane surface of the oxygen probe and the respiratory membranes of the organism, then the apparent critical DOT will be higher than the actual value (Figure 1). Such gradients can occur due to a number of circumstances; in poorly stirred solutions where liquid has laminar rather than turbulent flow then the gradient could become significant especially in cultures with very high respiration rates. For small discrete microorganisms such as bacteria or yeast under turbulent flow conditions, the diffusion gradient in the culture fluid can be assumed to be negligible. Where microbial cells form clumps, flocs or conglomerates, then a diffusion gradient will be set up between the outside of the clump and the inside so that the exterior DOT must be raised to a much higher level than the true critical value in order to saturate the respiratory system of the cells on the inside of the aggregate. Pellet type growth in fungi will generate this effect as will be formation of loose mycelial mass in liquid culture. In solid phase cultures respiration rates will readily become oxygen limited through diffusion gradients unless the cells are very loosely packed. For animal and plant cell cultures, it is generally not possible to create turbulent conditions as the shear forces are likely to disrupt the cells. Thus, in this case the DOT must be kept well above the true critical value to compensate for the diffusion gradient that will exist between the bulk of the liquid and the stagnant layer near the cell wall. Diffusion layers may also be set up by the accumulation of polysaccharide or slime layers on the outside of the cell.

14.3.5   Responses of Growing Microorganisms

14.3.5.1   Respiration rate

The generalized model for response in respiration rate described above is that normally obtained for cell suspensions in a non-growing state. However, more complicated responses have been obtained in cultures grown in chemostats where the organism has a longer time to adapt to the change in DOT imposed upon it. Cells which are growing or at least able to turn over protein can alter their enzyme complement with time in response to environmental changes. Changes in DOT often induce enzyme synthesis. Where the enzyme induced is rate limiting for respiration rate, such as a dehydrogenase or a cytochrome, then the net effect will be an increased respiration rate in response to the change in DOT. For example, in nitrogen fixing *Azotobacter* cultures the respiration rate of growing cultures increases almost linearly with DOT from zero to air saturation (Dalton and Postgate, 1968; Nagai *et al.*, 1971). This appears to be a protective response which prevents oxygen damage to the nitrogenase system. In such a case there will be no clear critical DOT value in growing cells.

In chemostat culture of *Klebsiella aerogenes* the respiration rates actually increased as the DOT was lowered below the critical value of about 600 Pa (Harrison, 1976b). Similar responses have been reported for other organisms, for instance the obligate aerobe *Haemophilus parainfluenzae* grown in batch culture, *E. coli* grown in chemostat culture and *Torula utilis* grown in chemostat culture under conditions of glucose limitation (see Harrison, 1972a).

The above discussion applies to *in situ* respiration rates as measured during growth and represents a respiration rate limited by a growth factor, be it a nutrient or oxygen. The term 'potential' respiration rate has been coined to express the respiration rate measured with cell suspensions removed from the restriction imposed during growth and supplied with excess oxygen and substrate. Changes in potential respiration rate will reflect changes in the enzyme complement or activity of the microorganisms. In glucose-limited chemostat cultures of *K. aerogenes* and *E. coli* both the *in situ* and the potential respiration rates were higher under oxygen-limited than oxygen-excess conditions (Table 2). In the case of *K. aerogenes* the potential respiration rate was highest in cells grown anaerobically. However, in cultures of the halophilic Gram-negative bacterium *B. natriegens* the *in situ* respiration rate fell as the DOT was reduced below the critical level (500 Pa) while the potential respiration rate increased. When the DOT was increased to more than 12×10^3 Pa, the critical potential respiration rate fell even though there was no change in the *in situ* rate (Table 2). This observation indicates that even though the organism may not demonstrate a direct metabolic response during the change in oxygen tension above the critical level, there must be changes in the enzymes constituents of the cell over this range.

Table 2 Response of *in situ* (glucose-limited) and Potential Respiration Rates (q_{O_2}) to DOT in *Escherichia coli*, *Klebsiella aerogenes* and *Beneckea natriegens*[a]

DOT range (Pa $\times 10^3$)	q_{O_2} (mmol g^{-1} h^{-1})[b] in situ	Potential
E. coli		
0.2–23	5.0	5.8
0.0–0.2	8.8	16.0
0.0	—	9.0
K. aerogenes		
0.2–23	5.6	9.1
0.0–0.2	7.2	12.7
0.0	—	22.4
B. natriegens		
12–20	8.0	10.0–11.0
0.5–12	8.5	14.0–15.5
0.0–0.5	5.0–8.5	16.0–21.0

[a] Harrison and Loveless , 1971a; Linton *et al.*, 1975.
[b] Where there is a change in q_{O_2} over the range of DOT the total range is shown.

14.3.5.2   Changes in cell constituents

Very many enzyme systems are either induced or repressed by oxygen. Generally, it has been found that the levels of cytochrome oxidase are increased as the dissolved oxygen tension falls below the critical value (Harrison, 1972a). This phenomenon has been shown to apply, in different studies, to cytochromes a_1, a_3, d and o, and to occur in obligate aerobes as well as in facultative organisms and in both yeast and bacteria (Harrison, 1976b). This is presumably an adaptive response: possibly a larger content of cytochromes at oxygen tensions below the K_m for the cytochrome oxidase enables organisms to prevent cytochromes from becoming limited for respiration. Generally, fermentative pathways are repressed in the presence of oxygen and, conversely, aerobic pathways may be repressed in the absence of oxygen. For example, in the absence of oxygen, *Klebsiella pneumoniae* dissimilates glycerol by the sequential action of NAD-dependent glycerol dehydrogenase and ATP-dependent dihydroxyacetone kinase to give DHAP which enters the glycolytic pathway. In parallel, glycerol is converted to 3-hydroxypropionaldehyde (3-HP) by coenzyme B_{12}-dependent glycerol dehydratase and the essential NAD is regenerated from the action of NADH-dependent 3-HP reductase. The product, 1,3-propanediol, is excreted into the medium. None of these four anaerobic enzymes is present during aerobic growth and catabolite repression has been implicated as a controlling factor (Forage and Lin, 1982). In the presence of oxygen, *K. pneumoniae* dissimilates glycerol by the sequential action of ATP-dependent glycerol kinase and FAD-linked glycerol 3-phosphate dehydrogenase again yielding DHAP. The enzymes of this aerobic pathway are repressed anaerobically but the mechanism for this is not known (Ruch *et al.*, 1974).

Molecular biology will provide some insights into the control of enzyme production by oxygen. Exposure of *K. pneumoniae* to oxygen causes repression of the nitrogen fixation (*nif*) genes and inactivation of nitrogenase. The *nif* L gene product is now known to be the repressor protein activated by this switch in metabolism but whether oxygen *per se* or some metabolite produced aerobically activates the protein is not known (Hill *et al.*, 1981; Buchanan-Wollaston *et al.*, 1981). When the admission of oxygen to an anaerobic culture renders the existing metabolic pathways inefficient, redundant or potentially toxic, key enzymes in those pathways may be specifically inactivated. For example, the citrate lyase of *Rhodopseudomonas gelatinosa* is inactivated by citrate lyase deacetylase and reactivation can occur through a separate enzyme catalysing the reverse reaction (Giffhorn and Gottshalk, 1975). Enzymes involved in the use of alternative electron acceptors to oxygen, such as respiratory nitrate reductase, are strongly repressed by oxygen so that nitrate does not act as an electron acceptor in the presence of oxygen. In facultative anaerobic yeasts, functional mitochondria are not formed under anaerobic conditions.

Non-respiratory oxygen-requiring enzymes such as oxygenases often require the presence of oxygen for induction. However, considerable variation in response to DOT has been found. Decane monooxygenase in a *Pseudomonas* sp. was induced to the greatest extent at low DOT values (Harrison *et al.*, 1969) while glucose oxidase production by *Aspergillus niger* increased with increasing DOT (Zetelaki and Vas, 1968). Cholesterol oxidase induction in *Nocardia rhodochrous* reached a maximum at DOT values equivalent to 35–45% air saturation and fell off both at lower and higher DOT values (Buckland *et al.*, 1976).

Other cell constituents besides enzymes change with dissolved oxygen, for example the amino acid pool of yeast has been shown to vary with dissolved oxygen tension in continuous cultures (Dawson, 1965). The fatty acid composition of yeast alters both qualitatively and quantitatively with DOT, there being a tendency for an increase in the unsaturated fatty acid component as oxygen is increased (Babij *et al.*, 1969).

14.3.5.3 Changes in metabolic products

As a broad generalization it can be stated that reduced or fermentative metabolic products are favoured by DOT levels below the critical value and their formation is repressed at DOT values above the critical value. This is not surprising as the role of such products is as alternative electron sinks to molecular oxygen. Thus production of ethanol, acetate, formate, lactate and butanediol are associated with anaerobic growth. Oxidized products of metabolism are generally generated at DOT values above the critical value. Thus citrate and pyruvate are associated with aerobic growth. Among other products favoured by aerobiosis are most commercially produced antibiotics, amino acids and enzymes. Products favoured by intermediate DOT values between zero and the critical value appear to be rare and of the cases reported many could be attributed to a secondary effect of growth inhibition at low DOT. There is at least one instance of an extracellular metabolite favoured by very high DOT values: *K. aerogenes* formed a bright yellow flavine maximally at DOT values more than twice that of air saturation (Harrison *et al.*, 1969).

A good deal of work has been carried out comparing product formation in aerated and unaerated batch cultures but very little of this has concerned the quantitative changes in metabolic products which occur with facultative organisms in response to dissolved oxygen tensions. *Klebsiella aerogenes* produces a wide variety of products at different stages of reduction. Harrison and Pirt (1967) followed the fermentation products in a chemostat culture of *K. aerogenes* in which DOT was monitored. In these studies it was not possible to correlate most of the changes in metabolic pattern with specific dissolved oxygen tensions as most changes took place at DOT values below the sensitivity of membrane probes. However, it was found that the response was certainly not an all-or-nothing effect in that different products responded to different extents to O_2 limitation. The response to DOT was affected by other environmental factors, especially pH. Restriction of oxygen supply can also cause a build-up of partially oxidized products. For instance, methanol has been found to accumulate under conditions of oxygen-limited growth in methane-utilizing cultures.

There are no simple rules for predicting microbial response to DOT in terms of metabolic products. Often restriction of oxygen supply will lead to the accumulation of intermediates of oxidative metabolism. However, reduced products of metabolism can also accumulate under aerobic conditions when they are subject to other regulatory influences. Thus ethanol is produced by yeast under aerobic conditions with glucose repression at rates equal to or greater than that obtained under anaerobic conditions (Moss *et al.*, 1971).

14.3.5.4 Transient responses to changes in DOT

Clearly, there are profound differences in the enzyme content and respiratory mechanisms of cells exposed to different dissolved oxygen regimes, and there must be a complex of transient states as microorganisms adapt to changes in DOT. Unfortunately, little attention has been paid to these transient states and consequently information is scarce. Studies on the response of *K. aerogenes* to a change from anaerobic to aerobic conditions in chemostat culture revealed three stages (Harrison and Loveless, 1971b): (i) a short period of increased acetate production and low carbon dioxide production immediately following reaeration of an aerobic culture; (ii) a subsequent period of high respiration rate with low yield coefficients for glucose and oxygen; and (iii) after about 8 h, a steady aerobic state is reached.

The high respiration rate in the second stage (lasting between 0.5 and 2.5 h after reaeration) represents a degree of uncoupling between growth and energy conservation. The net effect of this is a reduced yield of cells from glucose compared with aerobic growth during the transition. If a culture is subjected to repeated anaerobic/aerobic cycles the overall effect would be a growth efficiency much lower than that predicted by averaging the efficiency of the anaerobic and aerobic periods.

Such transient effects have important implications for industrial processes. Very large fermentation vessels cannot be ideally homogenous. Thus, cells circulating in a vessel where the *mean* DOT value is just above the critical value will be subjected to alternating low and high DOT values. The effect of this on metabolism may be difficult to predict from steady-state studies alone.

14.3.6   Control of DOT

DOT is one of the most difficult parameters to control in fermentations. Because of the low solubility of oxygen, its uptake by microorganisms is often rapid compared with the dissolution rate so that the DOT changes rapidly with even small changes in the respiration rate or oxygen transfer coefficient ($K_L a$ value). Also the response times of membrane electrodes are often slow compared with the rate of change of DOT in a highly active microbial culture. Therefore, sophisticated electronic control devices are essential for maintaining a constant DOT value. Fortunately, most commercial fermentations are favoured either by anaerobic or fully aerobic conditions. For the former it suffices to maintain the DOT at zero and for the latter at a value above the critical, *i.e.* between 1×10^2 and 1×10^5 Pa. Power input to aerate cultures is one of the most significant cost factors in commercial fermentation processes, so that ideally for aerobic processes the DOT value should be kept as close as possible to the critical value without dropping below it. However, the critical value is usually difficult to define accurately and, unless the culture is very well mixed, there is danger of anaerobic products forming in large fermenters so that a wide safety margin is generally given.

DOT can be simply monitored using membrane electrodes. Although these are not ideally sensitive at low DOT values, this is not usually a handicap in aerobic fermentations where the DOT is maintained above 100 Pa. Changes in DOT in response to feedback from the probe can be brought about by changing the air flow rate through the fermentation or by increasing the impeller speed. Often a combination of both is used. There are limitations to both as other parameters such as CO_2 concentration and shear force may be altered significantly.

In commercial fermentations, the power input for aeration and agitation is often fixed at a maximum throughout the process. In this case if a fermentation is in danger of becoming O_2 limited it is often necessary to slow down a nutrient feed to the culture to decrease O_2 demand and maintain aerobic conditions.

More sophisticated DOT control than is presently normally employed is certainly a prospect for the future. As this becomes available no doubt many more cases will be found of fermentations which are favoured by DOT values balanced accurately at very low values below the critical.

14.4   REDOX POTENTIAL

Redox potential can be defined in a simple system consisting of an oxidizable and reducible compound by the relationship:

$$E_h = E_0 + \frac{RT}{nF} \, ln \left[\frac{\text{(oxidant)}}{\text{(reductant)}} \right] \tag{2}$$

where E_h is the potential referred to the normal hydrogen electrode, E_0 is the standard potential of the system at 30 °C when the activities of all reactants are unity, R is the gas constant, T is the absolute temperature, F is the Faraday constant and n is the number of electrons transferred in the reaction. The E_h of a system will be dependent on the pH of the solution and thus the concept only has meaning in reversible systems where equilibrium is attainable. Individual redox couples in a complex system thus may have a well defined potential but for a conglomerate mixture of very different systems, some virtually irreversible and not in equilibrium with each other, the concept of redox potential cannot be strictly applied. Such a mixture would aptly describe most microbial cultures.

A platinum probe will respond to any system which gives an electrochemically reversible reaction at the electrode surface provided the rate of reaction is rapid compared with the electron drain or supply from the measuring electrode. For a rapidly reacting system such as Fe^{2+}/Fe^{3+}, measurements can be obtained with a platinum electrode at ion concentrations as low as 10^{-5} M. It is important to note that the electrode potential is independent of the capacity of the system.

Although oxygen does not influence a platinum probe in pure water, in solutions other reactions may occur which react at the probe surface and render it sensitive to dissolved oxygen levels. Moreover, in the presence of oxygen, all redox systems tend towards the fully oxidized state at a rate which will be dependent on the oxygen tension. Therefore, in complex systems a redox probe mostly reflects the dissolved oxygen tension. From equation (2), if oxygen is regarded as the oxidant, the output of a redox probe will be an exponential function of the dissolved oxygen tension; there is approximately a 60 mV decrease in redox potential per 90% decrease in DOT. At a low dissolved oxygen tension, very small changes in dissolved oxygen will be accompanied by relatively large changes in redox potential.

Measurements of E_h in growing microbial cultures cannot express a quantitation of reducing potential, because the system is not in equilibrium. Redox probes can be used to obtain empirical relationships between oxygen and metabolic changes at levels of dissolved oxygen below the sensitivity of membrane probes but at these low dissolved oxygen levels soluble metabolic products are formed which will certainly alter the relationship between DOT and E_h. Under anaerobic conditions a redox probe may be used as an empirical monitor of the reducing capacity, but, unless the E_h of the culture is carefully poised, inaccuracies are inevitable. If the main substrates or products of a microbial culture have a high activity at the electrode surface they may swamp all interfering reactions so that the redox potential as measured by the probe is a true representation of the state of that particular redox couple. In iron-reducing bacteria a platinum probe may be used to follow and control the balance between Fe^{3+} and Fe^{2+} ions. Similarly, in sulfate-reducing cultures a platinum probe may serve to monitor the sulfide concentration.

To avoid any ambiguities arising from readings produced by a platinum redox probe, it would seem preferable, wherever possible, to measure specific redox couples whose interaction with the cell is known. For example, in the case of cultures donating electrons to nitrate or nitrite ions the measurement of these compounds in the culture would best define the environmental changes which impinge on the organisms. Alternatively, the redox potential of intracellular components may be measured. Possibly, this would be the most revealing monitor of all because it is likely that through such components the cell response to external environment is instigated.

The NAD/NADH ratio represents a key redox potential for driving cellular oxidation/reduction reactions. A method for monitoring NADH in continuous culture has been developed (Harrison and Chance, 1970).

14.5 RESPONSES TO CARBON DIOXIDE

14.5.1 Requirement for Carbon Dioxide

There is an absolute requirement for CO_2 for growth of heterotrophic microorganisms on simple salts media with glucose or other sugars as the sole carbon source. CO_2 is required to replenish C_4 compounds drained from the TCA cycle during growth. In this case CO_2 is fixed *via* anaplerotic pathways. The affinity for CO_2 by such pathways is usually very high and the requirement is usually easily satisfied by CO_2 generated during respiration. However, at the initiation of growth, or if the culture is ventilated too vigorously, CO_2 limitation can occur. This is commonly manifested by a long lag phase after inoculation which can be avoided either by supplying CO_2 or TCA cycle metabolic intermediates such as citric acid.

14.5.2 Inhibition by Carbon Dioxide

There is little information on inhibition by CO_2 *per se*. Certainly most organisms appear to be unaffected by CO_2 partial pressures up to 0.2 atm equivalence (2×10^4 Pa). High CO_2 partial pressures in excess of 1 atm (10^5 Pa), may be reached in tall air-lift fermenters. It seems that cultures must be able to adapt to such high CO_2 partial pressures. CO_2 at high partial pressures will, however, exert an effect through depression of pH or chemical reaction with other compounds in the culture fluid. Carbonates may be formed which precipitate out minerals essential for growth.

14.6 WATER ACTIVITY

14.6.1 Introduction

The response of microorganisms to osmotic stress is most conveniently expressed in terms of water activity (a_w) which is a measure of the amount of free water in the system and is defined as the ratio of vapour pressure of water in solution (p_s^*) to that of pure water (p_w^*). Water activity may be altered through matrix effects caused by adsorption of water molecules at the interfaces of biphasic systems (as in solid state and polysaccharide synthesizing fermentations), or through osmotic effects brought about by solute–solvent interactions. Water activity is much simpler to measure than osmotic pressure and is related to solute concentration by Raoult's law:

$$a_w = n_w/(n_w + n_s) \tag{3}$$

where n_w and n_s are the molar concentrations of solvent and solute respectively. In low solute strength solutions, such as general culture media, a_w approximates to 1.0 and tonicity (osmolality) rather than water activity is the more correct measurement. Osmophiles are organisms which grow readily in environments with a high osmotic pressure (*i.e.* low water activity), in contrast to osmotolerant species which survive in such environments which are suboptimal for growth. Halophiles are special examples of osmophilic organisms and are discussed in detail below.

14.6.2 Water Relations in Bacteria

Gram-positive bacteria usually have higher internal osmotic pressures than Gram-negative organisms. For example, *Escherichia coli* and *Staphylococcus aureus* have osmotic pressures of 4–8×10^5 Pa and 20–25×10^5 Pa respectively (Rose, 1976). The difference between these groups of bacteria can be quite large and may in part reflect the greater mechanical rigidity of the Gram-positive cell wall structure.

Individual bacterial types are generally able to grow only over a very narrow range of water activity although bacterial growth is found at a_w values of 0.83–0.99. The minimum water activity for growth of Gram-negative bacteria is generally higher than that for Gram-positive bacteria reflecting the lower intracellular osmotic pressures. For example, *Pseudomonas fluorescens* and *Salmonella typhimurium* have minimum a_w values of 0.92 and 0.97 respectively whereas *Micrococcus* spp. *and S. aureus* have minimum a_w values of 0.83 and 0.86 respectively (Rose, 1976). These latter values are equivalent to, or in many cases less than, the minimum activities tolerated by many yeasts and a number of fungi.

Cold osmotic shock treatment is used to release periplasmic enzymes from Gram-negative bacteria. Although Gram-positive bacteria do not have a periplasmic space *per se*, similar treatment is useful in dissociating cell-bound forms of exoenzymes.

14.6.3 Water Relations in Yeasts and Fungi

Many yeasts tolerate minimum water activities similar to those of Gram-positive bacteria. The filamentous fungi, however, range from osmosensitive to the extremely osmophilic species important in food spoilage. For instance, *Botrytis cinerea* and *Rhizopus stolonifer* require relatively high water activities whereas *Xeromyces bisporus* isolated from prunes has a minimum a_w value of 0.61 (Pitt and Christian, 1968). Extreme osmophiles, such as the yeast *Saccharomyces rouxii*, are important as contaminants of sugar syrups, and are able to survive in low water activity conditions because of their ability to synthesize large quantities of polyols. These, being hydrophilic, increase the intracellular concentration of bound water and so lower the osmotic pressure difference across the plasmalemma. *Saccharomyces rouxii* can accumulate arabitol intracellularly up to a level equivalent to 18.5% (w/w) of the cell dry weight. Such compounds, termed compatible solutes, are also important in the halophilic organisms.

The internal osmotic pressure of apically growing fungal hyphae is an important factor in their extension process (see Trinci and Oliver, Volume 1, Chapter 11). In hyphal growth localized apical weakening of the structural wall polymers allows the plasmalemma to swell because of the high cytoplasmic osmotic pressures. Subsequently, rerigidification of the polymers occurs by further chitin synthesis and the net result is cell elongation. Interestingly, many fungi when removed from one solution and placed in another of higher osmotic pressure (but which does not cause hyphal bursting) stop apical growth, either transiently or permanently, and proliferate lateral branches. In others, adverse osmotic pressures may induce differentiation of the biomass into a plethora of sporing bodies.

Variations in growth kinetics may often be the first observed response to high solute concentrations. For instance, the maximum specific growth rate and yield from glucose for a glucose-limited chemostat culture of *Saccharomyces cerevisiae* decreased as the concentration of sodium chloride increased from 0.25 to 1.5 M (Watson, 1970).

14.6.4 Halotolerance and Halophily

This terminology relates to a group of organisms, predominantly bacteria and algae, which survive or grow optimally in the low water activity environment created specifically by NaCl. Microorganisms capable of surviving in concentrations of NaCl as high as 4.5 M but which grow optimally under low solute conditions are termed halotolerant. Such organisms include *Staphylococcus epidermidis*, *S. aureus* and salt tolerant yeasts and fungi. These organisms show a reduced growth yield from the available energy sources at high NaCl concentrations since a considerable energy expenditure is required to maintain the physiologically required intracellular solute concentrations. Halophilic organisms, in contrast, grow optimally at high salt concentrations. Thus moderate halophiles, including many marine microorganisms, grow optimally in 0.2–0.5 M saline solutions whereas extreme halophiles such as halobacteria and halococci (red halophiles) grow optimally in saline concentrations from 2.5 M to saturated solutions. The degree of halotolerance and halophily exhibited by a microorganism often depends on other environmental parameters of which temperature is the most critical. For example, *Vibrio costicola* grows in an NaCl concentration of 0.5–4.0 M at 30 °C, however, if the temperature is lowered to 20 °C growth can occur at solute concentrations as low as 0.2 M (Kushner, 1978).

Microbial adaptation to high salt environments has occurred by the evolution of physiological processes which function optimally at elevated internal solute concentrations. In the extreme halophile *Halobacterium halobium*, for example, the K^+ content may amount to 30–40% (w/w) of the cell dry weight (Gochnauer and Kushner, 1971). In the few halophiles studied, examples of enzymes requiring high solute levels for activity are known. However, in other organisms growing under halophilic conditions enzymes are not salt-insensitive and in these there is accumulation of compatible solutes in the cell to raise the intracelluar osmotic pressure. A specific illustration of this is seen in the halophilic green alga *Dunaliella viridis* which can grow over a wide range of NaCl concentrations up to saturation. The continued function of its salt-sensitive enzymes is achieved by the accumulation of glycerol at varying concentrations depending on the external salt concentrations. The glycerol probably protects enzymes from inhibition by high salt concentrations by providing functional hydroxyl groups (Brown, 1976).

14.7 EFFECTS OF pH

14.7.1 Introduction

The pH may be defined either in terms of hydrogen ion concentration (H^+) or activity a_{h^+} which are related:

$$pH = -\log a_{h^+} = -\log f(H^+) \tag{4}$$

For dilute solutions, which include most culture media, $f = 1$ so that hydrogen ion activity and concentration can be taken as interchangeable for all practical purposes. It is worth noting, however, that when considering intracellular pH or pH at membrane surfaces, this assumption is unlikely to be valid.

The pH value of a solution, of course, profoundly affects the chemistry of that solution in almost every aspect; it also directly affects the redox potential of a solution, therefore the intracellular pH will influence any reactions taking place outside the cell, the ionic state of polar molecules and perhaps the structure of complex molecules.

Enzyme kinetics are altered by changes in pH both in terms of K_m and V_{max} values, therefore any enzyme reaction in contact with the external medium, whether truly extracellular or in the periplasm, will be influenced by culture pH. Structures such as membranes in contact with the exterior are also subject to chemical changes in response to pH. Not surprisingly, microorganisms and animal and plant cells have evolved membranes that tend to exclude hydrogen ions so protecting the vital cell reactions from even small fluctuations in external pH.

14.7.2 Cellular-level Responses

14.7.2.1 Intracellular pH

As most microorganisms can tolerate a wide range of pH in the external environment, it is clear that the intracellular pH must be regulated to within narrow limits. In fact the intracellular pH values of all microorganisms, including the extreme acidophiles and alkalophiles, are similar and close to neutrality. The acidophile *Thermioplasma acidophila*, an organism with no cell wall, maintains an internal pH between 6.3 and 6.8 despite changes in extracellular pH from 2.0 to 6.0 (Hsung and Haug, 1975). This hydrogen ion gradient is apparently maintained by direct exclusion of H^+ because neither cell death nor uncoupling of energy metabolism led to a change in intracellular pH. However, other acidophiles depend on energy-driven protein exclusion to maintain internal pH at a constant value (Yamazaki *et al.*, 1973). At the other end of the scale, *Bacillus alcalophilus* maintains an internal pH close to neutral while growing at pH 10–11.5 by means of an energy dependent Na^+/H^+ antiporter which retrieves protons extruded during respiration (Mandel *et al.*, 1980).

14.7.2.2 Effects of pH on membrane function

According to the widely accepted Mitchel hypothesis of energy generation by proton motive force, a major function of the cell and mitochondrial membranes is to maintain a proton (pH) and charge gradient. For this purpose the cell membrane must have a very limited permeability to protons. At first consideration it would seem that as the intracellular pH is constant, then changes in the extracellular pH value must affect the proton translocation efficiency across the cell membrane and thus the generation of ATP. However, bacteria appear to be able to maintain their metabolic efficiency over a wide range of pH values. This is explained by the membrane potential value which adjusts to maintain the same proton motive force:

$$\Delta p = \Delta \psi - 60\Delta pH \tag{5}$$

where Δp = proton motive force; $\Delta \psi$ = membrane potential in mV; ΔpH = pH gradient across the membrane and temperature = 30 °C. Thus as the extracellular pH shifts towards acidity the membrane potential must fall for Δp to remain constant. However, for extracellular pH values above 9.0 and an internal pH around neutrality, the membrane potential required would be 350 mV. It is doubtful whether the cell membrane could retain its integrity at this charge intensity and the mechanism by which alkalophiles maintain such high pH gradients is not understood.

If the cell membrane becomes permeable to protons then the cell cannot generate ATP; respiration becomes uncoupled and the growth efficiency falls. Ultimately the cell will die if membrane function is disrupted as substrate transport fails and ion gradients cannot be maintained.

14.7.2.3 Effects of pH on uptake of substrate

(i) Passive diffusion

The rate of uptake of polar substrate molecules by passive diffusion will depend on the state of ionization of the molecule and therefore on the pH of the medium. The substrate molecule must enter in the undissociated state as the cell membrane has low permeability to charged compounds. Therefore, passive diffusion across membranes of acid substrates will be favoured by pH values above the pK of the substrate and the uptake of basic substrates by pH values below the pK.

(ii)   Active transport

For substrates taken into the cell by active transport driven by chemiosmosis, the favoured ionic state depends on the transport mechanism, which varies for different substrates. For example, in *Staphylococcus aureus* lysine is transported as a cation by a uniporter driven by the membrane potential, while glutamate is transported as an anion by a symporter driven by the pH gradient. The neutral amino acid, isoleucine, is taken up by a symporter driven by the total proton motive force (Hamilton, 1977). Therefore, the external pH will influence the availability of such substrates at low concentrations. Commonly, more than one transport system is available for uptake of each substrate and different transport systems may be important at different external pH values. Data on the kinetic effects of external pH on substrate uptake in actively growing cells are very sparse.

(iii)   Extracellular enzyme-dependent uptake

Intracellular enzymes are, of course, protected from environmental pH changes but all reactions dependent on extracellular enzymes will be sensitive to cultural pH value. Where substrate utilization is dependent on extracellular enzyme activity, either free in solution or in the periplasmic space, then external pH must exert a direct influence on substrate utilization. In the extreme case, the pH optima and range of extracellular enzymes will be crucial to an organism's ability to grow at a particular pH. For instance, the secretion of alkaline proteases is necessary for growth of organisms on proteins at high pH and the possession of acid cellulases may extend the pH range of cellulolytic fungi.

14.7.2.4   Effects of pH on products of metabolism

That the extracellular pH has a strong influence on the pathways of metabolism and products generated by microorganisms is well known and exploited in the fermentation industry. The mechanism by which metabolism is regulated is, however, not well understood. Often the response to pH in terms of metabolic product seems to be a logical adaptation by the microorganisms. For instance, *Klebsiella aerogenes* produces neutral products such as ethanol and 2,3-butanediol when growing at acidic pH but a preponderance of organic acids at pH values near or above neutrality (Table 3). However, acid-tolerant organisms will continue to produce acid products even at low pH values thereby, perhaps, diminishing competition from the non-acid tolerant organisms. *Aspergillus niger* produces citric acid at pH values below 3.0 while *Thiobacillus ferrooxidans* will generate sulfuric acid to pH values below 2.0.

Table 3   Changes in Fermentation Products Formed by *Klebsiella aerogenes* with Changes in Culture pH[a]

pH	Fermentation products[b]
5.0	Butanediol
6.0	Butanediol, ethanol
7.0	Butanediol, ethanol, lactate
8.0	Acetate, formate, lactate, butanediol, ethanol

[a] Harrison and Pirt, 1967.   [b] In order of quantity produced.

14.7.2.5   Effects of pH on cell morphology and structure

Dramatic changes in cell size and shape have been observed in response to pH. *Penicillium chrysogenum* altered its morphology from long hyphae to short vacuolated cells on raising the pH from 6.0 to 7.0 (Pirt and Callow, 1959). Whether such changes are a direct response to pH or to other chemical changes induced by pH is unclear. However, it can be expected that those parts of the cell in contact with the external liquid respond or adapt to changes of pH. There have been few studies of such responses but the composition of the cell wall of phosphate-limited *Bacillus* was found to change with pH in that teichuronic acid was replaced by teichoic as the pH of the culture liquid was altered from alkaline to acid (Ellwood and Tempest, 1972). Cell membrane lipids of yeast also vary with pH. No doubt if more studies were made it would be found that adaptations of cell wall and cell membrane components to pH changes are common especially among organisms with a wide range of pH tolerance.

14.7.2.6 Effects of pH on the chemical environment

The pH will exert indirect effects upon cell metabolism through changes in the chemical environment. Substrate availability may be affected by changes in the solubility of the substrate and, in the case of polar compounds, the degree of dissociation. Product concentration and state will be similarly affected. A product extruded by simple diffusion may be 'trapped' outside the cell by precipitation or dissociation allowing a higher extracellular concentration to be accumulated despite relatively low intracellular concentrations.

14.7.2.7 Effects of pH on flocculation and adhesion

The clumping or flocculation of microbial cells is an important phenomenon in industrial processes. Should flocculation occur during an active phase of fermentation or growth then the production of surface area and creation of diffusion gradients may impair the process. Harvesting or separation of cells by centrifugation, however, may be facilitated by flocculation. The activated sludge waste treatment process depends on the formation of compact, quick settling flocs of microorganisms for the separation of the returned sludge from effluent. In general, flocculation is strongly influenced by the pH of the suspending medium.

The charge on the cell surface is a function of the pH of the suspending medium and bacterial cultures have been shown to have a distinct 'pK_a' value at which the overall cell charge is neutral (Wilkinson and Hamer, 1974). Microorganisms in liquid suspension are acted upon by two opposing forces: van der Waals' attraction forces tend to bring the cells together and like charges on the cells (usually negative) repel and keep the cells apart. When the pH is equal to the pK_a value the cells can make contact and are held together by van der Waals' forces so the culture may flocculate. Flocculation can, in some instances, be reversed by readjustment of pH but more often the cells, once flocculated, become difficult to separate.

Adhesion of microorganisms to inert surfaces depends on a similar mechanism to flocculation. The organisms must first make contact with the surface and can only do this if there is no surface charge or if the charge on the microorganism and surface is opposite. Again, pH influences the surface charge on the microorganism and thus the tendency to adhere to inert surfaces. Wall growth on fermenters has been observed to be dependent on pH (Wilkinson and Hamer, 1974).

14.7.3 Optimum pH Values for Growth

The expression 'optimum pH' requires qualification as the optimum pH for growth-rate may be different from that for growth yield and entirely different from the optimum for product formation. In general, microorganisms demonstrate a broad optimum for growth rate and yield compared with the 'peak' value for temperature optima. Changes of pH of 0.5–1.0 units around the optimum for growth usually has little effect on growth rate or efficiency but may affect product formation.

Although there are organisms that can tolerate extremes of acid and others that tolerate extremes of alkalinity, there are none that can tolerate both. In general, microorganisms will tolerate and grow over pH ranges of 2–3 units and species that grow at pH values below neutrality are the more numerous. Typical pH response curves arc shown in Figure 2. Usually the slope of decrease in specific growth rate is more gradual on the acid than the alkaline side. A general statement is often made that fungi favour acid pH and bacteria more neutral conditions. While this is a useful guide and is often exploited in designing selective media, it must be borne in mind that many important genera of bacteria such as *Thiobacillus, Acetobacter* and *Lactobacillus* are acidophilic while pathogenic fungi thrive under neutral or alkaline pH conditions. The implication of the broad plateau shown in the pH response curves (Figure 2) is that, for most purposes, pH does not have to be poised within narrow limits. Many successful industrial fermentations have functioned very efficiently using the buffering capacity of the media for pH control rather than more precise electronic feed-back methods. However, although the optimum pH range for growth may be broad, sudden changes within this range may cause temporary pertubations in metabolism. Microorganisms may need to adapt their function to cope with a change in hydrogen ion concentration and if this change is too rapid the response may lag behind or over-shoot.

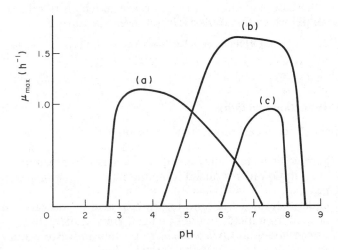

Figure 2 Typical responses of maximum growth rate (μ_{max}) to pH for (a) *Aspergillus niger*, (b) *Klebsiella aerogenes*, (c) *Methylococcus capsulatus*

The varied pH ranges of microorganisms have been exploited in many complex traditional fermentations where a succession of species dominance is required to complete the fermentation, for example the ancient process for the fermentation of soy-sauce is highly complicated and precise in its microbial ecology; the process depends on the salt tolerant *Pediococcus soyae* carrying out a homofermentative process on a koji and brine mixture, thus depressing the pH sufficiently to allow yeast to grow.

14.7.4 Causes of pH Changes in Cultures

Microbial growth and metabolism inevitably lead to a change in the hydrogen ion balance of the culture medium which, depending on the buffering capacity, will lead to a change in pH. In considering fermentation control strategy it is as well to consider the mechanism of pH changes.

14.7.4.1 Product formation

Most commonly, in simple salts media, microbial growth leads to a fall in pH of the medium. Complete aerobic metabolism of carbohydrates in simple media leads to the extrusion of protons and CO_2 which tends to depress culture pH while the products of carbohydrate fermentation are usually neutral or acid. Acid products can generate pH values below 2.5 in the industrial processes for citric, acetic and gluconic acids. Alkaline products of fermentation are much less common although some basic amino acids such as lysine and arginine are produced commercially. However, a rise in pH is often caused by the release of ammonia from deamination reactions during growth on complex, proteinaceous substrates.

14.7.4.2 Nutrient uptake

If the main carbon or energy source supplied to the culture is taken up as an anion not in association with a proton, then the pH of the medium must decrease. However, the uptake and utilization of organic acids as nutrients is not always accompanied by a fall in pH. When added to a glucose-limited continuous culture of *K. aerogenes*, acetate and succinate cause the pH to fall while citrate causes a rise in pH. When ammonium salts are supplied to a culture as sole nitrogen source, then uptake of ammonia leads to a fall in pH. Utilization of nitrate ion as a nitrogen source, on the other hand, causes a rise in pH, so that if both NH_4^+ and NO_3^- are used simultaneously the hydrogen concentration can be kept in balance.

Use of complex carbon and nitrogen sources, such as proteins and polypeptides, will generally lead to a rise in pH as excess nitrogen is released as ammonia. Urea is often employed as a cheap source of nitrogen. In this case, the effect of rapid urease activity is to release ammonia and CO_2 causing an initial rise in pH which then falls as the ammonia is utilized:

$$CO(NH_2)_2 + H_2O \rightarrow CO_2 + 2NH_3 \tag{6}$$

$$2NH_3 + H_2O \rightarrow 2NH_4^+ + 2H^+ \tag{7}$$

14.7.4.3 Oxidation/reduction reactions

The overall chemical effect of carbohydrate respiration will of course be the generation of carbonic acid and a decrease in pH. However, where the final electron acceptor is an inorganic molecule other than O_2, there will be a tendency for pH to increase. For example, where nitrate is the electron acceptor for respiration it is reduced to ammonia by *K. aerogenes* or even to N_2 gas by *Micrococcus denitrificans*.

Oxidation of inorganic molecules will tend to produce acid conditions. *Thiobacillus ferrooxidans* generates sulfuric acid from metal sulfides and can grow at pH values approaching 1.0. The nitrifying action of *Nitrosomonas* and *Nitrobacter*, important in waste water treatment, oxidizes ammonia through to nitrate thereby depressing the pH value.

14.7.4.4 Change in buffering capacity

Changes in pH can be brought on by a depletion of buffering capacity where the main buffering effect is by molecules that can act as substrate. The most commonly used buffer is phosphate which is generally supplied in great excess to that required for growth so that depletion does not occur. Organic acids such as citrate can be used as buffers and these may be used as both carbon and energy sources leading to rapid depletion of the buffering capacity. This effect of buffer depletion can be exploited where a rapid change in pH is required during the course of a fermentation.

14.7.5 Control of pH

14.7.5.1 By means of a buffer

This is the traditional method of moderating pH changes in cultures and is still most effective where: (i) there is a broad optimum range of pH for the fermentation; (ii) the fermentation is not strongly acidogenic or alkalogenic; (iii) the rate of fermentation is not rapid; (iv) the pH optimum of the fermentation is within the range of physiologically inert buffers.

The commonly used buffers and their ranges are summarized in Table 4.

Table 4 Buffers Commonly Used in Microbial Cultures

Buffer	Useful pH range	Comments
Phosphate (Na_2HPO_4/KH_2PO_4)	5.8–9.5	Only where phosphate can be in excess
Tris(tris(hydroxymethyl)aminomethane: HCl	7.2–9.1	Used for intracell systems; not necessarily inert to cell metabolism
Citric acid/HCl	1.0–5.0	Citrate chelates metals and may
Citric acid/NaOH	5.0–6.7	make them unavailable, also
Citric acid/Na_2HPO_4		citrate may be utilized as substrate.
Acetate/acetic acid	3.6–5.6	Only where acetate does not affect metabolism.

The strength of buffers that can be employed is limited by any inhibitory or toxic effects of the buffer and by osmotic tolerance of the organisms. Even buffers which are relatively inert physiologically, such as phosphate, may influence cell metabolism at very high concentrations. The methanol dehydrogenase of methane oxidizing bacteria, for example, can be inhibited by high phosphate concentrations (Tonge *et al.*, 1974). The excessive use of inert chemical buffers which are not converted to product is also to be avoided in industrial fermentations as an unnecessary extra cost. In addition, a high concentration of a chemical buffer is likely to add to the difficulty and cost of product recovery.

14.7.5.2 By balancing metabolism

We have already seen that pH changes in the culture medium can be caused by utilization of substrates and generation of products. If sufficient is known of the nature of these changes then it may be possible to design a balanced medium such that there is no net change in pH. For instance, in the culture of aerobic organisms on simple salts and glucose media the greatest part of acid production is caused by ammonia uptake and, to a lesser extent, by the oxidation of glucose to water and CO_2. By using a mixture of nitrate and ammonium salts in a nitrogen-limited culture (it is only under these circumstances that both nitrate and ammonia will be used) it would be possible to balance acid production by the rise in pH caused by nitrate utilization. In a more complex system, the pH can be monitored by an electrode and appropriate substrate fed in at a rate which maintains a constant pH. For example, growth of actinomycetes on complex media will tend to give a rise in pH as glucose becomes limiting and deamination of polypeptides and amino acids is induced. Glucose could be fed to the culture to keep the pH at a preset level. This may also have the effect of keeping the metabolic rate constant.

Control of pH by balancing the metabolism requires a high degree of knowledge of the process either through understanding or empirical experiment. The advantage should be that it avoids wasteful use of buffer and can be used to regulate metabolic rate. The disadvantage is the inflexibility of the system in that it would be difficult to alter pH in a prescribed way or to vary pH and metabolic rate independently.

14.7.5.3 By feedback control

Undoubtedly the most flexible means of regulating the pH of a culture is by the addition of acid or alkali in response to a measured deviation. Such control does not necessarily involve complex electronic circuitry and was employed long before the invention of sterilizable pH electrodes. In its crudest form the process was controlled at set (and not necessarily very frequent) intervals by estimating the pH of samples by titration and then adding a calculated amount of acid or, more frequently, alkali. Often an excess of titrant was added so the interval between additions could be lengthened. Adequate control could often be attained by this method due to the buffering capacity of the medium and the broad optimum in pH range of the process.

There are pH electrodes available now which can be sterilized by autoclaving and the calibration of which is stable for several hundred hours. Likewise, great advances have been made in pH control circuitry from the simplest on/off reaction to PID controllers and direct digital computer control. In general, the rate of pH change in cultures is slow compared with response times of electrodes and control circuits so that, for many purposes, simple controllers are adequate. However, modern computer-linked control provides infinite options for imposing accurate changes of pH through the course of a fermentation.

The chemicals used to adjust pH must be suited to the particular fermentation process. To compensate for acid production, NaOH is preferred to KOH in industrial fermentations because it costs far less. $Ca(OH)_2$ is even less expensive than NaOH and will be used where possible but $CaCO_3$ precipitation occurs at high pH values and, at lower pH values calcium salts of organic acids may precipitate causing inhomogeneity and increased viscosity of the culture. Under some circumstances the precipitation of an organic acid product as a calcium salt may be advantageous as it would prevent product inhibition. Sodium is generally tolerated by most microorganisms, particularly in the presence of potassium. However, sodium can antagonize potassium uptake when present in very large excess. Na^+ is taken up into the cell by diffusion and may replace K^+ which has to be pumped into the cell against a gradient at the expense of energy. A further energy drain is caused by the necessity to pump Na^+ out of the cell against a gradient. This energy drain can become severe if the Na^+ extracellular concentration is very high (1 mM) and K^+ supply is limiting. This can be prevented in fermentations with a strong alkali demand by adjusting pH with a mixture of KOH and NaOH.

For fermentations with a strong alkali demand, even if Na^+ antagonism of K^+ uptake is prevented, the osmolarity or sodium content may eventually inhibit the fermentation. This can be avoided by the use of ammonia as titrant by arranging for the rate of ammonia utilization as nitrogen source to approximately balance the demand as alkali. If the system is not in balance then nitrogen supply may become limiting for the process or ammonia may reach toxic concentrations. Where the system can be balanced, ammonia may be the preferred alkali for pH control on the basis of cost and because of its dual role.

The preferred acid for pH control is usually sulfuric. Hydrochloric acid can also be used but is generally avoided in industrial plants because of the severe corrosion problems that may be created. Neither sulfate nor chloride ions are inhibitory to most microorganisms at the concentrations normally attained. Acid demand of fermentations is generally small. Phosphoric acid may also be used but generally would be too costly for industrial processes. Organic acids can be employed but would only be economical where their use is demanded by particular circumstances. For instance, if it can be used as substrate in the fermentation, acetate might be employed to boost yield. Lactic acid may be used to lower the pH of a food or feed fermentation where it can also act as a preservative and added nutrient.

Despite its importance as an influence on microbial metabolism and growth, pH has been somewhat neglected in terms of control strategies. This can perhaps be attributed to the fact that very often quite crude control of pH within broad ranges was adequate for most experimental and industrial purposes. However, our increasing knowledge of microbial physiology and the availability of sophisticated computer-linked monitoring and control systems should lead to a more rapid advance in pH manipulation for industrial fermentations. In future we can expect changes, even very small ones, to be used as specific and quantifiable indications of culture metabolic activity and for control systems to respond to these in a very flexible manner. For instance, small pH changes may be used to monitor a nutrient availability and the supply of this can be precisely geared to demand. Knowledge of metabolism should enable the use of well balanced nutrient systems that avoid use of inert buffers and excess alkali and acids as titrants. Computer control enables any regime of pH changes to be imposed during the course of the fermentation and, as programmes and knowledge become more sophisticated, these changes themselves can be adapted during the course of fermentation. It should be a simple matter to control the pH of a fermentation initially at the optimum for growth and then to impose a pH change to induce enzymes for product formation, then perhaps change pH again to optimize formation and extrusion of the product, and yet again at the end of the process to flocculate the cells or precipitate the product. However, at present the sophistication of the control activity and software available far exceeds our knowledge of the organisms relation to pH changes.

14.8 TEMPERATURE

14.8.1 Cellular-level Responses

In contrast to pH, ionic composition and water activity, the internal temperature of a microorganism must be equal to that of its environment and, as for any chemical reaction, all microbial activity is sensitive to environmental temperature.

14.8.1.1 Temperature ranges for growth

Microorganisms are known which can grow above 90 °C or at less than 0 °C provided that liquid water and nutrients are available but no microorganism will grow at both ends of this range. The growth of most microbes is restricted to a 20–40 °C span and each species or strain has a characteristic minimum temperature (T_{min}), optimum temperature (T_{opt}) and maximum temperature (T_{max}). At T_{min} and T_{max}, the growth rate becomes zero and at T_{opt} growth occurs at its maximum rate. T_{opt} is usually only 5–10 °C below T_{max} (Figure 3).

The values T_{opt}, T_{min} and T_{max} are known as the cardinal temperatures and have been used to classify microorganisms into various categories. However, because individual authors set their own limitations on each of these categories and because it has become apparent that the ranges of T_{min} to T_{max} for microorganisms form an overlapping continuum, for the purposes of this chapter, only three categories will be recognized; these are: (a) psychrophiles: organisms which grow well at low temperatures (T_{opt} usually less than 20 °C) and may be able to grow below 0 °C provided liquid water is available; (b) mesophiles: organisms which grow well at moderate tem-

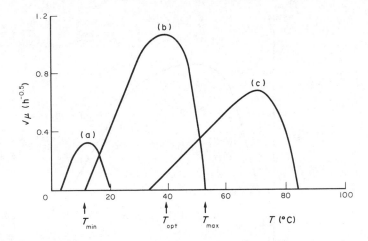

Figure 3 Temperature-dependent growth profiles for the psychrophile *Vibrio psychroerythrus* ATCC 27364 (a), the mesophile *Bacillus subtilis* (b) and the thermophile *Thermus aquaticus* (c). The original data from Mohr and Krawiec (1980) were used with permission and fitted with the μ relationship of Ratkowsky *et al.* (1983). The T_{min}, T_{opt} and T_{max} values for *B. subtilis* are marked

peratures (T_{opt}, 20–40 °C; these constitute the largest class of organisms studied to date); (c) thermophiles: organisms which grow well at high temperatures (T_{opt} may be greater than 60 °C).

14.8.1.2 Responses of growth rate to temperature

The Arrhenius equation describes the dependence of a chemical reaction upon temperature:

$$K = a_1 \exp(-E/RT) \tag{8}$$

where K is the reaction rate, R the gas constant, T the absolute temperature, a_1 a constant termed the frequency factor and E a constant termed the activation energy. A plot of $\ln K$ against $1/T$ should give a straight line with slope $-E/R$.

Attempts have been made to treat microbial growth as a simple chemical reaction by fitting growth data to the Arrhenius equation which becomes:

$$\mu = a_2 \exp(-E_g/RT) \tag{9}$$

where μ is the specific growth rate (h^{-1}), a_2 a constant and E_g a constant known as the temperature characteristic or the activation energy for growth. A plot of $\ln \mu$ against $1/T$ should give a straight line with slope $-E_g/R$ at least in the region from T_{min} to T_{opt}. In practice, Arrhenius plots give more than one straight line for some organisms in the region between T_{min} and T_{opt} (Figure 4), indicating a more complex response. These changes in slope most likely reflect changes in E_g which might be expected considering the complexity of growth and cellular composition. Nevertheless, values of E_g have been calculated for a number of organisms exhibiting simple Arrhenius profiles and they range from about 40 to 240 kJ mol^{-1} (Mohr and Krawiec, 1980).

More recently, Casolari (1981) has treated growth as a function of the collision frequency of a cell with water molecules having energy equal to or greater than E_u, the energy suitable for growth. Cell growth may thus be described by the equation:

$$X_0 = X_t^{1-ut} \tag{10}$$

where X_0 is the initial concentration of cells and X_t the concentration of cells after t hours growth. The temperature dependent parameter u is derived from the Maxwell distribution so that:

$$u = 18^{-1} N_0 \exp(-E_u/RT) \tag{11}$$

where N_0 is Avogadro's number. Knowing X_0, X_t, T and t, E_u can be calculated. E_u values of 125, 138 and 151 kJ mol^{-1} were recorded for the psychrophile *Vibrio marinus* MP1, the mesophile *Escherichia coli* and the thermophile *Bacillus stearothermophilus* respectively (Casolari, 1981).

Using this approach, the T_{max} of an organism can be predicted as the temperature at which the

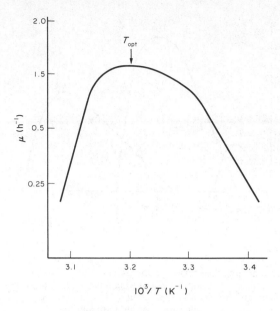

Figure 4 Arrhenius plot of temperature-dependent growth of *Bacillus subtilis*. (Data taken from Mohr and Krawiec, 1980)

doubling time equals the time taken to kill 50% of the population. So in terms of absolute temperature:

$$T_{max} = 0.009(2E_d - E_u) \tag{12}$$

where E_d is the lethal energy (see Section 14.8.1.3).

A good approximation of the growth curve from T_{min} to T_{opt} has been obtained from the relationship:

$$\sqrt{\mu} = b(T - T_{min}) \tag{13}$$

where b is a constant (Ratkowsky *et al.*, 1982). A model for the complete growth curve is:

$$\sqrt{\mu} = b(T - T_{min})[1 - \exp c(T - T_{max})] \tag{14}$$

where c is a further constant (Ratkowsky *et al.*, 1983). Where a set of pairs of μ and T are known, the values of T_{min}, T_{opt} and T_{max} can be determined by non-linear regression analysis. Although this relationship provides a good empirical fit to the data, there is as yet no obvious physiological basis for it.

14.8.1.3   *Effects of temperature on cell death*

Above T_{opt}, the specific growth rate rapidly declines. This decline is due to a combination of metabolic disruption and cell death. The relationship between the concentration of cells surviving after t minutes at temperature T, (X_t), and the initial concentration of cells (X_0) is usually expressed as an exponential function:

$$X_t = X_0 \exp(-kt) \tag{15}$$

where k is a death-rate parameter which increases with temperature. Casolari (1981) has suggested the following relationship:

$$X_0 = X_t^{1+Mt} \tag{16}$$

where M is the frequency of collision between the cells and water molecules having an energy greater than that leading to cell death (E_d). From the Maxwellian distribution:

$$M = \exp(103.7293 - 2E_d/RT) \tag{17}$$

The concentration of survivors (X_t) depends on X_0, T, t and E_d. As with E_u, E_d values increase according to the thermal characteristics of the microorganism ranging from about 130–160 kJ mol^{-1} for vegetative microorganisms (bacteria, yeasts and fungi) to values around 165 kJ mol^{-1} for bacterial spores.

14.8.1.4 Effects of temperature on cellular components

Marked differences in cellular composition have been shown for psychrophiles, mesophiles and thermophiles. Furthermore, adaptive changes have been shown in mesophiles shifted from temperatures at the lower end of their growth range to temperatures at the higher end giving some indication of the cellular factors which are important in determining T_{min} and T_{max}.

(i) Membranes

Membrane function requires fluidity in the phospholipid bilayer. Membrane phospholipids with unsaturated fatty acids melt at lower temperatures than those with saturated fatty acids so regulation of membrane composition gives microorganisms the ability to maintain optimum membrane fluidity at any temperature in their growth range. This process is known as homeoviscous adaptation. In the mesophiles *E. coli* and *B. subtilis* the proportion of unsaturated fatty acids incorporated into the phospholipids increases as the growth temperature is decreased (Table 5). As would be predicted, psychrophiles have a greater proportion of unsaturated or anteisobranched fatty acyl groups which are shorter in length than those from closely related mesophiles and thermophiles. On the other hand, the latter group contains a greater proportion of long straight chain saturated or isobranched fatty acids (McElhaney, 1976). The extreme acidothermophiles, which are members of the Archaebacteria, have phospholipids with ether linkages and a proportion of lipid chains cross-linked through the membrane, this unusual structure presumably increases thermal stability of the membrane.

Table 5 Effect of Growth Temperature on the Amounts of Major Fatty Acids of *Escherichia coli*[a]

Fatty acid	Growth temperature (°C)			
	10	25	35	43
Saturated	%	%	%	%
Myristic	3.9	3.8	4.7	7.7
Palmitic	18.2	27.6	31.7	48.0
Unsaturated	%	%	%	%
Hexadecenoic	26.0	23.2	23.3	9.2
Octadecenoic	37.9	35.5	24.6	12.2

[a] Marr and Ingraham, 1962

The membrane composition of *Acholeplasma laidlawii* can be manipulated experimentally and the gel to liquid-crystalline phase transition temperature of the membrane correlates with T_{min} values for growth down to 8 °C. Below this temperature some other cellular factor becomes limiting. In contrast, T_{opt} and T_{max} were less affected, but membrane composition will affect T_{max} if the motion of fatty acyl chains in the centre of the phospholipid bilayer increases to give a leaky membrane or affects the conformation of essential membrane proteins (McElhaney, 1976).

(ii) DNA

Temperature has little effect on the chromosomal content of wild type cells. Although increasing the G+C:A+T ratio of DNA increases the strand separation (melting) temperature of DNA, there is no evidence to suggest that thermophiles as a group have a higher G+C content than mesophiles or psychrophiles. Within the genus *Bacillus* however, the thermophilic members were found to have a higher G+C:A+T ratio (54%) than the mesophilic members (45%) giving a melting temperature of 92 °C compared to 88 °C (Stenesh *et al.*, 1968). These values are unlikely to influence T_{max}.

Increased growth temperature often results in the loss of plasmids from a proportion of the population, a process known as curing. The naturally occurring *Proteus* plasmids pRtsl (Terawaki *et al.*, 1967) and pR401 (Coetzee *et al.*, 1972) have plasmid-encoded temperature sensitive replication functions so a step up in growth temperature from 25 °C to 42 °C will result in rapid curing of the cells. However, temperature changes can be turned to advantage by genetic engineering procedures. Uhlin *et al.* (1979) constructed a mutant plasmid which loses control of its replication above 35 °C such that the copy number increases continuously. Eventually plasmid DNA may represent about 75% of the total DNA content of the cells. If the gene coding for a desired protein product is present on the plasmid and induced at the same time as the temperature shift the gene product will be made in large quantity. Inhibition of cell growth and loss of viability occurs after some 2–3 h as the metabolic load becomes too great (Uhlin and Clark, 1981).

(iii) RNA

As with DNA, there is no evidence that the base composition of rRNA, mRNA or tRNA affects stability to a degree that could account for T_{max} values of microorganisms. Nonetheless, the RNA content of cells is affected by temperature. The ratio of RNA to protein increases as temperature is decreased in continuous cultures of *Candida utilis* and *Klebsiella aerogenes*. This change reflects a decrease in the efficiency of protein synthesis (Alroy and Tannenbaum, 1973). To compensate, more ribosomes are produced and the cell volume increases to accommodate these changes (Hunter and Rose, 1972; Bull and Brown, 1979).

(iv) Proteins

As the temperature is increased the rate of protein denaturation is usually increased and the variation of denaturation rate with temperature obeys the Arrhenius law. In some cases, thermal denaturation is a simple unimolecular process, the denaturation kinetics being first order with respect to both concentration and time, but in others it may be complex and variable. For example, the denaturation of pepsin increases from first order to fifth order with respect to both concentration and time as the pepsin concentration is decreased (Casey and Laidler, 1951).

The thermal denaturation of proteins can be monitored as a loss of catalytic or biological function which is often irreversible. So, it is likely that the limits of growth for some organisms could be set by inactivation of one or more cellular components. Malcolm (1969) showed that *in vitro* inactivation of the glutamyl-tRNA synthetase and prolyl-tRNA synthetase of the psychrophile *Micrococcus cryophilus* coincided with its T_{max} of 25 °C. Similarly, T_{max} values for the psychrophiles *Candida gelida* and *Clostridium* strain 69 correlate with the stability of pyruvate carboxylase and triosephosphate isomerase respectively. A corollary of the thermal inactivation of proteins is that microbial metabolism can alter with temperature. Many microorganisms develop particular growth requirements as the temperature is increased above their T_{opt} values suggesting that specific enzyme syntheses or the enzymes themselves are temperature sensitive. For example, at temperatures near T_{max}, *E. coli* develops requirements for glutamate and nicotinamide (Ware, 1951) whilst *Saccharomyces cerevisiae* develops requirements for pantothenate and increased NaCl (Begue and Lichstein, 1963). Furthermore, temperature sensitive (ts) mutants can be obtained which exhibit narrowed growth ranges due to cold sensitivity or heat sensitivity of essential metabolic components such as glycolytic enzymes, ribosomal proteins or nucleic acid synthesizing enzymes.

Whilst some proteins from psychrophiles and mesophiles may be intrinsically heat stable, all proteins from thermophiles must be capable of adequate biological function at high temperature. In general, the proteins from thermophiles exhibit thermostability *in vitro* as well as *in vivo* although some, such as the glyceraldehyde 3-phosphate dehydrogenase from *Bacillus coagulans* are much less stable in isolation than would be expected from the T_{opt} (55 °C) of the organism. This protein (and presumably others) becomes stabilized in the polyelectrolyte milieu of the cell, an effect that can be mimicked by the addition of 10% NaCl *in vitro* (Amelunxen and Murdock, 1978).

In summary, only in a few cases have major factors contributing to the growth range limitations of microorganisms been identified. In most microorganisms no factor may be solely responsible for T_{max} or T_{min}, but the interaction of changes in membrane composition, DNA expression, protein synthesis and protein stability may bring about cessation of growth at defined temperatures.

14.8.2 Cultural Effects of Temperature

14.8.2.1 *Responses to temperature shifts*

The extent of cultural response to a change from one temperature to another depends both on the magnitude and on the rate of such a change. Where bacteria are subjected to rapid temperature elevations of about 10 °C there is usually a lag period of a few hours during which time synthesis of RNA (mainly rRNA) and proteins occurs. In contrast, a small temperature increase of one or two degrees or a larger increase occurring over an extended time period is accommodated immediately by the cell. This is because there is a slight surplus of growth capacity (macromolecular content, enzymatic activity, respiratory activity and transport) in cultures growing at a given constant temperature. A lag phase does not occur in response to a decrease in temperature, rather a finite period is involved where the now surplus growth capacity is reduced in response to the lower demand through turnover and breakdown of cellular constituents.

Continuous culture studies with *K. aerogenes* (Topiwala and Sinclair, 1971) have shown that microbial response to temperature approximates to a double Arrhenius-type curve thereby indicating that organisms have a sensitive temperature zone over which small variations in temperature induce large changes in metabolic rates and growth. As with other environmental parameters, the type and magnitude of temperature-induced metabolic responses also depends on the prevailing environmental conditions and particularly on medium composition and pH. Clearly, temperature control has to be considered closely in processes where temperature shifts can alter the culture disposition to growth and/or product formation.

14.8.2.2 Effects on substrate utilization

Temperature is particularly important in affecting the conversion efficiency of substrate into cell mass where the substrate is the carbon/energy source. With *K. aerogenes* (Topiwala and Sinclair, 1971) and *Hansenula polymorpha* (Snedecar and Cooney, 1974) growing under glucose and methanol limitations respectively, cell yield was observed to be inversely related to growth temperature. At temperatures above the optimum for growth, extra energy is utilized for maintenance purposes because of the faster turnover of proteins and nucleic acids and so less energy is available for growth-associated functions.

Oxygen availability tends to decrease with temperature due to a lower solubility and faster consumption rates. In *Aspergillus nidulans* for example, a temperature increase from 23 to 37 °C resulted in a concomitant increase in q_{O_2} from 1.54 to 3.24 (Rowley and Pirt, 1972), presumably due to higher maintenance rates. Temperature may affect substrate availability through solubility changes or by enhancing chemical interaction rates which lead to substrate–chemical complexes. Conversely, polymeric substrates such as starch and cellulose may be degraded more readily at higher temperatures as a result of higher specific enzyme activities or because of the induction of thermophilic enzymes. Certainly many of the fungal cellulases are thermophilic in nature. Temperature induced changes in kinetic parameters such as substrate affinity and yield are further exemplified by studies on *S. cerevisiae* which showed that when the temperature was raised from 25 to 38 °C the substrate affinity more than doubled whilst the yield decreased from 0.23 to 0.20 g g^{-1} (Jones and Hough, 1970). Concomitant with these changes were the accumulation of ethanol, pyruvate and 2-oxoglutarate in the medium at 38 °C compared to 25 °C.

14.8.2.3 Effects on product formation

The optimal temperature for growth may not be that best suited to product formation, especially where the product is predominantly non-growth associated as in the case of many of the secondary metabolites. Additionally, different producing strains may have different growth and production optima and so in any mutation-selection programme independent temperature optimization of each strain for both growth and subsequent product formation is often necessary. In penicillin production, for example, the mycelial growth phase of *Penicillium chrysogenum* is optimal near 30 °C whilst idiophasic penicillin production is greatest at about 20 °C. Thus, process optimization must consider imposed temperature changes which should be made when the culture is in a suitable physiological state. To achieve this an understanding of the biochemistry of product formation may be useful. The maximum rate of penicillin biosynthesis correlates with a minimum rate of lactose utilization which is favoured during penicillin production by the lower temperature (Woodruff, 1961).

Temperature modulation of metabolite overproduction is also known. In *Ashbya gosypii*, riboflavin overproduction occurs optimally at 28 °C although growth is maintained at 37 °C. In this case a deregulation of enzyme synthesis occurs at 28 °C leading to the overproduction (Demain, 1972). In streptomycin production by *Streptomyces griseus*, an increase as small as 1 °C at the threshold temperature results in an 80% reduction in antibiotic production (Dulaney, 1951).

The desired productivity must often be a compromise between those factors enhancing biosynthesis and excretion (*e.g.* enhanced enzyme activity; less integral membrane function) and those conducive to product stability, all of which are temperature-dependent. The net accumulation of product in the medium or intracellularly is the difference between total biosynthesis and product decay so that product stability in relation to temperature must be defined. Products may decay spontaneously, or through enzymatic attack, or as in the case of penicillins, through a combination of both. In all cases, temperature can markedly affect the decay rates, and may do so over a very narrow temperature range. The rate constants for penicillin decay follow an Arrhenius-type relationship with an activation energy of 50–63 kJ mol^{-1} (Constantinides *et al.*, 1970). Clearly the timing of a fermentation course has to be considered in relation to the effect of temperature on metabolic rates and product.

14.8.2.4 *Heat generation*

Generally, fermentations require an initial input of heat when operating above ambient conditions in order to provide optimum conditions for inoculum growth and thus to minimize the lag phase. However, during growth and product formation only a fraction of the energy available from complete catabolism of the substrate is retained within the biomass; the remainder (some 70% of alcohol fermentations) is dissipated as heat.

Thermodynamically, catabolic reactions are exothermic and consequently energy is lost as heat during growth. Greatest heat production comes from the respiratory system. The transfer of a pair of electron equivalents from NADH to molecular oxygen releases about -222 kJ mol^{-1} of energy. Maximally one NADH$_2$ molecule will bring about the synthesis of 3 ATP molecules from ADP, each with a free energy of formation of $+30.5$ kJ mol^{-1}. Thus energy coupling *via* the electron transport chain is at best only about 40% efficient, the other 60% being dissipated as heat. Subsequent utilization and hydrolysis of ATP to ADP and AMP releases more heat. Overall heat generation also depends on the type of the catabolism involved. For instance the complete oxidation of glucose to $CO_2 + H_2O$ yields -2868 kJ mol^{-1} whereas anaerobic fermentation to lactate generates only -197 kJ mol^{-1} of available energy. Only 2 mol of ATP are formed in the fermentation of glucose to lactate which corresponds to only 31% of the energy available from the reaction.

For aerobic growth where the only products formed are biomass, CO_2 and water, the heat evolution plus the heat of combustion of the biomass formed approximates to the heat of combustion of the substrate consumed. Heat production, therefore, is a stoichiometric function of biomass and product yield and may be used to monitor the course of both growth and product formation. Two approaches to such measurements are possible: (1) dynamic calorimetry which uses the entire fermenter as an adiabatic calorimeter, and (2) measurements made on the heat content of the cooling water (Wang *et al.*, 1979).

The importance of understanding heat evolution lies in the need for fermentations to be precisely temperature regulated in order to maintain optimal metabolic rates. Calculations on the rate and magnitude of heat evolution can be used to design cooling systems of the necessary capacity. Cooling requirements are often the largest single cost in a fermentation.

14.9 SHEAR

14.9.1 Generation of Shear

In submerged liquid fermentations agitation is required to provide adequate mixing of the contents and sufficient mass and heat transfer rates. The mechanical force which brings about relative movement of fluid molecules is known as the shear stress and the structural deformation and breakdown in the physical properties of the stressed fluid is known as the fluid shear rate. Shear rate is known to be a function of the impeller diameter and impeller tip velocity in stirred tank reactors. For Newtonian fluids, such as cultures of single-celled bacteria and yeasts, the viscosity is not altered by varying degrees of mixing although it is inversely related to temperature. However, the relationship is complicated in non-Newtonian broths such as those with dense mycelial biomass or those producing extracellular polysaccharides such as xanthan gums. In these latter fermentations viscosity is a shear dependent parameter and high mixing rates are necessary to create sufficient fluid shear for adequate mixing, mass and heat transfer.

14.9.2 Effects of Shear on Filamentous Fungi

Dense mycelial broths such as those used in industrial fermentations exhibit pseudoplastic rheological behaviour, that is the greatest changes in shear rate occur at the lower shear stress values. Additionally, a small dilution of the broth often leads to a much greater decrease in the apparent viscosity and hence a much reduced energy requirement for mixing. Shearing forces may improve broth rheology by restricting the average hyphal length. In the case of *Penicillium chrysogenum* impeller shear is thought to be responsible for a maintained steady state number of mycelial fragments in a glucose-limited chemostat culture (Pirt and Callow, 1959). The precise rheological response of a given filamentous fungus to mixing is known to vary with the environmental conditions, especially temperature and osmotic potential. With the latter parameter, changes in the broth osmotic pressure causes concomitant changes in the hyphal turgor pressure which in turn varies hyphal flexibility. Hyphal rigidity is important in forming stabilized flocs as well as in resisting shear damage. Where such flocs aggregate to form a rigid matrix, fungal pellets can be formed. Mycelial pelleting results in a very heterogeneous culture physiology as a consequence of poor nutrient, oxygen and by-product penetration through the pellets. However, in some fermentations the pellet morphology is preferred since it reverts the broth to a Newtonian rheology and thereby greatly reduces the power input for mixing and hence the energy costs of the fermentation. Furthermore post-fermentation separation of the biomass is simplified.

14.9.3 Effects of Shear on Protozoa and Animal Cells in Culture

Since protozoa and animal cells lack a protective cell wall they are considerably more susceptible to shear damage. Indeed, the shear sensitivity of some animal cell cultures is so great that the shearing planes caused by migration of air bubbles is often sufficient to cause cell death. For cell culture, mechanical agitators producing axial liquid flow such as marine-type propellers or vibromixers are more suitable than the flat blade turbines of conventional fermentation systems. Suitably designed airlift reactors with bulk flow created only by the introduction of a small air stream can provide a sufficiently low shear system for suspension culture of animal cell lines. Other methods of minimizing shear are: by increasing the serum concentration of the medium; by adding non-ionic surface active polymers; by supplying oxygen through head space aeration or through membrane permeation or as pre-oxygen saturated growth media. Animal cells in particular, but also some protozoa, have a strong surface attachment affinity, thereby making suspension cultures difficult to achieve. These may be grown on small solid particles in which case particle shape and size is important in minimizing shear damage; solid beads cause cell attrition by rubbing together in dynamic systems whereas small wafer like particles minimize shear contact and maximize cell contact area.

14.9.4 Effects of Products on Shear Rate

The formation of macromolecular products, particularly polysaccharides such as xanthan and pullulan, increase broth viscosity so that higher power inputs are needed to maintain constant heat and mass transfer rates. At high power inputs some polymer solutions exhibit a much reduced apparent viscosity (shear-thinning) so that a constant, high shearing rate is required to maintain mixing without increasing power input. Where highly viscous products are the aim of a fermentation, the impeller design may need to be modified to optimize mixing rather than gas transfer. For this, a marine type impeller may be more suitable than a flat-bladed turbine.

14.10 GENERAL CONTROL STRATEGIES

The response of microorganisms to their environment is potentially extremely complicated. The response to each environmental parameter can occur at several levels of cell regulation from allosteric control of enzymes to control of transcription and even selection of more 'fit' genotypes. Also there is simultaneous interaction between the various environmental factors and many aspects of the cell structure and metabolism. A change in temperature will have a direct effect on any cell enzyme system as well as on cell membranes but may also have indirect effects through changes in availability of nutrients. Maintenance of steady state conditions is relatively simple especially in continuous culture systems; often all that is required is that the important environmental parameters of substrate supply, pH, temperature and DOT are kept constant by feedback control. Generally the response times of a culture will be relatively slow compared with that of electronic monitoring probes and controls. The difficulty arises in controlling systems which are subjected to rapid transients. Such systems may become more common in industrial processes as understanding of microbial physiology increases, as it is likely that we shall attempt to direct microbial metabolism by imposing such transients.

One likely application of transients will be the application of a temperature jump to cultures of microorganisms containing temperature-sensitive copy number plasmids. This will enable cultures to be run at a temperature which favours efficient growth and then to be switched to product generation by raising the temperature to increase plasmid copy number and thus expression of a desired product. Accurate control of such transients will require a more detailed knowledge of the underlying regulatory systems of cell metabolism. Even in so-called steady-state systems it is very common, if not inevitable, that there are underlying oscillations. These have been reviewed by Harrison and Topiwala (1974). In brief, oscillations can result from the control equipment through overshoot, or arise spontaneously as a result of cellular feedback mechanisms. On scaling-up fermentations it is difficult to avoid subjecting microbial cells to changing conditions as they are moved around in a large fermentation vessel which is inevitably not perfectly homogeneous. Even small discontinuities in the control of the environment may be amplified through metabolism to have a significant effect on a culture. An example is the sensitivity of cultures of methanol utilizing organisms to small oscillations in the rate of methanol feed (Harrison and Topiwala, 1974). Unless methanol is fed into a culture at a very even rate, the effect of even small oscillations which expose the cells periodically to an excess of methanol, is to significantly depress the yield coefficient even though on average, methanol is limiting for growth. In fact, an oscillation in feed-rate of only 30 s periodicity will cause more than a 10% loss in yield (Meers, 1973), an effect which would be significant and costly on a commercial scale SCP plant.

If scale-up of processes is to be achieved reliably then ideally the organisms' response to the environment both in steady and transient states should be well understood. At present it must be said that there is not a thorough understanding of steady-state responses to important environmental parameters and studies on transient and oscillating changes are still in their infancy.

14.11 REFERENCES

Ackrell, B. A. C. and C. W. Jones (1971). The respiratory system of *Azotobacter vinelandii*—Part 2: Oxygen effects. *Eur. J. Biochem.*, **20**, 29–35.
Alroy, Y. and S. R. Tannenbaum (1973). The influence of environmental conditions on the macromolecular composition of *Candida utilis. Biotechnol. Bioeng.*, **15**, 239–256.
Amelunxen, R. E. and A. L. Murdock (1978). Microbial life at high temperatures: mechanisms and molecular aspects. In *Microbial Life in Extreme Environments*, ed. D. J. Kushner, pp. 217–218. Academic, New York.
Ashby, R. E. and D. E. F. Harrison (1980). Studies on the induction and turnover of citrate-oxidising capacity in *Klebsiella aerogenes* using chemostat culture. *J. Gen. Microbiol.*, **120**, 465–473.
Babij, T., F. J. Moss and B. J. Ralph (1969). Effects of oxygen and glucose levels on lipid composition of yeast *Candida utilis* grown in continuous culture. *Biotechnol. Bioeng.*, **11**, 593–603.
Begue, W. J. and H. C. Lichstein (1963). Growth requirements of *Saccharomyces cerevisiae* as a result of incubation at increased temperature. *Proc. Soc. Exp. Biol. Med.*, **114**, 625–628.
Brooks, J. D. and J. L. Meers (1973). The effect of discontinuous methanol addition on the growth of a carbon-limited culture of *Pseudomonas. J. Gen. Microbiol.*, **77**, 513–519.
Brown, A. D. (1976). Microbial water stress. *Bacteriol. Rev.*, **40**, 803–846.
Buchanan-Wollaston, V., M. C. Cannon, J. L. Beynon and F. C. Cannon (1981). Role of the *nif* A gene product in the regulation of *nif* expression in *Klebsiella pneumoniae. Nature (London)*, **294**, 776–778.
Buckland, B. C., M. D. Lilly and P. Dunnill (1976). The kinetics of cholesterol oxidase synthesis by *Nocardia rhodocrous. Biotechnol. Bioeng.*, **18**, 601–621.
Bull, A. T. and C. M. Brown (1979). Continuous culture applications to microbial biochemistry. In *Microbial Biochemistry*, ed. J. R. Quayle, pp. 177–226. University Park Press, Baltimore.
Casey, E. J. and K. J. Laidler (1951). The kinetics and mechanism of the heat inactivation of pepsin. *J. Am. Chem. Soc.*, **73**, 1455–1457.
Casolari, A. (1981). A model describing microbial inactivation and growth kinetics. *J. Theor. Biol.*, **88**, 1–34.
Chen, W. L. and P. C. Charalampous (1969). Mechanism of induction of cytochrome oxidase in yeast. 1. Kinetics of induction and evidence for accumulation of cytoplasmic and mitochondrial precursors. *J. Biol. Chem.*, **244**, 2767–2776.
Coetzee, J. N., N. Datta and R. W. Hedges (1972). R-factors from *Proteus rettgeri. J. Gen. Microbiol.*, **72**, 543–552.
Cole, J. A. (1976). Microbial gas metabolism. *Adv. Microb. Physiol.*, **14**, 1–92.
Constantinides, A., J. L. Spencer and E. L. Gaden (1970). Optimisation of batch fermentation process. 1. Development of mathematical models for batch penicillin fermentations. *Biotechnol. Bioeng.*, **12**, 803–830.
Dalton, H. and J. R. Postgate (1968). Effect of oxygen on growth of *Azotobacter chroococcum* in batch and continuous cultures. *J. Gen. Microbiol.*, **54**, 463–473.

Dawson, P. S. S. (1965). The intracellular amino acid pool of *Candida utilis* during batch and continuous flow cultures. *Biochim. Biophys. Acta*, **111**, 51–66.

Demain, A. E. (1972). Cellular and environmental factors affecting synthesis and excretion of metabolites. *J. Appl. Chem.*, **22**, 345–362.

Dulaney, E. L. (1951). Process for production of streptomycin. *US Pat.* 2 571 693.

Ellwood, D. C. and D. W. Tempest (1972). Influence of culture pH on the content and composition of teichoic acids in the walls of *Bacillus subtilis*. *J. Gen. Microbiol.*, **73**, 395–402.

Forage, R. G. and E. C. C. Lin (1982). *dha* System mediating aerobic and anaerobic dissimilation of glycerol in *Klebsiella pneumoniae* NCIB 418. *J. Bacteriol.*, **151**, 591–599.

Giffhorn, F. and G. Gottshalk (1975). Inactivation of citrate lyase from *Rhodopseudomonas gelatinosa* by a specific deacetylase and inhibition of this inactivation by L-(+)-glutamate. *J. Bacteriol.*, **124**, 1052–1061.

Gochnauer, M. B. and D. J. Kushner (1971). Potassium binding, growth and survival of an extremely halophilic bacterium. *Can. J. Microbiol.*, **17**, 17–23.

Gregory, E. M. and I. Fridovich (1973). Oxygen toxicity and the superoxide dismutase. *J. Bacteriol.*, **114**, 1193–1197.

Gregory, E. M., F. J. Yost and I. Fridovich (1973). Superoxide dismutases of *Escherichia coli*: intracellular localisation and function. *J. Bacteriol.*, **115**, 987–991.

Hamilton, W. A. (1977). Energy coupling in substrate and group translocation. In *Microbial Energetics*, ed. B. A. Haddock and W. A. Hamilton, pp. 185–216. Cambridge University Press, Cambridge.

Harrison, D. E. F. (1972a). Physiological effects of dissolved oxygen tension and redox potential on growing populations of microorganisms. *J. Appl. Chem.*, **22**, 417–440.

Harrison, D. E. F. (1972b). A study of the effect of growth conditions on chemostat-grown *Klebsiella aerogenes* and kinetic changes of a 500 nm absorption band. *Biochim. Biophys. Acta*, **275**, 83–92.

Harrison, D. E. F. (1976a). The measurement of dissolved oxygen in continuous fermentations. In *Measurement of Oxygen*, ed. H. Degn, I. Balslev and R. Brook, pp. 53–64. Elsevier, Amsterdam.

Harrison, D. E. F. (1976b). The regulation of respiration rate in growing bacteria. *Adv. Microb. Physiol.*, **14**, 243–313.

Harrison, D. E. F. and B. Chance (1970). Fluorimetric technique for monitoring changes in the level of reduced nicotinamide nucleotides in continuous cultures of microorganisms. *Appl. Microbiol.*, **19**, 446–450.

Harrison, D. E. F. and J. E. Loveless (1971a). The effect of growth conditions on respiratory activity and growth efficiency in facultative anaerobes grown in chemostat culture. *J. Gen. Microbiol.*, **68**, 35–43.

Harrison, D. E. F. and J. E. Loveless (1971b). Transient responses of facultatively anaerobic bacteria growing in chemostat culture to a change from anaerobic to aerobic conditions. *J. Gen. Microbiol.*, **68**, 45–52.

Harrison, D. E. F. and S. J. Pirt (1967). The influence of dissolved oxygen concentration on the respiration and glucose metabolism of *Klebsiella aerogenes* during growth. *J. Gen Microbiol.*, **46**, 193–211.

Harrison, D. E. F. and H. H. Topiwala (1974). Transient and oscillatory states of continuous culture. *Adv. Biochem. Eng.*, **3**, 167–219.

Harrison, D. E. F., D. G. MacLennan and S. J. Pirt (1969). Responses of bacteria to dissolved oxygen tension. In *Fermentation Advances*, ed. D. Perlman, pp. 117–144. Academic, New York.

Hill, S., C. Kennedy, E. Kavanagh, R. B. Goldberg and R. Hanau (1981). Nitrogen fixation gene (*nif* L) involved in oxygen regulation of nitrogenase synthesis in *K. pneumoniae*. *Nature (London)*, **290**, 424–426.

Hunter, K. and A. H. Rose (1972). Influence of growth temperature on the composition and physiology of microorganisms. *J. Appl. Chem.*, **22**, 527–540.

Hsung, J. C. and A. Haug (1975). Intracellular pH of *Thermoplasma acidophila*. *Biochim. Biophys. Acta*, **389**, 477–482.

Jones, R. C. and J. S. Hough (1970). The effect of temperature on the metabolism of baker's yeast growing on continuous culture. *J. Gen. Microbiol.*, **60**, 107–116.

Kubitschek, H. E. (1970). *Introduction to Research with Continuous Cultures*. Prentice Hall Inc., New Jersey.

Kushner, D. J. (1978). Life in high salt and solute concentrations: Halophilic bacteria. In *Microbial Life in Extreme Environments*, ed. D. J. Kushner, pp. 317–368. Academic, New York.

Linton, J. D., D. E. F. Harrison and A. T. Bull (1975). Molar growth yields, respiration and cytochrome patterns of *Beneckea natriegens* when grown at different medium dissolved oxygen tensions. *J. Gen. Microbiol.*, **90**, 237–246.

Linton, J. D., D. E. F. Harrison and A. T. Bull (1976). The effect of rate of respiration on sensitivity to cyanide and carbon monoxide in *Beneckea natriegens* grown in batch and continuous culture. *FEBS Lett.*, **64**, 358–363.

Malcolm, N. L. (1969). Molecular determinant of obligate psychrophily. *Nature (London)*, **221**, 1031–1033.

Mandel, K. G., A. A. Guffanti and T. A. Krulwich (1980). Monovalent cation/proton antiporters in membrane vesicles from *Bacillus alcalophilus*. *J. Biol. Chem.*, **225**, 7391–7396.

Marr, A. G. and J. L. Ingraham (1962). Effect of temperature on the composition of fatty acids in *E. coli*. *J. Bacteriol.*, **84**, 1260–1267.

McElhaney, R. N. (1976). The biological significance of alterations in the fatty acid composition of microbial membrane lipids in response to changes in environmental temperature. In *Extreme Environments: Mechanisms of Microbial Adaptation*, ed. M. R. Heinrich, pp. 255–281. Academic, New York.

Mohr, P. W. and S. Krawiec (1980). Temperature characteristics and Arrhenius plots for nominal psychrophiles, mesophiles and thermophiles. *J. Gen. Microbiol.*, **121**, 311–317.

Moss, F. J., P. A. D. Rickard, F. E. Bush and P. Caigen (1971). The response by microorganisms to steady state growth in controlled concentrations of oxygen and glucose. 1. *Candida utilis*. *Biotechnol. Bioeng.*, **11**, 561–580.

Nagai, S., Y. Nishizawa, M. Onodera and S. Aiba (1971). Effect of dissolved oxygen on growth yield and aldolase activity in continuous culture of *Azotobacter vinelandii*. *J. Gen. Microbiol.*, **66**, 197–203.

Novick, A. and L. Szilard (1950). Experiments with the chemostat on spontaneous mutations of bacteria. *Proc. Natl. Acad. Sci. USA*, **36**, 708–719.

Pirt, S. J. and D. S. Callow (1959). Continuous flow culture of the filamentous mould *Penicillium chrysogenum* and the control of its morphology. *Nature (London)*, **184**, 307–310.

Pitt, J. I. and J. H. B. Christian (1968). Water relations of xerophilic fungi isolated from prunes. *Appl. Microbiol.*, **16**, 1853–1858.

Ratkowsky, D. A., J. Olley, T. A. McMeekin and A. Ball (1982). Relationship between temperature and growth rate of bacterial cultures. *J. Bacteriol.*, **149**, 1–5.

Ratkowsky, D. A., R. K. Lowry, T. A. McMeekin, A. N. Stokes and R. E. Chandler (1983). Model for bacterial culture growth rate throughout the entire biokinetic temperature range. *J. Bacteriol.*, **154**, 1222–1226.

Rose, A. H. (1976). Osmotic stress and microbial survival. In *The Survival of Vegetative Microbes*, ed. T. R. G. Gray and J. R. Postgate, pp. 155–182. Cambridge University Press, Cambridge.

Rowley, B. I. and S. J. Pirt (1972). Melanin production by *Aspergillus nidulans* in batch and chemostat cultures. *J. Gen. Microbiol.*, **72**, 553–563.

Ruch, F. E., J. Lengeler and E. C. C. Lin (1974). Regulation of glycerol metabolism in *Klebsiella aerogenes*. *J. Bacteriol.*, **119**, 50–56.

Sapshead, L. M. and J. W. T. Wimpenny (1970). The effect of oxygen and nitrate on respiratory systems in *Micrococcus denitrificans*. *J. Gen. Microbiol.*, **63**, xiv–xv.

Snedecar, B. and C. L. Cooney (1974). Thermophilic mixed culture of bacteria utilising methanol for growth. *Appl. Microbiol.*, **27**, 1112–1117.

Stenesh, J., B. A. Roe and T. L. Snyder (1968). Studies of the deoxyribonucleic acid from mesophilic and thermophilic bacteria. *Biochim. Biophys. Acta*, **161**, 442–454.

Terawaki, Y., H. Takayasu and T. Akiba (1967). Thermosensitive replication of a kanamycin resistance factor. *J. Bacteriol.*, **94**, 687–690.

Tonge, G. M., C. J. Knowles, D. E. F. Harrison and I. J. Higgins (1974). Metabolism of one-carbon compounds: cytochromes of methane- and methanol-utilising bacteria. *FEBS Lett.*, **44**, 106–110.

Topiwala, H. and C. G. Sinclair (1971). Temperature relationships in continuous culture. *Biotechnol. Bioeng.*, **13**, 795–813.

Uhlin, B. E. and A. J. Clark (1981). Overproduction of the *Escherichia coli rec* A protein without stimulation of its proteolytic activity. *J. Bacteriol.*, **148**, 386–390.

Uhlin, B. E., S. Molin, P. Gustafsson and K. Nordstrom (1979). Plasmids with temperature dependent copy number for amplification of cloned genes and their products. *Gene*, **6**, 91–106.

Wang, E. I. C., C. L. Cooney, A. L. Demain, P. Dunhill, A. E. Humphrey and M. D. Lilly (1979). *Fermentation and Enzyme Technology*. Wiley, New York.

Ware, G. C. (1951). Nutritional requirements of *Bacterium coli* at 44 °C. *J. Gen. Microbiol.*, **5**, 880–884.

Watson, T. G. (1970). Effects of sodium chloride on steady state growth and metabolism of *Saccharomyces cerevisiae*. *J. Gen. Microbiol.*, **64**, 91–99.

Wilkinson, T. G. and G. Hamer (1974). Wall growth in mixed bacterial cultures growing on methane. *Biotechnol. Bioeng.*, **16**, 251–260.

Woodruff, H. B. (1961). Antibiotic production as an expression of environment. In *Microbial Reaction to Environment*, ed. G. G. Meynell and H. Gooder, pp. 317–342. Cambridge University Press, Cambridge.

Yamazaki, Y., N. Koyama and Y. Nosoh (1973). On the acidostability of an acidophilic thermophilic bacterium. *Biochim. Biophys. Acta*, **314**, 257–260.

Zetelaki, K. and K. Vas (1968). The role of aeration and agitation in the production of glucose oxidase in submerged culture. *Biotechnol. Bioeng.*, **10**, 45–49.

15

Mixed Culture and Mixed Substrate Systems

A. T. BULL
University of Kent at Canterbury, UK

15.1 INTRODUCTION

The science of microbiology has, for the greater part, developed from the study of pure or monospecies cultures despite the knowledge that microorganisms in natural environments exist, with very few exceptions, as complex, multispecies communities. The modern fermentation industry similarly has been dominated by the pure culture approach, again despite the fact that numerous traditional food, beverage and waste treatment processes evolved from the empirical deployment of mixed cultures. In recent years the properties of mixed cultures have attracted increasing attention. This interest has been stimulated by various events: (i) the need to understand how environmental pollution affects indigenous microfloras; (ii) the introduction of specific microorganisms into environments for purposes such as pest and pathogen control, enhancing soil fertility and plant growth, pollution treatment, and mineralization of crop residues in the field; (iii) the realization that the establishment and virulence phases of many diseases of man and animals are the consequence of mixed infections; and (iv) the exploitation of defined mixed cultures for a range of biotechnological purposes, including single cell protein production and bioconversions.

The significance of substrate mixtures for the growth and metabolism of microorganisms again has been appreciated only in fairly recent times. This situation also is rather surprising considering that most natural populations are presented with mixtures of carbon and other nutrients and that many commercial fermentations are based on complex mixtures of energy, carbon, nitrogen or other substrates. The serious study of mixed substrate utilization grew from Monod's analysis of bacterial growth on two-sugar mixtures and his original description of the phenomenon of diauxie.

The early emphasis on pure culture microbiology was due to several compelling reasons. First, eighteenth and nineteenth century mycology was much concerned with the elucidation of reproductive mechanisms and morphogenesis in fungi. Second, two of the major problems of twentieth century bacteriology related to so-called pleomorphism in bacteria and, most significantly, to the identity of the causal agents of infectious diseases. For all of these investigations the availability of pure cultures was a *sine qua non* and the development of enabling methodologies led to the blossoming of microbiology. These developments have been reviewed recently (Bull and Quayle, 1982; Bull and Slater, 1982a). The properties and commercial exploitation of mixed cultures has been constrained by various conceptual and practical difficulties, not least the mistaken belief that multispecies populations are inherently unstable. This chapter surveys the current state and potential of mixed culture processes and the utilization of mixed substrates.

15.2 MIXED CULTURES

15.2.1 Methods of Study

The chemostat type of continuous culture probably is the best experimental system for studying population interactions. Such cultures can be operated with or without feedback control: the former enables closely approximate steady state populations to be analysed, whereas the latter cultures are characterized by oscillatory behaviour which is more aptly described as stable rather than steady state. Chemostat culture enables environmental factors, either singly or in combination, to be altered systematically, and defined perturbations to be imposed on the mixed culture. Both of these manipulations can reveal the salient features of the mechanism(s) of species interaction. Thus, pulsing chemostat cultures of a methane-utilizing community with methanol confirmed that the role of a *Hyphomicrobium* species was to scavenge methanol which was produced by and was inhibitory towards the methane-oxidizer (see Section 15.2.7.2).

Multistage chemostats, in parallel or in series, also provide useful tools for interaction studies. For example, Cappenberg (1975) and Parker (1966) fed separate chemostat populations of single species into a combined final stage chemostat vessel to analyse the interactions of methanogenic and sulfate-reducing bacteria, and dental plaque bacteria, respectively. In the former case, commensalism based on the assimilation by the methanogens of acetate produced by the sulfate-reducers was confirmed. The application of a bidirectional flow of culture medium in a multistage chemostat allows growth and species interactions to be examined in opposing gradients of solutes or other environmental variables. Such a system has been termed a gradostat by Wimpenny and has been exploited extensively by his group (Wimpenny, 1982). The incorporation of membranes within continuous cultures also is particularly useful for exploring metabolic interactions between species. Finally, one other feature of continuous cultivation which is germane to interaction studies is the ability to make investigations over a very long time scale. This has special relevance to assessment of population and process stability and to the analysis of genetic interactions (see Section 15.2.6).

The analysis of microbial interaction and communities *in situ* is relevant to several aspects of environmental biotechnology (see Section 15.2.9). Fry (1982) provides details of the appropriate methodology.

15.2.2 Enrichment of Mixed Cultures

Continuous-flow enrichment techniques (chemostat and turbidostat types) have proved to be very powerful for organism selection, although they have not yet been widely adopted for this purpose (Veldkamp, 1970; Harrison, 1978; Bull, 1981). Continuous-flow enrichment almost invariably selects stable mixed cultures despite the fact that chemostat theory predicts a final population comprising only a single species. This apparent anomaly is explicable in terms of interactions additional to competition occurring between species. Enrichment protocols should be established so that they closely simulate the operating conditions of the intended processes in which the organisms are to be used (Linton and Drozd, 1982). The isolation of *n*-alkane-utilizing mixed cultures at different temperatures illustrate the extent to which variation in just one enrichment parameter can affect community composition and performance (Table 1).

Table 1 Effect of Temperature on the Enrichment of *n*-Hexadecane Communities[a]

T (°C)	μ_{max} (h^{-1})	Yield (g biomass g hexadecane^{-1})	Community ratio A	B	C
25	0.69	1.02	30	70	0
35	0.69	1.02	50	50	0
45	0.57	0.93	35	60	5
55	0.23	0.43	40	40	20
65	0.12	0.15	30	40	30

[a] From Sukatch and Johnson, 1972.

An extensive discussion of methods for enriching, isolating and analysing microbial communities can be found in Parkes (1982).

15.2.3 Analysis of Two-species Systems

The usual categorization of two-species interactions is that given in Figure 1. Interactions are defined in terms of the signs of the effects of species j upon species i and *vice versa*. Thus, a_{ij} +, 0 or − represent positive (beneficial), neutral, or negative (harmful) effects of species j upon i, depending on whether the population size of i increases, is unaffected, or decreases (Bull and Slater, 1982b). Figure 1 reveals six basic types of interaction: ++, mutualism; −−, competition; 00, neutralism; +0,0+, commensalism; +−, −+, parasitism, predation; 0−,−0, amensalism. The study of these various types of interaction has been very uneven and, in particular, neutralism and amensalism have received little attention. It must also be emphasized that the interactions between two species as defined in Figure 1 need not be the same under all conditions. Neither is the interaction between two species necessarily of a single type. Thus, cases of competitive commensalism and antagonistic mutualism, for example, have been described. Recent detailed discussions of two-species interactions from experimental and theory viewpoints can be found in Bull and Slater (1982b) and Culver (1981).

		Effects of species j on i (a_{ij})		
		+	0	−
	+	++	+0	+−
Effects of species i on j (a_{ji})	0	0+	00	0−
	−	−+	−0	−−

Figure 1 Matrix of interactions between microbial species i and j.

The two-species interactions just described provide the bases of community structure and for evolution and the extension of two-species interactions to three or more species generates greater complexity and additional biological phenomena. As Culver (1981) has remarked, there are two qualitatively distinct types of complexity deriving from three or more species interactions: (i) a consequence of increased dimensionality of a community with linear, pairwise interactions; and (ii) a consequence of multiway interactions. Linear interactions can introduce new complexities such as the stabilization of unstable interactions (*e.g.* the interaction of the predator *Tetrahymena pyriformis* on populations of *Azotobacter vinelandii* and *Escherichia coli* competing for glucose; Jost *et al.*, 1973), while non-linear interactions may produce shifts between competition, mutualism and predation depending on density or age structure of the populations.

15.2.4 Analysis of Multispecies Communities

A microbial community is simply a group of species which are often found living together; consortium and association are frequently used synonyms for the term community. Microbial communities, both in nature and in laboratory systems, possess a number of basic characteristics including successional development, homeostasis and evolution. We have delimited microbial communities on mechanistic bases as outlined in Table 2 (Slater, 1981; Bull and Slater, 1982b). The reader is referred to Slater (1981) for examples of these community classes.

Table 2   Classes of Microbial Communities based on Biological Mechanisms[a]

Class 1:	Structure based on provision of specific nutrients by different members of the community
Class 2:	Structure based on the alleviation of growth inhibition, including removal of metabolites which are inhibitory to the producer species (see also hydrogen transfer communities)
Class 3:	Structure based on interactions which result in the alteration of individual population growth parameters thereby producing more competitive and/or efficient community performance
Class 4:	Structure based on a combined metabolic activity not expressed by individual populations alone
Class 5:	Structure based on cometabolism
Class 6:	Structure based on hydrogen transfer reactions
Class 7:	Structure based on the interaction of several primary species

[a] From Bull and Slater (1982b).

15.2.5 Kinetics of Mixed Cultures

The development of mathematical models of microbial interactions, as of any other phenomenon, can have heuristic and predictive objectives, and the latter will be especially valuable in the optimization of biotechnological processes. Again the treatment of interactions from a mathematical position is very uneven. Most models of competition produce a stable equilibrium point or exclusion of one species in a pair-wise population; most models of predation produce limit cycles or damped oscillations; models of mutualism produce stable equilibria (weak interaction), and unbounded growth or extinction (strong interaction). Bazin (1981) provides a useful starting point for the modelling of multispecies systems.

15.2.6 Genetic Interactions

Microorganisms have relatively small genetic contents, a feature which limits their ability to adapt. However, microorganisms, particularly bacteria, possess a wide range of mechanisms for the exchange of genetic material, *viz*: transposition of DNA, plasmid–plasmid, plasmid–chromosome, plasmid–phage, phage–chromosome, and phage–phage interactions (Reanney *et al.*, 1982). Consequently, it is feasible to view microbial communities as forming genetically open networks in which, at least theoretically, information originating in one organism can pass to any other in the system. The most probable vectors for widespread dispersion of genes in communities are transposons and plasmids. However, it must be remembered that most of our understanding comes from pairwise interactions and that such widespread pooling of genes in communities is, at present, very largely conjectural. The experimental evidence for genetic interactions within microbial communities is evaluated in recent reviews by Reanney *et al.* (1982) and Williams (1982).

15.2.7 Mixed Culture Processes

Mixed cultures, of both the spontaneous non- or ill-defined and the deliberate defined types, have attracted renewed attention in the last few years for a wide diversity of applications. The number of traditional mixed culture processes is extremely large and in contrast to those few, well-understood and deliberately constituted mixed cultures. In this section a selection of processes will be included to illustrate the advantages and potential of mixed cultures in biotechnology. A number of recent publications can be consulted for detailed information (Harrison, 1978; Haas *et al.*, 1980; Bushell and Slater, 1981; Bull and Slater, 1982c).

15.2.7.1 Spontaneous mixed culture processes

Traditional microbial processes such as food and beverage fermentations and waste treatment, are almost invariably the consequences of mixed microbial action. In some fermented foods the essential microorganisms have been identified and the biochemical transformations which they bring about have been defined. In most fermented food processes, however, little elucidation has been made either at the microbiological or the biochemical levels. The same is frequently the case with other mixed processes such as waste treatment or metal leaching.

(i) Fermented foods and beverages

The development by man of fermented foods and beverages has been important for several reasons and such development has occurred in nearly all human cultures (Steinkraus, 1982; Wang and Hesseltine, 1982). The purposes of such processes include: (1) improvement of nutritional value and organoleptic properties of basic foodstuffs: fermentation frequently leads to an upgrading of the protein and vitamin contents of food materials, can enhance the texture and appearance (*e.g.* colour), and may reduce the cooking time of the product; (2) detoxification and improved digestibility: the often quoted example of cassava fermentation to the Nigerian dietary staple gari leads to the removal of a toxic cyanogenic glycoside; (3) preservation of foods such as vegetables, milk, meat and fish, most frequently by the fermentative production of lactic acid; and (4) production of intoxicant delights such as wines, beers and dairy materials. The number of known fermented foods and beverages is enormous but Steinkraus (1982) has conveniently distinguished four basic types.

> (a) *Development of meat-like flavours*. Japanese shoyu and miso are examples of such foods produced from the fermentation of soy beans, with or without the addition of wheat, rice or barley. These fermentations are based on the koji process, *i.e.* saccharification and hydrolysis of the starch and protein in the raw materials with species of *Aspergillus*, and subsequent fermentation by lactic acid bacteria and yeasts.
>
> (b) *Development of meat-like textures*. The best known products of this type are meat analogues or vegetable cheeses called tempe. Here soy beans are processed by fungi of the genus *Rhizopus* to form a firm bean cake (Indonesian tempe). Gram-negative bacteria present in the fermentation enrich the product with vitamin B_{12}.
>
> (c) *Alcoholic fermentations*. Most alcoholic fermentations lead to beverages but sweet-and-sour pastes such as lao chao (China) and tape ketan (Indonesia) are produced from starchy materials in mixed mould–yeast fermentations. The plethora of alcoholic beverages will be familiar to most microbiologists and include beers, wines, saké, pulque and very many others. Modern beers are mainly brewed with monocultures although top fermenters tend to comprise mixtures of yeasts. Kvass, Belgian lambic and geuze beers and African Kaffir beer all involve the participation of several yeasts and lactic acid bacteria. The traditional wine fermentations also depend on mixed yeast populations with least-alcohol tolerant species being successively displaced by more tolerant species (Goswell and Kunkee, 1977). Certain wine processes also involve the activities of lactic acid bacteria. The pulque fermentation of agave juice involves lactic acid bacteria and ethanol production by yeast and bacterium (*Zymomonas mobilis*). Saké fermentation also is complex: the koji step is followed by the development of nitrate-reducing bacteria, lactic acid bacteria and finally the mash is fermented by yeast.
>
> (d) *Acid fermentations*. Such fermentations usually produce lactic acid and foods, as a result, become much less susceptible to spoilage. Acid fermentation is one of the most effective and cheapest routes to food preservation (Raa, 1981). The range of substrates for acid food fermentations is wide, and includes fresh vegetables (sauerkraut), cereals (sour-dough bread), cereal gruels (ogi), pulses (dosa), milk (koumiss), fish and meats (Wood, 1981; Steinkraus, 1982). The fermentations are mediated by a comparable diversity of microorganisms: various lactic bacteria in association with other bacteria, yeast and moulds.

(ii) Waste treatment

Waste treatment processes are considered in great detail in Volume 3 of *Comprehensive Biotechnology* and several recent reviews deal with particular aspects of the population biology of such systems (anaerobic digestion, Hobson, 1981; aerobic waste treatment, Somerville, 1981; Curds, 1982). This material will not be examined here, rather one feature of the anaerobic digestion of wastes will be outlined in order to illustrate the tightly integrated nature of many of these waste degrading populations.

The degradation of polymeric and simple organic compounds to methane and carbon dioxide in anaerobic digesters is mediated by three bacterial populations (Mah, 1982): (1) chemoheterotrophic, fermentative H_2-producers: these bacteria hydrolyse polysaccharides and other polymers and ferment the hydrolysis products to volatile fatty acids, H_2 and CO_2; lactate and pyruvate also serve as fermentation substrates for these species; (2) chemoheterotrophic, non-fermentative, acetogenic, obligate proton reducers: certain fatty acids and alcohols are metabolized by these bacteria with the formation of acetate, H_2 and CO_2; and (3) methanogens: these strictly anaerobic bacteria enable an overall flow of electrons in the system by oxidizing H_2, reducing CO_2 and producing methane. They also metabolize the acetate produced in the system with the formation of CH_4 and CO_2. In the mixed culture environment of the digester (and similar environments such as the rumen), electron flow is channelled to proton reduction and H_2 becomes the major electron sink. This flow is possible thermodynamically only when H_2 is kept at very low concentrations, a situation which is realized by the activity of the methanogens. Interspecific hydrogen transfer is the name given to this process (see Wolin, 1982 for a detailed description). Such a strong physiological interdependence between different species leads to the establishment of very tight syntrophic associations. The classic case is that of *Methanobacillus omelianskii*, an ethanol-metabolizing methanogen described by H. A. Barker in 1936 and which, 30 years later, was shown to be a tight association of two distinct bacteria (Bryant *et al.*, 1967). The so-called S-organism oxidizes ethanol to acetate and hydrogen:

$$C_2H_5OH + H_2O \rightarrow CH_3CO_2^- + 2\,H_2 + H^+$$

The methanogen *sensu stricto* (*Methanobacterium bryantii*) reduces CO_2 by H_2

$$4\,H_2 + HCO_3^- \rightarrow CH_4 + 2\,H_2O + OH^-$$

Analysis of microbial interactions in anaerobic digesters has lagged behind that of the rumen. However, this situation is being redressed steadily and the information which is emerging on community structure and interactions is essential for the design and operation of new generations of methane digesters.

(iii)   Metal winning

The diversity of microorganisms in metal sulfide leaching operations is great and the processes are much more complex in microbiological terms than was initially imagined. Rather few observations have been made of leaching environments but a review of such studies (Norris and Kelly, 1982) reveals the occurrence of sulfur and iron oxidizing bacteria (*Thiobacillus, Leptospirillum, Metallogenium, Gallionella, Leptothrix*), heterotrophic bacteria, yeasts, filamentous fungi, algae and protozoa. The microbial interactions that occur in these environments and the consequences of such interactions for leaching efficiency are known only scantily. Analysis of laboratory model systems has revealed a variety of interactions.

Certain interactions may catalyse the cyclic transformation of inorganic substrates. Thus, iron-(III) produced by iron-oxidizing bacteria can be reduced by *Thiobacillus ferrooxidans* and *T. thiooxidans via* anaerobic sulfur oxidation to iron(II). Such recycling of iron species could be significant in the anaerobic removal of sulfur coating from metal sulfide ores in the oxygen deplete centres of large leach dumps. Mixed populations also may effect the leaching of ores that are recalcitrant to the component monopopulations. For example, whereas *T. ferrooxidans* can grow on pyrite (FeS_2), this ore does not support the growth either of *T. thiooxidans* (a sulfur oxidizer) or of *Leptospirillum ferrooxidans* (an iron oxidizer). However, coculture of *T. thiooxidans* and *L. ferrooxidans* produces pyrite breakdown at rates higher than those achieved by *T. ferrooxidans* alone (Norris and Kelly, 1978). The leaching activity of *T. ferrooxidans* itself may be potentiated in mixed culture with such mixotrophs as *T. acidophilus* and *T. organoparus*. The basis of the latter interaction is probably the removal of organic acids, such as pyruvate, excreted by *T. ferrooxidans* and which become growth inhibitory if allowed to accumulate. Finally the association of heterotrophic microorganisms with metal-leaching thiobacilli has been shown to improve leaching rates. The study of Tsuchiya *et al.* (1974) on the leaching of a copper–nickel concentrate is often quoted. These authors showed that a mixture of *T. ferrooxidans* and the acid-tolerant, nitrogen-fixing *Beijerinkia lacticogenes* was much more efficient at leaching than either monoculture. Increased availability of nitrogen in this latter system is the likely basis of the effect but this was not proved unequivocally and other interactions might also be significant. More recently Aitkeldieva (1982) has described the extraction of copper by a complex community of bacteria dominated by heterotrophs (>90%).

Most metal leaching operations currently employ large dumps but concentrates will probably be leached in heap and vat systems. In each of these systems the introduction of high activity strains is an attractive possibility for improving leaching rates. A few trials of this sort have been made. Genetically labelled *T. ferrooxidans* inoculated into a copper leach dump comprised over half the ferrobacillus population after two years but its high leaching capacity was not maintained (Groudev *et al.*, 1978). A detailed understanding of microbial interactions in leaching processes clearly is necessary to ensure the competitiveness of introduced strains and the maintenance of their activities.

Microorganisms can also be used to recover metals from industrial effluents either *via* accumulation, precipitation or the trapping of colloidal or particulate materials. Such recovery is made in stream, lagoon or activated sludge systems and in each, mixed microbial populations are involved. Metal accumulation in waste water treatment plants has been examined quite extensively (Cheng *et al.*, 1975; Nelson *et al.*, 1981) but the microbial communities and their dynamics are largely unknown. In addition the metal accumulating properties of some mixed cultures of known composition have been reported. The silver-accumulating community of *Pseudomonas maltophilia*, *Staphylococcus aureus* and a coryneform bacterium was more resistant to silver ions than a pure culture of *Ps. maltophilia* while the two other species were highly sensitive (Charley and Bull, 1979). Futhermore, accumulation of silver by the community (316 mg silver g^{-1} dry weight) was much greater than that by *Ps. maltophilia* (182 mg silver g^{-1} dry weight). Similarly, a 10-membered copper-accumulating community of bacteria isolated from activated sludge had a tolerance of copper superior to that of pure cultures of any of the constituent species (Dunn and Bull, 1983).

15.2.7.2 Defined mixed cultures

The number of defined, deliberately constituted mixed cultures deployed in the fermentation industry is small but introductions have been made into such diverse areas as the treatment of xenobiotic-containing wastes, single cell protein production, bioconversions and food fermentations.

(i) Yoghurt production

The use of mixed cultures in the dairy industry is well founded. The yoghurt fermentation, for example, is based on a protocooperative interaction between *Lactobacillus bulgaricus* and *Streptococcus thermophilus*, *i.e.* a mutualistic relationship that is not obligatory. This mixed culture is very stable under chemostat conditions and stability is mediated *via* product (lactic acid) inhibition. This stability has prompted various attempts to develop a process for continuous yoghurt manufacture. One successful development at the Netherlands Institute for Dairy Research (see Driessen, 1981 for review) is based on a two-stage fermentation. The first stage comprises the prefermentation of milk in a stirred tank reactor, and the second a plug flow coagulation. The total residence time from milk to product is only 3 h.

(ii) Single Cell Protein (SCP)

A number of mixed culture processes have been researched and commercialized for SCP production from carbohydrate, methane, natural gas and methanol feedstocks. The development of processes based on the latter three feedstocks are particularly interesting in the present context. The frequently observed instability of methylotroph monocultures on methane and the mixed alkane content of natural gases were two potent stimuli for the mixed culture R and D. Culture instability resulting from the accumulation of methanol in the broth can be relieved by the coculture with methanol-scavenging bacteria such as *Hyphomicrobium*. Similarly, the products of C_2 to C_5 *n*-alkane cometabolism by methane-utilizing bacteria are removed by heterotrophic species in mixed cultures, again relieving potential growth inhibition of the methylotroph and increasing the overall biomass yield. The advantages claimed for mixed cultures in SCP production include increased yield, specific growth rate, culture stability, foam prevention, resistance to contamination, and effective assimilation of mixed substrates (Harrison, 1978). A recent critical appraisal of this field has been made by Linton and Drozd (1982).

(iii)   Xenobiotic degradation

The increasing rate and level of introduction of synthetic chemicals into the environment in the past few decades has constituted an entirely novel challenge to the metabolic capabilities of microorganisms (Slater and Bull, 1982). Studies of the biodegradation of such chemicals have revealed that mixed microbial populations are frequently implicated (Bull, 1980). The intervention of mixed cultures may lead to enhanced rates of degradation, detoxification of higher concentrations of growth-inhibitory compounds, and the degradation of chemicals that prove recalcitrant in the face of monoculture attack (see Bull and Brown, 1979, for a review). The biodegradation of the organophosphorus insecticide parathion convincingly illustrates the latter point. *Pseudomonas stutzeri* synthesizes parathion hydrolase which cleaves the insecticide to *p*-nitrophenol and diethyl thiophosphonate; this is a cometabolic transformation and *p*-nitrophenol is growth-inhibitory to *Ps. stutzeri*. *Pseudomonas aeruginosa* cannot hydrolyse parathion but can grow on *p*-nitrophenol as the sole source of carbon. In complex mixed cultures containing both species of *Pseudomonas*, parathion is completely hydrolysed (Munnecke and Hsieh, 1974). Subsequently crude enzyme preparations from the mixed culture and parathion hydrolase from *Ps. stutzeri* have been developed for localized decontamination of parathion and related compounds (Munnecke, 1978).

(iv)   Biotransformations

For certain multistep biotransformations, a single-stage fermenter using a mixed microbial population may provide a viable processing option. Lee and his colleagues (1969) were among the first to exploit such cultures for the double transformation of steroids (*e.g.* 9α-fluorohydrocortisone to Δ^1-dehydro-16α-hydroxy-9α-fluorohydrocortisone with a single stage mixed culture of *Arthrobacter simplex* and *Streptomyces roseochromogenus*). An added advantage of this system was the repression of 20-ketoreductase which catalysed undesirable side reactions in monoculture fermentations. The progress of the mixed culture fermentation is strongly influenced by pH. Recently, Yoshida *et al.* (1981) have modelled the mixed batch culture and determined an optimal profile of pH change for the transformation.

15.2.8   Contamination and Degradation

Within the fermentation industry mixed culture situations arise inadvertently *via* contamination and strain degeneration.

15.2.8.1   *Contamination*

The risk of contamination is one that the fermentation industry has to live with perpetually and sterilization regimes are designed to meet the requirements of specific processes. Linton and Drozd (1982) remark that the antibiotics industry adopts sterilization regimes that ensure 99% freedom from contamination, whereas in the canning industry the probability of contamination has to be reduced to 1 in 10^{12}. Contamination is a much more serious event in time-independent continuous production than in batch fermentations which could be completed before contaminants reach significant proportions. Yeast-contaminant interactions have been studied quite extensively in the brewing process and, in particular, the implications of cosedimentation of yeast and bacteria have been recognized. A recent review of this latter topic is given by White and Kidney (1981). Other traditional fermentations such as those based on milk, are particularly susceptible to bacteriophages (Lawrence and Thomas, 1979, should be consulted for information).

Returning to large scale continuous fermentations, it has been suggested that it will be technically difficult and costly to attempt absolute monoculture operation, especially where recycling of process water is practised (Linton and Drozd, 1982, but see Smith, 1980, for a discussion of sterile auditing in production scale SCP continuous cultures). One means of circumventing this problem is to develop defined mixed cultures for operation under septic conditions (*e.g.* BP *n*-paraffin process; Laine, 1974) or aseptic conditions (*e.g.* Shell approach to SCP production from *n*-alkanes and methanol; Harrison 1978; Linton and Drozd, 1982). The comparative resistance of mono- and mixed-culture fermentations to contamination has received little experimental testing. The work of Rokem *et al.* (1980) demonstrated in impressive fashion that mixed cultures of the methylotrophic *Pseudomonas* C with heterotrophic soil bacteria withstood deliberate attempts to contaminate with *Escherichia coli*, *Salmonella typhimurium* or *Staphylococcus aureus*. In contrast, monocultures of the *Pseudomonas* C were susceptible to this contamination.

Finally, it should be pointed out that judicious selection of organisms can enable monoculture continuous fermentations to be operated as septic protected fermentations. For example, Rodriguez (Rodriguez, 1982; Bull, 1983) working in this laboratory selected by chemostat enrichment a strain of *Aspergillus terreus* which had maximum protein productivity at 45 °C and pH 3.5 when grown on starch. Under these conditions, continuous cultivation for SCP production was completely protected against massive challenge by soil microorganisms.

15.2.8.2 Industrial fermentations with unstable strains

Strain degeneration, particularly in terms of the loss of desired synthesis, is a phenomenon familiar to industrial microbiologists, and the selection kinetics of spontaneous, non-productive mutants has been treated by several authors. Much of the early work with continuous cultures was related to the selection of mutants (see Malek and Fencl, 1966). In more recent years the stability of genetically engineered organisms during batch and continuous culture has attracted attention. A number of experimental studies have been made of plasmid stability in batch and chemostat cultures and the effect that instability has on the competitiveness. The accumulated data reveal that plasmid maintenance (whole or partial) and copy numbers are strongly influenced by the growth environment. The presence of plasmids usually, but not invariably, leads to a slower growing recombinant in mixed culture with the faster growing plasmid-free or plasmid-modified revertants (Imanaka and Aiba, 1981). Recently Ollis (1982) has examined the implications of plasmid instability for product formation in batch and continuous cultures and has suggested various strategies for process operation under such circumstances, including selective cell recycling.

15.2.9 Environmental Biotechnology

A very rapidly expanding area of biotechnology involves the introduction of selected microorganisms into the environment for a diversity of purposes: improved plant growth (*e.g. Rhizobium* and mycorrhizal inoculants), improved nitrogen status of soil (*e.g.* blue green algal inoculants), enhanced silage production, pollutant dispersal, mineral leaching, degradation of crop residues *in situ*. The success of such operations is highly dependent on the competitiveness of introduced organisms *vis-à-vis* the indigenous microflora and an appreciation of mixed population dynamics cannot be neglected.

15.3 MIXED SUBSTRATES

In the present context, the term mixed substrate is most usefully defined in the manner proposed by Harder and Dijkhuizen (1982): 'the presence of a multiplicity of sources of nutrients that serve a similar physiological function'. The term embraces all types of nutrient–carbon sources, energy sources, nitrogen sources, phosphorus sources, *etc.*, and both homologous and non-homologous compounds. Culture media for industrial fermentations frequently are formulated so that they contain complex mixtures of nutrients, and mixed carbon sources in particular have been exploited for yield improvement of antibiotics. The great majority of biological waste treatment processes involve the dissimilation of mixed substrates. Finally, the growth of microorganisms in natural ecosystems usually occurs on low concentrations of diverse mixed substrates; indeed, as Harder and Dijkhuizen (1982) point out, studies in the last few years have revealed microorganisms that appear to be specially adapted to growth under mixed substrate-limited conditions. These latter organisms have been identified only by making continuous culture enrichments with mixed substrate limitation.

The data in Table 3 exemplify mixed substrate growth situations that have been investigated but it should be emphasized that the greatest attention has been paid to mixed carbon source utilization.

Table 3 Mixed Substrate Growth[a]

Substrate mixture	Physiological function	Organism
Glucose + lactose Acetate + oxalate Glucose + p-hydroxybenzoate	Mixed organic carbon and energy sources	Heterotrophs
Succinate + H_2/CO_2	Mixed organic/inorganic carbon and energy sources	Facultative chemolithotrophs
H_2 + CO	Mixed inorganic energy sources	Chemolithotrophs (carboxydobacteria)
CO_2 + acetate	Mixed carbon sources	Photoautotrophs
$O_2 + NO_3^-$ $CH_2(NH)_2 + NO_3^-$	Mixed electron acceptors	Denitrifiers Hyphomicrobium
$NH_4^+ + NO_3^-$	Mixed nitrogen sources	Numerous
$PO_4^{3-} + PO_3^{3-}$	Mixed phosphorus sources	—

[a] Based on Harder and Dijkhuizen, 1982.

15.3.1 Patterns of Mixed Substrate Utilization

Growth of microorganisms on two-substrate mixtures in batch culture elicits a number of different patterns of utilization (Figure 2). Substrate utilization usually is sequential or simultaneous. The classic case of mixed substrate utilization is that of diauxie. Monod's (1942) study of this phenomenon focused on two-sugar mixtures of which glucose/lactose utilization by *Escherichia coli* is particularly well known and understood (Figure 2a). Diauxic growth describes the situation in which the two substrates interfere with the utilization of each other such that two distinct cycles of growth occur, each at the expense of just one of the substrates. Thus, glucose interferes with the utilization of lactose in the above example: only when glucose is completely exhausted are enzymes synthesized which allow growth on lactose to proceed. A lag phase is observed between the two growth cycles.

Many other instances of diauxie have been studied since Monod's pioneering work and the phenomenon is not restricted to sugar mixtures. *Pseudomonas oxalaticus*, for example, exhibits diauxic growth on mixtures of organic acids. Whatever the sequence of substrate utilization in diauxic growth, the substrate utilized secondarily invariably supports a lower specific growth rate. It became generally accepted, therefore, that when presented with a choice of substrates, microorganisms grew on one at a time, with the so-called 'richer' substrate being used first. However, it has been realized subsequently that diauxie is not the only pattern of mixed substrate utilization.

Cases are known in which sequential utilization of two substrates occurs without an intervening lag phase. In some situations the switch from the 'richer' to the 'poorer' substrate is accompanied by a reduction in the specific growth (Figure 2b) as occurs when *Kluyveromyces fragilis* is cultured on a defined mixture of glucose and sorbitol, or on coconut water waste (Smith and Bull, 1976). In this example the specific growth rates on glucose and sorbitol were 0.40 and 0.08 h^{-1} respectively and on coconut water 0.58 h^{-1} (glucose plus fructose) and 0.11 h^{-1} (sorbitol). There are yet other cases of sequential substrate utilization known where the specific growth rate on the two substrates is essentially identical (Figure 2c). The growth of *Escherichia coli* on glucose plus galactose illustrates this latter situation.

Finally, the utilization of mixed substrates may proceed simultaneously (Figure 2d). Glucose and fructose utilization by *Kluyveromyces fragilis* is a case in point.

Fermentation media quite often have mixed substrate complexities greater than two, particularly where crude, natural materials are used. Rather few data are available in the published literature on the patterns of multisubstrate assimilation. Some relevant data have been provided from the study of food yeast production from coconut water waste, referred to above. This natural fermentation feedstock contains sucrose, glucose, fructose and sorbitol in various proportions depending on the age of the nut, and on palm variety. The pattern of substrate utilization in batch culture is shown in Figure 3. Growth at the expense of sugars produced a single exponential

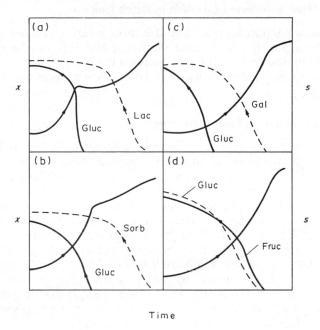

Figure 2 Patterns of dual substrate (carbon) utilization in batch cultures: (a) glucose–lactose, *Escherichia coli*; (b) glucose–sorbitol, *Kluyveromyces fragilis*; (c) glucose–galactose, *E. coli*; (d) glucose–fructose, *K. fragilis*. x, biomass and s, substrate concentrations

growth phase in which the specific growth rate was 0.39 to 0.41 h^{-1}; subsequent growth occurred on sorbitol at a predictably lower growth rate (0.06 to 0.08 h^{-1}).

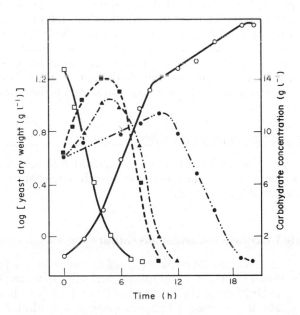

Figure 3 Batch culture of *Kluyveromyces fragilis* on a glucose–fructose–sucrose–sorbitol mixture: dry weight (○), glucose (■), fructose (▲), sucrose (□), sorbitol (●). The apparent increase in sorbitol concentration during the first 12 h of culture represents an interference by fructose of the sorbitol assay (Smith and Bull, 1976, with permission)

It is important to realize that different microorganisms may respond very differently to a given combination of substrates. Harder and Dijkhuizen (1982) also emphasize this point: 'A combination that gives diauxic growth in one organism may not in another . . . a preferred substrate for one organism may be a secondary substrate for another'.

15.3.2   Control of Mixed Substrate Utilization in Batch Culture

The control of substrate utilization is mediated at three levels: (i) *via* the regulation of substrate transport; (ii) *via* the regulation of enzyme synthesis; and (iii) *via* the regulation of enzyme activity. Harder and Dijkhuisen (1982) have discussed the control of mixed substrate utilization in terms of these latter regulatory mechanisms and only a few examples will be included here.

15.3.2.1   Control by regulation of substrate transport

Utilization of sugar mixtures by enteric bacteria may be controlled by interference with the phosphoenolpyruvate phosphotransferase system. Several studies have been made of glucose–fructose assimilation by *Escherichia coli*. Here it is believed that glucose blocks the formation of fructose phosphate by interacting with the membrane-located enzyme II component of the phosphotransferase system. The use of glucose and pyruvate by this bacterium can proceed simultaneously, but when growth is made on the mixture, reciprocal inhibition of the uptake rates occurs (Hegewald and Knorre, 1978). Thus glucose uptake is inhibited by the accumulation of acetyl-coenzyme A as a result of pyruvate metabolism. Subsequently acetyl-CoA stimulated oxaloacetate synthesis affects the phosphoenolpyruvate pool size thereby influencing glucose uptake. Organic acids also inhibit glucose uptake by pseudomonads. For example, when *Pseudomonas fluorescens* is grown under conditions of glucose excess, periplasmic dehydrogenases oxidize the sugar, and gluconate plus 2-oxogluconate accumulate extracellularly (Dawes *et al.*, 1976). Glucose transport is severely inhibited by these two organic acids.

15.3.2.2   Control by regulation of enzyme synthesis

The glucose–lactose diauxie is one of the most thoroughly revealed cases of mixed substrate catabolism being controlled *via* enzyme synthesis. Glucose inhibition of lactose utilization by *Escherichia coli* is a multivalent phenomenon encompassing catabolite repression of enzymes of the *lac* operon, inducer exclusion, and catabolite inhibition of extant lactose enzymes. Very many cases of catabolite repression of secondary substrate utilization are known that involve mixed carbon and energy compounds but such a regulatory mechanism is not restricted to these latter substrates. An interesting case, quoted by Harder and Dijkhuizen (1982) in their review, relates to methylamine utilization by *Arthrobacter*. In the presence of ammonia diauxic utilization of glucose and methylamine occurs with glucose being the preferred substrate. However, when growth occurs in the absence of ammonia, glucose and methylamine are utilized simultaneously, *i.e.* catabolite repression is alleviated when the secondary substrate is the source of nitrogen.

15.3.2.3   Control by regulation of enzyme activity

Allosteric modulation and adenylation are two significant mechanisms by which the activity (inhibition) of enzymes is affected. As indicated above such a mechanism constitutes part of the glucose-regulated utilization of lactose by *E. coli*.

15.3.3   Mixed Substrate Utilization in Continuous Culture

Mixed carbon utilization by chemostat populations of microorganisms is affected by the nature of the substrates and the dilution rate, and preferential and simultaneous utilization patterns have been reported (Harder and Dijkhuizen, 1976; Bull and Brown, 1979). Simultaneous utilization of carbon and energy substrates is a widespread phenomenon at low dilution rates in carbon limited chemostats. This observation suggests that simultaneous substrate utilization is of common occurrence in nutrient poor environments, a thesis which has been developed in a more recent review by Harder and Dijkhuizen (1982). The latter authors argue convincingly that organisms exist in Nature that are specially adapted to mixed nutrient limitation. Thus organisms that have the capacity to utilize low concentrations of multiple substrates concurrently probably flourish in environments where two or more substrates are turned over at similar rates. Harder and Dijkhuizen show that this situation pertains for certain chemolithotrophic and methylotrophic bacteria and suggest that it may be more widely occurring.

In continuous culture the utilization of mixed carbon sources is frequently subject to distinct on–off control. The data in Table 4 express the point of on–off control in terms of the relative dilution rate for a number of organism–substrate combinations. The relative dilution rate provides a measure of the physiological state of the organism. It can be seen that control of mixed carbon utilization is a function of the mixture of substrates and the organism (*cf.* fructose–glucose). The regulatory mechanisms that operate under these conditions are comparable to those described for batch cultures. Catabolite repression has been studied in this context and its regulation of sucrose (invertase) and acetamide (aliphatic amidase) utilization in mixed substrate chemostats has been reported by Smith and Bull (1976) and Clarke *et al.* (1968) respectively.

Table 4 Utilization of Mixed Carbon Sources in Carbon-limited Chemostats

Substrate combination[a]	Organism	On–off control in terms of W[b]
Fructose–glucose	*Pseudomonas fluorescens*	< 0.20
	Escherichia coli	0.45
	Saccharomyces fragilis	0.75
Lactose–glucose	*Escherichia coli*	0.72
Oxalate–acetate	*Pseudomonas oxalaticus*	0.43
Sorbitol–glucose	*Saccharomyces fragilis*	0.07
Sucrose–glucose	*Saccharomyces fragilis*	0.39

[a] Utilization of the first substrate is prevented by the second when $D > W$.
[b] W is the relative dilution rate, D/μ_{max} (Herbert, 1976).

15.3.3.1 Double substrate limited growth

In the context of chemostat culture, limiting substrate denotes that substrate which controls or limits the specific growth rate of an organism. Brief consideration will be made at this juncture of the possibility of more than one substrate functioning in such a rate limiting capacity.

A useful conceptual approach to double substrate limited growth has been presented by Bader (1978, 1982) in terms of interactive and non-interactive models. The interactive model assumes that both substrates are available in limiting concentrations such that both affect the growth rate. Such a situation might arise if a product P_1, which is required for growth, is synthesized from substrate S_1 and involves substrate S_2 as a cofactor (Figure 4a). If the concentrations of S_1 and S_2 are present at half μ_{max} concentrations, only one half of the enzyme is functioning at half its maximum rate and the rate of P_1 synthesis is only one quarter maximum. Thus, under these conditions, both substrates affect product formation and, consequently, growth. Similarly substrates S_1 and S_2 might be combined by enzyme E_4 to produce P_1 which again is required for growth (Figure 4b). Here some interaction always occurs but the extent is dependent upon the relative rate constants.

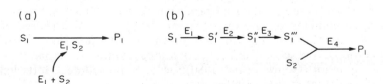

Figure 4 Interactive model of double substrate limited growth

Non-interactive models assume that the specific growth rate can only be limited by a single substrate at any one time. Thus, μ will be equal to the lowest value predicted from separate single substrate systems under otherwise identical conditions. In the non-interactive model the two substrates are metabolized by independent pathways (Figure 5) and growth is limited by the product present in shortest supply. The two pathways may not be completely independent but their interaction may be very small.

Mathematical modelling of double substrate limited growth can be based on various familiar growth kinetics (Monod, Blackman, exponential) and Bader (1982) has compared interactive and non-interactive approaches in each of these terms. Non-interactive models are discontinuous

$$S_1 \xrightarrow{E_1} P_1 \qquad S_2 \xrightarrow{E_2} P_2$$

Figure 5 Non-interactive model of double substrate limited growth

functions of the transition line from one substrate limitation to the other; these models also predict higher values of μ in the region where the dimensionless substrate concentrations S_1/K_{s1} and S_2/K_{s2} are both small (Figure 6). In contrast interactive models are continuous functions.

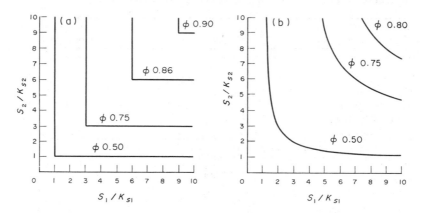

Figure 6 Relative specific growth rate, ϕ, as a function of two dimensionless substrate concentrations for (a) non-interactive and (b) interactive models based on Monod kinetics (Bader, 1982)

From an operational point of view interactive models are desirable. They can be developed to include multiple substrates and their structure is such that substrates present in non-limiting concentrations are eliminated. Unfortunately only a few experimental studies have been made of double substrate limitation and for substrates that interact only very weakly (*e.g.* vitamin B_{12} and phosphate) the non-interactive model gives the best fit. It is probable that the interactive and non-interactive models approach one or the other depending on the nature of the substrate combinations. True double substrate limitation of growth rate is predicted to be a rare event (see Bader, 1982 for an assessment) and very careful experimental control will be essential to achieve it. However, it is feasible to identify conditions for double substrate limitation and for certain fermentation processes, *e.g.* secondary metabolites, this may be highly desirable.

15.3.3.2 *Efficiency of growth on mixed substrates*

As indicated above, the metabolism of mixed substrates may occur in a totally independent manner. However, several cases are known where metabolism of mixed substrates is significantly interactive: it is quite common to find one substrate being used as a source of carbon and a second substrate as a source of energy. In the following list the first substrate in the pair functions as the carbon source, the second as an energy source: mannitol–methanol (van Verseveld *et al.*, 1979); glucose–ethanol (Geurts *et al.*, 1980); acetate–thiosulfate (Gottschal and Kuenen, 1980); glucose–formate (Linton *et al.*, 1981).

Linton and Stephenson (1978) observed a strong correlation between the energy content of a substrate and growth yield for substrates with heats of combustion up to *ca.* 46 kJ g^{-1} substrate carbon, or reductance degrees of *ca.* 4 electrons C-mol^{-1} substrate. Above these values Y_s^{max} was independent of the energy content. Thus, when growth is made on highly reduced substrates such as ethanol, methanol or hydrocarbons, not all the available energy can be utilized for biomass synthesis, *i.e.* the system is carbon limited. Subsequently, Linton and his colleagues (Linton *et al.*, 1981) showed that the assimilation of glucose carbon was significantly enhanced in the presence of formate. The addition of 30 mM formate to a glucose-limited chemostat culture of *Beneckea natriegens* increased the biomass field *from glucose* from 84 to 93 g mol^{-1}, formate alone did not support growth and in mixtures with glucose its metabolism provided only energy.

A detailed quantitative analysis of the above phenomenon has been provided by van Verseveld *et al.* (1979) for the growth of *Paracoccus denitrificans* on mannitol and methanol. The $Y_{-\text{man}}^{\max}$ values for two and three energy conservation sites are 95.3 and 109.7 g dry weight mol^{-1} mannitol respectively while the increase resulting from methanol catabolism is equivalent to $[N + N(M/R_{\text{m}})]150.2$, where M is the ratio methanol : mannitol, N is the proportion of mannitol assimilated in the absence of methanol, R_{m} is the methanol-equivalent having the energy content of one mannitol-equivalent, and 150.2 is the weight of one mole biomass. For two and three sites of energy conservation the calculated values of R_{m} are 5.88 and 5.46 respectively and these become 3.05 and 2.01 for a $Y_{-\text{man}}^{\max}$ of 150.2. Calculations of this sort are valuable in the design of processes for biomass production from reduced substrates and enable optimal substrate ratios to be determined.

More complex patterns of energy metabolism and carbon assimilation have been revealed and are exemplified by the work of Harder and his colleagues on acetate–formate utilization by *Pseudomonas oxalaticus* (see Harder and Dijkhuizen, 1982). Unlike *B. natriegens*, *P. oxalaticus* can grow on formate as the sole source of carbon and energy but in batch culture on a mixture of acetate and formate the formate only serves as a source of energy. When formate is added to the medium feed of an acetate limited chemostat (D, 0.10 h^{-1}, $S_{R(\text{acetate})}$, 30 mM) the steady state biomass concentration increased; under these conditions the metabolic fate of formate is concentration dependent. (a) At formate concentrations of 40 mM or less, formate functioned only as an ancillary source of energy for acetate assimilation: at 40 mM formate the biomass concentration increased by 52%. (b) At formate concentrations greater than 40 mM ribulose-1,5-bisphosphate carboxylase became progressively derepressed and autotrophic assimilation of CO_2 proceeded *via* the energetically expensive Calvin cycle. The extra synthesis of biomass that occurred under these conditions of high energy expenditure was very small (less than 2% when formate was increased from 40 to 90 mM). An exactly parallel situation is that of *Thiobacillus* A2 growing on acetate and thiosulfate in a chemostat. At thiosulfate : acetate ratios of less than 2 the Calvin cycle enzymes are repressed and energy derived from thiosulfate oxidation allows a greater proportion of acetate carbon to be assimilated.

15.4 COMETABOLISM

The cometabolic transformation of substrates does not necessarily conform to the strict definition of mixed substrate utilization given at the beginning of Section 15.3. However, the phenomenon is usefully introduced, albeit superficially, at this point because of its relevance to substrate transformations in Nature, in waste treatment operations and for industrial purposes.

The studies that focused attention on what is now referred to as cometabolism were made by Foster and his colleagues with hydrocarbon utilizing bacteria (Foster, 1962). For example, growth on a substrate such as methane enabled the non-growth substrates ethane and propane to be oxidized to products including acetaldehyde and acetate, and propionate and acetone, respectively. The methanotroph could not grow at the expense of ethane or propane, or of any of the oxidation products. Foster coined the term *cooxidation* to describe these latter metabolic transformations. Subsequently the definition was broadened to include reactions other than oxidation, *e.g.* hydrolysis, dehalogenation, for which the wider term *cometabolism* was introduced.

In 1972 Horvath assembled the information then available that purported to support the concept of cometabolism and from his survey the following criteria emerged as being necessary to describe a transformation as cometabolic: (i) the cosubstrate alone cannot sustain growth of the organism; (ii) products accumulate stoichiometrically from the cosubstrate; (iii) transformation of the cosubstrate is accompanied by increased oxygen consumption; and (iv) transformation of the cosubstrate occurs *via* the adventitious action of extant enzymes.

Unfortunately so much of the putative evidence in support of cometabolism failed to meet these criteria that Hulbert and Krawiec (1977) challenged the basic concept and disputed the metabolic novelty of such transformations. This criticism was timely and has prompted a reappraisal of the phenomenon. Many microbiologists argue that the term cometabolism should be perpetuated but its definition be clarified. Dalton has recently provided such a redefinition (see Dalton and Stirling, 1982) of cometabolism: 'the transformation of a non-growth substrate in the obligate presence of a growth substrate or another transformable compound'. Dalton and Stirling add two riders to their definition: (1) the term non-growth substrate should be restricted for compounds that do not support cell division; and (2) that the transformation of non-growth substrates in the absence of another substrate should be simply termed *fortuitous metabolism* and not viewed as a novel event. Thus, microorganisms which could utilize endogenous materials, such as glycogen and poly-β-hydroxybutyrate, might be able to transform non-growth substrates in the absence of exogenous energy. The redefinition of these terms has been prompted by research on methane oxidizing bacteria and the study of methane monooxygenases (Figure 7). In Figure 7 XH_2 is a reducing agent such as $NADH_2$. Thus, a methylotroph (1) can grow on methane, (2) can cometabolize carbon monoxide if methane is available to generate XH_2, and (3) can fortuitously metabolize ethane without the requirement of methane catabolism to generate XH_2.

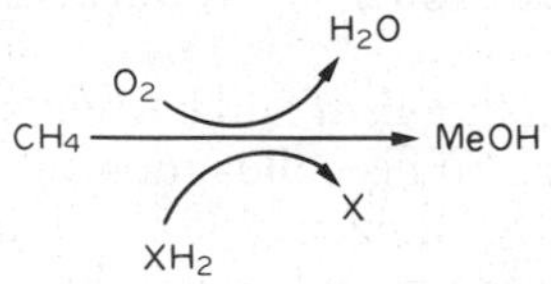

Figure 7 Bacterial oxidation of methane

Cometabolism can be viewed as the consequence of broad substrate specificity of enzymes or 'substrate ambiguity' as it has been termed by Dalton and Stirling (1982). It can be argued that an organism is advantaged when the cometabolic substrate is toxic and cometabolism leads to detoxification. The *quid pro quo* for broad enzyme specificity is the production of toxic cometabolic products (*e.g.* hydrolysis of parathion by *Pseudomonas stutzeri* to yield toxic non-metabolizable *p*-nitrophenol), and, as Dalton and Stirling (1982) have commented, cometabolism can be detrimental when an organism is faced with multiple substrates that can be metabolized by a single enzyme and metabolism of the cosubstrate is more kinetically favourable than the growth substrate. Thus, the K_m and V_{max} for methane (growth substrate) and CO (cosubstrate) metabolism of methane-oxidizing bacteria are 15 and 2.7 μM and 100 and 85 nmol mg^{-1} min^{-1} respectively.

15.4.1 Cometabolism in the Environment

Cometabolism has been studied especially vigorously in the context of the biodegradation of xenobiotic chemicals in the environment (Bull, 1980). The contribution of cometabolic degradation of xenobiotics may be significant but has rarely been quantified. In salt marsh ecosystems, for example, cometabolism has been considered to be primarily responsible for malathion degradation. The cometabolic products derived from xenobiotics can produce a variety of effects (Bull, 1980): (1) provision of growth substrates for other members of the microflora, *e.g.* *p*-chlorophenylacetic acid from DDT, *p*-nitrophenol from parathion; (2) generation of products toxic to the producer organism, *e.g.* 1-naphthol from carbaryl, diazene derivatives from arylhydrazines; (3) increased recalcitrance, *e.g.* transformation of propanil to 3,4-dichloroaniline followed by peroxidase condensation to the corresponding diazo compound.

The ubiquitous nature and abundance of methane-oxidizing bacteria suggests that they may have a significant role in the cometabolic removal of carbon monoxide from the environment. These bacteria may also be important in the fortuitous or cometabolic transformation of pollutants such as ammonia and halogenated alkanes (Dalton and Stirling, 1982).

15.4.2 Technological Potential

The cometabolic activities of microorganisms have been used extensively for the biotransformation of steroids (see Miller, Volume 3, Chapter 6) and the transformation of many other compounds, particularly aromatics, is well documented. In recent years increasing attention has been given to the cometabolic versatility of methanotrophic bacteria and commercial exploitation of the methane monooxygenase (see Higgins *et al.*, 1980).

15.5 REFERENCES

Aitkeldieva, S. A. (1982). Extraction of copper from Dzhezkazgan ore by an association of heterotrophic bacteria. *Izv. Akad. Nauk SSSR, Ser. Biol.*, **4**, 46–51. (*Chem. Abstr.* 1982, **97**, 159 325).
Bader, F. G. (1978). Analysis of substrate limited growth. *Biotechnol. Bioeng.*, **20**, 183–199.

Bader, F. G. (1982). Kinetics of double-substrate limited growth. In *Microbial Population Dynamics*, ed. M. J. Bazin, pp. 1–32. CRC Press Inc., Boca Raton, FL.

Bazin, M. J. (1981). Mixed culture kinetics. In *Mixed Culture Fermentations*, ed. M. E. Bushell and J. H. Slater, pp. 25–51. Academic, London.

Bryant, M. P., E. A. Wolin, M. J. Wolin and R. S. Wolfe (1967). *Methanobacillus omelianskii*, a symbiotic association of two species of bacteria. *Arch. Mikrobiol.*, **59**, 20–31.

Bull, A. T. (1980). Biodegradation: some attitudes and strategies of microorganisms and microbiologists. In *Contemporary Microbial Ecology*, ed. D. C. Ellwood, J. N. Hedger, M. J. Latham, J. M. Lynch and J. H. Slater, pp. 107–136. Academic, New York.

Bull, A. T. (1981). Strategies in microbial process optimization. In *Global Impacts of Applied Microbiology 6th International Conference*, ed. S. O. Emejuaiwe, O. Ogunbi and S. O. Sanni, pp. 623–645. Academic, London.

Bull, A. T. (1983). Continuous culture for production. In *Basic Biology of New Developments in Biotechnology*, ed. A. Hollaender, A. I. Laskin and P. Rogers, pp. 405–438. Plenum, New York.

Bull, A. T. and C. M. Brown (1979). Continuous culture applications to microbial biochemistry. In *International Review of Biochemistry*, ed. J. R. Quayle, vol. 21, pp. 177–226. University Park Press, Baltimore, MD.

Bull, A. T. and J. R. Quayle (1984). New dimensions in microbiology: an introduction. *Philos. Trans. R. Soc. London, Ser. B*, **297**, 447–457.

Bull, A. T. and J. H. Slater (1982a). Historical perspectives on mixed cultures and communities. In *Microbial Interactions and Communities*, ed. A. T. Bull and J. H. Slater, vol. 1, pp. 1–12. Academic, London.

Bull, A. T. and J. H. Slater (1982b). Microbial interactions and community structure. In *Microbial Interactions and Communities*, ed. A. T. Bull and J. H. Slater, vol. 1, pp. 13–44. Academic, London.

Bull, A. T. and J. H. Slater (1982c). *Microbial Interactions and Communities*. Academic, London.

Bushell, M. E. and J. H. Slater (1981). *Mixed Culture Fermentations*. Academic, London.

Cappenberg, Th. E. (1975). A study of mixed continuous cultures of sulphate-reducing and methane producing bacteria. *Microb. Ecol.*, **2**, 60–72.

Charley, R. C. and A. T. Bull, (1979). Bioaccumulation of silver by a multispecies community of bacteria. *Arch. Microbiol.*, **123**, 239–244.

Cheng, M. H. (1975). Heavy metals uptake by activated sludge. *J. Water Poll. Control Fed.*, **47**, 362–372.

Clarke, P. H., M. A. Houldsworth and M. D. Lilly (1968). Catabolite repression and the induction of amidase synthesis by *Pseudomonas aeruginosa* 8602, in continuous culture. *J. Gen. Microbiol.*, **51**, 225–234.

Culver, D. C. (1981). Introduction to the theory of species interactions. In *The Fungal Community. Its Organization and Role in the Ecosystem*, ed. D. T. Wicklow and G. C. Carrol, pp. 281–194. Dekker, New York.

Curds, C. (1982). The ecology and role of protozoa in aerobic sewage treatment processes. *Annu. Rev. Microbiol.*, **36**, 27–46.

Dalton, H. and D. I. Stirling (1982). Co-metabolism. *Philos. Trans. R. Soc. London, Ser. B*, **287**, 481–496.

Dawes, E. A., M. Midgley and P. H. Whiting (1976). Control of transport systems for glucose, gluconate and 2-oxogluconate, and of glucose metabolism in *Pseudomonas aeruginosa*. In *Continuous Culture 6: Applications and New Fields*, ed. A. C. R. Dean, D. C. Ellwood, C. G. T. Evans and J. Melling, pp. 195–207. Ellis Horwood, Chichester.

Driessen, M. F. (1981). Protocooperation of yoghurt bacteria in continuous culture. In *Mixed Culture Fermentations*, ed. M. E. Bushell and J. H. Slater, pp. 99–120. Academic, London.

Dunn, G. M. and A. T. Bull (1983). Bioaccumulation of copper by a defined community of activated sludge bacteria. *Eur. J. Appl. Microbiol. Biotechnol.*, **17**, 30–34.

Foster, J. W. (1962). Hydrocarbons as substrates for microorganisms. *Antonie van Leeuwenhoek*, **28**, 241–274.

Fry, J. C. (1982). The analysis of microbial interactions and communities *in situ*. In *Microbial Interactions and Communities*, ed. A. T. Bull and J. H. Slater, vol. 1, pp. 103–152. Academic, London.

Geurts, T. G. E., H. E. de Koek and J. A. Roels (1980). A quantitative description of growth of *Saccharomyces cerevisiae* CBS426 on a mixed substrate of glucose and ethanol. *Biotechnol. Bioeng.*, **22**, 2031–2043.

Goswell, R. W. and R. E. Kunkee (1977). Fortified wines. In *Alcoholic Beverages, Economic Microbiology*, ed. A. H. Rose, vol. 1, pp. 478–535. Academic, London.

Gottschal, J. C. and J. G. Kuenen (1980). Mixotrophic growth of *Thiobacillus* A2 in acetate and thiosulphate as growth-limiting substrates in the chemostat. *Arch. Microbiol.*, **126**, 33–42.

Groudev, S. N., F. N. Genchev and S. A. Gaiderjiev (1978). Observations on the microflora in an industrial dump leaching copper operation. In *Metallurgical Applications of Bacterial Leaching and Related Microbiological Phenomena*, ed. L. E. Murr, A. E. Torma and J. A. Brierley, pp. 253–274. Academic, New York.

Haas, C. N., H. R. Bungay and M. L. Bungay (1980). Practical mixed culture processes. *Annu. Rep. Ferment. Processes*, **4**, 1–29.

Harder, W. and L. Dijkhuizen (1976). Mixed substrate utilization. In *Continuous Culture 6: Applications and New Fields*, ed. A. C. R. Dean, D. C. Ellwood, C. G. T. Evans and J. Melling, pp. 297–314. Ellis Horwood, Chichester.

Harder, W. and L. Dijkhuizen (1982). Strategies of mixed substrate utilization in microorganisms. *Philos. Trans. R. Soc. London, Ser B*, **297**, 459–480.

Harrison, D. E. F. (1978). Mixed cultures in industrial fermentation processes. *Adv. Appl. Microbiol.*, **24**, 129–164.

Hegewald, E. and W. A. Knorre (1978). Kinetics of growth and substrate consumption of *Escherichia coli* ML30 on two carbon sources. *Z. Allg. Mikrobiol.*, **18**, 415–426.

Herbert, D. (1976). Stoicheiometric aspects of microbial growth. In *Continuous Culture 6: Applications and New Fields*, ed. A. C. R. Dean, D. C. Ellwood, C. G. T. Evans and J. Melling, pp. 1–30. Ellis Horwood, Chichester.

Higgins, I. J., D. J. Best and R. C. Hammond (1980). New findings in methane-utilizing bacteria highlight their importance in the biosphere and their commercial potential. *Nature (London)*, **286**, 561–564.

Hobson, P. N. (1981). Microbial pathways and interactions in the anaerobic treatment processes. In *Mixed Culture Fermentations*, ed. M. E. Bushell and J. H. Slater, pp. 53–79. Academic, London.

Horvath, R. S. (1972). Microbial cometabolism and the degradation of organic compounds in nature. *Bacteriol. Rev.*, **36**, 146–155.

Hulbert, M. H. and S. Krawiec (1977). Co-metabolism : a critique. *J. Theor. Biol.*, **69**, 287–291.

Imanaka, T. and S. Aiba (1981). A perspective on the application of genetic engineering: stability of recombinant plasmid. *Ann. N.Y. Acad. Sci.*, **369**, 1–14.

Jost, J. L., J. F. Drake, A. G. Fredrickson and H. M. Tsuchiya (1973). Interactions of *Tetrahymena pyriformis, Escherichia coli, Azotobacter vinelandii*, and glucose in a minimal medium. *J. Bacteriol.*, **113**, 834–840.

Laine, B. M. (1974). What proteins cost from oil. *Hydrocarbon Process.*, November, 139–142.

Lawrence, R. C. and T. D. Thomas (1979). The fermentation of milk by lactic acid bacteria. In *Microbial Technology: Current State, Future Prospects*, ed. A. T. Bull, D. C. Ellwood and C. Ratledge, pp. 187–219. Cambridge University Press, Cambridge.

Lee, B. K., D. Y. Ryu, R. W. Thoma and W. E. Brown (1969). Induction and repression of steroid hydroxylases and dehydrogenases in mixed culture fermentations. *J. Gen. Microbiol.*, **55**, 145–153.

Linton, J. D. and J. W. Drozd (1982). Microbial interactions and communities in biotechnology. In *Microbial Interactions and Communities*, ed. A. T. Bull and J. H. Slater, vol. 1, pp. 357–406. Academic, London.

Linton, J. D. and R. J. Stephenson (1978). A preliminary study on growth yields in relation to the carbon and energy content of various organic growth substrates. *FEMS Microbiol. Lett.*, **3**, 95–98.

Linton, J. D., K. Griffiths and M. Gregory (1981). The effects of mixtures of glucose and formate on the yield and respiration of *Beneckea natriegens*. *Arch. Microbiol.*, **129**, 119–122.

Mah, R. A. (1982). Methanogenesis and methanogenic partnerships. *Philos. Trans. R. Soc. London, Ser. B*, **297**, 153–170.

Malek, I. and Z. Fencl (1966). *Theoretical and Methodological Basis of Continuous Culture of Microorganisms*. Publishing House of Czechoslovak Academy of Sciences, Prague.

Monod, J. (1942). *Recherche sur la Croissance des Cultures Bacteriennes*. Herman, Paris.

Munnecke, D. M. (1978). Detoxification of pesticides using soluble or immobilised enzymes. *Process Biochem.*, **13**, 1–4.

Munnecke, D. M. and D. P. H. Hsieh (1974). Microbial decontamination of parathion and *p*-nitrophenol in aqueous media. *Appl. Microbiol.*, **28**, 212–217.

Nelson, P. O., A. K. Chung and M. C. Hudson (1981). Factors affecting the fate of heavy metals in the activated sludge process. *J. Water Poll. Control Fed.*, **53**, 1323–1333.

Norris, P. R. and D. P. Kelly (1978). Dissolution of pyrite (FeS_2) by pure and mixed cultures of some acidophilic bacteria. *FEMS Microbiol. Lett.*, **4**, 143–146.

Norris, P. R. and D. P. Kelly (1982). The use of mixed microbial cultures in metal recovery. In *Microbial Interactions and Communities*, ed. A. T. Bull and J. H. Slater, vol. 1, pp. 443–474. Academic, London.

Ollis, D. F. (1982). Industrial fermentations with (unstable) recombinant cultures. *Philos. Trans. R. Soc. London, Ser B*, **297**, 171–183.

Parker, R. B. (1966). Continuous culture system for ecological studies of microorganisms. *Biotechnol. Bioeng.*, **8**, 473–488.

Parkes, R. J. (1982). Methods for enriching, isolating, and analyzing microbial communities in laboratory systems. In *Microbial Interactions and Communities*, ed. A. T. Bull and J. H. Slater, vol. 1, pp. 45–102. Academic, London.

Raa, J. (1981). Biochemistry of microbial fish spoilage and preservation by lactic acid bacteria and added acid. In *Global Impacts of Applied Biology 6th International Conference*, ed. S. O. Emejuaiwe, O. Ogunbi and S. O. Sanni, pp. 3–16. Academic, London.

Reanney, D. C., W. P. Roberts and W. J. Kelly (1982). Genetic interactions among microbial communities. In *Microbial Interactions and Communities*, ed. A. T. Bull and J. H. Slater, pp. 287–322. Academic, London.

Rodriguez, J. A. (1982). Ph.D. Thesis, University of Wales Institute of Science and Technology.

Rokem, J. S., I. Goldberg and R. I. Mateles (1980). Growth of mixed cultures of bacteria on methanol. *J. Gen. Microbiol.*, **116**, 225–232.

Slater, J. H. (1981). Mixed cultures and microbial communities. In *Mixed Culture Fermentations*, ed. M. E. Bushell and J. H. Slater, pp. 1–24. Academic, London.

Slater, J. H. and A. T. Bull (1982). Microbial communities: biodegradation. *Philos. Trans. R. Soc. London, Ser. B*, **297**, 575–597.

Smith, M. E. and A. T. Bull (1976). Studies of the utilization of coconut water waste for the production of the food yeast *Saccharomyces fragilis*. *J. Appl. Bact.*, **41**, 81–95.

Smith, S. R. L. (1980). Single cell protein. *Philos. Trans. R. Soc. London, Ser. B*, **290**, 341–354.

Somerville, H. J. (1981). Mixed cultures in aerobic waste treatment. In *Mixed Culture Fermentations*, ed. M. E. Bushell and J. H. Slater, pp. 81–97. Academic, London.

Steinkraus, K. H. (1982). Fermented foods and beverages: the role of mixed cultures. In *Microbial Interactions and Communities*, ed. A. T. Bull and J. H. Slater, vol. 1, pp. 407–442. Academic, London.

Sukatch, D. A. and M. V. Johnson (1972). Bacterial cell production from hexadecane at high temperatures. *Appl. Microbiol.*, **23**, 543–546.

Tsuchiya, H. M., N. C. Trivedi and M. L. Schuler (1974). Microbial mutualism in ore leaching. *Biotechnol. Bioeng.*, **16**, 991–995.

van Verseveld, H. W., J. P. Boon and A. H. Stouthamer (1979). Growth yields and the efficiency of oxidative phosphorylation of *Paracoccus denitrificans* during two-(carbon) substrate-limited growth. *Arch. Microbiol.*, **121**, 213–223.

Veldkamp, H. (1970). Enrichment cultures of prokaryotic organisms. In *Methods in Microbiology*, ed. J. R. Norris and D. W. Ribbons, vol. 3A, pp. 305–361. Academic, London.

Wang, H. L. and C. W. Hesseltine (1982). Oriental fermented foods. In *Prescott and Dunn's Industrial Microbiology*, ed. G. Reed, 4th edn., pp. 492–538. Avi Publishing, Westport, CT.

White, F. H. and E. Kidney (1981). Yeast bacterium interactions in the brewing industry. In *Mixed Culture Fermentations*, ed. M. E. Bushell and J. H. Slater, pp. 121–135. Academic, London.

Williams, P. A. (1982). Genetic interactions between mixed microbial populations. *Philos. Trans. R. Soc. London, Ser. B*, **297**, 631–639.

Wimpenny, J. W. T. (1982). Response of microorganisms to physical and chemical gradients. *Philos. Trans. R. Soc. London, Ser. B*, **297**, 497–515.

Wolin, M. J. (1982). Hydrogen transfer in microbial communities. In *Microbial Interactions and Communities*, ed. A. T. Bull and J. H. Slater, vol. 1, pp. 323–356. Academic, London.

Wood, B. J. B. (1981). The yeast–Lactobacillus interaction: a study in stability. In *Mixed Culture Fermentations*, ed. M. E. Bushell and J. H. Slater, pp. 137–150. Academic, London.

Yoshida, T., M. Sueki, H. Taguchi, S. Kulprecha and N. Nilubol (1981). Modelling and optimization of steroid transformation in a mixed culture. *Annu. Rep. ICME, Osaka*, **4**, 1–13.

16

Animal and Plant Cell Cultures

R. E. SPIER
University of Surrey, Guildford, UK
and
M. W. FOWLER
University of Sheffield, UK

16.1   INTRODUCTION

Often overshadowed by other perhaps more glamorous aspects of biotechnology, animal and plant cell culture systems have nonetheless undergone major development in recent years. Such has been the progress that there are already animal cell processes in commercial usage, and a number of plant cell systems are rapidly approaching the same situation. In this review we have attempted to compare and contrast the two culture systems. The comparison includes the types of product which may be obtained from animal and plant cells, the nature of the cell systems, their growth characteristics and nutrient requirements, and finally moves to scale-up and process development. We have attempted to pinpoint areas where particular effort will be needed if industrial scale operation is to be achieved, especially in the plant cell area, and also to point out where progress may be hampered by major gaps in our knowledge of cellular physiology and biochemistry.

It should be noted that we have not dealt with those aspects of plant tissue culture concerned with micropropagation and the development of new varieties. Instead we have confined ourselves to plant cell culture and natural product synthesis to allow a more useful comparison with animal systems.

16.2   HISTORICAL PERSPECTIVES

Although in many ways 'infant biotechnologies' the first authenticated reports of the successful culture of animal and plant cells away from the parent body appeared about a century ago. Arnold demonstrated as early as 1880 that leucocytes would divide outside the body if bathed in an appropriate biological fluid, *e.g.* lymph or serum. A little later Haberlandt reported that plant cells from a variety of species could be maintained in a simple nutrient solution, but in contrast to Arnold's work no cell division was observed. Cell division in plant cell cultures was not in fact observed until after the turn of the century. In the early years of this century progress in animal cell culture was rapid. A particularly notable point being the development by Harrison in 1907, and subsequent extension by Carrel in 1913, of the hanging drop method of culture where a piece of tissue was bathed in lymph while being held on the underside of a coverslip which was itself sealed onto a hollowed out microscope slide. New media were developed and major progress made in reducing bacterial contamination which up to this time had been a particular problem. With plant cells, once Haberlandt was able to induce division in cultures, progress was also quite rapid. A large number of species were brought into culture and a wide range of nutrient media developed. A common feature of animal and plant cells is that they both have complex nutrient requirements which include not only a wide range of inorganic salts but also complex organic supplements. In the case of the majority of animal cell cultures a particular requirement was, and in many cases still is, foetal calf serum. The equivalent for plant cells is coconut milk; however, thanks to careful study of media formulation it has been found possible in many cases to do without coconut milk and today the majority of plant cell cultures are grown on defined, but not optimized, media. With animal cells a particular incentive towards the development of serum-free media has been the difficulty, and increasing cost, of obtaining mycoplasma-free foetal calf serum. A major problem of complex organic supplements such as foetal-calf serum and coconut milk is, of course, their inherent variability depending upon the source. This in turn leads to problems of reproducibility of data and so provides an additional incentive towards defined media.

During the period 1940 to 1950 there were many reports of both animal and plant cells from different species being grown successfully in culture, not just as a monolayer or a surface culture but also, and particularly in the case of plant cells, in liquid culture. At the end of the 1940s the first major application of animal cells was also to be seen: the production of polio virus in cells cultured from primate neural and neural kidney tissues. The production of virus material and antibodies are probably the major areas of interest for commercial application of animal cells and more will be said of this later. The first application of plant cells to produce commercially viable products did not occur till much later, in the 1970s, with the supply of an RNA restriction enzyme, 3',5'-phosphodiesterase derived from tobacco cells.

During the 1960s and 1970s developments in both technologies continued apace. Japanese workers showed that it was possible to grow plant cells in vessels of up to 20 m^3 capacity (see Fowler, 1982b) and certain animal cells were grown in suspension culture in vessels of up to 3 m^3 (see Spier, 1985) without facing what had been the major problem of surface adhesion and wall growth. Classical microbial techniques have now been applied to both cell systems and they have been grown in batch, semi-continuous and continuous culture. Lest this account should appear to be becoming euphoric let us stress that key problems still lie ahead. Simplistically these may be listed principally under the headings of low growth rate and biomass productivity, and often low specific product yield. We have also recently seen the beginnings of successful vector development for recombinant DNA technology and the cloning, particularly of animal genes, into microorganisms. These are all challenges which will have to be met in the not too distant future if animal and plant cell technologies are to prosper.

16.3 PRODUCTS AND POTENTIALS

In general terms the product potentials of animal and plant cells are rather different and will be treated separately except for a few initial comments. Both groups, however, tend to fall into the category of high added value, low volume markets.

Animal cell products are typically of high molecular weight and proteinaceous in nature. They cover a variety of enzymes, hormones and probably the most important area, animal vaccines. By contrast plant products are typically of low to medium molecular weight and cover a tremendous variety of chemical structure and biological/chemical activity. Certain plant proteins are potential targets but these are possibly in the minority when compared with animal systems.

It has to be said that neither animal nor plant cell systems will ever compete with generalized microbial biomass systems, *e.g.* for the production of single cell protein, or organic acid syntheses. Instead we must direct our attention towards those products which if not unique to animal and plant systems, are at least too expensive or difficult to derive from either microbial sources or through synthetic organic chemistry.

16.3.1 Animal Cells

Although mainly proteinaceous in nature, the range of products from animal cells is extensive (see Table 1), and likely to increase as the technology develops.

Table 1 Products of Animal Cells and Animal Cell Cultures

Immunobiologicals	Insecticides
Prophylactics (virus vaccines)	Enzymes
Monoclonal antibodies	Hormones
Immunobiological regulators	Whole cells

16.3.1.1 Immunobiologicals

There are three main product groups under this heading, virus vaccines (prophylactics), monoclonal antibodies, and immunobiological regulator materials (interleukines, lymphokines).

(i) Virus vaccines

Virus vaccines constitute a very important area for animal cell process technology and a number are already in production. In addition a number of others are in the development phase and approaching commercialization (Tables 2 and 3). The ideal virus vaccine should have the following characteristics: (a) it should be given as single dose; (b) it should provide lifelong immunity; and (c) it should be safe in use. Some vaccines meet all of these criteria. Typically they are the live virus vaccines such as smallpox, measles and rubella.

Table 2 Animal Cell Virus Products for Human Use

Existing	Prospective	Future
Mumps	Hepatitis A	Common cold
Measles	Hepatitis B	Some cancers
Rubella	Herpes simplex type 1	
Rabies	Herpes simplex type 2	
Yellow fever	Cytomegalovirus	
Polio	Varicella-zoster	
Influenza	Respiratory syncytial virus	
Adenovirus	Rift Valley fever	
Smallpox		

Table 3 Animal Cell Virus Products for Veterinary Use

Present	Present	Present	Present
Foot and mouth disease	Rabies	Canine contagious hepatitis	Contagious ecthyema
Marek's disease	Calf scours	Transmissible gastro-enteritis	Infectious bursal disease
Newcastle disease	Canine distemper		Loupiny ill
Infectious bovine rhinotracheitis	Canine and feline panleucopenia	Contagious pustular dermatitis	Laryngotracheitis
Parainfluenza 3	Ibaraki	Blue tongue	Rift Valley fever
Rinderpest	Fowl pox	Feline infectious enteritis	Pseudorabies
Bovine viral diarrhoea	Hog cholera		Ephemeral fever
Future	*Future*	*Future*	*Future*
African swine fever	Fish rhabdovivus	Equine influenza	Equine encephalitis

(ii)   Monoclonal antibodies

Perhaps one of the most discussed areas of the 'new' biotechnology products is monoclonal antibodies. The key breakthrough in this area was the demonstration that hybrid cell lines could be induced to grow and divide continuously and at the same time synthesize a monospecific or monoclonal antibody.

Applications of monoclonal antibodies are of course diverse, ranging from vaccines, to uses in diagnostic assay systems and preparative procedures. As diagnostic agents monoclonal antibodies are presently used to determine well over 150 drugs, toxins, vitamins and various other biological materials. The two key assay systems so far developed around antibodies, include radioimmunoassay (RIA) and enzyme linked immunoassay (ELISA). Both provide a very accurate means of determining and measuring very small amounts of material. Affinity chromatography is a good example of a preparative technique which is a good vehicle for the use of monoclonal antibodies in a preparative mode. It is already in use to separate interferons.

(iii)   Immunoregulator materials

A whole range of immunoregulatory factors have been identified in recent years, of which the interferons are possibly the best known. The group also includes substances such as thymosin, serum thymic factor and interleukin II. Although much development remains to be done in this area, it is undoubtedly one which will grow rapidly in the future.

16.3.1.2   Insecticides

Insecticides derived from viruses lethal for the insects they uniquely and specifically infect are of key importance for two reasons: they can specifically remove the target insect pest and leave helpful insects unharmed and they do not constitute a hazardous environmental pollutant. The baculoviruses have been particularly selected for use in this mode and have been applied on a number of occasions to the elimination of plant debilitating insects. Animal cell culture systems have been used to produce such viruses and have a great potential here.

16.3.1.3 Enzymes

The potential range of enzymes which could usefully be produced from animal cell systems is restricted and probably amounts to no more than about 10 enzymes (Table 4). All of these enzymes may also be targets for genetic engineering. However, problems of glycosylation, post translational modification and excretion from the host organism may all restrict such a development.

Table 4 Potential Enzyme Products from Animal Cells

Asparaginase	Factor VII	Hyaluronidase	Trypsin
Collagenase	Factor VIII	Pepsin	Tyrosine hydroxylase
Cytochrome P450	Factor X	Rennin	Urokinase

16.3.1.4 Hormones

Many animal hormones are relatively small polypeptides (20–30 amino acid residues) and are best produced by chemical synthesis. Some, however, are much larger (50–200 amino acid residues) and may also in some cases by glycosylated. In these situations cell cultures of the key hormone synthesizing organ could provide an important source of a number of hormones. Some examples are given in Table 5.

Table 5 Potential Hormone Products from Animal Cells

Interstitial cell stimulating hormone or luteinizing hormone
Follicle stimulating hormone
Chorionic hormone
Erythropoetin

16.3.1.5 Whole cells

It was stated earlier that it is unlikely that animal cells will ever compete with microbial cells for bulk biomass products such as single cell protein. There are, however, potential applications in the production of large amounts of animal cell material. These include the production of bulk cell material for enzyme and hormone extraction, toxicological testing, bioproduct assay and screening systems for potentially useful materials (*e.g.* anticancer agents).

16.3.2 Plant Cells

The plant kingdom is characterized by a great diversity of chemical components, often structurally complex and difficult to synthesize by conventional organic chemistry. This diversity of structure is matched by an equally wide range of properties and applications; these latter are listed in Table 6, with specific examples in Table 7.

Table 6 Application Areas of Plant Cell Natural Products

Pharmaceuticals	Perfumes
Food additives (flavours, aromas, colourants)	Enzymes
Agrochemicals	Speciality chemicals

Table 7   Plant Products and Associated Industries

Product	Plant	Activity	Industry
Anthocyanins	*Populus nigra*	Pigment	Food
Attar of roses	*Rosa* sp.	Perfume	Cosmetics
Codeine	*Papaver somniferum*	Analgesic	Pharmaceutical
Digoxin	*Digitalis lanata*	Cardiatonic	Pharmaceutical
Diosgenin	*Dioscorea deltoidea*	Antifertility agent	Pharmaceutical
Jasmine	*Jasminum* sp.	Perfume	Cosmetics
L-Dopa	*Vicia fabon*	Anticholinergic	Pharmaceutical
Morphine	*Papaver somniferum*	Analgesic	Pharmaceutical
Pyrethrin	*Chrysanthemum* sp.	Insecticide	Agriculture
Quinine	*Cinchona ledgeriana*	Antimalarial	Pharmaceutical
Quinine	*Cinchona ledgeriana*	Bittering agent	Food
Scopolamine	*Datura stramonium*	Antihypertensive	Pharmaceutical
Vincrystine	*Catharanthus roseus*	Antileukaemic	Pharmaceutical

16.3.2.1   Pharmaceuticals

For centuries plants have made a major contribution to medicine. In spite of major developments in synthetic organic chemistry and microbially derived drugs, the plant kingdom still contributes some 25% of all prescribed medicines, representing a multibillion dollar market worldwide. In Table 8 the 'top ten' drugs derived from plant sources are listed. Two points can be made from this list: first note the range of chemical structures involved, and second the wide range of pharmacological activities, ranging from antifertility agents to analgesics and antihypertensives. Closely allied to medicinals are narcotics, stimulants and plant derived poisons. The narcotics occupy a position midway between the medicinal agents and poisons. At low and carefully controlled concentrations they are usually extremely efficacious, *e.g.* codeine. At higher concentrations, or in continual usage they can be extremely dangerous leading to problems of addiction or even death, *e.g.* heroin and cocaine. Stimulants represent a somewhat different area and on the whole do not cause the major addiction problems associated with the narcotics. The most well known stimulant is caffeine.

Table 8   Ten Most-prescribed Drugs from the Plant Kingdom

Medicinal agent	Activity	Plant source
Steroids from diosgenin	Antifertility agents	*Dioscorea deltoidea*
Codeine	Analgesic	*Papaver somniferum*
Atropine	Anticholinergic	*Atropa belladonna* L.
Reserpine	Antihypertensive	*Rauwolfia serpentina* L.
Hyoscyamine	Anticholinergic	*Hyoscyamus niger* L.
Digoxin	Cardiatonic	*Digitalis lanata* L.
Scopolamine	Anticholinergic	*Datura metel* L.
Digitoxin	Cardiovascular	*Digitalis purpurea* L.
Pilocarpine	Cholinergic	*Pilocarpus jabonandi*
Quinidine	Antimalarial	*Cinchona ledgeriana*

Plant poisons have a long history of use both malevolent and beneficial. In a beneficial sense(!) they are still used by tribes in Africa and South America for hunting food, curare being a good example. A number of plant poisons are powerful neurotoxins, ricin from the castor bean being such a substance.

16.3.2.2   Food additives

Under this heading come flavours, aromas and colourants. A wide range of flavours and aromas is derived from the plant kingdom, including peppermint, onion flavour and vanilla. Many are complex chemically and require careful blending and composition. It is doubtful whether plant cell culture will contribute a great deal to this area in the immediate future where the desired product has a complex composition.

There is increasing pressure being placed upon the food industry by various governmental agencies to move from synthetic to natural colourants. Unfortunately many natural pigments are unstable when extracted from the plant and their potential application is therefore questionable.

16.3.2.3 Agrochemicals

Insecticides and plant growth regulators constitute the two key agrochemical sectors. The pyrethroids are possible the most successful insecticides yet developed. They are extremely effective without have the obvious side-effects of the organophosphates and DDT. The natural pyrethroids are derived from the Kenyan chrysanthemum and have to compete with synthetic products developed by scientists at Rothamsted Experimental Station. This is an area which shows major potential for the future.

Plant growth regulators cover the auxins, kinins and giberellins. All are capable of being produced through organic synthesis and their production through plant cell culture on a commercial scale is questionable.

16.3.2.4 Perfumes

A range of perfumes is produced using plant extracts in whole or part, 'attar of roses' possibly being the best known example together with jasmine. A particular point of interest with plant cell cultures is the possibility of producing novel perfumes, an area as yet relatively unexplored. Many plant-derived perfume materials are complex in nature and one must question whether it will be possible to reproduce such complex mixtures and blends using culture systems.

16.3.2.5 Enzymes

The range of enzyme targets from plant cells is restricted and relates to those having properties distinct from microbial systems. Examples may be seen in papain, bromelin and allinase, the onion flavour enzyme. Plant enzyme technology suffers from a lack of knowledge of the key biosynthetic pathways in plants and awaits developments in this area.

16.3.2.6 Speciality chemicals

There are relatively few speciality chemicals which are derived from plants and which could not be more effectively synthesized from microbial cells or by conventional organic chemistry. Some examples may occur in the area of the steroids and isoprenoids, but these are restricted.

16.3.2.7 Biomass applications of plant cell cultures

As with animal cell cultures there is no question of plant cell cultures competing with microbial systems for biomass products such as protein. However, there are one or two possible specialized applications including tobacco biomass, tea biomass (particularly decaffeinated) and fruit puree, which might develop in the not too distant future.

16.3.3 Cell Culture and Product Synthesis

Animal and plant cell culture may contribute to the production of animal and plant products in a number of ways; these include: (i) an alternative route to the synthesis of established products; (ii) a route of synthesis to novel products difficult to obtain from intact animals and plants in sufficient quantity to meet market demand; and (iii) as a source of enzymes for biotransformation processes converting low value substrates into high value products. In addition plant cell cultures may themselves be a source of novel substances of industrial interest in their own right.

It is against this background of present animal and plant products and potential applications that the following detailed sections on developments in cell culture technology must be assessed.

16.4   THE NATURE OF ANIMAL AND PLANT CELLS IN CULTURE

Animal and plant cells are, of course, both eukaryotic in nature and in discussions of the biotechnology of eukaryotic cells it is usual to associate the two. Fungi are omitted as they are generally dealt with in sections on antibiotic or alcohol production and, as such, the biotechnology which exploits them is generally on a scale of more than 10 000 l units. Animal and plant cells in culture, however, are presently used for low volume, high value products (see Section 16.3) and at scales of operation of up to 8000 l (cultures of baby hamster kidney (BHK) cells used for the production of foot and mouth disease (FMD) virus and lymphoblastoid cells for α-interferon) and 200–1000 l for plant cells for the bioconversion of β-methyldigitoxin to β-methyldigoxin (Barz, 1981). The largest reported plant cell culture was of 20 000 l of tobacco cells (Takayama *et al.*, 1977).

While it is usual to think of eukaryotic cells as being biologically similar, in the context of the technology of the culture of those cells and the generation of products from them, it is also necessary to consider the differences between the two types of cells. When attention is given to such differences it is easy to overlook the similarities. To prevent such a distortion a listing of the similarities and differences is given in Table 9, while Figure 1 depicts the way in which such structural differences appear.

Table 9   Structural Similarities and Differences between Commonly Cultured Animal and Plant Cells

Characteristic	Aminal cell	Plant cell
Nuclear membrane	+	+
Complex chromosomes	+	+
Mitochondria	+	+
Chloroplasts (protoplasts)	−	+
Cell membrane	+	+
Cell wall	−	+
Vacuole	−	+
Storage granules	−	+
Microvillae	+	−
Microtubule cytoskeleton	+	?
Endoplasmic reticulum	+	+
Departure from sphericity	−	+
Lysozomes	+	−

16.4.1   Cell Culture Initiation

In the first instance both animal and plant cell cultures are derived from complex organisms. Specific organs (lung and kidney from animals, leaf, stem or root from plants) each composed of numerous tissues are excised under conditions designed to ensure that the material removed is as free of exogenous contamination as possible. For animal cell cultures this requires the use of sterilized instruments, a work station which bathes the processed material in a laminar stream of filter-sterilized air and where obviously contaminated material is used (whole embryonated eggs) the outside of the material is cleansed with a weak solution of iodine. The surface sterilization of tissues excised from plants is effected by washing the tissue in 5% (w/v) sodium hypochlorite, 2% (w/v) mercury(II) chloride or 80% (w/v) ethanol for varying lengths of time after which the sterilizing agent is removed by washing the tissues repeatedly in sterile distilled water. The effectiveness of such techniques is enhanced by adding a wetting agent, such as Triton X100 or Tween. When such techniques are used it is important to use a regimen which does not damage the cells which are to form the tissue culture. Antibiotics are also occasionally used to control contaminating organisms for the initiation of cell cultures even though it is recognized that they may affect subsequent cell growth (Litwin, 1979). The use of seeds for initiation of plant cell cultures is directly analogous with the use of embryonated eggs for the initiation of animal cell cultures. In such circumstances it is more practical to use more vigorous techniques to clean the outer surfaces of the material preparatory to the removal of the tissues required for culture development (Fowler, 1983b).

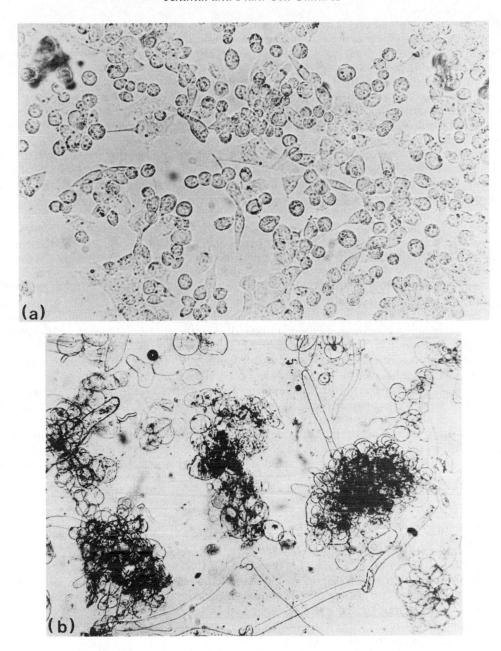

Figure 1 Photomicrographs of animal and plant cells to illustrate differing morphology: (a) animal cells, BHK magnification × 60; (b) plant cells *Catharanthus roseus*, magnification × 100

16.4.2 Culture Development

Present techniques for the production of plant cell cultures from excised tissues involve the siting of the explanted material on to the surface of an agar gel which has been formulated to contain those necessary nutrients and hormones which have been found to promote cell division and growth of the excised material (see Section 16.5 for details of the media used). It should be noted at this point that the hormonal component of such a medium working on tissues of particular origins can give rise to the production of differentiated organs, *viz.* leaf, stem or root (Kurz and Constabel, 1979). Such differentiation is not conducive to the generation of cell cultures for technological exploitation but rather serves to remind workers of the totipotency of tissues excised from plants and indeed use may be made of such systems for the production of seedlings and for the cloning of particular plants for cash crops. (This has been done successfully for the commercial oil palm, *Elaeis guineensis*). Under the appropriate conditions of medium composition, light, temperature (generally between 20–30 °C) cell growth occurs on the surface of the explant in an undifferentiated form and produces a callus. Such outgrowths from the central mass occur preferentially at particular points which gives a nodular appearance to the material. Also, unlike the animal cell system to be described, the cells do not migrate on to the surface of the agar and grow away from the original material, rather they stick together as a collection of clumps.

Systems quite similar to those described above for plant cells were used historically for the cultivation of animal cells. Lumps of animal tissue were placed on a slide or a dish and irrigated with a medium held to be suitable for the promotion of cell growth. Indeed, cells did divide, but unlike the plant cells which remained bonded together, the dividing cells of the animal explants moved away from the central cell mass and began to form a cell sheet on the surface of the supporting substratum. Such outgrowths were amongst the first signs that animal cells could be grown outside the body (see Willmer, 1935 for review).

The development of such outgrowth type cultures to the monodisperse cultures which are used widely today took place in 1916 (Rous and Jones, 1916). Rous and Jones found that they could prepare single cells by the application of the proteolytic enzyme trypsin to excised tissue and thereby generate single cells. This technique is now commonly used for the production of primary cultures of animal cells. For such cultures excised tissues are cut up into small (*ca.* 2 mm^3) pieces with scissors and then they are agitated in fluted conical flasks for up to 120 min at 37 °C in a dilute solution of trypsin (typically 0.25% w/v), in buffered saline. The resulting suspension is clarified by passing it through presterilized muslin and then washed in the centrifuge. The cells are finally resuspended in growth medium and distributed into bottles. Once the cells settle on to the glass base of the bottle, they attach and spread out. Once spread, they divide and, under the appropriate conditions of cells/surface area at the planting time, they spread out to form a contiguous monolayer of cells over the surface of the bottle. This process takes about 4 d.

As in the case of plant cell cultures, conditions can be modulated to promote the growth of a particular cell type. If no special precautions are taken, the fastest growing cell predominates; parenchyma for plants and fibroblasts for animals. While it is relatively easy to modulate plant cell cultures using simple synthetic hormones, animal cell cultures have to be carefully handled and often require multiple selective passaging in different culture systems, sometimes including live animals, to produce a cell culture which does not overgrow with fibroblasts (Sato and Buonassisi, 1964).

16.4.3 Secondary Cultures

Two fundamentally different types of secondary culture may be derived from established explants (calluses) or primary cultures. On the one hand the cultures may be simply divided and a secondary culture similar to the original culture can be produced. Alternatively, methods can be selected which promote the growth of cells in bulk suspension cultures similar to those commonly used for the production of yeasts, bacteria, algae and, when practicable, protozoa.

16.4.3.1 Culture replication

For plant cells the process of replicating the primary cultures is simple. The growing callus is merely removed from the surface of the agar and is then cut into a number (four to six) of smaller pieces which are replaced onto the surface of fresh agar in new flasks. With animal cells the process of culture replication involves the removal of the cells from the (normally) glass or plastic substratum to which they are attached as a confluent monolayer. This is effected by removing the spent growth medium and after washing the cell sheet with a buffered isotonic saline solution a second solution is added which loosens the attachment between the cells and the substratum. Such solutions may contain chelating agents (EDTA) and/or proteolytic enzymes such as trypsin, collagenase or pronase. Exposure times of 10–30 min at either room temperature or 37 °C are usual, after which medium containing an antiproteolytic enzyme agent (serum) is added to the container. After vigorously shaking the bottle, the cell suspension may be decanted, washed by centrifugation and diluted 3 to 10 times for planting new bottle cultures. These latter cultures are termed secondary cultures.

By repeating the procedures described above, both animal and plant cells may be kept viable in an *in vitro* environment indefinitely. There are, however, some difficulties. With the plant cells it has been thought that such passaging is the only way to maintain particular cell lines *in vitro*, as techniques for the long-term storage of such cells have not been fully established. More recently workers in Nottingham (L. Withers, personal communication) and Tubingen (W. Alfermann, personal communication) have been able to show that by the use of cryoprotectants such as DMSO or glycerol, plant cells frozen under controlled conditions (typically 1 °C min^{-1}), may be kept at the temperature of the vapour phase of liquid nitrogen (about −130 °C) for extended periods of time. Such procedures may also be used for storing animal cells where a commonly used storage medium would be composed of basal medium (50%), serum (25%) and cryoprotectant (25%) (Bolt *et al.*, 1979). In this way cells may be stored at concentrations of 10^7–10^8 cells ml^{-1} in volumes of between 1 ml and 20 ml.

16.4.3.2 *Industrially useful cell cultures*

The primary cultures of animal cells as described in the previous section are widely used for the production of viral vaccines for human use (Stones, 1981). Indeed, the required processes for the production of measles, Rubella and mumps vaccines are founded on primary cells grown in thousands of individual bottle cultures. Such multiple processes, however, are technologically inelegant, inefficient and in the case of plant cell product-generating systems uneconomic.

Developments in animal cell culture have taken two paths. Cells which are capable of dividing between 50–100 times, termed diploid cell strains, have been characterized extensively for the more reliable and controlled production of vaccines. Two cell strains in particular have achieved prominence for human vaccine production, namely the WI38 from the Wistar Institute in Philadelphia and the MRC5 strain of the Medical Research Council, London. A second cell type which is capable of indefinite replication, designated a cell line, is now under intense scrutiny as a possible source of human vaccines. This cell line, VERO, derived from a monkey kidney achieved some prominence as a substratum for the production of polio vaccine (Petricciani *et al.*, 1979).

The repeated passaging of animal and plant cells *in vitro* generally leads to a form of dedifferentiation. On the plant cell side this takes the form of loss of totipotency, *i.e.* individual cells may no longer be able to be cultured so as to regenerate a complete plant. With animal cells the situation is described in different terms. Here one is looking at the phenomenon of 'transformation' which presently is not fully understood. This state is analogous (and may be similar to) the observed changes when normal body cells become carcinogenic. They divide in a manner which is out of the control of the normal body regulatory systems. Many methods have been used to dilineate the 'degree' of transformation which a given cell culture expresses. Tests dependent on growth in soft agar, shape of cells when grown as monolayers and the growth of cells as multilayers have been used as indicators of transformation (Pastan and Willingham, 1978). Clearly the more transformed a cell culture the more dangerous it would be for the production of products for human use. While this may be an intellectually comfortable position to adopt, it may not be justified experimentally.

16.4.3.3 *Substrate independent cultures*

Rapidly growing cells derived from calluses or monolayer cell sheets can often be 'adapted' to grow free from a supporting substratum in 'homogeneous' suspension culture. The methods used to initiate such cultures for plant and animal cells are similar. Calluses are removed from the surface of agar cultures and are resuspended in a fluid medium in a conical flask held on a gyrotatory or reciprocating shaker. By swirling the culture at 100–150 r.p.m. cells which are loosely attached to the callus become detached and if suitably nurtured they begin to divide, thus establishing a suspension culture (Fowler, 1983b). Similar methods are used for preparing suspension cultures of animal cells. In the latter cases a 'spinner' flask is used. This consists of a vertical cylindrical bottle with a rotating drive magnet which prevents the 'denser than medium' cells from settling out. An example of such a procedure is that used by Capstick *et al.* (1962) for the production of a suspension cell line of BHK 21 C 13 cells.

While the production of plant cell suspension cultures is relatively routine, the production of the animal cell counterparts is not as common. Generally, transformed animal cells lend themselves to such procedures more readily than less transformed cells. This has resulted in a few generally used cell lines such as BHK, HeLa (originally a cancer cell derived from a Miss Helen

Lane), mouse L cells, lymphoblastoid cells and more recently hybridoma cells. The latter two cell lines are derived from cells of the lymphatic system of humans or mice, which have either been deliberately transformed or have been generated by fusion with an already transformed cell.

Subsequent sections of this chapter will describe the scale-up of these suspension cultures and also the scale-up of the immobilized cell cultures where the cells are grown on or in solid support materials either out of necessity (they will not grow at all in free suspension) or because product formation and recovery is preferentially promoted when the cells are held in a stationary configuration.

16.4.4 Individuality of Cell Lines in Relation to their Productivity

Although individual plant cells may be induced to grow as differentiated tissue, the productivity of such cells or the lines derived from them has been shown to be determined by the productivity of the plant from which they were derived. This has been most clearly established in the case of nicotine synthesis by different mutants of the plant *Nicotiana tabacum* (Fowler, 1983a) in which a single point mutation was shown to control the production of the alkaloid nicotine in cell cultures derived from the mutant. Also, selective excision of the more highly pigmented cells from a *Catharanthus roseus* callus culture led to the establishment of a suspension culture of predominantly highly pigmented cells.

The production of plant cell clones in tissue culture has proved difficult. Rather in the manner of animal cells it would seem that a certain minimal cell density is required for facile growth (Rein and Rubin, 1967). Such circumstances are generally circumvented either by using a medium which is 'conditioned,' that is cells have already been grown in it, or by the use of 'feeder layers' of cells (Takahashi and Okada, 1970). In the latter case a stationary culture of cells is established and its further replication prevented by X-ray irradiation. The cells which are to be established as clones are either grown directly on such a feeder layer or on the surface of an agar gel which has been formed over the feeder cell layer (Kurz and Constabel, 1979). Animal cells too express unique genotypes when cloned. Indeed, a prime factor which defines the productivity of the BHK suspension cell system for the manufacture of foot and mouth disease (FMD) vaccine is the particular cell population used for this operation. Recent investigations have shown that uncloned populations of BHK cells are made up of cells whose individual productivities may differ by 10–100 fold. When the cells from such mixed populations are cloned, cell cultures are derived which better reflect the productivity of the isolated cell (Clarke and Spier, 1980).

16.5 CULTURE MEDIA

Two different types of culture medium may be defined. The first type is the one used for the most rapid production of new cells: the growth medium. This, with some modifications, is used to initiate cultures. A second kind of medium may be used to maintain cultures in a non-growing and non-dividing state. Under such conditions specific chemicals may be added to induce product formation or virus seeds may be added for virus amplification.

16.5.1 Growth Media

The primary components of such media for both animal and plant cells are: (i) water; (ii) inorganic salts; (iii) trace elements; (iv) vitamins; (v) buffers; (vi) carbon and energy sources; (vii) nitrogen sources (defined and undefined); (viii) growth factors (defined and undefined); and (ix) other additives.

16.5.1.1 Water

While water seems to be one of the best defined materials in nature, as far as cell culture is concerned it can on occasion defy analytical examination (Girard, 1977). Well substantiated experimentation which enables growth failures to be unequivocally attributed to the water quality are rare. However, when water supplies may originate from rivers, reservoirs, wells and reprocessing stations or variable mixtures of all these sources at different times of the year, it is as well to consider water production and processing carefully.

Generally, demineralization and/or distillation are all that is necessary to produce water which will support the growth of plant and animal cells (Kurz and Constabel, 1979; Spier, 1985). Storage of distilled water can present problems. Bacterial counts of 0.5×10^6 ml^{-1} can be obtained from stored distilled water. Such water may be toxic for animal cells as bacterial toxins which may be destroyed on heating (during sterilization) are not degraded by the normal processes of filtration used to sterilize animal cell culture media (Konowalchuck and Speis, 1979). Thus, it becomes accepted practice to store distilled water at 80 °C which has the additional advantage that such water is regarded as being free of pyrogens and therefore suitable for the production of pharmacological products.

Where doubt about water quality persists, melted ice obtained from deep within a Greenland glacier or water prepared by a high pressure distillation process at over 200 °C or even produced by the burning of pure hydrogen in oxygen may be used. In such cases it is vital to use cells whose viability has been well established and which have been stored as aliquots of a successful cell culture.

Such strictures on water quality should not deter investigators from embarking on the generation of cell cultures as water derived from poorly specified sources has been used to grow BHK cells for FMD vaccines in Indonesia, Sri Lanka, Thailand and India.

16.5.1.2 Inorganic salts

Both cell types require isotonic solutions for optimal growth. The ions which form the bulk of such media are salts of the common acids. However, the comparison of the salts used for animal cell cultures has little else in common with those used for plant cells (Table 10).

Table 10 Inorganic Salts Used for Cell Culture Media

Salt	Animal cells (g l^{-1})	Plant cells (g l^{-1})
$CaCl_2 \cdot 2H_2O$	185–264	150–440
$Fe(NO_3)_3 \cdot 9H_2O$	0.08–0.1	—
$FeSO_4 \cdot 7H_2O$	—	27.8
KCl	320–400	—
KNO_3	—	1900–2500
KH_2PO_4	60	170–340
$MgCl_2 \cdot 6H_2O$	170	—
$MgSO_4 \cdot 7H_2O$	161–242	250–370
NaCl	5100–8000	—
$NaHCO_3$	350–3700	—
$NaH_2PO_4 \cdot 2H_2O$	100–1500	134
NH_4NO_3	—	1200–1650

[a] Range of values of 13 different variants of Eagle's medium (Flow Laboratories Catalogue, 1979).

Ammonium ions are held to be toxic for animal cells whereas they are used as a source of nitrogen for plant cells and while both potassium and sodium chlorides appear in large proportions in animal cell media, they do not seem necessary for plant cell growth. Inorganic materials are generally formulated as buffers so that phosphates and carbonates can be included in different proportions to obtain a particular pH. This is generally in the range 7–7.5 for animal cells and 5.5–5.8 for plant cells. It is also interesting to note the use of iron(II) sulfate as a source of iron for plant cells, while the more oxidized iron(III) nitrate is used for animals.

16.5.1.3 Trace elements

A medium lacking in small quantities of unusual inorganic ions would be unsuitable for any eukaryotic cell culture. At the level at which such materials are included in the final formulation (<1 μg l^{-1}) they may almost be regarded as impurities. Indeed, the water and the inorganic materials used in larger quantities may contain adequate amounts of such elements. Thus for most animal cell cultures additional trace elements are not routinely added, although specialized media may be supplemented with such materials (Weymouth, 1972). Plant cells on the other hand seem to be more demanding in this respect. The standard formulations referred to in Table 10 include such ions as I^-, BO_3^{3-}, Mn^{2+}, Zn^{2+}, MO_4^{2-}, Cu^{2+} and Co^{2+} in quantities ranging between 2.5 μg l^{-1} to 20 mg l^{-1} (Kurz and Constabel, 1979).

16.5.1.4 Vitamins

Vitamins are required by both plant and animal cells in culture. A comparison of the vitamins used in the different culture systems and an indication of the concentrations required is presented in Table 11. It should be noted that small molecules which become incorporated into coenzymes have been included. The levels at which these vitamins are used are similar in media designed for the growth of either cell type; a minor difference seems to show in the slightly greater range of materials used for the animal cell systems.

Table 11 Vitamins Used for Cell Culture Media[a]

Vitamin	Animal cells (mg l^{-1})	Plant cells (mg l^{-1})
Nicotinamide	1–4	—
Nicotinic acid	—	0.5–1
Pyridoxine–HCl (B$_6$)	—	0.5–1
Pyridoxal–HCl (B$_6$)	1–4	—
Thiamine–HCl (B$_1$)	1–4	0.1–10
Riboflavin (B$_2$)	0.1–0.4	—
Vitamin B$_{12}$	1.36	—
Folic acid	1.6	—
DL-Pantothenic acid	1.6	—
Choline chloride	1.6	—

[a] Flow Laboratories' Catalogue, 1979.

16.5.1.5 Buffers

The choice of a buffer system is critical to particular operational parameters. Alternatives which are available are those based on mixed alkali phosphates, bicarbonate, sodium succinate/succinic acid, barbiturates, HEPES and TRIS (Eagle, 1974). It would seem that plant cell cultures do not generally require a stable pH for maximal growth and the small variations which occur are not detrimental. This feature illustrates an interesting difference between animal and plant cells which is that the latter incorporate the medium nutrients in a manner which is in balance with the overall cellular composition. Animal cells, however, tend to partially metabolize the generally used carbon source (glucose) and generate, as a waste product, lactic acid. For the latter systems bicarbonate buffers are chosen as pH may be controlled either by modulating gas flows (CO_2 and/or air) through the culture, which on the small scale presents fewer problems in controlling possible sources of contamination, or by the addition of more concentrated solutions of Na_2CO_3 or HCl. It should be noted that addition of considerable quantities of such materials (as could happen in a poorly controlled system where acid was added to neutralize excess alkali and *vice versa*) the tonicity of the medium could change considerably due to a build up of NaCl. Some animal cell lines are sensitive to the HCO_3^- concentration in the culture fluid (McLimans, 1972) or some investigators may wish to be able to measure CO_2 in which case independent buffer systems are used.

16.5.1.6 Sources of energy and carbon

Much effort has been expended on the determination of the most appropriate source of energy for culturing cells. Most animal cell systems thrive on glucose at concentrations of between 1–4 g l^{-1}. Often additional carbon sources such as pyruvate, succinate and inositol are included as supplements. Where a less than maximal rate of cell growth is required, galactose can be substituted for glucose (Baugh *et al.*, 1970). Plant cells extend the range of useful energy sources to such materials as sucrose (the most preferred source of carbon), fructose, lactose, maltose, trehalose and starch. A recent review highlights the growth rate and efficiency of conversion of such substrates into cell mass (Fowler, 1982a). The metabolism of sucrose by the plant cells is interesting as it is known that plant cells possess an invertase in the cell wall which hydrolyzes the sucrose to glucose plus fructose prior to the entry of those substances into the cell *via* the plasmalemma. The extent of this reaction is determined by the origin of the cell line as well as other physiological factors.

Although plant cell lines make chlorophyll even though they are grown in the dark, the application of light to such cultures does not appreciably improve their ability to thrive in a fully autotrophic manner. Perhaps the most significant demonstration of the existence of chloroplast enzymes (carboxydismutase or ribulose 1,5-diphosphate carboxylase) may be inferred from the observation that overgassing cultures with air lowers growth rates while gassing at the same rates with additional CO_2 in the gas stream restores cell productivity (Fowler, 1983b). This indicates that some of the required carbon is most efficiently introduced into the metabolic system via the pentose phosphate pathway.

A further difference between animal and plant cells is highlighted in this area as plant cells use glucose directly for the production of the cellulose cell wall, a feature absent from animal cells. Also, older plant cell cultures excrete a variety of, as yet, undefined carbohydrate polymers whose stickiness plays a considerable role in determining the technology which has to be used to cultivate such cells. Such polymers cause the viscosity of cultures to rise to appreciable levels, the development of sizeable cell clumps of between 2–200 cells per clump and the attachment of cells to the vessel wall, to probes and to the foam which is generated (by the same polymers?) at the culture/air space interface. This latter accumulation of cells forms a crust on the top of the culture which generates stalactites which grow down into the reactor. The detachment of such structures and their subsequent demise also influences the operation of such systems.

16.5.1.7 *Nitrogen sources*

There are two types of nitrogen sources which can supply eukaryotic cells with the necessary material to build their proteins and nucleic acids; one such source may be clearly defined chemically while the second is a complex mixture of unknown nitrogen containing materials.

(i) *Defined nitrogen sources*

The major proportion of the nitrogen required for the growth of plant cells is derived from the inorganic ions NH_4^+ and NO_3^-. Minor sources of nitrogen such as glycine, glutamine or perhaps complex mixtures of undefined proteins and glycoproteins (see Section 16.5.1.8) can also be used by plant cells. Animal cells, on the other hand, do not prosper when presented with NH_4^+ ions, indeed every effort is made to prevent the build-up of such material as a waste product (Griffiths and Pirt, 1967; Birch and Cartwright, 1982). Also, nitrate cannot be readily reduced by animal cells to provide a source of nitrogen at a sufficient state of reduction.

It is well recognized that animals, and therefore the cell cultures derived from them, need a small number of 'essential' amino acids for growth. These materials, in addition to some nonessential amino acids, provide the nitrogen necessary for protein synthesis and cell growth. Glycine, serine, cysteine, proline, glutamate and glutamine are the commonly added non-essential amino acids. A number of recent studies on amino acid utilization by cultures of BHK cells (Arathoon and Telling, 1981) and on human diploid fibroplasts (Litwin, 1977) show that much work remains to be done for the formulation of a medium in which the parameters of cost, simplicity and efficacy are fully optimized.

In spite of the difficulties many investigators have developed media in which animal cells can grow which contain only defined materials. Several facets of such cell growth should be clearly perceived: (i) in order that a cell may be said to be able to grow on a defined medium, it should be passaged on that medium at least three times in succession to use up any stored, undefined, growth factors which may have accumulated in the cells; (ii) during the adaptation to the defined medium the population which has survived the transition may have been preferentially selected from the more heterogeneous population used in the first instance; and (iii) cells which lend themselves to growth in defined media tend to be of the more transformed variants (see Section 16.4 above) and may not, therefore, possess the characteristics of the less transformed cell from which they were derived. For example, Barteling (1982) succeeded in adapting BHK cells to grow on a defined medium, but was unable to grow his strain of foot and mouth disease virus to a sufficiently satisfactory level in them. On the other hand, Tomei and Issel (1975) claimed that they were able to adapt BHK cells to a chemically defined medium and also to be able to produce immunogenic amounts of foot and mouth disease virus from such cells. This latter work has, as yet, remained unexploited. The present situation is one of flux and industrially profitable processes based on such cells and media have yet to be fully established.

(ii)   Undefined nitrogen sources

Undefined proteinaceous materials have been used in eukaryotic cell cultures since their inception in the early years of this century. Their function has been to provide for cells an *in vitro* environment which, at least in biochemical terms, provides the kind of assortment of molecules which is experienced in a larger differentiated living organism. Early plant cell culturists commonly used the milk of coconuts and extracts of yeast, corn or malt (Kurz and Constabel, 1979) while their animal cell counterparts used sera derived from a variety of animals. In addition to these, basic brews, peptones, milk whey, hydrolysates of soya, bacteria, lactalbumin and other inexpensive protein concoctions have found relatively permanent places in, at least, animal cell culture systems.

Not only do such proteins provide sources of nitrogen which may result in less expensive media as they can substitute for refined amino acids (Mizrahi and Avihoo, 1976; Keay, 1975), but they also serve as sources of growth promoters, cell stabilizers in large scale cultures (Zolleto and Gagliardi, 1977), trypsin inhibitors and attachment factors for substratum-dependent cultures (Temin *et al.*, 1974). The problem inherent in the use of such materials is that they vary from batch to batch and they therefore have to be carefully controlled by realistic tests before acceptance by the cell technologist (Bolt *et al.*, 1979) and, once accepted, it is implicit that adequate stocks of stored material are on hand in the event that the current batches of material are found unsuitable (animal sera store well if frozen at $-20\,°C$). A second problem which is more pertinent to these latter materials is that the more powerful the growth promoter (foetal serum is a powerful growth promoter) the more likely it is that the material is contaminated with undesirable viruses and mycoplasma (Molander *et al.*, 1972; Horodniceanu, 1977). Such materials may also be expensive. Foetal calf serum, depending upon the supply and demand, varies in cost between £60 and £100 l^{-1} and, as it is used at levels of 5–10% (v/v) of culture it is an item whose necessary presence can determine the economic feasibility of a process. Other sera, often made in-house from blood derived from a local slaughterhouse, can cost less than £2 l^{-1}. Such sera are used in bulk for the large scale production of foot and mouth vaccines and other veterinary vaccines.

16.5.1.8   Growth factors

Defined chemicals, or undefined single or mixtures of chemicals, which when added to a formulated medium convert it into a medium in which cells replicate faster, are called growth factors. A listing of various growth factors is presented in Table 12. It is obvious from a cursory survey of this table that, for the most part, plant cell cultures are able to thrive with fewer undefined molecules in the medium and generally only require an auxin for cell elongation and a cytokinin (kinetin) for cell replication (Dougall, 1980). The situation is complicated by two additional features. Firstly, the response of a particular cell culture to these hormones is a consequence of the relative amounts of auxin and cytokinin. It has been shown (Skoog and Miller, 1957) that by altering the proportions of these two factors the resulting culture either will be a disorganized mass of cells, a callus, or differentiated stem or root type tissue. The second complicating factor is that the cells, as they grow, also produce and excrete both auxins and cytokinins into the culture fluids. This could lead to uncontrolled developments in culture systems; it also explains the fact that below a threshold cell concentration certain cells do not thrive unless heavily supplemented with less well defined growth factors (see Section 16.5.1.7). Once the plant cells have been adapted to the culture system, most investigators prefer to use media which are devoid of undefined elements as this improves their reliability and decreases their costs. Animal cell culture technologists are not so fortunate.

Media generally used for the production of animal cells either have a large number of defined growth factors or a few undefined complex mixtures which serve to enhance growth (Table 13). Indeed, the use of mixtures of such complex materials can lead to either simple, autoclavable media (Keay, 1975) or cheap media (Mizrahi and Avihoo, 1976) for animal cell cultures. Many of the growth factors for animal cells would be regarded as being specific. It is of interest to note that

Table 12 Common Growth Factors Used in Cell Cultures

Factor	Animal cells (g l^{-1})	Plant cells (g l^{-1})
Indoleacetic acid	—	1–30
Naphthaleneacetic acid	—	1
Kinetin	—	0.02–10
2,4-Dichlorophenoxyacetic acid	—	0.1–1.0
6-Benzyladenine	—	0.2–1.5
Zeatin	—	1.0–5.0
Giberellic acid[a]	—	0.5–10
Insulin	0.1–10	—
Glucagon	0.05–5	—
Follicle stimulating hormone	0.05–0.5	—
Luteinizing hormone	0.5–2	—
Prostaglandin E$_1$ and F$_2$	0.001–0.01	—
Triiodothyronine	0.001	—
Parathyroid hormone	0.001	—
Somatomedin C	0.001	—
Hydrocortisone	0.003–0.03	—
Progesterone	0.0003–0.03	—
Estradiol	0.0003–0.003	—
Testosterone	0.0003–0.003	—

[a] Giberellic acid includes at least four different chemical species.

many of the factors which are used to control cell growth in tissue culture are hormones and that it is possible to consider the mixtures of undefined growth factors as materials which contain as yet unidentified hormones. Such hormones can act (a) by alteration of activity of an existing enzyme or (b) by alteration of particular gene activity or (c) by alteration of cell permeability which in turn affects the level of activity of other cellular components. It is clear that the use of such materials to modulate the behaviour of cells in culture is of prime importance in controlling the productivity of cell cultures.

16.5.1.9 Other ingredients

Various materials are added to cell cultures to change their foaming properties, their viscosity and the frictional effects which sometimes interfere with cell growth on the large scale in stirred tank rectors (STRs) (see Section 16.6 below). A list of such materials has been compiled in Table 13. Again, plant cell cultures require few, if any, such materials and indeed for animal cell cultures addition of serum often obviates the need for some of the more esoteric components (PVP, methylcellulose).

It is often suggested that the widespread development of cell culture systems was contingent to the discovery of antibiotics. Primary cultures of animal cells are certainly promoted by such materials; however, cell lines of well worked cells can be grown with confidence in the absence of antibiotics. For a production operation, however, the cost of a periodic failure has to be offset against the cost of the antibiotic and most production operations retain such protection.

16.5.2 Maintenance Media

Media which promote cell viability but which do not cause cell growth or cell replication are called maintenance media. Such media are held to enhance the productivity of cultured cells because, it is argued, non-growing or stationary cells may be made to produce an end product in greater yields if this is the only metabolic activity which is left available to such cells. Maintenance media may be complex yet they can also be as simple as a buffered isotonic salt solution. In animal cell cultures such media may be used for the production of viruses for vaccines, although it is generally found that higher productivities result from more complex media; this is probably due to a small amount of cell growth which occurs in a growth factor free growth medium.

Both animal and plant cell cultures express the common feature that under appropriate conditions the nature and concentrations of hormones presented to the culture can influence differentiation so that particular genes become active with the result that whole biosynthetic pathways

Table 13 Cell Culture Additives

Additive	Putative function of additive
Pluronic F68	Shear protectant
Tween 80	Detergent–fat solubilizer
Antibiotics	
Penicillins	Bacteriostat–bacteriocide
Streptomycin	Bacteriostat–bacteriocide
Oxytocin	Bacteriostat–bacteriocide
Neomycin	Bacteriostat–bacteriocide
Gentamycin	Bacteriostat–bacteriocide
Midland Silicone M/S emulsion	
(Q 10321)	Prevents foaming
Methyl cellulose	Shear protectant–viscosity modifier
Thiolated gelatin	Shear protectant–viscosity modifier
Antifungals	
Mycostatin	Fungicide–fungistat
Fungizone	Fungicide–fungistat
Polyvinylpyrrollidone	Viscosity modifier
Dextrans	Viscosity modifier
Phenol red	pH indicator
Nucleosides/nucleotides	Nucleic acid precursors
Derivatives of nucleic acids	Nucleic acid precursors

are turned on or off (Kurz and Constabel, 1979; Janss *et al.*, 1980). This feature of animal cells has not generally been exploited commercially; for plant cells, however, it provides the basis of a cell culture production system which can compete economically with material prepared from plants grown in the field (Fowler, 1983a).

Biotransformations by fungi is an economic way of producing unusual steroids from more readily available steroid progenitors (Prescott and Dunn, 1959). Likewise, plant cell cultures may be used to good effect for the production of valuable digoxin from the low value digitoxin (Fowler, 1983b). Other examples of such biotransformations from precursors presented to plant cells in a maintenance culture are the production of the alkaloid dictamine from the precursor 4-hydroxy-2-quinolone, anthraquinone production promoted by the precursor *o*-succinylbenzoic acid and DL-tropic acid conversion into alkaloids by cell cultures of *Datura stramonium* (Kurz and Constabel, 1979). That plant cell cultures may be used for such biotransformations is due to the nature of the materials synthesized by the plant cell rather than the existence of appropriate technological systems. Animal cells do not generally accumulate or synthesize the immense range of low molecular weight chemicals that are so popular in the pharmaceutical, food, cosmetic and colouring industries. Rather, the animal cell produces larger molecules or low molecular weight variants of the basic polymers which make up the animal cell (*e.g.* proteins–peptides; complex carbohydrates–dextrans). Thus the design of maintenance media with particular precursor materials is unlikely to play a major role in animal cell technology.

16.6 CELL CULTURE TECHNOLOGIES

Although there has been relatively little contact between animal and plant cell technologists, the methods which they both use to produce and exploit cultured cells have more common features than would appear at first glance. There are three modes of operation. (a) Cells may be held on a surface in which state medium is moved past the relatively static cells. (b) Cells may be attached to carrier particles and the particles themselves may be set in motion relative to the container. The second mode involves trapping the cells within a gel matrix. The cell-containing gel may be fashioned into small beads or be held on supporting structures in the form of sheets. In this case also it is possible to irrigate the cell system by causing medium to flow across a static bed of cell-containing particles. (c) The third system is that termed bulk suspension culture. Under such conditions the cells are not attached to a surface nor are they embedded in an artificial particle; rather they are caused to be held in a more or less homogeneous distribution throughout the medium and they are kept in such a state by some means of mixing, agitation or shaking. In plant cell culture, as opposed to animal cell systems, the cells have a greater tendency to form clumps; however, while this presents problems due to microenvironmental inhomogeneity, it generally does not detract from the overall performance of the suspended culture system.

16.6.1 Cellular Characteristics which Influence the Choice of Cell Culture Technology

The technology used for the large scale cultivation of animal and plant cells reflects in many ways the unique properties of those cells. The culture contents should be as homogeneous as possible which requires adequate mixing. Yet, whereas in almost all the prokaryotic systems, the method of mixing is not critical, for animal and plant cells it assumes greater importance.

16.6.1.1 Mixing

Plant cells are large cells, 20–150 μm diameter, and in culture they can, as a result of having a cell wall, assume elongate forms (L/D between 1–30). Such cells are capable of withstanding tensile strain yet they are sensitive to relatively slight shear stresses and straining (Fowler, 1983b). Numerous investigators (Tanaka, 1981; Fowler, 1982b) have shown that when culture systems containing plant cells are agitated too vigorously (by rotating shake flasks above 150 r.p.m. or stirring cultures with turbine impellers at similar speeds) a decrease in cell growth rate results. There are, however, exceptions to this general observation as was found in the author's laboratory (Fowler, 1982b). Efforts to obtain homogeneous systems for plant cells are made more difficult because (a) daughter cells tend to stay together following cell division, and (b) particularly towards the end of the growth phase, the cells excrete more polysaccharide whose 'stickiness' enhances the formation of clumps. The plant cell technologist has then to steer between the Scylla of allowing clumps to grow to a size when the cells in the centre die or liberate toxic materials and the Charybdis of over-agitation which decreases cell viability through physical factors.

Animal cells are smaller (10 μm diameter) than plant cells, they lack a cell wall and do not generally make sticky polysaccharides. They assume a spherical configuration and seem to be more shear resistant in that suspension cultures may be agitated at several hundred r.p.m. with flat bladed turbines without detrimental effects on cell viability. Such considerations do not pertain to the situation where animal cells are grown on the surface of microcarriers, as it has been shown that under such conditions the shear regimen of the system can cause decreases in cell density and viability (Sinskey *et al.*, 1981).

The observations quoted above indicate that mixing systems which necessarily generate hydrodynamic shear forces influence cell physiology and growth. However, the careful quantification of such influences, independent of effects on associated parameters (oxygen levels), still needs to be done.

16.6.1.2 Aeration

Plant cells require 1×10^{-12} mol O_2 h^{-1} cell^{-1}, whereas animal cells use 0.06–0.2×10^{-12} mol O_2 h^{-1} cell^{-1}. Plant cell cultures run at cell concentrations of 1–25 g dry weight l^{-1}; for animal cells the figure is of the order of 0.25–0.5 g dry weight l^{-1}, for suspension cultures, although cells may be held at densities 10–50 times higher than this in cultures with a solid phase. For such reasons oxygen transfer rates for plant cell cultures have to be higher than those of animal cells, but are not as high as those required for the growth of bacteria or yeasts (Table 14). While data are available for the influence of the oxygen transfer coefficient ($K_L a$) on the growth and product generation activities of plant cells, animal cell technologists prefer to consider levels of dissolved oxygen (Spier and Griffiths, 1983). This difference of approach could reflect the difficulties of measuring medium dissolved oxygen levels in plant cell cultures as cells would stick to electrode surfaces, and in the later stages of the culture when the viscosity increases dramatically the flow of fluid past the end of the probe is insufficient to obtain a meaningful reading.

Table 14 Typical Oxygen Transfer Coefficients for Cell Growth in Various Systems[a]

System	$K_L a$ (h^{-1})
Animal cells, suspension culture 2×10^6 cells ml^{-1} (0.5 g l^{-1})	1–25
Plant cells, suspension culture (10–15 g l^{-1})	20–30
Bacterial cell cultures (10–20 g l^{-1})	100–1000
Yeast cell cultures (10–30 g l^{-1})	100–1000

[a] From Spier and Griffiths (1983).

16.6.1.3 Doubling times

Both animal and plant cells in culture have long doubling times. The shortest doubling time in a chemostat type animal cell culture has been reported as 8 h (Tovey, 1985) while times of between 18 and 48 h are regarded as more general (Griffiths and Pirt, 1967). Similarly, plant cell doubling times are between 25 and 100 h (Fowler, 1983a). The clear implication of such low growth rates is that were an organism with a lower doubling time to gain access to a culture then it would overrun it in a very short time. Thus, while it is possible to do experiments in bacterial systems in hours or a few days under conditions where contamination detection is difficult and effort has to be expended to show that the culture is impure, for animal and plant cell culture bacterial and yeast contamination is obvious and considerable attention has to be given to the design and operation of systems which can be worked free from the possibility of contamination by faster growing organisms.

(i) Sterilization of media

Plant cell culture media are more simple than those used for the growth of animal cells and in general may be sterilized by heating. Animal cell culture media are generally filter sterilized either *via* a cascade of membrane filters or through depth filters. Such filters have to be pre-washed after steam sterilization. As filtration is a 'probabalistic' process, care has to be taken to ensure that the upstream contaminant loading of the filter is such that the probability of a contaminant gaining access to the sterile side of the filter is vanishingly low.

It is well known that overheating media containing sucrose (present in most plant cell media) causes caramelization. The production of heat-sterilized sucrose-containing solutions may be effected either by sterilization in a flow-through heat exchanger or by sterilizing the sucrose separately and then adding it to the medium by aseptic transfer.

The sterilization of serum for animal cell cultures often presents problems. The method used is filtration and the material is generally presterilized before it is added into a medium formulation. A cascade of filter elements with decreasing porosities is used and by careful selection of the filter materials the losses of serum can be kept within tolerable limits. Once filtered, it may be stored frozen at $-20\ °C$ for long periods of time with undetectable changes in its growth promoting performance.

(ii) Sterilization of equipment

Equipment is sterilized conventionally by steam at 10–15 p.s.i. resulting in temperatures of between 110 and 120 °C. Sterilization using ethylene oxide, UV light, γ-radiation, or bleaches (hypochlorite solutions) are not commonly used because of lack of reliability (penetration of clumps of detritus in improperly cleaned vessels) or because they are difficult to implement and leave toxic residues. Fully automated (one button) sterilization systems are becoming more reliable, though they remain an expensive alternative to a properly monitored and controlled process operator implemented sterilization routine. Attention to detail in such procedures leads to the reliable and repeatable sterilization of process vessels, and chemostat cultures of animal and plant cells have been run for times in excess of two months.

16.6.1.4 Cell stickiness

The formation of cell clumps, morula or meringues (Fowler, 1983a) in plant cell cultures presents problems as the scale of the culture increases. Foams are also generated which aggravate the deleterious effects of clumping. As larger particles have higher sedimentation rates (providing that they do not generate gas which becomes entrapped with the bolus) more mechanical energy has to be applied to the culture to achieve the maximum degree of homogeneity. The crusts and wall growths which form decrease the operational efficiency of probes and can lead to diffficult problems when it comes to material transfers through pipes and valves in addition to the difficulties of cleaning vessels prior to sterilization. Animal cells are also sticky, but not to the extent of forming crusts. They do, however, form clumps (particularly in microcarrier cultures: see below) and they can form layers over probe surfaces though this is less of a problem as the cultures tend to be agitated more vigorously than their plant counterparts. Foams which are produced in aerated cultures are the result of the presence of serum in the medium rather than the properties of the cells or their exudates. While such foams cause practical problems (they immobilize a proportion of the cells at the gas/liquid interface) it is possible in the animal cell systems which require relatively little aeration (Section 16.6.1.2) to decrease the generation of foam by the use of a stainless steel mesh caged aeration system (Spier and Whiteside, 1983).

16.6.2 Immobilized Cell Systems

The production and exploitation of cells in an immobilized form leads to numerous technological advantages: (i) the cells may be held in a fraction of the overall reactor volume, thus enabling materials to be produced at higher concentrations and thereby facilitating downstream processing; (ii) the fluids which are made to wash over the cells may be changed readily without recourse to what is often a difficult cell/medium separation system (particularly if the products have to be maintained in an uncontaminated and viable state); (iii) the time needed to change from one fluid to the next can be very short which can be a critical factor when promoters and inhibitors have to be used in a critical time sequence, and as such systems can be washed easily the removal of the preceding reaction fluids can be achieved with efficacy; (iv) cells in such a configuration can be used as flow-through biocatalysts (more appropriate to plant cell systems at present, yet there are animal cell systems which could be developed for detoxification roles); (v) in such systems the cells are exposed to less physically demanding environments (medium mixing and oxygenation can be achieved in a separate vessel thus preventing the exposure of the cells to high shear forces) which leads to cultures which can be maintained in a non-dividing yet product generating phase for extensive periods of time (weeks to months), (vi) in plant cell systems, the cells stick to the measurement probes. When the cells are immobilized in one vessel it becomes practical to use the probes in a second, medium reservoir vessel, and thus achieve control of more process parameters.

The following two sections will describe two kinds of immobilized cell systems; firstly those in which the cells are grown on a solid (or relatively solid gel) surface and, secondly, systems which hold the cells immobile by restraining them within the matrix of a particle, gel, mesh or sponge.

16.6.2.1 *The growth and exploitation of cells grown on the surface of a supporting solid substratum*

Plant cell cultures are initiated on the surface of agar gels (Fowler, 1983b) but once this callus formation stage has passed they are either used in non-supported technologies or, if they are maintained on the agar surface, they are exploited in differentiated forms. This technology is useful for propagation of plants whose seed formation–germination cycle presents problems under controlled conditions. 'Rare' orchids may be propagated by such cell culture techniques (Fowler, 1983a). Animal cell cultures, too, begin their careers on glass or plastic surfaces (except for cell cultures derived from overt carcinomas). However, unlike the plant cell systems which, apart from the uses cited above, have not developed further, the animal cell surface-dependent culture systems have been developed in a wide variety of conformations and have been used commercially in most of those configurations. The remaining part of this section will, therefore, deal exclusively with the kinds of technologies which have been used to culture anchorage-dependent animal cells (for review see Spier, 1980).

Two fundamentally different approaches to growing animal cells are presently practiced. The one, wherein the scale-up of the process involves the increase in the number (or multiplicity) of units is defined as a multiple process; the second, wherein the scale-up of the process involves building a larger unit, is called a unit process. Advantages of multiple processes are: (i) if one unit is contaminated other units may be saved; (ii) scale-up does not alter system parameters, only management practices; (iii) the equipment used is simpler, more robust, easier to understand, trouble-shoot and replace; (iv) experiments are easy to conduct and thus small volumes of reagents are used. Lots of different formulations, times and conditions can be examined. Advantages of unit processes are: (a) they are labour and space saving and generally more cost efficient; (b) as 'all the eggs are in one basket', it is practicable to monitor and control the process conditions so as to yield better quality product with higher reliabilities; (c) systems lend themselves to higher degrees of automation; (d) fewer aseptic connections are required which leads to better overall freedom from contamination and, in reverse, better disease or biological 'security' to the environment outside the reactor. Within these two broad definitions there are a large number of variants as depicted in Figure 2. Indeed, there are innumerable subvariants which are finding their way into the literature in a steady trickle.

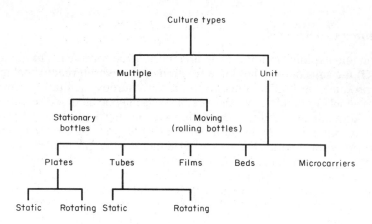

Figure 2   Immobilized cell culture systems for surface based cultures

(i)   Multiple processes

Animal cells are commonly grown in stationary cultures as described in the early sections of this chapter, on the inner surface of the base of a flattened bottle. Bottle materials can range from glass [soda or Pyrex (borosilicate)] to polystyrene (sterilized by γ-radiation and rendered suitable for the growth of animal cells by proprietary treatments; however, see Spier, 1980). Polycarbonate (expensive) and polypropylene (robust and sterilizable by heat at 120 °C) have also been used. The bottle sizes may vary from small test cultures of about 10–25 ml to bottles of 5 l capacity. As a rule of thumb, about 100 000–750 000 cells may grow on 1 cm^2 surface and, in so doing, generate 0.5 ml of spent growth medium. This relationship is a more or less constant feature of animal cell culture systems.

Dynamic bottle systems : roller bottles may be as small as a few hundred centilitres and as large as 10 l. The shape of the bottle can vary enormously from rectangular or square cross-section to elongate closed-off tubes two metres long and 10–15 cm diameter. Special racks, jigs, formers and supports are used to adapt the odd-shaped bottles to the rig which induces them to roll. The advantages they have over stationary bottles is that (i) all the internal surface of the bottle is available for cell growth; (ii) aeration is more efficient in a system in which the liquid is in movement and most of the cells are only covered by a thin film of liquid for most of the time; (iii) some people have produced more cells for less medium; and (iv) it is a more concentrated and hence space saving method. However, it is also (i) more expensive, (ii) liable to break down and vibrate, (iii) has to be carefully levelled, and (iv) introduces a complicating, additional variable, namely rotation speed.

(ii)   Unit processes

In Figure 2 five different sorts of unit process system have been depicted. Plate systems may be seen to be derived from stationary bottle systems and the propagator designed by Abbot Laboratories in the early 1960s was a 'plurality' of stacked plates held in a steel framework round which medium was circulated using an air lift pump. This was an awkward piece of equipment to use; the plastic version (multitary system) designed and built in the Nunc Laboratories has been used widely and has been highly mechanized for β-interferon production (Pakos and Johansson, 1982). It is an expensive alternative.

Rotating plates have also seen some applications. The plates may be vertical [commercialized by Biotec (titanium plate) or Connault Labs (glass plates)] or inclined to the horizontal (Birch *et al.*, 1981). Such systems tend towards overcomplex mechanical devices and scale-up changes too many factors. Weight, cost and the maintenance of parallelicity of the plates also present problems, as does inflexibility of the surface area to system volume ratio. Furthermore, it is often mechanically difficult to use both sides of the plates.

Systems which combine a multiplicity of tubes in a tank and either rotate the tubes in the medium or pump the medium through or around the tubes have become available in a number of forms recently. Small diameter tubes made of complex mixtures of materials which are generally used in filtration/reverse osmosis concentration systems, have been promoted by their manufacturers as animal cell culture systems. They have advantages in that it is possible to remove from the cells small molecules which pass to the other side of the membrane (detoxification or end-product negative-feedback removal). Yet they suffer from the disadvantage that cells grow in three dimensions and can block off the tubes resulting in 'dead' areas, also it is difficult to ensure that all tubes are experiencing the same conditions as different tubes become occluded to different degrees. These difficulties were reflected in Monsanto's second patent (1976) for such equipment, so that recently expressed optimism for such systems (Feder and Tolbert, 1983) may be unjustified.

Using, as the basis for their system, a multiplicity of wider bore glass tubes held between engineered end plates, Chemap and others have attempted to obtain the advantages of rolling bottle systems in a unit process configuration. Such systems have been scaled-up to 200 l and in spite of the mechanical and physical problems they have been shown to work. The main factors which count against them are cost, complexity (especially when they have to be disassembled and re-assembled as for cleaning) and inflexibility, for though they are more flexible than static systems they have to operate at least half-full for all the tubes to be irrigated by the various culture media. The extension of plates into films is inevitable and many spiral film propagators have been built. Melanex has been the film of choice, although others may be used. On the small scale (less than 2 l medium volume) such systems perform well, but larger systems run into problems of retaining homogeneity, medium circulation and a constant interleaf gap. A novel modification of the film into bags which can be wound up on to a roll has been used by Instrumentation Laboratories. The advantages claimed for this system are based on facile gas transfer which leads to good O_2 control and easy control of pH (using gaseous CO_2). Again, such systems are mechanically complex and costly and the gas control obtained easily in such systems may just as easily be obtained otherwise (see below).

The latter two systems described in this section have been more widely used and will be considered under separate headings.

(a) Beds. Columns containing a wide variety of packing materials have been the basis of many successful mass and heat transfer operations in the chemical industry. Such systems may also be used to propagate animal cells (Spier, 1980). The packing materials which have been used include glass, stainless steel, polystyrene and silicates moulded into various forms (spheres, jacks, static mixers, sintered glass matrix, springs, loops, spirals, raschig rings). The bed is normally held in a vessel separate from the vessel holding the bulk of the medium and the medium is circulated over the bed by airlift, centrifugal, gear, peristaltic and diaphragm pumps. This results in a simple, relatively inexpensive, robust technology using familiar substrates in a fairly flexible system. It scales-up well and can be used for a wide variety of cell types. Its chief disadvantage is that it is difficult to obtain a sample of the immobilized cells and it is not homogeneous. However, measurements on samples of the process fluids can be used to great effect to obtain control of the biological performance of the system (Whiteside and Spier, 1981).

Recent developments on the plant cell side (Rosevear and Lambe, 1982) have begun to study the effect of immobilizing plant cells on the surface provided by matrixes formed by nylon meshes, polyphenylene oxide porous particles and polyurethane sponges.

(b) Microcarriers. Microcarriers were offered to the biotechnological community in 1967 by the pioneering work of Van Wezel (1967). In 1982 the first 1000 l culture using such materials was reported (Montagnon and Fanget, 1982). Numerous variants of the microcarrier concept have been assessed. Microcarriers have been made from polystyrene, filter paper fibres, gelatin, chitosan (Nilsson and Mosbach, 1980) and glass, but the ones which have achieved most prominence are those derived from a 0.2 mm dextran bead substituted in different ways with defined amounts of diethylaminoethyl groups. The additional coating of such beads with nitrocellulose (Celloidin), serum or more recently collagen has made them more amenable to use. This type of culture system is homogeneous and can be run in simple reactors (for the most part modified fermenters). However, the more sophisticated microcarriers are costly, culture conditions have to be very carefully defined and controlled particularly in the early stages, the removal of process fluids from such systems can present problems because of the high liquid hold-up and finally the systems are shear sensitive (Sinskey *et al.*, 1981), and whereas oxygenation of such cultures has presented problems it is thought that generating the bubbles within a steel mesh cage which cannot be penetrated by the microcarriers may solve this problem (Spier and Whiteside, 1983). This system has much to offer yet in its present form it does not satisfy all needs.

16.6.2.2 *The growth of animal and plant cells immobilized within a confining matrix*

Most of the advantages which accrue to cultures immobilized on the surface of support materials also pertain to matrix based cultures (Section 16.6.2.1). There are, however, two problem areas. The first involves the process of generating the matrix containing the cells which could damage or alter the cell character, while the second concerns the structural inhomogeneity introduced by creating concentration gradients of (a) nutrients and (b) toxic by-products between the extraparticular environment and the centre of the particle. A third feature which is found in some compositions is that either cell growth or the leaching out of calcium or other cations causes the particles to lose their integrity on prolonged incubations (Rosevear and Lambe, 1982). In spite of such considerations, much successful immobilization work has been done in the laboratory although this has not yet been exploited at the pilot or production plant scale.

Most non-cancer derived animal cells attach to glass, metal and treated plastic surfaces readily, so that the requirement for immobilization has been met by the systems described above (Section 16.6.2.1). However, in recent times there has developed a need to use monoclonal antibody producing cells in the immobilized state and as these cells do not stick readily to surfaces, as they are derived from a myeloma cell whose cancer cell-like growth characteristics dominate the hybrid in which it is incorporated, alternative methods of immobilization have to be used.

The growth of plant cells on surfaces, however, is not very controllable: they form calluses or clumps and as they do not have the power of locomotion they do not, as do the animal cells, spread out in a beneficent monolayer over the surface of the provided substratum. For this reason much more effort has been expended on the immobilization of plant cell cultures within a confining matrix. Indeed, the technology developed for such plant cell systems may now find an application in the animal cell domain.

(i) *Gel entrapment systems*

Gels have been the most widely used system for entrapping cells. They have high porosities which decrease diffusional problems, they do not subject embedded cells to surface forces and they are sufficiently accommodating to allow a limited amount of cell division or cell size change (Rosevear and Lambe, 1982). The gels may take the form of sheets or beads. The latter are the preferred configuration as they may be stacked in columns or gently agitated in shake flasks.

The method of gel formation involves the mixing of a suspension of the cells to be entrapped with a 2–3% solution of sodium alginate (with additional nutrients). This is then injected into a second solution containing polyvalent cations, of which the most commonly used is calcium, at 50–100 mM concentrations. After about 1 h a stable bead (diameter 3–6 mm) forms. For analytical purposes the bead may be dissolved using chelating agents (EDTA, citrate, polyphosphate). Other gels have been used, agarose, carrageenin and other cross-linking agents. Thus acrylamide (5%) supplemented with xanthan gum or alginates has been used as gel sheets or sheets reinforced with open weave cotton cloth to yield systems in which cell viability has been retained.

(ii) *Applications of entrapped cells*

To date the work on entrapped cells has centred upon the retention of viability and the exploration of the possible products that could be produced from such systems, their yield, quality and productivity. The kinds of products (or processes) which have been looked at are summarized in Table 15. One area of considerable interest for animal cell biotechnologists is the possibility of providing insulin deficient patients with insulin producing cells which respond to glucose levels in the tissue fluids and so naturally augment defective pancreatic functions. Such systems could be a solution to the repeated injection or controlled release of the complex, yet technologically sophisticated, pump/microprocessor systems are under development to administer insulin to diabetics.

The exploitation of entrapped plant cells for biotransformations looks promising and the production of alkaloids and other fine and pharmaceutically active chemicals is under intensive investigation. It is too early to say yet which cell/product system will be found in the industrial exploitation situation, but it is clear that new materials and new processes to make them are an outcome which can be foreseen with a high probability.

16.6.3 Dynamic Cell Systems

Systems in which cells are grown in a dynamic state may be subdivided in a number of different ways. (i) Level of complexity would yield a grouping of: (a) simple systems consisting of a mag-

Table 15 Possible Products and Processes Based on
Entrapped Animal and Plant Cells

Cell type	Product or process
Animal cells	
Pancreatic β-cells	Insulin production
Hepatoma	Bilirubin glucuronide
Adipocyte cells	Controllable metabolism of lipids
Hybridoma	Antibody production
Plant Cells	
Digitalis lanata	Hydroxylation of cardiac glycosides
Daucus carota	Digitoxigenin→periplogenin
Morinda citrifolia	Biosynthesis of anthraquinones
Catharanthus	Ajmalicine production
Catharanthus	Serpentine production
Catharanthus	Indole alkaloids
Solanum aviculare	Glycoalkaloid production
Beetroot	Betacyanin pigments

netically driven stirrer in a bottle; (b) systems with a sterilize and service-in-place capability; (c) sophisticated computer monitored and controlled batch systems; and (d) computer monitored and controlled continuous systems. (ii) Agitation system differences would group impeller mixed systems differently from gas bubble mixed systems which would again be different from shaken or vibrated systems. An additional mixing system can be provided by an externally sited pumping mechanism. (iii) Systems in which the cells are held in one vessel and, by a variety of means, medium held in a second vessel is made to circulate through the cell-holding vessel (perfusion systems). Such systems are more readily achieved for immobilized cell systems but can be implemented with dynamic cell systems also.

While a full consideration of the different kinds of fermenters has been adequately dealt with in other chapters of this work, this section will concentrate on those fermenter designs in which most animal and plant cell biotechnologists have an interest. There designs have in outline been summarized in Figure 3. The objectives of the designs have been: (i) to achieve the lowest power input necessary to obtain a homogeneous culture wherein a sample taken at any place from the bulk of the culture is representative of the culture as a whole; (ii) to compensate for an increase in the size of the culture vessel (with a consequent change in the culture surface area/volume ratio) in such a way as to retain a constant microenvironment for cell growth, division and product expression; (iii) to achieve the above two conditions without creating a situation where the hydrodynamic forces disrupt cells or cause losses of cell viability; (iv) to give the flexibility of being able to change one process parameter while other parameters are held at contant levels; (v) to achieve a design which is as simple, robust, inexpensive and as versatile as possible for the cells and process specified—such vessels are easy to trouble-shoot, replacement of worn parts is simple, they adapt to changing economic and technical circumstances and offer the prospect of interchangeability of parts; and (vi) to achieve other features such as quietness of operation with reliability and replicability as desirable additional characteristics.

From Figure 3 it is clear that there are three fundamentally different reactor types, *viz.* those whose mixing is governed solely by the admission of air, those whose mixing is mainly determined by a rotating blade, impeller or turbine, and hybrid reactors which are a combination of the two previous types. For animal and plant cell cultures, the reactors are generally not designed to be efficient oxygen transfer systems, rather they aim at decreasing the severity of the hydrodynamic shear forces, which perhaps explains the profusion of designs available. From an operational point of view, the differences are subtle and may not be evident until attempts are made to scale them up. So, at small scale, any design will probably work but at larger scales there are considerable advantages in working with efficient units.

16.6.3.1 Air driven systems

The largest fermenter in the world, built by ICI for single cell protein production, is a 1000 m^3 air driven reactor (Figure 3, configuration 2). It is held to be efficient at oxygen transfer though as a mixing system it may have deficiencies as the carbon substrate has to be introduced through 1500 individually controlled inlet units. Variants of this system have been used for some time for

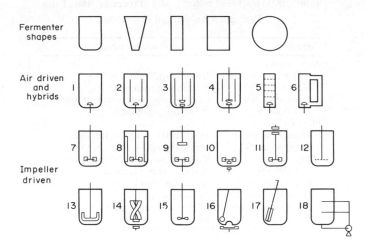

Figure 3 Alternative fermenter configurations. Configurations 8, 9 and 10 are industry standard models. Other configurations are variously dealt with in Spier (1985) and Fowler (1983b) and references therein

the culturing of plant cells (Fowler, 1982b). The air lift loop reactor (Figure 3, configuration 6) has been scaled-up to 100 l volumes and has been shown to function well up to cell concentrations of 20 g dry weight l^{-1}. Above such concentrations when the viscosity of the system increases, considerably more energy is required for mixing and other forms of culture vessel are required. Simple draught tube reactors (Figure 3, configuration 2) of up to 10 l volumes are also widely used. The advantage of such reactors is that they are mechanically simple, having no moving parts and generate substantial oxygen transfer rates. Hydrodynamic shear forces are low, which tends to lead to clump formation but cell damage is not incurred. The scale up of such systems has been shown to be practicable for bacteria (although with a considerable increase in complexity) yet with sticky clumped cells difficulties will obviously occur.

An air driven system (Figure 3, configuration 5) based on work in the plant cell area has recently been used with success for the growth of hybridoma cells. The direct comparison of such a system with an impeller agitated counterpart has yet to be reported. Such air driven systems under the guise of bubble column reactors, airlift reactors or Waldorf fermenters, have been used to cultivate BHK cells, lymphoblastoid cells and hybridoma cell lines. These systems are limited by the formation of foams or are adversely affected by the antifoam agents, yet under suitable low serum situations this technology has a role to fulfil.

16.6.3.2 Impeller and air driven systems

Once the air bubbles have been formed, there is a little control of their eventual development in the air-alone driven systems. Adaptations of this type of reactor have been made either to improve the mixing by using a screw impeller, or to improve the aeration by using a turbine impeller to break down the bubbles as they egress from the aeration discharge point and thereby dramatically increase the oxygen transfer coefficients (Martin, 1980; Figure 3). Vessels with internal structures are difficult to clean (particularly after a contaminated run) and are thereby more complex than is necessary. Furthermore, changes in the vessel size alter more parameters than the surface area to volume ratio which makes the reestablishment of the bubble flow patterns of the small scale units a trying exercise.

16.6.3.3 Impeller mixed systems

In Figure 3 there are sketches of impeller driven systems. This should not be taken to be an exclusive survey, indeed it is presented to show the range and variability which can be obtained from changes in this one area. Such variation could mean that it is an unsuccessful system and that new variants are needed to accommodate the weaknesses of the older versions; on the other hand the variations show the systems' basic flexibility, adaptability and capability to be tailored to particular applications as they arise. Like the tower fermenter (Figure 3, configuration 2) the impeller mixed fermenters have been built to capacities in excess of 200 m³ without the additional complication of creating homogeneity by multiple subdivisions of the substrate addition system.

These systems are characterized by a number of simple and complex parameters. The shape and size of the impeller determines the pump/shear ratio. This is not a measurable number but loosely characterizes the system. Thus large diameter impellers moving at low speeds have very high ratios as the shear action is generally low, whereas small diameter impellers rotating rapidly generate high shear forces but are less effective pumping systems. Thus this ratio determines the hydrodynamic characteristics of the vessel contents and for the shear sensitive cells dealt with in this chapter it is a major determinant of system performance. A second characteristic which relates to the pump/shear ratio is the power/volume ratio. This is often used as a parameter to hold constant when scaling up. It is dependent on the r.p.m. (r.p.m.³) and impeller (6 or 8 blade turbine) diameter (D^5). Thus the power per unit volume can be increased by increasing the diameter of the impeller at constant r.p.m. However, there is a corresponding increase in impeller tip speed which indicates that a regimen with higher shear forces is generated. For animal cell systems, scale-up technologists often hold the impeller tip speed as the constant factor and allow power/volume or other parameters to float.

Air is admitted into the impeller mixed reactor through a number of alternative devices. An open ended tube is common and specialized sinters or ceramics with different porosities are used. Annuli with perforations may also be practicable.

The number of parameters which can be varied with the impeller system are higher than those afforded by the air-alone driven systems. Such versatility may or may not be required for particular production systems, yet it is clear that as the scale of operation increases the demands made on the system for mixing become greater, which in turn requires that the opportunity to have an independent control of the hydrodynamic forces becomes a necessity rather than a luxury.

All of the designs depicted in Figure 3 have been used for the growth of either animal or plant cells. For animal cell systems turbine impellers are generally used at the larger scales of operation (4000 l) while at the smaller scales (<200 l) almost any system will provide a workable capability.

16.7 FINAL COMMENTS

We have attempted to provide a basic but reasonably detailed entrée into the field of cell culture. Space has not allowed us to delve into a number of areas which relate to and are of obvious importance to developments in cell culture. These areas include recombinant DNA technology, enhancement of specific product and biomass yield, the physiology and biochemistry of cultured cells and the recovery and processing of products. All of these are dealt with to a greater or lesser degree in various references given in the text. Readers wishing to have further information on these areas are referred to Spier (1985) and Fowler (1983b).

16.8 REFERENCES

Arathoon, W. R. and R. C. Telling (1981). Uptake of amino acids and glucose by BHK21 clone 13 suspension cell during cell growth. Joint ESACT/IABS meeting on the use of heteroploid and other cell substrates for the production of biologicals. Heidelberg 1981. *Dev. Biol. Standards*, **50**, 145–154.

Baugh, C. L., R. W. Lecher and A. A. Tytell (1970). The effect of pH on the propagation of the diploid cell W1–38 in galactose medium. *J. Cell. Physiol.*, **70**, 225–228.

Birch, J. R. and T. Cartwright (1982). Environmental factors influencing the growth of animal cells in culture. *J. Chem. Technol. Biotechnol.*, **32**, 313–317.

Birch, J. R., T. Cartwright and J. A. Ford (1981). Verfahan zum Zuchten von Tienschen Zellen und mit Einen Plantten-stapel Versehene Zellzuchtungsapparatur. *Deutesches Patentant*, **3**, 031–6761.

Bolt, K., J. Clarke and R. E. Spier (1979). The use of frozen BHK 21C13 cells to control the biological parameters for cell and foot and mouth disease virus growth. 2nd General Meeting ESACT, Paris 1978, *Dev. Biol. Standards*, **42**, 47–53.

Capstick, P. B., R. C. Telling, W. G. Chapman and D. L. Stewart (1962). Growth of a cloned strain of hampster kidney cells in suspended cultures and their susceptibility to the virus of foot and mouth disease. *Nature (London)* **195**, 1163–1166.

Clarke, J. B. and R. E. Spier (1980). Variation in the susceptibility of the BHK populations and cloned cell lines to three strains of foot and mouth disease virus. *Archiv. Virol.*, **63**, 1–9.

Dougall, D. K. (1980). Nutrition and metabolism. In *Plant Tissue Culture as a Source of Biochemicals*, ed. E. J. Staba, pp. 22–58. CRC Press, Boca Raton, FL.

Eagle, H. (1974). Some effects of environmental pH on cellular metabolism and function. In *Control of Proliferation in Animal Cells. Cold Spring Harbor Conference on Cell Proliferation*. ed. B. Clarkson and R. Baserga, pp. 1–11. Cold Spring Harbor Laboratory, New York.

Feder, J. and W. R. Tolbert (1983). Large scale cultivation of mammalian cells. *Sci. Am.*, **248**, 24–32.

Fowler, M. W. (1982a). Substrate utilisation by plant-cell cultures. *J. Chem. Technol. Biotechnol.*, **32**, 338–346.

Fowler, M. W. (1982b). The large scale cultivation of plant cells. *Prog. Ind. Microbiol.*, **17**, 207–229.

Fowler, M. W. (1983a). Commercial applications and economic aspects of mass plant cell culture. In *Plant Biotechnology*, ed. S. H. Mantell and H. Smith, pp. 3–37. Cambridge University Press, Cambridge.

Fowler, M. W. (1983b). Plant-cell cultures: fact and fantasy. *Biochem. Soc. Trans.*, **11**, 23–28.

Girard, H. (1977). Problems encountered in large scale cell production plants. In *Cell Culture and its Applications*, ed. R. T. Acton and J. D. Lynn, pp. 111–128. Academic, New York.

Griffiths, J. B. and S. J. Pirt (1967). The uptake of amino acids by mouse cells (strain LS) during growth in batch culture: the influence of cell growth rate. *Proc. R. Soc. London, Ser. B*, **168**, 421–438.

Horodniceanu, F. (1977). The problem of contaminants in the sera used for cell culture—a short review. *Proceedings of the First General Meeting of ESACT, Amsterdam*, ed. R. Spier and A. L. van Wezel, pp. 23–26. Instituut voor de Volksgezondheid, Bilthoven, Netherlands.

Janss, D. H., E. A. Hillman, L. B. Malan-Shibley and T. L. Ben (1980). Methods for the isolation and culture of normal human breast epithelial cells. *Methods Cell Biol.*, **21b**, 107–134.

Keay, L. (1975). Autoclavable low cost serum free cell culture media: the growth of L cells and BHK cells on peptones. *Biotechnol. Bioeng.*, **17**, 745–764.

Konowalchuck, J. and J. I. Speis (1979). Response of various cell lines to *Escherichia coli* toxic products. *Can. J. Microbiol.*, **25**, 335–339.

Kurz, W. G. W. and F. Constabel (1979). Plant cell suspension cultures and their biosynthetic potential. In *Microbial Technology*, ed. H. J. Peppler and D. Perlman, pp. 389–416. Academic, London.

Litwin, J. (1977). Necessary amino acids for the growth of human diploid fibroblasts. In *Proceedings of the First General Meeting of ESACT, Amsterdam*, ed. R. Spier and A. L. van Wezel, pp. 9–14. Instituut voor de Volksgezondheid, Bilthoven, Netherlands.

Litwin, J. (1979). A survey of various media and growth factors used in cell cultivation. Second General Meeting ESACT, Paris 1979. *Dev. Biol. Standards*, **42**, 37–45.

Martin, S. M. (1980). Mass culture systems for plant cell suspensions. In *Plant Tissue as a Source of Biochemicals*, ed. E. J. Staba, pp. 150–166. CRC Press, Boca Raton, FL.

McLimans, W. F. (1972). The gaseous environment of the mammalian cell in culture. In *Growth, Nutrition and Metabolism of Cells in Culture*, ed. G. H. Rothblat and V. J. Quitofalo, pp. 137–140. Academic, London.

Mizrahi, A. and A. Avihoo (1976). A simple and cheap medium for BHK cells in suspended cultures. *J. Biol. Standards*, **4**, 51–56.

Molander, C. W., A. J. Knavief, C. W. Boone, A. Paley and D. T. Imagawa (1972). Isolation and characterisation of viruses from foetal calf serum. *In vitro*, **7**, 168–173.

Monsanto Company (1976). *US Pat.*, 1 514 906.

Montagnon, B., J. C. Vincent-Falquet and B. Fanget (1983). Thousand litre scale microcarrier culture of vero cells for killed poliovirus vaccine: promising results. 5th meeting of ESACT, Copenhagen, 1982. *Dev. Biol. Standards*, **46**, 37–42.

Nilsson, K. and K. Mosbach (1980). Preparation of immobilised animal cells. *FEBS Lett.*, **118**, 145–150.

Pakos, V. and V. A. Johansson (1983). Large scale production of human fibroblast interferon in a multitray battery system. 5th meeting of ESACT, Copenhagen, 1982. Abstracts.

Pastan, I. and M. Willingham (1978). Cellular transformations and the morphologic phenotype of transformed cells. *Nature (London)*, **274**, 645–650.

Prescott, G. C. and C. G. Dunn (1959). The microbiological transformation of steriods. In *Industrial Microbiology*, 3rd edn., pp. 723–761. McGraw-Hill, New York.

Rein, A. and H. Rubin (1967). Effects of local cell concentrations upon the growth of chick embryo cells in tissue culture. *Exp. Cell Res.*, **49**, 666–678.

Rosevear, A. and C. A. Lambe (1982). Immobilised plant and animal cells. *Top. Enzyme Ferment. Biotechnol.*, **7**, 13–37.

Rous, P, and F. S. Jones (1916). A method for obtaining suspensions of living cells from fixed tissues and for the planting out of individual cells. *J. Exp. Med.*, **23**, 549–555.

Sato, G. and V. Buonassisi (1964). Hormone synthesis in dispersed cell cultures. In *Retention of Fundamental Differentiation in Cultured Cells*, ed. V. Defendi, *The Wistar Institute Symposium Monograph 1*, Philadelphia, pp. 27–34.

Sinskey, A. J., R. J. Fleischaker, M. A. Tyo, D. J. Giard and D. I. C. Wang (1981). Production of cell products: virus and interferon. *Ann. N.Y. Acad. Sci.*, **369**, 47–59.

Skoog, F. and C. O. Miller (1957). Chemical regulation of growth and organ formation in plant tissues cultured *in vitro*. *Symp. Soc. Exp. Biol.*, **11**, 118.

Spier, R. E. (1980). Recent developments in the large scale cultivation of animal cells in monolayers. *Adv. Biochem. Eng.*, **14**, 119–162.

Spier, R. E. (1985). Processes and products dependent on cultured animal cells. In *Basic Biotechnology*, ed. J. D. Bu'Lock and B. Kristianson, in press. Academic, London.

Spier, R. E. and B. Griffiths (1983). An examination of the data and concepts germane to the oxygenation of cultured animal cells. 5th meeting of ESACT, Copenhagen, 1982. *Dev. Biol. Standards*, **46**, 81–92.

Spier, R. E. and J. P. Whiteside (1983). The description of a device which facilitates the oxygenation of microcarrier cultures. 5th meeting of ESACT, Copenhagen, 1982. *Dev. Biol. Standards*, **46**, 151–152.

Stores, P. B. (1981). Viral Vaccines. In *Essays in Applied Microbiology*, ed. J. R. Norris and M. H. Richmond, pp. 1–32. Wiley, Chichester.

Takahashi, K. and T. S. Okada (1970). An analysis of 'conditioned medium' upon the cell culture at low density. *Dev. Growth Different*, **12**, 65–77.

Takayama, T. and T. Misato (1977). Effect of cultural conditions on the growth of *Agrostemna githago* cells in suspension culture and the concomitant production of an anti plant virus substance. *Physiol. Plant.*, **41**, 313–320.

Tanaka, H. (1981). Technological problems in cultivation of plant cells at high density. *Biotechnol. Bioeng.*, **23**, 1203–1218.

Temin, H. M., R. W. Pierson, Jnr. and N. C. Dulak (1974). The role of serum in the control of multiplication of avian and mammalian cells in culture. In *Control of Proliferation in Animal Cells. Cold Spring Harbor Conference on Cell Proliferation*, ed. B. Clarkson and R. Baserga, vol. 1, pp. 50–81. Cold Spring Harbor Laboratory, New York.

Tomei, L. D. and C. J. Issel (1975). Growth and immunogenicity of foot-and-mouth disease virus and baby hamster kidney cells adapted to and continuously grown in a serum-free chemical defined medium. *Biotechnol. Bioeng.*, **17**, 765–778.

Tovey, M. G. (1985). The cultivation of animal cells in continuous flow culture. In *Animal Cell Biotechnology*, ed. R. E. Spier and B. Griffiths, in press. Academic, New York.

van Wezel, A. L. (1967). Growth of cell-strains and primary cells on micro-carriers in homogeneous culture. *Nature (London)*, **216**, 64–65.

Weymouth, C. (1972). Construction of tissue culture media. In *Growth, Nutrition and Metabolism of Cells in Culture*, ed. G. H. Rothblat and V. J. Cristofalo, vol. 1, pp. 11–47. Academic, New York.

Whiteside, J. P. and R. E. Spier (1981). The scale-up from 0.1 to 100 litres of a unit process system based on 3 mm diameter glass spheres for the production of four strains of FMDV from BHK monolayer cells. *Biotechnol. Bioeng.*, **23**, 551–565.

Willmer, E. N. (1935). The growth and differentiation of normal tissues in artificial media. *Tissue Culture*. Methuen, London.

Zolleto, R. and G. Gagliardi (1977). New media and their advantages in the production of suspended cells and foot-and-mouth disease virus. *Dev. Biol. Standards*, **35**, 27–31.

CHEMICAL AND BIOCHEMICAL FUNDAMENTALS

17
Introduction

H. DALTON
University of Warwick, Coventry, UK

17.1 INTRODUCTION

Although microbes have been exploited for their products, particularly in the food and beverage industries for many centuries and the pharmaceutical and effluent treatment industries in the present century, a new and exciting stimulus to their exploitation has recently appeared. This has been due, in part, to the development over the last decade of recombinant DNA technology and the dramatic increase in oil prices which has engendered an evaluation of the potential of microbes to produce a wide variety of useful products.

We are quite clearly in the midst of a biological revolution in which it is now possible to seriously consider the 'construction' of microorganisms that might utilize a specific low cost substrate to make a desirable end-product. It is often quoted that we are only limited in the variety of products that can be made this way by our own imagination. Whilst there is a grain of truth in this it is also obvious that much still has to be understood about the microbe before we can be truly assertive about such a statement. Recombinant DNA technology has certainly shown that novel substrates can be utilized and novel products can be made by microbes, but in many instances this may not be necessary since judicious and careful screening procedures can turn up organisms with the desired properties (Dixon, 1983).

One of these desirable properties is the ability to utilize readily available cheap resources as starting substrates for bioconversion processes. Therefore the first nine chapters of this section are devoted to the metabolism of such substrates since it is through our understanding of how simple compounds can be transformed to more complex ones that we can begin to appreciate the extremely diverse abilities of microorganisms and discover how we might be able to harness these abilities to make valuable products.

Glucose, which in many parts of the world is an abundant source of carbon produced enzymically from starch, sucrose or even cellulose, is metabolized by aerobic and anaerobic organisms alike. Aerobic metabolism of glucose (Drs. Payton and Haddock, Chapter 18), particularly by prokaryotes such as *Escherichia coli*, generally results in complete oxidation to carbon dioxide and water in a highly regulated manner yielding few biotechnologically useful products. However, such organisms are important hosts for genetic material derived from other organisms, both prokaryotic and eukaryotic, which is expressed to produce either valuable proteins which are excreted or enzymes which may alter the metabolism of the organisms to produce new products.

Anaerobic glucose metabolism (Professor Morris, Chapter 19), on the other hand, results in the production of a variety of partially oxidized products and such activities in *Clostridium acetobutylicum* were, until recently, being exploited industrially in South Africa. It is also clear that our understanding of how glucose metabolism is regulated in the anaerobes could lead to several more products being produced on a commercial scale in the future. Studies in recent years on a subject not covered by our contributors have suggested that excretion of partially oxidized products is not restricted to the anaerobes, for example Vollbrecht (1980) and Vollbrecht and El Nawawy (1980) have found that aerobic organisms grown under restricted oxygen supply excrete a wide variety of potentially valuable metabolites and that the type and amount of product depends on oxygen availability. Furthermore many of these acids and alcohols are not normal metabolites of aerobes but are produced *via* the action of enzymes previously not thought to be present in strictly aerobic bacteria. Schlegel and Vollbrecht (1980) and Steinbuchel and Schlegel (1983) have suggested that the genetic information which codes for these fermentation enzymes reside in so called 'silent genes' (Riley and Anilionis, 1978) which are switched on when the environmental conditions, such as anoxia, favour their expression. Clearly there is now the impetus to search for more of these serendipitous activities which may even obviate the need to clone into microbes certain genes to perturb the natural metabolic balance within the cells to produce desirable products.

Methane and its oxidation product methanol (Dr. Stirling and Professor Dalton, Chapter 20) are already used as a 'biotechnological' substrate for single cell protein production but the organisms which grow on these substrates may well be used either for their specialized enzyme activities (*e.g.* bioconversions *via* methane monooxygenase) or as recipients of cloned eukaryotic genes (*e.g. Methylophilus methylotrophus* grown on methanol). Furthermore these substrates are both valuable energy sources in their own right, and although methanol is only produced in small amounts biologically, methane is certainly produced in large quantities from the anaerobic breakdown of organic material. It is only in recent years that the complex series of reactions implicating a wide variety of newly discovered cofactors involved in bacterial methanogenesis has been unravelled. Dr. Nagle and Professor Wolfe (Chapter 22) indicate the fascinating uniqueness of the structural and biochemical features of these Archaebacteria which could hold much promise, particularly in association with saccharolytic fermenters, in effecting extensive degradation of organic materials.

Hydrogen, both as an energy source for growth or as an end-product of metabolism, is one of our cheapest resources and the biochemical features of the hydrogenases involved in utilization and formation are reviewed by Professor Schlegel and Dr. Schneider (Chapter 23). The variety of strategies whereby carbon dioxide is utilized by autotrophs and heterotrophs is given by Dr. Dijkhuizen and Professor Harder (Chapter 21) and again serves to illustrate the diversity of mechanisms involved in the metabolism of a single compound. The importance of carboxylation reactions in the overall metabolism of microbes is now an established fact and the inclusion of this substrate in media as a growth initiator must not be overlooked by 'fermentologists' if early rapid growth in culture systems is required.

Alkanes are excellent growth substrates for many microbes and the yeasts and moulds in particular accumulate very high quantities of lipid and lipid-related compounds from the longer chain substrates (Dr. Boulton and Professor Ratledge, Chapter 24). Although the commercial promise for these lipids for technical purposes has not been fulfilled it is quite probable that a number of the more specialized uses could justify further development.

Many aromatic compounds, on the other hand, are considered as serious environmental hazards which are produced in ever increasing amounts by industry. Certainly it is possible to isolate organisms, either in pure culture or as consortia, capable of growth on such compounds which result in their complete removal from the environment. The wide variety of biological reactions involved in the transformation and subsequent mineralization of aromatic compounds are described by Professor Dagley (Chapter 25).

If microbes are to be successfully exploited in the production of new materials it is axiomatic that our understanding of the biochemical principles, to which all processes are subject, be clearly understood. Therefore the remaining chapters of this section have been chosen to encompass many of these principles. Enzyme kinetics (Dr. Cornish-Bowden, Chapter 27) and the mechanisms of enzyme catalysis (Dr. Suckling, Chapter 28) have always been the cornerstone of biochemistry and are now, or should be, an essential part of the repertoire of all biotechnologists who are using enzymes *per se* in a catalytic process, be they immobilized or freely dispersed. Bacterial respiration (Dr. Jones, Chapter 26) and microbial photosynthesis (Dr. Dow and Mr. Kelly, Chapter 30) are both concerned with the energetics of microbial systems. They are important to several types of biotechnological processes, namely biomass production, degradation of waste materials and metabolic production. If one wishes to maximize the efficiency of microbes either in conversion of substrate to biomass or substrate to products it is quite clear that different energetic considerations will apply; such aspects are considered in these two chapters. Although the chapters on production of metabolites (Dr. Neijssel and Professor Tempest, Chapter 32, and Professor Demain and Dr. Britz, Chapter 33) may have a titular similarity, the authors have adopted completely different approaches. Professor Demain considers how metabolism is internally regulated through the complex array of control circuits which ensure that the cell operates at maximum efficiency and then proceeds to indicate how one can perturb metabolism, largely by genetic modifications, to produce primary metabolites. Dr. Neijssel and Professor Tempest, on the other hand, indicate how by simple manipulation of the environmental conditions, such as pH, temperature, nutrient levels, *etc.*, it is possible to produce, often in large amounts, an impressive array of useful metabolites. The underlying physiological and biochemical and genetic features which control growth-associated metabolite production are obviously closely interwoven and we are now beginning to understand how these interact in a few examples presented in these chapters. One hopes that these approaches may pave the way for an understanding of how secondary (or non-growth associated) metabolites are induced in the biotechnologically important actinomycetes, a subject area that is not covered in this volume but is covered from a practical viewpoint in Volume 3, Section 1.

At the present time about 12 extracellular enzymes are produced in bulk quantity which have a variety of applications in the food and detergent industries (see Volume 3, Section 3). Our present understanding of the regulation of extracellular enzyme synthesis is fragmentary largely because exoenzyme synthesis is often a transient phenomenon being restricted in most cases to the idiophase and also because the molecular mechanisms involved are poorly understood. Dr. Priest (Chapter 31) considers the various proposed mechanisms of enzyme secretions and then outlines our current understanding of how these enzymes are synthesized.

Finally Dr. Hall (Chapter 29) considers, largely from a practical point of view, how enzymes may be produced, *in vivo*, which have completely novel functions. Such an approach stems from earlier observations that bacteria could be forced by selection to acquire the ability to metabolize novel substrates through the increased affinity and synthesis of pre-existing enzymes which normally had a very weak activity towards the new substrate. These sorts of studies in conjunction with rDNA technology could be very important if we wish to tap some of these unforeseen reservoirs within the microbial genome and produce organisms to meet specific industrial needs.

17.2 REFERENCES

Dixon, B. (1983). *Biotechnology*, **1**, 45.
Schlegel, H. G. and D. Vollbrecht (1980). *J. Gen. Microbiol.*, **117**, 475–481.
Steinbuckel, A. and H. G. Schlegel (1983). *Eur. J. Biochem.*, **130**, 329–334.
Riley, M. and A. Anilionis (1978). *Annu. Rev. Microbiol.*, **32**, 519–560.
Vollbrecht, D. (1980). *Biotechnol. Lett.*, **2**, 49–54.
Vollbrecht, D. and M. A. El Nawawy (1980). *Eur. J. Appl. Microbiol. Biotechnol.*, **9**, 1–8.

18

Aerobic Metabolism of Glucose

M. A. PAYTON and B. A. HADDOCK
Biogen SA, Geneva, Switzerland

18.1 INTRODUCTION

The aerobic metabolism of glucose by microorganisms has been the subject of a large number of extensive reviews and many of these can be found listed in the bibliography to this chapter. It is beyond the scope of this review to describe in detail how a large number of microbes metabolize glucose and completely oxidize this hexose, in theory at least, into CO_2 and H_2O. We will instead restrict this chapter to a discussion of a number of general points concerning aerobic glucose metabolism and illustrate these, where appropriate, with specific examples from the prokaryotes and lower eukaryotes.

It is important to distinguish between the terms aerobic, anaerobic, oxidative and fermentative glucose metabolism. These will be defined more fully with reference to specific examples in a later section. It will suffice for now to state that not all oxidative metabolism takes place under aerobic conditions and an aerobic environment, *i.e.* the presence of oxygen, does not always necessarily ensure oxidative or respiratory metabolism. We shall, however, in this chapter be primarily concerned with the aerobic oxidative mode of glucose catabolism; the fermentative modes of glucose catabolism are dealt with in detail in the following chapter.

The most obvious fundamental difference between oxidative and fermentative metabolism of glucose is that in addition to the ATP generated by substrate level phosphorylation, which is common to both, oxidative or respiratory metabolism offers the alternative generation of ATP by oxidative phosphorylation coupled *via* the electron transport chain ultimately to oxygen. This additional energy as ATP allows greater biosynthetic rates resulting usually in higher yields of microbial biomass per unit weight of glucose consumed.

In essence the general scheme of the aerobic breakdown of glucose can most conveniently here be considered in three phases. First, glucose enters the cell and appears inside as either free glucose or in a phosphorylated form. Second, the organism catabolizes the glucose or glucose phosphate so produced to the level of triose as pyruvate. The route taken between hexose or hexose phosphate and pyruvate differs from organism to organism but essentially follows one of the pathways illustrated in Figure 1. It should be stressed that almost all organisms use more than one of these pathways to produce pyruvate, usually two and in some cases, *e.g. Thiobacillus* A2, at least three. The relative flux through pathways which operate in parallel in any single microorganism can vary depending on such factors as the redox state of the cell, or changes in environmental conditions or in substrate levels.

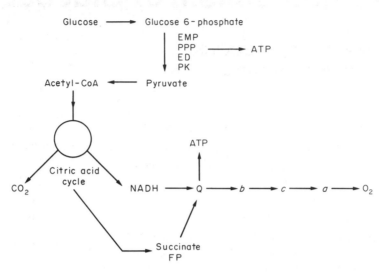

Figure 1   The three phases of aerobic glucose metabolism

The third phase of glucose breakdown shown in Figure 1 is the entry of the triose, pyruvate, into the citric acid cycle for total oxidation to CO_2 and H_2O. This point represents a key difference between oxidative and fermentative glucose metabolism. In fermentative metabolism the pyruvate is either itself a sink for reducing equivalents or is converted to a molecule which can act as one, the corresponding reduced products being the familiar 'typical' or 'characteristic' fermentation products such as ethanol and lactic acid. In oxidative metabolism, the role of electron sink is performed by the redox carriers comprising the electron transport chain ultimately coupled to oxygen, with the concomitant generation of the extra ATP mentioned above. Thus, intermediates, such as pyruvate produced from the breakdown of glucose, do not have to serve as organic electron acceptors. This makes pyruvate relatively more available for oxidation through the citric acid cycle. It is not surprising, therefore, that enzymes which compete for intracellular pyruvate are key regulatory points in organisms capable of both oxidative and fermentative glucose catabolism.

18.2 TRANSPORT OF GLUCOSE INTO MICROORGANISMS

The discoveries which signalled the existence of specific transport systems in microorganisms have been extensively reviewed (Kornberg 1972, 1976; Silhavy *et al.*, 1978; Saier, 1979). One important observation was the phenomenon of crypticity, that is the finding that whole cells of certain microbes were unable to utilize a particular sugar, though extracts from the same cells contained all the enzymes necessary for the metabolism of that sugar (Cohen and Monod, 1957). Also, a particular mutant of *E. coli*, while growing normally on glucose supplied internally by the breakdown of maltose, failed to grow on externally supplied glucose (Douderoff *et al.*, 1949). These and other discoveries, in particular with *E. coli* (Kaback, 1968), led to the biochemical and genetic analysis of other transport systems in different microorganisms. Such studies have identified two major classes of transport systems. The first, termed active transport or the permease system (Simoni and Postma, 1975), does not result in chemical alteration of the sugar on transport and can concentrate sugars against a gradient. This process requires energy, which according to Mitchell (1966) and others (Rosen, 1978; Wilson, 1978) is provided by membrane-bound ATPase activity coupled to the electrochemical (proton) gradient (Wilson *et al.*, 1976). The second major class of transport system is the 'group translocation' system and involves the appearance inside the cell of a chemically altered (phosphorylated) sugar (Rogers and Yu, 1962; Kundig, Ghosh and Roseman, 1961). The best studied system of this type is the phosphotransferase or PTS (Kundig *et al.*, 1966) system in organisms such as *E. coli*, which instead of membrane bound ATPases uses the energy derived from the hydrolysis of a phosphate donor, in this case phosphoenolpyruvate (PEP) (Romano *et al.*, 1970) to effect transport. The glucose appears within the cell in a phosphorylated form. In bacteria growing under anaerobic fermentative conditions the PTS system offers a number of advantages over the permease system. Not only does it allow for the conservation of ATP, but it also allows for a tight coupling of transport with glycolytic flux and the regulation of a number of other processes (for a review see Dills *et al.*, 1980). The precise role and occurrence of the PTS and PTS-like systems has been previously reviewed (Romano *et al.*, 1970; Dills *et al.*, 1980 and references therein) and is also described in some detail in the following chapter. However, such systems are primarily found in obligate anaerobes and in facultative anaerobes growing fermentatively on glucose, in particular in those organisms using the Embden–Meyerhof pathway (EMP) of glucose catabolism (see later), and play a lesser role in glucose uptake under aerobic oxidative conditions. In fact, according to Dills *et al.* (1980) a general rule amongst prokaryotes seems to be that obligate aerobes which use the EMP mode of glucolysis (see later) possess a PTS system, whereas obligate aerobes which use an alternative pathway, such as certain bacilli, lack the PTS. Similarly, obligate and facultative anaerobes which metabolize glucose *via* the EMP use a PTS system, whilst glucolysis *via*, for example, the Entner–Douderoff or phosphoketolase pathways is not found accompanying a PTS.

Amongst aerobically-used systems, active transport mechanisms involving permeases of differing specificities predominate. Even disregarding the presence of PTS systems, many microorganisms are still faced with a choice of more than one permease system for glucose. For example, amongst the fungi, *Neurospora crassa* (Klingmüller and Huh, 1972; Scarborough, 1973) and *Corprinus lagopus* (Moore and Devadatham, 1975) seemingly possess both high and low affinity hexose transport systems to cope optimally with a wide range of external glucose concentrations.

The relative utilization of competing glucose uptake pathways, however, is, as with most metabolic processes in microbes, not absolutely defined and a greater level of complexity can be imposed by the simultaneous option of a PTS system. A number of control mechanisms which affect this have been documented (Dills *et al.*, 1980) and it seems clear that at any one time a microbe may actually transport glucose by more than one system.

There has been some confusion in the literature concerning the eukaryotic facultative anaerobic yeast *Saccharomyces cerevisiae*. Early reports (Van Steveninck, 1968, 1969) suggested yeast had a PTS-type system whereby glucose was phosphorylated on transport. This fitted well with thoughts that facultative anaerobes which metabolized glucose primarily *via* the EMP (as yeast does, see later) possessed a PTS. However, more recent observations suggest that yeast does *not* possess a PTS and phosphorylation-on-transport is not achieved (Romano, 1982), though the interrelationship of alternative routes for glucose transport in this organism are still not fully understood (Cirillo, 1981).

18.3 FURTHER METABOLISM OF INTRACELLULAR GLUCOSE OR GLUCOSE PHOSPHATE

Following the intracellular accumulation of glucose or glucose 6-phosphate, most microorganisms are faced with a choice of four major pathways of subsequent metabolism to produce triose, usually as pyruvate. These are: (i) the Embden–Meyerhof pathway (EMP); (ii) the hexose monophosphate or pentose phosphate pathway (PPP); (iii) the Entner–Douderoff pathway (ED); and (iv) the phosphoketolase pathway (PK). The main features of these four major pathways are outlined below.

The *EMP* (Figure 2a) comprises the major pathway of glucose degradation in the Enterobacteriaceae (Doelle, 1975) and has the net overall reaction:

$$\text{glucose} + 2\,\text{ATP} + 2\,\text{NAD} \rightarrow \text{pyruvate} + 4\,\text{ATP} + 2\,\text{NADH}$$

Thus there is an overall net gain of 2 ATP and 2 NADH. This pathway is frequently found in fermentative organisms and forms the basis of the 'mixed acid' type fermentation of *E. coli*. (see Morris, Volume 1, Chapter 19).

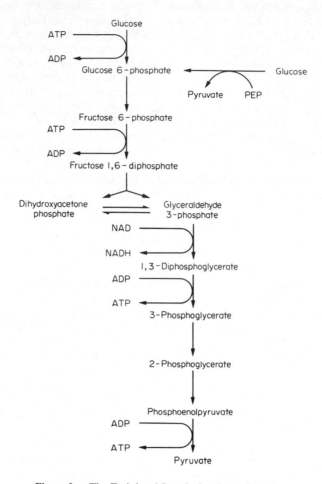

Figure 2a   The Embden–Meyerhof pathway (EMP)

The *PPP* (Figure 2b) is frequently found together with either the EMP or the ED (see below) and involves, like the EMP, initial conversion of glucose to glucose 6-phosphate. This stage is the point of deviation from the EMP. The glucose 6-phosphate is metabolized to glucono-Δ-lactone 6-phosphate rather than to fructose 6-phosphate as in the EMP. Further metabolism involves a series of interconversions.

The *ED* pathway (Figure 2c) was first discovered in *Pseudomonas saccharophila* (Entner and Douderoff, 1952) and is now known to form probably the major pathway of glucose metabolism in most *Pseudomonads* (Ornston, 1971) and is the only glucolytic pathway used by *Zymomonas* species (for review see Swings and De Ley, 1977). The first reactions of this pathway are similar if not identical to those of the PPP but the branch-point comes at 6-phosphogluconate, where instead of dehydrogenation, the intermediate is dehydrated to form 2-keto-3-deoxy-6-phospho-gluconate (KDPG).

The *PK* pathway (Figure 2d) is really a variation of the PPP and is found in organisms such as *Bifidobacteria, Lactobacilli* (Heath *et al.*, 1958) and *Thiobacillus* (Greenley and Smith, 1979). Phosphoketolases appear able to act on hexose phosphates or pentose phosphates (Doelle, 1975) according to Figure 2, with the production of acetyl phosphate and erythrose 4-phosphate (if a hexose phosphoketolase reaction cleaving fructose 6-phosphate) or acetyl phosphate plus glycer-aldehyde 3-phosphate (if a pentose phosphoketolase reaction cleaving xylulose 5-phosphate).

Before passing on to a discussion of these four pathways we should mention that organisms which do not phosphorylate glucose on transport (*i.e.* most prokaryotic organisms which use a glucolytic pathway other than the EMP and most eukaryotes) must first phosphorylate glucose to glucose 6-phosphate before further metabolism can occur. This reaction is performed by a specific glucokinase or a less specific hexokinase. A review of microbial hexokinases can be found in Doelle (1975). The hexokinase reaction has been extensively studied in the yeast *Saccharomyces cerevisiae* and this organism possesses three enzymic activities which convert glucose to glucose 6-

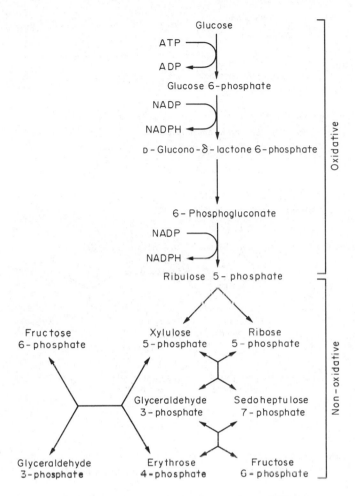

Figure 2b　The pentose phosphate pathway (PPP)

phosphate: hexokinase P1, hexokinase P2 and glucokinase, the structural genes for which have all been identified (Maitra and Lobo 1981).

18.3.1　Occurrence and Distribution of the Various Glucolytic Pathways

Although the majority of carbohydrate degradation pathways have intracellular glucose first converted to glucose 6-phosphate, a notable exception to this occurs where the unphosphorylated glucose is oxidized directly. In such organisms, usually a glucose dehydrogenase, and sometimes also a glucose oxidase, convert glucose directly to gluconate (Kitos *et al.*, 1958) which can then enter the pentose phosphate pathway (PPP) *via* gluconokinase which produces 6-phosphogluconate. One organism which does this rather than always metabolize intracellular glucose first to glucose 6-phosphate is *Pseudomonas aeruginosa*. At low oxygen tension in chemostats, this organism metabolizes glucose primarily *via* a phosphorylative pathway involving transport of glucose *via* an inducible glucose transport system which is permease-mediated and not of the PTS-type (see earlier). Subsequent metabolism of this is *via* glucose 6-phosphate into the Entner–Douderoff pathway. Under high oxygen tensions and at higher growth rates, however, glucose is first metabolized extracellularly by direct oxidation (Whiting *et al.*, 1976). This latter route is used primarily oxidatively, while the phosphorylative route is essential for anaerobic growth (Mitchell and Dawes, 1982). The detailed switch from one pathway to another has been recently described in some detail (Hunt and Phibbs, 1983) and the 'glucolytic options' now believed to exist for *Ps. aeruginosa* are shown in Figure 3.

An almost analogous system to the choice of glucolytic pathway possessed by *Ps. aeruginosa* is

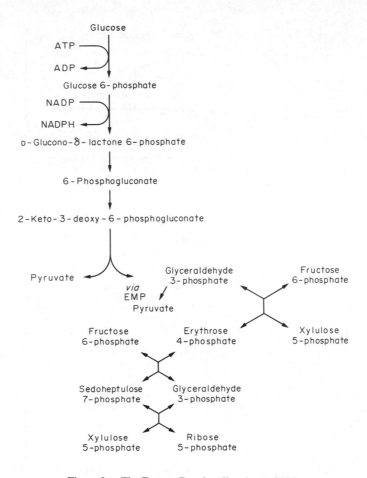

Figure 2c   The Entner–Douderoff pathway (ED)

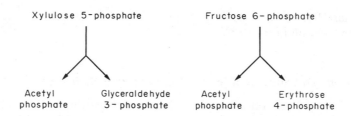

Figure 2d   The pentose phosphoketolase (PK) and hexose phosphoketolase reactions

found in *Gluconobacter oxydans* (Olijve and Kok, 1979a, 1979b), where a direct oxidative route *via* membrane-bound dehydrogenases to form gluconate is found along with a parallel PPP (not ED pathway as in *Ps. aeruginosa*). The organism responds to low growth rates by channelling up to 50% of glucose *via* the PPP, whereas at higher growth rates, the oxidative route is favoured with a concomitant accumulation of gluconate and ketogluconate in the medium (Olijve and Kok, 1979b).

Apart from exceptions such as those mentioned above, the major modes of glucolysis involve metabolism *via* glucose 6-phosphate. Recent results show that most microorganisms metabolize glucose by more than one pathway. As illustrated and discussed by Wood and Kelly in some detail (1980) even the old concept that most microorganisms possess one major pathway and one minor pathway of glucose metabolism is certainly an over-simplification. It appears, on the contrary, that most microbes can probably not only significantly vary the flux through two parallel pathways, but can probably additionally induce a third or even fourth pathway. For example, it is

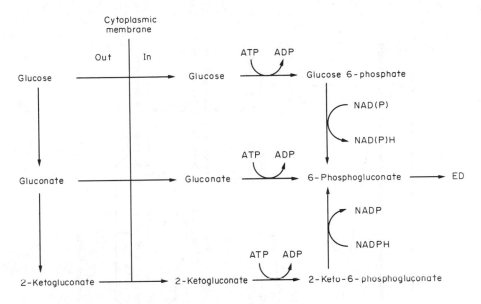

Figure 3 The initial stages of glucose catabolism in *Pseudomonas aeruginosa*

frequently documented that *E. coli* possesses two pathways for glucose dissimilation, the EMP and the PPP (see Doelle, 1975). However, *E. coli* actually possesses an inducible Entner–Douderoff pathway which allows gluconate-grown cells to metabolize glucose and gluconate independently *via* the EMP and the ED (Eisenberg and Dobrogosz, 1967).

Against this background of the current knowledge of just how variable the glucolytic pathways which a single organism can use are, it would be foolish now to state categorically the maximum number of pathways a particular microorganism appears to possess. However, certain specific examples of control of relative flux have been well documented and some of these are described below and a brief survey of the pathways believed simultaneously to operate in a range of microorganisms is shown in Table 1. In addition, the four major pathways have been condensed into a single scheme (Figure 4) to illustrate potential cross-over points and overlaps between the various glucolytic routes.

Table 1 The Distribution of the Four Major Routes of Glucolysis amongst a Variety of Microorganisms

Organism	Pathway Characteristic enzyme	EMP Phospho-fructokinase	PPP 6-Phospho-gluconate dehydrogenase	ED KDPG aldolase	PK Phosphoketolase
Zymomonas	—	—	+	—
Ps. aeruginosa	(+)	+	+	—
Ps. douderoffi	(+)	—	+	—
Gluconobacter oxydans	(+)	+	—	—
E. coli	+	+	+	—
S. typhimurium	+	+	—	—
B. subtilis	+	+	—	—
Rhodopseudomonas capsulata	+	—	+	—
Thiobacillus novellus	(+)	(+)	—	+
Thiobacillus A2	+	+	+	—
Bifidobacterium bifidum	—	—	—	+
S. cerevisiae	+	+	—	—
A. nidulans	+	+	—	—
Rhodotorula gracilis	(+)	+	—	—
Candida utilis	+	+	—	—
Penicillium chrysogenum	+	+	—	—

If we first consider the distribution of the EMP sequence of reactions then we see it is widespread and many common fermentation pathways such as the ethanologenic fermentation of *Saccharomyces cerevisiae* have this pathway as their basis. However, though energetically a

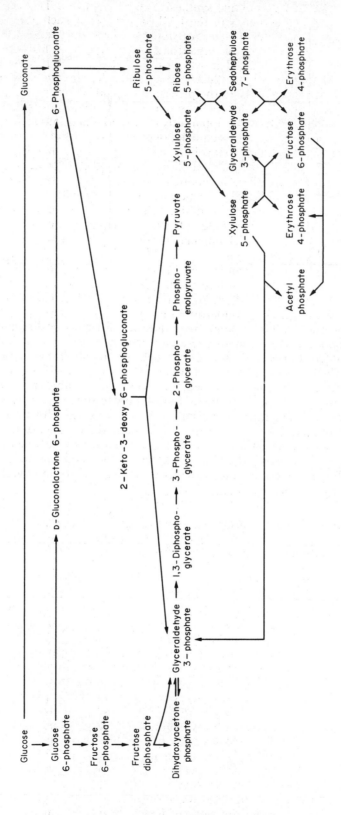

Figure 4   The interrelationship between the major pathways of glucose metabolism

favourable pathway, the EMP must be accompanied by the PPP in order that generation of pentose phosphates for purine and pyrimidine synthesis can be achieved. This is illustrated by the simultaneous use of the PPP together with the EMP in organisms such as *E. coli*, *Salmonella typhimirium* and in lower eukaryotes such as *Saccharomyces cerevisiae* and *Aspergillus nidulans* (McCullough *et al.*, 1977). The PPP in this organism can account for as little as a few percent of glucose metabolism under glucose-limited growth (Bainbridge *et al.*, 1971) or up to 40% at higher growth rates (Carter and Bull, 1969). As illustrated by McCullough *et al.* (1977) the function of the PPP in this organism is, in addition to the provision of pentose intermediates, the supply of NADPH by the oxidative part of the PPP for subsequent use in other biosynthetic reactions.

In some microorganisms the PPP can be apparently the predominant mode of glucolysis, though since pyruvate is not an intermediate in this sequence of reactions, some of the reactions of the EMP are required to convert glyceraldehyde 3-phosphate to pyruvate. Thus, microorganisms which possess the PPP as their major route of glucose catabolism must necessarily possess the EMP also.

One organism which uses the PPP as its major route of glucose catabolism is the obligately aerobic yeast *Rhodotorula gracilis*. Labelling experiments (Höfer *et al.*, 1971) suggest that 20% of added glucose is metabolized through the oxidative PPP (that is glucose to ribulose 5-phosphate *via* two dehydrogenases, glucose-6-phosphate dehydrogenase and 6-phosphogluconate dehydrogenase) while the remaining 80% is channelled through the non-oxidative part of the PPP (transaldolase and transketolase reactions).

Thus, for their respective reasons, *neither* the EMP *nor* the PPP can form the *sole* mode of glucolysis for any microorganism, except where the necessary intermediates not produced by the microbe's own metabolism are provided externally (see Doelle, 1975). This is in contrast to those organisms which possess the Entner–Douderoff pathway where it is possible to use this pathway alone for glucose catabolism and this is apparently the case in species of the genus *Zymomonas*; ethanol fermentations catalysed by *Zymomonas* species are described in more detail in Morris (Volume 1, Chapter 19). However, by no means all organisms which possess the ED pathway have this pathway operating in isolation. For example, *Ps. aeruginosa* as we have already seen, possesses the ED pathway as an alternative mode of glucolysis, as does *Rhodopseudomonas capsulata*, where a very interesting separation of two potentially competing pathways is seen. In this microorganism glucose is metabolized *via* the ED pathway (Eidels and Preiss, 1970). However, fructose grown cells, although they contain normal levels of ED enzymes, metabolize this hexose *via* the EMP (Conrad and Schlegel, 1977). The reason for this appears to be that fructose is initially phosphorylated to fructose 1-phosphate *via* a PTS system, rather than to fructose 6-phosphate, and thus avoids immediate conversion to glucose 6-phosphate (*via* phosphoglucose isomerase), the entry point into the ED pathway (see earlier). This is a variation on a commonly-found dichotomy of the EMP found in *E. coli* (Fraenkel, 1968), *B. subtilis* (Gay and Rapoport, 1970) and *Enterobacter aerogenes* (Hanson and Anderson, 1968) where glucose is metabolized to glucose 6-phosphate, thence to fructose 6-phosphate and to pyruvate *via* the EMP, while fructose, though still metabolized ultimately by the EMP, enters *via* fructose 1-phosphate which is converted to fructose diphosphate by a novel fructose-1-phosphate kinase (Ferenci and Kornberg 1971, 1973).

The ED pathway is found also amongst a large number of pseudomonads (Quayle, 1977) and was thought at one time to be characteristic (Lynch *et al.*, 1975). It was also thought that glucose was catabolized exclusively *via* this pathway in these organisms, though it is now known that for pseudomonads in general (Sawyer *et al.*, 1977) and for *Ps. douderoffi* (Baumann and Baumann, 1975) in particular, EMP reactions can occur and contribute significantly to overall glucolysis.

The fourth and final pathway which microorganisms may possess for glucose degradation is the phosphoketolase pathway. As outlined above, phosphoketolase pathways are essentially variations of the PPP, where phosphoketolases can cleave either hexose or pentose phosphate. Both types of phosphoketolase reaction are seen in *Bifidobacterium bifidum* fermentation and are described in the next chapter. An unusual pathway of glucolysis is also found in *Thiobacillus novellus* (Greenley and Smith, 1979) which possesses a phosphoketolase reaction which cleaves xylulose 5-phosphate into acetyl phosphate and glyceraldehyde 3-phosphate (Figure 4). The xylulose 5-phosphate is generated by the operation, in part, of a PPP and the glyceraldehyde 3-phosphate thus generated is further metabolized to pyruvate by a partial EMP.

One of the best studied examples of coexistence of multiple pathways and the relative fluxes through the pathways has been from the work of Kelly and his colleagues on *Thiobacillus* A2 (Wood and Kelly, 1978; Wood, Kelly and Thurston, 1977; Wood and Kelly, 1980). This organism possesses simultaneously three pathways, the EMP, the PPP and the ED pathway, which all operate during growth on glucose, though the ED pathway is usually the major glucolytic route accounting for between 68% and 90% of glucose flux (Wood and Kelly, 1980). In addition, phosphoketolase activity could be detected, though at very low levels.

The relative flux through these three major pathways is discussed in detail by Wood and Kelly (1980) and the reader is directed to that paper for a concise and illuminating discussion of the complexity of variable flux through coexisting glucolytic pathways.

18.3.2 The Fate of Pyruvate

As mentioned earlier, there are in all microorganisms a number of competing enzymes which use pyruvate as substrate. A number of these are shown in Figure 5. The partition of pyruvate *via* different pathways can, in fermentative organisms, be the crucial step in determining the final end-products of fermentation (see Morris, Volume 1, Chapter 19). In addition, in facultative anaerobes the switch from oxidative to fermentative growth can dictate whether pyruvate is oxidized through the citric acid cycle or diverted into, say, lactate or other fermentation products such as ethanol or acetate. The most common fate of pyruvate during aerobic glucose metabolism, however, is its conversion to acetyl-CoA, a reaction catalysed by the pyruvate dehydrogenase complex (PDH). This multienzyme complex consists of three different subunits in the bacteria *E. coli* (Koike, *et al.*, 1963), *Azotobacter vinelandii* (Bresteus *et al.*, 1972) and *Streptococcus faecalis* (Reed *et al.*, 1958), and the eukaryotes *Saccharomyces cerevisiae* and *Kluyveromyces lactis* (Wais *et al.*, 1973), and *Neuspora crassa* (Harding *et al.*, 1970). The components of the complex are: pyruvate dehydrogenase (E_1), dehydrolipoyl acetyltransferase (E_2) and the flavoprotein dihydrolipoyl dehydrogenase (E_3). The E_3 component is shared both by the PDH complex, and the analogous 2-oxoglutarate dehyrogenase complex, a key regulatory enzyme in the citric acid cycle.

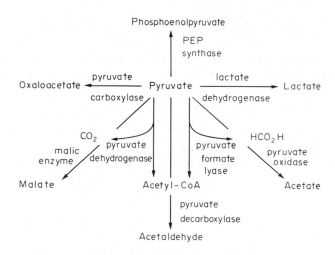

Figure 5   Alternative enzymes using pyruvate as substrate

In obligate aerobes such as the fungus *Aspergillus nidulans*, the normal fate of pyruvate is *via* the PDH complex to acetyl-CoA, and thence into the citric acid cycle (Payton *et al.*, 1977; McCullough *et al.*, 1977). The only significant competing pathway is the constant (and necessary, see later) removal of pyruvate in the anaplerotic reaction catalysed by pyruvate carboxylase. In facultative anaerobes, however, the situation becomes decidedly more complex. For example, in *E. coli*, there is a direct alternative to the PDH complex provided by pyruvate-formate lyase (PFL; Knappe *et al.*, 1968). This enzyme complex seems, under anaerobic conditions, to perform an analogous function to that of the PDH complex under aerobic conditions, but whereas the products of the PDH reaction are acetyl-CoA and reduced NAD, the products of the PFL are acetyl-CoA and formate. Formate can conveniently be thought of here as a product of using the CO_2 generated by the decarboxylation of pyruvate (C_3) to acetyl-CoA (C_2) as an acceptor for reducing equivalents, rather than using NAD. Under aerobic conditions, the NADH produced would normally be reoxidized *via* the electron transport chain, but under anaerobic conditions the generation of additional NADH (the EMP already generates 2 NADH per mole of glucose catabolized to 2 pyruvate molecules) would necessitate a major reshuffling of fermentative metabolism, since organisms growing fermentatively have to balance not only carbon, but also reducing equivalents, in order to survive (see next chapter). It is therefore not surprising that *E. coli* growing anaerobically, though it possesses some residual PDH complex activity (Henning *et al.*, 1969), inhibits the reaction by using NADH as a feed-back inhibitor of enzyme activity (Hansen and Henning, 1966; Pascal *et al.*, 1981). Also, in *E. coli* another major fate of pyruvate is its conversion to D-lactate *via* NAD-linked lactate dehydrogenase, probably more accurately termed pyruvate reductase since it operates unidirectionally (pyruvate to lactate) *in vivo* (Tarmy and Kaplan 1968). NAD-lactate dehydrogenase in *E. coli* is constitutively produced, though formed at noticeably higher levels during anaerobic growth (Doelle *et al.*, 1981), and can produce lactate during aerobic glucolysis under appropriate conditions such as growth on high glucose concentrations or in mutants of *E. coli* deficient in normal electron transport chain activity (Cox and Gibson, 1974; Clarke and Payton, unpublished data). Finally, *E. coli* has a pyruvate oxidase system which directly decarboxylates pyruvate to acetate (Williams and Hager, 1966; Mather *et al.*, 1982). The ability of mutants of *E. coli* lacking PDH activity (see Langley and Guest, 1977) to grow, albeit poorly (Chang and Cronan, 1983), aerobically on glucose medium can be explained by the operation of the pyruvate oxidase producing acetate, and subsequent reutilization of acetate as carbon source. This hypothesis is strengthened by the inability of *pox* (pyruvate oxidase deficient) *pdh* (PDH-deficient) double mutants to grow on glucose under identical conditions (Chang and Cronan, 1983). However, the precise physiological role of pyruvate oxidase is, at present, uncertain.

An interesting example of competition for pyruvate is provided by *Saccharomyces cerevisiae* where in addition to the anaplerotic pyruvate carboxylase (Haarasilta and Oura, 1975), lactate dehydrogenase, PDH and pyruvate decarboxylase (PDC) compete for intracellular pyruvate. The most interesting point of regulation appears to be whether pyruvate is metabolized to acetyl-CoA and hence oxidized by the citric acid cycle or whether to acetaldehyde *via* the PDC and subsequently reduced to ethanol. In terms of the balance of reducing equivalents, the PDC can clearly be seen to be optimal for glucose fermentation. However, both the PDH and PDC are produced constitutively in many strains of *S. cerevisiae* (Polakis and Bartley, 1965; Lancashire *et al.*, 1981). Polakis and Bartley (1965) argued that the relative K_m values for pyruvate assured that under aerobic conditions the majority of pyruvate was channelled *via* PDH, and under anaerobic conditions *via* PDC. The failure of *pdc* (pyruvate decarboxylase deficient; Lancashire *et al.*, 1981) mutants to grow anaerobically or aerobically on high glucose concentrations (and thus fermentatively) suggest the key role of this enzyme under such conditions. The reason for the failure of *pdc* mutants to grow aerobically on high glucose concentrations (Lancashire *et al.*, 1981) suggests that, though the PDH could theoretically cope with the conversion of pyruvate into acetyl-CoA, either the acetyl-CoA cannot be efficiently metabolized *via* the citric acid cycle due to glucose-repression of cycle activity (see later) or cannot be sufficiently converted into ethanol. Alternatively, the additional NADH generated by the PDH complex activity may 'overload' the reducing equivalent balance, particularly since the electron transport chain activity under such growth conditions would be reduced (Fiechter, 1981).

Apart from the C_3 units removed by anaplerotic reactions, the majority of pyruvate in microorganisms under oxidative aerobic growth conditions is metabolized *via* acetyl-CoA and then into the citric acid cycle. In obligate aerobes, the citric acid cycle is the final respiratory pathway in which oxidation of C_2 yields NADH which subsequently donates electrons to the electron transport chain, for the ultimate conservation of energy as ATP produced by oxidative phosphorylation. In addition, the cycle also provides key intermediates for biosynthesis of amino acids and porphyrins. The reactions of the cycle are shown in Figure 6. In eukaryotes, these reactions are mitochondrial, located either in the matrix or membrane-bound (Ernster and Kuylenstierna, 1969). Thus, eukaryotes must first transport pyruvate produced by the cytoplasmic glucolytic pathways into the mitochondrion before the PDH complex can produce acetyl-CoA for subsequent oxidation.

The initial entry of C_2 into the cycle is effected by a condensation reaction where oxaloacetate + acetyl-CoA is converted to citrate (Figure 6). This reaction is catalysed by citrate synthase. In prokaryotes, this enzyme, as the first enzyme of a complex series of reactions, is a point of regulation. In Gram-negative bacteria, citrate synthase is inhibited by reduced NADH, whereas in Gram-positive bacteria, such as *B. subtilis*, citrate synthase is regulated by levels of 2-oxoglutarate and ATP (Flechtner and Hanson, 1969, 1970). This is now known to correlate with the possession by Gram-positive bacteria of a much smaller enzyme (Weitzman and Dunmore, 1969) lacking sites for allosteric inhibition by NADH. In addition, AMP can relieve the NADH-

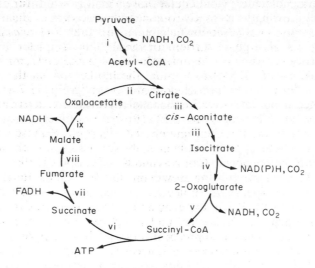

Figure 6 The citric acid cycle: i, pyruvate dehydrogenase; ii, citrate synthase; iii, aconitate hydratase; iv, isocitrate dehydrogenase; v, 2-oxoglutarate dehydrogenase; vi, succinyl-CoA synthetase; vii, succinate dehydrogenase; viii, fumarate hydratase; ix, malate dehydrogenase

mediated inhibition of citrate synthases from obligately aerobic Gram-negative bacteria (Flechtner and Hanson, 1970).

The 'fine-control' regulation of the various enzymes of the citric acid cycle has been described previously (Krebs, 1970; Sanwal, 1970; Doelle, 1975). For example, isocitrate dehydrogenase represents another key regulatory point for microorganisms which possess a *glyoxylate* cycle (Kornberg, 1966), a pathway used as an anaplerotic reaction during growth primarily on C_2 compounds. Most bacteria possess an NADP-linked enzyme, while that in eukaryotes is generally NAD-linked (Sanwal and Stachow, 1965; Cennamo *et al.*, 1970). However, this is not strictly true since in a number of fungi (Sanwal and Stachow, 1965; Kornberg and Prior, 1951; Ramakrishran and Martin, 1955) both nucleotide-dependent activities can be found, and in the obligate aerobe *Aspergillus nidulans*, only the NADP-lined enzyme and not the NAD-linked one can be detected (Hankinson and Cove, 1974; McCullough *et al.*, 1977).

The multienzyme complex 2-oxoglutarate dehydrogenase occupies as important a regulatory position in the citric acid cycle, especially in facultatively anaerobic microorganisms, as its glucolytic analogue PDH. The levels of overall activity of the complex are subject to change depending on growth conditions. Indeed, for *E. coli* not only the 2-oxoglutarate dehydrogenase complex (Amarasingham and Davis, 1965) but also isocitrate dehydrogenase, citrate synthase and aconitase (Hespell, 1976; Wimpenny and Cole, 1967) are induced by growth under aerobic conditions and repressed by anaerobic growth. One of the best studied systems of regulation affecting 2-oxoglutarate dehydrogenase has been in *E. coli*, where under aerobic conditions the high activity of the complex allows the citric acid cycle to function *in toto*, effecting potential complete oxidation of C_2 units to CO_2 and H_2O. However, under anaerobic conditions, the activity is insufficient to allow the cyclic operation of the series of oxidative reactions and under such conditions the 'cycle' operates as a branched pathway (see later and Morris, Volume 1, Chapter 19). A further interesting deviation from the normal 'aerobic' reactions of the citric acid cycle is also found in *E. coli*, where under anaerobic conditions the succinate dehydrogenase reaction is replaced by a fumarate reductase (Kroger, 1978). These effects of oxygen tensions on enzymes of the citric acid cycle, however, are not unique to facultative microorganisms. For example, oxygen availability is known to affect levels of the citric acid cycle enzymes citrate synthase, isocitrate dehydrogenase and malate dehydrogenase in *Azotobacter beijerinckii* (Jackson and Dawes, 1976) and *Pseudomonas aeroginosa* (Mitchell and Dawes, 1982). The more complex regulation of citric acid cycle enzymes found in facultative microbes is outside the scope of this chapter and can be found elsewhere (Stouthamer, 1978).

18.4 ANAPLEROTIC REACTIONS

In any consideration of glucose metabolism and the participation of the citric acid cycle, the role of the cycle in the generation of intermediates for biosynthesis cannot be over-emphasized.

The net 'drain' of cycle intermediates into cellular biosynthesis would naturally result in the eventual depletion of metabolites, particularly C_4 compounds. Such intermediates are clearly necessary for the further condensation of additional acetyl-CoA into the cycle *via* citrate synthase using oxaloacetate, seemingly as a 'catalytic carrier' for the C_2 unit. In order to maintain sufficient levels of C_4 intermediates to ensure uninterrupted flux through the cycle, microorganisms possess anaplerotic pathways (Kornberg, 1966) to 'top-up' the levels of C_4 in the cycle. The common anaplerotic mechanisms during growth on glucose are shown in Figure 7 and involve the direct carboxylation of C_3 to C_4. Two of these reactions involve carboxylate pyruvate while the remaining three involve carboxylate phosphoenolpyruvate (PEP). All of these activities are never found together (Kornberg, 1967) and it must be remembered that some of these enzymes, particularly malic enzyme, may have a more important role in anabolic processes. In the majority of bacteria, and particularly the Enterobacteriaceae (Canovas and Kornberg, 1965, 1966), PEP carboxylase is the major anaplerotic enzyme active during growth on glucose. In most fungi, however, pyruvate carboxylase is the major enzyme. For example, in *S. cerevisiae* (Haarasilte and Oura, 1975) and *A. nidulans* (Skinner and Armitt, 1972), the enzyme is formed constitutively, though apparently is required only during glycolysis. However, under such conditions, the activity is absolutely essential for growth on glucose, since mutants of *A. nidulans* lacking pyruvate carboxylase (Skinner and Armitt, 1972) fail to grow on glucose as sole carbon source.

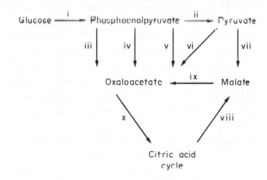

Figure 7 Common anaplerotic pathways during growth on glucose: i, glucolysis; ii, pyruvate kinase; iii, PEP carboxylase; iv, PEP carboxykinase; v, PEP carboxytransferase; vi, pyruvate carboxylase; vii, malic enzyme; viii, fumarate hydratase; ix, malate dehydrogenase; x, citrate synthase

18.5 THE ELECTRON TRANSPORT CHAIN

Eukaryotes differ from prokaryotes in that in the former the electron transport chain is mitochondrial and thus conveniently couples with the mitochondrial citric acid cycle. The components of the obligately aerobic electron transport chain can be represented as shown in Figure 8, and usually terminate with a cytochrome *c* oxidase. However, even in the presence of inhibitors of this oxidase, such as cyanide, many organisms such as trypanosomes and *Aspergillus nidulans* (Gunatilleke, Arst and Scazzochio, 1976) possess cyanide-insensitive alternative electron transport pathways. For a more detailed discussion of eukaryotic electron transport chains, the reader is directed to Borst and Grivell (1978). A discussion of the final coupling of cytochrome oxidase to oxygen can be found in Brunori and Wilson (1982) and references therein.

Though there are fundamental similarities between the electron transport systems of prokaryotes and eukaryotes, there clearly will be differences (Garland and Haddock, 1977). In prokaryotes, the situation appears more complex, though the membrane bound electron transport components of bacteria effect the same net result of proton translocation across a semi-permeable membrane. Reviews of the variety of components found in a variety of bacterial electron transport systems can be found in Haddock and Jones (1977) and Jones (Volume 1, Chapter 26) and will not be repeated here. In addition to oxygen under normal aerobic conditions, a variety of other acceptors can be used such as fumarate (Krüger, 1978) or NO_3^-, NO_2^-, SO_4^{2-} and CO_2 (Thauer *et al.*, 1977). Some bacteria which are capable of synthesizing a high potential cytochrome *c* such as *Paracoccus denitrificans* and *Rhodopseudomonas capsulata* are capable of producing equivalent yields of ATP by oxidative phosphorylation to those found for mitochondria (*i.e.* 3 ATP per NADH oxidized). Other bacteria, however, cannot synthesize such a cytochrome and thus generate only 2 ATP per NADH oxidized (*e.g. E. coli*; Jones, 1977). The assessment

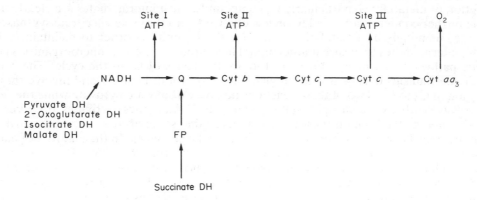

Figure 8   General scheme for the eukaryotic mitochondrial electron transport chain (DH = dehydrogenase)

of the efficiency of energy conservation in bacteria by oxidative phosphorylation has been discussed previously (Haddock 1980) as has the coupling of the chain to prokaryotic membrane ATPases (Downie *et al.*, 1979). In addition to a vast array of branched chains and alternative cytochromes (Haddock and Jones, 1977) the bacteria possess terminal oxidases different from *a*-type oxidases found in mitochondria. For example, a cytochrome oxidase type-*o* has been identified in *Acetobacter suboxydans* and also in *Vitreoscilla*, an aerobic myxobacterium. Also, a *d*-type cytochrome has been identified in *Azotobacter vinelandii* (Kauffmann and van Gelder, 1973), *E. coli* (Edwards *et al.*, 1981) and in *Pseudomonas aeroginosa* grown in the absence of oxygen, when it is using nitrate or nitrite as terminal electron acceptor (Greenwood *et al.*, 1978). A discussion of the relative structures, occurrence and physiological role of the various cytochrome oxidases can be found in Poole (1982).

It will become obvious that once again because of the availability of the appropriate mutants (Haddock, 1977; Cox and Gibson, 1974) *E. coli* provides a model system for examining the control and regulation of a variety of cytochrome components and a brief summary of the regulation of electron transport in this facultative anaerobe is included here.

E. coli synthesizes a variety of redox carriers, which constitute the electron transport chain, in response to varying growth conditions (Ashcroft and Haddock, 1975; Jones, 1977), and these number at least nine. During aerobic growth with O_2 as terminal electron acceptor, each mole of NADH oxidized produces 2 ATP (see above), one from the section involving the NADH dehydrogenase reaction and the other from the section containing cytochromes. As illustrated recently by Doelle *et al.* (1981), the proton motive force of the electron transport chain under aerobic conditions appears to reverse the role of ATPase to result in ATP synthesis. Anaerobically the reverse of this would occur, the substrate ATP in this case would be provided by substrate-level phosphorylation.

Aerobically, the cytochrome chain contains three *b*-type cytochromes which appear distinct from the *b*-cytochromes found in anaerobic growth (Reid and Ingledew, 1979) and the pathway terminates with cytochrome *o* oxidase (Castor and Chance, 1959), which is the only oxidase formed under aerobic conditions during exponential growth (Reid and Ingledew, 1979). However, when cells reach the stationary phase, cytochromes *a*, a_1 and *d* can be found (Shipp, 1972) though the role of a_1 remains a mystery. More recently, Hackett and Bragg (1983) have shown that when *E. coli* is grown anaerobically in the presence of nitrate as terminal acceptor, in addition to the *b*-556, *b*-555, *b*-562 and cytochrome *o* characteristic of aerobic growth, additional *b*-type cytochromes b^{fdh} (formate dehydrogenase) and b^{nr} (nitrate reductase) are formed. However, for a thorough description of anaerobic cytochromes consult Hackett and Bragg (1983a, 1983b) and references therein. The constitution and order of the terminal cytochrome chain in *E. coli* growing aerobically now appears to be:

$$\text{ubiquinone} \rightarrow b\text{-556} \rightarrow b\text{-555} \rightarrow b\text{-562} \rightarrow o \rightarrow O_2$$

18.6   THE REGULATION OF AEROBIC GLUCOSE METABOLISM

It would be impossible to generalize on the mechanisms of regulation of glucose metabolism in microorganisms, since each organism has its own peculiarities. In obligate aerobes the key points of regulation in glucolysis tend to be feed-back inhibition mechanisms to regulate the flow of carbon through the glucolytic pathway, and the theoretical end product of aerobic metabolism, ATP, plays a key role in this regulation (Doelle, 1975). In addition, as we have seen previously, even organisms which normally grow aerobically such as *Ps. aeruginosa* and *A. nidulans* respond to changes in oxygen concentration by varying the relative flux through alternative pathways of glucolysis and in the case of *Ps. aeruginosa* by producing altered levels of enzymes of the citric acid cycle. Another variable parameter which can affect glucose metabolism, at least in *A. nidulans*, is the provision of nitrate instead of ammonia as nitrogen source where nitrate stimulates glucolysis *via* the PPP, probably to produce additional NADPH for nitrate reduction (Hankinson and Cove, 1974; McCullough *et al.*, 1977).

Undoubtedly the most complex systems of regulation of glucose catabolism occur in facultative anaerobes, where superimposed on the normal 'housekeeping' regulation of flux through specific reactions of the aerobic glucolytic pathway are many varied responses involving the switch from oxidative to fermentative growth. In the introduction to this chapter we alluded to the distinctions between aerobic *versus* anaerobic and oxidative *versus* fermentative growth. Recently, for facultatively anaerobic bacteria such as *E. coli* (Doelle *et al.*, 1981) and for *Saccharomyces cerevisiae* (Fiechter *et al.*, 1981) it has been suggested that the terms oxidative and fermentative metabolism can be misleading since when originally coined, these appeared analogous to aerobic and anaerobic metabolism respectively. However, recent work has shown that this comparison is no longer tenable and to illustrate this we will go on to consider the regulation of glucose metabolism in yeast (*Saccharomyces cerevisiae*) and in *E. coli*.

18.6.1 Regulation of Glucose Catabolism in *Saccharomyces cerevisiae*

The facultatively anaerobic eukaryote *Saccharomyces cerevisiae* has been much studied as a model system for the regulation of glucose catabolism, largely because of its obvious biotechnological importance. In addition to studies on the regulation of flux through glucolysis and the control points therein, a number of investigators have concluded that the major control of yeast glucose metabolism is mediated through two phenomena: the Crabtree effect (Crabtree, 1929) and the Pasteur effect (De Deken, 1966).

The Crabtree effect is defined as the repression of normal respiration by sugars (usually glucose), even under fully aerobic conditions (De Deken, 1966), while the Pasteur effect, on the other hand, is the effect of oxygen on fermentation. The Crabtree effect is expressed in a large number of enzymic steps in *Saccharomyces cerevisiae* and almost certainly involves a very complex series of reactions involving ultimately, under high glucose concentrations, a gradual depletion of activity of the citric acid cycle enzymes, citrate synthase, aconitase, fumarase and malate dehydrogenase, and the respiratory chain. The net result of this 'shut-down' of normal glucose respiration is a switch to ethanol production by 'fermentation'. It should be noted here, however, that not all yeasts behave in exactly the same way, and this has led to a classification of yeasts into Crabtree-positive and Crabtree-negative yeasts (see Wöhrer *et al.*, 1981). The precise mechanism underlying the Crabtree effect is still not understood (Barford *et al.*, 1981) but may involve either catabolite-repression phenomena mediated by a system involving cyclic AMP (cAMP) and a cAMP receptor protein (see Peterkofsky, 1977) or a catabolite inactivation model involving proteolysis as a method of controlling protein turnover. A detailed account of the observations indicating the latter model are to be found in Fiechter *et al.* (1981) and references therein. Essentially, the regulatory mechanisms operate by feed-back control of key enzymes such as hexokinase (regulated by levels of glucose 6-phosphate), phosphofructokinase (regulated anaerobically by ATP and aerobically by ATP and citrate) and isocitrate dehydrogenase. When ATP is plentiful and the AMP : ATP ratio is low, the AMP-activated isocitrate dehydrogenase slows down.

18.6.2 Regulation of Glucose Metabolism in *E. coli*

We have seen how yeast is capable of growing on glucose in essentially four modes, aerobically or anaerobically and additionally either glucose repressed or not (Fiechter *et al.*, 1981). The same is true for *E. coli*; the four modes are shown in Table 2 and the conditions which result in each. For the purpose of the current review, a detailed description of the effect of oxygen on glucose metabolism is unnecessary. However, a comprehensive account of this phenomenon can be found (Doelle, 1975) and appears to involve no, or very little, regulation of phosphofructokinase (*E. coli* has two, an allosteric and a non-allosteric enzyme; Doelle *et al.*, 1981) in the bacterial Pasteur effect but rather an as yet poorly understood interaction with glucose transport *via* the PTS. Recently, a study has revealed the differences between aerobic and anaerobic growth in terms of the appearance and disappearance of particular proteins (Smith and Neidhart 1983a, 1983b). In addition a very interesting role of a potential regulatory protein (produced by the *Fnr* gene) which controls the synthesis of certain proteins required for anaerobic respiration has been identified by Guest and his colleagues (Shaw and Guest, 1981, 1982) and may be an analogue of the catabolite gene activator protein (CAP) as judged by nucleotide sequence homology (Shaw *et al.*, 1983).

Table 2   The Four Different Patterns of Growth Displayed by *Escherichia coli* Grown under Various Conditions

	Aerobic	*Anaerobic*
Oxidative	Growth on low glucose concentrations at low growth rates with oxygen as terminal electron acceptor	Growth on low glucose concentrations in the absence of oxygen using fumarate or nitrate as terminal electron acceptor
Fermentative	Growth on high concentrations of glucose at high growth rates; oxygen present but metabolic intermediates are an alternative electron acceptor	Growth on glucose in the absence of an externally supplied electron acceptor; the use of metabolic intermediates as electron acceptors leads to the production of characteristic fermentation products

The Crabtree effect, strictly defined, probably does not exist in *E. coli* (Doelle *et al.*, 1981), however both glucose concentration and specific growth rate do affect the switch from oxidative to fermentative growth in the presence of oxygen. The model proposed by Doelle (1981) to explain this phenomenon is as follows: high glucose concentrations or high specific rate of glucose utilization do not affect oxidative phosphorylation *per se*, but repress oxidative phosphorylation site I in the electron-transport chain (Haddock and Jones, 1977) involving the initial entry of reducing equivalents generated by the citric acid cycle (except entry *via* succinate dehydrogenase, which being flavoprotein linked by-passes site I) mediated by NADH dehydrogenase. The subsequent accumulation of NADH then causes the production of 'fermentation' products.

Under this form of 'glucose repression', an aerobic glucolysis can proceed *via* both respiration and fermentation, and, at higher specific growth rates, a reduction in one of the terminal oxidases cytochrome *d* (see earlier) is also seen together with a repression of citric acid cycle activity (Doelle *et al.*, 1981).

In summary, both *Saccharomyces cerevisiae* and *E. coli* apparently display a Pasteur and Crabtree effect. In yeast the Pasteur effect intimately involves the regulation of phosphofructokinase, whereas in bacteria, or at least in *E. coli*, this is not so. Also, it becomes clear that in defining the Crabtree effect, one must distinguish whether glucose represses biomass formation or can cause fermentation to occur in the presence of oxygen. In yeast, the glucose-mediated repression of biomass formation is seen, but in bacteria glucose repression results in only a shift to fermentative metabolism with no essential net change in ATP production and should perhaps, according to Doelle *et al.* (1981), best be considered as an effect both of specific growth rate and of glucose concentration.

18.7   POSTSCRIPT

In this chapter we have seen how the aerobic metabolism of glucose can be achieved by a number of microorganisms. The choices of pathways which each organism adopts to convert glucose or glucose phosphate to the level of triose as pyruvate have been outlined. There is, however, considerable variation in the relative flux through alternative pathways.

The presence of an electron-transport chain coupled ultimately to oxygen as electron acceptor facilitates the generation of more ATP than can be produced under fermentative conditions and thus aerobically-grown organisms typically have higher cell yields than anaerobically-grown cells.

The regulation of the switch from oxidative to fermentative growth under aerobic conditions is, as we have seen, complex and by reference to yeast and to *Escherichia coli* we outlined the current major thoughts on this topic, particularly with reference to the so-called Pasteur and Crabtree effects.

ACKNOWLEDGEMENT

Thanks are due to Deborah Hughes for help with research for this review.

18.8 REFERENCES

Amarasingham, C. R. and B. D. Davis (1965). Regulation of α-ketoglutarate dehydrogenase formation in *Escherichia coli*. *J. Biol. Chem.*, **240**, 3664–3668.
Ashcroft, J. R. and B. A. Haddock (1975). Synthesis of alternative membrane and redox carriers during aerobic growth of *Escherichia coli* in the presence of potassium cyanide. *Biochem. J.*, **148**, 349–352.
Bainbridge, B. W., A. T. Bull, S. J. Pirt, B. I. Rowley and A. P. J. Trinci (1971). Biochemical and structural changes in non-growing maintained and autolysing cultures of *Aspergillus nidulans*. *Trans. Br. Mycol. Soc.*, **56**, 371–385.
Barford, J. P., P. M. Jeffery and R. J. Hall (1981). The Crabtree effect in *Saccharomyces cerevisiae*—primary control mechanism or transient. In *Advances in Biotechnology. I. Scientific and Engineering Principles*, ed. M. Moo-Young, C. W. Robinson and C. Vezina, pp. 255–260. Pergamon, Oxford.
Baumann, L. and P. Baumann (1975). Catabolism of D-fructose by *Pseudomonas douderoffii*. II. Properties of 1-phosphofructokinase and 6-phosphofructokinase. *Arch. Microbiol.*, **105**, 241–248.
Borst, P. and L. A. Grivell (1978). The mitochondrial genome of yeast. *Cell*, **15**, 705–723.
Brunori, M. and M. T. Wilson (1982). Cytochrome oxidase. *Trends Biochem. Sci.*, **7**, 295–299.
Canovas, J. L. and H. L. Kornberg (1965). Fine control of phosphopyruvate carboxylase activity in *Escherichia coli*. *Biochim. Biophys. Acta*, **96**, 169–172.
Canovas, J. L. and H. L. Kornberg (1966). Properties and regulation of phosphopyruvate carboxylase activity in *Escherichia coli*. *Proc. R. Soc. Ser. B, London*, **165**, 189–205.
Carter, B. L. A. and A. T. Bull (1969). Studies on fungal growth and intermediary carbon metabolism under steady and nonsteady state conditions. *Biotechnol. Bioeng.*, **11**, 785–804.
Castor, L. N. and B. Chance (1959). Photochemical determination of the oxidases of bacteria. *J. Biol. Chem.*, **234**, 1587–1592.
Cennamo, C., L. Razzoli and F. Ferrari (1970). Kinetic behaviour of NAD$^+$-specific isocitrate dehydrogenase from bakers yeast and inhibition by AMP. *Ital. J. Biochem.*, **19**, 100–110.
Chang, Y. Y. and J. E. Cronan, Jr. (1983). Genetic and biochemical analysis of *Escherichia coli* strains having a mutation in the structural gene (poxB) for pyruvate oxidase. *J. Bacteriol.*, **154**, 756–762.
Cirillo, V. P. (1981). Unresolved questions on the mechanism of glucose transport in bakers yeast. In *Current Developments in Yeast Research: Advances in Biotechnology*, ed. G. G. Stewart and I. Russell, pp. 299–304. Pergamon, Oxford.
Cohen, G. N. and J. Monod (1957). Bacterial permeases. *Bacteriol. Rev.*, **21**, 169–194.
Conrad, R. and H. G. Schlegel (1977). Different degradation pathways for glucose and fructose in *Rhodopseudomonas capsulata*. *Arch. Microbiol.*, **112**, 39–40.
Cox, G. B. and F. Gibson (1974). Studies on electron transport and energy linked reactions using mutants of *Escherichia coli*. *Biochim. Biophys. Acta*, **346**, 1–25.
Crabtree, H. G. (1929). Observations on the carbohydrate metabolism of tumours. *Biochem. J.*, **23**, 536–545.
De Deken, R. H. (1966). The Crabtree effect: a regulatory system in yeast. *J. Gen. Microbiol.*, **44**, 149–157.
Dills, S. S., A. Apperson, M. R. Schmidt and M. H. Saier, Jr. (1980). Carbohydrate transport in bacteria. *Microbiol. Rev.*, **44**, 385–418.
Doelle, H. W. (1975). In *Bacterial Metabolism*. Academic, New York.
Doelle, H. W. (1981). New developments in the elucidation of the mechanisms of the Pasteur and Crabtree effects in bacteria. In *Advances in Biotechnology. I. Scientific and Engineering Principles*, ed. M. Moo-Young, C. W. Robinson and C. Vezina, pp. 249–254. Pergamon, Oxford.
Doelle, H. W., K. N. Ewings and N. W. Hollywood (1982). Regulation of glucose metabolism in bacterial systems. In *Advances in Biochemical Engineering*, ed. A. Fiechter, vol. 23, pp. 1–35. Springer-Verlag, Berlin.
Douderoff, M., W. Z. Hassid, E. W. Putman, A. L. Potter and J. Lederberg (1949). Direct utilization of maltose by *Escherichia coli*. *J. Biol. Chem.*, **179**, 921–934.
Downie, J. A., F. Gibson and G. B. Cox (1979). Membrane adenosine triphosphatases of prokaryotic cells. *Annu. Rev. Biochem.*, **48**, 103–131.
Edwards, C., S. Beer, A. Siviram and B. Chance (1981). Photochemical action spectra of bacterial *a* and *o*-type oxidases using a dye laser. *FEBS Lett.*, **128**, 205–207.
Eidels, L. and J. Preiss (1970). Carbohydrate metabolism in *Rhodopseudomonas capsulata*: enzyme titers, glucose metabolism and polyglucose polymer synthesis. *Arch. Biochem. Biophys.*, **140**, 75–89.
Eisenberg, R. C. and W. J. Dobrogosz (1967). Gluconate metabolism in *Escherichia coli*. *J. Bacteriol.*, **93**, 941–949.
Entner, N. and M. Douderoff (1952). Glucose and gluconic acid oxidation of *Pseudomonas saccharophila*. *J. Biol. Chem.*, **196**, 853–862.
Ernster, L. and B. Kwylenstierna (1969). Stucture, composition and function of mitochondrial membranes. *FEBS Symp.*, **17**, 5–31.
Ferenci, T. and H. L. Kornberg (1971). Pathway of fructose utilization by *Escherichia coli*. *FEBS Lett.*, **13**, 127–130.
Ferenci, T. and H. L. Kornberg (1973). The utilization of fructose by *Escherichia coli*. Properties of a mutant defective in fructose 1-phosphate kinase activity. *Biochem. J.*, **132**, 341–347.

Fiechter, A. (1981). Regulatory aspects in yeast metabolism. In *Advances in Biotechnology. I. Scientific and Engineering Principles*, ed. M. Moo-Young, C. W. Robinson and C. Vezina, pp. 261–266. Pergamon, Oxford.

Fiechter, A., G. F. Fuhrmann and O. Kappeli (1981). Regulation of glucose metabolism in growing yeast cells. *Adv. Microb. Physiol.*, **22**, 123–183.

Flechtner, V. R. and R. S. Hanson (1969). Course and fine control of citrate synthase from *Bacillus subtilis*. *Biochim. Biophys. Acta*, **184**, 252–262.

Flechtner, V. R. and R. S. Hanson (1970). Regulation of the tricarboxylic acid cycle in bacteria. A comparison of citrate synthases from different bacteria. *Biochim. Biophys. Acta*, **222**, 253–264.

Fraenkel, D. G. (1968). The phosphoenolpyruvate initiated pathway of fructose metabolism in *Escherichia coli*. *J. Biol. Chem.*, **243**, 6458–6463.

Garland, P. B. and B. A. Haddock (1977). Microbes and mitochondria. *Biochem. Soc. Trans.*, **5**, 479–484.

Gay, P. and G. Rapoport (1970). Etude des mutants dépourvus de fructose 1-phosphate kinase chez *Bacillus subtilis*. *C. R. Acad. Sci. (Paris)*, **271B**, 374–377.

Greenley, D. E. and D. W. Smith (1979). A novel pathway of glucose catabolism in *Thiobacillus novellus*. *Arch. Microbiol.*, **122**, 257–261.

Greenwood, C., D. Barber, S. R. Parr, E. Antonioni, M. Brunori and A. Colosimo (1978). The reaction of *Pseudomonas aeruginosa* cytochrome c-551 oxidase with oxygen. *Biochem. J.*, **173**, 11–17.

Gunatilleke, I. A. U. N., H. N. Arst, Jr. and C. Scazzocchio (1976). Three genes determine the carboxin-sensitivity of mitochondrial succinate oxidation in *Aspergillus nidulans*. *Genet. Res. Camb.*, **26**, 297–305.

Haarasilta, S. and E. Oura (1975). On the activity and regulation of anaplerotic and gluconeogenic enzymes during the growth process of bakers yeast. The biphasic growth. *Eur. J. Biochem.*, **52**, 1–7.

Hackett, N. R. and P. D. Bragg (1983a). Membrane cytochromes of *Escherichia coli* grown aerobically and anaerobically with nitrate. *J. Bacteriol.*, **154**, 708–718.

Hackett, N. R. and P. D. Bragg (1983b). Membrane cytochromes of *Escherichia coli* chl mutants. *J. Bacteriol.*, **154**, 719–727.

Haddock, B. A. (1977). The isolation of phenotypic and genotypic variants for the functional characterisation of bacterial oxidative phosphorylation. *Symp. Soc. Gen. Microbiol.*, **27**, 95–120.

Haddock, B. A. and C. W. Jones (1977). Bacterial respiration. *Bacteriol. Rev.*, **41**, 47–99.

Haddock, B. A. (1980). Microbial energetics. *Philos. Trans. R. Soc. London, Ser. B*, **290**, 329–339.

Hankinson, O. and D. J. Cove (1975). Regulation of mannitol 1-phosphate dehydrogenase in *Aspergillus nidulans*. *Can. J. Microbiol.*, **21**, 99–101.

Hansen, R. G. and U. Henning (1966). Regulation of pyruvate dehydrogenase activity in *Escherichia coli* K12. *Biochim. Biophys. Acta*, **122**, 355–358.

Hanson, T. E. and R. L. Anderson (1968). Phosphoenolypyruvate dependent formation of D-fructose 1-phosphate by a four-component phosphotransferase system. *Proc. Natl. Acad. Sci. USA*, **61**, 269–276.

Harding, R. W., D. F. Caroline and R. P. Wagner (1970). The pyruvate dehydrogenase complex from the mitochondrial fraction of *Neurospora crassa*. *Arch. Biochem. Biophys.*, **138**, 653–661.

Heath, E. C., J. Hurwitz, B. L. Horecker and A. Ginsberg (1958). Pentose fermentation by *Lactobacillus plantarum*. I. The cleavage of xylulose 5-phosphate by phosphoketolase. *J. Biol. Chem.*, **231**, 1009–1023.

Henning, U., W. Busch, G. Deppe and R. Marek (1969). Pyruvate dehydrogenase synthesis in *Escherichia coli* K12. *FEBS Symp.*, **19**, 19–28.

Hespell, R. B. (1976). Glycolytic and tricarboxylic acid cycle enzyme activities during intraperiplasmic growth of Bdellovibrio bacteriovorus on *Escherichia coli*. *J. Bacteriol.*, **128**, 677–680.

Hofer, M., K. Brand., K. Deckner and J. U. Becker (1971). Importance of the pentose phosphate pathway for D-glucose catabolism in the obligatory aerobic yeast *Rhodotorula gracilis*. *Biochem. J.*, **123**, 855–863.

Hunt, J. C. and P. V. Phibbs, Jr. (1983). Regulation of alternate peripheral pathways of glucose catabolism during aerobic and anaerobic growth of *Pseudomonas aeruginosa*. *J. Bacteriol.*, **154**, 793–802.

Jackson, F. A. and E. A. Dawes (1976). Regulation of the tricarboxylic acid cycle and poly-β-hydroxybutyrate metabolism in *Azotobacter beijerinckii* grown under oxygen or nitrogen limitation. *J. Gen. Microbiol.*, **97**, 303–312.

Jones, C. W. (1977). Aerobic respiratory systems in bacteria. *Symp. Soc. Gen. Microbiol.*, **27**, 23–59.

Kaback, H. R. (1968). The role of the phosphoenolpyruvate–phosphotransferase system in the transport of sugars by isolated membrane preparations of *Escherichia coli*. *J. Biol. Chem.*, **243**, 3711–3724.

Kauffman, H. F. and B. F. van Gelder (1973). The respiratory chain of *Azotobacter vinelandii*. I. Spectral properties of cytochrome d. *Biochim. Biophys. Acta*, **305**, 260–267.

Kitos, P., Ch. Wang, B. A. Mohler, T. E. King and V. H. Cheldelin (1958). Glucose and gluconate dissimilation in *Acetobacter suboxydans*. *J. Biol. Chem.*, **233**, 1295–1298.

Klingmüller, W. and H. Huh (1972). Sugar transport in *Neurospora crassa*. *Eur. J. Biochem.*, **25**, 141–146.

Knappe, J., J. Schacht, W. Möckel, T. Höpner, H. Vetter and R. Edenharder (1969). Pyruvate formate-lyase reaction in *Escherichia coli*. The enzymatic system converting an inactive form of the lyase into the catalytically active enzyme. *Eur. J. Biochem.*, **11**, 316–327.

Koike, M., L. J. Reed and W. F. Carroll (1963). β-Keto acid dehydrogenation complexes. IV. Resolution and reconstitution of the *Echerichia coli* pyruvate dehydrogenase complex. *J. Biol. Chem.*, **238**, 30–39.

Kornberg, A. and W. E. Pricer (1951). Di and triphosphopyridine nucleotide isocitric dehydrogenases in yeast. *J. Biol. Chem.*, **189**, 123–136.

Kornberg, H. L. (1966a). The role and control of the glyoxylate cycle in *Escherichia coli*. *Biochem. J.*, **99**, 1–11.

Kornberg, H. L. (1966b). Anaplerotic sequences and their role in metabolism. *Essays Biochem.*, **2**, 1–31.

Kornberg, H. L. (1972). Nature and regulation of hexose uptake by *Escherichia coli*. In *The Molecular Basis of Biological Transport*, ed. J. F. Woessner and F. Huijing, pp. 157–180. Academic, New York.

Kornberg, H. L. (1976). Genetics in the study of carbohydrate transport by bacteria. *J. Gen. Microbiol.*, **96**, 1–16.

Krebs, H. A. (1970). Rate control of the tricarboxylic acid cycle. *Adv. Enzyme Regul.*, **8**, 335–353.

Kröger, A. (1978). Fumarate as terminal acceptor of phosphorylative electron transport. *Biochim. Biophys. Acta*, **505**, 129–145.

Kundig, W., S. Ghosh and S. Roseman (1964). Phosphate bound to histidine in a protein as an intermediate in a novel phosphotransferase system. *Proc. Natl. Acad. Sci. USA*, **52**, 1067–1074.

Kundig, W., F. D. Kundig, B. Anderson and S. Roseman (1966). Restoration of active transport of glycosides in *Escherichia coli* by a component of a phosphotransferase system. *J. Biol. Chem.*, **241**, 3243–3246.

Lancashire, W. E., M. A. Payton, M. J. Webber and B. S. Hartley (1981). Petite negative mutants of *Saccharomyces cerevisiae*. *Mol. Gen. Genet.*, **81**, 409–410.

Langley, D. and J. R. Guest (1977). Biochemical genetics of the α-keto dehydrogenase complexes of *Escherichia coli* K12: isolation and biochemical properties of deletion mutants. *J. Gen. Microbiol.*, **99**, 263–276.

Lynch, W. H., J. MacLeod and M. Franklin (1975). Effect of temperature on the activity and synthesis of glucose-catabolising enzymes in *Pseudomonas fluorescens*. *Can. J. Microbiol.*, **21**, 1560–1572.

Maitra, P. K. and Z. Lobo (1981). Genetics of glucose phosphorylation in yeast. In *Advances in Biotechnology. Current Developments in Yeast Research*, ed. G. G. Stewart and I. Russell, pp. 293–297. Pergamon, Oxford.

Mather, M., R. Blake, J. Koland, H. Schrock, P. Russel, T. O'Brien, L. P. Hager, R. B. Gennis and M. O'Leary (1982). *Escherichia coli* pyruvate oxidase. Interation of a peripheral membrane protein with lipids. *Biophys. J.*, **37**, 87–88.

McCullough, W., M. A. Payton and C. F. Roberts (1977). Carbon metabolism in *Aspergillus nidulans*. In *Genetics and Physiology of* Aspergillus nidulans, ed. J. E. Smith and J. A. Pateman, pp. 97–129. Academic, New York.

Mitchell, C. G. and E. A. Dawes (1982). The role of oxygen in the regulation of glucose metabolism, transport and the tricarboxylic acid cycle in *Pseudomonas aeruginosa*. *J. Gen. Microbiol.*, **128**, 49–59.

Mitchell, P. (1966). Chemiosmotic coupling in oxidative and photosynthetic phosphorylation. *Biol. Rev. Camb. Philos. Soc.*, **41**, 455–502.

Moore, D. and M. S. Devadatham (1975). Distribution of mutant sites in the *Ftr* cistron depends upon the medium used for selection. *Mol. Gen. Genet.*, **138**, 81–84.

Olijve, W. and J. J. Kok (1979a). Analysis of growth of *Gluconobacter oxydans* in glucose containing media. *Arch. Microbiol.*, **121**, 283–290.

Olijve, W. and J. J. Kok (1979b). An analysis of the growth of *Gluconobacter oxydans* in chemostat cultures. *Arch. Microbiol.*, **121**, 291–297.

Ornston, L. N. (1971). Regulation of catabolic pathways in *Pseudomonas*. *Bacteriol. Rev.*, **35**, 87–116.

Pascal, M. C., M. Chippaux, A. Abou-Jaoudé, H. P. Blaschowski and J. Knappe (1981). Mutants of *Echerichia coli* K12 with defects in anaerobic pyruvate metabolism. *J. Gen. Microbiol.*, **124**, 35–42.

Payton, M. A., W. McCullough, C. F. Roberts and J. R. Guest (1977). Two unlinked genes for the pyruvate dehydrogenase complex in *Aspergillus nidulans*. *J. Bacteriol.*, **129**, 1222–1226.

Peterkofsky, A. (1977). Regulation of *Escherichia coli* adenylate cyclase by phosphorylation–dephosphorylation. *TIBS*, **2**, 12–14.

Polakis, E. S. and W. Bartley (1965). Changes in the enzyme activities of *Saccharomyces cerevisiae* during aerobic growth on different carbon sources. *Biochem. J.*, **97**, 284–297.

Poole, R. K. (1982). The oxygen reactions of bacterial oxidases. *Trends Biochem. Sci.*, **7**, 32–34.

Ramakrishnan, C. V. and S. M. Martin (1955). Isocitric dehydrogenases in *Aspergillus niger*. *Arch. Biochem. Biophys.*, **55**, 403–407.

Reed, L. J., F. R. Leach and M. Koike (1958). Studies on a lipoic acid-activating system. *J. Biol. Chem.*, **232**, 123–142.

Reid, G. A. and W. J. Ingledew (1979). Characterisation and phenotypic control of the cytochrome content of *Escherichia coli*. *Biochem. J.*, **182**, 465–472.

Rogers, D. and S-H. Yu (1962). Substrate specificity of a glucose permease of *Escherichia coli*. *J. Bacteriol.*, **84**, 877–881.

Romano, A. H. (1982). Facilitated diffusion of 6-deoxy-D-glucose in bakers yeast: evidence against phosphorylation-associated transport of glucose. *J. Bacteriol.*, **152**, 1295–1297.

Romano, A. H., S. J. Eberhard, S. L. Dingle and I. D. McDowell (1970). Distribution of the PEP : glucose phosphotransferase system in bacteria. *J. Bacteriol.*, **104**, 808–813.

Rosen, B. P. (ed.) (1978). *Bacterial Transport*. Decker, New York.

Saier, M. H., Jr. (1979). The role of the cell surface in regulating the internal environment. In *The Bacteria*, ed. S. R. Sokatch and L. N. Ornstein, pp. 168–227. Academic, New York.

Sanwal, B. D. (1970). Allosteric controls of amphibolic pathways in bacteria. *Bacteriol. Rev.*, **34**, 20–39.

Sanwal, B. D. and C. S. Stachow (1965). Allosteric activation of nicotinamide adenine dinucleotide specific isocitrate dehydrogenase: a possible regulatory protein. *Biochim. Biophys. Acta*, **96**, 28–44.

Sawyer, M. H., P. Baumann, L. Baumann, S. M. Berman, J. L. Canovas and R. H. Berman (1977). Pathways of D-fructose catabolism in species of *Pseudomonas*. *Arch. Microbiol.*, **112**, 49–55.

Scarborough, G. A. (1973). Transport in *Neurospora*. *Int. Rev. Cytol.*, **34**, 103–121.

Shaw, D. J. and J. R. Guest (1981). Molecular cloning of the *Fnr* gene of *Escherichia coli* K12. *Mol. Gen. Genet.*, **181**, 95–100.

Shaw, D. J. and J. R. Guest (1982). Amplification and product identification of the *Fnr* gene of *Escherichia coli*. *J. Gen. Microbiol.*, **128**, 2221–2228.

Shaw, D. J., D. W. Rice and J. R. Guest (1983). Homology between *CAP* and *Fnr*, a regulator of anaerobic respiration in *Escherichia coli*. *J. Mol. Biol.*, **166**, 241–247.

Shipp, W. S. (1972). Cytochromes of *Escherichia coli*. *Arch. Biochem. Biophys.*, **150**, 459–472.

Silhavy, T. J., T. Ferenci and W. Boos (1978). Sugar transport systems in *Escherichia coli*. In *Bacterial Transport*, ed. B. P. Rosen, pp. 127–169. Decker, New York.

Simoni, R. D. and P. W. Postma (1975). The energetics of bacterial active transport. *Annu. Rev. Biochem.*, **44**, 523–554.

Skinner, V. M. and S. Armitt (1972). Mutants of *Aspergillus nidulans* lacking pyruvate carboxylase. *FEBS Lett.*, **20**, 16–18.

Smith, M. W. and F. C. Neidhardt (1983a). Proteins induced by anaerobiosis in *Escherichia coli*. *J. Bacteriol.*, **154**, 336–343.

Smith, M. W. and F. C. Neidhardt (1983b). Proteins induced by aerobiosis in *Escherichia coli*. *J. Bacteriol.*, **154**, 344–350.

Stouthamer, A. H. (1978). Energy yielding pathways. In *The Bacteria*, ed. L. N. Ornston and J. R. Sokatch, vol. 6, pp. 389–462. Academic, New York.

Swings, J. and J. DeLey (1977). The biology of *Zymomonas*. *Bacteriol. Rev.*, **41**, 1–46.

Tarmy, E. M. and N. O. Kaplan (1968). Kinetics of *Escherichia coli* B: D-lactate dehydrogenase and evidence for pyruvate-controlled change in conformation. *J. Biol. Chem.*, **243**, 2587–2596.

Thaver, R. K., K. Jungerman and K. Decker (1977). Energy conservation in chemotrophic anaerobic bacteria. *Bacteriol. Rev.*, **41**, 100–180.

Van Dijken, J. P. and J. R. Quayle (1977). Fructose metabolism in four *Pseudomonas* species. *Arch. Microbiol.*, **114**, 281–286.

Van Steveninck, J. (1968). Transport and transport-associated phosphorylation of 2-deoxy-D-glucose in yeast. *Biochim. Biophys. Acta*, **54**, 386–394.

Van Steveninck, J. (1969). The mechanism of transmembrane glucose transport in yeast. Evidence for phosphorylation associated with transport. *Arch. Biochem. Biophys.*, **130**, 244–252.

Vicente, M. and J. L. Canovas (1973). Glucolysis in *Pseudomonas putida*: physiological role of alternative routes from the analysis of defective mutants. *J. Bacteriol.*, **116**, 908–914.

Wais, U., U. Gillmann and J. Ullrich (1973). Regulation, Isolierung und Charakterisierung Pyruvat-Dehydrogenase Komplexes aus Hefe. *Hoppe-Seyler's Z. Physiol. Chem.*, **345**, 206.

Weitzman, P. D. J. and P. Dunmore (1969). Citrate synthases: allosteric regulation and molecular size. *Biochim. Biophys. Acta*, **171**, 198–200.

Whiting, P. H., M. Midgley and A. E. Dawes (1976). The regulation of transport of glucose, gluconate and 2-oxygluconate and of glucose catabolism in *Pseudomonas aeruginosa*. *Biochem. J.*, **154**, 659–668.

Williams, F. R. and L. P. Hager (1966). Crystalline flavin pyruvate oxidase from *Escherichia coli*. 1. Isolation and properties of the flavoprotein. *Arch. Biochem. Biophys.*, **116**, 168–176.

Wilson, D. B. (1978). Cellular transport mechanisms. *Annu. Rev. Biochem.*, **47**, 933–965.

Wilson, D. M., J. F. Alderete, P. C. Maloney and T. H. Wilson (1976). Protomotive force as the source of energy for adenosine 5'-triphosphate synthesis in *Escherichia coli*. *J. Bacteriol.*, **126**, 327–337.

Wimpenny, J. W. T. and J. A. Cole (1967). The regulation of metabolism in facultative bacteria. 3. The effect of nitrate. *Biochim. Biophys. Acta*, **148**, 233–242.

Wöhrer, W., L. Forstenlehner and M. Röhr (1981). Evaluation of the Crabtree effect in different yeasts grown in batch and continuous culture. In *Advances in Biotechnology. Current Developments in Yeast Research*, ed. G. G. Stewart and I. Russell, pp. 405–410. Pergamon, Oxford.

Wood, A. P. and D. P. Kelly (1978). Triple catabolic pathways for glucose in a fast-growing strain of *Thiobacillus* A2. *Arch. Microbiol.*, **117**, 309–310.

Wood, A. P. and D. P. Kelly (1980). Carbohydrate degradation pathways in *Thiobacillus* A2 grown on various sugars. *J. Gen. Microbiol.*, **120**, 333–345.

Wood, A. P., D. P. Kelly and C. F. Thurston (1977). Simultaneous operation of three catabolic pathways in the metabolism of glucose by *Thiobacillus* A2. *Arch. Microbiol.*, **113**, 265–274.

19

Anaerobic Metabolism of Glucose

J. G. MORRIS
The University College of Wales, Aberystwyth, UK

19.1 INTRODUCTION

Since many species of obligately anaerobic bacteria can grow on glucose as sole source of carbon and energy, it is evident that biosynthesis of all of their organic components can be accomplished anaerobically from building blocks supplied by the catabolism of the carbohydrate. Aerobic bacterial growth is therefore chiefly notable for the opportunity it provides an organism to conserve free energy *via* aerobic respiration wherein molecular oxygen serves as the terminal oxidant for a membrane-associated series of electron transfer reactions which impel the active expulsion of protons from the cell. The ability to sustain high rates of respiration provides the aerobe with a flexible means of disposing of reducing equivalents (mainly as NAD(P)H) which, in turn, enables it to operate pathways such as the tricarboxylic acid cycle which generate reduced pyridine nucleotides and reduced flavin in relatively high yields. Under appropriate conditions therefore an aerobically respiring organism can completely oxidize glucose to carbon dioxide and water. The same is true, in principle, of those anaerobic bacteria which conserve free energy *via* some form of anaerobic respiration in which an oxidant other than dioxygen (*e.g.* nitrate, sulfate, fumarate) serves as the terminal electron acceptor for a membrane-integrated respiratory chain (Haddock and Jones, 1977). Thus although products of partial oxidation of glucose may accumulate when the sugar is provided in excess and the supply of terminal oxidant is limited, it is to its various modes of fermentation that we must look for the more distinctive products of anaerobic glucose metabolism.

19.2 FERMENTATION PROCESSES

A fermentation is a sequence of metabolic reactions whose chief purpose is the production of ATP entirely *via* substrate-level phosphorylation (SLP) reactions. Thus the most evident constraint that is placed on the routes that may be taken to catabolize any fermentation substrate is that at least one intermediate must be a suitable substrate for such an ATP-generating SLP reaction. The 'high energy' intermediates that are, in fact, produced in the course of carbohydrate fermentations are limited in number, consisting in the main of 1,3-diphosphoglycerate, phosphoenolpyruvate and acyl phosphates such as acetyl phosphate or butyryl phosphate. Each of these substances displays a sufficiently high phosphate transfer potential to be able to phosphorylate ADP to yield ATP in a spontaneous (exergonic) reaction catalysed by the appropriate cytoplasmic enzyme.

Since it is often the case that the substrate only enters the fermentative pathway after it has been converted into some suitable derivative (phosphate or CoA ester) by reactions which consume ATP, it follows that for the fermentation to give a net yield of ATP, more ATP must be produced by SLP reactions than was initially expended to modify/activate the fermentation substrate. Even so, the net yield of ATP from fermentative processes is not large, and obligately fermentative microbes are not well suited to the generation of 'single cell protein' because of their relatively low efficiency of conversion of substrate into biomass. Thus for each mol of glucose utilized in the ethanol fermentation of glucose by yeast only 2 mol of ATP is generated, whilst in ethanol fermentation by *Zymomonas mobilis* even this small net ATP yield is halved (1 mol ATP formed per mol glucose consumed). Furthermore, by no means all of the ATP that is formed is available for biosynthetic purposes, for a substantial fraction will be consumed in processes which sustain the viability of the existing biomass (maintenance energy requirements). It follows that a relatively small fraction of the carbon supplied by the fermented substrate will find its way into biomass. The major part will, in fact, accumulate in the so-called 'end-products of fermentation' which, in the main, consist of compounds formed by reduction of the various metabolites that serve as the terminal organic electron acceptors in the fermentation.

This highlights the second major constraint imposed upon any fermentation that is a balanced oxidation–reduction process. In brief, this is that in the course of the metabolic transformations which constitute the fermentation pathway, metabolites must be produced that can serve as adequate electron sinks. Reduction of these leads to reoxidation (and hence recycling) of the soluble electron carrier(s) that link the primary electron donating and terminal electron accepting reactions.

The homolactic fermentation of glucose that is practised by most species of *Streptococcus* and by several species of *Lactobacillus* well illustrates these features. In this fermentation each glucose molecule is converted into two molecules of pyruvate by the Embden–Meyerhof pathway of glycolysis. The course of glucose metabolism *via* this pathway (Figure 1) is so contrived that amongst the intermediates formed are two of the crucial SLP substrates, *viz.* 1,3-diphosphoglycerate and phosphoenolpyruvate. The ATP-generating reactions catalysed by phosphoglycerylkinase and pyruvate kinase are thus the key free energy-conserving reactions of this fermentation yielding in all 4 mol ATP per mol glucose consumed. Since two molecules of ATP are utilized in forming first glucose 6-phosphate and then fructose 1,6-diphosphate (in preparation for the crucial aldolase-catalysed split of the hexose diphosphate into two molecules of triose phosphate) the net yield of ATP in this fermentation is only 2 mol per mol of glucose utilized. A crucial step in the fermentation is the dehydrogenation of glyceraldehyde 3-phosphate which is associated with the uptake of inorganic phosphate to yield the first of the SLP substrates. This reaction, catalysed by triose phosphate dehydrogenase, utilizes NAD^+ as the electron acceptor, and maintenance of the fermentation is dependent on the regeneration of NAD^+ from the NADH so produced. This is accomplished by reduction of pyruvate to lactate. The coupling of triose phosphate oxidation to pyruvate reduction is highlighted when the fermentation pathway is rewritten as in Figure 2a, which demonstrates how inflexible is this interdependence in the absence of any alternative oxidant capable of regenerating NAD^+ from NADH. Figure 2b again illustrates the course of the homolactic fermentation but now in a manner which emphasizes that it is a linear (unbranched) fermentation pathway whose end product (lactate) and whose efficiency of free energy conservation (2 mol ATP per mol glucose utilized) are wholly predictable and invariable. This contrasts with the greater flexibility of a branched fermentation pathway such as that outlined in Figure 3, wherein each branch may be associated (or not) with the production of a distinctive SLP reaction substrate or electron acceptor. Such a fermentation will yield a mixture of end products whose proportions can be expected to vary in response to changes in cultural conditions. In turn this argues for the existence of indigenous control mechanisms that can normally regulate the flow of metabolites through the competing branch routes. By this means the efficiency of free energy conservation (*i.e.* as ATP production per mole of substrate utilized) can be adjusted to meet the demands made by the prevailing growth conditions.

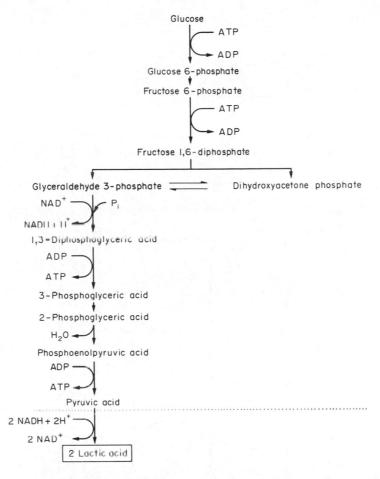

Figure 1 The Embden–Meyerhof pathway of glycolysis (anaerobic conversion of glucose to pyruvate), with additional terminal reduction of pyruvate to lactate

Bacterial fermentations of glucose afford several examples of such branched fermentation pathways. In many of these it is pyruvate that stands at the major branch point, the proportions in which the various constitutents are present in the terminal end-product mixture thus mirroring the division of the glycolytically generated pyruvate between the competing branches. It is the great diversity of fates that can befall pyruvate in bacteria that, in large measure, accounts for the wealth of products that can arise from the fermentation of glucose. This is evident in Figure 4 which summarizes the major enzyme catalysed reactions to which pyruvate is prone, and the terminal products that might, directly or indirectly, be formed thereby. Some of these reactions lead to the generation of an SLP reaction substrate and hence to additional ATP production, *e.g.* production of acetyl-CoA followed by formation of acetyl phosphate and eventually of acetate. Others serve primarily as a means of disposing of reducing power and of regenerating NAD^+, *e.g.* the routes which lead to the formation of ethanol, lactate, or 2,3-butanediol. It is even possible for both roles to be served by the same route, as in those which form butyrate or propionate. An additional complication is introduced by the possible formation of the same product from pyruvate but by different mechanisms. Thus in yeasts and a few bacteria including *Zymomonas mobilis* and *Sarcina ventriculi* acetaldehyde is produced (as a precursor of ethanol) *via* the action of a pyruvate decarboxylase. In other, heterofermentative, bacteria whose several end products include ethanol, this is generally formed from acetaldehyde produced by reduction of acetyl-CoA. Similarly, acetyl-CoA can be formed from pyruvate in several different ways. A single

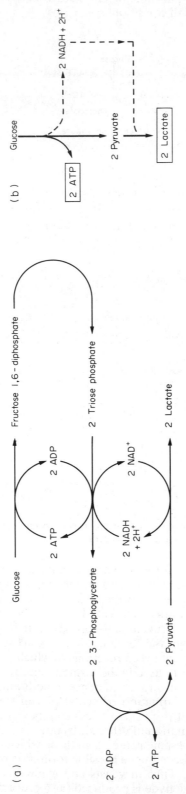

Figure 2 Homolactic fermentation of glucose *via* the Embden–Meyerhof pathway: (a) emphasizing the coupling between triose phosphate (as primary electron donor) and pyruvate (as terminal electron acceptor); and (b) illustrating the fact that this is a linear fermentation pathway with an invariable product (2 mol of lactate mol^{-1} glucose utilized) and an invariable net yield of ATP *via* substrate level phosphorylation reactions (2 mol ATP mol^{-1} glucose fermented)

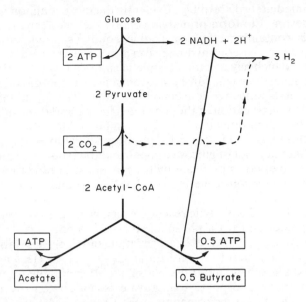

Figure 3 Outline scheme of a branched glucose fermentation pathway in which the possibility exists of variation both in the ratio of individual products (acetate, butyrate, H₂) and in the efficiency of free energy conservation (net ATP yield; see also Figures 9 and 10)

bacterium may contain the enzymes which catalyse more than one of these acetyl-CoA-yielding reactions, the prevailing cultural conditions often then determining which takes precedence.

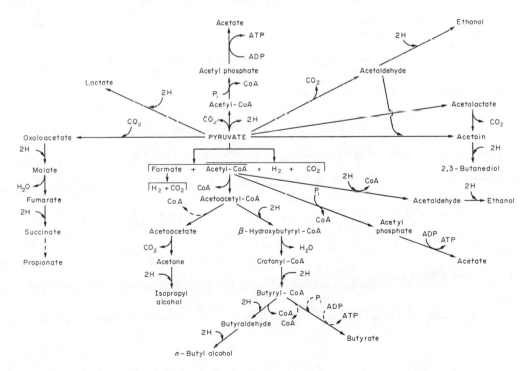

Figure 4 A composite diagram illustrating the major fates of pyruvate in various bacterial fermentations of glucose

Not only the fate of pyruvate but also the route whereby this key metabolite is formed from glucose may differ from organism to organism, so contributing to the diversity of microbial means of fermenting glucose. These can conveniently be classified into six main types summarized in Table 1. We shall consider each of these modes of glucose fermentation hereafter, but some common principles are immediately discernible. Evidently the most common terminal electron acceptors are aldehydes (destined to form primary alcohols), ketones (giving rise to secondary alcohols) and compounds containing ethylenic double bonds (yielding saturated carbon chains). Proton reduction to generate gaseous hydrogen can also serve as a means of disposing of reducing power in those bacteria which have a redox couple of suitably low potential (*e.g.* ferredoxin) and an appropriate hydrogenase enzyme. In homoacetogens, carbon dioxide serves as an electron acceptor in the course of its reductive conversion to acetate. The formation of pyruvate from glucose necessarily involves degradation of the substrate glucose and three major lytic routes are followed: (a) the fructose diphosphate route (*i.e.* Embden–Meyerhof pathway) in which the enzymes phosphofructokinase and fructose-1,6-bisphosphate aldolase play key roles in producing 2 mol of triose phosphate per mol of glucose catabolized; the triose phosphate (as glyceraldehyde 3-phosphate) is then converted to pyruvate by the reactions forming the remainder of the E–M pathway (Figure 1); these reactions generate 1 mol of NADH and 1 mol of ATP per mol of triose phosphate so oxidized; (b) the Entner–Doudoroff pathway, wherein dehydrogenation of glucose 6-phosphate precedes a $C_6 \rightarrow 2 \times C_3$ lytic reaction which produces an equimolar mixture of pyruvate and triose phosphate; and (c) a phosphoketolase-dependent pathway again commencing with glucose 6-phosphate but in which dehydrogenation and decarboxylation reactions precede a phosphorolytic reaction. This therefore consists of a $C_5 \rightarrow C_2 + C_3$ cleavage which yields an equimolar mixture of acetyl phosphate and triose phosphate. It is notable that in both (b) and (c) the further conversion of triose phosphate to pyruvate is undertaken by the same sequence of enzymes that accomplishes this transformation in the Embden–Meyerhof pathway. This testifies to the crucial importance of this ATP-generating segment of the E–M pathway which has been highly conserved during evolution.

Table 1 Major Types of Glucose Fermentation by Microorganisms

		Fermentation products	
Type of fermentation	*Representative organism*	*Major*	*Minor*
Lactic			
Homofermentation	*Streptococcus bovis*	Lactate	—
Heterofermentation	*Leuconostoc mesenteroides*	Lactate, ethanol, CO_2	—
Ethanolic			
Homofermentation	*Saccharomyces cerevisiae*	Ethanol, CO_2	—
	Zymomonas mobilis	Ethanol, CO_2	—
	Thermoanaerobacter ethanolicus	Ethanol, CO_2	(Lactate, H_2)
Heterofermentation	*Clostridium thermocellum*	Ethanol, acetate, CO_2, H_2	Lactate
	Sarcina ventriculi	Ethanol, CO_2	Acetate, H_2
	Thermoanaerobium brockii	Ethanol, lactate, CO_2	Acetate, H_2
Butyric			
Acidogenic	*Clostridium pasteurianum*	Butyrate, acetate, CO_2, H_2	—
Solventogenic	*Clostridium acetobutylicum*	Acetone, butanol, CO_2	Acetate, butyrate, H_2
Mixed acid			
Heterofermentation	*Escherichia coli*	Lactate, acetate, formate, ethanol, CO_2, H_2	Succinate
Butanediol	*Enterobacter aerogenes*	2,3-Butanediol, acetoin, ethanol, CO_2	Formate, acetate, H_2
Propionic	*Propionibacterium shermanii*	Propionate, acetate, CO_2	Succinate
Homoacetic	*Clostridium thermoaceticum*	Acetate	—

19.3 MODES OF GLUCOSE FERMENTATION

19.3.1 Fermentations that Yield Lactic Acid

The sequence of reactions that constitute the Embden–Meyerhof pathway of glycolysis has been outlined in Figure 1, which shows how two molecules of pyruvate are formed from one molecule of glucose with the concurrent reduction of two molecules of NAD^+ and the net production of two molecules of ATP. Figure 2 further illustrates how, in a *homolactic* fermentation, the necessary reoxidation of the NADH is accomplished by the reduction of both molecules of pyruvate to lactate. In such a fermentation lactate would account for at least 85% of the glucose carbon that was utilized. This is in marked contrast to any *heterolactic* fermentation wherein, although lactate would be a major product, it would not account for more than 50% of the glucose that was consumed, the remainder appearing chiefly as ethanol plus carbon dioxide with, possibly, some acetate. In obligately heterolactic species of bacteria, glycolysis does not proceed by the Embden–Meyerhof pathway but by a totally different route in which the enzyme phosphoketolase plays the key role (Figure 5). In this 'phosphoketolase pathway', glucose 6-phosphate is dehydrogenated and decarboxylated before cleavage. The substrate of the phosphoketolase is thus a pentulose phosphate (xylulose 5-phosphate) and the products of the lytic reaction are glyceraldehyde 3-phosphate plus acetyl phosphate. The triose phosphate is converted to pyruvate by the same sequence of reactions which is employed for this purpose in the Embden–Meyerhof pathway. Reduction of the pyruvate to lactate regenerates the NAD^+ reduced by the triose phosphate dehydrogenase in a nicely balanced manner. However, the two molecules of NADH generated by the initial oxidation of glucose 6-phosphate through 6-phosphogluconate to ribulose 5-phosphate plus CO_2, must also be reoxidized. This is ensured by conversion of the acetyl phosphate to acetyl-CoA which is then reduced *via* acetaldehyde to ethanol. The net yield of ATP in this ethanol plus lactate fermentation is only 1 mol per mol of glucose utilized. It is thus interesting to note that when a pentose, rather than a hexose, is consumed in a heterolactic fermentation based on the phosphoketolase pathway, the necessity to dispose of reducing power generated prior to the phosphoketolase reaction no longer exists. In this situation the acetyl phosphate can be converted to acetate with concomitant generation of ATP by the transphosphorylation reaction catalysed by acetate kinase. This acetate plus lactate fermentation is consequently more efficient in its conservation of free energy, since 2 mol of ATP are now formed per mol of pentose utilized. Acetate may also be produced on occasions in the course of a heterolactic fermentation of glucose when some alternative to acetaldehyde is available as a suitable oxidant for the regeneration of NAD^+. If this were an exogenous (alien) electron acceptor, acetate production would signal a proportionately enhanced yield of ATP from glucose utilization. However, in that more common variant of heterolactic fermentation of glucose in which acetate production occurs at the expense of the reduction of endogenously produced dihydroxyacetone phosphate to yield glycerol, the net yield of ATP would be less than in the 'normal' ethanol plus lactate fermentation.

Some lactic acid bacteria (*e.g. Streptococcus faecalis*) form only L(+)-lactate, others (*e.g. Lactobacillus delbrueckii*) produce D(−)-lactate, whereas some (*e.g. Lactobacillus plantarum*) accumulate DL-lactate. When DL-lactate is produced, this is generally due to the organism's possession of both L(+)- and D(−)-specific, NADH-dependent lactate dehydrogenases, though a lactate racemase has been found in a minority of species, for example in *Lactobacillus curvatus*. These fermentative, NADH-dependent lactate dehydrogenases might more properly be termed pyruvate reductases, for such is their function. The oxidative lactate dehydrogenases of bacteria, are in contrast, not dependent on NAD^+. The properties of a wide variety of bacterial lactate dehydrogenases have been studied in considerable detail (Garvie, 1980), largely because of the importance of lactic acid bacteria in the dairy industry and in the preparation of silage. An important finding (Wolin, 1964) was that the L(+)-lactate dehydrogenases of a number of lactic acid bacteria have an absolute or near-absolute requirement for fructose 1,6-diphosphate (FDP). This is the case for most species of *Streptococcus*, for several species of *Lactobacillus* including *Lb. casei*, *Lb. curvatus* and *Lb. xylosus*, and a few other bacteria including *Staphylococcus epidermidis* and *Actinomyces viscosus*. The FDP-dependent L(+)-lactate dehydrogenase of *Lb. casei* is additionally strongly inhibited by both ADP and ATP. Generally speaking, the L(+)-lactate dehydrogenase present in those bacteria which produce DL-lactate is not activated by FDP.

This susceptibility of the terminal L(+)-lactate dehydrogenase to allosteric control mediated by the intracellular concentration of a key intermediate of the Embden–Meyerhof pathway can help to explain, in some instances at any rate, why certain lactic acid bacteria respond to changes in environmental conditions by a shift from a homolactic to a heterolactic mode of glucose fermentation. By altering the initial pH of the growth medium from a slightly acidic to a slightly alkaline pH, lactic acid production by cultures of *Streptococcus faecalis* subsp. *liquefaciens* was decreased by 50% with consequent additional accumulation of ethanol, acetate and formate. Similar findings were made with *Streptococcus durans* and *Streptococcus thermophilus*. When either *Streptococcus mutans* or *Streptococcus sanguis* was grown in glucose-limited chemostat culture, lactic

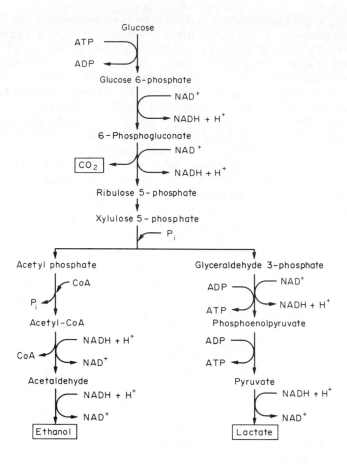

Figure 5 The phosphoketolase pathway of glucose metabolism in obligately heterolactic bacteria

acid production was almost totally suppressed, even though these organisms undertook a homolactic fermentation of glucose in media in which the carbohydrate was present in plentiful supply. Yet *Streptococcus bovis* and *Streptococcus salivarius* did not respond to deprivation of glucose in a like manner (see London, 1976). In cells of *Streptococcus mutans* JC2 grown on a limiting concentration of glucose and forming ethanol, acetate and formate together with some lactate, L(+)-lactate dehydrogenase was present at the same specific activity as in homolactic fermenting cells grown with excess glucose. However, the intracellular concentration of FDP was very low in the glucose-deprived cells and this was deemed sufficient to explain the shift from lactate production by these organisms. Since the L(+)-lactate dehydrogenase of *Streptococcus bovis* is also dependent on allosteric activation by FDP, it is not easy to see how this organism manages to sustain its homolactic fermentation even when glucose provision is growth rate limiting. Perhaps it is unable to utilize its Embden–Meyerhof pathway-derived pyruvate by any alternative reductive route (Garvie, 1980).

Streptococcus cremoris and *Streptococcus lactis* which are favourite 'starter' organisms in the dairy industry, noted for their high yield of lactate with only small amounts of secondary products such as acetate or diacetyl, were initially considered to be invariable homolactic fermenters. However, five out of six tested strains growing in glucose-limited chemostat culture switched from homolactic to heterolactic fermentation as the dilution rate was decreased (Lawrence and Thomas, 1979). The intracellular concentration of FDP remained high in the heterolactic fermenting cells and when these were suspended in a medium containing excess glucose they immediately reverted to a homolactic fermentation. The L(+)-lactate dehydrogenase is, however, not the only fermentative enzyme whose activity is regulated in these organisms. A regulating role has also been assigned to the pyruvate kinase of homolactic streptococci, for the activity of this enzyme is stimulated by glucose 6-phosphate, fructose 6-phosphate, fructose 1,6-diphosphate and glyceraldehyde 3-phosphate (Lawrence and Thomas, 1979).

It is therefore plain that a number of cultural conditions, including shift to a more alkaline pH, or substrate limitation, can cause some lactic acid bacteria to shift from a homolactic fermentation to a form of heterolactic fermentation in which the route of glycolysis (Embden–Meyerhof pathway) need not change but in which some of the pyruvate that is formed is metabolized in an alternative manner to yield the characteristic C_2 and C_1 products. This is to be distinguished from the anaerobic glucose metabolism of the obligate heterolactic fermenter which lacks fructosebisphosphate aldolase and which perforce must undergo glycolysis *via* the phosphoketolase pathway (Figure 5).

An interesting variant of the phosphoketolase route operates in *Bifidobacterium bifidum* (Scardovi, 1981). In this rather more complicated pathway (Figure 6), phosphoketolases are implicated at two steps. A fructose-6-phosphate phosphoketolase catalyses the cleavage of fructose 6-phosphate to give erythrose 4-phosphate and acetyl phosphate. The tetrose phosphate together with fructose 6-phosphate, *via* the anaerobic pentose cycle enzymes transketolase and transaldolase, give rise to xylulose 5-phosphate. This is split by xylulose 5-phosphate phosphoketolase to yield glyceraldehyde 3-phosphate and (additional) acetyl phosphate. The end products of the fermentation are thus acetate plus lactate in a theoretical ratio 3:2, though alternative metabolism of some of the intermediary pyruvate can alter the fermentation balance, sometimes quite substantially (Lauer and Kandler, 1976). Though unusual, this route of glucose fermentation is particularly productive of ATP, with 2.5 mol being generated per mol of glucose utilized.

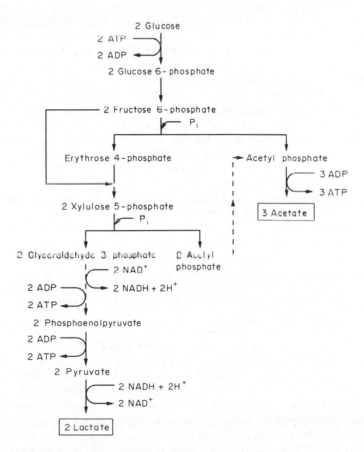

Figure 6 The pathway of glucose fermentation in *Bifidobacterium bifidum*

It should be noted that even bacteria which ferment glucose exclusively *via* the Embden–Meyerhof pathway and which may not possess glucose-6-phosphate dehydrogenase and 6-phosphogluconate dehydrogenase, are still able to synthesize the pentose sugars that they require for biosynthetic purposes. The crucial enzyme generally employed for his purpose is transketolase which can catalyze the interaction of fructose 6-phosphate with glyceraldehyde 3-phosphate to yield xylulose 5-phosphate plus erythrose 4-phosphate. In this way tetrose phosphate required, for example, for the synthesis of aromatic amino acids, is simultaneously manufactured (Fraenkel and Vinopal, 1973).

19.3.2   Fermentations that Yield Ethanol

The classical homofermentation of glucose to ethanol plus carbon dioxide is traditionally associated with species of yeasts such as *Saccharomyces cerevisiae* and *Saccharomyces uvarum* (formerly *Sacch. carlsbergensis*). In their metabolism of glucose, pyruvate is produced by the Embden–Meyerhof pathway and is then decarboxylated to yield acetaldehyde in an irreversible reaction catalysed by a thiamine pyrophosphate-dependent pyruvate decarboxylase. The acetaldehyde then serves as the terminal electron acceptor which effects the reoxidation of NADH generated by the triose phosphate dehydrogenase and is itself reduced to ethanol. As in the homolactic fermentation of glucose, formation of ethanol by this route is accompanied by a net yield of 2 mol ATP per mol of glucose consumed.

Although the ethanolic fermentation is itself a wholly anaerobic process, growth and reproduction of the yeast cells necessitates the synthesis of some cellular components (ergosterol, nicotinic acid, unsaturated fatty acids) by mechanisms in which molecular oxygen is an essential reactant. Thus yeast cultures cannot be grown anaerobically unless these additional nutrients are supplied preformed in the culture medium. It is for this reason that low concentrations of oxygen are generally maintained in commercial scale yeast fermentations. However, at higher oxygen concentrations, the rate of utilization of glucose is substantially diminished (the so-called Pasteur effect). This effect has been largely attributed to 'feedback' allosteric inhibition of the phosphofructokinase enzyme by ATP (Sols, 1976). Studies of *Sacch. cerevisiae* growing anaerobically on glucose in chemostat culture have shown that, in addition to ethanol, substantial quantities of glycerol may also be formed together with some pyruvate (Fiechter *et al.*, 1981). Glycerol production is also a notable accompaniment of ethanol production during the early exponential phase of growth of *Sacch. cerevisiae* in anaerobic batch culture when glucose concentrations are high. It has its origin in the alternative reduction of dihydroxyacetone phosphate to α-glycerol phosphate which is then dephosphorylated to yield glycerol. This was the basis of the glycerol production effected by 'Neuberg's bisulfite fermentation' which was employed during World War 1 for the production of glycerol (Wood, 1961). Addition of sodium sulfite to the fermentation medium caused acetaldehyde to be 'trapped' as its bisulfite addition product thus rendering it unavailable as an electron acceptor. Ethanol formation was not totally suppressed, nor could it be, for ATP could still only be generated during the conversion of triose phosphate to pyruvate. In another variant of glucose fermentation by yeast that was investigated by Neuberg, when an alkaline pH was maintained glycerol was again produced at the expense of ethanol production. In these circumstances, acetaldehyde accumulated and underwent a dismutation to yield equal quantities of acetate and ethanol. Indeed, this appeared to be the normal route of glucose fermentation in another yeast, *Zygosaccharomyces acidifaciens*. Both sugar tolerance and ethanol tolerance varies between different species of yeasts and can influence the choice of industrially suitable strains. Some of these can tolerate up to 22 to 35% (w/v) glucose, whilst the most ethanol tolerant strains, found amongst species of *Saccharomyces* and *Schizosaccharomyces*, can grow in media containing a maximum ethanol concentration of about 10% (w/v) whilst continuing to ferment at ethanol concentrations up to about 20% (w/v).

Though pyruvate decarboxylase is rarely present in bacteria it is found in certain species. Thus in *Sarcina ventriculi* this enzyme competes for pyruvate with the pyruvate ferredoxin oxidoreductase that generates acetyl-CoA. Since this acetyl-CoA, besides giving rise to acetyl phosphate and acetate may be reduced with NADH to yield acetaldehyde, the organism can produce ethanol from pyruvate by two distinct routes. Which takes precedence is influenced by the prevailing culture pH, more acid pH values (<pH 5) favouring the activity of the pyruvate decarboxylase with resultant increased ethanol production.

Another bacterium which possesses a pyruvate decarboxylase is *Zymomonas mobilis*. This organism, formerly called *Pseudomonas lindneri*, is responsible for the fermentation which produces the Mexican alcoholic beverage known as pulque. It can tolerate higher initial sugar concentrations than yeasts and has been found to produce up to 10% (w/v) ethanol from a 25% (w/v) glucose medium. It is thus a serious candidate for the industrial scale production of ethanol (Rogers *et al.*, 1980). *Zymomonas mobilis* accomplishes glycolysis, however, not by the Embden–Meyerhof pathway, but by an alternative route known as the Entner–Doudoroff pathway (Figures 7 and 8). In this pathway, glucose 6-phosphate is dehydrogenated to yield 6-phosphogluconate which is then dehydrated to give 2-oxo-3-deoxy-6-phosphogluconate in a reaction catalysed by a dehydratase which requires Fe^{2+} ions and glutathione for maximal activity. The 2-oxo-3-deoxy-6-phosphogluconate is then cleaved (by the action of an aldolase) to yield one molecule of pyruvate and one molecule of glyceraldehyde 3-phosphate. Again this triose phosphate is converted to additional pyruvate by the terminal reactions of the Embden–Meyerhof pathway. Since only 1 mol of triose phosphate is produced from each mol of glucose consumed, the net yield of ATP (1 mol ATP per mol glucose fermented) is only half that achieved by the Embden-–Meyerhof pathway. Perhaps in consequence of this relatively inefficient utilization of hexose the Entner–Doudoroff pathway, though quite common in Gram-negative bacteria with a purely respiratory metabolism, rarely operates in the fermentation of glucose.

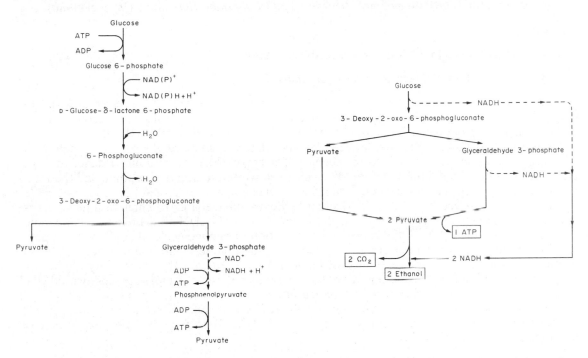

Figure 7 The Entner–Doudoroff pathway of glycolysis

Figure 8 Schematic diagram of the route of glucose fermentation (based on the Entner–Doudoroff pathway) in the ethanol-producing bacterium *Zymomonas mobilis*

Other bacteria which form ethanol from glucose do so not with the aid of a pyruvate decarboxylase but by the reduction of acetyl-CoA. This might be formed from pyruvate by oxidative decarboxylation such as is accomplished by the pyruvate:ferredoxin oxidoreductase found in species of *Clostridium, Ruminococcus, etc.* or by cleavage catalysed by a pyruvate formate lyase. Alternatively the acetyl-CoA may be produced from acetyl phosphate generated by a phosphoketolase (Figure 5). Thus ethanol is a common product of many mixed acid fermentations of glucose, providing a useful means for the disposal of excess reducing power and generally therefore being accumulated at the expense of acetate.

Ethanol production is a particularly frequent occurrence in carbohydrate fermentations accomplished by thermophilic anaerobes; indeed, it has been suggested that its volatility at the normal growth temperature of these thermophiles makes it a particularly apt fermentation product, its loss by distillation diminishing the likelihood that it will accumulate to toxic concentrations in the culture medium (Wiegel, 1980). For example, *Bacillus stearothermophilus* was found only to produce ethanol at growth temperatures in excess of 55 °C. *Clostridium thermocellum*, which is noted for its ability to grow and digest cellulose anaerobically at 60 °C, ferments glucose to yield ethanol plus acetate. *Clostridium thermohydrosulfuricum* similarly ferments glucose during its anaerobic growth at 69 °C to produce ethanol with lesser amounts of acetate, lactate, CO_2 and H_2. Improved yields of ethanol were obtained when, in batch culture, the pH was allowed to decrease from pH 7.5 to pH 6.9. A similarly mixed fermentation of glucose is undertaken by *Thermoanaerobium brockii* growing at 66 °C at pH 7.5 in a medium containing a small concentration of yeast extract, with about 1 mol of ethanol being produced per mol of glucose consumed (Zeikus, 1980). Interestingly, in media rich in yeast extract, *T. brockii* produces more lactate than ethanol. The anaerobic thermophile which produces the greatest proportional yield of ethanol (1.8 mol per mol glucose utilized) is however *Thermoanaerobacter ethanolicus* which grows at 69 °C over a broad range of pH values (pH 5.5 to 8.5). This nearly homoethanolic fermentation (with acetate, lactate and H_2 as the minor by-products) makes it likely that this organism will be intensively studied as a promising candidate for biotechnological exploitation (Ljundgahl *et al.*, 1981). Interest in these thermophilic ethanologens is not restricted to their fermentation of sugars in pure culture. *Thermoanaerobacter ethanolicus* can also ferment starch or cellobiose and in mixed culture with *Clostridium thermocellum* enhanced utilization of cellulose was obtained with nearly 1.5 mol of ethanol being formed per mol of glucose residue, *i.e.* a substantially greater yield of ethanol than the 0.84 mol mol^{-1} achieved by *Cl. thermocellum* alone (Wiegel, 1980).

19.3.3 Fermentations that Yield Acetic and Butyric Acids

19.3.3.1 Straightforward acidogenic fermentations

The anaerobic fermentation of glucose which yields acetic and butyric acids together with CO_2 plus H_2 has been well studied in species of butyric clostridia such as *Clostridium pasteurianum* (Figure 9). In this branched fermentation pathway, pyruvate is formed by the usual Embden–Meyerhof pathway and is then oxidatively decarboxylated to give acetyl-CoA in a reaction for which ferredoxin serves as the electron acceptor. The acetyl-CoA thus produced stands at a crucial branch point of metabolism; either (1) it can be converted to acetyl phosphate and thence to acetate in a two-step reaction sequence which generates one molecule of ATP, or (2) two molecules of acetyl-CoA can be condensed together to yield acetoacetyl-CoA which becomes the first substrate in a sequence of reactions in which two molecules of NADH are oxidized and one molecule of butyrate formed, with the concurrent production of one molecule of ATP. Were it possible for all the acetyl-CoA to be converted to acetate, then the fermentation of 1 mol of glucose would be accompanied by the generation of 4 mol of ATP. If, on the other hand, butyrate was the only fatty acid product of the fermentation, then the consumption of 1 mol of glucose would yield only 3 mol of ATP. In practice, when *Cl. pasteurianum* grows anaerobically in batch culture with excess glucose as the sole source of carbon and energy, acetate and butyrate are formed in such proportions (equation 1) that approximately 3.3 mol of ATP are generated per mol of glucose consumed (Thauer *et al.*, 1977).

$$\text{glucose} + 0.6\,H_2O \rightarrow 0.6\,\text{acetic acid} + 0.7\,\text{butyric acid} + 2\,CO_2 + 2.6\,H_2 \tag{1}$$

This particular ratio of acetate to butyrate in the fermentation products is sustained by regulation of a key cytoplasmic enzyme which indirectly determines what fraction of the NADH generated by glycolysis will be reoxidized *via* reduction of acetyl-CoA to butyrate. This enzyme is NADH-ferredoxin oxidoreductase which catalyses reduction of ferredoxin by NADH, the reduced ferredoxin joining that generated by the action of pyruvate-ferredoxin oxidoreductase. The reduced ferredoxin may thereafter be oxidized by a proton-reducing hydrogenase. In this manner 'excess' reducing power is disposed of as gaseous hydrogen (which is why, in equation 1, 2.6 H_2 is generated in place of the 2.0 H_2 which could maximally be formed *via* pyruvate oxidation alone). Thauer and his colleagues (see Thauer *et al.*, 1977) discovered that the activity of the NADH-ferredoxin oxidoreductase was stimulated by acetyl-CoA and inhibited by CoA. This responsiveness of NADH-ferredoxin oxidoreductase to the prevailing acetyl-CoA/CoA ratio means that the fate of acetyl-CoA at one branch point in the fermentation pathway is intimately linked to the fate of NADH at another. Similar controls operating on the NADH-ferredoxin oxidoreductases of other saccharolytic clostridia have been reported, as has their possession of reduced ferredoxin-NADP oxidoreductases. It is thus easy to see how by the provision of an alien electron acceptor which can rapidly reoxidize NADH, or again by coculture with a hydrogen utilizing anaerobe (such as *Desulfovibrio* or a methanogen), the fermentation balance of an organism such as *Clostridium pasteurianum* can be shifted in the direction of enhanced acetate formation and diminished butyrate production (with an increased specific yield of ATP). Given the reversibility of the reaction catalysed by hydrogenase, it is also likely that growth in an atmosphere enriched in H_2 would favour production of the more reduced end product (*i.e.* butyrate) at the expense of acetate.

Such behaviour is not limited to anaerobes producing acetate and butyrate. Zeikus (1980) has explained how in thermophilic ethanol-producing anaerobes, such as *Clostridium thermocellum* and *Thermoanaerobium brockii*, different fermentation product ratios can be determined by the specific activities and regulatory properties of the catabolic enzymes that control their fermentative electron flow; also, how it is that such electron flow(s) can be modified or even reversed by the addition of exogenous electron donors or acceptors.

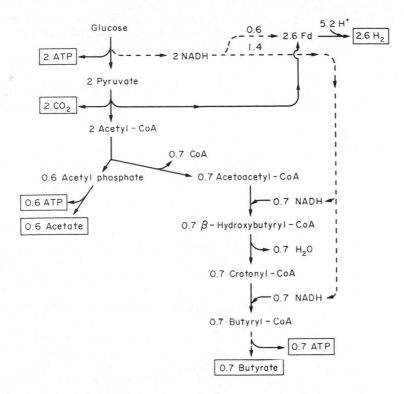

Figure 9 Acidogenic (acetic plus butyric) fermentation of glucose by the obligately anaerobic bacterium *Clostridium pasteurianum*; note that, of the 2 mol of NADH generated by the fermentation of 1 mol of glucose, 0.6 mol is reoxidized by ferredoxin in the reaction catalysed by an NADH-ferredoxin oxidoreductase whose activity responds to the prevailing intracellular ratio of acetyl-CoA/CoA. The remaining 1.4 mol of NADH is available for reoxidation *via* the conversion of 1.4 mol of acetyl-CoA to 0.7 mol of butyrate

19.3.3.2 The acetone plus butanol fermentation

Certain saccharolytic clostridia are noted for their ability to 'over reduce' the acid products of glucose catabolism and so accumulate ethanol in place of acetate, and butanol in place of butyrate. Thus strains of *Clostridium beijerinckii* generally produce butanol, whilst *Clostridium thermosaccharolyticum*, which forms butyrate during exponential growth in batch culture, produces ethanol in place of butyrate during the course of subsequent sporulation. A fermentation that has a distinguished record in having actually been usefully exploited on an industrial scale (with molasses or maize grain as feedstock) is the solventogenic fermentation yielding acetone and butanol that is accomplished by *Clostridium acetobutylicum* (Spivey, 1978). The fermentation pathway employed is shown in Figure 10, which demonstrates that it differs from the acidogenic fermentation of *Clostridium pasteurianum* in containing two additional branch points. The first affords an alternative fate for acetoacetyl-CoA which, in addition to being subject to reduction to β-hydroxybutyryl-CoA, may now interact with acetate in a reaction catalysed by a CoA transferase to give acetoacetate which is then subject to decarboxylation to yield acetone. The second enables butyryl-CoA, as an alternative to yielding butyrate, to be reduced to butyraldehyde and thence to butanol.

During anaerobic growth of *Clostridium acetobutylicum* in batch culture on excess glucose, the fermentation classically proceeds in two phases. The first consists of an acidogenic fermentation yielding acetic and butyric acids with H_2 plus CO_2. As a consequence of the production of the fatty acids, the pH of the medium decreases until, towards the end of the normal exponential phase of growth, it achieves a value in the region of pH 5–4.8. At this 'break point' the character of the fermentation changes as acetone and butanol begin to be produced. Formation of these neutral solvents in place of the acids moderates the fall in pH, which might actually rise slightly as some portion of the previously accumulated fatty acids is converted into acetone and butanol. Recent investigations of this shift from acidogenic to solventogenic fermentation have been car-

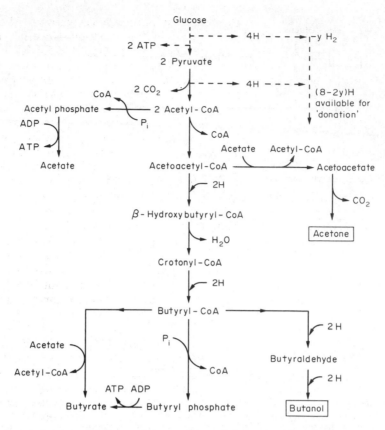

Figure 10 Solventogenic (acetone plus butanol) fermentation of glucose by *Clostridium acetobutylicum*; no attempt is made in this diagram to account for the actual proportions in which the individual end-products are formed

ried out with *Cl. acetobutylicum* growing at various rates and at fixed pH values in continuous flow culture both in chemostat culture (Bahl *et al.*, 1982) and in turbidostat culture (Gottschal and Morris, 1982). Strain differences have been revealed, with one strain producing solvents only at pH values below pH 4.7 (optimally at pH 4.3) and another strain undertaking solventogenesis at pH 5.5 but not tolerating a pH more acidic than pH 4.6. With this latter strain it was found that a low pH alone was insufficient to trigger solventogenesis which, however, occurred when the medium was supplemented with a mixture of acetic and butyric acids at a suitably high concentration (Gottschal and Morris, 1981). When supplied with butyric acid at a concentration above the threshold value this strain would, in batch culture at pH 5, commence solvent production from the outset. It remains to be seen whether this is a more general phenomenon, whilst the actual mechanism of endogenous regulation of the fermentation which contrives the switch from acid production to solvent formation has yet to be determined. Undoubtedly this might include both coarse controls (repression and/or depression of synthesis of enzymes) and fine controls which are likely to operate *via* allosteric regulation of the activities of key (especially branch point) enzymes. Regulation of the flow of electrons to proton reduction (H_2 generation) will also be involved and studies of the NAD(P)H-ferredoxin oxidoreductases of *Clostridium acetobutylicum* have been undertaken with this in mind.

In a variant of this type of fermentation, further reduction of acetone yields isopropyl alcohol (in strains of *Clostridium butyricum*) whilst under certain conditions, including deprivation of iron, substantial quantities of lactate may be formed by various saccharolytic clostridia including *Clostridium perfringens*.

19.3.4 Mixed Acid Fermentation of Glucose with Production of Formate

In enterobacteria such as *Escherichia coli* or *Salmonella typhimurium* the anaerobic metabolism of glucose yields a mixture of acids (acetic, lactic, formic with some pyruvic and succinic) plus substantial evolution of gas (carbon dioxide with hydrogen). In other enterobacteria (*e.g.* species of *Shigella*) much more formate accumulates with consequential minimal production of gas. Glycolysis is accomplished chiefly by the Embden–Meyerhof pathway but with some consumption of glucose (15% in *E. coli*) *via* the hexose monophosphate pathway. The rate of glycolysis is in large measure controlled at two points in the Embden–Meyerhof pathway. The phosphofructokinase is activated by AMP or ADP and is inhibited by PEP (as a form of 'feedback control'). Furthermore, *E. coli* possesses two pyruvate kinases one of which is activated by fructose 1,6-diphosphate and the other by AMP. The pyruvate so formed is subject to several fates including the direct reduction to lactate which oxidizes 1 mol NADH per mol pyruvate reduced, or carboxylation and reduction to yield succinate by a route (*via* oxaloacetate, malate and fumarate) wherein 2 mol NADH are oxidized per mol pyruvate utilized. The acetic and formic acids have their origin in the cleavage of pyruvate by a pyruvate formate lyase and the formate is further liable to be cleaved to yield CO_2 plus H_2 by the action of a formate hydrogenlyase. The proportion of formate that is liable to yield H_2 in this manner is highly variable, depending in large measure on the prevailing pH, with the intracellular specific activity of formate hydrogenlyase being much increased in organisms grown at a low pH. Indeed, the culture pH substantially influences the products accumulated in these enterobacterial glucose fermentations, the general trend being towards the production of a higher proportion of neutral products as the pH is lowered. When ethanol is produced, it is generally formed by the reduction of acetyl-CoA. It has been reported that even during wholly anaerobic growth of *Escherichia coli* on glucose this organism may obtain additional ATP from anaerobic respiration in which fumarate serves as terminal electron acceptor. Some succinate could accumulate in consequence, but it should be remembered that this will be the product of a special membrane-associated fumarate reductase which is not to be confused with the succinate dehydrogenase responsible for the intracytoplasmic reduction of fumarate.

A biotechnologically interesting variation of the mixed acid fermentation is practised by bacteria such as *Enterobacter aerogenes* which produce acetoin (acetylmethylcarbinol). This product may, in turn, either be oxidized to give diacetyl or further reduced to yield 2,3-butanediol. In *Enterobacter aerogenes* the acetoin is produced from two molecules of pyruvate in a thiamine pyrophosphate-dependent reaction sequence which progresses *via* α-acetolactate (Figure 11). Again, the prevailing culture pH influences the mix of end-products formed by this fermentation. At alkaline pH values, acidic products such as lactate, formate and acetate predominate, whilst at an acidic pH (below pH 6.3) the neutral end-products, 2,3-butanediol, acetoin and ethanol, are produced. Other bacteria which produce 2,3-butanediol by anaerobic metabolism of glucose include *Bacillus polymyxa*, *Bacillus subtilis* and *Serratia marcescens*.

It may be noted that in lactic acid bacteria (such as *Leuconostoc citrovorum* and *Streptococcus diacetilactis*) any diacetyl that is produced is formed directly from the interaction of acetyl-CoA with 2-hydroxyethyl-TPP and is then subject to reduction to give acetoin.

19.3.5 Fermentation of Glucose Yielding Propionate

Bacteria which are notable for their production of propionic acid from glucose (*e.g.* species of *Propionibacterium*) also form some acetate and carbon dioxide in this fermentation. Indeed, to achieve a balanced oxidation–reduction process the simplest net conversion would be that shown in equation (2) and would proceed by the pathway outlined in Figure 12 to generate 2.66 mol ATP per mol of glucose utilized. The basis for propionate production in this pathway is the carboxylation of pyruvate to give oxaloacetate and the further conversion of this dicarboxylic acid through succinate and succinyl-CoA to methylmalonyl-CoA and thence to propionyl-CoA. Free energy is subtly conserved by effecting the conversion of pyruvate to oxaloacetate by the donation of a carboxyl group from the carboxybiotin complex that is generated in the penultimate decarboxylation of methylmalonyl-CoA. Further free energy is conserved by the employment of a CoA transferase to generate succinyl-CoA from succinate at the expense of conversion of propionyl-CoA to propionate. From a biochemical standpoint, the most intriguing reaction is the rearrangement of succinyl-CoA to give methylmalonyl-CoA which is catalysed by a B_{12}-dependent mutase. However, it cannot be presumed that whenever propionate is produced in the course of a bacterial fermentation it will have been formed by this route. For example, in the propionate-yielding fermentation of lactate that is undertaken by *Clostridium propionicum* or *Bacteroides ruminicola* propionyl-CoA has its origin in acrylyl-CoA formed from lactyl-CoA by the action of a dehydratase.

$$3 \text{ glucose} \rightarrow 4 \text{ propionate}^- + 2 \text{ acetate}^- + 6H^+ + 2 CO_2 + 2 H_2O \qquad (2)$$

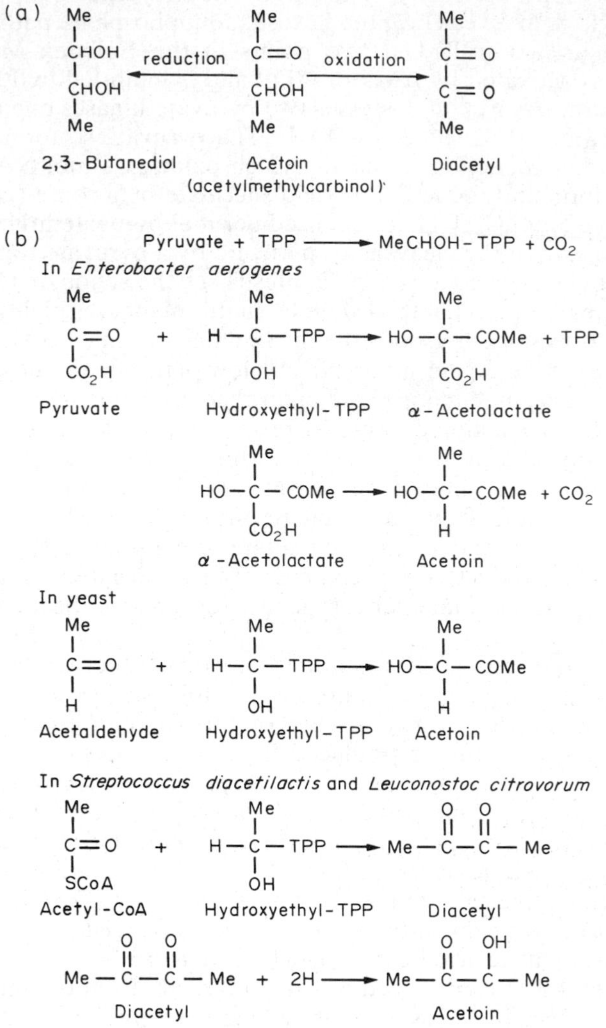

(a)

$$\underset{\text{2,3-Butanediol}}{\overset{\text{Me}}{\underset{\text{Me}}{\text{CHOH—CHOH}}}} \xrightarrow{\text{reduction}} \underset{\underset{\text{(acetylmethylcarbinol)}}{\text{Acetoin}}}{\overset{\text{Me}}{\underset{\text{Me}}{\text{C=O—CHOH}}}} \xrightarrow{\text{oxidation}} \underset{\text{Diacetyl}}{\overset{\text{Me}}{\underset{\text{Me}}{\text{C=O—C=O}}}}$$

(b)   Pyruvate + TPP ⟶ MeCHOH–TPP + CO_2

In *Enterobacter aerogenes*

$$\underset{\text{Pyruvate}}{\overset{\text{Me}}{\underset{\text{CO}_2\text{H}}{\text{C=O}}}} + \underset{\text{Hydroxyethyl-TPP}}{\overset{\text{Me}}{\underset{\text{OH}}{\text{H—C—TPP}}}} \longrightarrow \underset{\alpha\text{-Acetolactate}}{\overset{\text{Me}}{\underset{\text{CO}_2\text{H}}{\text{HO—C—COMe}}}} + \text{TPP}$$

$$\underset{\alpha\text{-Acetolactate}}{\overset{\text{Me}}{\underset{\text{CO}_2\text{H}}{\text{HO—C—COMe}}}} \longrightarrow \underset{\text{Acetoin}}{\overset{\text{Me}}{\underset{\text{H}}{\text{HO—C—COMe}}}} + CO_2$$

In yeast

$$\underset{\text{Acetaldehyde}}{\overset{\text{Me}}{\underset{\text{H}}{\text{C=O}}}} + \underset{\text{Hydroxyethyl-TPP}}{\overset{\text{Me}}{\underset{\text{OH}}{\text{H—C—TPP}}}} \longrightarrow \underset{\text{Acetoin}}{\overset{\text{Me}}{\underset{\text{H}}{\text{HO—C—COMe}}}}$$

In *Streptococcus diacetilactis* and *Leuconostoc citrovorum*

$$\underset{\text{Acetyl-CoA}}{\overset{\text{Me}}{\underset{\text{SCoA}}{\text{C=O}}}} + \underset{\text{Hydroxyethyl-TPP}}{\overset{\text{Me}}{\underset{\text{OH}}{\text{H—C—TPP}}}} \longrightarrow \underset{\text{Diacetyl}}{\overset{\text{O O}}{\text{Me—C—C—Me}}}$$

$$\underset{\text{Diacetyl}}{\overset{\text{O O}}{\text{Me—C—C—Me}}} + 2H \longrightarrow \underset{\text{Acetoin}}{\overset{\text{O OH}}{\text{Me—C—C—Me}}}$$

Figure 11   Microbial production and metabolism of acetoin: (a) acetoin may be oxidized to yield diacetyl, or reduced to give 2,3-butanediol; (b) three alternative routes for the biosynthesis of acetoin from pyruvate

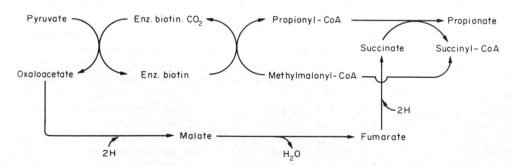

Figure 12   Route of reduction of pyruvate to propionate in species of *Propionibacterium*; when glucose is fermented the pyruvate and 'reducing power' in the form of NADH are generated by glycolysis *via* the Embden–Meyerhof pathway

An additionally interesting feature of the anaerobic metabolism of glucose in *Propionibacterium shermanii* is the ability of this organism to employ inorganic pyrophosphate in place of ATP in certain of its fermentation reactions. These include the formation of fructose 1,6-diphosphate in a reaction catalysed by a pyrophosphate-utilizing phosphofructokinase (equation 3). The inorganic pyrophosphate so consumed can be made in the reaction catalysed by the enzyme PEP-carboxytransphosphorylase (equation 4).

$$\text{fructose 6-phosphate} + PP_i \rightleftharpoons \text{fructose 1,6-diphosphate} + P_i \qquad (3)$$

$$PEP + P_i + CO_2 \rightleftharpoons \text{oxaloacetate} + PP_i \qquad (4)$$

19.3.6 Homoacetic Fermentation of Glucose

The fermentative pathway of glucose utilization by an anaerobic homoacetogen such as *Clostridium thermoaceticum* is outlined in Figure 13. By this route the bacterium is theoretically able to ferment 1 mol of glucose to give 3 mol of acetate. Glycolysis is accomplished by the Embden–Meyerhof pathway with generation of the usual amounts of NADH and ATP. Half of the resultant pyruvate is consumed in a transcarboxylation reaction involving a methylcorrinoid. The remainder of the pyruvate is presumed to be metabolized *via* a pyruvate-ferredoxin oxidoreductase to give acetyl-CoA, CO_2 and reduced ferredoxin. Acetyl-CoA is converted to acetate *via* acetyl phosphate and hence with the concurrent production of one molecule of ATP. The organism lacks hydrogenase and thus 'excess reducing power' cannot be disposed of as H_2. Instead, carbon dioxide serves as terminal electron acceptor in a complex sequence of reactions which ultimately leads to the production of acetate. The mechanism of the $C_1 + C_1$ condensation reaction has only recently been elucidated. The methyl group of the acetate has its origin in CO_2 and is formed by its reduction through a formate–tetrahydrofolate–corrinoid pathway which generates a methylcorrinoid protein. The methyl group is then combined with a 'C_1' unit which can have its origin either in the carboxyl group of pyruvate or in carbon monoxide (a nickel-containing carbon monoxide dehydrogenase being a key component of the multienzyme complex which catalyses this novel sequence of acetate-forming reactions). The important role of trace elements in the metabolism of *Clostridium thermoaceticum* is further emphasized by the finding that its NADPH-dependent formate dehydrogenase is a selenium tungsten, non-haem iron sulfur protein!

In this rather complex fermentation (Figure 13) oxidation of triose phosphate generates NADH and oxidation of pyruvate produces reduced ferredoxin. Conversion of CO_2 to formate is accomplished at the expense of oxidation of NADPH, whilst $FADH_2$ is the likely electron donor employed by the methylene tetrahydrofolate reductase. Thus the path of electron flow in this organism is particularly complicated and the possibility cannot yet be excluded that additional ATP synthesis might be accomplished in the course of some form of anaerobic 'respiratory' electron transport (Ljungdahl *et al.*, 1981).

19.4 REGULATION OF FERMENTATION PROCESSES

19.4.1 Cytoplasmic Fermentation Pathways

Several examples have already been provided of the manner in which the rate of flow of metabolites through a fermentation pathway may, in principle, be regulated. Control may be exercised on the synthesis of key enzymes, as in the case of the induced synthesis of acetoacetate decarboxylase in solventogenic cells of *Clostridium acetobutylicum*. Even more frequently the activities of existing enzymes can be modulated by the prevailing concentrations of key intermediates in the fermentation pathway, *e.g.* stimulation of the L(+)-lactate dehydrogenase of a homolactic streptococcus by fructose 1,6-diphosphate. When an organism provides more than one pathway for anaerobic glucose metabolism, the proportion of the glucose which is consumed by one pathway may be regulated by events in the other. Thus, *Streptococcus faecalis* besides operating the lactate yielding Embden–Meyerhof pathway also contains the hexose monophosphate pathway enzymes which oxidatively decarboxylate hexose phosphate to give pentulose phosphate. Glucose 6-phosphate thus occupies a pivotal position in glucose dissimilation in this organism, serving as it does as substrate for both of these pathways. Yet anaerobically the organism employs the hexose monophosphate pathway solely for biosynthetic purposes and the requirements of the organism are such that a much greater proportion of the glucose should be channelled *via* the ATP-yielding Embden–Meyerhof pathway than *via* the hexose monophosphate pathway to produce biosynthetic precursors (pentose phosphates, erythrose phosphate and NADPH). The necessary partition is nicely accomplished by the agency of the intracellular concentration of fructose 1,6-diphosphate, for this intermediate in the Embden–Meyerhof pathway both (1) activates the 'fermentative' lactate dehydrogenase and (2) inhibits the activity of the hexose monophosphate pathway

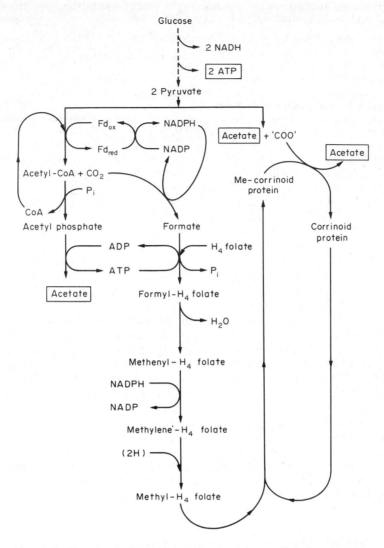

Figure 13 Homoacetic fermentation of glucose in *Clostridium thermoaceticum*

enzyme 6-phosphogluconate dehydrogenase. Whereas the phosphofructokinase of *Streptococcus faecalis* (unlike that of *Escherichia coli*) is not inhibited by ATP, inhibition of glucose 6-phosphate isomerase by 6-phosphogluconate accomplishes a comparable regulatory role by preventing the accumulation of fructose 6-phosphate. Under conditions when the demand for ATP is high, these controls channel glucose 6-phosphate preferentially through the Embden–Meyerhof pathway (Brown and Wittenberger, 1971).

Many other examples could be given of the ways in which by regulation of the activities of the cytoplasmic enzymes of fermentation, differential usage of 'competing' pathways and changes in overall rate of substrate consumption can be determined. However, there are two additional potential 'bottlenecks' in a fermentation which may control its overall rate, namely the rate of uptake of the fermentation substrate and the rate(s) of efflux of the fermentation products (Wood, 1981).

19.4.2   Transport of Glucose into Fermentative Bacteria

Active transport of sugars into bacterial cells can be accomplished by several mechanisms, each of which involves the participation of specific transport proteins located in the cytoplasmic membrane. These are of two main types, *viz.* those which vectorially translocate the unchanged substrate molecules and those which transport the substrate in a modified form. As an example of the former, glucose can be actively accumulated by *Escherichia coli via* the 'shock-sensitive', binding, protein-dependent β-methylgalactoside transport system. In other bacteria, glucose uptake is impelled by the transmembrane proton (electrochemical) gradient, being accomplished by cotransport with protons. Indeed, in *E. coli*, glucose 6-phosphate may also be cotransported with protons by the agency of an inducible hexose phosphate transport system. However, in the majority of facultative and obligate anaerobes which metabolize glucose by the Embden–Meyerhof pathway, uptake of glucose is accomplished by a 'group translocation mechanism' *via* a phosphotransferase system (PTS). Transport of glucose is associated with its concomitant phosphorylation in a process wherein phosphoenolpyruvate (PEP) serves as the primary phosphate donor. The result is the entry of glucose into the cytoplasm in the form of glucose 6-phosphate. The PTS system (Figure 14) consists of a number of cooperative enzyme components. Two of these are non-sugar specific proteins (enzyme I and HPr) that initiate the sequence of phosphorylation reactions and are relatively loosely associated with the cytoplasmic membrane (as peripheral proteins). The sugar-specific proteins are enzymes II and III of which enzyme II is integrated into the cytoplasmic membrane. The mechanism of the reaction sequence is such that phosphate is not transferred directly from PEP to the sugar but is sequentially transferred down the phosphate transfer chain of the PTS, first from PEP to enzyme I then to HPr and then to the sugar-specific enzyme III (Saier, 1977). It is the membrane integrated enzyme II complex that serves as the appropriate sugar recognition component of the system, so that an enzyme II^{glc} binds glucose for a PTS system, whose enzyme I and HPr components could additionally be concerned with the uptake of other sugars. The distribution of the glucose PTS system in fermentative bacteria has been surveyed (Romano *et al.*, 1979). It was found to operate in all homofermentative lactic acid bacteria that metabolize glucose by the Embden–Meyerhof pathway but in none of a group of heterofermentative species of *Lactobacillus* and *Leuconostoc* which ferment glucose *via* the phosphoketolase pathway. It was also not possessed by *Zymomonas mobilis* which, as we have seen, metabolizes glucose by the Entner–Doudoroff pathway. It was thus concluded that the glucose-PTS is seemingly restricted to bacteria which ferment glucose by the Embden–Meyerhof pathway.

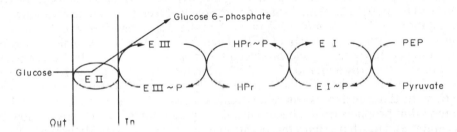

Figure 14 Diagrammatic representation of the phosphoenolpyruvate (PEP)-dependent glucose phosphotransferase system; the enzyme II component is shown as spanning the bacterial cell membrane; externally supplied glucose is transported into the bacterial cytoplasm in a modified form (*i.e.* as glucose 6-phosphate)

Although a fructose-PTS has been discovered in a number of strictly aerobic bacteria, and also in several photosynthetic bacteria, the metabolism of fructose in these organisms also proceeds either exclusively, or in part, *via* the Embden–Meyerhof pathway. An interesting example is provided by *Rhodopseudomonas sphaeroides* which lacks fructose 6-phosphate kinase and so cannot metabolize glucose *via* the Embden–Meyerhof pathway (it is, instead, glycolysed *via* the Entner–Doudoroff pathway). However, the fructose-PTS of this organism produces fructose 1-phosphate for which there is a specific kinase which catalyses its conversion to fructose 1,6-diphosphate and so allows it to be metabolized *via* the Embden–Meyerhof pathway.

It is reasonable to suppose that the correlation of (1) reliance on a glucose-PTS transport system with (2) subsequent utilization of the glucose 6-phosphate *via* the Embden–Meyerhof pathway, can be explained by the greater net yield of PEP that is obtained from this glycolytic pathway than from any alternative route of glucose metabolism. Two advantages can be perceived in this arrangement. First, it provides for a tight linkage between the transport of glucose and its subsequent metabolism, conferring upon the intracellular concentration of PEP an important role in regulating glucose utilization (at source). Second, it allows for the conservation of ATP (though not for enhanced conservation of free energy overall) since the product of the transport event is glucose 6-phosphate which can directly be partitioned between catabolic and anabolic pathways. However, it should be noted that some anaerobes which operate a complete

Embden–Meyerhof pathway do not take up glucose by a PTS mechanism, *e.g. Bacteroides the-taiotaomicron*. Furthermore, possession of a glucose-PTS by an anaerobic bacterium does not mean that all sugars will necessarily be taken up by the PTS mechanism. Thus, whilst *Clostridium pasteurianum* transports glucose and fructose by a PTS system, galactose is taken up by a proton symport mechanism.

At least five regulatory mechanisms have been elucidated which appear to account for the controlled rates of uptake of sugars *via* the various transport systems and *inter alia* the preferences expressed by bacteria presented with a choice of sugars (Dills *et al.*, 1980). Amongst these control mechanisms is the inhibition of uptake of one sugar substrate by another, due to competition of the corresponding enzyme II complexes for the common phosphoryl protein of the PTS. Sugar uptake may also be sensitive to the intracellular concentrations of sugar phosphate, whilst the activity of a proton motivated sugar transport system may be responsive to the transmembrane proton electrochemical gradient (PMF) in either a positive or a negative sense.

19.4.3 Efflux of Fermentation Products

Gaseous products (CO_2 and H_2) and relatively lipophilic products of low molecular weight are able to diffuse freely out of the producer organism into the surrounding medium along their concentration gradients. It is likely that this is the mechanism of efflux of solvents such as ethanol, butanol and acetone and again (and especially at more acid culture pH values) of short chain fatty acids such as formic, acetic, propionic and butyric acids, which permeate the cytoplasmic membrane in their undissociated forms. However, an active efflux process might be required to expel other more hydrophilic compounds to which the cell membrane is not freely permeable. This has been shown to be the case with the lactate produced by fermentation of glucose by *Streptococcus cremoris* (Otto *et al.*, 1980). Lactate is cotransported with protons from this organism, in a process which above pH 6.2 effects net export of protons. The outward transport of lactate at high external pH values thus contributes to the maintenance of a transmembrane electrochemical gradient of protons, thus 'sparing' some fermentatively-generated ATP which would otherwise have to be utilized for this purpose. The result is that the efficiency of free energy conservation by the homo-lactic fermetation in *Strep. cremoris* is greater than that represented by the 2 mol ATP per mol glucose associated with the operation of its Embden–Meyerhof pathway. This is confirmed by the greater than expected biomass yield produced by *Strep. cremoris* growing in chemostat culture under conditions of energy limitation (*i.e.* glucose limitation). It will be interesting to see whether the active efflux of any other products can similarly contribute to free energy conservation in fermentation. In those instances wherein specific transport proteins are employed to discharge such fermentation products from a bacterial cell the possibility exists for this outflow to be regulated. Whether, and to what extent, this occurs in practice has yet to be determined.

19.5 EFFICIENCY OF BIOMASS GENERATION BY ORGANISMS WHICH FERMENT GLUCOSE

It was initially thought that the yield of biomass per mol of substrate fermented (molar growth yield) should prove to be easily predictable once the pathway of fermentation had been discovered and the associated net yield of ATP (mol mol^{-1} substrate utilized) had been deduced. The example provided by the free energy conservation additionally attributable to lactate excretion by *Strep. cremoris* demonstrates that the calculation of the actual 'ATP equivalents' produced by an organism may not, in all cases, be so straightforward as was once imagined. This is even more true of the organism which is capable of supplementing ATP derived from substrate level phosphorylation with ATP generated by anaerobic electron transport-linked phosphorylation (Stouthamer, 1980). Furthermore, complications may arise, even in 'true' fermentations from subtle changes in end-product balances being induced by changes in environmental conditions. Thus, in the presence of nitrate serving as a non-respiratory but alien electron acceptor the acidogenic fermentation of glucose by *Clostridium perfringens* produced more acetate and less butyrate than in the absence of nitrate, with a consequential higher net specific yield of ATP reflected in an elevated biomass yield on a limiting concentration of glucose.

Even greater uncertainty is introduced into any attempt to predict the biomass derivable from the fermentation of a given amount of glucose, by the fact that a substantial fraction of the free energy conserved by a bacterium is then utilized in tasks other than (net) biosynthesis. The magnitude of this fraction is not constant but may vary with the growth rate of the organism and, for other reasons, with the prevailing culture conditions. In obligately fermentative anaerobes, membrane energization is sustained by hydrolysis of ATP at the membrane-located, proton motive ATP hydrolase (H^+-ATPase). This process probably accounts for a substantial part of the conserved free energy which is not thereafter used for the generation of new biomass. However, it is also clear that bacteria may possess several means of disposing of ATP when this is generated at a rate that exceeds immediate requirements. Aside from the temporary accumulation of some endogenous energy reserve polymer (Dawes and Senior, 1973), ATP might be consumed by the operation of some futile metabolic cycle (Neijssel and Tempest, 1976) or, more directly, by ATP hydrolysis at the membrane ATPase which is 'uncoupled' from proton expulsion. This last mechanism is an example of 'slippage' which may, in principle, occur in diverse ways within proton motivated energy-coupling systems so that divergence between theoretical and actual values of Y_{ATP}^{max} (the maximum biomass yield mol^{-1} of ATP generated in the fermentation) may, in large measure, be ascribed to diversion of ATP equivalents to partially abortive energization of the bacterial cell membrane. In some cases, at least, the magnitude of such slippage could be influenced by environmental conditions. Recognition of such a possibility counsels caution when any attempt is made to assess relative efficiencies of various fermentations on the basis solely of net ATP yield without regard to the fate of that ATP.

19.6 POSTSCRIPT

We have considered in this chapter pathways of anaerobic glucose metabolism as they present themselves in various species of microorganisms when each is grown in pure culture on glucose as sole source of carbon and energy. In nature such monocultures are, of course, somewhat exceptional. In processes of anaerobic digestion, glucose is frequently produced by the catabolism of some carbohydrate polymer and thereafter is metabolized by a mixed community of microbes. The fermentation product of one organism might prove a suitable substrate for growth of another; for example, the lactate produced by a streptococcus might then be fermented by a species of *Veillonella* or anaerobically respired by a species of *Desulfovibrio*. Such syntrophic associations can form the basis of stable consortia of anaerobic microbes (see Bull and Quayle, 1982). A particularly important example of syntrophy amongst anaerobes is provided by the close microbial associations forged by interspecies transfer of hydrogen (Wolin, 1982). Avid scavenging of H_2 by hydrogen-utilizing bacteria (methanogens, sulfate reducers, *etc.*) can significantly affect the fermentations accomplished by hydrogen-producing anaerobes with which they are cocultured. Efficient removal of gaseous hydrogen tends to 'short circuit' the normal electron flow in the hydrogen producer, with the result that it yields products that are more highly oxidized than those that it forms in the absence of the hydrogen scavenger. Thus, whereas in monoculture *Clostridium cellobioparum* ferments glucose to produce ethanol, formate, acetate, CO_2 and H_2, when it is grown in coculture with a H_2-utilizing anaerobe its glucose fermentation gives much acetate but a greatly diminished yield of ethanol. Conversely, growth of species of anaerobes which possess a 'reversible' hydrogenase in the presence of elevated concentrations of hydrogen can be expected to enhance the yield of their more reduced products of fermentation.

The following review articles are commendable sources of further information on anaerobic metabolism of glucose by pure and mixed cultures of microbes and the prospects for biotechnological exploitation of some of these processes: Bu'Lock, 1979; Esser and Schmidt, 1982; Flickinger, 1980; Gottschalk and Andreesen, 1979; Jones *et al.*, 1981; Mah, 1982; Tewes and Thauer, 1980.

19.7 REFERENCES

Bahl, H., W. Andersch, K. Braun and G. Gottschalk (1982). Effect of pH and butyrate concentration on the production of acetone and butanol by *Clostridium acetobutylicum* grown in continuous culture. *Eur. J. Appl. Microbiol. Biotechnol.*, **14**, 17–20.

Brown, A. T. and C. L. Wittenberger (1971). Mechanism for regulating the distribution of glucose carbon between the Embden–Meyerhof and hexose monophosphate pathways in *Streptococcus faecalis. J.Bacteriol.*, **106**, 456–467.

Bull, A. T. and J. R. Quayle (1982). New dimensions in microbiology: an introduction. *Philos. Trans. R. Soc. London, Ser. B*, **297**, 447–457.

Bu'Lock, J. D. (1979). Industrial alcohol. In *Microbial Technology: Current State, Future Prospects, 29th Symposium of the Society for General Microbiology*, ed. A. T. Bull, D. C. Ellwood and C. Ratledge, pp. 309–325. Cambridge University Press, Cambridge.

Dawes, E. A. and P. J. Senior (1973). The role and regulation of energy-reserve polymers in micro-organisms. *Adv. Microb. Physiol.*, **10**, 135–266.

Dills, S. S., A. Apperson, M. R. Schmidt and M. H. Saier, Jr. (1980). Carbohydrate transport in bacteria. *Microbiol. Rev.*, **44**, 385–418.

Esser, K. and U. Schmidt (1982). Alcohol production by biotechnology. *Process Biochem.*, **17**, 46–49.

Fiechter, A., G. F. Fuhrmann and O. Käppeli (1981). Regulation of glucose metabolism in growing yeast cells. *Adv. Microb. Physiol.*, **22**, 123–183.

Flickinger, M. C. (1980). Current biological research in conversion of cellulosic carbohydrates into liquid fuels: how far have we come? *Biotechnol. Bioeng.*, **22**, Suppl. 1, 27–48.

Fraenkel, D. G. and R. T. Vinopal (1973). Carbohydrate metabolism in bacteria. *Annu. Rev. Microbiol.*, **27**, 69–100.

Garvie, E. I. (1980). Bacterial lactate dehydrogenases. *Microbiol. Rev.*, **44**, 106–139.

Gottschal, J. C. and J. G. Morris (1981). The induction of acetone and butanol production in cultures of *Clostridium acetobutylicum* by elevated concentrations of acetate and butyrate. *FEMS Microbiol. Lett.*, **12**, 385–389.

Gottschal, J. C. and J. G. Morris (1982). Continuous production of acetone and butanol by *Clostridium acetobutylicum* growing in turbidostat culture. *Biotechnol. Lett.*, **4**, 477–482.

Gottschalk, G. and J. R. Andreesen (1979). Energy metabolism in anaerobes. In *Microbial Biochemistry*, ed. J. R. Quayle, vol. 21, pp. 85–115. University Park Press, Baltimore.

Haddock, B. A. and C. W. Jones (1977). Bacterial respiration. *Bacteriol. Rev.*, **41**, 47–99.

Jones, R. P., N. Pamment and P. F. Greenfield (1981). Alcohol fermentation by yeasts—the effect of environmental and other variables. *Process Biochem.*, **16**, 42–49.

Lauer, E. and O. Kandler (1976). Mechanismen der variation des verhältnisses acetat/lactat bei der vergärung von glucose durch bifidobakterien. *Arch. Microbiol.*, **110**, 271–277.

Lawrence, R. C. and T. C. Thomas (1979). The fermentation of milk by lactic acid bacteria. In *Microbial Technology*: *Current State, Future Prospects*, 29th Symposium of the Society for General Microbiology, ed. A. T. Bull, D. C. Ellwood and C. Ratledge, pp. 187–219. Cambridge University Press, Cambridge.

Ljungdahl, L. G., F. Bryant, L. Carreira, T. Saiki and J. Wiegel (1981). Some aspects of thermophilic and extreme thermophilic anaerobic microorganisms. In *Trends in the Biology of Fermentations for Fuels and Chemicals*, ed. A. Hollaender *et al.*, pp. 397–419. Plenum, New York.

London, J. (1976). The ecology and taxonomic status of the lactobacilli. *Annu. Rev. Microbiol.*, **30**, 279–301.

Mah, R. A. (1982). Methanogenesis and methanogenic partnerships. *Philos. Trans. R. Soc. London, Ser. B*, **297**, 599–616.

Neijssel, O. M. and D. W. Tempest (1976). The role of energy-spilling reactions in the growth of *Klebsiella aerogenes* NCTC 418 in aerobic chemostat culture. *Arch. Microbiol.*, **110**, 305–311.

Otto, R., A. S. M. Sonnenberg, H. Veldkamp and W. N. Konings (1980). Generation of an electrochemical proton gradient in *Streptococcus cremoris* by lactate efflux. *Proc. Natl. Acad. Sci. USA*, **77**, 5502–5506.

Rogers, P. L., K. J. Lee and D. E. Tribe (1980). High productivity ethanol fermentations with *Zymomonas mobilis*. *Process Biochem.*, **15** (6), 7–11.

Romano, A. H., J. D. Trifone and M. Brustolon (1979). Distribution of the phosphoenolpyruvate: glucose phosphotransferase system in fermentative bacteria. *J. Bacteriol.*, **139**, 93–97.

Saier, M. H., Jr (1977). Bacterial phosphoenolpyruvate: sugar phosphotransferase systems: structural, functional and evolutionary interrelationships. *Bacteriol. Rev.*, **41**, 856–871.

Scardovi, V. (1981). The genus *Bifidobacterium*. In *The Prokaryotes*, ed. M. P. Starr *et al.*, vol. 2, pp. 1951–1961. Springer–Verlag, Berlin.

Sols, A. (1976). The Pasteur effect in the allosteric era. In *Reflections on Biochemistry*, ed. A. Kornberg *et al.*, pp. 199–206. Pergamon, Oxford.

Spivey, M. J. (1978). The acetone, butanol, ethanol fermentation. *Process Biochem.*, **13**, 2–4, 25.

Stouthamer, A. H. (1980). Electron transport-linked phosphorylation in anaerobes. In *Anaerobes and Anaerobic Infections*, ed. G. Gottschalk, N. Pfennig and H. Werner, pp. 17–29. Gustav Fischer Verlag, Stuttgart.

Tewes, F. J. and R. K. Thauer (1980). Regulation of ATP-synthesis in glucose fermenting bacteria involved in interspecies hydrogen transfer. In *Anaerobes and Anaerobic Infections*, ed. G. Gottschalk, N. Pfennig and H. Werner, pp. 97–104. Gustav Fischer Verlag, Stuttgart.

Thauer, R. K., K. Jungermann and K. Decker (1977). Energy conservation in chemotrophic anaerobic bacteria. *Bacteriol. Rev.*, **41**, 100–180.

Wiegel, J. (1980). Formation of ethanol by bacteria. A pledge for the use of extreme thermophilic anaerobic bacteria in industrial ethanol fermentation processes. *Experientia*, **36**, 1434–1446.

Wolin, M. J. (1964). Fructose 1,6-diphosphate requirement of streptococcal lactic dehydrogenases. *Science*, **146**, 775–777.

Wolin, M. J. (1982). Hydrogen transfer in microbial communities. In *Microbial Interactions and Communities*, ed. A. T. Bull and J. H. Slater, vol. 1, pp. 323–356. Academic, London.

Wood, W. A. (1961). Fermentation of carbohydrates and related compounds. In *The Bacteria*, ed. I. C. Gunsalus and R. Y. Stanier, vol. 2, pp. 59–149. Academic, New York.

Wood, W. A. (1981). Basic biology of microbial fermentation. In *Trends in the Biology of Fermentations for Fuels and Chemicals*, ed. A. Hollaender *et al.*, pp. 3–17. Plenum, New York.

Zeikus, J. G. (1980). Chemical and fuel production by anaerobic bacteria. *Annu. Rev. Microbiol.*, **34**, 423–464.

20

Aerobic Metabolism of Methane and Methanol

D. I. STIRLING
Celanese Research Corporation, Summit, NJ, USA
and
H. DALTON
University of Warwick, Coventry, UK

20.1 INTRODUCTION

Over the past 15 years considerable effort has been expended on the understanding of how microorganisms grow on methane and methanol. Much of this interest has been engendered by industry, which has actively pursued the use of these substrates for single cell protein (SCP) production using either bacteria or yeasts. Although methane-based SCP processes have been abandoned in the West in favor of methanol processes, a number of operations are still being actively pursued in the Soviet Union.

One of the main reasons for developing SCP processes based on these C_1 compounds has been the ready availability of the substrates. Methane, an end product of anaerobic degradation of organic material by microorganisms, is being constantly produced and released into the atmosphere. It has been estimated (Ehhalt, 1976) that such sources produce around 528 to 812×10^{12} g methane per year which is matched by the output from natural gas wells. Furthermore, there are very large proven resources of methane gas under the Rocky Mountains in the USA as well as some speculative reports (see Paterson, 1978) that non-biogenic resources of methane gas inside the earth could sustain our present level of hydrocarbon consumption for one million years!

Methanol, on the other hand, is made chemically from synthesis gas by several processes. The preferred method is by the ICI low pressure process at around 1000 psi and 250 to 300 °C using a copper, aluminum–zinc catalyst (equation 1). The source of synthesis gas is usually methane although virtually anything containing carbon and hydrogen can be partially oxidized to carbon monoxide and hydrogen. At present, the worldwide supply and demand for methanol are in balance at around 12 million tonnes per annum but forecasts (see Cameron, 1982) indicate that demand for methanol may well increase by 63% by 1990. This demand could, for example, be met by converting much of the methane gas, which is flared off at the well site in many oil-producing countries, into methanol although there are many considerations (mostly economic) to take into account before this becomes a reality.

$$2\,H_2 + CO \rightarrow MeOH \qquad\qquad (1)$$
$$\text{synthesis gas}$$

It is quite clear, however, that both methane and hence methanol will be available in large amounts for many years to come and consequently will provide a cheap starting material for a number of biotechnological processes. Despite the unquestionable availability and relative cheapness of C_1 substrates, it is only in recent years that active research into the biology of methane- and methanol-utilizing microbes has been undertaken. Much of this has been presented recently in a very comprehensive book on the subject (Anthony, 1982) and several review articles (Colby *et al.*, 1979; Wolfe and Higgins, 1979; Higgins *et al.*, 1981) and a symposium proceedings (Dalton, 1981). The reader is referred to these articles for a detailed overview of the subject. Consequently, this chapter will only attempt to bring together what is known about these organisms if one is to be able to exploit their activities either as a source of single cell protein or as biocatalysts for converting cheap substrates into more expensive products.

20.2 DEFINITIONS

The general term given to microorganisms that use as their sole carbon source compounds more reduced than carbon dioxide which contain no carbon–carbon bonds is methylotroph (Colby and Zatman, 1972). Those organisms that can grow on methane as a sole carbon source are methanotrophs. Clearly, methanol-utilizing organisms are methylotrophs and methane-utilizers which will also grow on methanol are both methanotrophs and methylotrophs. This definition of methylotrophs also encompasses organisms that grow on C_1 compounds such as methylamine, formaldehyde, formate, formamide and carbon monoxide as well as compounds such as dimethylether, di- and tri-methylamine and trimethylamine *N*-oxide which contain more than one carbon atom but have no carbon–carbon bonds.

20.3 ECOLOGY AND DISTRIBUTION OF METHANE AND METHANOL UTILIZERS

All the methane oxiders and the vast majority of methanol-utilizing organisms are Gram-negative aerobes. The only exceptions to this are the Gram-positive methanol utilizers *Bacillus cereus* MM-33-1 (Akiba *et al.*, 1970), *Streptomyces* 239 (Kato *et al.*, 1974) and *Mycobacterium vaccae* (Loginova and Trotsenko, 1977) and the anaerobic methanol-utilizing budding bacterium *Hyphomicrobium* (Hirsch and Conti, 1964) which uses nitrate as its terminal electron acceptor when growing in the absence of oxygen.

20.3.1 Methane Utilizers

The ability to use methane as a sole carbon and energy source is restricted to the prokaryotes although there have been some fascinating reports from Wolf and Hanson (1979, 1980) and Wolf

(1981) that five yeast strains from the *Rhodotorula* and *Sporobolomyces* genera would grow on methane in a minimal medium with generation times of between 48 and 170 h. The methane oxidation rate of these cells was between 0.3 and 3 nmol CH_4 oxidized min^{-1} mg dry weight $cell^{-1}$. The corresponding figures for a typical obligate methanotroph are about 4 h and 200 nmol CH_4 oxidized min^{-1} mg dry weight $cell^{-1}$, respectively. The poor growth rates coupled with the reduced growth yields when attempts were made to grow the yeasts in liquid volumes greater than 15 ml do not make these organisms very attractive candidates for further study and, to the authors' knowledge, there have not been any other reports of yeasts capable of growth on methane. There have, however, been several reports of other eukaryotes being capable of methane oxidation but neither the report on green algae (Enebo, 1967) nor fungi (Zajic *et al.*, 1969) have been substantiated by other research groups.

As a result of the studies by Whittenbury *et al.* (1970), it is quite clear that the obligate methane-utilizing bacteria can be readily isolated from a wide variety of habitats. The obligate methane-utilizers were capable of growth on C_1 substrates only (generally methane and methanol) and were incapable of growth on substrates containing more than one carbon atom. Only a few strains grow well on methanol as isolated, but 'training' to tolerate increasing concentrations of the substrate is possible for several strains by the method of Hou *et al.*, (1979). They are capable of growth between 20 and 45 °C although one strain has been isolated which grows at 62 °C (Malashenko, 1976). All of the strains appear to have a fairly complex internal membrane structure and form either exospores or *Azotobacter*-like cysts. A classification scheme, based on morphological and biochemical characteristics, has been proposed (Whittenbury and Dalton, 1981) which delineates the organisms into three groups (Table 1). For the purposes of this chapter the classification scheme of Colby *et al.* (1979) will be presented since this also included a subgroup of facultative methanotrophs (Table 1). Since this classification scheme was proposed it has been reported (Murrell and Dalton, 1983) that the three types of methane oxidizers can also be differentiated by their mode of inorganic nitrogen metabolism and these results are incorporated in the table. Apart from *Methylococcus capsulatus* and *Methylomonas methanica* which appear in Bergey's Manual, none of the names of these organisms can be regarded as legitimate and should, therefore, be referred to in quotation marks.

Table 1 Tentative Classification Scheme for Methane-oxidizing Bacteria[a]

Determinants	Type I		Type II	
Membrane arrangement	Bundles of vesicular disks		Paired membranes around cell periphery	
Resting stages	Cysts (*Azotobacter*-like)		Exospores or lipid cysts	
Major carbon assimilation pathway	RMP (3-hexulosephosphate synthase)		Serine pathway (hydroxypyruvate reductase + 3-hexulosephosphate synthase)	
TCA cycle	Incomplete (2-oxoglutarate dehydrogenase negative)		Complete	
Nitrogenase	Some +		+	
Predominant fatty acid C chain length	16		18	
	Subgroup A	*Subgroup B*	*Subgroup obligate*[b]	*Subgroup facultative*[c]
Autotrophic CO_2 fixation	−	+		
DNA base ration (%G + C)	50–54	62.5	62.5 + (where tested)	
Isocitrate dehydrogenase	NAD or NADP dependent	NAD dependent	NADP dependent	
Cell shape	Rod and coccus	Coccus	Rod and vibrio	
Growth at 45 °C	Some +	+	−	
Presence of glutamate dehydrogenase when grown on ammonia	Present	absent (uses alanine dehydrogenase)	Absent (uses GS/GOGAT)	
Examples	*Methylomonas methanica* and *Methylomonas albus*	*Methylococcus capsulatus*	*Methanomonas methanooxidans*, *Methylosinus trichosporium* (both obligate) and *Methylobacterium organophilum* (facultative)	

[a] Not all strains classifiable into Type I and Type II have been shown to possess all the biochemical characteristics outlined in this scheme.
[b] Use methanol and formaldehyde as carbon and energy source, but not C_2+ compounds.
[c] Use variety of organic compounds, *e.g.* glucose, as carbon and energy source.

Following the elegant studies by Whittenbury on isolation procedures for the obligate methane utilizers, Patt *et al.* (1974) isolated a facultative methane-oxidizing bacterium from Lake Mendota, Wisconsin, USA which was also able to grow much better on a variety of multicarbon substrates. The organism was aptly named '*Methylobacterium organophilum*' XX but readily lost the ability to grow on methane after subculture in methanol or multicarbon substrates. This characteristic suggested that the ability to grow on methane could be explained by the possession of a plasmid and, indeed, a plasmid of molecular weight 8×10^7 was identified in methane grown on cells which was not present in cells grown on other substrates (see Schilling, 1978, in Hanson, 1980). Since 1974, workers at Exxon, New Jersey (Patel *et al.*, 1978) and Seattle, Washington (Lynch *et al.*, 1980) have also reported the isolation of facultative methanotrophs. However, very recently the existence of these facultative methanotrophs has been seriously questioned by the Seattle workers themselves (Lidstrom-O'Connor *et al.*, 1983). The facultative methanotrophic culture named as '*Methylobacterium ethanolicum*' H4-14 has now been shown to consist of a stable mixture of two methylotrophic organisms; one an obligate methanotroph considered to be a '*Methylocystis*' species (strain POC) and the other a *Xanthobacter* species (strain H4-14) which grows on $H_2 + CO_2$, methanol and multicarbon substrates. The POC strain was shown to contain three plasmids although their function is unknown. Their presence in POC only could explain why such plasmids were detected when other 'facultative methanotrophs' were grown on methane but not when grown on multicarbon substrates; under the latter conditions, the methanotroph would be present in very low amount and the plasmids would be virtually undetectable and could, therefore, give the false impression that the enzyme responsible for the initial oxidation of methane to methanol was plasmid encoded. It remains to be seen whether the other facultative methanotrophic cultures are axenic or mixed populations.

20.3.2 Methanol Utilizers

The methanol utilizers can be found in a wide variety of taxonomic groupings including the Gram-negative and Gram-positive bacteria, yeasts, filamentous fungi and the actinomycetes. Consequently, it has become convenient to distinguish them from one another largely in terms of their physiological and biochemical activities. In this respect, one can classify the methanol utilizers into two basic categories: the obligate methanol utilizers and the facultative methanol utilizers (Byrom, 1981). Anthony (1982) has included what Colby and Zatman (1975) called restricted facultative methylotrophs (Type M) into the obligate group because the Type M organisms are able to grow only very slowly on a few multicarbon compounds and are therefore, for all intents and purposes, obligate in nature. Furthermore, the less restricted facultative methylotrophs (Type L) of Colby and Zatman (1975) were included in the facultative methylotrophs because they grew well on a limited range of multicarbon compounds.

In many respects, the obligate methanol utilizers are similar to the methanotrophs but differ in their inability to use methane as a carbon source for growth, their lack of complex internal membrane structures, inability to fix N_2 and lack of differentiated resting bodies. The obligate methylotrophs are, in fact, very similar to one another in that they are aerobic, non-pigmented Gram-negative rods and are motile by means of a single polar flagellum. They all assimilate carbon *via* the ribulose monophosphate pathway (see Section 20.9) and have a (G + C) ratio of between 52% and 56%. However, it is possible, based on DNA/DNA hybridization experiments, to distinguish two groups within the obligate methanol utilizers (Byrom, 1981). It has been suggested that the name *Methylomonas* be reserved for obligate methane utilizers (Yordy and Weaver, 1977) and that the obligate methanol utilizers be given the generic names *Methylophilus* and *Methylobacillus* depending on their DNA hybridization group (Byrom, 1981).

The facultative methylotrophs are a much more diverse group of organisms since they include both Gram-negative, Gram-positive and Gram-variable representatives and show a variety of morphologies, physiologies and pigmentation groups. The Gram-negative methylotrophs can be subdivided into several groups. The pink pigmented facultative methylotrophs (PPFMs) which assimilate methanol *via* the serine pathway form red pigments giving rise to their characteristic pink colonies on solid media. They are aerobic large rods, catalase- and oxidase-positive and usually motile by a single polar flagellum. Most of them have been tentatively classified as *Pseudomonas* species but many of their features are not typical of this genus (Green and Bousefield, 1981). The non-pigmented organisms which have also been classified under the *Pseudomonas* genus differ from the PPFMs only in lack of pigmentation. It appears that the easiest way to isolate the PPFMs is to use methanol as the carbon source whereas the non-pigmented forms are readily isolated on methylated amines although some will grow on methanol.

One group of Gram-negative facultative methylotrophs, the stalked hyphomicrobia, have the characteristic of anaerobic growth on methanol using nitrate as the terminal electron acceptor. In this respect, they share a common feature with *Paracoccus denitrificans* (Cox and Quayle, 1975) although the mode of carbon fixation differs in that *Hyphomicrobium* uses a serine pathway to assimilate formaldehyde whereas *Paracoccus* completely oxidizes methanol to carbon dioxide and fixes the carbon *via* the Calvin cycle.

There are a number of facultative methylotrophs that grow on methanol or formate by oxidizing the substrate to carbon dioxide and then refixing the CO_2 *via* the ribulosebisphosphate carboxylase-driven Calvin cycle. These include the phototrophic bacteria of the genus *Rhodopseudomonas* which can grow on ethanol either anaerobically in the light (Quayle and Pfennig, 1975) or aerobically in the dark (Pfennig, 1979) as well as a wide variety of non-photosynthetic autotrophs such as *Thiobacillus novellus* (Chandra and Shethna, 1977) *Microcyclus aquaticus* (Loginova *et al.*, 1978) and *Blastobacter viscosus* (Loginova and Trotsenko, 1979).

The Gram-positive organisms that grow on methanol are rather poorly documented and little biochemical information is available on them. These include *Bacillus cereus* M-33-1 (Akiba *et al.*, 1970), *Streptomyces* 239 (Kato *et al.*, 1974) and *Mycobacterium vaccae* (Loginova and Trotsenko, 1977).

Of the eukaryotes that can grow on methanol, only three species of mycelial fungi have been described: *Gliocladium deliquescens* (Sakaguchi *et al.*, 1975), *Paecilomyces varioti* (Sakaguchi *et al.*, 1975) and *Trichoderma lignorum* (Tye and Willets, 1973). The attention paid to these eukaryotes pales in comparison with studies on the methanol-utilizing yeasts which have been reviewed quite comprehensively by Sahm (1977), Tani *et al.*, (1978), van Dijken *et al.*, (1981) and Anthony (1982). The first isolation of a methanol yeast was reported by Ogata *et al.*, (1969) which was ascribed to the genus *Kloeckera* sp. 2201 but may be a strain of *Candida boidinii* (Komagata, 1981). Since then, six more genera of yeast have been shown to have methylotrophic representatives but most studies have been restricted to *Candida*, *Hansenula*, *Pichia* and *Torulopsis*.

The mechanism of methanol oxidation in yeast differs markedly from that observed in bacteria in that specialized compartments within the yeast cell called peroxisomes (DeDuve and Baudhuin, 1966) are the site of the initial oxidation of methanol to formaldehyde (van Dijken et al., 1975; Fukui et al., 1975; Roggenkamp *et al.*, 1975) (see Section 20.6) which is then excreted and assimilated in the cytosol. Peroxisomes, of course, are not peculiar to the methanol-utilizing yeasts, but have also been found in alkane-utilizing yeasts and are fairly widespread amongst the eukaryotes (DeDuve, 1969). They contain, within their matrix, hydrogen peroxide-producing reactions which, in the case of the methanol-utilizing yeasts, is a consequence of the enzymic conversion of methanol to formaldehyde by the alcohol oxidase. The hydrogen peroxide generated in the reaction (see Section 20.5) is then catalytically decomposed to oxygen and water by catalase, thereby preventing hydrogen peroxide from inactivating sensitive enzymes. The peroxisomes and their enzymes are virtually absent from cells grown on glucose and only appear when grown with methanol. An interesting exception to this is found in the brown rot fungus *Poria contigua* which is capable of lignin degradation and when grown on a glucose medium will synthesize a peroxisome-containing methanol oxidase (Bringer *et al.*, 1979; van Dijken *et al.*, 1981) as well as formaldehyde and formate dehydrogenases but is incapable of growth on methanol. Presumably, the enzymes are present to break down methanol (which can arise from lignin degradation) to carbon dioxide. The yeasts also differ markedly from their bacterial counterparts in their mode of carbon assimilation and this will be discussed further in Section 20.9.

20.4 MORPHOLOGY OF METHANOTROPHS

At the ultrastructural level, some of the most interesting features to be observed are in the methane-utilizing bacteria. In this group, it appears at first sight that the complex intracytoplasmic membranes first observed by Davies and Whittenbury (1970) and Whittenbury *et al.* (1970) in the various obligate methane oxidizers may be involved in methane oxidation *per se* since the facultative methanotrophs did not possess membranes when grown on glucose (Patt and Hanson, 1978; Lynch *et al.*, 1980). However, because some of the facultative methane oxidizers have been shown to be mixed cultures, the interpretation of these results must now be treated with caution. Furthermore, cells grown on methanol (which might be expected to serve as a suitable control) have also been reported to contain membranes (Davies and Whittenbury, 1970; Linton and Vokes, 1978; DeBoer and Hazeu, 1972; Galachenko and Suzina, 1977; Best and Higgins, 1981). However, since such cells usually possess methane-oxidizing activity due to the fact that methanol is a substrate for the methane monooxygenase (Colby *et al.*, 1977), then these observations are not surprising and an unequivocal demonstration of membrane involvement in methane oxidation has yet to be shown.

The methane oxidizers can be divided into two types based on a variety of morphological, biochemical and physiological characters. One of the important features, from the morphological point of view, is the arrangement of the intracytoplasmic membranes. In Type I organisms, the membranes are evenly distributed throughout the cytoplasm in arrays consisting of stacked disc-shaped vesicle (Figure 1; Proctor *et al.*, 1969; Davies and Whittenbury, 1970; Smith and Ribbons, 1970; DeBoer and Hazeu, 1972) whereas in the Type II organisms the membranes were paired and often randomly distributed throughout the cytoplasm (Davies and Whittenbury, 1970). In some sections, the membranes were found at the periphery of the cell (Figure 2).

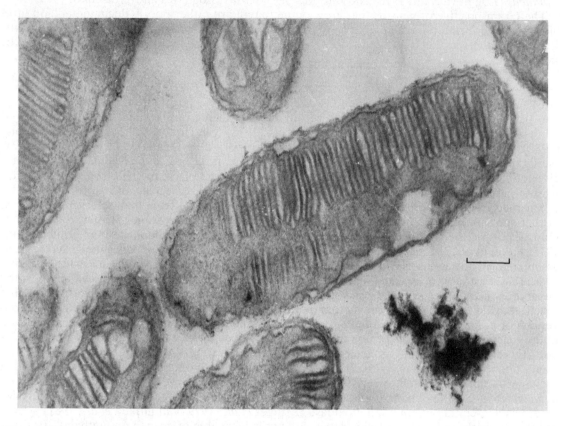

Figure 1 Electron micrograph of a thin section of the Type I organism '*Methylobacter capsulatus*' strain Y (bar marker represents 0.2 μm)

Other morphological features observed in the methane-oxidizers were the exospores and cysts (Whittenbury *et al.*, 1970). Desiccation- and heat-resistant exospores were formed by the Type II '*Methylosinus*' species usually as batch cultures entered the stationary phase. The spores appeared as rounded bodies at the tapered end of the cell which was subsequently budded off. Freshly formed spores would germinate within two to three days but older, heat-treated and dried spores germinated after one or two weeks. Cysts were formed by a number of strains and were of two distinct types: the 'lipid' cyst and an '*Azotobacter*-type' cyst. The 'lipid' cyst formed by '*Methylocystis parvus*' appeared during the stationary phase and was associated with an increasing complexity of the cell wall and a disappearance of the complex intracytoplasmic membrane system seen in the vegetative organisms. The '*Azotobacter*-type' cyst was observed in '*Methylobacter*' species and was manifest by an increase in size, refractivity and roundness. Neither cyst type was heat-resistant but they were able to withstand desiccation for one week in the case of the 'lipid' cyst and up to eighteen months for the '*Azotobacter*-type' cysts.

20.5 METHANE OXIDATION

The enzyme methane monooxygenase (MMO) is responsible for the initial O_2 and NAD(P)H-dependent hydroxylation of methane to form methanol. The first report of cell extracts capable of

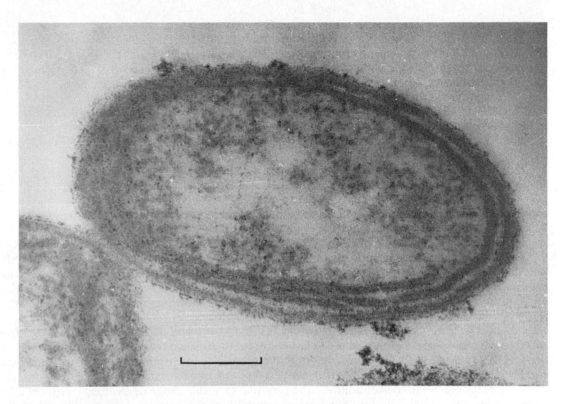

Figure 2   Electron micrograph of a thin section of the type II organism '*Methylosinus sporium*' strain 12 (bar marker represents 0.2 μm)

effecting this oxidation was in 1970 by Ribbons and Michalover who observed methane stimulation of oxygen uptake by membrane fractions isolated from *Methylococcus capsulatus* (Texas). Subsquently, Ferenci (1974) and Colby *et al.* (1975) reported that carbon monoxide and bromomethane respectively would serve as substrates for the particulate MMO from *Methylomonas methanica*. A particulate preparation was also observed in '*Methylosinus trichosporium*' by Tonge *et al.* (1975) and the first report of a soluble MMO in the Bath strain of *Methylococcus capsulatus* appeared in the following year (Colby and Dalton, 1976). The best characterized system to date is the soluble system from *M. capsulatus* (Bath). The soluble MMO system was found to be a complex of three proteins which could be resolved by DEAE cellulose chromatography and purified by chromatographic techniques (Colby and Dalton, 1978; Dalton, 1980; Woodland and Dalton, 1984). Component A of the complex has a molecular weight of about 210 000 and comprises three subunits of 54 000, 42 000 and 17 000 in equal amounts. The native protein contains both non-heme iron (2.3 mol mol^{-1}) and zinc (0.2–0.5 mol mol^{-1}) and no acid labile sulfur (Woodland and Dalton, 1984). From electron paramagnetic resonance (EPR) studies, it appears that the non-heme iron in component A is involved in substrate binding (Dalton, 1980; Woodland and Dalton, 1984) and that the iron may be present in a previously uncharacterized environment. Component C is an iron–sulfur flavoprotein of molecular weight 44 000 which catalyzes the reduction of a variety of acceptors (Colby and Dalton, 1979) including component A. Component B is a colorless protein of about 15 000 molecular weight, the role of which is uncertain but is necessary for reconstituted activity of the complex. Based on these studies of the reconstituted complex, a scheme of electron transfer between the MMO components has been proposed (Dalton, 1980) and is shown in Figure 3.

The enzyme from the Type II methanotroph, '*Methylosinus trichosporium*' OB3b, was originally purified by Tonge *et al.* (1977) after removal of two of the components from membrane fractions but this procedure no longer gives an active enzyme (Higgins *et al.*, 1981). In our hands, the MMO from '*M. trichosporium*' OB3b was soluble and one of the components even showed complementation of reactivity with components B and C from *M. capsulatus* (Bath) (Stirling and Dalton, 1979). Recently, these differences in properties of the OB3b enzyme between the Warwick and Kent groups were rationalized by Scott *et al.* (1981a, 1981b). They found the extracts of '*M. trichosporium*' OB3b had a particulate MMO activity when grown under oxygen-limiting

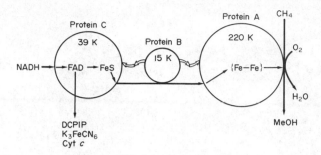

Figure 3 Pathway of electron transfer between proteins A, B and C of the MMO complex

conditions either in shake flasks or chemostat culture (see note added proof, Scott *et al.*, 1981a) but when grown under methane-limiting conditions the enzyme was soluble. These changes could also be correlated with the presence of an extensive intracytoplasmic membrane system (O_2-limited) or with a vesicular morphology with only a few membranes (CH$_4$-limited, NO$_3$-limited). How these findings account for the difference in properties between the enzymes of the two groups is not clear to the present authors since cultures of '*M. trichosporium*' OB3b in the Stirling and Dalton (1979) paper were grown under oxygen-limiting conditions and yielded extracts which had activity associated with the soluble proteins.

Recent work by the Warwick group does suggest that a switch between soluble and particulate activity can be observed in chemostat cultures of both *M. capsulatus* (Bath) and '*M. trichosporium*' OB3b in response to the oxygen regime and copper concentration in the growth medium (Stanley, Prior, Leak and Dalton, 1983). Under oxygen excess (methane-limiting) conditions, the enzyme is particulate under all conditions and sensitive to a wide range of electron transport inhibitors. Under oxygen-limiting (methane excess) conditions the enzyme is particulate when the concentration of copper is high, but soluble and insensitive to electron transport inhibitors when the copper concentration is low. A further complication arises in the latter instance (low copper) since at low cell densities the enzyme is particulate and does suggest that if copper is present largely in excess of requirements (as it would be at low cell densities) then the enzyme will be particulate no matter what the oxygen regime is. Why the copper level should regulate this switchover is unclear at present but it may implicate certain copper-containing proteins as electron carriers being involved with the MMO when the enzyme is in the particulate form, possibly in a manner similar to the observed *Pseudomonas* AM1 when growing on the C_1 substrate methylamine (Tobari and Harada, 1981).

Probably one of the most interesting biotechnological features of the MMO enzyme is its ability to insert an atom of oxygen into a wide range of substrates both *in vitro* and *in vivo*. The first reported instance of this appeared from the Warwick group (Colby *et al.*, 1977) in which they showed that *n*-alkanes, halogenated alkanes, *n*-alkenes, ethers, carbon monoxide, methanol and alicyclic, aromatic and heterocyclic hydrocarbons could be oxygenated to a variety of products by the soluble enzyme from *M. capsulatus* (Bath). Ammonia was then added to the list (Dalton, 1977) and subsequently extracts from '*M. trichosporium*' OB3b and *M. methanica* (Stirling *et al.*, 1979) were also shown to effect these bioconversions, although the *M. methanica* enzyme was more restricted in its range of substrates (Table 2). The biotransformations have also been observed in whole cells of '*M. trichosporium*' OB3b (Higgins *et al.*, 1979) and *M. capsulatus* (Bath) (Stirling and Dalton, 1979) and the application of these discoveries to biotechnology is given in Section 20.14. Subsequent to these discoveries, the Exxon group have also observed cell-free soluble (Patel *et al.*, 1982), cell-free particulate (Patel *et al.*, 1979) and whole cell (Hou *et al.*, 1980) activities in their facultative methanotroph '*Methylobacterium*' CRL-26.

20.6 METHANOL OXIDATION

Methanol oxidation to formaldehyde by methylotrophic microorganisms is catalyzed by either an NAD$^+$-independent dehydrogenase (bacteria) or oxidase (yeast). The ubiquitous bacterial methanol dehydrogenase (EC 1.1.99.8) was initially described in *Pseudomonas* M27 by Anthony and Zatman (1964a, 1964b, 1965). To demonstrate enzyme activity in cell extracts, phenazine methosulfate was required as electron acceptor plus ammonia or methylamine as activator. This

Table 2 Examples of Substrate Oxidations by Extracts of Three Methane Oxidizing Bacteria

Substrate	Products	M. capsulatus	M. trichosporium	M. methanica
		Specific activity ($nmol^{-1} min^{-1}$ mg protein^{-1})		
Methane	Methanol, methanal	84	51	73
Ethane	Ethanol, ethanal, acetate	68	33	87
Pentane	Pentan-1-ol, pentan-2-ol, *n*-pentanal	73	33	17
Hexane	Hexan-1-ol, hexan-2-ol, hexanal	40	25	0
Heptane	Heptan-1-ol, heptan-2-ol, heptanal	27	19	0
Chloromethane	Methanal	84	29	121
Dichloromethane	Carbon monoxide	82	26	43
Trichloromethane	Carbon dioxide	35	27	0
Bromomethane	Methanal	66	n.d.[a]	94
Methanol	Methanal	246	47	
Dimethyl ether	Methanol, methanal	30	26	65
Diethyl ether	ethanol, ethanal	45	144	27
Methyl formate	Methanal, formate	29	n.d.[a]	n.d.[a]
Carbon monoxide	Carbon dioxide	61	n.d.[a]	27
Ethylene	Epoxyethane	148	59	158
Propene	Epoxypropane	83	54	81
Cyclohexane	Cyclohexanol	62	20	0
Benzene	Phenol, hydroquinone	62	31	0
Toluene	Benzyl alcohol	53	13	0
Pyridine	Pyridine *N*-oxide	29	19	0
Styrene	Styrene epoxide	47	20	0

[a] n.d. not determined

in vitro activity had an optimum at around pH 9. It was found that the enzyme catalyzed the oxidation of a wide range of primary alcohols, hence the alternative name of primary alcohol dehydrogenase.

Since the original description of methanol dehydrogenase from *Pseudomonas* M27 (Anthony and Zatman, 1964a, 1964b, 1965, 1967a, 1967b), the same enzyme has been isolated and characterized in approximately 30 other methylotrophic bacteria (Anthony, 1982). They all have similar properties to those described above, in addition to showing a very low K_m for methanol. The majority of enzymes are dimers of two identical subunits of molecular weight 60 000, notable exceptions being *Methylomonas methanica* and *Methylosinus sporium* which contain monomeric enzymes of molecular weight 60 000 (Patel *et al.*, 1978; Patel and Felix, 1978).

In general, the substrate specificity of the enzyme is restricted to primary alcohols, with the affinity for the enzyme decreasing with increasing carbon chain length. Hydrated aldehydes are usually effective *in vitro* substrates, but whether this operates *in vivo* has yet to be determined. It has been reported that certain secondary alcohols are substrates for the enzymes isolated from *Methylobacterium organophilum* (Wolf and Hanson, 1978), *Pseudomonas* C (Goldberg, 1976) and *Diplococcus* PAR (Bellion and Wu, 1978).

Although 1,2-propanediol is not a substrate for pure methanol dehydrogenase, recent studies with *Pseudomonas* AM1 have indicated that the enzyme can catalyze the oxidation of the diol *in vivo*. It is thought that the diol is oxidized to lactic acid *via* lactaldehyde (Bolbot and Anthony, 1980).

In 1967, Anthony and Zatman described the *in vitro* fluorescence characteristics of methanol dehydrogenase and concluded that it possesses a novel prosthetic group, possibly a pteridine. This prosthetic group has now been characterized fully by two groups of workers independently (Salisbury *et al.*, 1979; Duine *et al.*, 1980) and determined to be a novel orthoquinone compound named pyrroloquinolinequinone (PQQ). The catalytic role of PQQ in the oxidation of methanol by methanol dehydrogenase has been discussed by Forrest *et al.* (1980) and DeBeer *et al.* (1983). PQQ has now been shown to be the prosthetic group of a range of dehydrogenases named quinoproteins and not restricted to methanol dehydrogenase as first thought (Duine *et al.*, 1979; Ameyama *et al.*, 1980). These various quinoproteins include glucose, amine, alcohol and aldehyde dehydrogenases.

The *in vivo* electron acceptor from methanol dehydrogenase is thought to be cytochrome *c* (see Anthony, 1982). This conclusion is based on the reduction of cytochrome *c* by methanol, catalyzed by anaerobically prepared methanol dehydrogenase (Duine *et al.*, 1979; O'Keefe and Anthony, 1980).

The oxidation of methanol to formaldehyde in methylotrophic yeasts differs greatly from the process in bacteria. In yeasts, the reaction occurs in organelles called peroxisomes and no energy

(NAD(P)H or ATP) is derived from the reaction. The oxidation is catalyzed by a non-specific alcohol oxidase, with oxygen as electron acceptor, yielding formaldehyde and hydrogen peroxide as products. The enzyme has been isolated from a number of yeasts and all appear to have very similar properties (Anthony, 1982). They are composed of eight subunits and range from 500 000 to 700 000 in molecular weight. The prosthetic group is flavin adenine dinucleotide. Optimum pH is 7.5 to 9.5 and optimum temperature 30–38 °C. The substrate specificity is restricted to primary alcohols up to about C_5, formaldehyde and oxygen.

High levels of catalase are induced in methylotrophic yeast during growth on methanol (Tani *et al.*, 1978a, 1978b). This enzyme removes the toxic hydrogen peroxide generated by the action of methanol oxidase. This can be achieved by two mechanisms; normal catalatic action generating water and oxygen or peroxidation where methanol is oxidized to formaldehyde using hydrogen peroxide as the oxidant. As the overall stoichiometries for both methanol oxidation reactions are the same, it has proved difficult to determine which is the preferred role of catalase in methylotrophic yeast. The localization of methanol oxidase and catalase activity in peroxisomes apparently protects the cell from the potentially harmful effects of hydrogen peroxide.

The structure and function of these microbodies are discussed in detail by Anthony (1982).

The differences between the complete oxidation of methanol to CO_2 in bacteria and yeast can be seen in Figures 4 and 5.

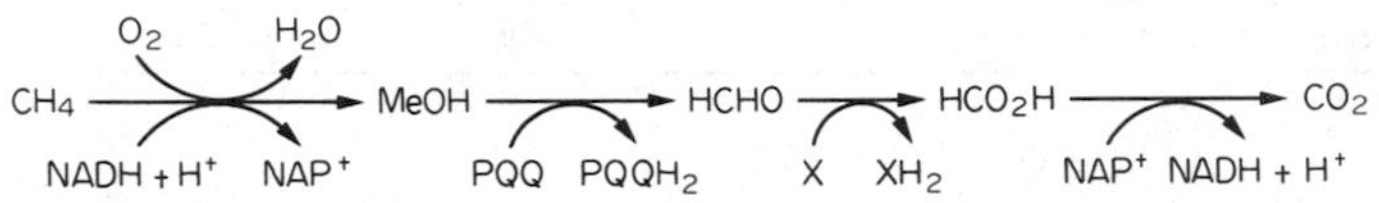

Figure 4   The complete oxidation of methane to CO_2 in methylotrophic bacteria

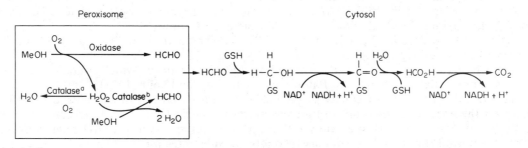

Figure 5   The complete oxidation of methanol to CO_2 in methylotrophic yeasts: catalase[a], catalytic activity; catalase[b], peroxidative activity; GSH glutathione

20.7   FORMALDEHYDE OXIDATION

The complete oxidation of formaldehyde to carbon dioxide in methylotrophic organisms results from successive dehydrogenase action or from a cyclic series of reactions involving C_1-assimilation enzymes.

In bacteria, formaldehyde dehydrogenases fall into two basic groups: $NAD(P)^+$-dependent and $NAD(P)^+$-independent enzymes. The various types detected have been discussed in detail by Stirling and Dalton (1978), Zatman (1981) and Anthony (1982). In certain methylotrophs, the ability to concomitantly reduce pyridine nucleotide with the oxidation of formaldehyde to formate can impact greatly on potential cell growth yield (see Section 20.10.)

The formaldehyde dehydrogenase, apparently ubiquitous in methylotrophic yeasts, is NAD^+-dependent and requires reduced glutathione for activity. The actual substrate for the oxidation is *S*-hydroxymethylglutathione, formed spontaneously from formaldehyde and reduced glutathione. The product generated from its oxidation is *S*-formylglutathione and not free formate. Kato *et al.* (1979) and Anthony (1982) provide a detailed description of yeast formaldehyde dehydrogenase.

Strom *et al.* (1974) and Colby and Zatman (1975) proposed a cyclic scheme for the complete oxidation of formaldehyde to CO_2 involving enzymes of the ribulose monophosphate pathway plus 6-phosphogluconate dehydrogenase (Figure 6). Two molecules of reduced pyridine nucleotide are generated by each turn of the cycle. Some bacteria using the ribulose monophosphate pathways for formaldehyde assimilation have little or no formaldehyde or formate dehydrogenase activity and, hence, rely entirely on this cyclic route for formaldehyde dissimilation. One obligate methylotroph, *Methylococcus capsulatus* (Bath) has the enzymic capability to oxidize formaldehyde to CO_2 by either successive dehydrogenation steps or *via* the above cyclic route (Stirling, 1978).

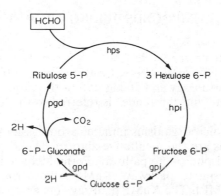

Figure 6 Cyclic pathway for oxidation of formaldehyde in bacteria: hps, hexulosephosphate synthase; hpi, hexulose-phosphate isomerase; gpi, glucose-6-phosphate isomerase; gpd, glucose-6-phosphate dehydrogenase; pgd, 6-phosphogluconate dehydrogenase

An analogous cyclic scheme of formaldehyde dissimilation is potentially active in methylotrophic yeast although the necessary oxidase and dehydrogenase functions for the direct oxidation to CO_2 are also present. The cyclic series of reactions are catalyzed by enzymes of the dihydroxyacetone cycle of formaldehyde assimilation (see Section 20.9) plus hexose-phosphate isomerase, glucose-6-phosphate dehydrogenase and 6-phosphogluconate dehydrogenase (Figure 7). As with the bacterial cyclic dissimilation scheme, two molecules of reduced pyridine nucleotide are formed, but in yeast this is exclusively NADPH. Hence, it is thought that the cycle is only operative to produce reductant for biosynthesis and not for the generation of energy.

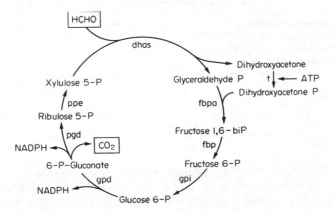

Figure 7 DHA dissimilation cycle of formaldehyde acidation in yeasts: dhas, dihydroxyacetone synthase; t, triokinase; fbpa, fructosebisphosphate aldolase; fbp, fructose bisphosphatase; gpi, glucose-6-phosphate isomerase; gpd, glucose-6-phosphate dehydrogenase; pgd, 6-phosphogluconate dehydrogenase; ppe, pentose phosphate

20.8 FORMATE OXIDATION

Formate is oxidized in methylotrophic bacteria by a soluble, NAD^+-dependent dehydrogenase which is specific for formate. In addition to this enzyme, a membrane-bound, $NAD(P)^+$-independent dehydrogenase has been detected in methylotrophs (Dijkhuizen, 1978, 1979; Rodionov and Zakharova, 1980). The distribution of these enzymes in a variety of bacteria plus their specific activities are detailed by Zatman (1981).

An analogous NAD^+-dependent formate dehydrogenase is found in methylotrophic yeasts. In *Hansenula polymorpha*, the enzyme appears to first hydrolyze *S*-formylglutathione (the product of formaldehyde oxidation) before oxidizing the enzyme-bound formate to CO_2 (van Dijken *et al.*, 1976).

20.9 ASSIMILATION OF C_1 COMPOUNDS DURING GROWTH ON METHANE AND METHANOL

This complex area of C_1 biochemistry has been covered by a number of excellent reviews (Quayle and Ferenci, 1978; Colby *et al.*, 1979; Quayle, 1980; Anthony, 1982). For this reason, the detailed biochemistry, enzymology and details of how the various assimilation pathways were elucidated will be omitted here and the reader is referred to the above papers for coverage of these aspects.

Nearly all methylotrophic microorganisms initiate carbon assimilation at the level of formaldehyde. The major exceptions are the facultative autotrophs/phototrophs which assimilate carbon dioxide *via* the ribulose bisphosphate pathway. Formaldehyde is assimilated by one of three pathways: ribulose monophosphate pathway (RuMP), the serine pathway and the xylulose monophosphate pathway (XuMP). The XuMP pathway appears to be exclusive to the methanol-utilizing yeast. Methylotrophic bacteria (other than autotrophs) can be split into two groups based on carbon assimilation pathways, *i.e.* organisms which use the RuMP pathway for C_1 assimilation and the organisms which use the serine pathway.

Table 3 shows the distribution of C_1 assimilation pathways within the various groups of microorganisms growing on methane and methanol. Interestingly, there have been no facultative methanotrophs isolated which use the RuMP pathway which is more energetically favorable than the serine pathway (see Section 20.10).

Table 3 Distribution of C_1 Assimilation Pathways in Organisms Capable of Growth on CH_4 and MeOH

	RuBP	*RuMP*	*Serine*	*XuMP*
Facultative autotroph/phototroph	+	−	−	−
Obligate methanotroph	−	+	+	−
Obligate methanotroph (Type X)	(+)[c]	+	(+)[c]	−
Facultative methanotroph	−	−	+	−
Obligate methylotroph[a]	−	+	−	−
Facultative methylotroph[a]	+	+	+	−
Methylotrophic yeast[a]	−	−	−	+
Methanotrophic yeast	(+)[c]	−	−	−

[a] In this table, methylotroph refers to organism capable of growth on methanol but not methane.
[b] This group contains only one organism, *Methylococcus capsulatus* (Bath).
[c] (+) Circumstantial evidence for operative pathway.

This is in contrast to the obligate methanol-utilizing organisms which appeared to use the RuMP pathway exclusively. Of the facultative methylotrophs (not capable of growth on methane), the vast majority are serine pathway organisms, with few organisms using other assimilation routes. No C_1 assimilation pathway has been ascribed to the methane yeasts as yet but preliminary ^{14}C labeling studies suggest that carbon is assimilated *via* CO_2 (Wolf, 1981). Depending on the definition of autotrophy (see Whittenbury and Kelly, 1977; Quayle and Ferenci, 1978; Zatman, 1981), the organism *Paracoccus denitrificans* could either be classed as a facultative autotroph or facultative methylotroph. The organism oxidizes methanol to CO_2 and fixes the CO_2 *via* the ribulose bisphosphate pathway, the energy for which is derived from the initial oxidative reactions. Hence, depending on definition, this microbe fulfils the requirements for two groups.

To determine the C_1 assimilation pathway of a methylotrophic organism, investigators usually assay for a number of key enzymes for each possible pathway. The presence of the key enzymes for a particular pathway is often accepted as evidence for the operation of that assimilation route. However, as has been pointed out in several instances, the presence of one of two enzymes of a pathway is not conclusive proof that the pathway in question is operative (Whittenbury *et al.*, 1975; Malashenko, 1976; Trotsenko, 1976; Bamforth and Quayle, 1977; Taylor, 1977; Taylor *et al.*, 1981).

20.9.1 Ribulose Bisphosphate Pathway

As this pathway is uncommon in methylotrophs and potentially excluded depending on the definition of methylotrophy, the reader is referred to Anthony (1982) for a detailed description of the pathway. The two key enzymes of the cycle are ribulosebisphosphate carboxylase, involved in the initial fixation of CO_2, and phosphoribulokinase which generates the ribulose bisphosphate for the above carboxylation. After three revolutions of the cycle, the net result is the condensation of three molecules of CO_2 to give one molecule of glyceraldehyde 3-phosphate (equation 2).

$$3\,CO_2 + 6\,NAD(P)H + 6\,H^+ + 9\,ATP \rightarrow \text{glyceraldehyde 3-phosphate} + 6\,NAD(P)^+ + 9\,ADP + 8\,P_i \tag{2}$$

20.9.2 Ribulose Monophosphate Pathway

This pathway is responsible for the cyclic condensation of three molecules of formaldehyde to produce either one molecule of pyruvate or one molecule of dihydroxyacetone phosphate. There are four potential variants of this cycle dependent on the cleavage and rearrangement reactions: KDPG aldolase/transaldolase variant, FBP aldolase/sedoheptulose bisphosphatase variant, KDPG aldolase/sedoheptulose bisphosphatase variant and FBP aldolase/transaldolase variant. Only the first two variants are well represented in described methylotrophs. The KDPG aldolase variants give rise to pyruvate as an end product whereas the FBP aldolase variants produce dihydroxyacetone phosphate.

The KDPG aldolase/transaldolase cycle is shown in Figure 8. This variant of the pathway is used mainly in obligate methylotrophs. A summary of the overall reactions is given in equation (3).

$$3\,HCHO + NAD^+ + 1\,ATP \rightarrow \text{pyruvate} + NADH + H^+ + 1\,ADP + 1\,P_i \tag{3}$$

The FBP aldolase/sedoheptulose bisphosphatase cycle is shown in Figure 9. Facultative methylotrophs are the major utilizers of this variant. The summary of the overall reactions involved is given in equation (4).

$$3\,HCHO + 2\,ATP \rightarrow \text{dihydroxyacetone phosphate} + 2\,ADP + 2\,P_i \tag{4}$$

There are two key enzymes of the RuMP pathway, 3-hexulose-phosphate synthase, the enzyme which catalyzes the initial aldol condensation between formaldehyde and ribulose 5-phosphate; secondly, 3-hexulosephosphate isomerase which isomerizes 3-hexulose 6-phosphate (condensation product) to fructose 6-phosphate.

20.9.3 Serine Pathway

This pathway effects the condensation of two molecules of formaldehyde with one of CO_2 to give one molecule of 3-phosphoglycerate. The cycle, shown in Figure 10, uses amino acid and carboxylic acid intermediates instead of the carbohydrate carriers in the RuMP pathway. There are a number of enzymes crucial to the operation of the cycle: serine transhydroxymethylase (formaldehyde/glycine condensing enzyme), serine-glyoxylate aminotransferase (performs two functions, Figure 10) hydroxypyruvate reductase, glycerate kinase, malyl-CoA lyase (crucial to the regeneration of the C_2 carrier glycine).

Two variants of the serine pathway exist and the difference revolves around the regeneration of

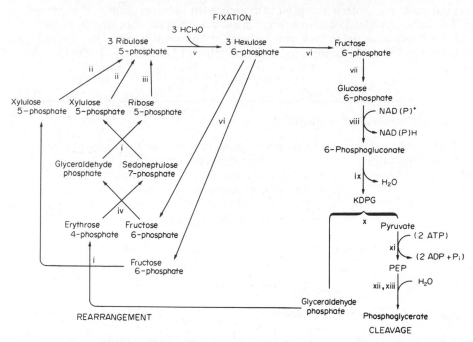

Figure 8 The ribulose monophosphate (RuMP) cycle of formaldehyde assimilation (KDPG aldolase/transaldolase variant). This variant occurs predominantly in obligate methylotrophs. i, transketolase; ii, pentosephosphate epimerase; iii, pentosephosphate isomerase; iv, transaldolase; v, hexulosephospate synthase; vi, hexulosephosphate isomerase; vii, glucosephosphate isomerase; viii, glucosephosphate dehydrogenase; ix, phosphogluconate dehydrase; x, 2-keto-3-deoxy-6-phosphogluconate (KDPG) aldolase; xi, PEP synthetase or equivalent enzymes; xii, enolase; xiii, phosphoglyceromutase

the second molecule of glyoxylate from acetyl-CoA. Acetyl-CoA is formed from the cleavage of malyl-CoA to glyoxylate and acetyl-CoA. The isocitrate lyase[+] variant uses a glyoxylate cycle to oxidize acetyl-CoA to glyoxylate, with isocitrate lyase being the key enzyme. The method by which the organisms with no isocitrate lyase activity, *i.e.* IL[-], catalyze the oxidation of acetyl-CoA to glyoxylate remains an enigma in C_1 biochemistry. In 1980, Korstee proposed a homoisocitrate–glyoxylate cycle involving the homologous intermediates of the glyoxylate cycle (Korstee, 1980, 1981). However, substantial doubt has now been cast upon these findings since other workers have been unable to repeat either the same experiments or equivalent experiments in other IL[-] serine pathway strains (Bellion *et al.*, 1981). An interesting observation made with the IL[-] serine organism *Protaminobacter ruber* was the ability of cell extracts to catalyze the cleavage of mesaconyl-CoA to propionyl-CoA and glyoxylate (Ueda *et al.*, 1981). It is possible, at least for this organism, that this reaction is involved in the oxidation of acetyl-CoA to glyoxylate. A solution to this perplexing problem would be most helpful, especially as the vast majority of methylotrophs utilizing the serine pathway are IL[-].

A summary of the overall reactions in one cycle of the IL[+] serine pathway is given in equation (5).

$$2\,HCHO + 1\,CO_2 + 2\,NADH + 2\,H^+ + 3\,ATP \rightarrow 3\,\text{phosphoglycerate}$$
$$+ 2\,NAD^+ + 3\,ADP + 3\,P_i \tag{5}$$

Until the IL[-] serine pathway is finally characterized, no summary equation can be written. The only probable difference will be in the form of reductant generated during the conversion of acetyl-CoA to glyoxylate.

20.9.4 Xylulose Monophosphate Pathway

This pathway is found exclusively in methanol-utilizing yeasts. It is responsible for the cyclic condensation of three molecules of formaldehyde yielding a molecule of triose phosphate (Figure 11). The only enzyme which operates exclusively in this pathway is the initial formaldehyde/xylu-

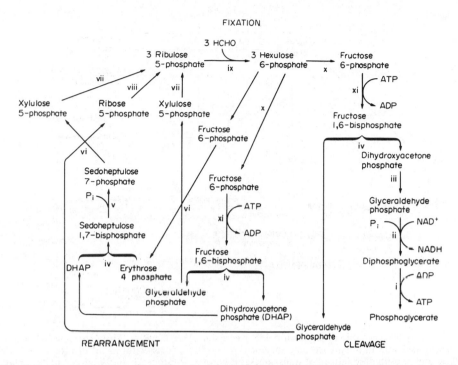

Figure 9 The ribulose monophosphate (RuMP) cycle of formaldehyde assimilation (fructosebisphosphate aldolase/sedo-heptulose bisphosphatase variant). This variant occurs predominantly in facultative methylotrophs. i, phosphoglycerate kinase; ii, glyceraldehydephosphate dehydrogenase; iii, triosephosphate isomerase; iv, aldolase; v, sedoheptulose bis-phosphatase; vi, transketolase; vii, pentosephosphate epimerase, viii, pentosephosphate isomerase; ix, hexulosephos-phate synthase; x, hexulosephosphate isomerase; xi, phosphofructokinase

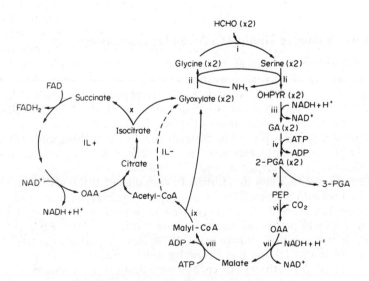

Figure 10 Serine pathway. i, serine transhydroxymethylase (EC 2.1.2.1); ii, serineglyoxylate aminotransferase; iii, hydroxypyruvate reductase (EC 1.1.1.29); iv, glycerate kinase (EC 2.7.1.31); v, phosphopyruvate hydratase (EC 4.2.1.11); vi, phosphoenolpyruvate carboxylase (EC 4.1.1.31); vii, malate dehydrogenase (EC 1.1.1.37); viii, malate thiokinase (EC 6.2.1.-); ix, malyl-CoA lyase (EC 4.1.3.24); x, isocitrate lyase (EC 4.1.3.1); ——, unknown reactions; OHPYR, hydroxypyruvate; GA, glycerate; PGA, phosphoglycerate; PEP, phosphoenolpyruvate; OAA, oxaloacetate.

Net reaction (IL+): $2\,HCHO + CO_2 + FAD + 2\,NADH_2 + 3\,ATP \rightarrow 3\,PGA + 2\,ADP + 2\,NAD + FADH_2$

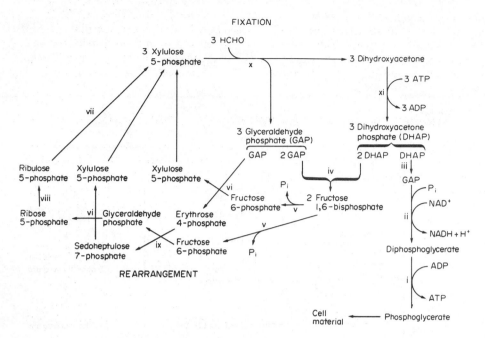

Figure 11 The dihydroxyacetone cycle of formaldehyde assimilation in yeast. The rearrangement part of this cycle involves transaldolase but an alternative rearrangement of the triose phosphates might involve sedoheptulosebisphosphate aldolase and phosphatase instead of transaldoase. i, phosphoglycerate kinase; ii, glyceraldehydephosphate dehydrogenase; iii, triosephosphate isomerase; iv, fructosebisphosphate aldolase; v, fructose bisphosphatase; vi, transketolase; vii, pentosephosphate epimerase; viii, pentosephosphate isomerase; ix, transaldolase; x, dihydroxyacetone synthase; xi, triokinase

lose phosphate condensing enzyme, dihydroxyacetone synthase. The remaining enzymes are all involved in carbohydrate or alcohol assimilation. The summary equation for the cycle is given in equation (6).

$$3\,HCHO + 3\,ATP \rightarrow \text{dihydroxyacetone phosphate} + 3\,ADP + 3\,P_i \qquad (6)$$

20.9.5 Organisms Containing Multiple C_1 Assimilation Pathways

To date, no methylotrophic organism has been shown to simultaneously operate more than one C_1 assimilation pathway. However, there are an ever-increasing number of reports of certain organisms possessing key enzyme activities from alternate pathways. This data has presented problems of classification with respect to the original definition of Type 1 and Type 2 methanotrophic organisms, *e.g. Methylococcus capsulatus* (Bath) (see Table 1; Whittenbury and Dalton, 1981).

Pseudomonas gazotropha is reported to utilize the ribulose bisphosphate (RuBP) pathway during growth on CO but assimilates methanol *via* the serine pathway (Romanova and Nozhevnikova, 1977). *Paracoccus denitrificans* and *Thiobacillus novellus* both contain constitutive hydroxypyruvate reductase activity in addition to enzymes of the RuBP pathway (Bamforth and Quayle, 1977; Chandra and Shethna, 1977). A range of RuMP pathway organisms, *Methylococcus capsulatus* (both Texas and Bath strains), *Methylomonas methanica* and a *Streptomyces* species have all been shown to possess hydroxypyruvate reductase activity plus other enzymes of the serine pathway (Salem *et al.*, 1973; Strom *et al.*, 1974; Reed, 1976; Kato *et al.*, 1977).

Methylococcus capsulatus (Bath) has been the most studied RuMP pathway organism with regard to potential ancillary pathways of carbon assimilation. In addition to the presence of the serine pathway enzymes (Reed, 1976; Whittenbury *et al.*, 1976), this organism has been shown to contain ribulose-bisphosphate carboxylase/oxygenase and phosphoribulokinase activities (Taylor, 1977; Taylor *et al.*, 1981). Although the organism can assimilate CO_2 over long periods of time in the presence of a suitable energy source, *e.g.* formate, no autotrophic growth can be demonstrated (Stanley and Dalton, 1982). RuBP carboxylase activity was not detected in a wide range of methanotrophs, therefore it does not appear to be a prerequisite for growth on methane.

It is postulated that in the absence of FBP aldolase the presence of RuBP carboxylase/oxygenase in this organism may be to provide an alternative cleavage pathway for the synthesis of 3-phosphoglycerate during growth on methane (Quayle, 1979; Stanley and Dalton, 1982). A number of intriguing questions arise from studying organisms which possess auxiliary C_1 assimilation enzymes. Are these pathways used for scavenging carbon under carbon limitation? Are they involved in fixing carbon in times of energy excess? Are these enzymes simply vestiges of pathways lost through evolution?

20.10 ENERGETICS OF GROWTH ON METHANE AND METHANOL

It is possible to predict the growth yield of an organism growing on any given substrate. First, a summary equation is formulated for the assimilation of the substrate into cellular material. Next, the equation is amended by adding additional carbon substrate whose complete oxidation would supply other energy and/or reductant required for biosynthesis. This additional carbon substrate will vary depending on the P/O ratio (moles of ATP formed per electron pair transferred to oxygen) for each pair of electrons generated during its complete oxidation. Hence, growth yields can be predicted for a given substrate and P/O ratio.

The growth yield on typical heterotrophic substrates with all nutrients in excess is predominantly ATP-limited for most bacteria (Anthony, 1982), *i.e.* a hypothetical increase in the supply of ATP would increase the growth yield. This is a reflection of the level of reduction of the substrate, *i.e.* if the compound is more oxidized than cellular material then growth will be ATP-limited. Growth on a more reduced substrate tends to be carbon-limited. Therefore, for growth on methane at least, and perhaps methanol, one would predict the growth yield to be carbon-limited.

The stoichiometries of growth on methane and methanol for all types of methylotrophs have been calculated by several workers (Harrison et al., 1972; van Dijken and Harder, 1975; Barnes *et al.*, 1976; Harder and Van Dijken, 1976; Anthony, 1978, 1980, 1982). For a detailed discussion of the concept of predicting cell yield, the mechanisms of constructing assimilation equations and a list of the assumptions used to generate such equations, the reader is referred to the above papers.

Table 4 lists the summary of assimilation equations for five typical methylotrophs. It is assumed that methane oxidation, catalyzed by methane monooxygenase, has a requirement for NAD(P)H, which is supported by the current literature (see methane oxidation). Both the organisms growing on methane will be verging on NAD(P)H limitation. This is a result of the NAD(P)H requirement for the initial hydroxylation of methane and the fact that methanol oxidation is not NAD(P)$^+$-linked. The limitation is further aggravated in the serine pathway methanotroph because of the NAD(P)H requirement for C_1 assimilation, resulting from having to refix CO_2. Of the three methanol-utilizing strains, only the serine pathway organism is likely to be NAD(P)H-limited and only when the formaldehyde dehydrogenase is not NAD(P)$^+$-linked. The only methylotroph which could be considered carbon-limited would be the RuMP methanol utilizers. The methanol-utilizing yeasts with the XuMP pathway are effectively ATP-limited as no ATP is derived from the oxidation of methanol to formaldehyde by the methanol oxidase. The potential limitations of growth are summarized in Table 5 with the proviso that P/O ratios can affect the conclusion. In general, high P/O ratios increase the tendency to carbon or NAD(P)H limitations.

Thus, although methane and methanol are reduced substrates, the growth yields of most organisms are still NAD(P)H or ATP-limited. This is a reflection of the unusual oxidizing enzymes involved in C_1 metabolism and not the respective thermodynamics.

20.11 REGULATION OF METHANE AND METHANOL ASSIMILATION

Genetic regulation studies have been almost exclusively restricted to facultative methylotrophic organisms, in particular serine pathway bacteria and methanol-utilizing yeasts.

Based on comparative specific activities of three serine pathway enzymes, it was suggested as far back as 1972 that assimilation enzymes of *Pseudomonas* AM1 may be coordinately expressed (Dunstan *et al.*, 1972). However, it is only recently that any biochemical and genetic evidence for such regulation has emerged. In five facultative serine pathway methylotrophs it has been shown that C_1 assimilation enzymes are inducible. Two of the organisms showed apparent coordinate

Table 4   Comparative Energetics of Growth for Five Methylotrophs

Growth Substrate	Assimilation pathway	Stoichiometry of C_1 assimilation
Methane	RuMP/KDGP variant	$12\ CH_4 + 13.5\ NAD(P)H + 37\ ATP + 3\ NH_3 \rightarrow 306\ g\ cell\ material^a + 12\ PQQH_2{}^b$
Methane	Serine (IL$^+$)	$12\ CH_4 + 21.5\ NAD(P)H + 41\ ATP + 3\ NH_3 \rightarrow 306\ g\ cell\ material + 12\ PQQH_2{}^b + 4\ YH_2{}^b + 4\ FPH_2{}^b$
Methanol	RuMP/KDGP variant	$12\ MeOH + 1.5\ NAD(P)H + 37\ ATP + 3\ NH_3 \rightarrow 306\ g\ cell\ material + 12\ PQQH_2$
Methanol	Serine (IL$^+$)	$12\ MeOH + 9.5\ NAD(P)H + 41\ ATP + 3\ NH_3 \rightarrow 306\ g\ cell\ material + 12\ PQQH_2 + 4\ YH_2 + 4\ FPH_2$
Methanol	XuMP	$12\ MeOH + 1.5\ NAD(P)H + 37\ ATP + 3\ NH_3 \rightarrow 306\ g\ cell\ material$

[a] Each equation represents the synthesis of 306 g cell material ($C_4H_8O_2N_3$).
[b] $PQQH_2$, YH_2 and FPH_2 are the reductants formed from methanol dehydrogenase, formaldehyde dehydrogenase and succinate dehydrogenase, respectively.

Table 5   Summary of Potential Growth
Limitations for Five Methylotrophs

Organism	Potential limitation
CH$_4$/RuMP	NAD(P)H
CH$_4$/serine	NAD(P)H
MeOH/RuMP	Carbon or ATP[a]
MeOH/serine	NAD(P)H or ATP[b]
MeOH/XuMP	ATP

[a] Dependent on P/O ratio of PQQH.
[b] Dependent on whether formaldehyde dehydrogenase is NAD(P)H$^+$ linked.

induction of all the assimilation enzymes. Three of the strains showed no repression of C_1 assimilation enzymes during concomitant growth on methanol and succinate, one showed a number of enzymes repressed, and all the enzymes assayed were repressed in the first (O'Connor and Hanson, 1977; McNerney and O'Connor, 1980; Lynch *et al.*, 1980; O'Connor, 1981). To determine the existence of operonic control within these strains would require genetic analysis but to date this has only proved possible with one of the organisms, *Methylobacterium organophilum* XX. Transformation was first reported in this strain in 1977 (O'Connor *et al.*) and was used to map a number of C_1 assimilation mutations. Based on gene linkage and enzyme levels within the mutants, a model was proposed for regulation of a number of methanol utilization genes (Figure 12) (O'Connor *et al.*, 1977; O'Connor and Hanson, 1977, 1978; O'Connor, 1981). For a detailed description of the mutants, the reader is referred to the above articles. The model depicts the genes coding for all the methanol assimilation enzymes including methanol dehydrogenase in a single operon. The gene coding for cytochrome *c* (required for the electron transfer from methanol dehydrogenase) is located distal to the main operon but is apparently coordinately regulated with the assimilation operon. The genes appeared to be induced by methanol itself as a mutant-lacking methanol dehydrogenase was still able to produce high levels of all the other enzymes in the presence of methanol.

Most of the regulation studies using methylotrophic yeasts have been concerned with the methanol dissimilation enzymes, showing increased synthesis of all the enzymes involved during

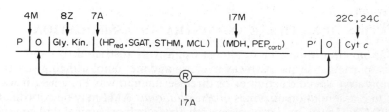

Figure 12   Model for genetic loci of C_1 enzymes in *M. organophilum* XX: Gly.Kin., glycerate kinase; HP$_{red}$, hydroxypyruvate reductase; SGAT, serine–glyoxylate aminotransferase; STHM, serine transhydroxymethylase; MCL, malyl-CoA lyase; MDH, methanol dehydrogenase; PEP$_{carb}$, phosphoenolpyruvate carboxylase

growth on methanol (Kato *et al.*, 1974; Eggli and Sahm, 1978; Eggli *et al.*, 1980; Egli, 1982; Egli *et al.*, 1982a). Recent mixed substrate culture studies (Egli *et al.*, 1982b, 1982c) are beginning to address the problem of the biochemical regulation of C_1 assimilation enzymes within methylotrophic yeasts but as yet no genetic studies have been reported.

One key regulatory step common to all methylotrophic organisms is the control over the fate of formaldehyde generated within the cell. There must be a rigid biochemical regulation to determine how much formaldehyde is assimilated or dissimilated. In bacteria possessing the serine pathway, the formaldehyde enters a cycle as methylene tetrahydrofolate and thereby provides a possible point for regulation. In organisms with the RuMP pathway, free formaldehyde is the substrate for assimilation. Indeed, in *Methylococcus capsulatus* (Bath), there are four potential fates for formaldehyde. It can be assimilated via the RuMP cycle or dissimilated by one of three possible routes: *via* NAD^+-linked formaldehyde dehydrogenase, *via* methanol dehydrogenase or *via* the dissimilatory RuMP route for formaldehyde oxidation (Stirling, 1978). Therefore, in this organism there are two possible branch points of metabolism, one at the level of free formaldehyde, the other at the level of 6-phosphogluconate.

Very little work has been done on the regulation of synthesis of methane monooxygenase (MMO) in either facultative or obligate methanotrophs. It appears for facultative methanotrophs that MMO activity is repressed during growth on methanol or multicarbon compounds (O'Connor, 1981). MMO activity has also been reported to be inducible in some obligate methanotrophs upon growth on methanol but this is not surprising considering methanol is a substrate for MMO (Colby *et al.*, 1977). The regulation of MMO synthesis is not yet clear. It has been reported that certain obligate methylotrophs show no MMO activity when grown on methanol (Patel *et al.*, 1980). However, for many other obligate methanotrophs, MMO activity was present in methanol grown cells (Linton and Vokes, 1978; Harder *et al.*, 1979; Best and Higgins, 1981) and some workers regard this as evidence for constitutive synthesis of MMO (Higgins *et al.*, 1981). This is probably a misconception as it has been demonstrated that methanol is indeed a substrate for MMO (Colby *et al.*, 1977; Stirling and Dalton, 1979) and therefore could induce synthesis of the enzyme. Recent culture studies using *Methylococcus capsulatus* (Bath) suggest that MMO is in fact inducible, as no oxygenase activity is detectable in cells grown continuously on formaldehyde (Dalton, unpublished observations).

Clearly, the area of biochemical and genetic regulation in methylotrophs is one which has been neglected until recently and requires more effort in the future.

20.12 GENETIC ANALYSIS OF METHYLOTROPHS

Until recently, the genetic analysis of any organism depended on the isolation of a range of phenotypic mutants. These mutants normally take the form of amino acid/vitamin auxotrophies or expression of antibiotic resistance and are generated by various well-documented physical, chemical and biological techniques. However, chromosome mapping of the obligate methylotroph *Methylophilus methylotrophus* AS1 has recently been achieved without the normally prerequisite mutant isolations (see below).

The isolation of stable metabolic mutants of methylotrophic bacteria depends greatly on the nature of their methylotrophy, *i.e.* obligate or facultative. Workers have experienced little difficulty in isolating a variety of stable phenotypic mutants with facultative methylotrophs whereas obligate methylotrophs have proved somewhat recalcitrant to mutant isolation procedures.

The facultative methylotroph, *Pseudomonas* AM1, was the first methylotroph to be analyzed by mutant characterization (Heptinstall and Quayle, 1970). Many workers have used mutants of this strain not only to study the assimilation and dissimilation of methanol (Anthony, 1982) but as potential recipients for *in vitro* and *in vivo* cloning experiments (Gautier and Bonewald, 1980; Dunstan, 1982; Nunn *et al.*, 1983). Mutant studies with the facultative methylotroph *Methylobacterium organophilum* XX have also proved useful in understanding the genetic and enzyme regulation of methanol assimilation in this strain (O'Connor and Hanson, 1977; O'Connor and Hanson, 1978; O'Connor, 1981a, 1981b).

A considerable amount of time and effort has been spent by a number of workers in an attempt to isolate metabolic mutants of obligate methylotrophs. A variety of repair-dependent and -independent mutagens failed to increase the mutation frequency of a number of obligate methanotrophs (Harwood *et al.*, 1972; Williams *et al.*, 1977; Warner *et al.*, 1983). *N*-Methyl-*N'*-nitro-*N*-nitrosoguanidine effected a slight increase but any auxotrophs isolated were extremely unstable. It was suggested that either excision is the dominant repair mechanism within these organisms or that repair is error-proof. This combined with the inherent selection and isolation problems associated with organisms dependent on methane or methanol for growth was presented as the reason for the inability to isolate mutants (Williams *et al.*, 1977). The reported toxicity of exogenous amino acids and other multicarbon compounds in methanotrophs (Eccleston and Kelly, 1972, 1973) could also exacerbate the problem. There is also the possibility that a lack of the appropriate permeases may prevent the uptake of exogenous compounds, as has been suggested for the obligate methanol-utilizer *Methylophilus methylotrophus* AS1 (Brammer, 1981). These problems can be overcome as demonstrated by the recent isolation of stable auxotrophic and other metabolic mutants of *Methylophilus methylotrophus* AS1 (Windass *et al.*, 1980; Brammer, 1981). Also, *nif* mutants have been obtained in an obligate methylotroph using a marker exchange technique (Toukdarien and Lindstrom, 1983).

Once a range of suitable strains have been selected, genetic analysis of an organism requires transfer of genetic information between strains. Transformation, conjugation and transduction are three mechanisms which mediate such exchanges. As no transducing bacteriophages have been isolated for a methylotroph, transduction has not been demonstrated. Transformation (chromosomal DNA) has been used successfully to initiate genetic studies of the facultative methanotroph *Methylobacterium organophilum* XX (O'Connor and Hanson, 1978). However, transformation (plasmid DNA) could not be demonstrated in a variety of other methylotrophs, both facultative and obligate (Gautier and Bonewald, 1980; Windass *et al.*, 1980).

Conjugation has proved by far the most useful technique for transfer of genetic information in methylotrophs. Transfer of plasmid DNA to methylotrophs, both facultative and obligate, has been demonstrated on numerous occasions (Jeyaseelen and Guest, 1979; Paraskeva, 1979; Warner *et al.*, 1980; Gautier and Bonewald, 1980; Windass *et al.*, 1980; Holloway, 1981). There are reports showing the transfer of chromosomal markers (Holloway, 1981; Dunstan, 1982) using enhanced chromosomal mobilizing (ECM) plasmids such as R68.45. Although conjugation and chromosome mobilization have been demonstrated in an obligate methylotroph (Holloway, 1981), effective genetic analysis of chromosomal markers in such an organism still remains a problem due to the difficulties in isolating a range of suitable mutants. For the obligate methylotroph *Methylophilus methylotrophus* AS1 this problem was cleverly overcome by the use of plasmid primes (plasmid/chromosome hybrids) (Moore *et al.*, 1983).

It had previously been shown that primes carrying chromosomal DNA fragments from a number of genera could complement phenotypic mutants of *Pseudomonas aeruginosa* to produce the wild-type phenotype (Johnson *et al.*, 1978; Holloway *et al.*, 1980). Using the ECM plasmid pMO 172, a library of prime plasmids carrying fragments of *M. methylotrophus* AS1 chromosome were constructed. The *M. methylotrophus* AS1 genes carried on these primes were identified by transferring the plasmids to mutant *Pseudomonas aeruginosa* strains and scoring for complementation of linked markers. In this way, four groups of linked markers were identified allowing the construction of a preliminary map of the *Methylophilus methylotrophus* AS1 chromosome (Moore *et al.*, 1983). Hence, when suitable mutant strains of a particular methylotroph are not available, mapping by complementation could provide a method for genetic analysis.

A number of genes associated with methanol assimilation have been cloned by recombinant DNA techniques using facultative methylotrophic bacteria (Gautier and Bonewald, 1980; Nunn *et al.*, 1983). In both cases, the functional clones were identified by complementation of suitable mutants of the same organism. A number of genes, both prokaryotic and eukaryotic, have been successfully cloned in *Methylophilus methylotrophus* AS1 (Windass *et al.*, 1980; Hennam *et al.*, 1982; De Maeyer *et al.*, 1982) demonstrating that this organism is a suitable host for the cloning and expression of foreign DNA (see Section 20.14).

Plasmid DNA has been isolated from a variety of facultative and obligate methylotrophs (Warner and Higgins, 1977; Hanson, 1980; Stirling, unpublished observations). No phenotypic trait has been ascribed to any of these plasmids although it has been suggested that a plasmid found in *Methylobacterium organophilum* XX grown on methane could encode for methane monooxygenase (Hanson, 1980). If these cryptic plasmids prove to be functionally uninteresting, they may still prove useful in providing replicons for cloning purposes.

20.13   COMETABOLISM—THE ROLE OF METHANOTROPHS IN THE ENVIRONMENT

Cometabolism and cooxidation are terms which have been used interchangeably in the literature to describe the phenomenon where complex organic compounds are partially degraded or completely mineralized by various consortia of microorganisms (Horvath, 1972; Alexander,

1981). No one organism could utilize these potentially recalcitrant molecules as sole source of carbon and energy and all required a cosubstrate for growth. The terms cooxidation (Foster, 1962) and cometabolism (Jensen, 1963) were, for many years, used ambiguously in the literature and this led to the need for such terms being questioned. Hulbert and Krawiec (1977) suggested that the transformation of non-growth substrates by microorganisms did not constitute novel metabolic events but merely reflected the non-specific nature of existing anabolic and catabolic pathways. However, due to the obvious ecological importance of such biological phenomena in the environment, it was proposed that the term cometabolism be retained but more clearly defined (Stirling and Dalton, 1979b). Cometabolism was redefined as 'the transformation of a non-growth substrate in the obligate presence of a growth substrate or another transformable compound'. The term non-growth substrate describes compounds that are unable to support cellular replication (as opposed to increase in biomass). It was further suggested that the transformation of non-growth substrates in the absence of another substrate did not constitute a novel metabolic event and should therefore be simply referred to as fortuitous metabolism, oxidation, dehalogenation, *etc.* (Stirling and Dalton, 1979b; Dalton an Stirling, 1982).

Methanotrophs are not only proficient in demonstrating such metabolic phenomena in the laboratory but potentially play a major role in the environment as a result of these ancillary reactions. Many mono- and di-oxygenases show what is considered to be a broad substrate specificity but no oxygenase described exhibits such a wide and diverse substrate specificity as methane monooxygenase. As described earlier, this enzyme can oxygenate an enormous range of *n*-alkanes, substituted alkanes, alkenes and ethers plus aromatic, alicyclic and heterocyclic compounds (Colby *et al.*, 1977; Stirling *et al.*, 1979; Stirling and Dalton, 1980). The enzyme appears to have an obligate requirement for NAD(P)H, the reductant necessary for substrate activation (Colby *et al.*, 1975; Colby and Dalton, 1976; Stirling *et al.*, 1979; Stirling and Dalton, 1979a, 1979b; Hou *et al.*, 1980; Scott *et al.*, 1981). The versatility of methane monooxygenase (MMO) means that methanotrophs can potentially oxidize a wide variety of hydrocarbons commonly found in the environment. Table 6 shows the oxidation of a number of such compounds by whole-cell suspensions of *Methylococcus capsulatus* (Bath). A few substrates, *e.g.* haloalkanes, required no cosubstrate for oxidation. As none of the compounds tested supported growth, these reactions are examples of fortuitous oxidation. The oxidation of the remaining substrates can be classed as cometabolic events due to the requisite presence of a cosubstrate (formaldehyde).

Table 6 Oxidation of Various Compounds by Resting Cell Suspensions of *Methylococcus capsulatus* (Bath)[a]

Substrate	*Oxidation rate* [nmol min^{-1} (mg dry cell weight)$^{-1}$] No Formaldehyde	+ 4 mM Formaldehyde	Product
Chloromethane	170	170	Not determined
Bromomethane	88	88	Not determined
Dimethyl ether	7	125	Not determined
Diethyl ether	0	16	Not determined
Carbon monoxide	0	520	Carbon dioxide
Ethane	0	25	Acetaldehyde
Propane	0	4	Propionaldehyde
Ethene	4	22	Epoxyethane
Propene	6	22	Epoxypropane
1-Butene	0	23	1,2-Epoxybutane
cis-2-Butene	0	27	*cis*-2,3-Epoxybutane, *cis*-buten-1-ol
trans-2-butene	0	52	*trans*-2,3-Epoxybutane, *trans*-2-buten-1-ol

[a] From Stirling and Dalton, 1979.

The nature of these oxidations can be explained by determining the supply of reductant for MMO activity. The products of the fortuitous oxidations were generally further metabolized yielding NAD(P)H and therefore did not require an exogenous source of reductant. Other examples of methanotroph cometabolism have been reported where cosubstrates were required for MMO oxidations (Ferenci, 1974; Ferenci *et al.*, 1975; Stirling and Dalton, 1976; Stirling, 1978; Higgins *et al.*, 1979; Hou *et al.*, 1980; Stirling and Dalton, 1981).

Other methanotrophs, in particular Type 2 organisms, are capable of a much wider range of fortuitous MMO oxidation than those described for *Methylococcus capsulatus* (Bath) (Higgins *et al.*, 1979). This is thought to be a reflection of the ability of Type 2 methanotrophs to accumulate large levels of poly-β-hydroxybutyrate as an endogenous reserve during growth under carbon excess (Dalton and Stirling, 1982). These organisms, upon mobilization of this endogenous reserve, will generate available reductant for potential MMO activity. If it were possible to grow a Type 1 methanotroph under conditions where a reserve polymer was synthesized then perhaps more compounds could be oxidized fortuitously by these strains. This appears to be the case for *Methylococcus capsulatus* (Bath) (Leak and Dalton, unpublished observations).

In the environment it appears that fortuitous metabolism and cometabolism of microorganisms play a major role in the degradation of complex hydrocarbons. For example, it is very difficult to isolate a single organism that can utilize unsubstituted cycloparaffins as sole source of carbon and energy (Imelik, 1948; Stirling *et al.*, 1977). However, these compounds are readily metabolized by microbial consortia in the soil (Beam and Perry, 1973, 1974). The initial functionalization of potentially recalcitrant molecules is almost certainly due to the promiscuity of some non-specific enzyme, *e.g.* oxygenase, dehalogenase. These modified compounds can now be further degraded by enzymes in the same organism or another member of the consortium. Methylotrophs possess MMO which can instigate the degradation of a variety of potentially recalcitrant or toxic compounds.

One such potentially toxic compound that is present in the atmosphere in vast quantities is carbon monoxide. The major sources of CO produced in the environment are anthropogenic and photochemical with microbiological processes contributing only a small proportion (Seiler and Schmidt, 1976). Of this enormous quantity of CO produced annually (approximately 10^{15} g; Uffen, 1983), it has been estimated that between 10% (Heichel, 1973) and 50% (Liebl and Seiler, 1976) is removed by microorganisms. It was a long thought that the organisms responsible were the aerobic, CO-utilizing carboxydobacteria (Zavarzin and Nozhevnikova 1977). However, recent studies comparing the kinetics of CO consumption by axenic cultures of carboxydobacteria and natural soil samples concluded that these organisms cannot contribute significantly to the removal of atmospheric CO at the soil interface (Conrad *et al.*, 1981). Indeed, it has been suggested that the fortuitous and cometabolic oxidations of CO are the major microbial degradative processes (Bartholomew and Alexander, 1979; Uffen, 1981).

There is a wealth of evidence supporting the involvement of MMO in the oxidation of CO by methanotrophs (Ferenci, 1974; Ferenci *et al.*, 1975; Hubley, 1975; Stirling and Dalton, 1976; Tonge *et al.*, 1977; Colby *et al.*, 1977; Stirling, 1978). Therefore, it is possible that these organisms play a significant role in the removal of CO from the atmosphere. This theory is supported when kinetic data for CO oxidation by carboxydobacteria and methanotrophs are compared (Table 7). The K_m for CO appears to be much lower (eight-fold) and the oxidation rates can be almost two-fold higher in methanotrophs. This biochemical evidence is supported by reports that the number of methanotrophs present in soils and lake sediments can be in the region of 10^6 and 10^8 organisms per gram (Adamse *et al.*, 1972; Whittenbury *et al.*, 1976). Thus, it appears that the fortuitous oxidation/cometabolism of methane-oxidizing bacteria could play a significant role in the biotransformation of potential pollutants in the environment.

Table 7 Kinetic Parameters for Carbon Monoxide Oxidation[a] (Data from Ferenci *et al.*, 1975; Conrad *et al.*, 1981; Stirling & Dalton, 1979)

	K_m (μm)	V_{max} ($nmol\ min^{-1}\ mg^{-1}$)
Ps. carboxydovorans	21	226[b]
Ps. methanica	2.7	85
M. capsulatus (Bath)	—	520

[a] Ferenci *et al.*, 1975; Stirling and Dalton, 1979; Conrad *et al.*, 1981; Dalton and Stirling, 1982.
[b] Assuming that protein content is 50% of dry mass.

Does the possession of an extremely versatile monooxygenase present a methanotroph with an evolutionary advantage in the environment? This has been a question of some debate (Higgins *et al.*, 1980; Stirling and Dalton, 1981). It was originally suggested that the wide substrate specificity of MMO resulted in the purely fortuitous oxidation of a wide variety of compounds, some of which are commonly found in the environment (Stirling and Dalton, 1979b). Contrary to these conclusions, it was proposed that there has been selective pressure for the development of a broad specificity MMO and that this enzyme initiates useful supplementary metabolism providing survival value to the organisms (Higgins *et al.*, 1980). The rationale for such claims was based upon evidence that obligate Type 2 methanotrophs can partially degrade phenylalkanes to short-chained alkanoic acids. β-Oxidation was thought to be the mechanism responsible and therefore acetate would be generated as a product. This acetate could be potentially assimilated into cellular material although it cannot be used as a sole source of carbon and energy.

Although it is possible that phenylalkanes could generate a small amount of assimilable carbon under optimal conditions, it is improbable that this 'scavenging' process provides sufficient selective pressure to maintain a broad substrate specificity MMO, certainly not in Type 1 methanotrophs which have an incomplete TCA cycle.

The lack of specificity shown by other enzymes usually results in wasteful metabolism, *e.g.* ribulosebisphosphate carboxylase/oxygenase activity, nitrogenase/hydrogen evolution. This evidence would tend to argue against retention of broad substrate specificity. Most of the potential biotransformations of methanotrophs initiated by MMO provide no NAD(P)H as a by-product. Therefore, the NAD(P)H required for MMO activity has to be supplied from within the cell. This must be extremely detrimental to the cell's overall energy balance, particularly in light of the fact that the growth of many methanotrophs is limited by the NAD(P)H supply (see Section 20.10). Many oxidations catalyzed by MMO form potentially toxic products that cannot be further metabolized, *e.g.* epoxides from alkenes, highly reactive halides from haloalkanes (Stirling and Dalton, 1980; Dalton and Stirling, 1982). This argues against retaining a catholic MMO.

All the methane-oxidizing organisms tested possess the ability to form elaborate resting cells (cysts or exospores) to ensure survival. These mechanisms surely afford more protection to unfavorable conditions than possession of a promiscuous MMO. Indeed, these resting stages may have evolved to protect the cells from the potentially harmful actions of MMO such as decribed above. Obviously, more experimental work is required to determine if MMO can indeed supply carbon or energy from non-growth substrate oxidations under environmental conditions.

20.14 INDUSTRIAL APPLICATIONS OF METHYLOTROPHS

The potential of methylotrophs for commercial exploitation can be divided into four areas: (1) the production of single cell protein (SCP), (2) the production of extracellular metabolites, (3) the production of both intracellular and extracellular polymers and (4) biocatalysis.

Methylotrophs have been considered excellent candidates for SCP production for some time. A number of companies are using or looking to use methylotrophs for their SCP process, *e.g.* Imperial Chemical Industries (ICI), Hoechst AG, Mitsubishi Gas Chemical Co., Norprotein and Phillips Petroleum Co. With the exception of Phillips Petroleum Co., they all use methylotrophic bacteria in their processes in preference to the yeasts as they have the advantage of higher growth rates, yields and protein content. Despite this, methylotrophic yeasts have been and are being considered as candidates for SCP production because their nucleic acid content is lower than bacteria and they have a much greater potential for consumer acceptance.

Obligate methanotrophic bacteria were seriously considered by Shell and others as the chosen organisms for SCP production. Although methane is cheap and plentiful it presents a number of problems. Methane has a low solubility and is explosive in nature. There is also a high oxygen requirement for growth of methanotrophs and lower growth yields compared to methanol-utilizing bacteria. Natural gas may also contain an appreciable amount of higher alkanes, *e.g.* ethane, propane and the cometabolites of these compounds can present toxicity problems (Drozd and McCarthy, 1981).

Hence, up until now the organism of choice has been an RuMP methanol-utilizing bacterium. These organisms have potentially the highest growth yield coefficient and are amongst the fastest growing methylotrophic organisms. The only commercial continuous process for SCP production using a methylotroph is the ICI 'Pruteen' plant at Billingham, England. The organism used in this process is the obligate methanol-utilizing bacterium *Methylophilus methylotrophus* AS1. A novel pressure cycle fermenter (1.5×10^6 l volume) was constructed and has been in operation for over two years. For a detailed description of the process and fermenter design, the reader is referred to two review articles, Senior and Windass (1980) and Smith (1981). It is hoped that 70 000 tonnes of SCP will be produced annually with the present process.

The growth yield coefficient of *Methylophilus methylotrophus* is to some extent ATP-limited (see Section 20.10). Therefore, any increase in ATP produced from the oxidation of methanol or decrease in ATP consumption would potentially increase cell yield. By the skillful use of genetic engineering, workers at ICI have managed to improve both these parameters. First, they managed to eliminate the energetically inefficient glutamine synthetase (GS)/glutamate synthase pathway (GOGAT) by mutation. This was replaced by the more efficient glutamate dehydrogenase (GDH) previously cloned from *Escherichia coli* (Windass *et al.*, 1980). They have also managed to replace the methanol dehydrogenase and oxidase genes of *Methylophilus methylotrophus* with an NAD^+-linked alcohol dehydrogenase gene from *Bacillus stearothermophilus* by similar genetic engineering techniques (Beardsmore *et al.*, 1982). The modified organism has an improved energy efficiency for carbon conversion (10–30%) and can be grown on higher methanol concentrations (5–500 mmol).

Methylotrophs have been shown to accumulate a number of extracellular metabolites, *e.g.* amino acids, vitamins. Most work in this area has been concentrated on the production of amino acids by methylotrophic bacteria. The production of L-glutamate was first reported using the strain *Methanomonas methylovora* which accumulated up to 1 g l^{-1} (Kouno *et al.*, 1972; Oki *et al.*, 1973). Subsequently, yields of 2–12 g l^{-1} were reported (Nakayama *et al.*, 1976).

There have been many reports of efforts to overproduce serine by adding exogenous glycine to facultative serine pathway methylotrophs (Keune *et al.*, 1976; Tani *et al.*, 1978a, 1978b; Morinaga *et al.*, 1981a, 1981b). Conversion rates of up to 70% have been achieved (Morinaga *et al.*, 1981b). Methylotrophic yeasts have been reported to accumulate amino acids during growth on methanol (Sahm, 1977). *Hansenula polymorpha* has been shown to accumulate tryptophan metabolites (Denenu and Demain, 1981a, 1981b).

Vitamin B_{12} has been produced during growth on methanol by a number of pink facultative methylotrophic bacteria (Tanaka *et al.*, 1974; Toraya *et al.*, 1976; Nishio *et al.*, 1977a). Only small quantities have been excreted, up to 3 mg l^{-1} under suitable growth conditions.

The intracellular polymer poly-β-hydroxybutyrate (PHB) is synthesized by a number of methylotrophs. A recent process patent by ICI (Powell *et al.*, 1980) has shown that *Methylobacterium organophilum* can accumulate 30% or more of its dry weight as PHB under optimum conditions. It is hoped that this compound can be used as a precursor for chemical polymer synthesis. Microbial polysaccharides have been shown to be useful for a number of applications, particularly in enhanced oil recovery and the food industry. Extracellular polysaccharide production by methylotrophs has been reported in both methane-utilizing bacteria (Hou *et al.*, 1977; Hida *et al.*, 1983) and in methanol-utilizing bacteria (Tam and Finn, 1977; Kanamaru *et al.*, 1982). Polysaccharide production by the obligate methylotroph *Methylomonas mucosa* has been the subject of a number of patents (Tannahill and Finn, 1975; Finn *et al.*, 1975).

The biocatalytic capabilities of methylotrophs have proved to be of potential commercial importance. A number of enzymic and biological assay systems have been devised using the biochemical capacity of methylotrophs. The specific microestimation of a variety of alkylamines can be determined using methylotrophic enzymes. Primary amine dehydrogenase from methylamine grown *Pseudomonas* AM1 was used to determine a number of primary amines (Large *et al.*, 1969; Cook and Anthony, 1978). Trimethylamine dehydrogenase isolated from *Hyphomicrobium vulgare* was used to determine amounts of trimethylamine as low as 10 nmol (Large and McDougall, 1975). A marine methanol-utilizing bacterium was used to assay for vitamin B_{12} in sea water and proved to be more specific and ten times more sensitive than the conventional assay (Yamamoto *et al.*, 1979). A number of methylotrophic alcohol-oxidizing enzymes have been suggested for the commercial production of aldehydes and ketones, *i.e.* methanol dehydrogenase, alcohol oxidase and secondary alcohol dehydrogenase (see Anthony, 1982).

Methylotrophic organisms and their enzymes have been used as potential candidates for both electroenzymological processes and biofuel cells. For a detailed discussion of the various types of reactors and processes envisaged, the reader is referred to Higgins and Hill (1979) and Higgins *et al.* (1980b). An enzymic fuel cell based on methanol dehydrogenase has been demonstrated (Plotkin *et al.*, 1981) which could also be used as a sensitive assay system for primary alcohols. An enzymic reactor incorporating methane monooxygenase was proposed (Higgins *et al.*, 1980b) in which the reductant for the oxidation reactions was supplied directly from the cathode in the form of electrons instead of the natural reductant NADH. This requirement for NADH puts a number of commercially very promising bioconversions by MMO in jeopardy.

After the discovery of the wide substrate specificity of MMO (Colby *et al.*, 1977), it was apparent that a number of hitherto difficult and low-yield chemical syntheses could be catalyzed by MMO. This led to a flurry of activity with regard to patents in this area (Dalton *et al.*, 1977; Higgins, 1978a, 1978b, 1979; Hou *et al.*, 1979a, 1979b, 1979c). One transformation of commercial importance is the epoxidation of propene to 1,2-epoxypropane. It has been shown that whole cells of obligate methylotrophs can catalyze this conversion quantitatively and excrete the product (Stirling, 1978; Stirling and Dalton, 1979b; Dalton, 1980; Hou *et al.*, 1979, 1980; Dalton and Stirling, 1982).

The supply of reductant (NADH) for this and other MMO transformations could prove critical for the economic feasibility of potential commercial processes. This, plus the inherent instability of the enzyme, probably precludes an isolated enzyme process. Under certain conditions, the reductant could be generated in whole cells from internal polymer reserves (see Section 21.13). Alternatively, a cosubstrate such as methanol could be used as a source of reducing power for the oxidations. It seems certain that the versatile oxygenase, methane monooxygenase, will soon be put to use in a commercial process, thus securing a position of methylotrophs as one of the most important groups of industrial microorganisms.

20.15 REFERENCES

Adamse, A. D., J. Hoeks, J. A. M. DeBont and J. F. Kessel (1972). Microbial activities in soil near natural gas leaks. *Arch. Mikrobiol.*, **83**, 32–51.

Akiba, M. M., H. Ueyama, M. Seki and T. Fukimbara (1970). Identification of lower alcohol-utilizing bacteria. *J. Ferment. Technol.*, **48**, 323–328.

Alexander, M. (1981). Biodegradation of chemicals of environmental concern. *Science*, **211**, 132–138.

Ameyma, M., K. Matshita, Y. Ohno, E. Shinagawa and O. Adachi (1981). Existence of a novel prosthetic group, PQQ, in membrane bound electron transport chain lined, primary dehydrogenase of oxidative bacteria. *FEBS Lett.*, **130**, 179–183.

Anthony, C. (1978). The prediction of growth yields in methylotrophs. *J. Gen. Microbiol.*, **104**, 41–104.

Anthony, C. (1980). Methanol as substrate: theoretical aspects. In *Hydrocarbons in Biotechnology*, ed. D. E. F. Harrison, I. J. Higgins and R. Watkinson, pp. 35–58. Heyden, London.

Anthony, C. (1982). *The Biochemistry of Methylotrophs*. Academic, London.

Anthony, C. and L. J. Zatman (1964a). The microbial oxidation of methanol I. Isolation and properties of *Pseudomonas* M27. *Biochem. J.*, **92**, 609–613.

Anthony, C. and L. J. Zatman (1964b). The microbial oxidation of methanol II. The methanol-oxidizing enzyme of *Pseudonomas* M27. *Biochem. J.*, **92**, 614–621

Anthony, C. and L. J. Zatman (1965). The microbial oxidation of methanol. The alcohol dehydrogenase of *Pseudomonas* M27. *Biochem. J.*, **96**, 808–812.

Anthony, C. and L. J. Zatman (1967a). The microbial oxidation of methanol. Purification and properties of the alcohol dehydrogenase of *Pseudomonas* M27. *Biochem. J.*, **104**, 453–959.

Anthony, C. and L. J. Zatman (1967b). The microbial oxidation of methanol. The prosthetic group of the alcohol dehydrogenase of *Pseudomonas* M27. *Biochem. J.*, **104**, 960–969.

Bamforth, C. W. and J. R. Quayle (1977). Hydroxypyruvate reductase activity in *Paracoccus denitrificans*. *J. Gen. Microbiol.*, **101**, 259–267.

Barnes, L. J., J. W. Drozd, D. E. F. Harrison and G. Hamer (1976). Process considerations and techniques specific to protein production from natural gas. In *Microbial Production and Utilization of Gases*. ed. H. G. Schlegel, G. Gottschalk and N. Pfennig, pp. 301–315. Akademie der Wissenschaften, Gottingen,

Bartholomew, G, W. and M. Alexander (1979). Microbial metabolism of carbon monoxide in culture and in soil. *Appl. Environ. Microbiol.*, **37**, 932–937.

Beam, H. W. and J. J. Perry (1973). Co-metabolism as a factor in microbial degradation of cycloparaffinic hydrocarbons. *Arch. Mikrobiol.*, **91**, 87–90.

Beam, H. W. and J. J. Perry (1974). Microbial degradation of cycloparaffinic hydrocarbons *via* co-metabolism and commensalism. *J. Gen. Microbiol.*, **82**, 163–169.

Beardsmore, A. J., S. H. Collins, K. A. Powell and P. J. Senior (1982). *Eur. Pat. Appl.* 82 302 608.3.

Bellion, E., J. A. Bolbot and T. D. Lash (1981). Generation of glyoxlate in methylotrophic bacteria. *Curr. Microbiol.*, **6**, 367–372.

Bellion, E. and G. T. S. Wu (1978). Alcohol dehydrogenases from a facultative methylotrophic bacterium. *J. Bacteriol.*, **135**, 251–258.

Best, D. J. and I. J. Higgins (1981). Methane-oxidising activity and membrane morphology in a methanol-grown obligate methanotroph, *Methylosinus trichosporium* OB3b. *J. Gen. Microbiol.*, **125**, 73–84.

Bolbot, J. A. and C. Anthony (1980). The metabolism of 1,2-propanediol by the facultative methylotroph *Pseudomonas* AM1. *J. Gen. Microbiol.*, **120**, 245–254.

Brammer, W. J. (1981). Possible applications of modern genetic techniques to SCP organisms. In *Microbial Growth on C1 Compounds*, ed. H. Dalton, pp. 312–316. Heyden, London.

Bringer, S., B. Sprey and H. Sahm (1979). Purification and properties of alcohol oxidase from *Poria contigua*. *Eur. J. Biochem.*, **101**, 563–570.

Byrom, D. (1981). Taxanomy of methylotrophs: a reappraisal. In *Microbial Growth on C1 Compounds*, ed. H. Dalton, pp. 278–284. Heyden, London.

Cameron, S. (1982). Wood alcohol has come a long way. *Financial Times*, London, March 3, 1982.

Chandara, T. S. and Y. I. Shethna (1977). Oxalate, formate, formamide and methanol metabolism in *Thiobacillus novellus*. *J. Bacteriol.*, **131**, 389–398.

Chida, K., G. J. Shen, T. Kodama and Y. Minoda (1983). Acidic polysaccharide production from methane by a new methane-oxidizing bacterium H–2. *Agric. Biol. Chem.*, **47**, 275–280.

Colby, J. and H. Dalton (1976). Some properties of a soluble methane monooxygenase from *Methylococcus capsulatus* strain Bath. *Biochem. J.*, **157**, 495–497.

Colby, J. and H. Dalton (1978). Resolution of the methane monooxygenase of *Methylococcus capsulatas* (Bath) into three components. Purification and properties of Component C, a flavoprotein. *Biochem. J.*, **171**, 461–468.

Colby, J. and H. Dalton (1979). Characterization of the second prosthetic group of the flavoenzyme NADH-acceptor reductase (Component C) of the methane monooxygenase from *Methylococcus capsulatus* (Bath). *Biochem. J.*, **177**, 903–908.

Colby, J., H. Dalton and R. Whittenbury (1979). Biological and biochemical aspects of microbial growth on C_1 compounds. *Annu. Rev. Microbiol.*, **33**, 481–517.

Colby, J., D. I. Stirling and H. Dalton (1977). The soluble methane monooxygenase of *Methylococcus capsulatus* (Bath). Its ability to oxygenate *n*-alkanes, *n*-alkenes, ethers and alicyclic, aromatic and heterocyclic compounds. *Biochem. J.*, **165**, 394–402.

Colby, J. and L. J. Zatman (1972). Hexose phosphate synthase and tricarboxylic acid cycle enzymes in bacterium 4B6, on obligate methylotroph. *Biochem. J.*, **128**, 1373–1376.

Colby, J. and L. J. Zatman (1975a). Tricarboxylic acid cycle and related enzymes in restricted facultative methylotrophs. *Biochem. J.*, **148**, 505–511.

Colby, J. and L. J. Zatman (1975b). Enzymological aspects of the pathways of trimethylamine oxidation and C_1 assimilation in obligate methylotrophs and restricted faculatative methylotrophs. *Biochem. J.*, **148**, 513–520.

Conrad, R., O. Meyer and W. Seiler (1981). Role of carboxydobacteria in consumption of atmospheric carbon monoxide by soil. *Appl. Environ. Microbiol.*, **42**, 211–215.

Cook, R. J. and C. Anthony (1978). The ammonia and methylamine active transport system of *Aspergillus nidulans*. *J. Gen. Microbiol.*, **109**, 265–274.

Cox, R. B. and J. R. Quayle (1975). The autotrophic growth of *Micrococcus denitrificans* on methanol. *Biochem. J.*, **150**, 569–571.

Dalton, H. (1977). Ammonia oxidation by the methane oxidizing bacterium *Methylococcus capsulatus* strain Bath. *Arch. Microbiol.*, **114**, 273–279.

Dalton, H. (1980). Transformations by methane monooxygenase. In *Hydrocarbons in Biotechnology*, ed. D. E. F. Harrison, I. J. Higgins and R. Watkinson, pp. 85–98. Heyden, London.

Dalton, H., J. Colby and D. I. Stirling (1977). *Br. Pat* 1 603 864.

Dalton, H. and D. I. Stirling (1982). Co-metabolism. *Philos. Trans. R. Soc. London, Ser. B*, **297**, 481–496.

Davies, S. L. and R. Whittenbury (1970). Fine structure of methane and other hydrocarbon utilizing bacteria. *J. Gen. Microbiol.*, **61**, 227–232.

DeBeer, R., J. A. Duine, J. Frank and J. Westerling (1983). The role of pyrroloquinoline semiquinone forms in the mechanism of action of methanol dehydrogenase. *Eur. J. Biochem.*, **130**, 105–109.

De Boer, W. E. and W. Hazeu (1972). Observations on the fine structure of a methane-oxidizing bacterium. *Antonie van Leeuwenhoek*, **38**, 33–47.

De Duve, C. (1969). Evolution of the peroxisome. *Ann. N.Y. Acad. Sci.*, **168**, 369–381.

De Duve, C. and P. Baudhuin (1966). Peroxisomes (microbodies and related particles). *Physiol. Rev.*, **46**, 323–357.

DeMaeyer, E., D. Skup, K. S. N. Prasad, J. DeMaeyer-Guignard, B. Williams, P. Meacock, G. Sharde, D. Pioli, J. Hennam, W. Schuch and K. Atherton (1982). Expression of a chemically synthesized human 1 interferon gene. *Proct. Natl. Acad. Sci. USA*, **79**, 4256–4259.

Denenu, E. O. and A. L. Demain (1981a). Enzymatic basis for the overproduction of tryptophan and its metabolities in *Hansenula* mutants. *Appl. Environ. Microbiol.*, **42**, 497–501.

Denenu, E. O. and A. L. Demain (1981b). Relationship between genetic deregulation of *Hansenula polymorpha* and production of tryptophan metabolites. *Eur. J. Appl. Microbiol. Biotechnol.*, **13**, 202–207.

Dijkhuizen, L., M. Knight and W. Harder (1978). Metabolic regulation in *Pseudomonas oxalaticus* OX1. Autotrophic and heterotrophic growth on mixed substrates. *Arch. Microbiol.*, **116**, 77–83.

Dikjhuizen, L., J. W. C. Timmerman and W. Harder (1979). A pyridine nucleotide-independent membrane-bound formate dehydrogenase in *Pseudomonas oxalaticus* OX1. *FEMS Microbiol. Lett.*, **6**, 53–56.

Drozd, J. W. and P. W. McCarthy (1981). Mathematical model of microbial hydrocarbon oxidation. In *Microbial growth on C_1 Compounds*, ed. H. Dalton, pp. 360–390. Heyden, London.

Duine, J. A., J. Frank and L. G. DeRuiter (1979b). Isolation of a methanol dehydrogenase with a functional coupling to cytochrome *c*. *J. Gen. Microbiol.*, **115**, 523–526.

Enebo, L. (1967). A methane-consuming green alga. *Acta Chem. Scand.*, **21**, 625–632.

Ferenci, T. (1974). Carbon monoxide-stimulated respiration in methane-utilizing bacteria. *FEBS Lett.*, **41**, 94–98.

Ferenci, T., T. Strom and J. R. Quayle (1975). Oxidation of carbon monoxide and methane by *Pseudomonas methanica*. *J. Gen. Microbiol.*, **91**, 79–91.

Finn, R. K., A. L. Tannahill and J. E. Laptewicz (1975). *US Pat.* 3 923 782.

Forrest, H. S., S. A. Salisbury and C. G. Kilty (1980). A mechanism for the enzymic oxidation of methanol involving methoxatin. *Biochem. Biophys. Res. Commun.*, **97**, 248–251.

Foster, J. W. (1962). Hydrocarbons as substrates for microorganisms. *Antonie van Leewenhoek*, **28**, 241–274.

Fukui, S., S. Kawamoto, S. Yasuhara, A. Tanaka, M. Osumi and F. Imaizumi (1975). Microbody of methanol-grown yeasts. Localization of catalase and flavin-dependent alcohol oxidase in the isolated microbody. *Eur. J. Biochem.*, **59**, 561–566.

Galachenko, V. F. and N. E. Suzina (1977). Peculiarities of membrane apparatus organization of methanotrophic bacteria. In *Microbial Growth on C_1 Compounds. Abstracts of the Second International Symposium on Microbial Growth on C_1 Compounds*, ed. G. K. Skryabin, M. V. Ivanov, E. N. Kondratjeva, G. A. Zavarzin, Yu. A. Trotsenko and A. I. Nesterov, pp. 12–15. USSR Academy of Sciences, Pushchino, USSR.

Gautier, F. and R. Bonewald (1980). The use of pladmid R11–62 and derivatives for gene cloning in the methanol-utilizing *Pseudomonas* AM1. *Mol. Gen. Genet.*, **178**, 375–380.

Goldberg, I. (1976). Purification and properties of a methanol-oxidizing enzyme in *Pseudomonas* C. *Eur. J. Biochem.*, **63**, 233–240.

Green, P. N. and I. J. Bousefield (1981). The taxonomy of the pink-pigmented facultatively methylotrophic bacteria. In *Microbial Growth on C_1 Compounds*, ed. H. Dalton, pp. 92–104. Heyden, London.

Hanson, R. S. (1980). Ecology and diversity of methylotrophic organisms. *Adv. Appl. Microbiol.*, **26**, 3–39.

Harder, W. and J. P. van Dijken (1976). Theoretical considerations on the relation between energy production and growth of methane-utilizing bacteria. In *Microbial Production and Utilization of Gases*, ed. H. G. Schlegel, G. Gottschalk and N. Pfennig, pp. 403–418. Akademie der Wissenschaften, Gottingen.

Harrison, D. E. F., H. H. Topiwala and G. Hamer (1972). Yield and productivity in single cell protein production from methane and methanol. In *Fermentation Technology Today*, ed. G. Turui, pp. 491–495. Society of Fermentation Technology, Kyoto.

Harwood, J. H., E. Williams and B. W. Bainbridge (1972). Mutation of the methane oxidizing bacterium, *Methylococcus capsulatus. J. Appl. Bacteriol.*, **35**, 99–108.

Haas, D. and B. W. Holloway (1978). Chromosome mobilization by the R plasmid R68.45: a tool in *Pseudomonas* genetics. *Mol. Gen. Genet.*, **152**, 229–237.

Heichel, G. H. (1973). Removal of carbon monoxide by field and forest soils. *J. Environ. Qual.*, **2**, 419–423.

Hennam, J. E., A. E. Cunningham, G. S. Sharpe and K. T. Atherton (1982). Expression of eukaryotic coding sequences in *Methylophilus methylotrophus. Nature (London)*, **297**, 80–83.

Heptinstall, J. and J. R. Quayle (1970). Pathways leading to and from serine during growth of *Pseudomonas* AM1 on C_1 compounds or succinate. *Biochem. J.*, **117**, 563–572.

Higgins, I. J. (1978a). *Br. Pat. Appl.* 78 19 712.

Higgins, I. J. (1978b). *Br. Pat. Appl.* 78 35 123.

Higgins, I. J., D. J. Best and R. C. Hammond (1980). New findings in methane-utilizing bacteria highlight their importance in the biosphere and their commercial potential. *Nature (London)*, **286**, 561–564.

Higgins, I. J., D. J. Best, R. C. Hammond and D. Scott (1981). Methane-oxidizing microorganisms. *Microbiol. Rev.*, **45**, 556–590.

Higgins, I. J. and H. A. O. Hill (1979). Microbial generation and interconversion of energy sources. *Symp. Soc. Gen. Microbiol.*, **29**, 359–378.

Higgins, I. J., R. C. Hammond, E. Plotkin, H. A. O. Hill, K. Vosaki, M. J. Eddowes and A. E. G. Cass (1980b). Electroenzymology and biofuel cells. In *Hydrocarbons in Biotechnology*, ed. D. E. F. Harrison, I. J. Higgins and R. Watkinson, pp. 181–194. Heyden, London.

Higgins, I. J., R. C. Hammond, F. S. Sariaslani, D. Best, M. M. Davies, S. E. Tryhorn and F. Taylor (1979). Biotransformation of hydrocarbons and related compounds by whole organism suspension of methane-grown *Methylosinus trichosporium* OB3b. *Biochem. Biophys. Res. Commun.*, **89**, 671–677.

Hirsch, P. and S. F. Conti (1964). Biology of budding bacteria. Enrichment isolation and morphology of *Hyphomicrobium spp. Arch Mikrobiol.*, **48**, 339–357.

Holloway, B. W. (1979). Plasmids that mobilize bacterial chromosomes. *Plasmid*, **2**, 1–19.

Holloway, B. W. (1981). The application of *Pseudomonas*-based genetics to methylotrophs. In *Microbial Growth on C_1 compounds*, ed. H. Dalton, pp. 317–324. Heyden, London.

Holloway, B. W., C. Crowther, P. Royle and M. Nayudu (1980). R-plasmids and bacterial chromosome transfer. In *Antibiotic Resistance, Transportation and other Mechanisms*, ed. S. Mitsuhashi, *et al.*, pp. 19–27. Avicenum, Prague.

Horvath, R. S. (1972). Microbial co-metabolism and the degradation of organic compounds in nature. *Bacteriol. Rev.*, **36**, 146–155.

Hou, C. T., A. I. Laskin and R. N. Patel (1979). Growth and polysaccharide production by *Methylocystis parvus* OBBP on methanol. *Appl. Environ. Microbiol.*, **37**, 800–804.

Hou, C. T., R. N. Patel and A. Laskin (1979a). *Br. Pat. Appl.* 79 13 054.

Hou, C. T., R. N. Patel and A. Laskin (1979b). *Br. Pat. Appl.* 79 13 063.

Hou, C. T., R. N. Patel, A. Laskin and B. Barnabe (1979a). Microbial oxidation of gaseous hydrocarbons: epoxidation of C_2 to C_4 n-alkenes by methylotrophic bacteria. *Appl. Environ. Microbiol.*, **38**, 127–134.

Hou, C. T., R. N. Patel, A. Laskin, N. Barnabe and I. Marczak (1979b). Identification and purification of a nicotinamide adenine dinucleotide dependent secondary alcohol dehydrogenase from C_1-utilizing microbes. *FEBS Lett.*, **101**, 179–183.

Hou, C. T., R. N. Patel, A. I. Laskin and N. Barnabe (1980). Microbial oxidation of gaseous hydrocarbons: oxidation of lower n-alkenes and n-alkanes by resting cell suspensions of various methylotrophic bacteria, and the effect of methane metabolites. *FEMS Microbiol. Lett.*, **9**, 267–270.

Hou, C. T., R. N. Patel, A. I. Laskin, N. Barnabe and I. Marczak (1979). Microbial oxidation of gaseous hydrocarbons: production of methyl ketones from their corresponding secondary alcohols by methane- and methanol-grown microbes. *Appl. Environ. Microbiol.*, **38**, 135–142.

Hubley, J. H. (1975). Ph.D. Thesis. University of Edinburgh.

Hulbert, M. H. and S. Krawiec (1977). Co-metabolism: a critique. *J. Theor. Biol.*, **69**, 287–291.

Hyder, S. L., A. Mayers and M. L. Cayer (1979). Membrane modulation in a methylotrophic bacterium *Methylcoccus capsulatas* (Texas) as a function of growth substrate. *Tissue Cell*, **11**, 597–610.

Imelik, B. (1948). Oxydation de cyclohexane par *Pseudomonas aeruginosa. C. R. Hebd. Seances Acad. Sci.*, **226**, 2082–2083.

Jensen, R. A. (1963). Carbon nutrition of some microorganisms decomposing halogen-substituted aliphatic acid. *Acta Agric. Scand.*, **13**, 404–412.

Jeyaseelan, K. and J. R. Guest (1979). Transfer of antibiotic resistance to facultative methylotrophs with plasmid R68.45. *FEMS Microbiol Lett.*, **6**, 87–89.

Johnston, A. W. B., S. M. Setchell and J. E. Beringer (1978). R-primes in *Rhizobium. J. Gen. Microbiol.*, **104**, 209–218.

Kanamaru, K., T. Hieda, Y. Kwamuro, Y. Mikami, Y. Obi and T. Kisaki (1982). Isolation and characterization of a *Hyphomicrobium* species and its polysaccharide formation from methanol. *Agric. Biol. Chem.*, **46**, 2411–2417.

Kato, N., H. Sahm and F. Wagner (1979). Steady state kinetics of formaldehyde dehydrogenase and formate dehydrogenase from a methanol-utilizing yeast *Candida bioidinii. Biochem. Biophy. Acta*, **566**, 12–20.

Kato, N., Y. Tani and K. Ogata (1974). Enzyme system for methanol oxidation in yeasts. *Agric. Biol. Chem.*, **38**, 675–677.

Karo, N., K. Tsuji, Y. Tani and K. Agata (1974). A methanol-utilizing actinomycete. *J. Ferment. Technol.* **52**, 917–920.

Kato, N., K. Tsuji, H. Ohashi, Y. Tani and K. Ogata (1977). Two assimilation pathways of C_1-compounds in *Streptomyces* sp. no. 239 during growth on methanol. *Agric. Biol. Chem.*, **41**, 29–34.

Kuene, H., H. Sahm and F. Wagner (1976). Production of L-serine by the methanol utilization bacterium, *Pseudomonas* 3ab. *Eur. J. Appl. Microbiol.*, **2**, 175–184.

Komagata, K. (1981). Taxonomic studies of methanol-utilizing yeasts. In *Microbial Growth on C_1 Compounds*, ed. H. Dalton, pp. 301–311. Heyden, London.

Kortstee, G. J. J. (1980). The homoisocitrate–glyoxylate cycle in pink, facultative methylotrophs. *FEMS Microbiol. Lett.*, **8**, 59–65.

Kortsee, G. J. J. (1981). The second part of the ICL⁻ pathway. In *Microbial Growth on C_1 Compounds*, ed. H. Dalton, pp. 211–219. Heyden, London.

Kuono, K., T. Oki, A. Kitai and A. Osaki (1972). *US Pat.* 3 663 370.

Nakayama, K., M. Kobata, Y. Tanaka, T. Nomura and R. Katsumata (1976). *US Pat.* 3 939 042.

Nishio, N., Y. Tsuchiya, M. Hayashi and S. Nagai (1977a). A fed-batch culture of methanol-utilizing bacteria with pH stat. *J. Ferment. Technol.*, **55**, 151–155.

Nunn, D. N., G. L. Fulton and M. L. Lidstrom (1983). Cloning by complemention of the genes for three C_1 specific enzymes from the methylotroph *Pseudomonas* AM1. *Abstr. Annu. Meet. ASM 1983*, 170.

O'Connor, M. L. (1981a). Regulation and genetics in facultative methylotrophic bacteria. In *Microbial Growth on C_1 Compounds*, ed. H. Dalton, pp. 294–300. Heyden, London.

O'Connor, M. L. (1981b). Extension of the model concerning linkage of genes coding for C-1 related functions in *Methylobacterium organophilum*. *Appl. Environ. Microbiol.*, **41**, 437–441.

O'Connor, M. L. and R. S. Hanson (1977). Enzyme regulation in *Methylobacterium organophilum*. *J. Gen. Microbiol.*, **101**, 327–332.

O'Connor, M. L. and R. S. Hanson (1978). Linkage relationship between mutants of *Methylobacterium organophilum* impaired in their ability to grow on one-carbon compounds. *J. Gen. Microbiol.*, **104**, 105–111.

O'Connor, M. L., A. E. Wopat and R. S. Hanson (1977). Genetic transformation in *Methylobacterium organophilum*. *J. Gen. Microbiol.*, **98**, 265–272.

Ogata, K., H. Nishikawa and M. Ohsugi (1969). A yeast capable of utilizing methanol. *Agric. Biol. Chem.*, **33**, 1519–1520.

O'Keefe, D. T. and C. Anthony (1980). The interaction between methanol dehydrogenase and the autoreducible cytochromes *c* of the facultative methylotroph *Pseudomonas* AM1. *Biochem. J.*, **190**, 481–484.

Oki, T., A. Kitai, K. Kudno and A. Ozaki (1973). Production of L-glutamic acid by methanol-utilizing bacteria. *J. Gen. Appl. Microbiol.*, **19**, 79–83.

Paraskeva, C. (1979). Transfer of kanamycin resistance mediated by plasmid R68.45 in *Paracoccus denitrificans*. *J. Bacteriol.*, **139**, 1062–1064.

Patel, R. N. and A. Felix (1976). Microbial oxidation of methane and methanol: crystallization and properties of methanol dehydrogenase from *Methylosinus sporium*. *J. Bacteriol.*, **128**, 413–424.

Patel, R. N., C. T. Hou and A. Felix (1978a). Microbial oxidation of methane and methanol: isolation of methane-utilizing bacteria and characterization of a facultative methane-utilizing isolate. *J. Bacteriol.*, **136**, 352–358.

Patel, R. N., C. T. Hou and A. Felix (1978b). Microbial oxidation of methane and methanol: crystallization of methanol dehydrogenase and properties of holo- and apo-methanol dehydrogenase from *Methylomonas methanica*. *J. Bacteriol.*, **133**, 641–649.

Patel, R. N., C. T. Hou, A. I. Laskin and A. Felix (1982). Microbial oxidation of hydrocarbons: properties of a soluble methane mono-oxygenase from a facultative methane-utilizing organism, *Methylobacterium* sp. strain CRL-26. *Appl. Environ. Microbiol.*, **44**, 1130–1137.

Patel, R. N., C. T. Hou, A. I. Laskin, A. Felix and P. Derelanko (1979). Microbial oxidation of gaseous hydrocarbons. II. Hydroxylation of alkanes and epoxidation of alkenes by cell-free particulate fractions of methane-utilizing bacteria. *J. Bacteriol.*, **139**, 675–679.

Patel, R. N., C. T. Hou, A. I. Laskin, A. Felix and P. Derelanko (1980). Microbial oxidation of gaseous hydrocarbons: production of secondary alcohols from corresponding *n*-alkanes by methane-utilizing bacteria. *Appl. Environ. Microbiol.*, **39**, 720–726.

Paterson, D. (1978). Methane from the bowels of the earth. *New Scientist*, 896–898.

Patt, T. E., G. C. Cole, J. Bland and R. S. Hanson (1974). Isolation and characterization of bacteria that grow on methane and organic compounds as sole sources of carbon and energy. *J. Bacteriol.*, **120**, 955–964.

Patt, T. E. and R. S. Hanson (1978). Intracytoplasmic membrane, phospholipid and sterol content of *Methylobacterium organophilum* cells grown under different conditions. *J. Bacteriol.*, **134**, 636–644.

Plotkin, E. V., I. J. Higgins and H. A. O. Hill (1981). Methanol dehydrogenase bioelectrochemical cell and alcohol detector. *Biotechnol. Lett.*, **3**, 187–192.

Powell, K. A., B. A. Collinson and K. R. Richardson (1980). Eur. Pat. Appl. 80 300 432.4.

Proctor, H. M., J. R. Norris and D. W. Ribbons (1969). Fine structure of methane-utilising bacteria. *J. Appl. Bacteriol.*, **32**, 118–121.

Quayle, J. R. (1980). Microbial assimilation of C_1 compounds. *Biochem. Soc. Trans.*, **8**, 1–10.

Quayle, J. R. (1980). Aspects of the regulation of methylotrophic metabolism. *FEBS Lett.*, **117**, K16–K27.

Quayle, J. R. and T. Ferenci (1978). Evolutionary aspects of autotrophy. *Microbiol. Rev.*, **42**, 251–273.

Quayle, J. R. and N. Pfennig (1975). Utilization of methanol by *Rhodospirillacaea*. *Arch. Microbiol.*, **102**, 193–198.

Reed, H. L. (1976). Ph.D. Thesis, University of Warwick.

Ribbons, D. W. and J. L. Michalover (1970). Methane oxidation by cell-free extracts of *Methylococcus capsulatus*. *FEBS Lett.*, **11**, 41–44.

Rodionov, Y. V. and E. V. Zakharova (1980). Two pathways of formate oxidation in methylotrophic bacteria. *Biochem. (USSR)*, **45**, 654–661.

Roggenkamp, R., H. Sahm, W. Hinkelmann and F. Wagner (1975). Alcohol oxidase and catalase in peroxisomes of *Candida boidinii*. *Eur. J. Biochem.*, **59**, 231–236.

Romanova, A. K. and A. N. Nozhevnikova (1977). Assimilation of one-carbon compounds by carboxydobsacteria. In *Microbial Growth on C_1 Compounds*, ed. G. K. Skryabin *et al.*, pp. 109–110. USSR Academy of Sciences, Puschino.

Sahm, H. (1977). Metabolism of methanol by yeasts. *Adv. Biochem. Eng.*, **6**, 77–103.

Sakaguchi, K., R. Kurane and M. Murata (1975). Assimilation of formaldehyde and other C_1 compounds by *Gliocladium deliquescens* and *Paecilomyces varioti*. *Agric. Biol. Chem.*, **39**, 1695–1702.

Salem, A. R., A. J. Hacking and J. R. Quayle (1973). Cleavage of malylcoenzyme A into acetylcoenzyme A and glyoxylate by *Pseudomonas* AM1 and other C_1 unit-oxidising bacteria. *Biochem. J.*, **136**, 89–96.

Salisbury, S. A., H. S. Forrest, W. B. T. Cruse and O. Kennard (1979). A novel coenzyme from bacterial dehydrogenases. *Nature (London)*, **280**, 843–844.

Scott, D., D. J. Best and I. J. Higgins (1981). Intracytoplasmic membranes in oxygen-limited chemostat cultures of *Methylosinus trichosporium* OB3b: biocatalytic implications of physiologically balanced growth. *Biotechnol. Lett.*, **3**, 641–644.

Scott, D., J. Brannan and I. J. Higgins (1981). The effect of growth conditions on intracytoplasmic membranes and methane mono-oxygenase activities in *Methylosinus trichosporium* OB3b. *J. Gen. Microbiol.*, **125**, 63–72.

Seifert, E., and N. Pfennig (1979). Chemoautotrophic growth of *Rhodopseudomonas* species with hydrogen and chemotrophic utilization of methanol and formate. *Arch. Microbiol.*, **122**, 177–182.

Seiler, W. and V. Schmidt (1976). The role of microbes in the cycle of atmospheric trace gases, especially hydrogen and carbon monoxide. In *Microbial Production and Utilization of Gases*, ed. H. G. Schlegel, G. Gottschalk and N. Pfennig, pp. 35–46. Akademie der Wissenschaften, Gottingen.

Senior, P. J. and J. Windass (1980). The ICI single cell protein process. *Biotechnol. Lett.*, **2**, 205–210.

Smith, S. R. L. (1981). Some aspects of ICI's single cell protein process. In *Microbial Growth on C_1 compounds*, ed. H. Dalton, pp. 342–348. Heyden, London.

Smith, U. and D. W. Ribbons (1970). Fine structure of *Methanomonas methanooxidans*. *Arch. Mikrobiol.*, **74**, 116–122.

Stanley, S. H. and H. Dalton (1982). Role of ribulose-1,5-bisphosphate carboxylase/oxygenase in *Methylococcus capsulatus* (Bath). *J. Gen. Microbiol.*, **128**, 2927–2935.

Stanley, S. H., S. D. Prior, D. J. Leak and H. Dalton (1983). Copper stress underlies the fundamental change in intracellular location of methane mono-oxygenase in methane-oxidizing organisms: studies in batch and continuous cultures. *Biotechnol. Lett.*, **5**, 487–492.

Stirling, D. I. (1978). Ph.D. Thesis, University of Warwick.

Stirling, D. I., J. Colby and H. Dalton (1979). A comparison of the substrate and electron donor specificities of the methane mono-oxygenases from three strains of methane-oxidizing bacteria. *Biochem. J.*, **177**, 361–364.

Stirling, D. I. and H. Dalton (1976). Cometabolism by an obligate methanotrophic bacterium. *Methylococcus capsulatus*. *Proc. Soc. Gen. Microbiol.*, **4**, 31.

Stirling, D. I. and H. Dalton (1978). Purification and properties of an $NAD(P)^+$-linked formaldehyde dehydrogenase from *Methylococcus capsulatus* (Bath). *J. Gen. Microbiol.*, **107**, 19–29.

Stirling, D. I. and H. Dalton (1979a). Properties of the methane mono-oxygenase from extracts of *Methylosinus trichosporium* OB3b and evidence for its similarity to the enzyme from *Methylococcus capsulatus* (Bath). *Eur. J. Biochem.*, **96**, 205–212.

Stirling, D. I. and H. Dalton (1979b). The fortuitous oxidation and cometabolism of various carbon compounds by wholecell suspensions of *Methylococcus capsulatus* (Bath). *FEMS Microbiol. Lett.*, **5**, 315–318.

Stirling, D. I. and H. Dalton (1980). Oxidation of dimethyl ether, methyl formate and bromomethane by *Methylococcus capsulatus* (Bath). *J. Gen. Microbiol.*, **116**, 277–283.

Stirling, D. I. and H. Dalton (1981a). Fortuitous oxidations by methane-utilizing bacteria. *Nature (London)*, **291**, 169–170.

Stirling, D. I. and H. Dalton (1981b). Electron donors for methane mono-oxygenase *in vivo*. Poster presented at 3rd International Symposium on Microbial Growth on C_1 Compounds. Sheffield, UK.

Stirling, L. A., R. J. Watkinson and I. J. Higgins (1977). Microbial metabolism of alicyclic hydrocarbons: isolation and properties of a cyclohexane-degrading bacterium. *J. Gen. Microbiol.*, **99**, 119–125.

Strøm, T., T. Ferenci and J. R. Quayle (1974). The carbon assimilation pathways of *Methylococcus capsulatus*, *Pseudomonas methanica* and *Methylosinus trichosporium* (OB3b) during growth on methane. *Biochem. J.*, **144**, 465–476.

Tam, K. T. and R. K. Finn (1977). Polysaccharide formation by a *Methylomonas*. In *Extracellular Microbial Polysaccharides*, ed. P. A. Sandford and A. Voskin, pp. 58–80. American Chemical Society, Washington, DC.

Tanaka, A., Y. Ohya, S. Shimizu and S. Fukui (1974). Production of vitamin B_{12} by methanol-assimilating bacteria. *J. Ferment. Technol.*, **52**, 921–924.

Tani, Y., T. Kanagawa, A. Hanpongkittikun, K. Ogata and H. Yamada (1978b). Production of L-serine by a methanolutilizing bacterium. *Arthrobacter globiformis* SK-200. *Agric. Biol. Chem.*, **42**, 2275–2279.

Tani, Y., N. Kato and H. Yamada (1978a). Utilization of methanol by yeasts. *Adv. Appl. Microbiol.*, **24**, 165–186.

Tannahill, A. L. and R. K. Finn (1975). *US Pat.* 3 878 045.

Tatra, P. K. and P. M. Goodwin (1982). Conjugative transfer of chromosomal markers in the facultative methylotroph *Pseudomonas* AM1 using plasmid R68.45. *Soc. Gen. Microbiol. Quart.*, **9**, M7.

Taylor, S. C. (1977). Evidence for the presence of ribulose-1,5-bisphosphate carboxylase and phosphoribulokinase in *Methylococcus capsulatus* (Bath). *FEMS Microbiol. Lett.*, **2**, 305–307.

Taylor, S. C., H. Dalton and C. S. Dow (1981). Ribulose-1,5-bisphosphate carboxylase/oxygenase and carbon assimilation in *Methylococcus capsulatus* (Bath). *J. Gen. Microbiol.*, **122**, 89–94.

Tobari, J. and Y. Harada (1981). Amicyanin: an electron acceptor of methylamine dehydrogenase. *Biochem. Biophys. Res. Commun.*, **101**, 502–508.

Tonge, G. M., D. E. F. Harrison and I. J. Higgins (1977). Purification and properties of the methane mono-oxygenase enzyme system from *Methylococcus trichosporium* OB3b. *Biochem. J.*, **161**, 333–344.

Tonge, G. M., D. E. F. Harrison, C. J. Knowles and I. J. Higgins (1975). Properties and partial purification of the methane-oxidising enzyme system from *Methylosinus trichosporium*. *FEBS Lett.*, **58**, 293–299.

Toraya, R., B. Yongsmith, S. Honda, A. Tanaka and S. Fukui (1976). Production of vitamin B_{12} by a methanol-utilizing bacterium. *J. Ferment. Technol.*, **54**, 102–108.

Toukdarian, A. E. and M. F. Lidstrom (1983). Isolation and characterization of *nif* mutants in the obligate methanotroph *Methylosinus* AT1. *Abstr. Annu. Meet. ASM 1983*, 189.

Trotsenko, Y. A. (1976). Isolation and characterization of obligate methanotrophic bacteria. In *Microbial Production and Utilization of Gases*, ed. H. G. Schlegel, G. Gottschalk and N. Pfennig, pp. 329–336. Akademie der Wissenschaften, Gottingen.

Tye, R. and A. Willets (1973). Fungal growth on methanol. *J. Gen. Microbiol.* **77**, 1P.

Ueda, S., K. Sato and S. Shimizu (1981). Gloxylate formation from mesaconyl-CoA and its related reactions in a methanol-utilizing bacterium, *Protaminabacter ruber*. *Agric. Biol. Chem.*, **45**, 823–830.

Uffen, R. L. (1981). Metabolism of carbon monoxide. *Enzyme Microbiol. Technol.*, **3**, 197–206.

van Dijken, J. P. and W. Harder (1975). Growth yield of microorganisms on methanol and methane. A theoretical study. *Biotechnol. Bioeng.*, **17**, 15–30.

van Dijken, J. P., W. Harder and J. R. Quayle (1981). Energy transduction and carbon assimilation in methylotrophic yeasts. In *Microbial Growth on C_1 compounds*, ed. H. Dalton, pp. 191–210. Heyden, London.

van Dijken, J. P., G. J. Oostra-Demkes, R. Otto and W. Harder (1976). *S*-formylglutathione: The substrate of formate dehydrogenase in methanol-utilizing yeasts. *Arch. Microbiol.*, **111**, 77–83.

van Dijken, J.P., M. Veenhuis, C. A. Vermeulen and W. Harder (1975). Cytochemical localization of catalase activity in methanol-grown *Hansenula polymorpha*. *Arch. Microbiol.*, **105**, 261–267.

Warner, P. J. and I. J. Higgins (1977). Examination of obligate and facultative methylotrophs for plasmid DNA. *FEMS Microbiol. Lett.*, **1**, 339–342.

Warner, P. J., J. W. Drozd and I. J. Higgins (1983). The effect of amino acids analogues on growth of an obligate methanotroph, *Methylosinus trichosporium* OB3b. *J. Chem. Technol. Biotechnol.*, **338**, 29–34.

Warner, P. J., I. J. Higgins and J. W. Drozd (1980). Conjugative transfer of antibiotic resistance to methylotrophic bacteria. *FEMS Microbiol. Lett.*, **7**, 181–185.

Whittenbury, R., J. Colby, H. Dalton and H. L. Reed (1976). Biology and ecology of methane oxidizers. In *Microbial Production and Utilization of Gases*, ed. H. G. Schlegel, G. Gottschalk and N. Pfennig, pp. 281–293. Akademie der Wissenschaften, Gottingen.

Whittenbury, R. and H. Dalton (1981). The methylotrophic bacteria. In *The Prokaryotes*, ed. M. P. Starr, H. Stlp, H. G. Truper, A. Balows and H. G. Schlegel, pp. 894–902. Springer-Verlag, Berlin.

Whittenbury, R. and D. P. Kelly (1977). Autotrophy: a conceptual phoenix. *Symp. Soc. Gen. Microbiol.*, **27**, 119–149.

Whittenbury, R., K. C. Phillips and J. F. Wilkinson (1970). Enrichment, isolation and some properties of methane-utilising bacteria. *J. Gen. Microbiol.*, **61**, 205–218.

Williams, E., M. A. Shimmin and B. W. Bainbridge (1977). Mutation in the obligate methylotrophs *Methylococcus capsulatus* and *Methylomonas albus*. *FEMS Microbiol. Lett.*, **2**, 293–296.

Windass, J. D., N. J. Worsey, E. M. Pioli, D. Pioli, P. T. Barth, K. T. Atherton, E. C. Dart, D. Byrom, K. Powell and P. J. Senior (1980). Improved conversion of methanol to single-cell protein by *Methylophilus methylotrophus*. *Nature* (*London*), **287**, 396–401.

Wolf, H. J. (1981). Biochemical characterization of methane-oxidizing yeast. In *Microbial Growth on C_1 Compounds*, ed. H. Dalton, pp. 202–210. Heyden, London.

Wolf, H. J. and R. S. Hanson (1978). Alcohol dehydrogenase from *Methylobacterium organophilum*. *Appl. Environ. Microbiol.*, **36**, 105–114.

Wolf, H. J. and R. S. Hanson (1979). Isolation and characterisation of methane utilising yeasts. *J. Gen. Microbiol.*, **114**, 187–194.

Wolf, H. J. and R. S. Hanson (1980). Identification of methane-utilising yeasts. *FEMS Microbiol. Lett.*, **7**, 177–179.

Woodland, M. P. and H. Dalton (1984). Purification and properties of component A of the methane mono-oxygenase from *Methylococcus capsulatus* (Bath). *J. Biol. Chem.*, **259**, 53–60.

Yamamoto, M., R. Okamoto and T. Invi (1979). Application of a marine methanol-utilizing bacterium for bioassay of vitamin B_{12} in sea water. *J. Ferment. Technol.*, **57**, 400–407.

Yordy, J. R. and T. C. Weaver (1972). *Methylobacillus*: a new genus of obligately methylotrophic bacteria. *Int. J. Sys. Bacteriol*, **27**, 247–255.

Zajic, E. J., B. Bolesky and A. Wellman (1969). Growth of *Graphium* sp. on natural gas. *Can. J. Microbiol*, **15**, 1231–1236.

Zatman, L. J. (1981). A search for patterns in methylotrophic pathways. In *Microbial Growth on C_1 Compounds*, ed. H. Dalton, pp. 42–54. Heyden, London.

Zavarzin, G. A. and A. N. Nozhevnikova (1971). Aerobic carboxydobacteria. *Microbial Ecol.*, **3**, 305–326.

21

Microbial Metabolism of Carbon Dioxide

L. DIJKHUIZEN and W. HARDER
University of Groningen, The Netherlands

21.1 INTRODUCTION

One of the most important contributions of microbes to natural ecosystems is in the recycling of carbon compounds. Various microorganisms play a role in the degradation of organic matter which ultimately results in the production of carbon dioxide. Although the concentration of this important constituent of the lithosphere is generally low, sediments, oceans and the atmosphere are enormous reservoirs of this compound. CO_2 holds a key position in the cycle of carbon on Earth and sustained life of heterotrophic organisms is strictly dependent on a proper balance between its production during mineralization and subsequent assimilation into organic compounds, as performed by green plants in the process of photosynthesis. This assimilation of CO_2 into organic compounds, however, is not only catalyzed by phototrophic eukaryotes but also by various groups of autotrophic prokaryotes.

This chapter reviews the process of CO_2 fixation in bacteria. These organisms differ widely in the amount of CO_2 required for growth. At one extreme, one finds the autotrophic bacteria mentioned above, which are able to synthesize all their cell material from CO_2, while at the other, there are so-called heterotrophic organisms which, generally in their intermediary metabolism, use this compound in a limited number of carboxylation reactions only. Although in earlier studies CO_2 incorporation by heterotrophs was often masked by its production in catabolic reactions, such studies were greatly facilitated by the introduction of radioactive tracers. Over the years this has resulted in the elucidation of a variety of mechanisms of CO_2 fixation and the properties of the enzymes involved have been investigated in numerous studies. This has led to the recognition that CO_2 is in fact one of the essential nutrients: the complete absence of CO_2 itself, or a source for its generation, generally has detrimental effects on the growth capacities of most, if not all microbes.

In the following sections the metabolic significance of CO_2 fixation in bacteria is discussed. Although this includes a discussion of the properties of some of the enzymes involved in the various carboxylation reactions, emphasis is placed on the physiological role of CO_2 fixation and therefore on the metabolic pathways in which these reactions function.

21.2 AUTOTROPHIC CARBON DIOXIDE FIXATION

Autotrophic bacteria, by definition, can use CO_2 as the major source of cell carbon (Rittenberg, 1969). To enable growth with CO_2 as the sole carbon source these organisms require special metabolic reactions in which this one-carbon substrate is converted into organic molecules containing carbon–carbon bonds. The latter compounds can subsequently be used for the synthesis of all the components of the cell *via* the common pathways of intermediary metabolism (Quayle, 1972).

In autotrophic bacteria three different pathways of CO_2 fixation are operative. The first to be recognized as such was a cyclic pathway, the reductive pentose phosphate or Calvin cycle (see Calvin, 1962), the net effect of which is the synthesis of a C_3-compound from CO_2. At the moment there is ample evidence that the Calvin cycle is involved in CO_2 assimilation in almost all autotrophic bacteria (Cole, 1976; Ohmann, 1979), with the exception of small groups of phototrophic bacteria and non-phototrophic anaerobes. In these phototrophic bacteria, which belong to the genus *Chlorobium*, the reductive carboxylic acid cycle (Evans *et al.* 1966), although highly disputed over the years, provides an alternative to the Calvin cycle. This cycle, which essentially is a reversal of the oxidative tricarboxylic acid cycle, effects the synthesis of acetyl-CoA from CO_2. The third mechanism of autotrophic CO_2 fixation to be considered here is found in non-phototrophic anaerobic bacteria. The situation in autotrophic representatives of this group is not completely clear yet, but significant progress has been made in recent years. Evidence is accumulating that in the methanogenic bacterium *Methanobacterium thermoautotrophicum* an acyclic CO_2 fixation pathway is operative. The first known fixation product of this pathway is acetyl-CoA which, apparently, is synthesized directly from (bound) C_1-units *via* an as yet unknown mechanism (Fuchs and Stupperich, 1982).

Before embarking on a more detailed description of these three pathways of autotrophic CO_2 fixation, the general approaches used in their elucidation, although outside the scope of this chapter, will be briefly considered. Over the years it has become clear that a number of methods are specifically useful in obtaining information about the sequence of reactions in a metabolic pathway (Dagley and Chapman, 1971). Most importantly, these include both short-term and long-term labelling experiments with radioactive tracers (Fuchs and Stupperich, 1982; Lindley *et al.*, 1980; Quayle, 1971), in order to identify early products and the ultimate fate and distribution of label in products of the pathway, and the demonstration that the necessary enzymes are present in sufficiently high activities to account for the observed growth rate of the organism in media containing the substrate under investigation. When possible, further confirmation should be sought in mutant studies and by studying the effect of specific inhibitors of enzymes in the proposed pathway. It has to be emphasized that only a combination of the methods mentioned above can provide unequivocal evidence for the functioning of a specific metabolic pathway.

21.2.1 The Calvin Cycle

Radioisotopic experiments with a number of autotrophic bacteria showed that 3-phosphoglycerate was the most heavily labelled early product of $^{14}CO_2$ fixation. Subsequent enzymological studies revealed that in these bacteria CO_2 is assimilated *via* the Calvin cycle which in design is similar to the one found in plants (Figure 1). Two enzymes are specifically involved in the operation of this cycle. The first one, phosphoribulokinase (ATP: D-ribulose-5-phosphate 1-phosphotransferase, EC 2.7.1.19) catalyzes the phoshorylation of ribulose 5-phosphate to ribulose 1,5-bisphosphate. The latter compound is the acceptor molecule in the actual CO_2 fixing reaction, which is catalyzed by the second enzyme, D-ribulose-1,5-bisphophate carboxylase (3-phospho-D-glycerate carboxylase [dimerizing], EC 4.1.1.39). The products of this reaction, two molecules of 3-phosphoglycerate, are converted to glyceraldehyde 3-phosphate and fructose 6-phosphate, which involves enzymes of the glycolytic pathway, followed by regeneration of ribulose 5-phosphate, *via* enzymes of the pentose phosphate pathway in so-called rearrangement reactions. The net effect of this cyclic pathway is the production of a C_3-compound from C_1-units with the following stoichiometry:

$$3\,CO_2 + 9\,ATP + 6\,NADH_2 \rightarrow \text{glyceraldehyde 3-phosphate} + 9\,ADP + 6\,NAD + 8\,P_i \qquad (1)$$

Equation (1) clearly shows that fixation of CO_2 *via* the Calvin cycle, in order to produce

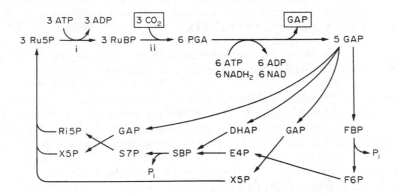

Figure 1 The Calvin cycle: the enzymes specifically involved in its operation are: i, phosphoribulokinase; and ii, ribulose-1,5-bisphosphate carboxylase. Abbreviations: Ru5P, ribulose 5-phosphate; RuBP, ribulose 1,5-bisphosphate; PGA, 3-phosphoglycerate; GAP, glyceraldehyde 3-phosphate; FBP, fructose 1,6-bisphosphate; F6P, fructose 6-phosphate; DHAP, dihydroxyacetone phosphate; E4P, erythrose 4-phosphate; X5P, xylulose 5-phosphate; SBP, sedoheptulose 1,7-bisphosphate; S7P, sedoheptulose 7-phosphate; Ri5P, ribose 5-phosphate

organic matter, requires the expenditure of considerable amounts of both ATP and $NADH_2$. It is therefore not surprising that the molar growth yields of autotrophic bacteria are low. Nevertheless, various representatives of this group are able to grow rapidly and doubling times as low as two hours have been reported for autotrophic growth. This indicates that rather effective mechanisms for energy generation must exist in these organisms. Although these autotrophic bacteria have a common mechanism for CO_2 assimilation, they are rather diverse with respect to the energy sources used for growth. As shown in Table 1 three groups can be distinguished. These energy sources are light in case of the phototrophic bacteria, whereas an inorganic compound, *e.g.* hydrogen, thiosulfate, and even carbon monoxide (see Schlegel and Meyer, 1981) or an organic C_1-compound (methanol, methylamine, formate; see Loginova and Trotsenko, 1979) can function as such in chemolithotrophic and methylotrophic bacteria, respectively. On the basis of the energy source used for autotrophic growth a number of bacteria can be classified in more than one group. This applies for instance to *Paracoccus denitrificans* which, although generally recognized as a hydrogen bacterium, is also able to grow with thiosulfate or methanol (see Bowien and Schlegel, 1981). A second example is found among those representatives of the purple non sulfur bacteria which are able to use methanol as an electron donor during photoautotrophic growth (Douthit and Pfennig, 1976; Quayle and Pfennig, 1975).

Table 1 Classification of Bacteria on the Basis of the Energy Sources Used for the Assimilation of Carbon Dioxide *via* the Calvin Cycle

Class	Energy source	Representatives
Phototrophs	Light	Purple (non)sulfur bacteria, cyanobacteria
Chemolithotrophs	Inorganic compounds	Thiobacilli, hydrogen bacteria, nitrifying bacteria, carbon monoxide oxidizing bacteria
Methylotrophs	Organic C_1-compounds	*Paracoccus denitrificans*, *Pseudomonas oxalaticus*, *Achromobacter* 1L, *Pseudomonas* 8, *Mycobacterium* 50

21.2.1.1 Molecular structure and properties of RuBPCase

Studies on the mechanism and control of CO_2 assimilation *via* the Calvin cycle have focused mainly on the enzyme involved in the actual CO_2-fixing step, ribulose-1,5-bisphosphate carboxylase (RuBPCase). This has resulted in the accumulation of a wealth of information on the properties of this enzyme. Little information, however, is available on the properties of the other enzyme specific for the Calvin cycle, phosphoribulokinase (see below). RuBPCase probably is the most abundant protein in nature (Ellis, 1979). It constitutes a major fraction of the total protein in extracts of both plant leaves (Fraction I) and autotrophic bacteria. The most likely explanation for the synthesis of these large amounts of RuBPCase protein is that the enzyme is a rather sluggish catalyst, with a low molecular activity of around 1100 min^{-1} (specific activity ~ 2 μmol min^{-1} mg protein^{-1}), as observed with a variety of enzymes purified to homogeneity from both plant and bacterial sources. Its purification is relatively easy, particularly when a sucrose density gradient is used (Goldthwaite and Bogorad, 1971; Tabita and McFadden, 1974a). Over the years several improvements have been added to the original purification protocol (Brown *et al.*, 1981; Covey and Taylor, 1980) which allow a rapid isolation of highly purified active enzyme in high yields. A detailed study of RuBPCases from various sources showed that the enzyme, in most cases, has an M_r of around 550 000 (McFadden and Purohit, 1978; McFadden, 1978). Dissociation of these enzymes with sodium dodecyl sulfate or urea, consistently revealed the presence of two types of subunits, namely large (M_r *ca.* 55 000) and small (M_r *ca.* 15 000). Electron microscopic studies of the enzymes isolated from *Nicotiana tabacum* (Baker *et al.*, 1975) and *Alcaligenes eutrophus* (see Bowien and Schlegel, 1981) subsequently showed a quaternary structure for the native enzymes of eight large and eight small subunits (L8S8). It has been established that in all the phototrophic eukaryotes investigated the enzyme has this L8S8 structure. However, this is not the case with all the bacterial RuBPCases.

In the past many conflicting results on the subunit composition of these enzymes have been reported. In some instances it was initially found that the enzymes consisted only of large subunits, whereas later it was shown that certain techniques used during enzyme purification (*e.g.* acid precipitation) removed the small subunits. Other reports indicated the presence of heterogeneous large subunits within one enzyme, which in some, but not all, cases, has been attributed to the action of proteolytic enzymes. Although it has become clear that most of the bacterial RuBPCases also are of the L8S8 type, ample evidence is available that the enzyme present in the purple non-sulfur bacterium, *Rhodospirillum rubrum* (molecular weight 114 000) is a dimer of large subunits (L2) only (Tabita and McFadden, 1974b). Surprisingly, the absence of small subunits does not seem to have a profound effect on the catalytic properties of this enzyme (McFadden, 1978), which poses the question what the actual function of the small subunits is (see below). Convincing evidence for the presence of similar small enzymes in other bacteria is lacking at the moment. A recent report on the properties of RuBPCase purified from the cyanobacterium *Synechococcus* sp. (Andrews *et al.*, 1981) showed that its originally proposed hexameric structure (L6S6), based on a molecular weight of the undissociated enzyme of 430 000, as measured by pore-gradient electrophoresis, was no longer tenable. Equilibrium sedimentation studies of the *Synechococcus* enzyme instead revealed a molecular weight of 530 000, which is within the range for L8S8 enzymes, and such an octameric structure was confirmed by electron microscopic studies. In the light of these results the L6S6 structure proposed for some other bacterial RuBP-Cases (Lawlis *et al.*, 1979; Taylor and Dow, 1980; Taylor *et al.*, 1980) certainly also requires a reevaluation. Finally, in some members of the *Rhodospirillaceae*, in particular *Rhodopseudomonas sphaeroides*, *Rhodopseudomonas capsulata* (see Gibson and Tabita, 1979) and *Rhodopseudomonas blastica* (C. S. Dow, personal communication) two RuBPCases with different molecular weights have been reported. In *Rps. sphaeroides* these enzymes have an L8S8 (molecular weight 556 000) and L6 structure (molecular weight 290 000), respectively, and a detailed structural and immunological analysis of the large subunits of these two enzymes showed that they are in fact distinct proteins (see Tabita, 1981).

Ribulosebisphosphate carboxylase is a bifunctional enzyme which catalyzes both the carboxylation (RuBPCase) and oxygenolysis (RuBPOase) of ribulose bisphosphate (Jensen and Bahr, 1977). Both reactions proceed *via* the 2,3-enediol form of RuBP. Its carboxylation leads, *via* a relatively unstable C_6 intermediate, to the production of two molecules of 3-PGA. Oxygen addition on the other hand results in the production of one molecule of 3-PGA and 2-phosphoglycolate each. Since the latter compound is a precursor of glycolate, the substrate for photorespiration, a process that opposes photosynthesis and therefore decreases plant productivity, the relationship between the RuBPOase and RuBPCase activities has drawn considerable attention. In these reactions CO_2 and O_2 compete with each other in a linearly competitive manner which indicates that they have affinity for the same site on the enzyme. Thus, the amount of glycolate produced by autotrophic organisms is dictated by the relative concentrations of oxygen and carbon dioxide in the gas atmosphere, as has been most elegantly demonstrated in continuous culture experiments with the chemolithotrophic bacterium *Thiobacillus neapolitanus* (Cohen *et al.*, 1979). Numerous attempts to reduce the RuBPOase activity specifically, *e.g.* by the use of inhibitors (see Jensen and Bahr, 1977) or the isolation of mutationally altered species of the enzyme in *A. eutrophus* (Andersen, 1979), invariably also resulted in a comparable loss in RuBPCase activity. The observation that every RuBPCase studied thus far, including those from anaerobically-grown phototrophic bacteria (McFadden and Tabita, 1974) shows oxygenase activity as well has been of considerable support for the conclusion that this activity in fact is an inevitable consequence of the active site chemistry of the enzyme molecule. Oxygenase activity has also been demonstrated with the L2 enzyme of *Rsp. rubrum* (Storrø and McFadden, 1981) and the L6 enzyme of *Rps. sphaeroides* (see Tabita, 1981), which, therefore, indicates that this phenomenon is not dependent on the presence of small subunits.

The kinetics of the reactions catalyzed by RuBPCase/RuBPOase purified from various sources have been subject to numerous investigations. The results of these studies clearly demonstrated that maximal activities of RuBPCase/RuBPOase are only obtained after a prolonged preincubation of the enzyme in the presence of CO_2 and Mg^{2+} (Figure 2). After the formation of an active ternary complex the enzyme can react with RuBP and CO_2 (or O_2). In this respect the order of substrate addition is highly important because preincubation of the enzyme with RuBP inhibits its subsequent activation by CO_2 and Mg^{2+} (see Jensen and Bahr, 1977; Lorimer, 1981). Interestingly, both the activation and the catalytic process, in the case of the carboxylase reaction, require CO_2 as the active species, rather than HCO_3^-, but involve different sites on the protein molecule. The requirement for preactivation of the enzyme provides also an explanation for the confusing kinetic data and wide range of apparent $K_m(CO_2)$ values reported in the earlier literature. In recent studies evidence has been obtained that this activation process is not restricted to enzymes with an L8S8 structure, *e.g.* spinach chloroplast RuBPCase (Lorimer *et al.*, 1976), but also affects the L2 enzyme in *Rsp. rubrum* (Whitman *et al.*, 1979) and the L6 enzyme in *Rps. sphaeroides* (Gibson and Tabita, 1979).

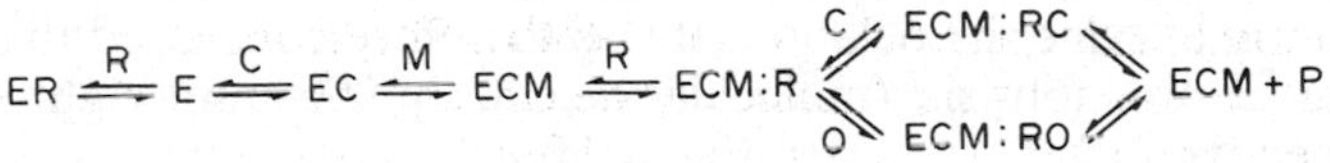

Figure 2   Schematic representation of the kinetics of activation and catalysis of RuBPCase/RuBPOase: abbreviations: E, enzyme; R, RuBP; C, CO_2; M, metal; ECM, active ternary complex; O, oxygen; P, product. (Modified after Tabita, 1981)

It is now generally recognized that the large subunit contains both the catalytic and the activation site. This is also clearly indicated by the occurrence and the properties of bacterial enzymes lacking the small subunits as described above. Based on a quantitative analysis of the overall amino acid composition of large subunits it was initially thought that these polypeptides were homologous from *Rsp. rubrum* (but see also Akazawa *et al.*, 1978) through higher plants (see McFadden and Purohit, 1978). The small subunits on the other hand appeared very dissimilar in composition and even non-homologous among each other. Obviously, the most direct approach in these studies would be a comparison of the primary structures of enzymes from the various sources. Although the complete amino acid sequences of the small subunits of pea and spinach and the large subunit of maize RuBPCase have been determined (see Lorimer, 1981), such studies are still hampered by the lack of sufficient information on the primary structures of bacterial RuBPCases. It has to be expected, however, that considerable progress in this field will be made in the near future, especially in view of the increasing efforts of various research groups to clone bacterial genes coding for RuBPCase. Meanwhile, some of the amino acids within the catalytic site of large subunit spinach RuBPCase, two lysyl and two cysteinyl residues, have been identified *via* an affinity labelling procedure (Schloss *et al.*, 1978). A comparison of the amino acid sequences near the catalytic site of the spinach enzyme with the comparable segments of the large subunit of maize RuBPCase showed a striking homology (see Lorimer, 1981). Absence of homology, however, was apparent between the cysteine-containing sequences of plant (maize or spinach) RuBPCases and the tryptic peptides of the *Rsp. rubrum* enzyme which contain cysteinyl residues. Thus, although this bacterial enzyme has many properties in common with RuBPCases from other sources (see above), this lack of homology indicates that at least its catalytic site is different. For the moment the intriguing question of whether non-homology between plant and bacterial RuBPCases is a general phenomenon cannot be answered since further sequence data are lacking. It is quite possible that the *Rsp. rubrum* enzyme represents a rather exceptional case, a view which is strongly supported by its unusual L2 structure. Interestingly, earlier studies on the immunological properties of this enzyme already showed that antibodies against this enzyme failed to cross-react with the enzymes from other sources, including plants and bacteria. This, however, might also be caused by a masking of antigenic determinants on the large subunits from the various sources by the small subunits present in these enzymes (see McFadden and Tabita, 1974).

Since the large subunit of RuBPCase contains both the catalytic and the activation sites, the function of the small subunits has remained obscure over the years. It was thought that the small subunits had some regulatory role, for instance by modulating the action of effectors upon the large catalytic subunit (Jensen and Bahr, 1977). The observation that in cells of *Rps. sphaeroides* both an L8S8 and an L6 enzyme are synthesized prompted Gibson and Tabita (1979) to further investigate the role of the small subunits by a comparison of the properties of the two purified enzymes. The results of their studies showed that these enzymes indeed differ considerably in various aspects of catalysis and activation. The presence of small subunits in the holoenzyme seemingly provided a conformation with, most importantly, a higher affinity for CO_2 and an increased rate of activation upon addition of CO_2 and Mg^{2+}. It should be noted here that similar differences have been observed between the *Rsp. rubrum* L2 enzyme and the spinach chloroplast L8S8 enzyme (see Tabita, 1981).

21.2.1.2 Phosphoribulokinase

Compared to RuBPCase, relatively little information is available on the structure and catalytic properties of phosphoribulokinase (PRK), the second enzyme specific for the Calvin cycle. Recently, this enzyme has been purified to homogeneity from two bacteria, *Rps. capsulata* and *A. eutrophus*, by means of affinity chromatography and shown to have a molecular weight of 220 000 (Tabita, 1980) and 256 000 (Siebert *et al.*, 1981), respectively. The molecular structure of PRK in these organisms is simple compared to that of RuBPCase since, in both cases, a single size subunit was observed upon enzyme dissociation. It was therefore concluded that these enzymes are oligomers consisting of probably six (molecular weight 36 000) and eight subunits (molecular weight 33 000), respectively.

An investigation of the regulatory properties of these enzymes, and earlier studies mainly performed with partially purified preparations from various sources, have shown that phosphoribulokinase, generally, is strongly activated by $NADH_2$, whereas AMP and PEP may act as inhibitors. This indicates that PRK most likely is a target enzyme for the *in vivo* control of the rate of CO_2 fixation, in keeping with its function in replenishing the CO_2 acceptor molecule RuBP. Interestingly, PRK from oxygen-evolving photosynthetic organisms, both prokaryotic (*Anabaena* sp.) and eukaryotic (*Chlorella sorokiniana*), is not affected by $NADH_2$ and this has also been reported for spinach chloroplast PRK (see Tabita, 1981).

21.2.1.3 Carboxysomes

Many autotrophic bacteria, including thiobacilli, nitrifying bacteria and cyanobacteria, contain polyhedral inclusion bodies (Shively, 1974). In *T. neapolitanus* (Shively *et al.*, 1973), *Nitrobacter agilis* (Shively *et al.*, 1977) and *Anabaena cylindrica* (Codd and Stewart, 1976) the presence of RuBPCase in these organelles has been demonstrated and they were therefore designated carboxysomes. Since in these organisms RuBPCase is present in the cytosol of the cell as well, the question was raised whether carboxysomal RuBPCase is the active species in CO_2 fixation. In a search for the physiological role of carboxysomes, Beudeker *et al.* (1980) reported that the number of these organelles and their volume fraction increased in cells of *T. neapolitanus* when the organism was grown in continuous cultures under CO_2 limitation, compared to thiosulfate or ammonium limitation. Moreover, in this organism the soluble and particulate RuBPCases were immunologically identical (Beudeker *et al.*, 1981). The subsequent isolation of carboxysomes from *T. neapolitanus*, followed by their ultrasonic disintegration, in fact revealed the presence of all relevant Calvin cycle enzymes in these organelles (Beudeker and Kuenen, 1981). Thus, in this organism a complete system for autotrophic CO_2 fixation is present both in soluble and organelle-bound form, which makes it difficult to understand of what advantage this might be. Based on the observation that malate dehydrogenase and aspartate aminotransferase activities are present in the carboxysomes as well, which suggests a shuttle mechanism for $NADH_2$, it has been speculated that this is a way to provide the Calvin cycle in these organelles with sufficiently high concentrations of $NADH_2$ (Beudeker and Kuenen, 1981). Experimental support for the suggested role of an $NADH_2$-shuttle was obtained by the demonstration that addition of malate, but not of $NADH_2$, to intact carboxysomes indeed increased (by 30%) the rate of RuBP-dependent $^{14}CO_2$

fixation. Especially in organisms which have to generate $NADH_2$, *via* reversed electron transport, such a mechanism could be of importance and may explain the presence of carboxysomes in some chemolithotrophic bacteria. However, the presence of Calvin cycle enzymes, in addition to RuBPCase, in carboxysomes of *T. neapolitanus* (Cannon and Shively, 1982) or other chemolithotrophs has yet to be confirmed by studies in independent laboratories.

21.2.1.4 Regulation of ribulose-1,5-bisphosphate carboxylase and phosphoribulokinase synthesis

Those members of the phototrophic and chemolithotrophic bacteria which are unable to utilize organic compounds as growth substrates (Smith and Hoare, 1977) show a limited ability to regulate the synthesis of the Calvin cycle enzymes. The facultative or versatile autotrophs on the other hand clearly adapt their enzymic machinery to changes in the environment. These organisms are therefore best suited for studies of the molecular mechanisms involved in the regulation of the synthesis of Calvin cycle enzymes. During growth of these organisms under 'autotrophic' conditions, the enzymes specifically involved in the Calvin cycle, RuBPCase and PRK, generally reach the highest activities. Dependent upon the organism under study and the nature of the organic compounds involved, addition of metabolizable organic substrates to such cultures often results in repression or even complete switch-off of the synthesis of these enzymes (Bowien and Schlegel, 1981; Dijkhuizen *et al.*, 1978; Matin, 1978; Tabita, 1981). In these experiments it was also observed that the specific activities of RuBPCase and PRK varied in a comparable manner which indicates that their synthesis is regulated coordinately. This, most likely, implies that the structural genes coding for these enzymes are located on a single operon whereby the same molecular signal governs the synthesis of both proteins. In view of these similarities it is tempting to speculate that a common mechanism is involved in the regulation of RuBPCase and PRK synthesis in these various facultative autotrophs.

In experiments with *P. oxalaticus*, which is able to grow autotrophically on formate, evidence was obtained that formate (or CO_2) does not act as an inducer for RuBPCase synthesis in this organism. In contrast, growth of the organism at low dilution rates in a carbon-limited continuous culture on mixtures of formate with oxalate or acetate, or on oxalate alone, resulted in a drastic decrease in the degree of repression normally exerted by these heterotrophic substrates on RuBPCase synthesis (Dijkhuizen and Harder, 1979a, 1979b). From these observations it was concluded that the synthesis of Calvin cycle enzymes in *P. oxalaticus* is regulated by a repression/derepression mechanism. The exact nature of the repressor molecule(s) involved in this mechanism is still unclear, but possible candidates are indicated by the observation that a situation in which very low concentrations of intermediary metabolites occur in the cells, *i.e.* low dilution rates in a carbon-limited continuous culture (Herbert *et al.*, 1956), seems to be required for derepression. In our view the most likely explanation for these phenomena is that 3-PGA, the first product of the Calvin cycle, and/or closely related metabolites, act as end-product repressors thus controlling the synthesis of the enzymes specific for this biosynthetic pathway (Dijkhuizen and Harder, 1984). Repression/derepression has also been proposed by Slater and Morris (1973) as a mechanism for control of RuBPCase synthesis in *Rsp. rubrum*.

Clearly, a complete elucidation of the regulatory mechanisms involved at the molecular level can only be obtained by the analysis of an appropriate set of mutants blocked in autotrophic CO_2 fixation. Although some years ago the genetic analysis of autotrophic organisms was still something of a pipe dream (Cole, 1976) such studies are now under way in various research groups (Srivastava *et al.*, 1982; Tabita, 1981).

21.2.2 The Reductive Carboxylic Acid Cycle

The second pathway involved in CO_2 fixation in autotrophic bacteria is the reductive carboxylic acid cycle (Evans *et al.*, 1966). Evidence for the functioning of this pathway in *Chlorobium limicola* forma *thiosulfatophilum*, a green sulfur bacterium (Kondratieva, 1979) able to grow anaerobically in the light on CO_2 as the sole carbon source and reduced sulfur compounds, *e.g.* thiosulfate, as the electron donor, was twofold. Firstly, a rapid incorporation of $^{14}CO_2$ into amino acids, especially glutamate, was observed in labelling experiments which is not consistent with the operation of a Calvin cycle (see above), but indicated that other carboxylation reactions were taking place. Secondly, in addition to phosphoenolpyruvate carboxylase and isocitrate dehydrogenase, activities of two unusual carboxylating enzymes, namely pyruvate synthase (EC 1.2.7.1;

equation 2) and α-ketoglutarate synthase (EC 1.2.7.3; equation 3) were detected in this organism. The latter two enzymes employ the low redox potential of reduced ferredoxin for the reductive carboxylation of acetyl-CoA and succinyl-CoA, respectively.

$$\text{acetyl-CoA} + CO_2 + \text{reduced ferredoxin} \rightarrow \text{pyruvate} + \text{CoA} + \text{oxidized ferredoxin} \qquad (2)$$

$$\text{succinyl-CoA} + CO_2 + \text{reduced ferredoxin} \rightarrow \alpha\text{-ketoglutarate} + \text{CoA} + \text{oxidized ferredoxin} \qquad (3)$$

Through the action of α-ketoglutarate synthase and isocitrate dehydrogenase, essential steps in the oxidative citric acid cycle can be reversed which allows, in one turn of the reversed cycle, the synthesis of one molecule of acetyl-CoA from two molecules of CO_2. These obervations led to the proposal that the reductive carboxylic acid cycle (Figure 3) is operating in members of the genus *Chlorobium*. After the conversion of acetyl-CoA into pyruvate, catalyzed by pyruvate synthase (Quandt *et al.*, 1978), synthesis of trioses and sugars proceeds *via* gluconogenetic enzymes, whereas phosphoenolpyruvate carboxylase functions in replenishing those intermediates of the cycle that are withdrawn for biosynthetic functions. Figure 3 shows one of the various possible ways to draw the sequence of reactions in this cycle. We prefer this scheme because it shows that any intermediate of the cycle can be withdrawn for biosynthetic purposes and therefore be regarded as the end-product. The representation chosen here also allows a comparison with the Calvin cycle and, from the stoichiometry (equation 4) it can be seen that the energy costs of the reductive carboxylic acid cycle are relatively low.

$$3\,CO_2 + 5\,\text{ATP} + 3\,\text{NAD(P)H}_2 + \text{FADH}_2 + 2\,\text{Fd}^{\text{red}} \rightarrow$$
$$\text{glyceraldehyde 3-phosphate} + 5\,\text{ADP} + 3\,\text{NAD(P)} + \text{FAD} + 2\,\text{Fd}^{\text{ox}} \qquad (4)$$

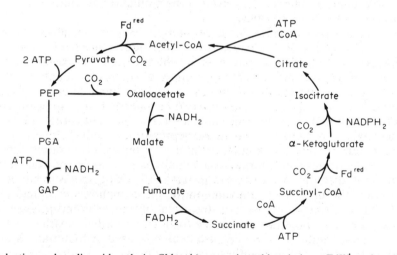

Figure 3 The reductive carboxylic acid cycle in *Chlorobium* species: abbreviations: Fd^{red}, reduced ferredoxin; PEP, phosphoenolpyruvate; CoA, coenzyme A. (Modified after Quayle, 1972)

Originally it was thought that the Calvin cycle operated in these organisms as well and therefore the reductive carboxylic acid cycle was only assigned an auxiliary role, mainly in the production of precursors for amino acids, porphyrins and lipids. In recent years, however, the presence of either of these CO_2-assimilating pathways in *C. limicola* has been a matter of considerable controversy. Sirevåg (1974), for instance, was unable to detect activities of the two enzymes specific for the Calvin cycle, RuBPCase and PRK in cell-free extracts of *C. limicola* strains Tassajara and 8327. Tabita *et al.* (1974), on the other hand, reported the presence of RuBPCase, albeit in low activities, in *C. limicola* strain Tassajara. Following this, Buchanan and Sirevåg (1976) analyzed the actual product(s) of the carboxylation reaction(s) in this organism and showed that these were not 3-phosphoglycerate, the expected product of RuBPCase activity, but rather glutamate, aspartate and malate. Obviously, the absence of the Calvin cycle in this organism would necessitate the presence of an alternative pathway for autotrophic CO_2 fixation. Up to that time the *in vivo* operation of the reductive carboxylic acid cycle had remained doubtful, mainly because of failure to demonstrate the presence of citrate lyase, one of the key reactions in this cycle, in sufficiently high activities (Beuscher and Gottschalk, 1972). As shown by Ivanovsky *et al.* (1980) this was probably due to the fact that in *C. limicola* strains 1C and L, citrate lyase is ATP-dependent (EC

4.1.3.8) and incubation of cell-free extracts of these organisms with citrate, ATP and CoA clearly resulted in the production of both acetyl-CoA and oxaloacetate. Unequivocal evidence for the operation of a reductive carboxylic acid cycle, and not the Calvin cycle in autotrophically growing cells of *C. limicola* was subsequently obtained by Fuchs *et al.* (1980a, 1980b). In labelling experiments with ^{14}C-pyruvate they showed that this compound rather than 3-phosphoglycerate, the product of RuBPCase activity when present, was used for carbohydrate synthesis. In addition, the labelling patterns obtained with specifically labelled ^{14}C-propionate were fully consistent with the conversion of this compound *via* succinyl-CoA, α-ketoglutarate and citrate, followed by cleavage of citrate into oxaloacetate and acetyl-CoA. The question whether the reductive carboxylic acid cycle also functions in organisms outside the genus *Chlorobium* cannot be answered with certainty. Although ATP-dependent citrate lyase activity was not detectable in various purple bacteria (Ivanovsky *et al.*, 1980), it remains possible that the cycle operates in other strictly anaerobic autotrophs. Its presence in aerobic organisms is unlikely in view of the oxygen sensitivity displayed by the two ferredoxin dependent enzymes, pyruvate synthase and α-ketoglutarate synthase.

21.2.3 The Anaerobic Non-phototrophic Autotrophs

Many non-phototrophic obligate anaerobes are able to grow with CO_2 as the sole carbon source and H_2 as the energy source. They are therefore true chemolithoautotrophic bacteria. The majority of these bacteria are methanogens which, characteristically, convert most of the CO_2 fixed into CH_4 in order to obtain energy (Zeikus, 1977). This peculiar and interesting type of anaerobic metabolism has drawn considerable attention over the years and will be dealt with by Nagle and Wolfe, Volume 1, Chapter 22. In this section we will concentrate on the mechanism of cell carbon synthesis from CO_2 in these organisms.

A detailed investigation of the physiology of methanogenic bacteria became possible only a number of years ago by the introduction of special cultivation techniques for these strictly anaerobic organisms (Wolfe and Higgins, 1979). Since most of these studies have concentrated on the process of methane formation from CO_2 and its role in energy generation, relatively little is known about the biochemical pathways of cell carbon synthesis. The best studied methanogen in this respect is *Methanobacterium thermoautotrophicum* which is able to grow rapidly (at 65 °C) in a mineral medium on H_2/CO_2. Initially, the general feeling was that autotrophic CO_2 assimilation in methanogens would also involve the ubiquitous Calvin cycle or, possibly, the reductive carboxylic acid cycle. Attempts to detect RuBPCase in this organism failed, however, and short-term labelling experiments with [14]CO_2 showed that amino acids, *i.e.* alanine, glutamate and aspartate, rather than 3-PGA were early products of CO_2 fixation (Daniels and Zeikus, 1978). This indicated that the Calvin cycle is not operating in this organism. Studies by Taylor *et al.* (1976) showed that synthesis of cell carbon from formaldehyde, which as a formyl derivative might be an intermediate in the process of reduction of CO_2 to CH_4, *via* the C_1-assimilation pathways present in methylotrophic bacteria (Quayle, 1980, Stirling and Dalton, Volume 1, Chapter 20), is also unlikely to occur. This conclusion, based on long-term labelling experiments with various ^{14}C-labelled compounds, was further supported by failure to detect appreciable activities of the enzymes specific for the serine pathway (hydroxypyruvate reductase) or the ribulose monophosphate pathway (hexulose phosphate synthase) in *M. thermoautotrophicum*. Finally, the possible functioning of a reductive carboxylic acid cycle in this organism was investigated in long-term labelling experiments with, for instance, U-^{14}C-succinate (Fuchs and Stupperich, 1978). In these experiments, however, label from succinate only appeared in glutamate, *via* a reductive carboxylation of succinyl-CoA, but not in alanine or aspartate, which indicated that pyruvate and oxaloacetate were not synthesized from α-ketoglutarate *via* isocitrate and citrate. These observations forced Fuchs and Stupperich (1978) to exclude also the reductive carboxylic acid cycle as a mechanism of autotrophic CO_2 assimilation in *M. thermoautotrophicum*.

Since neither the Calvin cycle nor the reductive carboxylic acid cycle, or any other known C_1-assimilation pathway appeared to be functional in this methanogenic bacterium, the operation of a third pathway of autotrophic CO_2 fixation became evident. At present the precise formulation of this pathway is not known. Its basic design (Figure 4), however, has been elucidated most elegantly in long-term labelling experiments in which the fate of radioactive tracers incorporated by growing cells was investigated (see Fuchs and Stupperich, 1982). The first known CO_2-fixation product is acetyl-CoA but the mechanism of its synthesis from CO_2 still remains to be established. Since regeneration of acetyl-CoA from a compound which itself is synthesized from acetyl-CoA and CO_2 could be excluded on the basis of the labelling experiments mentioned above, this apparently does not involve a cyclic pathway but rather a direct synthesis of acetyl-CoA from, most likely, bound C_1 units. The subsequent synthesis of precursors for cell material involves, as shown in Figure 4, the activities of three CO_2-fixing enzymes. Activities of two of these CO_2 fixing enzymes, pyruvate synthase (EC 1.2.–.–; equation 5) and α-ketoglutarate synthase (EC 1.2.–.–; equation 6), are also present in other strictly anaerobic bacteria (see above). The enzymes in this methanogenic bacterium, however, differ in that they require reduced factor 420 (F_{420}) rather than ferredoxin as a coenzyme. F_{420} is a coenzyme commonly found in methanogens (see Nagle and Wolfe, Volume 1, Chapter 22) and, in turn, is reduced in a reaction catalyzed by F_{420}-dependent hydrogenase. The third CO_2-fixing enzyme, phosphoenolpyruvate carboxylase, catalyzes the synthesis of oxaloacetate. This compound is the starting point for a series of reactions which ultimately result in the production of α-ketoglutarate and glutamate. A measurement of the specific activities of these various enzymes in cell-free extracts of *M. thermoautotrophicum* revealed that these are high enough to account for their proposed function and, therefore, provided additional support for the pathway shown in Figure 4 (Fuchs and Stupperich, 1982; Jansen *et al.*, 1982).

$$\text{acetyl-CoA} + CO_2 + \text{reduced } F_{420} \rightarrow \text{pyruvate} + \text{CoA} + \text{oxidized } F_{420} \tag{5}$$

$$\text{succinyl-CoA} + CO_2 + \text{reduced } F_{420} \rightarrow \text{α-ketoglutarate} + \text{CoA} + \text{oxidized } F_{420} \tag{6}$$

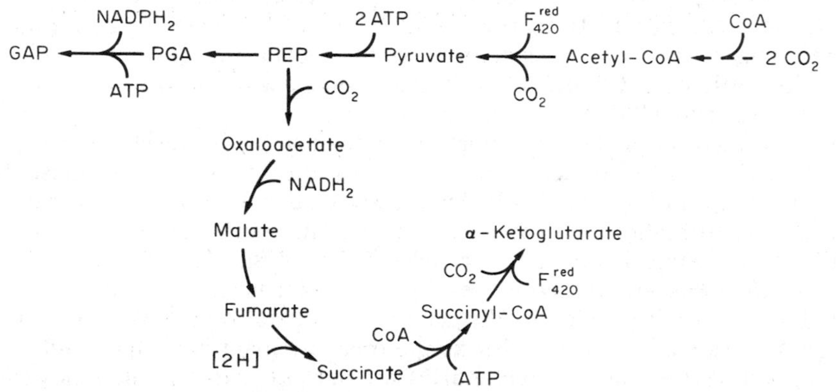

Figure 4 The autotrophic CO_2 assimilation pathway in *Methanobacterium thermoautotrophicum*. $---$ reaction sequence not known; F_{420}^{red}, reduced factor 420. (Modified after Fuchs and Stupperich, 1982)

At the moment it remains unknown how widespread the pathway of autotrophic CO_2 fixation proposed for *M. thermoautotrophicum* is among other methanogens. In fact, the pathway of CO_2 assimilation in most obligate anaerobic autotrophs has not been studied systematically yet. Such studies are, for instance, lacking for the only autotrophic representatives of the acetogenic bacteria which have been described so far, namely *Clostridium aceticum* and *Acetobacterium woodii* (see Braun *et al.*, 1981; Braun and Gottschalk, 1981). Although the methanogens and these two acetogens share the property of autotrophic growth on H_2/CO_2, they differ not only in the end-products of their anaerobic metabolism but also in many other aspects (see Wolfe and Higgins, 1979). Since the acetogenic bacteria are devoid of one of the coenzymes typical for methanogens, namely F_{420}, it can be predicted that, at least in detail, they will differ in their mechanism of autotrophic CO_2 fixation.

21.3 HETEROTROPHIC CARBON DIOXIDE FIXATION

Based on the contribution of CO_2 fixation to their cell carbon synthesis microorganisms can be divided in two groups, namely autotrophs and heterotrophs. Characteristically, in autotrophic bacteria all the organic compounds required for growth are initially derived from CO_2 which is fixed *via* one of the special metabolic pathways described above. Heterotrophic bacteria on the other hand are dependent on the presence of an organic compound as the carbon source and, generally in their intermediary metabolism, incorporate CO_2 *via* a variety of carboxylation reactions, which involve specific acceptor molecules. Although this may lead to the conclusion that in these organisms CO_2 fixation plays a minor role, its importance is clearly indicated by the observation that their incubation in mineral media, under conditions which ensure a total absence of CO_2, generally results in failure to grow (see Wimpenny, 1969). Thus, the presence of CO_2 is apparently also essential for an unimpaired metabolism of heterotrophic bacteria. Analogous to the situation with auxotrophic mutants, however, heterotrophs may be able to grow in the absence of CO_2 provided that the products of the carboxylation reactions are supplied, for instance by using complex media. Over the years the nature of many of these CO_2-fixing enzymes has been elucidated (Scrutton, 1971). These studies also led to the recognition that several of these enzymes may not only fulfil an essential role in heterotrophic bacteria, in which they are the only CO_2-fixing steps, but also in the intermediary metabolism of autotrophic bacteria. The precise mechanisms of the reactions catalyzed by these enzymes, the involvement of biotin and their regulatory properties (see Scrutton, 1971; Wood and Utter, 1965), will not be discussed here. Instead, the physiological role of a number of these so-called 'heterotrophic' CO_2-fixing enzymes is reviewed.

The contribution of 'heterotrophic' CO_2 fixation to carbon metabolism in a specific bacterium species is generally dependent on the growth conditions, *i.e.* medium composition, employed. In addition differences may be observed in the metabolic role of the CO_2-fixing enzymes present in different bacteria, in a reflection of the specific metabolic properties of these organisms. Table 2 lists some of the enzymes involved in heterotrophic CO_2 fixation and their most relevant physiological function.

Table 2 Examples of Enzymes Involved in Heterotrophic Carbon Dioxide Fixation

Enzyme	Function	Enzyme	Function
Pyruvate carboxylase	Anaplerotic	Malate enzyme	Malate oxidation
PEP carboxylase	Anaplerotic	Acetyl-CoA carboxylase	Fatty acid synthesis
PEP carboxytransphos-phorylase	Anaplerotic	Carbamylphosphate synthetase	Arginine and pyrimidine biosynthesis
PEP carboxykinase	Gluconogenetic	Methylmalonyl-CoA-pyruvate transcarboxylase	Propionate fermentation

During growth on carbohydrates or other carbon sources whose initial degradation does not result in the production of an intermediate of the tricarboxylic acid cycle, all organisms need so-called anaplerotic reactions (Kornberg, 1966) whose function is to replenish intermediates of this cycle which have been withdrawn for biosynthetic purposes. Some of these anaplerotic enzymes convert C_3-compounds into C_4-dicarboxylic acids *via* carboxylation reactions. The most widespread of these CO_2-fixing enzymes are pyruvate carboxylase (EC 6.4.1.1; equation 7) and PEP carboxylase (EC 4.1.1.31; equation 8), whereas PEP carboxytransphosphorylase (EC 4.1.1.38; equation 9) is present in propionibacteria. Similar interconversions of C_3- and C_4-compounds are catalyzed by the enzymes PEP carboxykinase (EC 4.1.1.32; equation 10) and malate enzyme (EC 1.1.1.40; equation 11). The metabolic role of the latter enzymes, however, is in general different from that of the carboxylases mentioned above. Although PEP carboxykinase is the only enzyme involved in the production of oxaloacetate from PEP in *Bacteroides* species (see Gottschalk and Andreesen, 1979), this probably represents a rather exceptional case. Mutant studies in for instance *E. coli* (Kornberg, 1966) showed that these two enzymes do not have an anaplerotic role. Apparently, *in vivo* these enzymes function as decarboxylases rather than carboxylases. PEP carboxykinase is therefore generally considered to be a gluconogenetic enzyme since during growth on acetate or malate, it decarboxylates oxaloacetate into PEP, a compound which is the precursor for carbohydrate synthesis. Malate enzyme on the other hand has an important role during growth on substrates such as malate. Through the action of this enzyme, malate is converted into pyruvate which, amongst others, is used for acetyl-CoA synthesis, thus enabling the further metabolism of malate *via* the TCA cycle.

$$\text{pyruvate} + CO_2 + \text{ATP} \rightarrow \text{oxaloacetate} + \text{ADP} + P_i \tag{7}$$

$$\text{PEP} + CO_2 \rightarrow \text{oxaloacetate} + P_i \tag{8}$$

$$\text{PEP} + CO_2 + P_i \rightarrow \text{oxaloacetate} + PP_i \tag{9}$$

$$\text{oxaloacetate} + \text{ATP} \rightarrow \text{PEP} + CO_2 + \text{ADP} \tag{10}$$

$$\text{L-malate} + \text{NADP} \rightarrow \text{pyruvate} + CO_2 + \text{NADPH}_2 \tag{11}$$

Two of the enzymes listed in Table 2 have a direct biosynthetic function. One of these, acetyl-CoA carboxylase (EC 6.4.1.2), catalyzes the carboxylation of acetyl-CoA to malonyl-CoA (equation 12), which is the first step in the process of fatty acid synthesis. The subsequent synthesis of fatty acids with a longer carbon chain proceeds from malonyl-CoA and acetyl-CoA *via* a cyclic series of reactions which involve so-called acyl carrier proteins (Gottschalk, 1979). The second example of a carboxylating enzyme with an important biosynthetic function is carbamylphosphate synthetase (EC 2.7.2.5). This enzyme catalyzes the production of carbamyl phosphate from CO_2, ATP and the amido group of L-glutamine (equation 13). Carbamyl phosphate is one of the immediate precursors for both arginine and pyrimidine synthesis. In the biosynthetic pathway leading from glutamate to arginine, it reacts with ornithine to form citrulline, a reaction catalyzed by ornithine carbamyltransferase. In the pathway leading to pyrimidine synthesis on the other hand, aspartate transcarbamylase catalyzes its condensation with aspartate to form carbamyl aspartate (Gottschalk, 1979).

$$\text{acetyl-CoA} + CO_2 + \text{ATP} \rightarrow \text{malonyl-CoA} + \text{ADP} + P_i \tag{12}$$

$$\text{L-glutamine} + CO_2 + 2\,\text{ATP} + H_2O \rightarrow \text{carbamyl phosphate} + \text{L-glutamate} + 2\,\text{ADP} + P_i \tag{13}$$

In the anaerobic fermentation characteristic for propionibacteria, pyruvate is reduced to propionate *via* the so-called succinate–propionate pathway (see Gottschalk and Andreesen, 1979). In this pathway, leading *via* oxaloacetate, succinate and methylmalonyl-CoA to propionyl-CoA, an unusual carboxylation reaction plays a most important role. In one of the later steps of the pathway methylmalonyl-CoA is decarboxylated to propionyl-CoA. This decarboxylation, however, does not result in CO_2 production. Instead, CO_2 remains enzyme-bound, in a complex with biotin, and is used for the carboxylation of pyruvate to oxaloacetate in one of the earlier steps of the pathway. Both these reactions are catalyzed by one and the same enzyme which has been given the appropriate name methylmalonyl-CoA-pyruvate transcarboxylase (EC 2.1.3.1; equation 14). The complex mechanism of this enzymic reaction offers one of the most intriguing examples of the important role of biotin in many of the enzymes involved in heterotrophic CO_2 fixation (see Wood and Utter, 1965).

$$\text{methylmalonyl-CoA} + \text{pyruvate} \rightarrow \text{oxaloacetate} + \text{propionyl-CoA} \tag{14}$$

So many different CO_2-fixing enzymes are known at the moment that it is impossible to describe them here in any detail. Two further mechanisms of CO_2 fixation, however, will be briefly referred to. In anaerobic bacteria various ferredoxin-dependent enzymes are involved in the reductive carboxylation of CoA esters to the corresponding α-keto acids. Examples of these reactions are for instance pyruvate synthase and α-ketoglutarate synthase, enzymes which fulfil not only an important function in autotrophic (see above) but also in heterotrophic anaerobes (see Gottschalk and Andreesen, 1979; Kondratieva, 1979).

The second mechanism of CO_2 fixation to be considered here is the involvement of CO_2 in the synthesis of acetate in clostridia. In *Clostridium formicoaceticum* and *C. thermoaceticum* hexoses are almost completely fermented to acetate (homoacetate fermentation). In this process one molecule of hexose is fermented to three molecules of acetate. One molecule of acetate is directly derived from pyruvate but the other two are produced in a complicated series of reactions in which the reducing equivalents formed during hexose degradation are used for the reduction of CO_2 to a methyl derivative (Gottschalk and Andreesen, 1979; Ljungdahl and Andreesen, 1976). CO_2 fixation in these organisms is initiated by its reduction to formate, catalyzed by a formate dehydrogenase which is clearly different from the enzyme present in formate-oxidizing organisms in that it contains tungsten and selenium instead of molybdenum. Formate is subsequently reduced to methyl tetrahydrofolate, followed by transfer of the methyl group to a corrinoid protein containing B_{12} as a coenzyme. Finally, two molecules of acetate are produced in a transcarboxylation reaction between the methylcorrinoid protein and the carboxyl group of the second molecule of pyruvate generated in hexose degradation (Ljungdahl and Wood, 1969). In the purine-fermenting species *C. acidiurici*, *C. cylindrosporum* and *C. purinolyticum*, corrinoid proteins apparently play no major role in acetate synthesis (see Dürre and Andreesen, 1982). In these organisms methylene tetrahydrofolate is converted to glycine (and tetrahydrofolate) in a CO_2-fixing reaction catalyzed by glycine synthase. Acetate is subsequently produced *via* serine and pyruvate or directly, in *C. purinolyticum*, by glycine reductase.

21.4 SUMMARY

Based on the contribution of CO_2 fixation to cell carbon synthesis, bacteria can be divided into two groups, namely heterotrophs and autotrophs. Heterotrophic bacteria are dependent on the presence of an organic carbon source for growth and, generally in their intermediary metabolism, incorporate CO_2 *via* a variety of (trans)carboxylation reactions. In autotrophic bacteria, on the other hand, CO_2 is the major source of cell carbon. In these organisms the operation of three different pathways specifically involved in CO_2 fixation has become manifest. The most widespread of these is the Calvin cycle which in bacteria is similar in design to the one found in plants. In phototrophic bacteria belonging to the genus *Chlorobium*, CO_2 is assimilated *via* the reductive carboxylic acid cycle. Thus far evidence for the functioning of this cycle in other autotrophic bacteria is lacking. Finally, evidence is accumulating that in *Methanobacterium thermoautotrophicum*, a methanogenic bacterium, a third pathway of autotrophic CO_2 fixation is operating. The first known product of CO_2 fixation in this organism is acetyl-CoA which, apparently, is not synthesized in a cyclic pathway, but directly from (bound) C_1-units *via* an as yet unknown mechanism. Most of the autotrophic representatives of the non-phototrophic obligate anaerobes, *e.g.* methanogenic and acetogenic bacteria, however, have not been studied systematically yet with respect to their pathway of autotrophic CO_2 fixation. It therefore remains to be elucidated whether other CO_2-fixing pathways exist in nature. In addition, considerable attention will be focused in future research on the evolutionary relationship between the various C_1-assimilation pathways (Quayle and Ferenci, 1978) which have become established and the organisms in which they function.

ACKNOWLEDGEMENT

We are grateful to Marry Pras for her invaluable help in the preparation of the manuscript.

21.5 REFERENCES

Akazawa, T., T. Takabe, S. Asami and H. Kobayashi (1978). Ribulose bisphosphate carboxylases from *Chromatium vinosum* and *Rhodospirillum rubrum* and their role in photosynthetic carbon assimilation. In *Photosynthetic Carbon Assimilation*, ed. H. W. Siegelman and G. Hind, pp. 209–226. Plenum, New York.

Andersen, K. (1979). Mutations altering the catalytic activity of a plant-type ribulose bisphosphate carboxylase/oxygenase in *Alcaligenes eutrophus*. *Biochim. Biophys. Acta*, **585**, 1–11.

Andrews, T. J., K. M. Abel, D. Menzel and M. R. Badger (1981). Molecular weight and quaternary structure of ribulose bisphosphate carboxylase from the cyanobacterium *Synechococcus* sp. *Arch. Microbiol.*, **130**, 344–348.

Baker, T. S., D. Eisenberg, F. A. Fiserling and L. Weisman (1975). The structure of form I crystals of D-ribulose-1,5-diphosphate carboxylase. *J. Mol. Biol.*, **91**, 391–399.

Beudeker, R. F. and J. G. Kuenen (1981). Carboxysomes: 'Calvinosomes'? *FEBS Lett.*, **131**, 269–274.

Beudeker, R. F., G. C. Cannon, J. G. Kuenen and J. M. Shively (1980). Relations between D-ribulose-1,5-bisphosphate carboxylase, carboxysomes and CO_2-fixing capacity in the obligate chemolithotroph *Thiobacillus neapolitanus* grown under different limitations in the chemostat. *Arch. Microbiol.*, **124**, 185–189.

Beudeker, R. F., G. A. Codd and J. G. Kuenen (1981). Quantification and intracellular distribution of ribulose-1,5-bisphosphate carboxylase in *Thiobacillus neapolitanus*, as related to possible functions of carboxysomes. *Arch, Microbiol.*, **129**, 361–367.

Beuscher, N. and G. Gottschalk (1972). Lack of citrate lyase—the key enzyme of the reductive carboxylic acid cycle in *Chlorobium thiosulfatophilum* and *Rhodospirillum rubrum*. *Z. Naturforsch.*, *Teil B*, **27**, 967–973.

Bowien, B. and H. G. Schlegel (1981). Physiology and biochemistry of aerobic hydrogen-oxidizing bacteria. *Annu. Rev. Microbiol.*, **35**, 405–452.

Braun, K. and G. Gottschalk (1981). Effect of molecular hydrogen and carbon dioxide on chemo-organotrophic growth of *Acetobacterium woodii* and *Clostridium aceticum*. *Arch. Microbiol.*, **128**, 294–298.

Braun, M., F. Mayer and G. Gottschalk (1981). *Clostridium aceticum* (Wieringa), a microorganism producing acetic acid from molecular hydrogen and carbon dioxide. *Arch. Microbiol.*, **128**, 288–293.

Brown, H. M., L. H. Bowman and R. Chollet (1981). An improved purification protocol for ribulose 1,5-bisphosphate carboxylase from *Chromatium vinosum*. *FEMS Microbiol. Lett.*, **12**, 105–109.

Buchanan, B. B. and R. Sirevåg (1976). Ribulose 1,5-diphosphate carboxylase and *Chlorobium thiosulfatophilum*. *Arch. Microbiol.*, **109**, 15–19.

Calvin, M. (1962). The path of carbon in photosynthesis. *Science*, **135**, 879–889.

Cannon, G. C. and J. M. Shively (1982). Enzymatic composition of carboxysomes from *Thiobacillus neapolitanus*. *Abstracts of the 13th International Congress of Microbiology*, p. 133. American Society for Microbiology, Washington.

Codd, G. A. and W. D. P. Stewart (1976). Polyhedral bodies and ribulose 1,5-diphosphate carboxylase of the blue-green alga *Anabaena cylindrica*. *Planta (Berlin)*, **130**, 323–326.

Cohen, Y., I. de Jonge and J. G. Kuenen (1979). Excretion of glycolate by *Thiobacillus neapolitanus* grown in continuous culture. *Arch. Microbiol.*, **122**, 189–194.

Cole, J. A. (1976). Microbial gas metabolism. *Adv. Microb. Physiol.*, **14**, 1–92.

Covey, S. N. and S. C. Taylor (1980). Rapid purification of ribulose 1,5-bis(phosphate) carboxylase from *Rhodomicrobium vannielii*. *FEMS Microbiol. Lett.*, **8**, 221–223.

Dagley, S. and P. J. Chapman (1971). Evaluation of methods used to determine metabolic pathways. In *Methods in Microbiology*, ed. J. R. Norris and D. W. Ribbons, vol. 6A, pp. 217–268. Academic, New York.

Daniels, L. and J. G. Zeikus (1978). One-carbon metabolism in methanogenic bacteria: analysis of short-term fixation products of $^{14}CO_2$ and $^{14}CH_3OH$ incorporated into whole cells. *J. Bacteriol.*, **136**, 75–84.

Dijkhuizen, L. and W. Harder (1979a). Regulation of autotrophic and heterotrophic metabolism in *Pseudomonas oxalaticus* OX1: growth on mixtures of acetate and formate in continuous culture. *Arch. Microbiol.*, **123**, 47–53.

Dijkhuizen, L. and W. Harder (1979b). Regulation of autotrophic and heterotrophic metabolism in *Pseudomonas oxalaticus* OX1: growth on mixtures of oxalate and formate in continuous culture. *Arch. Microbiol.*, **123**, 55–63.

Dijkhuizen, L., M. Knight and W. Harder (1978). Metabolic regulation in *Pseudomonas oxalaticus* OX1. Autotrophic and heterotrophic growth on mixed substrates. *Arch. Microbiol.*, **116**, 77–83.

Dijkhuizen, L. and W. Harder (1984). Current views on the regulation of autotrophic carbon dioxide fixation via the Calvin cycle in bacteria. *Antonie van Leeuwenhoek*, **50**, 473–487.

Douthit, H. A. and N. Pfennig (1976). Isolation and growth rates of methanol utilizing Rhodospirillaceae. *Arch. Microbiol.*, **107**, 233–234.

Dürre, P. and J. R. Andreesen (1982). Pathway of carbon dioxide reduction to acetate without a net energy requirement in *Clostridium purinolyticum*. *FEMS Microbiol. Lett.*, **15**, 51–56.

Ellis, R. J. (1979). The most abundant protein in the world. *TIBS*, **4**, 241–244.

Evans, M. C. W., B. B. Buchanan and D. I. Arnon (1966). A new ferredoxin-dependent carbon reduction cycle in a photosynthetic bacterium. *Proc. Natl. Acad. Sci. USA*, **55**, 928–934.

Fuchs, G. and E. Stupperich (1978). Evidence for an incomplete reductive carboxylic acid cycle in *Methanobacterium thermoautotrophicum*. *Arch. Microbiol.*, **118**, 121–125.

Fuchs, G. and E. Stupperich (1982). Autotrophic CO_2 fixation pathway in *Methanobacterium thermoautotrophicum*. *Zbl. Bakt. Hyg.*, *I. Abt. Orig. C3*, 277–288.

Fuchs, G., E. Stupperich and R. Jaenchen (1980a). Autotrophic CO_2 fixation in *Chlorobium limicola*. Evidence against the operation of the Calvin cycle in growing cells. *Arch. Microbiol.*, **128**, 56–63.

Fuchs, G., E. Stupperich and G. Eden (1980b). Autotrophic CO_2 fixation in *Chlorobium limicola*. Evidence for the operation of a reductive tricarboxylic acid cycle in growing cells. *Arch. Microbiol.*, **128**, 64–71.

Gibson, J. L. and F. R. Tabita (1979). Activation of ribulose 1,5-bisphosphate carboxylase from *Rhodopseudomonas sphaeroides*: probable role of the small subunit. *J. Bacteriol.*, **140**, 1023–1027.

Goldthwaite, J. J. and L. Bogorad (1971). A one-step method for the isolation and determination of leaf ribulose-1,5-diphosphate carboxylase. *Anal. Biochem.*, **41**, 57–66.

Gottschalk, G. (1979). *Bacterial Metabolism*. Springer-Verlag, Berlin.

Gottschalk, G. and J. R. Andreesen (1979). Energy metabolism in anaerobes. In *International Review of Biochemistry*, ed. J. R. Quayle, vol. 21, pp. 85–115. University Park Press, Baltimore.

Herbert, D., R. Elsworth and R. C. Telling (1956). The continuous culture of bacteria: a theoretical and experimental study. *J. Gen. Microbiol.*, **14**, 601–622.

Ivanovsky, R. N., N. V. Sintsov and E. N. Kondratieva (1980). ATP-linked citrate lyase activity in the green sulfur bacterium *Chlorobium limicola* forma *thiosulfatophilum*. *Arch. Microbiol.*, **128**, 239–241.

Jansen, K., E. Stupperich and G. Fuchs (1982). Carbohydrate synthesis from acetyl-CoA in the autotroph *Methanobacterium thermoautotrophicum*. *Arch. Microbiol.*, **132**, 355–364.

Jensen, R. G. and J. T. Bahr (1977). Ribulose 1,5-bisphosphate carboxylase-oxygenase. *Annu. Rev. Plant Physiol.*, **28**, 379–400.

Kondratieva, E. N. (1979). Interrelation between modes of carbon assimilation and energy production in phototrophic purple and green bacteria. In *International Review of Biochemistry*, ed. J. R. Quayle, vol. 21, pp. 117–175. University Park Press, Baltimore.

Kornberg, H. L. (1966). Anaplerotic sequences and their role in metabolism. In *Essays in Biochemistry*, ed. P. N. Campbell and G. D. Greville, vol. 2, pp. 1–31. Academic, New York.

Lawlis, V. B., Jr., G. L. R. Gordon and B. A. McFadden (1979). Ribulose 1,5-bisphosphate carboxylase/oxygenase from *Pseudomonas oxalaticus*. *J. Bacteriol.*, **139**, 287–298.

Lindley, N. D., M. J. Waites and J. R. Quayle (1980). A modified pulse-labelling technique for the detection of early intermediates in microbial metabolism: detection of (^{14}C)dihydroxyacetone during assimilation of (^{14}C)methanol by *Hansenula polymorpha*. *FEMS Microbiol. Lett.*, **8**, 13–16.

Ljungdahl, L. G. and J. R. Andreesen (1976). Reduction of CO_2 to acetate in homoacetate fermenting clostridia and the involvement of tungsten in formate dehydrogenase. In *Microbial Production and Utilization of Gases*, ed. H. G. Schlegel, G. Gottschalk and N. Pfennig, pp. 163–172. Akademie der Wissenschaften zu Göttingen, Goltze KG, Göttingen.

Ljungdahl, L. G. and H. G. Wood (1969). Total synthesis of acetate from CO_2 by heterotrophic bacteria. *Annu. Rev. Microbiol.*, **23**, 515–538.

Loginova, N. V. and Y. A. Trotsenko (1979). Autotrophic growth on methanol by bacteria isolated from activated sludge. *FEMS Microbiol. Lett.*, **5**, 239–243.

Lorimer, G. H. (1981). The carboxylation and oxygenation of ribulose 1,5-bisphosphate: the primary events in photosynthesis and photorespiration. *Annu. Rev. Plant Physiol.*, **32**, 349–383.

Lorimer, G. H., M. R. Badger and T. J. Andrews (1976). The activation of ribulose-1,5-bisphosphate carboxylase by carbon dioxide and magnesium ions. Equilibria, kinetics, a suggested mechanism, and physiological implications. *Biochemistry*, **15**, 529–536.

Matin, A. (1978). Organic nutrition of chemolithotrophic bacteria. *Annu. Rev. Microbiol.*, **32**, 433–468.

McFadden, B. A. (1978). Assimilation of one-carbon compounds. In *The Bacteria*, ed. L. N. Ornston and J. R. Sokatch, vol. 6, pp. 219–304. Academic, New York.

McFadden, B. A. and K. Purohit (1978). Chemosynthetic, photosynthetic and cyanobacterial ribulose bisphosphate carboxylase. In *Photosynthetic Carbon Assimilation*, ed. H. W. Siegelman and G. Hind, pp. 179–207. Plenum, New York.

McFadden, B. A. and F. R. Tabita (1974). D-Ribulose-1,5-diphosphate carboxylase and the evolution of autotrophy. *Biosystems*, **6**, 93–112.

Ohmann, E. (1979). Autotrophic carbon dioxide assimilation in prokaryotic microorganisms. In *Photosynthesis II. Encyclopedia of Plant Physiology*, ed. A. Pirson and M. H. Zimmermann, vol. 6, pp. 54–67. Springer-Verlag, Berlin.

Quandt, L., N. Pfennig and G. Gottschalk (1978). Evidence for the key position of pyruvate synthase in the assimilation of CO_2 by *Chlorobium*. *FEMS Microbiol. Lett.*, **3**, 227–230.

Quayle, J. R. (1971). The use of isotopes in tracing metabolic pathways. In *Methods in Microbiology*, ed. J. R. Norris and D. W. Ribbons, vol. 6B, pp. 157–183. Academic, New York.

Quayle, J. R. (1972). The metabolism of one-carbon compounds by microorganisms. *Adv. Microb. Physiol.*, **7**, 119–203.

Quayle, J. R. (1980). Microbial assimilation of C_1 compounds. *Biochem. Soc. Trans.*, **8**, 1–10.

Quayle, J. R. and T. Ferenci (1978). Evolutionary aspects of autotrophy. *Microbiol. Rev.*, **42**, 251–273.

Quayle, J. R. and N. Pfennig (1975). Utilization of methanol by Rhodospirillaceae. *Arch. Microbiol.*, **102**, 193–198.

Rittenberg, S. C. (1969). The role of exogenous matter in the physiology of chemolithotrophic bacteria. *Adv. Microb. Physiol.*, **3**, 159–195.

Schlegel, H. G. and O. Meyer (1981). Microbial growth on carbon monoxide, formate and hydrogen: a biochemical assessment. In *Microbial Growth on C_1 Compounds*, ed. H. Dalton, pp. 105–115. Heyden, London.

Schloss, J. V., C. D. Stringer and F. C. Hartman (1978). Identification of essential lysine and cysteinyl residues in spinach ribulosebisphosphate carboxylase/oxygenase modified by the affinity label *N*-bromoacetylethanolamine phosphate. *J. Biol. Chem.*, **253**, 5705–5711.

Scrutton, M. C. (1971). Assay of enzymes of CO_2 metabolism. In *Methods in Microbiology*, ed. J. R. Norris and D. W. Ribbons, vol. 6A, pp. 479–541. Academic, New York.

Shively, J. M. (1974). Inclusion bodies of prokaryotes. *Annu. Rev. Microbiol.*, **28**, 167–187.

Shively, J. M., F. L. Ball and B. W. Kline (1973). Electron microscopy of the carboxysomes (polyhedral bodies) of *Thiobacillus neapolitanus*. *J. Bacteriol.*, **116**, 1405–1411.

Shively, J. M., E. Bock, K. Westphal and G. C. Cannon (1977). Icosahedral inclusions (carboxysomes) of *Nitrobacter agilis*. *J. Bacteriol.*, **132**, 673–675.

Siebert, K., P. Schobert and B. Bowien (1981). Purification, some catalytic and molecular properties of phosphoribulokinase from *Alcaligenes eutrophus*. *Biochim. Biophys. Acta*, **658**, 35–44.

Sirevåg, R. (1974). Further studies on carbon dioxide fixation in *Chlorobium*. *Arch. Mikrobiol.*, **98**, 3–18.

Slater, J. H. and I. Morris (1973). Photosynthetic carbon dioxide assimilation by *Rhodospirillum rubrum*. *Arch. Mikrobiol.*, **88**, 213–223.

Smith, A. J. and D. S. Hoare (1977). Specialist phototrophs, lithotrophs and methylotrophs: a unity among a diversity of procaryotes? *Bacteriol. Rev.*, **41**, 419–448.

Srivastava, S., M. Urban and B. Friedrich (1982). Mutagenesis of *Alcaligenes eutrophus* by insertion of the drug-resistance transposon Tn5. *Arch. Microbiol.*, **131**, 203–207.

Storrø, I. and B. A. McFadden (1981). Glycolate excretion by *Rhodospirillum rubrum*. *Arch. Microbiol.*, **129**, 317–320.

Tabita, F. R. (1980). Pyridine nucleotide control and subunit structure of phosphoribulokinase from photosynthetic bacteria. *J. Bacteriol.*, **143**, 1275–1280.

Tabita, F. R. (1981). Molecular regulation of carbon dioxide assimilation in autotrophic microorganisms. In *Microbial Growth on C_1 Compounds*, ed. H. Dalton, pp. 70–82. Heyden, London.

Tabita, F. R. and B. A. McFadden (1974a). One-step isolation of microbial ribulose-1,5-diphosphate carboxylase. *Arch. Mikrobiol.*, **99**, 231–240.

Tabita, F. R. and B. A. McFadden (1974b). D-Ribulose-1,5-diphosphate carboxylase from *Rhodospirillum rubrum*. II. Quaternary structure, composition, catalytic and immunological properties. *J. Biol. Chem.*, **249**, 3459–3464.

Tabita, F. R., B. A. McFadden and N. Pfennig (1974). D-Ribulose 1,5-bisphosphate carboxylase in *Chlorobium thiosulfatophilum* Tassajara. *Biochim. Biophys. Acta*, **341**, 187–194.

Taylor, S., H. Dalton and C. Dow (1980). Purification and initial characterization of ribulose 1,5-bisphosphate carboxylase from *Methylococcus capsulatus* (Bath). *FEMS Microbiol. Lett.*, **8**, 157–160.

Taylor, S. C. and C. S. Dow (1980). Ribulose-1,5-bisphosphate carboxylase from *Rhodomicrobium vannielii*. *J. Gen. Microbiol.*, **116**, 81–87.

Taylor, G. T., D. P. Kelly and S. J. Pirt (1976). Intermediary metabolism in methanogenic bacteria. In *Microbial Production and Utilization of Gases*, ed. H. G. Schlegel, G. Gottschalk and N. Pfennig, pp. 173–180. Akademie der Wissenschaften zu Göttingen, Goltze KG, Göttingen.

Whitman, W. B., M. N. Martin and F. R. Tabita (1979). Activation and regulation of ribulose bisphosphate carboxylase-oxygenase in the absence of small subunits. *J. Biol. Chem.*, **254**, 10184–10189.

Wimpenny, J. W. T. (1969). Oxygen and carbon dioxide as regulators of microbial growth and metabolism. In *Microbial Growth*, ed. P. M. Meadow and S. J. Pirt, pp. 161–197. Cambridge University Press, Cambridge.

Wolfe, R. S. and I. J. Higgins (1979). Microbial biochemistry of methane—a study in contrasts. In *International Review of Biochemistry*, ed. J. R. Quayle, vol. 21, pp. 267–353. University Park Press, Baltimore.

Wood, H. G. and M. F. Utter (1965). The role of CO_2 fixation in metabolism. In *Essays in Biochemistry*, ed. P. N. Campbell and G. D. Greville, vol. 1, pp. 1–27. Academic, New York.

Zeikus, J. G. (1977). The biology of methanogenic bacteria. *Bacteriol. Rev.*, **41**, 514–541.

22

Methanogenesis

D. P. NAGLE, JR.* and R. S. WOLFE
University of Illinois, Urbana, IL, USA

22.1 INTRODUCTION

Recently the study of biological methanogenesis has attracted wide interest because of the vital role it plays in the anaerobic digestion of organic matter, the importance of methane for energy needs and chemicals, and because of the wealth of biochemical and biological knowledge to be found in methanogenic bacteria. Methane is found in the atmosphere at 1.4 p.p.m. Over 80% of this atmospheric component results from the recent anaerobic decay of organic matter. As much as 1–2% of the carbon fixed by plants per year becomes methane. The remainder of atmospheric methane is from geological or human-related sources. There is a dynamic flux of methane in the atmosphere; the content turns over about every five years, mainly by chemical oxidation (Ehhalt and Schmidt, 1978).

The purpose of this article is to summarize recent microbiological and biochemical studies of the microorganisms that function as biocatalysts in methanogenesis. There are a number of reviews, some very current, which cover the literature in intensive detail (Anthony, 1982; Balch *et al.*, 1979; Barker, 1956; Mah *et al.*, 1977; Mah and Smith, 1981; Mah *et al.*, 1981; Mah, 1982; McInerney and

*Now at University of Oklahoma.

Bryant, 1981; Prévot, 1980; Stadtman, 1967; Taylor, 1982; Wolfe, 1971; Wolfe and Higgins, 1979; Wolfe, 1980; Zeikus, 1977. The search of the literature for this article ended 30 November, 1982).

The abbreviations used in this chapter are as follows: HS-CoM, coenzyme M (2-mercaptoethanesulfonic acid); MeS-CoM, 2-methylthioethanesulfonic acid; BV, benzylviologen; factor F_{420}, (*N*-L-lactyl-γ-glutamyl)-L-glutamic acid phosphodiester of 7,8-didemethyl-8-hydroxy-5-deazariboflavin 5'-phosphate; CDR factor, carbon dioxide reduction factor; FAD, flavin adenine dinucleotide; FAF, formaldehyde activating factor; MP, methanopterin; MV, methylviologen; cTHMP, carboxytetrahydromethanopterin; OAA, oxaloacetate; PEP, phosphoenolpyruvate.

22.1.1 Natural Ecological Systems

Biological formation of methane occurs under strict, anaerobic conditions. The habitats in which this process takes place (Mah *et al.*, 1977; McInerney and Bryant, 1981; Stadtman, 1967; Taylor, 1982; Wolfe, 1971; Wuhrman, 1982) range from freshwater and marine sediments, peat bogs, and sewage digesters, to the rumen (Hungate, 1982), caecum and intestinal tract of higher animals as well as to the gut of insects. Organic matter is degraded in steps from polymers to sugars, to fatty acids, to acetate, hydrogen and CO_2, and finally to CH_4. As shown in Figure 1 three groups of organisms participate in this process. The methanogenic bacteria play a vital role in 'pulling' anaerobic fermentations to completion (McInerney and Bryant, 1981; Mah, 1982; Mah *et al.*, 1977; Taylor, 1982), two factors being involved. (i) Methane is a poorly soluble, non-toxic, volatile end product; by simple mass action the equilibrium of the overall degradation is displaced. (ii) H_2 is a product of some oxidative reactions upstream and some of the hydrogen producing organisms are strongly inhibited by it; even at very low partial pressures of H_2, growth of these cells is prohibited for thermodynamic reasons. Methanogens are effective scrubbers of hydrogen, and by removing H_2 permit growth of certain H_2-producing acetogenic organisms, and the degradation of reduced molecules. Fifteen years ago Wolin named this interaction 'interspecies hydrogen transfer' (Mah *et al.*, 1977; Mah, 1982; McInerney and Bryant, 1981; Wolfe, 1980). The importance of this process, a crucial concept in bacterial ecology, is demonstrated by two examples. First, the classical example is *Methanobacillus omelianski*, a symbiotic association of two organisms which together convert ethanol to methane. The ethanol-oxidizing organism (the 'S organism'), which produces hydrogen, is not able to grow on ethanol in the absence of the methanogen because it is strongly inhibited by the hydrogen that it produces. Second, the acetogenic, anaerobic fatty acid-oxidizing bacteria, which are obligate proton reducers (*e.g. Syntrophomonas wolfeii*, a butyrate oxidizer), are obligate syntrophs. They have not been grown successfully outside of coculture with an anaerobic hydrogen utilizer (Mah, 1982; McInerney and Bryant, 1981). In addition to driving the fermentation, the methanogens are important in influencing the distribution of end products. By inclusion of a methanogen with an organism that is fermenting a sugar, for example, the profile of end-products can shift dramatically from compounds such as lactate or propionate to acetate or methane ensuring more extensive degradation (Mah, 1982; McInerney and Bryant, 1981).

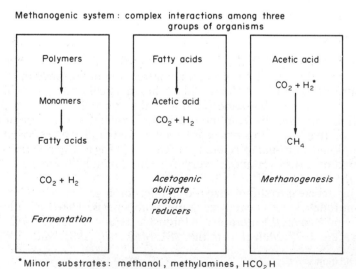

Figure 1   Simplified scheme of microbial groups involved in anaerobic biodegradation

22.1.2 Biomass conversion

Methanogenesis from complex organic matter in digesters has been harnessed with success on small and large scales for many years. With increasing demand for energy, the process may become commercially viable as a source of CH_4 (Lettinga, 1981; Melchior *et al.*, 1982; Pfeffer, 1982). In addition, it may be possible to use methane as precursor for methanol and other chemicals (Ghisalba and Heinzer, 1982; Zeikus, 1980). The digestion of manures, domestic sewage, and certain organic wastes may yield material which has superior properties as a fertilizer, an important item in the economic balance sheet.

Knowledge of the requirements of bacterial consortia which perform degradations, and of the physiology and metabolism of the individual species involved, will be important in process optimization to achieve predictable and controlled, large scale, methanogenic fermentations. Factor F_{420}, the fluorescent deazaflavin found in high amounts in methanogens, has been exploited in monitoring digesters (Melchior *et al.*, 1982).

22.2 METHANOGENIC ORGANISMS

22.2.1 Species

The methanogens were recognized as a coherent bacterial group on the basis of their physiology and limited growth substrates by Barker (1956). These non-spore forming organisms are among the most strict anaerobes known. Morphology includes cocci, rods, spirilla, and sarcina (Balch *et al.*, 1979; Wolfe, 1980; Zeikus, 1977). New and simplified techniques for growth of methanogens have allowed workers to obtain more than 20 species in pure culture. Balch *et al.* (1979) arranged the methanogens into three orders. Order I, Methanobacteriales, contains species of the genera *Methanobacterium* and *Methanobrevibacter*. Order II, Methanococcales, contains the methanococci, notably *M. voltae* (Whitman *et al.*, 1982) and *M. vannielii* (Stadtman, 1967). Order III, Methanomicrobiales, is made up of two families. The first family includes the genera *Methanomicrobium*, *Methanogenium* (marine isolates) and *Methanospirillum*. The second family is made up of the versatile species *Methanosarcina barkeri* and *Methanothrix soehngenii* (Stackebrandt *et al.*, 1982). These species, together with '*Methanococcus mazei*' (which actually belongs in the genus *Methanosarcina*, C. Woese, personal communication) are the only methanogens to utilize acetate. *Methanosarcina barkeri* and '*Methanococcus mazei*' utilize methanol and methylamines as well.

The new phylogenetic treatment of the methanogens was based upon the work pioneered by Woese, in which the partial nucleotide sequence of the 16S ribosomal RNA (which has a conserved function in all organisms) provides quantitative data on the relatedness of species (Fox *et al.*, 1980). The three orders of methanogens are distantly related, differing from each other as much as enterobacteria, bacilli, or cyanobacteria differ from each other.

22.2.2 Methanogens as Archaebacteria

By the criterion of ribosomal RNA relatedness at the sequence level, the methanogens are even more distantly related to typical bacteria than they are to each other. Since methanogens are more closely related to halophiles and thermoacidophiles than any of these groups are to typical bacteria, Woese and coworkers (Fox *et al.*, 1980) grouped the methanogens, halophiles and thermoacidophiles into a new kingdom, the Archaebacteria. Numerous properties in addition to the ribosomal RNA sequences emphasize the biochemical differences between the archaebacterial methanogens and Eubacteria. These properties appear as part of Sections 22.3–22.5.

One of these unique properties is the lipid structure of methanogens (Langworthy *et al.*, 1982). Methanogen lipids, like all archaebacterial lipids, do not contain glycerol ester bonded fatty acids. As shown in Figure 2 methanogen lipids contain ether linkages, and the chains are phytanyl or biphytanyl (**1**). In addition, a significant portion of methanogen lipid is made up of the diglyceryl tetraethers (**2**) (which cannot form a bilayer) and squalene (Langworthy *et al.*, 1982). The phospholipids are derived from (**1**) and (**2**). The mode of biosynthesis of the ether linkage is an unsolved question. Recently it has been pointed out that the isoprenoid chains of sediments and petroleum may have been derived from methanogens (or other archaebacteria) (Chappe *et al.*, 1982).

A second major difference between methanogens and eubacteria is the cell wall structure.

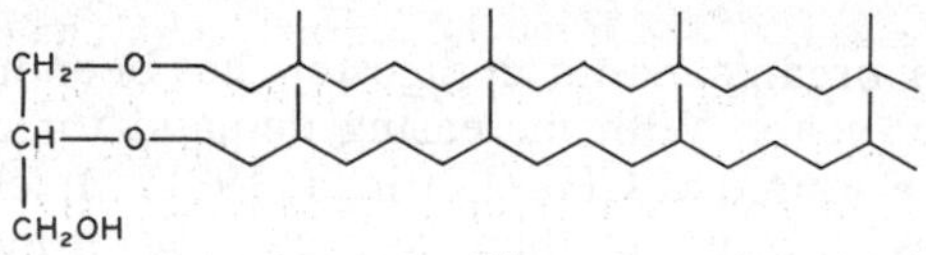

Figure 2 Glycerol lipids of methanogens: diphytanyl glycerol diether (**1**); dibiphytanyl diglycerol tetraether (**2**)

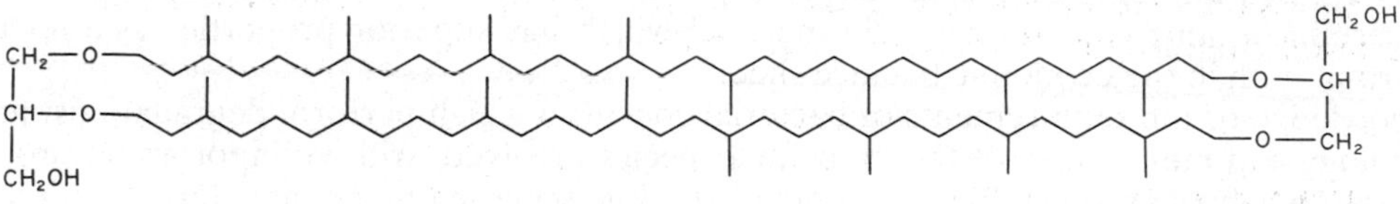

Work by Kandler and his group has uncovered a series of previously unknown cell wall types which include glycopeptide cell walls, peptide cell walls with no amino sugars, and pseudomurein. The latter contains L-amino acids exclusively and talosamine (a sugar previously unknown in Nature). The type of cell wall structure is consistent within each order of methanogens. As might be suspected, methanogens are resistant to antibiotics which interrupt eubacterial cell wall synthesis (*e.g.* penicillin, cycloserine; Kandler, 1982; Konig *et al.*, 1982).

The nucleic acids of methanogens differ from those of eubacteria in several ways. The genome size is small, roughly one-third that of *Escherichia coli*. The transfer RNA modification pattern is unlike that found in eubacteria and eukaryotes. The ribosomal protein pattern is unlike that of eubacteria, and the RNA polymerase subunit pattern (8–11 subunits) is not like the well-known eubacterial pattern, (4 subunits) but is more like that of the eukaryotic enzyme (Fox *et al.*, 1980).

Finally, several coenzymes which methanogens utilize in central metabolic roles are quite different from familiar coenzymes (Section 22.4). These novel cofactors highlight the seminal biochemical differences that exist between methanogens and eubacteria.

22.2.3 Growth of Methanogens

The methanogens have evolved a very simple way of making a living. The substrates which they are able to utilize are limited to those shown in reactions 1–8 of Table 1 (Thauer *et al.*, 1977; Wolfe, 1980). With one or two exceptions, methanogens are able to oxidize H_2 as the sole source of energy. A number of species use formate. Acetate is known to be utilized only by *Methanothrix soehngenii*, *Methanosarcina barkeri* and '*Methanococcus mazei*'. The latter organisms also grow on methanol and methylamines (Hippe *et al.*, 1979). Carbon monoxide was utilized by several species, but the small amount of growth obtained was at a rate of only 1% of that seen for H_2 and CO_2 (Daniels *et al.*, 1977).

Table 1 Standard Free Energy Changes of Methanogenic Reactions and Related Reactions

	Reaction	$kJ/reaction^a$	kJ/CH_4
(1)	$HCO_3^- + 4 H_2 + H^+ \longrightarrow CH_4 + 3 H_2O$	−135.6	−135.6
(2)	$4 HCO_2^- + H_2O + H^+ \rightarrow CH_4 + 3 HCO_3^-$	−130.1	−130.1
(3)	$4 MeOH \longrightarrow 3 CH_4 + HCO_3^- + H_2O + H^+$	−314.6	−104.9
(4)	$4 MeNH_3^+ + 3 H_2O \longrightarrow 3 CH_4 + 4 NH_4^+ + HCO_3^- + H^+$	−224.9	−75.0
(5)	$2 Me_2NH_2^+ + 3 H_2O \longrightarrow 3 CH_4 + 2 NH_4^+ + HCO_3^- + H^+$	−219.6	−73.2
(6)	$4 Me_3NH^+ + 9 H_2O \longrightarrow 9 CH_4 + 4 NH_4^+ + 3 HCO_3^- + 3 H^+$	−668.6	−74.3
(7)	$MeCO_2^- + H_2O \longrightarrow CH_4 + HCO_3^-$	−31.0	−31.0
(8)	$4 CO + 5 H_2O \longrightarrow CH_4 + 3 HCO_3^- + 3 H^+$	−196.7	−196.7
(9)	$ATP + H_2O \longrightarrow ADP + P_i$	−31.8	
(10)	$CO_2(g) + H_2O \longrightarrow HCO_3^- + H^+$	+4.8	
(11)	$HCO_3^- + H_2 \longrightarrow HCO_2^- + H_2O$	−1.3	
(12)	$HCO_2^- + H_2 + H^+ \longrightarrow H_2CO + H_2O$	+23.2	
(13)	$H_2CO + H_2 \longrightarrow MeOH$	−44.8	
(14)	$MeOH + H_2 \longrightarrow CH_4 + H_2O$	−112.5	

^{a} Free energy changes were calculated from free energies of formation reported by Thauer *et al.* (1977).

Prior to 1974 study of methanogens was hindered by difficulty in growing them. It is now possible to culture these strictest of anaerobes with facility, by means of methods which employ pressurized gas atmospheres (Balch *et al.*, 1979), a modification of the classical Hungate technique. Growth in both liquid medium and on plates is straightforward. Methanogens offer unique genetic challenges for the investigator both to manipulate genes which are involved in methanogenesis as well as to understand the mechanics of expression of archaebacterial genes. To date a genetic system has not been defined for these organisms; mutants or phage have not been reported. However, expression of methanogen DNA and proteins in *Escherichia coli* has been obtained (Bollschweiler and Klein, 1982; Reeve *et al.*, 1982).

22.3 METABOLISM

22.3.1 Methane from Carbon Dioxide

Results of early physiological studies on reduction of CO_2 to CH_4 were summarized in elegant fashion by Barker (1956), who proposed that CO_2 was fixed to a one-carbon carrier. In a stepwise manner, the one-carbon unit was reduced to CH_4 but was not freely diffusable at any oxidation level, *i.e.* as a formate, formaldehyde or methanol. The following three sections report the fleshing out of Barker's original model in ensuing years. The key to this work was the preparation of active crude extracts of methanogenic cells that were able to carry out CH_4 formation from CO_2. The early studies with crude extracts uncovered a requirement for ATP in reduction of CO_2 to CH_4. The requirement was catalytic, about 17 mol of CH_4 being produced per mol of ATP added. ATP could be removed once the enzyme system was activated, without interrupting CH_4 synthesis (Gunsalus and Wolfe, 1978; Whitman and Wolfe, 1983). When vesicles of *M. thermoautotrophicum* were studied (Sauer *et al.*, 1980) a requirement for ATP was not observed but stimulation was obtained. Probably ATP was being synthesized by the membrane vesicles.

Further work uncovered MeS-CoM (**4**; Figure 7; Section 22.4.1) as the C_1 carrier at the terminal step of CO_2 reduction. Gunsalus and Wolfe (1977) then found that CO_2 reduction to CH_4 was dependent on the presence of MeS-CoM (not HS-CoM). The data which led to the discovery of this effect, dubbed the RPG effect after the initials of its discoverer, is shown in Figure 3. As the figure indicates, for every MeS-CoM molecule added, about 11 CO_2 molecules were reduced to methane. Other substances which yield MeS-CoM (*e.g.* serine, formaldehyde and pyruvate) also permit CO_2 reduction to CH_4 by extracts (Romesser and Wolfe, 1982b). From these findings, the RPG effect was conceptualized as illustrated in Figure 4, where the energy derived from methyl group reduction is conserved for CO_2 activation. The identity of this activated state of the system is not known and is controversial. It may result from a conformational or allosteric change, or from a phosphorylated or adenylylated component, be it enzyme or cofactor (Kell *et al.*, 1981; Romesser and Wolfe, 1982b). It could be a radical state of the coenzyme M product (or of an enzyme) which is available for reaction with CO_2 after CH_4 leaves (Romesser and Wolfe, 1982b), or it may be an energized state of the membrane (Kell *et al.*, 1981). Some of these alternatives are testable.

No evidence for an RPG effect was found in extracts of *Methanosarcina barkeri*. HS-CoM, but not MeS-CoM was required for CH_4 production from MeOH or $H_2 + CO_2$ in these extracts, and MeS-CoM was not a precursor of CH_4 in these experiments (Hutten *et al.*, 1981). Likewise Romesser and Wolfe (1982b) found no RPG effect in *M. barkeri* and only a negligible effect in *M. formicicium* or *M. ruminantium*. Failure to observe the effect in preparations of these organisms may be related to the difficulty in preparing a soluble system. Perhaps the methylreductase system is embedded in the membrane where the active site is not freely accessible to added MeS-CoM.

In addition to ATP, MeS-CoM, and enzymes, three coenzymes are required for the reduction of CO_2 to CH_4 by extracts: carbon dioxide reduction factor (CDR factor, Section 22.4.5), methanopterin (Section 22.4.6), and the recently discovered formaldehyde activating factor (FAF; Section 22.4.7). Carboxytetrahydromethanopterin has been proposed as the first intermediate in CO_2 reduction (Vogels *et al.*, 1982; Section 22.4.6). FAF is needed for methanogenesis from formaldehyde, that is, from the $HOCH_2-$ level of oxidation (Section 22.4.7). The role of $HOCH_2S$-CoM is not so clear now, and it is not certain that it is the one-carbon carrier at the formaldehyde level of oxidation. In Figure 5, the current state of knowledge about the pathway from CO_2 to CH_4 is summarized.

Chemical inhibitors of methanogenesis include: chlorinated compounds, ranging from $CHCl_3$ to DDT; analogues of HS-CoM (bromo- and chloro-ethanesulfonic acids, Section 22.4.1); and

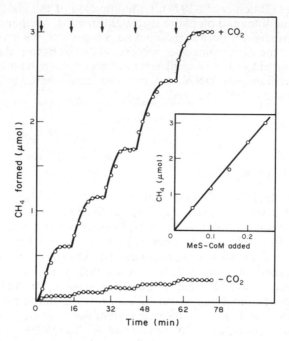

Figure 3 Stimulation of CH₄ formation from CO₂ *in vitro* by addition of MeS-CoM. Crude extract from *M. thermoauto-trophicum* was incubated under hydrogen or hydrogen + CO₂ (80:20, v:v). At times shown by arrows, 50 nmol of MeS-CoM was added anaerobically. The inset demonstrates that the amount of CH₄ formed after each addition equalled 12 times the amount of MeS-CoM added. (Reproduced from Gunsalus and Wolfe, 1977 with permission of the publisher)

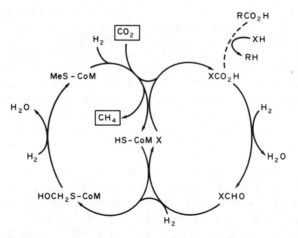

Figure 4 Scheme for activation of CO₂ concomitant with reduction of MeS-CoM. As MeS-CoM is reduced to CH₄, CO₂ is activated for another cycle of stepwise reduction. RCO₂H represents pyruvate, which, together with serine and formaldehyde (all precursors of MeS-CoM) also supports the effect. (Reprinted from Romesser and Wolfe, 1982b with permission of the publisher)

membrane active ionophores, which are uncouplers of energy metabolism (Gunsalus and Wolfe, 1978). Study of these inhibitors may be useful in elucidating biochemical routes involved in carbon metabolism. It is important to recognize the role such chemicals may play as inhibitors in digesters.

The alternative substrates for methanogenesis (Table 1) are less well studied than H₂ and CO₂. Formate, used by some methanogens, notably *Methanococcus vannielii* and *Methanobacterium formicicium*, is oxidized to CO₂ and hydrogen by formate dehydrogenase, an enzyme found in high levels in formate utilizing strains. In *M. vannielii*, it was demonstrated that formate dehydrogenase occurred in two forms, one of which contained selenocysteine (Jones and Stadtman, 1981). The reaction of the enzyme was shown to be linked to NADPH production *via* the novel

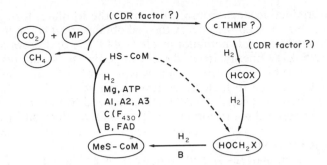

Figure 5 Current model of the cycle from CO_2 to CH_4. The simplified cycle shows the involvement of the terminal step in the activation of CO_2, as well as the known one-carbon carriers and known enzymes (Adapted from Romesser and Wolfe, 1982b; Vogels *et al.*, 1982; Wolfe, 1980; and Escalante-Semerena, Leigh, Nagle and Wolfe, unpublished results)

electron carrier, the deazaflavin factor F_{420} (Section 22.4.2). In complementary work, the formate dehydrogenase from *M. formicicum* was shown to contain a molybdenum cofactor, similar to that of nitrate reductase and sulfite oxidase (Schauer and Ferry, 1982). The enzyme clearly differs in many respects from that found in typical bacteria.

Acetate is utilized as a source of energy by *Methanosarcina*, '*Methanococcus mazei*', and *Methanothrix* (reaction 7, Table 1). Since the methane formed arises from the methyl carbon, without loss of protons, the mechanism for methanogenesis from acetate must involve only transfer, energy being obtained by methyl group reduction (Mah *et al.*, 1981; Wolfe and Higgins, 1979). Methanol and methylamines utilized by *Methanosarcina*, are presumably direct precursors to MeS-CoM, since no protons are lost from the carbon atom.

22.3.2 MeS-CoM Methylreductase

The terminal step of methane formation, MeS-CoM reduction to CH_4, has been the most studied reaction in methanogenesis (equation 1).

$$\text{MeS-CoM} + H_2 \xrightarrow[\text{A,B,C}]{\text{MgATP}} CH_4 + \text{HS-CoM} \qquad (1)$$

Components A, B, and C, required for the *in vitro* reaction were resolved by anaerobic chromatography of extracts of *M. thermoautotrophicum* (Gunsalus and Wolfe, 1980). Component B is an oxygen-labile coenzyme of unknown structure and function (Section 22.4.3). Component C, a yellow, oxygen stable protein, was purified to homogeneity, and shown to have an $\alpha_2\beta_2\gamma_2$ subunit structure (Ellefson and Wolfe, 1981). The protein appears to be the site at which methyl reduction takes place. Each mole of component C has bound to it two moles of factor F_{430}, the nickel-containing tetrapyrrole (Section 22.4.4; Ellefson *et al.*, 1982). Factor F_{430} from the protein appears to contain coenzyme M as a bound ligand (Keltjens *et al.*, 1982). The chemistry performed by this previously unknown, protein-bound coenzyme will be important to the understanding of MeS-CoM reduction, and possibly to the rest of the CO_2 to CH_4 pathway.

Component A had been shown to contain hydrogenase activity, and was quite unstable; recently, it has been stabilized. By anoxic chromatography, it has been resolved into four separate components, three proteins named A1, A2, and A3, and the fourth a coenzyme, FAD (Nagle and Wolfe, 1983). Component A1 contains the F_{420}-reducing hydrogenase activity. A2 and A3 are oxygen-labile. None of the protein fractions is homogeneous, and the functions they perform in the MeS-CoM reaction are unknown. The role of FAD is not clear, but may be related to the F_{420}-reducing hydrogenase, which is known to contain FAD (Jacobson *et al.*, 1982). Several major problems which remain to be solved include the nature of electron transport from H_2 to the sites of reduction, the ATP-dependent activation phenomenon, the role of components A1, A2 and A3 in the cycle of CO_2 reduction, the harnessing of energy derived from equation (1), and the role of FAD in the MeS-CoM methylreductase reaction.

22.3.3 Cellular Energy from Hydrogen Oxidation and Carbon Dioxide Reduction

One of the major unanswered questions concerning methanogenesis is how the cell obtains energy from the cycle shown in Figure 4. If the process of methanogenesis from CO_2 and H_2 is viewed as a series of reduction steps from CO_2 through the formate, formaldehyde, and methyl levels of oxidation (Table 1), then the energy the cell obtains must come from the two exergonic steps, between formaldehyde and methanol levels (−44.8 kJ) and between the methanol level and methane (−112.5 kJ). Energy from these exergonic steps supports ATP synthesis. In addition, free energy must be conserved in some way to provide for the endergonic steps of CO_2 activation and reduction to the formaldehyde level (Kell *et al.*, 1981; Stadtman, 1967). Arguments have been presented that conclude ATP is synthesized by electron transport phosphorylation, and not by substrate level phosphorylation. The chain of electron carriers has not yet been identified, but does not appear to consist of quinones or cytochromes (Thauer *et al.*, 1977; Kell *et al.*, 1981; Anthony, 1982; Wolfe, 1980).

Energy coupling in methanogens is still somewhat of a 'black box'. Recent advances have begun to illuminate the problem. The important concept is that the process involves membranes. When hydrogen was added to whole membrane preparations, a proton gradient was formed by hydrogenase (inside acid), and ATP synthesis was driven (Doddema *et al.*, reviewed in Kell *et al.*, 1981). Many methanogens possess intracellular vesicles, which may be the site at which coupling occurs. By histochemical methods, both the ATPase and the hydrogenase of *M. thermoautotrophicum* have been shown to be bound to intracellular membranes (Kell *et al.*, 1981). Other studies on the proton gradient in vesicle preparations (Jarrell and Sprott, 1981; Sauer *et al.*, 1981) supported the idea that ATP synthesis resulted from energized membranes. In addition, the results from experiments with uncouplers on whole cells suggested that the energy-yielding proton gradient was across internal membrane structures. By a different approach, the isotope discriminations against heavy protons during CH_4 synthesis by cells in heavy water confirm that an intracellular compartment is the site of one-carbon reduction (Spencer *et al.*, 1980).

The requirement for ATP by *in vitro* systems that produce CH_4 from either CO_2 or MeS-CoM (Sections 22.3.1 and 22.3.2) is not easily explained. Results of Whitman and Wolfe (1983) demonstrated that ATP is not required to set up a membrane potential, as previously suggested (Kell *et al.*, 1981). ATP hydrolysis appears to take place, and there is the possibility that a short-lived adenylated or phosphorylated intermediate (cofactor or enzyme) may be involved.

22.3.4 Hydrogenases

Methanogens that grow on hydrogen and CO_2 use hydrogenase as a source of electrons for methanogenesis; a primary electron acceptor is the deazaflavin, factor F_{420} (Section 22.4.2). Methanogen hydrogenases have shown new and interesting properties which have proven to be valuable in the study and understanding of well-known hydrogenases. These properties are summarized in Table 2 and the references therein. Two better known hydrogenases, of *Clostridium pasteurianum*, an anaerobe, and *Alcaligenes eutrophus*, an aerobe, are presented for comparison. Certain methanogen hydrogenases (like the *Alcaligenes* enzyme and unlike the clostridial enzyme) are stable to oxygen, unless the protein is reduced; then the enzyme becomes irreversibly inactivated by oxygen (*e.g.* Jacobson *et al.*, 1982). Extracts from most species appear to have two forms of hydrogenase; one is able to reduce factor F_{420}, while the other reduces only synthetic dyes. The latter may be an artifact of fractionation. FAD, nickel, and selenocysteine have been found in different F_{420}-reducing hydrogenases (Table 2). The nickel atom has now been shown to be EPR active (Albracht *et al.*, 1982; Kojima *et al.*, 1982). The EPR signal corresponds to one seen by Lancaster in whole cells, originally postulated to result from a nickel atom. Sparked by the finding of nickel in hydrogenase, Friedrich *et al.* (1982) tested the purified hydrogenase from *Alcaligenes eutrophus* and found it to contain nickel; the hydrogenase of *Desulfovibrio gigas* also has been shown to possess EPR-active nickel (Cammack *et al.*, 1982). The redox role of the nickel atom, the FeS centers and the FAD ligand as well as the role of selenocysteine in the hydrogenase reaction will be of great interest to students of electron transfer.

22.3.5 Carbon Fixation and Anabolism in Methanogens

An important question is the means by which methanogens, in large part autotrophic organisms, fix cell carbon from CO_2. The pathway for cell carbon assimilation is not based on the familiar ribulose bisphosphate cycle (Calvin cycle) nor the serine pathway, but is a new type, based on the citric acid cycle (Fuchs and Stupperich, 1982; Weimer and Zeikus, 1979). Results of labelling studies and determinations of enzyme activity in two species indicate that acetate is

Table 2 Hydrogenases of Methanogens

Species	Substrate	MW ($\times 10^{-3}$)	Subunits	Fe, S (mol^{-1})	Other properties	Reference
M. thermoautotrophicum strain ΔH	F_{420}	500	—	—	—	a
M. thermoautotrophicum strain ΔH	F_{420}	170	$\alpha_2\beta_2\gamma$	33, 24	2 FAD, 2.5 Ni, Ni EPR	b
M. thermoautotrophicum, strain Marburg	BV	60	—	—	1 Ni, Ni EPR	c
Methanobacterium, strain G2R	BV, MV, *etc.*	900	—	—	Membrane bound	d
Methanococcus vannielii	F_{420}	340	—	—	3.8 Selenocysteine	e
Alcaligenes eutrophus	NAD, BV, *etc.*	205	$\alpha_2\beta_2$	12, 12	1–2 Ni, 2 FMN	f, g
Clostridium pasterianum	Ferredoxin	60	α	12, 12	—	g

[a] Gunsalus and Wolfe (1980). [b] Kojima *et al.* (1983) and Jacobson *et al.* (1982). [c] Albracht *et al.* (1982). [d] McKellar and Sprott (1979). [e] Yamazaki (1982). [f] Friedrich *et al.* (1982). [g] Adams *et al.* (1981).

synthesized *de novo* from CO_2 in a mechanism similar to that of the acetogenic clostridia. Oxaloacetate is made *via* pyruvate, and at this step, significant differences between the two species are revealed, as shown in the pathways of Figure 6. *M. thermoautotrophicum* (Fuchs and Stupperich, 1982) appears to utilize a partial, reverse citric acid cycle in which α-ketoglutarate is made from succinate, as shown on the left of the figure (solid arrows). *Methanosarcina barkeri*, on the other hand, synthesizes α-ketoglutarate from citrate (Weimer and Zeikus, 1979) as shown on the right (dashed arrows). The components of a complete citrate cycle exist among the two methanogens, though neither organism has a complete cycle. Cell constituents are made from the cycle intermediates as follows: alanine from pyruvate, aspartate from oxaloacetate, glutamate from α-ketoglutarate, and sugars from pyruvate. In non-autotrophic methanogens, for example *Methanococcus voltae*, which requires isoleucine and leucine, these amino acids are a significant source of cell carbon. Even the autotrophic species will utilize exogenous acetate, and spare *de novo* synthesis.

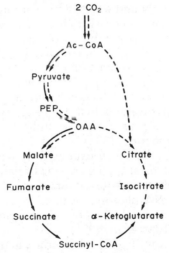

Figure 6 Pathway of CO_2 fixation into cell-carbon in two methanogens. *M. thermoautotrophicum* (———) (Fuchs and Stupperich, 1982); and *Methanosarcina barkeri* (– – – –) (Weimer and Zeikus, 1979). The pathway is also reviewed in Balch *et al.*, 1979

22.4 COENZYMES OF METHANOGENS

22.4.1 Coenzyme M

The novel cofactor HS-CoM (2-mercaptoethanesulfonic acid, **3**; Figure 7) and derivatives MeS-CoM (**4**; Figure 7) and (–S-CoM)$_2$ (the disulfide), are found only in methanogens. However, HS-CoM was used therapeutically as a mucolytic long before it was discovered in nature (*Merck Index*, 9th edn.). HS-CoM and derivatives are growth factors (vitamins) for the methanogen *M. ruminantium*; the organism provides the basis for a sensitive bioassay. The discovery, elucidation of structure, and the role of CoM in the terminal step of methanogenesis have been

amply reviewed (Wolfe, 1980). A question that remains is whether MeS-CoM is also an intermediate in fixation of cell carbon. The recent finding that CoM is a covalent part of protein-bound
factor F_{430} (Section 22.4.4) may provide a key to the role it plays in reaction mechanisms.

$$HSCH_2CH_2SO_3^-$$

(3)

$$MeSCH_2CH_2SO_3^-$$

(4)

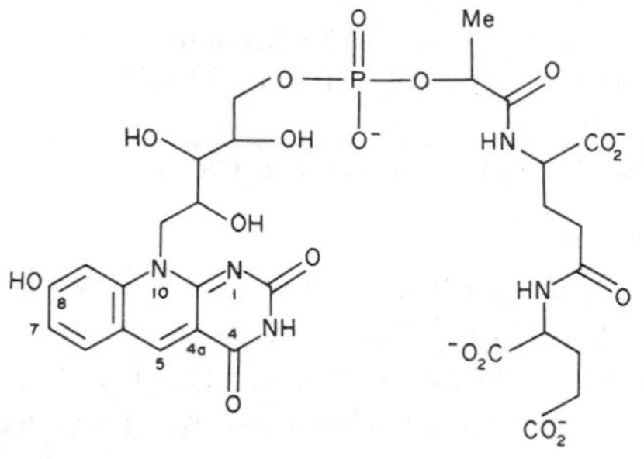

(5)

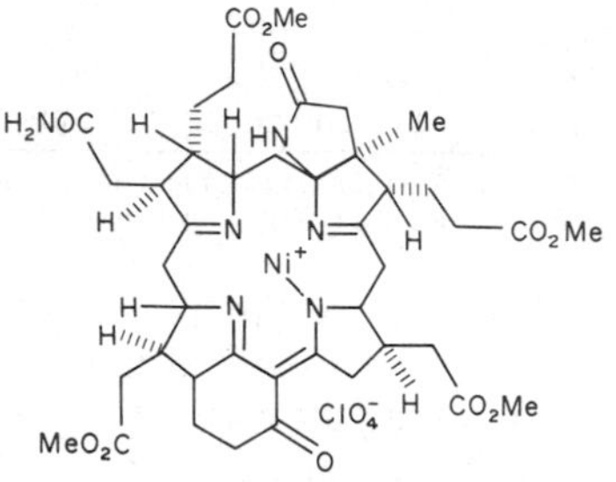

(6)

Figure 7 Structures of novel coenzymes from methanogens: coenzyme M (**3**); methyl-coenzyme M (**4**); factor F_{420} (**5**); $F_{430}M$ (**6**), the chemically methylated chromophore of factor F_{430} (Pfalz *et al.*, 1982)

22.4.2 Factor F_{420}

Factor F_{420} is the fluorescent compound which along with methanopterin gives oxidized methanogen cells their blue-green fluorescence (Eirich *et al.*, 1982; Doddema and Vogels, cited
therein). The structure (**5**; Figure 7) is notable in three respects: (1) it is a 7,8-didemethyl-8-
hydroxy-5-deazaflavin, unlike the typical 7,8-dimethylflavin; (2) the carbon at C-5, replacing N-5
in familiar flavins, limits the molecule to function only as a 2-electron carrier; and (3) the side-
chain is unique. The role F_{420} plays in methanogen metabolism is much like that played by ferredoxin in eubacteria. It functions in oxidoreduction reactions, accepting electrons from hydrogenase, formate dehydrogenase, pyruvate dehydrogenase, pyruvate synthase, α-ketoglutarate
synthase, and NADP-oxidoreductase, as shown by studies on a variety of methanogen species
(Eirich *et al.*, 1982; Fuchs and Stupperich, 1982).

It is now known that factor F_{420} has been exploited by organisms other than methanogens,
although in other roles. In the first instance, the enzyme of light-activated DNA repair in *Streptomyces griseus* contains F_{420} as chromophore (Eker *et al.*, 1981). This finding, together with
knowledge of action spectra of DNA photorepair in certain cyanobacteria, a green alga, and a
midge, led to the prediction that the involvment of F_{420} as a chromophore in DNA photorepair is
more general (Eirich *et al.*, 1982; Eker *et al.*, 1981).

The second newly uncovered role for F_{420} in eubacteria is in the biosynthesis of chlortetracycline
by *S. aureofasciens*. The terminal oxidation in synthesis of chlortetracycline requires Cosynthetic
Factor 1, now known to be identical to the fragment of F_{420} containing the deazaflavin system and
ribityl side chain, called FO (McCormick and Morton, 1982). These discoveries provide for the
exciting possibility that 8-hydroxy-5-deazaflavin coenzymes may be more widespread in nature
than previously suspected.

22.4.3 Component B of the MeS-CoM Methylreductase System

The formation of CH_4 from MeS-CoM (equation 1) by cell-free extracts of methanogens
requires a heat-stable, oxygen-labile molecule of 1000 daltons called component B (Section
22.3.2). The activity of this compound was found by Gunsalus and Wolfe (1980) and is now
known to occur only in methanogen cells (Tanner, PhD Thesis, University of Illinois, 1982).
Neither the structure nor the function of this unique compound is known at present.

22.4.4 Factor F_{430}

The yellow compound F_{430}, found in significant concentrations in methanogens, was shown to contain nickel (Diekert *et al.*, 1980; Whitman and Wolfe, 1980). A study of the biosynthesis of this molecule by Thauer *et al.* suggested that it was a tetrapyrrole compound, and the structure of the chromophore was determined by Pfalz *et al.* (1982). Structure (**6**; Figure 7) represents a chemically methylated derivative of F_{430}, called $F_{430}M$. The molecule is unique in that the chromophore portion (**7**; Figure 8) is a cross between the isobacteriochlorin nucleus of siroheme (**8**; Figure 8) and the corrin nucleus of the cobalamins (**9**; Figure 8). The cyclohexanone and 5-membered lactam modifications have not been seen previously in tetrapyrroles.

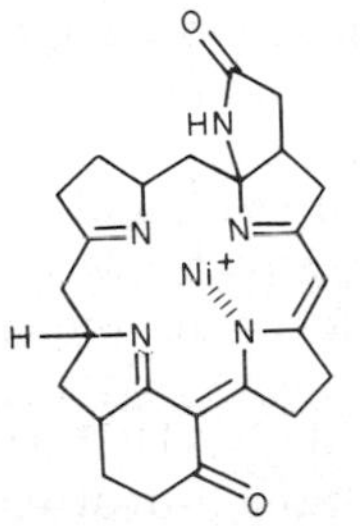

(7) 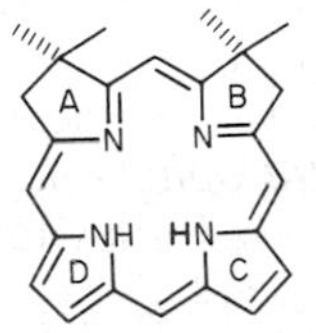   **(8)** 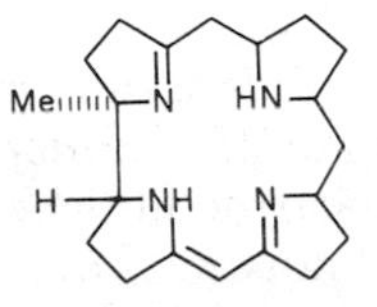   **(9)**

Figure 8 Comparison of the porphinoid ring systems of parent heterocycles: $F_{430}M$ parent (**7**); isobacteriochlorin parent (**8**); cobalamin parent (**9**)

Component C, the yellow protein of the MeS-CoM methylreductase system (Section 22.3.2) was shown to contain factor F_{430}, in the form called coenzyme MF_{430}. Results of nuclear magnetic resonance experiments indicated (Vogels *et al.*, 1982) and results of bioassays for coenzyme M (Section 22.4.1) confirmed that coenzyme M was associated with F_{430} (Keltjens *et al.*, 1982). Interpretation of the nuclear magnetic resonance data led to the suggestion that the sulfhydryl moiety of HS-CoM is positioned near the nickel atom in one axial position. The significance of this suggested arrangement is not yet known, but would be consistent with the known function of the coenzyme M sulfhydryl as one carbon carrier, and the possible reducing chemistry of the nickel tetrapyrrole. How the structure relates to the other steps on the redox pathway is not known.

22.4.5 CDR Factor

Dialysis of extracts of *M. thermoautotrophicum* resolves a small, heat-stable molecule which is required for reduction of CO_2 to CH_4 by the extract. The molecule, carbon dioxide reduction factor (CDR factor) is not required for reactions at the formaldehyde or methyl levels of oxidation (Romesser and Wolfe, 1982a). The structure of the molecule has not been determined, but it is not a familiar compound, nor is it methanopterin (Section 22.4.6) and it is needed in addition to methanopterin to support CH_4 production from CO_2 (Leigh and Wolfe, 1983).

22.4.6 Methanopterin and Related Compounds

A blue-fluorescent molecule with an absorbance maximum at 342 nm isolated from *M. thermoautotrophicum* was named factor F_{342} (Wolfe, 1980). F_{342} is now believed to be a degradation product of methanopterin (Bo); a tentative structure proposed for methanopterin (Vogels *et al.*, 1982) contains a pterin nucleus, glutamic acid, talosamine (see Section 22.2.2) and a cyclohexane-triol. Methanopterin is the parent of a compound called YFC (yellow fluorescent compound). Daniels and Zeikus (1978) showed that after short pulses with radioactive CO_2, methanogen cells contain labelled YFC; this compound gives rise to radioactive methane when incubated with extracts under H_2. Tentative structures have been proposed for YFC (Keltjens and Vogels, 1981; Vogels *et al.*, 1982), which is now called carboxy-5,6,7,8-tetrahydromethanopterin (cTHMP), a possible intermediate at the carboxy level of oxidation in the CO_2 reduction cycle (Figures 4 and

5). Both methanopterin and CDR factor are required for the crude extract system to reduce CO_2 to CH_4 (Leigh and Wolfe, 1983).

22.4.7   Other Coenzymes

Recently, a compound called formaldehyde activating factor (FAF) has been resolved from *M. thermoautotrophicum*. The compound is required for production of CH_4 from formaldehyde. It is not equivalent to other coenzymes from eubacteria or methanogens (Escalante-Semerena, PhD Thesis, University of Illinois, 1983); the structure has not been elucidated.

Two compounds, called F_A (λ_{max} 245 nm) and F_C (λ_{max} 252 and 257 nm) are small molecules of unknown structure and function (Keltjens and Vogels, 1982) although they contain sulfur at the sulfide level of oxidation. One or both of these have been proposed as one-carbon carriers at one of the levels of oxidation between formyl and methyl, but no evidence to support this has been published (Keltjens and Vogels, 1981; Vogels *et al.*, 1982).

Methanomicrobium mobile, a rumen methanogen, requires a small, oxygen-sensitive molecule, found in boiled extracts of *M. thermoautotrophicum*, for growth. Mobile factor, a compound of unknown structure, appears to fit the criteria for a vitamin–coenzyme relationship (Tanner, PhD Thesis, University of Illinois, 1982).

More familiar coenzymes found in methanogens include NAD(P), FAD (Lancaster, cited in Jacobson *et al.*, 1982), B_{12} (the factor III derivative, with the 5-hydroxybenzimidazole replacing the dimethylbenzimidazole of familiar cobalamins (Pol *et al.*, 1982, and references therein), and a compound which appears to be equivalent to the molybdenum cofactor of xanthine oxidase and nitrate reductase (Schauer and Ferry, 1982).

22.5   SUMMARY

Methanogenesis by microorganisms accounts for about 1% of the carbon fixed by photosynthesis. The process is vital in supporting the anaerobic breakdown of organic material. Methanogenesis can be harnessed to produce usable energy. Methanogens make their living by forming CH_4 from one or more of the following substrates: acetate, formate, methanol, methylamines, or carbon dioxide and hydrogen. Carbon monoxide is a minor substrate. On structural and biochemical levels methanogens are strikingly different from typical bacteria, and thus are grouped in the kingdom Archaebacteria.

Carbon dioxide is reduced to CH_4 at the expense of H_2, in a stepwise fashion, with CO_2 bound to newly described one-carbon carriers. Cellular energy is obtained from this process. Acetate is a major precursor of methane in nature. The methyl carbon becomes methane without loss of protons. The mechanism by which the cell obtains energy from the fermentation of acetate is not known.

Carbon is fixed by these cells in a newly discovered fashion, *via* an incomplete, reverse tricarboxylic acid cycle, with major variations in the details between species.

Methanogens contain certain unique coenzymes. Among these, coenzyme M, the one-carbon carrier, factor F_{420}, the 8-hydroxy-5-deazaflavin electron carrier, and factor F_{430}, the nickel tetrapyrrole which functions in the terminal step of methane production are of unusual biochemical interest.

Methanogenesis is a process most exciting to biologists, from ecologists to physical biochemists. Methanogens are biocatalysts, which may be exploited fruitfully by biotechnologists. They provide useful fuel gas, chemical feedstock, and perform a vital service in the anaerobic digestion of organic matter.

22.6   ADDENDUM

The structure of CDR Factor, a 2,4-disubstituted furan (Leigh *et al.*, 1984), and the role of tetrahydromethanopterin as a one-carbon carrier (Escalante-Semerena *et al.*, 1984) have been determined.

ACKNOWLEDGEMENTS

The work from this laboratory was supported by grants from the National Institutes of Health, National Science Foundation and Department of Energy. DPN was supported by NSF and NIH PHS Fellowships.

22.7 REFERENCES

Adams, M. W. W., L. E. Mortenson and J.-S. Chen (1981). Hydrogenase. *Biochim. Biophys. Acta*, **594**, 105–176.

Albracht, S. P. J., E.-G. Graf and R. K. Thauer (1982). The EPR properties of nickel in hydrogenase from *Methanobacterium thermoautotrophicum*. *FEBS Lett.*, **140**, 311–313.

Anthony, C. (1982), Methanogens and methanogenesis. In *Biochemistry of Methylotrophs*, pp. 269–327. Academic, London.

Balch, W. E., G. E. Fox, L. J. Magrum, C. R. Woese and R. S. Wolfe (1979). Methanogens: reevaluation of a unique biological group. *Microbiol. Rev.*, **43**, 260–296.

Barker, H. A. (1956). *Bacterial Fermentations*, pp. 1–27. Wiley, New York.

Bollschweiler, C. and A. Klein (1982). Polypeptide synthesis in *Escherichia coli* directed by cloned *Methanobrevibacter arborphilus* DNA. *Zbl. Bakt. Hyg.*, *I Abt. Orig. C3*, 101–109.

Cammack, R., D. Patil, R. Aguirre and E. C. Hatchikian (1982). Redox properties of the ESR-detectable nickel in hydrogenase from *Desulfovibrio gigas*. *FEBS Lett.*, **142**, 289–292.

Chappe, B., P. Albrecht and W. Michaelis (1982). Polar lipids of archaebacteria in sediments and petroleums. *Science*, **217**, 65–66.

Daniels, L. and J. G. Zeikus (1978). One-carbon metabolism in methanogenic bacteria: analysis of short-term fixation products of $^{14}CO_2$ and $^{14}CH_3OH$ incorporated into whole cells. *J. Bacteriol.*, **136**, 75–84.

Daniels, L., G. Fuchs, R. K. Thauer and J. G. Zeikus (1977). Carbon monoxide oxidation by methanogenic bacteria. *J. Bacteriol.*, **132**, 118–126.

Diekert, G., B. Klee and R. K. Thauer (1980). Nickel, a component of factor F_{430} from *Methanobacterium thermoautotrophicum*. *Arch. Microbiol.*, **124**, 103–106.

Ehhalt, D. H. and U. Schmidt (1978). Sources and sinks of atmospheric methane. *Pure Appl. Geophys.*, **116**, 452–464.

Eirich, L. D., G. D. Vogels and R. S. Wolfe (1982). An unusual flavin, coenzyme F_{420}. In *Flavins and Flavoproteins*, ed. V. Massey and C. H. Williams, pp. 435–441. Elsevier/North Holland, New York.

Eker, A. P. M., R. H. Dekker and W. Berends (1981). Photoreactivating enzyme from *Streptomyces griseus*-IV. On the nature of the chromophoric cofactor in *S. griseus* photoreactivating enzyme. *Photochem. Photobiol.*, **33**, 65–72.

Ellefson, W. L. and R. S. Wolfe (1981). Component C of the methylreductase system of *Methanobacterium*. *J. Biol. Chem.*, **256**, 4259–4262.

Ellefson, W. L., W. B. Whitman and R. S. Wolfe (1982). Nickel-containing Factor F_{430}: chromophore of the methylreductase of *Methanobacterium*. *Proc. Natl. Acad. Sci. USA*, **79**, 3707–3710.

Escalante-Semerena, J. C., K. L. Rinehart, Jr. and R. S. Wolfe (1984). *J. Biol. Chem.*, in press.

Fox, G. E., E. Stackebrandt, R. B. Hespell, J. Gibson, J. Maniloff, T. A. Dyer, R. S. Wolfe, W. E. Balch, R. S. Tanner, L. J. Magrum, L. B. Zablen, R. Blakemore, R. Gupta, L. Bonen, B. J. Lewis, D. A. Stahl, K. R. Luchrsen, K. N. Chen and C. R. Woese (1980). The phylogeny of prokaryotes. *Science*, **209**, 457–463.

Friedrich, C. G., K. Schneider and B. Friedrich (1982). Nickel in the catalytically active hydrogenase of *Alcaligenes eutrophus*. *J. Bacteriol.*, **152**, 42–48.

Fuchs, G. and E. Stupperich (1982). Autotrophic CO_2 fixation pathway in *Methanobacterium thermoautotrophicum*. *Zbl. Bakt. Hyg.*, *I Abt. Orig. C3*, 277–288.

Ghisalba, O. and F. Heinzer (1982). Methanol from methane — a hypothetical microbial conversion compared with the chemical process. *Experientia*, **38**, 218–223.

Gunsalus, R. P. and R. S. Wolfe (1977). Stimulation of CO_2 reduction to methane by methyl coenzyme M in extracts of *Methanobacterium*. *Biochem. Biophys. Res. Commun.*, **76**, 790–795.

Gunsalus, R. P. and R. S. Wolfe (1978). ATP activation and properties of the methyl coenzyme M reductase system in *Methanobacterium thermoautotrophicum*. *J. Bacteriol.*, **135**, 851–857.

Gunsalus, R. P. and R. S. Wolfe (1980). Methyl coenzyme M reductase from *Methanobacterium thermoautotrophicum*: resolution and properties of the components. *J. Biol. Chem.*, **255**, 1891–1895.

Hippe, H., D. Caspari, K. Fiebig and G. Gottschalk (1979). Utilization of trimethylamine and other N-methyl compounds for growth and methane formation by *Methanosarcina barkeri*. *Proc. Natl. Acad. Sci. USA*, **76**, 494–498.

Hungate, R. E. (1982). Methane formation and cellulose digestion — biochemical ecology and microbiology of the rumen ecosystem. *Experientia*, **38**, 189–192.

Hutten, T. J., M. H. DeJong, B. P. H. Peeters, C. Van Der Drift and G. D. Vogels (1981). Coenzyme M derivatives and their effects on methane formation from carbon dioxide and methanol by cell extracts of *Methanosarcina barkeri*. *J. Bacteriol.*, **145**, 27–34.

Jacobson, F. S., L. Daniels, J. A. Fox, C. T. Walsh and W. H. Orme-Johnson (1982). Purification and properties of an 8-hydroxy-5-deazaflavin-reducing hydrogenase from *Methanobacterium thermoautotrophicum*. *J. Biol. Chem.*, **257**, 3385–3388.

Jarrell, K. F. and G. D. Sprott (1981). The transmembrane potential and intracellular pH in methanogenic bacteria. *Can. J. Microbiol.*, **27**, 720–728.

Jones, J. B. and T. C. Stadtman (1981). Selenium-dependent and selenium-independent formate dehydrogenases of *Methanococcus vannielii*: separation of the two forms and characterization of the purified selenium-independent form. *J. Biol. Chem.*, **256**, 656–663.

Kandler, O. (1982). Cell wall structures and their phylogenetic implications. *Zbl. Bakt. Hyg.*, *I Abt. Orig. C3*, 149–160.

Kell, D. B., H. J. Doddema, J. G. Morris and G. D. Vogels (1981). Energy coupling in methanogens. In *Microbial Growth on C_1 Compounds*, ed. H. Dalton, pp. 159–170. Heyden, London.

Keltjens, J. T. and G. D. Vogels (1981). Novel coenzymes of methanogens. In *Microbial Growth on C_1 Compounds*, ed. H. Dalton, pp. 152–158. Heyden, London.

Keltjens, J. T., W. B. Whitman, C. G. Caerteling, A. M. van Kooten, R. S. Wolfe and G. D. Vogels (1982). Presence of coenzyme M derivatives in the prosthetic group (coenzyme MF_{430}) of methyl coenzyme M reductase from *Methanobacterium thermoautotrophicum*. *Biochem. Biophys. Res. Commun.*, **108**, 495–503.

Kojima, N., J. A. Fox, R. P. Hausinger, L. Daniels, W. H. Orme-Johnson and C. Walsh (1983). Paramagnetic centers in the nickel-containing, deazaflavin-reducing hydrogenase from *Methanobacterium thermoautotrophicum*. *Proc. Natl. Acad. Sci. USA*, **80**, 1300–1304.

König, H., R. Kralik and O. Kandler (1982). Structure and modifications of pseudomurein in *Methanobacteriales*. *Zbl. Bakt. Hyg.*, *I. Abt. Orig. C3*, 179–191.

Langworthy, T. A., T. G. Tornabene and G. Holzer (1982). Lipids of Archaebacteria. *Zbl. Bakt. Hyg.*, *I. Abt. Orig. C3*, 228–244.

Leigh, J. A. and R. S. Wolfe (1983). Carbon dioxide reduction factor and methanopterin: two coenzymes required for carbon dioxide reduction to methane by *Methanobacterium*. *J. Biol. Chem.*, **258**, 7536–7540.

Leigh, J. A., K. L. Rinehart, Jr. and R. S. Wolfe (1984). *J. Am. Chem. Soc.*, **106**, 3636–3640.

Lettinga, G. (1981). Anaerobic digestion for energy saving and production. In *Energy from Biomass*, ed. W. Palz, P. Chartier and D. O. Hall, pp. 270–278. Applied Science Publishers, London.

Mah, R. A. (1982). Methanogenesis and methanogenic partnerships. *Philos. Trans. R. Soc. London, Ser. B*, **297**, 599–616.

Mah, R. A. and M. R. Smith (1981). The methanogenic bacteria. In *The Prokaryotes: A Handbook on Habitats, Isolation, and Identification of Bacteria*, ed. M. P. Starr, H. Stolp, H. G. Truper, A. Balows, and H. G. Schlegel, vol. I, pp. 948–977. Springer-Verlag, New York.

Mah, R. A., M. R. Smith, T. Ferguson and S. Zinder (1981). Methanogenesis from H_2–CO_2, methanol and acetate by *Methanosarcina*. In *Microbial Growth on C_1 Compounds*, ed. H. Dalton, pp. 131–142. Heyden, London.

Mah, R. A., D. M. Ward, L. Baresi and T. L. Glass (1977). Biogenesis of methane. *Annu. Rev. Microbiol.*, **31**, 309–341.

McCormick, J. R. D. and G. O. Morton (1982). Identity of cosynthetic factor of *Streptomyces aureofaciens* and fragment FO from coenzyme F_{420} of *Methanobacterium* species. *J. Am. Chem. Soc.*, **104**, 4014–4015.

McInerney, M. J. and M. P. Bryant (1981). Review of methane fermentation fundamentals. In *Fuel Gas Production from Biomass*, ed. D. L. Wise, pp. 19–46. Chemical Rubber Co. Press, Inc., West Palm Beach, FL.

McKellar, R. C. and G. D. Sprott (1979). Solubilization and properties of a particulate hydrogenase from *Methanobacterium* strain G2R. *J. Bacteriol.*, **139**, 231–238.

Melchior, J.-L., R. Binot, I. A. Perez, H. Naveau and E.-J. Nyns (1982). Biomethanation: its future development and the influence of the physiology of methanogenesis. *J. Chem. Technol. Biotechnol.*, **32**, 189–197.

Nagle, D. P., Jr. and R. S. Wolfe (1983). Component A of the methyl coenzyme M methylreductase system of *Methanobacterium*: resolution into four components. *Proc. Natl. Acad. Sci. USA*, **80**, 2151–2155.

Pfaltz, A., B. Jaun, A. Fässler, A. Eschenmoser, R. Jaenchen, H. H. Gilles, G. Diekert and R. K. Thauer (1982). Factor-F430 from methanogenic bacteria: structure of the porphinoid ligand system. *Helv. Chim. Acta*, **65**, 828–865.

Pfeffer, J. T. (1982). Engineering, operation and economics of methane gas production. *Experientia*, **38**, 201–205.

Pol, A., C. van der Drift and G. D. Vogels (1982). Corrinoids from *Methanosarcina barkeri*: structure of the α-ligand. *Biochem. Biophys. Res. Commun.*, **108**, 731–737.

Prévot, A. R. (1980). Recherches recentes sur les bactéries méthanogènes. *Bull. Inst. Pasteur (Paris)*, **78**, 217–265.

Reeve, J. N., N. J. Trun and P. T. Hamilton (1982). Beginning genetics with methanogens. In *Genetic Engineering of Microorganisms for Chemicals*, ed. A. Hollaender, pp. 233–244. Plenum, New York.

Romesser, J. A. and R. S. Wolfe (1982a). CDR Factor, a new coenzyme required for carbon dioxide reduction to methane by extracts of *Methanobacterium thermoautotrophicum*. *Zbl. Bakt. Hyg.*, *I Abt. Orig. C3*, 271–276.

Romesser, J. A. and R. S. Wolfe (1982b). Coupling of methyl coenzyme M reduction with carbon dioxide activation in extracts of *Methanobacterium thermoautotrophicum*. *J. Bacteriol.*, **152**, 840–847.

Sauer, F. D., J. D. Erfle and S. Mahadevan (1980). Methane formation by the membranous fraction of *Methanobacterium thermoautotrophicum*. *Biochem. J.*, **190**, 177–182.

Sauer, F. D., J. D. Erfle and S. Mahadevan (1981). Evidence for an internal proton gradient in *Methanobacterium thermoautotrophicum*. *J. Biol. Chem.*, **256**, 9843–9848.

Schauer, N. L. and J. G. Ferry (1982). Properties of formate dehydrogenase in *Methanobacterium formicicum*. *J. Bacteriol.*, **150**, 1–7.

Spencer, R. W., L. Daniels, G. Fulton and W. H. Orme-Johnson (1980). Product isotope effects on *in vivo* methanogenesis by *Methanobacterium thermoautotrophicum*. *Biochemistry*, **19**, 3678–3683.

Stackebrandt, E., E. Seewaldt, W. Ludwig, K.-H. Schleifer and B. A. Huser (1982). The phylogenetic position of *Methanothrix soehngenii* elucidated by a modified technique of sequencing oligonucleotides from 16S rRNA. *Zbl. Bakt. Hyg.*, *I Abt. Orig. C3*, 90–100.

Stadtman, T. C. (1967). Methane fermentation. *Annu. Rev. Microbiol.*, **21**, 121–142.

Taylor, G. T. (1982). The methanogenic bacteria. In *Progress in Industrial Microbiology 16*, ed. M. J. Bull, pp. 231–329. Elsevier, New York.

Thauer, R. K., K. Jungermann and K. Decker (1977). Energy conservation in chemotrophic anaerobic bacteria. *Bacteriol. Rev.*, **41**, 100–180.

Vogels, G. D., J. T. Keltjens, T. J. Hutten and C. Van der Drift (1982). Coenzymes of methanogenic bacteria. *Zbl. Bakt. Hyg.*, *I. Abt. Orig. C3*, 258–264.

Weimer, P. J. and J. G. Zeikus (1979). Acetate assimilation pathway of *Methanosarcina barkeri*. *J. Bacteriol.*, **137**, 332–339.

Whitman, W. B. and R. S. Wolfe (1980). Presence of nickel in factor F430 from *Methanobacterium bryantii*. *Biochem. Biophys. Res. Commun.*, **92**, 1196–1201.

Whitman, W. B. and R. S. Wolfe (1983). Activation of the methylreductase system from *Methanobacterium bryantii* by ATP. *J. Bacteriol.*, **154**, 640–649.

Whitman, W. B., E. Ankwanda and R. S. Wolfe (1982). Nutrition and carbon metabolism of *Methanococcus voltae*. *J. Bacteriol.*, **149**, 852–863.

Wolfe, R. S. (1971). Microbial formation of methane. *Adv. Microb. Physiol.*, **6**, 107–146.

Wolfe, R. S. (1980). Respiration in methanogenic bacteria. In *Diversity of Bacterial Respiratory Systems*, ed. C. J. Knowles, vol. I, pp. 161–186. Chemical Rubber Company Press, Inc., Boca Raton, FL.

Wolfe, R. S. and I. J. Higgins (1979). Microbial biochemistry of methane — a study in contrasts. In *International Review of Biochemistry*, ed. J. R. Quayle, vol. 21, pp. 267–353. University Park Press, Baltimore.

Wuhrmann, K. (1982). Ecology of methanogenic systems in nature. *Experientia*, **38**, 193–198.

Yamazaki, S. (1982). A selenium containing hydrogenase from *Methanococcus vannielii*. *J. Biol. Chem.*, **257**, 7926–7929.

Zeikus, J. G. (1980). Chemical and fuel production by anaerobic bacteria. *Annu. Rev. Microbiol.*, **34**, 423–464.

Zeikus, J. G. (1977). The biology of methanogenic bacteria. *Bacteriol. Rev.*, **41**, 514–541.

23
Microbial Metabolism of Hydrogen

H. G. SCHLEGEL and K. SCHNEIDER
Universität Göttingen, Federal Republic of Germany

23.1 INTRODUCTION: THE ROLE OF HYDROGEN IN THE BIOSPHERE

Hydrogen is the most abundant element in the Sun where it plays a key role in energy generation. Together with oxygen it is the key element in the biological energy cycle on the Earth. In all organic matter hydrogen atoms are bound to carbon, nitrogen, sulfur and other elements.

In this chapter the term hydrogen is used to designate both hydrogen atoms in chemical compounds and molecular hydrogen (= gaseous hydrogen = dihydrogen = H_2). Gaseous hydrogen is produced as well as consumed by living organisms. However, while the conversion of hydrogen is widely distributed among microorganisms it is virtually absent from higher organisms. Large amounts of hydrogen are produced in the biosphere. Only a small proportion of this hydrogen reaches the atmosphere as such. The lower part of the atmosphere, called the troposphere, which extends up to a height of 10–12 km, contains about 2.2×10^9 tons H_2 (Figure 1). Compared to the amount of oxygen, 1.1×10^{15} tons, the amount of H_2 is very small. Its average concentration is 0.55 p.p.m. by volume of air. Due to this low mixing ratio of 0.55 p.p.m., hydrogen is designated as a trace gas. The concentration mentioned is the steady state concentration. Hydrogen is not conserved in the atmosphere as are, for example, the noble gases Ar, Ne and Kr which accumulated in the course of the Earth's history. It is rather continuously produced and decomposed in the troposphere. In this hydrogen cycle, biological processes play a significant role (Schlegel, 1974; Conrad and Seiler, 1980).

Hydrogen is produced as well as consumed by microorganisms in the presence and in the absence of oxygen (under both oxic and anoxic conditions). Both aerobic and anaerobic organisms

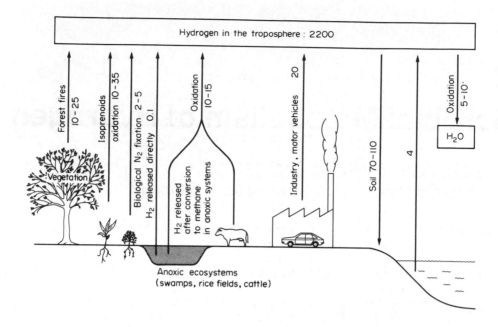

Figure 1 The hydrogen cycle (figures represent million tons; Conrad and Seiler, 1980, modified)

are involved. Figure 2 presents a survey of the metabolic processes accompanied by either evolution or consumption of hydrogen. The enzymes catalyzing these processes are the hydrogenases; they are discussed in Section 23.2. The organisms involved in the formation of hydrogen will be dealt with in Section 24.3.1, and those organisms able to utilize hydrogen will be discussed in Section 23.3.2. The biotechnological processes, which are in use or lend themselves for various applications are reviewed in Section 23.4.

This chapter only presents a brief systematic survey on the metabolism of hydrogen. For more

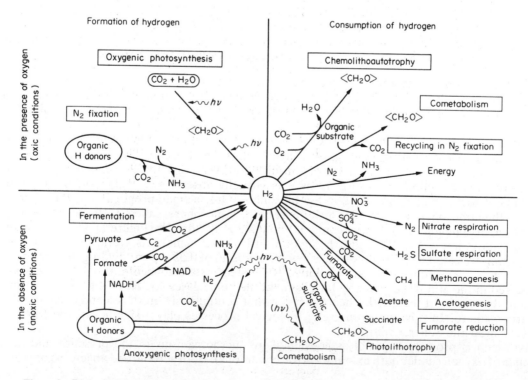

Figure 2 Biological processes of hydrogen formation and consumption (Conrad and Seiler, 1980, modified)

detailed presentations the reader may consult the proceedings volumes of symposia on 'Microbial Production and Utilization of Gases' (Schlegel *et al.*, 1976) and on 'Hydrogenases: Their Catalytic Activity, Structure and Function' (Schlegel and Schneider, 1978) as well as several recent review articles (Adams *et al.*, 1981; Zajic *et al.*, 1978). Reviews on the various groups of bacteria will be mentioned where appropriate. For reasons of brevity the original literature before 1978 is quoted in rare cases only.

23.2 ENZYMES CATALYZING THE EVOLUTION AND OXIDATION OF HYDROGEN

The enzymes catalyzing the formation and the oxidation of hydrogen are collectively called hydrogenases. The enzyme reaction is represented by the equation

$$H_2 \rightleftharpoons 2H^+ + 2e^-$$

All hydrogenases contain iron and acid-labile sulfur arranged in iron–sulfur clusters. As became known recently many hydrogenases, perhaps all, contain one or more nickel atoms per enzyme molecule. At least one hydrogenase contains FMN as a prosthetic group, another FAD. In spite of the many similarities between the hydrogenases their catalytic and physicochemical properties vary widely.

Hydrogenases can be differentiated with respect to their position in electron transport systems and their localization in the cell. The natural electron donor/acceptor is known only for the soluble, cytoplasmic or loosely bound periplasmic enzymes. For the membrane-bound hydrogenases this information is incomplete or lacking. Details of types and properties of hydrogenases are presented in the review articles of Adams *et al.* (1981) and Schlegel and Schneider (1978). A compilation of the papers on function and structure of hydrogenases has been published by Yagi (1981). Here a short survey will be presented of those hydrogenases which have been purified and partially characterized. For the reader's convenience nitrogenase is included because it is responsible for H₂ evolution by many bacteria. Assay systems are discussed in various reviews mentioned above and by Krasna (1978, 1979), van Dijk *et al.* (1981), and van Dijk and Veeger (1981).

23.2.1 H₂:Ferredoxin Oxidoreductase

The redox potential of the couple $2H^+/H_2$ is $E_0' = -413$ mV. Physiological electron carriers providing electrons at this potential are the ferredoxins. The ferredoxins present in the cytoplasm of many anaerobic bacteria readily release electrons if the appropriate hydrogenase is present.

$$H_2 + Fd(ox) \rightleftharpoons Fd(red) + 2H^+$$

This type of hydrogenase, H₂:ferredoxin oxidoreductase, has been isolated from *Clostridium pasteurianum* and characterized. It is present in the soluble, cytoplasmic fraction of the cells and catalyzes both the evolution and uptake of hydrogen. This type of hydrogenase has been described as an extraordinary oxygen-sensitive enzyme. It consists of a single polypeptide chain of a molecular mass of 60 000 and contains three [4Fe + 4S] clusters, only one of which is apparently involved in the oxidation–reduction reaction during catalysis. The natural electron carriers are ferredoxins and flavodoxins. *In vitro* these carriers can be substituted by viologen dyes such as benzyl- or methyl-viologen. These artificial carriers can be reduced by dithionite. In the presence of dithionite, hydrogen is evolved continuously. *In vivo*, electrons are supplied by the oxidation of pyruvate mediated by pyruvate:ferredoxin oxidoreductase or by the oxidation of NADH mediated by NADH:ferredoxin oxidoreductase.

The natural function of the ferredoxin-linked hydrogenase is the disposal of excess electrons in the form of H₂. The enzyme is present in the clostridia and many other strictly anaerobic, fermentative bacteria such as *Ruminococcus albus*, *Butyrivibrio fibrisolvens* and *Megasphaera elsdenii*, as well as facultatively anaerobic bacteria such as *Bacillus polymyxa*.

23.2.2 H₂:Ferricytochrome c_3 Oxidoreductase

The hydrogenase of sulfate-reducing bacteria is linked to cytochrome c_3 ($E_0' = -205$ mV) and is located in the periplasmic space or in the cytoplasmic membrane. Its molecular mass and location differ among species and strains. The H_2:ferricytochrome c_3 oxidoreductase catalyzes the reversible reaction

$$H_2 + \text{cytochrome } c_3(\text{ox}) \rightleftharpoons \text{cytochrome } c_3(\text{red}) + 2H^+$$

In *Desulfovibrio vulgaris* (Hildenborough) hydrogenase consists of a single polypeptide chain of a molecular mass of 50 000 and contains 12 iron and 12 labile sulfur atoms, whereas the hydrogenases of *Desulfovibrio vulgaris* (Miyazaki strain) and *Desulfovibrio gigas* consist of two non-identical subunits of molecular mass about 60 000 and 30 000. For the hydrogenase of *D. vulgaris* (Miyazaki strain) only eight atoms each for iron and labile sulfur have been determined. The enzymes from *D. gigas* and *D. desulfuricans* were recently found to contain 2 [4Fe+4S] clusters, 1 [3Fe+xS] cluster and nickel as redox-active components (Teixeira *et al.*, 1983; Krüger *et al.*, 1982).

The natural functions of the hydrogenases of sulfate-reducing bacteria appear to be both the uptake and the evolution of H_2 depending on the growth conditions. Hydrogen consumption is linked with the sulfite reductase system, hydrogen evolution with the oxidation of pyruvate.

$$\text{pyruvate} \rightarrow \text{Fd}_I \rightarrow \text{cyt } c_3 \rightarrow \text{Fd}_{II} \rightarrow \text{SO}_3^{2-}$$
$$\downarrow\uparrow$$
$$H_2$$

Properties of hydrogenase and its role in sulfate-reducing bacteria have been recently reviewed by Le Gall *et al.* (1982b). This type of hydrogenase may be present in other bacteria also, for example in phototrophic bacteria.

23.2.3 H_2:NAD^+ Oxidoreductase

A hydrogenase directly linked to nicotinamide dinucleotide (NAD) has so far been found only in the lithoautotrophic aerobic hydrogen-oxidizing bacteria, more specifically in *Alcaligenes eutrophus*, *A. ruhlandii* and *Nocardia opaca*. It catalyzes the reversible reaction

$$H_2 + NAD^+ \rightleftharpoons NADH + H^+$$

As the redox couple $NAD/NADH_2$ has a potential of $E_0' = -320$ mV the equilibrium of the reaction favors the reduction of NAD. The *in vitro* production of H_2 from $NADH_2$ has been demonstrated (Schneider and Schlegel, 1976; Pinchukova and Varfolomeev, 1980).

The H_2:NAD^+ oxidoreductases have been isolated and extensively studied from *A. eutrophus* strain H16 and from *Nocardia opaca* strain 1b. They are soluble cytoplasmic enzymes and have very similar structural properties. Both are tetramers and consist of four non-identical subunits which have, in the case of the *Nocardia* hydrogenase, molecular masses of 64 000, 56 000, 31 000 and 27 000 (Schneider *et al.*, 1984a). The NAD-linked hydrogenases apparently contain 1 [2Fe + 2S] cluster, one high potential [3Fe+xS] or [4Fe+4S] cluster, two low potential [4Fe + 4S] clusters, one FMN molecule and two nickel atoms (Schneider and Schlegel, 1978; Schneider *et al.*, 1979; Friedrich *et al.*, 1982; Schneider *et al.*, 1984b). Studies on the enzyme of *N. opaca* revealed that this type of hydrogenase is composed of two different subunit dimers. Only one of them, in which the nickel and the high potential Fe–S cluster are localized, shows hydrogenase activity, *e.g.* with methylviologen as electron carrier. The other dimer behaves like an NADH dehydrogenase and exhibits high diaphorase activity (Schneider *et al.*, 1984b).

The NAD reducing hydrogenase of *A. eutrophus* is reversibly inactivated and stabilized by oxygen in the absence of reducing compounds. In the simultaneous presence of oxygen, H_2 and $NADH_2$ superoxide radicals are formed by which the enzyme is irreversibly inactivated (Schneider and Schlegel, 1981). The natural function of the enzyme is apparently the reduction of NAD thus providing reducing power for the fixation of carbon dioxide *via* the Calvin cycle. It can also channel electrons into the electron transport chain.

23.2.4 H_2:Coenzyme F_{420} Oxidoreductase

The majority of methanogenic bacteria can grow on hydrogen and carbon dioxide. The hydrogenases of these bacteria seem to differ from each other; however, in general they are able to reduce coenzyme F_{420}, derivatives of F_{420}, flavins or artificial dyes, but not NAD or NADP.

$$H_2 + \text{coenzyme } F_{420}(\text{ox}) \rightleftharpoons \text{coenzyme } F_{420}(\text{red})$$

Coenzyme F_{420} is a derivative of 8-hydroxy-5-deazaflavin and acts in an analogous fashion to the flavins: it is reduced by accepting two hydrogen atoms. It has a low redox potential ($E_0' = -340$ mV) and reacts with hydrogenase and various dehydrogenases from methanogenic bacteria.

For the enzyme of *Methanobacterium thermoautotrophicum* molecular masses of 60 000 (Fuchs *et al.*, 1978) and 170 000 (Jacobson *et al.*, 1982) have been reported. For the latter protein a composition of three distinct subunits with molecular masses of 40 000, 31 000 and 26 000 in the ratio 2:2:1 has been described. Hydrogenases from methanogenic bacteria generally tend to form aggregates. Molecular masses up to 1.3×10^6 were reported (McKellar and Sprott, 1979; Jacobson *et al.*, 1982; Yamazaki, 1982). Recently FAD as well as nickel have been identified as components of hydrogenase from *M. thermoautotrophicum* (Graf and Thauer, 1981; Albracht *et al.*, 1982; Jacobson *et al.*, 1982; Kojima *et al.*, 1983). 33–43 iron atoms have been determined for this enzyme; however in contrast to other hydrogenases EPR signals from Fe–S clusters were detectable only with the reduced, not with the oxidized enzyme (Kojima *et al.*, 1983). From *Methanococcus vannielii* a selenium-containing hydrogenase was purified to near homogeneity. Selenocysteine is the chemical form of selenium in the enzyme (Yamazaki, 1982). The studies on the hydrogenases of methanogenic bacteria are rendered difficult by their strong or extreme oxygen sensitivity.

The natural function of the hydrogenases is to channel hydrogen into the reductive pathways for the assimilation of CO_2 for cell material synthesis and for the formation of methane. Apparently, from coenzyme F_{420} the hydrogen is mediated *via* NADP into biosynthetic pathways or is provided directly to the last step of methane formation, in which coenzyme MF_{430} is involved (Keltjens *et al.*, 1982).

23.2.5 Membrane-bound Hydrogenases

Unlike the soluble, cytoplasmic hydrogenases, the hydrogenases located in the cytoplasmic membrane have not been studied in great detail. With a few exceptions their physiological function is to take up hydrogen and to feed electrons into electron transport systems (respiratory chain, photosynthetic electron transport). Artificial electron carriers, preferentially dyestuffs, were used to measure their activity.

$$H_2 + acceptor(ox) \rightleftharpoons acceptor(red)$$

Their natural electron acceptors are still a matter of speculation. The quinones and cytochromes of the *b*- and *c*-type are possible candidates. The high redox potential (E_0' of about 0 mV) of these hydrogen or redox carriers, respectively, does not favor the involvement of these hydrogenases in hydrogen production. Therefore, these enzymes are sometimes called 'unidirectional' hydrogenases. On the other hand, many cytochromes turned out to have variable redox potentials depending on the protein environment (conformation) surrounding the prosthetic porphyrin compounds. The E_0' may be as negative as -350 mV. Therefore, it is too early to make predictions on the reversible nature of a hydrogenase reaction.

Membrane-bound hydrogenases have been found in Enterobacteriaceae, in the majority of aerobic hydrogen-oxidizing bacteria, almost all nitrogen-fixing bacteria, and anoxygenic and oxygenic phototrophic bacteria. They appear to be iron–sulfur proteins, and recent evidence suggests that at least some enzymes require nickel for activity. A few examples may illustrate that the membrane-bound hydrogenases differ from each other in size, subunit composition, number of iron–sulfur centers, sensitivity to oxygen, specific activity, K_m values, orientation in the membrane and a number of other properties (see review by Vignais *et al.*, 1981, and Adams *et al.*, 1981). The characteristics of the enzymes have been obtained by studying the solubilized, purified proteins. The properties of the isolated enzymes in the aqueous assay system may differ from those *in situ* in the membrane.

The hydrogenase from *Paracoccus denitrificans*, a lithoautotrophic hydrogen-oxidizing bacterium, is a tetramer of four identical subunits of a molecular mass of 63 000 for each subunit. This is in the presence of the detergent Triton X-100 which was used for solubilization. When detergent is removed the enzyme forms aggregates of two such tetramers. The enzyme will catalyze the reduction of benzyl- and methyl-viologen, methylene blue, ferricyanide and cytochrome *c*; it does not reduce NAD(P) or spinach ferredoxin (Sim and Vignais, 1978; Sim and Sim, 1979).

The membrane-bound hydrogenase from *Alcaligenes eutrophus* is a dimer of two non-identical subunits of molecular masses 67 000 and 32 000, and contains seven to nine atoms each of iron and acid-labile sulfur and approximately one nickel atom (Schneider *et al.*, 1983). In contrast to other types of membrane-bound hydrogenases, which are only paramagnetic in the oxidized state, the enzyme of *A. eutrophus* showed a complex EPR spectrum in the reduced state indicating the presence of two interacting low potential [4Fe+4S] clusters (Schneider *et al.*, 1983). It reduces methylene blue and, at a low rate, cytochrome *c* (Schink and Schlegel, 1978, 1979, 1980; Weiss and Schlegel, 1980). The enzyme is stable under air (Schink and Probst, 1980).

The membrane-bound hydrogenase from *Rhizobium japonicum* consists of a single polypeptide chain with a molecular mass of 63 000. It will reduce methylene blue and other redox carriers, but not however methylviologen. The *Rhizobium* hydrogenase in its isolated, solubilized state is a very oxygen-sensitive enzyme (Arp and Burris, 1979).

The heterogeneity within the membrane-bound hydrogenases may further be illustrated by mentioning the number of subunits and the molecular mass of enzymes from other bacteria: *Escherichia coli*, $2 \times 56\ 000$ (Adams and Hall, 1978, 1979b); *Proteus mirabilis*, $2 \times 63\ 000$ plus $2 \times 33\ 000$ (Schoenmaker *et al.*, 1979); *Rhodospirillum rubrum*, $1 \times 65\ 000$ (Adams and Hall, 1977, 1979a); *Thiocapsa roseopersicina*, $1 \times 47\ 000$, $1 \times 25\ 000$ (Gogotov, 1978); *Chromatium*, $2 \times 50\ 000$ (Gitlitz and Krasna, 1975; Strekas *et al.*, 1980).

23.2.6 Formate Hydrogenlyase

In species of Enterobacteriaceae hydrogenase is part of a membrane-bound multienzyme complex called formate hydrogenlyase. The enzyme system is involved in the evolution of hydrogen from formate

$$HCO_2H \rightarrow CO_2 + H_2$$

The reaction is, however, more complex and includes partially unknown electron carriers (Bernhard and Gottschalk, 1978; Schoenmaker *et al.*, 1978).

In vitro the hydrogenase reaction can be linked to benzylviologen. The hydrogenase has been located in the membrane fraction. The enzyme solubilized from *E. coli* has a molecular mass of about 100 000 and consists of two identical subunits with molecular masses of 56 000 and 33 000 and has been found to contain 12 atoms each of iron and labile sulfur (Adams and Hall, 1979b). In the hydrogen-evolving Enterobacteriaceae the hydrogenase is involved in H_2 formation from formate. However, the enzyme can equally well channel electrons into the electron transport chain and thus be linked to the fumarate reductase system (Bernhard and Gottschalk, 1978a; Lütgens and Gottschalk, 1982).

23.2.7 Nitrogenase

Nitrogenase is present in all nitrogen-fixing bacteria. It catalyzes the reduction of nitrogen to ammonia. However, it can also reduce several other compounds which have a triple bond in common, *e.g.* C_2H_2 to C_2H_4. In addition, it reduces protons to hydrogen. Electrons are supplied to the nitrogenase *via* ferredoxins or flavodoxins, respectively.

Nitrogenase consists of two subunits both of which are iron–sulfur proteins. One subunit contains molybdenum in addition. The enzyme, as well as the process of N_2 fixation, is extremely oxygen-sensitive.

H_2 formation by nitrogenase is irreversible and depends on a supply of large amounts of ATP in addition to electrons: the stoichiometry between H_2 production and N_2 fixation is summarized by the following overall equation:

$$8H^+ + 8e^- + N_2 \rightarrow 2\,NH_3 + H_2$$

H_2 formation appears to be an inevitable process and an intrinsic property of nitrogenase. Such a process may help to protect nitrogenase from oxygen damage by augmentation of respiratory protection in many organisms. Most aerobic nitrogen-fixing bacteria contain a membrane-bound hydrogenase which recycles the hydrogen and helps to trap the oxygen at the cell envelope (Dixon, 1972; Ruiz-Argüeso *et al.*, 1979; Robson and Postgate, 1980). Nitrogenase is involved in the production of hydrogen by phototrophic bacteria, the purple and green bacteria, the cyanobacteria, the root nodules of leguminous plants and various free-living N_2-fixing bacteria. In nature only the hydrogen produced by the rhizobia in the root nodules escapes to the atmosphere (see Section 23.3.1.2).

23.3 ORGANISMS INVOLVED IN THE CONVERSION OF HYDROGEN

Many microorganisms are able both to produce and utilize hydrogen depending on the existing conditions. Others can only utilize hydrogen.

23.3.1 Hydrogen-producing Microorganisms

23.3.1.1 *Anaerobic conditions*

Under anaerobic conditions in anoxic environments *via* several metabolic processes an excess of electrons is produced. If the respective enzymes, hydrogenases, and the electron-carriers are present, these electrons can be dumped taking advantage of protons as electron acceptors. The reduction of protons results in the production of molecular hydrogen. The hydrogen dissolved in the suspending medium can either be used by accompanying microorganisms or is evolved in gaseous form. There are several physiological groups of bacteria which evolve hydrogen under anaerobic conditions. These are fermentative bacteria, sulfate-reducing bacteria, and phototrophic bacteria.

(i) *Fermentation and fermentative bacteria*

Fermentation is a metabolic process aiming at regeneration of ATP, the metabolic energy currency, and at supplying metabolites for biosynthetic processes without involvement of oxygen (Gottschalk, 1981). Generally the organic substrate is cleaved and part of the cleavage products are used for energy generation. The conversions resulting in ATP regeneration are oxidative reactions. The oxidized carbon is usually released in the form of carbon dioxide. The electrons released in these oxidation reactions are transferred to carriers such as nicotinamide dinucleotides (NAD) or ferredoxins (Fd). The reduced carriers ($NADH_2$) are reoxidized by transferring the electrons to protons or other cleavage products or their derivatives which are thereby reduced and then excreted from the cells. The fermentation of carbohydrates and various other substances leads to the formation of the following products: ethanol, 2-propanol, 2,3-butanediol, *n*-butanol, formate, acetate, lactate, propanoate, butyrate, succinate, capronoate, acetone, carbon dioxide and hydrogen, either as sole products or as a mixture. According to the quantitatively prevailing or characteristic products various types of fermentation are differentiated such as ethanol fermentation, lactate fermentation, propionate fermentation, formate fermentation, butyrate fermentation, acetate fermentation and others (Figure 3).

In the lactic acid fermentation pyruvate is directly reduced to lactate. No hydrogen is produced when ethanol, lactate or propionate are the sole fermentation products (Figure 3). In other fermentations pyruvate is oxidized to acetyl-CoA. This can be converted to acetyl phosphate and give rise to ATP regeneration and excretion of acetate. In most instances the formation of acetyl-CoA is accompanied by the formation of hydrogen.

There are two types of biochemical reaction leading from pyruvate to hydrogen. One, which is typical for the clostridia (Gottschalk *et al.*, 1981) or other bacteria forming products such as acetate or butyrate, employs pyruvate:ferredoxin oxidoreductase. This enzyme catalyzes the reaction:

$$\text{pyruvate} + \text{CoA} + 2\,\text{Fd(ox)} \rightarrow \text{acetyl-CoA} + 2\,\text{Fd(red)} + CO_2$$

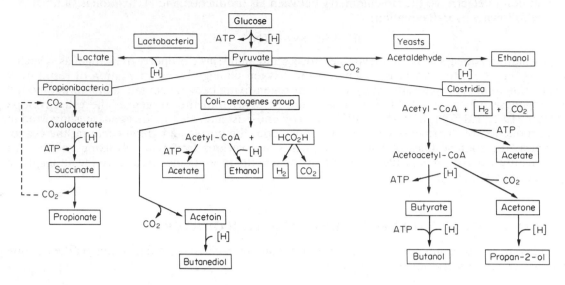

Figure 3   The products of glucose fermentation by various groups of fermentative microorganisms

Hydrogen is formed when oxidized ferredoxin is regenerated by ferredoxin:H_2 oxidoreductase according to the equation:

$$2\,Fd(red) \rightarrow 2\,Fd(ox) + H_2$$

The second type of H_2-evolving reaction is typical for the Enterobacteriaceae and related bacteria (Brenner, 1981). This fermentation is known as mixed acid fermentation, and formate is the characteristic product. The enzyme pyruvate formate lyase catalyzes the reaction:

$$\text{pyruvate} + CoA \rightarrow \text{acetyl-CoA} + \text{formate}$$

Hydrogen is formed when formate is cleaved by a hydrogenlyase enzyme system to give carbon dioxide and hydrogen:

$$\text{formate} \rightarrow CO_2 + H_2$$

   In several types of fermentation acetyl-CoA is converted to acetoacetyl-CoA. This intermediate serves as an electron acceptor for $NADH_2$ and is converted to yield butyrate, n-butanol, acetone, 2-propanol or capronoate. Of course, the synthesis of acetoacetyl-CoA prevents the bacterium from regenerating ATP from acetyl-CoA and therefore results in a low cell yield. Some fermentative bacteria contain an enzyme, $NADH_2$:ferredoxin oxidoreductase which catalyzes the reaction:

$$NADH_2 + 2\,Fd(ox) \rightarrow NAD + 2\,Fd(red)$$

As shown above hydrogen can be evolved from Fd(red) by mediation of a hydrogenase. The reduction of ferredoxin by $NADH_2$ enables bacteria to produce hydrogen even from reducing power which was primarily provided on the level of NAD. However, it may be remarked that the reduction of Fd by $NADH_2$ is a thermodynamically unfavorable reaction. This means that the reaction proceeds only if the reduced ferredoxin is reoxidized by hydrogenase, and hydrogen is evolved. As the hydrogenase reaction is reversible, hydrogen formation is inhibited by hydrogen accumulated in the environment. Consequently, hydrogen evolution originating from reducing power on the level of $NADH_2$ depends on continuous removal of H_2 from the environment. The removal of H_2 occurs readily in mixed cultures of H_2-evolving and H_2-consuming bacteria such as for example *Ruminococcus albus* and *Vibrio succinogenes* or methanogenic bacteria (Figure 4).
   The reactions discussed explain the formation of hydrogen by the Enterobacteriaceae, the clostridia and physiologically related bacteria. In pure cultures only reducing power derived from pyruvate *via* ferredoxin can be released as H_2. In appropriate mixed cultures both the reducing power provided as $NADH_2$ and as Fd(red) or formate can be released as H_2. Obviously, in mixed cultures the hydrogen does not appear in the form of gas bubbles, but only in dissolved form at

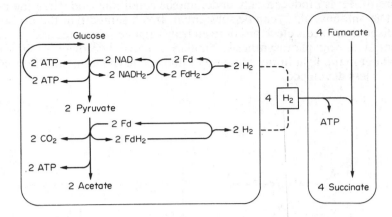

Figure 4 Glucose fermentation by *Ruminococcus albus* (left) and utilization of the hydrogen produced by *Vibrio succinogenes* (right), both grown in coculture; the nutritional symbiosis is an example of interspecies hydrogen transfer

low concentration (10^{-6} M) in the suspending medium due to its immediate utilization by anaerobic hydrogenotrophic bacteria such as the methanogenic, acetogenic and sulfate-reducing bacteria. In the presence of these bacteria and the corresponding acceptor compounds (CO_2 or sulfate) the reducing power released during fermentations as H_2 finally gives rise to the production of methane, acetate or hydrogen sulfide, respectively. Locations where these fermentations take place are the rumen of cattle, aquatic sediments, swamps, rice fields, waterlogged soils, and sewage digesters. The presently highly recommended biogas production from waste takes advantage of the processes described. In the ecosystems mentioned the fermentation substrates are cellulose, hemicelluloses and other poly- and oligo-saccharides. These substances are the leftovers of the vegetable food of animals, and plant litter; in aquatic ecosystems they settle down and are included into the sediment. The subsequent processes from cellulosic materials to methane and hydrogen sulfide are described as the anaerobic food chain (Figure 5).

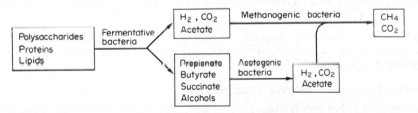

Figure 5 The anaerobic food chain

In pure culture many of the fermentative bacteria can be used to produce H_2 (Zajic *et al.*, 1978). Among the eukaryotes only a few protozoa are known to perform a fermentative type of metabolism and to produce hydrogen. These are *Tritrichomonas foetus* (Müller, 1980), an inhabitant of the human vagina, and *Dasytricha ruminantium* (Lloyd *et al.*, 1982) inhabitant of the rumen of cattle. These ciliates catalyze a kind of clostridia-type fermentation and are characterized by special cell organelles, hydrogenosomes, which harbor the enzyme system involved in the formation of H_2.

(ii)   Anoxygenic photosynthesis and phototrophic bacteria

The anoxygenic phototrophic bacteria (Stanier *et al.*, 1981) are well known under their previous name 'purple bacteria', derived from their red, brown and pinkish colors. They comprise at least three groups, the non-sulfur purple bacteria (Athiorhodaceae or Rhodospirillaceae), the sulfur purple bacteria (Thiorhodaceae or Chromatiaceae) and the green sulfur bacteria (Chlorobiaceae). Although their ability to evolve and consume hydrogen has been known for more than 50 years it is not possible to present a unified concept of the hydrogen metabolism of these bacteria. The majority of purple and green bacteria can produce H_2, some at high rates, either in the light or in the dark.

Photoproduction of hydrogen by the Rhodospirillaceae is evidently a nitrogenase-dependent reaction (Figure 6). H_2 is produced only under anoxic conditions and when the nitrogen sources limit growth. The amounts of H_2 can be substantial. It is assumed that the production of hydrogen serves to dispose of excess electrons derived from organic substrates and provided at a highly negative potential during photosynthesis. Similarly, active hydrogen evolution occurs when *Chlorobium* is kept in the light in the presence of thiosulfate. Among the Chromatiaceae, hydrogen production is less developed.

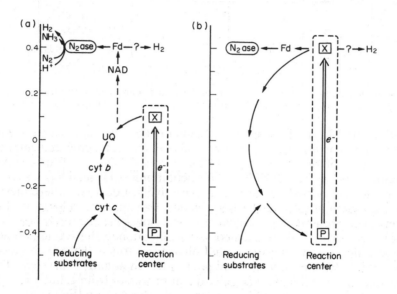

Figure 6 Schemes of photosynthetic electron transport in (a) Rhodospirillaceae and (b) Chlorobiaceae within the redox potential diagram. Abbreviations: cyt b, cytochrome b; Fd, ferredoxin; N_2ase, nitrogenase; P, pigments (bacteriochlorophyll a) as the electron donors of the photosynthetic reaction center; UQ, ubiquinone; X, electron acceptor of the reaction center

In the dark hydrogen is also produced, especially in the presence of pyruvate as carbon substrate. In *Rhodospirillum rubrum* the formate hydrogenlyase and the pyruvate formate lyase systems are apparently involved.

(iii) Oxygenic phototrophic bacteria (cyanobacteria)

The cyanobacteria (blue-green algae) under normal growth conditions in the light perform a plant-type photosynthesis, just like the higher plants (Stanier *et al.*, 1981; Rippka *et al.*, 1981). They are able to use water as the primary electron donor. Whereas the purple bacteria have only one photosystem and produce electrons at a low negative level (E_0' of about -250 mV), the cyanobacteria and green plants have two photosystems and generate a strong reductant (E_0' of about -500 mV). The acceptor of the photoenergized electrons is probably an iron–sulfur protein (compound X in Figure 7). It provides the electrons for CO_2 fixation, nitrate and sulfate reduction as well as (in the case of the N_2-fixing cyanobacteria) nitrogen fixation. Ferredoxin serves as an electron carrier.

Most, if not all, of the cyanobacteria possess the genetic information for the nitrogenase system and N_2 fixation (Gordon, 1981). However, the expression of the genes, *i.e.* the formation of the respective proteins, requires special conditions. Deficiency of combined nitrogen is a prerequisite for nitrogenase formation in all cyanobacteria. If this requirement is satisfied, the heterocystous cyanobacteria form heterocysts, special cells designed for N_2 fixation. Non-heterocystous cyanobacteria require both nitrogen deficiency and anoxic conditions for derepression of the formation of nitrogenase. In this respect non-heterocystous cyanobacteria are at a disadvantage because in the light, photosynthetic oxygen evolution occurs in the same cells in which nitrogenase has to be formed. Therefore, in these bacteria, for nitrogenase formation and its function, low light intensities or inhibitors (DCMU) to suppress oxygen evolution by photosystem II are required. In the majority of cyanobacteria hydrogen evolution is solely due to the function of nitrogenase (Bothe, 1982; Bothe *et al.*, 1978, 1980). The filamentous heterocystous cyanobacteria such as *Anabaena* species and *Nostoc muscorum* can produce hydrogen in the light. This is due to the

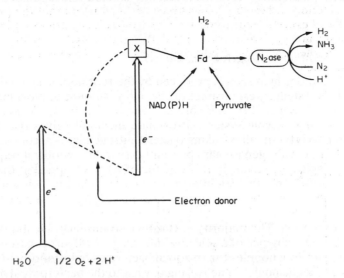

Figure 7 Scheme of photosynthetic electron transport in green plants and cyanobacteria; the nitrogenase (N₂ase) system is present in the cyanobacteria only (abbreviations as in Figure 6)

separation of CO_2 fixation, sugar synthesis and O_2 evolution (in the vegetative cells) and N_2 fixation and H_2 evolution (in the heterocysts). N_2 fixation is not or scarcely impaired by photosynthetic O_2 evolution of neighbouring cells. The non-heterocystous filamentous cyanobacteria can be used for H_2 production when they are periodically exposed to light and dark conditions. In the light they evolve O_2 and accumulate polysaccharides as storage materials. If they are then exposed to anoxic dark conditions, nitrogenase is formed and the storage materials are used to provide electrons for N_2 fixation and H_2 evolution.

As with all N_2-fixing bacteria, nitrogenase is apparently always accompanied by membrane-bound hydrogenases. These hydrogenases may also contribute to H_2 evolution in cyanobacteria. In the halophilic *Oscillatoria limnetica*, photoproduction of H_2 is due to a hydrogenase (Belkin and Padan, 1978). However, it is still controversial whether cyanobacteria contain only membrane-bound hydrogenases, which mainly take up hydrogen and which evolve hydrogen to a minimal extent or due to an artificial, solubilized state (Bothe *et al.*, 1980; Eisbrenner *et al.*, 1981), or whether cyanobacteria contain an additional soluble hydrogenase that is responsible for H_2 production (Tel-Or and Packer, 1978; Tel-Or *et al.*, 1978; Hallenbeck and Benemann, 1978; Peschek, 1979). Due to its importance, hydrogen metabolism in cyanobacteria has been repeatedly reviewed recently (Bothe *et al.*, 1980; Lambert and Smith, 1981).

(iv) Oxygenic green algae

Green algae are eukaryotes and perform oxygenic photosynthesis just as do the higher plants or the cyanobacteria. Unlike the latter group, green algae are not able to produce nitrogenase and fix nitrogen. As discovered 40 years ago, green algae release hydrogen after they have been adapted to anoxic conditions. Hydrogen production has been repeatedly confirmed, and green algae such as *Scenedesmus, Ankistrodesmus, Chlorella* and *Chlamydomonas* were shown to produce H_2 in the light and in the dark. The adaptation process for maximal H_2 evolution varies from species to species and requires from 30 min to 4 h or more. Adaptation can involve either the activation of a constitutive enzyme or *de novo* hydrogenase synthesis.

In the dark, green algae form only minute amounts of H_2. Hydrogen evolution is stimulated by organic substrates. The pattern of the accompanying products (CO_2, ethanol, acetate, glycerol) indicates that H_2 evolution is due to fermentative glucose (polysaccharide) breakdown.

Hydrogen evolution by anaerobically adapted green algae is stimulated by light. Photoproduction of H_2 ceases at high light intensities, obviously due to the production of O_2 to which H_2 evolution and hydrogenase are extremely sensitive. Two mechanisms for H_2 photoproduction have been proposed. One involves photolysis of water coupled to electron transport through photosystems I and II (Pow and Krasna, 1979; Rosen and Krasna, 1980). The alternative pathway involves metabolic breakdown of carbohydrates feeding electrons into photosystem I only. In both cases electrons are raised to the level of at least $-500\,mV$ and are apparently transferred *via* ferredoxin to hydrogenase. The possibilities have been recently studied and discussed (Bamberger *et al.*, 1982).

23.3.1.2 Aerobic conditions: nitrogen-fixing bacteria

Under aerobic conditions hydrogen is produced by the root nodule symbiosis of plants such as the leguminoses. This system is responsible for the major amount of biogenic H_2 released to the atmosphere (Conrad and Seiler, 1980a; Eisbrenner and Evans, 1983). As mentioned above (Section 23.2.6) the N_2-fixing enzyme system nitrogenase inevitably gives rise to the formation of hydrogen in addition to NH_3. In all N_2-fixing bacteria nitrogenase is accompanied by membrane-bound hydrogenase. The hydrogenase has apparently several functions: it traps the hydrogen and recycles it into the respiratory chain. In this way the oxygen is trapped at the cytoplasmic membrane and does not diffuse into the cell interior where the highly oxygen sensitive nitrogenase is located. The hydrogenase thus saves energy as well as protecting nitrogenase from inactivation by oxygen (Dixon, 1972, 1978).

Rhizobia are an exception. The majority of rhizobial strains lack the ability to form hydrogenase, apparently due to the absence of a selective advantage for hydrogenase-containing strains in nature. Within the plant root nodules the rhizobial bacteroids enjoy another O_2-protective mechanism, the presence of leghemoglobin delivering oxygen to the bacteroids at a controlled low rate and partial pressure. In the nodule microhabitat, O_2 protection by hydrogenase is consequently dispensable. The release of H_2 from fields of leguminoses is, therefore, plausible. Hydrogen evolution from the nodules (rhizothamnia) of the alder tree (*Alnus glutinosa*) has also been shown (Benson *et al.*, 1980).

23.3.2 Hydrogen-consuming Organisms

Hydrogen can serve as an electron donor for many microorganisms. Most of them can grow on hydrogen as the sole electron source, others depend on H_2 as an accessory electron source. All hydrogen-consuming organisms perform a respiratory type of metabolism, *i.e.* ATP generation occurs *via* electron transport phosphorylation. There are anaerobic and aerobic hydrogen-utilizing bacteria.

23.3.2.1 Hydrogen utilization by anaerobes

The bacteria which utilize hydrogen under anaerobic conditions are characterized by the ability to generate metabolic energy through 'anaerobic respiration'. This generalizing term points to the involvement of electron transport phosphorylation which was previously considered typical for aerobic, respiratory organisms only. The group of bacteria performing an anaerobic respiratory metabolism comprises the nitrate-reducing denitrifying bacteria, the sulfate- and sulfur-reducing bacteria, the methanogenic and acetogenic (carbonate-reducing) bacteria and those bacteria which generate energy *via* the fumarate reductase pathway. Representative bacteria are listed in Table 1. While the denitrifying bacteria, *e.g. Paracoccus denitrificans*, are facultative aerobes and can use oxygen as electron acceptors, the other groups (Table 1) are obligate anaerobes and their basic metabolism is fermentative. Most of these organisms are able to use organic substrates as electron donors. One or more bacterial species of each group can utilize hydrogen as an alternative source of reducing power, and the majority of methanogenic bacteria are apparently obligately bound to H_2 as electron donor.

Table 1 Hydrogen Consumption by Bacteria with Electron Transport Phosphorylation under Anaerobic Conditions

Reductive process	Representative organisms
$2\,NO_3^- + 5\,H_2 + 2\,H^+ \rightarrow N_2 + 6\,H_2O$	*Paracoccus denitrificans*
$SO_4^{2-} + 4\,H_2 \rightarrow S^{2-} + 4\,H_2O$	*Desulfovibrio vulgaris*
$S^0 + H_2 \rightarrow S^{2-} + 2\,H^+$	*Campylobacter* (saprophytic)
$CO_2 + 2\,H_2 \rightarrow CH_4 + 2\,H_2O$	*Methanobacterium*
$2\,CO_2 + 4\,H_2 \rightarrow MeCO_2H + 2\,H_2O$	*Acetobacterium*
Fumarate $+ H_2 \rightarrow$ succinate	*Vibrio succinogenes*

In their ecosystems the sulfate-, sulfur-, or carbonate-reducing bacteria are closely associated with the hydrogen-producing fermentative bacteria. They inhabit anoxic environments such as sediments and the intestinal tract of animals, especially the rumen of cattle. In anoxic marine sediments competitive relationships exist between methanogenic and sulfate-reducing bacteria, and sulfate is the preferential electron acceptor (Figure 8).

(i) Nitrate-reducing denitrifying bacteria

Growth with H_2 and nitrate is confined to *Paracoccus denitrificans*. During growth gas is consumed and evolved as indicated in Table 1. There are several strains differing by the regulation of hydrogenase formation and various minor characteristics (Nokhal and Schlegel, 1980, 1983).

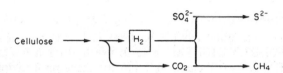

Figure 8 Competitive relationships between sulfate and carbon dioxide for accepting hydrogen produced in the anaerobic food chain

(ii) Sulfate-reducing bacteria

All sulfate-reducing bacteria which have been studied are able to utilize hydrogen as electron donor and form hydrogenase constitutively. H_2 serves as an energy source for growth of *Desulfovibrio* if either complex organic materials or acetate + CO_2 are available for the synthesis of cell material. While several strains of *Desulfovibrio desulfuricans*, *D. gigas* and *D. vulgaris* have been studied in great detail (recent comparative study by Brandis and Thauer, 1981) other recently isolated sulfate reducers which are able to utilize fatty acids (Widdel and Pfennig, 1981, 1982; Widdel *et al.*, 1983; Nethe-Jenchen and Thauer, 1984; Pfennig and Biebl, 1981) have still to be investigated.

(iii) Methanogenic bacteria

The majority of the hydrogen produced in nature is trapped by the methanogenic bacteria. The methanogens support the preceding fermentation steps of the anaerobic food chain by consuming the hydrogen and converting it to methane which is a volatile product inert to most anaerobic decomposition processes. Most of the methanogenic bacteria (review by Balch *et al.*, 1979; Mah and Smith, 1981; Taylor, 1982; Thauer and Morris, 1984) can utilize H_2 and depend upon it. The activation of H_2 is linked to dissimilatory and assimilatory reduction of carbon dioxide.

There are many methanogenic bacterial species, and presently at least seven genera are recognized (Balch *et al.*, 1979). Due to the recognition of basic differences with respect to metabolic traits, coenzymes, the composition of the cytoplasmic membrane, cell wall, ribosomes and polymerases compared to all other bacteria they are grouped as 'Archaebacteria' and gain much attention. Therefore, in spite of the difficulties in handling these fastidiously anaerobic bacteria research is making fast progress.

(iv) Acetogenic bacteria

Although a bacterium growing on H_2 and CO_2 and producing acetate, *Clostridium aceticum*, was discovered more than 50 years ago, progress in studying the acetogenic H_2-utilizing bacteria was slow. The original isolate of *C. aceticum* was revived recently (Braun *et al.*, 1981), and a new strain was isolated (Adamse, 1980). Related bacteria such as *Clostridium thermoaceticum, C. formicoaceticum* and *C. thermoautotrophicum* (Wiegel *et al.*, 1981) as well as *Acetobacterium woodii* (Balch *et al.*, 1977) and other species conducting assimilatory and dissimilatory reduction of CO_2 by H_2 and producing acetate have been described: *Acetobacterium wieringae* (Braun and Gottschalk, 1982) and *Acetogenium kivui* (Leigh *et al.*, 1981). The natural habitats of these bacteria are apparently sewage sludge and lake bottom sediments. No acetogenic bacteria were detected in rumen samples (Braun *et al.*, 1979). Metabolic studies are making fast progress (Thauer and Morris, 1984).

(v)   Fumarate-reducing bacteria

While the sulfate- and the nitrate-reducing bacteria are well delineated physiological groups, the heading 'fumarate-reducing bacteria' covers a heterogeneous group of organisms having a special kind of electron transport phosphorylation in common. The membrane-bound fumarate reductase is the only system of electron transport phosphorylation in obligately fermentative bacteria. Utilization of H_2 as one of the suitable electron donors for the reduction of fumarate to succinate has been demonstrated in *Desulfovibrio gigas*, *Proteus rettgeri*, *Escherichia coli*, *Bacteroides fragilis*, *Vibrio succinogenes* and others. In these bacteria the low potential electrons from H_2 are transferred *via* an electron transport chain to fumarate as terminal acceptor (Bernhard and Gottschalk, 1978b; Lütgens and Gottschalk, 1982); the succinate formed is either directly excreted or converted to propionate (for reviews see Kröger, 1977; Gottschalk, 1981).

23.3.2.2   Hydrogen utilization by phototrophs

(i)   Anoxygenic phototrophs

The majority of anoxygenic phototrophic bacteria (Rhodospirillaceae, Chromatiaceae and Chlorobiaceae; see Section 23.3.1.1(ii)) are able to use hydrogen for anaerobic growth in the light. Hydrogen is the most universal electron donor for these phototrophs. From hydrogen and other electron donors such as H_2S, S, thiosulfate or organic acids, electrons can be channelled into the photosynthetic reaction center (photosystem) but cannot reduce NAD(P) or ferredoxins directly. H_2 is activated by membrane-bound hydrogenases. Growth with H_2 and O_2 in the dark seems to be confined to only a few representatives of the Rhodospirillaceae. *Rhodopseudomonas capsulata* and *R. acidophila* were shown to grow just like the aerobic hydrogen-oxidizing bacteria (see Section 23.3.2.3) when they are incubated in the dark in the presence of H_2 plus less than 2% (vol/vol) O_2 (Siefert and Pfennig, 1979).

(ii)   Cyanobacteria

Hydrogen can be used by several species of cyanobacteria after adaptation and under special conditions. H_2 can be oxidized in the dark or used for photoreduction of CO_2 in the light. However, adaptation conditions and light intensities vary from species to species. Extended growth on H_2 either aerobically in the dark or anaerobically in the light has not been reported. The literature has been reviewed (Bothe *et al.*, 1978, 1980; Lambert and Smith, 1981).

(iii)   Green algae

Anaerobically adapted green algae can use H_2 for various reductive processes (CO_2, O_2, nitrate) in the light. These have been studied using the strains indicated above (Section 23.3.1.1(iv)) with the aim of elucidating the electron transport components involved (Kessler, 1978; Erbes and Gibbs, 1981). Extended growth on H_2 has not been reported.

23.3.2.3   Hydrogen utilization by the aerobic hydrogen-oxidizing bacteria

The aerobic hydrogen-oxidizing bacteria are a physiologically well defined group. They are chemolithoautotrophs, *i.e.* they are able to gain energy by oxidizing hydrogen with oxygen as electron acceptor and to fix CO_2 *via* the Calvin cycle. The overall equation for gas consumption is

$$6\,H_2 + 2\,O_2 + CO_2 \rightarrow <CH_2O> + 5\,H_2O$$

They are facultative lithoautotrophs; they can also grow on one of a large variety of organic substrates. Their metabolism is strictly respiratory (Probst, 1980).

The physiological group comprises about 25 species belonging to nine bacterial genera (Bowien and Schlegel, 1981). Taxonomically it is a very heterogeneous group; there are Gram-negative and Gram-positive bacteria, mesophiles and thermophiles; some can alternatively grow on carbon monoxide, other C_1-compounds or thiosulfate as electron donors; others can fix nitrogen. In essence, hydrogen-oxidizing bacteria are normal respiratory bacteria equipped with two additional enzyme systems: hydrogenase(s) and the components of the Calvin cycle. With respect to the hydrogenases three types of bacteria have to be differentiated: the majority of the hydrogen-oxidizing bacteria contain only a membrane-bound hydrogenase which does not reduce NAD; a few species (*Alcaligenes eutrophus*, *A. ruhlandii*, *A. hydrogenophilus*) contain in addition to the memrane-bound enzyme a cytoplasmic, NAD-reducing hydrogenase; a few strains (*e.g. Nocardia opaca*) have the latter enzyme only.

The hydrogen-oxidizing bacteria can be easily isolated from soil, mud, compost heaps and water (Aragno and Schlegel, 1981). Their habitats are apparently those locations which are accessible by air and in close contact with anoxic microenvironments or root nodules releasing hydrogen. However, studies on hydrogen oxidation in soil are just starting (Conrad and Seiler, 1979, 1980b).

Alcaligenes eutrophus is a very robust, reliable bacterium. It has been proposed for single cell protein production and is being used to produce the bacterial lipid poly-β-hydroxybutyric acid on an industrial scale (King, 1982). Furthermore, it serves as a model organism for the physiological group of chemolithoautotrophs. The genome of many hydrogen-oxidizing bacteria is divided into two parts, the chromosome and a plasmid (Gerstenberg *et al.*, 1982). In *A. eutrophus* the latter carries information for the ability to oxidize hydrogen (Friedrich *et al.*, 1981b; Andersen *et al.*, 1981).

At least three species, *Alcaligenes latus* (Malik *et al.*, 1981), *Xanthobacter autotrophicus* (Wiegel *et al.*, 1978) and *X. flavus* (Malik and Claus, 1979) can fix nitrogen and thus can grow in the absence of combined nitrogen provided the oxygen concentration is low. Other N_2-fixing hydrogen-oxidizing bacteria have been described (Malik and Schlegel, 1981; Pedrosa *et al.*, 1980, 1982). Chemolithoautotrophic nitrogen fixation was reviewed (Dalton, 1980).

A few species of the hydrogen-oxidizing bacteria, such as *Pseudomonas carboxydovorans*, *P. carboxydohydrogena*, *Comamonas compransoris* among others can use carbon monoxide as an alternative electron donor (Zavarzin, 1978; Meyer and Schlegel, 1983). The stoichiometry of the gases consumed during growth is represented by the equation:

$$7\,CO + 2.5\,O_2 + H_2O \rightarrow [CH_2O] + 6\,CO_2$$

For growth on CO molybdenum is required as a microelement. The CO-oxidizing enzyme is a molybdo-flavoprotein (Meyer, 1982).

23.4 THE POTENTIAL USE OF HYDROGENASES AND HYDROGEN IN BIOTECHNOLOGY

The biotechnological aspects of research on hydrogenases concern the potential use of hydrogenases as catalysts either for hydration reactions or for the evolution of hydrogen. The variety of possible practical applications of hydrogenases has recently been reviewed by Klibanov (1983) Hydrogenases can be used to regenerate reduced coenzymes such as nicotinamide dinucleotides, flavins or ferredoxins. Coenzyme regeneration by hydrogen *via* hydrogenases is advantageous because by-products are not formed. Respective methods using dissolved or immobilized enzymes are being developed. Furthermore, the bacterial hydrogenases lend themselves to the production of hydrogen from water and light using *in vitro* systems of broken chloroplasts or artificial membranes (Rao and Hall, 1979).

Progress has also been made in the production of hydrogen from (i) organic waste materials in light using the anoxygenic phototrophic bacteria (Jouanneau *et al.*, 1980; Kim *et al.*, 1981); or (ii) water using cyanobacteria (Bothe, 1982; Lambert and Smith, 1981) or green algae (Pow and Krasna, 1979) as catalysts.

Finally the aerobic hydrogen-oxidizing bacteria offer themselves for biomass production provided that inexpensive hydrogen and carbon dioxide are available. Single cell protein (Schlegel, 1969; Schlegel and Lafferty, 1971) and poly-β-hydroxybutyric acid (King, 1982) are the main products. Advantage can be drawn from the simple nutritional demands and easy handling of the respective bacteria.

23.5 REFERENCES

Adams, M. W. W. and D. O. Hall (1977). Isolation of the membrane-bound hydrogenase from *Rhodospirillum rubrum*. *Biochem. Biophys. Res. Commun.*, **77**, 730–737.
Adams, M. W. W. and D. O. Hall (1978). Solubilization and partial purification of the membrane-bound hydrogenase of *Escherichia coli*. *Biochem. Soc. Trans.*, **6**, 1339–1341.
Adams, M. W. W. and D. O. Hall (1979a). Properties of the solubilized membrane-bound hydrogenase from the photosynthetic bacterium *Rhodospirillum rubrum*. *Arch. Biochem. Biophys.*, **195**, 288–299.

Adams, M. W. W. and D. O. Hall (1979b). Purification of the membrane-bound hydrogenase of *Escherichia coli*. *Biochem. J.*, **183**, 11–22.

Adams, M. W. W., L. E. Mortenson and J. S. Chen (1981). Hydrogenase. *Biochim. Biophys. Acta*, **594**, 105–176.

Adamse, A. D. (1980). New isolation of *Clostridium aceticum* (Wieringa). *Antonie van Leeuwenhoek J. Microbiol. Serol.*, **46**, 523–531.

Albracht, S. P. J., E.-G. Graf and R. K. Thauer (1982). The EPR properties of nickel in hydrogenase from *Methanobacterium thermoautotrophicum*. *FEBS Lett.*, **140**, 311–313.

Andersen, K., R. C. Tait and W. R. King (1981). Plasmids required for utilization of molecular hydrogen by *Alcaligenes eutrophus*. *Arch. Microbiol.*, **129**, 384–390.

Aragno, M. and H. G. Schlegel (1981). The hydrogen oxidizing bacteria. In *The Prokaryotes. A Handbook on Habitats, Isolation and Identification of Bacteria*, ed. M. P. Starr, H. Stolp. H. G. Trüper, A. Balows and H. G. Schlegel, vol. II, pp. 865–893. Springer-Verlag, Berlin.

Arp, D. J. and R. H. Burris (1979). Purification and properties of the particulate hydrogenase from the bacteroids of soybean root nodules. *Biochim. Biophys. Acta*, **570**, 221–230.

Balch, W. E., S. Schoberth, R. S. Tanner and R. S. Wolfe (1977). *Acetobacterium*, a new genus of hydrogen-oxidizing, carbon dioxide-reducing, anaerobic bacteria. *Int. J. Syst. Bacteriol.*, **27**, 355–361.

Balch, W. E., G. E. Fox, L. J. Magrum, C. R. Woese and R. S. Wolfe (1979). Methanogens: reevaluation of a unique biological group. *Microbiol. Rev.*, **43**, 260–296.

Bamberger, E. S., D. King. D. L. Erbes and M. Gibbs (1982). H_2 and CO_2 evolution by anaerobically adapted *Chlamydomonas reinhardtii* F–60. *Plant Physiol.*, **69**, 1268–1273.

Belkin, S. and E. Padan (1978). Hydrogen metabolism in the facultative anoxygenic cyanobacteria (blue-green algae) *Oscillatoria limnetica* and *Aphanothece halophytica*. *Arch. Microbiol.*, **116**, 109–111.

Benson, D. R., D. J. Arp and R. H. Burris (1980). Hydrogenase in actinorhizal root nodules and root nodule homogenates. *J. Bacteriol.*, **142**, 138–144.

Bernhard, T. and G. Gottschalk (1978a). The hydrogenase of *Escherichia coli*, purification, some properties and the function of the enzyme. In *Hydrogenases: Their Catalytic Activity, Structure and Function*, ed. H. G. Schlegel and K. Schneider, pp. 199–208. Goltze KG, Göttingen.

Bernhard, T. and G. Gottschalk (1978b). Cell yields of *Escherichia coli* during anaerobic growth on fumarate and molecular hydrogen. *Arch. Microbiol.*, **116**, 235–238.

Bothe, H. (1982). Hydrogen production by algae. *Experientia*, **38**, 59–64.

Bothe, H., E. Distler and G. Eisbrenner (1978). Hydrogen metabolism in blue-green algae. *Biochimie*, **60**, 277–289.

Bothe, H., G. Neuer, I. Kalbe and G. Eisbrenner (1980). Electron donors and hydrogenase in nitrogen-fixing microorganisms. In *Annual Proceedings of the Phytochemical Society of Europe*, ed. W. D. P. Stewart and J. R. Gallon, vol. 18, pp. 83–112. Academic, London.

Bowien, B. and H. G. Schlegel (1981). Physiology and biochemistry of aerobic hydrogen-oxidizing bacteria. *Annu. Rev. Microbiol.*, **35**, 405–452.

Brandis, A. and R. K. Thauer (1981). Growth of *Desulfovibrio* species on hydrogen and sulphate as sole energy source. *J. Gen. Microbiol.*, **126**, 249–252.

Braun, M., S. Schoberth and G. Gottschalk (1979). Enumeration of bacteria forming acetate from H_2 and CO_2 in anaerobic habitats. *Arch. Microbiol.*, **120**, 201–204.

Braun, M., F. Mayer and G. Gottschalk (1981). *Clostridium aceticum* (Wieringa), a microorganism producing acetic acid from molecular hydrogen and carbon dioxide. *Arch. Microbiol.*, **128**, 288–293.

Braun, M. and G. Gottschalk (1982). *Acetobacterium wieringae* sp. nov., a new species producing acetic acid from molecular hydrogen and carbon dioxide. *Zbl. Bakt. Hyg.I. Abt.*, *Orig. C3*, 368–376.

Brenner, D. J. (1981). Introduction to the family Enterobacteriaceae. In *The Prokaryotes. A Handbook on Habitats, Isolation and Identification of Bacteria*, ed. M. P. Starr, H. Stolp, H. G. Trüper, A. Balows and H. G. Schlegel, vol. II, pp. 1105–1127. Springer-Verlag, Berlin.

Cammack, R., D. Patil, R. Aguirre and E. C. Hatchikian (1982). Redox properties of the ESR-detectable nickel in hydrogenase from *Desulfovibrio gigas*. *FEBS Lett.*, **142**, 289–292.

Conrad, R. and W. Seiler (1979). The role of hydrogen bacteria during the decomposition of hydrogen by soil. *FEMS Microbiol. Lett.*, **6**, 143–145.

Conrad, R. and W. Seiler (1980). The role of microbiological processes for the atmospheric hydrogen cycle. *Forum Mikrobiologie*, **4**, 219–225.

Conrad, R. and W. Seiler (1980a). Contribution of hydrogen production by biological nitrogen fixation to the global hydrogen budget. *J. Geophys. Res.*, **85**, 5493–5498.

Conrad, R. and W. Seiler (1980b). Decomposition of atmospheric hydrogen by soil microorganisms and soil enzymes. *Soil Biol. Biochem.*, **13**, 43–49.

Dalton, H. (1980). Chemoautotrophic nitrogen fixation. In *Nitrogen Fixation. Annual Proceedings of the Phytochemical Society of Europe*, ed. W. D. P. Stewart and J. R. Gallon, vol. 18, pp. 177–195. Academic, London.

Dixon, R. O. D. (1972). Hydrogenase in legume root nodule bacteroids: occurrence and properties. *Arch. Mikrobiol.*, **85**, 193–201.

Dixon, R. O. D. (1978). Nitrogenase-hydrogenase interrelationships in Rhizobia. *Biochimie*, **60**, 233–236.

Eisbrenner, G., P. Roos and H. Bothe (1981). The number of hydrogenases in Cyanobacteria. *J. Gen. Microbiol.*, **125**, 383–390.

Eisbrenner, G. and H. J. Evans (1983). Aspects of hydrogen metabolism in nitrogen-fixing legumes and other plant–microbe associations. *Annu. Rev. Plant Physiol.*, **34**, 105–136.

Erbes, D. L. and M. Gibbs (1981). Kinetics of the oxyhydrogen reaction in the presence and absence of carbon dioxide in *Scenedesmus obliquus*. *Plant Physiol.*, **67**, 129–132.

Friedrich, B., E. Heine, A. Finck and C. G. Friedrich (1981a). Nickel requirement for active hydrogenase formation in *Alcaligenes eutrophus*. *J. Bacteriol.*, **145**, 1144–1149.

Friedrich, B., C. Hogrefe and H. G. Schlegel (1981b). Naturally occurring genetic transfer of hydrogen-oxidizing ability between strains of *Alcaligenes eutrophus*. *J. Bacteriol.*, **147**, 198–205.

Friedrich, C. G., K. Schneider and B. Friedrich (1982). Nickel in the catalytically active hydrogenase of *Alcaligenes eutrophus*. *J. Bacteriol.*, **152**, 42–48.

Fuchs, G., J. Moll, P. Scherer and R. Thauer (1978). Activity, acceptor specificity and function of hydrogenase in *Methanobacterium thermoautotrophicum*. In *Hydrogenases: Their Catalytic Activity, Structure and Function*, ed. H. G. Schlegel and K. Schneider, pp. 83–92. Goltze KG, Göttingen.

Gerstenberg, C., B. Friedrich and H. G. Schlegel (1982). Physical evidence for plasmids in autotrophic, especially hydrogen-oxidizing bacteria. *Arch. Microbiol.*, **133**, 90–96.

Gitlitz, P. H. and A. I. Krasna (1975). Structural and catalytic properties of hydrogenase from *Chromatium*. *Biochemistry*, **14**, 2561–2567.

Gogotov, I. N. (1978). Relationships in hydrogen metabolism between hydrogenase and nitrogenase in phototrophic bacteria. *Biochimie*, **60**, 267–275.

Gordon, J. K. (1981). Introduction to the nitrogen-fixing prokaryotes. In *The Prokaryotes. A Handbook on Habitats, Isolation and Identification of Bacteria*, ed. M. P. Starr, H. Stolp, H. G. Trüper, A. Balows and H. G. Schlegel, vol. I, pp. 781–794. Springer-Verlag, Berlin.

Gottschalk, G. (1981). The anaerobic way of life of prokaryotes. In *The Prokaryotes. A Handbook on Habitats, Isolation and Identification of Bacteria*, ed. M. P. Starr, H. Stolp, H. G. Trüper, A. Balows and H. G. Schlegel, vol. II, pp. 1415–1424. Springer-Verlag, Berlin.

Gottschalk, G., J. R. Andreesen and H. Hippe (1981). The genus *Clostridium* (nonmedical aspects). In *The Prokaryotes. A Handbook on Habitats, Isolation and Identification of Bacteria*, ed. M. P. Starr, H. Stolp, H. G. Trüper, A. Balows and H. G. Schlegel, vol. II, pp. 1767–1803. Springer-Verlag, Berlin.

Graf, E.-G. and R. K. Thauer (1981). Hydrogenase from *Methanobacterium thermoautotrophicum*, a nickel-containing enzyme. *FEBS Lett.*, **136**, 165–169.

Hallenbeck, P. C. and J. R. Benemann (1978). Characterization and partial purification of the reversible hydrogenase of *Anabaena cylindrica*. *FEBS Lett.*, **94**, 261–264.

Jacobson, F. S., L. Daniels, J. A. Fox, C. T. Walsh and W. H. Orme-Johnson (1982). Purification and properties of an 8-hydroxy-5-deazaflavin-reducing hydrogenase from *Methanobacterium thermoautotrophicum*. *J. Biol. Chem.*, **257**, 3385–3388.

Keltjens, J. T., W. B. Whitman, G. Caerteling, A. M. van Kooten, R. S. Wolfe and G. D. Vogels (1982). Presence of coenzyme M derivatives in the prosthetic group (coenzyme MF_{430}) of methylocoenzyme M reductase from *Methanobacterium thermoautotrophicum*. *Biochem. Biophys. Res. Commun.*, **108**, 495–503.

Kessler, E. (1978). Hydrogenase in green algae. In *Hydrogenases: Their Catalytic Activity, Structure and Function*, ed. H. G. Schlegel and K. Schneider, pp. 415–422. Goltze KG, Göttingen.

King, P. P. (1982). Biotechnology. An industrial view. *J. Chem. Technol. Biotechnol.*, **32**, 2–8.

Klibanov, M. (1983). Biotechnological potential of the enzyme hydrogenase. *Process Biochem.*, **18**, 13–16.

Kojima, N., J. A. Fox, R. P. Hausinger, L. Daniels, W. H. Orme-Johnson and C. Walsh (1983). Paramagnetic centers in the nickel-containing, deazaflavin-reducing hydrogenase from *Methanobacterium thermoautotrophicum*. *Proc. Natl. Acad. Sci. USA*, **80**, 378–382.

Krasna, A.I. (1978). Oxygen-stable hydrogenase and assay. In *Methods in Enzymology*, ed. S. P. Colowick and N. O. Kaplan, vol. 53, pp. 296–314. Academic, New York.

Krasna, A. I. (1979). Hydrogenase: properties and applications. *Enzyme Microb. Technol.*, **1**, 165–172.

Kröger, A. (1977). Phosphorylative electron transport with fumarate and nitrate as terminal hydrogen acceptors. In *Microbial Energetics*, ed. B. A. Haddock and W. A. Hamilton, pp. 61–93. Cambridge University Press, Cambridge.

Krüger, H. J., B. H. Huynh, P. O. Ljungdahl, A. V. Xavier, D. V. DerVartanian, I. Moura, H. D. Peck, Jr., M. Teixeira, J. J. G. Moura and J. LeGall (1982). Evidence for nickel and a three-iron center in the hydrogenase of *Desulfovibrio desulfuricans*. *J. Biol. Chem.*, **257**, 14 620–14 623.

Lambert, G. R. and G. D. Smith (1981). The hydrogen metabolism of cyanobacteria (blue-green algae). *Biol. Rev.*, **56**, 589–660.

LeGall, J., P. O. Ljungdahl, I. Moura, H. D. Peck, Jr., A. V. Xavier, J. J. G. Moura, M. Teixera, B. H. Huynh and D. V. DerVartanian (1982a). The presence of redox-sensitive nickel in the periplasmic hydrogenase from *Desulfovibrio gigas*. *Biochem. Biophys. Res. Commun.*, **106**, 610–616.

LeGall, J., J. J. G. Moura, H. D. Peck, Jr. and A. V. Xavier (1982b). Hydrogenase and other iron-sulfur proteins from sulfate-reducing and methane-forming bacteria. In *Iron-Sulfur Protein*, ed. T. G. Spiro, vol. 4, pp. 183–208. Wiley, New York.

Leigh, J. A., F. Mayer and R. S. Wolfe (1981). *Acetogenium kivui*, a new thermophilic hydrogen-oxidizing acetogenic bacterium. *Arch. Microbiol.*, **129**, 275–280.

Lloyd, D., J. Williams, N. Yarlett and A. G. Williams (1982). Oxygen affinities of the hydrogenosome-containing protozoa *Tritrichomonas foetus* and *Dasytricha ruminantium*, and two aerobic protozoa, determined by bacterial bioluminescence. *J. Gen. Microbiol.*, **128**, 1019–1022.

Lütgens, M. and G. Gottschalk (1982). Cell and ATP yields of *Citrobacter freundii* growing with fumarate and H_2 or formate in continuous culture. *J. Gen. Microbiol.*, **128**, 1915–1921.

Mah, R. A. and M. R. Smith (1981). The methanogenic bacteria. In *The Prokaryotes. A Handbook on Habitats, Isolation and Identification of Bacteria*, ed. M. P. Starr, H. Stolp, H. G. Trüper, A. Balows and H. G. Schlegel, vol. I, pp. 948–977. Springer-Verlag, Berlin.

Malik, K. A. and D. Claus (1979). *Xanthobacter flavus*, a new species of nitrogen-fixing hydrogen bacteria. *Int. J. Syst. Bacteriol.*, **29**, 283–287.

Malik, K. A., C. Jung, D. Claus and H. G. Schlegel (1981). Nitrogen fixation by the hydrogen-oxidizing bacterium *Alcaligenes latus*. *Arch. Microbiol.*, **129**, 254–256.

Malik, K. A. and H. G. Schlegel (1981). Chemolithoautotrophic growth of bacteria able to grow under N_2-fixing conditions. *FEMS Microbiol. Lett.*, **11**, 63–67.

McKellar, R. C. and G. D. Sprott (1979). Solubilization and properties of a particulate hydrogenase from *Methanobacterium* strain G2R. *J. Bacteriol.*, **139**, 231–238.

Meyer, O. (1982). Chemical and spectral properties of carbon monoxide: methylene blue oxidoreductase. *J. Biol. Chem.*, **257**, 1333–1341.

Meyer, O. and H. G. Schlegel (1983). Biology of aerobic carboxydotrophic bacteria. *Annu. Rev. Microbiol.*, **37**, 277–310.

Müller, M. (1980). The Hydrogenosome. In *The Eukaryotic Microbial Cell, Society for General Microbiology Symposium No. 30*, ed. G. W. Gooday, D. Lloyd and A. P. J. Trinci. Cambridge University Press, Cambridge.

Nethe-Jenchen, R. and R. K. Thauer (1984). Growth yields and saturation constant of *Desulfovibrio vulgaris* in chemostat culture. *Arch. Microbiol.*, **137**, 236–240.

Nokhal, T.-H. and H. G. Schlegel (1980). The regulation of hydrogenase formation as a differentiating character of strains of *Paracoccus denitrificans*. *Antonie van Leeuwenhoek J. Microbiol. Serol.*, **46**, 143–155.

Nokhal, T.-H. and H. G. Schlegel (1983). Taxonomic study of *Paracoccus denitrificans*. *Int. J. Syst. Bacteriol.*, **33**, 26–37.

Pedrosa, F. O., J. Döbereiner and M. G. Yates (1980). Hydrogen-dependent growth and autotrophic carbon dioxide fixation in *Derxia*. *J. Gen. Microbiol.*, **119**, 547–551.

Pedrosa, F. O., M. Stephan, J. Döbereiner and M. G. Yates (1982). Hydrogen-uptake hydrogenase activity in nitrogen-fixing *Azospirillum brasilense*. *J. Gen. Microbiol.*, **128**, 161–166.

Peschek, G. A. (1979). Evidence for two functionally distinct hydrogenases in *Anacystis nidulans*. *Arch. Microbiol.*, **123**, 81–92.

Pfennig, N. and H. Biebl (1981). The dissimilatory sulfur-reducing bacteria. In *The Prokaryotes. A Handbook on Habitats, Isolation and Identification of Bacteria*, ed. M. P. Starr, H. Stolp, H. G. Trüper, A. Balows and H. G. Schlegel, vol. I, pp. 941–947. Springer-Verlag, Berlin.

Pinchukova, E. E. and S. D. Varfolomeev (1980). Reversible oxidation–reduction of NAD by hydrogen catalyzed by soluble hydrogenase from *Alcaligenes eutrophus* Z-1. *Biokhimiya*, **45**, 1405–1411.

Pow, Th. and A. I. Krasna (1979). Photoproduction of hydrogen from water in hydrogenase-containing algae. *Arch. Biochem. Biophys.*, **194**, 413–421.

Probst, I. (1980). Respiration in hydrogen bacteria. In *Diversity of Bacterial Respiratory Systems*, ed. C. J. Knowles, vol. II, pp. 160–181. CRC Press, Boca Raton, FL.

Rao, K. K. and D. O. Hall (1979). Hydrogen production from isolated chloroplasts. In *Photosynthesis in Relation to Model Systems*, ed. J. Barber, pp. 299–329. Elsevier Biomedical Press, Amsterdam.

Rippka, R., J. B. Waterbury and R. Y. Stanier (1981). Isolation and purification of cyanobacteria: some general principles. In *The Prokaryotes. A Handbook on Habitats, Isolation, and Identification of Bacteria*, ed. M. P. Starr, H. Stolp, H. G. Trüper, A. Balows and H. G. Schlegel, vol. I, pp. 212–220. Springer-Verlag, Berlin.

Rosen, M. M. and A. I. Krasna (1980). Limiting reactions in hydrogen photoproduction by chloroplasts and hydrogenase. *Photochem. Photobiol.*, **31**, 259–265.

Schink, B. and H. G. Schlegel (1978). Hydrogen metabolism in aerobic hydrogen-oxidizing bacteria. *Biochimie*, **60**, 297–305.

Schink, B. and H. G. Schlegel (1979). The membrane-bound hydrogenase of *Alcaligenes eutrophus*. I. Solubilization, purification, and biochemical properties. *Biochim. Biophys. Acta*, **567**, 315–324.

Schink, B. and I. Probst (1980). Competitive inhibition of the membrane-bound hydrogenase of *Alcaligenes euthrophus* by molecular oxygen. *Biochem. Biophys. Res. Commun.*, **95**, 1563–1568.

Schink, B. and H. G. Schlegel (1980). The membrane-bound hydrogenase of *Alcaligenes eutrophus*: II. Localization and immunological comparison with other hydrogenase systems. *Antonie van Leeuwenhoek J. Microbiol. Serol.*, **46**, 1–14.

Schlegel, H. G. (1969). From electricity *via* water electrolysis to food. In *Fermentation Advances*, ed. D. Perlman, pp. 807–832. Academic, New York.

Schlegel, H. G. (1974). Production, modification, and consumption of atmospheric trace gases by microorganisms. *Tellus*, **26**, 11–20.

Schlegel, H. G. (1984). Studies on the regulation and genetics of enzymes of *Alcaligenes eutrophus*. In *Aspects of Microbial Ecology and Metabolism*, ed. G. A. Codd, pp. 187–209. Academic, New York.

Schlegel, H. G. and R. M. Lafferty (1971). Novel energy and carbon sources. A. The production of biomass from hydrogen and carbon dioxide. *Adv. Biochem. Eng.*, **1**, 143–168.

Schlegel, H. G. and K. Schneider (1978). Introductory report: distribution and physiological role of hydrogenases in microorganisms. In *Hydrogenases: Their Catalytic Activity, Structure and Function*, pp. 15–44. Goltze KG, Göttingen.

Schlegel, H. G., G. Gottschalk and N. Pfennig (eds.) (1976). *Microbial Production and Utilization of Gases*, Goltze KG, Göttingen.

Schneider, K. and H. G. Schlegel (1976). Purification and properties of soluble hydrogenase from *Alcaligenes eutrophus* H16. *Biochim. Biophys. Acta*, **452**, 66–80.

Schneider, K. and H. G. Schlegel (1978). Identification and quantitative determination of the flavin component of soluble hydrogenase from *Alcaligenes eutrophus*. *Biochem. Biophys. Res. Commun.*, **84**, 564–571.

Schneider, K. and H. G. Schlegel (1981). Production of superoxide radicals by soluble hydrogenase from *Alcaligenes eutrophus* H16. *Biochem. J.*, **193**, 99–107.

Schneider, K., D. S. Patil and R. Cammack (1983). ESR properties of membrane-bound hydrogenases from aerobic hydrogen bacteria. *Biochim. Biophys. Acta*, **748**, 353–361.

Schneider, K., H. G. Schlegel and K. Jochim (1984a). Effect of nickel on activity and subunit composition of purified hydrogenase from *Nocardia opaca* 1b. *Eur. J. Biochem.*, **138**, 533–541.

Schneider, K., R. Cammack and H. G. Schlegel (1984b). Content and localization of FMN, Fe–S clusters and nickel in the NAD-linked hydrogenase of *Nocardia opaca* 1b. *Eur. J. Biochem.*, **142**, 75–84.

Schneider, K., R. Cammack, H. G. Schlegel and D. O. Hall (1979). The iron–sulfur centres of soluble hydrogenase from *Alcaligenes eutrophus*. *Biochim. Biophys. Acta*, **578**, 445–461.

Schoenmaker, G. S., L. F. Oltmann and A. H. Stouthamer (1978). Hydrogenase of *Proteus mirabilis*. In *Hydrogenases: Their Catalytic Activity, Structure and Function*, ed. H. G. Schlegel and K. Schneider, pp. 209–219. Goltze KG, Göttingen.

Schoenmaker, G. S., L. F. Oltmann and A. H. Stouthamer (1979). Purification and properties of the membrane-bound hydrogenase from *Proteus mirabilis*. *Biochim. Biophys. Acta*, **567**, 511–521.

Siefert, E. and N. Pfennig (1979). Chemoautotrophic growth of *Rhodopseudomonas* species with hydrogen and chemotrophic utilization of methanol and formate. *Arch. Microbiol.*, **122**, 177–182.

Sim, E. and R. B. Sim (1979). Hydrodynamic parameters of the detergent-solubilised hydrogenase from *Paracoccus denitrificans*. *Eur. J. Biochem.*, **97**, 119–126.

Sim, E. and P. M. Vignais (1978). Hydrogenase activity in *Paracoccus denitrificans*. Partial purification and interaction with the electron transport chain. *Biochimie*, **60**, 307–314.

Stanier, R. Y., N. Pfennig and H. G. Trüper (1981). Introduction to the phototrophic prokaryotes. In *The Prokaryotes. A Handbook on Habitats, Isolation, and Identification of Bacteria*, ed. M. P. Starr, H. Stolp, H. G. Trüper, A. Balows and H. G. Schlegel, vol. I, pp. 197–211. Springer-Verlag, Berlin.

Strekas, T., B. C. Antanaitis and A. I. Krasna (1980). Characterization and stability of hydrogenase from *Chromatium*. *Biochim. Biophys. Acta*, **616**, 1–9.

Taylor, G. T. (1982). The methanogenic bacteria. In *Progress in Industrial Microbiology*, ed. M. J. Bull, vol. 16, pp. 231–329. Elsevier, Amsterdam.

Teixeira, M., I. Moura, A. V. Xavier, D. V. DerVartanian, J. LeGall, H. D. Peck, Jr., B. H. Huynh and J. J. G. Moura (1983). *Desulfovibrio gigas* hydrogenase: redox properties of the nickel and iron–sulfur centers. *Eur. J. Biochem.*, **130**, 481–484.

Tel-Or, E., L. W. Luijk and L. Packer (1978). Hydrogenase in N_2-fixing cyanobacteria. *Arch. Biochem. Biophys.*, **185**, 185–194.

Tel-Or, E. and L. Packer (1978). The hydrogenase of *Anabena cylindrica* and *Nostoc muscorum* in heterocysts and vegetative cells. In *Hydrogenases: Their Catalytic Activity, Structure and Function*, ed. H. G. Schlegel and K. Schneider, pp. 371–380. Goltze KG, Göttingen.

Thauer, R. K. and J. G. Morris (1984). Metabolism of chemotrophic anaerobes: old views and new aspects. In *The Microbe 1984: Part II Prokaryotes and Eukaryotes*, ed. D. P. Kelly and N. G. Carr. Society for General Microbiology, Cambridge University Press.

van Dijk, C., S. G. Mayhew and C. Veeger (1981). An analysis of activity determinations in a series of coupled redox reactions with special reference to hydrogenase. *Eur. J. Biochem.*, **114**, 201–207.

van Dijk, C. and C. Veeger (1981). The effects of pH and redox potential on the hydrogen production activity of the hydrogenase from *Megasphaera elsdenii*. *Eur. J. Biochem.*, **114**, 209–219.

Vignais, P. M., M. F. Henry, E. Sim and D. B. Kell (1981). The electron transport system and hydrogenase of *Paracoccus denitrificans*. *Curr. Top. Bioenerget.*, **12**, 115–196.

Weiss, A. R. and H. G. Schlegel (1980). Reduction of horse heart cytochrome *c* by the membrane-bound hydrogenase of *Alcaligenes eutrophus*. *FEMS Microbiol. Lett.*, **8**, 173–176.

Widdel, F. and N. Pfennig (1981). Studies on dissimilatory sulfate-reducing bacteria that decompose fatty acids. I. Isolation of new sulfate-reducing bacteria enriched with acetate from saline environments: description of *Desulfobacter postgatei* gen. nov. sp. nov.. *Arch. Microbiol.*, **129**, 395–400.

Widdel, F. and N. Pfennig (1982). Studies on dissimilatory sulfate-reducing bacteria that decompose fatty acids. II. Incomplete oxidation of propionate by *Desulfobulbus propionicus* gen. nov. sp. nov.. *Arch. Microbiol.*, **131**, 360–365.

Widdel, F., G. W. Kohring and F. Mayer (1983). Studies on dissimilatory sulfate-reducing bacteria that decompose fatty acids. III. Characterization of the filamentous gliding *Desulfonema limicola* gen. nov. sp. nov., and *Desulfonema magnum*, sp. nov. *Arch. Microbiol.*, **134**, 286–294.

Wiegel, J., D. Wilke, J. Baumgarten, R. Opitz and H. G. Schlegel (1978). Transfer of the nitrogen-fixing hydrogen bacterium *Corynebacterium autotrophicum* Baumgarten *et al.* to *Xanthobacter* gen. nov.. *Int. J. Syst. Bacteriol.*, **28**, 573–581.

Wiegel, J., M. Braun and G. Gottschalk (1981). *Clostridium thermoautotrophicum* species novum, a thermophile producing acetate from molecular hydrogen and carbon dioxide. *Curr. Microbiol.*, **5**, 255–260.

Yagi, T. (1981). Function and structure of hydrogenases. *Nat. Sci.*, **32**, 29–83.

Yamazaki, S. (1982). A selenium containing hydrogenase from *Methanococcus vannielii*. *J. Biol. Chem.*, **257**, 7926–7929.

Zajic, J. E., N. Kosaric and J. D. Brosseau (1978). Microbial production of hydrogen. In *Advances in Biochemical Engineering*, ed. T. K. Ghose, A. Fiechter and N. Blakebrough, vol. 9, pp. 57–109. Springer-Verlag, Berlin.

Zavarzin, G. A. (1978). *Hydrogen Bacteria and Carboxydobacteria*. USSR Academy of Sciences, Nauka Publishing House, Moscow.

24

Biosynthesis of Fatty Acids and Lipids

C. A. BOULTON and C. RATLEDGE
University of Hull, UK

24.1 INTRODUCTION

Lipids are components of all living cells. They are characterized solely by their solubility in an organic solvent such as chloroform. As this is their only common property, lipids encompass a wide range of molecules from fatty acids to terpenes, and will even include such materials as chlorophylls and many quinones. The full range of lipids is too great to be covered in a short chapter such as this and the reader wishing to know more is referred to the general introductory text of Gurr and James (1980) and also, if their German is good, to the major monograph by Thiele (1979). Microbial lipids are not well served with recent books though there are many reviews on individual topics to which we shall draw the reader's attention in the appropriate place. Our intention in this chapter will be primarily, but not exclusively, focused on the fatty acids and on the fatty acyl containing lipids. As microorganisms growing on alkanes are a prominent part of biotechnology, and because alkanes can influence very considerably the nature of the lipid within the microbial cell, we shall also be discussing the methods whereby microorganisms are able to utilize these materials.

However, let us stress that this is a chapter in which the biochemistry and the relevance of lipids will be covered. For readers wishing to know about the exploitable aspects of microbial oils and fats we must refer them to Volume 2, Chapter 53 of this work. For reasons of space, as well as because of their specialized nature, we have not included algae or protozoa in this chapter. Both groups of organisms produce a wide array of lipids, sometimes quite complex.

24.1.1 Nomenclature

The current nomenclature of lipids was recommended in 1976 by the IUPAC–IUB Commission on Biochemical Nomenclature. This document was published in various journals in 1977; reprints of it can be obtained from Dr. W. E. Cohn, NAS–NRC Office of Biochemical Nomenclature, Biology Division, Oak Ridge National Laboratory, PO Box 7, Oak Ridge, TN 37830, USA.

The main points in the recommendations were that the terms mono-, di- and tri-glyceride be replaced by mono-, di- and tri-acylglycerol. A typical triacylglycerol, tristearoylglycerol, is given in Figure 1. To designate the configuration of other fatty acyl derivatives of glycerol, the carbon atoms must be stereospecifically numbered (the *sn* system); thus *sn*-glycerol 3-phosphate is to be used, not L-glycerol 3-phosphate. Consequently triacylglycerols with different fatty acyl groups on each carbon atom should be described, for example, as *sn*-1,3-dipalmitoyl-2-oleoylglycerol.

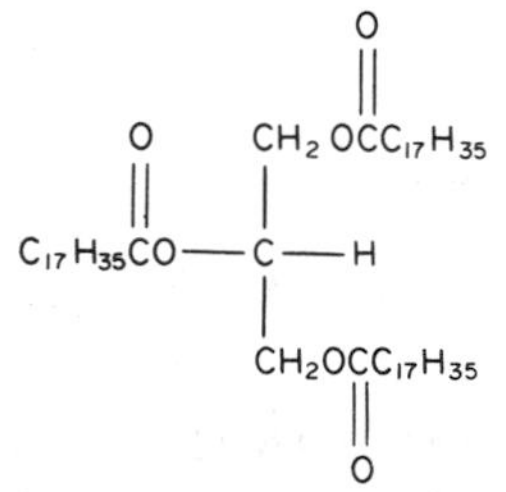

Figure 1 Tristearoylglycerol

As will now be appreciated, the existing trivial names for fatty acids (see Table 1) are still permissible in the new nomenclature system. The numbering system for fatty acids is particularly important as it is a frequently used shorthand system. Where the possibility of isomeric forms exists, as in the presence of unsaturation in the acyl chain, the position of the double bond must be specified. Thus 18:1(9), or $18:1\Delta^9$, refers to oleic acid though strictly this should be specified as a *cis* bond (*c*9) so that it is distinguished from the *trans* isomer (*t*9), which would be elaidic acid. Table 1 lists most of the common fatty acids found in microorganisms together with their numerical symbol.

Those lipids which contain phosphoric acid, either as a mono- or di-ester, are referred to as phospholipids. The term phosphoglyceride again should not be used but glycerophospholipid instead. A typical phospholipid is shown in Figure 2, this is 3-*sn*-phosphatidylcholine. The full

Table 1 Some Common Fatty Acids of Microorganisms

Systematic name	Trivial name	Nomenclature
Saturated Me(CH$_2$)$_n$CO$_2$H		
$n = 10$ Dodecanoic acid	Lauric	12 : 0
$n = 12$ Tetradecanoic acid	Myristic	14 : 0
$n = 14$ Hexadecanoic acid	Palmitic	16 : 0
$n = 16$ Octadecanoic acid	Stearic	18 : 0
$n = 18$ Eicosanoic acid	Arachidic	20 : 0
Unsaturated Me(CH$_2$)$_n$(CH$_2$CH=CH)$_x$(CH$_2$)$_m$CO$_2$H		
$n = 4, m = 7, x = 1$ 9-Hexadecenoic acid	Palmitoleic	16 : 1 (9c)
$n = 6, m = 7, x = 1$ 9-Octadecenoic acid	Oleic	18 : 1 (9c)
$n = 4, m = 9, x = 1$ 11-Octadecenoic acid	*cis*-Vaccenic	18 : 1 (11c)
$n = 3, m = 7, x = 2$ 9, 12-Octadecadienoic acid	*cis,cis*-Linoleic	18 : 2 (9c,12c)
$n = 0, m = 7, x = 3$ 9,12,15-Octadecatrienoic acid	α-Linolenic	18 : 3 (9c,12c,15c)
$n = 3, m = 4, x = 3$ 6,9,12-Octadecatrienoic acid	γ-Linolenic	18 : 3 (6c,9c,12c)
$n = 3, m = 3, x = 4$ 5,8,11,14-Eicosatetraenoic acid	Arachidonic	20 : 4 (5c, 8c, 11c, 14c)
Branched MeCH(Me)(CH$_2$)$_n$CO$_2$H (isosaturated, ω-branched)		
$n = 12$ 14-Methylpentadecanoic acid	Isopalmitic	br-16 : 14 Me
MeCH$_2$CH(Me)(CH$_2$)$_n$CO$_2$H (anteisosaturated, ω-1-branched)		
$n = 13$ 15-Methylheptadecanoic acid	Anteisostearic	br-16 : 15 Me
$n = 14$ 10-Methyloctadecanoic acid	Tuberculostearic	br-19 : 10 Me
Cyclopropane Me(CH$_2$)$_3$CH–CH(CH$_2$)$_n$CO$_2$H		
CH$_2$		
$n = 7$ 9,10-Methylenehexadecanoic acid	—	cyc-17 : 0
$n = 9$ 11,12-Methyleneoctadecanoic acid	Lactobacillic	cyc-19 : 0
Hydroxy Me(CH$_2$)$_n$CHOH(CH$_2$)$_m$CO$_2$H		
$n = 8, m = 7$ 9-Hydroxy-12-*cis*-octadecenoic acid	Ricinoleic	9-OH 18 : 1
$n = 10, m = 1$ 3-Hydroxytetradecanoic acid	—	3-OH 14 : 0
$n = 11, m = 0$ 2-Hydroxytetradecanoic acid	—	2-OH 14 : 0

systematic name would be 1,2-diacyl-*sn*-glycero-3-phosphocholine. The previous trivial name for this compound, lecithin, is no longer preferred.

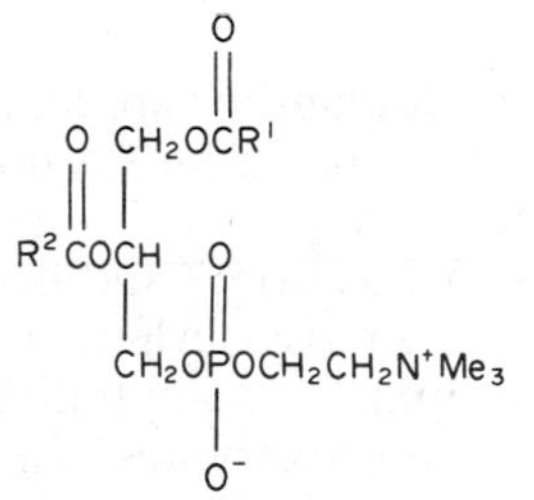

$$O$$
$$\|$$
$$O\ \ CH_2OCR^1$$
$$\|\ \ |$$
$$R^2COCH\ \ \ \ O$$
$$|\ \ \ \ \ \|$$
$$CH_2OPOCH_2CH_2N^+Me_3$$
$$|$$
$$O^-$$

Figure 2 3-*sn*-Phosphatidylcholine; R^1 and R^2 may be any long-chain alkyl group

For glycolipids, which designates any compound containing one or more monosaccharide residues linked by a glycosyl linkage to a lipid part, no radical changes were recommended. A typical glycolipid, diacylgalactosylglycerol which has the systematic name 1,2-diacyl-3-β-D-galactosyl-*sn*-glycerol, is given in Figure 3.

Other lipids, such as sterols, carotenoids, quinones and other terpenoid lipids, were not encompassed by this IUPAC–IUB recommendation.

24.1.2   Relevance and Importance of Lipids

Lipids, in particular the phospholipids, are the principal components of all membranes both of the cytoplasm and of the individual organelles, such as the mitochondrion, within the cell. Within

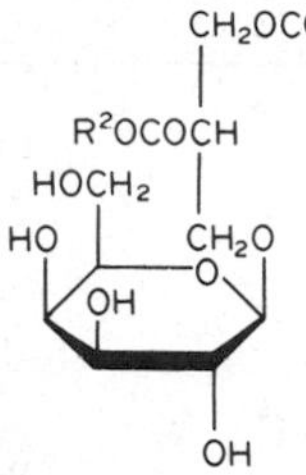

$$CH_2OCOR^1$$

Figure 3   1,2-Diacyl-3-β-D-galactosyl-*sn*-glycerol (1,2-diacyl-*sn*-glycerol 3-β-D-galactoside)

these lipid bilayers can be found other lipid molecules such as sterols, and even carotenoids. Lipids can act in other structural capacities and are frequently found as integral parts of bacterial cell envelopes. Here the lipid is usually conjugated to a polysaccharide, for example, as in lipid A which is found in the outer membrane of Gram-negative bacteria. A useful collection of reviews on the membrane lipids of prokaryotes has recently appeared (Razin and Rottem, 1982).

Besides structural roles, lipids fulfil several functional roles. They can act as storage compounds of which the triacylglycerols (Figure 1) are the prime, but not sole, example. They can confer hydrophobicity on various protein molecules making them less soluble in water and with a tendency, therefore, to associate with membranes. In this capacity we have some of the components of the respiratory chain; other components, such as quinones, are of course lipids themselves. Lipids may also be produced by microorganisms to solubilize water-insoluble substrates being used to support its growth. Here the lipid, which is often a glycolipid, is an extracellular product and forms an emulsion, or micelle, with a substrate such as an alkane. (This is dealt with in greater detail in Section 24.6.)

Although we will not be covering such materials as the chlorophylls and associated pigments of the photosynthetic apparatus, nevertheless it should be remembered that these compounds too are lipids.

24.2   LIPID COMPOSITION OF MICROORGANISMS

24.2.1   General Survey

24.2.1.1   Bacteria

The lipid types which occur in microorganisms are many and varied. With bacteria, the complex array of lipids varies from fatty acids, from C_6 up to very complex multi-branched C_{88} molecules as are found in the mycobacteria, through various acylglycerols, fatty alcohols, glycosulfo-, phospho- and peptido-lipids to wax esters and hydrocarbons. There are also glycophospholipids, ornithine-containing lipids, glycosylacylglycerols, phosphonolipids, ether lipids, *etc*. (see Goldfine, 1972). Many unusual lipid types are found in the Archaebacteria (primitive bacteria) which is a new subdivision of the prokaryotes. The variety of types of lipid, and even of individual fatty acids, is now so great that it has become possible to use lipids as an important aid in the taxonomy of bacteria. The reader wishing to know more about the individual bacterial lipids in this context is referred to the detailed reviews of Lechevalier (1977, 1982).

Although it is not possible even to attempt to condense this data in a chapter such as this, or even to give a tabulated presentation of data due to the wide variation there is in lipid types, the following points may be useful.

Bacteria contain straight chain saturated fatty acids such as are found ubiquitously in nature (see Table 1). Monounsaturated fatty acids are of one or two types depending on their route of synthesis (Section 24.5.3): these are the *cis*-vaccenic acid type, which is the most common, or the oleic acid type. Bacteria with a few exceptions do not contain di- or poly-unsaturated fatty acids. They do, however, contain branched chain acids of both the ω- and ω-1 type as well as tuberculostearic acid: 10-methylstearic acid (br-19:10 Me), where the methyl group is in the mid-section of the molecule. Multibranched fatty acids occur in the mycobacteria. Cyclopropane and cyclopropene fatty acids (Table 1) occur in many bacteria.

Hydroxy fatty acids are of two main types: the α (or C-2) and the β (or C-3) hydroxy types. The latter type is found extensively in the Actinomycetales. In a few bacteria hydroxy, branched chain fatty acids have been reported.

Bacteria do not synthesize sterols though they do produce a variety of carotenoids and various isopentyl containing materials including menaquinones and chlorophylls.

Whilst bacteria contain some tri-, di- and mono-acylglycerols this is not a widespread characteristic as triacylglycerols are usually lipid storage molecules and, in general, bacteria do not store very much lipid. An unusual variation occurs with various Archaebacteria where glycerol ethers occur (Figure 4).

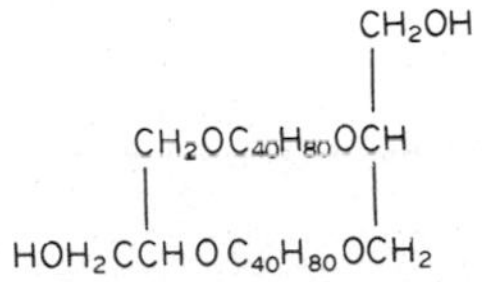

Figure 4 Diglycerol tetraether lipid from *Thermoplasma acidophilum* (from Langworthy, in Shilo, 1980)

Phospholipids occur in all living cells as part of the cell membrane. Bacteria are no exception. The common ones found in bacteria include phosphatidylethanolamine, though the methylated derivatives mono-, di- and tri-methylethanolamine (which is phosphatidylcholine, see Figure 2) are confined to the Gram-negative bacteria and the actinomycetes. Phosphatidylserine also is restricted to the Gram-negative bacteria. Phosphatidylinositol mannosides (Figure 5) are found in actinomycetes and also in some yeasts. These compounds may contain up to six mannose residues and there may be one or two further acyl groups esterified both to the inositol ring and (if there are two acyl groups) to the 2-O-mannose ring.

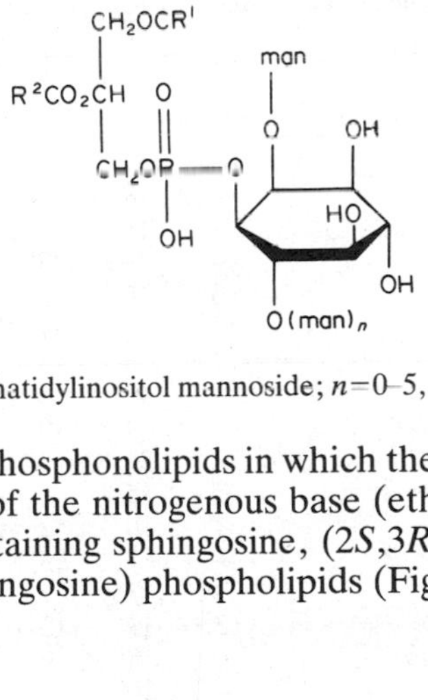

Figure 5 Phosphatidylinositol mannoside; $n=0$–5, man = mannose

Unusual phospholipids include phosphonolipids in which the phosphorus atom (see Figure 2) is directly linked to the first carbon of the nitrogenous base (ethanolamine or choline). Sphingolipids, which are phospholipids containing sphingosine, (2*S*,3*R*,4-*erythro*)-2-amino-4-octadecane-1,3-diol, and ceramide (*N*-acylsphingosine) phospholipids (Figure 6) occur in a variety of Gram-negative anaerobes.

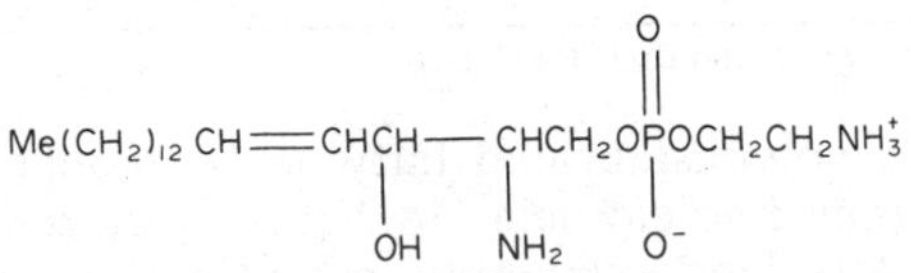

Figure 6 Ceramide phosphoethanolamine

Conjugated lipids within bacteria include a variety of complex glycolipids the most predominant of which are the *sn*-1,2-diacylglycerols glycosidically linked to various carbohydrates at the *sn*-3 position. Glucose, galactose, mannose and glucuronic acid are the main sugars in these molecules. Simpler acylated sugars, of one sugar + one fatty acyl group, are also to be found in numerous bacteria although the dimycolyltrehalose glycolipid found in mycobacteria is a molecule of some 188 carbon atoms, the mycolic acids being complex α-branched C_{88} fatty acids (see Minnikin, 1982).

Waxes (esters of fatty acids and fatty alcohols) have been found in a few genera of bacteria: *Acinetobacter*, *Micrococcus* and *Clostridium*. However, the polyester of β-hydroxybutyrate (Figure 7) is more widely distributed with some 23 bacteria genera being listed by Dawes and Senior (1973). Additions to the list can be expected.

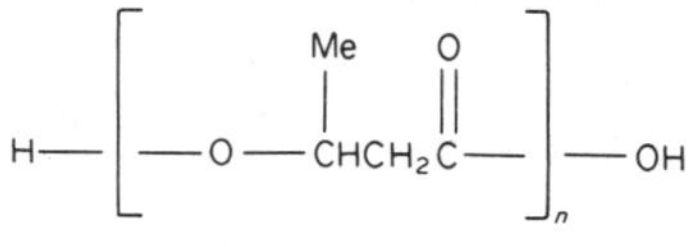

Figure 7 Poly-β-hydroxybutyrate; n = 5000 approx.

24.2.1.2 Yeasts

The reader wishing to know more about the detailed aspects of yeast lipids is referred to the excellent reviews by Hunter and Rose (1971), Rattray *et al.* (1975) and Brennan *et al.* (1974).

In general, the complex array of lipids seen in bacteria is not repeated in the yeasts. The range of fatty acids is much more constrained, there being no branched chain fatty acids and only a few species, mainly of *Torulopsis*, which produce hydroxy fatty acids. Fatty acids (see Table 2) are usually of 16 and 18 carbon atoms; few reports of fatty acids longer than C_{22} have been given. Yeasts being eukaryotic organisms synthesize their unsaturated fatty acids by the oleic acid route rather than the *cis*-vaccenic pathway (see Section 24.5.3) and oleic acid usually, if not always, is the most abundant fatty acid.

Table 2 Fatty Acids of Selected Yeasts[a]

Organism	Fatty acid (relative % w/w)						
	14:0	*16:0*	*16:1*	*18:0*	*18:1*	*18:2*	*18:3*
Candida utilis	—	11	—	1	24	39	25
C. utilis	—	18	—	—	42	34	—
Cryptococcus terricolus	—	18	1	6	60	12	2
Debaryomyces hansenii	—	24	3	8	50	3	2
Hansenula anomala	3	35	2	36	3	—	3
Kluyveromyces polysporus	11	13	51	1	21	—	—
Lipomyces lipofer	1	36	3	21	44	5	—
Lipomyces starkeyi	—	40	6	5	44	4	—
Pichia fermentans	—	9	19	—	22	30	16
Pullaria pullulans	—	31	3	9	42	13	1
Rhodosporidium toruloides	1	25	1	13	46	12	2
Rhodotorula graminis	1	30	2	12	36	15	4
Saccharomyces carlsbergensis	—	7	58	2	34	—	—
S. cerevisiae	1	13	37	5	44	—	—
S. fragilis	3	19	12	3	27	25	10
Schizosaccharomyces pombe	0	8	0	0	89	0	0
Trichosporon cutaneum	—	30	—	13	46	11	—
Trigonopsis variabilis	—	19	34	—	23	22	—

[a] Erwin (1972) Weete (1980), and Ratledge (1982a).

Yeasts, however, produce polyunsaturated fatty acids, though *Saccharomyces cerevisiae* is unable to produce either *cis,cis*-linoleic acid, 18:2 (9,12), or α-linolenic acid, 18:3 (9,12,15), which occur in most other yeasts. Under anaerobic conditions, it can no longer produce oleic acid and this must be provided in the growth medium as an essential nutrient.

It can be said that all the fatty acids which have been identified in yeasts have also been found in the lipids of plants. The fatty acids are, as in all microorganisms, not found in a free form. They are esterified usually in the form of triacylglycerols, phospholipids, *etc*. Triacylglycerols are the principal form of storage lipid in yeasts, as well as in most other eukaryotic organisms. The fatty acid distribution on the glycerol is similar to that found with plant oils with there being an almost complete absence of C_{16} and C_{18} saturated acids in the *sn*-2 position (Ratledge, 1982a). Little work has been carried out concerning differentiation of acyl groups attached to the *sn*-1 and *sn*-3 positions: Haley and Jack (1974) using the triacylglycerols from *Lipomyces lipofer* were able to show that different acyl groups (mainly 16:0, 18:1 and 18:2) were attached in the two positions but as far as we are aware no other work in this direction has been done.

The phospholipids of yeasts have been well researched. In order of usual abundance, the major ones are: phosphatidylcholine, phosphatidylethanolamine, phosphatidylinositol and phosphatidylserine. Cardiolipin (Figure 8) and lysophosphatidylcholine (which has the acyl group at *sn*-2 removed and a free hydroxyl instead—see Figure 2) have been isolated from *Saccharomyces cerevisiae*.

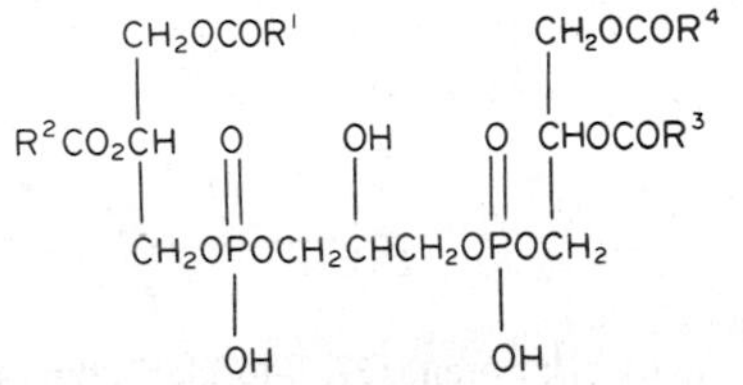

Figure 8 Cardiolipin

Sterols, which are not usually found in the prokaryotes, are present in some abundance in yeasts. The commonest two are ergosterol (**1**; Figure 9) and zymosterol (**2**; Figure 9) though there are numerous other minor sterols. The sterols may occur free or esterified through the 3-hydroxy group to a fatty acid. Squalene, the C_{30} hydrocarbon precursor of sterols, has been isolated from yeasts.

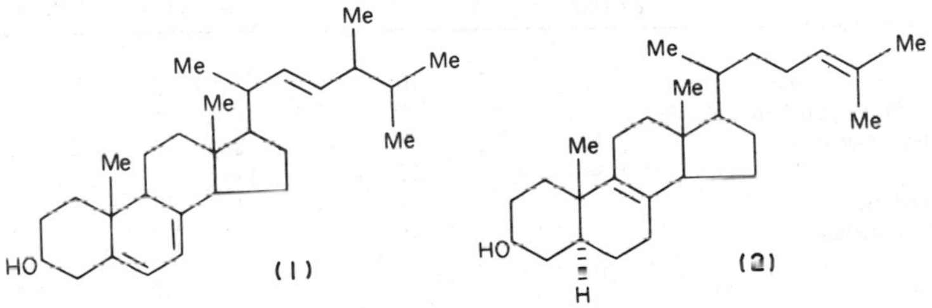

Figure 9 Yeast sterols: ergosterol (**1**); zymosterol (**2**)

Yeasts, particularly *Rhodotorula*, *Cryptococcus* and *Sporobolomyces*, produce a variety of carotenoids (see Goodwin, 1980) which may be either yellow (β-carotene; **3**; Figure 10) or red (torularhodin; **4**; Figure 10) thus giving the culture a coloured appearance. The carotenoid astaxanthin (**5**; Figure 10) is responsible for giving *Phaffia rhodozyma* a flamingo-pink colour. Numerous other carotenoids abound as do various polyprenols, such as farnesol, which arise from intermediates of the carotenoid biosynthetic pathway.

Sphingolipids (see Figure 6) of the ceramide type have been reported to be present in small amounts in yeasts. Glycolipids also occur. Both types of lipid have been reported as occurring in large amounts as extracellular lipids. Acylated derivatives of mannose, arabitol, xylitol, and sophorose have been reported (see Spencer *et al.*, 1979). The acyl moieties, however, are not usually simple fatty acids but 16:0 (3-HO), 18:0 (3-HO), 22:0 (13-HO), 18:0 (17-HO) and a number of polyhydroxy fatty acids such as 8,9,13-trihydroxydocosanoic acid which has varying degrees of acetylation and acylation.

24.2.1.3 Fungi

Specialized reviews covering the lipids of fungi are those of Brennan *et al.* (1974), Brennan and Losël (1978), Erwin (1973), Wassef (1977) and Weete (1980). In general it can be said that there is a greater diversity of lipid types and of range within each lipid type than in yeasts. With respect to

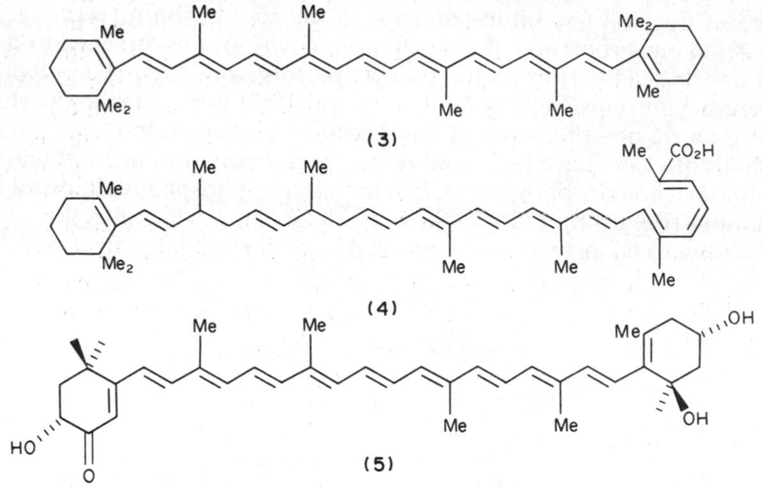

Figure 10 Carotenoids: β-carotene (**3**); β-torularhodin (**4**); astaxanthin (**5**)

fatty acids (see Table 3) there are members of the genus *Entomophthora* in which shorter chain fatty acids, C_8 to C_{12}, can be found in some abundance. Longer chain acids, C_{22} and C_{24} and even longer, have been reported in some genera. Branched chain fatty acids occur in *Conidiobolus* species but apparently not elsewhere.

Table 3 Fatty Acids of Selected Moulds[a]

	Fatty acid (relative % w/w)								
Organism	12:0 or shorter	14:0	16:0	16:1	18:0	18:1	18:2	18:3	20:4
Phycomycetes									
Conidiobolus denaesporus	1	15	10	5	3	15	2	2[b]	9
Cunninghamella echinulata	—	1	19	2	7	55	11	6[b]	—
Entomophthora coronata	40	31	9	—	2	14	2	1[b]	1
Mucor mucedo	—	1	17	1	11	31	33	6[b]	—
Rhizopus arrhizus	—	1	18	4	11	30	16	—	—
Saprolegnia parasitica	—	7	18	2	3	20	14	3[b]	10
Ascomycetes									
Aspergillus niger	—	2	22	3	5	7	46	11	1
Botyritis cinerea	—	3	19	1	4	11	16	42	—
Fusarium moniliforme	—	2	14	—	11	30	42	1	—
Microsporium gypseum	—	—	17	—	8	10	64	—	—
Penicillium chrysogenum	—	3	13	—	12	19	43	6	—
P. notatum	—	—	20	3	6	14	54	2	—
Stilbella thermophila	—	2	43	2	14	25	14	—	—
Basidiomycetes									
Agaricus campestris	3	2	12	—	6	3	63	—	—
Fomes sp.	1	2	12	2	3	4	70	4	—
Tricholoma nudum	—	1	29	2	13	33	35	26	—

[a] Erwin (1972), Weete (1980) and Ratledge (1982a).　[b] γ-Linolenic acid, 18:3 (6,9,12).

Polyunsaturated fatty acids are usually present in some abundance in fungal lipids. Besides *cis, cis*-linoleic acid, 18:2 (9,12), the triply unsaturated linolenic acid can also be found in most species. Two isomers of this acid occur: α-linolenic acid, 18:3 (9,12,15) which is present in all species except those belonging to the Phycomycetes order in which the γ-isomer, 18:3 (6,9,12), is found instead. There have been no reports of the simultaneous occurrence of both these fatty acids. *Entomophthora* species have been reported as containing 20:3 (8,11,14) and 20:4 (5,8,11,14) though it would not be unreasonable to assume that these acids could not be found by close inspection of other fungal lipids.

Hydroxy fatty acids, including ricinoleic acid (12-hydroxyoleic acid), have been reported in several species of *Claviceps*. However these organisms, the cause of ergot in rye, only produce this acid when grown in its sclerotial form and this is difficult, though not impossible, to achieve in the laboratory. Morris and Hall (1966) reported the growth of *Claviceps purpurea* in the laboratory to give 42% of its total fatty acids as ricinoleic acid and Waiblinger and Gröger (1972) have reported up to 62% of the total fatty acids of the same organism grown in submerged culture. Although the ricinoleic acid is esterified to glycerol the hydroxyl groups within the triacylglycerol are not free but are esterified with a non-hydroxylated fatty acid giving an estolide structure. Epoxy and dihydroxy fatty acids have also been identified in the lipid from *Claviceps* spp.

Fungi invariably contain phospholipids of the same types, and in approximately the same order of abundance, as are present in yeasts (see Section 24.2.1.2). A wide range of sterols has been recovered from fungi (see Brennan *et al.*, 1974); these include cholesterol, stigmasterol and derivatives of these lipids. Sphingolipids, including cerosides and sphingoglycolipids, have been recovered from many species, with a large number of individual variations upon the basic theme. Glycolipids similar to those found in *Torulopsis* yeasts (see above) but based on cellobiose or erythritol rather than sophorose have been identified in the lipids of *Ustilago maydis* and *U. nuda*. Numerous carotenoids also occur (Goodwin, 1980).

24.2.2 Oleaginous Microorganisms

Whilst there is a great deal of interest amongst chemists and microbiologists in the lipids of microorganisms, mainly from the point of chemical novelty and of taxonomic usefulness, there is increasing interest in these materials by biotechnologists as useful oils and fats in their own right. Microbial oils may be considered either as alternative sources to existing plant oils or as sources of novel materials and these are discussed in greater detail in Volume 3, Chapter 53.

24.3 PATTERNS OF LIPID ACCUMULATION

It has long been known that many eukaryotic microorganisms will increase their lipid content if they become depleted of a nutrient provided that the supply of carbon to the cell stays plentiful. This applies in particular to oleaginous organisms but also to many non-oleaginous species. The course of lipid accumulation follows a biphasic pattern in batch cultures (Figure 11a). In the first phase, when all nutrients are present in excess, there is a period of balanced growth during which the lipid content of cells stays approximately constant. This ends when a nutrient, usually nitrogen, is exhausted. There then follows an interim period of about six hours during which time the nitrogen pool (mainly amino acids) within the cells becomes diminished. When the cell is completely devoid of any further utilizable nitrogen, protein and nucleic acid biosynthesis cease though the existing cell machinery continues to take up glucose, metabolize it and convert it into lipid. There is therefore a continued build up of lipid in the second phase without there being much, if any, increase in the total cell population.

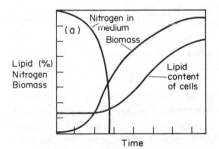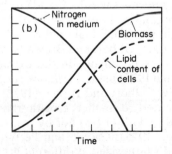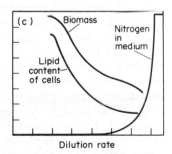

Figure 11 Patterns of lipid accumulation in eukaryotic microorganisms: (a) in a typical oleaginous organism growth in batch-culture; (b) in an organism in which lipid accumulation parallels the growth rate; (c) in an oleaginous organism grown in continuous culture

With some of the slower growing moulds, a different pattern has been occasionally reported where the rate of lipid accumulation appears to coincide with the growth rate (Figure 11b).

In continuous culture (Figure 11c), lipid accumulation can still be achieved provided the culture is run under nitrogen-limited conditions and at a dilution rate (specific growth rate) which is sufficiently slow to allow the organism time to assimilate the carbon which is available to it within the fermenter. The accumulation of lipid in yeasts studied in continuous culture appears to be attributable not to the rate of lipid biosynthesis (more correctly, the specific rate of lipid biosynthesis expressed as grams of lipid synthesized per gram of lipid-free cell weight per hour) increasing as the growth rate falls, but because the rate of protein and nucleic acid synthesis, *etc.* falls. Thus the rate of lipid synthesis stays approximately constant but assumes a greater proportion of the cell's activity as the growth rate declines. Lipid accumulation, at least in yeasts, is therefore not linked to growth.

Lipid accumulation can be triggered by other nutrient limitations: magnesium-, phosphate- and even iron-limited cultures have been reported as stimulating lipid accumulation (Gill *et al.*, 1977) but these effects do not seem to be clear cut as with nitrogen limitation (see also Section 24.4.7).

24.4 FACTORS INFLUENCING LIPID BIOSYNTHESIS

There are many factors which affect the composition of the lipid within a cell and may, coincidentally, also cause a rise or fall in the total amount of lipid within a cell. The greatest single influence on the lipid composition of a microorganism undoubtedly comes when the growth substrate is changed to a hydrocarbon or fatty acid (Ratledge, 1977, 1979). The ability of a paraffin to determine the fatty acyl chain lengths within various lipid classes is so great that this will be covered separately in Section 24.6.

As almost any change in the growth condition of a microorganism can bring about a change in lipid composition, it is prudent of the investigator to be aware of the sequential train of events which are set in motion when a single parameter is changed in a batch culture. For example, the oft-quoted example of temperature affecting the degree of unsaturation of the fatty acyl groups of the lipid (see Section 24.4.4) has to take into account that lowering the temperature will slow down the growth rate of the organism and simultaneously will increase the amount of oxygen dissolved in the medium. Changes in both these conditions could then influence the metabolic status of the cells which might result in the pH of the culture falling (or rising). How then is the effect of temperature to be interpreted? The only answer is to use a chemostat where each growth condition can be changed *independently* of all the others. Where this has not been done, one must then bear in mind that it may not be possible to give unequivocal interpretations of the results without a great deal of work.

General reviews covering this aspect in some detail have been compiled for bacteria by Lechevalier (1982), and by Rattray *et al.* (1975) and Hunter and Rose (1971) for yeasts and by Weete (1980) for fungi in general.

24.4.1 Growth Rate

The influence of the growth rate on lipid accumulation in oleaginous microorganisms is a major determinant on the amount of lipid which is built up within the cells (see Figure 11c) though in non-oleaginous eukaryotic organisms the effect tends to be less evident. However, because the majority of the lipid in these organisms is associated with the various membranes of the cell, there are many reports of changes in the individual lipid components with varying growth rates. With bacteria, which are not oleaginous, Gilbert and Brown (1978), for example, showed with *Pseudomonas aeruginosa* that there was no change in the extractable lipid content of cells grown at between 0.04 and 0.35 h^{-1}. Gill (1975) using *Ps. fluorescens* and Calcott and Petty (1980) using *E. coli* have commented similarly. Gilbert and Brown (1978) did find the proportion of phospholipids in *Ps. aeruginosa* decreased at higher growth rates whereas the neutral lipid content increased. This was not found by Gill (1975) nor by researchers who have also used *E. coli* as a model organism (Günter *et al.*, 1975). Calcott and Petty (1980), whilst not commenting on the actual amount of phospholipid synthesized at different growth rates, did reach broadly the same conclusions as Gilbert and Brown that the individual components of the phospholipids were subject to striking alterations in relative proportions at different growth rates.

24.4.2 Substrate Concentration

Most work on the effect of substrate concentration has been carried out with yeasts; there seems to be little information concerning bacteria. Glucose has been the usual substrate whose concentration has been varied though other substrates including methanol (see Section 24.4.3) have been examined (Rattray and Hambleton, 1980). Glucose, with its metabolism through the Embden–Meyerhof pathway, poses the most problems for eukaryotic microorganisms which can be divided into two groups according to whether they are Crabtree positive or Crabtree negative. In the former group, an increase in the concentration of glucose (or any other carbohydrate metabolized *via* the glycolytic pathway) results in a repression of the synthesis of oxidative (respiratory) enzymes and is manifested by decreased oxygen uptake coupled with accumulation of a metabolically reduced metabolic intermediate such as ethanol. *Saccharomyces cerevisiae* is a typical Crabtree positive yeast; an increase in glucose concentration therefore brings about increased ethanol production even under aerobic conditions. The metabolic changes are probably associated with a decrease in mitochondrial components which leads to a general decrease in total fatty acids, glycerophospholipids and sterol esters (Brown and Johnson, 1970; Johnson *et al.*, 1972).

In Crabtree negative yeasts, such as *Candida utilis* and *Saccharomyces fragilis*, there is an increase in lipid accumulated as the glucose concentration is increased (Johnson *et al.*, 1972; Babij *et al.*, 1969). The increase in lipid is mainly triacylglycerol and, even though the total lipid content of such cells may still be only 20%, this serves to illustrate that modest lipid accumulation can still be achieved even in non-oleaginous species.

24.4.3 Growth Substrate

Apart from alkanes and other hydrocarbons (see Section 24.6) which have been briefly mentioned as profoundly influencing lipid composition, there are numerous reports of different growth substrates not only on changing the amount of lipid accumulated but also the fatty acids to be found within the lipid. With respect to the oleaginous microorganisms, the little information there is has already been reviewed (Ratledge, 1982a). With regard to the non-oleaginous organisms, perhaps the most interesting developments have occurred when methanol has been used as a substrate, as methanol is now being widely contemplated as a feedstock in many fermentation processes.

The pathway of methanol assimilation has been described elsewhere in this volume (Chapter 20). With methanol-utilizing bacteria (see Table 4), those which are obligate methylotrophs (*e.g.* *Methylomonas methylotrophus*) contain mainly C_{16} fatty acids plus phosphatidylglycerol whereas facultative methylotrophs contain C_{18} fatty acids plus phosphatidylethanolamine (Byrom, 1981). With the latter organisms, which can grow on substrates other than methanol, there does not appear to be any significant change in the fatty acids produced by the organism when grown on methanol or another substrate (see Table 4).

Table 4 Fatty Acids of Methanol-utilizing Microorganisms

	Substrate	*Lipid* (%)	*14:0*	*16:0*	*16:1*	*18:0*	*18:1*	*cyc-16:0*	*cyc-18:0*	*HO-14:0*	*HO-16:0*
Bacteria											
Methylomonas methylotrophus[a]	Methanol	6.5	1.5	43	49	0.8	0.8	0.8	0.8	—	3
Protaminobacter ruber[b]	Methanol	—	—	5	11	6	78	—	—	1	—
	Nutrient broth	—	—	3	8	8	77	—	—	2	—
Pseudomonas extorquens[b]	Methanol	—	—	7	9	8	75	—	—	2	—
	Nutrient broth	—	—	3	10	5	79	—	—	1	—

	Substrate	*Lipid* (%)	*14:0*	*16:0*	*16:1*	*18:0*	*18:1*	*18:2*	*18:3*
Yeasts									
Candida guilliermondii	Methanol	12.3	—	14	7	1	53	24	2
	Glucose	22.2	—	20	9	3	57	9	1
Candida boidinii[d]	Methanol	6	tr	19	18	2	30	28	—
	Glucose	3.5	1	19	20	5	11	41	—
Hansenula polymorpha[d]	Methanol	1.4	tr	15	3	5	38	36	—
	Glucose	1.2	tr	16	6	8	25	41	—

[a] Byrom and Ousby (1975.)   [b] Ikemoto *et al.* (1978).   [c] Jigami *et al.* (1979).   [d] Rattray and Hambleton (1980).

With methanol-utilizing yeasts, which are all facultative methylotrophs, greater changes in the fatty acids have been noted between glucose-grown cells and methanol-grown ones than occur with bacteria (see Table 4). Rattray and Hambleton (1980), who examined the lipid composition of two methanol-grown yeasts, commented on the considerable modifications in phospholipid composition which occurred when cells grew on methanol rather than glucose: phosphatidyl-serine and phosphatidylinositol increased at the expense of phosphatidylcholine.

24.4.4 Temperature

In general, the degree of unsaturation of the fatty acids of biological systems increases as the growth temperature is decreased and there are many reports which have confirmed this with both prokaryotic and eukaryotic microorganisms. The vast majority of this work has unfortunately been carried out using batch cultivation techniques where the effects of temperature on growth rate and oxygen tension have tended to be ignored. However when studies have been carried out with chemostats the correlation of degree of unsaturation with the temperature has been confirmed in many instances with, for example, *Pseudomonas fluorescens* (Gill, 1975) and various *Candida* spp. (Brown and Rose, 1969; McMurrough and Rose, 1971, 1973). But not all organisms show this pattern of behaviour; little significant difference has been recorded in the degree of fatty acid unsaturation with *Saccharomyces cerevisiae* between 30 and 15 °C (Hunter and Rose, 1972); with *Candida* sp. no. 107 from 19 to 32.8 °C (Hall and Ratledge, 1977); or with *Rhodotor-ula gracilis* from 22.5 to 30.0 °C (Ratledge and Hall, 1979). However in the closely monitored behaviour of *S. cerevisiae*, Hunter and Rose (1972) showed that there was an increased content of triacylglycerols and glycerophospholipids at lower temperatures. Gill (1975) also noted changes in the relative proportions of individual phospholipids in *P. fluorescens* at various temperatures.

The changes which do occur in the degree of unsaturation or even in the shortening of the chain length of the fatty acids, which has the same effect of increasing of the fluidity of the membrane lipids at lower temperatures, are explicable in terms of the changes which occur either in the inducibility of the various desaturases at low growth temperatures or in the activity of the fatty acid synthetase (see Fulco, 1974; Okuyama *et al.*, 1979).

Above 50 °C many constraints to microbial life become evident. The membrane lipids of ther-mophilic bacteria, which are capable of growth at 95 °C and survival at perhaps 105 °C, are con-siderably different from those organisms which have a much lower tolerance to temperature. For details of the lipids of such organisms the reader is referred to the several chapters on this subject which are to be found in various books edited by Heinrich (1976), Kushner (1978) and Shilo (1979).

24.4.5 Oxygen

In eukaryotes, oxygen is required for the conversion of stearic acid to oleic acid and thence to linoleic and linolenic acids (see Section 24.5.3). Whilst there is some little evidence to show that deprivation of oxygen to a culture does lead to a decline in the amount of polyunsaturated fatty acids being produced (see Ratledge, 1982a), this need not always be the case as the outcome of oxygen deprivation will depend upon the relative affinities of the fatty acid desaturase and, prob-ably, cytochrome oxidase. If the latter has the weaker affinity, growth of the organism will decline *before* there is any effect manifested upon the synthesis of unsaturated fatty acids. As there is only one step in the biosynthesis of fatty acids which requires ATP (acetyl-CoA carboxy-lase, see Section 24.5.1), and none which require oxygen directly, it may be expected that oxygen deprivation would not decrease the physical amount of lipid being produced by a microorganism. Only the biomass yield would be affected. This has been confirmed in at least one instance (Hall and Ratledge, 1977).

24.4.6 pH and Salinity

It is only at the extreme of pH and salinity that effects on lipid composition become strikingly evident. These are genotypic changes rather than phenotypic. The lipids of acidophilic alkanophi-lic and halophilic organisms are ably reviewed by several authors in the books edited by Heinrich (1976), Kushner (1978) and Shilo (1979). Ordinarily the membranes of other microorganisms are not capable of modification to allow them to survive in such environments and consequently the changes which are seen in these organisms tend to be minimal.

24.4.7 Other Factors

As has been mentioned (Section 24.3), the influence of limiting the amount of nitrogen available to a culture can serve to increase the amount of lipid accumulated in oleaginous microorganisms as well as in many non-oleaginous ones. Limitations of other nutrients such as Mg, PO_4, *etc.* may, or may not, bring about similar increases in lipid accumulation but they may bring about many changes in lipid composition independently of their effect on lipid accumulation. For example, phosphate limitation of growth of *Pseudomonas diminuta* results in the partial replacement of acidic phospholipids by acidic glycolipids (Minnikin *et al.*, 1974). With *Saccharomyces cerevisiae* sterol esters and triacylglycerol declined though without any significant change occurring in the amounts of phosphilipids (Ramsay and Douglas, 1979). Observations concerning the effect of both phosphate and magnesium limited growth on the polar lipids of *Bacillus subtilis* have been made by Minnikin and Abdolrahimzadeh (1974).

Deprivation of essential vitamins on lipid metabolism has been studied in a few instances. The best described effect is that of inositol-deficiency with certain strains of *S. carlsbergensis* and *S. cerevisiae* which then leads to an increase in cell lipid. It is considered that the effect is brought about by an increase in the activity of acetyl-CoA carboxylase though the exact mechanism of how this links to inositol-deficiency is still not clear (Daum *et al.*, 1977). Thiamine deficiency in the same strain of *S. carlsbergensis* on the other hand led to a decline in content of all lipids: sterol esters, acylglycerols and all glycerophospholipids (Nishikawa *et al.*, 1977).

24.5 LIPID BIOSYNTHESIS

24.5.1 Synthesis of Saturated Fatty Acids

Saturated fatty acids are synthesized from acetyl-CoA by the concerted action of two complex enzyme systems, acetyl-CoA carboxylase and fatty acid synthetase. The properties of these two enzyme systems are described below. Readers requiring more detailed information are referred to the comprehensive reviews of Volpe and Vagelos (1976), Bloch and Vance (1977) and Lynen (1980).

24.5.1.1 *Acetyl-CoA carboxylase*

Acetyl-CoA carboxylase is frequently considered to catalyze the first committed step in the synthesis of fatty acids from acetyl-CoA. Structurally it is complex: three functional components have been identified, namely, biotin carboxyl-carrier protein (BCCP), biotin carboxylase and carboxyl transferase. Thus, the enzyme catalyzes the biotin-dependent carboxylation of acetyl-CoA, to give malonyl-CoA, in a two-stage reaction dependent on the presence of ATP:

$$\text{biotin-BCCP} + HCO_3^- + H^+ + ATP \xrightarrow{\text{biotin carboxylase}} \text{carboxybiotin-BCCP} + ADP + P_i$$

$$\text{carboxybiotin-BCCP} + \text{acetyl-CoA} \xrightarrow{\text{carboxyl transferase}} \text{biotin-BCCP} + \text{malonyl-CoA}$$

The enzyme, particularly in eukaryotic organisms, is widely regarded as catalyzing the principal regulatory step in lipogenic pathways and a number of mechanisms have been proposed. These include activation by tricarboxylic acid cycle intermediates, inhibition by long chain fatty acyl-CoA esters and reversible phosphorylation. The allosteric activation of acetyl-CoA carboxylation by citrate has been reported for many mammalian enzymes (Volpe and Vagelos, 1976) and although it has also been reported for certain yeast acetyl-CoA carboxylases this has also been disputed. Long chain acyl-CoA esters (FACEs) inhibit acetyl-CoA carboxylase from numerous sources and may represent a plausible feed-back control mechanism although certain authors have urged caution pointing out that these molecules are powerful detergents and possibly disrupt enzyme structure non-specificity (see Sumper, 1974).

In bacteria it has been suggested that acetyl-CoA carboxylase may be regulated by variations in the cellular concentration of ppGpp (guanosine 3',3',5',5'-tetraphosphate). Membrane phospholipids constitute the major lipid fraction in bacteria and, therefore, lipogenesis may be closely linked with growth. Using stringent *E. coli* strains, it has been demonstrated that when growth is limited by amino acid starvation the intracellular concentration of ppGpp increases. Since this metabolite inhibited the carboxyl-transferase component of acetyl-CoA carboxylase it was suggested that this formed the mechanism by which the rate of phospholipid synthesis was coregulated with the growth rate (Bloch and Vance, 1977).

24.5.1.2   *Fatty acid synthetase*

Malonyl-CoA, derived from the acetyl-CoA carboxylase reaction, is used to synthesize long chain saturated fatty acids in a series of reactions also requiring NADPH and acetyl-CoA, catalyzed by the fatty acid synthetase complex. The structure of the complex varies according to the source (see below), however, the sequence of reactions are basically similar in all organisms:

Acetyl transacylation (priming reaction):
$$\text{MeCOSCoA} + \text{HS-enzyme} \rightarrow \text{MeCOS-enzyme} + \text{CoASH} \tag{1}$$

Malonyl transacylation:
$$\text{HO}_2\text{CCH}_2\text{COSCoA} + \text{HS-enzyme} \rightarrow \text{HO}_2\text{CCH}_2\text{COS-enzyme} + \text{CoASH} \tag{2}$$

β-Ketoacyl synthase condensation:
$$\text{HO}_2\text{CCH}_2\text{COS-enzyme} + \text{MeCOS-enzyme} \rightarrow \text{HS-enzyme} + \text{CO}_2 + \text{MeCOCH}_2\text{COS-enzyme} \tag{3}$$

β-Ketoacyl reduction:
$$\text{MeCOCH}_2\text{COS-enzyme} + \text{NADPH} \rightarrow \text{MeCHOHCH}_2\text{COS-enzyme} + \text{NADP}^+ \tag{4}$$

β-Hydroxyacyl dehydration:
$$\text{MeCHOHCH}_2\text{COS-enzyme} \rightarrow \text{H}_2\text{O} + \text{MeCH}{=}\text{CHCOS-enzyme} \tag{5}$$

2,3-*trans*-Enoylacyl reduction:
$$\text{MeCH}{=}\text{CHCOS-enzyme} + \text{NAD(P)H} \rightarrow \text{MeCH}_2\text{CH}_2\text{COS-enzyme} + \text{NAD(P)} \tag{6}$$

As the end-product of equation (6) now becomes the substrate in equation (3), where it reacts with a further molecule of malonyl-S-enzyme, the net result of each cycle of reactions is the introduction of two additional carbon atoms on to the initial acetyl group. These additional carbon atoms are themselves derived from the acetyl-CoA substrate of acetyl-CoA carboxylase. After seven turns of the cycle, palmitoyl-coenzyme A is synthesized with the following stoichiometry:

$$\text{MeCOSCoA} + 7\,(\text{HO}_2\text{CCH}_2\text{COSCoA}) + 14\,\text{NADPH} + 14\text{H}^+ \rightarrow$$
$$\text{Me(CH}_2)_{14}\text{COSCoA} + 7\,\text{CO}_2 + 7\,\text{CoASH} + 14\,\text{NADP}^+ + 6\,\text{H}_2\text{O}$$

In bacteria, except for the mycobacteria, the components of the fatty acid synthetase are separable (termed type II) and the component to which the growing acyl chain is attached *via* a thioester linkage, is a small molecular weight ($\sim$10 000 daltons) protein termed the acyl carrier protein (ACP). Thus in equations (1) to (6) above, ACP replaces the designated 'enzyme'.

The nature of the final products from the complex are governed to a large extent by the substrate specificity of the β-ketoacyl-enzyme (or ACP) synthase. For example, propionyl-CoA may replace acetyl-CoA in the initial priming reaction with the consequence that odd-numbered saturated fatty acids would be synthesized. Similarly, palmitoyl-CoA is the end-product in many systems because this compound is not itself a substrate for the β-ketoacyl-enzyme synthase enzyme; thus further elongation is blocked.

The fatty acid synthetases of eukaryotic microorganisms, and of the mycobacteria, are structurally and functionally organized into undissociable complexes. These synthetases, termed type I, consist of multiple copies of only two multifunctional polypeptides. Acyl carrier protein has not been isolated from type I synthetases although its presence has been inferred by the detection of 4'-phosphopantotheine which is the prosthetic group of ACP. Type I fatty acid synthetases are considered to be more highly evolved than the type II variety since intermediates are not released. Full details of the nature of the various fatty acid synthetase types may be obtained from the reviews given at the beginning of this section. Additional information on the mycobacteria, which contain both type I and type II synthetases, may be found in the reviews of Ratledge (1982b) and Minnikin (1982).

24.5.2   Origin of Acetyl-CoA

24.5.2.1   *Bacteria*

Reactions in bacteria leading to the generation of acetyl-CoA, and those in which it is utilized, are summarized in Figure 12. The principal sites of regulation of the flux of metabolism between oxidative and lipogenic pathways appear to be at the levels of citrate synthase and acetyl-CoA carboxylase. Thus, NADH and ATP may be regarded as end-products of the tricarboxylic acid cycle and it is suggested that variations in the concentrations of these metabolites govern the activity of citrate synthase and consequently regulate the oxidation of acetyl-CoA (Weitzman and Danson, 1976). The mechanisms by which bacterial acetyl-CoA carboxylases might be regulated have been discussed in Section 24.5.1.1.

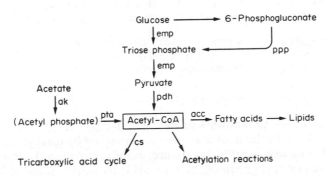

Figure 12 Illustration of the amphibolic nature of acetyl-CoA in bacterial metabolism: pdh, pyruvate dehydrogenase; ppp, pentose phosphate pathway; emp, Embden Meyerhof glycolytic pathway; ak, acetyl kinase; pta, phosphotransacetylase; acc, acetyl-CoA carboxylase; cs, citrate synthase (coupling with oxaloacetate)

24.5.2.2 *Eukaryotic microorganisms*

In eukaryotes, acetyl-CoA is produced in the mitochondria, *via* the action of pyruvate dehydrogenase complex, and since the inner mitochondrial membrane is relatively impermeable to this metabolite, mechanisms must exist for its transport into the cytosol, the site of fatty acid biosynthesis.

Three possibilities exist. (a) Acetyl-CoA might be hydrolysed to acetate, which can pass freely through the mitochondrial membrane, and then be regenerated by a cytosolic acetyl-CoA synthetase. (b) Acetyl groups might be transported across the mitochondrial membrane in the form of acetylcarnitine, $Me_3N^+CH_2CH(COMe)CH_2 CO_2H$. (c) Acetyl groups may be transferred into the cytosol in the form of carbon atoms 1 and 2 of citrate. Citrate is accessible to the mitochondrial membrane and in the cytosol acetyl-CoA may be regenerated by ATP:citrate lyase: citrate + ATP + CoASH → acetyl-CoA + oxaloacetate + ADP + P_i.

Recent evidence has shown that the ATP:citrate lyase pathway is only operative in oleaginous microorganisms (Boulton and Ratledge, 1981). In non-oleaginous organisms the carnitine-mediated system is probably predominant (Kohlhaw and Tan-Wilson, 1977).

24.5.3 Biosynthesis of Unsaturated Fatty Acids

Monounsaturated fatty acids are synthesized by either aerobic or anaerobic mechanisms.

The aerobic pathway is the most common being found in yeast, protozoa, algae, some bacteria and mammals. Invariably the double bond is introduced into the 9 position of the fatty acid, thus palmitoleic, 16:1(*c*9), and oleic, 18:1(*c*9), acids are common products. The double bond is introduced into the saturated fatty acid by a specific monooxygenase in which NADPH is the coreductant. The reaction involves an electron transport chain usually featuring either flavoprotein or iron–sulfur protein components.

In *Eubacteriales* monounsaturated fatty acids are synthesized by a modified fatty acid synthetase. The modification takes the form of a branch point at the dehydration stage (equation 5, in Section 24.5.1.2). The products at the branch points are the normal *trans*-2-enoyl-ACP and, by an isomerization reaction, *cis*-3-enoyl-ACP. The former may be used to synthesize saturated fatty acids as indicated in the scheme, whilst the latter may be elongated with C_2 units (from malonyl-CoA) to produce a longer chain *cis*-unsaturated fatty acid (see Figure 13).

Polyunsaturated fatty acids, which occur only in eukaryotes, are synthesized from saturated or

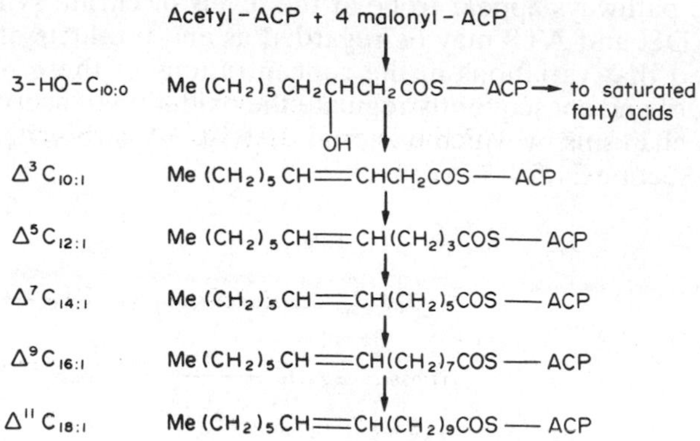

Acetyl – ACP + 4 malonyl – ACP

$3-HO-C_{10:0}$ Me $(CH_2)_5$ CH_2 CHCH$_2$COS——ACP → to saturated
                                                   fatty acids

                                      |
                                      OH

$\Delta^3 C_{10:1}$ Me $(CH_2)_5$ CH═CHCH$_2$COS——ACP

$\Delta^5 C_{12:1}$ Me $(CH_2)_5$ CH═CH$(CH_2)_3$COS——ACP

$\Delta^7 C_{14:1}$ Me $(CH_2)_5$ CH═CH$(CH_2)_5$COS——ACP

$\Delta^9 C_{16:1}$ Me $(CH_2)_5$ CH═CH$(CH_2)_7$COS——ACP

$\Delta^{11} C_{18:1}$ Me $(CH_2)_5$ CH═CH$(CH_2)_9$COS——ACP

Figure 13   Biosynthesis of unsaturated fatty acids of the *cis*-vaccenic acid series in bacteria

monounsaturated precursors *via* pathways involving both chain elongation and desaturases of the monooxygenase type. Usually the double bonds are separated by methylene groups (*i.e.* —CH═CHCH$_2$CH═CH—) and new double bonds are usually introduced between existing double bonds and the terminal methyl group of the fatty acid. However, some algae and Phycomycete moulds can also desaturate towards the terminal carboxyl group.

25.5.4   Biosynthesis of Other Fatty Acids

Branched chain fatty acids are synthesized either by the insertion of methyl groups into unsaturated fatty acids or from branched chain amino acid precursors. An example of the latter mechanism is the synthesis of D-12-methyltetradecanoic acid from isoleucine (Figure 14).

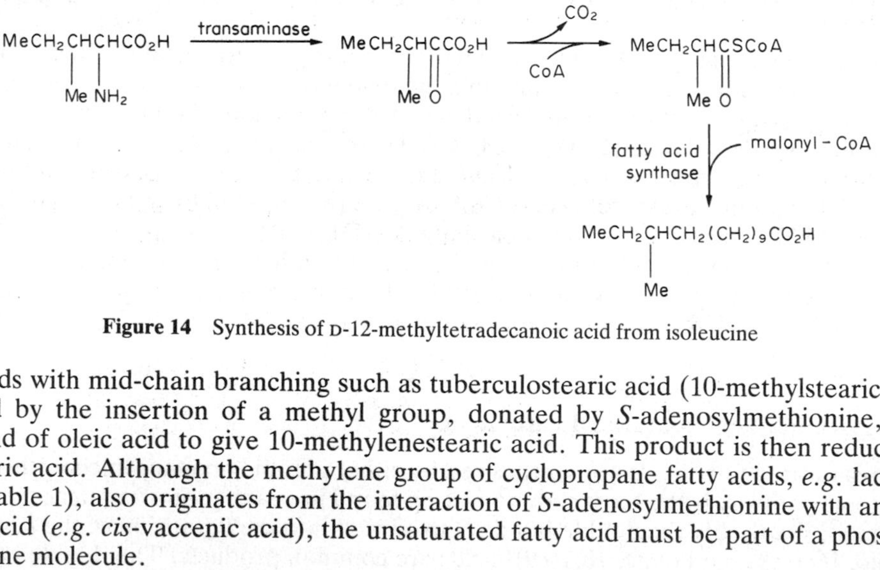

MeCH$_2$CHCHCO$_2$H $\xrightarrow{\text{transaminase}}$ MeCH$_2$CHCCO$_2$H $\xrightarrow[\text{CoA}]{\text{CO}_2}$ MeCH$_2$CHCSCoA
       | | | || | ||
       Me NH$_2$ Me O   Me O

                                                                    fatty acid ⟍ malonyl – CoA
                                                                    synthase

                                              MeCH$_2$CHCH$_2$(CH$_2$)$_9$CO$_2$H
                                                     |
                                                     Me

Figure 14   Synthesis of D-12-methyltetradecanoic acid from isoleucine

Fatty acids with mid-chain branching such as tuberculostearic acid (10-methylstearic acid) are synthesized by the insertion of a methyl group, donated by *S*-adenosylmethionine, into the double bond of oleic acid to give 10-methylenestearic acid. This product is then reduced to 10-methylstearic acid. Although the methylene group of cyclopropane fatty acids, *e.g.* lactobacillic acid (see Table 1), also originates from the interaction of *S*-adenosylmethionine with an unsaturated fatty acid (*e.g. cis*-vaccenic acid), the unsaturated fatty acid must be part of a phosphatidylethanolamine molecule.

Hydroxy fatty acids are synthesized as by-products of the various pathways of fatty acid oxidation. Thus, 3-hydroxy fatty acids arise as products of the β-oxidation sequence of fatty acid degradation (see Figure 20). Similarly, α-oxidation yields 2-hydroxy acids (see Section 24.6. 5.2, p. 478).

24.5.5   Biosynthesis of Lipids from Fatty Acids

It is beyond the scope of this article to give an account of the biosynthesis of all the lipid types occurring in microorganisms; however, some of the more important storage and structural lipids are discussed.

24.5.5.1 Triacylglycerols

Triacylglycerols constitute the major storage lipid in eukaryotic microorganisms and the immediate precursors for *de novo* biosynthesis are fatty acyl-CoA esters and *sn*-glycerol 3-phosphate. This pathway is illustrated in Figure 15.

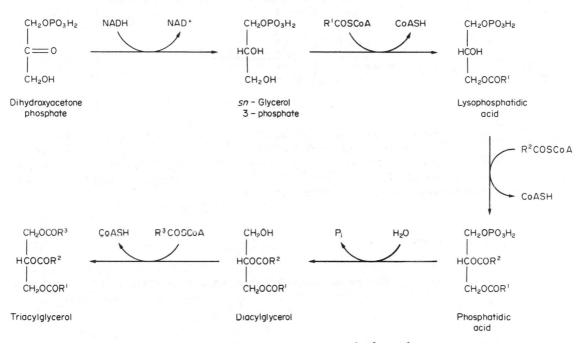

Figure 15 *De novo* biosynthesis of triacylglycerols, where R^1, R^2 and R^3 are long alkyl chains

24.5.5.2 Phospholipids

Phospholipids are synthesized from phosphatidic acid in reactions involving cytidine nucleotides, which serve as carriers of either alcohol groups or phosphatidic acid itself. The pathways leading to the formation of some common phospholipids are shown in Figure 16.

24.5.5.3 Waxes

The mechanisms by which true waxes (*i.e.* esters of long chain fatty acids with long chain primary alcohols) are synthesized are uncertain. Three possibilities have been postulated: (1) the reversal of a wax esterase activity; (2) transfer of acyl groups from a phospholipid; (3) transfer of acyl groups from acyl-CoA.

24.5.5.4 Poly-β-hydroxybutyrate

This polymer forms the major energy store in many bacteria (Dawes and Senior, 1973) and it is synthesized, from acetyl-CoA, as shown in Figure 17. There is no suggestion of the involvement of ACP.

24.6 MICROBIAL METABOLISM OF ALKANES AND FATTY ACIDS

24.6.1 Alkane-utilizing Organisms

A wide variety of yeasts, fungi and bacteria possess the ability to utilize alkanes and comprehensive lists may be obtained from the reviews of Klug and Markovetz (1971), Shennan and Levi (1974) and Bemmann and Tröger (1975).

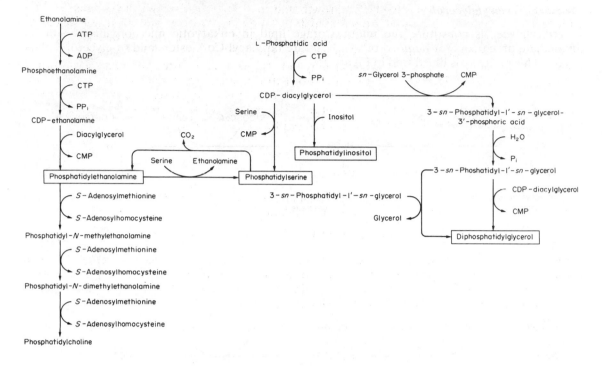

Figure 16   Biosynthetic pathways for some common phospholipids

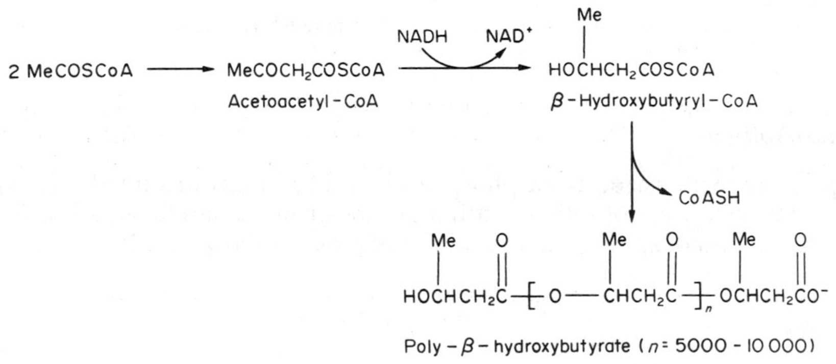

Figure 17   Poly-β-hydroxybutyrate synthesis

24.6.2   Uptake of Alkanes

The limited solubility of alkanes in aqueous media necessitates that specific mechanisms exist for their uptake and subsequent transport to the site of further metabolism. Two hypotheses have been proposed:
   (1) uptake *via* direct contact between cell surfaces and alkane droplets; and
   (2) 'pseudo'-solubilization of the alkane prior to uptake.
Evidence for the first hypothesis is provided by the observation that many alkane-utilizing microorganisms preferentially adhere to hydrocarbons forming agglomerates, termed flocs. In contrast, non-utilizers do not show this phenomenon. Growth of many organisms on alkanes induces the synthesis of cell wall lipopolysaccharides conferring a lipophilic nature to the cell surface and this is responsible for the affinity of the cell for hydrocarbons. Meisel *et al.* (1976) and Fukui and Tanaka (1981) have both reported the presence of channels and slime-like outgrowths in the cell walls of alkane-grown *Candida tropicalis* and *C. albicans*. These authors considered

that alkanes become attached to the outgrowth and then migrate through the channels to the endoplasmic reticulum, the site of the initial hydroxylation (see Section 24.6.3). It remains to be established whether hydrocarbon uptake can be described as a true active transport process (Käppeli and Fiechter, 1976, 1981; Lindley and Heydeman, 1983). The second and most favoured hypothesis proposes that alkanes may only be assimilated after an initial 'solubilizing' step and that this is achieved by the synthesis of extracellular surfactants (see Section 24.5.3). Thus, the large hydrocarbon droplets may be regarded simply as substrate reservoirs from which submicroscopic droplets are formed and utilized. Alkane in this latter form is termed accommodated (Einsele *et al.*, 1975) or pseudosolubilized (Goma *et al.*, 1973).

24.6.3 Mechanisms of Alkane Oxidation

Following uptake, alkanes are subjected to an initial oxidation step. In the majority of microorganisms this takes the form of monoterminal oxidation.

$$Me(CH_2)_nMe \rightarrow Me(CH_2)_nCH_2OH \rightarrow Me(CH_2)_nCHO \rightarrow Me(CH_2)_nCO_2H$$

Monoterminal oxidation of the alkane to the corresponding primary alcohol is catalyzed by a complex hydroxylase in which a variety of electron-carrier systems are involved. In various species of *Pseudomonas*, a monooxygenase system involving rubredoxin, NADH-rubredoxin reductase and ω-hydroxylase has been described (Figure 18).

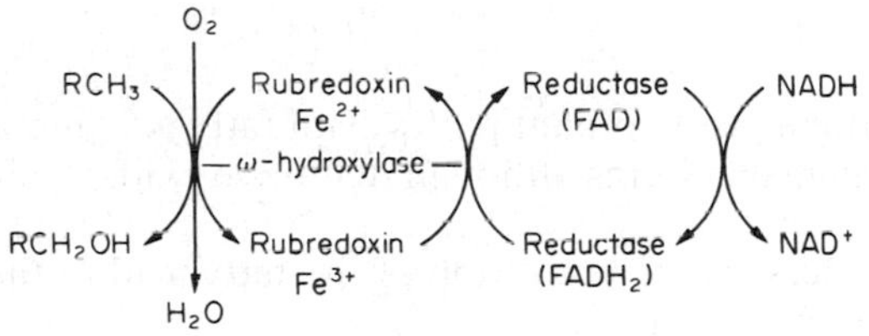

Figure 18 Oxidation of alkanes in some *Pseudomonas* spp.

Another system has been detected in certain *Corynebacteria* and also in the microsomal fraction of alkane-utilizing yeasts. This system utilizes cytochrome P-450 as the terminal oxidase and an NADPH-cytochrome P-450 reductase as the electron transfer component (Figure 19).

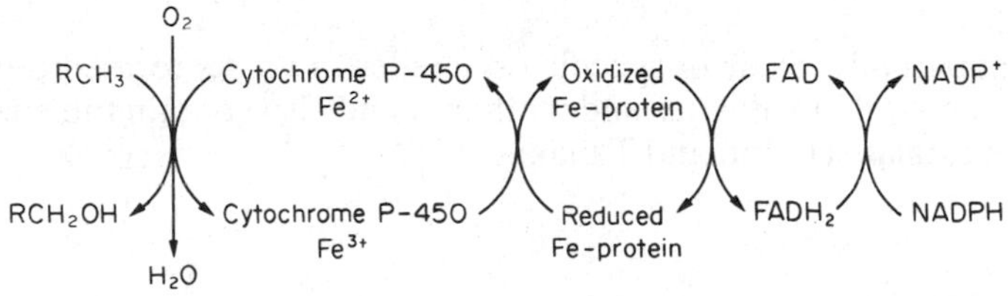

Figure 19 Oxidation of alkanes in some yeasts and *Corynebacteria* spp.

The ability to degrade alkanes at the subterminal carbon atoms has been reported for a number of microorganisms, though it seems to be most common in the moulds (see Boulton and Ratledge, 1984).

$$MeCH_2(CH_2)_nCH_2Me \rightarrow MeCH_2(CH_2)_nCH(OH)Me \rightarrow MeCH_2(CH_2)_nCOMe \rightarrow$$
$$Me(CH_2)_{n-2}CH_2OH + MeCO_2H$$

Diterminal oxidation of *n*-alkanes has also been reported to occur in some bacteria and yeasts (Boulton and Ratledge, 1984). The principal pathway is a two-stage process in which one methyl group is oxidized to give the corresponding fatty acid then in the second stage the ω-methyl group is oxidized in a similar fashion:

$$Me(CH_2)_nMe \rightarrow Me(CH_2)_nCH_2OH \rightarrow Me(CH_2)_nCHO \rightarrow Me(CH_2)_nCO_2H \rightarrow$$
$$HOH_2C(CH_2)_nCO_2H \rightarrow OHC(CH_2)_nCO_2H \rightarrow HO_2C(CH_2)_nCO_2H$$

Recent work has indicated that oxidation may also proceed with the diol as an intermediate. A list of extracellular dioics produced by yeasts may be obtained from the recent review of Rehm and Reiff (1981).

24.6.4 Oxidation of Primary Alcohols to Fatty Acids

The higher alcohols, produced as a result of alkane oxidation, are ultimately oxidized to the corresponding fatty acids *via* the aldehyde. These reactions are catalyzed by specific long-chain alcohol dehydrogenases, which are usually NAD^+-dependent, and aldehyde dehydrogenases. The enzyme systems involved are inducible by alkanes, higher alcohols or aldehydes. Both the alcohol and aldehyde dehydrogenases tend to show activity only with long chain substrates (Lebeault and Azoulay, 1971). The intracellular location of the enzymes seems to vary according to the organism with mitochondrial, cytosolic and membrane-bound types being reported (Ratledge, 1977).

24.6.5 Metabolism of Fatty Acids Derived from Alkanes

There are several ways in which long chain fatty acids, either derived from alkanes or directly introduced to a microorganism, can be further metabolized.

24.6.5.1 β-Oxidation

It is widely assumed that the predominant pathway of fatty acid metabolism is one involving β-oxidation as described in animal systems although definitive evidence for its occurrence in microorganisms is still lacking.

The initial step in this process is the activation of the fatty acid to the corresponding acyl-CoA ester by acyl-CoA synthetase:

$$RCO_2H + CoA + ATP \rightarrow RCO\text{-}SCoA + ADP + P_i$$

Mishina *et al.* (1978) reported that *Candida lipolytica* contained two long chain acyl-CoA synthetases, one of which was induced by growth on oleate. This latter enzyme had a broader specificity than the former and these authors suggested that it was involved in fatty acid degradation, whereas the former type was involved in acyl-CoA synthesis for lipogenesis.

A fatty acid β-oxidation system reported to occur in various yeasts (Fukui and Tanaka, 1981) is illustrated (Figure 20).

These reactions apparently occur in specific organelles in yeasts termed peroxisomes. Peroxisomes are induced by growth on alkanes and methanol and also contain the enzymes of the glyoxylate cycle as well as catalase (Fukui and Tanaka, 1979).

24.6.5.2 α-Oxidation

The first intermediate in the α-oxidation pathway is the formation of an α-hydroxy fatty acid, which in turn may be oxidatively decarboxylated:

$$RCH_2CO_2H \rightarrow RCHOHCO_2H \rightarrow RCHO\ (+ CO_2) \rightarrow RCO_2H$$

Evidence for this pathway in most organisms is slight; however, further information may be obtained from the reviews of Finnerty and Makula (1975) and Ratledge (1978).

24.6.6 Microbial Products Derived from Alkanes

A wide variety of useful products may be derived from alkane-grown microorganisms. It is beyond the scope of this article to give a complete account of all the end-products, although these include the biomass itself as single cell protein (SCP), amino acids, organic acids, carbohydrates, nucleic acids, vitamins, coenzymes and antibiotics. In this article only the nature of lipid products

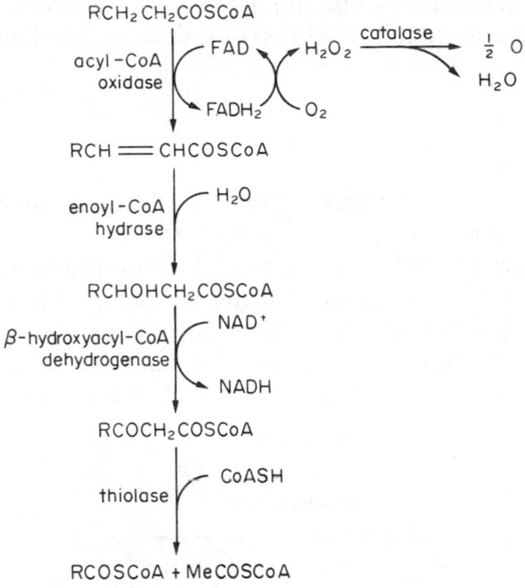

Figure 20 Fatty acid β-oxidation in peroxisomes of yeasts

derived from alkanes will be discussed. However, for further information on non-lipid products the reader should consult the review of Fukui and Tanaka (1981).

In general, microorganisms cultivated on alkanes have a higher lipid content when compared with that observed for non-hydrocarbon substrates. This holds true when allowances are made for growth rate differences and the fact that many microorganisms can accommodate unchanged alkane which would, of course, coextract with lipid. The ability to accumulate greater than usual quantities of intracellular lipid or lipid-related compounds is particularly evident in the yeasts and moulds. In addition, various classes of lipid products may be produced extracellularly by all types of alkane-utilizing microorganisms.

24.6.6.1 Fatty alcohols and aldehydes

The immediate intermediates of alkane oxidation, namely higher alcohols and aldehydes, do not usually accumulate in significant quantities. However, by careful strain screening or by metabolic manipulation or by the selection of appropriately blocked mutants the accumulation of higher alcohols has been achieved. The alcohols, which usually appear in the medium, may be primary or secondary depending on the pathway of alkane oxidation operating.

Aldehydes are usually only present at low concentrations and this is thought to be due to the fact that the rate of aldehyde oxidation to the acid is more rapid than the rate of alcohol oxidation. Thus, aldehyde accumulation is not favoured.

24.6.6.2 Fatty acids

The fatty acid profile of any given microorganism is usually a reflection of the alkane upon which it is cultivated (Ratledge, 1977, 1980). Thus, alkanes from C_{14} to C_{18} may be assimilated with little change in the chain length. This is true of both odd and even chain length alkanes. Alkanes with chain lengths shorter than C_{14} are not accommodated with such ease, as is reflected by the observation that high proportions of these fatty acids are exceptional in microorganisms. It seems that short-chain alkanes are elongated. However, growth on short-chain odd-number alkanes (C_{11} to C_{13}) usually results in a fatty acid profile with over 50% even-chain alkanes (C_{16} and C_{18}). Whether this is a consequence of elongation or *de novo* synthesis from acetyl-CoA, derived from alkane degradation, is not clear.

It should be made clear that free fatty acids do not normally occur either intra- or extra-cellularly owing to their toxic nature. Instead they are esterified to either glycerol, phosphoglyceric acid derivatives, sugars, or fatty alcohols. In fact, an increase in the second of these categories, the phospholipids, is characteristic of microbial growth on alkanes. This is presumably a reflection of the dramatic increase in membrane biosynthesis associated with alkane-utilization.

24.6.6.3 Surfactants

The importance of surfactants in alkane uptake has already been discussed (Section 24.6.2). The nature of these compounds, which cover a wide range of chemical types, and their possible commercial importance have recently been reviewed (Cooper and Zajic, 1980).

Structurally, the compounds are lipids with the surfactant properties arising from a combination of polar and apolar moieties in a single molecule. Thus, the apolar moiety often consists of a long alkyl chain, whilst the polar portion may be supplied by a variety of groups, including sugars, alcohols and even proteins. Some typical examples are shown in Figure 21.

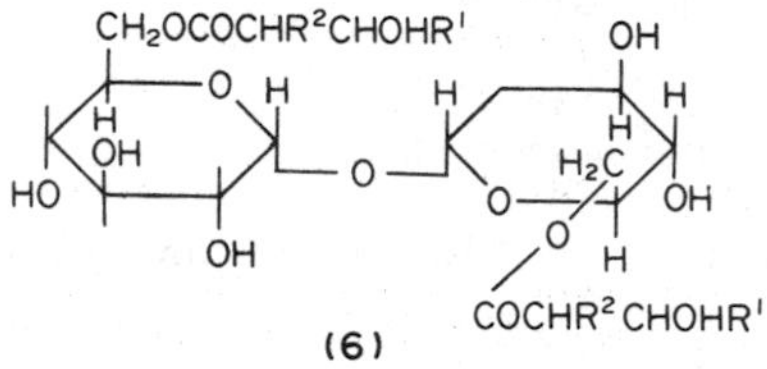

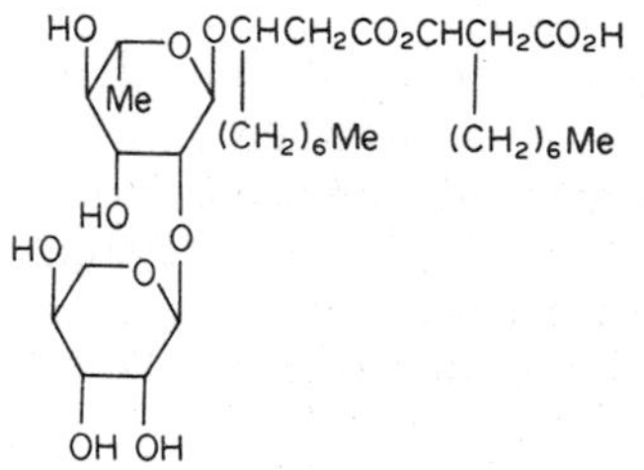

Figure 21 Examples of surfactants produced by microorganisms: (6) trehalose lipid (*Rhodococcus erythropolis*), where R^1 and R^2 are long alkyl chains; rhamnolipid (7) (*Arthrobacter paraffineus*); sophorolipid (8) (*Torulopsis bombicola*); surfactin (9) (*Bacillus subtilis*)

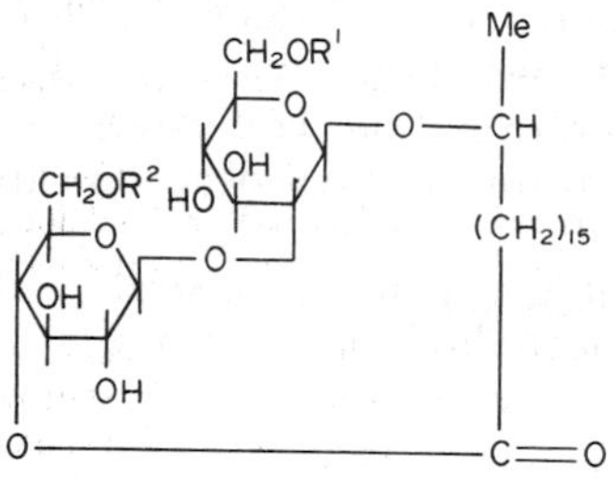

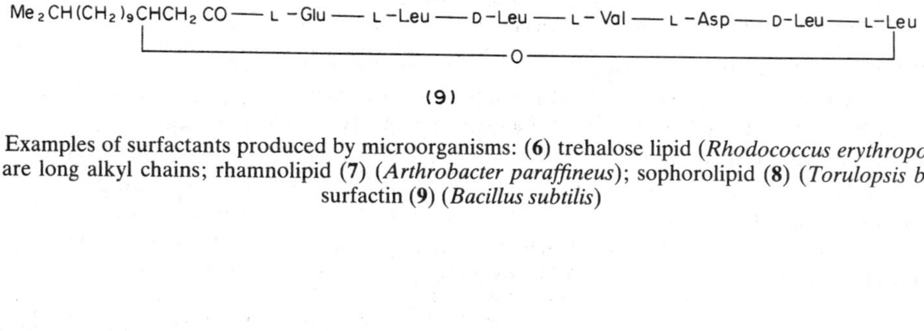

24.7 SUMMARY

The major lipid types, including the various fatty acids, of bacteria, yeasts and fungi are enumerated together with brief indications of the metabolic relevance of the various materials. The patterns of lipid accumulation are described for eukaryotic microorganisms growing in batch- and continuous culture. Information is provided concerning how the amount of lipid and its composition in an organism varies with its environment and growth conditions.

The biosyntheses of fatty acids, saturated, unsaturated, branched chain and hydroxylated, are described and include mention of how oleaginous microorganisms are able to accumulate high amounts of lipid. There is brief coverage of the synthesis of triacylglycerols, phospholipids and waxes.

The final section covers the utilization of alkanes and deals primarily with the mechanisms of oxidation and the ensuing pathways of metabolism. The various products which can arise from alkanes are enumerated.

24.8 REFERENCES

Babij, T., F. J. Moss and B. J. Ralph (1969). Effects of oxygen and glucose levels on lipid composition of yeast *Candida utilis* grown in continous culture. *Biotechnol. Bioeng.*, **11**, 593–603.

Bemmann, W. and R. Tröger (1975). Zur Möglichkeit der Kohlenwasserstoffverwertung von Pilzen. *Zbl. Bakt. Hyg. Abt. II*, **129**, 742–752.

Bloch, K. and D. Vance (1977). Control mechanisms in the synthesis of saturated fatty acids. *Annu. Rev. Biochem.*, **46**, 263–298.

Boulton, C. A. and C. Ratledge (1981). Correlation of lipid accumulation in yeasts with possession of ATP:citrate lyase. *J. Gen. Microbiol.*, **127**, 169–176.

Boulton, C. A. and C. Ratledge (1984). The physiology of hydrocarbon-utilizing microorganisms. *Top. Enz. Ferment. Biotechnol.* **9**, 11–77.

Brennan, P. J., P. F. S. Griffin, D. M. Lösel and D. Tyrrell (1974). The lipids of fungi. *Prog. Chem. Fats Lipids*, **14**, 49–89.

Brennan, P. J. and D. Lösel (1978). Physiology of fungal lipids: selected topics. *Adv. Microb. Physiol.*, **17**, 47–179.

Brown, C. M. and A. H. Rose (1969). Fatty acid composition of *Candida utilis* as affected by growth temperature and dissolved-oxygen tension. *J. Bacteriol.*, **99**, 371–378.

Brown, C. M. and B. Johnson (1970). Influence of the concentration of glucose and galactose on the physiology of *Saccharomyces cerevisiae* in continuous culture. *J. Gen. Microbiol.*, **64**, 279–287.

Byrom, D. and J. C. Ousby (1975). In *Proceedings of the International Symposium on Microbial Growth on C_1 Compounds*, ed. G. Terui, pp. 23–28. Society of Fermentation Technology, Tokyo.

Byrom, D. (1981). In *Microbial Growth on C_1 Compounds*, ed. H. Dalton, pp. 278–284. Heyden, London.

Calcott, P. H. and R. S. Petty (1980). Phenotypic variability of lipids of *Escherichia coli* grown in chemostat culture. *FEMS Microbiol. Lett.*, **7**, 23–27.

Cooper, D. G. and J. F. Zajic (1980). Surface-active compounds from microorganisms. *Adv. Appl. Microbiol.*, **26**, 229–251.

Daum, G., H. Glatz and F. Paltauf (1977). Lipid metabolism in an inositol-deficient yeast, *Saccharomyces cerevisiae*. The influence of temperature and anaerobiosis on the cellular lipid composition. *Biochim. Biophys. Acta*, **488**, 484–492.

Dawes, E. A. and P. J. Senior (1973). The role and regulation of energy reserve polymers in microorganisms. *Adv. Microb. Physiol.*, **10**, 136–266.

Einsele, A., A. Fiechter and H. P. Knoepfel (1972). Respiratory activity of *Candida tropicalis* during growth on hexadecane and on glucose. *Arch. Mikrobiol.*, **82**, 247–253.

Erwin, J. D. (1973). *Lipids and Biomembranes of Eukaryotic Microorganisms*. Academic, London.

Finnerty, W. R. and P. A. Makula (1975). Microbial lipid metabolism. *Crit. Rev. Microbiol.*, **4**, 1–40.

Forney, F. W. and A. J. Markovetz (1970). Subterminal oxidation of aliphatic hydrocarbons. *J. Bacteriol.*, **102**, 281–282.

Fukui, S. and A. Tanaka (1981). Metabolism of alkanes by yeasts. *Adv. Biochem. Eng.*, **19**, 217–237.

Fulco, A. J. (1974). Metabolic alterations of fatty acids. *Annu. Rev. Biochem.*, **43**, 215–241.

Gilbert, P. and M. R. W. Brown (1978). Influence of growth rate and nutrient limitation on the gross cellular composition of *Pseudomonas aeruginosa* and its resistance to 3- and 4-chlorophenol. *J. Bacteriol.*, **133**, 1066–1072.

Gill, C. O. (1975). Effect of growth temperature on the lipids of *Pseudomonas fluorescens*. *J. Gen. Microbiol.*, **89**, 293–298.

Gill, C. O., M. J. Hall and C. Ratledge (1977). Lipid accumulation in an oleaginous yeast, *Candida* 107, growing on glucose in a single-stage continuous culture. *Appl. Environ. Microbiol.*, **33**, 231–239.

Goldfine, H. (1972). Comparative aspects of bacterial lipids. *Adv. Microb. Physiol.*, **8**, 1–58.

Goma, G., A. Pareilleux and G. Durand (1973). Specific hydrocarbon solubilization during growth of *Candida lipolytica*. *Hakko Kogaku Zasshi*, **51**, 616–618.

Goodwin, T. W. (1980). *The Biochemistry of the Carotenoids*, 2nd edn., vol. 1. Chapman and Hall, London.

Günter, T., L. Richter and J. Schmalbeck (1975). Phospholipids of *Escherichia coli* in magnesium deficiency. *J. Gen. Microbiol.*, **86**, 191–193.

Gurr, M. I. and A. T. James (1975). *Lipid Biochemistry: An Introduction*. Chapman and Hall, London.

Haley, J. E. and R. C. Jack (1974). Stereospecific analysis of triacylglycerols and major phosphoglycerides of *Lipomyces lipoferus*. *Lipids*, **9**, 679–681.

Hall, M. J. and C. Ratledge (1977). Lipid accumulation in an oleaginous yeast (*Candida* 107) growing on glucose under various conditions in a one- and two-stage continuous culture. *Appl. Environ. Microbiol.*, **33**, 577–584.

Heinrich, M. R. (ed.) (1976). *Extreme Environments: Mechanisms of Microbial Adaptation*. Academic, New York.

Hunter, K. and A. H. Rose (1971). In *The Yeasts*, ed. A. H. Rose and J. S. Harrison, vol. 2, pp. 211–270. Academic, London.

Hunter, K. and A. H. Rose (1972). Lipid composition of *Saccharomyces cerevisiae* as influenced by growth temperature. *Biochim. Biophys. Acta*, **260**, 639–653.

Ikemoto, S., K. Katoh and K. Komagata (1978). Cellular fatty acid composition in methanol-utilizing bacteria. *J. Gen. Appl. Microbiol.*, **24**, 41–49.

Jigami, Y., O. Suzuki and S. Nakasato (1979). Comparisons of lipid composition of *Candida guilliermondii* grown on glucose, ethanol and methanol as the sole carbon source. *Lipids*, **14**, 937–942.

Johnson, B., S. J. Nelson and C. M. Brown (1972). Influence of glucose concentration on the physiology and lipid composition of some yeasts. *Antonie van Leeuwenhoek*, **38**, 129–136.

Käppeli, O. and A. Fiechter (1976). The mode of interaction between the substrate and cell surface of the hydrocarbon-utilizing yeast *Candida tropicalis*. *Biotechnol. Bioeng.*, **18**, 967–974.

Käppeli, O. and A. Fiechter (1981). Properties of hexadecane uptake by *Candida tropicalis*. *Curr. Microbiol.*, **6**, 21–26.

Klug, M. J. and A. J. Markovetz (1971). Utilization of aliphatic hydrocarbons by micro-organisms. *Adv. Microb. Physiol.*, **5**, 1–43.

Kohlaw, G. B. and A. Tan-Wilson (1977). Carnitine acetyltransferase: candidate for the transfer of acetyl groups through the mitochondrial membrane of yeast. *J. Bacteriol.*, **129**, 1159–1161.

Kushner, D. J. (ed.) (1978). *Microbial Life in Extreme Environments*. Academic, London.

Lebeault, J. M. and E. Azoulay (1971). Metabolism of alkanes by yeast. *Lipids*, **6**, 444–447.

Lechevalier, M. P. (1977). Lipids in bacterial taxonomy – a taxonomist's view. *CRC Crit. Rev. Microbiol.*, **5**, 109–210.

Lechevalier, M. P. (1982) In *CRC Handbook of Microbiology*, ed. A. I. Laskin and H. A. Lechevalier, 2nd edn., vol. 4, pp. 435–564. CRC Press, Boca Raton, FL.

Lindley, N. D. and M. T. Heydeman (1983). Uptake of vapour phase [^{14}C] dodecane by whole mycelia of *Cladosporium resinae*. *J. Gen. Microbiol.*, **129**, 2301–2305.

Lynen, F. (1980). On the structure of fatty acid synthetase of yeast. *Eur. J. Biochem.*, **112**, 431–442.

McMurrough, I. and A. H. Rose (1973). Effects of temperature variation on fatty acid composition of a psychrophilic *Candida* species. *J. Bacteriol.*, **114**, 451–452.

Meisel, M. N., G. A. Mednedova and T. M. Kozlova (1976). Cytological mechanisms of the assimilation of *n*-alkanes by yeasts. *Mikrobiologiya*, **45**, 844–851.

Minnikin, D. E., H. Abdolrahimzadeh and J. Baddiley (1974). Replacement of acidic phospholipids with acidic glycolipids in *Pseudomonas diminuta*. *Nature (London)*, **249**, 268–269.

Minnikin, D. and H. Abdolrahimzadeh (1974). Effect of pH on the proportions of polar lipids, in chemostat cultures of *Bacillus subtilis*. *J. Bacteriol.*, **120**, 999–1003.

Minnikin, D. (1982). In *The Biology of the Mycobacteria*, ed. C. Ratledge and J. L. Stanford, vol. 1, pp. 94–184. Academic, London.

Mishina, M., T. Kamiryo, S. Tashiro and S. Numa (1978). Separation and characterization of two long-chain acyl-CoA synthetases from *Candida lipolytica*. *Eur. J. Biochem.*, **82**, 347–354.

Morris, L. J. and S. W. Hall (1966). The structure of the glycerides of ergot oils. *Lipids*, **1**, 188–196.

Nishikawa, Y., I. Nakamura, T. Kamihara and S. Fukui (1977). Effects of thiamine and pyridoxine on the lipid composition of *Saccharomyces cerevisiae* 4228. *Biochem. Biophys. Acta*, **486**, 483–489.

Okuyama, H. and M. Saito (1979). Regulation by temperature of the chain length of fatty acids in yeast. *J. Biol. Chem.*, **254**, 12281–12284.

Ramsay, A. M. and J. L. Douglas (1979). Effect of phosphate limitation of growth on the cell-wall and lipid composition of *Saccharomyces cerevisiae*. *J. Gen. Microbiol.*, **110**, 185–191.

Ratledge, C. (1977). In *Developments in Biodegradation of Hydrocarbons*, ed. R. J. Watkinson, pp. 1–46. Applied Science, London.

Ratledge, C. (1980). In *Hydrocarbons in Biotechnology*, ed. D. E. F. Harrison, I. J. Higgins and R. J. Watkinson, pp. 133–153. Heyden, London.

Ratledge, C. (1982a). *Prog. Ind. Microbiol.*, **16**, 119–206.

Ratledge, C. (1982b). In *The Biology of the Mycobacteria*, ed. C. Ratledge and J. L. Stanford, vol. 1, pp. 53–93. Academic, London.

Ratledge, C. and M. J. Hall (1979). Accumulation of lipid by *Rhodotorula glutinis* in continous culture. *Biotechnol. Lett.*, **1**, 115–120.

Rattray, J. B. M., A. Schibeci and D. K. Kidby (1975). Lipids of yeasts. *Bacteriol. Rev.*, **39**, 197–231.

Rattray, J. B. M. and J. E. Hambelton (1980). The lipid components of *Candida boidinii* and *Hansenula polymorpha* grown on methanol. *Can. J. Microbiol.*, **26**, 190–195.

Razin, S. and S. Rottern (eds.) (1982). Membrane lipids of prokaryotes. *Current Topics in Membranes and Transport*, vol. 17. Academic, New York.

Rehm, H. J. and I. Reiff (1981). Mechanisms and occurrence of microbial oxidation of long chain alkanes. *Adv. Biochem. Eng.*, **19**, 175–215.

Shennan, J. L. and J. D. Levi (1974). The growth of yeasts on hydrocarbons. *Prog. Ind. Microbiol.*, **13**, 1–57.

Shilo, M. (ed.) (1979). *Strategies of Microbiol Life in Extreme Environments*. Verlag Chemie, Weinheim.

Spencer, J. F. T., D. M. Spencer and A. P. Tulloch (1979). In *Economic Microbiology*, ed. A. H. Rose, vol. 3, pp. 523–540. Academic, London.

Sumper, M. (1974). Control of fatty acid biosynthesis by long-chain acyl-CoA esters and by lipid membranes. *Eur. J. Biochem.*, **49**, 469–475.

Thiele, O. W. (1979). *Lipide, Isoprenoide mit Steroiden*. Georg Thieme Verlag, Stuttgart.

Volpe, J. J. and P. R. Vagelos (1976). Biosynthesis of fatty acids. *Physiol. Rev.*, **56**, 339–417.

Waiblinger, K. and D. Gröger (1972). On the production of ergoline alkaloids and fatty acids composition of the mycelium in submerged cultures of various *Claviceps* cultures. *Biochem. Physiol. Pflanzen*, **63**, 468–476.

Wassef, M. K. (1977). Fungal lipids. *Adv. Lipid Res.*, **15**, 159–232.

Weete, J. D. (1980). *Lipid Biochemistry of Fungi and Other Organisms*. Plenum, New York.

Weitzman, P. D. J. and M. J. Danson (1976). Citrate synthase. *Curr. Top. Cellul. Regulation*, **10**, 161–205.

25

Microbial Metabolism of Aromatic Compounds

S. DAGLEY
University of Minnesota, St. Paul, MN, USA

25.1 INTRODUCTION

25.1.1 Fission of the Benzene Nucleus

Next to glucosyl residues, the benzene ring is the most widely distributed unit of chemical structure in the biosphere, and to a large extent the continuous operation of the carbon cycle depends upon its rapid fission by microorganisms (Dagley, 1977, 1978). The physiological significance of each one of the vast array of aromatic compounds synthesized by plants is by no means clear, but lignin, a plant biopolymer more abundant than protein, has certainly played an essential part in evolution by serving as a structural material that allows plants to stand erect and compete effectively for solar energy. The ability of animals to degrade aromatic compounds is extremely restricted, being mainly confined to the amino acids phenylalanine, tyrosine and tryptophan, whereas it may be safely assumed that somewhere on Earth various microorganisms exist which, perhaps working with others in a consortium, are capable of degrading any given naturally occurring aromatic compound when conditions are favorable to growth. For this reason, the contribution of microorganisms to the functioning of the carbon cycle and the maintenance of life cannot be exaggerated.

The relative metabolic inertness of the benzene nucleus is due to its stable resonance structure. Microorganisms are exceptional in their ability to invest energy in reactions that reduce the resonance energy barrier presented by the nucleus and permit fission to occur; this investment is repaid by subsequent reactions that release energy. When conditions are anaerobic, resonance due to conjugation of unsaturated bonds is removed by reduction, whereas aerobic microorganisms make direct use of oxygen gas (dioxygen) to hydroxylate the benzene ring and so facilitate its fission by further reaction with dioxygen.

Three types of reaction sequences are available to anaerobes; they were reviewed by Evans (1977) and are illustrated for benzoic acid (1) in Figure 1. Each one of the three sequences is initiated by reduction of benzoic acid to cyclohexanecarboxylic acid (2; Figure 1). Reactions of pathway (a) in Figure 1 are catalyzed by pure cultures of certain soil pseudomonads: hydrogen atoms removed from metabolites are transferred anaerobically to nitrate, and generation of the ATP required to sustain growth is coupled to electron flow through a cytochrome chain. The photosynthetic organism *Pseudomonas palustris* utilizes benzoic acid anaerobically as sole source of carbon by the second pathway (b; Figure 1). The reactions are essentially those of β-oxidation of fatty acids. Dehydrogenation and addition of water followed by a second dehydrogenation occur as in sequence (a), but the substrates are the corresponding coenzyme A esters. Accordingly, the cyclohexanone ring is opened by thiolysis instead of hydrolysis, and another difference between the proposed pathways is the decarboxylation step (3 → 4; Figure 1) which is confined to pathway (a). Benzoic acid is also converted anaerobically into methane by consortia of microorganisms by a third pathway (c; Figure 1). Compounds (2), (3), (4) and (5) appear to be formed as intermediates, as they are in the case of organisms using nitrate as terminal electron acceptor; hydrolysis and dehydrogenation then give rise to free fatty acids (pathway c). In this sequence, however, it is necessary for a certain proportion of these products to serve as sources of acetate for accepting hydrogen using reactions catalyzed by methanogens in the consortium.

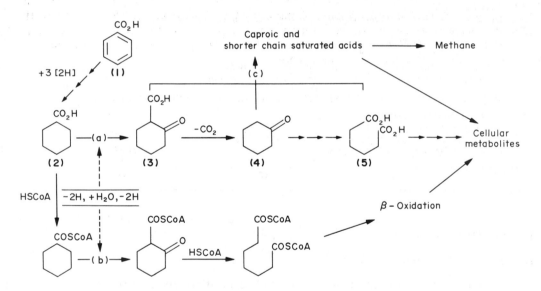

Figure 1 Three reaction sequences proposed for bacteria growing anaerobically at the expense of benzoic acid. In each case the benzene nucleus of benzoic acid (1) is first reduced to a cyclohexane structure (2). Sequence (a) is used by certain denitrifiers that transfer hydrogen atoms to nitrate ions; sequence (b) by the photosynthesizer *Pseudomonas palustris* which obtains cellular carbon by a process resembling β-oxidation of fatty acids; and sequence (c) by anaerobes that form short chain fatty acids which, by accepting transferred hydrogen atoms, are converted into methane by methanogenic bacteria present in the consortium

Anaerobic processes make an important contribution to the carbon cycle. Thus, Howarth and Teal (1979) showed that sulfate-mediated respiration of organic matter in a salt marsh was twelve times higher than that mediated by oxygen, while Strayer and Tiedje (1978) estimated that one-third of the total primary productivity of a freshwater lake in Michigan could be recovered as methane during summer. Furthermore, in addition to benzoic acid and phenol, the ligno-aromatic compounds ferulic acid, vanillin, cinnamic acid, protocatechuic acid and catechol can be converted anaerobically into methane by microbial consortia (Healy and Young, 1978, 1979). However, despite its importance, progress towards understanding biochemical details of anaerobic aromatic catabolism has been slow, due to the experimental difficulties attending the use of consortia of fastidious organisms that are extremely sensitive to oxygen. It may also be noted that lignin itself, as distinct from the above mentioned degradation products of small molecular weight, resists anaerobic attack, as do aromatic hydrocarbons which are a focus of much current interest. Accordingly, the remainder of this review will be concerned with aerobic aromatic metabolism.

Aerobic fission of the benzene ring is summarized in Figure 2 where the main points of cleavage are shown by the dotted lines. Features of interest are as follows. (i) When the ring opens, O_2 is 'fixed' in the reaction product. This is seen for two reactions of compound (3; Figure 2); mode of fission (a) is often described as meta cleavage and (b) as ortho cleavage, catalyzed by the enzymes protocatechuate 4,5-dioxygenase and protocatechuate 3,4-dioxygenase respectively. A third site of attack for compound (3) also falls into the category of meta fission; the enzyme, protocatechuate 2,3-dioxygenase, is found in certain bacilli (Crawford, 1975). (ii) There is no 'all purpose' dioxygenase. Even when one microorganism can open the benzene ring in several of the ways shown in Figure 2, a separate and distinct dioxygenase is elaborated for each occasion. (iii) Dioxygenases are nonheme iron proteins. The ortho cleaving enzyme protocatechuate 3,4-dioxygenase contains tightly-bound iron(III), is deep red in color, and the valency of the iron does not alter during the catalytic cycle. Protocatechuate 4,5-dioxygenase is colorless and is activated by iron(II), although this event is accompanied by a valence change (Wood, 1980). (iv) The presence of at least two hydroxyl groups suitably placed usually constitutes a prerequisite for ring fission; often these groups are situated on adjacent carbons, but they may also be placed *para* to each other, as in gentisic and homogentisic acids, compounds (4) and (6) of Figure 2. Exceptions to this requirement for two hydroxyl groups have been reported but they are very uncommon. Crawford *et al.* (1979) isolated a *Bacillus* strain that utilized 5-chloro-2-hydroxybenzoate as source of carbon: the benzene ring was cleaved by a dioxygenase in a similar manner to that of gentisic acid, with chlorine taking the place of the hydroxyl at C-5 of gentisate. Kiyohara and Nadao (1977) reported that a 1-hydroxy 2-naphthoate 1,2-dioxygenase cleaved one of the benzene rings of a naphthalene structure bearing one carboxyl and one hydroxyl substituent. Que (1979) showed that 2-aminophenol was slowly attacked by catechol 2,3-dioxygenase but the physiological significance of the reaction is not clear.

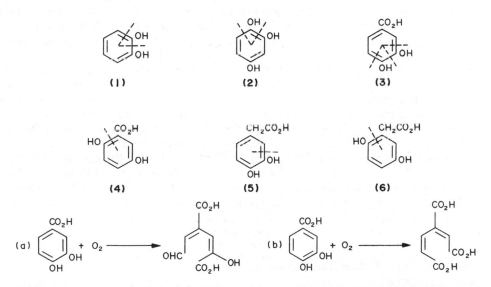

Figure 2 Fissions of the benzene nucleus catalyzed by dioxygenases. The main substrates for these enzymes are: catechol (1); hydroxyquinol (2); protocatechuic acid (3); gentisic acid (4); homoprotocatechuic acid (5); homogentisic acid (6). Fission of a benzene ring at each point indicated by a broken line has been described; separate dioxygenases are used. Oxygen is incorporated into each substrate, as shown for reactions catalyzed by: (a) protocatechuate 4,5-dioxygenase, (b) protocatechuate 3,4-dioxygenase

25.1.2 Preparation of the Nucleus for Aerobic Fission

In general, therefore, before it can be opened the benzene ring must carry two hydroxyl groups that are suitably oriented, and these must first be introduced if they are not already in place. This reaction, catalyzed by a flavoprotein monooxygenase (Dagley, 1975) involves a four electron reduction of dioxygen in which one atom of oxygen is added to the aromatic substrate and the second atom is reduced to water at the expense of a reduced pyridine nucleotide (Ballou, 1982):

$$\text{aryl—H} + O_2 + \text{NAD(P)H} + H^+ \rightarrow \text{aryl—OH} + \text{NAD(P)}^+ + H_2O$$

These enzymes are remarkably specific for their natural substrates: thus, 4-hydroxybenzoate 3-hydroxylase is a different enzyme from either 3-hydroxybenzoate 4-hydroxylase or 4-hydroxyphenylacetate 3-hydroxylase; however, they are not absolutely substrate specific, and chemically synthesized substrate analogs have been useful in kinetic studies.

Four examples to illustrate the diversity of flavoprotein aromatic monooxygenases are given in Figure 3. (1) 4-Hydroxybenzoate 3-hydroxylase has been studied intensively both kinetically and structurally, so that the amino acid sequence and many features of its X-ray crystal structure are known (Wierenga *et al.*, 1982). The enzyme is a dimer. Each monomer contains 394 amino acid residues and binds one FAD. Sites for binding NADH and 4-hydroxybenzoate (which causes a conformational change when bound) have been examined and a channel through which O_2 reaches the active site has been revealed. (2) 4-Hydroxyphenylacetate 1-hydroxylase catalyzes hydroxylation of the substrate with a simultaneous shift of the side chain to form homogentisic acid; this is one of the ring fission substrates central to aromatic catabolism (**6**; Figure 2). Hareland *et al.* (1975) have proposed a feasible mechanism for this unusual reaction. It may be noted that another enzyme is known that hydroxylates the same compound at C-3 of the benzene ring to give a different ring fission substrate, namely, homoprotocatechuic acid (Sparnins *et al.*, 1974), compound (**5**; Figure 2). (3) Salicylate hydroxylase, which has been purified (White-Stevens and Kamin, 1972) and is still receiving attention (Suzuki and Katagiri, 1982; Wang *et al.*, 1982), catalyzes formation of catechol from salicylate with simultaneous loss of carbon dioxide. A bacterial enzyme has been described (Kobayashi and Hayaishi, 1970) which oxidizes anthranilate (2-aminobenzoate) to catechol and ammonia, but the resemblance to salicylate hydroxylase is only superficial since the former enzyme consists of two proteins and incorporates both atoms of $^{18}O_2$ into catechol (Kobayashi *et al.*, 1964). (4) This equation describes the reaction of a different anthranilate hydroxylase, forming 2,3-dihydroxybenzoate instead of catechol. It is found in fungi (Subba Rao *et al.*, 1967) and yeast (Anderson and Dagley, 1981) rather than bacteria, and it is of interest insofar as the hydroxyl group at C-2 of 2,3-dihydroxybenzoate is derived from water and that at C-3 is derived from O_2 (Powlowski and Dagley, 1982). It is, therefore, a monooxygenase even though two atoms of oxygen are incorporated into the product. On the other hand, the enzyme system catalyzing reaction (5) of Figure 3 is classified as dioxygenase because both oxygens originate in O_2. Axcell and Geary (1975) showed that the system consists of a flavoprotein and two nonheme iron proteins: a similar combination operates for the hydroxylation of toluene, and the flavoprotein component has been purified by Subramanian *et al.* (1981). The outstanding investigations of Gibson and his group concerning the dihydrodiols formed from benzene, toluene, biphenyl, naphthalene and other hydrocarbons have encompassed determinations of absolute configuration where molecular asymmetry is involved, as is the case for toluene; references are given by Dagley (1978) to some of these studies, and to those of Reiner and Hegeman (1971) concerning a similar system that oxidizes benzoic acid. The dihydrodiols formed in these reactions are converted by NAD-linked dehydrogenases into catechols that serve as substrates for ring fission dioxygenases.

25.1.3 Reactions which Follow Ring Fission

Initial attack by a monooxygenase upon a benzene nucleus uses NADH that could otherwise be used to generate ATP. By contrast, ring fission reactions themselves neither consume nor generate energy. In this respect they resemble those reactions that follow, which simply involve breaking up the original carbon chain to form, eventually, metabolites of the Krebs cycle; only NAD-dependent dehydrogenases for the aldehydes in these sequences are potential sources of energy. It may be suggested that microorganisms used hydratases, hydrolases and aldolases early in evolution and that these enzymes, therefore, were available for metabolizing ring fission products when dioxygenases made their appearance at a time of increasing concentration of atmospheric dioxygen. Types of reactions that break carbon–carbon bonds after the benzene ring has opened are shown in Figure 4. Reaction (1) is catalyzed by an aldolase acting upon a substrate having hydroxyl and keto substituents in the 1,3 position: this is also a structural feature of fructose 1,6-bisphosphate, the substrate of the aldolase used in glycolysis and the most thoroughly investigated enzyme in this category. Reaction (2) shows hydrolysis of a 1,3-diketone, and reac-

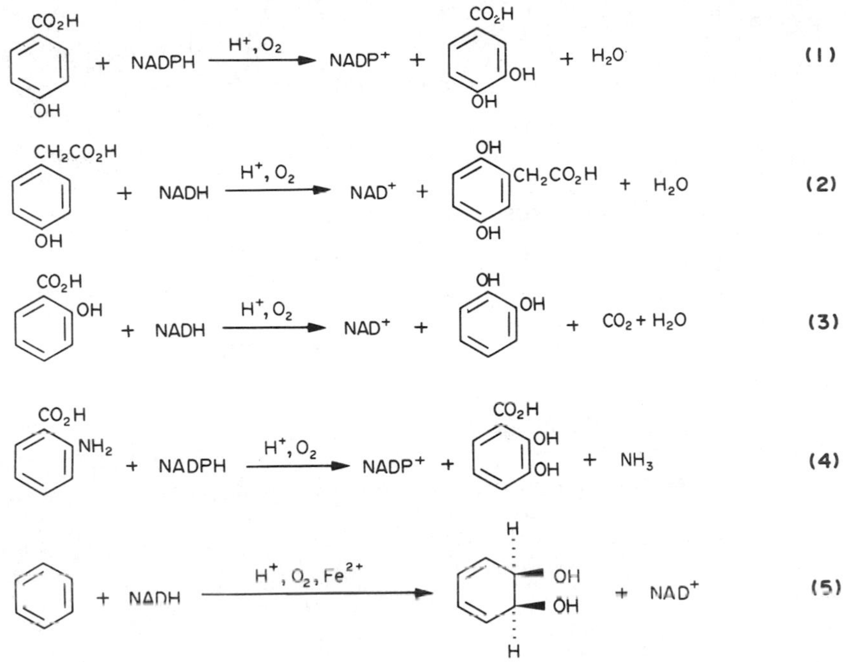

Figure 3 Examples of hydroxylations catalyzed by flavoproteins: (1) 4-hydroxybenzoate 3-hydroxylase (or 4-hydroxybenzoate 3-monooxygenase); (2) 4-hydroxyphenylacetate 1-hydroxylase; (3) salicylate hydroxylase; (4) anthranilate hydroxylase (deaminating) from eukaryotes; (5) benzene 1,2-dioxygenase

tion (3) is familiar as that catalyzed by the enzyme thiolase of β-oxidation of fatty acids; thioesterification confers carbonyl properties upon the carboxyl carbon of the original acid. Reaction (4) shows decarboxylation of a β-keto acid.

The substrates formed by fission of the benzene nucleus possess two ethylenic bonds in conjugation, and they must be prepared for degradation by reactions that introduce the structural features mentioned. Hydration, as in Figure 4, equation (5), gives the substrate of equation (1). At one stage, following ortho fission of the benzene ring of a catechol, a lactone is formed; upon hydrolysis (equation 6) this gives a β-keto acid, the thioester of which can be cleaved by coenzyme A as in equation (3). Equation (7) shows how isomerization of 2-hydroxy-*cis,cis*-muconic acid gives a substrate that decarboxylates when shifts of electrons occur, similar to those of equation (4): a vinyl analog of a β-keto acid is formed by this isomerization. Equation (8) accounts for the unusual reaction by which formate or acetate is released by hydrolysis from the meta fission products of catechol or 3-methylcatechol; isomerization gives a suitable substrate for a hydrolase.

25.2 PATHWAYS OF DEGRADATION

25.2.1 Meta Fission Pathways

These catabolic sequences are summarized in Figure 5. Enzyme a, catechol 2,3-dioxygenase, oxidizes catechol (1) to 2-hydroxymuconic semialdehyde which can be metabolized by two routes: hydrolytic or oxidative. As shown in Figure 4, reaction (8), hydrolysis releases formate and gives 2-hydroxy-penta-2,4-dienoic acid (8, Figure 5). The oxidative route involves prior isomerization before decarboxylation can occur, giving the same intermediate (8). Proof of isomerization was furnished by Collinsworth *et al.* (1973) who showed that when 4-methylcatechol (2; Figure 5) is degraded by these enzymes, compound (9; Figure 5) is a metabolite. This is the *cis*-isomer of 2-hydroxyhexa-2,4-dienoic acid: the *trans*-isomer which is not metabolized would be expected had not isomerization removed the double bond between C-4 and C-5, thereby permitting rotation at C-5. This would not occur by generating a methylene group at C-3 through keto–enol tautomerism, as is still shown occasionally in some proposed schemes.

The hydrolytic route serves for the degradation of 3-methylcatechol (4; Figure 5), 2,3-dihy-

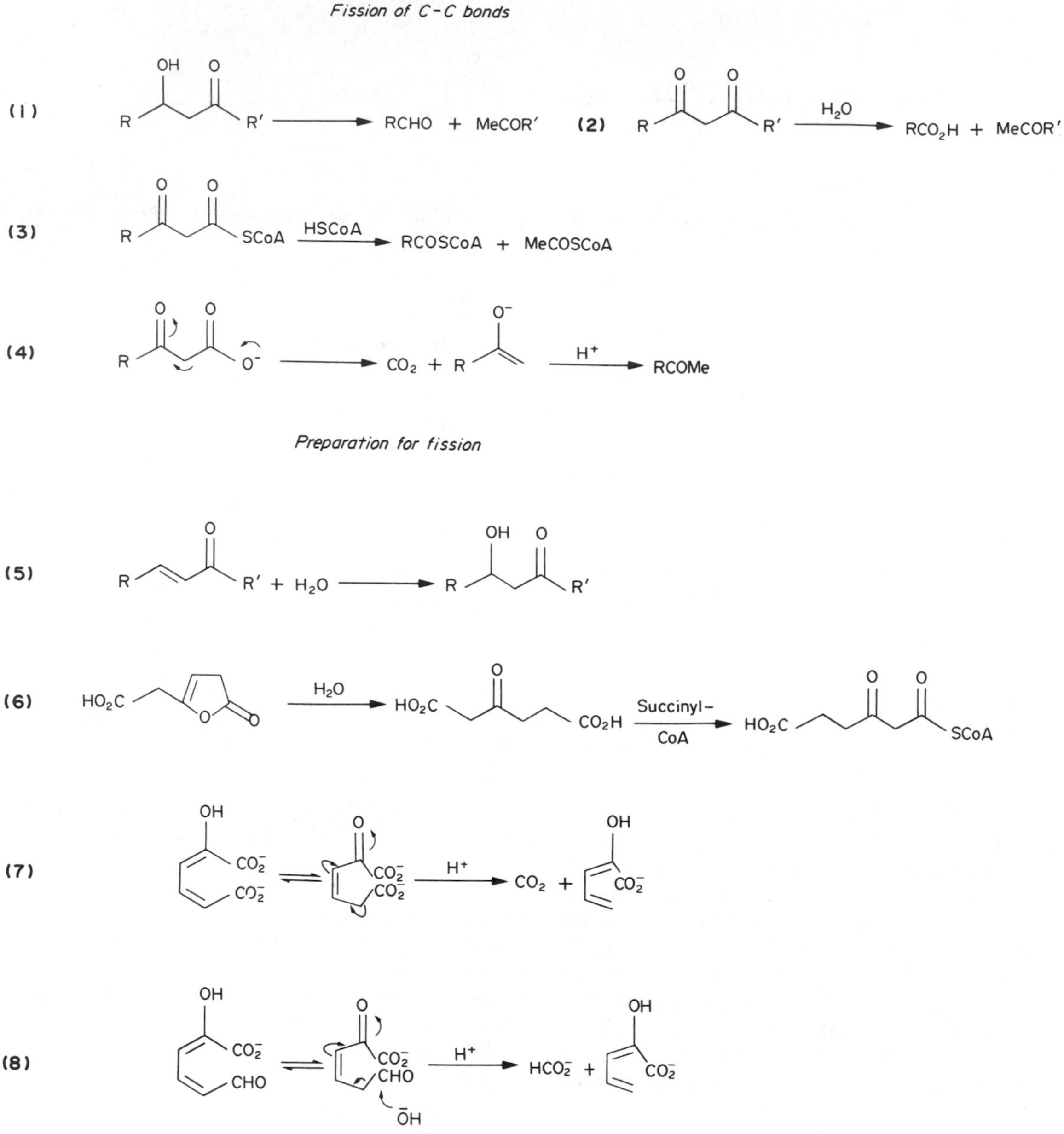

Fission of C–C bonds

(1) R...OH...O...R' ⟶ RCHO + MeCOR'   (2) R...O...O...R' —H₂O→ RCO₂H + MeCOR'

(3) R...O...O...SCoA —HSCoA→ RCOSCoA + MeCOSCoA

(4) R...O...O...O⁻ ⟶ CO₂ + R...O⁻ —H⁺→ RCOMe

Preparation for fission

(5) R...O...R' + H₂O ⟶ R...OH...O...R'

(6) HO₂C...lactone —H₂O→ HO₂C...O...CO₂H —Succinyl–CoA→ HO₂C...O...O...SCoA

(7) OH...CO₂⁻...CO₂⁻ ⇌ O...CO₂⁻...CO₂⁻ —H⁺→ CO₂ + OH...CO₂⁻

(8) OH...CO₂⁻...CHO ⇌ O...CO₂⁻...CHO...ŌH —H⁺→ HCO₂⁻ + OH...CO₂⁻

Figure 4 Types of reactions involved in breaking carbon–carbon bonds of substrates encountered in aromatic catabolism: (1) aldol fission; (2) hydrolysis of a 1,3-diketone; (3) thiolysis of the coenzyme A ester of a β-keto acid; (4) decarboxylation of a β-keto acid. After the benzene ring has been opened, substrates for the foregoing reactions are prepared by: (5) hydration of an ethylenic bond; (6) hydrolysis of a lactone to give a β-keto acid, followed by esterification with coenzyme A; (7) or (8) release of either CO_2 or formate from a benzene ring-fission product

droxy-β-phenylpropionate (**5**; Figure 5) and 2,3-dihydroxy-*p*-cumate (**6**; Figure 5). The latter is a degradation product of *Pseudomonas putida* when utilizing the hydrocarbon *p*-cymene (**7**; Figure 5) as source of carbon (DeFrank and Ribbons, 1977). Conversion of (**7**) to (**8**) entails several steps whereby methyl is oxidized to carboxyl and two hydroxyl groups are introduced by reactions that involve a *cis*-dihydrodiol (see Figure 3, reaction 5). A dioxygenase specific for 2,3-dihydroxybenzoates (enzyme d, Figure 5) opens the benzene ring, decarboxylation occurs and hydrolysis then liberates isobutyrate and forms compound (**8**) of the catechol meta fission sequence (Wigmore and Ribbons, 1980). 2,3-Dihydroxy-β-phenylpropionate (**5**; Figure 5) is a metabolite of cinnamic, hydrocinnamic and 2-hydroxyhydrocinnamic (melilotic) acids. After ring opening by a specific dioxygenase, enzyme c, succinate is liberated by hydrolysis. 3-Methylcatechol (**4**; Figure 5) is degraded by the same enzymes in *P. putida* as are used for catechol (**1**), although the structure of 3-methylcatechol precludes use of the oxidative pathway. Some bacilli

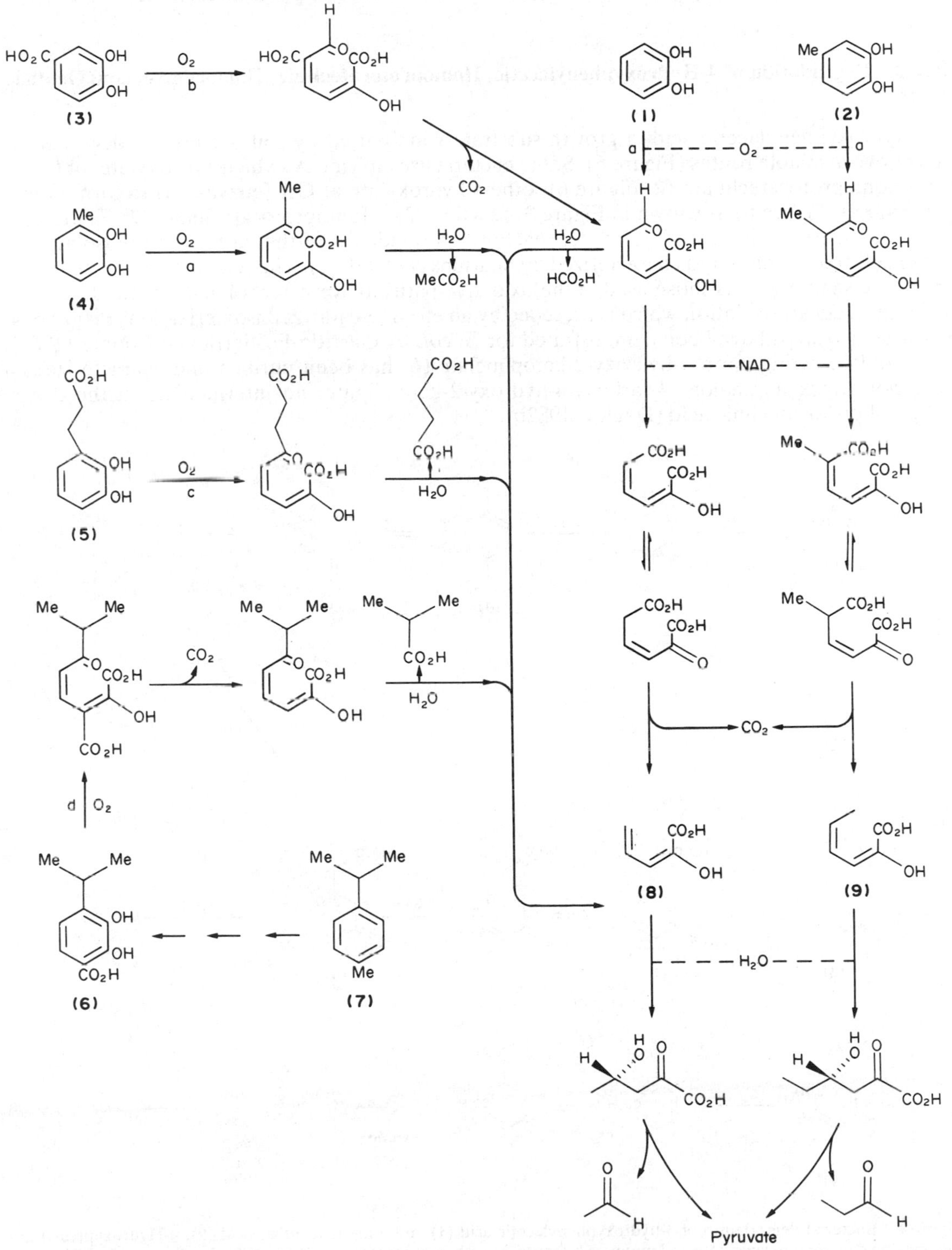

Figure 5 Meta fission pathways for degrading the following catecholic compounds: catechol (1); 4-methylcatechol (2); protocatechuic acid (3): this pathway is taken by certain bacilli, whereas that shown in Figure 7 is used by other bacteria; 3-methylcatechol (4); 2,3-dihydroxy-β-phenylpropionic acid (5); 2,3-dihydroxy-*p*-cumic acid (6), a metabolite of *p*-cymene (7). Three ring-fissions are catalyzed by the same enzyme, a, catechol 2,3-dioxygenase. Enzymes b, c and d are separate dioxygenases. Oxidative routes involving NAD are to the right, hydrolytic routes to the left

elaborate a 2,3-dioxygenase for protocatechuate (**3**; Figure 5) and are able to use catechol meta fission enzymes for its degradation (Crawford, 1975). The final reactions of these pathways involve stereospecific addition of water to compounds (**8**) and (**9**) giving the L(*S*)-hydroxyketo acids shown; aldol fission produces pyruvate plus acetaldehyde or propionaldehyde.

25.2.2 Degradation of 4-Hydroxyphenylacetic, Homoprotocatechuic, Homogentisic and Gentisic Acids

4-Hydroxyphenylacetic acid, a growth substrate widely used by soil bacteria, is degraded by one of two available routes (Figure 6). Some bacteria use enzyme A, which hydroxylates at C-3 to form homoprotocatechuate (**2**; Figure 6); others hydroxylate at C-1 (enzyme B) to form homogentisate (**3**; Figure 6) as shown in Figure 3, reaction (2). Homoprotocatechuate (**2**; Figure 6) is metabolized to succinate and pyruvate: reaction intermediates of the pathway were isolated by use of cell-free extracts and characterized by Sparnins *et al.* (1974). The reactions of this sequence are of the same types as those used in the oxidative pathway for catechol and 4-methylcatechol, including a decarboxylation which is preceded by an enzyme-catalyzed isomerization (Figure 6, **4**, **5**); these two steps have been demonstrated for *E. coli* by Garrido-Pertierra and Cooper (1981). The aldolase acting upon 4-hydroxy-2-ketopimelate (**6**) has been purified and characterized; it does not attack its analog, 4-carboxy-4-hydroxy-2-ketoadipate, an intermediate in the degradation of protocatechuic acid (Dagley, 1982b).

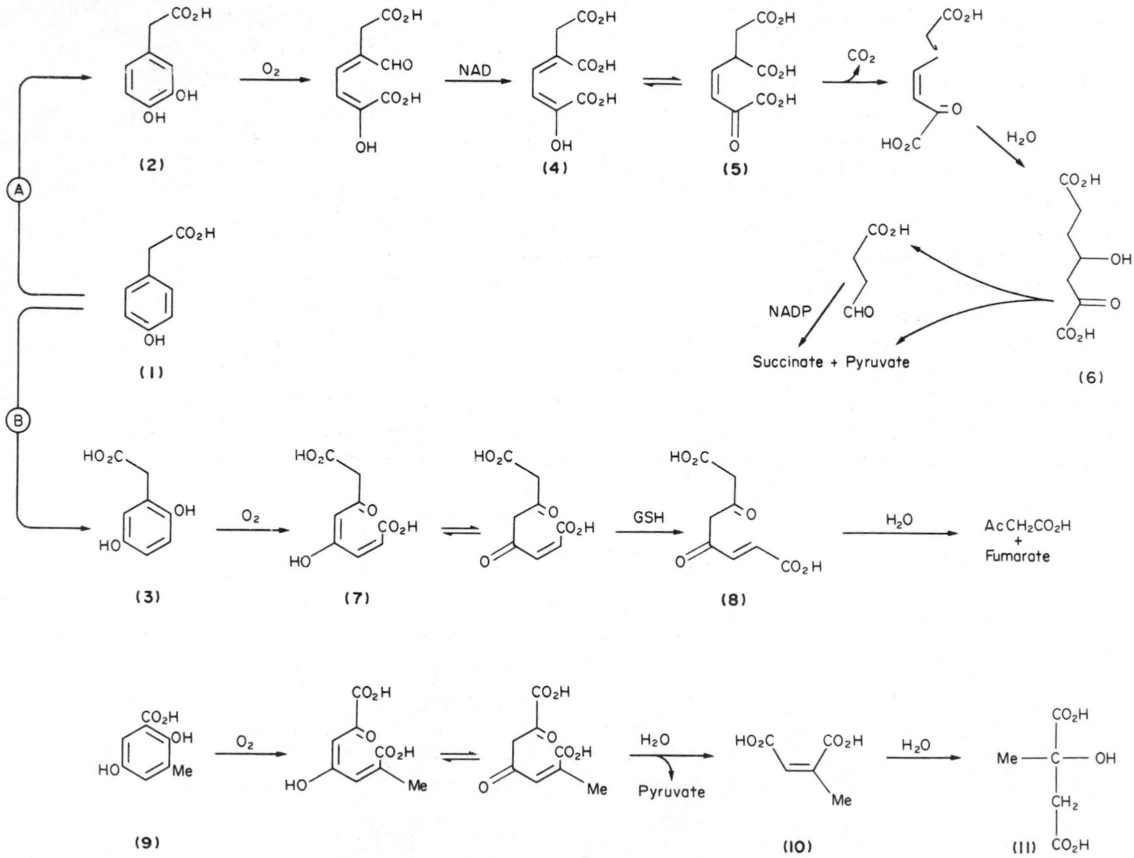

Figure 6 Bacterial degradation of 4-hydroxyphenylacetic acid (**1**) and 3-methylgentisic acid (**9**). 4-Hydroxyphenylacetate may be degraded to succinate and pyruvate by route A, with homoprotocatechuate (**2**) as ring-fission substrate; or by route B to acetoacetate and fumarate, involving homogentisate (**3**). 3-Methylgentisate is catabolized to pyruvate and D-citramalate (**11**)

Degradation of homogentisate (**3**; Figure 6) to fumarate and acetoacetate proceeds by reactions similar to those in mammals. Enzymic isomerization of maleylacetoacetate (**7**; Figure 6) to fumarylacetoacetate (**8**; Figure 6) is glutathione-dependent (ketonization of **7** is non-enzymic).

Some bacteria degrade gentisic acid by reactions similar to those used for homogentisate, while others lack the glutathione-dependent isomerization and so give rise, *via* maleic acid, to D-malate instead of the Krebs cycle intermediate L-malate; the other product formed is pyruvate. Such organisms possess an additional enzyme system for oxidizing D-malate (Hopper *et al.*, 1970). Some of them are able to degrade methyl-substituted gentisates when their lack of the isomerase, an enzyme that exhibits rather stringent substrate specificity, proves to be an advantage: maleyl-pyruvate hydrolase has broad specificity. This pathway is shown in Figure 6 for 3-methylgentisic acid (**9**), the product being D-citramalic acid (**11**). Compound (**11**) and other alkylmalic acids are degraded by ATP-coenzyme A-dependent aldolases (Hopper *et al.*, 1971). It may be noted that, following fission of the benzene nucleus, the carbon chain is broken according to reaction (2), Figure 4, namely by hydrolysis of a 1,3-diketone.

25.2.3 Protocatechuate 4,5-Dioxygenase

When non-fluorescent pseudomonads are grown with 4-hydroxybenzoate as carbon source the meta fission enzyme, protocatechuate 4,5-dioxygenase, is induced whereas these conditions result in the induction of protocatechuate 3,4-dioxygenase in fluorescent organisms (Stanier *et al.*, 1966). Until Maruyama (1979) identified the product as 2-oxopyran-4,6-dicarboxylic acid (**4**; Figure 7; also referred to as 2-pyrone-4,6-dicarboxylic acid) it was generally assumed that the aldehyde (**2**; Figure 7) formed by meta fission of protocatechuate is oxidized directly to the corresponding tricarboxylic acid (**7**; Figure 7). However, Kersten *et al.* (1982) suggested that the aldehyde undergoes oxidation in its hemiacetal form (**3**; Figure 7) and they found support for the degradative sequence of Figure 7 by purifying a hydrolase that is specific for 2-oxopyran-4,6-dicarboxylate and is strongly induced in non-fluorescent pseudomonads when they are grown with 4-hydroxybenzoate. Oxidation of a hemiacetal to a pyrone, followed by opening of the ring to give a hydroxy acid, is analogous to the well known pathway of glucose metabolism that involves oxidation of glucose 6-phosphate to 6-phosphogluconic acid lactone, with subsequent hydrolysis. The pyrone hydrolase of Figure 7 was absent from all the fluorescent pseudomonads examined by Kersten *et al.* (1982) with one exception, and in this particular organism the enzyme was induced by its substrate (**4**; Figure 7) and not by 4-hydroxybenzoate or protocatechuate: these compounds were degraded, in the usual manner for fluorescent pseudomonads, by ortho fission reactions which do not involve the hydrolase. After the pyrone (**4**; Figure 7) has been hydrolyzed by a non-fluorescent pseudomonad, one of the stereoisomers of the open chain tricarboxylic acid is then hydrated (Figure 7, reaction d) to give 4-carboxy-4-hydroxy-2-ketoadipate; this compound undergoes aldol fission (reaction e) yielding pyruvate and oxaloacetate. The aldolase, which has been purified and characterized (Tack *et al.*, 1972) does not attack its analog, a catabolite in homoprotocatechuate degradation (**6**; Figure 6). Protocatechuate 4,5-dioxygenase of non-fluorescents also attacks gallic acid (**8**) which is then degraded as shown in Figure 7.

25.2.4 Degradation of 3-*O*-Methylgallic Acid: Biological Formation of Methanol

The complex biopolymer, lignin, contains numerous benzenoid residues that carry methoxyl group substituents. Oxidative biodegradation of lignin by fungi releases methoxylated aromatic acids such as ferulic (3-methoxycinnamic), vanillic (4-hydroxy-3-methoxybenzoic) and syringic (3,5-dimethoxy-4-hydroxy-benzoic) acids. These compounds are good growth substrates for various soil bacteria that oxidize methoxyl to hydroxyl before opening the benzene nucleus. The demethylase systems contain both flavoprotein and iron–sulfur protein (Ribbons, 1970, 1971) and form an unstable intermediate that decomposes spontaneously with release of formaldehyde:

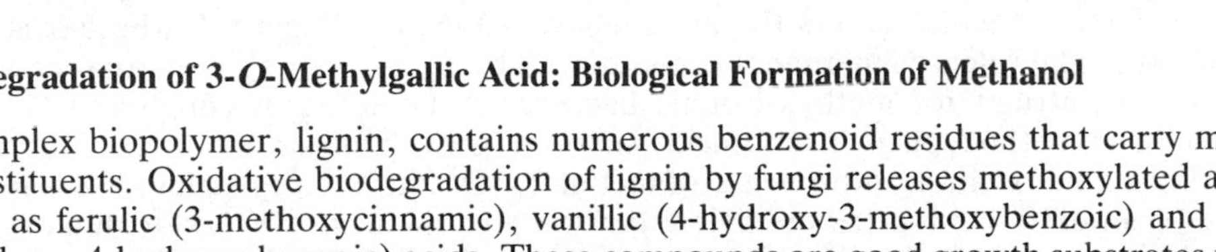

Strains of *Pseudomonas putida* capable of utilizing 3,4,5-trimethoxybenzoic and 3,4,5-trimethoxycinnamic acids for growth are readily selected from soil (Donnelly and Dagley, 1980, 1981a). In the case of 3,4,5-trimethoxycinnamic acid (**1**; Figure 8) the side chain substituent is

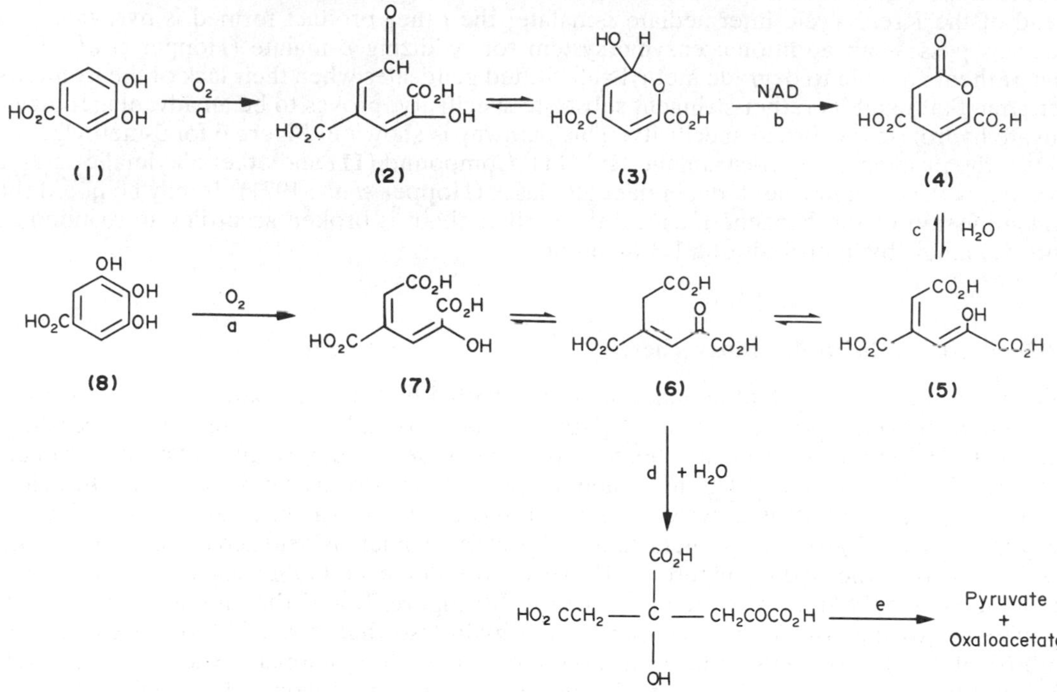

Figure 7 Degradation of protocatechuic (**1**) and gallic (**8**) acids by *Pseudomonas testosteroni*. Both substrates are attacked by the same enzyme, a, protocatechuate 4,5-dioxygenase. The ring-fission product (**2**) is oxidized to 2-oxopyran-4,6-dicarboxylate (**4**) by the NAD-dependent hydrogenase b with hemiacetal (**3**) as the suggested intermediate. Isomeric forms of open chain ring-fission products (**5**), (**6**) and (**7**) have not been individually characterized (Kersten *et al.*, 1982, with the permission of the American Society for Microbiology)

released, by reactions not yet fully documented, giving 3,4,5-trimethoxybenzoic acid (**2**; Figure 8). Methoxyl groups on two adjacent carbons of the benzene ring of this substrate are first oxidized to hydroxyls by the demethylase enzyme system mentioned, so that a ring-fission substrate, 3-*O*-methylgallate (**4**; Figure 8), now becomes available. It may be noted that CH_3 of any methoxyl group is bound to the benzene ring by an ether linkage resistant to hydrolysis; indeed, the routine chemical procedure for demethoxylation is treatment with hydrogen iodide. By the reaction sequence of Figure 8, however, the methoxyls at C-3 of compounds (**1**) and (**2**) become part of an ester grouping. Methanol is, therefore, readily released by hydrolysis, giving rise to a metabolite (**5**; Figure 8) which enters the meta fission pathway of Figure 7, where it is designated compound (**6**). The released methanol is not utilized by *P. putida*, and in natural habitats doubtless serves as nutrient for methylotrophic bacteria. A homolog of compound (**2**; Figure 8), namely 3,4,5-trimethoxyphenylacetic acid, supports growth of strains of *Arthrobacter* and is degraded in similar fashion with release of methanol (Donnelly and Dagley, 1981b). Due to the presence of an extra carbon (as methylene) in the growth substrate, 3-ketoglutarate is formed instead of oxaloacetate and is then decarboxylated to give acetoacetate, which enters the Krebs cycle as two molecules of acetyl-CoA.

The 3-*O*-methylgallate dioxygenase of *P. putida* (Figure 8) readily oxidizes gallate, but does not attack protocatechuate; it is not, therefore, a protocatechuate dioxygenase. By contrast, no separate enzyme is required by the non-fluorescent organisms, *P. testosteroni* and *P. acidovorans*, for metabolizing gallate: protocatechuate 4,5-dioxygenase and the gallate dioxygenase of non-fluorescents are synonymous (Figure 7). This is also shown in Figure 9: protocatechuate 4,5-dioxygenase, enzyme i, gives the open chain tricarboxylic acid from gallate. However, this same enzyme, acting upon 3-*O*-methylgallate, gives 2-oxopyran-4,6-dicarboxylic acid with expulsion of methanol (Kersten *et al.*, 1982). As a consequence, non-fluorescent pseudomonads when grown with 4-hydroxybenzoate are equipped to metabolize 3-*O*-methylgallate because they possess a hydrolase for the pyrone in addition to protocatechuate 4,5-dioxygenase. The reason why fluorescent pseudomonads such as *P. putida* require a special dioxygenase for both 3-*O*-methylgallate (enzyme iii; Figure 9) and gallate becomes clear when it is realized that they elaborate the 3,4-

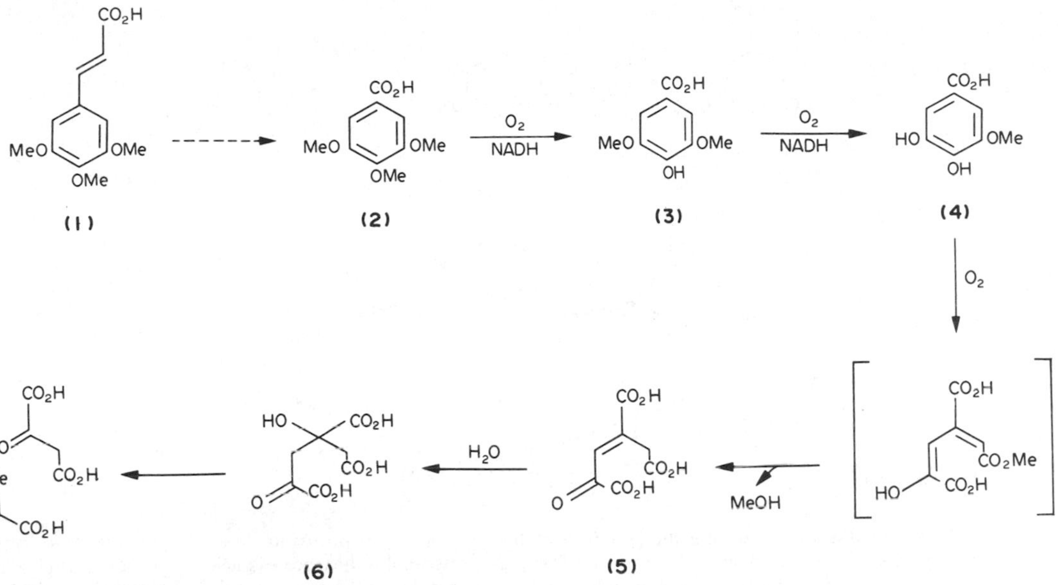

Figure 8 Degradation of 3,4,5-trimethoxycinnamic acid (**1**) and 3,4,5-trimethoxybenzoic acid (**2**) to give pyruvate and oxaloacetate. These catabolites support growth of *Pseudomonas putida* but methanol, which is also formed, is not utilized (Donnelly and Dagley, 1981a with the permission of the American Society for Microbiology)

but not the 4,5-dioxygenase for protocatechuate. Protocatechuate 3,4-dioxygenase (enzyme ii; Figure 9) forms 2-oxopyran-4,6-dicarboxylate from gallate, but no hydrolase is present in these fluorescents to open the pyrone ring, so that further degradation of gallate cannot occur; moreover, protocatechuate 3,4-dioxygenase does not attack 3-*O*-methylgallate. In summary, gallate and its *O*-methyl ether are bacterial metabolites abundant in nature. The fluorescent pseudomonad *P. putida* elaborates special dioxygenases for their degradation and releases methanol by hydrolysis, when it degrades 3-*O*-methylgallate, whereas the non-fluorescents, *P. acidovorans* and *P. testosteroni*, utilize protocatechuate 4,5-dioxygenase, expel methanol by pyrone formation, and open the pyrone ring by means of a substrate-specific hydrolase. A crucial factor in clarifying this area of bacterial catabolism by Kersten *et al.* (1982) was the establishment of spectral and chromatographic characteristics both for 2-oxopyran-4,6-dicarboxylic acid and for the product of enzymic and mild alkaline hydrolysis of the pyrone.

25.2.5 Ortho Fission Pathways

When catechol and protocatechuate are degraded by ortho fission reactions, the metabolic strategy is to place an oxygen atom in the appropriate position on the carbon chain by means of lactonization rather than hydration. Catechol (**2**; Figure 10) and protocatechuate (**3**; Figure 10) are metabolized by reactions of the same type. These are catalyzed by different enzymes in each sequence, and converge upon one compound, designated for convenience as β-ketoadipate enollactone. The lactone is hydrolyzed enzymically to β-ketoadipate which, as mentioned previously (Figure 4, reaction 6), is a suitable substrate for thiolytic fission when it has been esterified with coenzyme A. Hydroxyquinol (1,2,4-benzenetriol; **1**; Figure 10) is a catabolite of resorcinol in *P. putida* (Chapman and Ribbons, 1974) and *T. cutaneum* (Gaal and Neujahr, 1979), and of benzoate in *T. cutaneum* (Anderson and Dagley, 1980a). There is no need for lactonization to occur in this pathway since the oxygen of the third hydroxyl group of (**1**; Figure 10) subsequently appears in the carbonyl group of β-ketoadipate; this is formed when maleylacetic acid (**4**; Figure 10) is reduced by NADH. Excellent reviews are available of the β-ketoadipate pathways for catechol and protocatechuate, including aspects of regulation, in various bacterial species (Stanier and Ornston, 1973) and in nocardioform actinomycetes (Cain, 1981). Methyl group substituents when suitably placed in substrates are tolerated by enzymes of the gentisate or catechol meta fission pathways of *Pseudomonas*, and complete degradation is often thereby accomplished. This is

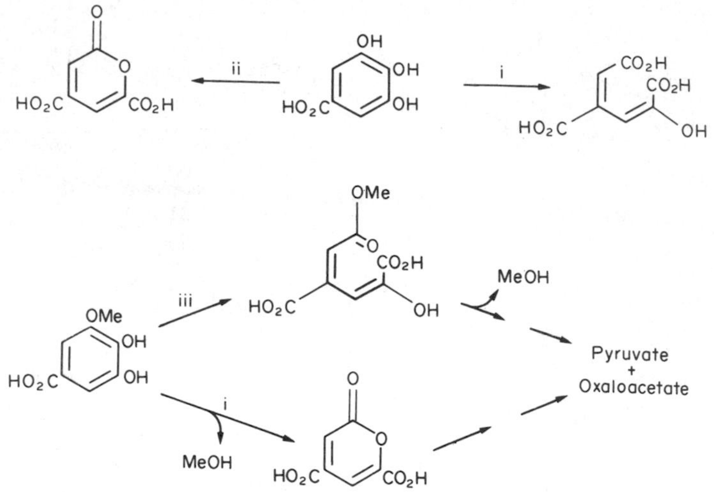

Figure 9 Action of dioxygenases on gallic and 3-*O*-methylgallic acids: (i) protocatechuate 4,5-dioxygenase forms the open chain ring-fission product from gallic acid, but forms 2-oxopyran-4,6-dicarboxylic acid from 3-*O*-methylgallic acid with elimination of methanol; (ii) protocatechuate 3,4-dioxygenase forms 2-oxopyran-4,6-dicarboxylic acid from gallic acid; the enzyme does not attack 3-*O*-methylgallic acid; (iii) a dioxygenase induced in 3,4,5-trimethoxycinnamic acid-grown *Pseudomonas putida* gives an open chain product from 3-*O*-methylgallic acid (see Figure 8) (Kersten *et al.*, 1982, with the permission of the American Society for Microbiology)

not the case for ortho fission pathways, for even if the benzene ring can be cleaved, a 'dead end' lactone is then formed, having the methyl group so placed as to prevent delactonization (Catelani *et al.*, 1971). However, the actinomycete *Gordona rubra* can degrade 4-methylcatechol to completion, forming a methyl-substituted lactone of different structure which is subject to hydrolysis (Miller, 1981). Further, the soil yeast *Trichosporon cutaneum* can grow at the expense of *m*- and *p*-cresols, using reactions that involve methyl-substituted catabolites of the *β*-ketoadipate pathway (Powlowski, 1983).

25.2.6 Separation of Pathways Used for Aromatic Catabolism by Bacteria

One general feature emerges, somewhat unexpectedly, from the foregoing survey. Repeatedly, pairs of substrates possessing very similar chemical structures, and undergoing reactions of the same type, are metabolized by separate, substrate-specific enzymes. Thus, gentisate and homogentisate are degraded in similar fashion, but each step in one pathway is catalyzed by a different enzyme from that used in the other. The enzymology of meta fission of protocatechuate is entirely separate from that of homoprotocatechuate, involving different dioxygenases and aldolases; and the ortho fission pathway for protocatechaute differs from both, with regard to reaction types as well as enzymes. This marked specialization of enzymic equipment appears from purely chemical considerations to be unnecessary, or indeed, wasteful. However, on reflection, it seems less surprising. As nutrients in the microbial world, aromatic compounds compare in importance to, say, carbohydrates in mammalian metabolism, and this has probably been so ever since O_2 became a significant constituent of the terrestrial atmosphere. Given a reaction sequence which, on thermodynamic grounds, offers a feasible route from an aromatic compound to the Krebs cycle, then it would be left to evolution to shape the structures of the enzymes required, and their substrate specificities. If it happened that gentisate was frequently encountered by evolving organisms, and homogentisate was not, ability to degrade the latter would not be advantageous and would not necessarily be acquired.

Some of the principal pathways used by bacteria are summarized, without detail, in Figure 11. A great number of naturally occurring benzenoid compounds are degraded to give, ultimately, one of the dihydric phenols shown. Sometimes the degradation product will not possess the two hydroxyls required for fission, and in such a case the ring-fission substrate will be generated from it by the action of a hydroxylase of the type described earlier (Figure 3). It has been mentioned that these flavoproteins usually exhibit stringent substrate specificities in accordance with their

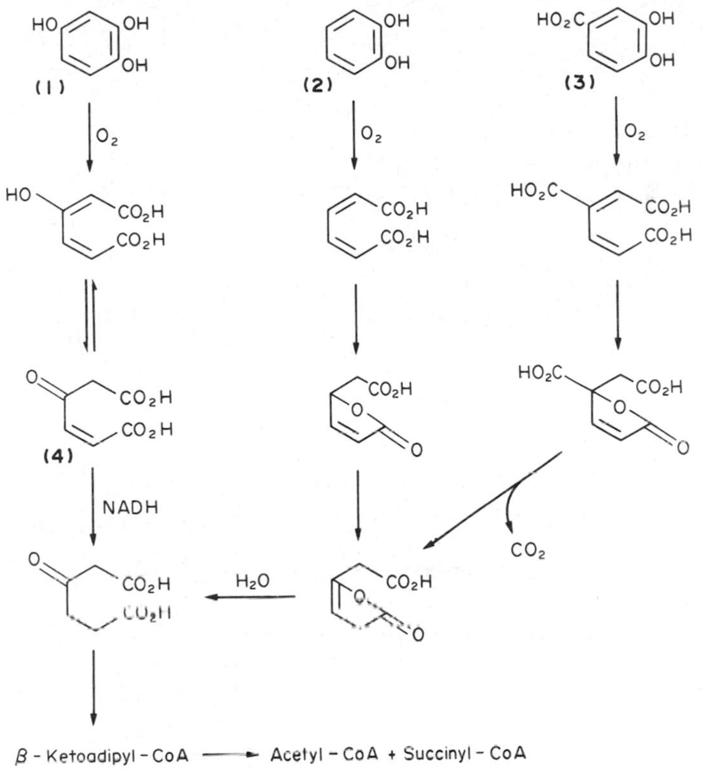

Figure 10 Ortho fission pathways for degrading hydroxyquinol (**1**), catechol (**2**) and protocatechuic acid (**3**) to give acetyl-CoA and succinyl-CoA

function, which is to select one catabolic route, and no other. Thus, two degradative routes for tyrosine are found in prokaryotes; in both, the amino acid undergoes, successively, transamination, decarboxylation and then oxidation to give 4-hydroxyphenylacetic acid. Many Gram-negative organisms hydroxylate this compound, with a simultaneous shift of the side chain, to form homogentisate which is then degraded (see reaction 2 of Figure 3). Alternatively, some Gram-positive organisms, such as certain bacilli, elaborate 4-hydroxyphenylacetate 3-hydroxylase so that their growth substrate, tyrosine, is degraded by the homoprotocatechuate route (Sparnins and Chapman, 1976). The 'choice' of pathways by some of the Gram-negative bacteria in the former category is made more striking by the fact that they possess all the enzymes required to oxidize homoprotocatechuate to CO_2, but do not possess the 3-hydroxylase needed to form this metabolite from 4-hydroxyphenylacetate. Separation of catabolic pathways is also emphasized by the fact that enzyme derepression mechanisms, as well as the enzymes themselves, are highly specific. For example, when a strain of *Acinetobacter* was grown with 4-hydroxybenzoate, the 3-hydroxylase was derepressed and the growth substrate was metabolized by the homoprotocatechuate pathway. When grown with 3-hydroxybenzoate, a 6-hydroxylase for this substrate was derepressed, as were all the enzymes for the gentisate route. It should be emphasized that bacteria, in general, cannot be divided into meta fission or ortho fission categories; thus, it is only when pseudomonads are tested after growth with one particular substrate, 4-hydroxybenzoate, that the generalization can be made that fluorescent organisms use ortho fission, specifically for protocatechuate, whereas non-fluorescent organisms use meta fission. When typical strains of *P. putida* are grown with phenol, the meta fission pathway is used for catechol but the ortho route for protocatechuate is taken by the same organisms when grown with 4-hydroxybenzoate.

25.2.7 Catabolism of Aromatic Compounds in *Trichosporon cutaneum*

By contrast with bacteria, information about the utilization of aromatic compounds by eukaryotes is scant, with one exception. The pioneering studies of Neujahr and her colleagues (Neujahr and Varga, 1970; Neujahr and Kjellen, 1978; Gaal and Neujahr, 1979) showed that the soil

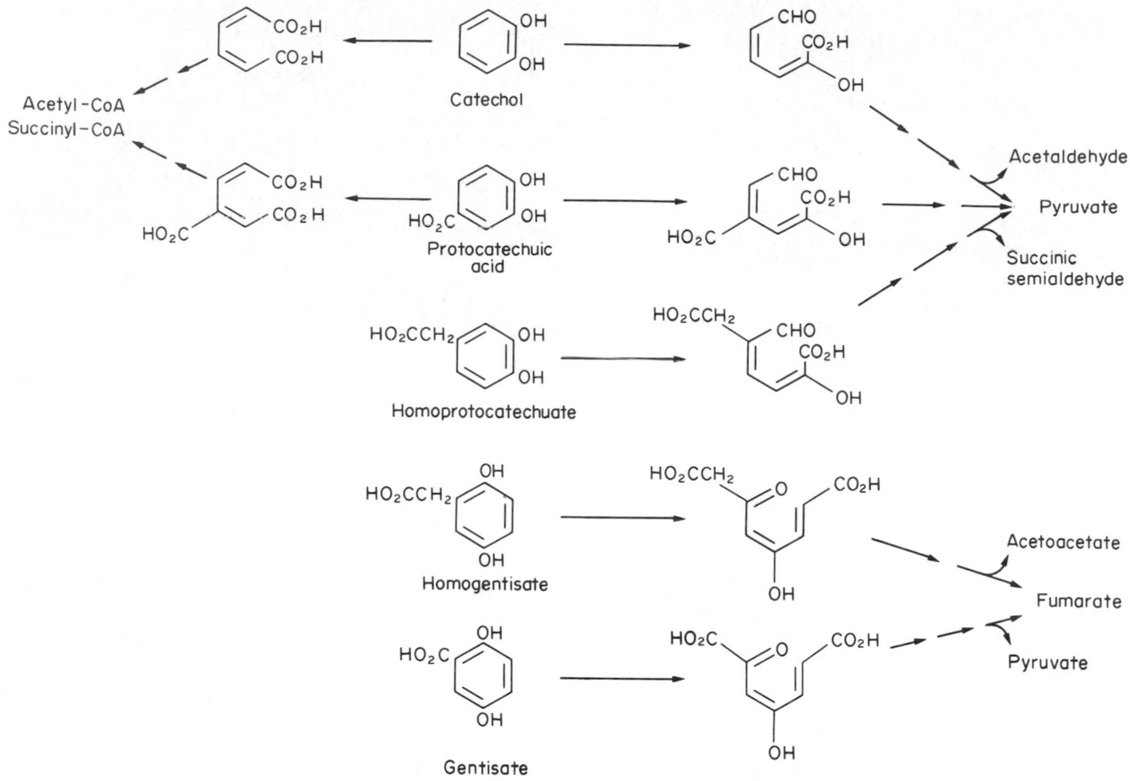

Figure 11 Summary of the principal pathways taken by bacteria that utilize organic compounds for growth. The various dihydric phenols are themselves intermediates in the degradation of a wide range of aromatic compounds. The catabolism of other important phenolic ring-fission substrates, such as hydroxyquinol, is described in the text, and a third degradative sequence for protocatechuate is given in Figure 5

yeast *T. cutaneum* displays catabolic versatility worthy of most pseudomonads; and as this work has progressed since that time, several contrasts with bacterial systems have emerged. For example, *T. cutaneum* lives under two restrictions by comparison with bacteria. It does not employ meta fission routes, being confined to using either the homogentisate or ortho fission pathways; further, it lacks dioxygenases for three dihydric phenols that are central to aromatic metabolism, namely protocatechuate, homoprotocatechuate and gentisate. These restrictions are surmounted in two ways: by hydroxylating a third time whenever two hydroxyls are not suitably placed for the types of ring-fission that are available to the organism, and by using enzymes of a slightly broader substrate specificity than those found in bacteria. These features are illustrated by the work of Anderson and Dagley (1980a) summarized in Figure 12. Thus, *T. cutaneum* hydroxylates 4-hydroxyphenylacetate (**11**; Figure 12) to give homoprotocatechuate (**12**; Figure 12) for which it possesses no dioxygenase, unlike many bacteria. Accordingly, a third hydroxyl is now inserted (reaction viii) to give a substrate which, in this organism, is accepted by the same dioxygenase that attacks homogentisate (**15**; Figure 12). As in bacteria, salicylate (**8**; Figure 12) is oxidatively decarboxylated to give catechol (**9**); however, salicylate hydroxylase from *T. cutaneum* also attacks gentisate (**7**; Figure 12), catalyzing reaction (iv) to give a ring-fission substrate, hydroxyquinol (**4**). In this way, by using a hydroxylase less stringent than its bacterial counterpart, gentisate can be oxidized to completion without the need of a special dioxygenase. As a catabolite of benzoate, hydroxyquinol assumes a greater importance in this scheme than in bacterial metabolism; three consecutive hydroxylations are used to convert benzoate (**1**) into (**4**; Figure 12). Many soil bacteria possess a single enzyme system that oxidizes benzoate, with loss of CO_2, to give the ring-fission substrate catechol, but this reaction is not used by *T. cutaneum*.

Another feature of aromatic metabolism in *T. cutaneum* suggests that some general assumptions made in this area of study should be reevaluated. In bacteria, enzyme derepression is quite specific: this factor is partly responsible for separation of the pathways of Figure 11. It also supplies the basis for the technique known as simultaneous adaptation (sequential induction) which Stanier (1947) developed during early studies in this area. Thus, if a suspension of benzoate-

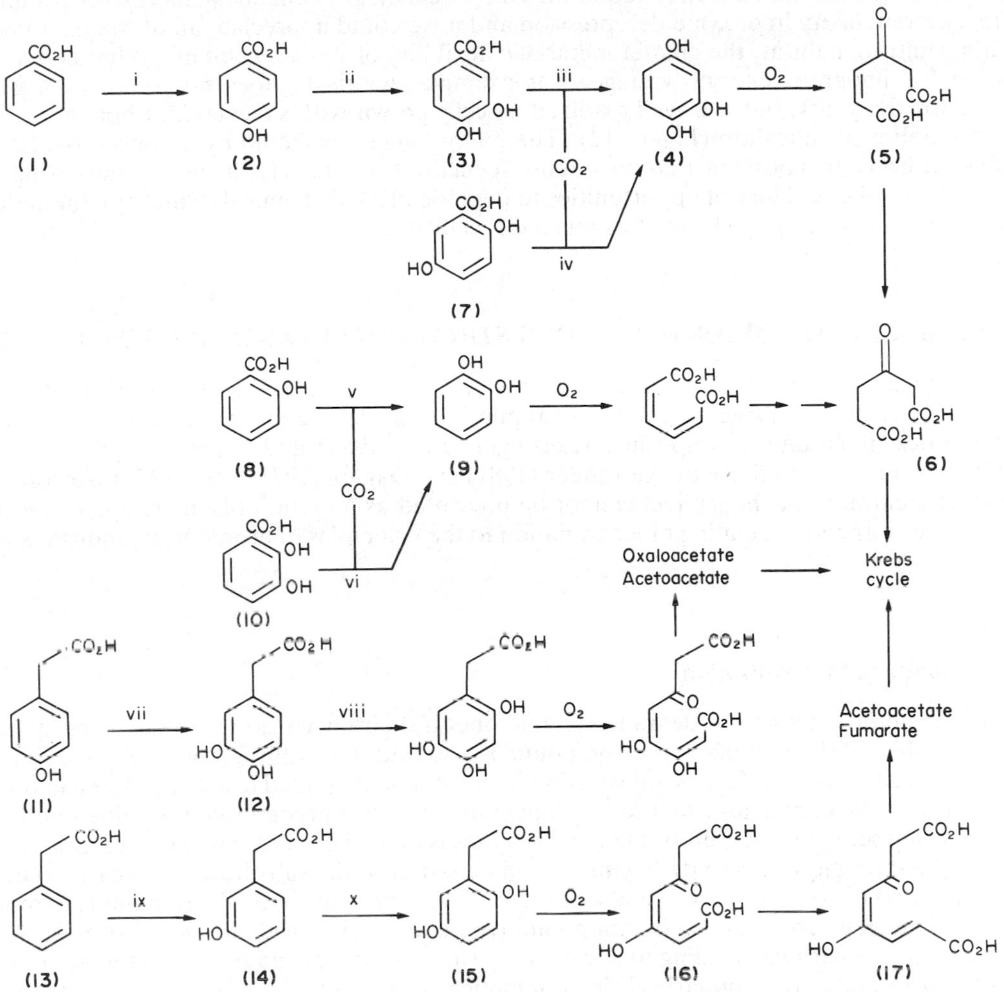

Figure 12 Degradation of the following aromatic acids by the soil yeast *Trichosporon cutaneum*: benzoic (**1**); gentisic (**7**); salicylic (**8**); phenylacetic (**13**); and 4-hydroxyphenylacetic (**11**). Cofactor requirements have been omitted; however, hydroxylases for reactions iii, iv, v, viii and x all required NADH, replaced less effectively by NADPH. Reactions ii and vii were catalyzed by hydroxylases with a specific requirement for NADPH. Reaction vi is a non-oxidative decarboxylation with no cofactor requirement; reactions iv and v are catalyzed by the same enzyme (Anderson and Dagley, 1980a, with the permission of the American Society for Microbiology)

grown bacteria is found to be adapted to oxidize gentisate, the cells do not usually attack catechol or protocatechuate which lie on feasible alternative pathways; and when this is observed, the investigator need no longer consider them as catabolites. However, when the yeast *T. cutaneum* is grown with benzoate it is found to be fully adapted to catechol, protocatechuate and gentisate; and for good measure also to 2-, 3- and 4-hydroxybenzoates. Simultaneous adaptation is a useful technique as a preliminary approach to delineating a bacterial catabolic route; but in the case of *T. cutaneum* it is evidently without value.

The old concept of survival of the fittest is sometimes thought of, currently, as a consequence of available energy being put to use by successful competitors in the most efficient manner possible. It is tempting for the investigator of novel microbial pathways to be guided by this principle, and to focus upon a suggested catabolic route that appears to be energetically the most economical. This approach is not always advisable. We have seen that *T. cutaneum* is given to hydroxylating three times while its bacterial competitors hydroxylate no more than twice. Hydroxylations are expensive in energy, as Anderson and Dagley (1980b) showed by microcalorimetry: each one involves evolution of energy as heat that would otherwise by assigned to assimilating cellular material. However, these extra hydroxylations, as we have seen, compensate for limitations in the range of the organism's dioxygenases. Regarding specificity of enzyme induction, the stringent control exercised by bacteria seems admirably economical. The organisms do not derepress enzymes that are not immediately required. But *T. cutaneum* remains a successful competitor despite a certain laxity in enzyme derepression and if we could appreciate all of the factors operating in its natural habitat, the greater metabolic flexibility of *T. cutaneum* might be seen as compensating for apparent energy wastage. For example, gentisate does not serve as a growth substrate for the yeast, but it is well oxidized by cells grown with salicylate, although it is not a direct catabolite of salicylate (Figure 12). The energy made available by gentisate oxidation is doubtless utilized. It is possible that the intense specialization of bacteria when catabolizing some compounds could entail loss of opportunities to degrade related chemicals which are immediately available to *T. cutaneum* by virtue of its relaxed specificity.

25.3 DEGRADATION OF AROMATIC INDUSTRIAL POLLUTANTS AND PESTICIDES

Technological societies have shown a growing concern for improving environmental quality, and this is reflected by a large output of recent publications devoted to the microbial degradation of xenobiotics, including a symposium (Leisinger *et al.*, 1981) and numerous reviews among which may be mentioned those by Alexander (1981) and Dagley (1975, 1977, 1978). Accordingly, a complete treatment of the subject cannot be presented as a section of one chapter, and comments will be confined to relating biodegradation to the principles of aromatic catabolism already outlined.

25.3.1 Complete Mineralization

An industrial product will be degraded by microbes if their enzymes can attack without serious restriction. Thus, if the manufactured compound is also found in nature, it will be biodegradable. If a natural product is modified by substitution it will still be degraded if the enzymes can tolerate such changes in the chemical structures of their substrates. For accommodating substituents, the most promising sequence on purely chemical grounds is meta fission of catechols (Figure 5), since the hydrolytic branch of this route begins with elimination of the substituent as a carboxylic acid (see degradation of compounds **4, 5** and **6** of Figure 5). After hydrolysis, the remaining carbons of the benzene nucleus are retained in compound (**8**; Figure 5) in each case, so beyond this point there is a common pathway leading to pyruvate and acetaldehyde, and thence to the Krebs cycle. Fates of substituents are summarized diagrammatically in Figure 13. Thus, in sequence (1), 3-methylcatechol is converted into acetate, pyruvate and acetaldehyde by the hydrolytic pathway, details of which were given in Figure 5, whereas in sequence (2), Figure 13, 4-methylcatechol gives CO_2, pyruvate and propionaldehyde by the oxidative pathway. Where there are two methyl substituents, as is the case when the catechol is formed by hydroxylating either 2,3- or 3,4-xylenol, sequence (3) shows that the products are acetate, pyruvate and propionaldehyde.

When the substituent group is phenyl in sequence (1; Figure 13) the corresponding catechol is 2,3-dihydroxybiphenyl which was found by Gibson *et al.* (1973) to be the substrate for a ring-fission dioxygenase elaborated by microorganisms capable of growth with biphenyl. They also showed that the hydrocarbon undergoes preliminary oxidation by a dihydroxylating dioxygenase system, giving *cis*-2,3-dihydro-2,3-dihydroxybiphenyl which is then dehydrogenated to the corresponding catechol. These reactions are similar to those converting benzene into catechol (see Figure 3, reaction 5). At low concentrations, biphenyl and other polycyclic hydrocarbons are distributed widely enough, at least in the United States, to permit ready isolation of microorganisms that can use them as carbon sources for growth. Since such compounds are formed when wood, or other organic matter, burns, they could have arisen from the sequence of forest fires that have occurred at regular intervals over the centuries even in uninhabited areas. It is probable, therefore, that mankind has always been exposed to trace amounts of these hydrocarbons, which exhibit carcinogenic properties when present at more elevated concentrations. Likewise, it may be recalled that the phenols and catechols, which are normal microbiol metabolites of aromatic compounds, also become toxic even to the organisms that degrade them, so that growth on phenol, for example, is not usually supported when the concentration exceeds 0.5 g l^{-1}. Ohmori *et al.* (1973) showed that 4-chlorobenzoic acid is formed by bacterial degradation of *p*-chlorobiphenyl, as Figure 13 predicts. Further substitution by chlorine, particularly in the *ortho* positions, tends to retard degradation: not only is hydroxylation sterically hindered thereby, but since the enzymically activated species of oxygen for hydroxylations is electrophilic, electron-withdrawing chlor-

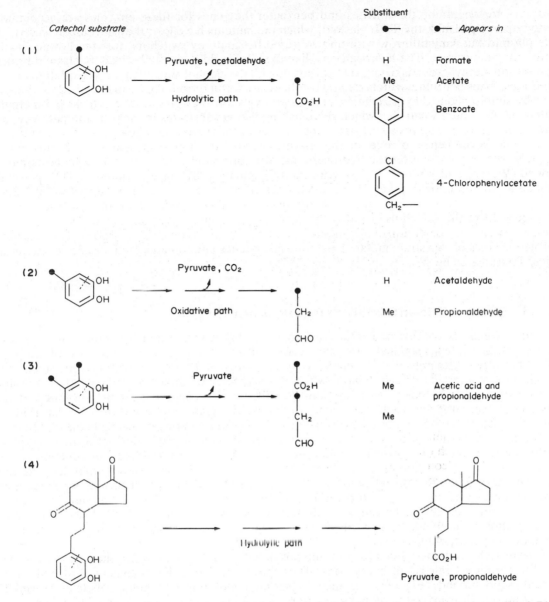

Figure 13 Summary of meta fission routes to show how the metabolic fate of a substituent group can be predicted. Ring-fission by dioxygenases is shown by broken lines; details of reaction sequences are given in Figure 5. Sequence (4) is a special example of (3) taken from studies of microbial catabolism of steroids by Gibson *et al.*, 1966 (Dagley, 1982a, with the permission of Academic Press)

ine substituents would be expected to deactivate the substrate. Polychlorinated biphenyls are notoriously persistent pollutants.

25.3.2 Catabolic Plasmids

When they function for the degradation of aromatic compounds that are both ubiquitous and abundant, it is likely that the enzymes catalyzing the main catabolic pathways, summarized in Figure 11, will be chromosomally encoded. But most aromatic hydrocarbons are not included in this category and we find, accordingly, that their catabolic enzymes are plasmid-borne and can be mobilized if and when hydrocarbons become available as nutrients. Thus, a strain of *P. putida* can use *m*- or *p*-toluic acids (3- and 4-methylbenzoic acids) and also the hydrocarbons toluene, *m*-xylene and *p*-xylene as single carbon sources. Mutants may be selected that no longer grow with these substrates but are still able to grow with unsubstituted benzoic acid, and when they do so, they use the enzymes of the ortho fission route. By contrast, the wild-type uses the meta fission enzymes for degrading the toluates and benzoate: the genes for these enzymes are accommodated upon a plasmid, the TOL plasmid, which the mutants have lost; they have been 'cured' of the plasmid but can still grow with unsubstituted benzoate by switching over to chromosomal, ortho fission enzymes. The hydrocarbons, like the toluic acids, are catabolized by plasmid-borne genes: one methyl group is oxidized to carboxyl, with the alcohol and aldehyde as oxidation intermediates. Thus benzoate, *m*-toluate and *p*-toluate are formed from the respective hydrocarbons, and by simultaneous dihydroxylation and decarboxylation they give rise to catechol, 3-methylcatechol and 4-methylcatechol which serve as ring-fission substrates in meta fission pathways as shown in Figure 5. The broad substrate specificities of these enzymes allow, apparently, one set of plasmid-borne genes to code for the enzymes needed to support growth with 12 substrates, namely, the particular aromatic hydrocarbons, alcohols, aldehydes, and carboxylic acids mentioned (Worsey and Williams, 1975). Kunz and Chapman (1981) have examined a TOL plasmid from another strain of *P. putida* that also encodes enzymes for meta fission catabolism of 1,2,4-trimethylbenzene and 3-ethyltoluene. The products were those predicted by applying the scheme of Figure 13 to the catechols given by the corresponding substituted benzoic acids; these were formed by enzymic oxidation of one methyl group in each of the two hydrocarbons. The plasmid-borne enzymes of this strain metabolized toluene, xylenes and toluates by the same reactions as those investigated by Worsey and Williams (1975).

25.3.3 Release of Halogen Substituents from the Benzene Nucleus

Many chemicals of environmental concern are halogen-substituted aromatics, and investigations of their fates in soil and water have generated an extensive, but characteristically empirical, literature. Attempts will be made to bring a few of those observations into focus by specifically relating halide elimination, which is a necessary preliminary for complete microbiol mineralization, to the operation of systems described in this chapter. First, halide ions may be released fortuitously, as is the case when 4-hydroxybenzoate 3-hydroxylase (see reaction 1, Figure 3) acts upon 3,5-difluoro-4-hydroxybenzoate to form 5-fluoroprotocatechuate with elimination of a stoichiometric amount of fluoride ion (Husain *et al.*, 1980). This is not the usual physiological function of the enzyme; halide release in this case is a fortuitous consequence of the mechanism of enzyme action. Second, and in contrast, the organism may employ an enzyme specially designed to release halide ions: halogen-substituted natural products are by no means as uncommon as was once supposed (see Dagley, 1975), moreover halogen-containing pesticides have been routinely spread for long periods over vast areas of agricultural land. This favors effective selection of those microorganisms that have acquired plasmids for the purpose of using these compounds as nutrients.

Enzymes that release methanol from methoxylated aromatic natural products can also serve for the fortuitous release of halide ions. Thus, protocatechuate 4,5-dioxygenase of non-fluorescent pseudomonads (Figure 9, enzyme i) expels methanol from 3-*O*-methylgallate and forms 2-oxopyran-4,6-dicarboxylate; 5-fluoro-, 5-chloro, and 5-bromo-protocatechuic acids undergo the same transformation with elimination of the respective halide ions. Since non-fluorescent pseudomonads typically possess 2-oxopyran-4,6-dicarboxylic acid hydrolase, some strains are able to grow with 5-chlorovanillic acid (5-chloro-4-hydroxy-3-methoxybenzoic acid) as their source of carbon: the methoxyl group is oxidized to hydroxyl, and chloride is then eliminated from 5-chloroprotocatechuate during pyrone formation. Elimination of halide can be understood in terms of the sequence proposed for conversion of 3-*O*-methylgallate into 2-oxopyran-4,6-dicarboxylate by protocatechuate 4,5-dioxygenase (Figure 14). It is proposed that, before leaving the enzyme, the ring-fission product is converted into its stereoisomer *via* two tautomeric changes: the enolic hydroxyl at C-2 is now in a position to attack C-6 and displace methanol. The leaving group could clearly be a halide ion instead of methoxyl.

In principle, halide might be released from C-6 by two other mechanisms. An acyl halide might be formed by, say, enzyme iii of Figure 9 which does not simultaneously catalyze the tautomeric changes of Figure 14 and therefore gives the open chain product instead of forming pyrone. The acyl halide so formed might then undergo nucleophilic attack by H_2O (hydrolysis), giving the free acid and liberating halide ion; but this reaction has not been reported. Conversely, a nucleophilic amino acid residue on the enzyme might attack at C-6, displacing halide and acylating the enzyme. This is more likely, and has in fact been observed (Knackmuss, 1981) for 3-halocatechols which exhibit typical 'suicide' kinetics with catechol 2,3-dioxygenase. Clearly, the most effective and safe way to expel halide from a ring-fission product is by internal nucleophilic attack that

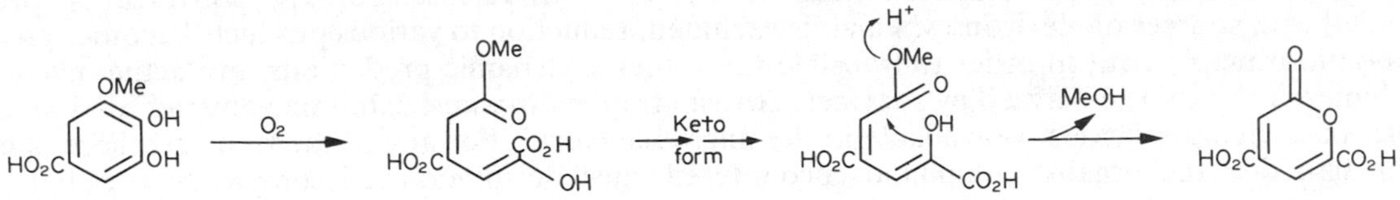

Figure 14 Sequence of chemical changes proposed for the conversion of 3-*O*-methylgallic acid into methanol and 2-pyrone-4,6-dicarboxylic acid, catalyzed by protocatechuate 4,5-dioxygenase. Intermediates are considered to remain bound to the enzyme throughout the course of the reaction

forms a new ring-system. Reactions of this type are also catalyzed by lactonizing enzymes in ortho fission sequences (Figure 10) and lactonization is, in fact, used to expel chloride ion in the bacterial degradation of 2,4-dichlorophenoxyacetic acid (2,4-D). However, this is not fortuitous elimination, because the lactonizing enzyme (a cycloisomerase) is not the same as that used in the catechol ortho fission route of Figure 10; also, a different type of lactone is formed, having a second double bond in the side chain substituent (Figure 15).

When 2,4-D is degraded by bacteria, the first reaction is a monooxygenase attack upon the side chain, resulting in its release as glyoxylate to form 2,4-dichlorophenol which is hydroxylated giving 3,5-dichlorocatechol. Ortho fission then gives a substituted muconic acid (Figure 15) from which one Cl⁻ ion is expelled by a cycloisomerase particularly active upon chlorinated substrates and specifically induced by growth of the organism on chloroaromatics. Hydrolysis of the lactone gives the 2-chloro derivative of maleylacetic acid which, as shown in Figure 10, is enzymically reduced to β-ketoadipate by NADH. An additional equivalent of NADH as electron donor is needed to accomplish this conversion in the case of 2-chloromaleylacetic acid, since chlorine must be eliminated as Cl⁻ ion (Chapman, 1979). Reductive dehalogenation of halogenobenzoates by anaerobic consortia has also been recently demonstrated (Suflita *et al.*, 1982). This observation has great promise for environmental studies.

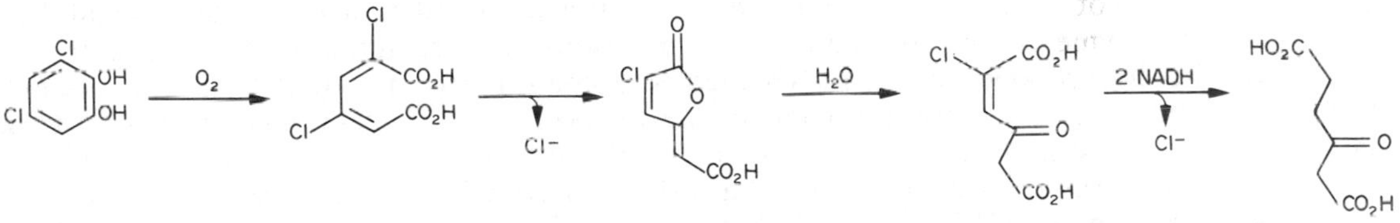

(1) **(2)** **(3)**

Figure 15 Bacterial degradation of 3,5-dichlorocatechol. The benzene ring of this compound, a catabolite of the herbicide 2,4-dichlorophenoxyacetic acid, is opened by ortho fission and chloride is expelled in lactonization. The lactone (**1**) is then hydrolyzed to 2-chloromaleylacetic acid (**2**) from which a second chloride ion is expelled upon reduction to β-ketoadipic acid (**3**). This compound is then metabolized as in Figure 10

25.3.4 Incomplete Degradation of Aromatics

Since all enzymes exhibit some degree of substrate specificity, features of chemical structure may prevent a compound from being attacked by an enzyme in a degradative pathway. Thus, *Pseudomonas testosteroni* grew with *p*-cresol (4-hydroxytoluene) as sole source of carbon but it did not grow when provided with structural analogs substituted with methyl groups at C-2 or C-3 of the nucleus (Dagley and Patel, 1957). When cells were induced to grow with *p*-cresol, however, these compounds were oxidized as far as the corresponding 4-hydroxymethylbenzoic acids, which therefore accumulated in culture fluids; but this partial oxidation of the substrate was insufficient to sustain growth. In this example further oxidation was prevented by the strict substrate specificity of 4-hydroxybenzoate 3-hydroxylase which does not tolerate substitutions at C-2 and C-3. This example was taken from work performed with a pure culture under laboratory conditions; in a natural habitat other organisms would doubtless take over and complete the degradative process where *P. testosteroni* left off.

Any organic compound tends to oxidize in air to CO_2 with release of energy; conversely, if provided with sources of electrons when air is excluded, reduction to various products becomes possible in principle. But in order to translate these thermodynamic predictions into actual events, changes have to be catalyzed by enzymes. A vast array of enzymes, exhibiting a very broad range of substrate specificities, is available in the microbial world. For any of them to act, there is no proviso that the organic compound encountered must be mineralized completely, or that its degradation must benefit the organism. Alexander (1981) has compiled a comprehensive list of reactions that bring about modifications of man-made compounds of environmental concern; essentially this list reflects the tolerance, or lack of it, of microbial enzymes for alterations in the structure of their natural substrates. But although some semblance of order emerges when it is realized that one pollutant may give rise to another by the action of enzymes that belong to well established categories, these products of microbial catabolism still remain bewildering in their diversity. This is a very different impression from that gained in the laboratory by a biochemical microbiologist working with pure cultures and single carbon sources, for the catabolic pathways he reveals are assembled with an economy that reflects their relevance to the life of the organism. By contrast, the microbial transformations of pesticides and other environmental chemicals are quite haphazard: they are random rather than purposeful. These compounds were not manufactured in the first place to serve as microbial substrates, but rather for use against target pest organisms and microbial enzymes will attack them only if they bear some structural resemblance to substrates they meet in nature. This contrasts with the integration of these enzymes into metabolic sequences, shaped by evolution for efficient degradation of a commonly encountered natural product.

Examples of partial degradation, causing compounds to accumulate that prove to be more hazardous than the original substrates, are furnished by investigations of the fates of aniline-based herbicides in soil. Fungal attack is initiated (Figure 16) by hydrolysis, using acylamidases. The acids, alcohols or amines thereby released may serve as fungal nutrients but the other products, substituted anilines, do not. The herbicide propanil (3,4-dichloropropionanilide; $R = C_2H_5$ in Figure 16, b with $X = Y = Cl$) has received particular attention. Bartha and Pramer (1967) reported that soil treated with 2000 p.p.m. propanil (a very high application) produced a major metabolite subsequently identified as 3,3',4,4'-tetrachloroazobenzene, formed by the condensation of two molecules of 3,4-dichloroaniline arising from propanil by hydrolysis. Other herbicides in this category, and various dichloroanilines and monochloroanilines have also been shown to be transformed into chlorine-substituted azobenzenes in soil. These compounds are structurally related to known potent carcinogens, and studies of their mode of formation from 4-chloroaniline and 3,4-dichloroaniline provide instructive examples of how chemically reactive and potentially hazardous intermediates can arise from microbial action. Kaufman *et al.* (1973) showed that the soil fungus *Fusarium oxysporum* Schlecht degraded 4-chloroaniline (**1**, Figure 16) to the following products, which were isolated and characterized: 4-chlorophenylhydroxylamine (**2**); 4-chloronitrosobenzene (**3**); 4-chloronitrobenzene (**4**); 4,4'-dichloroazoxybenzene (**5**); and 4,4'-dichloroazobenzene (**6**). It appears likely that microbial activation of the amino group of (**1**) by hydroxylation to give (**2**) and then (**3**) by oxidation, facilitated the condensation of (**1**) with (**3**) to give (**6**), and of (**2**) with (**3**) to give (**5**).

Two general reasons have been given why microorganisms initiate attack upon an aromatic compound: either as a prelude to its utilization as a nutrient, or because rather unspecific enzymes recognise features of chemical structure resembling those of natural substrates. There is a third reason which emerges from the work of Gibson and his colleagues; one example from an extensive literature must suffice. Cerniglia and Gibson (1979, 1980) showed that the filamentous fungus, *Cunninghamella elegans* oxidizes an environmental carcinogen, the hydrocarbon benzo[*a*]pyrene, to a complex mixture of polar products which are remarkably similar to those arising from detoxication reactions in higher organisms; moreover, sulfuric and glucuronic acid conjugates of these metabolites from *C. elegans* were also detected. It seems that these particular filamentous fungi do not regard aromatic hydrocarbons as nutrients to be degraded, but as xenobiotics to be detoxified; however, preliminary hydroxylation may well assist later degradation by bacteria. The behavior of *C. elegans* toward these hydrocarbons might be described as that of a minute, filamentous liver. In many respects, this may be useful material for medical research. It encourages the hope that *C. elegans* might provide future assistance in cancer research and toxicology similar to that given by streptococci and lactobacilli when vitamins and coenzymes were being discovered, or by *E. coli* in the days when the whole range of intermediary metabolism, mammalian and microbiol, was a ferment of research activity.

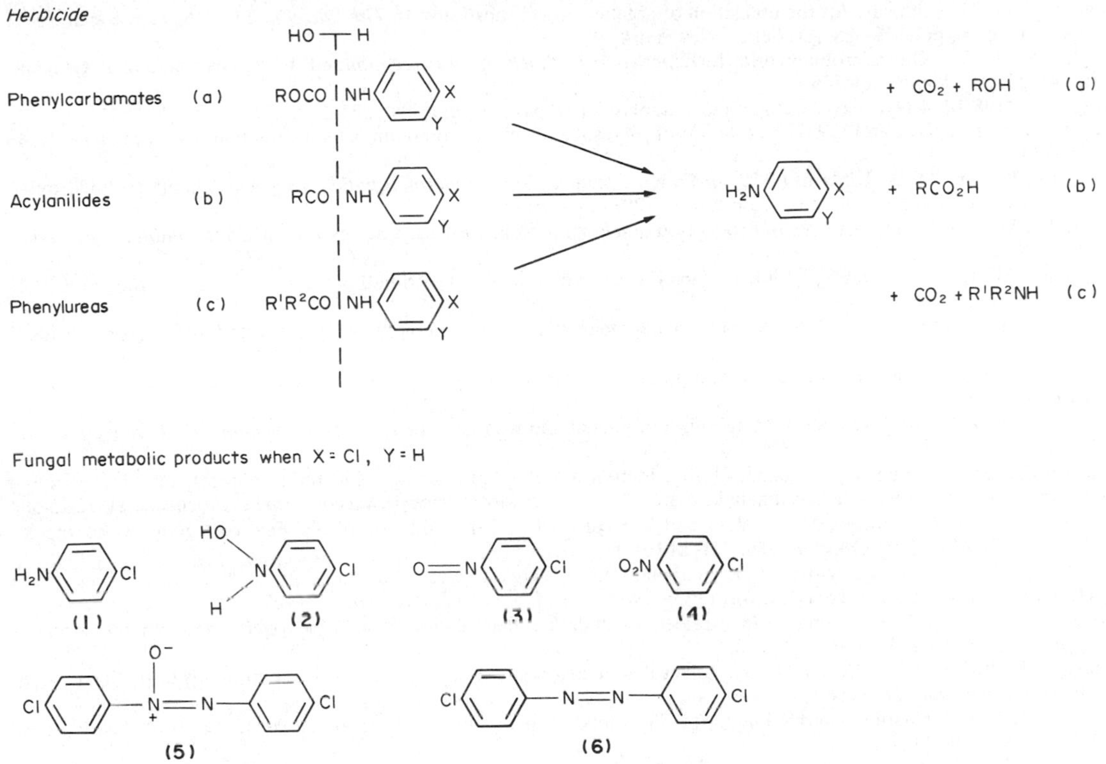

Herbicide

Phenylcarbamates (a)

Acylanilides (b)

Phenylureas (c)

$+ CO_2 + ROH$ (a)

$+ RCO_2H$ (b)

$+ CO_2 + R^1R^2NH$ (c)

Fungal metabolic products when X = Cl, Y = H

Figure 16 Enzymatic hydrolysis of some aniline-based herbicides. When 4-chloroaniline (1) is released, fungi can metabolize this compound to compounds (2) to (6) (Dagley, 1977, with the permission of Academic Press)

25.4 REFERENCES

Alexander, M. (1981). Biodegradation of chemicals of environmental concern. *Science*, **211**, 132–138.

Anderson, J. J. and S. Dagley (1980a). Catabolism of aromatic acids in *Trichosporon cutaneum*. *J. Bacteriol.*, **141**, 534–543.

Anderson, J. J. and S. Dagley (1980b). Heat evolution of microbial catabolism: effects of monooxygenases. *J. Bacteriol.*, **143**, 525–528.

Anderson, J. J. and S. Dagley (1981). Catabolism of tryptophan, anthranilate and 2,3-dihydroxybenzoate in *Trichosporon cutaneum*. *J. Bacteriol.*, **146**, 291–297.

Axcell, B. C. and P. J. Geary (1975). Purification and some properties of a soluble benzene-oxidizing system from a strain of *Pseudomonas*. *Biochem. J.*, **146**, 173–183.

Ballou, D. P. (1982). Flavoprotein monooxygenases. In *Flavins and Flavoproteins*, ed. V. Massey and C. H. Williams, pp. 301–310. Elsevier–North Holland, New York.

Bartha, R. and D. Pramer (1967). Pesticide transformation to aniline and azo compounds in soil. *Science*, **156**, 1617–1618.

Cain, R. B. (1981). Regulation of aromatic and hydroaromatic catabolic pathways in nocardioform actinomyces. In *Actinomycetes*, *Zbl. Bakt. Suppl. 11*, ed. K. P. Schaal and G. Pulverer. Gustav Fischer Verlag, Stuttgart.

Catelani, D., A. Fiecchi and E. Galli (1971). (+)-γ-Carboxylmethyl-γ-methyl-α-butenolide. A 1,2-ring-fission product of 4-methylcatechol by *Pseudomonas desmolyticum*. *Biochem. J.*, **121**, 89–92.

Cerniglia, C. E. and D. T. Gibson (1979). Oxidation of benzo[a]pyrene by the filamentous fungus *Cunninghamella elegans*. *J. Biol. Chem.*, **254**, 12 174–12 180.

Cerniglia, C. E. and D. T. Gibson (1980). Fungal oxidation of benzo[a]pyrene and (±)-*trans*-7,8-dihydroxy-7,8-dihydrobenzo[a]pyrene. *J. Biol. Chem.*, **255**, 5159–5163.

Chapman, P. J. (1979). Degradation mechanisms. In *Microbial Degradation of Pollutants in Marine Environments*, ed. A. W. Bourquim and P. H. Pritchard, pp. 28–66. US Environmental Protection Agency, Gulf Breeze, FL.

Chapman, P. J. and D. W. Ribbons (1974). Metabolism of resorcinylic compounds by bacteria: alternative pathways for resorcinol catabolism in *Pseudomonas putida*. *J. Bacteriol.*, **125**, 985–998.

Collinsworth, W. L., P. J. Chapman and S. Dagley (1973). Stereospecific enzymes in the degradation of aromatic compounds by *Pseudomonas putida*. *J. Bacteriol.*, **113**, 922–931.

Crawford, R. L. (1975). A novel pathway for degradation of protocatechuic acid in a *Bacillus* species. *J. Bacteriol.*, **121**, 531–536.

Crawford, R. L., P. E. Olson and T. D. Frick (1979). Catabolism of 5-chlorosalicylate by a *Bacillus* isolated from the Mississippi River. *Appl. Environ. Microbiol.*, **38**, 379–384.

Dagley, S. (1975). A biochemical approach to some problems of environmental pollution. *Essays Biochem.*, **11**, 81–138.

Dagley, S. (1977). Microbial degradation of organic compounds in the biosphere. *Surv. Prog. Chem.*, **8**, 121–170.

Dagley, S. (1978). Pathways for the utilization of organic growth substrates. In *The Bacteria*, ed. L. N. Ornston and J. R. Sokatch, vol. 6, pp. 305–388. Academic, New York.

Dagley, S. (1982a). Our microbial world. In *Experiences in Biochemical Perception*, ed. L. N. Ornston and S. G. Sligar, pp. 45–57. Academic, New York.

Dagley, S. (1982b). 4-Hydroxy-2-ketopimelate aldolase. *Methods Enzymol.*, **90**, 277–280.

Dagley, S. and M. D. Patel (1957). Oxidation of *p*-cresol and related compounds by a *Pseudomonas*. *Biochem. J.*, **66**, 227–233.

DeFrank, J. J. and D. W. Ribbons (1977). *p*-Cymene pathway in *Pseudomonas putida*: ring cleavage of 2,3-dihydroxy-*p*-cumate and subsequent reactions. *J. Bacteriol.*, **129**, 1356–1364.

Donnelly, M. I. and S. Dagley (1980). Production of methanol from aromatic acids by *Pseudomonas putida*. *J. Bacteriol.*, **142**, 916–924.

Donnelly, M. I. and S. Dagley (1981a). Bacterial degradation of 3,4,5-trimethoxycinnamic acid. *J. Bacteriol.*, **147**, 471–476.

Donnelly, M. I. and S. Dagley (1981b). Bacterial degradation of 3,4,5-trimethoxyphenylacetic and 3-ketoglutaric acids. *J. Bacteriol.*, **147**, 477–481.

Evans, W. C. (1977). Biochemistry of the bacterial catabolism of aromatic compounds in anaerobic environments. *Nature (London)*, **270**, 17–22.

Gaal, A. and H. Y. Neujahr (1979). Metabolism of phenol and resorcinol in *Trichosporon cutaneum*. *J. Bacteriol.*, **137**, 13–21.

Garrido-Pertierra, A. and R. A. Cooper (1981). Identification and purification of distinct isomerase and decarboxylase enzymes involved in the 4-hydroxyphenylacetate catabolic pathway of *Escherichia coli*. *Eur. J. Biochem.*, **117**, 581–584.

Gibson, D. T., R. L. Roberts, M. C. Wells and V. M. Kobal (1973). Oxidation of biphenyl by a *Beijerinckia* species. *Biochem. Biophys. Res. Commun.*, **50**, 211–219.

Hareland, W. A., R. L. Crawford, P. J. Chapman and S. Dagley (1975). Metabolic function and properties of a 4-hydroxyphenylacetate 1-hydroxylase from *Pseudomonas acidovorans*. *J. Bacteriol.*, **121**, 272–285.

Healy, J. B., Jr. and L. Y. Young (1978). Catechol and phenol degradation by a methanogenic population of bacteria. *Appl. Environ. Microbiol.*, **35**, 216–218.

Healy, J. B., Jr. and L. Y. Young (1979). Anaerobic biodegradation of eleven aromatic compounds to methane. *Appl. Environ. Microbiol.*, **38**, 84–89.

Hopper, D. J., P. J. Chapman and S. Dagley (1970). Metabolism of L-malate and D-malate by a species of *Pseudomonas*. *J. Bacteriol.*, **104**, 1197–1202.

Hopper, D. J., P. J. Chapman and S. Dagley (1971). The enzymic degradation of alkyl-substituted gentisates, maleates and malates. *Biochem. J.*, **122**, 29–40.

Howarth, R. W. and J. M. Teal (1979). Sulfate reduction in a New England salt marsh. *Limnol. Oceanogr.*, **24**, 999–1013.

Husain, M., B. Entsch, D. P. Ballou, V. Massey and P. J. Chapman (1980). Fluoride elimination from substrates in hydroxylation reactions catalyzed by *p*-hydroxybenzoate hydroxylase. *J. Biol. Chem.*, **255**, 4189–4197.

Kaufman, D. D., J. R. Plimmer and U. I. Klingebiel (1973). Microbial oxidation of 4-chloroaniline. *J. Agric. Food Chem.*, **21**, 127–132.

Kersten, P. J., S. Dagley, J. W. Whittaker, D. M. Arciero and J. D. Lipscomb (1982). 2-Pyrone-4,6-dicarboxylic acid, a catabolite of gallic acids in species of *Pseudomonas*. *J. Bacteriol.*, **152**, 1154–1162.

Kiyohara, H. and K. Nagao (1978). The catabolism of phenanthrene and naphthalene by bacteria. *J. Gen. Microbiol.*, **105**, 69–75.

Knackmuss, H.-J. (1981). Degradation of halogenated and sulfonated hydrocarbons. In *Microbial Degradation of Xenobiotics and Recalcitrant Compounds*, ed. T. Lesinger, A. M. Cook, R. Hutter and J. Nuesch, pp. 190–212. Academic, New York.

Kobayashi, S. and O. Hayaishi (1970). Anthranilic acid conversion to catechol. *Methods Enzymol.*, **17A**, 505–510.

Kobayashi, S., S. Kuno, N. Itada, O. Hayaishi, S. Kozuka and S. Oae (1964). O^{18} studies on anthranilate hydroxylase—a novel mechanism of double hydroxylation. *Biochem. Biophys. Res. Commun.*, **16**, 556–561.

Kunz, D. A. and P. J. Chapman (1981). Catabolism of pseudocumene and 3-ethyltoluene by *Pseudomonas putida* (*arvilla*) mt2: evidence for new functions of the TOL (pWWO) plasmid. *J. Bacteriol.*, **146**, 179–181.

Miller, D. J. (1981). Toluate metabolism in nocardioform actinomycetes: utilization of the enzymes of the 3-oxoadipate pathway for the degradation of methyl-substituted analogues. In *Actinomycetes, Zbl. Bakt. Suppl. 11*, ed. K. P. Schaal and G. Pulverer. Gustav Fischer Verlag, Stuttgart.

Neujahr, H. Y. and K. G. Kjellen (1978). Phenol hydroxylase from yeast. Reaction with phenol derivatives. *J. Biol. Chem.*, **253**, 8835–8841.

Neujahr, H. Y. and J. M. Varga (1970). Degradation of phenols by intact cells and cell-free preparations of *Trichosporon cutaneum*. *Eur. J. Biochem.*, **13**, 37–44.

Ohmori, T., T. Ikai, Y. Minoda and K. Yamada (1973). Utilization of polyphenyl and polyphenyl-related compounds by microorganisms, part I. *Agric. Biol. Chem.*, **37**, 1599–1605.

Powlowski, J. (1983). Ph.D. Thesis, University of Minnesota.

Powlowski, J. and S. Dagley (1982). Anthranilate hydroxylase (deaminating) from *Trichosporon cutaneum*. In *Flavins and Flavoproteins*, ed. V. Massey and C. H. Williams, pp. 339–341. Elsevier–North Holland, New York.

Que, L. (1978). Extradiol cleavage of *o*-aminophenol by pyrocatechase. *Biochem. Biophys. Res. Commun.*, **84**, 123–129.

Reiner, A. M. and G. D. Hegeman (1971). Metabolism of benzoic acid by bacteria. Accumulation of (−)-3,5-cyclohexadiene-1,2-diol-1-carboxylic acid by a mutant strain of *Alcaligenes eutrophus*. *Biochemistry*, **10**, 2530–2535.

Ribbons, D. W. (1970). Stoicheiometry of *O*-demethylase activity in *Pseudomonas aeruginosa*. *FEBS Lett.*, **8**, 101–104.

Ribbons, D. W. (1971). Requirement of two protein fractions for *O*-demethylase activity in *Pseudomonas testosteroni*. *FEBS Lett.*, **12**, 161–165.

Sparnins, V. L. and P. J. Chapman (1976). Catabolism of L-tyrosine by the homoprotocatechuate pathway in gram-positive bacteria. *J. Bacteriol.*, **127**, 363–366.

Sparnins, V. L., P. J. Chapman and S. Dagley (1974). Bacterial degradation of 4-hydroxyphenylacetic acid and homo-protocatechuic acid. *J. Bacteriol.*, **120**, 159–167.

Stanier, R. Y. (1947). Simultaneous adaptation: a new technique for the study of metabolic pathways. *J. Bacteriol.*, **54**, 339–348.

Stanier, R. Y. and L. N. Ornston (1973). The β-ketoadipate pathway. *Adv. Microb. Physiol.*, **9**, 89–149.

Stanier, R. Y., N. J. Palleroni and M. Doudoroff (1966). The aerobic pseudomonads: a taxonomic study. *J. Gen. Microbiol.*, **43**, 159–271.

Strayer, R. F. and J. M. Tiedje (1978). *In situ* methane production in a small, hypereutrophic, hard-water lake. Loss of methane from sediments by vertical diffusion and ebullition. *Limnol. Oceanogr.*, **23**, 1201–1206.

Subba Rao, P. V., K. Moore and G. H. N. Towers (1967). *O*-Pyrocatechuic acid carboxylase from *Aspergillus niger*. *Arch. Biochem. Biophys.*, **122**, 466–473.

Subramanian, V., T-N. Liu, W-K. Yeh, M. Narro and D. T. Gibson (1981). Purification and properties of NADH-ferredoxin (TOL) reductase, a component of toluene dioxygenase from *Pseudomonas putida*. *J. Biol. Chem.*, **256**, 2723–2730.

Suflita, J. M., A. Horowitz, D. R. Shelton and J. M. Tiedje (1982). Dehalogenation, a novel pathway for the anaerobic biodegradation of haloaromatic compounds. *Science*, **218**, 1115–1117.

Suzuki, K. and M. Katagiri (1982). Functional consequence of modifying an arginyl residue of salicylate hydroxylase. In *Flavins and Flavoproteins*, ed. V. Massey and C. H. Williams, pp. 301–310. Elsevier–North Holland, New York.

Tack, B. F., P. J. Chapman and S. Dagley (1972). Purification and properties of 4-hydroxy-4-methyl-2-oxoglutarate aldolase. *J. Biol. Chem.*, **247**, 6444–6449.

Wang, L-H., R. Y. Hamzah and S-C. Tu (1982). On the mechanism of salicylate hydroxylase: studies using deuterated substrates. In *Flavins and Flavoproteins*, ed. V. Massey and C. H. Williams, pp. 346–349. Elsevier–North Holland, New York.

White-Stevens, R. H. and H. Kamin (1972). Studies of a flavoprotein, salicylate hydroxylase. 1, Preparation, properties and the uncoupling of oxygen reduction from hydroxylation. *J. Biol. Chem.*, **247**, 2358–2370.

Wierenga, R. K., K. H. Kalk, J. M. van der Laan, J. Drenth, J. Hofsteenge, W. J. Weijer, P. A. Jekel, J. J. Beintema, F. Muller and W. J. H. van Berkel (1982). The structure of *p* hydroxybenzoate hydroxylase. In *Flavins and Flavoproteins*, ed. V. Massey and C. H. Williams, pp. 11–18. Elsevier–North Holland, New York.

Wigmore, G. J. and D. W. Ribbons (1980). *p*-Cymene pathway in *Pseudomonas putida*: selective enrichment of defective mutants by using halogenated substrate analogs. *J. Bacteriol.*, **143**, 816–824.

Wood, J. M. (1980). Recent progress on the mechanism of action of dioxygenases. In *Metal Ion Activation of Dioxygen*, ed. T. G. Spiro, pp. 165–180. Wiley, New York.

Worsey, M. J. and P. A. Williams (1975). Metabolism of toluene and xylenes by *Pseudomonas putida* (*arvilla*) mt-2: evidence for a new function of the TOL plasmid. *J. Bacteriol.*, **124**, 7–13.

26

Bacterial Respiration

C. W. JONES
University of Leicester, UK

26.1   INTRODUCTION

Respiration is characterized by the transfer of reducing equivalents (H, H^-, e^-) from an exogenous donor to an exogenous acceptor *via* a series of sequential oxidation–reduction reactions. These reactions are catalysed by spatially-organized redox carriers housed in a unit membrane. Overall, they liberate an amount of free energy which is determined by the difference between the redox potentials of the donor and acceptor couples, and therefore different types of respiration can be accompanied by widely differing energy yields.

Most species of chemoheterotrophic bacteria, and of course mitochondria from higher organisms, oxidize organic donors (*e.g.* NADH, succinate and, less frequently, substrates such as lactate, methanol or methylated amines) using molecular oxygen as the terminal acceptor. Some of these organisms can also catalyse anaerobic respiration, during which oxygen is replaced by alternative oxidants such as various oxyanions of nitrogen and sulfur, fumarate or iron(III). A fairly restricted number of chemolithotrophic species of bacteria oxidize inorganic donors such as various nitrogen or sulfur compounds, hydrogen or iron(II) at the expense of oxygen or, occasionally, other acceptors. This chapter will confine itself to aerobic respiration in chemoheterotrophic bacteria, since it is principally organisms of this type which will be used in biotechnology for processes such as the production of biomass, the treatment of organic waste materials or the overproduction of metabolites, in which energy transduction is of major significance.

The aerobic respiratory chains of bacteria contain a diverse complement of redox carriers which include flavoproteins, iron–sulfur proteins, quinones, cytochromes and copper proteins; the first two usually, but not always, comprise the primary dehydrogenases which catalyse the initial oxidation of the donor, whereas specialized autoxidizable cytochromes act as the oxidases which catalyse the terminal reduction of oxygen to water. The overall oxidation of the organic donor entails the transfer of two hydrogen atoms per atom of oxygen consumed. However, since the individual redox carriers which constitute the respiratory chain contain either organic centres (flavoproteins, quinones) or metal centres (iron–sulfur proteins, cytochromes, copper proteins) they catalyse the transfer of hydrogen atoms (or hydride ions) and electrons, respectively, and thus respiration between these two types of centres is accompanied by the uptake or release of one or more protons.

It is now generally accepted that energized protons play an intimate part in the mechanism of respiratory chain energy transduction in bacteria, *i.e.* the redox reactions of respiration generate a membrane-associated proton motive force (Δp) which is subsequently used to drive ATP synthesis, solute transport and other energy-dependent membrane reactions (Garland, 1977; Jones, 1977; Harold, 1978). Oxidative phosphorylation, and most examples of photosynthetic phosphorylation, is thus characterized by the presence of a proton motive redox system plus a proton motive ATP phosphohydrolase complex. These are embedded in an energy-coupling membrane which is impermeable to ions, particularly H^+ and OH^-, except *via* specific exchange –diffusion systems. There is, however, considerable controversy about the way in which the proton motive force is developed and used, and exactly where it is located relative to the membrane. A number of hypotheses currently seek to answer these questions, most importantly the chemiosmotic hypothesis and several versions of the localized proton hypothesis. Chemiosmosis envisages energy transduction as occurring *via* a proton current which circulates through a passive, insulating membrane and the adjacent bulk aqueous phases (Mitchell, 1966, 1979, 1981a; Figure 1a). The proton motive force is thus a delocalized, transmembrane electrochemical potential difference of H^+ ($\Delta\bar{\mu}_{H+}$; mV) which is variably composed of a transmembrane pH gradient (ΔpH) and a membrane potential ($\Delta\psi$) according to the equation:

$$\Delta\bar{\mu}_{H+} = \Delta\psi - \frac{2.303RT\Delta\text{pH}}{F}$$

In contrast, the various versions of the localized proton hypothesis consider the proton motive force to be intramembrane or transinterface (*i.e.* to lack an osmotic component, the transported H^+), or even to be transmembrane but restricted to relatively localized surface 'channels' which allow lateral proton transfer between the various types of proton translocating systems in the membrane (Williams, 1978a, 1978b, 1979; Kell, 1979; Kell *et al.*, 1981; see also Westerhoff *et al.*, 1984; Figure 1b). In each case, however, proton movement is under strict kinetic or diffusional control such that the protons enter the bulk aqueous phases only under rather unusual circumstances, such as in the presence of charge-neutralizing ions which collapse $\Delta\psi$; furthermore, it is likely that the proton motive force across or within the membrane is normally greater than that between the bulk aqueous phases (*i.e.* $\Delta p > \Delta\bar{\mu}_{H+}$). These more localized views of energy conservation thus endow the membrane itself with a major role in energy coupling and storage. In this chapter, chemiosmosis will generally be adopted as the working mechanism of energy transduction during bacterial respiration, although it should be noted that some data can be interpreted as well or better in terms of a more localized proton current.

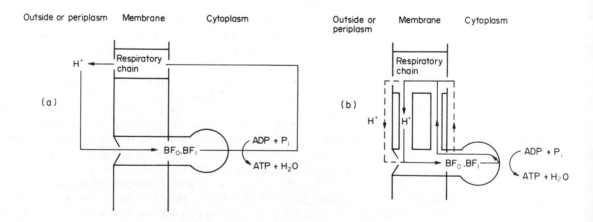

Figure 1 The pathway of the proton current during respiratory chain energy conservation: (a) chemiosmosis, and (b) two versions of the localized proton hypothesis (— intramembrane; – – –transmembrane with lateral surface conduction). Note that the ATP phosphohydrolase complex ($BF_0.BF_1$) could be replaced by a suitable proton motive force-dependent solute transport system

The activity of the proton motive respiratory chain is encapsulated by the equation:

$$DH_2 + 0.5 O_2 + z H^+_{(in)} \rightarrow z H^+_{(out)} + H_2O + D$$

where z is the $\rightarrow H^+/O$ quotient (g-ion H^+ per g-atom O) for the aerobic oxidation of the donor DH_2. Several possible mechanisms of redox-linked proton translocation have been formulated. Some of these represent direct mechanisms of energy coupling, based on the concept of vectorial group translocation, and include the proton motive redox arm and loop ($\rightarrow H^+/2e^- = 2$) and the proton motive redox cycle ($\rightarrow H^+/2e^- = 4$), all of which depend on the precise spatial and sequential organization in the membrane of the constituent organic and metal redox centres such that protons are released from the outer surface of the membrane and taken up at the inner surface. A fourth possible mechanism, the proton motive redox pump ($\rightarrow H^+/2e^- > 0$) exemplifies indirect coupling and envisages redox-linked conformational changes in the apoproteins of the redox carriers which lead to alterations in the pK_a values of appropriately located amino or carboxyl groups and hence to the release of protons from one side of the membrane and their uptake from the other side (the so-called vectorial, membrane Bohr effect); subsequent conformational changes allow outwardly-directed proton transfer across the membrane such that the appropriate surface groups are restored to their original protonation states (see Wikström *et al.*, 1981).

The synthesis of ATP at the expense of the proton motive force can be described by the equation:

$$H^+ + ADP^{3-} + HPO_4^{2-} + x\,H^+_{(out)} \rightarrow x\,H^+_{(in)} + ATP^{4-} + H_2O$$

where x is the $\rightarrow H^+/ATP$ quotient (g-ion H^+ per mol ATP synthesized). The overall stoichiometry of ATP synthesis during aerobic respiration, the ATP/O quotient, is thus equal to the overall $\rightarrow H^+/O$ quotient (*i.e.* the sum of the individual $\rightarrow H^+/2e^-$ quotients at each proton-translocating site in the respiratory chain) divided by the $\rightarrow H^+/ATP$ quotient. ATP synthesis is catalysed by the ATP phosphohydrolase complex, $BF_0.BF_1$; BF_1 is a hydrophilic, adenine nucleotide-containing polypeptide complex which is located on the cytoplasmic side of the membrane and which is responsible for catalysing the dehydration of ADP and inorganic phosphate, whereas BF_0 consists of several hydrophobic polypeptides which presumably span the membrane and hence facilitate the inwardly-directed movement of protons down their electrochemical gradient to the active site of BF_1 (Kozlov and Skulachev, 1977; Fillingame, 1980; Mitchell, 1981b; Maloney, 1982; Walker *et al.*, 1984).

The exact mechanism of ATP synthesis at the expense of energized protons remains to be elucidated, although evidence from kinetic and substrate-binding experiments increasingly supports some kind of alternating catalytic site mechanism involving cooperative interactions between up to three catalytic domains (Boyer, 1977; Cross, 1981). It is envisaged that during the latter process energized protons cause conformational changes in BF_1 such that the binding of ADP and phosphate at one site becomes tighter, whereas the binding of ATP at the second site is relaxed; ATP is then formed at the first site, the already synthesized ATP is released from the second site, and ADP and phosphate become loosely bound to the third site. This third-of-a-cycle is then effectively repeated twice more such that three molecules of ATP are synthesized and released during one complete cycle. It should be noted that the energized protons probably facilitate the binding of ADP and phosphate and the release of ATP, rather than the dehydration reaction *per se*.

26.2 THE GENERATION OF THE PROTON MOTIVE FORCE

26.2.1 Bacterial Respiratory Chains

The aerobic respiratory chains of chemoheterotrophic bacteria exhibit large variations in redox carrier composition. These variations occur both between different species of bacteria and within the same species following growth under different conditions (Haddock, 1977; Jones, 1977). Interspecies variations often involve the replacement of one or more redox carriers by others with related properties, a situation which is exemplified by the replacement of ubiquinone with menaquinone (both of which are hydrogen atom carriers, albeit with different standard redox potentials), by the replacement of, for example, cytochrome oxidases aa_3 and/or o by d, a_1 and o (all of which are electron carriers, albeit with different standard redox potentials and/or affinities for molecular oxygen) and by the replacement of one type of nicotinamide nucleotide transhydrogenase by another (both of which catalyse the reversible transfer of a hydride ion between NADPH and NAD^+, but which may or may not be proton motive; Table 1). Interspecies variations may alternatively or additionally reflect the uncompensated absence of a specific redox carrier such as cytochrome c or transhydrogenase.

Table 1   The Simplified Aerobic Respiratory Chain Compositions of Selected Chemoheterotrophic Bacteria[a]

Organism	Respiratory chain composition										
Paracoccus denitrificans	Th	Ndh	Q		*b*	[PQQ]	*c*	aa_3	(*o*)		
Alcaligenes eutrophus	Th	Ndh	Q		*b*		*c*	aa_3	(*o*)		
Methylophilus methylotrophus		Ndh	Q		*b*	[PQQ]	*c*	aa_3	(*o*)		
Micrococcus lysodeikticus	Th	Ndh		MK	*b*		*c*	aa_3	(*o*)		
Pseudomonas fluorescens	Th	Ndh	Q		*b*		*c*		*o*		
Azotobacter vinelandii		Ndh	Q		*b*		*c*		*o*	a_1	*d*
Escherichia coli	(Th)	Ndh	Q	(MK)	*b*				*o*		
Escherichia coli (oxygen-limited)	(Th)	Ndh	Q	MK	*b*				(*o*)	a_1	*d*

[a] Abbreviations: Th, nicotinamide nucleotide transhydrogenase; Ndh, NADH dehydrogenase; Q, ubiquinone; MK, menaquinone; PQQ, pyrroloquinoline quinone. Curved brackets indicate low concentration or activity. Note that PQQ is principally associated with the oxidation of methanol or methylamine, and hence is not part of the NADH oxidizing respiratory chain.

Intraspecies variations in redox carrier composition usually reflect changes in the growth environment such as the limited availability of essential nutrients, most strikingly that of oxygen, iron or sulfur. Thus oxygen-limited growth is often accompanied by the increased synthesis of alternative cytochrome oxidases which may have higher affinities for this acceptor (*e.g.* cytochrome oxidase *o* compared with aa_3, or *d* relative to aa_3 and/or *o*), and growth under iron- or sulfate-limited conditions is characterized by decreased levels of cytochromes and/or iron–sulfur proteins. Variations in redox carrier composition within a single species of bacterium can, of course, also be effected by genetic manipulation, although the ease with which this can be carried out differs dramatically between different organisms. Mutants have been reported which are deficient in ubiquinone, menaquinone or haem, or which lack the apoproteins of various dehydrogenases (most interestingly those catalysing the oxidation of NADH, succinate and methanol) or of various cytochromes (including cytochrome *c* and cytochrome oxidases aa_3 and *o*).

In addition to variations in redox carrier composition, considerable variations in the sequential organization of these carriers have been reported (Jurtshuk *et al.* 1975; Jones, 1977; Jones *et al.* 1977; John, 1981; Figure 2). All aerobic respiratory chains in bacteria exhibit extensive branching at the level of the primary dehydrogenases, and are thus able to oxidize a wide range of substrates with differing redox properties. In contrast, the terminal cytochrome system may be either linear or branched; the former are limited to a few organisms which contain only one functional cytochrome oxidase, whereas the branched systems utilize at least two oxidases and may bifurcate either immediately prior to the oxidases or at the level of cytochrome *b* (in which case one branch usually contains cytochrome *c* plus cytochrome oxidases aa_3 or *o*, and the other branch donates electrons directly to cytochrome oxidases *d*, aa_3 or *o*).

Considerably less is known about the spatial organization of the redox carriers in the cytoplasmic membrane. There is reasonably good evidence, however, that transhydrogenase and most primary dehydrogenases are intrinsic membrane proteins whose substrate-binding sites are on the cytoplasmic side of the membrane. In contrast, methanol dehydrogenase and methylamine dehydrogenase are peripheral membrane proteins which are only loosely attached to the periplasmic side of the membrane where they are closely associated with cytochrome *c* and, at least in some organisms, with a small copper protein of the azurin or amicyanin type. Ubiquinone and menaquinone are probably embedded within the interior of the membrane in accordance with their hydrophobic properties, but the scant evidence so far available for the various cytochrome oxidases is most simply interpreted as indicating that they are intrinsic membrane proteins which probably have their oxygen-binding sites on the cytoplasmic side of the membrane (see Ludwig, 1980; Poole, 1983).

26.2.2   Respiration-linked Proton Translocation

The sequential and spatial organization of the redox carriers clearly has considerable implications for respiration-linked proton translocation, particularly if the latter process occurs *via* group translocation mechanisms with fixed and predictable stoichiometries. Investigations into the sequence and topography of the respiratory chain components has thus been supplemented by detailed analysis of the ejection of protons, and to a lesser extent the secondary uptake of other ions such as K^+, during whole cell respiration. The results of such studies indicate that both the rate and the extent of acidification of the external medium is maximal only under conditions

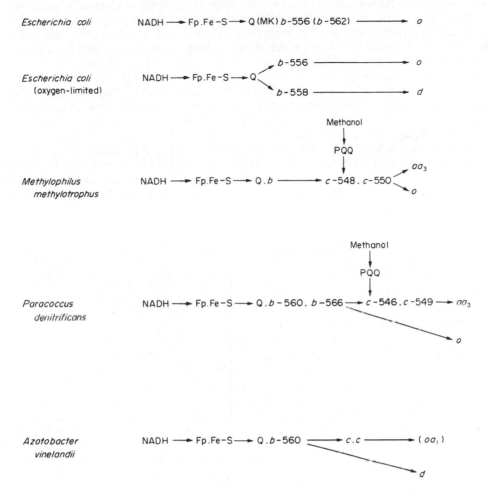

Figure 2 The aerobic respiratory chains of selected chemoheterotrophic bacteria: numbers in, for example, *b*-556 refer to wavelength maxima in low temperature reduced *minus* oxidized difference spectra, and curved brackets indicate low concentrations or activities

where the $\Delta\psi$ component of the proton motive force is collapsed (*e.g.* in the presence of a permeant anion such as thiocyanate, or of K^+ plus the ionophorous antibiotic, valinomycin). Under these conditions respiration is accompanied by the translocation of up to approximately eight protons per atom of oxygen consumed ($\rightarrow H^+/O = 8$, $\rightarrow$ charge/O = 8) according to the identity of the substrate and the redox carrier composition of the respiratory chain (Jones, 1977; Jones *et al.*, 1977; Stouthamer, 1980; Dawson and Jones, 1981; Vignais *et al.*, 1982).

During the oxidation of endogenous substrates, which are principally comprised of reduced nicotinamide nucleotides, the highest quotients are obtained with organisms that catalyse respiration *via* a respiratory chain which contains, amongst other components, a proton motive transhydrogenase, cytochrome *c* and, possibly, a proton-pumping cytochrome oxidase aa_3; two or more of these criteria are satisfied by *Paracoccus denitrificans* and *Bacillus stearothermophilus* ($\rightarrow H^+/O = 8$; $\rightarrow$ charge/O = 8). In contrast, the lowest quotients are exhibited by organisms such as *Escherichia coli* and some strains of *Bacillus megaterium* and *Bacillus subtilis*, whose respiratory chains lack several or all of these features ($\rightarrow H^+/O = 4$, $\rightarrow$ charge/O not reported). In each group of organisms the stoichiometry of respiration-linked proton or charge translocation is significantly decreased when endogenous substrates are replaced by a flavin-linked substrate such as succinate, and is further diminished, in appropriate organisms, when cytochrome *c*-linked substrates such as methanol or ascorbate-TMPD are oxidized.

It has been concluded, on the basis of these experiments, that bacterial respiratory chains contain up to four proton- (and/or charge-) translocating sites, *viz.* at the level of the energy-linked transhydrogenase (site 0; $\rightarrow H^+/2e^- = 2$, $\rightarrow$ charge/$2e^- = 2$), NADH dehydrogenase (site 1; $\rightarrow H^+/2e^- = 2$, $\rightarrow$ charge/$2e^- = 2$), the quinone–cytochrome *b* region (site 2; $\rightarrow H^+/2e^- = 4$, $\rightarrow$

charge/$2e^-$ = 2) and cytochrome c oxidase (site 3; → H$^+$/$2e^-$ ⩾0,→ charge/$2e^-$ ⩾2). Site 0 contains no detectable redox carriers and probably acts as some form of proton pump, whereas site 1 is either a loop or a pump, and site 2 is probably a quinone cycle involving the oxidized, semiquinone and reduced forms of Q or MK (in organisms lacking cytochrome c, site 2 is either a loop or a pump). According to the organism involved and the origin of the reducing power, site 3 can be an electron-transferring arm (→ H$^+$/$2e^-$ = 0, → charge/$2e^-$ = 2), a proton motive redox arm (as during the oxidation of methanol *via* the periplasmic methanol dehydrogenase; → H$^+$/$2e^-$ = 2, → charge/$2e^-$ = 2), or a combination of either type of redox arm and a proton pump (→ H$^+$/$2e^-$ = 2 or 4, → charge/$2e^-$ = 4; Figure 3). There is convincing evidence that respiration *via* sites 1, 2 and, if present, 3 generates a substantial proton motive force in whole cells (inside alkaline and electrically positive), whereas site 0 is able to conserve energy only under rather artificial conditions, *i.e.* when [NADPH][NAD$^+$] ≫ [NADP$^+$][NADH]. Indeed, redox potential considerations suggest that site 0 probably acts in reverse *in vivo*, catalysing the energy-dependent reduction of NADP$^+$ by NADH at the expense of the proton motive force generated by respiration through the other sites.

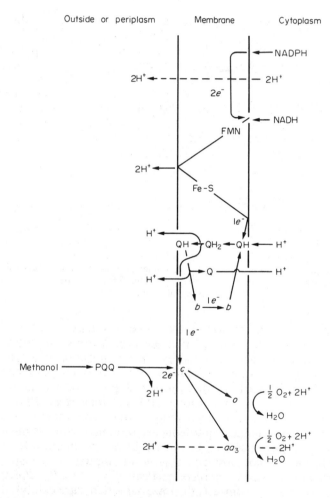

Figure 3 Respiration-linked proton translocation in aerobic chemoheterotrophic bacteria. See text for further details

26.2.3 The Proton Motive Force

There is now good evidence that aerobic respiration in chemoheterotrophic bacteria generates a substantial proton motive force which is variably composed of ΔpH and $\Delta\psi$ (cytoplasm alkaline and electrically negative). The values of ΔpH and $\Delta\psi$ during whole cell respiration have been determined principally, although not entirely, by measuring the transmembrane distribution of permeant marker molecules such as weak acids (ΔpH) and lipophilic cations ($\Delta\psi$), and more recently by phosphorus NMR techniques; it should be noted, therefore, that these methods measure the delocalized driving force ($\Delta\bar{\mu}_{H^+}$) which may only approach the true proton motive force (Δp) under certain rather unphysiological conditions (see Kell, 1979). The results indicate values for $\Delta\bar{\mu}_{H^+}$ of up to 197 mV during whole cell respiration under conditions where the dissipation of $\Delta\bar{\mu}_{H^+}$ by competing reactions such as ATP synthesis, solute transport or H^+ leakage is minimized.

In a perfectly coupled system at equilibrium the proton motive force can be predicted from the equation:

$$\Delta p = \frac{2\Delta E_h}{\rightarrow K^+/O}$$

where ΔE_h is the actual difference in redox potential between the donor and acceptor couples. However, since the value of $2\Delta E_h/ \rightarrow K^+/O$ is probably at least 285 mV for the oxidation of any respiratory chain substrate, even in bacteria which exhibit high stoichiometries of respiration-linked proton translocation, it must be concluded either that the measured proton motive force is significantly less than the true proton motive force or that the latter is not at equilibrium with the respiratory chain.

The composition of the proton motive force is largely determined, in the absence of exogenous ΔpH- or $\Delta\psi$-collapsing reagents, by the external pH. Thus in neutrophilic bacteria ΔpH is maximal at an external pH of 5.5 to 6.0, but falls to zero at pH 7.0 to 7.5, and even becomes reversed at more alkaline pH values; $\Delta\psi$ undergoes approximately converse changes. As a result of these as yet incompletely understood effects, both $\Delta\bar{\mu}_{H^+}$ and the internal pH of the organism are maintained relatively constant over quite a wide range of external pH values. This latter phenomenon is particularly important in acidophilic and alkaliphilic bacteria since it allows them to maintain their cytoplasm at a pH much closer to neutral than that of the external environment, *i.e.* approximately pH 6 compared with an external pH as low as 2 in acidophiles, and approximately pH 9 compared with an external pH as high as 12 in alkaliphiles.

26.3   THE UTILIZATION OF THE PROTON MOTIVE FORCE

The proton motive force generated by aerobic respiration in chemoheterogrophic bacteria is used to drive several physiologically important membrane functions including ATP synthesis, solute transport, reversed electron transfer and motility, of which the first two impose by far the highest demand on the overall energy budget of the cell and hence merit discussion here.

26.3.1   ATP Synthesis

The conversion of energy in the form of the membrane-associated proton motive force (Δp) into the free energy of hydrolysis of ATP within the cytoplasm of the cell (the phosphorylation potential, ΔG_p) is catalysed by the ATP phosphohydrolase complex, $BF_0.BF_1$. The physicochemical properties of this complex have been extensively investigated in *E. coli* and in the thermophilic bacterium PS3, which is probably closely related to *B. stearothermophilus*, and there is now almost general agreement that it consists of eight different polypeptide or proteolipid components (Kagawa, 1978; Fillingame, 1980; Walker *et al.*, 1984). Five of these hydrophilic polypeptides comprise BF_1 (probably in the ratio $\alpha_3\beta_3\gamma\delta\varepsilon$; MW approximately 380 000), whereas BF_0 contains up to three predominantly hydrophobic polypeptides or proteolipids (possibly in the ratio $a_2b_{2-3}c_{12}$; MW approximately 170 000). The extremely difficult problem of determining the functions of the individual subunits has been largely solved by a judicious mixture of physical reconstitution experiments and genetic analysis. The results indicate that the α and β subunits form the active site of the complex, possibly in association with the γ subunit, and that the δ and ε subunits, probably in association with the b subunit, form a proton-translocating stalk which attaches BF_1 to BF_0; proton movement through BF_0 probably occurs *via* a transmembrane channel composed of subunits c and a. The entire $BF_0.BF_1$ complex thus acts during respiration as a proton-injecting ATP synthetase in which protons traverse the membrane *via* BF_0 to reach the active site in BF_1 where ATP synthesis occurs.

There is good evidence that net synthesis of ATP occurs only when the value of $\Delta\bar{\mu}_{H^+}$, generated by respiration or by artificial means, exceeds a threshold value of 150 to 180 mV (see Maloney, 1982). It seems likely that the movement of protons to the active site of BF_1 is regulated by a proton motive force-dependent gating mechanism of some sort, possibly mediated *via* the $\gamma\delta\varepsilon$ subunits, which ensures not only that significant proton conductance occurs only when the proton motive force is sufficient to drive net ATP synthesis, but also that a large proton motive force can be maintained even at low respiration rates.

Although the details of the possible mechanism of ATP synthesis will not be considered in this chapter, it is interesting at this point to note that the alternating catalytic site mechanism outlined above (Section 26.1) is fully commensurate with the known structural properties of BF_1, in particular with the observations that the enzyme contains three loose (catalytic) and strong (regulatory) binding sites for adenine nucleotides on the β and α subunits respectively, and that the inactivation of a single β subunit in the enzyme leads to the complete inhibition of ATP synthesis. At the moment it is not known, however, how the proton motive force induces the required changes both in the conformation of the $\alpha_3\beta_3$ complex and in the affinities of the enzyme for ATP, ADP and phosphate which are essential to such a mechanism of ATP synthesis.

In contrast to the respiratory chain, where $\rightarrow H^+/O$ and $\rightarrow K^+/O$ quotients are relatively easy to measure, the determination of the $\rightarrow H^+/ATP$ quotient exhibited by the ATP phosphohydrolase complex has proved a much more difficult problem. Direct (kinetic) assays involving measurement of H^+ translocation concomitant with ATP synthesis or hydrolysis have largely proved unreliable for a number of reasons, both physiological and technical. Most attention has therefore been directed to the use of an indirect (thermodynamic) approach which is based on the parallel assay of the proton motive force and the phosphorylation potential, since under conditions where Δp is in equilibrium with the intracellular adenine nucleotide and phosphate pools,

$$\Delta p = \frac{\Delta G_p}{F(\rightarrow H^+/ATP)}$$

where ΔG_p, the phosphorylation potential, is defined as

$$\Delta G_p = \Delta G^\circ + 2.303RT\log\frac{[ATP]}{[ADP][P_i]} \text{ (kJ mol}^{-1})$$

In most species of bacteria so far examined in this way, $\Delta G_p/F$ varies from approximately 450 to 500 mV, whereas Δp (measured as $\Delta\bar{\mu}_{H^+}$) is usually in the range 140 to 197 mV, and hence the resultant $\rightarrow H^+/ATP$ quotients range upwards from approximately 2.3 according to the exact conditions of assay. If, as seems increasingly likely, the proton current is more localized than originally envisaged by chemiosmosis, $\Delta\bar{\mu}_{H^+}$ will be a variable underestimate of Δp and hence the $\rightarrow H^+/ATP$ quotients measured by this approach will be variable overestimates of the true (mechanistic) stoichiometry, *i.e.* the latter is probably very close to 2 g-ion H^+/mol ATP. Alternatively if Δp and ΔG_p are not in equilibrium, then ΔG_p will tend to be underestimated and thus the calculated $\rightarrow H^+/ATP$ quotients will also be low; in this respect it is worth noting that the unusually high values of ΔG_p measured in whole cells of *P. denitrificans* are indicative of an $\rightarrow H^+/ATP$ quotient of 3–4 g-ion H^+/mol ATP and a similarly high value has recently been reported *via* measurement of $\rightarrow H^+/O$ and P_i/O quotients in *M. methylotrophus*. It seems likely, therefore, that the $\rightarrow H^+/ATP$ quotient in bacteria is probably about 3.

26.3.2 Active Transport of Solutes

During active transport a solute moves across the cytoplasmic membrane from a region of low electrochemical potential to a region of high electrochemical potential, *i.e.* the solute is accumulated in one compartment relative to a second compartment at the expense of free energy. Thus, as described above, the release of free energy during respiration allows the active translocation of H^+ to generate a proton motive force (primary active transport), which in turn can drive the active accumulation or ejection of various other solutes *via* variably specific solute carriers (secondary active transport). These solutes do not undergo chemical modification during transport and include amino acids, inorganic ions, lactose and, in obligately aerobic bacteria, glucose. In addition, some examples of secondary active transport are known in which solute movement occurs at the expense of an electrochemical potential difference of sodium ions ($\Delta\bar{\mu}_{Na^+}$) rather than of protons (Hamilton, 1977; West, 1980).

Unlike ATP synthesis and hydrolysis (Section 27.3.1), which are dependent on the magnitude rather than the composition of the proton motive force, the active uptake or ejection of a particular solute is affected by both of these parameters. Firstly, the direction in which the solute moves during active transport is determined by the relative values of the opposing driving forces, *i.e.* the proton motive force and the transmembrane electrochemical potential difference of the solute ($\Delta\bar{\mu}_S$). When Δp exceeds $\Delta\bar{\mu}_S$ the result is a net accumulation or a net ejection of the solute until the opposing forces are equalized; in contrast, when $\Delta\bar{\mu}_S$ exceeds Δp the solute moves down its own concentration gradient until $\Delta\bar{\mu}_S$ and Δp are again equal. It should be noted, however, that there is increasing evidence that solute transport often attains a kinetic steady state rather than a thermodynamic equilibrium, *i.e.* $\Delta\bar{\mu}_S$ is often significantly lower than Δp (measured as $\Delta\bar{\mu}_{H^+}$) such that the accumulation ratio, or the exclusion ratio, of a particular solute is often lower than the value predicted solely from thermodynamic considerations. Secondly, since the directional properties of the proton motive force requires that the actively transported species is either neutral or positively charged, anions and neutral molecules enter in cotransport with one or more H^+ (H^+ symport), whereas cations are taken up alone (cation uniport). Similarly, surplus cations leave in counter transport with H^+ (H^+.cation antiport), whereas anions exit in cotransport with H^+ (H^+.anion symport). The active accumulation or ejection of a particular solute thus depends uniquely on either ΔpH (*e.g.* $2H^+$.Ca^{2+} antiport), $\Delta\psi$ (*e.g.* K^+ uniport) or the overall proton motive force, Δp (*e.g.* H^+.lactose symport). Similarly, the efflux of a metabolic end-product down its own concentration gradient may uniquely generate ΔpH and/or $\Delta\psi$ (Konings and Veldkamp, 1980).

The active accumulation of essential major nutrients or cofactors (*e.g.* the carbon source, NH_4^+, K^+) and the active ejection of surplus solutes (*e.g.* Na^+, Ca^{2+}) can consume up to a third of the free energy released during catabolism and respiration, although the precise amount obviously varies with the physiological properties of the organism, the nature of the growth substrate, and the conditions under which the organism is cultured. It has recently become clear, however, that some of this expenditure of energy can be saved as a result of the passive efflux of a metabolic end product, such as lactate, in cotransport with one or more protons, which thus generates a significant ΔpH or Δp. The homolactic fermentation of glucose by *Streptococcus* provides an extreme example of this phenomenon since the growth yield of the organism increases by up to 60% if the excreted lactate is continuously removed in order to maintain a high $\Delta\bar{\mu}_{lactate}$. This phenomenon is, however, of relatively little significance during the growth of aerobic bacteria since the excretion of metabolites other than carbon dioxide and water generally infers incomplete oxidation of the carbon substrate and hence the loss of a very much greater amount of free energy than can be salvaged *via* H^+.product symport.

The expenditure of energy on the active transport of solutes, in particular the ejection of Na^+, can be significantly minimized of course by growing organisms in a medium whose solute composition is maintained close to that of the cell cytoplasm; under such conditions a higher proportion of the available energy can be directed towards the biosynthesis of cell material. In a similar vein it has been shown that a mutation in a solute transport system which leads to a higher H^+.solute quotient, and hence to an increased ability to accumulate the solute, engenders a higher rate of growth when that solute is the limiting component in the growth medium (although the efficiency of growth is presumably somewhat diminished). Finally, it should be noted that the presence of high concentrations of weak acids such as acetate or formate in the growth medium, particularly at external pH values which are equal to or less than the pK_a values of these acids, leads to significant uncoupling of respiratory chain energy conservation and hence to the inhibition of cell growth; under these conditions the acid diffuses into the cell in the undissociated state but subsequently undergoes rapid and extensive dissociation such that the ΔpH component of the proton motive force is significantly collapsed.

26.4 BIOTECHNOLOGICAL ASPECTS OF BACTERIAL RESPIRATION

Respiratory chain energy conservation in aerobic bacteria is crucially important to two distinct types of biotechnological processes, *viz.* (*i*) biomass production and (*ii*) the treatment of waste materials and the overproduction of useful metabolites. Paradoxically, these can be said to have virtually opposite aims. Thus, during biomass production the aim is to maximize the efficiency with which cell material is synthesized at the expense of the carbon source and other essential nutrients, whereas during waste treatment or metabolite overproduction processes the aim is to maximize the efficiency with which the carbon source is degraded to suitable end products, in most instances with the minimum production of cell material. Since the respiratory chain provides the energetic link between the exergonic reactions of degradative metabolism and the endergonic reactions of biosynthesis, a thorough understanding of the features which determine the efficiency of respiratory chain energy conservation in bacteria is particularly important to these biotechnological processes.

26.4.1 Biomass Production

Biomass production is currently best, and most simply, exemplified by the production of bacterial single cell protein (SCP), *i.e.* the commercial production of an expensive animal, or ultimately human, feedstock from a relatively cheap carbon source (Kharatyan, 1978; Mateles, 1979; Drozd and Linton, 1981). It is thus necessary to select for this large scale process an organism, or a consortium of different organisms, which not only has the appropriate catabolic and biosynthetic properties, but which also can catalyse respiratory chain energy conservation with high efficiency such as to maximize the yield of ATP per molecule of donor oxidized and acceptor reduced. The effect of this is of course to minimize the consumption of both the donor and the acceptor, and also, perhaps less obviously but very importantly, to limit the production of heat. In general terms, therefore, the strategy adopted is to ensure that respiration releases the largest available amount of free energy and that as much of this as possible is conserved, initially in the generation of the proton motive force and subsequently in the synthesis of ATP and in the accumulation of essential nutrients.

There is little doubt that oxygen is the best terminal electron acceptor for SCP production since it is fairly cheap when provided as compressed air, it readily penetrates biological membranes, it has a high gas transfer coefficient, its reduction product is entirely non-toxic, and the redox potential of the oxygen/water couple is high ($E_0' = O_2/H_2O = +820$ mV). In contrast, several possible sources of reducing substrates have been considered including methane and other saturated hydrocarbons, methanol and ethanol. Early investigations were directed towards the use of hydrocarbons, mainly because of their relative cheapness and easy availability in natural gas and as by-products of the oil industry. Unfortunately, however, they have a limited solubility in water, and methane in particular is potentially highly explosive; furthermore, catabolism of the hydrocarbons occurs *via* an initial oxygenation reaction, for example to yield methanol from methane, which not only wastes both oxygen and reducing power but also generates a large amount of heat. Methane was therefore abandoned in favour of methanol, which is significantly cheaper than ethanol. In addition, methanol presents few hazards, readily permeates biological membranes (although, as it turns out, it does not need to do so) and is highly miscible with water; the redox potential of the formaldehyde/methanol couple is also relatively low (E_0' HCHO/ MeOH = -180 mV).

The one SCP process based on bacteria which is currently in large scale production, the ICI PRUTEEN process, thus involves the aerobic continuous culture of a methylotrophic bacterium, in this case *Methylophilus methylotrophus*, on a minimal salts medium plus methanol (Senior and Windass, 1980; Smith, 1981). The use of methanol does, however, have two potential disadvantages. Firstly, the product of methanol oxidation, formaldehyde, is rather toxic; this hazard can be overcome by growing cells under methanol-limited conditions (which also lead to the highest efficiencies of carbon conversion, *i.e.* of methanol carbon into cell carbon) such that much of the formaldehyde produced is immediately assimilated and the remainder is rapidly oxidized to carbon dioxide, with the concomitant release of reducing power mainly in the form of NAD(P)H. Secondly, although the redox potential of the formaldehyde/methanol couple is fairly low, and is thus commensurate with the oxidation of methanol *via* a methanol dehydrogenase with flavin or even NAD$^+$ as its prosthetic group/coenzyme (as during the oxidation of lactate by bacteria), all of the methanol dehydrogenases so far investigated in methylotrophic bacteria contain a novel pyrroloquinolinequinone prosthetic group (E_0' PQQ/PQQH$_2$ $\simeq$ +120 mV) and mostly donate reducing equivalents to the main respiratory chain at the level of cytochrome *c* (but see Duine *et al.*, 1984). The oxidation of methanol to formaldehyde is thus associated with energy coupling only at site 3 ($\rightarrow$ H$^+$/O = 2, $\rightarrow$ K$^+$/O = 2) and hence probably leads to a maximum ATP/O quotient of 1. In contrast to some other species of methylotrophic bacteria such as *Pseudomonas* AM1 (Anthony, 1981), cytochrome *c* appears to be an obligatory prerequisite for respiration from NAD(P)H in *M. methylotrophus*; the oxidation of the NAD(P)H generated by the further catabolism of formaldehyde is thus associated with energy coupling at all three sites ($\rightarrow$ H$^+$/O = 6, $\rightarrow$ K$^+$/O = 6) and hence with an ATP/O quotient $\geqslant$ 1.5 (Dawson and Jones, 1981; Jones *et al.*, 1984). It is interesting to note that although the respiratory chain of *M. methylotrophus* contains cytochrome oxidase *aa*$_3$ in addition to cytochrome oxidase *o*, there is currently no evidence to indicate that the cytochrome *aa*$_3$ has any capacity to pump protons (although such an activity could be obscured if cytochrome oxidase *o* consumed two protons from the periplasm rather than from the cytoplasm during the terminal reduction of an atom of oxygen to water).

The respiratory membrane of *M. methylotrophus*, like that of most mesophilic and neutrophilic bacteria so far investigated, is relatively impermeable to protons except *via* specific porter systems such as the ATP phosphohydrolase complex or the various proton-linked solute transport systems, *i.e.* little of the proton current is short-circuited by proton leakage through the membrane, at least when measured at the moderate to low levels of the proton motive force which pertain when such measurements are made.

It is clear, therefore, that the membrane-bound respiratory chain of *M. methylotrophus* is quite well suited to the commercial role which has recently been thrust upon this organism. The carbon conversion efficiency of methanol-limited cultures is therefore fairly high. There are, however, several features of the respiratory chain which are obviously less than perfect, and which it may be possible to improve upon at a later date by genetic manipulation. If such an approach proves successful (as has indeed been claimed with respect to alterations in the nitrogen assimilation system of this organism; Senior and Windass, 1980) the carbon conversion efficiency might be expected to rise significantly, and hence bring about associated decreases in the costs for methanol and energy (the latter being used mainly for cooling and air compression) which currently comprise approximately 59% and 20% respectively of the total operating costs of the process.

Finally it is interesting to note that although energy-limited growth conditions are required for the most efficient production of biomass in the form of SCP, the synthesis of bacterial biomass as exo- or endo-polysaccharide storage polymers (*e.g.* xanthan gums, alginates, poly-β-hydroxybutyrate) is optimal under conditions of carbon and energy excess, such as when the growth of the selected organism is limited by the availability of ammonia or sulfate (Dawes and Senior, 1973; Sutherland and Ellwood, 1979).

26.4.2 Waste Treatment and Metabolite Production

Sewage and industrial effluent processes have traditionally concentrated on the maximal degradation of the waste materials to simple, non-toxic end products; solids (sludge), mainly in the form of biomass, are then removed and the innocuous effluent is then recycled back into the original industrial process or disposed of *via* the usual routes (see Slater and Somerville, 1979). Such treatment processes are usually aerobic, oxygenation occurring relatively slowly in processes which entail the use of unstirred ponds or filter-beds, but rather more rapidly in systems such as the UNOX or FMC where air is replaced by oxygen, and very rapidly in highly agitated systems, such as that exemplified by the ICI deep shaft process which operates *via* a novel air-lift design (Hines *et al.*, 1975).

Almost by definition, such effluent treatment processes operate under conditions of excess carbon substrate(s) with either excess or limiting concentrations of oxygen and, very often, limiting concentrations of other essential nutrients such as phosphate or ammonia. These carbon-sufficient conditions inevitably lead to incomplete oxidation of the substrates, firstly because the cells appear unable closely to control the uptake and/or metabolism of the excess carbon source, and secondly because discrete bottlenecks occur in catabolic pathways due to the low availability of required cofactors or metabolites. The products of aerobic catabolism thus reflect both the identity of the carbon source and the nature of the growth-limiting nutrient, and commonly include acetate, lactate, ethanol, gluconate and 2-oxoglutarate, as well as carbon dioxide and water (Neijssel and Tempest, 1979). Most of these effluent treatment processes employ a consortium of bacterial species, which thus afford a wide range of respiratory chains with varied redox carrier compositions and electron transfer pathways, together with different affinities for various electron donors and molecular oxygen, and significant variations in membrane integrity. The overall efficiency of respiratory chain energy conservation in such a population is thus relatively low and this, allied to the metabolic imperfections outlined above plus a significant degree of uncoupling of respiration from energy conservation, leads to low efficiencies of bacterial growth and hence to the relatively low cell yields which have traditionally characterized such processes.

More recently, however, attention has been increasingly diverted from the traditional view that effluents are necessarily waste materials which need to be treated and discharged, and towards the idea that they could be used as potential feedstocks for the formation of biomass (either as single cell protein for animal feed or as high-quality fertilizer) or other bulk products of commercial importance (in particular methane for fuel). The production of biomass from effluents has so far been based upon the cultivation of yeast, algae or fungi, rather than bacteria, using carbon sources such as starch wastes or sulfite liquor. In contrast, methane production is principally a bacterial-based process, albeit an anaerobic fermentation during which carbon dioxide undergoes a complex eight electron reduction at the expense of hydrogen or suitable organic donors.

ACKNOWLEDGEMENTS

The author is indebted to Drs. M. J. Dawson and M. Carver for many stimulating discussions. Work carried out in the author's laboratory was supported financially by the University of Leicester, the Science and Engineering Research Council and ICI.

26.5 REFERENCES

Anthony, C. (1981). Electron transport in methylotrophic bacteria. In *Microbial Growth on C_1 Compounds*, ed. H. Dalton, pp. 220–230. Heyden, London.
Boyer, P. D. (1977). Conformational coupling in oxidative phosphorylation and photophosphorylation. *TIBS*, **2**, 38–40.
Cross, R. L. (1981). The mechanism and regulation of ATP synthesis by F_1-ATPases. *Annu. Rev. Biochem.*, **50**, 681–714.
Dawes, E. A. and P. J. Senior (1973). The role and regulation of energy reserve polymers in microorganisms. *Adv. Microb. Physiol.*, **10**, 135–266.
Dawson, M. J. and C. W. Jones, (1981). Chemiosmotic aspects of respiratory chain energy conservation in *Methylophilus methylotrophus*. In *Microbial Growth on C_1 Compounds*, ed. H. Dalton, pp. 251–257. Heyden, London.
Drozd, J. W. and J. D. Linton (1981). Single cell protein production from methane and methanol. In *Continuous Cultures of Cells*, ed. P. H. Calcott, vol. 1, pp. 113–141. C.R.C. Press, Boca Raton, FL.
Duine, J. A., J. Frank, J. A. Jongejan and M. Dijkstra (1984). Enzymology of the bacterial methanol oxidation step. In *Microbial Growth on C_1 Compounds*, ed. R. L. Crawford and R. S. Hanson, pp. 91–96. American Society for Microbiology, Washington.
Fillingame, R. H. (1980). The proton-translocating pumps of oxidative phosphorylation. *Annu. Rev. Biochem.*, **49**, 1079–1113.
Garland, P. B. (1977). Energy transduction in microbial systems. In *Microbial Energetics*, ed. B. A. Haddock and W. A. Hamilton, pp. 1–21. Cambridge University Press, Cambridge.
Haddock, B. A. (1977). The isolation of phenotypic and genotypic variants for the functional characterisation of bacterial oxidative phosphorylation. In *Microbial Energetics*, ed. W. A. Hamilton and B. A. Haddock, pp. 95–120. Cambridge University Press, Cambridge.
Hamilton, W. A. (1977). Energy coupling in substrate and group translocation. In *Microbial Energetics* ed. W. A. Hamilton and B. A. Haddock, pp. 185–216. Cambridge University Press, Cambridge.
Harold, F. M. (1978). Vectorial metabolism. In *The Bacteria*, ed. L. N. Ornston and J. R. Sokatch, vol. 6. pp. 463–521. Academic, New York.
Hines, D. A., M. Bailey, J. C. Ousby and F. C. Roesler (1975). The ICI deepshaft aeration process for effluent treatment. *Inst. Chem. Engineers Symp.*, **41**, D1–10.
John, P. (1981). Schematic representations of branched respiratory chains. *Trends Biochem. Sci.*, **6**, VIII–X.
Jones, C. W. (1977). Aerobic respiratory systems in bacteria. In *Microbial Energetics*, ed. B. A. Haddock and W. A. Hamilton, pp. 23–59. Cambridge University Press, Cambridge.
Jones, C. W., J. M. Brice and C. Edwards (1977). Bacterial cytochrome oxidases and respiratory chain energy conservation. In *Functions of Alternative Terminal Oxidases*, ed. H. Degn, D. Lloyd and G. C. Hill, FEBS Meeting, vol. 49, pp. 89–97. Pergamon, Oxford.
Jones, C. W., M. A. Carver, M. J. Dawson and R. A. Patchett (1984). Respiration and ATP synthesis in *M. methylotrophus*. In *Microbial Growth on C_1 Compounds*, ed. R. L. Crawford and R. S. Hanson, pp. 134–140. American Society for Microbiology, Washington.
Jurtshuk, P., T. J. Mueller and W. C. Acord (1975). Bacterial terminal oxidases. *CRC Crit. Rev. Microbiol.*, **3**, 399–468.
Kagawa, Y. (1978). Reconstitution of the energy transformer, gate and channel, subunit reassembly, crystalline ATPase and ATP synthesis. *Biochim. Biophys. Acta*, **505**, 45–93.
Kell, D. B. (1979). On the functional proton current pathway of electron transport phosphorylation; an electrodic view. *Biochim. Biophys. Acta*, **549**, 55–99.
Kell, D. B., D. J. Clarke and J. G. Morris (1981). On proton-coupled information transfer along the surface of biological membranes and the mode of certain colicins. *FEMS Microbiol. Lett.*, **11**, 1–11.
Kharatyan, S. G. (1978). Microbes as food for humans. *Annu. Rev. Microbiol.*, **32**, 301–327.
Konings, W. N. and H. Veldkamp (1980). Phenotypic responses to environmental change. In *Contemporary Microbial Ecology*, ed. D. C. Ellwood, J. N. Hedger, M. J. Latham, J. M. Lynch and J. H. Slater, pp.161–191. Academic, London.
Kozlov, I. A. and V. P. Skulachev (1977). H^+-adenosine triphosphatase and membrane energy coupling. *Biochim. Biophys. Acta*, **463**, 29–89.
Ludwig, B. (1980). Haem aa_3-type cytochrome c oxidases from bacteria. *Biochim. Biophys. Acta*, **594**, 177–189.
Maloney, P. C. (1982). Coupling between H^+ entry and ATP synthesis in bacteria. *Curr. Top. Membr. Transp.*, **16**, 175–193.
Mateles, R. J. (1979). The physiology of single-cell protein (SCP) production. In *Microbial Technology:Current State, Future Prospects*, ed. A. T. Bull, D. C. Ellwood and C. Ratledge, pp. 29–52. Cambridge University Press, Cambridge.
Mitchell, P. (1979). David Keilin's respiratory chain concept and its chemiosmotic consequences. In *Les Prix Nobel en 1978*, pp. 135–172. Nobel Foundation, Stockholm.

Mitchell, P. (1981a). Davy's electrochemistry: nature's protochemistry. *Chem. Br.* **17**, 14–23.

Mitchell, P. (1981b). Biochemical mechanism of proton motive phosphorylation in F_0F_1adenosine triphosphatase molecules. In *Mitochondria and Microsomes*, ed. C. P. Lee, G. Schatz and G. Dallner, pp. 427–457. Addison–Wesley, New York.

Neijssel, O. M. and D. W. Tempest (1979). The physiology of metabolite overproduction. In *Microbial Technology: Current State, Future Prospects*, ed. A. T. Bull, D. C. Ellwood and C. Ratledge, pp. 53–82. Cambridge University Press, Cambridge.

Poole, R. K. (1983). Bacterial cytochrome oxidases: a structurally and functionally diverse group of electron transfer proteins. *Biochim. Biophys. Acta*, **726**, 205–243.

Senior, P. J. and J. Windass (1980). The ICI single cell protein process. In *Biotechnology—A Hidden Past, A Shining Future* (13th International TNO conference), pp. 97–101. Netherlands Central Organisation for Applied Research.

Slater, J. H. and H. J. Somerville (1979). Microbial aspects of waste treatment with particular attention to the degradation of organic compounds. In *Microbial Technology: Current State, Future Prospects*, pp. 222–261. Cambridge University Press, Cambridge.

Smith, S. R. L. (1981). Some aspects of ICI's single cell protein process. In *Microbial Growth on C_1 Compounds*, ed. H. Dalton, pp. 342–348. Heyden, London.

Stouthamer, A. H. (1980). Bioenergetic studies on *Paracoccus denitrificans*. *Trends Biochem. Sci.*, **5**, 164–166.

Sutherland, I. W. and D. C. Ellwood (1979). Microbial exopolysaccharides—industrial polymers of current and future potential. In *Microbial Technology: Current State, Future Prospects*, ed. A. T. Bull, D. C. Ellwood and C. Ratledge, pp. 107–150. Cambridge University Press, Cambridge.

Vignais, P. M., M. F. Henry, E. Sim and D. B. Kell (1982). The electron transport system and hydrogenase of *Paracoccus denitrificans*. *Curr. Top. Bioenergetics*, **12**, 116–196.

Westerhoff, H. V., B. A. Melandri, G. Venturoli, G. F. Azzone and D. B. Kell (1984). Mosaic protonic coupling hypothesis for free energy transduction. *FEBS Lett.*, **165**, 1–6.

Walker, L. E., M. Saraste and N. J. Gay (1984). The *unc* operon; nucleotide sequence, regulation and structure of ATP synthase. *Biochim. Biophys. Acta*, **768**, 164–200.

West, I. C. (1980). Energy coupling in secondary active transport. *Biochim. Biophys. Acta*, **604**, 91–126.

Wikström, M., K. Krab and M. Saraste (1981). Proton-translocating cytochrome complexes. *Annu. Rev. Biochem.*, **50**, 623–655.

Williams, R. J. P. (1978a). The multifarious couplings of energy transduction. *Biochim. Biophys. Acta*, **505**, 1–44.

Williams, R. J. P. (1978b). The history and the hypotheses concerning ATP formation by energised protons. *FEBS Lett.*, **85**, 9–19.

Williams, R. J. P. (1979). Some unrealistic assumptions in the theory of chemiosmosis and their consequences. *FEBS Lett.*, **102**, 126–132.

27

Enzyme Kinetics

A. CORNISH-BOWDEN
University of Birmingham, UK

27.1 INTRODUCTION

Although chemical kinetics developed in the last century as an academic discipline, the parallel study of enzymes and enzyme kinetics had from the beginning a biotechnological flavour, and could almost be regarded as an offshoot of the brewing industry. The enzymes of alcoholic fermentation, such as invertase and the amylases, were among the first to be intensively studied, and the people who laid the foundations of enzymology around the turn of the century (O'Sullivan and Tompson, 1890; Brown, 1892, 1902; Henri, 1902, 1903; Sørensen, 1909; Michaelis and Menten, 1913) were for the most part involved to some degree with the brewing industry. Today, when pH has been absorbed into the mainstream of chemistry, it is easy to forget that it had its origins not in chemistry but in industrial enzymology and that Sørensen was employed by a brewery. This is not in fact very surprising, because in scientific brewing there is an acute need to be able to manage reactions that only occur under conditions that most chemists, even now, regard as absurdly mild: it is salutary to remember that the substance known to chemists as 'dilute hydrochloric acid' has a pH less than zero. Moreover, the great rapidity with which Michaelis built

521

upon Sørensen's work (Michaelis and Davidsohn, 1911) makes it clear that the ability to control pH by means of buffers satisfied a need that Sørensen was by no means unique in feeling.

After the time of Michaelis, brewing ceased to be a dominant theme in enzymology, probably because once the biotechnological problems that enzymology was then capable of solving had been solved and the solutions had become part of the established knowledge of the industry, further development of enzymology seemed to be academic. Today it is mainly enzymologists who are interested in enzymes for their own sake, but a much larger group of people require a basic understanding of enzyme kinetics in order to use enzymes for many practical purposes in medicine and in biotechnology.

27.2 KINETICS OF SINGLE-SUBSTRATE REACTIONS

27.2.1 The Michaelis–Menten Mechanism

Compared with simple uncatalysed reactions, the most striking characteristic of enzyme-catalysed reactions is that they display saturation. This means that as the concentration of substrate is increased the rate of reaction does not increase indefinitely but instead approaches a finite limit. The simplest explanation of this behaviour is that an enzyme-catalysed reaction must pass through at least two steps: first, enzyme E and substrate A must bind together to form an enzyme–substrate complex, EA; then conversion of substrate into product Z takes place within this complex and the product is released:

$$E + A \overset{fast}{\rightleftharpoons} EA \overset{slow}{\rightarrow} E + Z \tag{1}$$

This is called the Michaelis–Menten mechanism, and it forms the starting point for studying all enzyme-catalysed reactions. Michaelis and Menten (1913) assumed that the binding reaction was fast enough to be treated as an equilibrium (as indicated), but this assumption is unnecessary and essentially the same sort of kinetic behaviour is expected regardless of the relative rates of the two steps. It is also unnecessary to suppose that there are only two steps or that the reaction is irreversible: if the chemical conversion of A into Z occurs in a separate step from the release of Z, or if the whole process is reversible, the behaviour of the initial rate as a function of substrate concentration is still of the same kind as for the simplest form of the Michaelis–Menten mechanism shown in equation (1).

27.2.2 Steady-state Treatment

If we make no assumptions about the relative rates of the two steps or about the irreversibility of the whole process, the Michaelis–Menten mechanism may be written as follows:

$$E + A \underset{k_{-1}}{\overset{k_{+1}}{\rightleftharpoons}} EA \underset{k_{-2}}{\overset{k_{+2}}{\rightleftharpoons}} E + Z \tag{2}$$
$$e_0-x \quad a \qquad\qquad x \qquad\qquad z$$

where k_{+1}, k_{-1}, k_{+2} and k_{-2} are the rate constants for the various steps. Briggs and Haldane (1925) showed that this general mechanism led to the same sort of dependence of initial rate on substrate as the more restricted Michaelis–Menten mechanism. They argued that after an initial rapid rise in the concentration of the intermediate EA, it would be in a steady state in which its rate of formation was balanced by the rate of its conversion into products.

If the concentrations of the various species at any instant are as shown under equation (2), the steady-state assumption implies that dx/dt, the rate of change of intermediate concentration, is zero. However, this does not immediately lead to a simple result unless we make two further stipulations: the substrate concentration must be much larger than the enzyme concentration, so that we can ignore the distinction between free and total substrate concentrations, and the product Z is essentially absent during the period of measurement, so that the reverse reaction can be neglected. It is usually possible to ensure that these conditions are satisfied in practice, but if they are not a more complex analysis than what follows is required. When they are satisfied, we can write

$$\frac{dx}{dt} = k_{+1}(e_0 - x)a - k_{-1}x - k_{+2}x = 0 \tag{3}$$

which may be rearranged to give the steady-state concentration of EA:

$$x = \frac{k_{+1}e_0a}{k_{-1} + k_{+2} + k_{+1}a} \tag{4}$$

The rate of reaction is then $k_{+2}x$, *i.e.*

$$v = \frac{k_{+1}k_{+2}e_0a}{k_{-1} + k_{+2} + k_{+1}a} = \frac{k_{+2}e_0a}{[(k_{-1} + k_{+2})/k_{+1}] + a} \tag{5}$$

This is called the Michaelis–Menten equation.

27.2.3 Properties of the Michaelis–Menten Equation

An equation of the form of equation (5) can be derived for many mechanisms more complex than equation (2). Consequently it is usual to write the Michaelis–Menten equation in a way that does not imply more knowledge about the individual steps than one has, *i.e.* as

$$v = \frac{k_0e_0a}{K_m + a} \tag{6}$$

in which k_0 is called the catalytic constant and K_m is the Michaelis constant. In the special case where the mechanism is as simple as equation (2), they are equivalent to k_{+2} and $(k_{-1} + k_{+2})/k_{+1}$ respectively. The ratio k_0/K_m is no less important a parameter than k_0 and K_m, and the Nomenclature Committee of the International Union of Biochemistry (1982) has recommended that it be given the symbol k_A and the name specificity constant. (These recommendations are rather recent and not yet widely adopted.) The Michaelis–Menten equation can be expressed equivalently in terms of any pair out of k_0, K_m and k_A, *i.e.* instead of equation (6) we may write

$$v = \frac{k_Ae_0a}{1 + a/K_m} \tag{7}$$

or

$$v = \frac{k_0k_Ae_0a}{k_0 + k_Aa} \tag{8}$$

For partially purified or otherwise incompletely characterized enzymes it may be impossible to specify the enzyme concentration in meaningful units, and it is often convenient to replace k_0e_0 by V, a quantity known as the limiting velocity. (The name maximum velocity is also often used but is less appropriate because there is no maximum in the mathematical sense.) So equation (6) becomes

$$v = \frac{Va}{K_m + a} \tag{9}$$

The practical meanings of the parameters of the Michaelis–Menten equation may be seen from Figure 1, which shows the predicted dependence of v on a. At low concentrations the kinetics approximate to a first-order dependence on a (second-order overall), with k_A as second-order rate constant. At high values of a the rate approaches a limiting value of k_0e_0 or V. (The curve is drawn very inaccurately in many textbooks and may give the impression that V can be measured directly; Figure 1 should make it clear that v approaches V too slowly for this to be feasible.) The Michaelis constant is the concentration of substrate when the rate is half the limiting value, *i.e.* $a = K_m$ when $v = 0.5V$.

Mechanistic meanings can also be attached to the various parameters. The Michaelis constant is an inverse measure of the strength of substrate binding: although it is not in general a true thermodynamic equilibrium constant it does express the same kind of ratio, *i.e.* [E][A]/[EA], but under steady-state, not equilibrium, conditions. The catalytic constant measures the reactivity of the enzyme–substrate complex once formed, *i.e.* it measures the catalytic potential of the enzyme uncomplicated by considerations of binding. The specificity constant provides the best measure of the total effectiveness of an enzyme, as it takes account of both binding and chemical reactivity, and is the only parameter with the appropriate dimensions for comparison with second-order rate constants of model reactions. It also provides a direct measure of the specificity of an enzyme, as its name suggests. In the laboratory it is tempting to consider specificity by examining a second

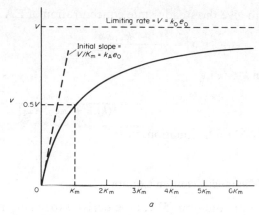

Figure 1 Plot of initial rate v against substrate concentration a for an enzyme obeying the Michaelis–Menten equation

reaction, *e.g.* the conversion of B into Y, in isolation, so that the second rate v_B can be expressed as follows (*cf.* equation 7):

$$v_B = \frac{k_B e_0 b}{1 + b/K_m^B} \qquad (10)$$

but this is not a meaningful way of considering physiological specificity. In the living organism specificity can only mean the ability of an enzyme to discriminate between two (or more) substrates that are available simultaneously, *i.e.* mixed together. Under such circumstances the rate v_B is

$$v_B = \frac{k_B e_0 b}{1 + (a/K_m^A) + (b/K_m^B)} \qquad (11)$$

and the rate of the first reaction, which can be written as v_A, is

$$v_A = \frac{k_A e_0 a}{1 + (a/K_m^A) + (b/K_m^B)} \qquad (12)$$

For the derivation of these equations, see Cornish-Bowden (1979). As both equations have the same denominator, it follows that the ratio of rates is simply

$$v_A/v_B = k_A a/k_B b \qquad (13)$$

i.e. it depends only on the concentrations and specificity constants of the two substrates. The catalytic constants and Michaelis constants are immaterial.

27.2.4 Determination of Michaelis–Menten Parameters

27.2.4.1 *Graphical methods*

Graphs provide the most convenient method for presenting the results of kinetic experiments, and they are often used as well for the actual analysis. It is rather important therefore that the graphs used should be capable of yielding satisfactory results. We have seen (Figure 1) that at realistic substrate concentrations the rate does not approach close enough to the limiting rate V for it to be found by direct measurement, or for the plot of v against a to give an acceptable method of determining any of the kinetic parameters. The second worst method is also, unfortunately, by far the most widely used. It involves plotting $1/v$ against $1/a$, which gives a straight line with slope $1/k_A e_0$, or $K_m/k_0 e_0$, and intercept $1/k_0 e_0$ on the ordinate axis (Figure 2), as may be seen by examining the equation that results from taking reciprocals of both sides of equation (8) or equation (6):

$$\frac{1}{v} = \frac{1}{k_A e_0 a} + \frac{1}{k_0 e_0} = \frac{K_m}{k_0 e_0 a} + \frac{1}{k_0 e_0} \qquad (14)$$

This is called a double-reciprocal plot. It has the very undesirable property of making it almost impossible to recognize by inspection which observations are less trustworthy than others, as may be seen from the great variation in the lengths of the error bars in Figure 2, each of which represents the same amount of uncertainty in the corresponding v. The double-reciprocal plot also tends to allow an apparently well fitting line to be drawn through poor observations, thereby suggesting greater precision than was actually achieved. Despite these defects (or perhaps because of them) the double-reciprocal plot remains in wide use.

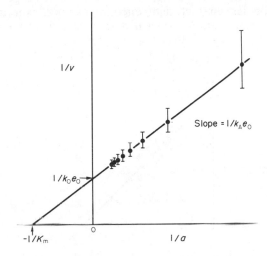

Figure 2 Plot of $1/v$ against $1/a$. The error bars represent errors of $\pm 0.05V$ in the initial rate v. This plot is commonly known as a double-reciprocal plot or a Lineweaver–Burk plot

A much better plot is given by the equation that results from multiplying every term in equation (14) by a:

$$\frac{a}{v} = \frac{1}{k_A e_0} + \frac{a}{k_0 e_0} = \frac{K_m}{k_0 e_0} + \frac{a}{k_0 e_0} \tag{15}$$

In this case a/v is plotted against a to give a straight line of slope $1/k_0 e_0$ and intercept $1/k_A e_0$ ($-K_m/k_0 e_0$) on the ordinate axis (Figure 3). The main reason for preferring this plot is that it gives a reasonably reliable indication of the quality of each observation, as may be seen from the small variation in the lengths of the error bars in Figure 3.

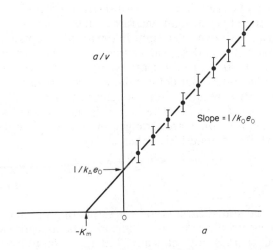

Figure 3 Plot of a/v against a. As in Figure 2, the error bars represent errors of $\pm 0.05V$ in v. This plot is sometimes called a Hanes plot

A third rearrangement of the Michaelis–Menten equation

$$v = k_0 e_0 - \frac{k_0 v}{k_A a} = k_0 e_0 - K_m(v/a) \tag{16}$$

shows that a plot of v against v/a is also a straight line, with slope $-k_0/k_A$ $(= -K_m)$ and intercept $k_0 e_0$ on the ordinate axis (Figure 4). Although the effect of experimental error on this plot is complicated by the presence of v in both coordinates, so that errors in v result in displacement of experimental points towards or away from the origin, the parameter values are in practice more trustworthy than those given by the double-reciprocal plot (Dowd and Riggs, 1965). The plot also has the advantages that it tends to make poor data look worse, it emphasizes any departure from Michaelis–Menten kinetics, and it includes the entire observable range of v from 0 to V on a finite range of graph paper. This last characteristic encourages good experimental design, because it makes it difficult for experimenters to conceal design defects either from themselves or from others.

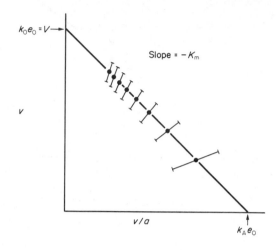

Figure 4   Plot of v against v/a. As in Figures 2 and 3, the error bars represent errors of $\pm 0.05V$ in v. This plot is called an Eadie–Hofstee plot

The plots that have been described are variously attributed to Woolf (1932), Hanes (1932), Lineweaver and Burk (1934), Eadie (1942), Hofstee (1952), and even to Scatchard (1948). In general it is preferable to use descriptive names that make it clear what is being plotted against rather than uninformative and confusing personal names that in some cases perpetuate historical inaccuracies.

A rather different method of plotting the Michaelis–Menten equation is illustrated in Figure 5. It is called a direct linear plot (Eisenthal and Cornish-Bowden, 1974), and unlike the other plots it represents each observation by a straight line rather than by a point. It depends on the fact that if axes labelled V and K_m are drawn, a straight line with intercepts v and $-a$ on these two axes respectively relates all pairs of V and K_m values that satisfy the particular observation for which the line is drawn. If several such lines are drawn, for several observations, the lines will all (ideally) intersect at a unique point that defines the values of V and K_m for the whole set of data. In practice, experimental error results in the appearance of a family of intersections, each of which defines one V and one K_m value. The best V is then defined as the median of all the V values, *i.e.* the middle value when they are arranged in rank order, and the best K_m the median of all the K_m values (Cornish-Bowden and Eisenthal, 1974).

27.2.4.2   Statistical methods

The most commonly used statistical method for estimating parameters is the method of least squares. Many modern electronic calculators include linear-regression facilities that provide least-squares fits to straight lines automatically, but these should not be used for any of the straight-line forms of the Michaelis–Menten equation, because they normally incorporate an assumption that each point on the straight line is equally reliable, whereas this assumption is most unlikely to be correct if the straight line is arrived at by transforming a non-linear equation. This comment applies to all transformed equations, but it is especially noticeable in Figure 2 that uniform errors in v generate grossly unequal errors in $1/v$.

The correct way of applying the least-squares method depends on information about the distri-

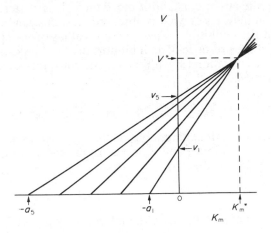

Figure 5 Direct linear plot. Each observation is represented by a line rather than a point, this line being a straight line with intercepts $-a$ on the K_m axis and v on the V axis. The point of intersection of the lines gives the coordinates of the values of K_m and V that fit the data

bution of experimental error that is tedious to acquire (see, *e.g.*, Storer *et al.*, 1975; Askelöf *et al.*, 1976) and is not usually available. If one is willing to assume that each v value is known with about the same percentage accuracy (*i.e.* that the coefficient of variation of v is constant), then it is appropriate to define a sum of squares SS as follows:

$$SS = \Sigma\left(\frac{v_i}{\hat{v}_i} - 1\right)^2 \tag{17}$$

where each $\hat{v}_i$ represents the rate calculated with any parameter values at the substrate concentration a_i that corresponds to the ith observed rate v_i, and the summation is carried out over all observations. We can make this expression a minimum (for details, see Cornish-Bowden, 1979) by defining best-fit parameter values as follows:

$$1/\hat{k}_A e_0 = \hat{K}_m/\hat{V} = \frac{\Sigma v_i^2 \Sigma v_i/a_i - \Sigma v_i^2/a_i \Sigma v_i}{\Sigma v_i^2/a_i^2 \Sigma v_i^2 - (\Sigma v_i^2/a_i)^2} \tag{18}$$

$$1/\hat{k}_0 e_0 = 1/\hat{V} = \frac{\Sigma v_i^2/a_i^2 \Sigma v_i - \Sigma v_i^2/a_i \Sigma v_i/a_i}{\Sigma v_i^2/a_i^2 \Sigma v_i^2 - (\Sigma v_i^2/a_i)^2} \tag{19}$$

Although these expressions give satisfactory results in most experiments it is important to realize that they embody assumptions that may not be correct. Their accuracy is sensitive to any variation in the precision of the observations, and they are especially sensitive to the presence of any outliers, or very bad observations, among the data. For discussion of ways of avoiding these assumptions, see Cornish-Bowden and Endrényi (1981). The considerations discussed therein led to the development of a Fortran program capable of fitting a wide variety of enzyme kinetic equations in the presence of various kinds of error behaviour. This is described and reproduced in full in Cornish-Bowden (1985).

27.3 INHIBITION

27.3.1 Types of Inhibition

A substance that causes an enzyme-catalysed reaction to proceed more slowly when it is present in the reaction mixture is called an inhibitor. Inhibition may result from interaction between inhibitor and enzyme, or between inhibitor and substrate, and it may be reversible by diluting or dialysing out the inhibitor or it may not. In this account only the most widely encountered case, linear reversible inhibition, will be considered.

Inhibition is often classified according to its presumed mechanism, but this is inadvisable, because it is often necessary to describe experimental results before their interpretation is clear, and in any case it is not desirable that a purely descriptive account of an experiment should be liable to change as one comes to understand it more fully. It is better, therefore, to define the different kinds of inhibition in terms of what is observed, and this can be done most conveniently by examining the effect of the inhibitor on the apparent values of the specificity constant k_A and the catalytic constant k_0. In the presence of an inhibitor, the kinetics can often be described by an equation of the same form as equation (8) with the kinetic constants k_A and k_0 replaced by apparent values k_A^{app} and k_0^{app} respectively:

$$v = \frac{k_0^{app} k_A^{app} e_0 a}{k_0^{app} + k_A^{app} a} = \frac{e_0 a}{(1/k_A^{app}) + (a/k_0^{app})} \tag{20}$$

In any linear inhibition, one or both apparent parameters are decreased by a factor of $(1 + i/K_i)$, where i is the concentration of inhibitor and K_i is a constant known as the inhibition constant. In competitive inhibition, only k_A^{app} is affected:

$$k_A^{app} = k_A/(1 + i/K_{ic}) \tag{21}$$

and $k_0^{app} = k_0$ is independent of i; in this case the inhibition constant K_{ic} is a competitive inhibition constant.

In mixed inhibition, k_A^{app} varies according to equation (21) and in addition k_0^{app} varies in accordance with a similar equation:

$$k_0^{app} = k_0/(1 + i/K_{iu}) \tag{22}$$

and the inhibition constant K_{iu} is called the uncompetitive inhibition constant. There is no particular reason for the competitive and uncompetitive inhibition constants to be equal, *i.e.* $K_{ic} = K_{iu}$, but when they are, the inhibition is called non-competitive. Non-competitive inhibition is neither common nor important, but it survives in many elementary textbooks from the early studies of Michaelis and Pechstein (1914). Mixed inhibition is sometimes called non-competitive inhibition even when K_{ic} and K_{iu} are unequal, but this usage is confusing and best avoided.

A less common, but theoretically important, kind of inhibition occurs when k_0^{app} varies according to equation (22) but $k_A^{app} = k_A$ is independent of i. This is called uncompetitive inhibition.

Table 1 summarizes the characteristics of linear inhibitors. It shows not only the effects on k_A^{app} and k_0^{app} but also those on K_m^{app}, the apparent value of the Michaelis constant. These are much less regular and memorable than the effects on k_A^{app} and k_0^{app} and for that reason do not provide a satisfactory basis for defining inhibition types. In general, effects on K_m^{app} are most easily remembered by regarding K_m as a derived quantity equivalent to k_0/k_A.

Table 1   Characteristics of Linear Inhibitors

Type of inhibition	k_A^{app}	k_o^{app}	K_m^{app}
Competitive	$\dfrac{k_A}{1 + i/K_{ic}}$	k_0	$K_m(1 + i/K_{ic})$
Mixed	$\dfrac{k_A}{1 + i/K_{ic}}$	$\dfrac{k_0}{1 + i/K_{iu}}$	$\dfrac{K_m(1 + i/K_{ic})}{1 + i/K_{iu}}$
Pure non-competitive[a]	$\dfrac{k_A}{1 + i/K_i}$	$\dfrac{k_0}{1 + i/K_i}$	K_m
Uncompetitive	k_A	$\dfrac{k_0}{1 + i/K_{iu}}$	$\dfrac{K_m}{1 + i/K_{iu}}$

[a] The qualifier 'pure' is to distinguish this case, for which $K_i = K_{ic} = K_{iu}$, from the more general case of mixed inhibition, which is referred to as non-competitive inhibition by some authors, though this latter usage is confusing and is not recommended.

27.3.2   Mechanisms of Linear Inhibition

Conceptually the simplest way in which competitive inhibition can arise is by binding of the inhibitor to the free enzyme in such a way as to prevent binding of substrate:

$$\mathrm{EI} \overset{K_{ic}}{\rightleftharpoons} \mathrm{E} + \mathrm{A} \rightleftharpoons \mathrm{EA} \rightarrow \mathrm{E} + \mathrm{Z}$$
$$+$$
$$\mathrm{I}$$

A classical example of this is the competitive inhibition of succinate dehydrogenase by malonate, which resembles the true substrate succinate sufficiently to bind tightly to the succinate-binding site but lacks the dimethylene group that would permit it to be dehydrogenated. Similar competitive inhibition occurs with many enzymes when the inhibitor is a structural analogue of the substrate. This mechanism is an example of dead-end inhibition, because the enzyme–inhibitor complex EI occurs off the reaction pathway and can break down only by reversing its formation.

A different way in which competitive inhibition can arise is as a case of product inhibition. It often happens that the step of a reaction in which product is released also regenerates the form of the enzyme to which the substrate binds: if this happens and the product-release step is reversible, the product acts as a competitive inhibitor.

Mixed inhibition arises when the inhibitor, which is usually not a structural analogue of the substrate in this case, binds both to the free enzyme and to the enzyme–substrate complex in such a way as to prevent the catalytic reaction from occurring. It is also common as a form of product inhibition, especially in reactions of more than one substrate; it typically occurs when the product and substrate bind to different forms of enzyme in the mechanism.

If a dead-end inhibitor binds to the enzyme–substrate complex but not to the free enzyme it brings about uncompetitive inhibition, but although this is formally the simplest case of uncompetitive inhibition it represents a somewhat unlikely sort of behaviour and is not common. Uncompetitive inhibition is encountered mainly as a form of product inhibition, but even then it is much less common than competitive and mixed inhibition except in reactions with three or more products.

27.3.3 Analysis of Inhibition Experiments

It is obvious from equations (21) and (22) that the type of inhibition and the values of the inhibition constants can be determined by measuring the apparent parameters of the Michaelis–Menten equation at various inhibitor concentrations. If K_{ic} is finite, *i.e.* in competitive and mixed inhibition, a plot of $1/k_A^{app}$ against i gives a straight line of slope $1/k_A K_{ic}$, and intercept $-K_{ic}$ on the i-axis; if K_{iu} is finite, *i.e.* in mixed and uncompetitive inhibition, a plot of $1/k_0^{app}$ against i gives a straight line of slope $1/k_0 K_{iu}$, and intercept $-K_{iu}$ on the i-axis.

As an alternative to this approach, it is common practice to examine the appearance of a plot of $1/v$ against i at each of several a values (Dixon, 1953). This plot is shown on the left-hand side of Figure 6 for each of the common types of linear inhibition. It will be seen that it does not provide a complete diagnosis of the inhibition type, as it does not distinguish between the two most common types, competitive and mixed inhibition. The missing information can be obtained by plotting a/v against i (Cornish-Bowden, 1974) as well, as shown on the right-hand side of Figure 6.

The theoretical basis for the plot of $1/v$ against i may be seen by considering the equation for mixed inhibition, obtained by substituting equations (21) and (22) in equation (20):

$$v = \frac{e_0 a}{\left[\frac{1 + i/K_{ic}}{k_A}\right] + \left[\frac{(1 + i/K_{iu})a}{k_0}\right]} \tag{23}$$

In reciprocal form,

$$\frac{1}{v} = \frac{1 + i/K_{ic}}{k_A e_0 a} + \frac{1 + i/K_{iu}}{k_0 e_0} \tag{24}$$

this is a linear function of i, so that a plot of $1/v$ against i is a straight line. If plots are made at two substrate concentrations a_1 and a_2, the point of intersection can be found by equating the two expressions for $1/v$:

$$\frac{1 + i/K_{ic}}{k_A e_0 a_1} + \frac{1 + i/K_{iu}}{k_0 e_0} = \frac{1 + i/K_{ic}}{k_A e_0 a_2} + \frac{1 + i/K_{iu}}{k_0 e_0} \tag{25}$$

If K_{ic} is finite, *i.e.* in competitive and mixed inhibition, this equation simplifies to $i = -K_{ic}$; if K_{ic} is infinite, as in uncompetitive inhibition, the equation has no solution, *i.e.* the lines are parallel. The characteristics are thus as shown on the left-hand side of Figure 6. The corresponding analysis for the plot of a/v against i follows from multiplying both sides of equation (24) by a and proceeding in the same way, to show that $i = -K_{iu}$ at the point of intersection for this plot.

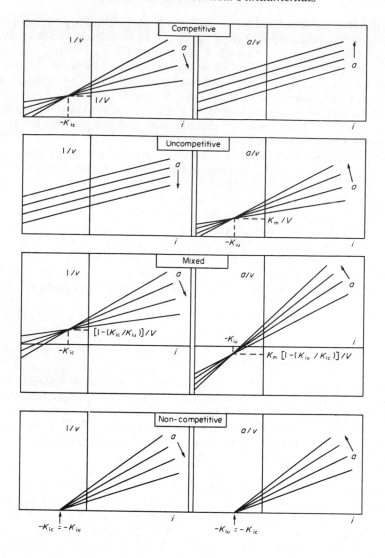

Figure 6 Determination of inhibition constants and diagnosis of inhibition type. On the left-hand side, plots of $1/v$ against i at various a values (Dixon, 1953) give lines that intersect to show the value of K_{ic}; on the right-hand side, plots of a/v against i at various a values (Cornish-Bowden, 1974) give the value of K_{iu}

27.4 REACTIONS WITH MORE THAN ONE SUBSTRATE

27.4.1 Michaelis–Menten Kinetics for a Two-substrate Reaction

A reaction $A + B \rightleftharpoons Y + Z$, with two substrates A and B, at concentrations a and b respectively, and two products Y and Z, can be said to obey Michaelis–Menten kinetics if the initial rate in the absence of products is given by an expression of the following form:

$$v = \frac{e_0}{\dfrac{1}{k_0} + \dfrac{1}{k_A a} + \dfrac{1}{k_B b} + \dfrac{1}{k_{AB} ab}} \tag{26}$$

This is a generalization of equation (8), and further generalizations to reactions with three or more substrates follows in an obvious way. In the presence of both products the behaviour becomes much more complex, because then the back reaction is possible and must be allowed for; but if only one product is present, *e.g.* Z at a concentration z, the numerator of the rate expression is unchanged, but additional terms of the form $z/k_{AB}^{Z}ab$, $z/k_{A}^{Z}a$ or z/k^{Z} appear in the denominator. These bring about various characteristic kinds of product inhibition.

Although equation (26) is apparently much more complex than equation (8), its relationship to the Michaelis–Menten equation may be seen by considering the effect of varying one substrate concentration while keeping the other constant. Thus, with constant b and variable a, it can be rearranged as follows:

$$v = \frac{e_0}{\left(\dfrac{1}{k_0} + \dfrac{1}{k_B b}\right) + \left(\dfrac{1}{k_A} + \dfrac{1}{k_{AB} b}\right)\dfrac{1}{a}} \tag{27}$$

This is of exactly the same form as equation (8) with k_0 and k_A replaced by apparent values k_0^{app} and k_A^{app} given by

$$\frac{1}{k_0^{app}} = \frac{1}{k_0} + \frac{1}{k_B b} \tag{28}$$

$$\frac{1}{k_A^{app}} = \frac{1}{k_A} + \frac{1}{k_{AB} b} \tag{29}$$

It follows, therefore, that when one substrate concentration is varied at a constant value of the other the kinetic behaviour is very similar to that observed in a one-substrate reaction.

27.4.2 Discrimination Between Mechanisms

One of the main reasons for studying the kinetics of a reaction with more than one substrate is that different mechanisms give rise to different kinetic behaviour, so one can use kinetic measurements to discriminate between mechanisms. A major division between two-substrate mechanisms concerns the question of whether both substrates must bind to the enzyme before any products are released, or whether the first product can be released before the second substrate is bound. In the former case, known as a ternary-complex mechanism, the rate equation in the absence of products is equation (26). In the latter case, the substituted-enzyme mechanism, the rate equation differs from equation (26) by missing the term $1/k_{AB}ab$ from the denominator:

$$v = \frac{e_0}{\dfrac{1}{k_0} + \dfrac{1}{k_A a} + \dfrac{1}{k_B b}} \tag{30}$$

The absence of this term has no effect on equation (28), so the dependence of k_0^{app} on the constant substrate concentration is the same for both mechanisms. But equation (29) collapses to $1/k_A^{app} = 1/k_A$: thus if k_A^{app} varies with b a ternary-complex mechanism is implied, whereas if it does not a substituted-enzyme mechanism is a possibility. In practice discrimination is more complex than this, because only the simplest cases of two-substrate mechanisms have been considered, but the principle applies in general: experiments can reveal the presence or absence of particular terms in the rate equation, which can be used to support or exclude particular mechanisms.

In the presence of products, added one at a time to prevent the back reaction from occurring, additional terms appear in the denominators of the expressions in equations (26) and (30). These allow product-inhibition experiments to be used to confirm discrimination between mechanisms. Moreover, the particular terms that appear depend on the order in which substrates are bound and products are released, so one can use product inhibition to show which substrate is which and which product is which.

27.5 EFFECTS OF pH AND TEMPERATURE

27.5.1 pH Dependence of Enzymes

Nearly all physiological processes occur in the neutral range of pH, and many enzymes are inactivated at pH values outside the range 5–9. Consequently, proper control of pH by means of buffers is an essential component of any enzyme investigation. In addition, by studying the way in which an enzyme responds to moderate variation in the pH one can obtain valuable information about its structure and mechanism of action.

Acids and bases have been defined in various ways to meet the needs of chemistry, but for studying enzyme behaviour the only definition that is of interest is that of Brønsted (1923): 'An acid is a species having a tendency to lose a proton, and a base is a species having a tendency to add on a proton'. It follows that deprotonation of any acid AH yields a species A^- that satisfies the definition of a base. Thus Brønsted acids and bases occur as pairs, and A^- is called the conjugate base of AH. The dissociation reaction can be written as follows:

$$AH \rightleftharpoons A^- + H^+$$

and an acid dissociation constant K_a can be defined as the equilibrium constant:

$$K_a = [A^-][H^+]/[AH] \tag{31}$$

As it is usually convenient to express the hydrogen ion concentration $[H^+]$ as its negative logarithm, $pH = -\log[H^+]$, it is correspondingly convenient to replace K_a by $pK_a = -\log K_a$, so

$$pK_a = pH - \log([A^-]/[AH]) \tag{32}$$

Typical pK_a values for the ionizing groups commonly found in proteins are given in Table 2. These are 'typical' rather than exact because the environment provided by a protein surface is so complex that actual values can vary substantially. In the active site of an enzyme pK_a values may be 'perturbed' by as much as 2 pH units. The groups that ionize in the vicinity of pH 7 are especially important in understanding the catalytic properties of enzymes. In particular, numerous enzymes have reactive histidine or cysteine side-chains in their active sites, even though these are among the less common of amino acids in protein structures generally.

Table 2 Ionizing Groups in Proteins

Name	Type of Group	pK_a	Name	Type of Group	pK_a
C-terminal	Carboxylate	3.4	N-terminal	Amine	7.5
Aspartate	Carboxylate	4.1	Cysteine	Thiol	8.3
Glutamate	Carboxylate	4.5	Tyrosine	Phenol	9.6
Histidine	Imidazole	6.3	Lysine	Amine	10.4

27.5.2 Ionization of a Dibasic Acid

It is not uncommon for an enzyme to display a bell-shaped pH dependence of the sort illustrated in Figure 7. One often encounters such curves in the literature that result from measurements of initial rate (or, worse, extent of reaction) under arbitrary conditions at various pH values. Such curves are virtually meaningless for understanding the catalytic properties of the enzyme, because the initial rate depends on several variables that need to be properly controlled. A minimum requirement, therefore, is for the Michaelis–Menten parameters to be determined at each pH.

If the catalytic constant k_0 or the specificity constant k_A varies with pH as shown in Figure 7 the variation can most easily be explained in terms of a requirement for two ionizing groups to exist in particular states of ionization, one protonated, the other deprotonated. These groups may be either on the enzyme or on the substrate, but for simplicity only the case of an enzyme that acts as a dibasic acid will be examined:

$$H_2E \overset{K_1}{\rightleftharpoons} HE^- + H^+ \overset{K_2}{\rightleftharpoons} E^{2-} + 2H^+$$

for which only the singly ionized form HE^- is active. The dissociation constants K_1 and K_2 define the concentrations of the three enzyme species at any H^+ concentration:

$$K_1 = [HE^-][H^+]/[H_2E] \tag{33}$$
$$K_2 = [E^{2-}][H^+]/[HE^-] \tag{34}$$

Then, if the total enzyme concentration is defined as e_0, we have

$$e_0 = [H_2E] + [HE^-] + [E^{2-}] = [HE^-](1 + [H^+]/K_1 + K_2/[H^+]) \tag{35}$$

and so

$$[HE^-] = e_0/(1 + [H^+]/K_1 + K_2/[H^+]) \tag{36}$$

It is clear from inspection of this result that $[HE^-]$ tends to zero if $[H^+]$ is very large or very small,

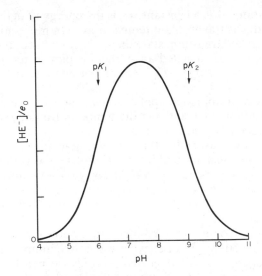

Figure 7 Bell-shaped pH dependence. The curve is calculated from equation (36) with $pK_1 = 6.0$, $pK_2 = 9.0$

and is appreciable only if $K_1 > [H^+] > K_2$. More detailed analysis shows that a plot of $[HE^-]$ against pH is a symmetrical bell-shaped curve of the form shown in Figure 7, with a maximum at $[H^+] = (K_1K_2)^{1/2}$, which is the same as pH = $(pK_1 + pK_2)/2$.

If pK_1 and pK_2 are well separated, with a difference $pK_2 - pK_1$ greater than about 4, there is a rather flat maximum with $[HE^-] \simeq e_0$, and the half-maximum points occur close to pH = pK_1 and pH = pK_2. This is, however, an exceptional circumstance because $pK_2 - pK_1$ is not usually as large as 4. Consequently $[HE^-]$ does not normally approach e_0 closely and the half-maximum pH values are further apart than the pK values. Nonetheless, the pK values may be calculated by a simple method suggested by Dixon (1979): define the difference between half-maximum pH values as $2 \log q$; then $pK_2 - pK_1 = 2 \log (q - 4 + 1/q)$. For example, suppose the maximum occurs at pH 6.3 and the half-maximum pH values are 5.0 and 7.6; then the width at half-height is 2.6, so $\log q = 1.3$, $q = 20$, and $pK_2 - pK_1 = 2 \log (20 - 4 + 1/20) = 2.41$. Hence $pK_1 = 5.1$, $pK_2 = 7.5$.

The pK values determined in this way are called molecular dissociation constants, because they refer to deprotonation of a complete molecule. A dibasic acid such as H_2E has two non-equivalent dissociable protons and can be deprotonated to two different singly deprotonated forms with two different group dissociation constants. As these two forms are unlikely to be equally responsible for the catalytic activity of the enzyme, one would ideally like to know their relative proportions and the relevant group dissociation constants, but this information cannot normally be obtained with certainty from pH–activity profiles, and interpretation of molecular dissociation constants as if they referred to particular groups is attended by some risk of error.

Plots of either the specificity constant k_A or the catalytic constant k_0 against pH may have the appearance of Figure 7. The pH dependence of k_A reflects the ionizing behaviour of the free enzyme or the free substrate (or a combination of the two), whereas that of k_0 reflects that of the enzyme–substrate complex. In practice, therefore, it is usually helpful to measure both parameters as a function of pH. The Michaelis constant K_m is affected by ionization of both free and bound enzyme and substrate, and is therefore liable to display complicated pH behaviour. The only time when it is directly helpful to consider the pH dependence of K_m is when it shows no variation in a pH range where k_0 and k_A vary: K_m is then the true thermodynamic dissociation constant of the enzyme–substrate complex (see Section 28.2.3), for reasons that have been discussed elsewhere (Cornish-Bowden, 1976).

27.5.3 Effect of Temperature on Enzymes

The effect of temperature on elementary chemical reactions is commonly considered in terms of the Arrhenius equation:

$$k = A \exp(-E_a/RT) \tag{37}$$

in which k is any rate constant, A is a constant, E_a is the energy of activation, $R = 8.31$. J mol^{-1} K^{-1} (the gas constant) and T is the absolute temperature. (In more advanced work a somewhat different approach known as the transition state theory is often used, but it will not be considered here.) A plot of log k against $1/T$ is called an Arrhenius plot: when the Arrhenius equation is obeyed it gives a straight line with slope $-E_a/2.3R$, and provides a convenient method of evaluating the activation energy.

In principle, the same sort of approach applies to enzyme-catalysed reactions, but in practice there are so many complications that studying the temperature dependence of an enzyme-catalysed reaction cannot be recommended, at least until the behaviour at one temperature is well understood. There are several reasons for this. In enzyme-catalysed reactions one rarely has a measure of an individual rate constant, and if log k is replaced as ordinate of the Arrhenius plot by log v the results are likely to depend in a complex way on several rate constants. In addition, one can normally study an enzyme-catalysed reaction over an excessively restricted temperature range, *e.g.* 5–45 °C (278–318 K), too little to allow the temperature dependence to be properly characterized.

In the older literature it was common for a temperature optimum to be reported, *i.e.* a temperature at which the catalytic activity of an enzyme was a maximum. Such a maximum is nearly always an artifact, a consequence of the fact that not only do nearly all reaction rates increase with temperature but also enzymes are denatured, often irreversibly, by excessive temperatures, *i.e.* they lose their structure and activity. Denaturation is often time-dependent, and so the temperature at which the rate is a maximum can vary in an arbitrary way with the experimental conditions.

27.6 DEVIATIONS FROM MICHAELIS–MENTEN KINETICS

27.6.1 Mixtures of Isoenzymes

Not all enzymes obey the Michaelis–Menten equation, equation (8), or its generalization to reactions of more than one substrate, equation (26). Indeed, some authors have argued that Michaelis–Menten kinetics occur very rarely, if at all (Hill *et al.*, 1977), though it must be added that their views have not received wide support. If the expected straight lines are not seen in the various plots discussed in Section 27.2.4.1, it follows that the Michaelis–Menten equation is not obeyed. The simplest explanation is that the enzyme is a mixture of isoenzymes, *i.e.* distinct enzyme species catalysing the same reaction. This possibility should always be examined before entertaining more exotic explanations of non-Michaelis–Menten behaviour.

In the presence of a mixture of isoenzymes the net rate is the sum of the rates due to the separate catalysts. Thus, if there are two isoenzymes, both obeying Michaelis–Menten kinetics, one at a concentration e_0 and with specificity constant k_A and catalytic constant k_0, the other at concentration e_0' with constants k_A' and k_0' respectively, the net rate is:

$$ v = \frac{k_0 k_A e_0 a}{k_0 + k_A a} + \frac{k_0' k_A' e_0' a}{k_0' + k_A' a} \tag{38} $$

Even though both components obey the Michaelis–Menten equation, equation (38) is not of Michaelis–Menten form and cannot be rearranged into the form of a straight line (except in the special case where the Michaelis constants are equal, *i.e.* $K_m = k_0/k_A = K_m' = k_0'/k_A'$). In general the sort of behaviour predicted by this equation (and generalizations with more than two components) is similar to that for negative cooperativity, which is discussed in Section 27.6.3.

It is not normally profitable to try to analyse the kinetics of reactions catalysed by mixtures. Instead one must attempt to separate the components and study the catalytic properties of the pure isoenzymes. Only when one is satisfied that one is working with a pure catalyst is it worthwhile considering the possibility that its properties may be due to negative cooperativity.

27.6.2 Substrate Inhibition

Many enzymes are subject to inhibition at high concentrations of their substrates. In the simplest case the enzyme–substrate complex EA is able to bind a second substrate molecule A to produce a complex EA$_2$ that is incapable of reaction. When this happens the rate v is not accurately described by the Michaelis–Menten equation but instead varies with substrate concentration *a* as follows:

$$v = \frac{k_0 k_A e_0 a}{k_0 + k_A a(1 + a/K_{iA})} \tag{39}$$

in which K_{iA} is the dissociation constant of the abortive complex EA_2 and k_0 and k_A are similar in meaning to the catalytic constant and specificity constant considered previously in Section 27.2.3. (The ratio k_0/k_A is sometimes loosely called the Michaelis constant and given the symbol K_m, but this is not recommended as it is illogical and misleading to refer to a Michaelis constant in a context where Michaelis–Menten kinetics are not obeyed.)

At low concentrations of substrate, if a/K_{iA} is negligible compared with 1, equation (39) simplifies to equation (8), the Michaelis–Menten equation, and k_0 and k_A can be estimated from the data in this range by the methods of Section 27.2.4. At high concentrations *v* is less than that predicted by the Michaelis–Menten equation and passes through a maximum at $a = (k_0 K_{iA}/k_A)^{0.5}$, thereafter decreasing.

27.6.3 Cooperativity and Negative Cooperativity

Enzymes that occupy key positions in metabolism, especially those catalysing steps at metabolic branch points, frequently do not obey Michaelis–Menten kinetics. This is because Michaelis–Menten kinetics do not provide sufficient sensitivity to changing conditions: for example, it is evident from equation (9) that an enzyme obeying Michaelis–Menten kinetics requires an 81-fold increase in substrate concentration (from $K_m/9$ to $9K_m$) to bring about an increase in *v* from 10% to 90% of *V*. The way in which enzymes overcome this problem is by means of a phenomenon known as cooperativity. This is a complex subject that has been discussed more fully elsewhere (Cornish-Bowden, 1979); here no attempt will be made to describe the underlying mechanisms of cooperativity but will simply indicate how it can be recognized.

The most obvious property of cooperative enzymes is that they typically display a sigmoid (S-shaped) plot of *v* against substrate concentration, as shown in Figure 8a. It is often convenient to quantify the degree of cooperativity by means of a plot of log $v/(V - v)$ against log *a*, which is known as a Hill plot. Here *V* is not strictly the same parameter as in equation (9), as equation (9) is not obeyed, but it has a similar meaning: it is the limiting rate obtained by extrapolation to saturating concentrations. For cooperative enzymes the limit is approached more closely than with Michaelis–Menten kinetics and so *V* can usually be found reasonably accurately by direct measurement. A typical Hill plot is shown in Figure 8b, Note that the points lie approximately but not exactly on a straight line, which has slope *h*, called the Hill coefficient. At the extremes this slope always approaches unity, but there is often an extensive almost-straight central region.

No equation for the Hill plot has been given here, because this plot is not based on a theoretically valid equation, but is simply an experimental convenience as it allows the degree of cooperativity to be expressed in a way that is widely understood. A value of $h = 1$ corresponds to Michaelis–Menten kinetics, and larger values to cooperativity: typical values are in the range 2 to 4, values greater than 4 being very rare. Some authors have interpreted *h* as a measure of the number of catalytic sites per molecule of enzyme: this has no theoretical basis and certainly gives a wrong result in nearly all cases.

Values of *h* less than 1 are also observed (though in this case it is usually difficult to get an accurate value of *V* for use in the Hill plot). The simplest explanation of this is that the enzyme is a mixture (Section 27.6.1), but it can also be a sign of negative cooperativity. This means that the enzyme is less sensitive to substrate concentration than one obeying Michaelis–Menten kinetics.

27.7 KINETICS OF IMMOBILIZED ENZYMES

Industrially useful catalysts should be stable, so that they can remain in use for an economically acceptable period, and solid, so that they are not washed away with the chemical components of the reaction catalysed. Enzymes as provided by nature are not usually either stable or solid, and a great deal of research has gone into finding ways of 'immobilizing' them to make them more useful for technology. At its crudest, immobilization can simply mean attachment of the enzyme to a solid support such as cellulose, either by covalent bonding or simply by physical attraction, so that the reaction mixture can pass through the catalyst without washing it away. But there are

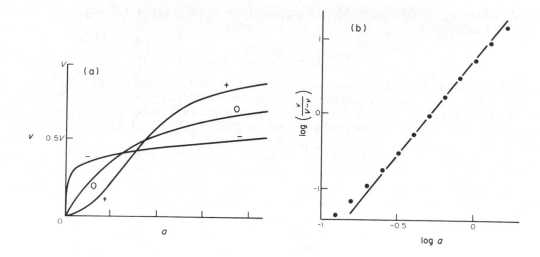

Figure 8 Deviations from Michaelis–Menten kinetics. (a) Plots of v against a. The curve labelled (0) is calculated from the Michaelis–Menten equation; the curve labelled (+) shows positive cooperativity; the curve labelled (−) shows negative cooperativity. (b) Hill plot of the positively cooperative case. An enzyme obeying Michaelis–Menten kinetics would give a straight line of slope 1 (not shown), and a slope less than 1 indicates negative cooperativity

many ways of immobilizing an enzyme, and not all of them bring it into the solid phase: for example, the enzyme may remain in a true liquid solution while trapped within a bead by a matrix with pores that are too small to allow passage of the enzyme molecule but large enough for small reactant molecules to diffuse through.

Rigorous treatment of the kinetics of immobilized enzymes is considerably more complex than for enzymes in solution (see, for example, Laidler and Bunting, 1973). However, for industrial purposes it is hardly necessary to have a rigorous analysis, provided one has a qualitative understanding of the ways that immobilized enzymes are likely to differ in their kinetic behaviour from enzymes in solution. Here only a brief account will be given, but the subject is treated much more extensively, at a practical rather than an academic level, by Trevan (1980).

There are three main reasons why the kinetics of an immobilized enzyme may differ from those of the same enzyme free in solution. First, the means of immobilizing it, especially if it is covalently attached, may alter its chemical properties. Second, the 'microenvironment', *i.e.* the composition and chemical characteristics of the solution in the immediate vicinity of the enzyme molecule may differ appreciably from those in the bulk phase. Third, diffusion of reactant molecules between the bulk phase and the region around the enzyme active site is slower than in free solution, and can be much slower.

The polymers used for immobilizing enzymes are often polyionic, *i.e.* they contain large numbers of charged groups. These groups have substantial effects on the partitioning of small ions between the bulk phase and the microenvironment of the enzyme, and one of the most important ions to be influenced in this way is the proton. If the polymer is polyanionic the negative charges will tend to concentrate protons above the concentration in the bulk phase, whereas a polycationic polymer has the opposite effect. As a result, the pH–activity profile of an immobilized enzyme may apparently be displaced along the pH axis from the curve seen in free solution. This is an apparent shift rather than a real one because it reflects a difference in pH between that in the vicinity of the enzyme and the measured value in the bulk phase, rather than a change in the properties of the enzyme. It is typically a consequence of inadequate buffering, and can be overcome if the medium is well buffered with a buffer that can equilibrate between the bulk phase and the microenvironment.

Diffusion of substrate molecules to the catalytic site of an enzyme can assume great importance when the enzyme is immobilized. Instead of the usual Michaelis–Menten formulation of the reaction considered in Section 27.2.1 (equation 1), we must allow for an extra step representing diffusion of substrate in the bulk phase, $A_{outside}$, to become substrate in the vicinity of the enzyme, A_{inside}:

$$A_{outside} \rightleftharpoons A_{inside} + E \rightleftharpoons AE \rightarrow E + Z \tag{40}$$

The qualitative effect of this extra process is that diffusion is typically first order over the whole concentration range, whereas the enzyme-catalysed reaction approximates to first order only at low substrate concentrations (see Figure 1). Consequently, at low concentrations of substrate in the bulk phase, the enzyme may be able to react with every substrate molecule that reaches it, so the reaction is diffusion-limited and the kinetic constants of the enzyme itself do not affect the observed kinetics. However, as the concentration increases the diffusion rate continues to increase proportionately and eventually reaches a point where the enzyme begins to be saturated and diffusion ceases to be limiting. It follows, therefore, that an immobilized enzyme typically shows apparently enzyme-independent kinetics at low concentrations but normal enzyme-dependent kinetics at high concentrations. From the practical point of view this means that the transition from first-order to zero-order dependence on the substrate concentration in the bulk phase is more abrupt than that seen in Figure 1 for a typical enzyme-catalysed reaction in free solution.

Diffusion limitation can have important effects on product inhibition (see Section 27.3.2) for similar reasons. In free solution the product is rapidly diluted by diffusion away from the enzyme, and its effect as an inhibitor may not be noticed until an appreciable fraction of reaction. If the enzyme is immobilized, the product may in contrast be unable to diffuse rapidly out of the microenvironment, and consequently its concentration may rise very rapidly in the vicinity of the enzyme, producing substantial inhibition even though its concentration averaged over the whole system remains negligible.

27.8 REFERENCES

Askelöf, P., M. Korsfeldt and B. Mannervik (1976). Error structure of enzyme kinetic experiments. *Eur. J. Biochem.*, **69**, 61–67.

Briggs, G. E. and J. B. S. Haldane (1925). A note on the kinetics of enzyme action. *Biochem. J.*, **19**, 338–339.

Brønsted, J. N. (1923). Einige Bemerkungen über die Begriff der Säuren und Basen. *Recl. Trav. Chim. Pays-Bas*, **42**, 718–728.

Brown, A. J. (1892). Influence of oxygen and concentration on alcoholic fermentation. *J. Chem. Soc. (Trans.)*, **61**, 369–385.

Brown, A. J. (1902). Enzyme action. *J. Chem. Soc. (Trans.)*, **81**, 373–388.

Cornish-Bowden, A. (1974). A simple graphical method for determining the inhibition constants of mixed, uncompetitive and non-competitive inhibitors. *Biochem. J.*, **137**, 143–144.

Cornish-Bowden, A. (1976). Estimation of the dissociation constants of enzyme–substrate complexes from steady-state measurements. *Biochem. J.*, **153**, 455–461.

Cornish-Bowden, A. (1979). *Fundamentals of Enzyme Kinetics*. Butterworths, London.

Cornish-Bowden, A. (1985). A Fortran program for robust regression of enzyme kinetic data. In *Techniques in Protein and Enzyme Biochemistry*, ed. K. F. Tipton. Elsevier, Amsterdam.

Cornish Bowden, A. and R. Eisenthal (1974). Statistical considerations in the estimation of enzyme kinetic parameters by the direct linear plot and other methods. *Biochem. J.*, **139**, 721–730.

Cornish-Bowden, A. and L. Endrényi (1981). Fitting of enzyme kinetic data without prior knowledge of weights. *Biochem. J.*, **193**, 1005–1008.

Dixon, H. B. F. (1979). Derivation of molecular pK values from pH-dependences. *Biochem. J.*, **177**, 249–250.

Dixon, M. (1953). The determination of enzyme inhibition constants. *Biochem. J.*, **55**, 170–171.

Dowd, J. E. and D. S. Riggs (1965). Comparison of estimates of Michaelis–Menten kinetic constants for various linear transformations. *J. Biol. Chem.*, **240**, 863–869.

Eadie, G. S. (1942). The inhibition of cholinesterase by physostigmine and prostigmine. *J. Biol. Chem.*, **146**, 863–869.

Eisenthal, R. and A. Cornish-Bowden (1974). The direct linear plot. *Biochem. J.*, **139**, 715–720.

Hanes, C. S. (1932). Studies on plant amylases. *Biochem. J.*, **26**, 1406–1421.

Henri, V. (1902). Théorie générale de l'action des quelques diastases. *C. R. Hebd. Séances Acad. Sci.*, **135**, 916–919.

Henri, V. (1903). *Lois Générales de l'Action des Diastases*. Hermann, Paris.

Hill, C. M., R. D. Waight and W. G. Bardsley (1977). Does any enzyme follow the Michaelis–Menten equation? *Mol. Cellul. Biochem.*, **15**, 173–178.

Hofstee, B. H. J. (1952). Specificity of esterases. *J. Biol. Chem.*, **199**, 357–364.

International Union of Biochemistry, Nomenclature Committee (1982). Symbolism and terminology in enzyme kinetics, Recommendations 1981. *Arch. Biochem. Biophys.*, **224**, 732–742 (1983); *Biochem. J.*, **213**, 561–571 (1983); *Eur. J. Biochem.*, **128**, 281–291 (1982).

Laidler, K. J. and P. S. Bunting (1973). *The Chemical Kinetics of Enzyme Action*, 2nd edn., pp. 383–412. Clarendon Press, Oxford.

Lineweaver, H. and D. Burk (1934). The determination of enzyme dissociation constants. *J. Am. Chem. Soc.*, **56**, 658–666.

Michaelis, L. and H. Davidsohn (1911). Die Wirkung der Wasserstoffionen auf das Invertin. *Biochem. Z.*, **35**, 386–412.

Michaelis, L. and M. L. Menten (1913). Die Kinetik der Invertinwirkung. *Biochem. Z.*, **49**, 333–369.

Michaelis, L. and H. Pechstein (1914). Über der verschiedenartige Natur der Hemmungen der Invertasewirkung. *Biochem. Z.*, **60**, 79–90.

O'Sullivan, C. and F. W. Tompson (1890). Invertase: a contribution to the history of an unorganized ferment. *J. Chem. Soc. (Trans.)*, **57**, 834–931.

Scatchard, G. (1948). The attractions of proteins for smaller molecules and ions. *Ann. N. Y. Acad. Sci.*, **51**, 660–672.

Sørensen, S. P. L. (1909). Etudes enzymatiques. *C. R. Trav. Lab. Carlsberg*, **8**, 1–168.
Storer, A. C., M. G. Darlison and A. Cornish-Bowden (1975). The nature of experimental error in enzyme kinetics. *Biochem. J.*, **151**, 361–367.
Trevan, M. D. (1980). *Immobilized Enzymes*. Wiley, Chichester.
Woolf, B. (1932). Cited by J. B. S. Haldane and K. G. Stern in *Allgemeine Chemie der Enzyme*, pp. 119–120. Steinkopff, Dresden.

28

Mechanisms of Enzyme Catalysis

K. E. SUCKLING
University of Edinburgh, UK

28.1 INTRODUCTION

Enzymes are attractive as potential industrial catalysts because they operate with high specificity and efficiency under mild conditions. These same properties have for many years inspired studies of the molecular events associated with enzyme catalysis by biochemists and chemists. It has been possible to identify and quantify many of the factors which lead to the large rate enhancements which enzymes are able to achieve over their non-enzymic counterparts (Jencks, 1975; Lipscomb, 1982). The extent to which these studies have been successful is shown by the fact that it is now possible to design and synthesize model enzymes which operate with efficiencies close to their biological prototypes (Tabushi, 1982). However, despite the close interaction of chemistry and biochemistry over several decades it is still true that much remains to be learnt about even the best understood enzyme and its catalytic mechanism (Warshel, 1981).

From the point of view of technological applications of enzymes it may be unnecessary to understand the mechanism of a given enzyme in detail, although this knowledge will often be important as a process is developed. Indeed, the mechanism of an enzyme-catalysed reaction may change in some respects when the enzyme is adapted to a process, for example when constrained in immobilized form.

Whilst a detailed understanding of the mechanism of a reaction catalysed by an enzyme which is used in a process may not be available or even essential, an understanding of the general characteristics of enzyme mechanisms provides a conceptual basis which can help in rationalizing the behaviour of an enzyme in a complex process. In this chapter we shall examine the stages by which an enzyme-catalysed reaction occurs, noting the various atomic and molecular phenomena which are important at each stage. We shall attempt to put all these concepts together by concluding with a discussion of an enzyme which has important industrial applications. For detailed accounts of the biochemical concepts outlined in this chapter the reader should refer to texts by Jencks (1969, 1975), Metzler (1977) or Walsh (1979) and to the reviews published in such serials as Annual Reviews in Biochemistry, Annual Reviews in Biophysics and Bioengineering, Advances in Enzymology, *etc.*

28.2   THE EVENTS IN AN ENZYME-CATALYSED REACTION

In any chemical reaction in which bonds are made and broken the sequence of events which occurs follows a regular scheme which is outlined in Table 1. The reagents initially diffuse together (1), and undergo changes in conformation, if these are necessary (2). The changes in covalent bonding now occur (3 and 4) and the products diffuse away from each other (5). This scheme includes the main interactions which take place, those between the reagents, but it neglects many other interactions, particularly those associated with the medium in which the reactions occur. Such interactions are especially significant in protein conformation and in enzyme catalysis. In reactions in solution, such as we are considering in biological processes, solvent molecules associate with reactants, products and the intermediates and transition states to varying degrees, stabilizing some and destabilizing others according to the extent to which they interact. The removal of such solvent effects by an enzyme, essentially causing the reaction to take place in a different, usually non-aqueous medium, can be an important factor in enzymic catalysis. This is because the weak non-covalent bonds, which are very important in biological systems, become relatively stronger in the absence of a polar solvent like water. It should be pointed out in passing that there is no reason why an enzyme-catalysed reaction should not be carried out in a solvent other than water, provided that the enzyme will work at a useful rate. A substrate like cholesterol, for example, is very insoluble in water, but it can be oxidized by cholesterol oxidase in the appropriate medium (Dunnill, 1978).

Table 1   Events in a Chemical Reaction in Solution

(1)	Reagents diffuse together
(2)	Conformational and orientation changes to reacting configuration of functional groups
(3)	Reorganization of electrons to form intermediate by way of a transition state
(4)	Breakdown of intermediate by way of another transition state with further electronic reorganization
(5)	Diffusion of products away from each other

The sequence of events outlined in Table 1 will obviously be more complex in an enzyme-catalysed reaction. Let us assume that the reaction takes place in aqueous solution and is catalysed by a soluble, non-particulate enzyme. Table 2 outlines the events which must occur, and this will provide the framework for our discussion of the mechanism of catalysis. In the initial binding steps (1) several possibilities may arise, depending upon the nature of the reaction. If more than one substrate is involved, the order in which the substrates bind may be important for the mechanism. As a consequence of the interaction between the enzyme and the substrate the whole assembly may change its conformation (2). Such a change, whilst leading to a lower overall energy state, as most chemical reactions do, may promote the subsequent events in the reaction. The stage which is most similar to the corresponding non-enzyme-catalysed reaction is stage (3) in which the covalent bonds are made and broken. The molecular details of these last events may be very similar in enzymic and non-enzymic catalysis. Once the bonding changes have occurred, the release of products requires a sequence of events which mirrors the initial steps. There will be further adjustments in conformation (4) and finally the products will dissociate from the enzyme and be resolvated by water (5). In principle each step of the sequence could be affected by the enzyme with one major exception. The enzyme cannot alter the rate at which the substrate molecules diffuse to meet it in solution, although the organization of a cell may well be able to do so (Park *et al.*, 1982). It is likely, therefore, that in many biotechnological processes the biological catalysts themselves will not be rate-limiting (Goldstein, 1976). Diffusion processes in the bulk phase and transport processes across barriers such as cell membranes, if present, may be much slower than the chemical changes which it is desired to effect. These very significant binding and dissociation steps are usually studied in detail by kinetic experiments. Enzyme kinetics has been discussed in Volume 1, Chapter 27 so here we will make use of the results of kinetic experiments without examining the methods used or the concepts on which they are based.

28.3   ENZYME MECHANISMS

The mechanism of an enzyme-catalysed reaction is a composite description of all the events that take place at a molecular and atomic level from the initial binding of substrates to the release of products. The picture is built up from studies of many different kinds (Suckling and Suckling, 1980a, 1980b) using all available chemical and physical methods. The use of many techniques is

Table 2 Events in an Enzyme-catalysed Reaction

(1)	Binding of substrate(s) to the active site of the enzyme
(2)	Conformational changes of enzyme or substrate or both as a result of binding
(3)	Changes in chemical bonding by way of transition states and intermediates (equivalent to steps 3 and 4 of Table 1)
(4)	Further conformational changes on formation of products
(5)	Release of products to the solvent water

necessary because each method has its own small window of vision which is distinct in some way from the others. Thus kinetic methods can identify the rate-determining steps, detect the order of binding of substrates and release of products and implicate certain functional groups in catalysis. Rapid-reaction techniques and pre-steady-state kinetics can detect conformational changes in the enzyme–substrate complex (Fersht, 1984).

These methods do not throw any light on the structures of the interacting species or how the functional groups come together during the bond making and breaking steps. Structural methods, of which X-ray crystallography and nuclear magnetic resonance (NMR) are now the most important (Steitz and Shulman, 1982), can fill in many of the gaps and by a combination of these approaches a detailed sequence of molecular events covering binding, conformational and electronic changes can be built up. A final refinement, applied to organic chemistry for many years, but now available for enzyme-catalysed reactions, is a theoretical analysis of the catalytic process based on classical or quantum mechanical theories of chemical reactions. This kind of study is leading to a critical examination of the relative importance of the various effects that operate in enzyme catalysis (Warshel, 1981; Pincus and Scheraga, 1981; Lipscomb, 1982).

The term mechanism is used here in a precise sense but it will often be found to be used rather loosely for a description of a reaction based on very limited data and on analogy with simpler chemical processes and some better-understood enzyme mechanisms. Care should always be taken to examine the experimental evidence on which a mechanism is based. Unfortunately for applications of enzymes in technology, many of the best understood enzymes do not catalyse reactions which have industrial importance. One exception to this is the class of enzymes known as serine proteases. This class includes mammalian enzymes such as trypsin and chymotrypsin and also bacterial enzymes like subtilisin which are widely included in domestic detergent preparations (Aunstrup, 1979).

28.3.1 Enzyme Kinetics

As the previous chapter has indicated, kinetic information is essential in the practical application of enzymes as well as a major base on which a molecular description of the reaction can be built. As part of a simple kinetic study one determines optimum parameters for the rate of the enzyme-catalysed reaction under study. Thus the optimum pH, substrate concentration and temperature can be found. These parameters may be very useful guides when developing a process based on a reaction studied in the laboratory but they may not be the appropriate ones for a scaled-up plant. A compromise using suboptimal pH or substrate concentration may be used for other reasons. Indeed the kinetic optima determined in solution may not even be relevant to the *in vivo* state itself, but they are essential for a defined and reliable study of an enzyme mechanism.

We shall assume that from kinetic studies we can obtain a scheme of the order of binding of substrates to an enzyme and of the release of products. We may also obtain evidence for the participation of certain functional groups in the reaction from the pH/rate dependence. For example, in glucose isomerase, which we shall discuss in more detail later, the maximum rate occurs at a pH of 6.35, which implicates a carboxyl group in the mechanism. With this information as a basis we can examine what other approaches can tell us about the remaining events in the scheme outlined in Table 2 and, as we proceed, attempt to develop a unified picture of an enzyme-catalysed reaction.

28.3.2 Binding of the Substrate to the Enzyme

The most significant step in an enzyme-catalysed reaction is probably the binding of the substrate. Indeed many important biological events, such as the transmission of a nerve impulse to a muscle (Conti-Tronconi and Raftery, 1982) are triggered by just such a simple binding event without any subsequent changes in covalent bonding at the binding site. It is at the binding stage that the specificity of the enzyme is expressed, and the molecular understanding of this is important in providing the structural basis for the chemical reactions which follow.

An enzyme is a unique polypeptide which folds in solution in a defined three-dimensional way. Usually the polar amino acid residues (such as lysine, arginine and glutamic acid) are found on the outer surface of the protein molecule which is in contact with the aqueous medium. Apolar residues (like phenylalanine, leucine and alanine) contribute to the conformation by clustering inside the folded polypeptide chain away from the solvent water. Certain areas in the structure can be found, often in clefts in the surface or at the area where two polypeptides associate together to give an active enzyme, where the arrangement of the amino acid side chains appears to have a distinct function (Rossman and Argos, 1981). These areas have often been revealed by X-ray diffraction studies and more recently by NMR (Steitz and Shulman, 1982). Such techniques allow one to demonstrate that the substrate of an enzyme binds in a specific orientation to a region of the protein which is known as the active site. A major goal, once a kinetic description of an enzyme-catalysed reaction is available, is to define the amino acid residues and the three-dimensional structure of the active site since this provides many clues about the remainder of the mechanism.

It should be emphasized that, although for simplicity most studies of enzyme mechanisms concentrate on events at the active site, the catalytic activity is a property of the whole assembly, enzyme, substrate and solvent. Important roles may be played by other parts of the ensemble at sites distal to the active site. This is especially true of the molecular mechanisms by which the rates of enzyme-catalysed reactions are thought to be controlled in the intact cell (Suckling and Suckling, 1980b). The specific binding of enzyme and substrate is usually reflected in complementary interactions between functional groups on the substrate and the amino acid side chains or metal ions or cofactors which may be found in the vicinity of the active site. Figure 1 illustrates some examples of such complementary interactions in carboxypeptidase A which hydrolyses the terminal peptide bond in polypeptides. The main bonding types are ionic bonds (between Arg-145 and the terminal carboxyl group of the substrate), hydrogen bonds (between the phenolic hydroxyl group of Tyr-248 and the peptide bonds of the substrate) and hydrophobic bonds (hydrophobic amino acid residues surrounding the terminal tyrosine residue of the substrate). Although individually these bonds are much weaker than covalent bonds, together they can lead to very strong overall interactions between protein and substrate. Often the strength of binding is increased by the presence of several subsidiary binding sites which allow further interactions of larger, usually polymeric substrates with the enzyme. Thus lysozyme has six sites for binding the carbohydrate rings of its polysaccharide substrate and the chemical reaction occurs at just one of these. There are similar binding subsites in the serine proteases.

Interactions such as these make a substantial contribution to the entropy of activation component of the rate equation for the reaction (see below). Specific non-covalent binding is the main additional factor which enzymes can bring to catalysis over non-enzymic catalysis in solution (Jencks, 1975).

As mentioned earlier, the initial binding step is one which cannot be catalysed by the enzyme itself. Unless some external form of organization is available the rate of binding will be limited by the rate of diffusion of the participating molecules in solution. It has been shown that the catalytic efficiency of some enzymes, for example triose phosphate isomerase, is limited only by this diffusion process: all other steps in the mechanism have evolved to occur at the optimum rate (Knowles and Alberry, 1977).

Some biological systems act faster than this and there must be ways, which are now beginning to be understood, of increasing the rate at which the enzyme and substrate come together (Jürss *et al.*, 1979). Nature adopts another strategy in metabolic sequences by setting up a biochemical production line in which several related catalytic activities are brought together. Such multienzyme complexes can consist of several associated polypeptides, or even only one, which carry a series of active sites. The substrate is directed from one site to the next without diffusing off into solution. Often the substrate remains covalently bound to the enzyme in a manner somewhat related to modern methods for the chemical synthesis of proteins or of nucleic acids. In this way the diffusion steps between enzymes are eliminated and the whole sequence of reactions can be regulated together (Suckling and Suckling, 1980b). Several artificial systems which incorporate these ideas have been prepared, for example malate dehydrogenase, citrate synthase and lactate dehydrogenase have been immobilized together in a functional form (Mosbach and Mattiason, 1976).

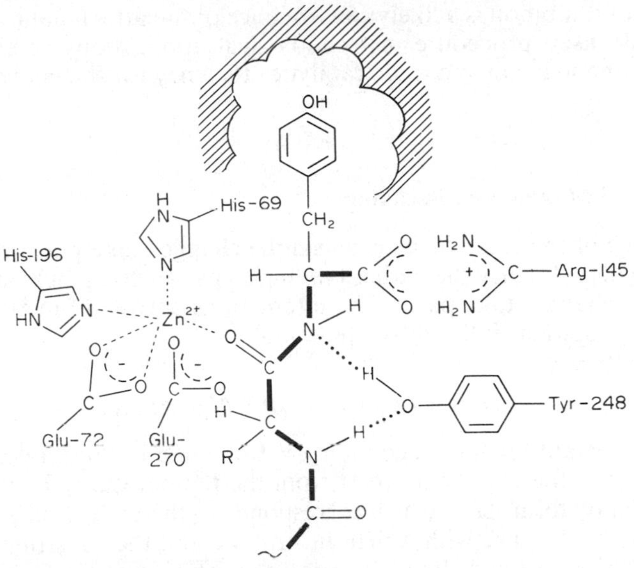

Figure 1 The enzyme–substrate complex of carboxypeptidase A. The aromatic ring of the terminal tyrosine residue of the peptide substrate is secured at the active site by the hydrophobic pocket (shaded). Ionic interactions link Arg-145 with the terminal carboxyl group of the peptide. Tyr-248 moves on binding of substrate to effect hydrogen bonds with two peptide linkages. The zinc ion coordinates with the carbonyl group of the peptide bond to be hydrolysed, polarizing it and activating it towards nucleophilic attack by water. This is assisted by Glu-270 acting as a general acid. 'Solid' bonds (■■■■) emphasize the polypeptide chain of the substrate. (Drawing based on Lipscomb, 1972)

28.3.3 Conformational Changes

Once the substrate has bound to the active site all may not yet be ready for covalent bond changes to take place. In a two-substrate reaction, for example, both substrates may be required to bind to different regions of the active site and this binding may take place in a defined order. Usually a kinetic study will help define these stages.

In many enzymes in the resting state, with no substrate bound, the amino acid side chains required for catalysis may not be in the correct orientation. The necessary change to the catalytic configuration may only take place when the substrate has bound. The enzyme uses part of the binding energy released by the enzyme–substrate interaction to make often small but critical changes in the orientation of functional groups for catalysis to occur. For example, in carboxypeptidase A several residues move as the substrate binds, especially tyrosine-248 (Figure 1). Hexokinase adjusts its conformation as its substrate, glucose, binds. Presumably this adjustment excludes water from the active site which might otherwise compete for the phosphate group which is transferred to the glucose from ATP by the enzyme. This is the classical example of what is known as an induced fit (Bennett and Steitz, 1978).

The best example of a conformational change in a protein in which much of the molecular detail is known and which has served as a prototype for many other systems is found not in an enzyme but in haemoglobin, the protein of red blood cells which binds oxygen. Although haemoglobin does not catalyse a chemical reaction, it shows many characteristics which are typical of enzymes. Haemoglobin consists of four subunits, each of which can bind a molecule of oxygen. As a molecule of oxygen binds to one subunit, changes take place in the conformation of the protein which alter the ease with which the binding of the next oxygen molecule occurs. This process, when it leads to an increasing ease of binding, is known as cooperativity, and it is of critical importance in the function of haemoglobin as it balances the binding of oxygen in the lungs and its release to the tissues (Perutz, 1978). Similar phenomena have been implicated in many enzyme-catalysed reactions and are often associated with the intracellular regulation of the rate of these reactions.

Conformational changes are among the most subtle and critical in an enzyme-catalysed reaction. If the conformational mobility of an enzyme is restricted, the catalysis may be markedly inhibited. This kind of inhibition is a likely consequence of the attachment of an enzyme to a solid support, a commonly used procedure in technological applications of enzymes (Wang *et al.*, 1979). In such cases one may have to trade catalytic efficiency for the convenience of an immobilized enzyme.

28.3.4 Covalent Bond Making and Breaking

At this central stage of the catalytic mechanism the changes take place which result in the formation of new molecules. Clearly the earlier binding steps can contribute significantly to the way in which the bonding changes take place. The interactions between functional groups that begin with binding now develop their full catalytic potential.

By transition state theory, the rate of a chemical reaction can be written as:

$$k = (k_B T/h)\exp(-\Delta H^{\ddagger}/RT + \Delta S^{\ddagger}/R)$$

where k = the rate constant for the reaction, k_B = Boltzmann's constant and h = Planck's constant. The main terms in the exponent, apart from the temperature, T, which is assumed to be constant, are an enthalpy term $\Delta H^{\ddagger}$, which corresponds to the activation energy, and an entropy term $\Delta S^{\ddagger}$, which reflects the ease with which the enzyme and the substrate come together. Thus many of the factors which we have already discussed which favour binding will be included in the entropy term. A favourable entropy change is achieved by bringing the appropriate functional groups in the substrate and enzyme together at the active site. This is similar to the phenomenon of neighbouring-group participation which is well known to organic chemists (Fersht and Kirby, 1980). In carboxypeptidase these groups are glutamate-270, tyrosine-248, arginine-145 and the zinc ion (Figure 1). The other consequence of the binding of the peptide substrate to carboxypeptidase may be to increase the strain in the peptide bond to be cleaved.

A number of mechanisms of catalysis in solution in non-enzymic systems are well known (Jencks, 1969) and most of these can be detected in enzyme-catalysed reactions. These include acid–base catalysis (involving both proton-transfer reactions and Lewis acids), covalent catalysis, in which a covalent intermediate is formed between the enzyme and the substrate, and stabilization of reactive intermediates such as carbenium ions and carbanions. These mechanisms all allow a reduction in the enthalpy of activation to be achieved.

The activation energy for a reaction could also be lowered by raising the ground state energy of the substrate with respect to the transition state. This could occur as a consequence of a conformational change forced on the substrate molecule by binding to the enzyme, as may occur in the reaction catalysed by lysozyme (Lipscomb, 1972). The removal of a polar substrate from the influence of the bulk solvent into an essentially non-aqueous active site may have a similar effect on the energetics of the reaction. Obviously a number of the catalytic mechanisms we have just outlined may participate in the mechanism of one particular enzyme. Whilst these effects can often be identified in the mechanism of an enzyme of interest, it is difficult to assess their relative importance. This is particularly so in the case of strain induced in the substrate molecule by binding to the enzyme (Knowles, 1977). Several of the theoretical approaches currently being applied to the analysis of enzyme-catalysed reactions are directed towards estimating the importance of each catalytic mechanism (Warshel, 1981; Pincus and Scheraga, 1981).

In the next few paragraphs we shall examine some of these mechanisms as they are found in enzyme-catalysed reactions. In carboxypeptidase A (Figure 1), the zinc ion acts as a general Lewis acid and polarizes the carbonyl group of the peptide bond to be hydrolysed, making it more susceptible to nucleophilic attack by glutamate-270, or, more probably, by water, which is itself hydrogen bonded to glutamate-270 (Breslow and Warnick, 1977). Tyrosine-248 also probably provides general acid catalysis.

A similar polarization of a peptide carbonyl group to make it more reactive is achieved by different means in the serine proteases such as trypsin, chymotrypsin and subtilisin. In these enzymes the carbonyl group is polarized by hydrogen bonding to the amino group of a peptide bond of the enzyme. Further hydrogen bonding between amino acid residues present at the active site, serine, imidazole and aspartate, leads to the development of a highly nucleophilic serine hydroxyl group which attacks the polarized carbonyl group of the peptide bond to be cleaved (Figure 2). This results in the formation of a substrate fragment which is covalently bound to the enzyme as an ester through the serine hydroxyl group. The acyl enzyme is hydrolysed in the subsequent stages of the reaction to yield free enzyme and product. In this mechanism we have examples of both acid–base and covalent catalysis. For an account of the details of this mechanism the reader should refer to the review by Steitz and Shulman (1982).

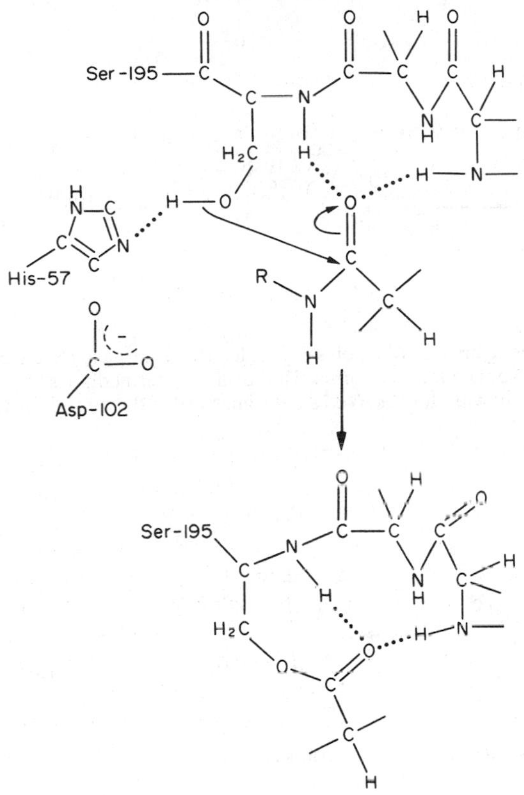

Figure 2 Some stages in the mechanism of serine proteases. The carbonyl group of the peptide bond to be hydrolysed is hydrogen bonded to two peptide amino groups of the enzyme. The hydroxyl group of Ser-195 is activated by hydrogen bonding to His-57 and attacks the polarized carbonyl group. This results in the formation of a tetrahedral intermediate (not shown) which decomposes into the acyl enzyme shown in the lower part of the figure and a peptide fragment is released. The covalently bound intermediate product still possesses a polarized carbonyl group and is hydrolysed by water to release the second peptide fragment from the enzyme. (Drawing based on Steitz and Shulman, 1982)

Ionic intermediates are common species in chemical reactions, particularly those occurring in polar solvents. Enzymes can help stabilize such intermediates by the provision of complementary functional groups in the correct orientation at the active site. Positively charged carbon species, usually known as carbonium ions but nowadays more properly referred to as carbenium ions, frequently arise in substitution reactions at carbon. In carbohydrates, which commonly undergo substitution reactions, the ring oxygen participates, stabilizing the positive charge and forming an oxonium ion. Figure 3 shows a possible mechanism for the hydrolysis of a polysaccharide in which an oxonium ion may be formed. Such a mechanism may be appropriate for glucoamylase, which is widely used in conjunction with glucose isomerase in the industrial production of fructose syrups from starch (Reilly, 1979).

In contrast to positively charged ions, enzymes are not so readily able to stabilize transient negatively charged species and so a range of coenzymes has evolved which allow carbanions and related species to be generated at enzyme active sites. Examples include coenzyme A, pyridoxal phosphate and thiamine pyrophosphate (see Metzler, 1977 or Suckling and Suckling, 1980 for a detailed discussion of the chemistry and biochemistry of coenzymes). Coenzyme A forms thiol esters with carboxylic acid substrates. Thiol esters are intrinsically more reactive than the oxygen analogues. Thus enolate anions form readily under base catalysis at enzyme active sites, allowing condensation reactions such as the formation of citrate from oxaloacetate and acetyl-coenzyme A to proceed (Figure 4). Coenzymes and also transition metal ions are important in oxidation–reduction reactions, which involve electron or formal hydride ion transfer. Space does not permit a

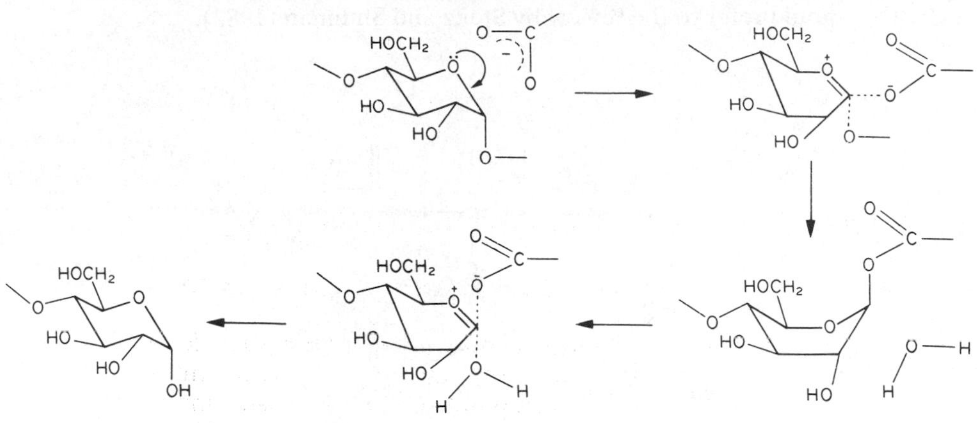

Figure 3 A mechanism for the hydrolysis of a polysaccharide. The negatively charged carboxyl group of the enzyme stabilizes the positively charged oxonium ion as it forms. The covalent intermediate shown may form, and this would be consistent with the observed stereochemistry of the hydrolysis reaction

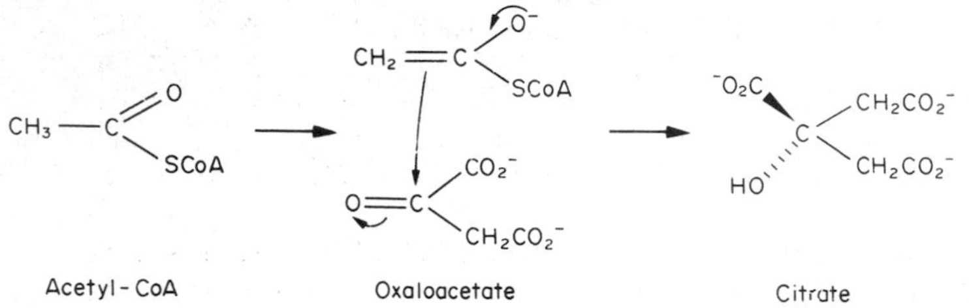

Figure 4 Formation of citrate from acetyl-CoA and oxaloacetate. The thiol ester, acetyl-CoA, readily forms an enolate ion on the enzyme active site. This ion acts as a nucleophile, adding to the carbonyl group of oxaloacetate in a stereospecific manner to form citrate. The stereospecificity of the reaction implies precise control of all stages by the enzyme

discussion of these or of other representative examples of each class of enzyme-catalysed reaction. A comprehensive discussion can be found in Walsh (1979) or Metzler (1977).

The mechanisms we have just considered all contribute to the central part of the enzyme-catalysed reaction. The final stages in which the products dissociate and the enzyme relaxes to its resting state to a large extent mirror the initial binding steps. Again diffusion may limit the reaction rate, making product dissociation the slow step. With all these factors in mind, let us examine the mechanism of an enzyme which is of great importance as an industrial catalyst, glucose isomerase.

28.4 GLUCOSE ISOMERASE

The production of fructose syrups from starch derived from maize or wheat by enzyme-catalysed processes is now a well established commercial operation. Starch is initially hydrolysed by amylase or glucoamylase (Reilly, 1979) and the resulting glucose enzymically isomerized into an equilibrium mixture of glucose and fructose (Antrim *et al.*, 1979). Much of the industrial use of glucose isomerase relies on an immobilized form of the enzyme.

In a sense the widely used name of the enzyme is incorrect since the natural substrate for the commercial glucose isomerases is xylose. Thus from the biochemical point of view the enzyme is strictly known as D-xylose ketol-isomerase, EC 5.3.1.5 and is derived from bacteria such as *Pseudomonas hydrophila* or *Lactobacillus pentosus*. The biological and industrial reactions catalysed are shown in Figure 5.

Xylose isomerase is widely used in preference to other enzymes with glucose isomerase activity because it is heat stable (temperatures of 45–65 °C can be used) and the addition of cofactors is

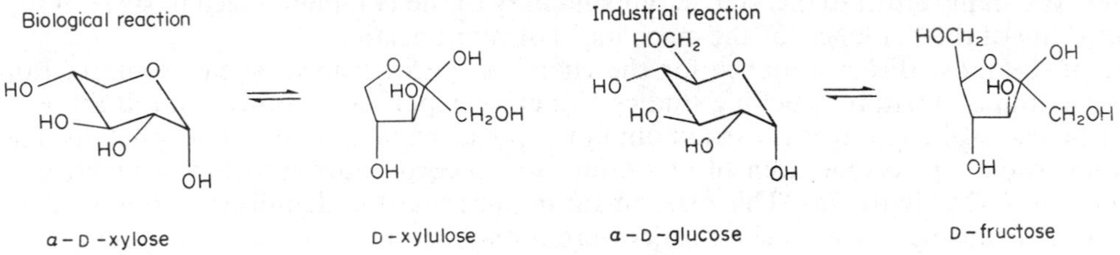

Biological reaction

Industrial reaction

a – D –xylose   D –xylulose   a – D –glucose   D –fructose

Figure 5   Reactions catalysed by xylose (glucose) isomerase, EC 5.3.1.5

unnecessary, particularly when the enzyme is used in the immobilized form. A comprehensive review of this enzyme and its technological application can be found in Antrim *et al.* (1979).

A number of xylose isomerases have been purified and their basic kinetic properties studied. pH optima of 6 to 8.5 have been reported. The K_m for xylose is about 100 fold less than that for glucose: a nearly molar concentration of glucose is required to achieve the half maximal rate of isomerization. Thus although glucose is clearly not the best substrate for this enzyme, its catalytic activity is sufficient for it to be used in an industrial process.

With the basic kinetic parameters of the reaction defined we can proceed to examine more details of the structure and mechanism of glucose isomerase. The reaction catalysed, that of aldose–ketose isomerization, is common to many enzymes, so as we proceed we will be able to draw on mechanistic analogy from a wide range of studies of related reactions (Rose, 1975).

Many glucose isomerases require a metal ion such as Co^{2+}, Mn^{2+}, Mg^{2+} or Cr^{2+} for activity. We shall need to account for this requirement in the mechanism. Other metal ions inhibit the reaction. It is probable that these ions displace the ions required for catalysis and produce an inactive enzyme. The inhibitory ions include Ag^+, Hg^{2+}, Cu^{2+}, Zn^{2+} and Ni^{2+}. It would obviously be important to exclude these ions from the solutions used and the vessels used in a plant designed for glucose isomerization.

Other inhibitors include the substrate analogue xylitol (Figure 6) which is a sugar alcohol and which presumably competes with the substrate for binding at the active site. This pattern of inhibition is similar to other sugar isomerases.

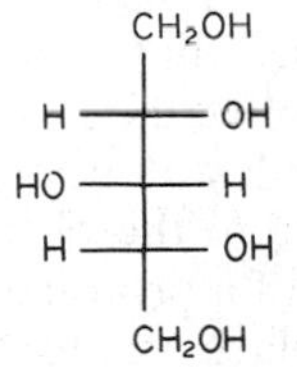

Figure 6   Xylitol, the sugar alcohol corresponding to xylose, the aldo-sugar which is the natural substrate for xylose isomerase. The reduced form of the natural substrate such as this is often found to be an inhibitor of sugar isomerases

The isomerization reaction involves the conversion of one substrate molecule into one product molecule and so we might expect the following simple kinetic scheme to apply to the steady state of the reaction:

$$E + glucose \rightleftharpoons E \cdot glucose \rightleftharpoons E \cdot fructose \rightleftharpoons E + fructose$$

Experiments with xylose isomerase were consistent with this scheme and the turnover number for D-xylose was found to be about 7000 min^{-1}, a relatively rapid reaction. As we mentioned before, the isomerization reaction is an equilibrium and this means that mechanistic studies can be carried out with the reaction running initially in either direction.

The kinetic situation is a little more complicated than this simple scheme suggests. Sugars can exist in a number of different anomeric forms in solution which are in equilibrium with each other. It is likely that the enzyme will only recognize one such form as the substrate, since the orientation of functional groups at the active site will be that required for reaction with that specific anomer. Several studies have shown that the α-D-glucopyranose form is the substrate for glucose isomerase (Figure 5). Since the isomerization reaction takes place on the ring-opened form of the sugar, an initial step in the catalysis of the reaction is likely to be the ring opening of the substrate. We shall return to this step in our summary of the complete mechanism, but first let us consider some of the evidence for the mechanism of isomerization.

Some of the most direct evidence for the chemical mechanism of sugar isomerization came from deuterium and tritium labelling studies. An early experiment showed that if the isomerization of fructose 6-phosphate was carried out in the presence of glucose-6-phosphate isomerase in deuterium oxide, up to one atom of deuterium was incorporated into the product, glucose 6-phosphate, at C-2 (Figure 7a). This experiment implies that the deuterium atom is added to an intermediate from the solvent. The results were thought to be consistent with the intermediate being an enediol. (Figure 7b).

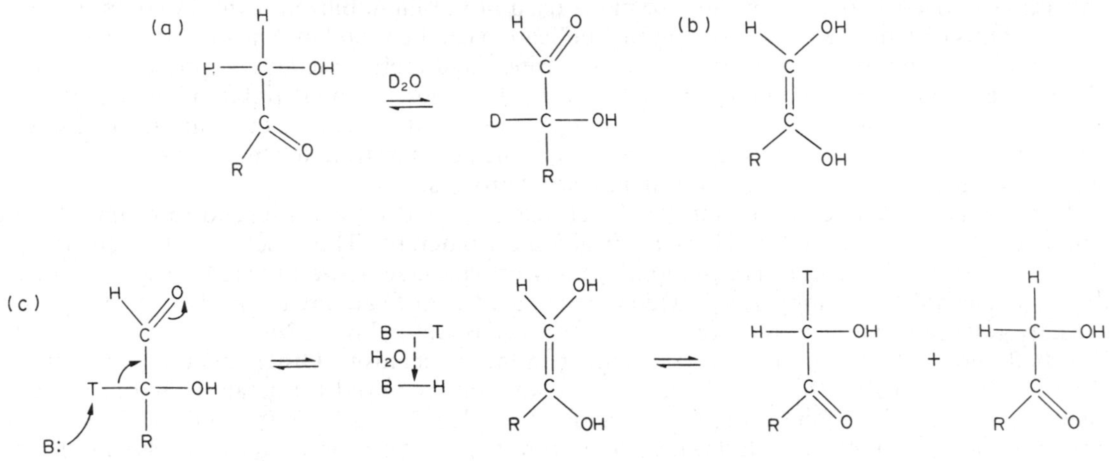

Figure 7 Isotopic evidence for the mechanism of sugar isomerases: (a) uptake of deuterium when the reaction is carried out in deuterated water; (b) the enediol intermediate; (c) intramolecular transfer of a proton to a basic group on the enzyme and part exchange of this proton with water during catalysis

The mechanism was elaborated using the results of a tritium labelling experiment. Here the isomerization was allowed to proceed using a substrate labelled with tritium at C-2 (Figure 7c). Tritium was found not only in the product but also in the solvent water. This was interpreted in the way shown in Figure 7c. It was proposed that a basic group on the enzyme would accept the proton (or tritium) from C-2 as the enediol formed. The enzyme-bound proton could then partially exchange with protons from the solvent water before being transferred back to the intermediate to form the product. This would lead to the observed distribution of radioactivity. In this way we have both intramolecular transfer of a proton and also partial exchange with the solvent. The extent to which this exchange takes place was found to vary with the isomerase studied. In reactions catalysed by xylose isomerase very little of the tritium label is lost to the solvent, but the fact that some measurable exchange does occur supports the view that the enediol intermediate is formed also in this case. It may be that the Mn^{2+} ion which is present in the enzyme coordinates water molecules associated with the enzyme rather tightly and so hinders the rate of exchange with solvent.

More detailed isotopic work showed that the label was transferred intramolecularly in a stereospecific way. Stereospecificity in a reaction implies a very precise orientation of reacting groups and is very characteristic of enzyme-catalysed reactions where stereochemical control is achieved by the geometry of the active site (see also Figure 4). The intermediate could thus be more precisely defined as a *cis*-enediol. The chemical mechanism, with its precise stereochemistry can be summarized as shown in Figure 8.

A number of questions arise from a plausible chemical mechanism such as this. The labelling pattern in substrate and product, for example, should be independent of the direction in which the reaction is initially run. This was shown to be the case for glucose-6-phosphate isomerase. We can also ask what enzyme functional groups are necessary for catalysis and in particular what functional group performs the role already defined in the mechanism, that of the base which transfers the proton to and from the enediol intermediate. We also need to be able to understand more precisely the role of the metal ion.

A possible function of the metal ion could be to polarize the carbonyl group of the aldo- or

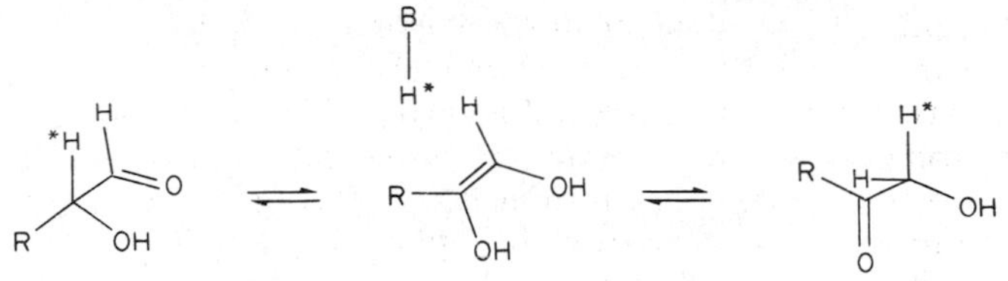

Figure 8 Stereochemistry of the isomerization reaction revealed by more detailed isotopic studies. The labelled proton H* is removed from the substrate to form the *cis*-enediol. Transfer of a proton back to the same face of the enediol leads to the stereospecifically labelled product shown

keto-sugar in a manner similar to the zinc ion in carboxypeptidase A (Section 28.3.4). NMR studies of the water molecules surrounding the manganese ion in xylose isomerase suggested that the substrate displaced some of the bound water when it entered the active site. The implication of this was that the metal ion might be close enough to the substrate for it to exert its Lewis acid function. Alternatively it was thought that the positively charged ion could stabilize the developing enolate ion which is in equilibrium with the enediol intermediate. More recent studies (see Antrim *et al.*, 1979) suggested in contrast that the metal ion was too remote from the substrate to allow any interaction. The main role of the metal ion in xylose isomerase may therefore be simply to help maintain the protein in its catalytically active configuration. The observation that the metal ion can be omitted under certain conditions when an immobilised xylose isomerase is used (Poulsen and Zittan, 1976) would tend to support this latter view.

The use of active site directed inhibitors allowed the importance of a carboxyl group in catalysis to be demonstrated for triosephosphate isomerase and glucose-6-phosphate isomerase. The 1-halo-3-hydroxyacetone phosphate (Figure 9a) inhibited triosephosphate isomerase and the epoxide analogue, 1,2-anhydro-D-mannitol 6-phosphate (Figure 9b), was shown to inhibit glucose-6-phosphate isomerase in a way consistent with attack of an enzyme carboxyl group at C-1 of the inhibitor (Rose, 1975). These studies were carried out with great attention to detail to ensure that the kinetics and pH/rate dependence of the inhibition were consistent with the proposed mechanism.

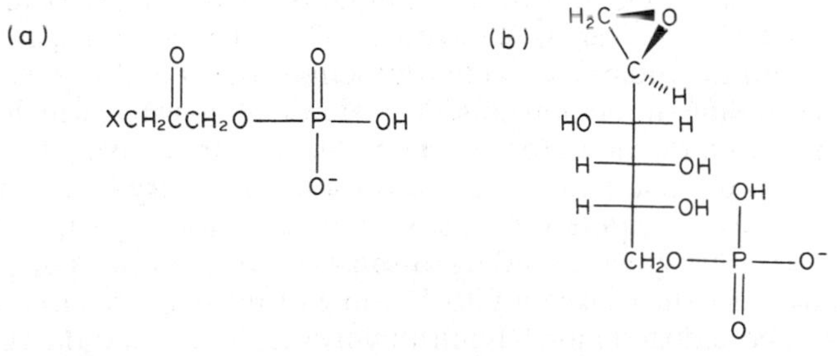

Figure 9 Active site directed inhibitors of (a) triosephosphate isomerase (X = halogen); and (b) glucose-6-phosphate isomerase

At this stage it is clear that we have a reasonable chemical understanding of the isomerization reaction and of the mechanism as it is seen by the substrate. However, structural details of the enzyme and of the active site are lacking. An X-ray crystallographic study of glucose-6-phosphate isomerase allowed a more comprehensive, but still incomplete, picture to be obtained (Shaw and Muirhead, 1976).

The mechanism, drawn here by analogy for α-D-glucose, takes into account all the chemical and other evidence which we have discussed (Figure 10). Since the exact amino acid residues which take part in catalysis at the active site are unknown, we can only indicate their mechanistic function. The crystallographic study led to one modification of the mechanism as we have described it so far. It was concluded that the carboxylate group modified by the epoxide substrate analogue (Figure 9b) does not catalyse the proton transfer associated with the enediol intermediate. Rather it was thought to assist in the ring opening of the substrate and in the proton transfers shown in Figure 10 during the isomerization.

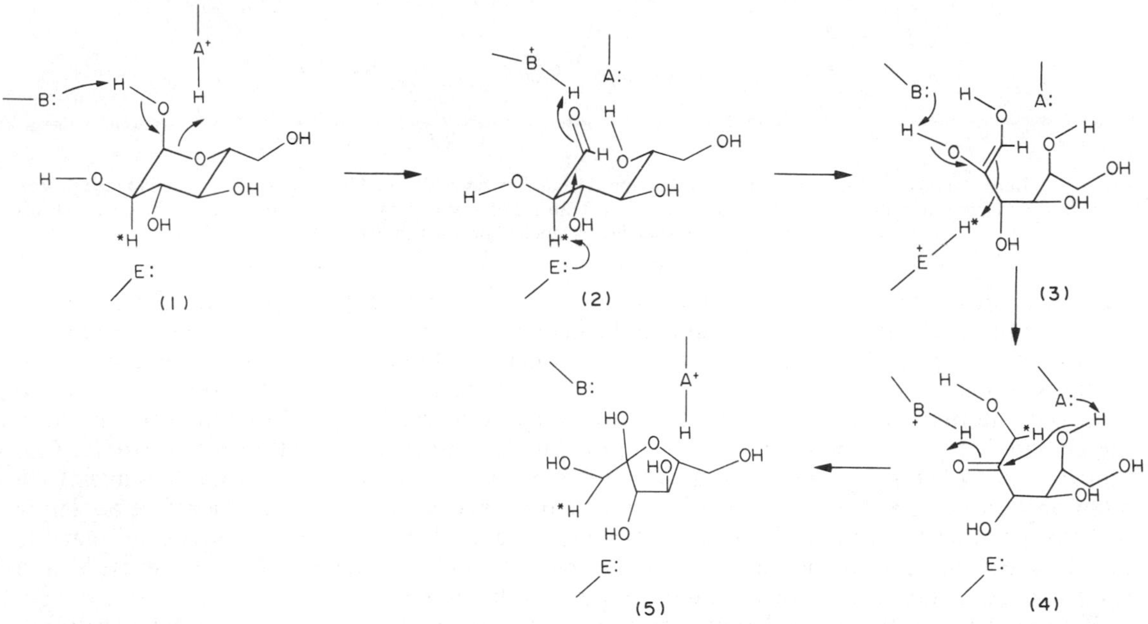

Figure 10 A possible mechanism for the isomerization of α-D-glucose. (1) Basic group B on the enzyme (possibly a carboxyl group) initiates ring opening of the α-pyranose form of glucose which is known to be the preferred substrate. This process is assisted by the acidic group HA⁺—. (2) A second basic group E accepts the labelled proton, H*, and the *cis*-enediol intermediate forms. (3) HB*— promotes the collapse of the intermediate and H* is transferred to C-1 of the substrate. (4) Groups A and B again cooperate in the ring closing of the intermediate to form the furanose product (5)

28.5   SUMMARY

An enzyme mechanism attempts to provide a rationalization of all the molecular events which occur during catalysis. The mechanism is based on a wide range of studies which can include kinetic measurements, structural studies by X-ray diffraction and the effects caused by a variety of inhibitors. The precise folding of the polypeptide chain of an enzyme, which may be observed by structural studies, leads to a defined area of the molecule, the active site, where the substrate binds and where catalysis occurs. The active site is complementary in structure to the substrate and, with cofactors if these are required, contains all the functional groups necessary for catalysis of the chemical reaction. This specific binding of substrate to enzyme is one of the factors which make enzymes attractive as industrial catalysts. Binding of substrate to enzyme makes use of non-covalent interactions. These include ionic bonds, hydrogen bonds and the so-called hydrophobic interactions. The protein structure itself is also dependent on the integrity of many non-covalent interactions and it is important that these remain unperturbed by manipulations of the enzyme, such as immobilization, if catalytic activity is to be retained.

The catalysis of the reaction by an enzyme takes place in a series of events which follow the initial binding of substrate(s) to the enzyme. The enzyme may catalyse any or all of these steps but it is unable to control the rate at which substrate diffuses onto the active site or at which product diffuses away from the enzyme. Thus some reactions in the cell and many reactions *in vitro* and in industrial plants may be limited in their rates by the rate of diffusion of small substrate and product molecules.

Catalysis is achieved by a variety of mechanisms. Perhaps the most important is the specific binding of the substrate(s) to the enzyme which causes the appropriate functional groups in the enzyme and substrate to be correctly oriented for catalysis. In addition, acid–base catalysis, covalent catalysis and stabilization of reactive intermediates or transition states contribute to the rate enhancement caused by the enzyme. The relative importance of these effects is difficult to assess in any specific case.

These considerations dictate that the conditions adopted in a biotechnological process based on an enzyme-catalysed reaction should be such that the reactive conformation of the enzyme should be minimally disturbed. The temperature and the pH of the medium should allow the enzyme to remain active for long periods. It does not follow, however, that the optimum conditions used in an *in vitro* mechanistic study are appropriate for a plant; a compromise set of conditions in which product yield and catalyst life are optimized together would be ideal.

28.6 REFERENCES

Antrim, R. L., W. Colilla and B. J. Schnyder (1979). Glucose isomerase production of high-fructose syrups. *Appl. Biochem. Bioeng.*, **2**, 97–155.
Aunstrup, K. (1979). Production, isolation and economics of extracellular enzymes. *Appl. Biochem. Bioeng.*, **2**, 27–69.
Bennett, W. S. and T. A. Steitz (1978). Glucose-induced conformational change in yeast hexokinase. *Proc. Natl. Acad. Sci. USA*, **75**, 4848–4852.
Breslow, R. and D. L. Wernick (1977). Unified picture of mechanisms of catalysis by carboxypeptidase. *Proc. Natl. Acad. Sci. USA*, **74**, 1303–1307.
Conti-Tronconi, B. M. and M. A. Raftery (1982). The nicotinic cholinergic receptor: correlation of molecular structure with functional properties. *Annu. Rev. Biochem.*, **51**, 491–530.
Dunnill, P. (1978). In *Enzymic and Non-enzymic Catalysis*, ed. P. Dunnill *et al.*, chap. 2. Horwood, Chichester.
Fersht, A. (1984). *Enzyme Structure and Mechanism*, 2nd edn., chap. 4. W. H. Freeman, San Fransisco.
Fersht, A. R. and A. J. Kirby (1980). Intramolecular catalysis and the mechanism of enzyme action. *Chem. Br.*, **16**, 136–142.
Goldstein, L. (1976). Kinetic behaviour of immobilised enzyme systems. *Methods Enzymol.*, **44**, 397–444.
Jencks, W. P. (1969). *Catalysis in Chemistry and Enzymology*. McGraw-Hill, New York.
Jencks, W. P. (1975). Binding energy, specificity and enzymic catalysis, *Adv. Enzymol.*, **43**, 219–410.
Jürss, R., H. Prinz and A. Maelicke (1979). NBD-5-acylcholine: fluorescent analog of acetylcholine and agonist at the neuromuscular junction. *Proc. Natl. Acad. Sci. USA*, **76**, 1064–1068.
Knowles, J. R. (1977). Why mechanistic enzymology? *Trends Biochem. Sci.*, **2**, N278–279.
Knowles, J. R. and W. J. Alberry (1977). Perfection in enzymic catalysis: the energetics of triose phosphate isomerase. *Acc. Chem. Res.*, **10**, 105–111.
Lipscomb, W. N. (1972). Three dimensional structures and chemical mechanisms of enzymes. *Chem. Soc. Rev.*, **1**, 319–336.
Lipscomb, W. N. (1982). Acceleration of reactions by enzymes. *Acc. Chem. Res.*, **15**, 232–238.
Metzler, D. E. (1977). *Biochemistry*. Academic, New York.
Mosbach, K. and B. Mattiason (1976). Multistep enzyme systems. *Methods Enzymol.*, **44**, 453–457.
Park, C. H., F. Y-H. Wu and C-W. Wu (1982). Molecular mechanism of promoter selection in gene transcription. *J. Biol. Chem.*, **257**, 6950–6956.
Perutz, M. F. (1978). Hemoglobin: structure and respiratory transport. *Sci. Am.*, **239**, 68–86.
Pincus, M. R. and H. A. Scheraga (1981). Theoretical calculations on enzyme–substrate complexes: the basis of molecular recognition and catalysis. *Acc. Chem. Res.*, **14**, 299–306.
Poulsen, P. B. and L. Zittan (1976). Continuous production of high-fructose syrup by cross-linked cell homogenates containing glucose isomerase. *Methods Enzymol.*, **44**, 809–822.
Reilly, P. J. (1979). Starch hydrolysis with soluble and immobilised glucoamylase. *Appl. Biochem. Bioeng.*, **2**, 185–207.
Rose, I. A. (1975). Mechanism of the aldose–ketose isomerase reactions. *Adv. Enzymol.*, **43**, 491–517.
Rossman, M. G. and P. Argos (1981). Protein folding. *Annu. Rev. Biochem.*, **50**, 497–532.
Shaw, P. J. and H. Muirhead (1976). Active site of glucose 6-phosphate isomerase. *FEBS Lett.*, **65**, 50–55.
Steitz, T. A. and R. G. Shulman (1982). Crystallographic and NMR studies of the serine proteases. *Annu. Rev. Biophys. Bioeng.*, **11**, 419–444.
Suckling, K. E. and C. J. Suckling (1980a). *Biological Chemistry*, chap. 11–12. Cambridge University Press, Cambridge.
Suckling, K. E. and C. J. Suckling (1980b). *Biological Chemistry* chap. 13. Cambridge University Press, Cambridge.
Tabushi, I. (1982). Cyclodextrin catalysis as a model for enzyme action. *Acc. Chem. Res.*, **15**, 66–72.
Walsh, C. (1979). *Enzymatic Reaction Mechanisms*. W. H. Freeman, San Fransisco.
Wang, D. I. C., C. L. Cooney, A. L. Demain, P. Dunnill, A. E. Humphrey and M. D. Lilly (1979). *Fermentation and Enzyme Technology*, chap. 13. Wiley, New York.
Warshel, A. (1981). Electrostatic basis of structure-function correlation in proteins. *Acc. Chem. Res.*, **14**, 284–290.

29

Enzyme Evolution

B. G. HALL
University of Connecticut, Storrs, CT, USA

29.1 INTRODUCTION

Because enzymes catalyze essentially all chemical reactions in biological systems, a significant amount of the effort in the field of biotechnology is devoted to manipulations designed to enhance the levels of specific enzyme activities within cells. Gene cloning, using recombinant DNA technology, has proven to be a powerful tool for such manipulations. Gene cloning lends itself to two primary manipulations: (i) increasing the amount of enzyme that is synthesized by increasing the number of copies of the gene that are present within a cell, which is usually accomplished by cloning a gene onto a high copy number plasmid; and (ii) allowing a cell to produce a new enzyme, which is accomplished by introducing a foreign gene that codes for the new enzyme into an appropriate plasmid. The power of the latter manipulation is that it potentially allows the genetic engineer to combine particularly desirable features of two different organisms. For example, an organism that is particularly efficient at producing a useful waste product (ethanol) might be unable to utilize a potentially available substrate (pentose sugars). One approach might be to attempt to clone bacterial genes for pentose utilization, then to introduce them into the appropriate yeast strain. A common problem with that approach is that foreign genes are often not expressed in the desirable organism. For instance, the rabbit β-globin genes are incorrectly initiated, and splice sites are not recognized in yeast (Beggs, 1980).

An altogether different problem arises when the desired enzyme activity is not known to exist in any organism. Engineering of microorganisms to detoxify specific synthetic chemical waste products is an example of a situation that might require a novel enzymatic activity. Because recombinant DNA technology cannot create new functions, but can only bring existing functions into new genetic environments, cloning does not provide a solution to the requirement for novel functions. Development of novel enzyme activities demands an understanding of the process by which new enzymatic functions evolve. That process is being investigated utilizing the approach of experimental, or 'directed' evolution. This chapter will explore several of these studies, then will suggest ways in which the results of those studies are applicable to the field of biotechnology.

29.2 CATABOLISM OF UNNATURAL SUGARS

29.2.1 Regulatory Mutations

Of the eight aldopentoses and four pentitols, three pentoses (D-xylose, L-arabinose and D-ribose) and two pentitols (ribitol and D-arabitol) are known to be common in nature. The remaining sugars are unknown or rare in nature, and are thus xenobiotic compounds. Pathways for the catabolism of the natural pentoses and pentitols were known, and nearly 20 years ago two groups began to study the means by which the soil bacterium *Klebsiella* could acquire the ability to utilize xylitol, a xenobiotic sugar that is now in common use as an artificial sweetener.

Learner *et al.* (1964) applied strong selection to *Klebsiella aerogenes* and isolated a mutant that could grow on xylitol. The mutant strain, designated strain X1, constitutively synthesized a high level of ribitol dehydrogenase (RDH), the enzyme which normally is inducible by ribitol and which converts ribitol to D-ribulose. The properties of the RDH were indistinguishable from those of wild-type RDH, suggesting that it was the constitutive synthesis of the RDH that permitted catabolism of xylitol. The demonstration that an independently isolated RDH constitutive mutant could grow on xylitol confirmed that hypothesis. From strain X1 they isolated a second mutant that could grow faster on xylitol, strain X2. The RDH from strain X2 proved to have a lower K_m for xylitol than the wild-type enzyme (Wu *et al.*, 1968). Continued selection for faster growth led to strain X3, which constitutively synthesized an arabitol permease that also had a specificity for xylitol (Wu *et al.*, 1968; see Figure 1).

Mortlock and his coworkers obtained very similar results when they selected mutants of *Klebsiella pneumoniae* var. *oxytoca* that could utilize xylitol. Again, the xylitol positive mutants synthesized RDH constitutively, and they showed that, *in vitro*, the wild-type enzyme (whether obtained from a wild-type strain or from the constitutive mutant) exhibited a low level of activity toward xylitol.

The above results were extended by Hartley's group (Hartley *et al.*, 1972; Rigby *et al.*, 1974), who subjected *Klebsiella* to long term selection in continuous culture. They found that, following the initial mutation that resulted in constitutive synthesis of RDH, a series of gene duplications occurred that led to increasing copies of the RDH gene and consequent faster growth. Eventually a strain was obtained that synthesized 20% of its protein as RDH. This strategy can be very successful in organisms that are refractory to *in vitro* DNA manipulations, however the resulting strains are quite unstable and the additional gene copies are rapidly lost upon withdrawal of the selective pressure for rapid utilization of the novel resource.

These examples point out the most common means by which microorganisms can acquire novel catabolic capabilities: *via* regulatory mutations that lead to constitutive synthesis of an enzyme that already had a specificity for the novel resource, but whose synthesis was not induced by the novel resource. We have also seen that subsequent mutations that improve the activity of the enzyme toward the novel resource can be obtained by selection for more efficient catabolism (see Figure 1).

29.2.2 Modular Pathways

In some cases acquisition of novel capabilities involves 'capturing' elements from more than one preexisting metabolic pathway. We saw that capture of an element of the arabitol catabolic pathway, the arabitol permease, resulted in improved growth on xylitol when the strain had already acquired an efficient, constitutive ribitol dehydrogenase. In effect, *Klebsiella* assembled an 'optimal' xylitol pathway from two modules, one an element of the ribitol pathway, the other, an element of the arabitol pathway. More complex examples of such modular construction of new pathways are presented by the acquisition of pathways for D-arabinose and for L-1,2-propanediol catabolism.

A single mutation is required for *Klebsiella* to grow on D-arabinose (reviewed in Mortlock, 1981). The mutation results in the constitutive synthesis of L-fucose isomerase, an enzyme of the L-fucose pathway that is capable of converting D-arabinose to D-Ribulose. D-Ribulose is an intermediate in the ribitol catabolic pathway, and is the natural inducer of the enzymes for that path-

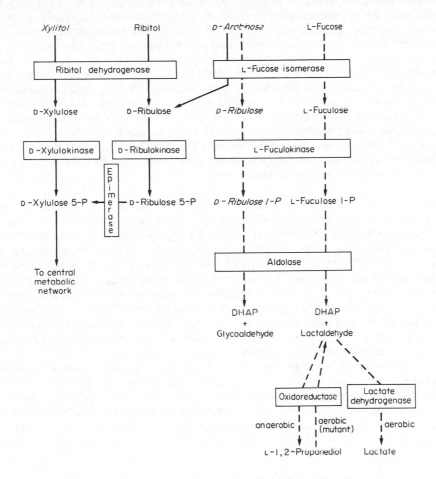

Figure 1 Metabolism of xylitol, D-arabinose and propanediol in *Klebsiella* and *E. coli*: novel substrates and intermediates are italicized; enzymes are enclosed in boxes; solid arrows show *Klebsiella* and *E. coli* C pathways and dashed arrows show *E. coli* K-12 pathways (Hall, 1983)

way. Thus, the first step in D-arabinose metabolism is carried out by an element of the L-fucose pathway, while subsequent steps are catalyzed by elements of the ribitol pathway.

In *E. coli* K-12 the above pathway for D-arabinose metabolism is not possible because *E. coli* K-12 does not possess a ribitol pathway. Nevertheless, mutants of *E. coli* K-12 capable of metabolizing D-arabinose were obtained by Mortlock's group (LeBlanc and Mortlock, 1971). The D-arabinose positive mutants of *E. coli* did not express the L-fucose pathway constitutively, instead a regulatory mutation had led to D-arabinose being recognized as an inducer of the L-fucose pathway. Again, L-fucose isomerase converted the D-arabinose to D-ribulose, but in the absence of a ribitol pathway the D-ribulose could not enter the central metabolic network as a pentulose 5-phosphate. Instead, the D-ribulose continued to be metabolized by the enzymes of the L-fucose pathway, resulting in the production of dihydroxyacetone phosphate + glycoaldehyde, a less energy efficient pathway than the pentose pathway employed by *Klebsiella*.

The L-fucose pathway has contributed to the evolution of yet another new pathway, that for catabolism of L-1,2-propanediol (Hacking and Lin, 1977; reviewed in Hall, 1982). *E. coli* K-12 is unable to utilize propanediol. When mutants capable of aerobic utilization of L-1,2-propanediol were isolated several clues suggested the involvement of the L-fucose pathway. First, propanediol is excreted by *E. coli* during anaerobic growth on L-fucose. Second, the L-1,2-propanediol positive mutant had become incapable of growth on L-fucose. Third, genetic analysis showed that the mutation that allowed L-1,2-propanediol utilization mapped in the same region as did the genes for fucose utilization. It was shown that during L-fucose fermentation an NAD-dependent dehydrogenase was induced that converted lactaldehyde to propanediol. In the wild-type strain that enzyme was not produced during aerobic L-fucose utilization. The L-1,2-propanediol positive mutant synthesized the lactaldehyde dehydrogenase constitutively, and that dehydrogenase was indistinguishable from the wild-type enzyme. Reversing the normal flow of the anaerobic pathway, the lactaldehyde dehydrogenase converted the L-l,2-propanediol to lactaldehyde, which was metabolized to pyruvate by two other enzymes of the L-fucose pathway. Based on its ability to catalyze the interconversion of lactaldehyde and L-1,2-propanediol in either direction, the lactaldehyde dehydrogenase is now called a propanediol oxidoreductase. The failure of the propanediol positive strain to utilize L-fucose was explained by the observation that the enzymes for the first three steps in L-fucose catabolism had become uninducible in the mutant. The number and naure of the mutations involved in the acquisition of L-1,2-propanediol utilization is unclear. Some of the enzymes of the L-fucose pathway, one of which is normally expressed only under anaerobic conditions, had become constitutive, while three other enzymes of the same pathway had become uninducible. It has been possible to isolate L-1,2-propanediol positive mutants that retain the ability to catabolize L-fucose, but these rapidly give rise to L-fucose negative mutants upon prolonged growth on L-1,2-propanediol.

The propanediol oxidoreductase has also provided an alternative pathway for the acquisition of xylitol utilization. Wu (1976) observed that the propanediol oxidoreductase could oxidize xylitol to D-xylulose, the first intermediate in the D-xylose catabolic pathway. Starting with the L-1,2-propanediol positive mutant, selection for growth on xylitol resulted in a mutant that expressed both the D-xylose pathway enzymes and a D-xylose transport system constitutively. Xylitol was able to enter the cell *via* the D-xylose transport system, was oxidized to D-xylulose by the propanediol oxidoreductase of the anaerobic L-fucose system, and was further metabolized by the enzymes of the D-xylose pathway. This is a particularly important example, because it illustrates the point that some new metabolic capabilities cannot be selected directly, but must be selected in a series of steps. Because it does not possess a ribitol pathway, and hence cannot synthesize an RDH, *E. coli* K-12 could not evolve the same pathway for xylitol utilization as did *Klebsiella* strains. Direct selection for xylitol utilization from a wild-type *E. coli* K-12 is not possible because no single mutation provides all of the necessary enzymatic activities required for xylitol metabolism. Constitutive expression of the D-xylose pathway alone does not permit xylitol metabolism because, although it can be taken up into the cell, xylitol has no entry to the D-xylose pathway. Constitutive expression of the propanediol oxidoreductase alone is not sufficient because no xylitol transport system is provided, and xylitol is not an inducer of the D-xylose pathway enzymes. In fact, no pentitol-utilizing mutants of *E. coli* K-12 have been directly selected as one-step mutants from wild-type strains.

29.3   EVOLUTION OF AN ALIPHATIC AMIDASE IN *PSEUDOMONAS*

Pseudomonas aeruginosa utilizes acetamide and propionamide which are hydrolyzed by an aliphatic amidase (*amiE* gene product) to yield ammonia and the corresponding aliphatic acids (reviewed in Clarke, 1978). The amidase is inducible by these substrates, but the fact that related substrates are either poor inducers, poor substrates, or both, prevents their use as carbon or nitrogen sources by this organism. In an elegant series of studies Clarke and her colleagues have directed the evolution of this system to greatly extend the range of substrates utilized by *Pseudomonas aeruginosa*. Clarke's experiments have yielded a series of mutants altered in both the properties of the aliphatic amidase (*amiE* mutants) and in the regulation of the amidase gene (*amiR* mutants).

Formamide is hydrolyzed by the amidase at about 20% of the rate at which acetamide is hydrolyzed, but it is not utilized as a nitrogen source because it is not an effective inducer of amidase synthesis (formamide cannot be used as a carbon source because the product of hydrolysis, formic acid, is not metabolized by *P. aeruginosa*). Selection for utilization of formamide as a nitrogen source yielded several classes of regulatory mutants (Figure 2). One class had become sensitive to formamide as an inducer, while the remaining classes synthesized amidase constitutively. Among the constitutives, two important classes could be distinguished: one resembled the wild-type in that it remained sensitive to butyramide as an anti-inducer, while the other had lost its sensitivity to butyramide. One strain, C11, exhibited a high constitutive rate of amidase synthesis that allowed it to utilize formamide, but was sensitive to butyramide inhibition. Because butyramide is a very poor substrate, being hydrolyzed at <2% of the acetamide rate, strain C11 was unable to grow on butyramide. Two different lines of mutants able to utilize butyramide were derived from strain C11. One mutant strain, CB4, resulted from a second regulatory mutation that rendered the regulatory protein insensitive to butyramide inhibition. That strain synthesized wild-type amidase, and was phenotypically indistinguishable from the single-step butyramide-resistant constitutive mutants that had been isolated directly from the wild-type strain. A second mutant, strain B6, synthesized an altered amidase with greatly increased activity toward butyramide. The amidase from strain B6 exhibited a 10 fold higher V_{max} and a six fold lower K_m for butyramide than wild-type enzyme.

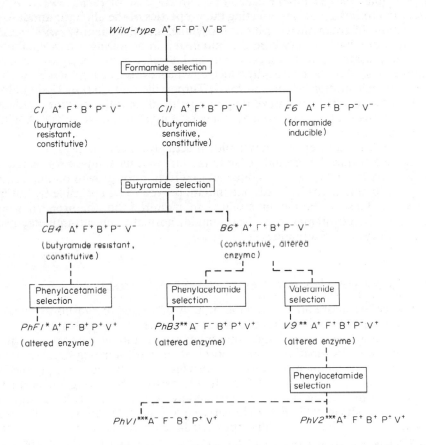

Figure 2 Evolution of the aliphatic amidase system in *Pseudomonas areuginosa*: selective conditions are in boxes; solid lines indicate mutations affecting the regulatory gene *amiR*, while dashed lines indicate mutations affecting the structural gene for the amidase enzyme, *amiE*; asterisks indicate the number of structural gene mutations present in the strain; growth phenotypes are indicated following the strain names; A=acetamide; F=formamide (as nitrogen source); B= butyramide; P=phenylacetamide (as nitrogen source); V=valeramide

Neither strain CB4 nor strain B6 was able to utilize phenylacetamide, and neither the wild-type enzyme nor the altered enzyme from strain B6 exhibited detectable activity toward phenylacetamide. Both strains yielded mutants capable of utilizing phenylacetamide, and in both cases the mutations altered the properties of the amidase itself. The amidase of strain PhB3, derived from strain B6, exhibited high activity toward phenylacetamide, but had lost its activity toward acetamide. The enzyme from strain PhF1, derived from strain CB4, also exhibited good activity toward phenylacetamide, and retained activity toward acetamide, but became quite thermolabile. Note that both of the phenylacetamide-utilizing mutants were the result of three mutations: PhF1 carried two mutations in the regulatory gene *amiR* and one mutation in the amidase gene, *amiE*, while PhB3 carried a single mutation in *amiR* and two mutations in *amiE*.

Strain B6 was also used to generate mutants capable of utilizing another analog, valeramide. The valeramide positive mutant, V9, possessed a second mutation in the amidase gene that permitted activity toward valeramide, but rendered the enzyme thermolabile and resulted in reduced activity toward propionamide. Again, mutant V9 carried one regulatory and two structural gene mutations.

Two phenylacetamide-utilizing mutants were isolated from strain V9. Both carried new mutations in the amidase gene, but the two mutations altered the properties of the amidase in strikingly different ways. The enzyme from strain PhV1 had a relative activity of 100 for phenylacetamide, but its activity had been reduced from 8 μmol min^{-1} mg^{-1} to 0.5 μmol min^{-1} mg^{-1} on butyramide. At the same time, its thermostability had returned to nearly the wild-type level. The enzyme from the sibling strain PhV2 had a relative activity of only 55 on phenylacetamide, but its activity toward butyramide was 2.09 μmol min^{-1} mg^{-1}. The PhV2 enzyme, however, exhibited the same thermolability as did the enzyme from strain V9.

This family of mutants has provided a striking example of the variety of means by which micro-organisms can acquire new metabolic capabilities. In the case of *Pseudomonas* the majority of mutations were structural mutations affecting the properties of the aliphatic amidase, rather than regulatory mutations affecting the conditions for synthesis or the quantity synthesized. It must be recognized, however, that in every case the initial step in acquiring a new catabolic capability involved a regulatory mutation. It appears that the interplay between regulatory and structural gene mutations is critical to the process by which enzymes evolve new catalytic capabilities.

In an explicit demonstration of this interplay, Turberville and Clarke (1981) deliberately isolated a regulated mutant that can utilize butyramide; recall that the mutant amidase capable of efficient butyramide hydrolysis was present in a constitutive strain, strain B6. A wild-type *amiR* allele was introduced into strain B6. Because the wild-type regulatory protein is not only insensitive to butyramide as an inducer, but butyramide is actually an anti-inducer, the resulting recombinant was unable to utilize butyramide despite the fact that its amidase was active toward that substrate. The recombinant strain was subjected to selection for growth on butyramide. Among 19 butyramide positive mutants isolated, one mutant proved to be inducible by butyramide. The properties of the amidase in the mutant strain were unaltered, but the properties of the regulatory protein were altered dramatically. Butyramide, normally an anti-inducer, had become a more powerful inducer than acetamide.

29.4 EVOLUTION OF A NEW β-GALACTOSIDASE IN *E. COLI*

The studies described above all adopted the strategy of forcing organisms to acquire the ability to metabolize truly novel resources, novel in the sense that the wild-type organism cannot use the resource. That strategy has the limitation that many new catabolic capabilities will require the evolution of a series of new functions: a transport system, multistep degradative system and possibly a mechanism for the elimination of toxic by-products. An alternative approach is to focus on a single, clearly defined new function that is already part of an existing and well characterized metabolic network. This strategy has been exploited by directing the evolution of a new β-galactosidase function in a strain of *E. coli* K-12 carrying a large deletion within the *lacZ* gene, the gene for β-galactosidase. That strain still expresses a functional lactose permease, and retains the ability to metabolize the breakdown products of lactose, *i.e.* glucose and galactose. Selection for mutants that can metabolize lactose has led to the development of a family of mutants that have reacquired the ability to utilize lactose and a variety of related β-galactoside sugars. Such a system has the advantage that it utilizes a genetically well characterized organism, making possible a level of genetic analysis and manipulation not available with more esoteric organisms. Similarly, both the general physiology of the organism and the specific biochemistry of lactose metabolism are already well understood. Finally, it was possible to take advantage of the variety of artificial substrates that had been developed for the purpose of investigating the original *lac* operon.

The strain from which the Lac$^+$ mutants were obtained carried a well characterized deletion that removed the middle third of the *lacZ* gene, thus no Lac$^+$ mutants could arise simply as the consequence of a back-mutation. The method for selecting lactose positive mutants from this strain is both simple and inexpensive. The strain is streaked onto lactose-TTC plates that contain broth, lactose, the fermentation indicator triphenyltetrazolium chloride (TTC), and isopropyl-thiogalactoside (IPTG). The broth supports the growth of the Lac$^-$ cells and allows them to form colonies on the medium. The sole purpose of the IPTG is to induce expression of the gene for lactose permease, and thus to permit the lactose that is present to enter the cell. The TTC permits lactose-fermenting colonies (white) to be easily distinguished from non-fermenting colonies (red). The lactose itself, of course, provides the selective function. Cells initially grow at the expense of the broth to form dark red colonies. As colonies reach a size of about 10^{10} cells they exhaust the broth in their immediate vicinity. Rare spontaneous Lac$^+$ mutants that arise are able to continue growing at the expense of the lactose, and eventually form outgrowths called papillae on the surface of the colonies. This process requires about two weeks, and because the papillae are white they are easily recognized. The mutant Lac$^+$ cells are easily purified by restreaking onto appropriate indicator plates. After two weeks of incubation virtually every colony on a selective plate exhibits several papillae. This selection scheme permits the rapid isolation of a large number of independent mutants by the simple expedient of picking only one papilla from each colony for purification and further study.

The selective method is of general applicability. By streaking cells on to plates that contain both broth and a novel resource, mutants have been selected with new catabolic capabilities in several systems. In *E. coli* a mutant was selected that was capable of fermenting cellobiose. In *Klebsiella* the selected mutant ferments lactose by an entirely new pathway involving a phosphorylating transport system and a phosphogalactosidase. In *Saccharomyces cerevisae* sucrose, maltose, and melibiose positive mutants have been selected.

The mutations that allow *lacZ* deletion strains of *E. coli* to grow on lactose are all in the genes of the ebg (evolved β-galactosidase) operon, located, far away from the lac operon, at 66 minutes on the *E. coli* K-12 map (reviewed in Hall, 1982). There are two structural genes in the ebg operon. *ebgA* specifies a 120 000 molecular weight polypeptide that is the subunit of a hexameric β-galactosidase enzyme called ebg enzyme. Ebg enzyme is unrelated to the classical β-galactosidase specified by the *lacZ* gene, since no serological cross reaction between the two enzymes has ever been detected. The wild-type, or 'unevolved', enzyme is called ebg° enzyme, the 'o' standing for 'original'; and the allele specifying ebg° enzyme is *ebgA*°. The designation *ebgA*$^+$ is reserved for mutant alleles specifying enzymes with greatly enhanced activity toward lactose. Much of this section will focus on the mutations that convert *ebgA*° to *ebgA*$^+$ alleles.

The second structural gene of the ebg operon is *ebgB*, a gene that specifies a 79 000 molecular weight protein of unknown function. *ebgB* is located distal to *ebgA* in the operon, but little more is known about that gene.

Expression of the ebg operon is subject to negative control (repression) by the product of the *ebgR* gene. The *ebgR* gene is located immediately adjacent to *ebgA*. *ebgR*$^-$ mutants are constitutive, synthesize about 5% of their soluble protein as ebg enzyme, and are recessive to *ebgR*$^+$. As has been the case for the other systems described above, mutations in the regulatory gene have proven to be critical to the process by which *lacZ* deletion strains of *E. coli* K-12 evolve lactose utilization.

It must be emphasized that the wild-type alleles of the genes, *ebgA*°, *ebgB*$^+$ and *ebgR*$^+$ exist in wild-type strains of *E. coli*. The evolutionary process does not create these genes, instead it modifies their properties in order to permit lactose utilization.

29.4.1 Properties of the Wild-type Proteins

The wild-type, or unevolved, ebg enzyme (ebg° enzyme) is virtually inactive toward lactose, but it does exhibit hydrolytic activity toward the synthetic substrate *O*-nitrophenyl-β-galactoside (ONPG). That activity is sufficient to permit accurate measurement of the amount of ebg° enzyme synthesized under a variety of conditions. The basal, or uninduced, level of enzyme is very low, corresponding to 3–5 molecules per cell, or about one transcription of the gene per cell division. Synthetic inducers such as IPTG or thiomethylgalactoside that are powerful inducers of the lac operon are not effective inducers of the ebg operon, nor are most of the β-galactoside sugars that have been tested. Lactose itself is the most effective inducer tested (see Table 1). Even lactose, which induces synthesis to 100 fold above the basal level, is not a particularly effective inducer when compared with the level of synthesis exhibited by a constitutive (*ebgR*$^-$) cell that produces no functional repressor. Such a cell makes ebg enzyme at 2000 times the basal level, or about 3% of the total cellular protein.

The ebg° enzyme has been purified to homogeneity (Hall, 1976 and 1981), and its catalytic constants determined on a variety of natural and synthetic substrates (Table 2). The activity of this enzyme toward lactose is extremely low. Comparison of the values in Table 2 with those obtained by Huber *et al.* (1976) for the *lacZ* β-galactosidase show that the V_{max} of ebg for lactose is only 0.019 of the V_{max} of the classical β-galactosidase (32 000 nmol min^{-1} mg^{-1}), while the K_m is 60 fold higher than that of the classical enzyme (2.5 mM lactose). Under physiological conditions this means that ebg enzyme is about 0.005 times as active as the *lacZ* β-galactosidase. It is therefore not surprising that even a strain that synthesizes ebg enzyme constitutively is unable to utilize lactose as a carbon source.

29.4.2 Evolution of Lactose Utilization

Campbell *et al.* (1973) isolated the first lactose-utilizing mutant of a *lacZ* deletion strain of *E. coli* and showed that it synthesized a new β-galactosidase enzyme which they designated EBG

Table 1   Induction of ebg Enzyme in the Unevolved ($ebgR^+$) Strain

Additions	Specific activity[a]	Additions	Specific activity[a]
Non-inducers		Weak inducers	
None	0.28	Methyl-β-galactoside	1.4
Isopropylthiogalactoside	0.30	Thiodigalactoside	1.4
Melibiose	0.20	Lactulose	3.0
Galactose	0.20	Galactosyl-D-arabinose	2.3
Lactobionic acid	0.20		
Glycerol-β-galactoside	0.20	Strong inducers	
β-Methylthiogalactoside	0.34	Lactose	29
Phenyl-β-galactoside	0.13	Lactose	33
Galacturonic acid	0.22		
		Constitutive strain	
		None	536

[a] Specific activities are in units per mg of soluble protein as determined in crude extracts.

Table 2   Properties of ebg Enzymes and Growth of Strains Synthesizing those Enzymes Constitutively[a]

Class[b]	Property[c]	Lactose	Substrate lactulose	Galactosyl-arabinose	Lactobionate
ebg^0 ($n = 1$)	V_{max}	620	270	52	No detectable activity
	K_m	150	180	64	
	Specificity	4.0	1.5	0.81	
	Growth rate	0	0	0	0
Class I ($n = 3$)	V_{max}	3566	69	185	No detectable activity
	K_m	22	34	14	
	Specificity	160	2.1	12.7	
	Growth rate	0.45	0	0.03	0
Class II ($n = 3$)	V_{max}	2353	1887	356	No detectable activity
	K_m	59	26	25	
	Specificity	40	73	14.4	
	Growth rate	0.19	0.26	0.02	0
Class IV ($n = 9$)	V_{max}	1461	430	737	105
	K_m	0.82	7.9	3.0	15.4
	Specificity	1800	55	244	6.7
	Growth rate	0.37	0.18	0.13	0
Class V ($n = 1$)	V_{max}	590	215	349	370
	K_m	0.69	6.5	4.9	3.0
	Specificity	850	33	70	123
	Growth rate	0.18	0.10	0.07	0.20

[a] Hall, 1978a, 1981.   [b] n = number of purified enzymes from independent mutants.   [c] Units: V_{max} (mg^{-1} of pure enzyme), K_m (mmol of substrate), growth rates (h^{-1}); standard errors of these mean values are less than 10% of the values.

enzyme. Their mutant strain had been obtained after five rounds of selection for improvement on lactose, and it was assumed to have resulted from five mutations. That assumption was later shown to have been in error, and it was determined that exactly two mutations were required for the evolution of lactose utilization (Hall, 1977; Hall and Clarke, 1977). One mutation must be in *ebgA* in order to increase the activity of ebg enzyme toward lactose. Such mutations to *ebgA*$^+$ occur spontaneously at a frequency of 2×10^{-9} per cell division (Hall, 1977). The other mutation must be in *ebgR* in order to increase the quantity of ebg enzyme synthesized during growth on lactose (Hall and Clarke, 1977).

29.4.3   Evolution of New Activities for ebg Enzymes

As shown in Table 2, ebg^0 enzyme exhibits very low activity toward the various disaccharides tested, and no activity at all toward lactobionic acid. With the exception of lactose, these four sugars (lactose, lactulose, galactosyl-D-arabinose, and lactobionic acid) are very poor inducers of ebg enzyme synthesis. To examine the relationship between properties of enzymes and growth phenotype without generating regulatory complications, we considered strains that synthesize ebg enzyme constitutively.

The ebg° strain was unable to utilize any of the four β-galactoside sugars. Table 2 shows that the specificity (V_{max}/K_m) of ebg for each of these sugars is very low.

When lactose-utilizing mutants were selected, two distinct classes were obtained based upon growth rates on the four sugars. Growth rates are very homogeneous within a class, and we have obtained well over 50 independent mutants of each class. Each of the lactose-utilizing mutants obtained from an ebg° strain could be unambiguously assigned to one of these two classes (Hall, 1978a).

Class I strains grow rapidly on lactose, very slowly on galactosylarabinose (doubling times in excess of 24 hours), and not at all on lactulose or lactobionic acid. Class II strains grow more slowly on lactose, very slowly on galactosylarabinose, not at all on lactobionic acid, but grow well on lactulose. The frequencies at which these two classes arise spontaneously is such that each is the consequence of a single point mutation (Hall, 1977). Ebg enzyme was purified to homogeneity from three independent Class I and three independent Class II mutants. Table 2 shows that the properties of these enzymes were entirely consistent with the growth properties of the strains from which they were obtained. The two classes exhibited entirely different catalytic constants for the sugars they hydrolyzed, and neither class exhibited detectable activity toward lactobionic acid. The Class I enzyme exhibited a 40 fold increase in specificity for lactose, while the Class II enzyme showed a 10 fold increase in lactose specificity. Similarly, the Class II enzyme showed a 35 fold increase in specificity for lactulose, while the Class I enzyme exhibited virtually no change in lactulose specificity. There is a rough correlation between growth rate and the specificity of the enzyme for its substrate. The correlation is not exact for two reasons: (1) the substrates are not equally good substrates for the permease, hence the physiological concentrations of the substrates within the cell are not identical; (2) *E. coli* are unable to utilize D-arabinose, thus cleavage of galactosyl-D-arabinose yields only a single metabolizable hexose sugar, while hydrolysis of the other substrates yields two metabolizable hexoses.

In a second set of experiments lactulose was substituted for lactose in the selective plates, and a set of lactulose positive mutants was obtained from the ebg° strain. All of these mutants proved to be Class II mutants based both upon growth phenotype and upon characterization of the enzymes they produced (Hall, 1978a and 1981).

Class II strains exhibited a metabolic capability absent in both the parental ebg° strain and in sibling Class I strains, *i.e.* the ability to utilize lactulose. A series of Class I strains were subjected to selection for lactulose utilization on lactulose-TTC plates. The mutants that were obtained were designated Class IV, and Table 2 shows that their properties were quite distinct from those of Class II strains. First, Class IV strains grew more slowly on lactulose than did Class II; second they grew more rapidly on lactose than did Class II strains. Third, and most surprising, Class IV strains grew at respectable rates on galactosylarabinose. The property of efficient galactosylarabinose utilization represented, at a biologically meaningful level, the acquisition of another new catabolic function.

Again, the properties of Class IV ebg⁺ enzymes *in vitro* correlated well with the growth properties. A dramatic decrease in the K_m for all substrates had occurred, in the case of lactose increasing the specificity to 450 times the value of ebg° enzyme. The real surprise was that although Class IV strains cannot utilize lactobionic acid, Class IV ebg enzyme was able to hydrolyze lactobionate *in vitro*. The failure of Class IV strains to grow on lactobionate is explained by the low specificity of the Class IV enzyme for that substrate, a specificity that is comparable to that of ebg° enzyme for lactose. Although the Class I and Class II mutations clearly conferred new catabolic capabilities on the cell, the Class IV enzyme was the first example of the acquisition of a completely new catalytic capability at the molecular level.

The observation that the double-mutant Class IV strains could grow well on galactosylarabinose suggested that it might be possible to select galactosylarabinose utilizing mutants directly from ebg°, or from Class II strains. Despite intensive efforts no galactosylarabinose-utilizing mutants have ever been selected from the ebg° strain. They were, however, obtained easily from Class II strains and arose spontaneously at a frequency of about 10^{-9}, the frequency expected for single point mutations. The growth properties and the properties of enzymes purified from those strains were indistinguishable from the properties of Class IV strains, and the strains were initially designated Class IVa. The inability to distinguish Classes IV and IVa at any phenotypic level suggested that they might be identical and simply have arisen by different evolutionary pathways, one *via* Class I, the other *via* Class IV. That implied that a Class IV strain simply carried a Class I and a Class II mutation together in the same *ebgA* gene. That hypothesis was tested by genetic analysis, and in a critical experiment it was shown that Class IV recombinants could be obtained in a cross between a Class I and a Class II strain (Hall and Zuzel, 1980). Enzymes purified from Class IV strains were indistinguishable regardless of the sequence of the two mutations that generated the Class IV allele, or whether the allele had been obtained by recombination (Hall, 1981). Figure 3 shows the evolutionary pathway for these enzyme classes.

The observation that the double-mutant Class IV ebg enzyme could hydrolyze lactobionic acid

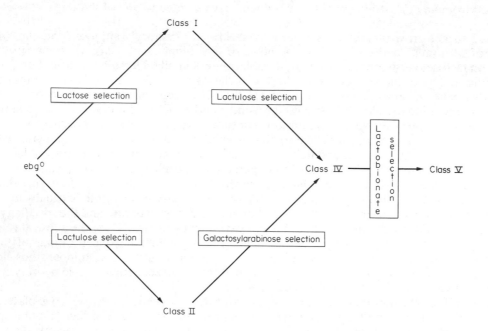

Figure 3 ebg evolutionary pathway; selective conditions are indicated in boxes

in vitro suggested that it might be possible to obtain mutants capable of growth on lactobionate. Such mutants were successfully selected, and arose spontaneously at a frequency of 10^{-9}. The enzyme was purified from one such mutant and its properties are shown in Table 2. It should be noted that the increase in activity toward lactobionic acid was accompanied by a decrease in activity toward the other substrates.

Three general observations flow from these studies. First, selection for utilization of a particular new sugar always leads to an increase in the enzyme's V_{max} and a decrease in its K_m for that substrate. The catalytic constants for the unselected substrates may be unaffected, may be improved, or may be worsened. Second, if there is detectable activity for a substrate *in vitro*, it is likely that the activity can be improved to a point that permits utilization of that substrate *in vivo*. From a practical point of view this means that if activity is detected *in vitro*, selection for growth will probably yield an improved enzyme. Third, and perhaps most important, many new functions cannot be obtained by direct selection. Consider these examples: galactosylarabinose utilization has never been successfully selected directly from the wild-type *ebgA* allele. Only by selecting first for either lactose or lactulose utilization, then selecting for galactosylarabinose utilization is it possible to obtain galactosylarabinose positive mutants. Similarly, repeated efforts were made to obtain lactobionic acid positive mutants from ebg°, from Class I and from Class II strains. None of those efforts were successful. Lactobionate positive mutants could be derived only from Class IV strains (Hall, 1978a). It is reasonable to conclude that galactosylarabinose utilization requires two specific mutations, and lactobionate utilization three mutations. Since these mutations arise at a rate of about 10^{-9} per cell division, the expectation is that only one in 10^{27} cells, or about one in 10^{12} kg of *E. coli* would by chance carry all three mutations simultaneously. Since, for substrates such as lactobionate, the direct route to the desired phenotype is unproductive, it is often necessary to select for utilization of closely related compounds first.

29.4.4 Evolution of the ebg Repressor

The lactose fermenting strain isolated by Campbell *et al.* (1973) synthesized ebg enzyme constitutively, and it was assumed that ebg enzyme arose from mutations in an unregulated operon.

When we isolated 34 independent *ebg*$^+$ mutants (Hall and Hartl, 1974), we found that only three of these were constitutive, the remainder synthesizing ebg enzyme only when induced by lactose. As further investigation showed that the unevolved (wild-type) ebg operon was also inducible by lactose (Hartle and Hall, 1974), it seemed likely that inducible *ebg*$^+$ mutants carried a wild-type repressor gene. Some time later, purification of both the wild-type and the mutant ebg enzymes (Hall, 1976) made it possible to measure the level of ebg enzyme synthesis very accurately. These measurements showed that, although the basal levels of enzyme synthesis were indistinguishable, the *ebg*$^+$ (evolved, lactose fermenting) strain synthesized four times as much ebg enzyme as did the unevolved strain (Table 3). Genetic analysis showed that the increased level of enzyme synthesis was the result of a mutation in the *ebgR* gene (Hall and Clarke, 1977). Estimates of the *in vivo* lactase activity were made based upon the K_m and V_{max} for lactose of the various ebg enzymes, upon the amount of enzyme synthesized during growth on lactose, and upon the physiological lactose concentration. A comparison of *in vivo* lactase activity with the growth rates on lactose showed that there was a nearly perfect correlation between these two parameters. More importantly, our equation predicted that there was a minimum level of lactase activity, amounting to about 5 nmol of lactose hydrolyzed per min per mg of soluble cell protein, below which no growth on lactose could occur. These measurements suggested that even the most active ebg enzyme, when synthesized at the level permitted by the wild-type repressor, would not produce that minimal level of activity. We constructed a recombinant strain that synthesized a Class I enzyme under the control of a wild-type repressor. Exactly as predicted, that strain was unable to grow on lactose (Hall and Clarke, 1977). Thus, careful physiological measurements made it clear that coevolution of the regulatory and structural genes of the ebg operon was obligatory for the evolution of lactose utilization.

Table 3 Inducers of ebg Enzyme Synthesis in the Unevolved Strain DS4680A and in the Inducible Evolved Strain A4[a]

| | Specific synthesis $\times$ 100[b] | |
Inducer	DS4680A	A4
None	0.0022	0.0030
Thiodigalactoside	0.012	0.015
Methylgalactoside	0.012	0.014
Lactulose	0.027	0.056
Lactose	0.26	1.23

[a] Data from Hall and Clarke, 1977. [b] Specific synthesis = the fraction of total soluble protein that is ebg enzyme.

The wild-type repressor is virtually insensitive to lactulose as an inducer (Table 1). The chance isolation of a strain carrying a Class II *ebgA* allele and a wild-type *ebgR* gene provided the opportunity to attempt to direct the evolution of the ebg repressor itself. Although the Class II enzyme produced by that strain was active toward lactulose, the fact that lactulose was such a poor inducer prevented the strain from growing on lactulose. Some 3000 lactulose fermenting mutants were obtained, but most of them proved to be constitutive mutants. Nine mutants inducible by lactose were obtained, but only three were characterized in detail. In each case, the repressor had not only become sensitive to lactulose as an inducer, but it had become hypersensitive to lactose as well (Hall, 1978b).

The ebg studies in many ways confirm the impressions generated by earlier studies, especially those of Clarke on the aliphatic amidase. Regulatory mutations are critical to the evolutionary process. Equally important are mutations that alter the catalytic properties of the enzymes in the catabolic pathway. Many new functions require a series of mutations that must be individually selected using compounds related to the specific new function. Lactobionate utilization, requiring one regulatory and three structural gene mutations, is a case in point from the ebg system. Similarly, phenylacetamide utilization requires one regulatory mutation and two structural gene mutations in the amidase system of *Pseudomonas aeruginosa*.

29.5 A PHOSPHOLACTASE SYSTEM IN *KLEBSIELLA*

Klebsiella normally possesses two lac operons, one on the chromosome and the other on a plasmid (Reeve, 1976). A mutant was obtained (Reeve, 1976) that had lost both of its lac operons and was consequently lactose negative. It was shown that this Lac strain synthesized a β-galactosidase that was inactive toward lactose, but whose synthesis was induced by lactose. Despite these similarities, that enzyme (β-galactosidase III) was shown to be unrelated to ebg enzyme (Hall and Reeve, 1977). Under intensive selective pressure a Lac$^+$ mutant of that strain was obtained. It was expected that lactose utilization would have arisen *via* mutations affecting the new β-galactosidase, but that was not the case. Instead, the mutant strain synthesized a phospho-β-galactosidase, an enzyme that was active only toward ONPG or lactose that was phosphorylated on C-6 of the galactose moiety (Hall, 1979). Several lines of evidence were developed that showed that the new mutant transported and phosphorylated lactose *via* a phosphoenolpyruvate dependent phosphotransferase system (Iami and Hall, 1981; Hall *et al.*, 1982). It was shown that the parental lactose negative strain also synthesized the phospho-β-galactosidase enzyme (Hall, 1979), and it appears likely that the normal function of that enzyme is to act as a phospho-β-glucosidase and to cleave phosphorylated cellobiose (Hall *et al.*, 1982). The critical step in the acquisition of lactose utilization appears to be a complex regulatory mutation that permits expression of the lactose transport system (Hall *et al.*, 1982).

Why did the system not respond to selection in the expected way by altering the β-galactosidase itself? The lactose negative strain possesses neither a β-galactoside transport system nor the immediate potential for such a system (Hall, 1980). In the absence of such a transport system, mutations that increase the activity of the β-galactosidase enzyme for lactose are of absolutely no benefit to the cell. This is an example of an instance in which ignorance of the physiology of the cell with respect to another step in the pathway (transport) resulted in the failure to apply selective pressure to the expected function. It is important for the biotechnologist to clearly recognize that one never applies selective pressure to a gene, only to a cell, and that the cell responds with its own best solution to the problem presented. The cell's solution is not always that desired by the experimenter.

29.6 DECRYPTIFYING EXISTING GENES

Evolution of new metabolic capabilities does not always involve point mutations. In some cases the desired information is present in an unexpressed, or cryptic, form. *E. coli*, including natural isolates as well as laboratory strains, is unable to utilize β-glucosides such as salicin and arbutin. The information for β-glucoside utilization is present, however, in all strains examined. A variety of mutations, including insertion of IS elements, can activate the *bgl* operon that specifies the transport and hydrolytic proteins necessary for β-glucoside metabolism (Reynolds *et al.*, 1981). When a new metabolic function arises that involves expression of several genes that were not expressed in the wild-type organism, it is likely that a set of genes has become decryptified.

An example of direct interest to the field of biotechnology comes from the work of Slater and his associates. The presence of halogenated compounds in the biosphere constitutes a major contemporary problem. While natural halogenated compounds do exist, the majority are generated by man and many constitute a significant environmental hazard. Slater has sought organisms that can degrade the herbicide Dalapon (2,2-dichloropropionic acid). *Pseudomonas putida* strain PP1 is unable to metabolize Dalapon. During long term selection in a chemostat a mutant arose that could utilize Dalapon as a sole carbon source (Senior *et al.*, 1976). The mutant strain, designated PP3, expressed two dehalogenases and two associated permeases. Expression of these genes was regulated and subjected to induction by Dalapon and by a related family of halogenated alkanoic acids (Slater *et al.*, 1979). The acquisition of the ability to express two dehalogenases and two transport systems simultaneously and in a regulated fashion strongly suggests that an entire set of genes was decryptified. Recent work (Slater *et al.*) suggests that the dehalogenase genes are located on transposable element, and that expression of those genes may be related to mobility of those elements.

29.7 CONCLUSIONS

Recent unpublished work from several laboratories is beginning to suggest that the physical location of some genes is critical to their expression. As microorganisms have adapted to specialized niches expression of blocks of genes may have been silenced by genetic rearrangements that placed the genes in unacceptable locations. It seems likely that familiar microorganisms possess a significant reservoir of unexpressed information. A thorough understanding of the physiology, the ecology and the taxonomy of familiar organisms will be necessary in order to develop a data base that will allow the biotechnologist to be able to predict the chances of selecting a particular new function in a particular organism. As the techniques of direct evolution are combined with those of recombinant DNA technology it will become increasingly possible to engineer organisms to meet specific needs of industry and society.

29.8 REFERENCES

Beggs, J. D., J. van den Berg, A. van Ooyen and C. Weissman (1980). Abnormal expression of chromosomal rabbit β-globin gene in *Saccharomyces cerevisae*. *Nature (London)*, **238**, 835–840.

Campbell, J. H., J. Lengyel and J. Langridge (1973). Evolution of a second gene for β-galactosidase in *Escherichia coli*. *Proc. Natl. Acad. Sci. USA*, **70**, 1841–1845.

Clarke, P. H. (1978). Experiments in microbial evolution. In *The Bacteria*, ed. L. N. Ornston and J. R. Sokatch, vol. VI, pp. 137–218. Academic, New York.

Hacking, A. J. and E. C. C. Lin (1977). Regulatory changes in the fucose system associated with the evolution of a pathway for the catabolism of propanediol in *Escherichia coli*. *J. Bacteriol.*, **130**, 832–838.

Hall, B. G. (1976). Experimental evolution of a new enzymatic function. Kinetic analysis of the ancestral (ebgᵒ) and evolved (ebg⁺) enzymes. *J. Mol. Biol.*, **107**, 71–84.

Hall, B. G. (1977). Number of mutations required to evolve a new lactase function in *Escherichia coli*. *J. Bacteriol.*, **129**, 540–543.

Hall, B. G. (1978a). Experimental evolution of a new enzymatic function. II. Evolution of multiple functions for EBG enzyme in *E. coli*. *Genetics*, **89**, 453–465.

Hall, B. G. (1978b). Regulation of newly evolved enzymes. IV. Directed evolution of the *ebg* repressor. *Genetics*, **90**, 673–691.

Hall, B. G. (1979). Lactose metabolism involving phospho-β-galactosidase in *Klebsiella*. *J. Bacteriol.*, **138**, 691–698.

Hall, B. G. (1980). Properties of β-galactosidase. III: implications for entry of galactosides into *Klebsiella*. *J. Bacteriol.*, **142**, 433–438.

Hall, B. G. (1981). Changes in the substrate specificities of an enzyme during directed evolution of new functions. *Biochemistry*, **20**, 4042–4049.

Hall, B. G. (1982). Evolution on a petri dish. The evolved β-galactosidase system as a model for studying acquisitive evolution in the laboratory. *Evol. Biol.*, **13**, 85–150.

Hall, B. G. (1983). Evolution of new metabolic functions in laboratory organisms. In *Evolution of Genes and Proteins*, ed. M. Nei and R. K. Koehn, pp. 234–257. Sinauer Associates, Sunderland, MA.

Hall, B. G. and N. D. Clarke (1977). Regulation of newly evolved enzymes. III. Evolution of the ebg repressor during selection for enhanced lactase activity. *Genetics*, **85**, 193–201.

Hall, B. G. and D. L. Hartl (1974). Regulation of newly evolved enzymes. I. Selection of a novel lactase regulated by lactose in *Escherichia coli*. *Genetics*, **76**, 391–400.

Hall, B. G., K. Imai and C. P. Romano (1982). Genetics of the lac-PTS system of *Klebsiella*. *Genet. Res. Camb.*, **39**, 287–302.

Hall, B. G. and E. C. R. Reeve (1977). A third β-galactosidase in a strain of *Klebsiella* that possesses two *lac* genes. *J. Bacteriol.*, **132**, 219–223.

Hall, B. G. and T. Zuzel (1980). Evolution of a new enzymatic function by recombination within a gene. *Proc. Natl. Acad. Sci. USA*, **77**, 3529–3533.

Hartl, D. L. and B. G. Hall (1974). Second naturally occurring β-galactosidase in *E. coli*. *Nature (London)*, **248**, 152–153.

Hartley, B. S., B. D. Burleigh, G. G. Midwinter, C. H. Moore, H. R. Morris, P. W. J. Rigby, M. J. Smith and S. S. Taylor (1972). Where do new enzymes come from? In *Enzymes: Structure and Function*, 8th FEBS Meeting, ed. J. Denreth, R. A. Oosterbaan and C. Veeger, vol. 29, pp. 151–176. North-Holland, Amsterdam.

Huber, R. E., G. Kurz and K. Wallenfels (1976). A quantitation of the factors which affect the hydrolase and transgalactosylase activities of β-galactosidase (*E. coli*) on lactose. *Biochemistry*, **15**, 1994–2001.

Imai, K. and B. G. Hall (1981). Properties of the lactose transport system in *Klebsiella* sp. strain CT-1. *J. Bacteriol.*, **145**, 1459–1462.

Learner, S. A., T. T. Wu and E. C. C. Lin (1964). Evolution of a catabolic pathway in bacteria. *Science*, **146**, 1313–1315.

LeBlanc, D. J. and R. P. Mortlock (1971). Metabolism of D-arabinose: a new pathway in *Escherichia coli*. *J. Bacteriol.*, **106**, 90–96.

Mortlock, R. P. (1981). Regulatory mutations and the development of new metabolic pathways by bacteria. *Evol. Biol.*, **14**, 205–267.

Mortlock, R. P., D. D. Fossitt, D. H. Petering and W. A. Wood (1965). Metabolism of pentoses and pentitols by *Aerobacter aerogenes*. III. Physical and immunological properties of pentitol dehydrogenases and pentulokinases. *J. Bacteriol.*, **89**, 129–135.

Reeve, E. C. R. (1976). The lactose system of *Klebsiella aerogenes* V9A. 5. Lac-permease defective mutants of two *Klebsiella* lac plasmids and their apparent reversion to wild type. *Genet. Res. Camb.*, **28**, 61–74.

Reynolds, A. E., J. Felton and A. Wright (1981). Insertion of DNA activates the cryptic *bgl* operon in *E. coli* K-12. *Nature (London)*, **293**, 625–629.

Rigby, P. W. J., B. D. Burleigh and B. S. Hartley (1974). Gene duplication in experimental enzyme evolution. *Nature (London)*, **251**, 200–204.

Senior, E., A. T. Bull and J. H. Slater (1976). Enzyme evolution in a microbial community growing on the herbicide Dalapon. *Nature (London)*, **262**, 476–479.

Slater, J. H., D. Lovatt, A. J. Weightman, E. Senior and A. T. Bull (1979). The growth of *Pseudomonas putida* on chlorinated aliphatic acids and its dehalogenase activity. *J. Gen. Microbiol.*, **114**, 125–136.

Slater, J. H., A. Weightman and B. G. Hall (1985). Dehalogenase genes of *Pseudomonas putida* PP3 are on chromosomally located transposable elements. *Mol. Biol. Evol.*, submitted.
Turberville, C. and P. H. Clarke (1981). A mutant of *Pseudomonas aeruginosa* PAC with an altered amidase inducible by
the novel substrate. *FEMS Microbiol. Lett.*, **10**, 87–90.
Wu, T. T. (1976). Growth of a mutant of *Escherichia coli* K-12 on xylitol by recruiting enzymes for D-xylose and L-1,2-
propanediol metabolism. *Biochem. Biophys. Acta*, **428**, 656–663.
Wu, T. T., E. C. C. Lin and S. Tanaka (1968). Mutants of *Aerobacter aerogenes* capable of utilizing xylitol as a novel
carbon source. *J. Bacteriol.*, **96**, 447–456.

30
Microbial Photosynthesis

D. J. KELLY and C. S. DOW
University of Warwick, Coventry, UK

30.1 INTRODUCTION

Prokaryotes capable of satisfying all, or part, of their cellular energy demands from light are the cyanobacteria ('blue-green algae') the purple and green eubacteria and the halobacteria. Within these groups there lies a wealth of structural and metabolic diversity; they are, in addition, of considerable ecological importance and have substantial economic potential. Our aim in this article is to describe the essential nature of microbial photosynthesis and to emphasize the distinctive characteristics of phototrophic microbes.

30.2 HISTORICAL BACKGROUND

The existence of microbes capable of utilizing light energy to drive the synthesis of cellular components was firmly established by the late 19th century. The classic work of Winogradsky

(1888) and Molisch (1907) elucidated the properties of the purple sulfur and purple non-sulfur bacteria respectively and it was these two groups of phototrophs which exerted a profound influence on the development of concepts relating to both the mechanism of photosynthesis ('light reactions') and the pathway of carbon assimilation ('dark reactions'). Nevertheless, until approximately 1930, research on photosynthetic bacteria was generally unpopular, this being primarily attributable to the early observation of Englemann (1883) that these microbes do not produce molecular oxygen, as green plants do. Consequently at a time when comparative biochemistry was gaining popularity, this observation seemed incompatible with the accepted 'unity of biochemical processes' and, as amiably recounted by Gest (1982), the 'pandemic of overly comparative biochemistry' which pervaded this period hindered a clear perception of the nature of microbial photosynthesis.

It was largely through the pioneering work of Van Niel in the 1930s on the physiology of purple sulfur bacteria that a clearer picture of their photosynthetic processes started to emerge. At this time the reduction of CO_2 to carbohydrate was seen as a central feature of all photosynthetic processes (Gest, 1982) and Van Niel's initial conclusion that purple sulfur bacteria could use H_2S but not H_2O as the H donor was summed up in the following equation, where H_2A represents the H donor:

$$CO_2 + 2\,H_2A \rightarrow [CH_2O] + H_2O + 2A \tag{1}$$

Later this equation was modified (Van Niel, 1935) to include H_2O as the direct H donor in all photosyntheses with differences in the nature of H_2A reflecting differences in the details of the process (Chl is chlorophyll):

$$4\,[Chl{\cdot}H_2O + h\nu \rightarrow Chl{\cdot}OH + H] \tag{2a}$$

$$CO_2 + 4H \rightarrow CH_2O + H_2O \tag{2b}$$

$$2\,[2\,Chl{\cdot}OH + H_2A \rightarrow 2\,Chl{\cdot}H_2O + 2A] \tag{2c}$$

$$CO_2 + 2\,H_2A + 4\,h\nu \rightarrow [CH_2O] + H_2O + 2A \tag{3}$$

Equation (2c) embodied the realization that simple organic compounds could act as H_2A, as in the purple non-sulfur bacteria, while equations (2a) and (2b) were seen as being similar in all types of photosyntheses. This example serves to illustrate how the basic concepts of microbial photosynthesis veered between two standpoints, either stressing fundamental differences or emphasizing unity.

From 1945 onwards, the idea began to emerge that light energy could be converted to chemical bond energy in the form of ATP and subsequently there was considerable interaction between those studying plant and microbial photosynthesis. The studies of Gest and Kamen (1948) showed that illumination caused a marked increase in P_i uptake by intact cells of both algae and bacteria and Levitt (1953) interpreted green plant photosynthesis as a photoelectric phenomenon in which light absorption by a chlorophyll molecule resulted in the expulsion of an electron. It was soon demonstrated by Frenkel (1954), with 'pigmented particles' isolated from *Rhodospirillum rubrum*, and Arnon *et al.* (1954), with isolated chloroplasts, that light-induced phosphorylation in the absence of supplied electron acceptors or donors could be reproduced *in vitro*. It was not until the 1960s that the concept of cyclic electron flow in anoxygenic photosynthesis began to emerge. At about the same time, Hill and Bendall (1960) proposed a non-cyclic, linear scheme of electron transport to explain oxygenic photosynthesis in green plants, thus exposing an apparent dichotomy between bacterial and higher plant and algal photosynthesis. However, Mitchell (1961) provided an explanation for the coupling of electron transport to phosphorylation applicable to both. Upon the realization that the so-called 'blue-green algae' had a prokaryotic cellular organization (Stanier and Van Niel, 1962) many prejudices were violated and the diversity of microbial photosynthesis became more widely appreciated. This was underlined in 1971 when Oesterhelt and Stoeckenius (1971) demonstrated that the halobacteria exhibited an unusual type of photosynthesis, radically different from any other known system, which was based upon bacteriorhodopsin mediated proton translocation.

To date, three broad patterns of microbial photosynthesis are known (Table 1). In the following sections, we give a comparative description of the general nature of cyanobacterial and eubacterial photosynthesis. Halobacterial photophosphorylation is briefly described separately in Section 30.5.

Table 1 Patterns of Microbial Photosynthesis

Group	Type of photosynthesis	Pigment associated with the primary photoact	Electron donors	Products	Carbon source
Eubacteria	Anoxygenic	Bacteriochlorophyll	H_2, H_2S, S, organic compounds	ATP + NAD(P)H	CO_2 + organics
	Oxygenic	Chlorophyll	H_2O (H_2S)[a]	ATP + NAD(P)H	CO_2, organics[c]
Archaebacteria	Halobacterial	Bacteriorhodopsin	—[b]	ATP	Organics

[a] Only certain cyanobacteria, *e.g. O. limnetica* under anaerobic conditions. [b] Electron donors do not participate. [c] Some cyanobacteria can grow photoheterotrophically.

30.3 GENERAL CHARACTERISTICS OF MICROBIAL PHOTOSYNTHESIS

30.3.1 Structure and Synthesis of Photosynthetic Pigments

30.3.1.1 *Chlorophylls and bacteriochlorophylls*

The chlorophylls are the pigments responsible for initiating primary photochemistry in the cyanobacteria and higher plants. The basic structure of chlorophyll (Figure 1) is remarkably similar to that of haem except that iron is replaced by magnesium as the chelated metal (transition metals would cause quenching of the initial photochemical reactions; Mauzerall, 1978) and an additional pentanone ring is present. Higher plants contain both chlorophylls *a* and *b*, the difference between these molecules lying solely in the nature of the substituent at position C-3 (Figure 1), whilst cyanobacteria contain only chlorophyll *a*.

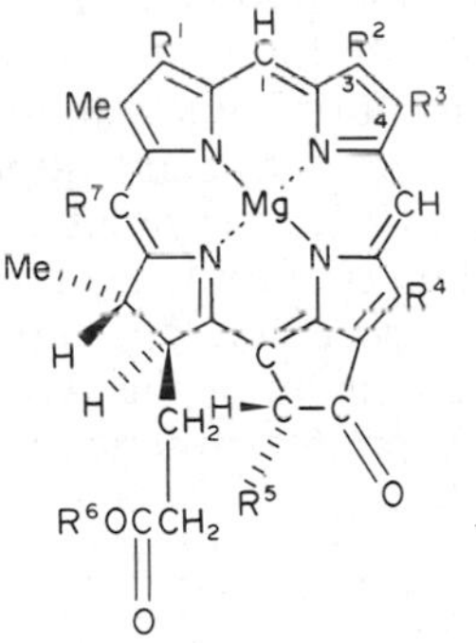

Figure 1 The chemical structure of chlorophyll and bacteriochlorophyll. The double bond between C-3 and C-4 is absent in Bchl *a* and *b*. R^1–R^7 represent the substituents characteristic of the different types of Bchl and Chl (Table 2)

To date, six types of bacteriochlorophyll (Bchl) have been identified from photosynthetic bacteria (Bchl *a*, *b*, *c*, *d*, *e* and *g*) and some of their characteristics are listed in Table 2. Bacteriochlorophyll *a* and chlorophyll *a* are virtually identical except that the former has a vinyl group and the latter a methoxy group at position C-1 (Figure 1). Each type of bacteriochlorophyll has a corresponding pheophytin, also of importance in primary photochemistry, in which the chelated Mg^{2+} is replaced by $2H^+$. The different types of bacteriochlorophyll are distinguished by the nature of their peripheral substituent groups and these have a profound effect upon the absorption spectrum of the molecule (Table 2). The *in vivo* longest wavelength absorption maximum is also strongly influenced by protein interactions and is always red-shifted when compared to its counterpart in organic solvents.

Traditionally, the taxonomy of the green and purple photosynthetic bacteria has always been linked primarily to the presence of a particular type of bacteriochlorophyll (Pfennig, 1978; Truper and Pfennig, 1978). Nevertheless only the biosynthesis of bacteriochlorophyll *a*, in *Rhodopseudomonas sphaeroides*, has been thoroughly investigated (Jones, 1978). The pathway

Table 2 Properties of Chlorophylls[a]

Chlorophyll	λ_{max} in vivo (nm)	Red shift in vivo (nm)	R^1	R^2	R^3	R^4	R^5	R^6	R^7
Bchl a	800–900	75–135	Ac	Me	Et	Me	CMe	Ph/Gg	—
b	1020–1040	230–246	Ac	Me	=CH	Me	CMe	Phy	—
c	745–760	85	CH(OH)Me	Me	Et	Et	—	Fa	Me
d	725–745	75	CH(OH)Me	Me	Et	Et	—	Fa	—
e	715–725	65	CH(OH)Me	Fo	Et	Et	—	Fa	Me
g	788	—	$CHCH_2$	Me	=CH	Me	CMe	Gg	—
Chl a	680–683	15–18	$CHCH_2$	Me	Et	Me	CMe	Phy	—

[a] R^1–R^7 indicate substituents in Figure 1. Abbreviations: Ac, acetyl; Me, methyl; Fo, formyl; Et, ethyl; CMe, carboxymethyl; Ph, phytyl; Gg, geranylgeranyl; Fa, farnesyl

begins with the condensation of glycine and succinyl-CoA to form 5-aminolevulinic acid (ALA) in a reaction catalyzed by ALA-synthetase. Ensuing reactions result in the formation of protoporphyrin IX from four pyrrole units and up to this point chlorophyll, bacteriochlorophyll and haem share the same intermediates. Insertion of iron into protoporphyrin IX results in haem synthesis, while Mg-protoporphyrin is the first intermediate on the pathway to Bchl formation. In bacteria, there is no evidence for a light dependent step in the formation of the final form of Bchl (Jones, 1978) as there is in chlorophyll formation in plants. The biosynthesis of Bchl c, d and e (the chlorobium chlorophylls) appears to branch from that of Bchl a and b at an early stage (Jones, 1978) and there is as yet little information on the synthesis of the newly identified Bchl g (Brockman and Lipinski, 1983) from *Heliobacterium chlorum* (Gest and Favinger, 1983). This molecule is of more than passing interest because several of the substituent groups are identical to those of chlorophyll a (Table 2).

30.3.1.2 Carotenoids

The carotenoid pigments are largely responsible for the colour of the green and purple bacteria. They form an integral part of the photochemical process by functioning in both light absorption and excitation transfer, in addition to extending the usable wavelength range of light and in protecting the cell against harmful photooxidation reactions. In these microbes carotenoids are usually aliphatic tetraterpenoids with eight isoprene units (Figure 2) synthesized along at least four different routes (Schmidt, 1978). In cyanobacteria, the dominant carotenoid is β-carotene, while those of the spirilloxanthin, spheroidene and okenone series are characteristic of the purple bacteria, and carotenoids of the isorenieratene series are found in the green sulfur bacteria (Schmidt, 1978). The halobacteria, members of the Archaebacteria (Fox *et al.*, 1980), contain bacterioruberins, which do not serve in photosynthetic reactions, but do protect the cells from the harmful effects of solar radiation.

30.3.1.3 The phycobilins

The cyanobacteria contain, in place of Chl b, a range of linear tetrapyrroles covalently bonded to proteins which determine both the colour of the organisms and the properties of their light-harvesting apparatus (Section 30.3.2.1). Structurally, phycobilins contain the uncommon ethylidene group which is also found in Bchl b and g (Figure 2).

30.3.2 The Initial Reactions—Primary Photochemistry and Electron Transport

Photosynthesis is a carefully regulated redox process which depends upon a precise spatial relationship of the components of the photosynthetic apparatus. The organization of all chlorophyll and bacteriochlorophyll based photosyntheses can be described in terms of four components: (i) a light-harvesting antenna system, (ii) the photochemical reaction centre, (iii) the electron transfer chain, and (iv) the proton-translocating ATPase. The reactions catalysed by these components are largely membrane bound, vectorial events directed at supplying a source of reducing power (NAD(P)H) and energy (ATP and a proton motive force, Δp) in order to drive carbon assimilation. In both anoxygenic and oxygenic photosynthesis, light harvested by the antenna pig-

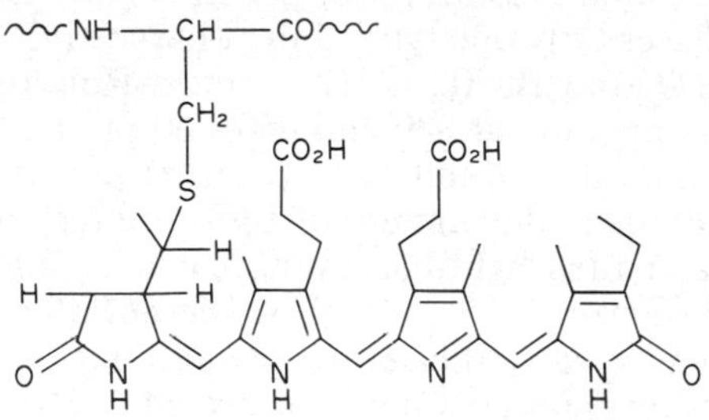

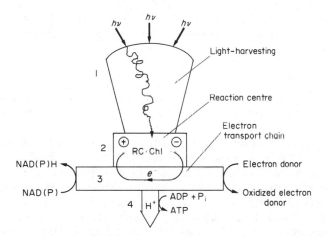

Figure 2 The structure of phycocyanobilin (**1**), the polypeptide bound phycobilin chromophore of phycocyanin, and spirilloxanthin (**2**), an aliphatic carotenoid commonly found in members of the Rhodospirillaceae

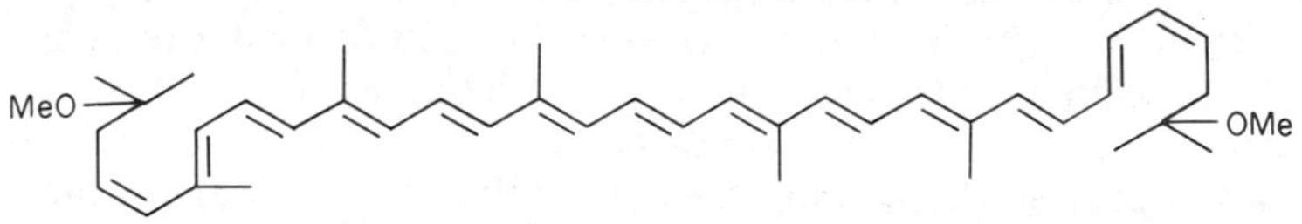

ments is transferred to a few specialized sites, the reaction centres (RC), where charge separation produces an oxidized form of chlorophyll and an electron which is transferred to the electron transport chain (Figure 3). However, the details of this process vary greatly in the different groups of phototrophic prokaryotes.

Figure 3 Schematic diagram of photosynthetic energy flow. Light harvested by the antenna system (1) arrives by a tortuous route at the reaction centre (2) as excitation energy. Here, charge separation and electron flow (3) are initiated, and subsequently exploited in the production of a proton motive force (PMF). ATP synthesis *via* the proton-translocating ATPase (4) then results in the dissipation of the PMF

30.3.2.1 *Light-harvesting*

In green and purple photosynthetic bacteria most of the bulk bacteriochlorophyll is photochemically inactive, that is, it does not undergo photooxidation when excited by light. Instead the energy is transferred randomly between adjacent pigment molecules until trapped by the reaction centre or lost as fluorescence or heat. The idea of a large number of pigment molecules acting in concert to provide a small number of reaction centres with energy is implicit in the concept of the photosynthetic unit, *i.e.* bulk bacteriochlorophyll associated with a reaction centre and electron transport chain. The details of light-harvesting in the Rhodospirillaceae are now reasonably clear due to refined spectroscopic techniques and the isolation of defined pigment–protein complexes which contain Bchl *a* non-covalently bound to low molecular weight (<14 000 M_r) proteins

(Feick and Drews, 1978; Sauer and Austin, 1978; Broglie *et al.*, 1980; Peters and Drews, 1983). *Rhodospirillum rubrum* produces only one type of light harvesting (LH) complex (Figure 4) with an absorption maximum at 890 nm ('B890', LHI) but most Rhodospirillaceae produce an accessory complex with absorption maxima at 800 and 850–860 nm ('B800 + 850', LHII) and which contains the bulk of the membrane bound Bchl (Lien *et al.*, 1973; Drews and Oelze, 1981; Thornber *et al.*, 1983). Variations in the amount of LHII are responsible for variation in the size of the photosynthetic unit and thus light-harvesting capacity under changing light conditions (Aagaard and Sistrom, 1972; Oelze, 1983a). In *R. sphaeroides* at least, energy is thought to be transferred with high efficiency between the carotenoid and Bchl *a* within LHII and it is generally assumed that further transfer to the reaction centre occurs *via* LHI (Monger and Parson, 1977; Feick *et al.*, 1980). This agrees with the organization of LHII into large 'pools' surrounding LHI–RC complexes *in vivo* (Monger and Parson, 1977). However fluorescence emission studies (Feick *et al.*, 1980) and chemical cross-linking data (Peters *et al.*, 1983) give some support for energy transfer direct from LHII to the reaction centre. In addition, there is evidence for the existence of other types of accessory LH complexes (Thornber *et al.*, 1983) but their function is as yet unclear.

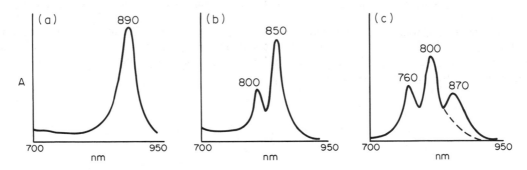

Figure 4 Typical absorption spectra of pigment–protein complexes from the Rhodospirillaceae. (a) The B890 LHI complex from *Rhodospirillum rubrum*. The corresponding LHI complex from *Rhodopseudomonas sphaeroides* is 'B875', while the accessory, LHII complex (b) has absorption maxima at 800 and 850 nm. (c) Absorption spectra of the *R. sphaeroides* reaction centre before (——) and after (– – – – –) photooxidation

The green phototrophic bacteria contain much larger photosynthetic units than those of the purple bacteria; up to 2000 molecules of Bchl *c*, *d* or *e* and 100 Bchl *a* molecules per reaction centre. The chlorobium chlorophylls are associated with protein in a vesicular structure, the chlorosome, appressed to the cytoplasmic membrane while two types of Bchl *a*–protein complexes are more closely associated with the reaction centre in the cytoplasmic membrane (Staehlin *et al.*, 1980; Olson, 1981). Evidence from fluorescence emission spectroscopy suggests that energy transfer occurs from Bchl *c*, *d* or *e* to Bchl *a* and then to the reaction centre which is consistent with this organization.

The distinctive light-harvesting system surrounding photosystem II in cyanobacteria consists of the phycobiliproteins, phycoerythrin (λ_{max} 565 nm), phycocyanin (λ_{max} 622 nm) and allophycocyanin (λ_{max} 652 nm) arranged in close proximity in membrane bound structures termed phycobilisomes (Section 30.4.3.1). Overlapping fluorescence and absorption bands allow efficient energy transfer from phycoerythrin to phycocyanin and thence to allophycocyanin and the reaction centre. Some cyanobacteria, for example *Fremyella diplosiphon*, can alter the composition of their phycobiliprotein complement in response to illumination by light of different wavelengths. Thus red light stimulates phycocyanin synthesis while phycoerythrin is stimulated by green light. This complementary chromatic adaptation ensures maximal efficiency of quantum absorption under differing light qualities.

30.3.2.2 *Charge separation and electron transport in anoxygenic photosynthesis*

That the photochemical reaction centre (RC) is a defined physical entity in bacteria was first proposed by Clayton (1963). This followed the observation that when the light-harvesting bacteriochlorophyll of a carotenoidless mutant (R-26) of *R. sphaeroides* was destroyed chemically, light-induced bleaching of a component in the absorption spectrum could still be observed. The subsequent isolation of the RC as a pigment–protein complex (Reed and Clayton, 1968) precipitated intense research on the mechanism of primary photochemistry.

Essentially the RC acts as an energy sink with Bchl in a specialized environment within which it can convert the energy of excitation received from the antenna into an electron transfer event. Isolated RCs of *R. sphaeroides* contain 3 polypeptides of M_r 28 000 (H), 24 000 (M) and 21 000 (L) in a 1:1:1 stoichiometry, 4 Bchl molecules, 2 bacteriopheophytin molecules, 2 ubiquinone molecules and one iron atom (Feher and Okamura, 1978). The absorption spectrum of the RC (Figure 4) shows a peak at 870 nm which undergoes reversible bleaching upon illumination in the presence of an electron acceptor or chemical oxidant. Evidence from circular dichroism and electron paramagnetic resonance spectroscopy indicates this peak to be composed of a dimer of Bchl molecules (P870) which are known as the 'special pair' (Norris and Katz, 1978). The remaining two Bchl molecules are responsible for the 800 nm peak, and the 760 nm peak is due to bacteriopheophytin.

Figure 5 illustrates the mechanism of energy transfer in the *R. sphaeroides* RC. After quantum absorption the P870 Bchl enters the singlet excited state and rapidly transfers an electron, probably *via* one of the 800 nm absorbing 'voyeur' Bchls, to a single bacteriopheophytin molecule. The initial charge separation is stabilized by electron transfer to the primary quinone acceptor molecule, Q_I. Further transfer to the secondary acceptor, Q_{II}, is dependent *in vitro* on the RC iron content. *In vivo*, electron transfer occurs *via* quinone and cytochrome *b* to a *c*-type cytochrome, which acts as the electron donor to P870, thus completing a cyclic electron transport chain (Figure 6). The efficiency of the system is such that only one quantum is needed to transfer one electron from the primary donor (P870) to the primary acceptor bacteriopheophytin and the accompanying redox change is nearly 1000 mV.

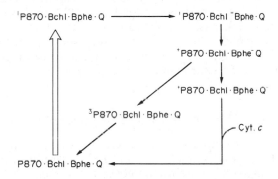

Figure 5 The mechanism of charge separation and electron flow within the *R. sphaeroides* reaction centre. ^{1}P870 represents the singlet excited state, and ^{3}P870 the triplet state, of the special pair dimer. Bchl represents a bacteriochlorophyll monomer absorbing at 800 nm. Bphe is bacteriopheophytin and Q, the primary quinone acceptor

Reaction centres have now been isolated from the Bchl *b* containing bacteria *R. viridis* and *Thiocapsa pfennigii* (Pucheu *et al.*, 1976; Seftor and Thornber, 1984) and although spectrally distinct, with P960 as the primary donor, they appear to operate in a similar manner to the RC of *R. sphaeroides*. In these Bchl *b* containing species, firmly bound *c*-type cytochromes are present in the RC preparations and recent chemical cross-linking studies have indicated a near-neighbour relationship *in vivo* (Peters *et al.*, 1984).

Energy conservation in cyclic electron transport is associated with electrogenic proton translocation to form a transmembrane proton motive force that can be used in ATP synthesis (Wraight, 1982). In purple photosynthetic bacteria this is associated with a cytochrome b–c_1 complex that has recently been isolated (Gabellini *et al.*, 1982; Yu *et al.*, 1984) and which is believed to be homologous to the ubiquinol-cytochrome *c* reductase from mitochondria and the cytochrome b_6–f(c) complex in green plants (Barber, 1984a). However, in addition to a source of ATP, photosynthesis must also satisfy the demand for reducing power in the form of NAD(P)H for biosynthetic purposes. In the purple bacteria the redox potential of the quinone acceptor ($Q_I/Q_I^- = -170$ mV) is insufficient to reduce NAD(P) directly and this must therefore occur by reversed electron transport driven by Δp (Figure 6). The reaction centres of the green sulfur bacterium *Chlorobium limicola f. thiosulfatophilum*, however, appear to contain an iron–sulfur primary acceptor with a much lower redox potential (-550 mV) making possible the direct reduction of NAD(P) by a non-cyclic pathway (Jennings and Evans, 1977). The electron donor to the RC is cytochrome

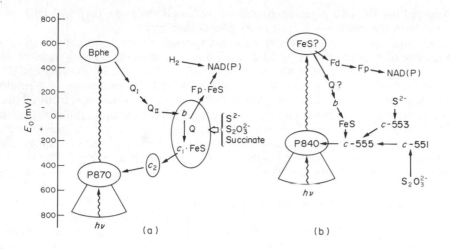

Figure 6 Electron transport in anoxygenic photosynthesis in (a) the Rhodospirillaceae and (b) the Chlorobiaceae. In (a) electrons entering the cytochrome b–c_1 segment can either return to the reaction centre, through cytochrome c_2, or reduce NAD(P) by 'reversed' electron transport through flavoprotein (Fp) and iron–sulfur centres (FeS). Electrons from succinate, or reduced sulfur compounds also enter the b–c_1 segment *via* specific dehydrogenases or reductases. In (b) the whole system operates at a significantly lower redox potential, allowing direct NAD(P) reduction *via* ferredoxin (Fd) that is not energy consuming

c-555 which receives electrons from sulfide or thiosulfate *via* specific reductases and cytochrome c-553 and c-551 (Figure 6). The primary donor in the RC, P840, supplies electrons to NAD(P) *via* ferredoxin and a flavoprotein reductase. Although much less is known about the mechanism of electron transport and primary photochemistry in the green sulfur bacteria than in Rhodospirilla-ceae, there are some similarities to non-cyclic electron transport in oxygenic photosynthesis and there is some evidence for cyclic flow through quinone and cytochrome b (Figure 6).

30.3.2.3 Charge separation and electron transport in oxygenic photosynthesis

Cyanobacterial oxygenic photosynthesis is supported by essentially the same mechanism as in higher plants and algae. Two photosystems, PSI and PSII, comprised of a reaction centre and associated light-harvesting pigment–protein complexes are present (Figure 7). Cyclic electron transport associated with PSI and non-cyclic electron transport associated with PSII both lead to the generation of Δp. The reaction centre of PSI contains P700 Chl a as a primary donor which transfers an electron to a pheophytin molecule and then to a series of iron–sulfur centres of rather low redox potential. NAD(P) reduction can occur by electron transport through ferredoxin or alternatively electrons can cycle through the plastoquinone 'pool' and the cytochrome b_6–f complex to the donor plastocyanin (Figure 7).

This system has functional analogies with that of the purple and green phototrophic bacteria and indeed some cyanobacteria such as *Oscillatoria limnetica* can photoassimilate CO_2 anoxygenically by using H_2S as the electron donor to PSI (Cohen *et al.*, 1975). Under normal conditions, however, H_2O is the ultimate electron donor, coupling with PSII, with the production of molecular oxygen. The details of this reaction in cyanobacteria are not clear but may involve the same type of mangano–protein complex recently isolated from higher plants (Barber, 1984b). The PSII reaction centre contains a P680 Chl a primary donor which receives excitation energy from allophycocyanin in the unique phycobiliprotein light-harvesting system. Electron transfer occurs *via* the plastoquinone pool, the cytochrome b_6–f segment and PSI to ferredoxin and hence to NAD(P). The energy that drives this non-cyclic system of NAD(P)H production is entirely solar in origin, rather like the process in *Chlorobium*, but unlike the Δp driven reversed electron transport in members of the Rhodospirillaceae.

Mechanisms which regulate the flow of electrons between the two photosystems in cyanobacteria operate at several levels. For example, light harvested by the Chl a molecules surrounding PSI is transferred almost exclusively to P700, while excitation transfer from the phycobiliprotein complex associated with PSII can activate both P700 and P680 (Lemasson *et al.*, 1973).

The essential role of PSII in cyanobacteria, as in higher plants, is to allow the use of water as an

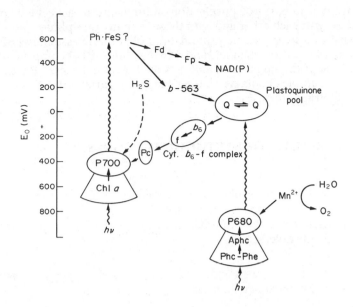

Figure 7 Electron transport in cyanobacterial, oxygenic photosynthesis. There are similarities with both the Rhodospirillacean system (cyclic electron transport through a cytochrome b–c(f) complex) and that of the Chlorobiaceae (direct NAD(P) reduction through ferredoxin–Fd) in the pathway surrounding photosystem I (*i.e.* P700 + Chl a). In some cyanobacteria, H_2S is an alternative electron donor to photosystem I. Photosystem II, *i.e.* P680 + allophycocyanin (Aphc) and phycocyanin (Phc) – phycoerythrin (Phe), supplies electrons to the plastoquinone pool which acts as a mobile redox carrier between photosystems. These electrons can then be used to reduce NAD(P) by a non-cyclic pathway.

electron donor to reduce pyridine nucleotide. For maximum efficiency of this process, *via* non-cyclic electron transport, the two photosystems must operate synchronously and this requires equal excitation rates of the two RCs. In higher plants the regulator of excitation transfer is phosphorylation of the PSII Chl a/b light-harvesting complex, which preferentially causes energy transfer to PSI. The redox state of the plastoquinone pool is thought to control the level of phosphorylation (Allen *et al.*, 1981). In cyanobacteria, there is evidence that electrons transferred to the plastoquinone pool from PSII are not homogeneously distributed to all PSIs (Kawamura and Fujita, 1982), suggesting that some PSI is not connected with electron flow from H_2O to NAD(P). How this is achieved is unclear but it is a potential regulatory mechanism to control electron movements between the photosystems.

30.3.2.4 ATP synthesis

Mechanisms to account for light driven ATP synthesis (photophosphorylation) are best understood in members of the Rhodospirillaceae. Very little is known of such processes in the green bacteria. Cyanobacterial photophosphorylation is rather similar to the process in higher plant chloroplasts, in that electron transport driven by both photosystems I and II can result in the synthesis of ATP *via* an ATPase with subunit composition and properties akin to those of the chloroplast. However, it is important to note that relatively few cyanobacteria have been studied with regard to their ATPase activity and in even fewer have measurements of the proton motive force (Δp) been undertaken.

In the Rhodospirillaceae it appears that the quinone system associated with the cytochrome b–c_1 complex sequentially binds and releases two protons per electron transferred from Q to cytochrome c_2, such that a transmembrane proton gradient from the inner to the outer bulk phases results. As postulated by Mitchell (1961) this sets up both a pH gradient (ΔpH) and membrane potential ($\Delta \psi$) which both contribute to the overall proton motive force (Δp). Direct measurements of ΔpH and $\Delta \psi$ in illuminated chromatophores of *R. sphaeroides* (Michels and Konings, 1978) and the effects of uncouplers and ionophores, underline the utility of the chemiosmotic mechanism in explaining photophosphorylation.

Energy conservation is mediated by the proton-translocating ATPase, just as in oxidative phosphorylation, and this in effect dissipates Δp with the concomitant formation of ATP from ADP +

P_i. The ATPases from a number of purple non-sulfur bacteria have been purified and show remarkable homology with each other and with those of many non-photosynthetic, heterotrophic bacteria. They consist of a hydrophobic protein–lipid oligomeric assembly (BF_0) which is membrane bound and acts as a proton channel, and a hydrophilic (BF_1) complex which is located on the cytoplasmic surface of the membrane and acts as the catalytic site (Baccarini-Melandri and Melandri, 1978). Each complex contains binding sites for a variety of ATPase inhibitors, such as oligomycin (BF_0) and efrapeptin (BF_1).

30.4 THE EUBACTERIAL PHOTOSYNTHETIC MICROBES

30.4.1 Introduction

With respect to the mechanism of photosynthetic energy transfer, the eubacteria (Fox *et al.*, 1980) fall into two groups. The anoxygenic phototrophs are of diverse evolutionary origin with many disparate properties, while the cyanobacteria, oxygenic phototrophs, are phylogenetically related and morphologically diverse.

30.4.2 The Anoxygenic Phototrophic Bacteria

30.4.2.1 The major groups

The anoxyphotobacteria (Gibbons and Murray, 1978) comprise two orders: the Rhodospirillales which contain the families Rhodospirillaceae (purple non-sulfur bacteria) and Chromatiaceae (purple sulfur bacteria), and the Chlorobiales which contain the Chlorobiaceae (green sulfur bacteria) and Chloroflexaceae (green non-sulfur bacteria). The two orders are well defined with regard to fine structure and photosynthetic pigments (Table 3). In the Rhodospirillaceae, the sole pigments are either Bchl *a* or *b*, which are contained in the cytoplasmic membrane or intracytoplasmic membranes (ICM) derived from it (Figure 8). In the Chlorobiales, the light-harvesting Bchls *c*, *d* or *e* are contained within specialized organelles (chlorosomes) attached to, but not continuous with, the cytoplasmic membrane (Figure 9).

Table 3 Characteristics of the Anoxyphotobacteria

Groups	DNA G + C (mol %)	Motility	Membrane glycolipids	Photosynthetic pigments	Arrangement of photosynthetic membranes
Anoxyphotobacteria					
Rhodospirillales					
Rhodospirillaceae	61–70	+	None	Bchl *a* or *b*	Vesicular, lamellate, stacks
Chromatiaceae	45–70	+	None	Bchl *a* or *b*	Usually vesicular
Chlorobiales					
Chlorobiacae	48–58	−	Monogalactosyl	Bchl *a* + *c*, *d*, *e*[a]	Chlorosomes
Chloroflexaceae	53–55	Gliding	Monogalactosyl	Bchl *a* + *c*, *d*[a]	Chlorosomes

[a] The light harvesting Bchl *c*, *d*, and *e* of the Chlorobiales is located in non-unit membrane bound chlorosomes which may consist of a galactolipid monolayer.

30.4.2.2 Development of the photosynthetic apparatus

In growing cultures the composition of the photosynthetic apparatus is carefully regulated by at least four major factors; light flux, oxygen tension, growth rate (nutrition) and temperature (Drews and Oelze, 1981; Oelze, 1983a). With strict photoanaerobes, the effect of environmental light flux is of major regulatory significance. In the Chlorobiaceae a decrease in light flux causes an increase in the ratio of light-harvesting Bchl *c* contained in the chlorosomes to reaction centre plus light-harvesting Bchl *a* contained in the cytoplasmic membrane, as well as an increase in the number, size and density of the chlorosomes. The effect of light intensity on the Rhodospirillales' (purple bacteria) photosynthetic apparatus is more pronounced and a variety of responses are known. *Rhodospirillum rubrum* produces a photosynthetic unit of constant size (RC:LHI Bchl ratio of 1:25–35). Light flux appears to control the specific Bchl content only *via* the amount of

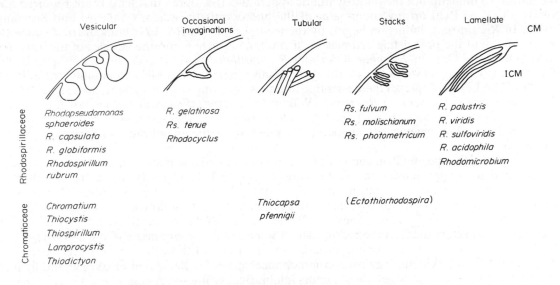

Figure 8 Intracytoplasmic membranes in the Rhodospirillales. The Ectothiorhodospiraceae have been designated as a separate family (Imhoff *et al.*, 1984)

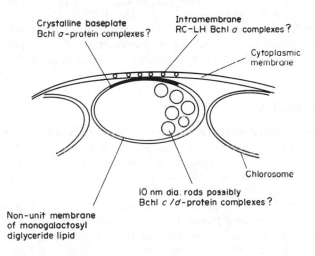

Figure 9 Schematic diagram of the chlorosome, the light-harvesting structure found in the Chlorobiaceae

ICM and the number of photosynthetic units (Drews and Oelze, 1981). In *Rhodopseudomonas sphaeroides*, *R. capsulata* and *R. palustris*, however, the major effect of a decrease in light intensity is to cause an increase in the amount of the accessory LHII Bchl–protein complex. As in *Rs. rubrum* the RC and LHI complexes are produced in constant proportion and so the size of the photosynthetic unit is increased (Aagaard and Sistron, 1972; Firsow and Drews, 1977; Schumacher and Drews, 1979). A similar pattern is seen in *Chromatium vinosum* but here a change in the composition as well as the amount of the accessory LH complex is evidenced by differences in the absorption spectra of high- and low-light grown cells. These changes can also be seen upon variation of the growth temperature (Mechler and Oelze, 1978). Such qualitative changes can be observed in some members of the Rhodospirillaceae (Thornber *et al.*, 1983) but their significance is unclear.

The response of facultative phototrophic bacteria to a change in oxygen tension is quite dramatic. The early studies of Cohen-Bazire *et al.* (1957) clearly showed that high oxygen tensions specifically repressed the formation of Bchl and thus prevented the formation of the photosynthetic apparatus. Growth of cultures under aerobic conditions in the dark therefore results in very low specific Bchl contents, few intracytoplasmic membranes (in most species) and a small photosynthetic unit. Transfer of such cultures to anaerobic-dark or anaerobic-light conditions results in the multistep induction of the photosynthetic apparatus. This shows that light is not required for the synthesis of Bchl or any component of the photosynthetic system (Aagaard and Sistrom, 1972). In *Rs. rubrum*, induction begins by the incorporation of RC–LHI units of fixed stoichiometry, first into the preexisting cytoplasmic membrane and then into the vesicles of the growing ICM (Oelze, 1983a). In *R. sphaeroides* and *R. capsulata* RC–LHI complexes are specifically inserted into invagination sites of the cytoplasmic membrane which eventually form ICM vesicles. The LHII complexes then assemble into these growing vesicles (Niederman *et al.*, 1979). A different pattern is seen in *Rs. tenue* (Wakim and Oelze, 1980) which does not form ICM. Instead, this species incorporates pigment–protein complexes homogeneously into the cytoplasmic membrane but changes its shape, that is increases the membrane surface area, upon induction.

In summary, the specific Bchl content of cultures of facultatively phototrophic bacteria is inversely related to oxygen tension in aerobic cultures and to light intensity in anaerobic cultures. However, the regulatory mechanisms underlying these changes are still unresolved. It has been known for some time that photosynthetic and respiratory electron flow are mediated by shared electron transport components (Jones, 1977) and it was postulated that light and oxygen act *via* common transmitters to affect the redox state of some cellular 'component' (Cohen-Bazire *et al.*, 1957) which effects control of Bchl synthesis. The existence of this effector has never been substantiated. Indeed, evidence against a common mechanism for the action of oxygen and light is the different kinetics with which these factors inhibit Bchl synthesis (Arnheim and Oelze, 1983a). Moreover, the RC–LHI and –LHII complexes show different inhibition kinetics with oxygen (LHII is more sensitive) but the same kinetics with light (Arnheim and Oelze, 1983b). In fact, differences in the ratio of the complexes under varying oxygen and light flux conditions can be explained largely in terms of variation in the rate of supply of the Bchl precursor, 5-aminolaevulinic acid (ALA), at least in *R. sphaeroides* H5 (Oelze, 1983b). The system assembling the B875 LHI complex has a higher affinity for Bchl than that assembling the B800 + 850 LHII complex so that at low rates of supply the former will always predominate. Coupled with the fact that oxygen, but not light, inhibits ALA synthetase, this explains the order of assembly upon a drop in pO_2.

30.4.2.3   Carbon metabolism

The major routes of carbon assimilation have been best studied in members of the Rhodospirillaceae and Chromatiaceae. Evidence from the incorporation of $^{14}CO_2$ into 3-phosphoglyceric acid during autotrophic CO_2 fixation, and the demonstration of all the key enzymes of the reductive pentose phosphate, or Calvin, cycle in cell-free extracts have led to the conclusion that this is the major route of CO_2 fixation in both families (Fuller, 1978, Sojka, 1978). An outline of the cycle is shown in Figure 10. It is important to note that the key reactions depend upon a continued supply of ATP and reducing power (NADPH) which must be obtained from light driven electron transport (Section 30.3.2.2). The enzyme ribulose bisphosphate carboxylase/oxygenase (RubisCo) catalyses the initial carboxylation of ribulosebisphosphate to form 3-phosphoglyceric acid which is required for several biosynthetic purposes. Consequently high activities of this enzyme can be detected in cell-free extracts of photoautotrophically grown cells. RubisCo is also present under photoheterotrophic growth conditions, but here its activity is dependent upon the phase of growth, the ease with which the carbon source is used by other pathways, such as the tricarboxylic acid cycle, and the amount of CO_2 generated during metabolism.

The properties of RubisCo have been well studied in certain members of the Rhodospirillaceae and are considered by Dijkhuizen and Harder (Volume 1, Chapter 21).

In addition to the Calvin cycle, a variety of ancillary carboxylation mechanisms exist which serve an anaplerotic role. An important example is phosphoenolpyruvate (PEP) carboxylase which functions in the formation of C_4 acids. The metabolic flexibility of members of the Rhodospirillaceae for growth on carbon sources other than CO_2 suggests the presence of an active tricarboxylic acid cycle and this has been confirmed in many species growing chemo- or photo-heterotrophically (Ormerod and Gest, 1962; Fuller, 1978). The photometabolism of fatty acids, amino acids, organic acids, alcohols, sugars and even aromatic and C_1 compounds is ample testament to this (Quayle and Pfennig, 1975; Sojka, 1978; Dutton and Evans, 1978) and contrasts with members of the Chromatiaceae which show a much narrower carbon source utilization spectrum. Some of the Rhodospirillaceae can grow under anaerobic conditions in the dark by 'fermentation', provided dimethyl sulfoxide (DMSO) or trimethylamine *N*-oxide (TMAO) are present as electron acceptors to maintain the redox balance. However, recent evidence suggests that elec-

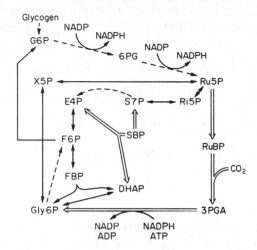

Figure 10 The reductive and oxidative pentose phosphate cycles. Reactions unique to the reductive (Calvin) cycle ($\Rightarrow$) are those involved in CO_2 fixation in most anoxyphotobacteria. Reactions unique to the dissimilatory, oxidative cycle ($--\rightarrow$) occur in some cyanobacteria. Reactions common to both are also shown ($\rightarrow$). G6P, glucose 6-phosphate; 6PG, 6 phosphogluconate; Ru5P, ribulose 5-phosphate; RuBP, ribulose bisphosphate, 3-PGA, 3-phosphoglycerate; Gly6P; glyceraldehyde 6-phosphate; X5P, xylulose 5-phosphate; E4P, erythrose 4-phosphate; S7P, sedoheptulose 7-phosphate; SBP, sedoheptulose 1,7-bisphosphate, F6P; fructose 6-phosphate; FBP, fructose 1,6-bisphosphate; DHAP, dihydroxyacetone phosphate; Ri5P, ribose 5-phosphate

tron flow to DMSO or TMAO results in the formation of a membrane potential which can be linked to ATP production (McEwan *et al.*, 1983) emphasizing again the plethora of energy generating pathways in this group.

Carbon metabolism in the green sulfur bacteria is less well understood. The evidence for CO_2 fixation *via* the Calvin cycle is far less compelling, as $^{14}CO_2$ incorporation results in the early labelling of glutamate and RubisCo cannot be detected in many strains of *Chlorobium*. This led Evans *et al.* (1966) to propose the existence of a reductive carboxylic acid cycle in which 1 mol of acetyl-CoA results from the fixation of 2 mol of CO_2 by the ferredoxin dependent enzymes, pyruvate and α-ketoglutarate synthetase. For a detailed discussion of the reductive carboxylic acid cycle, see Volume 1, Chapter 21 by Dijkhuizen and Harder.

30.4.3 The Cyanobacteria: Oxygenic Photosynthesis in a Diverse Prokaryotic Group

For 150 years, the cyanobacteria were regarded as a division of the algae and even now their nomenclature is still governed by the rules of the botanical code (see Stanier and Cohen-Bazire, 1977 for an account of the tortuous history of cyanobacteriology). They represent the largest single assemblage of photosynthetic prokaryotes and are amongst the most diverse and ubiquitous of all prokaryotic groups. Consequently we can only give a selective account of their properties as they relate to photosynthesis and the reader is recommended to consult the treatise by Carr and Whitton (1982) to obtain a balanced view of the biology of the group.

30.4.3.1 *Organization of the photosynthetic apparatus*

In most cyanobacteria the photosynthetic apparatus is localized on a system of intracytoplasmic membranes generally termed 'thylakoids'. These have some similarity to the thylakoids of the chloroplasts of higher plants except that they do not form extensive stacking regions. Although in most species the thylakoid system is confined to the periphery of the cell, they may also be scattered throughout the cytoplasm (*e.g. Chroococcidiopsis*) or confined to the central region (*e.g. Trichodesmium*). Unlike the intracytoplasmic membrane system in some anoxyphotobacteria, there is no evidence for a specific relationship between the cytoplasmic membrane and the thylakoid system: the two are topologically distinct.

In the unicellular cyanobacterium *Gloeobacter violaceus* (Rippka *et al.*, 1974) thylakoids are absent and the ensemble of light-harvesting phycobiliproteins are located in rod or bundle shaped phycobilisomes in a continuous layer attached to the cell membrane. In all other cyanobacteria the phycobiliproteins are contained within hemidiscoidal shaped phycobilisomes on the surface of the thylakoids (Cohen-Bazire and Bryant, 1982). The structure of these phycobilisomes reflects the order of excitation energy transfer between the biliprotein chromophores (Section 30.3.1.3). Thus, allophycocyanin forms a triangular core, of trimeric construction, surrounded by up to six peripheral rods consisting of disc-shaped hexamers of phycocyanin at the base and phycoerythrin at the apex (Figure 11). A number of polypeptides are present in the structure which appear to have roles in attaching the phycobilisome to the thylakoid and/or stabilizing interactions between the subunits (linker polypeptides).

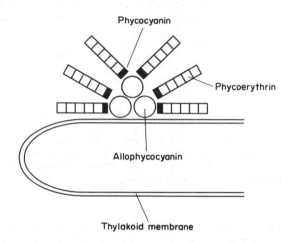

Figure 11 Structure of the phycobilisome. The hemidiscoidal shaped phycobilisomes consist of a core of allophycocyanin with peripheral rods of phycocyanin. In the example shown, the rods consist largely of phycoerythrin as would be the case in a chromatically adapting strain grown in green light

Several environmental factors, notably light fluency and quality, and nitrogen deprivation cause structural changes in the phycobilisomes. In those cyanobacteria capable of complementary chromatic adaptation (Section 30.3.2.1) changes in the ratio of phycoerthrin to phycocyanin cause changes in the size of the phycobilisomes (if phycoerythrin synthesis alone is stimulated in green light and repressed in red light, *e.g. Synechocystis* 6701) or in the composition of the peripheral rods (if both phycoerythrin and phycocyanin synthesis is modulated). There is also evidence that a decrease in light intensity can cause an increase in the length of the peripheral rods in some cyanobacteria (Cohen-Bazire and Bryant, 1982). Nitrogen starvation results in a rapid breakdown of biliproteins, mediated by an inducible, membrane associated protease. Under these conditions, chlorophyll *a* must assume the role of the primary light receptor for PSII and it has been suggested that the biliproteins may represent a major nitrogenous reserve material.

30.4.3.2 *Interrelationships between photosynthetic and chemosynthetic carbon metabolism in cyanobacteria*

It has become axiomatic that all cyanobacteria are aerobic photoautotrophs with an endogenous dark respiratory metabolism which is subject to severe inhibition by light (Stanier and Cohen-Bazire, 1977). However, although obligate photoautotrophy is widespread within the group, some species have the capacity to use a limited range of sugars in the dark, *i.e.* as facultative chemoheterotrophs, or in the light, *i.e.* as facultative photoheterotrophs.

That CO_2 is fixed *via* the reductive pentose phosphate or Calvin cycle in photoautotrophically grown cyanobacteria was first confirmed by Pelroy and Bassham (1972). Early labelling experiments showed 3-phosphoglyceric acid to be the first product, consistent with the presence of ribulosebisphosphate carboxylase. Subsequently, RubisCo was purified from a number of species and most found to be of the L8S8 type, although L8 and L4 types have been reported (Smith, 1982). Because of the aerobic nature of cyanobacterial photosynthesis, the oxygenase reaction of this carboxylase could lead to the potential loss of fixed carbon as glycollate, which is excreted into the medium. However cyanobacteria do possess a carbon recovery pathway which converts glycollate to 3-phosphoglycerate *via* tartronic semialdehyde (Codd and Stewart, 1973).

The chief stable product of light driven CO_2 fixation in cyanobacteria is glycogen. When light grown cells are transferred to the dark, there is a gradual decrease in their glycogen content. This is accompanied by an increase in the pool sizes of 6-phosphogluconate, fructose 1,6-bisphosphate and sedoheptulose 1,7-bisphosphate (Pelroy and Bassham, 1972) consistent with the operation of an oxidative pentose phosphate pathway. Much of the enzymatic machinery of the reductive pentose phosphate pathway (Figure 10) is also used in this pathway, with the notable exceptions of ribulosebisphosphate carboxylase and phosphoribulokinase. These enzymes are repressed under dark conditions, and instead glucose-6-phosphate and 6-phosphogluconate dehydrogenases are produced which are specific for NADP, rather than NAD. This suggests that NADPH is the principal pyridine nucleotide used for dark respiratory electron transport to molecular oxygen, possibly *via* some of the redox carriers used in photosynthetic electron flow (Stanier and Cohen-Bazire, 1977).

Although in principle it would be possible to link the oxidative pentose phosphate pathway with the tricarboxylic acid cycle as a terminal oxidation route, this is negated due to the lack of α-ketoglutarate dehydrogenase in cyanobacteria. The individual reactions thus serve a purely biosynthetic role, as in some of the anoxyphotobacteria, methylotrophs and chemolithotrophs.

Those cyanobacteria able to grow as facultative chemoheterotrophs in the dark can do so only slowly, and at the expense of a limited range of sugars, most commonly glucose. Again, the oxidative pentose phosphate pathway appears to be the major route for respiratory dissimilation and some of the intermediates are converted to glycogen for storage. All of the facultative chemoheterotrophs can also develop photoheterotrophically by using the appropriate carbon source in place of CO_2 in the light. The ability to grow in this manner can be demonstrated by rigorously excluding CO_2, or in its presence by using a light intensity too low to support CO_2 assimilation. Alternatively cultures can be treated with dichlorophenyldimethylurea (DCMU) which selectively blocks PSII and thus prevents generation of the reductant (NADPH) necessary for CO_2 assimilation. Nevertheless, under these conditions, the rate of ATP synthesis by cyclic photophosphorylation is high, allowing far better growth with the particular organic substrate than is obtained in the dark. Accordingly this technique can be used as a screening method for general carbon-source utilization (Rippka, 1972).

Whilst cyanobacterial heterotrophy is readily demonstrated in the laboratory, its significance in the carbon balance of natural populations is harder to assess (Smith, 1982). There is no doubt that cyanobacteria can exist in a variety of environments that are permanently dark, conditions under which chemoheterotrophic metabolism would be a necessity. Photoheterotrophic growth may conceivably be important at limiting light intensities, but what seems certain is that subtle adjustments to cyanobacterial metabolism can be made in response to ever fluctuating environmental conditions and that, as a group, these organisms are not limited to obligate photoautotrophy.

30.5 HALOBACTERIAL PHOTOPHOSPHORYLATION

The halobacteria are representatives of the archaebacterial group of prokaryotes (Fox *et al.*, 1980), which also includes the methanogens, thermoacidophiles and certain thermophilic sulfur oxidizers. This ancient assemblage, although bacterial in appearance, is no more related to the eubacteria than to the eukaryotic line of descent. It is therefore not surprising that novel bioenergetic systems have been discovered amongst them.

The natural environment of the halobacteria is the surface layer of hypersaline ponds where the salt concentration is at or near saturation and which are subject to intense solar radiation (Dundas, 1977). Energy is normally obtained by oxygen-linked respiration of amino acids mediated by a membrane-bound electron transport chain. However, oxygen solubility in saturated salt solutions is low and thus oxygen limitation would be a potential problem if it did not induce the formation of an alternative energy conserving system dependent on light. Under these conditions, a special chromoprotein, bacteriorhodopsin (bR), is produced which aggregates to form a crystalline 'purple membrane' within the preexisting cell membrane (Oesterhelt and Stoeckenius, 1971). Bacteriorhodopsin consists of a 26 000 M_r apoprotein with the chromophore, *i.e.* retinal (vitamin A aldehyde), covalently linked to a lysine residue *via* a Schiff base. Upon illumination, the photoreaction cycle of the all-*trans* retinal isomer leads to bleaching of the chromophore and a consequent deprotonation of the Schiff base (Figure 12).

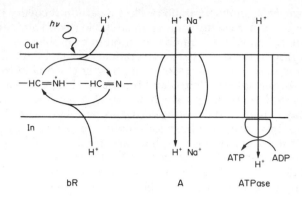

Figure 12 Photophosphorylation and ionic movements in the halobacteria. Deprotonation of the Schiff base in bacteriorhodopsin (bR) results in the net movement of protons from the cytoplasm to the external bulk phase across the cytoplasmic membrane. Inwardly directed proton translocation *via* the Na$^+$/H$^+$ antiporter (A) and the ATPase provides energy for the maintenance of ionic gradients and ATP synthesis respectively

Translocation of this proton to the external bulk phase results in the generation of both a pH gradient (ΔpH) and membrane potential ($\Delta\psi$) which both contribute to the proton motive force (Δp). The Δp is used to power motility, to generate ATP *via* the ATPase, to maintain ionic balances *via* Na$^+$/H$^+$ antiporters and to support active transport of amino acids. Recently, it has become clear that there is another light-driven ionic pump situated in the residual 'red membrane' of the halobacteria (Stoeckenius and Bogomolni, 1982). This has been termed halorhodopsin and acts as an inwardly directed Cl$^-$ ion-pump. It is therefore not directly involved in energy conservation.

In the context of microbial photosynthesis, it is appropriate to view the halobacterial system as a mechanism for the maintenance of a variety of essential ionic gradients across the cell membrane, some of which can lead to ATP synthesis. The halobacteria are not true phototrophs as their growth cannot occur solely at the expense of light energy but they have developed a means of supplementing their ATP pool by a method not found in any other biological group.

30.6 SUMMARY

In this chapter we have concentrated primarily on the mechanisms used by phototrophic microorganisms to generate ATP and reducing power to meet the requirements of cellular metabolism. The biochemistry and biophysics of the various processes are now rapidly being elucidated, although much of the research has been based on one organism, *Rhodopseudomonas sphaeroides*. As we have attempted to show there is considerable diversity, both structural and physiological, to be found in the phototrophic microbes. Consequently extrapolation of this data to other microbes can only be done with extreme caution.

Considerable effort is now being applied to membrane differentiation, topography and function. The restraining factor in advances in microbial photosynthetic research is the application and exploitation of genetic and molecular biological techniques. The fusion of these approaches, *i.e.* physiological and genetic, should rapidly enhance our understanding of the organisms and the process.

30.7 REFERENCES

Aagaard, J. and W. R. Sistrom (1972). Control of synthesis of reaction-center bacteriochlorophyll in photosynthetic bacteria. *Photochem. Photobiol.*, **15**, 209–225.
Allen, J. F., J. Bennett, K. E. Steinback and C. J. Arntzen (1981). Chloroplast protein phosphorylation couples plastoquinone redox state to distribution of excitation energy between photosystems. *Nature (London)*, **291**, 25–29.
Arnheim, K. and J. Oelze (1983a). Differences in the control of bacteriochlorophyll formation by light and oxygen. *Arch. Microbiol.*, **135**, 299–304.
Arnheim, K. and J. Oelze (1983b). Control by light and oxygen of B875 and B850 pigment–protein complexes in *Rhodopseudomonas sphaeroides*. *FEBS Lett.*, **162**, 61–63.
Arnon, D. J., M. B. Allen and F. R. Whatley (1954). Photosynthesis by isolated chloroplasts. *Nature (London)*, **174**, 394–396.

Baccarini-Melandri, A. and B. A. Melandri (1978). In *The Photosynthetic Bacteria*, ed. R. K. Clayton and W. R. Sistrom, pp. 615–628. Plenum, New York.

Barber, J. (1984a). Further evidence for the common ancestry of cytochrome *b–c* complexes. *Trends Biochem. Sci.*, **9**, 209–211.

Barber, J. (1984b). Has the mangano–protein of the water splitting reaction of photosynthesis been isolated? *Trends Biochem. Sci.*, **9**, 79–80.

Brockmann, H. Jr. and A. Lipinski (1982). Bacteriochlorophyll *g*. A new bacteriochlorophyll from *Heliobacterium chlorum. Arch. Microbiol.*, **136**, 17–19.

Broglie, R. M., C. N. Hunter, P. Delepelaire, R. A. Neiderman, N. H. Chua and R. K. Clayton (1980). Isolation and characterization of the pigment–protein complexes of *Rhodopseudomonas sphaeroides* by lithium dodecyl sulfate/polyacrylamide gel electrophoresis. *Proc. Natl. Acad. Sci. USA*, **77**, 87–91.

Carr, N. G. and B. A. Whitton (1982). *The Biology of Cyanobacteria*, p. 688. Blackwell, Oxford.

Clayton, R. K. (1963). Toward the isolation of a photochemical reaction centre in *Rhodopseudomonas sphaeroides. Biochim. Biophys. Acta*, **75**, 312–333.

Codd, G. A. and W. D. P. Stewart (1973). Pathways of glycollate metabolism in the blue-green alga *Anabaena cylindrica. Arch. Mikrobiol.*, **94**, 11–28.

Cohen, Y., E. Padan and M. Shilo (1975). Facultative anoxygenic photosynthesis in the cyanobacterium *Oscillatoria limnetica. J. Bacteriol.*, **123**, 855–861.

Cohen-Bazire, G. and D. A. Bryant (1982). In *The Biology of Cyanobacteria*, ed. N. G. Carr and B. A. Whitton, pp. 143–190. Blackwell, Oxford.

Cohen-Bazire, G., W. R. Sistrom and R. Y. Stanier (1957). Kinetic studies of pigment synthesis by non-sulphur purple bacteria. *J. Cell. Comp. Physiol.*, **49**, 25–68.

Drews, G and J. Oelze (1981). Organization and differentiation of membranes of phototrophic bacteria. *Adv. Microb. Physiol.*, **22**, 1–92.

Dundas, I. E. P. (1977). Physiology of Halobacteriaceae. *Adv. Microb. Physiol.*, **15**, 85–120.

Dutton, P. L. and W. C. Evans (1978). In *The Photosynthetic Bacteria*, ed. R. K. Clayton and W. R. Sistrom, pp. 719–728. Plenum, New York.

Englemann, T. W. (1883). *Bacterium photometricum*. Ein Beitrag zur vergleichenden Physiologie des Licht- und Farbensinnes. *Pflugers. Arch. Physiol.*, **30**, 95–124.

Evans, M. C. W., B. B. Buchanan and D. I. Arnon (1966). A new ferredoxin dependent carbon reduction cycle in a photosynthetic bacterium. *Proc. Natl. Acad. Sci. USA*, **55**, 928–934.

Feher, G. and M. Y. Okamura (1978). In *The Photosynthetic Bacteria*, ed. R. K. Clayton and W. R. Sistrom, pp. 349–386. Plenum, New York.

Feick, R. and G. Drews (1978). Isolation and characterization of light harvesting bacteriochlorophyll–protein complexes from *Rhodopseudomonas capsulata. Biochim. Biophys. Acta*, **501**, 499–513.

Feick, R., R. van Grondelle, C. P. Rijgersberg and G. Drews (1980). Fluorescence emission by wild-type and mutant strains of *Rhodopseudomonas capsulata. Biochim. Biophys. Acta*, **593**, 241–253.

Firsow, N. N. and G. Drews (1977). Differentiation of the intracytoplasmic membrane of *Rhodopseudomonas palustris* induced by variations of oxygen partial pressure or light intensity. *Arch. Microbiol.*, **115**, 299–306.

Fox, G. E., E. Stackebrandt, R. B. Hespell, J. Gibson, J. Maniloff, T. A. Dyer, R. S. Wolfe, W. E. Balch, R. J. Tanner, L. J. Magrum, L. B. Zablen, R. Blakemore, R. Gupta, L. Bonen, B. J. Lewis, D. A. Stahl, K. R. Luehrsen, K. N. Chen and C. R. Woese (1980). The phylogeny of prokaryotes. *Science*, **209**, 457–463.

Frenkel, A. W. (1954). Light induced phosphorylation by cell-free preparations of photosynthetic bacteria. *J. Am. Chem. Soc.*, **76**, 5568–5569.

Fuller, R. C. (1978). In *The Photosynthetic Bacteria*, ed. R. K. Clayton and W. R. Sistrom, pp. 691–706. Plenum, New York.

Gabellini, N., J. R. Bowyer, E. Hurt, B. A. Melandri and G. Hauska (1982). A cytochrome *b/c₁* complex with ubiquinol-cytochrome *c₂* oxidoreductase activity from *Rhodopseudomonas sphaeroides* G A. *Eur. J. Biochem.*, **126**, 105–111.

Gest, H. (1982). In *From Cyclotrons to Cytochromes*, ed. N. O. Kaplan and A. Robinson, pp. 305–321. Academic, New York.

Gest, H. and J. L. Favinger (1983). *Heliobacterium chlorum*, an anoxygenic brownish-green photosynthetic bacterium containing a 'new' form of bacteriochlorophyll. *Arch. Microbiol.*, **136**, 11–16.

Gest, H. and M. D. Kamen (1948). Studies on the phosphorus metabolism of green algae and purple bacteria in relation to photosynthesis. *J. Biol. Chem.*, **176**, 299–318.

Gibbons, N. E. and R. G. E. Murray (1978). Proposals concerning the higher taxa of bacteria. *Int. J. Syst. Bacteriol.*, **28**, 1–6.

Hill, R. and F. Bendall (1960). Function of the two cytochrome components in chloroplasts: a working hypothesis. *Nature (London)*, **186**, 136–137.

Imhoff, J. F., H. G. Truper and N. Pfennig (1984). Rearrangement of the species and genera of the phototrophic purple non-sulfur bacteria. *Int. J. Syst. Bacteriol.*, **34**, 340–343.

Jennings, J. V. and M. C. W. Evans (1977). The irreversible photoreduction of a low potential component at low temperatures in a preparation of the green photosynthetic bacterium *Chlorobium thiosulfatophilum. FEBS Lett.*, **75**, 33–36.

Jones, O. T. G. (1977). In *Microbial Energetics*, ed. B. A. Haddock and W. A. Hamilton, *Symposium of the Society for General Microbiology*, vol. 27, pp. 151–183. Cambridge University Press, Cambridge.

Jones, O. T. G. (1978). In *The Photosynthetic Bacteria*, ed. R. K. Clayton and W. R. Sistrom, pp. 751–778. Plenum, New York.

Kawamura, M. and Y. Fujita (1982). In *Photosynthetic Prokaryotes, Cell Differentiation and Function*, ed. G. C. Papageorgiou and L. Packer, pp. 127–145. Elsevier Biomedical, New York.

Lemasson, C., N. Tandeau de Marsac and G. Cohen-Bazire (1973). Role of allophycocyanin as a light-harvesting pigment in cyanobacteria. *Proc. Natl. Acad. Sci. USA*, **70**, 3130–3133.

Levitt, C. S. (1953). Photosynthesis as a photoelectric phenomenon. *Science*, **118**, 686–691.

Lien, S., H. Gest and A. San Pietro (1973). Regulation of bacteriochlorophyll synthesis in photosynthetic bacteria. *Bioenergetics*, **4**, 423–434.

Mauzerall, D. (1978). In *The Photosynthetic Bacteria*, ed. R. K. Clayton and W. R. Sistrom, pp. 223–231. Plenum, New York.

McEwan, A. G., S. J. Ferguson and J. B. Jackson (1983). Electron flow to dimethyl sulphoxide or trimethylamine *N*-oxide generates a membrane potential in *Rhodopseudomonas capsulata*. *Arch. Microbiol.*, **136**, 300–305.

Mechler, B. and J. Oelze (1978). Differentation of the photosynthetic apparatus of *Chromatium vinosum* strain D. I. The influence of growth conditions. *Arch. Microbiol.*, **118**, 91–97.

Michels, P. A. M. and W. Konings (1978). The electrochemical proton gradient generated by light in membrane vesicles and chromatophores of *Rhodopseudomonas sphaeroides*. *Eur. J. Biochem.*, **85**, 147–155.

Mitchell, P. (1961). Coupling of phosphorylation to electron and hydrogen transfer by a chemiosmotic type of mechanism. *Nature (London)*, **191**, 144–148.

Molisch, H. (1907). *Die Purpurbakterium nach neuen untersuchungen*, p. 95. Fischer Verlag, Jena.

Monger, T. G. and W. W. Parson (1977), Singlet–triplet fusion in *Rhodopseudomonas sphaeroides* chromatophores: a probe of the organization of the photosynthetic apparatus. *Biochim. Biophys. Acta*, **460**, 393–407.

Niederman, R. A., D. E. Mallon and L. C. Parks (1979). Membranes of *Rhodopseudomonas sphaeroides*. IV. Isolation of a fraction enriched in newly synthesised bacteriochlorophyll *a*–protein complexes. *Biochim. Biophys. Acta*, **555**, 210–220.

Norris, J. R. and J. J. Katz (1978). In *The Photosynthetic Bacteria*, ed. R. K. Clayton and W. R. Sistrom, pp. 397–418. Plenum, New York.

Oelze, J. (1983a). In *The Phototrophic Bacteria: Anaerobic Life in the Light*, ed. J. G. Ormerod, pp. 8–34. Blackwell, Oxford.

Oelze, J. (1983b). Control of the formation of bacteriochlorophyll complexes in *Rhodopseudomonas sphaeroides* mutant strain H5. *Arch. Microbiol.*, **136**, 312–316.

Oesterhelt, D. and W. Stoeckenius (1971). Rhodopsin-like protein from the purple membrane of *Halobacterium halobium*. *Nature (New Biol.)*, **233**, 149–152.

Olson, J. M. (1981). Chlorophyll organization in green photosynthetic bacteria. *Biochim. Biophys. Acta*, **594**, 33–51.

Ormerod, J. G. and H. Gest (1962). Hydrogen photosynthesis and alternative metabolic pathways in photosynthetic bacteria. *Bacteriol. Rev.*, **26**, 51–66.

Pelroy, R. A. and J. A. Bassham (1972). Photosynthetic and dark carbon metabolism in unicellular blue-green algae. *Arch. Mikrobiol.*, **86**, 25–28.

Peters, J. and G. Drews (1983). The B870 pigment–protein complex of *Rhodopseudomonas capsulata* contains two different pigment-binding polypeptides. *FEMS Lett.*, **17**, 235–237.

Peters, J., J. Takemato and G. Drews (1983). Spatial relationships between the photochemical reaction centre and the light-harvesting complexes in the membrane of *Rhodopseudomonas capsulata*. *Biochemistry*, **22**, 5660–5667.

Peters, J., W. Welte and G. Drews (1984). Topographical relationships of polypeptides in the photosynthetic membrane of *Rhodopseudomonas viridis* investigated by reversible chemical cross-linking. *FEBS Lett.*, **171**, 267–270.

Pfennig, N. (1978). In *The Photosynthetic Bacteria*, ed. R. K. Clayton and W. R. Sistrom, pp. 3–18. Plenum, New York.

Pucheu, N. L., N. L. Kerber and A. F. Garcia (1976). Isolation and purification of reaction centres from *Rhodopseudomonas viridis* NHTC 133 by means of LDAO. *Arch. Microbiol.*, **109**, 301–305.

Quayle, J. R. and N. Pfennig (1975). Utilization of methanol by Rhodospirillaceae. *Arch. Microbiol.*, **102**, 193–198.

Reed, D. W. and R. K. Clayton (1968). Isolation of a reaction centre fraction from *Rhodopseudomonas sphaeroides*. *Biophys. Biochem. Res. Commun.*, **30**, 471–475.

Rippka, R. (1972). Photoheterotrophy and chemoheterotrophy among unicellular blue-green algae. *Arch. Mikrobiol.*, **87**, 93–98.

Rippka, R., J. B. Waterbury and G. Cohen-Bazire (1974). A cyanobacterium which lacks thylakoids. *Arch. Mikrobiol.*, **100**, 419–436.

Sauer, K. and L. A. Austin (1978). Bacteriochlorophyll–protein complexes from the light-harvesting antenna of photosynthetic bacteria. *Biochemistry*, **17**, 2011–2019.

Schmidt, K. (1978). In *The Photosynthetic Bacteria*, ed. R. K. Clayton and W. R. Sistrom, pp. 729–750. Plenum, New York.

Schumacher, A. and G. Drews (1979). Effects of light intensity on membrane differentiation in *Rhodopseudomonas capsulata*. *Biochim. Biophys. Acta*, **547**, 417–428.

Seftor, R. E. B. and J. P. Thornber (1984). The photochemical reaction centre of the bacteriochlorophyll *b* containing organism *Thiocapsa pfennigii*. *Biochim. Biophys. Acta*, **764**, 148–159.

Smith, A. J. (1982). In *The Biology of the Cyanobacteria*, ed. N. G. Carr and D. A. Whitton, pp. 47–86. Blackwell, Oxford.

Sojka, G. A. (1978). In *The Photosynthetic Bacteria*, ed. R. K. Clayton and W. R. Sistrom, pp. 707–718. Plenum, New York.

Staehlin, L. A., J. R. Golecki and G. Drews (1980). Supramolecular organization of chlorosomes (chlorobium vesicules) and of their membrane attachment sites in *Chlorobium limicola*. *Biochim. Biophys. Acta*, **589**, 30–45.

Stanier, R. Y. and G. Cohen-Bazire (1977). Phototrophic prokaryotes: the cyanobacteria. *Annu. Rev. Microbiol.*, **31**, 225–274.

Stanier, R. Y. and C. B. Van Niel (1962). The concept of a bacterium. *Arch. Mikrobiol.*, **42**, 17–35.

Stoeckenius, W. and R. A. Bogomolni (1982). Bacteriorhodopsin and related pigments of Halobacteria. *Annu. Rev. Biochem.*, **51**, 587–616.

Thornber, J. P., R. J. Cogdell, B. K. Pierson and R. E. B. Seftor (1983). Pigment–protein complexes of purple photosynthetic bacteria: an overview. *J. Cell. Biochem.*, **23**, 159–169.

Truper, H. G. and N. Pfennig (1978). In *The Photosynthetic Bacteria*, ed. R. K. Clayton and W. R. Sistrom, pp. 19–27. Plenum, New York.

Van Niel, C. B. (1935). Photosynthesis of bacteria. *Cold Spring Harbor Symp. Quant. Biol.*, **3**, 138–150.

Wakim, B. and J. Oelze (1980). The unique mode of adjusting the composition of the photosynthetic apparatus to different environmental conditions by *Rhodospirillum tenue*. *FEMS Lett.*, **7**, 221–223.

Winogradsky, S. (1888). *Beitrage zur Morphologie und Physiologie der Bakterien*, vol. 1, *Zur Morphologie und Physiologie der schwefelbacterien*, p. 120. Verlag A. Felix, Leipzig.

Wraight, C. A. (1982). In *Photosynthesis*, ed. C. P. Govindjee, vol. I, pp. 17–61. Academic, New York.

Yu, L., Q. C. Mei and C. A. Yu (1984). Characterization of purified cytochrome b–c_1 complex from *Rhodopseudomonas sphaeroides* R-26. *J. Biol. Chem.*, **259**, 5752–5760.

31
Extracellular Enzymes

F. G. PRIEST
Heriot-Watt University, Edinburgh, UK

31.1 INTRODUCTION

Following a recession in the late 1960s largely caused by fear of side-effects from enzymes used in washing detergents, the enzyme industry has now recovered to such an extent that some 25 products are being marketed in bulk quantity. From an economic point of view the proteases are the most important industrial enzymes with several hundred tonnes being produced annually. The predominant applications are the use of alkaline serine protease from *Bacillus licheniformis* in detergents and rennet preparations from *Mucor* and *Endothea* strains in cheese manufacture. Amylolytic enzymes are used extensively in the starch-processing and allied food industries particularly with the success of high-fructose syrup produced by isomerization of glucose. These enzymes are principally α-amylases from *Bacillus amyloliquefaciens* and *B. licheniformis*, amyloglucosidase from *Aspergillus* strains, and glucose isomerase from various bacteria including *Streptomyces* species. Other major enzymes that are produced from microorganisms for industrial purposes are listed in Table 1 and have recently been reviewed (Aunstrup, 1978). Two related features are apparent in Table 1: industrial enzymes are almost invariably extracellular proteins in that they are prepared from the fermentation broth rather than from the cells and, in the case of the bacterial enzymes, the producer-organisms are Gram-positive with the sole exception of '*Klebsiella aerogenes*' (*K. pneumoniae*). For commercial fermentations, extracellular enzymes have two advantages; they can be produced in very high yield and they can be easily recovered by a series of simple fractionation and purification steps. Processing of intracellular enzymes involves additional complicating stages such as cell breakage, without damage to the product, and separation from other components of the cell, particularly nucleic acids.

Table 1 Enzymes Produced in Bulk Quantity and Some Representative Producer Microorganisms[a]

Enzyme	Producer organism	Enzyme	Producer organism
α-Amylase	*Aspergillus oryzae*	Lactase	*Aspergillus* spp.
	Bacillus amyloliquefaciens		*Kluyveromyces marxianus*
	Bacillus licheniformis		('*Saccharomyces fragilis*')
Amyloglucosidase	*Aspergillus niger*		*Kluyveromyces lactis*
	Rhizopus spp.	Microbial rennet	*Endothea parasitica*
Dextranase	*Penicillium* spp.		*Mucor miehei*
β-Glucanase	*Aspergillus niger*		*Mucor pusillus*
	Bacillus pumilus	Pectinase	*Aspergillus niger*
	Bacillus subtilis	Proteases	*Aspergillus niger*
Glucose isomerase	*Arthrobacter* spp.		*Aspergillus oryzae*
	Bacillus coagulans		*Bacillus amyloliquefaciens*
	Streptomyces spp.		*Bacillus licheniformis*
Hemicellulase	*Aspergillus* spp.		*Bacillus thermoproteolyticus*
Invertase	*Aspergillus* spp.	Pullulanase	*Klebsiella pneumoniae*
	Saccharomyces spp.		('*K. aerogenes*')

[a] Data collated from Aunstrup (1978) and Rose (1980).

The problem of defining exactly what is meant by the term extracellular has been discussed by Glenn (1976) who concluded that the most fundamental aspect of these molecules was that they were exported through the cytoplasmic membrane. This can conveniently be extended to molecules that are partially transported through the membrane (*i.e.* membrane-bound) but the final location of these proteins must relate to the structure of the cell wall. Bacteria may be divided into Gram-positive and Gram-negative types which reflects the chemical composition and structure of the cell walls (reviewed by Costerton *et al.*, 1974). Gram-positive bacteria have a relatively simple cell wall largely composed of a thick (200 Å) layer of peptidoglycan bounding the cytoplasmic membrane. This net-like molecule provides the rigidity of the cell wall and may be covered by protein or carbohydrate capsular material. Enzymes that have crossed the membrane may be temporarily restricted by the wall but generally they diffuse into the culture fluid. Some enzymes, on the other hand, are transported across the membrane but remain attached to the inner or outer surface (see Section 31.2.3). Such enzymes are sometimes partially or completely released upon removal of the peptidoglycan with lysozyme in isotonic medium (protoplasting) and have consequently been described as periplasmic. However, since the Gram-positive bacterium does not possess a periplasm this description can be misleading. The location of these enzymes can vary with growth conditions. For example the alkaline phosphatase of *B. licheniformis* appears to be firmly attached to the inner surface of the cytoplasmic membrane in young, exponential phase cells, but as the culture ages it becomes less tightly bound and an increased amount is released upon protoplasting (Glynn *et al.*, 1977). Similarly α-glucosidase in *B. licheniformis* is membrane-bound during exponential phase growth and later released to the culture fluid (unpublished results).

The Gram-negative cell wall is a complex structure comprising a cytoplasmic or inner membrane bounded by a thin layer of peptidoglycan, a periplasmic space and a second or outer membrane. The two membranes are connected across the periplasm at adhesion sites (Bayer, 1979). The Gram-negative equivalent of the extracellular enzyme from a Gram-positive cell is generally located in the periplasm and consequently extracellular proteins *sensu stricto* are rare in Gram-negative bacteria. Nevertheless they do occur particularly in pseudomonads, aeromonads and some enterobacteria. Indeed the enterotoxin of *Vibrio cholerae* is secreted into the surrounding medium. Most exoenzymes are, however, located in the periplasm which appears to play vital roles for cell growth and may account for as much as 20–40% of the total cell mass (Nikaido, 1979). It contains specific amino acid and sugar transport binding proteins and enzymes that degrade macromolecules or low molecular weight organic compounds as a source of nutrients. Moreover, transport proteins and enzymes are located in both the inner and outer membranes of the Gram-negative cell wall providing a system of considerable complexity. Since exoenzymes *sensu stricto* are a rarity in these bacteria they have little commercial application to date, the important exception being pullulanase (a 'leaky' periplasmic enzyme) production from *K. pneumoniae*.

Finally, the fungal cell wall can be likened to the Gram-positive bacterial wall in that it does not possess a periplasm surrounded by a hydrophobic outer membrane. Instead, it largely comprises α-1,3 and β-1,3 glucan and chitin with varying amounts of cellulose and protein (Rosenberger, 1976). It is by no means however a homogeneous mixture of its constituent polymers but appears to be a structured and complex assembly. Fungi elaborate a variety of polysaccharide hydrolases (Table 1). It appears that those enzymes that scavenge the environment for macromolecular nutrients are liberated from the cell but a second class of enzymes have substrates within the wall and are themselves tightly bound to the wall. Like the autolysins of bacteria, these are involved in growth and morphogenesis.

31.2 MECHANISM OF SECRETION

31.2.1 Membrane-bound Ribosomes

Current models for the transport of proteins across membranes derive from early studies with pancreas and liver cells. Electron micrographs of these cells revealed two populations of ribosomes, those attached to the inner surface of the endoplasmic reticulum and those free in the cytoplasm. It gradually became evident that secretory proteins are produced on membrane-bound polysomes and that newly synthesized proteins do not appear in the cytosol but are transferred directly into the lumen of the endoplasmic reticulum. These investigations led to the concept that exported proteins are cotranslationally translocated through the membrane of the endoplasmic reticulum (Palade, 1975).

In bacteria, the dense packing of the ribosomes obscures the visual demonstration of membrane-bound polysomes and evidence for a functional attachment has been sought in other ways. One approach has been to characterize the products from *in vitro* protein synthesis using purified membrane-bound and soluble ribosomes. Membrane-bound polysomes from *Escherichia coli* were shown to be rich in mRNAs for periplasmic and outer membrane proteins while cytoplasmic proteins were derived exclusively from soluble polysomes. This observation has since been extended to several bacteria (Davis and Tai, 1980) and in this laboratory we have shown that synthesis of the extracellular α-amylase and membrane-bound α-glucosidase of *B. licheniformis* occurs exclusively on membrane-bound polysomes. It would therefore seem likely that bacterial proteins are secreted cotranslationally. Smith *et al.* (1977) argued that if this were the case, the growing end of the polypeptide chain could be labelled as it emerged from the cytoplasmic membrane with a compound that could not penetrate the membrane. Demonstration of the other end of the polypeptide attached to the ribosome would confirm the process. *E. coli* spheroplasts were labelled with [^{35}S]acetylmethionyl methylphosphate sulfone which reacts with free amino groups but does not penetrate membranes. The spheroplasts were then lysed and the soluble and membrane-bound polysomes purified by sedimentation on a sucrose gradient. The membrane–polysome complexes contained a considerable proportion of the initial label which was evidently attached to the ribosome through peptidyl-tRNA since it was released by puromycin. Moreover, when protein synthesis was allowed to proceed on these ribosomes, alkaline phosphatase, a periplasmic enzyme, was identified amongst the products. These experiments have been refined and extended and cotranslational secretion by membrane-bound polysomes has been demonstrated in *E. coli* (alkaline phosphatase), *B. licheniformis* (penicillinase), *B. subtilis* (α-amylase) and *Corynebacterium diptheriae* (toxin) (reviewed by Davis and Tai, 1980).

Cotranslational secretion is therefore firmly established in both eukaryotes and prokaryotes but it must be emphasized that it is not exclusive. Post-translational mechanisms (see Section 31.2.4) operate in the biosynthesis of cytochrome *c*, other mitochondrial proteins and chloroplast and peroxisomal proteins (reviewed by Kreil, 1981), and in some bacterial systems, in particular toxin biosynthesis by *V. cholerae* (Nichols *et al.*, 1980).

31.2.2 Signal Hypothesis

It was suggested by Blobel and Sabatini (1971) that membrane-bound polysomes involved in protein secretion in eukaryotes might be attached to the rough endoplasmic reticulum (RER) through the NH$_2$-terminal region of the growing peptide chain. Evidence in support of this was presented by Milstein *et al*. (1972) in their studies on the translation of immunoglobulin light chain mRNA in cell-free systems. A larger precursor form of IgG light chain containing extra amino acids at the NH_2-terminus was synthesized in the reticulocyte lysate but when microsomal membranes were added only the authentic light chain was produced. Similar precursors for various secreted proteins were subsequently described using *in vitro* translation systems (see Harwood, 1980; Kreil, 1981 for complete lists).

These early findings were assembled into the signal hypothesis by Blobel and Dobberstein (1975). It was envisaged (Figure 1) that translation of all mRNAs begins on soluble ribosomes, thus eliminating any requirement for specialized ribosomes or initiation factors. The polypeptide chain would then elongate until some 50 to 80 amino acids had been assembled of which the first 10 to 40, which would be protruding from the larger ribosomal subunit, would 'signal' the beginning of the secretion process. The polysomes would be attracted to the RER by this signal sequence and attach to specific sites on the membrane. In the original formulation it was proposed that the emergence of the signal sequence recruited two or more receptor proteins to form a pore or tunnel in the membrane. The nascent protein was then envisaged to pass through this pore as the polypeptide chain was being extended. After the entire signal sequence had traversed the membrane, it would be removed by a specific protease allowing the remainder of the protein to be discharged through the membrane ultimately to assume its mature form. This model is based on data from a number of systems and since its proposal many aspects have been experimentally confirmed. In addition, specific binding proteins termed ribophorins that bind ribosomes tightly have been found in the RER membrane (Kreibich *et al*., 1978). It is suggested that these proteins are the site of attachment of polysomes engaged in the synthesis of excreted proteins. However, results that implicate specific ribosome binding are still controversial with contradictory results being obtained in different systems. Detailed discussion of the signal hypothesis has been given by Harwood (1980) and Kreil (1981).

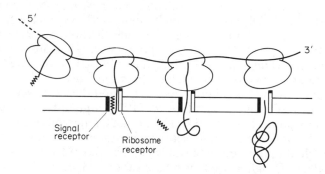

Figure 1 Schematic illustration of the signal hypothesis for the transport of proteins across membranes. As the signal peptide (zig-zag line) emerges from the large ribosomal subunit it recruits receptor proteins in the membrane which form a pore. Signal peptidase cleaves the signal peptide from the growing polypeptide chain allowing the protein to adopt its mature configuration as it is discharged from the membrane. Once translation is complete, the ribosome leaves the membrane and the receptor proteins are free to diffuse in the membrane (Blobel *et al*., 1979)

31.2.2.1 *Signal hypothesis in bacteria*

There is now overwhelming evidence that the signal hypothesis operates in the secretion of proteins from bacteria. Membrane-bound polysomes have been demonstrated (Section 31.2.1) and additional evidence in the form of larger precursors for secreted proteins was soon gathered in several systems, particularly *E. coli*. The demonstration of precursors *in vivo* was initially elusive, presumably because they were very short lived and rapidly processed. Consequently *E. coli* was treated with perturbants of the envelope structure which would allow accumulation of intermediates by inhibition at different stages of biosynthesis. When the membrane proteins of toluene-treated *E. coli* were characterized with antilipoprotein antiserum, two distinct peaks were observed after electrophoresis of the immunoprecipitate. One was identified as the authentic lipoprotein (an outer membrane protein), the other was larger by 18 or 19 amino acid residues but retained all the features of the lipoprotein and was designated prolipoprotein. Phenylethyl alcohol, tyrosylarginylmethyl ester and reduction in membrane fluidity by low temperature have similarly been used to inhibit maturation for the detection of precursor proteins in *E. coli* (Halegoua and Inouye, 1979).

Three other approaches have been used to demonstrate precursors of secreted proteins containing putative signal sequences in bacteria. DNA from specialized transducing phages carrying genes for secreted proteins have been used as templates for coupled transcription and translation. A higher molecular weight (by several thousand daltons) precursor for alkaline phosphatase from *E. coli* has been detected in this way. Precursors for several excreted proteins have been detected by using membrane-bound polysome preparations for protein synthesis and analysis of the products by electrophoresis. Both periplasmic (*e.g.* maltose-, leucine- and arabinose-binding proteins) and outer membrane (phage lambda receptor protein and *ompA* product) proteins of *E. coli* have been shown to possess larger precursor forms in this way (Kreil, 1981). Finally the precursors of the β-lactamases from *E. coli* and *B. licheniformis* have been identified by comparison of the nucleotide sequences of the genes with the amino acid sequences of the proteins (Kroyer and Chang, 1981).

31.2.2.2 Signal sequence structure

A comparison of the NH_2-terminal extensions found in bacterial proproteins with signal peptides of secretory polypeptides from eukaryotic cells reveals no principal differences. They are similar in length (18 to 29 amino acids), in distribution of polar and apolar amino acids and in the nature of the CO_2H-terminal residue. Moreover, in *E. coli* the composition of these polypeptides is similar irrespective of whether the mature protein is located in the inner or outer membrane or the periplasm, which suggests that the peptide acts as a true signal sequence required solely for transport of the protein across the cytoplasmic membrane.

The signal sequences of various proproteins have been compared in detail by Halegoua and Inouye (1979). The peptide extension can be divided into three main sections containing the following general features. The NH_2-terminal portion, section *I*, invariably has a basic character. It is of variable length (10 to 25 Å) and accounts for the differences in the length of individual signal peptides. Nevertheless, section *I* always terminates some 55 Å from the cleavage site. The first basic residue (lysine or arginine) invariably appears at the 16th or 17th position from the cleavage site and marks the start of section *II*, the hydrophobic portion. This length of high hydrophobicity lies between the basic section and the cleavage site and is about 50 Å long. Within section *II*, three subsections are apparent which are separated by glycine or proline residues at about positions 6 and 11. Finally, the amino acid residue at the cleavage site (section *III*) is generally glycine, alanine, serine or cysteine.

31.2.2.3 Function of signal peptide and translocation

Based on the general features of signal peptides described above, its possible mode of action in the translocation process has been proposed (DiRenzio *et al.*, 1978; Halegoua and Inouye, 1979). As the precursor protein is synthesized, the basic section *I* emerges from the ribosome and attaches to the inner surface of the cytoplasmic membrane through ionic interactions (Figure 2). The hydrophobic section *II* is then progressively inserted into the membrane by hydrophobic interaction with the lipid bilayer. Eventually the cleavage site (section *III*) is exposed to the outer surface of the membrane and cleavage occurs. The length of sections *II* and *III* (about 53 Å) is sufficient to extend across the membrane. The 'loop' model described above differs from the original signal hypothesis for eukaryotic cells in that there is no requirement for the association of membrane receptor proteins to form a tunnel. However, no direct evidence for either model is currently available.

The energy involved in the translocation of the protein across the membrane has been reviewed by Davis and Tai (1980). It has been suggested that chain elongation on the ribosome might push the peptide through the membrane but this would require that the ribosome is firmly attached to the membrane. Although this appears to be the case in eukaryotic cells, when membrane-bound polysomes from *E. coli* were treated with puromycin, 75 percent were released from the membrane. Since puromycin releases ribosomes from nascent polypeptide chains by formation of peptidylpuromycin this suggests that ribosomes are attached to the membrane solely by the nascent peptide chain. A second possibility is that the folding of the protruding polypeptide

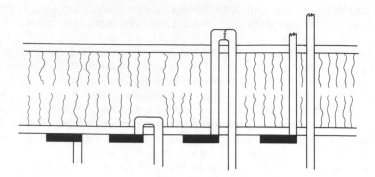

Figure 2 Loop model for the translocation of secretory proteins across the membrane. Solid portions represent the basic region of the signal peptide (section *I*), the following blank portion is the hydrophobic central region (section *II*). The cleavage site (section *III*) is followed by the protein itself. Section *I* attaches to the inner surface of the membrane and the hydrophobic region is progressively inserted into the membrane. When the cleavage site emerges it is attacked by signal peptidase allowing the protein to be vectorially transported (Halegoua and Inouye, 1979).

on the outside of the membrane draws the rest of the chain through the membrane. Finally, the favoured scheme is that the energy may be derived from the membrane itself, possibly from proton motive force.

31.2.2.4 *Processing of the precursor*

In the original hypothesis, the precursor protein is processed by removal of the signal peptide during or immediately after the synthesis of the protein. Studies in which precursor protein has been accumulated in *E. coli* by treatment with phenylethyl alcohol have shown that processing is not required for the translocation process itself, but is necessary for the proper assembly of outer membrane proteins into their correct location (Halegoua and Inouye, 1979).

The diversity of amino acid residues adjacent to the cleavage site in both eukaryotic and pro-karyotic precursor proteins suggests that there may be a multiplicity of 'signalase' processing enzymes. Certainly very substantial evidence exists that cleavage of the signal peptide from secretory proteins in eukaryotes is carried out by peptidase(s) located in the RER and most prob-ably occurs during translation prior to chain completion (reviewed by Harwood, 1980). In *E. coli*, processing activity associated with an outer membrane has been described for alkaline phospha-tase but for the procoat protein of bacteriophage f1 the activity was in the cytoplasmic mem-brane. This discrepancy has recently been clarified by the purification and characterization of indistinguishable leader (signal) peptidases from the inner and outer membranes of *E. coli* (Zwinzinski *et al.*, 1981).

31.2.2.5 *Gene fusion studies*

A novel approach to the study of protein localization in *E. coli* has used gene fusions in which the NH$_2$-terminal portion of β-galactosidase is replaced by the NH$_2$-terminal sequence of another protein. Such hybrid proteins retain β-galactosidase activity and can thus be located in the cell. Silhavy *et al.* (1979) have used this technique to obtain strains of *E. coli* in which the *lacZ* gene (β-galactosidase) is fused to three loci from the maltose (*mal*) utilization operon: *malF* which codes for a probable transport protein located in the cytoplasmic membrane, *malE* which codes for the periplasmic maltose binding protein and *lamB* which codes for the outer membrane protein which is essential for the penetration of maltodextrins into the cell and also serves as the recep-tor for phage λ. Gene fusions between *malF* and *lacZ* which contained a substantial portion of *malF* produced a hybrid protein that was tightly bound to the cytoplasmic membrane. However, *malF* – *lacZ* fusions that contained only a small amount of *malF* DNA produced a hybrid pro-tein that was located in the cytoplasm, the normal site for β-galactosidase. Similarly *lamB* – *lacZ* fusions containing only a small portion of the *lamB* gene produced a hybrid protein that was located in the cytoplasm but when a substantial portion of *lamB* DNA was present, the protein was transported to the outer membrane. However, with *malE* – *lacZ* fusions, the hybrid protein was located in the cytoplasm if a small portion of the binding protein was present but when a large portion of *malE* was fused to *lacZ* the resultant protein was not transported to the periplasm. Instead the protein was found to be associated with the cytoplasmic membrane and its accumulation there had deleterious effects resulting in cell death since the export of other proteins to the periplasm became blocked. These strains (and the *lamB – lacZ* fusions) show a maltose sensitive phenotype since transcription of the fused genes is induced by maltose. Mutations in the signal sequence can now be selected by their maltose resistant phenotype and hybrid proteins in such strains are located in the cytoplasm. Analysis of the signal sequences in these export-defective strains showed replacement of single apolar amino acids in section *II* of the signal peptide by an acidic or basic residue. These findings clearly demonstrate the importance of the central hydrophobic region of the signal peptide.

The similar maltose sensitive phenotype of *malE – lacZ* and *lamB – lacZ* has been explained by proposing that the *lamB* gene product (normally located in the outer membrane) is initially located in the inner membrane probably following dissociation of the ribosome from the membrane shortly after vectorial transport has been initiated. The protein is then transported to the outer membrane *via* the fusion points between the membranes. By adding only a little *lamB* DNA to β-galactosidase there was insufficient information for the correct translocation to take effect and the hybrid became lethally jammed. In strains containing a higher proportion of *lamB* DNA, the β-galactosidase was effectively transported to the outer membrane. These gene fusion studies indicate that information directing a protein to a non-cytoplasmic location is contained within the gene that codes for that protein. Moreover, it would seem that any protein can be directed to certain non-cytoplasmic locations if the correct genetic sequence is present.

31.2.3 Membrane Associated Intermediates

Membrane associated intermediates seem to be intimately involved in the secretion of some proteins by Gram-positive bacteria. The most extensively studied system is penicillinase synthesis by *B. licheniformis* (Lampen, 1978). This enzyme can be recovered in essentially two forms, as an extracellular hydrophilic protein of molecular weight around 29 000 and as a hydrophobic membrane-bound protein of 33 000. The hydrophobicity of this form was originally attributed to a covalently linked phospholipid moiety but, after much controversy, it would seem that it is glyceride attached by a thioethyl linkage to cysteine that confers this character (Smith *et al.*, 1981). Though the membrane-bound form can be converted to exoenzyme, kinetic studies suggest that it is not an obligatory intermediate. Recently, chain completion on membrane-bound ribosomes has revealed a third precursor form (molecular weight 36 000). The extra sequence on this molecule is evidently a hydrophobic signal sequence since the precursor was retained by decylagarose. Evidently secreted proteins can go through successive cleavages like the preproteins that have been recognized in eukaryotic cells as precursors of protein hormones.

Other, unidentified proteins have been observed to occur as highly hydrophobic intermediates (Nielson *et al.*, 1981; Smith *et al.*, 1981) and it appears that addition of glycerol esterified with long chain fatty acids may be a widespread phenomenon that confers upon polypeptide chains a hydrophobic domain sufficient to ensure attachment to the cell membrane.

31.2.4 Alternative Export Mechanisms; Post-translational Secretion

Cotranslational secretion of membrane-bound and extracellular proteins predicts that the protein will not occur within the cytoplasm of the cell. Several lines of evidence support this proposal for various extracellular enzymes including the amylases and proteases of *Bacillus* strains (reviewed by Priest, 1977) but there are exceptions. Considerable controversy has arisen over the biosynthesis of a phage coat protein localized in the cytoplasmic membrane of *E. coli*. Both co- and post-translational insertion and processing have been proposed for the precursor of M13 coat protein. However, *in vivo* studies have revealed that procoat is synthesized on soluble rather than membrane-bound polysomes and subsequently inserted into the cytoplasmic membrane where it is converted to coat protein (reviewed by Wickner, 1979).

Cholera toxin is an extracellular protein but can be detected in the cytoplasm of *Vibrio cholerae*. *In vitro* translation studies revealed that it is synthesized in larger precursor form solely by free ribosomes. It would seem that the toxin is initially located in the cytoplasm and exported post-translationally. Similarly various colicins and *E. coli* enterotoxin can be detected in the cytosol and may be synthesized on soluble ribosomes and subsequently exported (Nichols *et al.*, 1980).

On the basis of the synthesis of M13 coat protein the membrane trigger hypothesis was formulated which emphasized the ability of a membrane to 'trigger' the folding of a polypeptide chain into a membrane associated form. By analogy with the signal hypothesis, an NH_2-terminal extension on the protein, the leader peptide, modifies the folding pathway of the protein such that a configuration integrally associated with the membrane is formed. The leader sequence may be proteolytically removed at a later stage to allow the protein to function, and the protein itself can subsequently be released from the membrane to become extracellular (Wickner, 1979).

The signal hypothesis can also accommodate post-translational secretion after synthesis on free ribosomes. Blobel *et al.* (1979) suggest that after synthesis in the cytoplasm an excreted or membrane-bound enzyme would form a configuration in which the signal sequence is exposed. The signal sequence would thus be able to interact with membrane receptor proteins to cause their association to form a pore, as in the original hypothesis, through which the protein traverses the membrane in an unfolded form. This post-translational translocation and indeed cotranslational translocation could be interruped by an internal stop sequence in the protein which would result in a membrane-bound location.

31.2.5 Aspects of Enzyme Secretion in Fungi

It is important to consider briefly those aspects of enzyme secretion in fungi that differ from the simpler bacterial systems. Much of the information on enzyme secretion in eukaryotes has been obtained from studies of specialized animal secretory cells and, since fungal cells are eukaryotic, there has been a tendency to suggest that the pathway of protein secretion in both is similar. Since there has been little systematic study of protein secretion in fungi, the secretory pathway in animal cells will be outlined but it must be emphasized that many of the basic subcellular bodies involved in protein secretion in these cells are frequently absent from fungal cells (Ghosh, 1978).

Although the transport of the protein across the membrane is similar in prokaryotic and eukaryotic cells, in the latter the molecule is discharged into the cisternal space of the RER. The molecules are then transported to the smooth endoplasmic reticulum from which they are released in vesicles. The vesicles fuse with the 'cis' part of the Golgi apparatus in which post-translational modification may occur and are discharged from the 'trans' region of the Golgi apparatus in vesicles. These undergo concentration (condensing vacuoles) to form zymogen granules which, in a Ca^{2+} dependent fusion with the plasma membrane, release the enzyme from the cell.

The sequence of post-translational events in yeast glycoprotein secretion has recently been determined with the aid of temperature-sensitive mutants that affect the secretory apparatus. The results indicate a similar sequence of events to the animal cell scheme. Secretory proteins enter the RER where the initial steps of glycosylation occur. The material is transferred to Golgi-like structures in which further glycosylation occurs before the proteins are packaged in vesicles that are transported to the bud where they fuse with the plasma membrane (Novick *et al.*, 1981). Cytoplasmic vesicles have been noted in the apical tips of some filamentous fungi and the subapical zone has been shown to contain vesicles, RER, smooth surfaced cisternae and some Golgi apparatus. Frequently a specialized structure (Spitzenkörper) comprising a collection of vesicles and tubes occurs. On the other hand, an ultrastructural study of *Trichoderma viride* revealed no recognizable RER or Golgi structures suggesting that secretory pathways vary considerably not only between fungal and mammalian cells but also within the fungi.

31.3 REGULATION OF EXTRACELLULAR ENZYME SYNTHESIS

Many extracellular and periplasmic enzymes have no cellular substrate and it would appear that they have evolved as 'scavenger enzymes' which degrade polymeric material in the environment to provide the organism with assimilable nutrients. Their regulation will therefore be related to the control of endocellular, catabolic enzymes. However, some exported enzymes have cellular substrates (for example proteases, nucleases and cell wall lytic enzymes) and have been implicated in various cellular processes. This complicates their regulation and since most commercial enzymes conform to the first category, this chapter will deal almost exclusively with the catabolic exoenzymes.

31.3.1 Regulation of Protein Synthesis

Since the formulation of the operon concept for the regulation of lactose metabolism in *E. coli*, there has been a dramatic increase in our understanding of genetic control in prokaryotes and eukaryotes. An attempt to outline briefly our current notions in this area will be given since it must relate to the control of extracellular enzyme synthesis.

The regulation of protein synthesis in bacteria is principally exerted at the transcriptional level and groups of related structural and regulatory genes comprise operons that are under common control. These operons have a single promoter site to which RNA polymerase binds prior to initiating transcription. The frequency at which the genes are transcribed can therefore be controlled by a regulatory locus placed between the promoter and the structural genes. Negative control defines a regulatory locus that allows free passage of RNA polymerase from promoter to structural genes but is progressively closed by a specific protein repressor molecule. Positive control, on the other hand, comprises a regulatory region that by itself is closed, but is progressively opened by binding a specific activator protein. These two modes of regulation are known to control both inducible (catabolic) and repressible (anabolic) operons in prokaryotes (for review see Goldberger, 1979). For example, lactose and maltose utilization in *E. coli* are governed by negative and positive control respectively and the biosynthesis of several amino acids including lysine and arginine is regulated by negative control. Genetic evidence suggests that cysteine biosynthesis may be regulated by an activator (Savageau, 1979). In these cases, the regulator proteins, whether they be repressors or activators, are synthesized constitutively from their own promoter and are present in virtually constant concentration within the cell (classical regulation). It is also possible for the structural gene of the regulatory protein to lie within the operon itself. In such cases (autogenous regulation) the concentration of regulatory protein within the cell changes according to the level of induction or repression of the operon and results in varying functional effectiveness depending on the system. For example, in negatively controlled catabolic operons since the concentration of repressor will be amplified alongside the enzymes of the metabolic pathway during induction, this tends to buffer the system against fluctuations in the external concentration of inducer, and a constant presence of inducer is necessary to efficiently induce the operon. Models can be constructed for autogenous regulation of operons of any type; negative or positive, inducible or repressible and such systems have been extensively compared and reviewed by Savageau (1979).

Transcriptional control can also be exerted by varying the affinity of RNA polymerase for the promoter since the rate of transcription appears to be dependent on the rate of RNA polymerase binding. Some promoters are 'stronger' than others and different forms of RNA polymerase may recognize different classes of promoter (Travers *et al.*, 1981 and see Section 31.3.5.1). Moreover, some promoters require additional activator molecules to function effectively (Sections 31.3.4 and 31.3.5.1).

Three lines of evidence suggest that extracellular enzyme synthesis, at least in bacilli, is largely regulated at the transcriptional level. Antibiotic inhibition studies suggest that transient repression of amylase synthesis by glucose and of protease synthesis by amino acids in *B. amyloliquefaciens* operates at transcription. Second, hybridization analysis of total mRNA in *B. amyloliquefaciens* indicates an increase late in the exponential phase that is almost entirely due to a 25-fold increase in exoprotein mRNA (Brown and Coleman, 1975). It should be noted, however, that such an increase may be due to increased stability of the mRNA. Finally, genetic studies of the amylase operon in *B. subtilis* have revealed two loci; *amyE*, the structural gene for α-amylase and *amyR* a closely linked regulatory element. The *amyR* may be the promoter for *amyE* and consequently these findings suggest that promoter recognition and binding is a regulatory element in amylase synthesis (reviewed by Yamane and Maruo, 1980). Nevertheless, it is possible that part or all of *amyR* may be transcribed and have some translational control function.

Transcriptional control systems can only function effectively if the product (mRNA) is an unstable molecule that is rapidly eliminated from the cell thereby halting protein synthesis once the operon is switched off. Although this is generally the case in prokaryotes (average half-life for mRNA in *E. coli* is around 2.5 min at 37 °C) there are several examples of relatively stable mRNA (10–50 min). This aspect of translational control centres on excreted proteins and will be discussed in Section 31.3.6.

Genetic regulation in the eukaryotic cell is considerably more complex than that in bacteria. The primary transcript from the eukaryotic chromosome is not mRNA but high molecular weight RNA ranging from 2000–20 000 nucleotides in length. This heterogeneous nuclear RNA (hnRNA) is post-transcriptionally 'spliced' and intervening sequences (introns) removed to give rise to a translatable message. Although some mRNAs are derived from original transcripts of similar size to the finished mRNA, the majority of mRNAs are probably derived from transcription units considerably larger than the mRNA itself (reviewed by Crick, 1979). Moreover, both ends of the mRNA are modified. A methylated oligonucleotide structure (cap) is added to the 5'-end (Bannerjee, 1980) probably at the start of transcription. This appears to promote mRNA stability and is also required for efficient translation. The 3'-end contains a long stretch of adenine residues [poly(A)] which are added post-transcriptionally and are probably involved in stability and transport of the mRNA from the nucleus.

Two stages of gene expression have been extensively examined in eukaryotes for evidence of regulation, transcription and processing. The facility to clone unique DNA sequences in *E. coli* has enabled the production of DNA hybridization probes which can be used to measure the level of particular RNA molecules in cells. In this way the yeast cytochrome *c* gene has been shown to be under transcriptional control during catabolite repression by glucose and the magnitude of regulation provided by modulating mRNA levels was sufficient to account for observed changes in protein concentration (Zitomer *et al.*, 1979). Further evidence for transcriptional control in fungi has been reviewed by Arst and Bailey (1977). But it may be that processing of hnRNA is not automatic and the selection of sequences for export to the cytoplasm as mRNA may be an important mode of control. There are indications that mRNA populations may change extensively while the hnRNA is more or less constant suggesting that some genes are controlled by selection of sequences for processing and/or nucleocytoplasmic transport (for reviews see Darnell, 1979; Lewin, 1980). The control of protein synthesis in eukaryotes is therefore complicated and operates at several points covering both transcription and translation. Nevertheless it seems that those enzymes from fungi in which the applied microbiologist might be interested, are probably regulated at the transcriptional level. This generalization excludes enzymes associated with developmental changes for which it is likely that translational controls, particularly stabilization and destabilization of mRNA, are important.

31.3.2 Induction of Exoenzymes

Inducible catabolic enzymes are typified by the enzymes of the lactose utilization operon of *E. coli* and are synthesized at a barely detectable basal level in the absence of inducer. The inducing molecule (substrate, by-product from the metabolic pathway in question, or a product) interacts with the regulatory molecule, transcription is increased by removal of repressor or activation, and there is a dramatic amplification in the rate of protein synthesis from the operon. When the inducer is removed, transcription returns to the constitutive basal rate.

The efficient regulation of extracellular enzymes which digest polymeric substrates presents the microorganism with two problems which are not encountered in the regulation of intracellular enzymes. Firstly, for induction to occur the cell must be able to recognize the presence of a large and potentially unassimilable substrate. Secondly, for the enzyme to be secreted at appropriate rates into environments that may vary widely in their effects on the activity and stability of the enzyme, synthesis must be coupled to the external specific activity of the enzyme. It is consequently generally assumed that extracellular enzymes are secreted at a constitutive basal rate. In the presence of substrate, low molecular weight product(s) are produced that enter the cell and effect induction. However, to limit the secretion of exoenzyme, the accumulation of high concentrations of products can lead to 'self catabolite repression' (Tsuyumu, 1979). The scheme is supported by several physiological observations in a variety of systems including pectic acid lyase synthesis in *Erwinia carotovora*, which is induced by unsaturated digalacturonic acid, and α-amylase synthesis in some strains of *Bacillus licheniformis* and *B. stearothermophilus*, which is induced by maltooligosaccharides particularly maltotetraose. The cellulase complex of enzymes is induced in *Trichoderma reesei* by cellulose but low molecular weight products of cellulose degradation are poor inducers. It is thought that such molecules may be rapidly degraded to glucose and thus cause 'self catabolite repression' since sophorose, a β-1,2-disaccharide of glucose is less rapidly hydrolysed and induces the cellulase complex strongly. There are numerous reports of the necessity for polymeric substrate in the medium for maximum enzyme secretion by a variety of organisms but in most instances the low molecular weight effector molecules have not been identified (reviewed by Priest, 1977; Ingle and Erickson, 1978).

The basal level of inducible exoenzymes appears to vary between a barely detectable amount that is increased many thousand fold on induction, to a relatively high level that is best described as partially constitutive and which increases little upon induction. Many amylases from *Bacillus* strains are secreted in the absence of α-glucan but maximum yields are only obtained when the bacteria are grown on complex media containing starch or some hydrolysis product thereof, and this has been referred to as induction. However, lack of catabolite repression may be at least partially responsible for these observations and in the absence of rigorous experimentation, preferably using gratuitous inducers, it is difficult to distinguish catabolite derepression from induction (Ingle and Erickson, 1978). Finally, some exoenzymes are constitutive. The α-amylases of *B. amyloliquefaciens*, most strains of *B. licheniformis* and *B. subtilis* are synthesized irrespective of the growth medium and there is no firm evidence for induction by hydrolysis products from starch.

The genetic and molecular background to the induction of extracellular enzymes in bacteria and fungi is largely unknown. Nevertheless, there are no *a priori* reasons to think that the control of these enzymes should be markedly different from the mechanisms governing intracellular enzymes (reviewed by Priest, 1983). The most extensively studied inducible exoenzyme is the penicillinase of *B. licheniformis* (Imsande, 1978). Both the membrane-bound and extracellular forms of the enzyme derive from the same chromosomal gene (*penP*) which is controlled by a closely linked repressor gene (*penI*). These genes have been cloned into phage and plasmid vectors and recombinant plasmids have been used for complementation studies. These tests have confirmed that regulation of penicillinase biosynthesis is mediated through a repressor (Imanaka *et al.*, 1981). However, a second regulatory gene (effector or antirepressor) has also been implicated.

31.3.3 End-product Repression

It was mentioned in the last section that high external concentrations of hydrolysis products from the action of exoenzymes on their substrates can lead to 'self catabolite repression'. This phenomenon has long been recognized in protease secretion where it has been termed end-product repression.

Protease secretion in bacteria and fungi is usually constitutive but repressed by single or combinations of amino acids (Keay *et al.*, 1972). In *B. megaterium* isoleucine and threonine repress efficiently. Inhibitor studies with *B. amyloliquefaciens* suggest that the repression of protease synthesis by high concentrations of amino acids acts at the transcriptional level (Both *et al.*, 1972) but the molecular details are unknown.

An interesting variation is the alkaline protease, secreted by *Neurospora crassa*. This enzyme is induced when protein is supplied as the sole source of carbon, nitrogen or sulfur but repressed by high concentrations of amino acids. Induction is not a response to general starvation however, since absence of other metabolites does not lift the repression. Hanson and Marzluf (1975) have suggested a model for this system in which one of three positive control signals turns on protease transcription by interacting with individual operator regions which lie adjacent to the protease gene on the chromosome. These positive regulators are repressed by NH_4^+, cysteine and glucose but interaction of any one at its operator site is sufficient to activate transcription of the gene. This model does not take into account induction of the protease by protein, an interesting feature of this system.

It should be noted that enzymes other than proteases are subject to end-product repression, in particular various DNAases and RNAases are inhibited by high phosphate concentration as is alkaline phosphate (Priest, 1977).

31.3.4 Catabolite Repression

Catabolite repression is the permanent repression of inducible or constitutive enzyme synthesis that occurs in most microorganisms when growing rapidly on a readily utilized carbon source (*e.g.* glucose or glycerol). Moreover, if such a carbon source is added to a culture growing slowly on a poor carbon source there is a severe, transient repression of catabolite sensitive operons which subsequently assume their catabolite repressed rate of expression. Catabolite and transient repression play a major role in the regulation of extracellular enzyme synthesis in most, if not all, microorganisms (see also Section 31.3.2) and are responsible for the inhibition of proteases and nucleic acid hydrolysing enzymes as well as most carbohydrases. For example α-amylase and protease synthesis in *B. licheniformis* and *B. subtilis* is repressed by glucose and mannitol, while α-amylase from *B. amyloliquefaciens* is repressed by glucose (Ingle and Boyer, 1976). Extracellular enzymes from *Vibrio parahaemolyticus*, *Pseudomonas maltophilia*, *Staphylococcus aureus* and streptomycetes are similarly controlled by catabolite repression (Priest, 1977). In eukaryotes, the cellulases of *Trichoderma* species and *Myrothecium*, the β-glucanases of *Penicillium*, and protease synthesis by *N. crassa* are subject to catabolite repression (Eveleigh and Montenencourt, 1979).

Despite the widespread occurrence of catabolite repression, with the exception of *E. coli* the molecular details are poorly understood. In *E. coli* and other enterobacteria, adenosine 3',5'-cyclic monophosphate (cAMP) plays a central role in catabolite repression. Together with its receptor protein (CRP) sometimes referred to as the cAMP acceptor protein (CAP), the cAMP/CRP complex activates those operons (about 30 in *E. coli*) involved in the transport and utilization of fuel sources. Without this activation, even in the presence of inducers, transcription of catabolite sensitive operons is reduced to a very low level.

The CRP has two distinct domains, one that binds cAMP and a second that binds to DNA. In the presence of cAMP the protein undergoes an allosteric transition into an active conformation that binds to sites within promoters enabling RNA polymerase to bind and initiate transcription. The RNA polymerase binding site is some 30–50 nucleotides distal to the CRP binding region and the relevant DNA sequences in the *lac*, *ara* and *gal* operons of *E. coli* have been determined (Botsford, 1981).

During growth of *E. coli* on a rapidly metabolized carbon source, the intracellular concentration of cAMP is low and consequently the superfluous catabolic enzymes are not transcribed. There is evidence that the promoters of operons controlled by cAMP form a heirarchy with different affinities for the cAMP/CRP complex. Thus as the cell increases its cAMP concentration in response to carbon starvation, the cAMP concentration rises and more catabolic operons are activated in readiness for the relevant inducer molecules (Alper and Ames, 1978). Mutants of *E. coli* that lack CRP or adenyl cyclase (the enzyme that synthesizes cAMP from ATP) are unable to grow on most carbon sources except glucose or gluconate. However, several lines of evidence suggest that cAMP is not solely responsible for catabolite repression in *E. coli*, in particular a group of mutants has been isolated that is resistant to catabolite repression but contains a low level of cAMP. However, in some oxidative bacteria including *Pseudomonas aeruginosa* and *Pseudomonas putida*, cAMP is present but does not vary in concentration in response to growth conditions and does not appear to be involved in catabolite repression (Phillips and Mulfinger, 1981). Moreover, despite intensive searches, cAMP has not been detected in several *Bacillus* and *Lactobacillus* species although cyclic 3',5'-guanosine monophosphate has been found in *B. licheniformis* and *B. megaterium* (Botsford, 1981). The amounts are very small however, only a few molecules per cell, and it seems unlikely that it has any physiological significance. Certain highly phosphorylated nucleotides (adenosine tetraphosphate and related molecules) have been excluded from any major role in catabolite repression in *B. subtilis* since mutants, defective in the synthesis of these molecules behave as wild type in the presence of glucose (Dowds *et al.*, 1978). Although catabolite repression is a common phenomenon in prokaryotes, much remains to be discovered about the apparent diversity of the molecular mechanisms that have evolved.

The situation in eukaryotic microorganisms is no better. Cyclic AMP is postulated to operate in these organisms through activation of protein kinases which control the phosphorylation of various critical proteins and thereby modulate the activity of the proteins. Since cAMP is intimately involved in catabolite repression in some bacteria, yeast was examined in this connection. *Saccharomyces cerevisiae* (*carlsbergensis*) growing on galactose was found to contain 70% more cAMP than when it was growing on glucose and *Kluyveromyces marxianus* (*Saccharomyces fragilis*) growing on lactate contained more cAMP than when it was growing on glucose. However, there was no correlation between cAMP levels in *S. cerevisiae* and sensitivity of invertase to glucose repression, and the inducible *N*-acetylglucosamine catabolic pathway of *S. cerevisiae* could not be protected from catabolite repression by exogenous cAMP. Firm evidence that cAMP is not involved in catabolite repression in *S. cerevisiae* has recently been provided by Matsumoto *et al.* (1982) who isolated mutants able to grow on cAMP as adenine source (thereby ensuring penetration of the cell by the nucleotide). Catabolite repression of galactokinase synthesis in these mutants could not be relieved by cAMP. We therefore remain rather ignorant of the molecular mechanisms of this widespread phenomenon which occurs in most, if not all, microorganisms.

31.3.5 Patterns of Exoenzyme Synthesis

An aspect of exoenzyme synthesis that has been the cause of considerable controversy is the time course of enzyme secretion during batch culture growth. Growth of microorganisms in batch culture can be divided into the exponential growth phase or trophophase and the late exponential/stationary phase or idiophase. Extracellular enzyme secretion is generally associated with either the trophophase or the idiophase. The secretion of proteases and amylases by *B. subtilis* and *B. amyloliquefaciens* occurs almost exclusively after the end of exponential growth and the industrial microbiologist arranges the fermentation conditions to provide extensive growth followed by an extended idiophase, before sporulation is initiated and exoenzyme accumulation declines. In most strains of *B. licheniformis* and *Aspergillus oryzae*, on the other hand, amylase synthesis accompanies growth and the extended stationary phase is not required for maximum enzyme yield. Similarly, the various exoenzymes of *Bacillus polymyxa* (laminarinase, protease, amylase and xylanase) are secreted during growth and the rate of synthesis declines as the culture enters the stationary phase. It should be noted that these patterns can sometimes be changed by mutation or by variation in the growth medium. Mutation to increased yields often leads to synthesis earlier in the growth cycle and secretion of neutral protease by *B. megaterium* occurs throughout the growth cycle in minimal medium, but in complex medium it is repressed during growth and only synthesized in the stationary phase prior to sporulation. These few examples typify the patterns observed in various prokaryotes and eukaryotes for which there is a vast literature which has been reviewed elsewhere (Coleman *et al.*, 1975; Glenn, 1976; Priest, 1977; Ingle and Erickson, 1978).

An important implication of this variation in exoenzyme synthesis during batch culture growth concerns product formation during steady state growth in chemostat culture. Microorganisms growing in the chemostat are held in exponential phase and consequently only those products that are secreted during exponential growth can be maintained at constant levels in the culture fluid. The various exoenzymes of *B. licheniformis* and *B. stearothermophilus* are in this category and have been successfully studied in chemostats where evidence for induction and catabolite repression has been obtained (Wouters and Buysman, 1977; Davis *et al.*, 1980). It has been pointed out on various occasions that this technique should be valuable but has been used infrequently. This is presumably because most exoenzymes are idiophase products and, as such, it is impossible to obtain constant yields during steady state growth however low the dilution rate. Amylase and protease secretion by *B. amyloliquefaciens* is typical in that the initial yield is high (after starting with batch culture growth) but once continuous operation is initiated the yield fluctuates and ultimately declines. This can be partially overcome by the two-stage chemostat. This comprises a growth vessel into which fresh medium enters and from which culture is fed into a second vessel. The transferred culture is effectively 'shifted down' and it mimics entry into stationary phase in batch culture. Such systems have been successful for exoenzyme synthesis from *B. subtilis* and may have some commercial value (Fencl *et al.*, 1972).

Various theories have been suggested to explain the repression of exoenzyme synthesis in some bacteria during exponential phase growth of which two have received considerable attention.

31.3.5.1 *RNA polymerase modification*

Coleman *et al.* (1975) have suggested a model for the control of exoenzyme secretion that derives from their studies on *B. amyloliquefaciens* but has recently been extended to other bacteria including *Staphylococcus aureus*. Two separate effects are considered in this scheme. First, it is suggested that at the end of exponential growth there is a change in the transcriptional selectivity of RNA polymerase such that exoenzyme promoters are preferentially recognized. Second, the turnover of stable (ribosomal) RNA that occurs as the organism enters stationary phase provides a large 'pool' of nucleotide precursors which enhances transcription by the modified polymerase.

This change in the transcriptional selectivity of RNA polymerase was originally considered to be due to structural or compositional changes in the enzyme. Such modifications have been noted in *B. subtilis* and are thought to be partially responsible for gene selection during differentiation (Doi, 1977). However, it is unlikely that such changes are involved in derepression of all exoenzyme since mutants blocked at the initial stages of sporulation generally fail to modify their RNA polymerase and yet secrete normal quantities of amylase (reviewed by Priest, 1977). However, a recent analysis of the serine protease gene of *B. subtilis* has shown that the promoter sequence is recognized by RNA polymerase combined with a sigma factor associated with transcription of sporulation-specific genes (Wong *et al.*, 1984).

Nevertheless as Ingle and Erickson (1978) have pointed out these observations require more rigorous experimentation in order to reach any firm conclusion concerning RNA polymerase modification and derepression of exoenzymes during the idiophase as a general phenomenon.

RNA polymerase selectivity in *E. coli* can be modified by a variety of small molecules including fMet-tRNA, adenosine and guanine nucleotides (reviewed by Travers *et al.*, 1981). For example, *in vitro* experiments have demonstrated that ATP inhibits transcription of the *lac* operon but stimulates stable RNA synthesis. In both cases the effect can be reversed by 5'-ADP, 5'-AMP or adenosine tetraphosphate. At the end of exponential growth there are considerable changes in the concentrations of various nucleotides in bacilli (Hanson, 1976) and it may be that these are involved in the modulation of RNA polymerase activity such that exoenzyme promoters are preferentially selected.

Returning to the hypothesis for the regulation of extracellular enzyme synthesis proposed by Coleman *et al.* (1975), it was suggested that the enlarged pool of RNA precursors evident in *B. amyloliquefaciens* at the end of exponential growth could stimulate exoenzyme synthesis. There is no firm evidence for this proposal and current notions consider RNA polymerase binding to be the rate-limiting stage in transcription in bacteria so long as the supply of precursors is not limiting.

31.3.5.2 *Catabolite repression*

The alternative explanation for exoenzyme synthesis being restricted to the early stationary phase of growth involves catabolite repression (Schaeffer, 1969). Thus, during exponential growth the cells would be strongly catabolite repressed and exoenzyme synthesis minimal. Upon entering stationary phase, catabolite repression would be lifted and synthesis ensue. Although catabolite repression is undoubtedly involved in the control of exoenzyme synthesis in most microorganisms, it is difficult to reconcile it with this specific pattern of enzyme secretion. For example, amylase synthesis responds similarly to catabolite repression in both *B. licheniformis* and *B. subtilis* and yet, in the former, amylase is secreted during exponential phase while in the latter only during stationary phase. Consequently, although there is considerable experimental evidence for this scheme, until the molecular details of catabolic repression are more completely understood it will remain unproven.

31.3.6 Translational Control of Exoenzyme Synthesis in Bacteria

Although the primary control of enzyme synthesis is transcriptional (see Section 31.3.1) there is considerable evidence that relatively stable mRNAs for exported proteins offer a measure of translational control. One of the original observations in this context concerned late exponential phase cultures of *B. amyloliquefaciens* which had been washed and resuspended in fresh medium. Exoenzyme secretion by these cells was resistant to inhibition of transcription by rifampicin or actinomycin D for long periods (30–90 min) but was rapidly halted by translation inhibitors such as chloramphenicol. This suggested that *de novo* synthesis was taking place from a preformed message and similar findings were subsequently reported for a variety of bacilli and other bacteria including *Clostridium* and *Pseudomonas* (Priest, 1977). These findings are open to criticism since such inhibitor studies are notoriously non-specific, but the use of Ura⁻ mutants to block mRNA synthesis by uracil starvation substantiates this aspect of exoenzyme mRNA metabolism. Two explanations have been offered for these observations. Elliott's group (O'Connor *et al.*, 1978) favour the existence of a large pool of mRNA of normal stability. This accumulates in late exponential phase cells and is sufficient to sustain translation of exoenzymes in the absence of transcription for the observed period. Although this model accounts for many aspects of extracellular enzyme synthesis there are inconsistencies; in particular the mRNA decay kinetics following the inhibition of transcription are not compatible with a pool of mRNA which has accumulated due to limitation of translation sites. Indeed, the decay kinetics support the alternative explanation for the continued translation of exoenzymes in the absence of transcription, stable mRNA.

Although, as originally conceived, transcriptional regulation required very unstable mRNA, there are now many examples of relatively stable messages in prokaryotes. A recent analysis of the functional half-lives for mRNAs in *E. coli* gave estimates varying between 40 s and 20 min and the more stable messages were associated with cell envelope proteins (Pederson *et al.*, 1978). Indeed, the remarkable stability of the lipoprotein mRNA in *E. coli* has allowed its purification and characterization. The molecule comprises 322 bases of which some 88 bases are not translated. The nucleotide sequence of the 3'-end contains very stable stem and loop structures and it is thought that these secondary structures confer the stability to the molecule (Pirtle *et al.*, 1978).

Similarly the half-lives of some extracellular protein mRNAs from Gram-positive bacteria are longer than for cytoplasmic proteins. For example, the half life for extracellular protease mRNA from *B. megaterium* has been estimated at 6 to 7 min and for amylase mRNA from *B. licheniformis* around 8 min. This compares with average cytoplasmic mRNA half-lives of 4 and 2.5 min respectively (Priest, 1983). Moreover, a highly stable mRNA (half-life 58 min) has been implicated in toxin biosynthesis in *Clostridium perfringens* (Labbe and Duncan, 1977).

However, not all messages for extracellular proteins are so stable; penicillinase mRNA in *B. licheniformis* has a relatively short half-life of about 4 min (Imsande, 1978) and the half-life of levansucrase mRNA in *B. subtilis* is 2.1 min (Petit-Glatron and Chambert, 1981). Moreover, it is difficult to reconcile a stable mRNA with some aspects of exoenzyme synthesis. In particular enzyme secretion in washed cells is not invariably resistant to inhibitors of transcription and under certain conditions exported proteins behave in the same way as cytoplasmic proteins. The 'mRNA pool' hypothesis accommodates this by assuming depletion of the pool and coupled transcription and translation. The 'stable message' theory, on the other hand, has to invoke mRNA with a variable half-life. In conclusion, RNA metabolism associated with extracellular protein synthesis in bacteria demonstrates some unusual characteristics which will probably only be understood when hybridization probes are available to accurately determine mRNA levels and stabilities. It nevertheless seems likely that these peculiarities are associated with the mobilization of the mRNA from the chromosome to the translation site although no details of this process have yet been obtained.

31.3.7 Control of Secretion

Very little is known about the regulation, if any, that occurs during secretion. Processing enzymes constitute an obvious control point and there is evidence that modulation of a specific protease may regulate the processing and amount of penicillinase released from the membrane of *B. licheniformis* (Lampen, 1978). The leader peptidase gene from *E. coli* has been cloned into a multicopy plasmid which results in a 4 to 10-fold over-production of the enzyme in both the inner and outer membranes of the cell but the effect on protein export has not been reported in detail (Zwinzinski *et al.*, 1981).

Once released from the membrane, exoproteins may accumulate transiently beneath the wall of Gram-positive cells but ultimately diffuse into the environment. Similarly, Eveleigh and Montenencourt (1979) have suggested that release of the enzyme may be the rate limiting factor in secretion by fungi. In Gram-negative bacteria, on the other hand, most secreted proteins are retained in the periplasm. Those that do cross to the outer membrane may subsequently be released to the environment after processing by the relevant enzyme. An interesting example in this area concerns *Serratia marcescens* in which secretion of extracellular lipase can be enhanced by treating the cells with certain extracellular polysaccharides. It is thought that these may function by changing the conformation of the enzyme or detaching exolipase bound to the outside of the bacterium through steric exclusion (Winkler and Stuckman, 1979).

Numerous genetic studies have revealed pleiotropic mutations that affect the composition or structure of the cell envelope and simultaneously the rate or yield of protein export. Although it is generally assumed that these events are correlated this may not necessarily be the case. For example tunicamycin (an inhibitor of cell wall and glycoprotein synthesis) resistant mutants of *B. subtilis* secrete increased levels of α-amylase (but not protease nor other extracellular enzymes), but the hyperproductivity phenotype was segregated from antibiotic resistance when the characters were transferred into other strains (Yamane and Maruo, 1980). In other cases, the two features do seem to be correlated however; *B. subtilis* strains possessing the $SacU^h$, *amyB* or *pap* mutations (these are identical genotypes bearing different labels; Steinmetz *et al.*, 1976) over produce several extracellular enzymes including α-amylase, levansucrase and proteases. These bacteria are poorly, if at all, transformable, generally lack flagella and produce less autolysin which, in affecting the structure of the cell envelope, may be responsible for the complex phenotype (Ayusawa *et al.*, 1975). Similarly, in *Serr. marcescens*, oxacillin hypersensitive mutants have been isolated which have increased sensitivity to various antibiotics, detergents and dyes, form less red pigment and have low extracellular lipase activity. These various features have been traced to a deficiency in lipopolysaccharide synthesis and absence of a particular outer membrane protein (Winkler *et al.*, 1978).

Finally, several pleiotropic fungal mutants that have abnormal cell walls secrete altered amounts of extracellular enzyme. Examples include kabacidine resistant mutants of *Fusarium* which export high yields of alkaline protease (Nakao *et al.*, 1973) and a single mutation in *N. crassa* that derepresses α-amylase, glucoamylase, invertase and trehalase synthesis and concomitantly affects cell wall composition (Gratzner, 1972). It is possible, then, that an altered cell surface can lead to increased enzyme secretion which suggests that in the normal cell the wall constitutes a barrier that should not be discounted in studies of enzyme synthesis and secretion.

31.4 CONCLUSIONS AND SUMMARY

During the past decade, protein secretion in bacteria has been intensively studied and our knowledge of the process in *E. coli* now outstrips our understanding of eukaryotic systems. It would seem that the translocation of the protein across the membrane and the role of signal (leader) peptides in the process is essentially the same in both types of cell although the details may vary. The most striking confirmation of this similarity is the recent finding that preproinsulin can be secreted across the *E. coli* membrane and processed correctly to proinsulin.

With regard to the regulation of extracellular enzyme synthesis in bacteria and fungi, although much has been reported on the effect of the environment on enzyme production and factors such as catabolite repression, induction or end-product repression implicated, virtually nothing is known of the molecular mechanisms involved. This must be a major area where the application of the new genetic techniques will benefit both academic and industrial microbiology.

31.5 REFERENCES

Alper, M. D. and B. N. Ames (1978). Transport of antibiotics and metabolite analogs by systems under cyclic AMP control : positive selection of *Salmonella typhimurium cya* and *crp* mutants. *J. Bacteriol.*, **133**, 149–157.
Arst, H. N. Jr. and C. R. Bailey (1977). The regulation of carbon metabolism in *Aspergillus nidulans*. In *Genetics and Physiology of Aspergillus*, ed. J. E. Smith and J. A. Pateman, pp. 131–146. Academic, London.
Aunstrup, K. (1978). Enzymes of industrial interest, traditional products. In *Annual Reports on Fermentation Processes*, ed. D. Perlman and T. Tsao, vol. 2, pp. 125–154. Academic, London.
Ayusawa, C., Y. Yoneda, K. Yamane and B. Maruo (1975). Pleiotropic phenomena in autolytic enzyme(s) content, flagellation and simultaneous hyperproduction of extracellular α-amylase and protease in a *Bacillus subtilis* mutant. *J. Bacteriol.*, **124**, 459–469.
Bannerjee, A. K. (1980). 5′-Terminal cap structure in eukaryotic messenger ribonucleic acids. *Microbiol. Rev.*, **44**, 175–205.
Bayer, M. E. (1979). The fusion sites between outer membrane and cytoplasmic membrane of bacteria : their role in membrane assembly and virus infection. In *Bacterial Outer Membranes*, ed. M. Inouye, pp. 167–202. Wiley, New York.
Blobel, G. and D. D. Sabatini (1971). Ribosome interaction in eukaryotic cells. In *Biomembranes*, ed. L. A. Manson, vol. 2, pp. 193–195. Plenum, New York.
Blobel, G. and B. Dobberstein (1975). Transfer of proteins across membranes. II. Reconstitution of functional rough microsomes by heterologous components. *J. Cell Biol.*, **67**, 852–862.
Blobel, G., C. N. Chang, B. M. Goldman, A. H. Erickson and V. R. Lingappa (1979). Translocation of proteins across membranes: the signal hypothesis and beyond. *Symp. Soc. Exp. Biol.*, **33**, 9–36.
Both, G. W., J. L. McInnes, J. E. Hanlon, B. K. May and W. H. Elliott (1972). Evidence for an accumulation of messenger RNA specific for extracellular protease and its relevance to the mechanism of enzyme secretion in bacteria. *J. Mol. Biol.*, **67**, 199–207.
Botsford, J. L. (1981). Cyclic nucleotides in prokaryotes. *Microbiol. Rev.*, **45**, 620–642.
Brown, S. and G. Coleman (1975). Messenger ribonucleic acid content of *Bacillus amyloliquefaciens* throughout its growth cycle compared with *Bacillus subtilis* 168. *J. Mol. Biol.*, **96**, 354–362.
Coleman, G., S. Brown and D. A. Stormonth (1975). A model for the regulation of bacterial extracellular enzyme and toxin biosynthesis. *J. Theor. Biol.*, **52**, 143–148.
Costerton, J. W., J. M. Ingram and K. J. Cheng (1974). Structure and function of the cell envelope of Gram negative bacteria. *Bacteriol. Rev.*, **38**, 87–110.
Crick, F. (1979). Split genes and RNA splicing. *Science*, **204**, 264–271.
Darnell, J. E., Jr. (1979). Transcription units for mRNA production in eukaryotic cells and their DNA viruses. *Prog. Nucleic Acid Res. Mol. Biol.*, **22**, 327–353.
Davis, B. D. and P.-C. Tai (1980). The mechanism of protein secretion across membranes. *Nature (London)*, **283**, 433–438.
Davis, P. E., D. L. Cohen and A. Whitaker (1980). The production of α-amylase in batch and chemostat culture by *Bacillus stearothermophilus*. *Antonie van Leeuwenhoek*, **46**, 391–398.
DiRenzio, J. M., K. Nakamura and M. Inouye (1978). The outer membrane proteins of Gram negative bacteria: biosynthesis, assembly and functions. *Annu. Rev. Biochem.*, **47**, 481–532.
Doi, R. H. (1977). Role of ribonucleic acid polymerase in gene selection in prokaryotes. *Bacteriol. Rev.*, **41**, 568–594.
Dowds, B., L. Baxter and M. McKillen (1978). Catabolite repression in *Bacillus subtilis*. *Biochim. Biophys. Acta*, **541**, 18–34.

Eveleigh, D. E. and B. S. Montenencourt (1979). Increasing yields of extracellular enzymes. *Adv. Appl. Microbiol.*, **25**, 57–74.

Fencl, Z., J. Řičicia and J. Kodešová (1972). The use of the multi-stage chemostat for microbial product formation. *J. Appl. Chem. Biotechnol.*, **22**, 405–426.

Ghosh, B. K. (1978). Introduction to the ultrastructure of fungi. In *Handbook of Microbiology*, ed. H. A. Lechevalier and A. I. Laskin, pp. 11–69. CRC Press, Cleveland.

Glenn, A. R. (1976). Production of extracellular proteins by bacteria. *Annu. Rev. Microbiol.*, **30**, 41–62.

Glynn, J. A., S. D. Schafell, J. M. McNicholas and F. M. Hulett (1977). Biochemical localisation of alkaline phosphatase of *Bacillus licheniformis* as a function of culture age. *J. Bacteriol.*, **129**, 1010–1019.

Goldberger, R. F. (1979). Strategies of genetic regulation in prokaryotes. In *Biological Regulation and Development*, ed. R. F. Goldberger, vol. 1, pp. 1–18. Plenum, New York.

Gratzner, H. G. (1972). Cell wall alterations associated with the hyperproduction of extracellular enzymes in *Neurospora crassa*. *J. Bacteriol.*, **111**, 443–446.

Halegoua, S. and M. Inouye (1979). Biosynthesis and assembly of outer membrane proteins. In *Bacterial Outer Membranes*, ed. M. Inouye, pp. 67–113. Wiley, New York.

Hanson, M. A. and G. A. Marzluf (1975). Control of the synthesis of a single enzyme by multiple regulatory circuits in *Neurospora crassa*. *Proc. Natl. Acad. Sci. USA*, **72**, 1240–1244.

Hanson, R. S. (1976). Role of small molecules in regulation of gene expression and sporulation in bacilli. In *Spores VI*, ed. P. Gerhardt, H. L. Sadoff and R. N. Costilow, pp. 318–326. American Society for Microbiology, Washington, DC.

Harwood, R. (1980). Protein transfer across membranes: the role of signal sequences and signal peptidase activity. In *The Enzymology of Post-translational Modification of Proteins*, ed. R. B. Freedman and H. C. Hawkins, vol 1, pp. 3–52. Academic, London.

Imanaka, T., T. Tanaka, H. Tsunekawa and S. Aiba (1981). Cloning of the genes for penicillinase, *penP* and *penI* of *Bacillus licheniformis* in some vector plasmids and their expression in *Escherichia coli*, *Bacillus subtilis* and *Bacillus licheniformis*. *J. Bacteriol.*, **147**, 776–786.

Imsande, J. (1978). Genetic regulation of penicillinase synthesis in Gram positive bacteria. *Bacteriol. Rev.*, **42**, 67–83.

Ingle, M. B. and E. W. Boyer (1976). Production of industrial enzymes by *Bacillus* species. In *Microbiology 1976*, ed. D. Schlessinger, pp. 420–426. American Society for Microbiology, Washington, DC.

Ingle, M. B. and J. R. Erickson (1978). Bacterial amylases. *Adv. Appl. Microbiol.*, **24**, 257–278.

Keay, L., M. H. Moseley, R. G. Anderson, R. J. O'Connor and B. S. Wildi (1972). Production and isolation of microbial proteases. *Biotechnol. Bioeng. Symp.*, **3**, 63–92.

Kreibich, G., C. M. Freienstein, P. N. Pereyra, B. C. Urlich and D. D. Sabatini (1978). Proteins of rough microsomal membranes related to ribosome binding. II. Cross-linking of bound ribosomes to specific membrane proteins exposed at the binding sites. *J. Cell Biol.*, **77**, 488–505.

Kreil, G. (1981). Transfer of proteins across membranes. *Annu. Rev. Biochem.*, **50**, 317–348.

Kroyer, J. and S. Chang (1981). The promoter-proximal region of the *Bacillus licheniformis* penicillinase gene: nucleotide sequence and predicted leader peptide sequence. *Gene*, **15**, 343–347.

Labbe, R. G. and C. L. Duncan (1977). Evidence for stable messenger ribonucleic acid during sporulation and enterotoxin synthesis by *Clostridium perfringens* Type A. *J. Bacteriol.*, **124**, 843–849.

Lampen, J. O. (1978). Phospholipoproteins in enzyme excretion by bacteria. *Symp. Soc. Gen. Microbiol.*, **28**, 231–247.

Lewin, B. (1980). In *Gene Expression 2*, 2nd edn., pp. 752–755. Wiley, New York.

Matsumoto, K., I. Uno, A. Toh-E, T. Ishikawa and Y. Oshima (1982). Cyclic AMP may not be involved in catabolite repression in *Saccharomyces cerevisiae*: evidence from mutants capable of utilising it as a carbon source. *J. Bacteriol.*, **150**, 277–285.

Milstein, C., C. G. Brownlee, T. M. Harrison and M. B. Mathews (1972). A possible precursor of immunoglobulin light chain. *Nature (London)*, **239**, 117–120.

Nakao, Y., M. Suzuki, M. Kumo and K. Macjima (1973). Production of alkaline protease from *n*-paraffins by a kabicidin resistant mutant of *Fusarium* sp. *Agric. Biol. Chem.*, **37**, 1223–1224.

Nichols, J. C., P.-C. Tai and J. R. Murphy (1980). Cholera toxin is synthesized in precursor form on free polysomes in *Vibrio cholerae* 569 B. *J. Bacteriol.*, **144**, 518–523.

Nielson, J. B. K., M. P. Caulfield and J. O. Lampen (1981). Lipoprotein nature of *Bacillus licheniformis* membrane penicillinase. *Proc. Natl. Acad. Sci. USA*, **78**, 3511–3515.

Nikaido, H. (1979). Non specific transport through the outer membrane. In *Bacterial Outer Membranes*, ed. M. Inouye, pp. 361–408. Wiley, New York.

Novick, P., S. Ferro and R. Schekman (1981). Order of events in the yeast secretory pathway. *Cell*, **25**, 461–469.

O'Connor, R., W. H. Elliott and B. K. May (1978). Modulation of an apparent mRNA pool for extracellular protease in *Bacillus amyloliquefaciens*. *J. Bacteriol.*, **136**, 24–34.

Palade, G. E. (1975). Intracellular aspects of the process of protein synthesis. *Science*, **189**, 347–358.

Pederson, S., E. Reech and J. D. Friesen (1978). Functional mRNA half-lives in *Escherichia coli*. *Mol. Gen. Genet.*, **166**, 329–336.

Petit-Glatron, M.-F. and R. Chambert (1981). Levansucrase of *Bacillus subtilis*: conclusive evidence that its production and export are unrelated to fatty-acid synthesis but modulated by membrane-modifying agents. *Eur. J. Biochem.*, **119**, 603–611.

Philips, A. T. and L. M. Mulfinger (1981). Cyclic adenosine 3′,5′-monophosphate levels in *Pseudomonas putida* and *Pseudomonas aeruginosa* during induction and carbon catabolite repression of histidase synthesis. *J. Bacteriol.*, **145**, 1268–1292.

Pirtle, I. M., I. L. Pirtle and M. Inouye (1978). Homologous nucleotide sequences between prokaryotic and eukaryotic mRNAs : the 5′-end sequence of the mRNA of the lipoprotein of the *Escherichia coli* outer membrane. *Proc. Natl. Acad. Sci. USA*, **75**, 2190–2194.

Priest, F. G. (1977). Extracellular enzyme synthesis in the genus *Bacillus*. *Bacteriol. Rev.*, **41**, 711–753.

Priest, F. G. (1983). Enzyme synthesis : regulation and process of secretion in microorganisms. In *Microbial Enzymes and Biotechnology*, ed. W. M. Fogarty, pp. 319–366. Applied Science, London.

Rose, A. H. (1980) (ed.) *Microbial Enzymes and Bioconversions*. Academic, London.

Rosenberger, R. F. (1975). The cell wall. In *The Filamentous Fungi*, ed. J. E. Smith and D. R. Berry, vol. 2, pp. 328–344. Edward Arnold, London.

Savageau, M. A. (1979). Autogenous and classical regulation of gene expression: a general theory and experimental evidence. In *Biological Regulation and Development*, ed. R. F. Goldberger, vol. 1, pp. 57–108. Plenum, New York.

Schaeffer, P. (1969). Sporulation and the production of antibiotics, exoenzymes, and exotoxins. *Bacteriol. Rev.*, **33**, 48–71.

Silhavy, T. J., P. S. Bassford, Jr. and J. R. Beckwith (1979). A genetic approach to the study of protein localization in *Escherichia coli*. In *Bacterial Outer Membranes*, ed. M. Inouye, pp. 203–254. Wiley, New York.

Smith, W. P., P.-C. Tai, R. C. Thompson and B. D. Davis (1977). Extracellular labeling of nascent polypeptide traversing the membrane of *Escherichia coli*. *Proc. Natl. Acad. Sci. USA*, **74**, 2830–2834.

Smith, W. P., P.-C. Tai and B. D. Davis (1981). *Bacillus licheniformis* penicillinase : cleavage and attachment of lipid during cotranslational secretion. *Proc. Natl. Acad. Sci. USA*, **78**, 3051–3055.

Steinmetz, M., F. Kunst and R. Dedonder (1976). Mapping of mutations affecting synthesis of exoenzymes in *Bacillus subtilis*. *Mol. Gen. Genet.*, **148**, 281–285.

Travers, A. A., C. Kari and H. A. F. Mace (1981). Transcriptional regulation by bacterial RNA polymerase. *Symp. Soc. Gen. Microbiol.*, **31**, 111–130.

Tsuyumu, S. (1979). 'Self-catabolite repression' of pectate lyase in *Erwinia carotovora*. *J. Bacteriol.*, **137**, 1035–1036.

Wickner, W. (1979). The assembly of protein in biological membranes: the membrane trigger hypothesis. *Annu. Rev. Biochem.*, **48**, 23–45.

Winkler, U., K. B. Heller and B. Folle (1978). Pleiotropic consequences of mutations towards antibiotic-hypersensitivity in *Serratia marcescens*. *Arch. Microbiol.*, **116**, 259–268.

Winkler, U. K. and M. Stuckman (1979). Glycogen, hyaluronate and some other polysaccharides greatly enhance the formation of exolipase by *Serratia marcescens*. *J. Bacteriol.*, **138**, 663–670.

Wong, S. L., C. W. Price, D. S. Goldberg and R. H. Doi (1984). The subtilisin *E* gene of *Bacillus subtilis* is transcribed from a σ^{37} promoter *in vivo*. *Proc. Natl. Acad. Sci. USA*, **81**, 1184–1188.

Wouters, J. T. M. and R. J. Buysman (1977). Production of some extracellular enzymes by *Bacillus licheniformis* 749/C in chemostat cultures. *FEMS Microbiol. Lett.* **1**, 109–112.

Yamane, K. and B. Maruo (1980). *B. subtilis* α-amylase genes. In *Molecular Breeding and Genetics of Applied Microorganisms*, ed. K. Sakaguchi and M. Okanishi, pp. 117–123. Academic, New York.

Zitomer, R. S., D. L. Montgomery, D. L. Nichols and B. D. Hall (1979). Transcriptional regulation of the yeast cytochrome *c* gene. *Proc. Natl. Acad. Sci. USA*, **76**, 3627–3631.

Zwinzinski, C., T. Date and W. Wickner (1981). Leader peptidase is found in both the inner and outer membrane of *Escherichia coli*. *J. Biol. Chem.*, **256**, 3593–3597.

32

Overproduction of Microbial Metabolites

O. M. NEIJSSEL and D. W. TEMPEST
Universiteit van Amsterdam, The Netherlands

32.1 INTRODUCTION

One of the most interesting features of microorganisms is their enormous metabolic versatility. On the one hand there are some bacteria that can grow on a mixture of CO_2, H_2 and N_2 as the sole sources of carbon, energy and nitrogen, respectively. On the other hand, there are bacteria, such as the *Rickettsia* spp., that cannot grow independently from a host cell, presumably because of their complex nutritional requirements. Moreover, many microorganisms are capable of growing on a wide variety of organic chemicals such as hydrocarbons, phenols, halogenated compounds, pesticides, nylon precursors, *etc.* If we also take into account the fact that the chemical conversions effected by most microbes occur at moderate temperatures and at atmospheric pressure, and are usually stereospecific, it is understandable that these organisms have attracted much attention as providing possible alternatives to the classical organic chemical syntheses of certain compounds, which frequently are performed using expensive starting material and catalysts and at increased temperatures and pressures.

The sum of the chemical reactions carried out by a cell is called metabolism and it serves two major goals: (i) to keep the cell viable and intact (maintenance), and (ii) to effect synthesis of new cell material (growth). From this it follows that metabolite production must be functional in one or both of these two processes, and therefore will be regulated by the cell's demands. In other words one may assume that every metabolite produced by a particular microbe has a certain function. In the case of ATP, pyruvate, amino acids and other primary metabolites, this function is clear and well understood, but the functional significance of the production of other (secondary) metabolites, such as some antibiotics, mycotoxins, pigments, *etc.*, is at the moment unclear. In fact, some investigators think that these compounds serve no function at all and that their production is fortuitous. However, we feel that even if one cannot yet ascribe a certain physiological role to these compounds, it is not correct to tacitly assume that they serve no function at all in the survival and growth of the organism.

In this connection pigment production by *Pseudomonas fluorescens* provides a good example. Under certain growth conditions this organism produces a water soluble, yellow-green fluorescent pigment (pyoverdine), sometimes in quantities equal to the cellular dry weight (Meyer and Abdallah, 1978). In the past no clear physiological role could be assigned to this pigment. However, recently it has been shown that this compound is an iron chelator that is produced by the organism in environments in which the availability of iron is low. The production of pyoverdine allows the organism to scavenge the last traces of iron from such environments, which results in a growth advantage (Meyer and Hornsperger, 1978).

It is difficult to give a meaningful definition of metabolite overproduction. If a mutant organism produces a substance in higher amounts than the wild type (under identical growth conditions) then, clearly, one can call this overproduction. But when *Saccharomyces cerevisiae* produces ethanol from glucose in the absence of oxygen, it is not at all clear why this should be called *over-production*. However, for pragmatic reasons we will use the term overproduction for any accumulation or excretion of a substance. Thus, the production of pyoverdine by *Pseudomonas fluorescens* and ethanol by *Saccharomyces cerevisiae* is, in the context of this chapter, overproduction.

Many microorganisms always excrete metabolites as a consequence of their genetic constitution. Some of them exist in Nature: bacteria like the streptococci and clostridia lack the genes for a respiratory chain, therefore the reducing equivalents generated in the breakdown of the carbon source cannot be reoxidized by an external electron acceptor, such as oxygen or nitrate. Instead, these organisms use organic intermediates of their metabolism as electron acceptors and, once reduced, these compounds are excreted. Other organisms have acquired the property of overproduction by genetic manipulation in a laboratory. The genetic technique enables one to multiply genes, to select mutant genes coding for enzymes that are insensitive to inhibition, to delete genes coding for enzymes catalysing 'unwanted' side reactions, or, in the case of recombinant DNA, to insert genes not belonging to the natural gene content of a species, but coding for the synthesis of the desired product (*e.g.* human insulin in *Escherichia coli*). In principle, one can now optimize genetically a microbe for the production of a particular metabolite.

The conclusion could be drawn, then, that metabolite overproduction can be achieved almost exclusively by genetic means and that there is hardly any physiology involved! This is not true, and the purpose of this chapter is to show that by careful manipulation of the growth environment one can also elicit metabolite overproduction in microorganisms. We shall show that excretion of products of metabolism is part of a survival strategy of microbes in certain environments and is therefore a stable property. One example has already been mentioned: the production of pyoverdine by *Pseudomonas aeruginosa* under iron-limited growth conditions.

In many of the reported studies, use has been made of a chemostat. One should not conclude that metabolite overproduction by wild-type organisms is a phenomenon which only occurs in this type of culture, but since a study of the influence of the growth environment on microbial physiology is most effectively studied by using a chemostat, emphasis is put on these types of investigations.

32.2 EFFECTS OF NUTRIENT LIMITATION

The growth environment influences the behaviour of microorganisms to a great extent. Thus, phenomena like enzyme induction and enzyme repression are well known, though the phenotypic variability of the microbial cell extends much further and is far greater than often is realized. This sensitivity of cell physiology to the growth environment led Herbert (1961) to conclude: 'It is virtually meaningless to speak of the chemical composition of a microorganism without at the same time specifying the environmental conditions that produced it.' The key to understanding these changes in structure and behaviour is to be found in Nature: there microbes routinely face changes in temperature, pH, water activity, and, most frequently, changes in nutrient availability. During the course of evolution the organisms necessarily must have acquired a great flexibility in their adaptation to these environmental changes, for example a lowering of the growth temperature results in an organism with a higher unsaturated fatty acid content in its membranes, so that the membrane fluidity is maintained at its optimum (Ellar, 1978). Similarly, an increase of osmolarity of the medium is counterbalanced in some Gram-negative bacteria by an increased intracellular osmotic pressure due to the synthesis of glutamate and enhanced uptake of potassium (Tempest *et al*, 1970).

Nutrient limitation leads to a great number of adaptive responses (for a review, see Tempest and Neijssel, 1978). One interesting response in the context of this chapter is metabolite overproduction and a simple, perhaps rather trivial, example again is ethanol production by *Saccharomyces cerevisiae*. This yeast can grow aerobically in a minimal medium with glucose as the sole carbon source. Under these circumstances it produces little or no ethanol (we ignore here the Crabtree effect). When grown under anaerobic circumstances in the same medium (plus a small amount of essential growth factors) one observes a number of phenomena: a substantial amount of ethanol is now formed from glucose and the growth rate and yield of the cells is lowered considerably. The important point to note here is that, although the growth is less abundant, there still *is* growth. In contrast to obligate aerobes, *S. cerevisiae* can cope with the absence of oxygen, but the price that is paid is a lower efficiency of cell synthesis concomitant with metabolite overproduction. Thus, in this environment metabolite (ethanol) overproduction is essential for the survival of the microbe and is therefore a stable property. The importance of this example lies in the fact that the absence of oxygen creates a metabolic block in much the same way as a mutation in a gene coding for an enzyme would do, but whereas a back mutation to the wild-type property is possible, and sometimes difficult to prevent, the environmentally controlled metabolic block cannot disappear spontaneously.

A low availability of a particular nutrient (thus not a total absence) does not create a block but one or more bottlenecks, *i.e.* the metabolic flow through a pathway is diminished to a minimum rate. A typical example is the glucose metabolism of *Klebsiella aerogenes* growing in a fully aerobic *ammonia-limited* environment, when compared with a fully-aerobic glucose-limited environment (Table 1). Under glucose limitation the sole products of glucose metabolism are new cell material and carbon dioxide. The oxygen consumption is comparatively low and the yield of new cells is high. When grown under ammonia-limited conditions at the same rate, *K. aerogenes* consumes three times as much glucose and almost double the amount of oxygen: now metabolite overproduction (or overflow metabolism) is observed and acetate, pyruvate and particularly 2-oxoglutarate are excreted into the medium. Why does a low availability of ammonia cause this phenomenon? It is not due to an increased membrane permeability causing leakage of the intracellular pool, because with different growth limiting conditions, such as potassium limitation, other products are found in the extracellular fluids. A consideration of the way in which ammonia is metabolized provides the answer. In contrast to carbon-limited cells, which incorporate ammonia (present in high concentrations) *via* glutamate dehydrogenase, ammonia-limited cells consume ammonia *via* glutamine synthetase plus glutamate synthase (Tempest *et al.*, 1973). The glutamate formed is used subsequently for the synthesis of all other amino acids necessary for protein synthesis. Increased metabolism of the carbon source must provide a high, enzyme saturating, concentration of the key reactants ATP, NADPH and 2-oxoglutarate to enable the enzymes to react optimally with the trace levels of ammonia. This results in an intracellular accumulation and/or excretion of high molecular weight products of glucose metabolism (such as glycogen intracellularly and polysaccharide outside the cell) and low molecular weight substances to which the cell membrane is permeable, *i.e.* acetate, pyruvate and 2-oxoglutarate. It has to be stressed, however, that the permeability of the cell membrane plays a crucial role in this process. Whereas glycogen synthesis is observed generally in heterotrophic microbes growing in media with a high C/N ratio (Dawes and Senior, 1973), excretion of 2-oxoglutarate is not so common under these circumstances due to the fact that in some organisms the membrane is not permeable for this compound (Ghei and Kay, 1973; Neijssel and Tempest, 1979). This emphasizes again the importance of the choice of the organism for the production of a particular metabolite.

Table 1 Substrate Utilization Rates and Rates of Product Formation in Variously-limited Chemostat Cultures of *Klebsiella aerogenes* NCTC 418[a]

Limitation	Glucose	O_2	NH_3	Cells	CO_2	H_2O	HAc	Pyr	2-OG	GA	KGA	$(CH_2O)_n$	Prt
C	12	26	9	1	31	50	0	0	0	0	0	0	0
N	36	44	9	1	40	72	4	3.5	11	0	0	1.9	0
K	58	106	13	1	113	165	13	6.7	1.2	11	6.8	0	0.4
P	38	65	9	1	41	94	4.5	0	0	3.2	13.3	0.8	0

[a] Cells were grown in a mineral medium ($D = 0.17 \pm 0.1 h^{-1}$; pH 6.8 ± 0.1; 35 °C). All values are adjusted to a cell production rate of 1 kg h^{-1}. Extracellular polysaccharide, $(CH_2O)_n$, and extracellular protein are expressed in kg h^{-1}, whereas the other values are expressed in mol h^{-1}. HAc = acetate; Pyr = pyruvate; 2-OG = 2-oxoglutarate; GA = gluconate; KGA = 2-ketogluconate. Recalculated from the data of Neijssel and Tempest (1975) and Tempest and Neijssel (1978). The chemical composition of cell mass was taken from the data of Herbert (1976). The values of NH_3 and H_2O have been calculated so as to achieve an approximately ($\pm 5\%$) balanced chemical equation.

A different situation is encountered with potassium-limited growth, or more generally, metal ion-limited growth. In contrast to carbon, nitrogen or sulfur, potassium is present in the cell in the same chemical state as in the medium. Although some binding may occur to intracellular components, it is nevertheless clear that there will exist an intracellular pool of freely mobile potassium ions. In Gram-negative organisms such as *Klebsiella aerogenes* the concentration of this pool of K^+ is maintained at a value of at least 40 mM. In Gram-positive organisms the concentration of the pool of K^+ is even higher: 100–300 mM (Tempest *et al.*, 1966). The growth limiting potassium concentration outside the cell is immeasurably low. Thus, a substantial potassium ion gradient must exist across the cell membrane, which is built up by an energy-linked K^+ transport system, and, once established, also has to be maintained. It has been shown that if the supply of energy in the cell is suddenly interrupted, the pool K^+ will immediately leak from the cell. For this purpose *K. aerogenes* was grown under carbon-limited conditions but also at a low, but not growth limiting, concentration of K^+. When the culture was in a steady state, the medium supply to the culture was suddenly stopped by switching off the pump. This removal of the supply of the carbon and energy source (in this case glucose) created an immediate shortage in the supply of reducing equivalents to the respiratory chain, which resulted in a decrease of the rate of oxygen consumption. Since the membrane potential could no longer be maintained at the appropriate level, potassium started to leak from the cells (Figure 1). Switching the pump on again quickly restored the membrane potential and the original gradient (Hueting *et al.*, 1979). This shows that when growing in an environment with a low availability of K^+ the cell must maintain a high rate of supply of energy in the form of a membrane potential to take up K^+ and to keep it inside the cell. This is achieved with a high rate of carbon source metabolism, particularly *via* those pathways that produce large amounts of NAD(P)H. The data in Table 1 indicate that there is a substantial increase in TCA-cycle activity, since the production of CO_2 is increased by more than threefold when compared with carbon-limited growth. *K. aerogenes* also has available an extra pathway to generate NAD(P)H: a rapid oxidation of glucose to gluconic and 2-ketogluconic acid. These products are excreted, since there is no further use for them in metabolism.

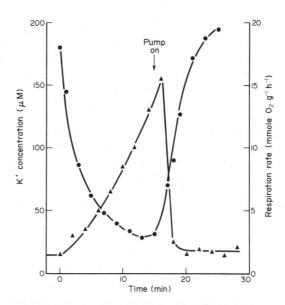

Figure 1 Influence of a step-down to zero dilution rate on the respiration rate (●) and extracellular K^+ concentration (▲) of a glucose-limited culture of *Klebsiella aerogenes* NCTC 418 (dilution rate = 0.4 h^{-1}; pH 6.8; temperature 35 °C). At $t=0$ the pump was turned off, after 15 min the pump was turned on again. Redrawn from Hueting *et al.* (1979)

This example shows that in this case metabolite overproduction is provoked not because of a bottleneck, but because of a process which we refer to as a 'metabolic drain'. The imposed limitation induces a process which demands a continuous supply of energy, which 'pulls' carbon source metabolism. Both processes, 'bottleneck' and 'drain', cause an increased rate of energy production, either in the form of a membrane potential or as ATP. The fate of this extra energy is clear only with potassium-limited cells. In ammonia-limited cells mechanisms must exist to maintain the energy status of the cell, reflected in the ATP/ADP ratio, in a physiological range. A detailed discussion of this energetic aspect of metabolite overproduction is given elsewhere (Tempest and Neijssel, 1980; Stouthamer, 1979). It can be concluded that almost always metabolite overproduction is accompanied by extra energy dissipation.

The examples above have been selected to illustrate some principles that dominate metabolite overproduction. It is important to realize that in many cases a combination of these two extremes is observed, for example with a phosphate limitation. On the one hand phosphate is metabolized into phospholipids, nucleic acids and nucleotides, on the other hand a pool of freely soluble phosphate must always be present in a microbe and plays an important role in the regulation of the energy flow *via* the reactions:

$$AMP + P_i \rightleftharpoons ADP + P_i \rightleftharpoons ATP$$

Thus it comes as no surprise that phosphate-limited cells of *K. aerogenes* behave in a way that is intermediate between cells subjected to ammonia and potassium limitations. They excrete polysaccharides, but also gluconic and 2-ketogluconic acids, and their oxygen consumption is higher than ammonia-limited cells, but lower than cells that are potassium-limited (Table 1). The metabolic bottleneck caused by phosphate limitation is, of course, any phosphate-requiring reaction, but the most important assimilatory route of phosphate is *via* the phosphorylation of the adenine nucleotides, thus the membrane bound ATPase is the major target of this limitation. If one applies the same reasoning that was used for ammonia limitation, *i.e.* a saturation of the enzyme system with all reactants other than the limiting substrate, it becomes clear immediately that an increased rate of respiration is fundamental to this process. Moreover, the oxidation of glucose to gluconate and 2-ketogluconate is an excellent way of producing reducing equivalents with a minimum involvement of phosphate.

In the synthesis of antibiotics and other so-called secondary metabolites, the nature and extent of nutrient limitation plays a crucial role. However, a detailed physiological explanation for these observations is generally lacking, because one does not know which role the secondary metabolites play in the survival of the producer strain, although cell differentiation (spore formation or mycelium formation) and the facilitation of uptake of metal ions have been implicated as processes under regulation by this sort of compound (Katz and Demain, 1977; Khoklov, 1982). High levels of glucose, ammonia (and other readily utilizable nitrogen sources), and phosphate have been described as factors that depress the synthesis of different antibiotics (for a review, see Demain, 1982), therefore carbon sources other than glucose are frequently used along with an insoluble nitrogen source (for example soya bean meal), for the production of antibiotics. A key regulatory factor is the extracellular concentration of phosphate. Concentrations of 10 mM or higher inhibit the synthesis of peptide antibiotics, polyene macrolides, tetracyclines and other compounds. In the case of streptomycin this is due to an accumulation of streptomycin phosphate, which is biologically inactive (Miller and Walker, 1969), but more generally phosphate seems to repress the synthesis of enzymes involved in antibiotic synthesis. The exact mechanism of phosphate repression is not yet understood. The involvement of several intracellular molecules that are known to regulate primary metabolism (*e.g.* cyclic AMP, glucose 6-phosphate, adenosine polyphosphates and guanosine polyphosphates) have been studied, but an unequivocal role for any of these compounds in mediating enzyme repression by phosphate has not been established (Demain, 1982).

A limitation on metal ions such a iron, zinc or manganese has been shown to promote the production of citrate by *Aspergillus niger* and other organisms (Kubicek and Röhr, 1977; Marchal *et al.*, 1977). It may be that the target enzyme for this type of limitation is aconitase, which needs an iron–citrate complex in order to react properly (Glusker, 1968) so that iron limitation would create a metabolic bottleneck at the level of this enzyme, but this does not explain the effect of zinc and manganese.

Apart from the nature of the growth limiting nutrient, its concentration is an important factor. This is a well known phenomenon in the production of antibiotics in batch cultures: production usually starts at the end of the log phase when the growth rate of the culture declines, yet it is not invariably true that the highest productivities are attained at the lowest growth rates. In principle, there are two other relationships possible between growth rate and product formation. In the first place, product formation can increase with increasing growth rate and near, or at, μ_{max} the production also is maximal. This can be illustrated by the data of Olijve (1978) obtained with *Gluconobacter oxydans*. When this organism was grown in a glucose-limited chemostat at pH 5.5, and at a growth rate of 0.05 h^{-1}, only 35% of the glucose was converted into gluconic, 2-ketogluconic and 5-ketogluconic acids. Increasing the dilution rate led to an increase in the conversion of glucose to these products, such that at a growth rate of 0.26 h^{-1} 98% of the glucose was found to be oxidized and subsequently excreted as gluconic acid (91%) and ketogluconic acids (7%). Similarly, the data of Rosenberger and Elsden (1960) with *Streptococcus faecalis* show the same trend (Table 2). They cultured this organism anaerobically in a defined medium in a chemostat. The table shows the specific rates of glucose consumption ($q_{glucose}$) and lactate production ($q_{lactate}$) for the different growth conditions, calculated from the authors' data. As can be seen, when the growth was limited by the glucose supply, the rate of glucose consumption by the organism was proportional to the dilution rate. For comparison, the data for tryptophan limitation also are given, and these show that here again, the nature of the growth limitation had a great influence. Like the ammonia-limited *K. aerogenes* cells, tryptophan-limited *S. faecalis* consumed glucose at a higher rate than did glucose-limited cells. This resulted in an increased rate of lactate production. It is interesting to note that these fermenting bacteria also must possess ATP dissipating reactions when growing under tryptophan-limited conditions, since the rate of ATP production (which is equal to the rate of lactate production) is higher under these growth conditions as compared with glucose-limited growth at the same rate.

Table 2 Specific Rates of Glucose Consumption and Lactate Production by Glucose-limited and Tryptophan-limited Chemostat Cultures of *Streptococcus faecalis*[a]

Limitation	Dilution rate (h^{-1})	$q_{glucose}$ (mol $kg^{-1} h^{-1}$)	$q_{lactate}$ (mol $kg^{-1} h^{-1}$)
Glucose	0.22	4.9	3.8
	0.31	7.5	6.8
	0.43	9.6	17.6
Tryptophan	0.22	12.9	13.4
	0.31	16.6	20.2
	0.43	16.1	30.6

[a] Calculated from the data of Rosenberger and Elsden (1960).

The proportionality between the rate of product formation and the growth rate can be explained as follows. An increased growth rate necessitates an increased rate of carbon source metabolism, particularly *via* the energetically most favourable pathways. In growing bacteria that consume oxygen, this leads to the well known, frequently linear, relationship between q_{O_2} and growth rate (Pirt, 1965). If metabolism of the carbon source is coupled with product formation, then this parameter will follow the same trend.

Another frequent observation is that the rate of product formation first increases with increasing growth rate and then at yet higher growth rates decreases; for instance, Matteo *et al.* (1976) found that the highest amounts of the two enzymes of gramicidin S biosynthesis were synthesized at growth rates well below the maximum. The same pattern has been observed with polysaccharide (Tempest, unpublished result) and 2,3-butanediol (Pirt and Callow, 1958) production by *K. aerogenes*. The explanation here is that product formation is now subject to regulation by at least two factors. Firstly, an increased rate of carbon source metabolism at higher growth rates will stimulate metabolite production (see above). If the growth limiting nutrient inhibits or represses product formation, its negative effect will become more and more apparent when the growth rate increases. Moreover, one has to realize that near μ_{max}, growth is virtually unrestricted and that all primary biosynthetic reactions occur at a fast rate. Thus, there will be a general shift of the carbon flow from secondary to primary pathways. Concluding this section one can state that metabolite overproduction is influenced by the following environmental factors: (i) carbon-sufficient growth conditions; (ii) the nature of the carbon source; (iii) the nature and concentration of the growth limiting nutrient; (iv) the regulation of carbon source metabolism; (v) the permeability of the cell membrane towards products; and (vi) the type of organism.

32.3 EFFECTS OF pH AND UNCOUPLERS OF OXIDATIVE PHOSPHORYLATION

The intracellular pH value of a microbe is thought to be maintained within certain limits so as to allow enzymes to work effectively. The external pH value, however, can vary markedly and when this occurs a pH gradient will be formed in much the same way as a potassium gradient under potassium-limited growth conditions. Many years ago Gale and Epps (1942) observed that bacteria seemingly possess compensation mechanisms that steer the extracellular pH towards a more neutral direction (*i.e.* neutralization reactions). Thus, if the pH value of the medium became alkaline, acidic products were formed and if the medium became acidic, then neutral or alkaline products were formed. In the microbes that were studied, *Escherichia coli* and *Micrococcus lysodeikticus*, this regulation of product formation was achieved by an alteration in the cells' enzymic constitution. At low pH values amino acid decarboxylases were formed, whereas at high pH values the synthesis of these enzymes was repressed and amino acid deaminases were formed instead. Table 3 shows the effect of medium pH value on product formation by different bacteria. From these data it is obvious that the compensation mechanism proposed by Gale and Epps (1942) seems to function generally. Particularly striking is the effect of medium pH value on the production of formic acid and 2,3-butanediol by cultures growing anaerobically. The synthesis of the enzymes responsible for the formation of these two products, α-acetolactate synthase and formate-hydrogen lyase, are regulated precisely in accord with the compensation mechanism: at low pH values both enzymes are formed and much CO_2, H_2 and 2,3-butanediol are produced, at high pH values their syntheses are repressed and formate is formed and the production of CO_2 and H_2 are very much diminished (Størmer, 1977; Wood, 1961).

Table 3 Effect of Medium pH Value on the Fermentation of Glucose by Different Organisms[a]

Organism	*Escherichia coli*[b]		*Aerobacter cloacae*[c]		*Klebsiella aerogenes*[d]		
pH value	6.0	7.8	5.0	7.2	6.10	6.85	7.80
Acetate	36.5	38.7	5.0	69.5	37.8	77.6	91.9
Acetoin	0.06	0.19	0	1.5	—	—	—
2,3-Butanediol	0.30	0.26	38.8	2.3	43.2	17.1	0.27
Butyrate	0	7.10	—	—	—	—	—
CO_2	88.0	1.75	—	—	170	159	106
Ethanol	49.8	50.5	61.8	67.5	41.8	55.9	55.4
Formate	2.43	86.0	—	—	0.21	6.85	37.2
Glycerol	1.42	0.32	—	—	—	—	—
Hydrogen	75	0.26	—	—	—	—	—
Lactate	79.5	70.0	0	3.6	3.38	3.24	14.3
Succinate	10.7	14.8	3.3	6.4	1.39	2.06	5.08

[a] — signifies not determined. All values are in mol per 100 mol of glucose fermented. [b] Data from Blackwood *et al.* (1956). [c] Data from Hernandez and Johnson (1967). [d] Data for an anaerobic potassium-limited culture (unpublished results of M. J. Teixeira de Mattos and P. Plomp).

From these and previous examples it becomes clear that the maintenance of gradients of protons and other ions requires energy and affects product formation, implicating a metabolic drain. Hence, if one can manipulate this drain in such a way that it is increased, the likely result will be an increased rate of production of metabolites and perhaps the synthesis of other desired products. A way to achieve this would be to increase the membrane's permeability towards these ions. Some antibiotics, like valinomycin and gramicidin, but also weak lipophilic acids (including phenols and other anionic aromatic compounds) have this effect. Significantly, these compounds are known collectively as uncouplers of oxidative phosphorylation (Hanstein, 1976), because they stimulate respiration of mitochondria and microorganisms without invoking a concomitant stimulation of ATP synthesis (thus, respiration is dissociated from phosphorylation). The general mechanism of action of these compounds is depicted in Figure 2. Because the compounds are lipophilic they seemingly can freely traverse the cell membrane and, depending on the selectivity of the compound, gradients of a particular ion are abolished. The cell can react to this functional change by increasing the pumping rate of the relevant ion, which causes an increased energy demand. If this pumping rate can be increased such that it exceeds the backflow of ions caused by the uncoupler, the cells will still be able to maintain the necessary gradient, albeit at the expense of extra energy. If the cells do not succeed in doing this they will cease to function unless they are cultivated in a very special environment. Thus, Harold and Van Brunt (1977) have shown that *Streptococcus faecalis*, which generates its proton gradient through ATP hydrolysis by the membrane bound ATPase, could grow seemingly normally in the presence of gramicidin and other uncouplers. Under these conditions there was no gradient of pH or of electrical potential across the plasma membrane and currents of H^+, K^+ and Na^+ were short-circuited. Growth required a rich medium, a slightly alkaline pH value (7.7), and a relatively high external concentration of K^+ (0.28 M). Clearly, in such a medium there is no need for the maintenance of gradients of the relevant nutrients, since their external concentration will be at least as high as the intracellular concentration necessary for growth. In commonly used culture media, however, a high concentration of an uncoupler could be fatally toxic and therefore has to be avoided.

Formic, acetic, propionic and butyric acids are common fermentation products and, because they are weak lipophilic acids, are also uncouplers particularly at pH values near to their respective pK_a values, where the concentration of the undissociated acid is substantial. This provides a

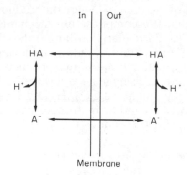

Figure 2 General mechanism of the dissipation of a cation gradient across a membrane by a weak lipophilic acid or an ionophore. H^+ = cation, A^- = lipophilic anion or ionophore. Note that not all ionophores are necessarily negatively charged

rationale for the compensation mechanism: production of these acids at low pH values would lead to an uncontrollable dysfunction of the cells and death of the culture. In this connection Gottschal and Morris (1981) have investigated whether this effect of acetate and butyrate caused the switch of fermentation pattern in cultures of *Clostridium acetobutylicum*. This organism ferments sugars to acetate and butyrate. If the fermentation is allowed to proceed without pH control, the pH drops initially from 7.0 to a value of about 4.5–5.0 and then increases again, due to the conversion of substantial amounts of butyrate and acetate into butanol and acetone. They found that at a low pH value relatively high concentrations of acetate and butyrate were required for the induction of this phenomenon. However, when another weak acid which was known to permeate the clostridial membrane (5,5-dimethyloxazolidine-2,4-dione) was added to the culture, it failed to induce the production of acetone and butanol, but stimulated butanol production only.

The concentration of a product in the culture fluids is, among other factors, dependent on the culture density. Since the uncoupling effect of the lower fatty acids increases with their concentration, a dense culture producing, for example, acetate will be affected to a greater extent than a more dilute culture. In industry the use of very dense populations of microorganisms is preferred. Thus, a physiological incompatibility can arise when one wishes to produce a substance in high concentration, which is itself toxic or whose production is accompanied by the production of other toxic compounds. A way to solve this problem is to remove the toxic compound from the fermentation broth. It was for this reason that dialysis culture was studied and it was shown by Landwall and Holme (1977) that in such a culture very dense populations of microorganisms could be grown without any inhibition of cell proliferation due to accumulated fatty acids. It seems probable that dialysis culture is a promising approach for some types of production processes, although immobilized cells are much more favoured today as an alternative to the usual culture methods.

Another way to circumvent the problem of toxicity of volatile fatty acids has been found in studies of the anaerobic digestion of waste waters. In principle, the generation of methane from organic material occurs in two steps: first the conversion of the organic material into volatile fatty acids by bacteria such as the clostridia, and subsequently the conversion of these acids into methane by the methanogenic bacteria. When this purification is carried out in one compartment, the system is very sensitive to sudden peaks in the availability of organic material (*i.e.* so-called 'shock loading'). When such a peak occurs it causes a dramatic increase in the concentration of the volatile fatty acids and a marked decrease in pH value. Under these circumstances growth of the methanogenic bacteria cannot occur and the culture stops forming methane, but produces only the volatile fatty acids. When, however, the anaerobic digestion was carried out in two fermenters (reactors), one for the conversion of organic material into volatile fatty acids, and the second fermenter carrying out methanogenesis from the waste of the first fermenter, this so-called two-phase system was shown to be considerably more stable (Cohen *et al.*, 1980, 1982). Also, one was able to choose the optimal physiological parameters for both processes independent of each other and thereby avoid sudden 'runaway' situations.

Addition of artificial uncouplers such as 2,4-dinitrophenol (DNP) to the culture medium of *K. aerogenes* had indeed a drastic effect on metabolite production, as is shown in Table 4. The oxygen consumption was increased as was the glucose consumption. Significantly, no polysaccharide was produced, but large amounts of gluconate, and in particular 2-ketogluconate, were formed.

This change in excretion pattern is consistent with the explanation given in the previous section for ammonia and potassium limitation, respectively. The presence of the uncoupler causes a metabolic drain on top of a metabolic bottleneck. Moreover, due to the increased demand for reducing equivalents by the respiratory chain, the $NAD(P)H/NAD(P)^+$ ratio will be lowered when compared with ammonia-limited cells in the absence of DNP and therefore the synthesis of polysaccharides will be suppressed. In carbon-limited cells at low growth rates (0.08 h^{-1}) the presence of DNP at a concentration of 0.25 mM caused excretion of low amounts of acetate, but a DNP concentration of 1 mM had a greater effect. Substantial amounts of acetate and D-lactate were produced and although the bacteria took up oxygen, their metabolism was similar to that of anaerobic cells, the ATP being generated by substrate level phosphorylation and respiratory energy seemingly being used to counteract the leakage of H$^+$ ions induced by DNP (data not shown, Neijssel, 1977).

Table 4 Substrate Utilization Rates and Rates of Product Formation in Ammonia-limited Chemostat Cultures of *Klebsiella aerogenes* Grown in the Presence and Absence of 2,4-Dinitrophenol[a]

DNP (mM)	Glucose	O_2	NH_3	Cells	CO_2	H_2O	HAc	Pyr	2-OG	GA	KGA	$(CH_2O)_n$
0	36	44	9	1	40	72	4	3.5	11	0	0	1.9
1	84	76	9	1	35	123	8	19	7	4	48	0

[a] Cells were grown in a mineral medium ($D = 0.17 \pm 0.01$ h^{-1}; pH 6.8 ± 0.1; 35 °C). All values are adjusted to a cell production rate of 1 kg h^{-1}. Extracellular polysaccharide, $(CH_2O)_n$, is expressed in kg h^{-1}, whereas the other values are expressed in mol h^{-1}. HAc = acetate; Pyr = pyruvate, 2-OG = 2-oxoglutarate, GA = gluconate, KGA = 2-ketogluconate. Recalculated from the data of Neijssel and Tempest (1975) and Neijssel (1977). The chemical composition of cell mass was taken from the data of Herbert (1976). The values of NH_3 and H_2O have been calculated so as to achieve an approximately ($\pm 5\%$) balanced chemical equation.

One can conclude that the pH value of the culture medium, and the possible uncoupling effect of some fermentation products, have a profound influence on product formation. In some cases this effect is advantageous and stimulates the synthesis of more or different compounds, in other cases, where the effect of the uncouplers is too toxic, their effect must be avoided by taking appropriate measures.

32.4 EFFECTS OF TEMPERATURE

Temperature is an important process parameter for at least two reasons: firstly, it has an influence on microbial physiology, and secondly the maintenance of an optimal process temperature is, in industry, one of the major factors in the economics of the process. There are a number of reports in the literature on the effect of temperature on the synthesis of metabolites. The biosynthesis of iron transport compounds is such an example. *Salmonella typhimurium* excretes a phenolate iron transport compound into the medium when grown at 31.0 °C. As the temperature of incubation is raised from 31.0 to 36.9 °C, the organism excretes less of this compound; a further rise of the temperature to 40.3 °C results in an inability of the organism to grow at that temperature unless it is supplemented with iron transport compounds. As expected, relatively high concentrations of iron in the medium suppress the synthesis of the phenolate iron transport compounds (Garibaldi, 1972). A similar observation was made with a fluorescent pseudomonad (Garibaldi, 1971).

The synthesis of other toxins is also under temperature control. *Aspergillus ochraceus* produces penicillic acid when grown at 10–20 °C, whereas ochratoxin A is produced at 28 °C (Ciegler, 1972). Dietrich *et al.* (1972) showed that enterotoxin formation by *Staphylococcus aureus* at 37 °C was four times higher than when the organism was grown at 25 °C, whereas the cell yield at 25 °C was about 16% higher than at 37 °C.

An interesting application of the effect of temperature on a fermentation has been made with the production of ethanol by *Bacillus stearothermophilus*. The organism was grown in an anaerobic chemostat in a complex medium with 3% glucose. The temperature was maintained at 65 °C and this relatively high temperature value caused a distillation of a major part of the ethanol that was dissolved in the culture's extracellular fluids. The total amount of ethanol produced depended on the dilution rate (it increased with D) but the amount of ethanol distilled depended almost entirely upon the rate of gas (N_2) flow through the medium. The level of ethanol in the distillate was only 5.3%, but this represented a 76-fold concentration of the alcohol produced by the culture and a consequent considerable reduction of its concentration in the culture fluids

(Atkinson *et al.*, 1975). This organism is obviously not suitable for producing ethanol, but these experiments suggest that alcohol production by thermophilic microorganisms may present a simple alternative for concentrating the product while simultaneously reducing its concentration within the culture, thus permitting the fermentation of normally unacceptably high levels of carbohydrate. In this case the accumulation of a potentially toxic product of metabolism was prevented by means of simple manipulation of one of the process parameters.

The examples that have been discussed in this section show that there is generally a good description of the effect of temperature on metabolite production, but that again a detailed understanding of the mechanism of this parameter is usually lacking.

32.5 CONCLUSIONS

In the preceding sections we have tried to formulate some general principles of the physiology of metabolite overproduction. These have been demonstrated with some examples in which the physiological significance of product formation was clear and understandable. At the same time we have already noted that in many cases, particularly with the production of secondary metabolites, a physiological understanding of this process is still lacking. It is therefore necessary to undertake more studies in this area, keeping in mind the lessons learned from similar studies on the production of primary metabolites. On the other hand, one must not overlook the fact that the actinomycetes, whose members are the most prolific producers of secondary metabolites, are very different indeed from *Escherichia coli* or *Klebsiella aerogenes*. One of the most prominent differences is the capacity of the Actinomycetales to show cell differentiation and, as noted previously, many authors have argued that production of secondary metabolites serves a function in this process. If this is true, this behaviour has consequences for choosing the best culture method. Cell differentiation means that there is a development of the organism in time and space, but a typical property of a well stirred batch culture is that it is homogenous (*i.e.* space independent) and a chemostat culture in steady state is both homogenous and time independent. For this reason Campbell *et al.* (1982) prefer to use solid cultures for studying the function of secondary metabolites. They consider the whole colony of a mycelium-forming microbe more or less as one organism and therefore use different types of solid culture which enable them to study aerial mycelium, vegetative mycelium and surrounding medium independently from one another. With two *Penicillium* spp. three patterns of formation and subsequent disposal within or around the colony emerged: (i) discharge into the medium, (ii) localization predominantly in the aerial mycelium and (iii) distribution between aerial and vegetative mycelia. Moreover, they were able to show that conidiophores of *Penicillium brevicompactum* that have a penicillus that is mature enough to be producing brevianamide rotate when illuminated by white or UV light. If no brevianamide was present, no rotation was observed.

In many cases, however, a stirred liquid culture is optimal and preferable for the production of substances. From a physiological point of view a chemostat culture is superior to a batch culture in that the former allows the experimenter to vary a single environmental parameter independently from the others, while in a batch culture these parameters are mostly interdependent. Using the chemostat technique, a vast amount of knowledge on the behaviour of microorganisms in nutrient-limited environments has been amassed and it is this sort of knowledge in particular which is essential for a better understanding of the physiology of product formation. This does not imply that genetic studies are irrelevant; on the contrary, these are also essential and have provided us with spectacular results in the last ten years, but a genetically engineered producer organism must be maintained in optimal condition during the production process, and it is for the control of this process that a good understanding of physiology will prove to be essential.

A good example of a successful combination of physiology and genetics is the production of single cell protein from methanol by ICI. In this process the organisms themselves are the desired product and, keeping in mind the considerations put forward in this chapter, it is not surprising to learn that a methanol-limited culture was chosen for cell production. Nevertheless, the investigators identified a potential source of methanol wastage due to the organisms' mode of ammonia assimilation. Instead of using glutamate dehydrogenase to catalyse the conversion of ammonia and 2-oxoglutarate into glutamate, this organism used the two-stage pathway involving glutamate synthetase and glutamate synthase, to carry out the same function. The latter pathway requires one extra molecule of ATP per molecule of glutamate formed. By means of genetic engineering this energetically expensive pathway was removed and the glutamate dehydrogenase gene of *Escherichia coli*, cloned into broad host-range plasmids, was introduced into the organism. The resultant organism was tested in a methanol-limited chemostat at a dilution rate of $0.2\ h^{-1}$. Subsequent enzyme assays and plasmid analysis showed that the *gdh* gene was stably maintained, even in the absence of antibiotic selection, and that the modified organism gave 4–7% higher carbon conversion than its unmodified precursor (Windass *et al.*, 1980). This example typifies the fruitful complementary use of physiology and genetics in optimizing a microbial process and sets the trend for future developments.

32.6 REFERENCES

Atkinson, A., D. C. Ellwood, C. G. T. Evans and R. G. Yeo (1975). Production of alcohol by *Bacillus stearothermophilus*. *Biotechnol. Bioeng.*, **17**, 1375–1377.

Blackwood, A. C., A. C. Neish and G. A. Ledingham (1956). Dissimilation of glucose at controlled pH values by pigmented and non-pigmented strains of *Escherichia coli*. *J. Bacteriol.*, **72**, 497–499.

Campbell, I. M., D. L. Doerfler, B. A. Bird, A. T. Remaley, L. M. Rosato and B. N. Davis (1982). Secondary metabolism and colony development in solid cultures of *Penicillium brevicompactum* and *Penicillium patulum*. In *Overproduction of Microbial Products*, ed. V. Krumphanzl, B. Sikyta and Z. Vanek, pp. 141–151. Academic, London.

Ciegler, A. (1972). Bioproduction of ochratoxin A and penicillic acid by members of the *Aspergillus ochraceus* group. *Can. J. Microbiol.*, **18**, 631–636.

Cohen, A., A. M. Breure, J. G. van Andel and A. van Deursen (1980). Influence of phase separation on the anaerobic digestion of glucose, I. Maximum COP-turnover rate during continuous operation. *Water Res.*, **14**, 1439–1448.

Cohen, A., A. M. Breure, J. G. van Andel and A. van Deursen (1982). Influence of phase separation on the anaerobic digestion of glucose, II. Stability and kinetic responses to shock loadings. *Water Res.*, **16**, 449–455.

Dawes, E. A. and P. J. Senior (1973). The role and regulation of energy reserve polymers in microorganisms. In *Advances in Microbial Physiology*, ed. A. H. Rose and D. W. Tempest, vol. 10, pp. 135–266. Academic, London.

Demain, A. L. (1982). Catabolite regulation in industrial microbiology. In *Overproduction of Microbial Products*, ed. V. Krumphanzl, B. Sikyta and Z. Vanek, pp. 3–20. Academic, London.

Dietrich, G. G., R. J. Watson and G. J. Silverman (1972). Effect of shaking speed on the secretion of enterotoxin B by *Staphylococcus aureus*. *Appl. Microbiol.*, **24**, 561–566.

Ellar, D. J. (1978). Membrane fluidity in microorganisms. In *Companion to Microbiology*, ed. A. T. Bull and P. M. Meadow, pp. 265–295. Longman, London.

Gale, E. F. and H. M. R. Epps (1942). The effect of the pH of the medium during growth on the enzymic activities of bacteria (*Escherichia coli* and *Micrococcus lysodeikticus*) and the biological significance of the changes produced. *Biochem. J.*, **36**, 600–618.

Garibaldi, J. A. (1971). Influence of temperature on the iron metabolism of a fluorescent pseudomonad. *J. Bacteriol.*, **105**, 1036–1038.

Garibaldi, J. A. (1972). Influence of temperature on the biosynthesis of iron transport compounds by *Salmonella typhimurium*. *J. Bacteriol.*, **110**, 262–265.

Ghei, O. K. and W. W. Kay (1973). Properties of an inducible C_4-dicarboxylic acid transport system in *Bacillus subtilis*. *J. Bacteriol.*, **114**, 65–79.

Glusker, J. P. (1968). Mechanism of aconitase action deduced from crystallographic studies of its substrates. *J. Mol. Biol.*, **38**, 149–162.

Gottschal, J. C. and J. G. Morris (1981). The induction of acetone and butanol production in cultures of *Clostridium acetobutylicum* by elevated concentrations of acetate and butyrate. *FEMS Microbiol. Lett.*, **12**, 385–389.

Hanstein, W. G. (1976). Uncoupling of oxidative phosphorylation. *Biochim. Biophys. Acta*, **456**, 129–148.

Harold, F. M. and J. Van Brunt (1977). Circulation of H^+ and K^+ across the plasma membrane is not obligatory for bacterial growth. *Science*, **197**, 372–373.

Herbert, D. (1961). The chemical composition of microorganisms as a function of their environment. In *Microbial Reaction to Environment, 11th Symposium of the Society for General Microbiology*, ed. G. G. Meynell and H. Gooder, pp. 391–416. Cambridge University Press, Cambridge.

Herbert, D. (1976). Stoicheiometric aspects of microbial growth. In *Continuous Culture 6: Applications and New Fields*, ed. A. C. R. Dean, D. C. Ellwood, C. G. T. Evans and J. Melling, pp. 1–30. Ellis Horwood, Chichester.

Hernandez, E. and M. J. Johnson (1967). Anaerobic growth yields of *Aerobacter cloacae* and *Escherichia coli*. *J. Bacteriol.*, **94**, 991–995.

Hueting, S., T. de Lange and D. W. Tempest (1979). Energy requirement for maintenance of the transmembrane potassium gradient in *Klebsiella aerogenes* NCTC 418: a continuous culture study. *Arch. Microbiol.*, **123**, 183–188.

Katz, E. and A. L. Demain (1977). The peptide antibiotics of *Bacillus*: chemistry, biogenesis, and possible functions. *Bacteriol. Rev.*, **41**, 449–474.

Khokhlov, A. S. (1982). Low molecular weight microbial bioregulators of secondary metabolism. In *Overproduction of Microbial Products*, ed. V. Krumphanzl, B. Sikyta and Z. Vanek, pp. 97–109. Academic, London.

Kubicek, C. P. and M. Röhr (1977). Influence of manganese on enzyme synthesis and citric acid accumulation in *Aspergillus niger*. *Eur. J. Appl. Microbiol.*, **4**, 167–175.

Landwall, P. and T. Holme (1977). Removal of inhibitors of bacterial growth by dialysis culture. *J. Gen. Microbiol.*, **103**, 345–352.

Marchal, R., O. Chaudé and M. Metche (1977). Production of citric acid from *n*-paraffins by *Saccharomycopsis lipolytica*: kinetics and balance of the fermentation. *Eur. J. Appl. Microbiol.*, **4**, 111–123.

Matteo, C. C., C. L. Cooney and A. L. Demain (1976). Production of gramicidin S synthetases by *Bacillus brevis* in continuous culture. *J. Gen. Microbiol.*, **96**, 415–422.

Meyer, J. M. and M. A. Abdallah (1978). The fluorescent pigment of *Pseudomonas fluorescens*: biosynthesis, purification and physicochemical properties. *J. Gen. Microbiol.*, **107**, 319–328.

Meyer, J. M. and J. M. Hornsperger (1978). Role of pyoverdine$_{pf}$, the iron-binding fluorescent pigment of *Pseudomonas fluorescens*, in iron transport. *J. Gen. Microbiol.*, **107**, 329–331.

Miller, A. L. and J. B. Walker (1969). Enzymatic phosphorylation of streptomycin by extracts of streptomycin-producing strains of *Streptomyces. J. Bacteriol.*, **99**, 401–405.

Neijssel, O. M. (1977). The effect of 2,4-dinitrophenol on the growth of *Klebsiella aerogenes* NCTC 418 in aerobic chemostat culture. *FEMS Microbiol. Lett.*, **1**, 47–50.

Neijssel, O. M. and D. W. Tempest (1975). The regulation of carbohydrate metabolism in *Klebsiella aerogenes* NCTC 418 organisms, growing in chemostat culture. *Arch. Microbiol.*, **106**, 251–258.

Neijssel, O. M. and D. W. Tempest (1979). The physiology of metabolite over-production. In *Microbial Technology: Current State, Future Prospects, 29th Symposium of the Society for General Microbiology*, ed. A. T. Bull, D. C. Ellwood and C. Ratledge, pp. 53–82. Cambridge University Press, Cambridge.

Olijve, W. (1978). Glucose metabolism in *Gluconobacter oxydans*. PhD thesis. University of Groningen, The Netherlands.

Pirt, S. J. (1965). The maintenance energy of bacteria in growing cultures. *Proc. R. Soc. London, Ser. B.*, **163**, 224–231.

Pirt, S. J. and D. S. Callow (1958). Exocellular product formation by microorganisms in continuous culture. I. Production of 2,3-butanediol by *Aerobacter aerogenes* in a single stage process. *J. Appl. Bacteriol.*, **21**, 188–205.

Rosenberger, R. F. and S. R. Elsden (1960). The yields of *Streptococcus faecalis* grown in continuous culture. *J. Gen. Microbiol.*, **22**, 726–739.

Størmer, F. C. (1977). Evidence for the regulation of *Aerobacter aerogenes* pH 6 acetolactate-forming enzyme by acetate ion. *Biochem. Biophys. Res. Commun.*, **74**, 898–902.

Stouthamer, A. H. (1979). The search for correlation between theoretical and experimental growth yields. In *International Review of Biochemistry*, ed. J. R. Quayle, vol. 21, pp. 1–47. University Park Press, Baltimore.

Tempest, D. W., J. W. Dicks and J. R. Hunter (1966). The interrelationship between potassium, magnesium and phosphorus in potassium-limited chemostat cultures of *Aerobacter aerogenes. J. Gen. Microbiol.*, **45**, 135–145.

Tempest, D. W., J. L. Meers and C. M. Brown (1970). Influence of environment on the content and composition of microbial free amino acid pools. *J. Gen. Microbiol.*, **64**, 171–185.

Tempest, D. W., J. L. Meers and C. M. Brown (1973). Glutamate synthase (GOGAT); a key enzyme in the assimilation of ammonia by prokaryotic organisms. In *The Enzymes of Glutamine Metabolism*, ed. S. Prusiner and E. R. Stadtman, pp. 167–182. Academic, London.

Tempest, D. W. and O. M. Neijssel (1978). Ecophysiological aspects of microbial growth in aerobic nutrient-limited environments. In *Advances in Microbial Ecology*, ed. M. Alexander, vol. 2, pp. 105–153. Plenum, New York.

Tempest, D. W. and O. M. Neijssel (1980). Growth yield values in relation to respiration. In *Diversity of Bacterial Respiratory Systems*, ed. C. J. Knowles, vol. 1, pp. 1–32. CRC Press, Boca Raton, FL.

Windass, J. D., M. J. Worsey, E. M. Pioli, D. Pioli, P. T. Barth, K. T. Atherton, E. C. Dart, D. Byrom, K. Powell and P. J. Senior (1980). Improved conversion of methanol to single-cell protein by *Methylophilus methylotrophus. Nature, (London)*, **287**, 396–401.

Wood, W. A. (1961). Fermentation of carbohydrates and related compounds. In *The Bacteria—a Treatise on Structure and Function*, ed. I. C. Gunsalus and R. Y. Stanier, vol. II, pp. 59–149. Academic, London.

33

Regulation of Metabolite Synthesis

M. L. BRITZ* and A. L. DEMAIN
Massachusetts Institute of Technology, Cambridge, MA, USA

33.1 INTRODUCTION

Primary metabolites are those microbial products whose synthesis is an integral part of the normal growth process and which are made during the exponential phase of growth. They include intermediates and end-products of anabolic metabolism which are normally used by the cell as building blocks for essential macromolecules (*e.g.* amino acids, nucleotides) or are converted to coenzymes (*e.g.* vitamins). Other primary metabolites result from catabolic metabolism; they are not used for building cellular constituents but their production is essential to allow growth to continue, as it is related to energy production and substrate utilization (*e.g.* citric acid, acetic acid and ethanol). Industrially, the most important metabolites are the amino acids, nucleotides, vitamins, solvents and organic acids, which are made by a diverse range of bacteria and fungi and have numerous uses in the food and chemical industries (Table 1). Many of these metabolites are made by microbial fermentation rather than chemical synthesis because the fermentations are economically competitive and produce the biologically useful isomeric forms. Several other industrially important chemicals have the potential of being made by microbial fermentations (*e.g.* glycerol and other polyhydroxy alcohols; Spencer and Spencer, 1978) but are presently synthesized cheaply as petroleum by-products. This is particularly relevant to solvent production where the popularity of the acetone–butanol fermentation and other analogous systems has had an erratic history which correlates with the availability and cost of petroleum (Hastings, 1978). There is now renewed interest in the microbial production of ethanol, organic acids (such as acetic acid), and other solvents as the cost of petroleum fluctuates and the necessity for eventual replacement fuels has become apparent. Although acetone and butanol can be regarded as 'marginal' primary metabolites, since their synthesis by *Clostridium acetobutylicum* is not an absolute requirement for growth (Ryden, 1958), they are considered here because of some recent novel developments which may elucidate some of the factors involved in regulation of their formation. Detailed descriptions of solvent fermentations will appear elsewhere in this series. This chapter

* Now at CSIRO, Clayton, Victoria, Australia.

aims at describing mechanisms of regulation relevant to primary metabolite synthesis, providing examples from the fermentation industry, and indicating approaches which have facilitated strain development for primary metabolite production.

Table 1 Some Primary Metabolites of Industrial Interest and Some Overproducing Species

Metabolite	Organism	Uses (present and future)
Glutamate	*Corynebacterium glutamicum, Brevibacterium* spp.	Food industry as flavor-accentuating agent
Lysine, threonine	*Corynebacterium glutamicum, Brevibacterium flavum, Escherichia coli*	Essential amino acids, added to supplement low grade protein
Guanylic, inosinic and xanthylic acids	*Bacillus subtilis, Corynebacterium glutamicum*	Food industry as flavor-accentuating agents
Citric acid	*Aspergillus niger, Candida* spp.	Acidulant in food industry; pharmaceuticals (effervescent powders); esters used as plasticizers
Itaconic acid	*Aspergillus terreus, Aspergillus itaconicus*	Synthetic fiber and resin manufacture; copolymers (*e.g.* styrene–butadiene)
Fumaric acid	*Rhizopus arrhizus, Rhizopus nigricans*	Plastics and food industries
Cyanocobalamin (vitamin B_{12})	*Propionibacterium shermanii, Pseudomonas denitrificans*	Food and feed supplement
Riboflavin (vitamin B_2)	*Eremothecium ashbyii, Ashbya gossypii*	Food and feed supplement
β-Carotene	*Blakeslea* spp., *Choanephora* spp.	Precursor of vitamin A; coloring agent in food industry; food supplement
Xanthan gum, dextran	*Xanthomonas campestris, Leuconostoc mesenteroides*	Food, pharmaceutical, textile industries, useful for thickening, stiffening and setting properties
Acetic acid	*Saccharomyces ellipsoideus* or *Saccharomyces cerevisiae,* plus *Acetobacter* spp.; *Clostridium* spp.	Vinegar, preservative in food industry; chemical feedstocks; polymer industry
Acetone, butanol	*Clostridium acetobutylicum*	Solvents in chemical industry; thinners; synthetic polymers.
Ethanol	*Saccharomyces cerevisiae, Zymomonas mobilis, Clostridium thermocellum* and other *Clostridium* spp.	Alcoholic beverages; solvent in chemical industry; fuel extender

33.2 REGULATION OF METABOLISM

Microbial metabolism is an integrated process which requires the coordinated activity of numerous enzymes involved in the assimilation of nutrients (sources of carbon, nitrogen, phosphorus, sulfur, and minerals) from the environment into cellular constituents. This is a conservative process which, under normal circumstances, does not expend energy or carbon for making compounds already available in the environment, nor does it overproduce components of intermediary metabolism. Coordination of metabolism functions to ensure that, at any particular moment, only the necessary enzymes, and the correct amount of each, are made. Once sufficient quantities of a particular material are made, the enzymes concerned with its formation are no longer synthesized and the activity of preformed enzymes is curbed by a number of specific regulatory mechanisms. Industrial synthesis of primary metabolites aims at circumventing these usually rigid regulatory mechanisms to create, in essence, inefficient microorganisms which overproduce products or intermediates of anabolic or catabolic metabolism. One way this can be achieved is by altering environmental conditions, which manipulates the functioning of metabolic enzymes and alters the organism's phenotype while not altering the genetic make-up of the organism. Microbes sense such changes in their environment and respond to these by changing their composition and metabolism. A good example of this is found in the citric acid fermentation of *Aspergillus niger* and similar filamentous molds (Miall, 1978). These organisms have an inherent capacity to excrete small amounts of some organic acids during aerobic growth. This phenomenon is exploited for the commercial manufacture of citric acid by providing fermentation conditions which amplify this metabolic peculiarity. In *A. niger*, glucose is converted to pyruvate by the Embden–Meyerhof–Parnas (EMP) pathway, thence to acetyl-coenzyme A, which enters the tricarboxylic acid (TCA) cycle by condensing with oxaloacetate to form citrate.

Normally, citrate is converted to *cis*-aconitate, and then to isocitrate, by the action of aconitate hydratase. This enzyme appears to require iron for its activity. Thus, controlling trace elements in the fermentation by use of ferrocyanide to chelate divalent cations and then supplementing the medium with essential trace elements leads to an inhibition of citrate breakdown and accumulation of this TCA cycle intermediate. A similar effect has been achieved with *Candida guilliermondii* which excretes large quantities of citric acid when cultured in the presence of metabolic inhibitors (such as sodium fluoroacetate, *n*-hexadecylcitric acid or *trans*-aconitic acid) which block the TCA cycle at the same point (see Miall, 1978). Although some strain improvement has been reported (Hannan *et al.*, 1973), suitably cultured natural isolates produce high yields of citric acid. In contrast, strain improvement is required absolutely in other systems to develop organisms with altered genotypes resulting in deregulation of metabolism. Indeed, it is usually the combination of a genetically modified microbe and growth under optimal environmental conditions which allows commercial synthesis of primary metabolites. Models of regulatory systems, and circumvention of these in strain development, are described in the following sections.

33.2.1 Induction

Some enzymes synthesized by microorganisms are present regardless of the growth medium, such as those which convert glucose to pyruvate. These enzymes are termed 'constitutive' to distinguish them from other enzymes which are formed to a large extent only when their substrate (or a closely-related compound) is present. Substrate analogues which are not attacked by the enzyme (gratuitous inducers) are often excellent inducers of enzyme synthesis. Induction allows rapid formation of enzymes when they are needed, for example when the organism is in an environment containing a specific compound (*e.g.* a polysaccharide, oligosaccharide or an amino acid) as the sole carbon and energy source, and thus prevents the wastage of energy and amino acids for making unnecessary enzymes. Although most inducers are substrates of catabolic enzymes, products function as inducers in certain cases; for example in some microorganisms, maltodextrins induce amylase, fatty acids induce lipase, urocanic acid induces histidase, and galacturonic acid induces polygalacturonase. Some coenzymes induce enzymes, as in thiamine induction of pyruvate decarboxylase.

The most thoroughly studied inducible enzyme system is that for lactose hydrolysis in *Escherichia coli*, which provided the basis of a model system for negative control of protein synthesis described originally by Jacob and Monod (1961). Synthesis of enzymic proteins is encoded by structural genes which are often grouped together on the genome next to regulatory loci known as the promoter and operator (Figure 1). The group of structural genes in association with operator and promoter regions is known as an operon. In the absence of inducer, structural genes are not transcribed by RNA polymerase (hence enzyme production is repressed) since this enzyme is either unable to bind to its attachment site on DNA (the promoter region) or, once bound, cannot move along the DNA to transcribe. This hindrance is brought about by the prior binding of a regulatory (repressor) protein to its attachment site on the operator locus. Negative control means that the regulatory protein coded by the regulator locus interferes with transcription. The regulator gene coding for the repressor protein need not be located near the operon it regulates. Induction of enzyme synthesis occurs when the repressor is not in contact with the operator region; this is brought about when the inducer is present in the medium (or synthesized by the cell) *i.e.* inducer binds to the repressor and removes it from the operator. Enzyme synthesis is then derepressed or 'turned on' when there is sufficient inducer to bind all of the repressor molecules in the cell. In the case of the *lac* operon in *E. coli*, about 10 molecules of repressor are made per regulator gene (Riggs *et al.*, 1970).

Another type of control mechanism exists where the product of the regulator gene functions in positive regulation of transcription. Positive control means that the regulatory protein coded by the regulator is necessary for transcription to occur. It occurs in *E. coli* for utilization of L-rhamnose, maltose and arabinose (see Rose, 1976). In the latter case, the product of the *ara* C (regulator) gene is a protein which combines with arabinose to form an activator molecule which acts at an initiator site in the operon. Transcription and translation are then free to occur and arabinose is utilized. The activator molecule therefore positively controls the reading of genes in the presence of substrate. Although induction has no role in controlling biosynthetic operons in *E. coli*, it does in other microorganisms. In *Pseudomonas putida*, tryptophan synthetase is induced by

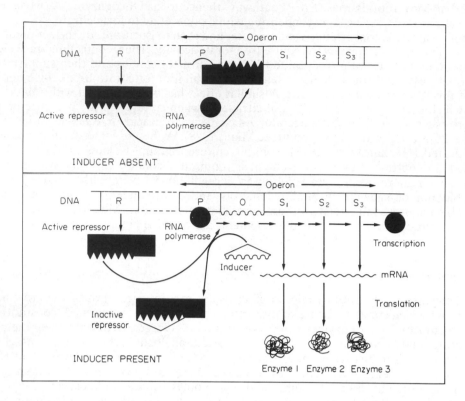

Figure 1 Negative control of enzyme synthesis by induction (see text): R = regulatory gene; P = promoter; O = operator; S_1, S_2, S_3 = structural genes; mRNA = messenger RNA

indole glycerophosphate (Proctor and Crawford, 1976) and the entire tryptophan branch is induced by chorismate in *Bacillus subtilis* (Kane and Jensen, 1970). Induction in fungi such as *Aspergillus nidulans* is mainly of the positive control type (Arst, 1981).

Regulatory mutations can be used to eliminate the dependence of enzyme formation on inducer. This occurs when the regulator gene is modified to produce a non-functional repressor, or mutation in the operator gene eliminates the ability of repressor to bind. Such mutants, which produce a normally inducible enzyme in the absence of inducer, are called 'constitutive' mutants. Many procedures have been devised to isolate mutants of *E. coli* constitutive for β-galactosidase synthesis which are based on the survival and growth of mutants under poor conditions of induction; wild-type cells, still requiring induction, are disadvantaged due to poor substrate utilization and become a smaller proportion of the population. Thus, maintaining *E. coli* in a chemostat with low concentrations of lactose selects for mutants which can efficiently use low levels of substrate, *i.e.* are no longer repressed in the absence of inducing substrate (Novick and Horiuchi, 1961). Similarly, growing mutagenized cells on a compound which is a good carbon source but a poor or non-inducer (*e.g.* phenyl-β-galactoside or 2-nitrophenyl-α-L-arabinoside for selection of constitutive β-galactosidase mutants) selects for those cells which use the substrate well and no longer require induction (Jacob and Monod, 1961; Jayaraman *et al.*, 1966). This method has recently been used to isolate constitutive maltase producers from *Saccharomyces italicus*; the non-inducing substrate used was sucrose. In addition to constitutivity, the mutant was a 2–3 fold hyperproducer (Schaefer and Cooney, 1982).

33.2.2 Nutritional Repression

Carbon 'catabolite' repression, like enzyme induction, is one of the conservative mechanisms which safeguard against waste of the cell's protein-synthesizing machinery, and operates when more than one utilizable substrate is present in the environment. The cell produces enzymes to catabolize the best carbon substrate present (usually glucose), *i.e.* the most rapidly assimilated carbon source; synthesis of enzymes utilizing other substrates is repressed until the primary substrate is exhausted. The repressed enzymes are usually inducible, although constitutive enzymes may also be involved, and very potent inducers can sometimes reverse catabolite repression. Catabolite repression is a phenomenon usually caused by glucose but in different organisms, other rapidly metabolized carbon sources can repress and, indeed, even repress glucose catabolism. An example of this occurs in *Pseudomonas aeruginosa*, where citrate is the preferred carbon source over glucose (Clarke and Lilly, 1969).

In enteric bacteria, catabolite repression is mediated by the intracellular levels of cyclic 3',5'-adenosine monophosphate (c'AMP). The main evidence for this has been provided by studies with *E. coli*. In *E. coli*, c'AMP stimulates the synthesis of a large number of enzymes and is necessary for synthesis of mRNA corresponding to all inducible operons. c'AMP binds to the promoter region *via* a specific binding protein (c'AMP receptor protein or CRP), a dimer of identical subunits and two separate domains. The N-terminus attaches to c'AMP and the C-terminus to DNA. Each CRP subunit binds one c'AMP molecule. After binding c'AMP, the CRP undergoes an allosteric transition to an active state in which it binds to specific portions of DNA, thus increasing the affinity of RNA polymerase to that particular promoter and thus the frequency of transcription (Pastan and Perlman, 1970; Botsford, 1981). Binding of CRP-c'AMP to its part of the promoter region is necessary for binding of RNA polymerase to its site on the promoter and thus initiation of transcription which occurs at a site 30 to 50 nucleotides distal to the CRP-c'AMP binding site. Promoters of different operons have different affinities for the complex (Piovant and Lazdunski, 1975). Thus, not all promoters are binding the complex and undergoing transcription initiation at the same time. During glucose assimilation, the intracellular concentration of c'AMP is depressed 1000-fold, whereas metabolism of a non-repressive carbon source has little effect on c'AMP levels. c'AMP reverses catabolite repression of many enzymes in *E. coli*. Mutants of *E. coli* which cannot make an effective c'AMP binding protein or adenylate cyclase (the enzyme converting ATP to c'AMP) fail to grow or grow poorly on lactose, glycerol or other sugars, whereas mutants lacking c'AMP phosphodiesterase (which degrades c'AMP to AMP) are insensitive to catabolite repression (Monard *et al.*, 1969). c'AMP may regulate enzyme action in yeasts, its presence having been described in *Saccharomyces cerevisiae*, *Kluyveromyces fragilis* and *Schizosaccharomyces pombe*. However, in some other organisms where catabolite repression occurs, for example in *Bacillus* species *P. aeruginosa*, *Arthrobacter crystallopoietes*, *Rhizobium meliloti* and anaerobic bacteria such as *Bacteroides fragilis*, c'AMP has not been detected (Siegel *et al.*, 1977) or has been shown not to play a role in carbon source repression of enzymes (Botsford, 1981). It is doubtful that c'AMP plays any role in carbon source repression in molds such as *A. nidulans* (Arst and Bailey, 1977).

Since carbon catabolite repression in *E. coli* appears to be caused not by a specific catabolite of glucose or other rapidly-assimilated carbon source but by transport of such compounds, we prefer the term 'carbon source' repression over 'carbon catabolite' repression. Transport systems known to inhibit adenylate cyclase include the PEP-phosphotransferase system, proton symport and facilitated diffusion (Botsford, 1981).

Carbon source repression can be avoided simply by not using the repressive carbon source (if non-repressive carbon sources are utilized and costs remain competitive) or by limiting the rate of growth by controlled feeding of the repressive carbon source. Production of cellulase by *Pseudomonas fluorescens* var. *cellulosa* is suppressed by glucose and galactose; growing this organism on mannose results in cells producing over 1500 times as much cellulase as cells grown on galactose. Similar relief from carbon source repression of cellulase production in this organism occurs by slow feeding on glucose, wherein cellulase production increases by almost 200-fold.

Alternatively, mutants can be selected which are resistant to carbon source repression. Many such mutants are apparently modified in their glucose catabolic pathway, which causes slower glucose utilization and derepression of a number of usually repressed enzymes. Mutation to resistance to carbon source repression involving specific enzymes can also be achieved by making this type of mutation obligatory for growth. This occurs, for example, when histidine is offered as the sole nitrogen source for growth of *Klebsiella aerogenes* on media containing glucose. Glucose usually represses production of the histidine-degrading enzymes. However, growth cannot occur unless these enzymes are produced so that histidine can provide nitrogen for cell growth, hence only derepressed mutants will grow under these conditions (Neidhardt, 1960). Another type of method relies on the fact that synthesis of normally repressed enzymes takes a particular period of time after the organisms have been transferred from a repressing carbon source. Such a lag in initiation of growth occurs when *E. coli* is transferred from glucose to a medium containing lactose, maltose, acetate, or succinate (Hsie and Rickenberg, 1967). Alternation of growth between glucose and, for example, succinate selects mutants resistant to carbon source repression, as only organisms which are making enzymes for succinate utilization during growth on glucose grow rapidly after transfer. These mutants have a shorter generation time on maltose, acetate or succinate than their repressible parents, which aids in the selection.

Nitrogen source repression is known by many other names such as nitrogen metabolite repression, nitrogen catabolite repression or ammonia repression. Simply, it is the repression of enzymes acting on nitrogenous substrates by high concentrations of ammonium or certain organic nitrogen compounds. Enzymes typically under such control are proteases, amidases, urease and amino acid-degrading enzymes. An enzyme of practical significance which is regulated by ammonium ion is glycine decarboxylase. This enzyme is crucial in the bioconversion of glycine to L-serine by *Nocardia butanica* (Tanaka *et al.*, 1981). Addition to the medium of the ammonium-trapping agent tribasic magnesium phosphate derepressed the enzyme two- to five-fold and increased serine production three- to eight-fold. Key enzymes which appear to be involved in the mechanism of nitrogen source repression are those of ammonium assimilation such as NADP-glutamate dehydrogenase and glutamine synthetase but it is not clear whether they are involved as regulatory proteins or simply as catalytic proteins; in the latter case, the pool sizes of one or more substrates and/or products of these enzymes would be critical in bringing about repression, *e.g.* glutamine or glutamate. A gene controlling nitrogen source repression (*are* A) has been identified in *Aspergillus* (for an excellent review, see Marzluff, 1981). In molds, the gene codes for a regulatory protein exerting positive control on transcription. The regulatory protein appears to be active under conditions of derepression (*e.g.* low ammonium supply) and inactive under repressive conditions (*e.g.* high ammonium supply). The actual intracellular effector appears to be glutamine rather than ammonia but more work is needed on this point. Derepressed mutants can often be obtained by selection for resistance to methylammonium, a toxic analogue of ammonium (Arst and Cove, 1969).

Nutrient repression of certain enzymes is also effected by high concentrations of phosphorus sources (especially orthophosphate) and sulfur sources (especially inorganic sulfate); for example nucleases and phosphatases are usually repressed by phosphate. Recently, it has been observed that isocitrate lyase, fructose diphosphate aldolase, NADP isocitrate dehydrogenase and malate dehydrogenase in *Neurospora* are repressed by phosphate (Savant *et al.*, 1982).

Phosphate suppresses the production of riboflavin by *Eremothecium ashbyii*. The mechanism appears to involve phosphate repression of flavin mononucleotide (FMN) hydrolase and stimulation of flavokinase production resulting in conversion of riboflavin to FMN (Mehta and Modi, 1981). Regulation of additional enzymes, probably prior to riboflavin, must also be involved since the total production of riboflavin and FMN is decreased by excess phosphate.

Phosphate-derepressed mutants can be selected by growth with a phosphate ester (*e.g.* β-glycerol phosphate) as the sole source of carbon in the presence of high phosphate concentrations (Torriani and Rothman, 1961). Proteases and sulfatases are often repressed by sulfate or sulfur amino acids.

33.2.3 Feedback Regulation

Induction and catabolite repression by carbon, nitrogen, sulfur and phosphorus sources are important in the context of regulating industrial extracellular enzymes such as amylases, proteases and cellulases (see Elander and Demain, 1981). They are also important considerations in primary metabolite regulation for two other reasons: (1) they may be involved in regulation of breakdown of a major substrate, which could be a macromolecule such as cellulose or starch and which might be a cheap carbon source, and hence determine the rate and efficiency of carbon flow down a particular pathway; and (2) efficient utilization of carbon sources presented as mixtures of carbohydrates, as in molasses, may be achieved only after carbon source repression is overcome or constitutive synthesis of degradative enzymes occurs. However, the most important mechanism responsible for regulation of biosynthetic enzymes concerned with making amino acids, nucleotides and vitamins is feedback regulation. This category of regulation functions at two levels: at the level of enzyme action (feedback inhibition) and at the level of enzyme synthesis (feedback repression). Both mechanisms act to adjust the rate of production of end-products to the rate of synthesis of macromolecules required by the cell at any particular moment.

In feedback inhibition, the final metabolite of a pathway, when present in sufficient quantities, inhibits the action of one or more early enzymes of the pathway (usually the first enzyme) to prevent further synthesis of intermediates and products of that pathway. The end-product need not resemble the enzyme substrate in size, shape or charge since it binds to a regulatory site on the protein which differs from the substrate binding site. Binding of the inhibitor alters the conformation of the enzyme and prevents attachment of the substrate to the catalytic site: such enzymes are termed 'allosteric' (other shape). Some allosteric enzymes are composed of two types of subunit, one being a catalytic subunit binding the substrate and another a regulatory subunit binding the inhibitor (*e.g.* aspartate transcarbamylase), whereas other allosteric enzymes have both regulatory and catalytic sites within the same polypeptide.

Feedback repression involves the 'turning off' of enzyme synthesis when sufficient amounts of the product have been made and it starts to accumulate. The mechanism of this process is shown in Figure 2. In the absence of protein repressor, RNA polymerase binds to the promoter region and then moves on to transcribe the structural genes coding for the enzymes of the pathway. The end-product of the pathway acts as a corepressor. The aporepressor specified by the regulator gene is inactive in the absence of its corepressor and is unable to bind to the operator. However, in the presence of corepressor, an active repressor is formed which binds to the operator to prevent transcription by RNA polymerase and hence prevents enzyme synthesis. Feedback repression is a widespread phenomenon and is particularly important in regulating pathways where only a limited number of molecules of product are needed per cell, as in the case of synthesis of vitamins.

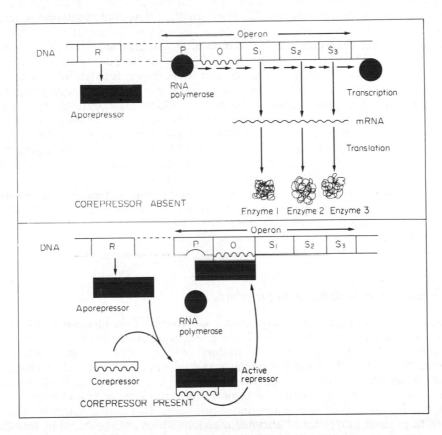

Figure 2 Feedback repression of enzyme synthesis (see text and legend to Figure 1)

Many of the amino acid biosynthetic pathways are regulated not by the amino acids themselves but by their charged tRNA molecules. Thus, whereas feedback repression is effected by amino acid end-products acting as corepressors interfering with transcription initiation, another type of control called attenuation involves charged tRNA and transcription termination, *i.e.* in the presence of an excess of charged tRNA, transcription is initiated but terminated before the first structured gene is transcribed (Kolter and Yanofsky, 1982).

Attenuation (also known as 'transcription termination control') is known to control six bacterial amino acid biosynthetic operons producing seven amino acids (Thr, Ile, Val, Trp, Leu, Phe, His). The *trp* operon actually employs both repression and attenuation control.

Attenuation requires the participation of RNA polymerase and charged tRNA. In operons controlled by attenuation, there is a 140 to 200 base pair leader sequence between the promotor and the first structural gene. The upstream portion of the leader region contains codons coding for a particular amino acid which is the product of the operon. Thus the *thr* operon contains eight Thr and four Ile codons (there is multivalent control by Thr and Ile); the *ilv* operon leader region contains several Ile, Leu and Val codons (there is multivalent control by Ile, Leu and Val).

The crucial event in attenuation is the manner in which the mRNA transcripts of the leader region forms base pairs. When an amino acid is in excess, *i.e.* when there is a high concentration of the particular charged tRNA available for translation of the early leader region, one type of mRNA base pairing occurs forming a 'terminator' structure. Under conditions of amino acid limitation and thus a deficiency in charged tRNA, another type of mRNA base pairing occurs; this structure is called an antiterminator. Simply, when a terminator is present, RNA polymerase is stopped before it transcribes the first structural gene. On the other hand, when an antiterminator forms, transcription of the operon occurs, followed by translation of the structural genes.

Feedback regulation is complicated by the existence of branched biosynthetic pathways, where a common metabolic sequence branches at one or more points to make two or more products. Without a more sophisticated regulatory mechanism, the possibility could arise in which shutting down the common pathway by one of the products could starve the cell of the other end-products. Many regulatory devices have evolved to avoid this problem; each type is applicable to either feedback inhibition or repression. (1) Isoenzymes: differential regulation is achieved when several enzymes catalyze the same reaction but each enzyme is sensitive to inhibition and/or repression by a different end-product. Regulation of the aspartic acid family of amino acids and the aromatic amino acids occurs by this means in *E. coli* (Umbarger, 1978); the former is shown in Figure 3. (2) Concerted (covalent, multivalent or cooperative) regulation: this occurs when the presence of more than one end-product is required to cause inhibition and/or repression of the one enzyme involved; a single end-product in excess produces little or no effect. Concerted feedback inhibition of aspartokinase by threonine and lysine occurs in *Corynebacterium glutamicum* (Nayakama *et al.*, 1966; Figure 3). (3) Cumulative regulation: each end-product in excess acts separately to cause some inhibition and/or repression but the presence of all inhibitors in excess is required for a total effect. The best example of this is glutamine synthetase in *E. coli*, where eight products regulate the activity of this enzyme (Stadtman, 1966).

Two major types of manipulation have been employed to relieve feedback regulation. The first type prevents accumulation of repressive or inhibitory concentrations of end-products in the cell, whereas the second alters the enzymes or the enzyme-forming systems so they are less sensitive to feedback effects. These are described in the following paragraphs.

33.2.3.1 *Limiting accumulation of end-products*

Substantial increases in enzyme production can be realized if the internal build-up of corepressors is limited. The most effective way of doing this is to obtain an auxotrophic (nutritional) mutant which fails to make the repressive product and requires it as a medium constituent for growth; limited feeding of the product partially starves the mutant of its end-product requirement and thus inhibitory concentrations never occur intracellularly. This principle allows accumulation of intermediates of biosynthetic pathways as well as end-products of branched pathways and has been applied to production of nucleotides and amino acids. In a simple pathway, where substrate is converted to product in a series of enzymic reactions which are regulated by intracellular concentrations of that product, intermediates are accumulated by auxotrophs requiring the final product due to (1) lack of feedback regulation on the earlier enzymes of the pathway, and (2) lack of conversion of the desired intermediate to the next (not necessarily final) product. In branched pathways, auxotrophy not only alleviates feedback regulation but allows rechannelling of carbon into other branches of the pathway, hence accumulation of other end-products. For example, production of lysine by *C. glutamicum* and *Brevibacterium flavum* is regulated *via* concerted feedback inhibition by threonine and lysine on a single aspartokinase (Figure 3). Mutation resulting in the removal of homoserine dehydrogenase yields auxotrophs requiring both methionine and threonine for growth. Limited feeding of threonine by-passes the concerted inhibition of aspartokinase, leading to excretion of over 50 g l^{-1} of lysine in culture fluids (Nakayama, 1976). Accumulation of lysine by these bacteria is aided by both the absence of lysine-degrading enzymes and of regulation in the lysine branch of this pathway.

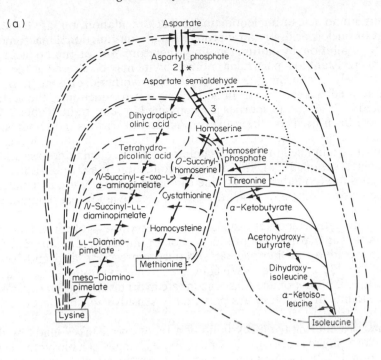

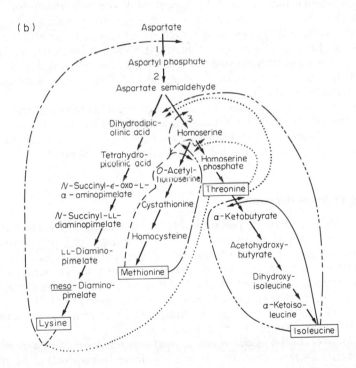

Figure 3 Comparison of feedback regulation of the aspartic acid family of amino acids in *Escherichia coli* and *Corynebacterium glutamicum*. The tight control in *E. coli* is shown in (a) and the loose control in *C. glutamicum* is shown in (b): ---------, lysine repression; ──-──, lysine inhibition; ─────, methionine repression; ────, methionine inhibition; ──────────, isoleucine repression; ─────, isoleucine inhibition; ──────, threonine + isoleucine repression; ·······, threonine inhibition. *Not shown is the additional control in *E. coli* in which concerted feedback repression of the second enzyme (aspartic semialdehyde dehydrogenase) is exerted by lysine, methionine plus threonine

33.2.3.2 *Feedback resistance mutations*

One of the most common ways to remove feedback regulation is to obtain mutants which are resistant to toxic structural analogues of the desired compound. Such antimetabolites substitute for their analogous amino acid or nucleotide in feedback regulation, but are either not incorporated into proteins or nucleic acids or cause production of non-functional macromolecules. Organisms sensitive to inhibition by antimetabolites therefore die of amino acid or nucleotide starvation. Mutants resistant to a particular antimetabolite may be altered at the level of enzyme synthesis (constitutively producing the appropriate biosynthetic enzymes), or may produce enzymes resistant to feedback inhibition. In both cases, overproduction of the metabolite results, and this is amplified if these two properties are combined in one strain. Table 2 lists analogues, many of which have been used to obtain overproduction of various metabolites. Occasionally, selection for resistance to antimetabolites results in gene amplification; for example the *his* operon was found to be amplified in certain 3-amino-1,2,4-triazolealanine resistant mutants of *S. typhimurium* (Anderson *et al.*, 1976). In this case, production of enzymes of the *his* operon was doubled. If analogues fail to inhibit growth, one can often get inhibition by changing the carbon or nitrogen source or by adding a detergent, *e.g.* polyoxyethylene stearylamine, to the medium (Yoshida *et al.*, 1981).

'Super-sensitivity' to antimetabolites can also cause oversynthesis of certain metabolites by directing carbon flow towards particular products. For example, in *Candida lipolytica*, mutants selected for super-sensitivity to fluoroacetate produce more citric acid and less isocitric acid (Akiyama *et al.*, 1973). Fluoroacetate, when converted to fluorocitrate, is toxic due to the latter's action as a competitive inhibitor of aconitase (aconitate hydratase). Mutant organisms which produce only small amounts of aconitase are particularly sensitive to fluoroacetate and no longer convert citrate to isocitrate.

Another approach to altering feedback inhibition relies on mistakes made by the cell in correcting errors. Auxotrophy may arise due to one or more changes in a biosynthetic enzyme which render it non-functional. When selecting prototrophic revertants (which no longer have the specific nutritional requirement) from auxotrophs, a portion of the revertant population will produce the original feedback-sensitive enzyme. However, most of the population produce a functional enzyme which is not identical to the original structure but during reversion has become resistant to feedback inhibition, for example auxotrophic mutation and reversion of an isoleucine requirement in a strain of *Hydrogenomonas* leads to overproduction and excretion of isoleucine. The enzyme affected is threonine deaminase; the enzyme in the revertant is much less inhibited by isoleucine (Reh and Schlegel, 1969).

Often, a desired product is degraded by the producing organism or converted into another metabolite. Correction of this problem is usually accomplished by mutating out the enzyme which acts on the product. Mutation of the glutamate producer *B. flavum* to inability to grow on glutamate improved glutamate production in four out of 75 mutants (Shiio *et al.*, 1982).

33.2.4 Additional Types of Regulation

Other types of regulation, possibly less important from the point of view of fermentation technology, include metabolic interlock, guanosine tetraphosphate (ppGpp) control and regulatory inactivation. In metabolic interlock (also known as 'cross-pathway regulation'), the product of one biosynthetic pathway interacts with a completely different biosynthetic pathway. This interaction has been observed between the histidine, leucine or methionine biosynthetic pathways and that for aromatic amino acids (Nester and Montoya, 1976; Jensen, 1969). Other examples are the interactions between the aspartic acid and the branched chain amino acid families (Patte *et al.*, 1972) and the purine and pyrimidine pathways (Jensen, 1979).

Control by ppGpp involves the effect of amino acid deficiency on a large number of physiological activities in bacteria. Deficiency of any amino acid leads to production of ppGpp from GTP. This intracellular effector redirects the cells' activities to correction of the amino acid deficiency. Thus, ppGpp inhibits synthesis of ribosomal and transfer RNA, fatty acids, lipids, nucleotides, peptidoglycan, glycolytic esters, carbohydrates and polyamines, and activates amino acid biosynthetic operons, proteases and protein turnover (Stephens *et al.*, 1975; Peterkofsky, 1976). The inhibition of stable RNA formation is carried out by ppGpp at the transcription level; it impedes transcription at promoters for stable RNA while stimulating transcription at other promoters, *e.g.* amino acid biosynthetic operons. ppGpp production is elicited by limitation of any aminoacyl-tRNA species (Gallant, 1979); it is produced by an enzyme (which is associated with ribosomes) in the presence of ribosomes, uncharged tRNA, GTP, ATP and Mg. The significance of ppGpp in eukaryotes is doubtful.

Regulatory inactivation refers to the selective inactivation of enzymes (Switzer, 1977) by two

Table 2 Analogues of Natural Metabolites

Metabolite	Analogue
Adenine	Benzimidazole, 2,6-diaminopurine
α-Alanine	α-Aminoethanesulfonic acid, glycine, α-aminoisobutyric acid, serine
D-Alanine	D-Cycloserine, O-carbamyl-D-serine, D-α-aminobutyric acid
β-Alanine	β-Aminobutyric acid, propionic acid, asparagine, D-serine
p-Aminobenzoic acid	Sulfonamide
Arginine	Canavanine, lysine, ornithine, homoarginine, arginine hydroxamate, D-arginine
Aspartic acid	Cysteic acid, β-hydroxyaspartic acid, diaminosuccinic acid, aspartophenone, α-methylaspartic acid, β-aspartic acid hydrazide, S-methylcysteine sulfoxide, β-methylaspartic acid, hadacidin
Asparagine	2-Amino-2-carboxyethanesulfonamide
Biotin	Avidin, dethiobiotin, α-dehydrobiotin
Cysteine	Allylglycine
Glutamic acid	Methionine sulfoxide, γ-glutamylethylamide, β-hydroxyglutamic acid, methionine sulfoximine, α-methylglutamic acid, γ-phosphonoglutamic acid, D-ethyl-γ-phosphonoglutamic acid, γ-fluoroglutamic acid
Glutamine	S-Carbamylcysteine, O-carbamylserine, O-carbazylserine, 3-amino-3-carboxypropanesulfonamide, N-benzylglutamine, azaserine, 6-diazo-5-oxonorleucine, γ-glutamyl hydrazide
Glycine	α-Aminomethanesulfonic acid
Guanosine	8-Azaxanthine, psicofuranine, decoyninine
Histidine	D-Histidine, imidazole, 2-thiazolealanine, 1,2,4-triazolealanine, 2-methylhistidine
Isoleucine	Leucine, methallylglycine, ω-dehydroisoleucine, 3-cyclopentene-1-glycine, 2-cyclopentene-1-glycine, O-methylthreonine, β-hydroxyleucine, valine, isoleucine hydroxamate, α-aminobutyric acid, β-hydroxynorvaline (α-amino-β-hydroxyvaleric acid)
Leucine	D-Leucine, α-aminoisoamylsulfonic acid, norvaline, norleucine, methallylglycine, α-amino-β-chlorobutyric acid, valine, δ-chloroleucine, isoleucine, β-hydroxynorleucine, β-hydroxy leucine, 3-cyclopentene-1-alanine, 2-amino-4-methylhexenoic acid, 5′,5′,5′-trifluoroleucine, 4-azaleucine, α-aminobutyric acid
Lysine	α-Amino-ε-hydroxycaproic acid, arginine, 2,6-diaminoheptanoic acid, oxalysine, 3-aminomethylcyclohexaneglycine, 3-aminocyclohexanealanine, trans-4-dehydrolysine, S-(β-aminoethyl)cysteine (thialysine), 4-azalysine, α-chlorocaprolactam
Methionine	Crotylalanine, crotylglycine, methoxinine, α-methyl-DL-methionine, norleucine, ethionine, methionine sulfoximine, threonine, selenomethionine
Niacin	Pyridine-3-sulfonic acid, 3-acetylpyridine, picolinic acid
Ornithine	α-Amino-δ-hydroxyvaleric acid, α-methylornithine
Phenylalanine	α-Amino-β-phenylethanesulfonic acid, tyrosine, phenylserine, cyclohexylalanine, O-aminophenylalanine, p-aminophenylalanine, fluorophenylalanines, chlorophenylalanines, bromophenylalanines, β-2-thienylalanine, β-3-thienylalanine, β-2-furylalanine, β-3-furylalanine, β-2-pyrrolealanine, 1-cyclopentene-1-alanine, 1-cyclohexane-1-alanine, 2-amino-4-methyl-4-hexenoic acid, S-(1,2-dichlorovinyl)cysteine, β-4-pyridylalanine, tryptophan, β-2-pyridylalanine, β-4-pyrazolealanine, β-4-thiazolealanine, p-nitrophenylalanine, 3-aminotyrosine, p-ethylphenylalanine, phenylalanine hydroxamate, tyrosine hydroxamate
Proline	Hydroxyproline, 3-methylproline, 3,4-dehydroproline, azetidine-2-carboxylic acid
Pyridoxine	Isoniazid
Riboflavin	7-Methyl-8-trifluoromethyl-10-(1′-D-ribityl)isoalloxazine
Serine	α-Methylserine, homoserine, threonine, isoserine
Thiamine	Pyrithiamine, bacinaethrin
Threonine	Serine, β-hydroxynorvaline (α-amino-β-hydroxyvaleric acid), β-hydroxynorleucine
Tryptophan	Methyltryptophans, naphthylalanines, indoleacrylic acid, naphthylacrylic acid, β-(2-benzothienyl)alanine, styrylacetic acid, indole, α-amino-β-3-(indazole)propionic acid (tryptazan), fluorotryptophans, 7-azatryptophan, tyrosine hydroxamate, p-ethylphenylalanine, p-aminophenylalanine, phenylalanine hydroxamate, tryptophan hydroxamate
Tyrosine	Aminotyrosine, fluorotyrosines, p-aminophenylalanine, m-nitrotyrosine, β-(5-hydroxy-2-pyridyl)alanine, 3-methyltryptophan, fluorophenylalanines; D-tyrosine, tyrosine hydroxamate, p-ethylphenylalanine
Uracil	5-Fluorouracil, 5-fluorouridine, 2-amino-4-methylpyrimidine
Valine	α-Aminoisobutanesulfonic acid, α-aminobutyric acid, norvaline, leucine, isoleucine, methylglycine, β-hydroxyvaline, ω-dehydroalloisoleucine

different mechanisms. In modification inactivation, the enzyme remains intact but its physical state is changed or it is covalently modified. Covalent modifications include phosphorylation of a specific serine or threonine residue, nucleotidylation of a specific tyrosine residue, ADP-ribosylation of an arginine residue, methylation of a glutamate or aspartate carboxyl group, acetylation of an ε-amino group of a lysine residue, or tyrosinolation of a protein terminal carboxyl group (Chock *et al.*, 1980). In degradative inactivation, at least one peptide bond is broken; such breakages may represent the first step in protein turnover. It is carried out by proteases which are restricted from non-selective action by confinement in vacuoles or by protease inhibitors. Regulatory inactivation usually occurs after the exponential phase of growth, especially after exhaustion of a source of carbon or nitrogen. This inactivation serves to prevent futile cycles of metabolism, to destroy enzymes no longer needed, or to divert branch point metabolism from one branch to another.

33.3 PERMEABILITY CONSIDERATIONS

The integrity of cell surfaces is important in primary metabolite production for two reasons; first, in terms of the ability of the cell to excrete the oversynthesized product and, second in determining the tolerance and hence survival of an organism making a toxic product such as ethanol or butanol. The most striking example of the former is found in glutamate oversynthesis by certain coryneform bacteria, which typically are naturally auxotrophic for biotin. In these organisms, glucose is converted to α-ketoglutarate *via* the EMP pathway and the TCA cycle, thence to glutamate by reductive amination. It was once thought that glutamate overproducers lacked the TCA cycle enzyme, α-ketoglutarate dehydrogenase, which would normally compete with NADP-glutamate dehydrogenase. However, this conclusion was the result of the enzyme's lability during preparation of extracts (Shiio and Ujigawa-Takeda, 1980). The preferred metabolic flux of α-ketoglutarate towards glutamate is due to inhibition of α-ketoglutarate dehydrogenase by *cis*-aconitate and oxalacetate and repression by citrate. During growth on glucose, the overproducers accumulate glutamate intracellularly until saturation is reached (about 50 mg g^{-1} dry weight). Accumulation then ceases, due to feedback regulation, unless the permeability barrier is altered to facilitate exit of the amino acid. This feedback effect is due to inhibition by glutamate of NADP-glutamate dehydrogenase, leading to an accumulation of α-ketoglutarate. The α-ketoglutarate cooperates with aspartate to inhibit phosphoenolpyruvate (PEP) carboxylase covalently, thus shutting down the source of C_4 units to the TCA cycle. Glutamate also represses PEP carboxylase (Shiio and Ujigawa-Takeda, 1979). Avoidance of feedback regulation is achieved by biotin limitation or by addition of agents which alter cell surface integrity, such as penicillin or fatty acid derivatives. Membrane permeability can also be altered for glutamate excretion by restricting the supply of glycerol to glycerol auxotrophs (Nakao *et al.*, 1972). Addition of penicillin during logarithmic growth triggers glutamate excretion, decreasing intracellular levels to 5 mg g^{-1} dry weight. Excretion continues for 40 to 50 h, since lysis does not occur in these bacteria, and concentrations of glutamate in fermentation broths reach over 100 g l^{-1}. Since biotin is important as a cofactor in lipid synthesis, biotin limitation during glutamate fermentation causes marked changes in the lipid composition of cell envelopes. Nakao *et al.* (1973) noted that membranes of these biotin-limited organisms were deficient in phospholipids, and that penicillin treatment, while not inhibiting phospholipid synthesis, caused excretion of phospholipids, presumably by interfering with their incorporation into membranes. Biotin levels used are critical for glutamate production from glucose: 1 to 5 μg l^{-1} of biotin results in glutamate accumulation whereas 15 μg l^{-1} increases growth rates and organic acids such as lactate are made instead of glutamate. High biotin concentrations are desirable for production of other amino acids such as lysine and threonine, as glutamate remains intracellular to serve as the major source of nitrogen (*via* transamination) for these other amino acids. Ammonia assimilation in these organisms occurs *via* amination of α-ketoglutarate, catalyzed by NADP-glutamate dehydrogenase.

Biotin also appears to have some role in the lysine fermentation, as Tosaka *et al.* (1979) reported that lysine production increased from 34 to 42 g l^{-1} by a strain of *Brevibacterium lactofermentum* when biotin levels were increased from 50 to 500 μg l^{-1}. Carbon dioxide liberation and accumulation of alanine, valine and leucine decreased. Hanel *et al.* (1981) have shown a relationship between the nature of cell surface lipids in *C. glutamicum* and excretion of lysine: oxygen limitation increased intracellular lysine pools three-fold while decreasing the proportion of oleic acid in membranes.

Table 3 lists some amino acid processes developed by mutation. Many of the cultures have multiple mutations often of a different class.

Production of solvents by fermentation is often limited by sensitivity of the producing organism to the product, for example ethanol concentrations of 10 to 12% (w/v) are toxic for yeast used in brewing. Toxicity probably results from effects of ethanol acting on the cell surface and on enzymes involved in ethanol production and/or other cellular functions. Millar *et al.* (1982) investigated the stability of the glycolytic enzymes of yeasts and *Zymomonas mobilis*, finding that denaturation is unlikely to play a direct part in ethanol sensitivity but that ethanol may be responsible for retarding the rates of some EMP pathway reactions. The lipid composition of the plasma membrane has been shown to influence the inhibitory effect of ethanol on transport of sugars and

Table 3 Amino Acid Production by Mutants

Amino acid	Microorganism	Type of mutation Auxotophic requirement	Type of mutation Analogue resistance	Other characteristics	Potency $(g\,l^{-1})$
L-Arginine	*Brevibacterium flavum*	Guanine	2-Thiazolealanine	—	35
L-Glutamate	*Corynebacterium glutamicum*	Biotin	—	—	100
L-Histidine	*Serratia marcescens*	—	1,2,4-Triazolealanine, 2-methylhistidine, 6-methylproline	Histidase-negative, uroca-nase-negative,	22
L-Isoleucine	*Brevibacterium flavum*	—	Thialysine, hydroxy-norva-line, ethionine	—	33
L-Leucine	*Brevibacterium lactofermentum*	Methionine, isoleucine	2-Thiazolealanine	—	28
L-Lysine	*Brevibacterium flavum*	Threonine	—	Threonine-sensitive	75
I-Methionine	*Ustilago maydis*	Leucine	—	—	6
L-Ornithine	*Corynebacterium glutamicum*	Arginine	—	—	77
L-Phenylalanine	*Arthrobacter paraffineus*	Tyrosine	—	—	15
L-Proline	*Corynebacterium glutamicum*	Purine, isoleucine, histidine, ornithine	—	—	42
L-Threonine	*Brevebacterium flavum*	—	Hydroxynorvaline	—	27
L-Tryptophan	*Corynebacterium glutamicum*	Phenylalanine, tyrosine	Tryptophan analogues	—	12
L-Tyrosine	*Corynebacterium* sp.	Phenylalanine	—	—	19
L-Valine	*Brevibacterium lactofermentum*	Isoleucine, methionine	2-Thiazolealanine	—	31

amino acids into yeasts (Thomas and Rose, 1979). It has also been suggested (Watson, 1982) that the phospholipid composition of membranes is important in ethanol tolerance, since cells whose membrane phospholipids are enriched in unsaturated fatty acids produce higher concentrations of ethanol (13 to 15.5%, w/v).

Production of acetone and butanol by *C. acetobutylicum* is limited by the sensitivity of the organism to butanol. A mutant of *C. acetobutylicum* which is altered in its cell surface as demonstrated by its resistance to autolysin and which produces less autolysin, was found to be more tolerant to butanol (Van der Westhuizen *et al.*, 1982). Solvent production by this mutant is now under investigation.

33.4 RECENT APPROACHES TO STRAIN CONSTRUCTION

Organisms used today for industrial production of amino acids and nucleotides have been developed by mutation and selection/screening programs which often started with organisms having some capacity to make the desired product but which required multiple mutations leading to deregulation in a particular biosynthetic pathway before high productivity can be obtained (see, for example, Kinoshita and Nakayama, 1978). The sequential mutations ensure that carbon sources are channelled efficiently to the appropriate products without significant deviation to other pathways or loss of carbon in by-products. These mutations presumably involve not only release of feedback controls but also enhancement of the formation of pathway precursors. This approach to strain improvement has been remarkably successful in producing organisms which make industrially significant concentrations of primary metabolites. However, some of the problems with this approach include (1) the necessity of screening large numbers of mutants for the rare combination of traits sequentially obtained which lead to overproduction, and (2) the vigor of the producing strain, which may be substantially weakened following several rounds of mutagenesis. More recent approaches are utilizing the techniques of modern genetic engineering to develop strains overproducing primary metabolites. The rationale for strain construction relies largely on the same principles of regulation discussed in our previous sections, but aims at assembling the appropriate characteristics (sometimes by means of *in vitro* recombinant DNA techniques) in genetic backgrounds which have not been subjected to repeated mutagenesis. This is particularly valuable in organisms with complex regulatory systems, where deregulation would involve many genetic alterations. It is not within the scope of this chapter to give a detailed description of these methods; excellent descriptions of these are found in the text of Old and Primrose (1981), with suggested background reading in Stent and Calendar (1978). In the following sections, we describe some approaches to strain construction resulting in amino acid hyperproducing strains of *E. coli* and other Gram-negative bacteria. These strains have not yet been used for commercial production of amino acids.

33.4.1 Amino acid Production by Genetically Engineered Strains of *E. coli* and Related Organisms

Production of a primary metabolite by deregulated organisms may inevitably be limited by the inherent capacity of the particular organism to make the appropriate biosynthetic enzymes, *i.e.* even in the absence of repressive mechanisms there may not be enough of the enzyme made to obtain high productivity. One way to overcome this is to increase the number of copies of structural genes coding for these enzymes, and another way often used in combination with this strategy is to increase the frequency of transcription which is related to the frequency of binding of RNA polymerase to the promoter region (Rosenberg and Court, 1979). The former can be achieved by incorporating the biosynthetic genes *in vitro* into a plasmid which, when placed in a cell by genetic transformation, will replicate into multiple copies; some 'amplifiable' plasmids, such as pBR322, can exist at 50 copies per cell. Increasing the frequency of transcription involves constructing a hybrid plasmid *in vitro* which contains the structural genes of the biosynthetic enzymes but lacks the regulatory genes (promoter and operator) normally associated with these. Instead, the structural genes are placed next to an efficiently and frequently read promoter and operator, and are now subject to regulation by these genes. One promoter which has been used to regulate the production of many genetically-engineered proteins is that of the *lac* operon. For example, Itakawa *et al.* (1977) constructed plasmids using pBR322 which contained the *lac* promoter, catabolite gene activator-protein binding site, the operator, the ribosome binding site (a sequence of DNA which when transcribed into mRNA is the position where ribosomes bind to start translation) and part or all of the β-galactosidase structural gene. Behind this was inserted the chemically synthesized gene for the hormone somatostatin. If the somatostatin gene was inserted directly behind the *lac* regulatory region, production of the hormone was not detected due to intracellular degradation of the polypeptide. However, placing the gene next to that for β-galactosidase protected the fused protein from degradation. A similar approach was used to construct strains of *E. coli* making human insulin (see Old and Primrose, 1981). Even if the *lac* promoter is still subject to regulation, when it is present on a multicopy plasmid there are sufficient copies of the control region to titrate out all the repressor produced by a single copy of the repressor gene on the chromosome. The ideal plasmid for primary metabolite synthesis would contain a regulatory region with a constitutive phenotype, preferably not subject to nutritional repression.

The plasmid pBR322 has been used recently as the cloning vector to construct strains of *E. coli* making 55 g l^{-1} of L-threonine, where the conversion of carbohydrate into amino acids exceeded 40% (Debabov, 1982). A description of the approaches used for strain construction is given in Debabov *et al.* (1981), which reports on strains making up to 20 g l^{-1} of threonine. The source of the structural genes for threonine biosynthesis was chromosomal DNA from one of two strains which were resistant to feedback regulation by threonine (due to resistance to the threonine analogue, β-hydroxynorvaline). Plasmids constructed by ligating fragments of donor chromosomal DNA with pBR322 were enzymatically modified to decrease plasmid size (allowing higher copy number per cell) and then transformed into a recipient which was auxotrophic for threonine and partially auxotrophic for isoleucine. Organisms carrying the appropriate hybrid plasmids were selected on the basis of drug resistance specified by pBR322 (ampicillin and tetracycline resistance or ampicillin resistance alone, depending on the enzymic treatment of DNA used to construct the hybrid plasmid) and prototrophy. As the block in isoleucine synthesis was only partial, any organisms overproducing threonine could make enough isoleucine to allow the organisms to grow on media not supplemented with isoleucine. This gave a selective advantage to organisms overproducing threonine, thus ensuring that only organisms maintaining the correct type of plasmid could survive during fermentation.

L-Tryptophan, like the aspartic acid family of amino acids, is an essential amino acid whose production by fermentation would be desirable. However, despite many attempts using different organisms (Table 4), the titers of L-tryptophan remain relatively low and the processes often require addition of expensive precursors, thus they do not warrant large scale industrial production on economic grounds. Most recent approaches to strain development for tryptophan production by *E. coli* have used *in vitro* recombinant DNA techniques (Aiba *et al.*, 1980, 1982). Biosynthesis of the aromatic amino acids in *E. coli* involves a common pathway converting glucose to chorismate which is the central branch point (Figure 4). Regulation involves three isoenzymes of DAHP synthase which are feedback regulated by tyrosine, tryptophan and phenylalanine; repression of the phenylalanine-DAHP synthetase is multivalent, requiring tryptophan plus phenylalanine (Figure 4; Umbarger, 1978). All of the enzymes in the terminal pathway of tryptophan synthesis are repressed by tryptophan. In the work described by Aiba *et al.* (1982), a composite plasmid was constructed which contained the operon for tryptophan synthesis (*i.e.* the genes coding for the tryptophan branch in which chorismate is converted to tryptophan). This was transformed into a host cell which lacked the structural genes of this operon (*i.e.* was auxotrophic for tryptophan due to deletion of these genes), did not degrade tryptophan (*i.e.* lacked tryptophanase) and was constitutive with respect to tryptophan repression (*i.e.* it failed to make an active aporepressor specific for tryptophan). Deletion of the *trp* operon structural genes in the host organism ensured that the incoming plasmid-bound structural genes had no homology with host DNA, and would therefore not recombine into the host chromosome. Once in the host cell, the plasmid was mutagenized and selection for resistance to 5-methyltryptophan was performed, resulting in a plasmid carrying the *trp* operon whose anthranilate aggregate (anthranilate synthetase + PRA phosphoribosyltransferase) was resistant to inhibition by tryptophan. The host containing the mutated plasmid made 6.2 g l^{-1} of tryptophan. These authors reported that a moderate plasmid copy number (five to ten) is optimal for tryptophan production. Further, they pointed out that there is still room to improve tryptophan production, as the host strain was not auxotrophic for tyrosine or phenylalanine, nor was it altered in the regulation of carbon flow in the common aromatic amino acid pathway. This point is supported by Choi and Tribe (1982) who showed that under suitable growth conditions, a strain of *E. coli* (constructed by conventional techniques of genetic recombination) which lacked feedback repression and inhibition of the common aromatic amino acid pathway and of the phenylalanine branch made 8.7 g l^{-1} of phenylalanine.

Table 4 L-Tryptophan Production in Fermentation by Various Microorganisms

Microorganism	Titer (g l^{-1})	Productivity (g l^{-1} h^{-1})	Reference
Claviceps purpurea	1.5	0.010	Malin and Westhead, 1959
Escherichia coli	0.7	0.015	Kida and Matsushiro, 1965
Hansenula anomala	14	0.054	Ebihara *et al.*, 1969
Bacillus subtilis	6.2	0.128	Shiio *et al.*, 1973
Corynebacterium glutamicum	12	0.125	Hagino and Nakayama, 1975
Escherichia coli	1.3	0.080	Tribe and Pittard, 1979
Escherichia coli	6.2	0.229	Aiba *et al.*, 1982
Brevibacterium flavum	11.4	—	Shiio *et al.*, 1982

One of the major problems in using strains in which the desired characteristics are encoded by a plasmid is the difficulty in maintaining plasmids during fermentation. Plasmid instability in the absence of selective pressure leads to a dilution of the plasmid in the population and loss of the desired phenotype. Making the presence of the threonine-plasmid essential for survival of the threonine-overproducing strains mentioned above overcame this problem. Another way is to exert an antibiotic selection pressure during fermentation so that only organisms resistant to the antibiotic due to the presence of a plasmid-borne resistance gene can survive. This was the approach taken by Aiba *et al.* (1980) for tryptophan production, where the plasmid was maintained during batch fermentation by positive selection provided by tetracycline, resistance to which was encoded by the plasmid containing the *trp* operon. However, attempts to improve tryptophan yields in continuous culture resulted in the majority of the cells failing to make tryptophan (Dwivedi *et al.*, 1982); this could not be attributed to the loss of plasmid from cells, as most cells remained resistant to tetracycline. The authors suggest that the region of plasmid DNA coding for tryptophan production was highly susceptible to mutation, resulting in decreased enzyme synthesis.

An alternative approach to strain construction not involving recombinant DNA technology was reported by Komatsubara *et al.* (1979). Here, a strain of *Serratia marcescens* producing high

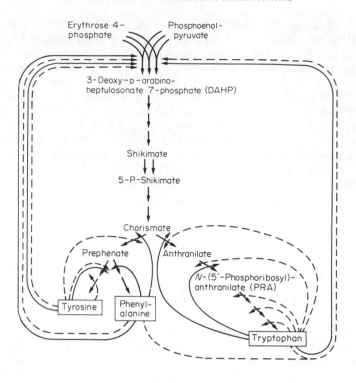

Figure 4 Feedback regulation of aromatic amino acid biosynthesis in *E. coli*. Solid lines represent inhibition, dashed lines represent repression

levels of threonine was constructed by transductional crosses which combined several feedback control mutations in one organism. Regulation of the aspartic acid family of amino acids in *S. marcescens* diverges slightly from that in *E. coli* (Figure 3). Two isoenzymes of aspartokinase are present, one of which is feedback inhibited by threonine and covalently repressed by threonine and isoleucine, and the second is feedback inhibited and repressed by lysine. The wild-type parent strain was first made auxotrophic for isoleucine by mutagenesis (eliminating threonine deaminase) and then mutated to eliminate the other threonine-degrading enzyme threonine dehydrogenase. Three classes of mutants were then obtained from the above strain as the source of genetic material for transduction: (1) strain HNr21, in which both the threonine-regulated aspartokinase and homoserine dehydrogenase are resistant to inhibition by threonine, which was selected on the basis of β-hydroxynorvaline resistance, (2) strain HNr59, also selected for β-hydoxynorvaline resistance, in which homoserine dehydrogenase is resistant to both inhibition and repression and the threonine-regulated aspartokinase is constitutively synthesized, and (3) strain AECr174, which is resistant to S-2-aminoethylcysteine, in which the lysine-regulated aspartokinase is resistant to feedback inhibition and repression. Since at least one of the three key enzymes in threonine synthesis was subject to regulation in each of these strains, each produced only modest amounts of threonine (4.1 to 8.7 g l^{-1}). Recombination of the three strains by transduction yielded a transductant which produced high levels of threonine (25 g l^{-1}), had aspartokinase and homoserine dehydrogenase activities which were resistant to feedback regulation by threonine and lysine, and was also a methonine bradytroph (leaky auxotroph). Transductional construction of strains producing histidine, arginine, urocanate, isoleucine and norvaline has also been reported (Komatsubara *et al.*, 1979, 1980).

33.4.2 Strain Construction in Other Species

E. coli and related enteric organisms have been used for constructing amino acid-producing strains since they are amenable to genetic manipulation and their biosynthetic pathways plus regulatory mechanisms are relatively well understood. However, use of such organisms industrially poses certain problems, including (1) potential pathogenicity (a relatively minor problem with laboratory-utilized and debilitated *E. coli* strains), (2) production of endotoxins and (3) susceptibility to bacteriophage infection during fermentation. As mentioned earlier, a further problem may be the instability of genetically engineered strains, but this may also be true for species other than *E. coli*. Most industrially useful organisms involved in primary metabolite synthesis lack some of these drawbacks, although bacteriophage attacks during fermentation have been noted for the glutamic acid-producing bacteria (see Momose *et al.*, 1976) and in the acetone–butanol fermentation. In the latter case, bacteriophage-resistant strains of *C. acetobutylicum* can be selected (Hastings, 1978). Strain construction by means other than the traditional induced mutation-selection methods, *i.e.* by transduction, transformation, or conjugation/plasmid transfer, has not occurred as such genetic systems were not developed for these organisms. Recently, some attention has been given to developing genetic systems in the coryneform bacteria and a preliminary report on their use in constructing improved strains of amino acid-producing bacteria has appeared. Momose *et al.* (1976) reported the isolation of bacteriophage mediating generalized transduction in *B. flavum*. The frequency of transduction was low (about 10^{-6}) and there have been no further reports on the use of this system for strain construction. Similarly, a plasmid in *B. lactofermentum* (ATCC 13869; strain number corrected by the authors, personal communication) has been detected and characterized with respect to size (3.7×10^7 daltons) and digestion with restriction endonucleases (Kaneko *et al.*, 1979). Genetic markers on this plasmid have not yet been identified nor has it been used in strain construction. A procedure for the production and regeneration of protoplasts of *B. flavum* and the subsequent fusion of protoplasts accompanied by genetic recombination was described by Kaneko and Sakaguchi (1979). Exponentially growing cultures were treated with 0.3 units of penicillin per ml followed by lysozyme treatment in hypertonic medium, resulting in the generation of osmotically sensitive protoplasts. The fusion frequencies (the number of cells fused in relation to the total protoplasts mixed) were low (5×10^{-6}) relative to frequencies seen for protoplast fusion in other Gram-positive organisms (see Hopwood, 1981), suggesting that the methods reported have yet to be optimized. Katsumata *et al.* (1981) also noted low fusion frequencies using this procedure. A modification of this method appears in a preliminary report by Santamaria *et al.* (1982), where low frequency intraspecific fusion occurred between strains of *B. lactofermentum* (2.3×10^{-6} fusants per *B. lactofermentum* cell) and *C. glutamicum* (3.1×10^{-3} fusants per *C. glutamicum* cell); recombinant clones with several combinations of genetic markers were noted but not specified. Here, cells were grown in glycine (0.1 to 10 mg ml^{-1}) and treated with sodium dodecyl sulfate (10 μg ml^{-1}) to produce protoplasts. The procedure of Kaneko and Sakaguchi (1979) for protoplast fusion in coryneform bacteria has been applied directly to construct amino acid-producing strains. Tosaka *et al.* (1982) reported that interspecific and intraspecific recombinants among mutants of *B. lactofermentum* and *B. flavum* occurred at frequencies of 10^{-3} to 10^{-5}; lysine auxotrophy was introduced into a lysine plus threonine producing strain of *B. lactofermentum* by protoplast fusion, resulting in an organism making only threonine. Also, a lysine-producing strain of *B. lactofermentum*, isolated after repeated mutagenesis, had poor glucose consumption rates; glucose utilization was improved three-fold by recombination with a wild-type strain. Such preliminary reports indicate an optimistic future for strain development by protoplast fusion techniques.

Genetics in some genera of organisms of industrial importance are relatively well understood, for example, in *S. cerevisiae* (Mortimer and Schild, 1980), *Bacillus subtilis* and *B. megaterium* and several *Streptomyces* species (see Hopwood, 1981). However, there are relatively few reports at present on the use of genetics in strain construction for primary metabolite synthesis. One exception is the use in *Saccharomyces* of protoplast fusion to construct novel yeast strains making more ethanol (Panchal *et al.*, 1982). As in the coryneform group of bacteria, genetics of other organisms of present or potential commercial interest are at an early stage of development. Allcock *et al.* (1982) reported production of protoplasts of *C. acetobutylicum* by treating mid-exponential cells grown in a medium containing 0.4% glycine with lysozyme, using 0.3 M sucrose as the osmotic stabilizer. Phage DNA could be introduced into such protoplasts by transformation, as indicated by the generation of mature phage particles on regeneration of cells (Reid *et al.*, 1982). Genetic studies in the Gram-negative, ethanol-producing bacterium *Z. mobilis* have been facilitated by genetic exchange between this organism and other Gram-negative species, which allows the introduction of plasmids into this organism (Skotnicki *et al.*, 1980). DNA can also be introduced by transformation, so that inter- and intra-specific genetic exchange is possible. These authors suggest that such genetic techniques can be used to improve growth on sucrose and extend the range of substrates which can be utilized for ethanol production (Skotnicki *et al.*, 1982). Other strain selection programmes with *Z. mobilis* have improved ethanol production in association with improved ethanol tolerance and the generation of flocculent strains (Lee *et al.*, 1982).

33.5 REFERENCES

Aiba, S., T. Imanaka and H. Tsunekawa (1980). Enhancement of tryptophan production by *Escherichia coli* as an application of genetic engineering. *Biotechnol. Lett.*, **2**, 525–530.

Aiba, S., H. Tsunekawa and T. Imanaka (1982). New approach to tryptophan production by *Escherichia coli*: genetic manipulation of composite plasmids *in vitro*. *Appl. Environ. Microbiol.*, **43**, 289–297.

Akiyama, S., T. Suzuki, Y. Sumino, Y. Nakao and H. Fukuda (1973). Induction and citric acid productivity of fluoroacetate-sensitive mutant strains of *Candida lipolytica*. *Agric. Biol. Chem.*, **37**, 879–884.

Allcock, E. R., S. J. Reid, D. T. Jones and D. R. Woods (1982). *Clostridium acetobutylicum* protoplast formation and regeneration. *Appl. Environ. Microbiol.*, **43**, 719–721.

Anderson, R. P., C. G. Miller and J. R. Roth (1976). Tandem duplications of the histidine operon observed following generalized transduction in *Salmonella typhimurium*. *J. Mol. Biol.*, **105**, 201–218.

Arst, H. N., Jr. (1981). Aspects of the control of gene expression in fungi. *Symp. Soc. Gen Microbiol.*, **31**, 131–160.

Arst, H. N., Jr. and C. R. Bailey (1977). The regulation of carbon metabolism in *Aspergillus nidulans*. In *Genetics and Physiology of Aspergillus*, ed. J. E. Smith and J. A. Pateman, pp. 131–146. Academic, London.

Arst, H. N., Jr. and D. J. Cove (1969). Methylammonium resistance in *Aspergillus nidulans*. *J. Bacteriol.*, **98**, 1284–1293.

Botsford, J. L. (1981). Cyclic nucleotides in procaryotes. *Microbiol. Rev.*, **45**, 620–642.

Chock, P. B., S. G. Rhee and E. R. Stadtman (1980). Interconvertible enzyme cascades in cellular regulation. *Annu. Rev. Biochem.*, **49**, 813–843.

Choi, Y. J. and D. E. Tribe (1982). Continuous production of phenylalanine using an *Escherichia coli* regulatory mutant. *Biotechnol. Lett.*, **4**, 223–228.

Clarke, P. H. and M. D. Lilly (1969). The regulation of enzyme synthesis during growth. *Symp. Soc. Gen Microbiol.*, **19**, 113–159.

Debabov, V. (1982). Construction of strains producing L-threonine. *Abstracts of 4th International Symposium on Genetics of Industrial Microorganisms, Kyoto, Japan*. p. 9.

Debabov, V. G., J. I. Kozlov, N. I. Zhdanova, E. M. Khurges, N. K. Yankovsky, M. N. Rozinov, R. S. Shakulov, B. A. Rebentish, V. A. Livshits, M. M. Gusyatiner, S. V. Mashko, V. N. Moshentseva, L. F. Kozyreva and R. A. Arsatiants. (1981). Method for preparing strains which produce amino acids. *US Pat.* 4 278 765.

Dwivedi, C. P., T. Imanaka and S. Aiba (1982). Instability of plasmid-harboring strain of *E. coli* in continuous culture. *Biotechnol. Bioeng.*, **24**, 1465–1468.

Ebihara, Y., H. Niitsu and G. Terui (1969). Fermentative production of tryptophan from indole by *Hansenula anomala*. *J. Ferment. Technol.*, **47**, 733–738.

Elvin, C. M. and H. L. Kornberg (1982). A mutant β-D-glucoside transport system of *Escherichia coli* resistant to catabolite inhibition. *FEBS Lett.*, **147**, 137–142.

Gallant, J. A. (1979). Stringent control in *E. coli*. *Annu. Rev. Genet.*, **13**, 393–415.

Hagino, H. and K. Nakayama (1975). L-Tryptophan production by analog-resistant mutants derived from a phenylalanine and tyrosine double auxotroph of *Corynebacterium glutamicum*. *Agric. Biol. Chem.*, **39**, 343–349.

Hanel, F., M. Hilliger and V. Gräfe (1981). Effect of oxygen limitation on cellular L-lysine pool and lipid spectrum of *Corynebacterium glutamicum*. *Biotechnol. Lett.*, **3**, 461–464.

Hannan, M. A., F. Rabbi, A. T. M. Faizur Rahman and N. Choudhury (1973). Analysis of some mutants of *Aspergillus niger* for citric acid production. *J. Ferment. Technol.*, **51**, 606–608.

Hastings, J. J. H. (1978). Acetone–butyl alcohol fermentation. In *Economic Microbiology*, ed. A. H. Rose, vol. 2, pp. 31–45. Academic, London.

Hopwood, D. A. (1981). Genetic studies with bacterial protoplasts. *Annu. Rev. Microbiol.*, **35**, 237–272.

Hsie, A. W. and H. V. Rickenberg (1967). Catabolite repression in *Escherichia coli*: the role of glucose-6-phosphate. *Biochem. Biophys. Res. Commun.*, **29**, 303–310.

Itakura, K., T. Hirose, R. Crea, A. D. Riggs, H. L. Heyneker, F. Bolivar and H. W. Boyer (1977). Expression in *Escherichia coli* of a chemically synthesized gene for the hormone somatostatin. *Science*, **198**, 1056–1063.

Jacob, F. and J. Monod (1961). Genetic regulatory mechanisms in the synthesis of proteins. *J. Mol. Biol.*, **3**, 318–356.

Jayaraman, K., B. Muller-Hill and H. V. Rickenberg (1966). Inhibition of the synthesis of β-galactosidase in *Escherichia coli* by 2-nitrophenyl-β-D-fucoside. *J. Mol. Biol.*, **18**, 339–343.

Jensen, R. A. (1969). Metabolic interlock. Regulatory interactions exerted between biochemical pathways. *J. Biol. Chem.*, **244**, 2816–2823.

Kane, J. F. and R. A. Jensen (1970). The molecular aggregation of anthranilate synthase in *Bacillus subtilus*. *Biochem. Biophys. Res. Commun.*, **41**, 328–333.

Kaneko, H. and K. Sakaguchi (1979). Fusion of protoplasts and genetic recombination of *Brevibacterium flavum*. *Agric. Biol. Chem.*, **43**, 1007–1013.

Kaneko, H., T. Tanaka and K. Sakaguchi (1979). Isolation and characterization of a plasmid from *Brevibacterium lactofermentum*. *Agric. Biol. Chem.*, **43**, 867–868.

Katsumata, R., K. Takayama and A. Furuya (1981). Preparation of genetic recombinant bacterial strains. *Jpn. Pat.* 56 109 587.

Kida, S. and A. Matsushiro (1965). The application of cellular regulatory theories to the tryptophan accumulation by *Escherichia coli*. *J. Ferment. Technol.*, **43**, 307–310.

Kinoshita, S. and K. Nakayama (1978). Amino acids. In *Economic Microbiology*, ed. A. H. Rose, vol. 2, pp. 209–261. Academic, London.

Kolter, R. and C. Yanofsky (1982). Attenuation in amino acid biosynthetic operons. *Annu. Rev. Genet.*, **16**, 113–134.

Komatsubara, S., M. Kisumi and I. Chibata (1979). Transductional construction of a threonine-producing strain of *Serratia marcescens*. *Appl. Environ. Microbiol.*, **38**, 1045–1051.

Komatsubara, S., M. Kisumi and I. Chibata (1980). Transductional construction of an isoleucine producing strain of *Serratia marcescens*. *J. Gen. Microbiol.*, **119**, 51–61.

Lee, J. M., M. L. Skotnicki and P. L. Rogers (1982). Kinetic studies on a flocculent strain of *Zymomonas mobilis*. *Biotechnol. Lett.*, **4**, 615–620.

Malin, B. and J. Westhead (1959). Production of L-tryptophan in submerged culture. *J. Biochem. Microbiol. Technol. Eng.*, **1**, 49–57.

Marzluf, G. A. (1981). Regulation of nitrogen metabolism and gene expression in fungi. *Microbiol. Rev.*, **45**, 437–461.

Mehta, H. B. and V. V. Modi (1981). Regulation of flavin levels by phosphate in flavinogenic *Eremothecium ashbyii*. *Biochem. Int.*, **2**, 181–186.

Miall, L. M. (1978). Organic acids. In *Economic Microbiology*, ed. A. H. Rose, vol. 2, pp. 47–119. Academic, London.

Millar, D. G., K. Griffiths-Smith, E. Algar and R. K. Scopes (1982). Activity and stability of glycolytic enzymes in the presence of ethanol. *Biotechnol. Lett.*, **4**, 601–606.

Momose, H., S. Miyashiro and M. Oba (1976). On the transducing phages in glutamic acid-producing bacteria. *J. Gen. Appl. Microbiol.*, **22**, 119–129.

Monard, D., J. Janacek and H. V. Rickenberg (1969). The enzymatic degradation of 3′,5′-cyclic AMP in strains of *E. coli* sensitive and resistant to catabolite repression. *Biochem. Biophys. Res. Commun.*, **35**, 584–591.

Mortimer, R. K. and D. Schild (1980). Genetic map of *Saccharomyces cerevisiae*. *Microbiol. Rev.*, **44**, 519–571.

Nakada, D. and B. Magasanik (1964). The roles of inducer and catabolite repressor in the synthesis of β-galactosidase by *Escherichia coli*. *J. Mol. Biol.*, **8**, 105–127.

Nakao, Y., T. Kanamaru, M. Kikuchi and S. Yamatodani (1973). Extracellular accumulation of phospholipids, UDP-*N*-acetylhexosamine derivatives and L-glutamic acid by penicillin-treated *Corynebacterium alkanolyticum*. *Agric. Biol. Chem.*, **37**, 2399–2401.

Nakao, Y., M. Kikuchi, M. Suzuki and M. Doi (1972). Microbial production of L-glutamic acid by glycerol auxotrophs. I. Induction of glycerol auxotrophs and production of L-glutamic acid from *n*-paraffins. *Agric. Biol. Chem.*, **36**, 490–496.

Nakayama, K. (1976). The production of amino acids. *Process Biochem.*, **11** (3), 4–9.

Nakayama, K., H. Tanaka, H. Hagino and S. Kinoshita (1966). Studies on lysine fermentation. 5. Concerted feedback inhibition of aspartokinase and the absence of lysine inhibition on aspartic semialdehyde–pyruvate condensation in *Micrococcus glutamicus*. *Agric. Biol. Chem.*, **30**, 611–616.

Neidhardt, F. C. (1960). Mutant of *Aerobacter aerogenes* lacking glucose repression. *J. Bacteriol.*, **80**, 536–543.

Nester, E. W. and A. Montoya (1976). Involvement of a histidine locus in tyrosine and phenylalanine synthesis in *Bacillus subtilis*. In *Microbiology 1976*, ed. D. Schlessinger, pp. 141–144. American Society for Microbiology, Washington, DC.

Novick, A. and T. Horiuchi (1961). Hyper-production of β-galactosidase by *Escherichia coli* bacteria. *Cold Spring Harbor Symp. Quant. Biol.*, **26**, 239–245.

Old, R. W. and S. B. Primrose (1981). *Principles of Gene Manipulation. An Introduction to Genetic Engineering*. 2nd edn. University of California Press, Berkeley. Blackwell Scientific Publications, Oxford.

Panchal, C. J., A. Harbison, I. Russell and G. G. Stewart (1982). Ethanol production by genetically modified strains of *Saccharomyces*. *Biotechnol. Lett.*, **4**, 33–38.

Pastan, I. and R. Perlman (1970). Cyclic adenosine monophosphate in bacteria. *Science*, **169**, 339–344.

Patte, J. C., M. Zuber and F. Borne (1972). Regulation of the synthesis of aspartokinase III of *Escherichia coli* by amino acids other than lysine. *Mol. Gen. Genet.*, **116**, 35–39.

Peterkofsky, A. (1976). Cyclic nucleotides in bacteria. *Adv. Cyclic Nucleotide Res.*, **7**, 1–48.

Piovant, M. and C. Lazdunski (1975). Different cyclic adenosine 3′,5′-monophosphate requirements for induction of β-galactosidase and tryptophanase. Effect of osmotic pressure on intracellular cyclic AMP concentrations. *Biochemistry*, **14**, 1821–1825.

Proctor, A. R. and I. P. Crawford (1976). Evidence for autogenous regulation of *Pseudomonas putida* tryptophan synthase. *J. Bacteriol.*, **126**, 547–549.

Reh, M. and H. G. Schlegel (1969). Die Biosynthese von Isoleucin und Valin in *Hydrogenomonas* H16. *Arch. Mikrobiol.*, **67**, 110–127.

Reid, S., E. Allcock, D. Jones and D. Woods (1982). Protoplast formation, regeneration and transformation in *Clostridium acetobutylicum*, Abstracts. *4th International Symposium on Genetics of Industrial Microorganisms, Kyoto, Japan*, p. 88.

Riggs, A. D., R. F. Newby and S. Bourgeous (1970). *lac* Repressor-operator interaction, 2. Effect of galactosides and other ligands. *J. Mol. Biol.*, **51**, 303–314.

Rose, A. H. (1976). *Chemical Microbiology*, 3rd edn. Plenum, New York.

Rosenberg, M. and D. Court (1979). Regulatory sequences involved in the promotion and termination of RNA transcription. *Annu. Rev. Genet.*, **13**, 319–353.

Ryden, R. (1958). Development of anaerobic fermentation processes: acetone-butanol. In *Biochemical Engineering*, ed. R. Steel, pp. 125–148. Heywood, London.

Santamaria, R., J. M. Mesas and J. F. Martin (1982). Genetic recombination by protoplast fusion in coryneform bacteria. *Abstracts, 4th International Symposium on Genetics of Industrial Microorganisms, Kyoto, Japan*, p. 60.

Savant, S., N. Parikh and H. S. Chhatpar (1982). Phosphate mediated regulation of some of the enzymes of carbohydrate metabolism in *Neurospora crassa*. *Experientia*, **38**, 310–312.

Schaeffer, E. J. and C. L. Cooney (1982). Production of maltase by wild-type and a constitutive mutant of *Saccharomyces italicus*. *Appl. Environ. Microbiol.*, **43**, 75–80.

Shiio, I., K. Ishii and K. Yokozeki (1973). Production of L-tryptophan by 5-fluorotryptophan resistant mutants of *Bacillus subtilis*. *Agric. Biol. Chem.*, **37**, 1991–2000.

Shiio, I., H. Ozaki and M. Mori (1982). Glutamate metabolism in a glutamate-producing bacterium, *Brevibacterium flavum*. *Agric. Biol. Chem.*, **46**, 493–500.

Shiio, I., S. Sugimoto and K. Kawamura (1982). Production of L-tryptophan by azaserine-resistant mutants of *Brevibacterium flavum*. *Agric. Biol. Chem.*, **46**, 1849–1854.

Shiio, I. and K. Ujigawa-Takeda (1979). Regulation of phosphoenolpyruvate carboxylase by synergistic action of aspartate and 2-oxoglutarate. *Agric. Biol. Chem.*, **42**, 2479–2485.

Shiio, I. and K. Ujigawa-Takeda (1980). Presence and regulation of α-ketoglutarate dehydrogenase complex in a glutamate-producing bacterium *Brevibacterium flavum*. *Agric. Biol. Chem.*, **44**, 1897–1904.

Siegel, L. S., P. B. Hyleman and P. V. Phibbs, Jr. (1977). Cyclic adenosine 3′,5′-monophosphate levels and activities of adenylate cyclase and cyclic adenosine 3′,5′-monophosphate phosphodiesterase in *Pseudomonas* and *Bacteroides. J. Bacteriol.*, **129**, 87–96.

Skotnicki, M. L., D. E. Tribe and P. L. Rogers (1980). R-plasmid transfer in *Zymomonas mobilis. Appl. Environ. Microbiol.*, **40**, 7–12.

Skotnicki, M. L., R. G. Warr, A. E. Goodman and P. L. Rogers (1982). Development of genetic techniques and strain improvement in *Zymomonas mobilis. Abstracts, 4th International Symposium on Genetics of Industrial Microorganisms, Kyoto, Japan*, p. 32.

Spencer, J. F. T. and D. M. Spencer (1978). Production of polyhydroxy alcohols by osmotolerant yeasts. In *Economic Microbiology*, ed. A. H. Rose, vol. 2, pp. 398–425. Academic, London.

Stadtman, E. R. (1960). Allosteric regulation of enzyme activity. *Adv. Enzymol.*, **28**, 41–154.

Stadtman, E. R. (1968). The role of multiple enzymes in the regulation of branched metabolic pathways. *Ann. N.Y. Acad. Sci.*, **151**, 516–530.

Stent, G. S. and R. Calendar (1978). *Molecular genetics. An Introductory Narrative*, 2nd edn. W. H. Freeman, San Francisco.

Stephens, J. C., S. W. Artz and B. N. Ames (1975). Guanosine 5′-diphosphate 3′-diphosphate (ppGpp): positive effector for histidine operon transcription and general signal for amino acid deficiency. *Proc. Natl. Acad. Sci. USA*, **72**, 4389–4393.

Switzer, R. L. (1977). The inactivation of microbial enzymes *in vivo. Annu. Rev. Microbiol.*, **31**, 135–157.

Tanaka, Y., S. Omura, K. Araki and K. Nakayama. (1981). Derepression of glycine decarboxylase synthesis by magnesium phosphate in *Nocardia butanica. Agric. Biol. Chem.*, **45**, 2661–2664.

Thomas, D. S. and A. H. Rose (1979). Inhibitory effect of ethanol on growth and solute accumulation by *Saccharomyces cerevisiae* as affected by plasma membrane lipid composition. *Arch. Microbiol.*, **122**, 49–55.

Torriani, A. and F. Rothman (1961). Mutants of *Escherichia coli* constitutive for alkaline phosphatase. *J. Bacteriol.*, **81**, 835–836.

Tosaka, O., H. Hirakawa and K. Takinami (1979). Effect of biotin levels on L-lysine formation in *Brevibacterium lactofermentum. Agric. Biol. Chem.*, **43**, 491–495.

Tosaka, O., M. Karasawa, S. Ikeda and H. Yoshii (1982). Genetic recombination in amino acid-producing bacteria by protoplast fusion. *Abstracts, 4th International Symposium on Genetics of Industrial Microorganisms, Kyoto, Japan*, p. 61.

Tribe, D. E. and J. Pittard (1979). Hyperproduction of tryptophan by *Escherichia coli*: genetic manipulation of the pathways leading to tryptophan formation. *Appl. Environ. Microbiol.*, **38**, 181–190.

Umbarger, H. E. (1978). Amino acid biosynthesis and its regulation. *Annu. Rev. Biochem..*, **47**, 533–606.

Van der Westhuizen, D. T. Jones and D. R. Woods (1982). Autolytic activity and butanol tolerance of *Clostridium acetobutylicum. Appl. Environ. Microbiol.*, **44**, 1277–1281.

Watson, K. (1982). Unsaturated fatty acid but not ergosterol is essential for high ethanol production in *Saccharomyces. Biotechnol. Lett.*, **4**, 397–402.

Yoshida, H., K. Araki and K. Nakayma (1981). L-Arginine production by arginine and analog-resistant mutants of microorganisms. *Agric. Biol. Chem.*, **45**, 959–963.

Appendix 1: Glossary of Terms

Because of the broad multidisciplinary nature of biotechnology, both beginners and specialists in this field may find this glossary useful. It covers terms often used in the relevant areas of the biological, chemical and engineering sciences. It is not intended to be exhaustive. The material was generated primarily from four sources: 'Commercial Biotechnology', Office of Technology Assessment, Congress of the US (1984); 'Advances in Biotechnology', M Moo-Young *et al.* (1981); *Pure and Applied Chemistry*, Vol. 54, No. 9, pp. 1743–1749 (1982); and 'Dictionary of Biochemistry', J. Stenish (1975).

Acclimatization: The biological process whereby an organism adapts to a new environment. For example, it describes the process of developing microorganisms that degrade toxic wastes in the environment.

Activated sludge: Biological growth that occurs in aerobic, organic-containing systems. These growths develop into suspensions that settle and possess clarification and oxidative properties.

Activation energy: The difference in energy between that of the activated complex and that of the reactants; the energy that must be supplied to the reactants before they can undergo transformation to products.

Active immunity: Disease resistance in a person or animal due to antibody production after exposure to a microbial antigen following disease, inapparent infection or inoculation. Active immunity is usually long-lasting.

Active transport: The movement of a solute across a biological membrane such that the movement is directed against the concentration gradient and requires the expenditure of energy.

Activity: A measure of the effective concentration of an enzyme, drug, hormone or some other substance. It is also the product of the molar concentration of an ionic solute and its activity coefficient.

Adsorption: The taking up of molecules of gases, dissolved substances or liquids at the surfaces of solids or liquids with which they are in contact.

Aerobic: Living or acting only in the presence of free-form oxygen, as in air.

Affinity chromatography: The use of compounds, such as antibodies, bound to an immobile matrix to 'capture' other compounds as a highly specific means of separation and purification.

Airlift fermenter: Vessel in which a bioconversion process takes place; the sparged gas is the only source of agitation. The presence of a draft tube inside a fermenter distinguishes this type of fermenter from a bubble column.

Alga (pl. **algae**): A chlorophyll-containing, photosynthetic protist; algae are unicellular or multicellular, generally aquatic and either eukaryotic or prokaryotic.

Amino acids: The building blocks of proteins. There are 20 common amino acids.

Amino acid sequence: The linear order of amino acids in a protein.

Amylase: An enzyme that catalyzes the hydrolysis of starch.

Anabolism: The phase of intermediary metabolism that encompasses the biosynthetic and energy-requiring reactions whereby cell components are produced. Also, the cellular assimilation of macromolecules and complex substances from low-molecular weight precursors (*cf.* Catabolism).

Anaerobic: Living or acting in the absence of free-form oxygen.

Anaerobic digestion: The energy-yielding metabolic breakdown of organic compounds by microorganisms that generally proceeds under anaerobic conditions and with the evolution of gas. The term is most often used to describe the reduction of waste sludges to less solid mass and for methane gas production.

Antibiotic: A specific type of chemical substance that is administered to fight infections, usually bacterial infections, in humans or animals. Many antibiotics are produced by using microorganisms; others are produced synthetically.

Antibody: A protein (immunoglobulin) produced by humans or higher animals in response to exposure to a specific antigen and characterized by specific reactivity with its complementary antigen (see also Monoclonal antibodies).

Antigen: A substance, usually a protein or carbohydrate, which, when introduced into the body of a human or higher animal, stimulates the production of an antibody that will react specifically with it.

Antiserum: Blood serum containing antibodies from animals that have been inoculated with an antigen. When administered to other animals or humans, antiserum produces passive immunity.

Aromatic compound: A compound containing a benzene ring. Many speciality and commodity chemicals are aromatic compounds.

Ascites: Liquid accumulations in the peritoneal cavity. Used as a method for producing monoclonal antibodies.

Asepsis: The prevention of access of microorganisms causing disease, decay or putrefaction to the site of a potential infection.

Aseptic: Of, or pertaining to, asepsis. In directed fermentation processes, the exclusion of unwanted (contaminating) organisms.

Assay: A technique that measures a biological response.

Attenuated vaccine: Whole, pathogenic organisms that are treated with chemical, radioactive, or other means to render them incapable of producing infection. Attenuated vaccines are injected into the body, which then produces protective antibodies against the pathogen to protect against disease.

Autolysis: The self-destruction of a cell as a result of the action of its own hydrolytic enzymes.

Autotrophic: Capable of self-nourishment (opposed to heterotrophic).

Axial dispersion: Mixing along the flow path of fluids during processing as in a bioreactor.

Bacteria: Any of a large group of microscopic organisms having round, rodlike, spiral or filamentous unicellular or noncellular bodies that are often aggregated into colonies, are enclosed by a cell wall or membrane, and lack fully differentiated nuclei. Bacteria may exist as free-living organisms in soil, water, organic matter, or as parasites in the live bodies of plants and animals.

Bacteriophage (or **phage**)/**bacterial virus**: A virus that multiplies in bacteria. Bacteriophage lambda is commonly used as a vector in rDNA experiments.

Batch processing: A method of processing in which a bioreactor, for example, is loaded with raw materials and microorganisms, and the process is run to completion, at which time products are removed (*cf.* Continuous processing).

Binary fission: Asexual division in which a cell divides into two, approximately equal, parts, as in the growth method of some single cells.

Biocatalyst: An enzyme, in cell-free or whole-cell forms, that plays a fundamental role in living organisms or industrially by activating or accelerating a process.

Biochemical: Characterized by, produced by, or involving chemical reactions in living organisms; a product produced by chemical reactions in living organisms.

Biochip: An electronic device that uses biological molecules as the framework for molecules which act as semiconductors and functions as an integrated circuit.

Bioconversion: A chemical conversion using a biocatalyst.

Biodegradation: The breakdown of substances by biological agents, especially microbes.

Biological oxygen demand (BOD): The oxygen used in meeting the metabolic needs of aerobic organisms in water containing organic compounds.

Biological response modifier: Generic term for hormones, neuroactive compounds and immunoactive compounds that act at the cellular level; many are possible targets for production with biotechnology.

Biologics: Vaccines, therapeutic serums, toxoids, antitoxins and analogous biological products used to induce immunity to infectious diseases or harmful substances of biological origin.

Biomass: Organic matter of biological origin such as microbial and plant material.

Biooxidation: Oxidation (the loss of electrons) catalyzed by a biocatalyst.

Biopolymers: Naturally occurring macromolecules that include proteins, nucleic acids and polysaccharides.

Bioprocess: Any process that uses complete living cells or their components (*e.g.* enzymes, chloroplasts) to effect desired physical or chemical changes.

Bioreactor: Vessel in which a bioprocess takes place; examples include fermenter, enzyme reactor.

Biosensor: A device, usually electronic, that uses biological molecules to detect specific compounds.

Biosurfactant: A compound produced by living organisms that helps solubilize compounds such as organic molecules (*e.g.* oil and tar) by reducing surface tension between the compound and liquid.

Biosynthesis: Production, by synthesis or degradation, of a chemical compound by a living organism.

Biotechnology: Use of biological agents or materials to produce goods or services for industry, trade and commerce; a multidisciplinary field.

Bubble column: A gas–liquid contacting vessel in which sparged gas is the only source of agitation.

Bubbly flow: Type of two-phase flow in which the gas phase is distributed in the liquid in the form of bubbles whose dimensions are small compared to the characteristic dimension of the flow cross-section, as in certain designs of bioreactor geometrics.

Budding: A form of asexual reproduction typical of yeast, in which a new cell is formed as an outgrowth from the parent cell.

Buffer: A solution containing a mixture of a weak acid and its conjugate weak base that is capable of resisting substantial changes in pH upon the addition of small amounts of acidic or basic substances, as used for some fermentation media.

Bulking: Increase in volumetric solids as in certain waste treatment bioreactors, which limits the weight concentration that a clarifier can handle.

Callus: An undifferentiated cluster of plant cells that is a first step in regeneration of plants from tissue culture.

Carbohydrate: An aldehyde or a ketone derivative of a polyhydroxy alcohol that is synthesized by living cells. Carbohydrates may be classified either on the basis of their size into mono-, oligo- and poly-saccharides, or on the basis of their functional group into aldehyde or ketone derivatives.

Carboxylation: The addition of an organic acid group (COOH) to a molecule.

Catabolism: The phase of intermediary metabolism that encompasses the degradative and energy-yielding reactions whereby nutrients are metabolized. Also, the cellular breakdown of complex substances and macromolecules to low-molecular weight compounds (*cf.* Anabolism).

Catalysis: A modification, especially an increase, in the rate of a chemical reaction induced by a material (*e.g.* enzyme) that is chemically unchanged at the end of the reaction.

Catalyst: A substance that induces catalysis; an agent that enables a chemical reaction to proceed under milder conditions (*e.g.* at a lower temperature) than otherwise possible. Biological catalysts are enzymes; some nonbiological catalysts include metallic complexes.

Cell: The smallest structural unit of living matter capable of functioning independently; a microscopic mass of protoplasm surrounded by a semipermeable membrane, usually including one or more nuclei and various nonliving products, capable alone, or interacting with other cells, of performing all the fundamental functions of life.

Cell culture: The *in vitro* growth of cells usually isolated from a mixture of organisms. These cells are usually of one type.

Cell differentiation: The process whereby descendants of a common parental cell achieve and maintain specialization of structure and function.

Cell fusion: Formation of a single hybrid cell with nuclei and cytoplasm from different cells (as in 'cell fusion technology'; see also Hybridoma.)

Cell line: Cells that acquire the ability to multiply indefinitely *in vitro* (especially in plant cell tissues).

Cellulase: The enzyme that digests cellulose to sugars.

Cellulose: A polymer of six carbon sugars found in all plant matter; the most abundant biological compound on earth.

Chemical clarification: Characterization of a wastewater process involving distinct operations: coagulation, flocculation and sedimentation.

Chemostat: An apparatus for maintaining microorganisms in the (exponential) phase of growth over prolonged periods of time. This is achieved by the continuous addition of fresh medium and the continuous removal of effluent, so that the volume of the growing culture remains constant.

Chemostat selection: Screening process used to identify microorganisms with desired properties, such as microorganisms that degrade toxic chemicals (see Acclimatization).

Chemotherapeutic agent: A chemical that interferes with the growth of either microorganisms or cancer cells at concentrations at which it is tolerated by the host cells.

Chemotherapy: The treatment of a disease by means of chemotherapeutic agents.

Chitin: A homopolysaccharide of *N*-acetyl-D-glucosamine that is a major constituent of the hard, horny exoskeleton of insects and crustaceans.

Chlorophyll: The green pigment that occurs in plants and functions in photosynthesis by absorbing and utilizing the radiant energy of the sun.

Chloroplasts: Cellular organelles where photosynthesis occurs.

Chromosome: A structure in the nucleus of eukaryotic cells that consists of one or more large double-helical DNA molecules that are associated with RNA and histones; the DNA of the chromosome contains the genes and functions in the storage and in the transmission of the genetic information of the organism.

Chromatography: A process of separating gases, liquids or solids in a mixture or solution by adsorption as the mixture or solution flows over the adsorbent medium, often in a column. The substances are separated because of their differing chemical interaction with the adsorbent medium.

Chromosomes: The rodlike structures of a cell's nucleus that store and transmit genetic information; the physical structures that contain genes. Chromosomes are composed mostly of DNA and protein and contain most of the cell's DNA. Each species has a characteristic number of chromosomes.

Clinical trial: One of the final stages in the collection of data for drug approval where the drug is tested in humans.

Clone: A group of genetically identical cells or organisms produced asexually from a common ancestor.

Cloning: The amplification of segments of DNA, usually genes.

Coagulation: The process whereby chemicals are added to wastewater resulting in a reduction of the forces tending to keep suspended particles apart.

Coagulation–flocculation aids: Materials used in relatively small concentrations which are added either to the coagulation and/or flocculation basins and may be classified as: oxidants, such as chlorine or ozone; weighting agents, such as bentomite clay; activated silica; and polyelectrolytes.

Coding sequence: The region of a gene (DNA) that encodes the amino acid sequence of a protein.

Codon: The sequence of three adjacent nucleotides that occurs in messenger RNA (mRNA) and that functions as a coding unit for a specific amino acid in protein synthesis. The codon determines which amino acid will be incorporated into the protein at a particular position in the polypeptide chain.

Coefficient of thermal conductivity: A physical parameter characterizing intensity of heat conduction in a substance; it is numerically equal to the conductive heat flux density due to a temperature gradient of unity.

Coenzyme: The organic molecule that functions as a cofactor of an enzyme.

Cofactor: The nonprotein component that is required by an enzyme for its activity. The cofactor may be either a metal ion (activator) or an organic molecule (coenzyme) and it may be attached either loosely or tightly to the enzyme; a tightly attached cofactor is known as a prosthetic group.

Colony: A group of contiguous cells that grow in or on a solid medium and are derived from a single cell.

Complementary DNA (cDNA): DNA that is complementary to messenger RNA; used for cloning or as a probe in DNA hybridization studies.

Compulsory licensing: Laws that require the licensing of patents, presumably to ensure early application of a technology and to diffuse control over a technology.

Constitutive enzyme: An enzyme that is present in a given cell in nearly constant amounts regardless of the composition of either the tissue or the medium in which the cell is contained.

Conjugation: The covalent or noncovalent combination of a large molecule, such as a protein or a bile acid, with another molecule. Also, the alternating sequence of single and double bonds in a molecule. Also, the genetic recombination in bacteria and in other unicellular organisms that resemble sexual reproduction and that entails a transfer of DNA between two cells of opposite mating type which are associated side by side.

Continuous processing: Method of processing in which raw materials are supplied and products are removed continuously, at volumetrically equal rates (*cf.* Batch processing).

Continuum: A medium whose discrete heterogeneous structure can be neglected, as in certain fluid flow problems.

Convective mass transfer: Mass transfer produced by simultaneous convection and molecular diffusion. The term is usually used to describe mass transfer associated with fluid flow and involves the mass transfer between a moving fluid and a boundary surface or between two immiscible moving fluids.

Convective transfer: The transfer of mass, heat or momentum in a medium with a nonhomogeneous distribution of velocity, temperature, or concentration; it is accompanied by the displacement of macroscopic elements through the medium.

Cosmid: A DNA cloning vector consisting of plasmid and phage sequences.

Corporate venture capital: Capital provided by major corporations exclusively for high-risk investments.

Crabtree effect: The inhibition of oxygen consumption in cellular respiration that is produced by increasing concentrations of glucose (see also Pasteur effect).

Critical dilution rate: Dilution rate at which wash-out conditions of cells in a bioreactor occurs (see Chemostat).

Cross flow filtration: Method of operating a filtration device whereby the processed material prevents undue build-up of filtered material on filter.

Crystalloid: A noncolloidal low-molecular weight substance.

Culture medium: Any nutrient system for the artificial cultivation of bacteria or other cells; usually a complex mixture of organic and inorganic materials.

Cyclic batch culture: Method of operating a bioreactor whereby a fill-and-dump approach retains enough biocatalyst to avoid need for re-inoculation. In cell cultures, relatively insignificant growth between fill and dump stages occurs.

Cytoplasm: The 'liquid' portion of a cell outside and surrounding the nucleus.

Cytotoxic: Damaging to cells.

Declining phase (or **Death phase**): The phase of growth of a culture of cells that follows the stationary phase and during which there is a decrease in the number (or the mass) of the cells.

Denitrification: The formation of molecular nitrogen from nitrate by way of nitrite.

Deoxyribonucleic acid (DNA): A linear polymer, made up of deoxyribonucleotide repeating units, that is the carrier of genetic information; present in chromosomes and chromosomal material of cell organelles such as mitochondria and chloroplasts, and also present in some viruses. The genetic material found in all living organisms. Every inherited characteristic has its origin somewhere in the code of each individual's DNA.

Diagnostic products: Products that recognize molecules associated with disease or other biological conditions and are used to diagnose these conditions.

Dialysis: The separation of macromolecules from ions and low-molecular weight compounds by means of a semipermeable membrane that is impermeable to (colloidal) macromolecules but is freely permeable to crystalloids and liquid medium.

Dicots (dicotyledons): Plants with two first embryonic leaves and nonparallel veined mature leaves. Examples are soybean and most flowering plants.

Diffusion boundary layer: Characterized by a transverse concentration gradient of a given component in a mixture; the effect of the gradient produces a transverse (mass transfer) of this component.

Diffusion coefficient: A physical parameter that appears as a proportionality coefficient with the gradient of concentration of a specified component in the equation which establishes the dependence of the mass diffusion flux density of the given component on the concentration gradients of all the components in the mixture. 'Self-diffusion coefficient' denotes a physical parameter which characterizes the diffusion of some molecules in the same medium with respect to others for a single-component medium.

Diffusional mass flux: Mass flux due to molecular diffusion.

Dilution rate: Reciprocal of the residence time of a culture in a bioreactor; given by the flow rate divided by bioreactor volume.

Dimensional analysis: Method of determining the number and structure of dimensionless groups consisting of variables essential to a given process on the basis of a comparison of the dimensions of these variables.

Diploidy (or **Diploid state**): The chromosome state in which each of the various chromosomes, except the sex chromosome, is represented twice.

Disclosure requirements: A patent requirement for adequate public disclosure of an invention that enables other people to build and use the invention without 'undue' experimentation.

Distal: Remote from a particular location or from a point of attachment.

DNA: Deoxyribonucleic acid (see above).

DNA base pair: A pair of DNA nucleotide bases. Nucleotide bases pair across the double helix in a very specific way: adenine can only pair with thymine; cytosine can only pair with guanine.

DNA probe: A sequence of DNA that is used to detect the presence of a particular nucleotide sequence.

DNA sequence: The order of nucleotide bases in the DNA helix; the DNA sequence is essential to the storage of genetic information.

DNA synthesis: The synthesis of DNA in the laboratory by the sequential addition of nucleotide bases.

Doubling time: The observed time required for a cell population to double in either the number of cells or the cell mass; it is equal to the generation time only if all the cells in the population are capable of doubling, have the same generation time, and do not undergo lysis (see Generation time).

Downstream processing: After bioconversion of materials in a bioreactor, the separation and purification of the product(s).

Drug: Any chemical compound that may be administered to humans or animals as an aid in the treatment of disease.

Dry weight: The weight of a sample from which liquid has been removed, usually by drying of filtered material.

Eddy mass diffusivity: A quantity characterizing the intensity of turbulent mass transfer of a particular component, as in intensely mixed bioreactors.

Electrophoresis: The movement of charged particles through a stationary liquid under the influence of an electric field. Electrophoresis is a tool for the separation of particles in both preparative and analytical studies of macromolecules. Separation is achieved primarily on the basis of the charge on the particles and to a lesser extent on the basis of the size and shape of the particles. Potentially useful in downstream processing.

Elution: The removal of adsorbed material from an adsorbent, such as the removal of a product from an enzyme bound in a chromatography column.

Emulsification: The process of making lipids, oils, fats, more soluble in water.

Endoenzyme: An enzyme that acts at random in cleaving molecules of substrate.

Endorphins: Opiate-like, naturally occurring peptides with a variety of analgesic effects throughout the endocrine and nervous systems.

Enrichment culture: A culture used for the selection of specific strains of an organism from among a mixture; such a culture favors the growth of the desired strain under the conditions used.

Enzyme: Any of a group of catalytic proteins that are produced by living cells and that mediate and promote the chemical processes of life without themselves being altered or destroyed.

Enzyme induction: The process whereby an inducible enzyme is synthesized in response to an inducer. The inducer combines with a repressor and thereby prevents the blocking of an operator by the repressor.

Escherichia coli (*E. coli*): A species of bacteria that inhabits the intestinal tract of most vertebrates. Some strains are pathogenic to humans and animals. Many nonpathogenic strains are used experimentally as hosts for rDNA.

Eukaryote: A cell or organism with membrane-bound, structurally discrete nuclei and well-developed cell organelles. Eukaryotes include all organisms except viruses, bacteria and blue-green algae (*cf.* Prokaryote).

Exoenzyme: An enzyme that acts by cleaving the ends of molecular chains in a substrate.

Exponential growth: The growth of cells in which the number of cells (or the cell mass) increases exponentially.

Export controls: Laws that restrict technology transfer and trade for reasons of national security, foreign policy or economic policy.

Facultative: Capable of living under more than one set of conditions, usually with respect to aerobic or anaerobic conditions (*e.g.* a facultative anaerobe is an organism or a cell that can grow either in the absence, or in the presence, of molecular oxygen).

Fatty acids: Organic acids with long carbon chains. Fatty acids are abundant in cell membranes and are widely used as industrial emulsifiers.

Fed-batch culture: As in 'cyclic batch culture' except that significant changes in the medium (*e.g.* cell growth) occur during the addition and/or removal of materials from bioreactor.

Feedback inhibition: A negative feedback mechanism in which a product of an enzymatic reaction inhibits the activity of an enzyme that functions in the synthesis of this product.

Feedstocks: Raw materials used for the production of chemicals.

Fermentation: A bioprocess. Fermentation is carried out in bioreactors and is used in various industrial processes for the manufacture of products such as antibiotics, alcohols, acids and vaccines by the action of living organisms (strictly speaking, anaerobically).

Film boiling: Boiling in which a continuous film of vapor that collapses periodically into the bulk of the liquid is formed on the heated surface, as in certain evaporation processes.

Flagellum: (pl. **flagella**): A threadlike, cellular extension that functions in the locomotion of bacterial cells and of unicellular eukaryotic organisms.

Flavin adenine dinucleotide (FAD): The flavin nucleotide, riboflavin adenosine diphosphate, which is a coenzyme form of the vitamin riboflavin, and which functions in dehydrogenation reactions catalyzed by flavoproteins.

Flocculating agent: A reagent added to a dispersion of solids in a liquid to bring together the fine particles into larger masses.

Flocculation: The agglomeration of suspended material to form particles that will settle by gravity, as in the 'tertiary' treatment of waste materials.

Food additive (or **Food ingredient**): A substance that becomes a component of food or affects the characteristics of food and, as such, is regulated, *e.g.* by the US Food and Drug Administration.

Forced convection: Motion of fluid elements induced by external forces, *e.g.* in a bioreactor by a mechanical stirrer.

Free convection: Motion of fluid elements induced by 'natural' forces, *e.g.* by density differences caused by concentration or temperature gradients.

Free-living organism: An organism that does not depend on other organisms for survival.

Fruiting body: A mass of vegetative cells which swarm together at the same stage of growth.

Fungus: Any of a major group of saprophytic and parasitic plants that lack chlorophyll, including molds, rusts, mildews, smuts and mushrooms.

Gamma globulin (GG): A protein component of blood that contains antibodies and confers passive immunity.

Gene: The basic unit of heredity; an ordered sequence of nucleotide bases, comprising a segment of DNA. A gene contains the sequence of DNA that encodes one polypeptide chain (*via* RNA).

Gene amplification: In biotechnology, an increase in gene number for a certain protein so that the protein is produced at elevated levels.

Gene expression: The mechanism whereby the genetic directions in any particular cell are decoded and processed into the final functioning product, usually a protein (see also Transcription and Translation).

Gene transfer: The use of genetic or physical manipulation to introduce foreign genes into host cells to achieve desired characteristics in progeny.

Generation time: The time required by a cell for the completion of one cycle of growth (see also Doubling time).

Genetic engineering: Loose term used to describe any gene manipulative technique, especially recombinant DNA techniques.

Genome: The genetic endowment of an organism or individual.

Genus: A taxonomic category that includes groups of closely related species.

Germ cell: The male and female reproductive cells; egg and sperm.

Germplasm: The total genetic variability available to a species.

Glycoproteins: Proteins with attached sugar groups.

Glycoside: A mixed acetal (or ketal) derived from the cyclic hemiacetal (or hemiketal) form of an aldose (or a ketose); a compound formed by replacing the hydrogen or the hydroxyl group of the anomeric carbon of the carbohydrate with an alkyl or aryl group.

Glycosylation: The attachment of sugar groups to a molecule, such as a protein.

Gram negative: Designating a bacterium that does not retain the initial Gram stain but retains the counterstain. Gram-negative bacteria possess a relatively thin cell wall that is not readily digested by the enzyme lysozyme, and in which the peptidoglycan layer is covered with lipopolysaccharide.

Gram positive: Designating a bacterium that retains the initial Gram stain and is not stained by the counterstain. Gram-positive bacteria generally possess a relatively thick and rigid cell wall that is readily digested by the enzyme lysozyme, and that consists of a layer of peptidoglycan.

Gram stain: A set of two stains (chemicals) that are used to stain bacteria; the staining depends on the composition and the structure of the bacterial cell wall.

Growth hormone (GH): A group of peptides involved in regulating growth in higher animals.

Heat flux: The quantity of heat that passes through an arbitrary surface per unit time.

Heat flux density: Heat flux per unit area.

Heat transfer: Spontaneous irreversible process of heat transmission in a space with a nonisothermal temperature field, as in the cooling or heating of bioreactors and ancilliary equipment.

Hemicellulose: A polymer of D-xylose that contains side chains of other sugars and that serves to cement plant cellulose fibers together.

Herbicide: An agent (*e.g.* a chemical) used to destroy or inhibit plant growth; specifically, a selective weed killer that is not injurious to crop plants.

Heterofermentative lactic acid bacteria: Lactic acid bacteria that produce in fermentation less than 1.8 moles of lactic acid per mole of glucose; in addition to lactic acid, these organisms produce ethanol, acetate, glycerol, mannitol and carbon dioxide (see also Homofermentative lactic acid bacteria).

Heterotroph: A cell or organism that requires a variety of carbon-containing compounds from animals and plants as its source of carbon, and that synthesizes all of its carbon-containing biomolecules from these compounds and from small inorganic molecules.

Heterotrophic: Pertaining to a regulatory enzyme in which the effector is a metabolite other than the substance of the enzyme.

High performance liquid chromatography (HPLC): A recently developed type of chromatography that is potentially important in downstream processing of bioreactor products.

Histone: A basic, globular, and simple protein that is characterized by its high content of arginine and lysine. Histones are found in association with nucleic acids in the nuclei of many eukaryotic cells.

Homofermentative lactic acid bacteria: Lactic acid bacteria that produce 1.8–2.0 moles of lactic acid per mole of glucose during fermentation.

Hormone: A chemical messenger found in the circulation of higher organisms that transmits regulatory messages to cells.

Host: A cell whose metabolism is used for growth and reproduction of a virus, plasmid or other form of foreign DNA.

Host–vector system: Compatible combinations of host (*e.g.* bacterium) and vector (*e.g.* plasmid) that allow stable introduction of foreign DNA into cells.

Hybrid: The offspring of two genetically dissimilar parents (*e.g.* a new variety of plant or animal that results from cross-breeding two different existing varieties; a cell derived from two different cultured cell lines that have fused).

Hybridization: The act or process of producing hybrids.

Hybridoma: Product of fusion between myeloma cell (which divides continuously in culture and is 'immortal') and lymphocyte (antibody-producing cell); the resulting cell grows in culture and produces monoclonal antibodies.

Hybridoma technology: See Monoclonal antibody technology.

Hydrolysis: Chemical reaction involving addition of water to break bonds.

Hydroxylation: Chemical reaction involving the addition of hydroxyl (OH) groups to chemical compounds.

Hypha (pl. **hyphae**): The filamentous and branched tube that forms the network which contains the cytoplasm of the mycelium of a fungus.

Immobilized enzyme or cell techniques: Techniques used for the fixation of enzymes or cells on to solid supports. Immobilized cells and enzymes are used in continuous bioprocessing in bioreactors and upstream or downstream processing of materials.

Immune response: The reaction of an organism to invasion by a foreign substance. Immune responses are often complex, and may involve the production of antibodies from special cells (lymphocytes), as well as the removal of the foreign substance by other cells.

Immunization: The administration of an antigen to an animal organism to stimulate the production of antibodies by that organism. Also, the administration of antigens, antibodies or lymphocytes to an animal organism to produce the corresponding active, passive or adoptive immunity.

Immunoassay: The use of antibodies to identify and quantify substances. The binding of antibodies to antigen, the substance being measured, is often followed by tracers such as radioisotopes.

Immunogenic: Capable of causing an immune response (see also Antigen).

Immunotoxin: A molecule attached to an antibody capable of killing cells that display the antigen to which the antibody binds.

Inducible enzyme: An enzyme that is normally either absent from a cell or present in very small amounts, but that is synthesized in appreciable amounts in response to an inducer in the process medium.

Interface: The boundary between two phases.

Interferons (Ifns): A class of glycoproteins (proteins with sugar groups attached at specific locations) important in immune function and thought to inhibit viral infections.

In vitro: Literally, in glass; pertaining to a biological reaction taking place in an artificial apparatus; sometimes used to include the growth of cells from multicellular organisms under cell culture conditions. *In vitro* diagnostic products are products used to diagnose disease outside of the body after a sample has been taken from the body.

In vivo: Literally, in life; pertaining to a biological reaction taking place in a living cell or organism. *In vivo* products are products used within the body.

Ionic strength: A measure of the ionic concentration of a solution.

Laminar flow: Fluid motion in which the existence of steady fluid particle trajectories can exist; in processing equipment, it represents relatively low levels of mixing intensities.

Isoelectric pH (isoelectric point): The pH at which a molecule has a net zero charge; the pH at which the molecule has an equal number of positive and negative charges, which includes those due to any ions bound by the molecule.

Isoelectrophoretic pH (isoelectrophoretic point): The pH at which the electrophoretic mobility of a protein is zero; this pH may coincide with the theoretical isoelectric pH of the protein, depending on the surface structure of the protein, the ionic strength, and the nature of the ionic double layer around the protein.

Lag phase: That phase of growth of a cell that precedes the exponential phase and during which there is only little or no growth.

Leaching: The removal of a soluble compound such as an ore from a solid mixture by washing or percolating.

Lignocellulose: The composition of woody biomass, including lignin and cellulose.

Lignolytic: Pertaining to the breakdown of lignin.

Lime: Various natural forms of the chemical compound CaO, as in hydrated lime and dolomitic lime.

Linker: A small fragment of synthetic DNA that has a restriction site useful for gene cloning, which is used for joining DNA strands together.

Lipase: An enzyme that catalyzes the hydrolysis of fats.

Lipids: A large, varied class of water-insoluble fat-based organic molecules; includes steroids, fatty acids, prostaglandins, terpenes and waxes.

Lipopolysaccharide: A water-soluble lipid–polysaccharide complex.

Liposome transfer: The process of enclosing biological compounds inside a lipid membrane and allowing the complex to be taken up by a cell.

Lymphocytes: Specialized white blood cells involved in the immune response; B lymphocytes produce antibodies.

Lymphokines: Proteins that mediate interactions among lymphocytes and are vital to proper immune function.

Lyophilization: The removal of water under vacuum from a frozen sample; a relatively gentle process in which water sublimes.

Lysis: The rupture and dissolution of cells.

Mass exchange: Mass transfer across an interface or a permeable wall (membrane) between two phases.

Mass flux: The mass of a given mixture component passing per unit time across any surface.

Mass flux density: Mass flux per unit area of surface.

Mass transfer: Spontaneous irreversible process of transfer of mass of a given component in a space with a nonhomogenous field of the chemical potential of the component. In the simplest case, the driving force is the difference in concentration (in liquids) or partial pressure (in gases) of the component. Other physical quantities, *e.g.* temperature difference (thermal diffusion), can also induce mass transfer.

Mass transfer coefficient: A quantity characterizing the intensity of mass transfer; it is numerically equal to the ratio of the mass flux to the difference of its mass fractions. For the case of mass transfer between a liquid medium and a gas, the mass fraction of a given component in the liquid is determined by phase equilibrium parameters (distribution coefficient) with allowance, if necessary, for resistance to transfer at the phase boundary *per se*.

Mass velocity: Mass flow rate across a unit area perpendicular to the direction of the velocity vector.

Mesophile: An organism that grows at moderate temperatures in the range 20–45 °C, and that has an optimum growth temperature in the range 30–39 °C.

Mesophilic: Of, or pertaining to, mesophiles.

Messenger RNA (mRNA): RNA that serves as the template for protein synthesis; it carries the transcribed genetic code from the DNA to the protein synthesizing complex to direct protein synthesis.

Metabolism: The physical and chemical processes by which chemical components are synthesized into complex elements, complex substances are transformed into simpler ones, and energy is made available for use by an organism.

Metabolite: Any reactant, intermediate or product in the reactions of metabolism.

Metallothioneins: Proteins, found in higher organisms, that have a high affinity for heavy metals.

Methanogens: Bacteria that produce methane as a metabolic product.

Microorganisms: Microscopic living entities; microorganisms can be viruses, prokaryotes (*e.g.* bacteria) or eukaryotes (*e.g.* fungi). Also referred to as microbes.

Microencapsulation: The process of surrounding cells with a permeable membrane.

Mitochondrion (pl. **mitochondria**): A subcellular organelle in aerobic eukaryotic cells that is the site of cellular respiration and that carries out the reactions of the citric acid cycle, electron transport, and oxidative phosphorylation. Mitochondria contain DNA and ribosomes, carry out protein synthesis, and are capable of self-replication.

Mixed culture: Culture containing two or more types of microorganisms.

Molecular diffusion: Mass transfer resulting from thermal motion. Concentration diffusion refers to molecular diffusion resulting from a nonhomogenous distribution of concentrations of components of a mixture.

Monoclonal antibodies (MAbs): Homogeneous antibodies derived from a single clone of cells; MAbs recognize only one chemical structure. MAbs are useful in a variety of industrial and medical capacities since they are easily produced in large quantities and have remarkable specificity.

Monoclonal antibody technology: The use of hybridomas that produce monoclonal antibodies for a variety of purposes. Hybridomas are maintained in cell culture or, on a larger scale, as tumors (ascites) in mice. Also referred to as 'hybridoma' technology.

Monocots (monocotyledons): Plants with single first embryonic leaves, parallel-veined leaves, and simple stems and roots. Examples are cereal grains such as corn, wheat, rye, barley and rice.

Monosaccharide: A polyhydroxy alcohol containing either an aldehyde or a ketone group; a simple sugar.

Multigenic: A trait specialized by several genes.

Multi-phase medium: A medium consisting of two or more single-phase portions with physical properties changing discontinuously (stepwise) at the boundaries of the medium, as in a gas–liquid dispersion used in aerobic fermentations.

Mutagenesis: The induction of mutation in the genetic material of an organism; researchers may use physical or chemical means to cause mutations that improve the production of capabilities of organisms.

Mutagen: An agent that causes mutation.

Mutant: An organism with one or more DNA mutations, making its genetic function or structure different from that of a corresponding wild-type organism.

Mutation: A permanent change in a DNA sequence.

Mycelium (pl. **mycelia**): The vegetative structure of a fungus that consists of a multinucleate mass of cytoplasm, enclosed within a branched network of filamentous tubes known as hyphae.

Myeloma: Antibody-producing tumor cells.

Myeloma cell line: Myeloma cells established in culture.

Natural convection: Free motion due to gravitational forces in a system with a non-homogeneous density distribution (see also Free convection).

Neurotransmitters: Small molecules found at nerve junctions that transmit signals across those junctions.

Newtonian fluid: A fluid, the viscosity of which is independent of the rate and/or duration of shear.

NIH Guidelines: Guidelines, established by the US National Institutes of Health, on the safety of research involving recombinant DNA.

Nitrogen fixation: The conversion of atmospheric nitrogen gas to a chemically combined form, ammonia (NH_3), which is essential to growth. Only a limited number of microorganisms can fix nitrogen.

Nodule: The anatomical part of a plant root in which nitrogen-fixing bacteria are maintained in a symbiotic relationship with the plant.

Nodulins: Proteins, possibly enzymes, present in nodules; function unknown.

Non-Newtonian fluid: A fluid, the viscosity of which depends on the rate and/or duration of shear.

Nucleate boiling: Boiling in which vapor is generated in the form of periodically produced and growing discrete bubbles.

Nucleic acids: Macromolecules composed of sequences of nucleotide bases. There are two kinds of nucleic acids: DNA, which contains the sugar deoxyribose, and RNA, which contains the sugar ribose.

Nucleoside: A glycoside composed of D-ribose or 2-deoxy-D-ribose and either a purine or a pyrimidine.

Nucleotide base: A structural unit of nucleic acid. The bases present in DNA are adenine, cytosine, guanine and thymine. In RNA, uracil substitutes for thymine.

Nucleus: In the biological sciences, a relatively large spherical body inside a cell that contains the chromosomes.

Oligomer: A molecule that consists of two or more monomers linked together, covalently or noncovalently.

Oligonucleotides: Short segments of DNA or RNA.

Optical density (absorbance): A measure of the light absorbed by a solution.

Organelle: A specialized part of a cell that conducts certain functions. Examples are nuclei, chloroplasts and mitochondria, which contain most of the genetic material, conduct photosynthesis and provide energy, respectively.

Organic compounds: Molecules that contain carbon.

Organic micropollutant: Low molecular weight organic compounds considered hazardous to humans or the environment.

Osmosis: The movement of water or another solvent across a semipermeable membrane from a region of low solute concentration to one of a higher solute concentration.

Osmotic pressure: The pressure that causes water or another solvent to move in osmosis from a solution having a low solute concentration to one having a high solute concentration; it is equal to the hydrostatic pressure that has to be applied to the more concentrated solution to prevent the movement of water (solvent) into it.

Oxidation ponds: Quiescent earthen basins that provide sufficient hydraulic hold-up time for the natural processes to effect removal and stabilization of organic matter.

Oxygen transfer rate (OTR): Mass transfer for oxygen solute as in fermentation medium.

Parasite: An organism that lives in or upon another organism from which it derives some or all of its nutrients.

Passive immunity: Disease resistance in a person or animal due to the injection of antibodies from another person or animal. Passive immunity is usually short-lasting (*cf.* Active immunity).

Pasteur effect: The inhibition of glycolysis and the decrease of lactic acid accumulation that is produced by increasing concentrations of oxygen.

Patent: A limiting property right granted to inventors by a government allowing the inventor of a new invention the right to exclude all others from making, using or selling the invention unless specifically approved by the inventor, for a specified time period in return for full disclosure by the inventor about the invention.

Pathogen: A disease-producing agent, usually restricted to a living agent such as a bacterium or virus.

Pectin: A polysaccharide that occurs in fruits and that consists of a form of pectic acid in which many of its carboxyl groups have been methylated.

Peptide: A linear polymer of amino acids. A polymer of numerous amino acids is called a *polypeptide*. Polypeptides may be grouped by function, such as 'neuroactive' polypeptides.

Permease: An enzyme that is instrumental in transporting material across a biological membrane or within a biological fluid.

Pharmaceuticals: Products intended for use in humans, as well as *in vitro* applications to humans, including drugs, vaccines, diagnostics and biological response modifiers.

Photorespiration: Reaction in plants that competes with the photosynthetic process. Instead of fixing CO_2, RuBPCase can utilize oxygen, which results in a net loss of fixed CO_2.

Photosynthesis: The reaction carried out by plants where carbon dioxide from the atmosphere is fixed into sugars in the presence of sunlight; the transformation of solar energy into biological energy.

Plasma: The liquid (noncellular) fraction of blood. In vertebrates, it contains many important proteins (*e.g.* fibrinogen, responsible for clotting).

Plasmid: An extrachromosomal, self-replicating, circular segment of DNA; plasmids (and some viruses) are used as 'vectors' for cloning DNA in bacterial 'host' cells.

Plug flow: Flow of materials in which there is no mixing in the direction of flow (see Axial dispersion).

Polymer: A linear or branched molecule of repeating subunits.

Polypeptide: A long peptide, which consists of amino acids.

Polysaccharide: A polymer of sugars.

Pool boiling: Boiling with convective (free) motion in a liquid volume whose dimensions in all directions are large compared to the breakaway diameter of the bubble.

Primary metabolite: Metabolite that is required for the function of the organism's life support system.

Proinsulin: A precursor protein of insulin.

Prokaryote: A cell or organism lacking membrane-bound, structurally discrete nuclei and organelles. Prokaryotes include bacteria and the blue-green algae. (*cf.* eukaryote).

Promoter: A DNA sequence in front of a gene that controls the initiation of 'transcription' (see below).

Prophylaxis: Prevention of disease.

Protease: Protein-digesting enzyme.

Protein: A polypeptide consisting of amino acids. In their biologically active states, proteins function as catalysts in metabolism and, to some extent, as structural elements of cells and tissues.

Protist: A unicellular or multicellular organism that lacks the tissue differentiation and the elaborate organization that is characteristic of plants and animals.

Protoplast fusion: The joining of two cells in the laboratory to achieve desired results, such as increased viability of antibiotic-producing cells.

Protozoa: Diverse forms of eukaryotic microorganisms; structure varies from simple single cells to colonial forms; some protozoa are pathogenic.

Psychrophile: An organism that grows at low temperatures in the range of 0–25 °C, and that has an optimum growth temperature in the range 20–25 °C.

Pure culture: A culture containing only microorganisms from one species.

Pyrogenicity: The tendency for some bacterial cells or parts of cells to cause inflammatory reactions in the body, which may detract from their usefulness as pharmaceutical products.

Recarbonation: Unit water treatment process in which carbon dioxide is added to a lime-treated water. Basic purpose is the downward adjustment of the pH of the water.

Recombinant DNA (rDNA): The hybrid DNA produced by joining pieces of DNA from different organisms together *in vitro* (*i.e.* in an artificial apparatus).

Recombinant DNA technology: The use of recombinant DNA for a specific purpose, such as the formation of a product or the study of a gene.

Recombination: Formation of a new association of genes or DNA sequences from different parental origins.

Reducing sugar: A sugar that will reduce certain inorganic ions in solution, such as the copper(II) ions of Fehling's or Benedict's reagent; the reducing property of the sugar is due to its aldehyde or potential aldehyde group.

Regeneration: In biological sciences, the laboratory process of growing a whole plant from a single cell or small clump of cells.

Regulatory sequence: A DNA sequence involved in regulating the expression of a gene.

Repressible enzyme: An enzyme, the synthesis of which is decreased when the intracellular concentration of specific metabolites reaches a certain level.

Resistance gene: Gene that provides resistance to an environmental stress such as an antibiotic or other chemical compound.

Respiration: The cellular oxidative reactions of metabolism, particularly the terminal steps, by which nutrients are broken down; the reactions which require oxygen as the terminal electron acceptor, produce carbon dioxide as a waste product, and yield utilizable energy.

Restriction enzymes: Enzymes that cut DNA at specific DNA sequences.

Ribosome: One of a large number of subcellular, nucleoprotein particles that are composed of approximately equal amounts of RNA and protein and that are the sites of protein synthesis in the cell.

Ribosomal RNA: The RNA that is linked noncovalently to the ribosomal proteins in the two ribosomal subunits and that constitutes about 80% of the total cellular RNA.

RNA: Ribonucleic acid (see also Messenger RNA).

RuBPCase (ribulosebisphosphate carboxylase): An enzyme that catalyzes the critical step of the photosynthetic CO_2 cycle.

Salting out: The decrease in the solubility of a protein that is produced in solutions of high ionic strength by an increase of the concentrations of neutral salts.

Saccharification: The degradation of polysaccharides to sugars.

Scale-up: The transition of a process from research laboratory bench scale to engineering pilot plant or industrial scale.

Secondary metabolite: Metabolite that is not required by the producing organism for its life-support system.

Semiconductor: A material such as silicon or germanium with electrical conductivities intermediate between good conductors such as copper wire and insulators such as glass.

Shake flask: A laboratory flask for culturing microorganisms in a shaker–incubator which provides mixing and aeration.

Shear rate: The variation in velocity within a flowing material, as in a bioreactor.

Single cell protein (SCP): Cells, or protein extracts, of microorganisms grown in large quantities for use as human or animal protein supplements. A misnomer for multicellular SCP products.

Single-phase medium: Continuous single- or multi-component medium whose properties in space can vary in a continuous manner with no phase boundaries, *e.g.* a gas or a liquid.

Slaking: Process of adding water to quicklime, or recalcined lime, to produce a slurry of hydrated lime.

Slant culture: A culture grown in a tube that contains a solid nutrient medium which was solidified while the tube was kept in a slanted position.

Slimes: Aggregations of microbial cells that pose environmental and industrial problems; may be amenable to biological control.

Sludge: Precipitated solid matter produced by water and sewage treatment or industrial problems; may be amenable to biological control.

Slug flow: Type of two-phase flow in which the gas phase flows in the form of large bubbles whose transverse dimensions are commensurate with the characteristic dimension of the flow cross section as in some pipeline operations.

Somaclonal variation: Genetic variation produced from the culture of plant cells from a pure breeding strain; the source of the variation is not known.

Species: A taxonomic subdivision of a genus. A group of closely related, morphologically similar individuals which actually or potentially interbreed.

Specific growth rate: The rate of growth of a population of microorganisms, per unit mass of cells.

Spectrometer: An instrument used for analyzing the structure of compounds on the basis of their light-absorbing properties.

Spheroplast: A bacterial cell that is largely, but not entirely, freed of its cell wall.

Spore: A dormant cellular form, derived from a bacterial or a fungal cell, that is devoid of metabolic activity and that can give rise to a vegetative cell upon germination; it is dehydrated and can survive for prolonged periods of time under drastic environmental conditions.

Starch: The major form of storage carbohydrates in plants. It is a homopolysaccharide, composed of D-glucose units, that occurs in two forms: amylose, which consists of straight chains, and in which the glucose residues are linked by means of alpha(1–4) glycosidic bonds; and amylopectin, which consists of branched chains, and in which the glucose residues are linked by means of both alpha(1–4) and alpha(1–6) glycosidic bonds.

Stationary phase: The phase of growth of a culture of microorganisms that follows the exponential phase and in which there is little or no growth.

Sterile: Free from viable microorganisms.

Sterilization: The complete destruction of all viable microorganisms in a material by physical and/or chemical means.

Steroid: A group of organic compounds, some of which act as hormones to stimulate cell growth in higher animals and humans.

Stirred tank bioreactor: Agitated vessel in which a bioprocess takes place; mixing is provided by the mechanical action of an impeller/agitator.

Storage protein genes: Genes coding for the major proteins found in plant seeds.

Strain: A group of organisms of the same species having distinctive characteristics but not usually considered a separate breed or variety. A genetically homogeneous population of organisms at a subspecies level that can be differentiated by a biochemical, pathogenic or other taxonomic feature.

Substrate: A substance acted upon, for example, by an enzyme.

Subunit vaccine: A vaccine that contains only portions of a surface molecule of a pathogen. Subunit vaccines can be prepared by using rDNA technology to produce all or part of the surface protein molecule or by artificial (chemical) synthesis of short peptides.

Surfactant: A substance that alters the surface tension of a liquid, generally lowering it; detergents and soaps are typical examples.

Symbiont: An organism living in symbiosis, usually the smaller member of a symbiotic pair of dissimilar size.

Symbiosis: In the biological sciences, the living together of two dissimilar organisms in mutually beneficial relationships.

Synchronous growth: Growth in which all of the cells are at the same stage in cell division at any given time.

Tangential flow filtration: See Cross-flow filtration.

Taxis: The movement of an organism in response to a stimulus.

Taxonomy: The scientific classification of plants and animals that is based on their natural relationships; includes the systematic grouping, ordering and naming of the organisms.

Therapeutics: Pharmaceutical products used in the treatment of disease.

Thermal diffusivity: Numerically equal to the ratio of the coefficient of thermal conductivity to the volumetric specific heat of a substance (see also Heat transfer).

Thermophile: An organism that grows at high temperatures in the range 45–70 °C (or higher temperatures) and that has an optimum growth temperature in the range 50–55 °C.

Thermophilic: Heat loving. Usually refers to microorganisms that are capable of surviving at elevated temperatures; this capability may make them more compatible with industrial biotechnology schemes.

Thermotolerant: Capable of withstanding relatively high temperatures (45–70 °C).

Thrombolytic enzymes: Enzymes such as streptokinase and urokinase that initiate the dissolution of blood clots.

Ti plasmid: Plasmid from *Agrobacterium tumefacciens*, used as a plant vector.

Tissue plasminogen activator (TPA): A hormone that selectively dissolves blood clots that cause heart attacks and strokes.

Totipotency: The capacity of a higher organism cell to differentiate into an entire organism. A totipotent cell contains all the genetic information necessary for complete development.

Toxicity: The ability of a substance to produce a harmful effect on an organism by physical contact, ingestion or inhalation.

Toxin: A substance, produced in some cases by disease-causing microorganisms, which is toxic to other living organisms.

Toxoid: Detoxified toxin, but with antigenic properties intact.

Transcription: The synthesis of messenger RNA on a DNA template; the resulting RNA sequence is complementary to the DNA sequence. This is the first step in gene expression (see also Translation).

Transfer RNA (tRNA): A low-molecular weight RNA molecule, containing about 70–80 nucleotides, that binds an amino acid and transfers it to the ribosomes for incorporation into a polypeptide chain during translation.

Transformation: In the biological sciences, the introduction of new genetic information into a cell using naked DNA.

Transistor: An active component of an electrical circuit consisting of semiconductor material to which at least three electrical contacts are made so that it acts as an amplifier, detector or switch.

Translation: In the biological sciences, the process in which the genetic code contained in the nucleotide base sequence of messenger RNA directs the synthesis of a specific order of amino acids to produce a protein. This is the second step in gene expression (see also Transcription).

Transposable element: Segment of DNA which moves from one location to another among or within chromosomes in possibly a predetermined fashion, causing genetic change; may be useful as a vector for manipulating DNA.

Turbulent flow: In the engineering sciences, fluid motion with particle trajectories varying chaotically (randomly) with time; irregular fluctuations of velocity, pressure and other parameters, non-uniformly distributed in the flow; indicates relatively high levels of mixing (*cf.* Laminar flow).

Vaccine: A suspension of attenuated or killed bacteria or viruses, or portions thereof, injected to produce active immunity (see also Subunit vaccine).

Vector: DNA molecule used to introduce foreign DNA into host cells. Vectors include plasmids, bacteriophages (virus) and other forms of DNA. A vector must be capable of replicating autonomously and must have cloning sites for the introduction of foreign DNA.

Viable: Describing a cell or an organism that is alive and capable of reproduction.

Viable count: The number of viable cells in a culture of microorganisms.

Virus: Any of a large group of submicroscopic agents infecting plants, animals and bacteria and unable to reproduce outside the tissues of the host. A fully formed virus consists of nucleic acid (DNA or RNA) surrounded by a protein or protein and lipid coat.

Viscosity: A measure of a liquid's resistance to flow.

Volatile fatty acids (VFAs): Mixture of acids, primarily acetic and propionic, produced by acidogenic microorganisms during anaerobic digestion.

Volatile organic compounds (VOCs): Group of toxic compounds found in ground water and that pose environmental hazards; their destruction during water purification may be done biologically.

Wash out: Condition at which the critical dilution rate is exceeded in a chemostat.

Wild-type: The most frequently encountered phenotype in natural breeding populations.

Yeast: A fungus of the family Saccharomycetacea that is used especially in the making of alcoholic liquors and fodder yeast for animal feeds, and as leavening in baking. Yeasts are also commonly used in bioprocesses.

Yield: For a general chemical reaction: the weight of product obtained divided by the theoretical amount expected. Also, for the isolation of an enzyme: the total activity at a given step in the isolation divided by the total activity at a reference step.

Appendix 2: Nomenclature Guidelines

Provisional list of symbols and units recommended for use in biotechnology by the IUPAC Commission on Biotechnology as reported in *Pure Appl. Chem.*, **54**, 1743–1749 (1982). This list is intended for use in conjunction with other recommendations on symbols, in particular *Manual of Symbols and Terminology for Physicochemical Quantities and Units* (Pergamon, 1979) and 'Letter Symbols for Chemical Engineering', *Chem. Eng. Prog.*, 73–80 (1978).

A2.1 GENERAL CONCEPTS

	Symbol	SI units	Other units
Activation energy	E	$J\,mol^{-1}$	$cal\,mol^{-1}$
for growth	E_g		
for death	E_d		
Area dimensions			
area per unit volume	a	m^{-1}	cm^{-1}
Linear dimension			
impeller diameter	D_i	m	m
tank diameter	D_t	m	m
liquid depth	D_l	m	m
width of baffle	D_b	m	m
Pressure	p	Pa	atm, bar
denote partial pressure with appropriate subscript, *e.g.* P_{O_2} for partial pressure of oxygen			
Ratio, in general	R		
for stoichiometric mass ratio, *e.g.* mass of substrate A consumed per mass of substrate B consumed	$R_{A/B}$		
for stoichiometric molar ratio, *e.g.* mole of substrate A consumed per mole of B consumed	$R_{MA/B}$		
Temperature			
absolute	T	K	K
general	t, T	°C	°C
Time	t	s	min, h
identify specific time periods by appropriate subscripts, *e.g.* t_d for			

653

	Symbol	SI units	Other units
doubling time, t_l for lag time, and t_r for replacement or mean residence time			

Volume dimensions

	Symbol	SI units	Other units
volume	V	m^3, L	L
identify by subscript, *e.g.* V_1 for volume of stage 1, *etc.*			

Yield, general mass ratio expressing output over input — Y

without further definition, Y refers to the mass conversion ratio in terms of g dry weight biomass per g mass of substrate used. It should be further defined by subscripts to denote other ratios, *e.g.* $Y_{P/S}$ and $Y_{P/X}$ for g mass of product per g mass of substrate and per g dry weight of biomass, respectively

Yield, growth mass ratio corrected for maintenance, where — Y_G

$$\frac{1}{Y} = \frac{1}{Y_G} + \frac{m}{\mu}$$

or

$$q_s = \frac{\mu}{Y_G} + m$$

	Symbol	SI units	Other units
Yield, molar growth	Y_{GM}	$kg\,mol^{-1}$	$g\,mol^{-1}$

kg biomass formed per mole of mass used, or further defined as above to denote other molar yields

A2.2   CONCENTRATIONS AND AMOUNTS

Concentration

Biomass*

	Symbol	SI units	Other units
total mass (dry wt. basis)	x	kg	g
mass concentration (dry wt. basis)	X	$kg\,m^{-3}$	$g\,L^{-1}$
volume fraction	ϕ		
total number	N		
number concentration	n	m^{-3}	
Substrate concentration mass or moles per unit volume	C_S	$kg\,m^{-3}$, $kmol\,m^{-3}$	$mg\,L^{-1}$, $mmol\,L^{-1}$
Product concentration mass or moles per unit volume	C_P	$kg\,m^{-3}$, $kmol\,m^{-3}$	$mg\,L^{-1}$, $mmol\,L^{-1}$
Gas hold-up volume of gas per volume of dispersion	ε_G		

* Note: because of the difficulty in expressing biomass (cells) in molar terms, a separate symbol (other than C) is recommended.

	Symbol	*SI units*	*Other units*
Inhibitor concentration mass or moles per unit volume	C_i	$\text{kg m}^{-3}, \text{kmol m}^{-3}$	$\text{mg L}^{-1}, \text{mmol L}^{-1}$
Inhibitor constant dissociation constant of inhibitor–biomass complex	K_i	kg m^{-3}	g L^{-1}
Saturation constant as in the growth rate expression $\mu = \mu_m C_S/(K_s + C_S)$	K_s	kmol m^{-3}	$\text{g L}^{-1}, \text{mmol L}^{-1}$
Total amount, *e.g.* mass or moles	C	kg, kmol	g, mol

A2.3 INTENSIVE PROPERTIES

	Symbol	*SI units*	*Other units*
Density, mass	ϱ	kg m^{-3}	g L^{-1}
Diffusivity, molecular, volumetric	D_v	$\text{m}^2\,\text{s}^{-1}$	$\text{cm}^2\,\text{s}^{-1}$
Enthalpy, mass, of growth heat produced per unit of dry weight biomass formed	H_X	J kg^{-1}	J g^{-1}
Enthalpy, molar, of substrate consumption or of product formation	H_S, H_P	J mol^{-1}	J mol^{-1}
Vapor pressure denote with appropriate subscript, *e.g.* $p_i^* = $ vapor pressure of material i	p^*	Pa	atm, bar
Viscosity, absolute	μ	Pa s	poise
Viscosity, kinematic	ν	$\text{m}^2\,\text{s}^{-1}$	$\text{cm}^2\,\text{s}^{-1}$

A2.4 RATE CONCEPTS

	Symbol	*SI units*	*Other units*
Death rate, specific $\delta = -(\text{d}n/\text{d}t)/n$	δ	s^{-1}	$\text{s}^{-1}, \text{h}^{-1}$
Dilution rate volume flow rate/culture volume	D	s^{-1}	$\text{h}^{-1}, \text{d}^{-1}$
Dilution rate, critical value at which biomass washout occurs in continuous flow culture	D_c	s^{-1}	$\text{h}^{-1}, \text{d}^{-1}$
Doubling time, biomass $t_d = (\ln 2)/\mu$	t_d	s	min, h
Flow rate, volumetric identify stream by appropriate subscript, *e.g.* A for air, G for gas, L for liquid, *etc.*	F	$\text{m}^3\,\text{s}^{-1}$	L h^{-1}
Growth rate, colony radial rate of extension of biomass colony on a surface	K_r	m s^{-1}	$\mu\text{m h}^{-1}$
Growth rate, maximum specific	μ_m	s^{-1}	$\text{h}^{-1}, \text{d}^{-1}$
Growth rate, specific $\mu = (\text{d}x/\text{d}t)/x$	μ	s^{-1}	$\text{h}^{-1}, \text{d}^{-1}$

	Symbol	SI units	Other units
Heat transfer coefficient			
individual	h	$\text{W m}^{-2}\,\text{K}^{-1}$	$\text{cal h}^{-1}\,\text{cm}^{-2}\,{}^\circ\text{C}^{-1}$
overall	U	$\text{W m}^{-2}\,\text{K}^{-1}$	$\text{cal h}^{-1}\,\text{cm}^{-2}\,{}^\circ\text{C}^{-1}$
Maintenance coefficient, substrate or non-growth term associated with substrate consumption as defined in yield relationship (see yield term)	m	s^{-1}	h^{-1}
Mass transfer coefficient (molar basis)			
Individual, area basis	k	$\text{kmol m}^{-2}\,\text{s}^{-1}$ (driving force)$^{-1}$	$\text{gmol h}^{-1}\,\text{cm}^{-2}$ (driving force)$^{-1}$
gas film	k_G	$\text{kmol m}^{-2}\,\text{s}^{-1}\,\text{kPa}^{-1}$	"
liquid film	k_L	m s^{-1}	"
Overall, area basis	K	"	"
gas film	K_G	"	"
liquid film	K_L	"	"
Individual, volumetric basis	ka	$\text{kmol m}^{-3}\,\text{s}^{-1}\,\text{kPa}^{-1}$	h^{-1}
gas film	$k_G a$	"	h^{-1}
liquid film	$k_L a$	s^{-1}	h^{-1}
Metabolic rate, maximum specific	q_m	s^{-1}	h^{-1}
Metabolic rate, specific	q	s^{-1}	h^{-1}

$q = (\mathrm{d}C/\mathrm{d}t)/X$
where C may be a substrate or product mass concentration. Subscripts may further define the rates, *e.g.* q_S, q_P, q_{O_2}, which are substrate utilization, product formation, and oxygen uptake rates, respectively

	Symbol	SI units	Other units
Mutation rate	w	s^{-1}	h^{-1}
Power	P	W	W
Productivity, mass concentration rate basis, use appropriate subscripts, *e.g.* r_X for biomass productivity and r_P for product productivity	r	$\text{kg m}^{-3}\,\text{s}^{-1}$	$\text{kg m}^{-3}\,\text{h}^{-1}$
Revolutions per unit time or stirring speed	N	s^{-1}	s^{-1}
Velocity	V	m s^{-1}	cm s^{-1}

V_s for superficial gas velocity $= F_G/\pi D_t^2$

V_i for impeller tip velocity $= \pi N D_i$

Subject Index